Gerhard Einsele
Werner Ricken
Adolf Seilacher (Eds.)

Cycles and Events in Stratigraphy

With 393 Figures

Springer-Verlag
Berlin Heidelberg New York
London Paris Tokyo
Hong Kong Barcelona
Budapest

Prof. Dr. GERHARD EINSELE
Dr. WERNER RICKEN
Prof. Dr. ADOLF SEILACHER

Geologisches Institut
Universität Tübingen
Sigwartstraße 10
7400 Tübingen, Germany

ISBN 3-540-52784-2 Springer-Verlag Berlin Heidelberg New York
ISBN 0-387-52784-2 Springer-Verlag New York Berlin Heidelberg

Library of Congress Cataloging-in-Publication Data. Cycles and events in stratigraphy / Gerhard Einsele, Werner Ricken, Adolf Seilacher, eds. p. cm. Includes bibliographical references and index. ISBN 3-540-52784-2 (alk. paper). – ISBN 0-387-52784-2 (U.S.: alk. paper) 1. Geology, Stratigraphic. 2. Geology – Periodicity. I. Einsele, Gerhard. II. Ricken, Werner. III. Seilacher, Adolf. QE651.C9 1991 551.7–dc20 91-11656

This work is subject to copyright. All rights are reserved, whether the whole or part of the material is concerned, specifically the rights of translation, reprinting, reuse of illustrations, recitation, broadcasting, reproduction on microfilms or in other ways, and storage in data banks. Duplication of this publication or parts thereof is only permitted under the provisions of the German Copyright Law of September 9, 1965, in its current version, and a copyright fee must always be paid. Violations fall under the prosecution act of the German Copyright Law.

© Springer-Verlag Berlin Heidelberg 1991
Printed in Germany

The use of registered names, trademarks, etc. in this publication does not imply, even in the absence of a specific statement, that such names are exempt from the relevant protective laws and regulations and therefore free for general use.

Typesetting: International Typesetters Inc., Makati, Philippines
32/3145-543210 – Printed on acid-free paper

Preface

Earth's history is recorded in her strata. Each stratum is a historical accretion event. It may retain evidence of millions of subevents, but normally the stratum is the lowest level of coherent historical resolution, and the succession of strata thus constitutes the succession of links in the chain of history, as recognized by Steno.

Descriptive and *chronological* stratigraphy are concerned with sets of strata grouped into lithic or biotic units: with recognizing, describing, and naming them, and with mapping them in space and time. This work has established the framework of historical geology.

New insights were reached when seismic profiling brought to view the subtleties in the geometric relationships of stratal packages, and when the *sequence stratigraphy* thus defined turned out to be widely correlated, suggesting that transgressive-regressive events were driven by a global dictator – eustatic oscillation. While the level to which this correlation may be relied upon is debated, and while the mechanisms continue to be controversial, sequence stratigraphy has emerged as a major new branch of our science.

Event stratigraphy in a narrower sense is concerned with the individual strata and their substance. Progress in geophysics, geochemistry, geobiology, and other fields is providing ever-improving means of reading the information inherent in the stratum. Normally, a lithic unit is composed of hundreds or thousands of strata of a few kinds, recording a regime in which certain events occurred over and over again. Regimes recording the recurrent deposition of turbidites or of tempestites find their match in presentday settings, but others do not: for example, formations in which sediments with normal bottom fauna alternate with black shales devoid of benthic fossils record regimes in which whole seas became anaerobic time after time – a kind of regime which nowadays occurs only at the local level.

Beyond this lie the rare events, such as strata showing global isotope anomalies or geochemical signatures such as enrichment in iridium, which bespeak environmental shocks of global extent, events of sorts beyond human experience (but not necessarily beyond our abilities to decipher). Event stratigraphy teaches us that the Earth has not always been as it is today; and event stratigraphy is thus challenged to decipher the nature, extent, and cause of such deviations.

In this endeavor, a special concern is that of timing. Oscillatory or more complex cyclic repetition are the rule in stratigraphic sequence. To what extent are these randomly timed, to what extent the products of programmed processes? The world of our experience is rigorously structured by the alternation of night and day and of summer and winter, imposed by our planet's orbital characteristics. Random variations such as those of war and peace and those of the stock market are superimposed. The annual cycle is imprinted in stratigraphy at the level of varves, and the question arises as to whether stratigraphic cyclicity at larger scales may also have been driven by larger-scale cycles, such as the Earth's precession, the variation in obliquity, the cycles of orbital eccentricity, and the oscillation and revolution of the Solar System in the Milky Way Galaxy.

There is no doubt about the existence of these cycles (indeed, there may be others), but there is diversity of opinion as to whether they have left their imprint in sediments, and whether this imprint can be sifted out from the stochastically generated oscillations and cycles, and from diagenetically superimposed features. This is the special concern of *Cyclostratigraphy*.

The response of Pleistocene glaciation to the orbital variations has already been firmly established, and Pleistocene stratigraphy is being tied increasingly to the detailed proxy record of global ice flux provided by the oceanic oxygen isotope record, with potential time resolution to the 10^4 year level. The possibility of achieving such time resolution in the more distant past opens new horizons to stratigraphy. Orbital forcing patterns through the ages would also shed much new light on the functioning of climatic and oceanic regimes, and might well provide new data on the changes in length of day and other orbital features.

Some of the questions about events, changing regimes, and cycles were asked long ago, but the focus on deciphering individual beds in successive strata and of visualizing the nature of such ancient regimes, by combined sedimentology and paleontology, is comparatively recent. It can now be cast in the framework of sequence stratigraphy, and is now drawing strength from increasing involvement of geochemistry and geophysics. The specter of the enormous numbers of strata in the record is being laid to rest by the computer. We stand on the threshold of a new stratigraphy.

The predecessor volume, *Cyclic and Event Stratification*, approached some of these problems by means of individual case studies. The success which it enjoyed led Publisher and Editors to plan this successor volume, with somewhat broadened scope and an attempt to provide more overview. The proof of a book is in the reading, and this one will be read extensively.

San Pedro, December 1990   Alfred G. Fischer

Cycles and Events in Stratigraphy follows a precursor volume, *Cyclic and Event Stratification*, which was published in 1982 as a result of a workshop held in Tübingen. That volume contained essentially case studies dealing with three major topics of marine environments: rhythmic marl-limestone sequences, sandy and calcareous tempestites, and bedding phenomena in black shales. Although translated also into Russian and Chinese, *Cyclic and Event Stratification* was out of print within a few years.

Since then, knowledge about, and general interest in bedding features and cyclic sequences have increased considerably. New concepts and techniques have become available. Encouraged by friends and colleagues, we therefore felt it appropriate to present the state of the art in a new book, instead of just preparing a revised edition of the first volume. *Cycles and Events in Stratigraphy* has a much wider approach than the first volume. A total of 62 experts present reviews and concepts of various aspects of stratification. Most of these papers were discussed during a second international meeting held in January 1989 in Tübingen.

The principal changes are twofold: the new book for the most part presents review articles illustrated with conceptual figures, and covers a much wider range of topics than did the former volume. It describes not only a larger variety of marine bedding features, including gravity mass flows, siliceous sediments, phosphorites, shallow water carbonate, and glacio-marine cycles, but also stratification phenomena in lacustrine sediments, coal cycles, and tephra layers on land and below the sea. Scales range from annual varves to larger, mainly sea-level controlled sedimentary sequences. Secondary effects, such as biological response and feedback mechanisms, trace fossil tiering, and diagenetic overprinting, are also presented, as well as special techniques in timing and correlating cyclic and event bedding phenomena. A limited number of chapters were reserved to address only particularly interesting case studies. Most of the contributions are newly written, only one chapter has been taken from the first volume, but in a completely revised version. In spite of changes in authorship and treatment of much wider aspects, the book maintains the old dualism between cyclic and event signals in the stratigraphic record and extends it to the level of sequence stratigraphy.

We gratefully acknowledge the excellent contributions by all of the authors, as well as the advice and help from countless other colleagues. The secretaries of the Geological-Paleontological Institute in Tübingen, Marlis Lupold, Margot Pilopp, and Helga Wörner, helped with the typing, and Hermann Vollmer did the drafting for not only our own, but other chapters as well. Werner Wetzel made some of the photographic reductions. Linda Hobert of Albany, NY, and Susanne Borchert of Tübingen, had the difficult task of editing the language for the non-native English speaking authors. Marie-Luise Starke compiled the tremendous number of references from all articles to a single list. From the very beginning of this project, we enjoyed the close and fruitful cooperation with Dr. Wolfgang Engel and the Springer-Verlag.

This book addresses graduate students as well as professionals in geology, sedimentology, stratigraphy, and paleontology. Presenting concept-oriented reviews, it is expected to close the gap between the conventional text book and the standard symposium volume published for a small insider group of specialists. We hope that, in spite of any shortcomings, this book will contribute to keeping alive the discussion on the exciting field of dynamic stratigraphy.

Tübingen, December 1990   G. Einsele, W. Ricken, A. Seilacher

Contents

Introduction
Cycles and Events in Stratigraphy – Basic Concepts and Terms
G. Einsele, W. Ricken, and A. Seilacher 1

Part I Structure of Individual Beds

Chapter 1 Rhythmic Stratification: the "Undistorted" Record of Periodic Environmental Fluctuations

1.1 Limestone-Marl Alternations – an Overview
G. Einsele and W. Ricken 23

1.2 Orbital Cyclicity in Mesozoic Strata
A.G. Fischer 48

1.3 Pelagic Black Shale-Carbonate Rhythms:
Orbital Forcing and Oceanographic Response
P.L. de Boer 63

1.4 Environmental Factors Controlling Cretaceous Limestone-Marlstone Rhythms
D.L. Eicher and R. Diner 79

1.5 Rhythmic Carbonate Content Variations in Neogene Sediments Above the Oceanic Lysocline
L. Diester-Haass 94

1.6 Carbonate Cycles in the Pacific:
Reconstruction of Saturation Fluctuations
J. Grötsch, G. Wu, and W.H. Berger 110

1.7 A Holistic Geochemical Approach to Cyclomania:
Examples from Cretaceous Pelagic Limestone Sequences
M.A. Arthur and W.E. Dean 126

1.8 Variation of Sedimentation Rates in Rhythmically Bedded
Sediments: Distinction Between Depositional Types
W. Ricken 167

1.9 Sedimentary Rhythms in Lake Deposits
C. R. Glenn and K. Kelts 188

Chapter 2 Event Stratification: Records of Episodic Turbulence

2.1 Events and Their Signatures – an Overview
A. Seilacher 222

2.2 Shallow Marine Storm Sedimentation –
the Oceanographic Perspective
D. Nummedal 227

2.3 Storm Deposition at the Bed, Facies, and Basin Scale:
the Geologic Perspective
A. Seilacher and T. Aigner 249

2.4 Taphonomic Feedback (Live/Dead Interactions) in the Genesis
of Bioclastic Beds: Keys to Reconstructing Sedimentary Dynamics
S. M. Kidwell 268

2.5 Fossil Lagerstätten:
a Taphonomic Consequence of Event Sedimentation
C. E. Brett and A. Seilacher 283

2.6 Secular Changes in Phanerozoic Event Bedding
and the Biological Overprint
J. J. Sepkoski Jr, R. K. Bambach, and M. L. Droser 298

2.7 Submarine Mass Flow Deposits and Turbidites
G. Einsele 313

2.8 Calcareous Turbidites and Their Relationship
to Sea-Level Fluctuations and Tectonism
G. P. Eberli 340

2.9 Fine-Grained Turbidites
D. J. W. Piper and D. A. V. Stow 360

2.10 Distinction of Tempestites and Turbidites
G. Einsele and A. Seilacher 377

Contents

2.11 Flash Flood Conglomerates
F. Pflüger and A. Seilacher 383

2.12 Tephra Layers and Tephra Events
H.-U. Schmincke and P. van den Bogaard 392

Chapter 3 The Diagenetic Overprint: Enhancement Versus Obliteration of Primary Signals in Calcareous Rocks

3.1 Diagenetic Modification of Calcareous Beds — an Overview
W. Ricken and W. Eder 430

3.2 Pressure-Dissolution and Limestone Bedding:
the Influence of Stratified Cementation
R. G. C. Bathurst 450

Chapter 4 Cherts and Phosphorites: Primary and Diagenetic Bedding in Special Environments

4.1 Rhythmic Bedding in Siliceous Sediments — an Overview
K. Decker 464

4.2 Compaction and Cementation in Siliceous Rocks
and Their Possible Effect on Bedding Enhancement
R. Tada 480

4.3 Stratification in Phosphatic Sediments:
Illustrations from the Neogene of California
K. B. Föllmi, R. E. Garrison, and K. A. Grimm 492

Chapter 5 Preservation and Biological Destruction of Laminated Sediments

5.1 Stratification in Black Shales:
Depositional Models and Timing — an Overview
A. Wetzel 508

5.2 Redox-Related Benthic Events
C. E. Savrda, D. J. Bottjer, and A. Seilacher 524

5.3 Biofacies Models for Oxygen-Deficient Facies
in Epicontinental Seas: Tool for Paleoenvironmental Analysis
B. B. Sageman, P. B. Wignall, and E. G. Kauffman 542

5.4 Anaerobic – Poikiloaerobic – Aerobic:
a New Facies Zonation for Modern and Ancient Neritic Redox Facies
W. Oschmann 565

5.5 Cyclical Deposition of the Plattenkalk Facies
Ch. Hemleben and N. H. M. Swinburne 572

5.6 Biolaminations – Ecological Versus Depositional Dynamics
G. Gerdes, W. E. Krumbein, and H.-E. Reineck 592

Part II Larger Cycles and Sequences

Introductory Remarks
G. Einsele and W. Ricken 611

Chapter 6 Sequences: Hierarchies, Causes, and Environmental Expression

6.1 The Stratigraphic Signatures
of Tectonics, Eustacy and Sedimentology – an Overview
P. R. Vail, F. Audemard, S. A. Bowman, P. N. Eisner, and C. Perez-Cruz   617

6.2 Asymmetry in Transgressive-Regressive Cycles in Shallow Seas
and Passive Continental Margin Settings
G. Einsele and U. Bayer 660

6.3 Condensed Deposits in Siliciclastic Sequences:
Expected and Observed Features
S. M. Kidwell 682

6.4 Biological and Evolutionary Responses
to Transgressive-Regressive Cycles
G. R. McGhee Jr, U. Bayer, and A. Seilacher 696

6.5 Lagoonal-Peritidal Sequences in Carbonate Environments:
Autocyclic and Allocyclic Processes
A. Strasser 709

6.6 A Basic Model for Lofer Cycles
J. Haas 722

6.7 Coal Cyclothems and Some Models for Their Origin
W. Riegel 733

6.8 Cycles, Rhythms, and Events
on High Input and Low Input Glaciated Continental Margins
R. Henrich 751

Chapter 7 Timing and Correlation

7.1 Time Span Assessment – an Overview
W. Ricken 773

7.2 High-Resolution Correlation: a New Tool in Chronostratigraphy
E. G. Kauffman, W. P. Elder, and B. B. Sageman 795

7.3 Varves, Beds, and Bundles in Pelagic Sequences
and Their Correlation (Mesozoic of SE France and Atlantic)
P. Cotillon 820

7.4 The Spectral Analysis of Stratigraphic Time Series
G. P. Weedon 840

7.5 Milankovitch Cycles and the Measurement of Time
W. Schwarzacher 855

References 865

Subject Index 945

List of Contributors

Aigner, Thomas: Geologisches Institut, Universität Tübingen, Sigwartstr. 10, 7400 Tübingen, FRG

Arthur, Michael A.: Department of Geosciences, The Pennsylvanian State University, 503 Deike Building, University Park, PA 16802, USA

Audemard, Felipe: Department of Geology and Geophysics, Rice University, Houston, Texas 77251, USA

Bambach, Richard K.: Department of Geological Sciences, Virginia Polytechnic Institute and State University, Blacksburgh, Virginia 24061, USA

Bathurst, Robin G. C.: Derwen Deg Fawr, Llanfair D.C., Ruthin, Clwyd, North Wales LL15 2SN, United Kingdom

Bayer, Ulf: Kernforschungsanlage, Institut für Erdöl und Organische Geochemie, 5170 Jülich, Postfach 1913, FRG

Berger, Wolfgang H.: Scripps Institution of Oceanography, University of California, San Diego, La Jolla, California 92093, USA

Bottjer, David J.: Department of Geological Sciences, University of Southern California, Los Angeles, California 90089, USA

Bowman, Scott A.: Department of Geology and Geophysics, Rice University, Houston, Texas 77251, USA

Brett, Carlton E.: Department of Geology, Hutchinson Hall, University of Rochester, Rochester, New York 14627, USA

Cotillon, Pierre: Département des Sciences de la Terre, Université Claude-Bernard Lyon 1, 27–43 Bd du 11 Novembre, 69622 Villeurbanne Cedex, France

Dean, Walter E.: U.S. Geological Survey, MS 939, Box 25046, Federal Center, Denver, Colorado 80225-0046, USA de Boer, Poppe L.: Comparative Sedimentology Division, Institute of Earth Sciences, Budapestlaan 4, 3508 TA, Utrecht, Netherlands

Decker, Kurt: Geologisch-Paläontologisches Institut der Universität, Universitätsstraße 7, 1010 Wien, Austria

Diester-Haass, Lieselotte: Fachrichtung Geographie, Universität des Saarlandes, 6600 Saarbrücken, FRG

Diner, Richard: Department of Geological Sciences, University of Colorado, Campus Box 250, Boulder, Colorado 80309, USA

Droser, Mary L.: Department of Geology, Oberlin College, Oberlin, Ohio 44074-1087, USA

Eberli, Gregor: University of Miami, RSMAS-MGG, 4600 Rickenbacker Cswy., Miami, FL 33149, USA

Eder, F. Wolfgang: Universität Göttingen, Geologisches Institut, Goldschmidtstraße 3, 3400 Göttingen, FRG

Eicher, Don L.: Department of Geological Sciences, University of Colorado, Campus Box 250, Boulder, Colorado 80309, USA

Einsele, Gerhard: Universität Tübingen, Geologisches Institut, Sigwartstraße 10, 7400 Tübingen, FRG

Eisner, Pablo N.: Department of Geology and Geophysics, Rice University, Houston, Texas 77251, USA

Elder, William P.: Geological Survey, Branch of Paleontology and Stratigraphy, Mail Stop 915, 345 Middlefield Road, Menlo Park, California 94025, USA

Fischer, Alfred G.: Department of Geological Sciences, University of Southern California, Los Angeles, California 90089, USA

Föllmi, Karl B.: Geologisches Institut, ETH Zürich, Sonneggstraße 5, 8092 Zürich, Switzerland

Garrison, Robert E.: Earth Sciences, Applied Sciences Building, University of California, Santa Cruz, California 95064, USA

Gerdes, Giesela: Institut für Chemie und Biologie des Meeres, Universität Oldenburg, Postfach 2503, 2900 Oldenburg, FRG

Glenn, Craig R.: Department of Geology and Geophysics, University of Hawaii, Honolulu, Hawai 96822, USA

List of Contributors

Grimm, Kurt A.: Earth Sciences, Applied Sciences Building, University of California, Santa Cruz, California 95064, USA

Grötsch, Jürgen: Institut für Paläontologie, Universität Erlangen-Nürnberg, Loewenichstraße 28, 8520 Erlangen, FRG

Haas, János: Central Office of Geology, 1051 Budapest, Arany János u. 25, Hungary

Hemleben, Christoph: Universität Tübingen, Geologisches Institut, Sigwartstraße 10, 7400 Tübingen, FRG

Henrich, Rüdiger: Geomar, Forschungszentrum für marine Geowissenschaften, Wischhofstraße 1-3, 2300 Kiel, FRG

Kauffman, Erle G.: Department of Geological Sciences, University of Colorado, Campus Box 250, Boulder, Colorado 80309, USA

Kelts, Kerry: Limnological Research Center, University of Minnesota, Minneapolis, MN 55455, USA

Kidwell, Susan M.: Department of Geophysical Sciences, University of Chicago, 5734 South Ellis Avenue, Chicago, Illinois 60637, USA

Krumbein, Wolfgang E.: Institut für Chemie und Biologie des Meeres, Universität Oldenburg, Postfach 2503, 2900 Oldenburg, FRG

McGhee, George R.: The State University of New Jersey, Rutgers, Department of Geological Sciences, Wright-Rieman Geological Laboratory, Busch Campus, New Brunswick, New Jersey 08903-201, USA

Nummedal, Dag: Department of Geology and Geophysics, Louisiana State University, Baton Rouge, Louisiana 70803-4104, USA

Oschmann, Wolfgang: Institut für Paläontologie, Universität Würzburg, Pleicherwall 1, 8700 Würzburg, FRG

Perez-Cruz, Guillermo: Department of Geology and Geophysics, Rice University, Houston, Texas 77251, USA

Pflüger, Friedrich: Universität Tübingen, Geologisches Institut, Sigwartstraße 10, 7400 Tübingen, FRG

Piper, David J. W.: Atlantic Geoscience Centre, Bedford Institute of Oceanography, P.O. Box 1006, Dartmouth, Nova Scotia, Canada, B2V 4A2

Reineck, Hans-Erich: Senckenberg Institut, Schleusenstraße 39a, 2940 Wilhelmshaven, FRG

Ricken, Werner: Universität Tübingen, Geologisches Institut, Sigwartstraße 10, 7400 Tübingen, FRG

Riegel, Walter: Universität Göttingen, Geologisches Institut, Goldschmidtstraße 3, 3400 Göttingen, FRG

Sageman, Bradley B.: The Pennsylvania State University, Department of Geosciences, 503 Deike Building, University Park, PA 16802, USA

Savrda, Charles E.: Department of Geology, Auburn University, 210 Petrie Hall-Thach Ave., Auburn, Alabama 36849-5305, USA

Schmincke, Hans-Ulrich: Geomar, Forschungszentrum für marine Geowissenschaften, Wischhofstraße 1–3, 2300 Kiel, FRG

Schwarzacher, Walther: Department of Geology, The Queen's University Belfast, Belfast BT7 1NN, Northern Ireland

Seilacher, Adolf: Universität Tübingen, Geologisches Institut, Sigwartstraße 10, 7400 Tübingen, FRG; and Yale University, Kline Geology Laboratory, P.B. 6666, New Haven, Connecticut 06520, USA

Sepkoski, J. John Jr.: Department of Geophysical Sciences, University of Chicago, 5734 South Ellis Avenue, Chicago, Illinois 60637, USA

Stow, Dorrik A. V.: The University of Nottingham, Department of Geology, University Park, Nottingham, NG7 2RD, United Kingdom

Strasser, André: Université de Fribourg, Institut de Géologie, Pérolles, 1700 Fribourg, Switzerland

Swinburne, Nicola H. M.: Department of Earth Sciences, The Open University, Walton Hall, Milton Keynes MK7 6AA, United Kingdom

Tada, Ryuji: University of Tokyo, Geological Institute, Faculty of Science Building 5, 7-3-1 Hongo, Tokyo 113, Japan

Vail, Peter R.: Department of Geology and Geophysics, Rice University, Houston, Texas 77251, USA

van den Bogaard, Paul: Geomar, Forschungszentrum für marine Geowissenschaften, Wischhofstraße 1–3, 2300 Kiel, FRG

Weedon, Graham P.: Department of Earth Sciences, University of Cambridge, Downing Street, Cambridge CB2 3EQ, United Kingdom

Wetzel, Andreas: Universität Basel, Geologisch-Paläontologisches Institut, Bernoullistraße 32, 4056 Basel, Switzerland

Wignall, Paul B.: University of Leicester, Leicester LEI 7RH, United Kingdom

Wu, G.: Scripps Institution of Oceanography, University of California, San Diego, La Jolla, California 92093, USA

Introduction

Cycles and Events in Stratigraphy – Basic Concepts and Terms

G. Einsele, W. Ricken, and A. Seilacher

1 Introduction

One of the most conspicuous features of sedimentary rocks is their stratification. Caused by various periodic and episodic processes, as well as by biologic and diagenetic overprints, such stratification commonly exhibits a distinctive kind of rhythmicity, due to regularly alternating beds or a repetition of larger units which are referred to as depositional cycles. Rhythmic and cyclic sequences occur worldwide in presumably every environmental and stratigraphic system. Several textbooks have summarized the knowledge and concepts of rhythmic stratification (e.g., Merriam 1964; Duff et al. 1967; Elam and Chuber 1972; Schwarzacher 1975; Einsele and Seilacher 1982; Berger et al. 1984), and many books on facies analysis deal with this subject. In the meantime, our knowledge and techniques have increased considerably, and more geologists, sedimentologists, stratigraphers, and paleontologists appear to be interested in rhythmic and cyclic stratification than in earlier years.

The basic subjects addressed in this book include: (1) The description of depositional processes and associated overprints forming various beds, which may be grouped into cyclic and episodic stratification types; and (2) the description of the ordering of beds into a hierarchical pattern of smaller cycles and larger sequences. However, before such concepts are presented in Parts I and II of this Volume, some very basic introductions to the subjects dealt with here and some clarifications regarding terminologies used are given in this first chapter. In the following, these introductory remarks start with comments on the types of beds treated in this book.

2 Types of Beds Addressed in this Book

The basic sedimentological unit addressed in this book is the laterally traceable bed (Fig. 1). Smaller and larger sedimentary structures are not discussed in this book, including small-scale ripples, larger trough cross beds, larger sand and carbonate bodies, etc. The typical bed addressed in this book is defined as a three-dimensional body of relatively uniform composition, as compared to the underlying or succeeding beds. Compositional characteristics include primary chemical-mineralogical and structural properties, biological aspects, and diagenetic structures. Bed thicknesses can vary from a few centimeters to several meters, representing various time spans ranging from seconds to several 100 ka. Thicknesses of 5 to 40 cm are common, but

Einsele et al. (Eds.)
Cycles and Events in Stratigraphy
©Springer-Verlag Berlin Heidelberg 1991

mean thicknesses seem to have increased through the Phanerozoic due to slowdown of orbital frequencies (Berger et al. 1987), and to an increase in the depth of bioturbational sediment mixing eliminating smaller beds (see Sepkoski et al. Chap. 2.6, this Vol.). The lateral extent of beds ranges from a few meters to thousands of km (see Kauffman et al. Chap. 7.2, this Vol.). Beds are separated from each other by thinner or thicker intercalations of differing composition or structure. The boundaries between beds and their intercalations, which usually are finer-grained and softer, form the upper and lower bedding planes. They may be described as gradational, sharp, wavy, and sutured. Bedding planes are of primary depositional and biological or diagenetic origin (see Bathurst Chap. 3.2, Tada Chap. 4.2, Ricken and Eder Chap. 3.1, all this Vol.). Subbedding planes are subtle, indistinct surfaces, which may grade laterally into distinctive bedding planes. Event beds often have a sharp lower (i.e.,

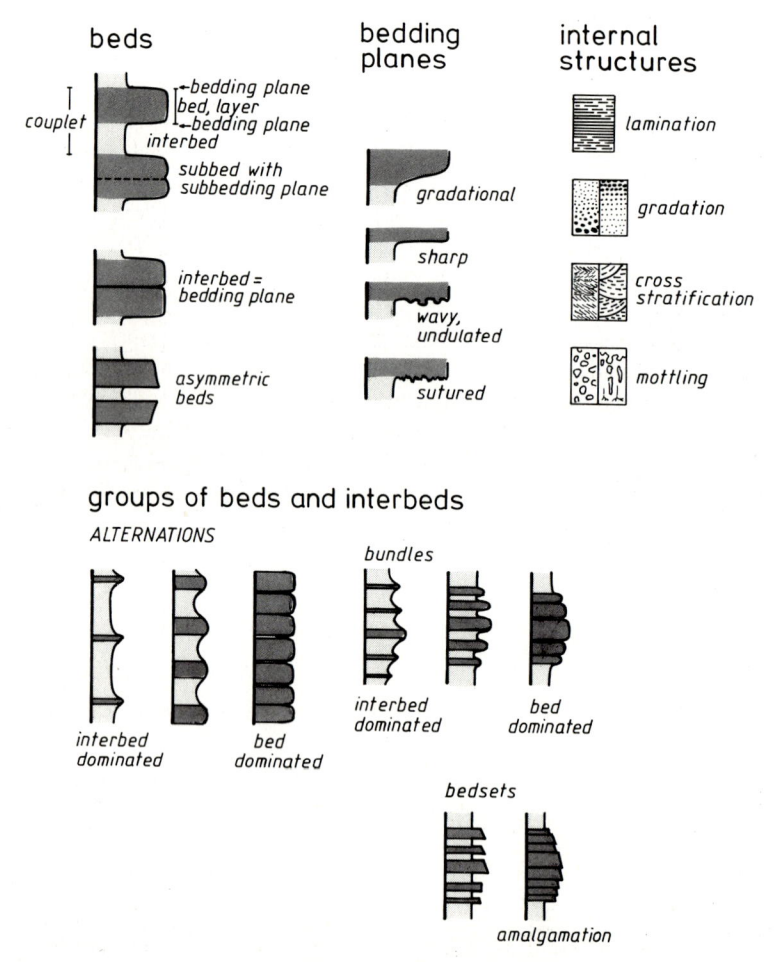

Fig. 1. Descriptive terms for bedding phenomena as described in this book

Einsele et al: Basic Concepts

the sole) and a gradational upper bedding plane (see Seilacher and Aigner Chap. 2.3, this Vol.). For a detailed description of beds and associated phenomena, the reader is referred to McKee and Weir (1953), Campbell (1967), Reineck and Singh (1980), as well as Collinson and Thompson (1982).

In an alternation of different beds, weathering-resistant beds are described as beds or layers, while less resistant beds may be called interbeds or interlayers (Fig. 1). A layer and a subsequent interlayer form a bedding couplet (Fischer and Schwar-

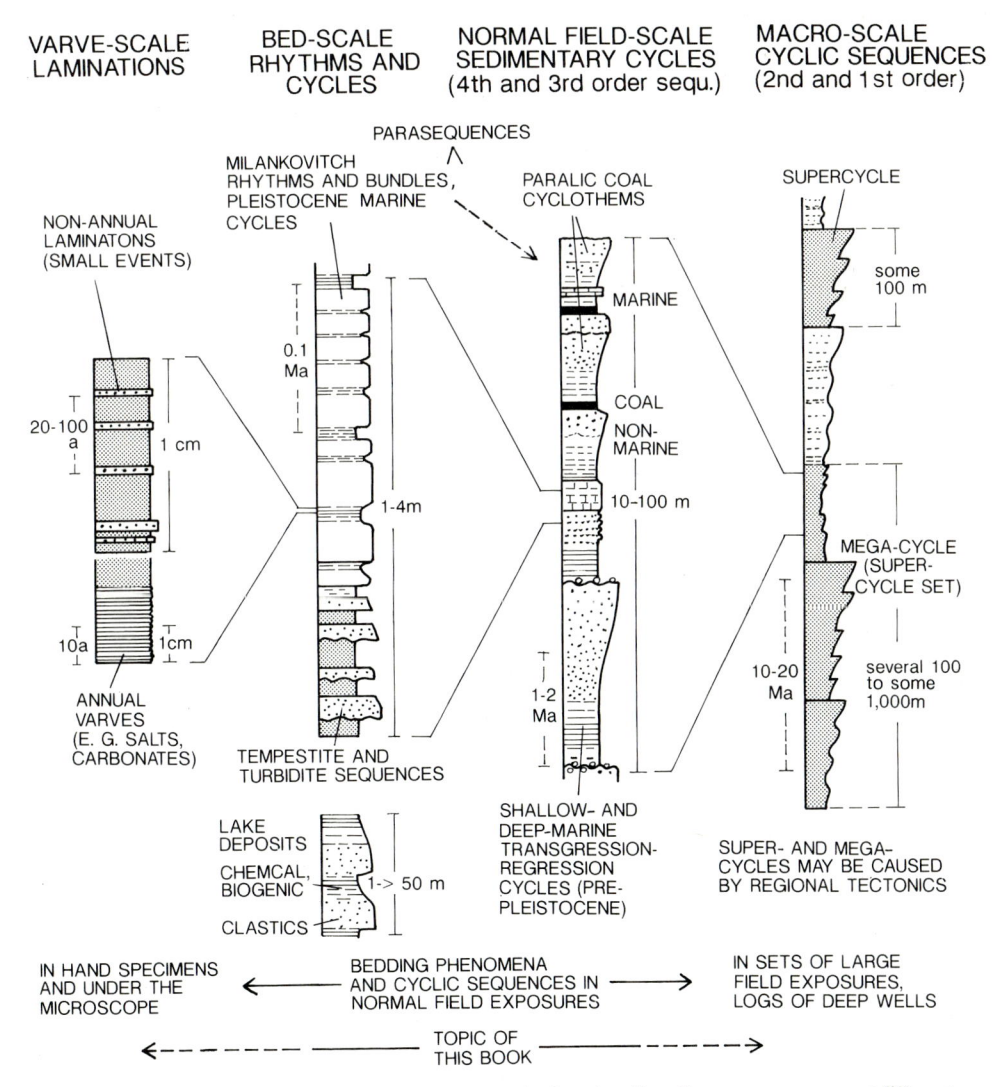

Fig. 2. Nomenclature used in this book to describe rhythmic and cyclic sedimentary sequences of different origins, thicknesses, and time periods in the field. Orders of cycles in parentheses after Vail et al. (1977c) and Haq et al. (1987). The term megacycle is here used for "supercycle set" in Haq et al. (1987)

zacher 1984). When the interlayer becomes very thin, as found in bed-dominated alternations, it may essentially represent a bedding plane.

Groups of beds and interbeds denote alternating or rhythmic bedding (see below). Alternations can be bed-dominated or interbed-dominated. Bedsets (Campbell 1967) or bundles (Schwarzacher 1975) represent several bedding couplets, separated by thicker interbeds (see Fig. 3g). They may be symmetrical, with the thickest layer in the middle, or asymmetrical, due to thinning-upward or fining-upward processes. Bedsets without interbeds are formed by the amalgamation of event layers.

3 Rhythmic Bedding and Sedimentary Cycles of Different Scales

Various sedimentary sequences of a regional or possibly even global nature may consist primarily of:

1. Two alternating bed types (succession AB, AB, etc.) which can be called *rhythmic or cyclic bedding*, or *rhythmic sequences* (see Fig. 3b-d).
2. Several different sediment types, at least three, forming a succession (e.g., ABC, ABC, etc.) which is repeated, known as *cyclic sequences* (see Fig. 3e-g).

Unfortunately, there is no general agreement on the terminology related to these two basic stratification types. This is mainly due to the fact that both rhythmic and cyclic sequences can be generated by entirely different processes, and that the geological time span required to deposit a single couplet or bundle in such rhythms or cycles may vary greatly. Therefore, none of the attempts to apply a general terminology to the genesis or time periods of rhythmic or cyclic bedding phenomena has received general accord. Probably the only practical way, at least as a first step, is a purely descriptive approach, which may later be complemented or replaced by a terminology based on genetic relationships, as soon as sufficient information is available.

A simple, primarily descriptive classification of rhythmic and cyclic phenomena is shown in Fig. 2, which is based mainly on the thicknesses of beds and larger sedimentary cycles. Rhythmic and cyclic sediments are subdivided into four groups, the varve-scale laminations, bed-scale rhythms and cycles, field-scale sedimentary cycles (including third- and fourth-order cycles and parasequences), and various orders of macroscale cyclic sequences (i.e., supercycles and megacycles, according to the Vail-Haq nomenclature, see below). Apart from the macroscale sequences, this classification can be easily applied in the field and in well logs without any information on genesis, sedimentation rates, and associated time spans. Varve-scale laminations or bed-scale alternations are well-known features, but none of them are the products of only one specific process in a certain environment, and each may represent quite different time periods. The field-scale represents the typical outcrop cycle, several meters to tens of meters thick. Many of these field-scale marine cycles are interpreted today as representing global and relative sea level changes on the order of 100 ka to several Ma. However in lacustrine environments, where overall sedimentation rates usually are much higher than in the deeper sea, cycles of the same thickness are often controlled by shorter-term climatic fluctuations on the order of 20 to 100 ka (see Glenn and Kelts Chap. 1.9, this Vol.). Climatic variations with time periods on this order are also believed to be the cause of many rhythmic, marine marl-limestone sequences. Macroscale cycles normally cannot be seen in single exposures. They comprise successions of considerable thickness (100 m up to several km) and represent long time periods (usually between 10 Ma and more than 100 Ma). They are often caused by long-term tectonic subsidence processes associated with rifting and subsequent spreading, which may lead to lithospheric flexure and persisting onlap of coastal sediments on continental margins (Watts et al. 1982; Sheridan 1987).

Vail et al. (1977c and Chap. 6.1, this Vol.) and Haq et al. (1987) proposed a subdivision of depositional sequences into cycles from a first to sixth order, which are primarily defined in terms of time periods. The first- and second-order cycles have a time period too long and thus a thickness too great to be seen in normal sections. Conspicuous cycles in the field are third-order sequences with a time range of 0.5 to 5 Ma. The fourth-order cycles have a time period on the order of 0.1 to 0.5 Ma, representing eustatic variations around the larger eccentricity cycles. Into this category fall the so-called parasequences of the Vail nomenclature as well as many features we are dealing with here. For example, some of the coal cycles seem to have this periodicity (Riegel Chap. 6.7, this Vol.). The fifth-order sea level variations represents the periodicities of the Milankovitch frequency band, such as glacio-eustatic cycles (0.01 to 0.1 Ma). Variations on these orders of magnitude include field-scale and bed-scale phenomena, as expressed for instance in peritidal carbonate cycles (see Strasser Chap. 6.5, this Vol.). Sea level changes with a higher frequency as the Milankovitch variations (shorter 10 ka) are denoted as sixth-order cycles (see Table 1 in Vail et al. Chap. 6.1, this Vol.).

The term *cyclothem* (e.g., Weller 1964) is widely used in North America, especially for coal-bearing sequences, particularly those of Pennsylvanian age (see Riegel Chap. 6.7, this Vol.). This term describes the "basic cycle", i.e., a package of lithologies representing the smallest cyclic unit of a sequence. The succession of different beds and interbeds in a cyclothem may be symmetric, subsymmetric, or completely asymmetric (compare Fig. 3e and f). The term cyclothem is a purely descriptive expression, as cyclothems may be caused by eustatic marine sea level fluctuations (e.g., Heckel 1986), episodic thrust loading, and by other mechanisms (de Klein and Willard 1989; Riegel Chap. 6.7, this Vol.). In seismic stratigraphy this term is replaced by *depositional sequence*, which is defined by its lower and upper boundaries as well as by lowstand, transgressive, and highstand deposits, or, if it is a smaller unit, by the term *parasequence* (van Wagoner et al. 1987; Vail 1987; Vail et al. Chap. 6.1, this Vol.). In this text, the terms depositional sequence or depositional cycle and, for a succession of such cycles, cyclic sequence, are mainly used.

4 Basic Types of Rhythmic Bedding and Sedimentary Cycles

Using the term sedimentary cycle may foster the opinion that such cycles represent equivalent time periods. This is not necessarily true, and in fact most workers use the term cycle not in this sense, but only as a convenient way to describe repeated successions of certain lithologies and facies types. In the case of many rhythmic

RHYTHMIC BEDDING

a STOCHASTIC "BEDDING"
TIME

"DEPOSITIONAL NOISE"

b EVENT STRATIFICATION,
TIME (LINEAR)

AB_3
AB_2
AB_1

DISCYCLIC, EPISODIC (NON-PERIODIC)
$AB_1 \neq AB_2 \neq AB_3$

c INCREASING THICKNESS OF BEDDING COUPLETS

- DUE TO EXPANDED TIME PERIOD (3 INSTEAD OF 1), OR
- PERMANENTLY HIGHER SEDIMENTATION RATE (1),(2)...

TIME (LINEAR)

± ENHANCED BY DIAGENETIC OVERPRINT AND/OR WEATHERING

STRICTLY CYCLIC (PERIODIC) TIME PERIOD FOR AB= const.

d INCREASING SED. RATE,
TIME (NON-LINEAR)

NORMAL FIELD-SCALE CYCLES

e SYMMETRIC
GRAIN SIZE

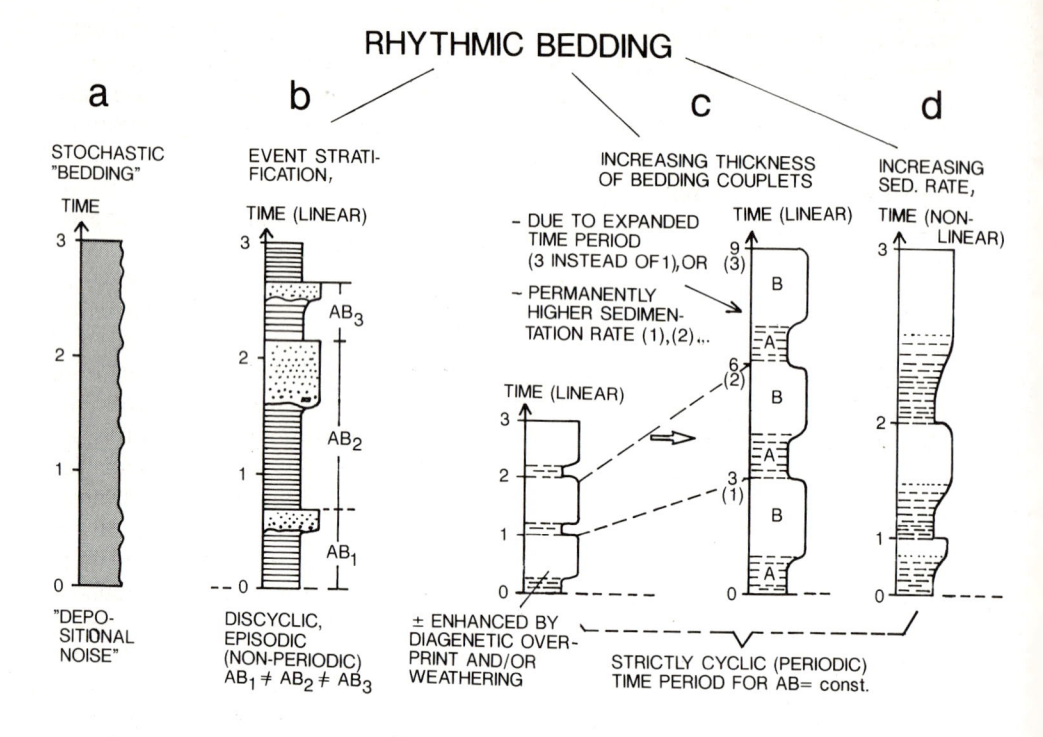

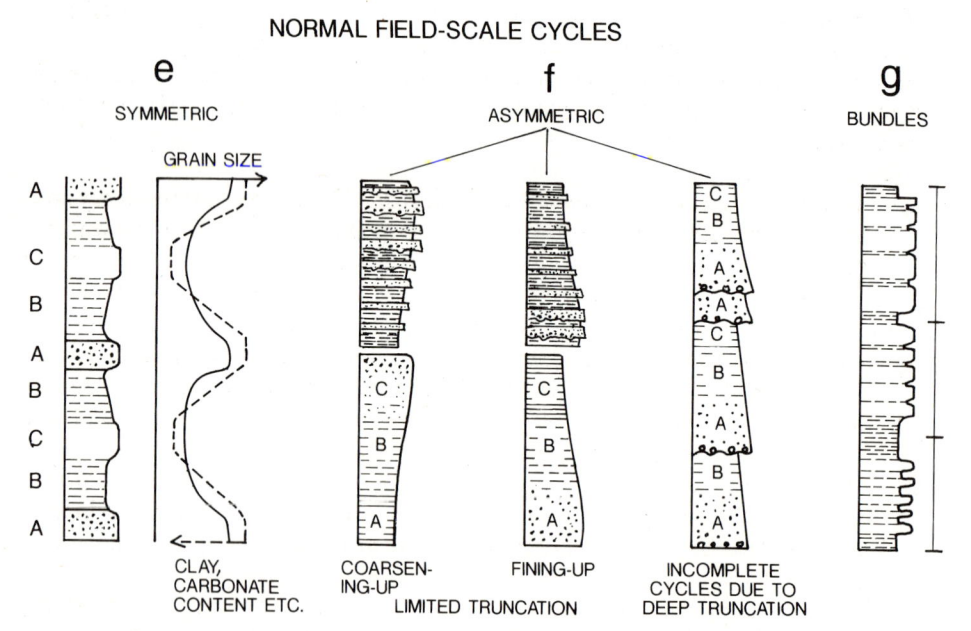

CLAY, CARBONATE CONTENT ETC.

f ASYMMETRIC

COARSENING-UP
LIMITED TRUNCATION

FINING-UP

INCOMPLETE CYCLES DUE TO DEEP TRUNCATION

g BUNDLES

Fig. 3a-g. Stochastic bedding (**a**) and different types of rhythmic (**b** through **d**) and cyclic (**e** through **g**) sedimentary successions. All sequences may be either strictly cyclic or periodic (**c** and **d**) or nonperiodic (discyclic, **b**). The thickness of corresponding bed types can change from cycle to cycle, either due to variations in the time period (**c**) of succeeding cycles or in the rates of sedimentation (**c** and **d**). Rhythmic bedding caused by depositional events (**b**) may show similar alternations as cyclic sequences, but they are always discyclic or nonperiodic sequences (e.g., turbidite sequences), it is clear that they are the result of frequently but irregularly recurring sedimentological events. Therefore some authors theoretically distinguish between *strictly cyclic or periodic* sequences with a regular time period (Fig. 3c and d), and *discyclic or nonperiodic* sequences (Fig. 3b) caused by irregular stratigraphic events (e.g., Dott 1988) or by recurrent processes associated with the depositional regime (i.e., autocyclic processes, see below). Smaller irregularities in the mode and rate of deposition are sometimes referred to as "depositional noise", resulting in purely *stochastic* sequences (Fig. 3a).

It is, however, very difficult to prove whether or not a given sedimentary cycle is really caused by a mechanism with a constant time period; numerical techniques, such as time series analyses, are important testing tools for this (see Weedon Chap. 7.4, Schwarzacher Chap. 7.5, both this Vol.). The most prominent examples of periodicity affecting depositional patterns are the Earth's orbital cycles of precession, obliquity, and eccentricity with periods of 21, 41, and about 100 and 400 ka (i.e., Milankovitch cycles, see Fischer Chap. 1.2, Einsele and Ricken Chap. 1.1, both this Vol.). If, as demonstrated in the Pleistocene oxygen isotope curve, some sediments are repeatedly affected by phenomena of a constant or quasi-constant recurrence time, they may be termed *periodites* (Einsele 1982). However, even in this case, the thicknesses of the single layers A and B may vary due to changing sediment composition (Fig. 3c) or fluctuating sedimentation rates (Fig. 3d).

Sedimentary cycles may also be *symmetric* or *asymmetric* (e.g., coarsening-upward or fining-upward sequences) as well as *complete* or *incomplete*, the latter commonly an effect of low subsidence values (Fig. 3f). The superposition of two or several periodically recurring processes can generate a succession of bundles (Fig. 3g, Schwarzacher and Fischer 1982; see Fischer Chap. 1.2, this Vol.).

Genetically, one can distinguish between two groups of mechanisms leading to rhythmic and cyclic sequences, originally defined by Beerbower (1964) to describe cyclic styles in fluvial systems:

1. *Autocyclic sequences* (or *autogenetic sequences*, Dott 1988) are primarily controlled by processes taking place in the sedimentary prism itself (e.g., within a basin or part of it, Fig. 4a). Their beds usually show only limited stratigraphic continuity. Prominent examples in this category are nonperiodic tempestites (i.e., storm events) and turbidite sequences, the migration and superposition of channel and lobe systems in fluvial environments as well as in the deep sea, or successions of coal seams produced by switching lobes of a subsiding delta plain (see Seilacher and Aigner Chap. 2.3, Einsele Chap. 2.7, Eberli Chap. 2.8, Piper and Stow Chap. 2.9, as well as Riegel Chap. 6.7, all this Vol.).
2. *Allocyclic sequences* (or *allogenetic sequences*) are mainly caused by variations external to the considered sedimentary system (e.g., the basin), such as climatic changes, tectonic movements in the source area, global sea level variations, etc. (Fig. 4b). Such processes often tend to generate cyclic phenomena of a larger lateral continuity and time period than autocyclic processes. The most characteristic effect of some of the allocyclic processes is that they operate simultaneously in different basins in a similar way. Thus it should be possible to correlate part of the allocyclic sequences over long distances and perhaps even from one basin to another.

It is often not possible to distinguish sharply between autocyclic and allocyclic processes, depending on the size of the depositional unit under consideration. Large sedimentary systems include more autocyclic mechanisms, while small sedimentary units are more affected by allocyclic mechanisms. Considering a whole basin, regional tectonics may affect both the drainage area outside the depositional basin as well as smaller tectonic structures within the proper basin floor. In addition, the occurrence and frequency of mainly autocyclic mass flows and turbidites is strongly affected by allocyclic eustatic sea level changes, etc. In fact, there are many depositional sequences displaying the results of both allocyclic and autocyclic phenomena (e.g., Riegel Chap. 6.7, Strasser Chap. 6.5, Kauffman et al. Chap. 7.2, all this Vol.).

5 Features and Recurrence Time of Cyclic and Discyclic Bedding

Rhythmic and cyclic bedding repetitions can be subdivided into two major groups:

1. Bedding variations by repeated *slow gradual changes* in deposition. As mentioned above, this type is called *cyclic bedding*, or when strictly periodic, *periodic bedding*. It is usually caused by allocyclic, e.g., orbital climatic phenomena. Typical examples are unlithified, chalk-marl sequences (see Part I.1, this Vol.).
2. Bedding variations by *abrupt changes* in sedimentation due to depositional events or episodes at random to quasi-periodic time intervals (*stochastic, episodic, and discyclic bedding*, see Part I.2, this Vol.). The most prominent examples of this group are tempestites and turbidites, usually originating from autocyclic mechanisms. Flooding episodes in alluvial plains (Pflüger and Seilacher Chap. 2.11, this Vol.) and repeated volcanic ashfalls into basins of normally accumulating sediments also belong to this group (Schmincke and van den Bogaard Chap. 2.12, this Vol.). Changes in the intensity or frequency of depositional events, however, may be also affected by allocyclic processes (Kauffman et al. Chap. 7.2, this Vol.).

Some characteristics of the entirely different processes generating rhythmic sequences are shown in Fig. 5. In the case of *cyclic bedding*, slow, gradual variations in the primary composition of the sediments and associated variations in sedimentation rates (Fig. 5a) lead to a smoothly changing, vertical sediment buildup with time; this is valid for both calcareous and siliceous bedding (see Decker Chap. 4.1, this Vol.). However, a small variation in primary composition, texture, or fabric may be sufficient to promote a secondary differentiation into accentuated beds and interbeds by diagenetic overprint (see below). When not interrupted by depositional events, the intensity of burrow mottling is continuous in succeeding beds, although there may occur slow variations in the number and type of bottom-dwelling species as well as planktonic organisms (see below). Anoxic intervals may further modify the alternating beds as well as their biological record.

The recurrence interval of the various cyclic deposits can theoretically be strictly periodic, quasi-periodic, or nonperiodic. The 11-year sun spot cycles found in varved salt deposits and lake sediments (see Glenn and Kelts Chap. 1.9, this Vol.) represent a strict periodicity. Milankovitch cycles, however, hold some complications.

a AUTOCYCLIC MECHANISMS

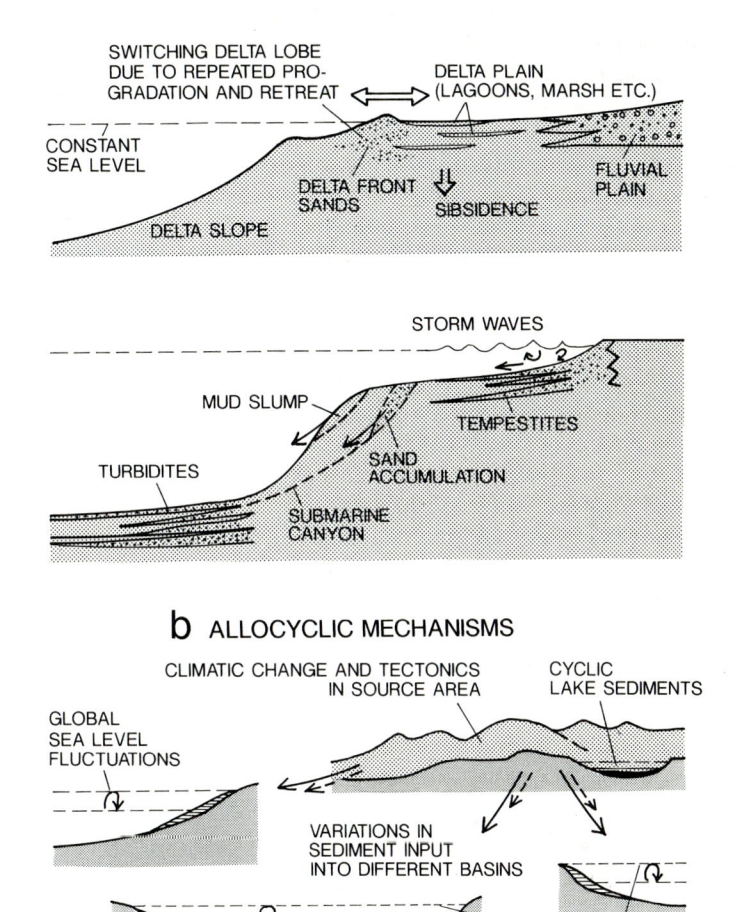

b ALLOCYCLIC MECHANISMS

Fig. 4a,b. Autocyclic (a) and allocyclic mechanisms (b) as a primary cause of the generation of rhythmic and cyclic sequences

Whereas, for example, the individual orbital parameters seem to have constant amplitudes and time periods, their complicated combined effects on climate and sediments tend to create a periodicity with considerably varying amplitudes and a changing importance of certain time intervals (see Einsele and Ricken Chap. 1.1, Fischer Chap. 1.2, both this Vol.). Cyclic sequences caused by longer-term, third-order sea level changes (see Part II.1 this Vol.) also show varying time periods within a certain order of magnitude. It is the superposition of several mechanisms of differing time periods and intensities which often renders it difficult to find out from the response of sediments whether there is a true periodic process in operation above the depositional background noise. For these reasons, many allocyclic phenomena in sediments appear to be quasi-periodic, and some even nonperiodic, rather than strictly periodic.

Alternations due to *episodic phenomena* (tempestites, turbidites) show a very irregular time curve of sediment buildup (Fig. 5c and d). Slow and more or less continuous vertical accumulation of fine-grained background sediment is interrupted irregularly by erosional and immediately succeeding depositional episodes. In the case of turbidite sequences (Fig. 5c), the background sediment, which is produced in the area of deposition itself (i.e., autochthonous sediment), is discontinuously interrupted by laterally transported, allochthonous turbidite sediment (see Eberli Chap. 2.8, Piper and Stow Chap. 2.9, Einsele Chap. 2.7, all this Vol.). Therefore, the rates of sedimentation also change abruptly. The bioturbated surface layer of the background sediment, normally migrating slowly upward, is episodically truncated and replaced by an event deposit. Recolonization of the event layer is achieved by specially adapted fauna burrowing from the new surface downward (Sepkoski et al. Chap. 2.6, this Vol.). As in cyclic sequences, there may be intervals of anoxic conditions. Whereas turbidites always contain material from distant sources, often mixed with incorporated basinal sediment, the material of tempestites is either entirely autochthonous or derived from nearby sources (Einsele and Seilacher Chap. 2.10, this Vol.).

Oxygen-deficient environments such as black shales, phosphatic sediments, and varved siliceous ooze may show gradual periodic variations and episodic bedding types (see Wetzel Chap. 5.1, Föllmi et al. Chap. 4.3, Decker Chap. 4.1, Glenn and Kelts Chap. 1.9, all this Vol.). This is related to the lateral influx of allochthonous sediment or storm-related winnowing of clays to produce bioclast-rich lags, as well as gradual periodic changes (Fig. 5b). Longer-term rhythms or cycles are often superimposed by shorter-term fluctuations, generating thin varve-scale (bio)laminations, or contain benthic and bioturbated intervals due to water mass mixing and turnover (see Savrda et al. Chap. 5.2, Sageman et al. Chap. 5.3, Oschmann Chap. 5.4, as well as Gerdes et al. Chap. 5.6, all this Vol.).

The recurrence time of the various types of quasi-periodic and non-periodic processes as well as events spans a wide scale (Fig. 6). Between the recurrence time of tidal cycles and the large-scale tectonic and eustatic processes lie 11 orders of magnitude. The quasi-periodic variations related to tidal, rotational, and orbital parameters have short repetitions and occur on a relatively small time interval ranging

Fig. 5a-d. Vertical sediment buildup-time curves (BT curves) to characterize different types of rhythmic bedding. **a** Cyclic or periodic bedding due to slow gradual changes in accumulation and composition of sediments. **b** and **c** Episodic bedding caused by repeated erosional and depositional events of different magnitudes at random time intervals. **d** Cyclic bedding (interval *A*) and episodic bedding (interval *B*) in oxygen-deficient environments, superimposed by varve-scale laminations. **a**, **b**, and **d** may also show anoxic or oxygenated intervals. B_{upw} zone of bioturbation migrating slowly upward; B_{dw} new community of bottom-dwelling organisms burrowing downward

Einsele et al: Basic Concepts

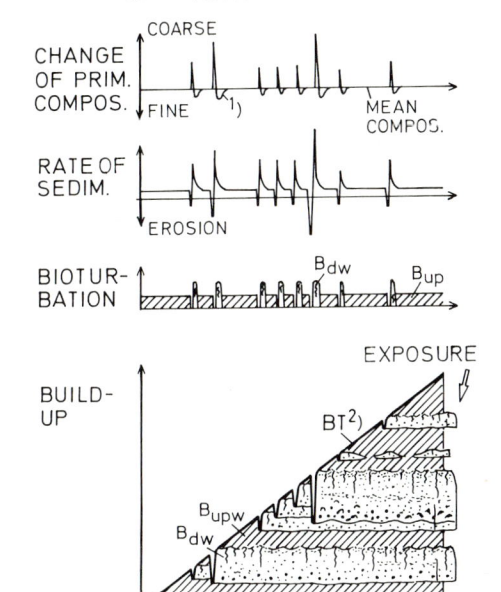

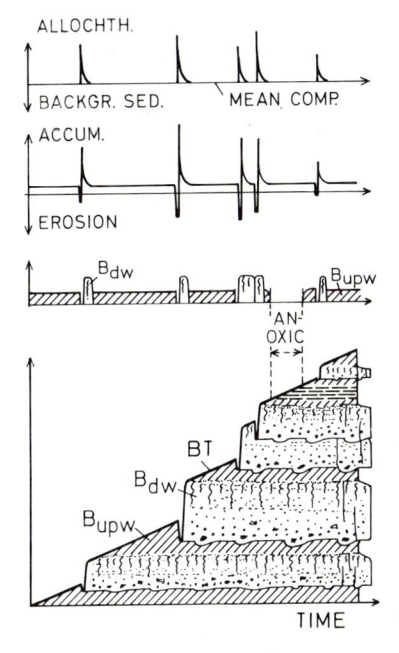

from hours to more than a million years (see Fig. 8 in Glenn and Kelts Chap. 1.9, this Vol.). Geologic events such as tempestites, turbidites, and volcanic eruptions generally have the same range of recurrence as the periodic variations. Thus, these events tend to disturb the periodic stratification. Both quasi-periodic signals and interrupting events take place during the background changes of the large-scale repetitions of global eustatic and tectonic processes. However, the shorter periods of these larger-scale processes overlap with the Milankovitch frequencies. This is observed for magnetic reversals and high-frequency eustatic variations.

Observations on major river floods and tsunamis show repetitions between 10 and 150 years, but the recurrence time of tempestites and turbidites in the rock record is one to two orders of magnitude longer. Measured in terms of human life span or the written historical record, these floods and tsunamis are large and rare, but not gigantic enough to be transmitted to the rock record (Dott 1988; Clifton 1988). The commonly found reappearance of events in turbidite and tempestite sequences is

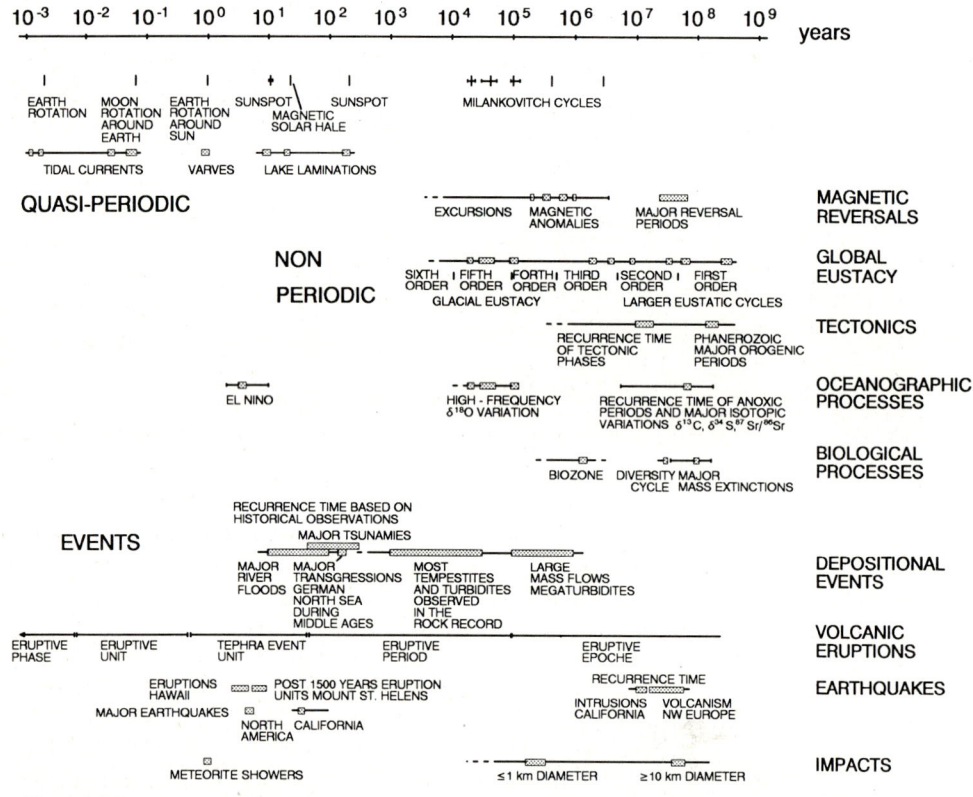

Fig. 6. Major recurrence time of quasi-periodic and nonperiodic processes as well as events. For explanation see text. Data is compiled according to Fischer Chap. 1.2, Glenn and Kelts Chap. 1.9, Schmincke and Boogard Chap. 2.12, as well as Vail et al. Chap. 6.1, (all this Vol.); Bolt (1978); Dott (1988); Fisher and Schmincke (1984); Haq et al. (1987); Holser (1984); Jacobs (1984); Reineck (1978); Sheridan (1987); Shoemaker (1984); van Andel (1985); as well as Ziegler (1982)

around or slightly below the Milankovitch frequencies. This means that orbitally controlled marl-limestone rhythms can be intercalated and modified by event beds. On the other hand, the dominant reappearance of events on the bed scale is within a larger time span than the smaller periodic signals from tidal processes, Earth rotation and sunspot activity (Fig. 6). Thus, successions with tidal current stratification, varves, and other laminations (e.g., in black shales, lithographic limestones, and deeper water evaporites) are usually little affected by tempestite and turbidite beds.

6 Biological Response

Biological processes require time. Therefore, reactions to gradual, cyclic fluctuations versus short episodic events are fundamentally different. Long term environmental changes on the order of Milankovitch cycles will be accommodated by shifts in the faunal and floral spectrum and species dominance. Physical events, in contrast, will be experienced as catastrophes, with some of the victims becoming smothered in tempestitic or turbiditic fossil lagerstätten (Brett and Seilacher Chap. 2.5, this Vol.).

A true response can begin only after the depositional event. Reestablishment of the original bottom community, for instance, implies the formation of a new tiered zone of burrows at the event bed surface, in which at least the deepest level is not obscured by subsequent tier ascension. In the case of turbidites, the post-event community also differs from the background fauna, because the import of detrital food changes the local trophic conditions for a short period (see Sepkoski et al. Chap. 2.6, as well as Einsele and Seilacher Chap. 2.10, this Vol.).

Similarly, post-event conditions may be different in storm-dominated settings, if an originally muddy substrate was replaced by a shell pavement of infaunal bivalves through winnowing. In this case the post-event community will be dominated by epifaunal encrusters that could not settle on the pre-tempestite mud bottoms. This taphonomic feedback (Kidwell Chap. 2.4, this Vol.) is not automatic, however, but depends on the overall state of the depositional system. In cases where long-term sedimentation and subsidence are in balance, the storm-suspended sediment will be redeposited in place, but as a graded bed. This means that the post-event bottom will be as muddy or even muddier than it was before. Only in a regressive situation will the muddy fraction be exported into deeper parts of the basin and leave the shell layer exposed for epifaunal settlement.

Another kind of taphomomic feedback may take place in a transgressive situation. Here the decisive factor is not a change in substrate consistency, but the general reduction of coarser-grained terrigenous sediment influx. Reduced rates of terrigenous sediment supply allow the establishment of sessile epifaunal organisms (such as some species of corals or oysters), who achieve stability through their body size. Since reclining or mud-sticking strategies favor excessively large, massive skeletons, they provide outsized bioclasts in a situation in which terrigenous sedimentation is reduced and very fine-grained. Measured in terms of life spans, recliners and mud stickers require not only low sedimentation rates, but also quiet water conditions. They will therefore be restricted to depth zones below the influence of normal waves, i.e., to regions near or below storm wave base. The rare storms which still reach them (at intervals of many years) will be too weak to transport the large shells, but may make them sink stratigraphically to the erosional base by winnowing away the muddy substrates, and then blanket them with the redeposited mud. Over significant periods of time, the combination of long-term production of outsized bioclasts (and of concretional "diaclasts" formed by early diagenetic processes) with episodic winnowing and burial, may eventually generate thick shell beds or mounds, even while overall fine-grained substrate conditions persisted (see Brett and Seilacher Chap. 2.5, this Vol.).

Epibenthic shell beds of this kind occur throughout the Phanerozoic record. They have been conveniently used as marker beds, because they tend to persist over considerable distances and because they commonly coincide with biozonal boundaries. Due to their coarser grain size and associated erosional features, such marker beds have traditionally been related to regressive phases. This picture changes if the coarse particles are not transported, but are produced in place by biological processes and enhanced by diagenesis, and if the sedimentary structures reflect short turbulence events rather than the background regime. Placing boundary shell beds into the ongoing transgression – rather than in the regressive peak – also solves another paradox, because now the faunal changes coincide with the periods in which the gates of the epicontinental seas opened and facilitated the immigration of new ammonites and other guilds (McGhee et al. Chap. 6.4, as well as Kauffman et al. Chap. 7.2, this Vol.).

Since the publication of the precursor of this book, perspectives have also changed with regard to black shale facies. In this realm obvious storm layers are rare or virtually absent, and "benthic events" provide the only accentuation. Bioturbation horizons and monospecific shelly laminae were appropriately assigned to short-term oxygenation events. This view may still be adequate in many cases, but the discovery of chemosymbiosis modifies the picture in an interesting manner. If the organisms in these horizons are not normal benthos, but required very specific redox and substrate conditions, their ecologic "window" may have opened also along a continuous redox gradient. In other words, the "events" may reflect a discontinuity in the ecologic spectrum rather than in depositional conditions (Savrda et al. Chap. 5.2, Sageman et al. Chap. 5.3, both this Vol.).

In summary, the biological responses of benthic communities are not an epiphenomenon, but an integral part of dynamic stratigraphy. By providing a gauge of time, the life spans of organisms allow us to estimate the time frame of sedimentary processes and the amount of fine sediment lost by winnowing. But in order to avoid misinterpretation, the paleontological data should be scrutinized with all the rigor of paleontological and taphonomic analysis.

7 Diagenetic Overprint

7.1 Basic Processes

Primary bedding features, including sedimentary structures, bedding planes and bedding rhythms, can be significantly modified by diagenetic overprints, particularly in carbonates and siliceous sediments (see Part I.4 and 5, this Vol.). Further alterations are caused when sections are exposed to weathering agents. Diagenesis includes various mechanical and chemical-mineralogical processes during burial history. Under certain conditions, even minor variations in primary composition and pore space are sufficient to cause significant diagenetic modifications. As a result, primary structures and bedding rhythms may be enhanced, modified, or in some cases even obliterated (Fig. 7).

The most important diagenetic processes affecting the original bedding rhythms include: mechanical compaction, cementation by precipitating additional cement in the available pore space, replacement of less stable components or sediment parts by new substances (e.g., chert nodules), displacement of soft sediment by internally growing crystals (e.g., pyrite and evaporites), dolomitization, dissolution, and pressure dissolution. Most of these processes are highly selective in respect to individual components and beds; thus one specific process operates at certain sites, while other processes occur elsewhere in the sediment. This has the following consequences regarding composition, compaction, and weathering:

Accentuation of primary compositional variations. In bedded sequences, selective cementation and pressure dissolution are the most important processes enhancing compositional variations. In calcareous rocks, they are predictably related to three major factors: (1) The amount of cement, which is practically equivalent to the volume of pore space available at the time of cementation (e.g., Lippmann 1955;

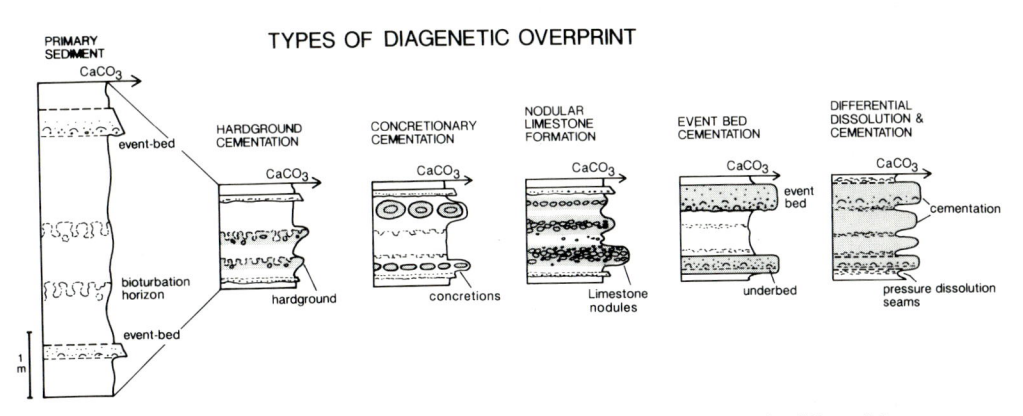

Fig. 7. Commonly found types of diagenetic overprints in calcareous rocks. Note the differential preservation of certain parts of the section, forming various types of bedding rhythms compared with the primary sediment (which may be chalk or marl, *left*)

Seibold 1962; Bathurst 1975). This pore space in turn is controlled by the primary porosity of the sediment and the degree of mechanical compaction (Ricken 1986). (2) The intensity of carbonate dissolution, which can be substantial when the carbonate system is open, but limited when it is closed. (3) The carbonate content and the mineralogical composition of the sediment, which affect the processes of mechanical compaction, dissolution, and cementation.

Differential compaction. This process is well known from marked differences in compaction between sandy and clayey sediments or rigid reef structures and surrounding fine-grained sediments. But differential compaction also results from carbonate or silicate mass redistribution, which is always associated with changes in porosity and composition. At the site of cementation, the pore space is filled with solid cement, thus inhibiting further compaction, while the surrounding, noncemented or dissolution-affected sediment is subjected to further compaction. Zones of particularly extensive dissolution taking place along seams, fitted fabrics, and stylolites lead to higher compaction than in mechanically compacted sediment (Bathurst Chap. 3.2, Tada Chap. 4.2, both this Vol.). As a consequence, beds, interbeds, and primary sedimentary structures become disproportionately preserved in the rock record (Fig. 7). Cemented parts essentially retain their original thickness, while other parts undergoing mechanical compaction or dissolution are relatively reduced in thickness (Ricken 1986). Examples for this type of differential compaction are drapings around concretions and thickness reduction of marly interbeds in marl-limestone alternations. It is obvious that differential compaction and the associated dissolution and cementation processes change the original bedding rhythms.

Modification by weathering. In weathered field exposures of calcareous rocks, carbonate content and weathering conditions determine the apparent proportions of marl and limestone beds (Seibold 1952). According to the concept of the weathering boundary (Einsele 1982), there is a certain carbonate content, e.g., 65 to 85%, above which the carbonate rock is resistant and appears as limestone; below this boundary the rock is more easily disintegrated and appears as marl. Carbonate oscillations in unweathered rocks above or below the weathering boundary are difficult to resolve in field sections, because they appear either as pure limestones or as marls (see Fig. 6 in Einsele and Ricken Chap. 1.1, this Vol.).

7.2 Types of Diagenetic Overprint

Hardgrounds. They are formed by early cementation at the sediment surface in various calcareous environments in areas of erosion and winnowing, i.e., on carbonate platforms, shelves, and continental slopes. When cemented hardgrounds occur repeatedly in a noncemented sequence, the stratification of such a sequence of hard beds is dominated by erosional surfaces (Fig. 7).

Concretion layers. Concretions of carbonate, silica, and phosphate generally form early in diagenesis near the sea floor, but they also grow at some depth below the sediment-water interface (Raiswell 1987). Concretions are frequently concentrated in certain layers which deviate from the normal sediment, e.g., by high organic carbon contents, high porosities, and low sedimentation rates (see Föllmi et al. Chap. 4.3, this Vol.). Such beds and thicker concretion zones are found to be parallel or sub-parallel to the primary bedding (Fig. 7, see Kaufmann et al. Chap. 7.2, this Vol.). In pure chalks, layers of concretionary flints may form an essentially diagenetic stratification, with a different number of beds compared to the primary bedding rhythm (Ricken and Eder Chap. 3.1, this Vol.). Nodular limestones are generated by early concretionary cementation of bioturbated carbonate muds and by pressure dissolution during the later stages of diagenesis.

Event bed cementation. Cementation of calcareous event beds is commonly found to pre-date that of the marly to shaly background sediment. This is because calcareous event beds usually contain reactive shallow water carbonates. Cementation of graded beds can also affect a thin layer below the event bed, thus forming an "underbed" (Eder 1971); occasionally an "overbed" is also observed. Both processes, the early cementation of proper event beds and the formation of underbeds, enhance the relative thickness and carbonate content of the resulting compound beds, while the non-event sediment is subjected to further compaction. As a consequence, the stratification of event beds is substantially intensified (Fig. 7).

Differential dissolution and cementation in layered sequences. Differential dissolution and cementation are amongst the most important diagenetic processes in bedded calcareous rocks (see Bathurst Chap. 3.2, Ricken and Eder Chap. 3.1, both this Vol.). It is thought that this phenomenon also occurs in siliceous rocks, where it contributes to final cementation (see Tada Chap. 4.2, this Vol.). Primarily at great burial depths, carbonate is dissolved in the interbeds and completely or partially reprecipitated as cement in the neighboring layers. As a result, original variations in composition are amplified, and thus the original bedding rhythm is enhanced due to differential compaction associated with the redistribution process. At high carbonate contents, bedding eventually becomes brick-like, consisting of limestone layers of a rather uniform thickness, alternating with thin, highly compacted and carbonate-depleted marl beds (Fig. 7). In skeletal carbonates, differential dissolution and cementation can be combined with an early phase of cementation, affecting bedded carbonates selectively (see Bathurst Chap. 3.2, this Vol.).

8 Use of Cycles and Events in Stratigraphic Correlation

In the interpretation of sedimentary sequences, the concepts of cyclic and episodic bedding are not exclusive; on the contrary, both phenomena may occur in the same section. The response of sediments to a regularly recurring cyclic signal is frequently superimposed by various event types. Such sequences are characterized by frequent changes in lithology, geochemical variations, and isotope shifts, and various successions of biological events (Kauffman 1988a; Kauffman et al. Chap. 7.2, this Vol.). Under appropriate conditions, all these phenomena can be traced over large distances within an epeiric basin or shelf basin and sometimes even between several of these basins.

The occurrence of both cyclic and episodic phenomena in a certain region can be used to refine stratigraphic correlation (Fig. 8). Biostratigraphic boundaries or datable volcanic ash layers provide the chronostratigraphic framework. Between these boundaries, short-period rhythms or cycles and the succession of various event beds permit the subdivision of large stratigraphic zones into relatively short subunits, thus providing a detailed succession of relative time lines within a basin fill. This method is referred to as "high resolution cyclic and event stratigraphy" (Kauffman et al. Chap. 7.2, this Vol.). In the pelagic realm, where the Milankovitch-type variations are much better preserved than in shallow seas, there is hope to develop a detailed cyclostratigraphy (Fischer et al. 1990).

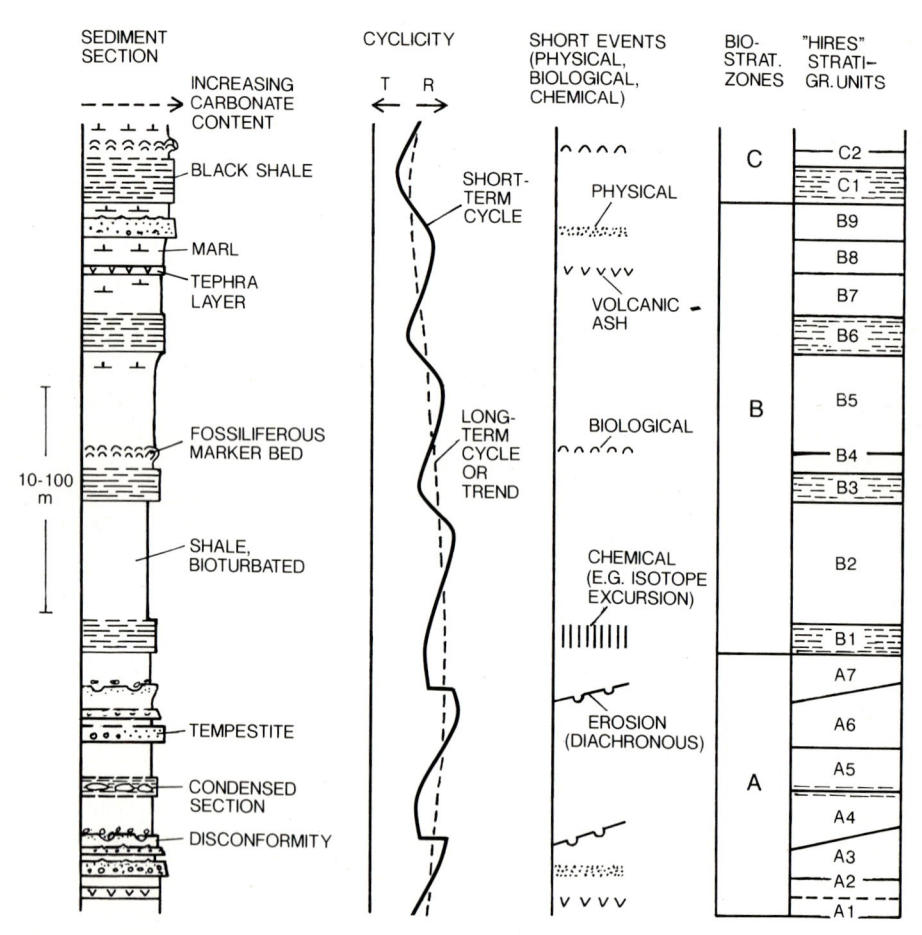

Fig. 8. Principle of high-resolution cyclic and event stratigraphy ("HIRES") within a chronostratigraphic framework provided by index fossils or radiometric age determination (e.g., tephra layers). (Based on Kauffman 1988a)

9 Conclusions

The most basic terms and processes are introduced which are addressed in more detail in the various parts of *Cycles and Events in Stratigraphy*. This chapter deals with definitions of beds, interbeds, rhythms, and the recurrence time of successions of cyclic or event beds in various environments, and the nomenclature of larger depositional cycles. The role of autocyclic and allocyclic processes in forming such sequences is discussed. Modifications of individual beds through biological responses or feedbacks as well as diagenetic overprints and their effects on bedding rhythms are briefly presented. Within a stratigraphic framework, short-term depositional cycles and event beds allow a considerable refinement of stratigraphy on a regional to possibly global scale.

Acknowledgements. The authors would like to acknowledge the fruitful discussions with the many friends and participants of the workshop on this rather difficult subject. We would especially like to thank our colleagues K. Föllmi, Santa Cruz; L. Hobert, Albany; B. Sageman, Boulder; and A. Strasser, Fribourg for reviewing this paper and providing their many helpful suggestions.

Part I
Structure of Individual Beds

Chapter 1 Rhythmic Stratification:
the "Undistorted" Record of Periodic Environmental Fluctuations

Chapter 2 Event Stratification:
Records of Episodic Turbulence

Chapter 3 The Diagenetic Overprint: Enhancement Versus Obliteration of Primary Signals in Calcareous Rocks

Chapter 4 Cherts and Phosphorites:
Primary and Diagenetic Bedding in Special Environments

Chapter 5 Preservation and Biological Destruction of Laminated Sediments

Chapter 1 Rhythmic Stratification

1.1 Limestone-Marl Alternation – an Overview

G. Einsele and W. Ricken

1 Introduction

Rhythmic carbonate sediments are a common and striking feature in the sedimentary record. They occur in a number of depositional settings, the most important of which are: pelagic to hemipelagic marine environments from the outer shelf to deep water above the CCD, shallow marine carbonate platforms and lagoons, and lacustrine basins. This chapter discusses only the first group. Since the pioneering work of Gilbert (1895), Milankovitch (1941), Emiliani (1955), and Hays et al. (1976), a great amount of new information has become available from investigations of both young and old oceanic sediments, especially through new coring and laboratory techniques applied to deep sea sediments. It seems appropriate to apply this knowledge to ancient rocks and develop some basic models for the generation of limestone-marl couplets and larger rhythmic sequences. Earlier summaries of this topic appear in various books (Duff et al. 1967; Schwarzacher 1975; Imbrie and Imbrie 1979; Einsele and Seilacher 1982; Berger et al. 1984) and in several review articles (e.g., Einsele 1982; Arthur et al. 1984b; Fischer 1986; Fischer et al. 1989, 1990).

2 Rhythmic Sedimentation: Orbital Forcing and Depositional Response

2.1 Orbital Parameters and Periodicities: Milankovitch Theory

Along with annual varves and sunspot cycles (see Kelts and Glenn Chap. 1.9, this Vol.), orbital variations of the Earth, i.e., Milankovitch cycles, are the most common type of astronomical forcing represented in sedimentary rocks (Fischer 1986 and Chap. 1.2, this Vol.). Milankovitch cycles have quasi-periodicities between 20 to 400 ka, resulting in seasonal effects in insolation (Milankovitch 1941; Imbrie and Imbrie 1979; Berger et al. 1984) that are amplified by various climatic-oceanic feedback systems and are then transformed into the sedimentary record (Fig. 1). Three major orbital cycles are observed:

1. Variations in eccentricity. The weakly elliptic orbit of the Earth is influenced by gravitational interactions with the other planets of the solar system, causing slight variations in eccentricity as the Earth in its orbit itself rotates around the sun. The eccentricity fluctuates with average quasi-periodicities of 100 and 410 ka, with major periods at 95, 100, 120, and 410 ka (Fig. 2; Berger and Tricot 1986).

Einsele et al. (Eds.)
Cycles and Events in Stratigraphy
©Springer-Verlag Berlin Heidelberg 1991

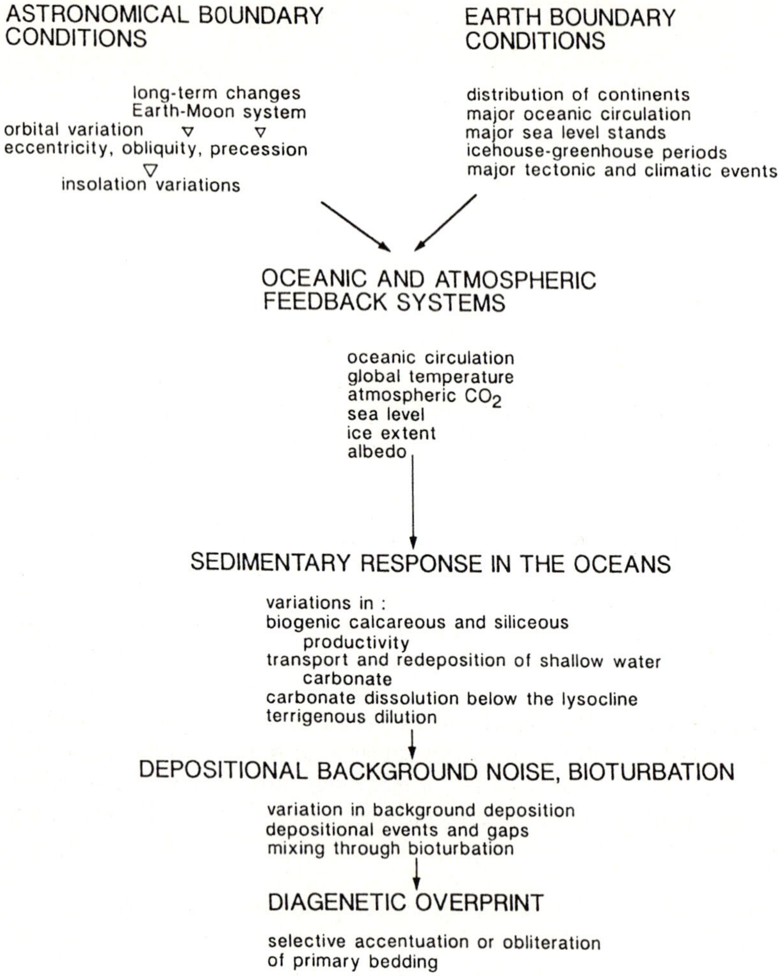

```
ASTRONOMICAL BOUNDARY                    EARTH BOUNDARY
CONDITIONS                               CONDITIONS

         long-term changes               distribution of continents
         Earth-Moon system               major oceanic circulation
orbital variation   ▽    ▽               major sea level stands
eccentricity, obliquity, precession      icehouse-greenhouse periods
         ▽                               major tectonic and climatic events
   insolation variations
                        ╲       ╱
                   OCEANIC AND ATMOSPHERIC
                   FEEDBACK SYSTEMS

                       oceanic circulation
                       global temperature
                       atmospheric CO₂
                       sea level
                       ice extent
                       albedo
                          ↓
              SEDIMENTARY RESPONSE IN THE OCEANS

                   variations in :
                   biogenic calcareous and siliceous
                       productivity
                   transport and redeposition of shallow water
                       carbonate
                   carbonate dissolution below the lysocline
                   terrigenous dilution
                          ↓
            DEPOSITIONAL BACKGROUND NOISE, BIOTURBATION

                   variation in background deposition
                   depositional events and gaps
                   mixing through bioturbation
                          ↓
                   DIAGENETIC OVERPRINT

                   selective accentuation or obliteration
                   of primary bedding
```

Fig. 1. Basic processes controlling rhythmic sedimentation related to orbital variations, Earth boundary conditions, oceanic-atmospheric feedbacks and responses, depositional noise, and diagenetic overprinting

2. Variations in obliquity or tilt: The angle of tilt between the Earth's axis and the ecliptic plane undergoes small variations between 22° and 24°30'; today the angle of tilt is 23°27'. This variation has a major quasi-periodicity of 41 ka, with minor components at 29 and 54 ka (Imbrie and Imbrie 1980).
3. Variations in precession. The Earth describes a precessional movement as a result of the gravitational pull of the sun on the tilted equatorial bulge. The average quasi-periodicity is 21.7 ka, with major periods of 19 and 23 ka, and with extreme variations between 14 and 28 ka (Berger and Tricot 1986). The climatically important "precession index", $e \sin \omega$, represents the interrelationship between precession and eccentricity (e), describing the precession of the equinox. The

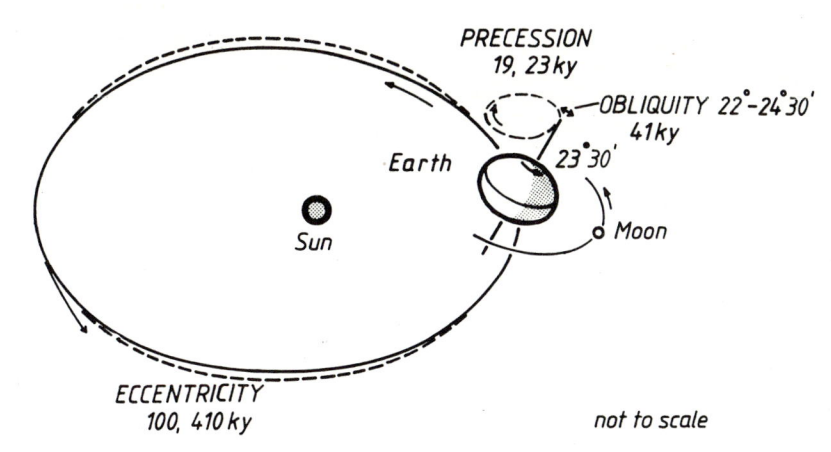

Fig. 2. Orbital parameters and their major frequencies

precession index varies with the period of precession and the amplitude of eccentricity (Fig. 3).

There is some evidence that orbital periodicities have slowed down over the course of the Earth's history. In the Paleozoic, shorter periodicities for the obliquity and precession are assumed, related to a smaller distance between the Earth and Moon and a higher rotation velocity of the Earth (Denis 1986). For the Lower Silurian (440 Ma B.P.), Berger et al. (1987) estimated that the mean periodicities for tilt and precession were only 60 and 80% of the present values, respectively. Thus, Milankovitch rhythms in the Paleozoic should contain more beds per equivalent thickness unit than their younger counterparts.

2.2 Amplification of Orbital Variations by Atmospheric and Oceanic Feedback Systems

When considering an entire year and integrating over all latitudes, tilt and precession do not result in any essential changes in received insolation (Imbrie and Imbrie 1980; Berger 1987). However, the composite effect of the three orbital parameters generates a fluctuating seasonality, with a net effect of generally ±10% or less (e.g., Ruddiman and McIntyre 1984; Berger and Tricot 1986; Fischer et al. 1989). Obliquity variations affect mainly the climate at the poles and intensify the contrast between the seasons. Times with higher obliquity are associated with a warmer polar summer and a colder polar winter. Fluctuations in the precession index are expressed by reversed seasonality for the two hemispheres of the Earth, and by orbital positions defined by the perihelion during winter or during summer (Imbrie and Imbrie 1979; see Fig. 2). Near the equator, received radiation has the periodicity of the eccentricity and precession, while radiation in polar regions is determined more by obliquity (Berger 1978b; Ruddiman and McIntyre 1984). The coincidence of ice buildup and decay in both hemispheres indicates a global energy redistribution in the late Pleistocene (Broecker 1984). It is not known whether this hemisphere coupling observed in the late Pleistocene also occurred during ice-free periods of the Earth.

Since Milankovitch cycles have only a small seasonal effect, they must be enhanced by climatic-oceanic feedback systems in order to produce a sedimentary signal (Kerr 1981; Ruddiman and McIntyre 1984; Fischer et al. 1989). The most important feedback systems amplifying orbital variations include the magnitude of continental ice buildup (Berger et al. 1984), the content of atmospheric CO_2 (Sundquist and Broecker 1985), shifting of monsoon climates and wet and dry belts (Barron et al. 1985; Prell and Kutzbach 1987), and changing of the oceanic circulation and of upwelling intensities (Sarnthein et al. 1988). These feedbacks and nonlinear threshold processes are modified by changing boundary conditions throughout the Earth's history, such as different land-sea distributions, eustatic sea level fluctuations, overall climatic-oceanic regimes, etc. (see Fig. 1). For example, continental ice buildup, with its albedo-temperature and moisture-ice mass feedback systems, amplifies the 100 ka eccentricity signal in the late Pleistocene (Fig. 4; Start and Prell 1984; Berger and Pestiaux 1985; Ruddiman et al. 1987).

All these atmospheric-oceanographic feedback systems enhance the weak orbital-climatic variations and translate the orbital signals into marl-chalk rhythms.

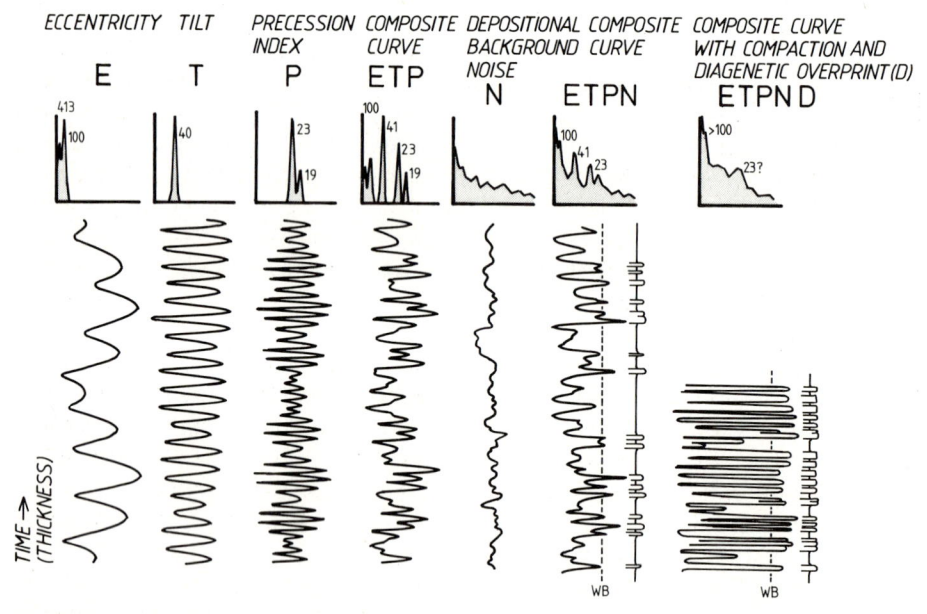

Fig. 3. Schematic diagram for the generation of rythmic bedding by superposition of various parameters, including orbital frequencies, depositional background noise, and diagenesis. For the ETPN and ETPND curves, simplified weathered sections are shown, assuming that only signals above the carbonate content of the weathering boundary (*WB*) are translated into limestone layers. Orbital curves E, T, and P with their power spectra adapted from Imbrie et al. 1984

Superimposed on these signals is depositional background noise; both signals are later subjected to differential compaction and diagenetic overprint (Fig. 1 and 3).

2.2.1 The Role of Atmospheric CO_2 and the Carbon Cycle

The formation of rhythmic marl-limestone alternations is closely associated with three feedback processes inherent in the global carbon cycle: (1) Amplification of orbital signals through changes in atmospheric CO_2 creating the greenhouse or icehouse effect; (2) biogenic carbonate production as related to climate-driven oceanic circulation and the recycling of nutrients; and (3) carbonate dissolution in the deep sea as a consequence of the CO_2 content of the bottom waters and undersaturation with respect to calcite. The complicated interactions among these processes are now becoming increasingly understood for the Pleistocene climatic cycles, but there is no doubt that they were also important in ice-free periods.

Measurements of CO_2 in polar ice cores have shown the existence of a fundamental link between atmospheric CO_2 and global temperature for the last climatic cycle (Delmas et al. 1980; Neftel et al. 1982; Oeschger et al. 1984; Barnola et al. 1987). In the Vostok ice core from Antarctica, the entrapped atmospheric CO_2 content varied closely with the overall marine $\delta^{18}O$ record (Imbrie et al. 1984), indicating that CO_2 essentially changed simultaneously with global ice volume and temperature (Fig. 5; Barnola et al. 1987). During the climatic minimum of the last glacial, atmospheric CO_2 was approximately 100 ppmv lower than in the succeeding, pre-industrial interglacial. A 21 ka periodicity of the CO_2 curve suggests that the carbon dioxide content of the atmosphere is associated with oceanic processes acting at low latitudes.

The role of atmospheric CO_2 in amplifying orbital forcing has been widely discussed (Shackleton and Pisias 1985; Genthon et al. 1987). There is growing evidence that the small atmospheric CO_2 reservoir is essentially controlled by the CO_2 content of oceanic surface water, and to the storage or release of CO_2 which is thought to influence carbonate preservation in the deep sea (Broecker and Peng 1984; Keir and Berger 1985). As shown by Peterson and Prell (1985b) and Farell and Prell (1987), the position of the lysocline oscillated during the late Quaternary in the equatorial Pacific by 500 m and in the central Indian Ocean by 1000 m, due to a complex interaction between carbonate accumulation and dissolution (Grötsch et al. Chap. 1.6, this Vol.). Several models for influencing the CO_2 reservoir in oceanic surface water have been suggested, including the storage of nutrients or carbonate carbon on the shelf (Broecker 1982; Broecker and Peng 1984; Shackleton et al. 1983; Berger and Keir 1984), changes in upwelling intensities (e.g., Siegenthaler and Wenk 1984; Vincent and Berger 1985; Sarnthein et al. 1987, 1988), and storage or release of carbon in the terrestrial biosphere (Shackleton 1977).

3 Limits to the Recognition of Limestone-Marl Rhythms in Field Exposures

In natural outcrops or weathered quarries, an alternation consisting of the two principal phases of "clay" and fine-grained carbonate may, under certain conditions,

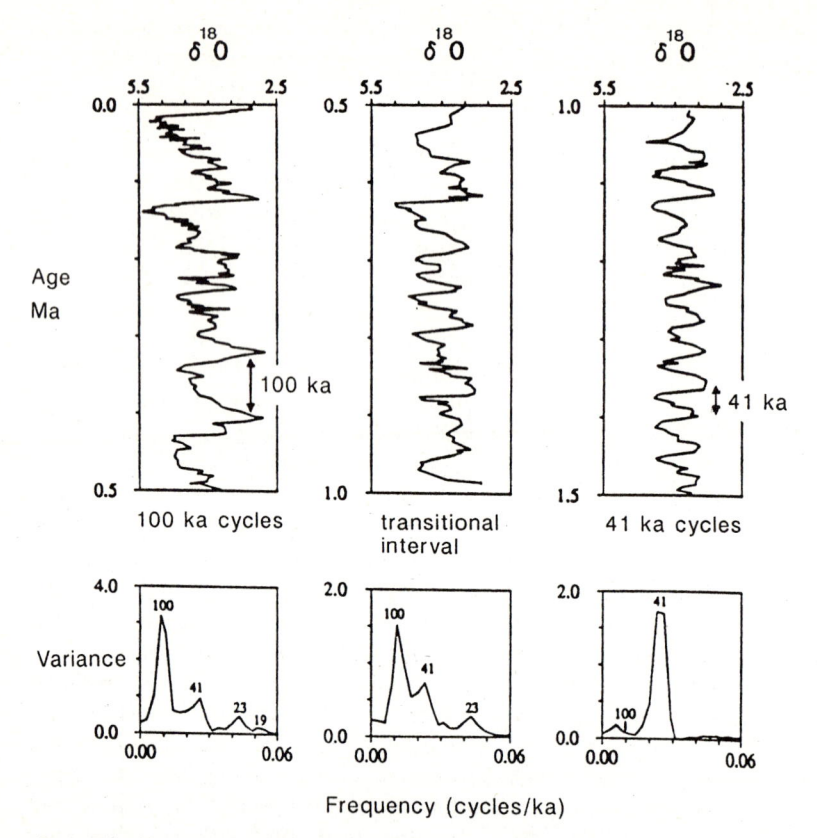

Fig. 4. Transition from the 41 ka obliquity signal to the 100 ka eccentricity signal in the middle Pleistocene documented by the $\delta^{18}O$ record for the North Atlantic, DSDP Site 607, for the last 1.5 Ma. (Ruddiman et al. 1987)

not show any distinct lithologic change or bedding features. In weathered lithified rocks, an observable transition from marlstones to limestones usually takes place when the carbonate content surpasses a certain threshold value (Seibold 1952). Above or below this limit, the material weathers either into layers of hard limestone or beds of less resistant marl or marly shale. A rhythmic sequence can be recognized in the field only if the alternating beds have a carbonate content fluctuating around this weathering boundary (Fig. 6). Oscillations in the carbonate content below or above this limit do not significantly change the field appearance of marlstone or limestone, respectively.

The carbonate content defining the weathering boundary depends on climate, exposure, and duration of the weathering process. It is further related to porosity and the type and grain size of the carbonate fraction. Generally, the carbonate content at the weathering boundary is somewhat lower for carbonate-poor alternations. For many typical marl-limestone alternations representing various climatic zones in Europe and North America, values for the weathering boundary between 65 and 85%

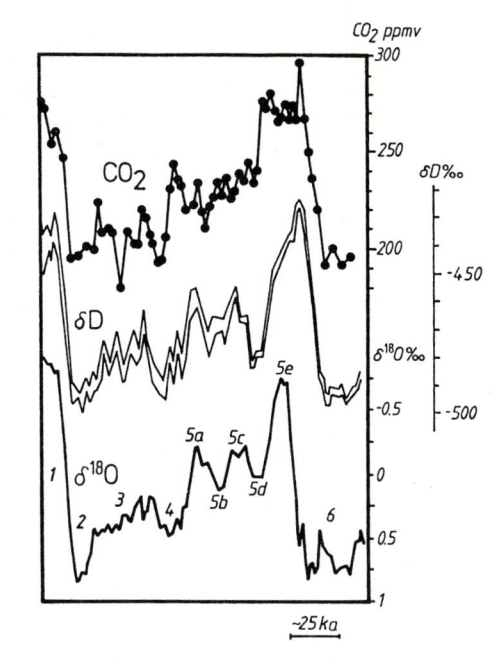

Fig. 5. Record of global atmsopheric CO_2 and δD for the last climatic cycle in the Vostock ice core, Antarctica, compared to the marine $δ^{18}O$ record. The δD curve is related to the local Antarctic temperature; the stacked $δ^{18}O$ curve reflects global ice volume. (After Jouzel et al. 1987 and Barnola et al. 1987)

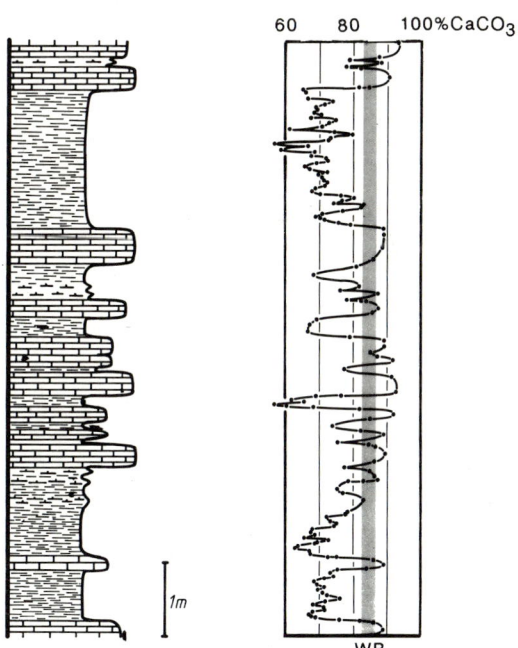

Fig. 6. The influence of weathering on the outcrop appearance of bedding rhythms. Comparison of a weathered limestone-marl sequence (*left*) with measured variations in carbonate content (*right*). Smaller carbonate variations below the carbonate content of the weathering boundary (*WB* around 85% $CaCO_3$) cannot be seen to form distinctive limestone and marl beds in the outcrop. Quarry face is approximately 20 years old. Upper Jurassic, southern Germany. (After Ricken 1986)

$CaCO_3$ are encountered (Seibold 1952; Cotillon et al. 1980; Arthur et al. 1984b; Pratt et al. 1985; Ricken 1986; Weedon 1986; Herbert and Fischer 1986; Erba 1988). Consequently, the mean carbonate accumulation rate must be considerably higher than the noncarbonate rate in order to obtain limestone-marl sequences in field exposures. Hence, this type of rhythm is strongly related to the rate of carbonate production, the ratio of carbonate and clay in marine environments, and to early and late diagenetic processes influencing the carbonate content. A threshold for the preservation of rhythmic marl-limestone alternations is the thickness of the surface layer completely mixed by bioturbation, which in deep-sea sediments is about 5 to 10 cm thick (Ekdale et al. 1984b; Broecker and Peng 1984; Dalfes et al. 1984). Beds thinner than this critical thickness can easily be destroyed by bioturbation.

4 Mechanisms Forming Pelagic Limestone-Marl Rhythms

In the following, the major processes forming marl-limestone alternations and black shale-carbonate alternations are described. These include variations in carbonate productivity, terrigenous dilution, carbonate dissolution, and the redox conditions of the bottom waters; later, the various types of rhythms will be discussed in terms of diagenetic overprinting. For each of these processes, simplified models are introduced which show the involved depositional and diagenetic variations, and the resulting bedding rhythms. The basic processes described herein must be regarded as end members of the frequently encountered composite types of alternations, where two or three parameters oscillate simultaneously. Methods for distinguishing such composite cycles are presented by Ricken (Chap. 1.8, this Vol.) by using characteristic carbonate-organic carbon relationships. Other types of cycles, such as color and scour cycles, are mentioned only briefly here, as they are already comprehensively treated by Bottjer et al. (1986) and Dean and Gardner (1986). It is assumed that the origin of scour rhythms by Milankovitch cycles is difficult to prove, because of the high number of stratigraphic gaps associated with this kind of repetition.

4.1 Periodic Fluctuations of Pelagic Carbonate Supply (Productivity Cycles)

Variation in surface water carbonate productivity leading to the formation of marl-limestone alternations is widely discussed for the Quaternary (Prell and Hays 1976; Pisias 1976; Adelseck and Anderson 1978) and pre-Quaternary (Cotillon 1985; Eicher and Diner 1985; Bottjer et al. 1986; Herbert and Fischer 1986; Pratt and King 1986; Tornaghi et al. 1989). Productivity changes are thought to be generally important for carbonate cycles with an entirely pelagic carbonate fraction, and which do not show any signs of varying carbonate dissolution and terrigenous dilution. However, such productivity changes are not easily demonstrated from the diversity of calcareous microfossils. The occurrence of calcispheres in the carbonate-rich bed and the covariance of carbonate and siliceous ooze accumulation point to productivity rhythms (see Eicher and Diner, as well as Decker Chap. 4.1, this Vol.).

A pure productivity cycle is characterized by a fluctuating supply of pelagic carbonate during a steady contribution of clay. A bedded sequence resulting from this mechanism is shown in Fig. 7a, where a homogeneous marly sediment with an assumed initial carbonate content of 60% is transformed into a sequence of alternating marl and limestone beds. In this example, the periodically increased carbonate production must be more than two times greater than the original production, in order to obtain a weathering-resistant limestone layer containing at least 75% carbonate, as mentioned above.

With periodically increasing carbonate production, the limestone beds become thicker stepwise, and alternations are more and more dominated by the carbonate-rich bed (Fig. 7a). For a production factor of 5, the ratio comparing bed thicknesses of limestones and marls is 3.4, and for a production factor of 10, it is 6.4. It can be seen from this simple test that an already high background carbonate production (in our example 60% of the total sediment) must be multiplied periodically by a factor of 10 to produce the common type of limestone-marl sequences with relatively thick limestone beds and thin marly interbeds. Hence, this type of cycle must be associated with large variations in sedimentation rates between beds and interbeds (see Ricken Chap. 1.8, this Vol.). However, smaller production factors are required when primary carbonate variations are enhanced by diagenetic carbonate redistribution, which leads to differential compaction between beds.

It remains doubtful whether the assumed fluctuations in carbonate production alone are sufficient to produce the common type of rhythmic limestone-marl sequences mentioned above. Only extreme oceanic environmental changes, i.e., shifting of upwelling cells and climatic belts, can cause large changes in productivity. Hence, cyclic variation in productivity seldom seems the sole factor controlling limestone-marl rhythms.

4.2 Periodic Fluctuation of Supply with Terrigenous Sediment (Dilution Cycles)

Periodic fluctuation in terrigenous dilution is thought to be a major process for calcareous depositional environments with a minor but oscillating terrigenous input, such as the outer shelf or epicontinental sea (see Fig. 10). Terrigenous input is through fluvial, eolian, or glacial processes, and thus is closely related to climatic changes influencing runoff and erosion on the continents (Gardner 1982; Pratt 1984; Bottjer 1986; Dean and Gardner 1986; Pratt and King 1986; Diester-Haass Chap. 1.5, this Vol.). In basins not far from land areas, the input of fluvial and eolian sediment in the silt and clay fraction can show considerable fluctuations in quantity and composition, depending on whether they are located in arid or humid climatic zones (Sarnthein 1978; Sirocco 1989). In cold phases of the Pleistocene, more siliciclastic material was transported into the oceans by ice-rafting and delivery of sediment-laden meltwater streams than in warmer intervals (Ruddiman and McIntyre 1984; Diester-Haass Chap. 1.5, this Vol.).

Terrigenous dilution is ideally modelled when a steady production of biogenic carbonate is diluted by fluctuating input of the detrital phase (Fig. 7b). Through this

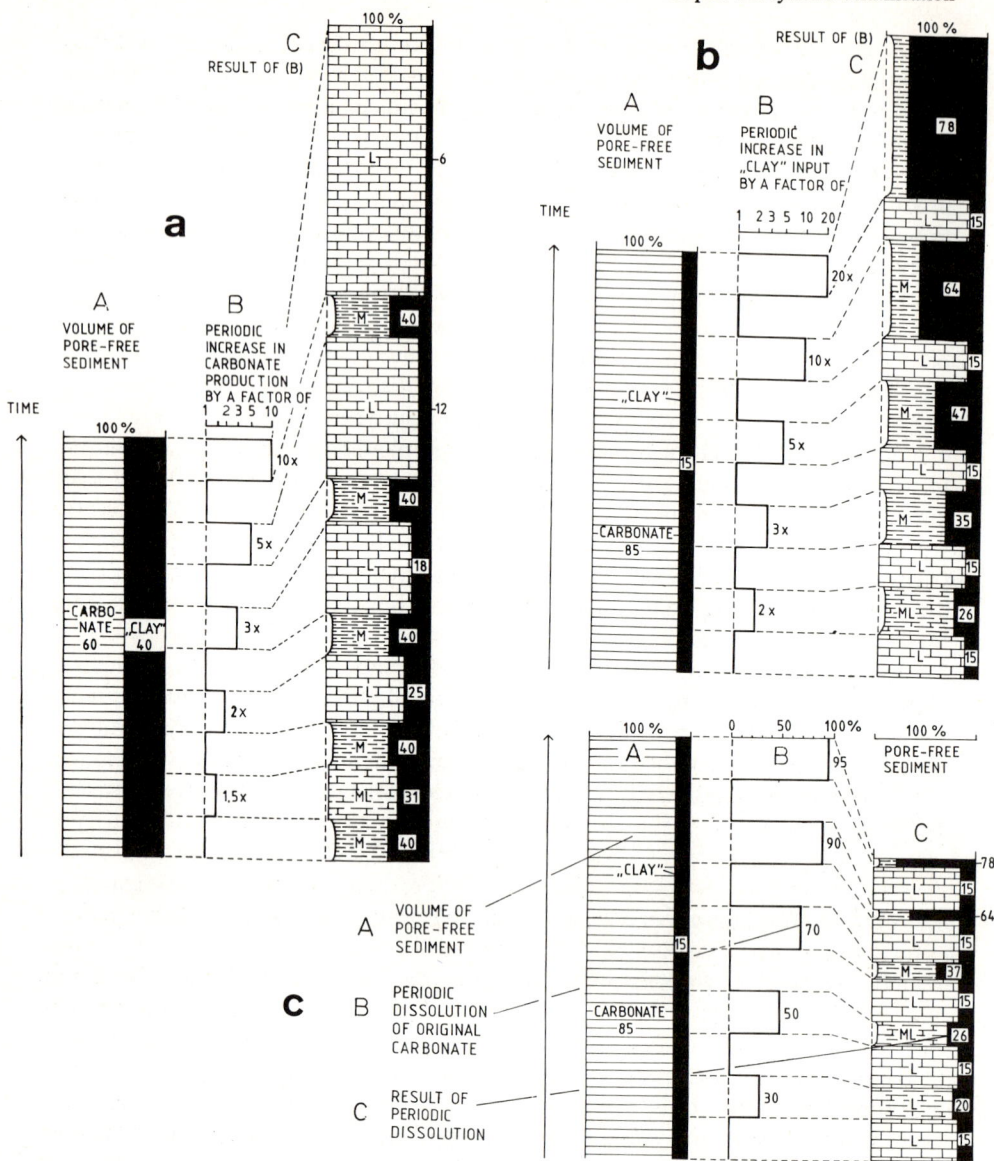

Fig. 7a-c. Simplified models for the developmental of limestone-marl (*L-M*) alternations assuming either variations in carbonate productivity, terrigenous dilution, or dissolution. *ML* marly limestone. **a** Variation in carbonate productivity. Transition of a homogeneous marly sediment column (*A*) into a limestone-marl succession (*C*) by periodically increased carbonate production. (*B*). Note the relation between the thickness of limestone layers and the carbonate production factor. **b** Variation in terrigenous dilution. Transition of a homogeneous clay-bearing carbonate sediment (*A*) into a limestone-marl succession (*C*) by periodically increased clay input (*B*). Note growing thickness of marl layers with increasing dilution factor, whereas the thickness of limestones remains constant. **c** Variation in carbonate dissolution. Transition of a homogeneous carbonate sequence (*A* containing 85% CaCO$_3$ and 15% clay) into a limestone-marl succession (*C*) by periodic dissolution of primary carbonate (*B* dissolution cycles). Note the decreasing thickness of individual marl beds with increasing dissolution

increasing noncarbonate input, the layers relatively rich in clay and silt will be thicker than the alternating carbonate-rich beds; therefore, this rhythm can be distinguished from alternations caused by variation in carbonate productivity or dissolution which have limestones beds thicker than marl beds (see Fig. 7a and b). In this model, a homogeneous sequence primarily consisting of 85% carbonate and 15% clay, i.e., the same original composition as assumed by Dean et al. (1981), is diluted by periodic augmentation of the noncarbonate input. Doubling of the clay and silt input leads to a marly limestone with 26% noncarbonate; marly interbeds develop when the dilution factor is greater than approximately 3.

4.3 Periodic Dissolution of Carbonate (Dissolution Cycles)

Dissolution of carbonate is most significant for sites situated within the lysocline and the CCD, a zone which for the present oceans is 1 to 1.5 km thick (Berger 1968; Thunell 1976; Berger et al. 1982; Arrhenius 1888). Additional dissolution is observed for sites above the lysocline where the sediments are relatively rich in organic matter which can be decomposed and thus provide aggressive CO_2 (Emerson and Bender 1981; Diester-Haass Chap. 1.5, this Vol.). Dissolution is quantified by applying various indexes based on benthic foraminiferal abundance and fragmentation of planktonic foraminifera (Berger 1968; Thunell 1976; Peterson and Prell 1985b; Grötsch et al. Chap. 1.6, Diester-Haass Chap. 1.5, both this Vol.). Also, the nannoplankton composition is utilized, as the susceptibility to dissolution differs for various species (Roth and Bowdler 1981).

Periodic and simultaneous dissolution of carbonate has been quoted by several authors as the main factor controlling fluctuating carbonate contents, although slightly different records of dissolution were found for Pleistocene sediments in the Pacific, Atlantic, and Indian Oceans. In the Pacific, carbonate dissolution is generally greater in warm climatic periods; thus, the total carbonate content of glacial periods is higher than that of the interglacial muds (Arrhenius 1952; Berger 1973; Thompson and Saito 1974; Luz and Shackleton 1975; Volat et al. 1980; Prell 1982; Farell and Prell 1987). In the central equatorial Pacific, large carbonate variations ranging from about 50 to 80% have been encountered at water depths between 4.4 and 4.8 km, reflecting climatic oscillations of the lysocline in the Pleistocene and Pliocene (Farell and Prell 1987). In contrast to the Pacific, the Atlantic seems to show the opposite pattern, with more dissolution during glacial periods and less dissolution and higher carbonate contents during the intergalcials (Gardner 1975; Be et al. 1976; Balsam 1981; Crowley 1983, 1985). This out of phase dissolution can be explained by different deep water production rates (Crowley 1985). In the Indian Ocean, a dissolution pattern intermediate between the Atlantic and Pacific types is observed (Peterson and Prell 1985b; see Fig. 11 in Grötsch et al. Chap. 1.6, this Vol.).

The wide distribution of Tertiary dissolution cycles is documented by various studies involving similar indices of faunal preservation, as used for Quaternary cycles (Kaneps 1973; Diester-Haass 1975; Gardner 1975; Keller 1980; Vincent et al. 1981; Dean et al. 1981; Diester-Haass and Rothe 1988). Oceanic dissolution cycles are also thought to be common in the Cretaceous, as the CCD was considerably shallower than it is today. Arthur et al. (1985c) showed that in the Middle Cretaceous the CCD was located at water depths of only 2 to 3 km, while for the Upper Cretaceous and Tertiary varying positions between 3.5 and 5 km are reported (Seibold and Berger 1982).

The effect of increasing dissolution for the development of carbonate-poor beds is shown for a model assuming a relatively high initial carbonate content of 85% (Fig. 7c). As Dean et al. (1977) have already pointed out, in this case dissolution of 50% of the carbonate is not sufficient to decrease the percentage carbonate below the weathering boundary to produce a marly interbed. Increasing dissolution leads to a thinning of the marly interbeds that alternate with thicker limestone beds. If 95% of the original carbonate is periodically dissolved, i.e., if there is a dissolution factor of 20, the limestone-marl thickness ratio becomes approximately 5. A value on this order of magnitude appears to be characteristic for many limestone-marl sequences; however, a smaller degree of early dissolution would be required in alternations with diagenetic modifications. Note that alternations formed by dissolution or productivity variations result essentially in the same bedding rhythm, with thicker limestone layers and thinner marl beds.

4.4 Calcareous Redox Cycles

Alternations composed of carbonate-rich beds and organic carbon-rich shales is a common bedding pattern in many black shale units and their facies transitions into marls and carbonates (see de Boer Chap. 1.3, Eicher and Diner Chap. 1.4, as well as Arthur and Dean Chap. 1.7, all this Vol.). Such very conspicuous alternations are found in environments that range from the epicontinental sea to the deep sea; they are especially common in the Atlantic Ocean in Aptian to Cenomanian time (e.g., Arthur et al. 1984b; Pratt 1984; Dean and Gardner 1985; de Boer 1986; Dean and Arthur 1986; Ogg et al. 1987). Light, bioturbated limestone beds alternate with dark, laminated, carbonate-poor interbeds rich in organic carbon. Varve-type lamination within the black shale bed and the specific tiering pattern of various burrow types in the limestones or the whole limestone-shale couplet indicate fluctuating oxygenation of the bottom waters, i.e., redox cycles (Savrda et al. Chap. 5.1, this Vol.). In the Cretaceous of the Atlantic ocean, redox cycles seem to be caused by a combination of varying input of terrestrial and marine organic matter, varying sea-surface carbonate production, and fluctuating oxygenation of bottom waters (Dean and Arthur 1986). Quite often, black shale beds show geochemical signals indicating a terrestrial origin for the organic matter, especially for the Cretaceous North Atlantic Ocean (Tissot et al. 1980; Summerhayes 1981). Several authors made the observation that many black shale beds contain redeposited organic-rich mud turbidites (Arthur et al. 1984b; also see Piper and Stow Chap. 2.9, this Vol.).

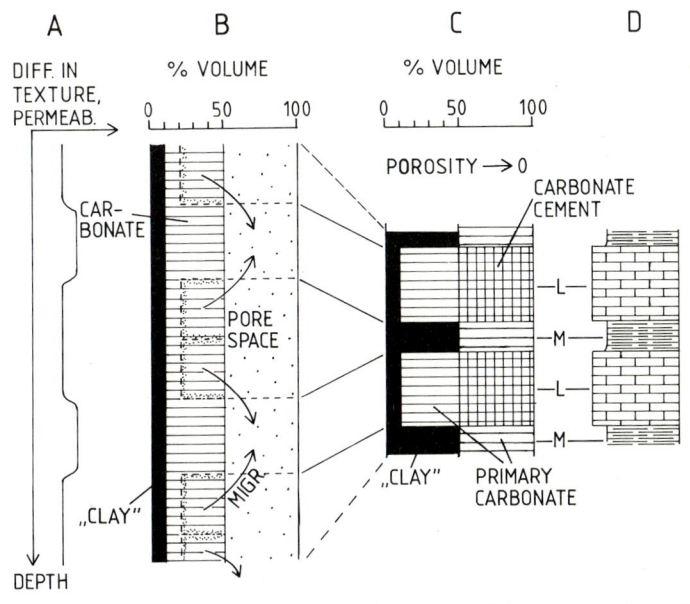

Fig. 8A-D. Model of limestone-marl couplets solely produced by diagenesis. *A* Primary fluctuations of texture and pore size. *B* Primary composition with a carbonate/clay ratio of 4:1 and reduction of original pore space to 50%. *Arrows* show the subsequent migration of carbonate from marl (*M*) to limestone (*L*) layers. *C* final composition. *D* field aspect of limestone-marl couplets. Limestone (*L*) consist of 40% primary carbonate, 50% cement, and 10% clay, but are not further compacted after *B*

4.5 Diagenetic Overprinting

Diagenetic overprinting is described by most authors as the enhancement of original carbonate and clay differences between beds by carbonate redistribution, which modifies and often augments the primary bedding rhythm (e.g., Arthur et al. 1984; Ricken 1986; Bathurst 1987). Diagenetic carbonate redistribution seems to occur within both open and closed diagenetic systems and under both low and high amounts of overburden (Bathurst Chap. 3.2, Ricken and Eder Chap. 3.1, both this Vol.). Some authors have even assumed that diagenetic processes are the only cause for generating marl-limestone alternations, as proposed by Sujkowski (1958) and Hallam (1986). In such a case, oblique bedding and a random distribution of limestone nodules within a marly matrix is thought to be created. Both phenomena, however, are very seldom observed on the scale of beds; thus, rhythmic unmixing of a completely homogeneous sediment appears unlikely. In nodular limestones, nodules are not randomly distributed; instead, they are commonly encountered in layers parallel to the general stratification.

Under extreme conditions, however, the diagenetic overprint may affect a sediment with an equally distributed carbonate content, but with variation in texture, pore space, and carbonate type between beds. A simple model describing a closed-system redistribution for such a case is based on the following assumptions (Fig. 8): (1) Prior to the onset of diagenesis, the original pore volume was reduced to 50% by mechanical compaction (Schlanger and Douglas 1974). (2) Thereafter, the remaining pore space in the subsequent limestone layers is filled with carbonate cement, and thus no further compaction can take place. (3) The carbonate for cementation of the limestone layers is obtained by carbonate dissolution only from the diagenetically evolving and compacting marly interbeds, while carbonate precipitation is restricted to the texturally different subsequent limestone beds.

The result of this simplified model is that the marl beds become considerably reduced in thickness, and that the alternation is finally dominated by the thickness of the limestone layers, thus augmenting the initial bedding rhythm. The carbonate content of the limestone layers has been increased to 90% by cementation, while marls have lost 75% of their original carbonate by "chemical compaction". This model may give a feeling for the quantities of mass transport involved and the augmentation of the bedding rhythm taking place in carbonate diagenesis. A more detailed description of the diagenetic overprint is given by Bathurst (Chap. 3.2) as well as Ricken and Eder (Chap. 3.1, both this Vol.).

5 Criteria for the Identification of Depositional and Diagenetic Processes in Marl-Limestone Alternations

1. Composition, trace elements and organic matter. Since the primary carbonate mineralogy is considerably affected by diagenesis, the composition of the noncarbonate fraction may be more suitable for distinguishing between primary and secondary effects. Clay mineral studies (e.g., Pratt 1984; Pratt and King 1986) or plots of element ratios of the insoluble residue, e.g., Al/Si or Na/K ratios (Arthur et al. 1985c; Arthur and Dean Chap. 1.7, this Vol.), give hints as to the climatic and erosional variation on adjacent land masses, when diagenetic clay mineral changes are moderate. Carbonate-bounded trace elements, such as Sr and Mg, can be used in Quaternary cycles to indicate changing aragonite and Mg calcite proportions in the sediment (Veizer 1983b). As coccoliths begin to experience overgrowth by carbonate cement, however, Sr and Mg values can provide information on the diagenetic exchange processes involved (Matter 1974). In alternations with high diagenetic overprint, Mg, Fe, and Mn become enriched in the marl layers by differential compaction (Wanless 1979; Ricken 1986). This process also greatly augments the organic carbon content in the more compacted marl beds, relative to the less compacted, cemented limestones (Ricken and Eder Chap. 3.1, this Vol.). In several alternations in the U.S. Western Interior, we found high compactional enrichment of organic carbon contents by factors of 2 to 10. For some Cretaceous alternations, it was shown that marls and limestones contain different types of organic matter as indicated on van Krevelen diagrams (Deroo et al. 1984). On the other hand, Pratt (1984) provided evidence that in redox cycles, metabolism of the organic substances in bioturbated and nonbioturbated beds is different and thus influences the type of organic matter that is preserved. For element correlation techniques of the clastic and the calcareous fraction as well as organic matter composition the reader is referred to the article by Arthur and Dean (Chap. 1.7, this Vol.).

2. *Isotopes.* Oxygen and carbon isotopes contained in the carbonate are difficult to interpret, because primary variations are small and diagenetic modifications are often large (Arthur and Dean Chap. 1.7, this Vol.). Quaternary $\delta^{18}O$ values show a global simultaneous variation due to the changing continental ice volume (see Fig. 5), but an opposing pattern relative to the carbonate content for many sites in the Atlantic and the Pacific is observed. Despite extreme climatic variations, only small isotopic changes in Quaternary cycles are recorded, as $\delta^{18}O$ and $\delta^{13}C$ values vary by 1 to 2‰ and 0.5‰, respectively (Imbrie et al. 1984; Shackleton et al. 1983). Bulk analyses of deep sea ooze show slightly positive values for both oxygen and carbon isotopes (Hudson 1977). For ice-free periods, $\delta^{18}O$ variations have been interpreted in terms of salinity changes (e.g., Pratt 1984), or changes in surface water temperatures (e.g., de Boer 1982a), while $\delta^{13}C$ variations are thought to reflect organic productivity and water stratification (Arthur et al. 1984b). Diagenetic processes such as increasing overburden and differential cementation-dissolution, may have various influences on the $\delta^{18}O$ values, including homogenization of oxygen values between marl and limestone layers, and both differential and equal shifts in limestone and marl layers. Diagenetic $\delta^{18}O$ shifts commonly result in negative values (e.g., Scholle 1977), however, after several km of overburden isotopes may again become more positive (Arthur et al. 1984b). The initial $\delta^{13}C$ contents can change to lighter values during diagenesis, as early carbonate cementation involves light organic matter. This is especially important when marl beds are rich in organic carbon, as in redox cycles.

3. *Fauna.* Changes in diversity and faunal preservation are an important tool for evaluating environmental differences for uncemented marl-limestone alternation (Eicher and Diner Chap. 1.4, Diester-Haass Chap. 1.5, as well as Grötsch et al. Chap. 1.6, all this Vol.). In carbonite-rich, cemented alternations that commonly have a marked diagenetic overprint, the microfossils are affected by differential carbonate dissolution and cementation. However, as diagenetic carbonate dissolution is concentrated in small zones and clay seams, leaving small parts of the marl bed unaffected, fairly preserved calcareous microfauna can sometimes still be obtained (Seibold and Seibold 1953; Darmedru et al. 1982; Eicher and Diner 1985; Erba 1988). More suitable for the detection of primary environmental variations seems the employment of the noncalcareous siliceous and organic microfauna (Darmedru et al. 1982; Herbert et al. 1986). Data denoting absolute fossil abundance may be ambiguous, because of alternating differences in sedimentation rates, compaction, and employed preparation methods.

4. *Structures.* In many alternations, depositional structures are not preserved due to complete bioturbation. Only redox cycles contain laminations indicating either varve-type variations or current activity (Robertson 1984). Some alternations in epicontinental seas show an asymmetric distribution of the bioturbation pattern, with gradually increasing burrow density and oxygen tolerance at the tops of the limestone layers, followed by a sudden drop in bioturbation at the transition to the laminated marl bed, which might be related to a systematic decrease in sedimentation rate (see Fig. 2 in Savrda et al. Chap. 5.2, this Vol.). If the rhythmic sequence is not completely destroyed by intensive burrowing, one can often observe burrows

in one type of the alternating beds filled by sediment from the following bed, documenting primary differences in composition (Hallam 1964). The types, number, and tiering patterns of burrows can be used as indicators of bottom water oxygenation (see Savrda et al. Chap. 5.2, Sageman et al. Chap. 5.3, as well as Oschmann Chap. 5,4, all this Vol.). The flattening of burrow diameters is higher in marl beds, indicating differential compaction between these beds and the limestone layers. In the marl layers, carbonate dissolution fabrics are encountered, including clay seams, fitted fabrics, and stylolites (see Bathurst Chap. 3.2, this Vol.).

6 Correlation and Timing of Limestone-Marl Rhythms

The hierarchical pattern of orbital-controlled bedding rhythms can be used to correlate them laterally over long distances, in environments that are only slightly affected by stratigraphic gaps and depositional background noise (see Kauffman et al. Chap. 7.2, this Vol.). Using various marker beds, a bed-by-bed correlation has been accomplished in several regions over tens to hundreds of km, e.g., in the upper Jurassic of southern Germany (von Freyberg 1966), in the lower Cretaceous of southeastern France (Cotillon et al. 1980), in the Upper Cretaceous of northern Germany (Seibertz 1979) and the Upper Cretaceous of the U.S. Western Interior, where the synchronism of beds can be excellently proved by the intercalation of numerous bentonites (Hattin 1985). Cotillon and Rio (1984) made an attempt to correlate lower Cretaceous DSDP sites from the Gulf of Mexico with sections in southeastern France (see Cotillon Chap. 7.3, this Vol.). In the future, there is hope that integrated approaches of correlation in special areas might lead to a detailed "cyclostratigraphy", with the maximum time resolution corresponding to the precession (Fischer et al. 1989).

The success of stratigraphic correlation and accurate time span determination is hampered by the intercalation and superposition of noncyclic deposition and by the occurrence of stratigraphic gaps (von Freyberg 1966; Thiede and Ehrmann 1986; Anders et al. 1987). Often, marl-limestone alternations are intercalated with widely distributed carbonate-mud turbidites and tempestites with bed thicknesses similar to that of typical marl-limestone alternations (Piper and Stow Chap. 2.9, this Vol.). After bioturbation mottling and diagenetic overprint, the event nature of these beds is difficult to recognize and to distinguish from Milankovitch-type stratification (Ricken 1986). Thus, determination of the time duration represented by a bedding couplet depends on the completeness of a cyclic sequence and on an objective method to distinguish cyclic phenomena from event beds and background noise.

1. The usual method for determining the time span of a bedding couplet is to date the bottom and top of a thick sequence which contains a great number of cycles, and to calculate the average cycle periodicity. The error for the timing of one cycle may become relatively high, because of the increasing probability that both changes in sedimentation rate and stratigraphic gaps are included when the considered time span is large (Sadler 1981; Anders et al. 1987; Schwarzacher Chap. 7.5, as well as Ricken Chap. 7.1, both this Vol.). Gaps and depositional

noise such as unrecognized calcareous mud turbidites, extinction of smaller beds by bioturbation (Herbert and Fischer 1986), and diagenetic welding of several layers (Ricken 1986), as well as weathering effects, are all factors which may greatly influence the timing of beds.
2. The hierarchy of the orbital cycles can be statistically distinguished from the quasi-stochastic background noise by applying Fourier and Walsh power spectra (Weedon 1989; Weedon Chap. 7.4, as well as Schwarzacher Chap. 7.5, both this Vol.). However, this is difficult to accomplish when one or two of the different Milankovitch frequencies are not regularly recorded in the sedimentary sequence (Fischer Chap. 1.2, this Vol.). In addition, power spectra are affected by gaps, variations in sedimentation rate, and differential compaction (Weedon Chap. 7.4, this Vol.).

For all these reasons, a precise timing of the limestone-marl periods is seldom possible, and the accuracy of the published data is difficult to evaluate. In Fig. 9 and Table 1, the mean periods of couplets and their sedimentation rates are summarized. The data represent fine-grained marl-limestone alternations and redox cycles from various regions which are Mesozoic and Cenozoic in age. Despite the fact that this data collection is erroneous and far from complete, it allows one to draw the following conclusions:

1. Despite considerable data scattering, the time period of the bedding couplets is approximately within the Milankovitch frequency band. Both Quaternary and pre-Quaternary marine cycles show major variations between 10 and 100 ka. This is a strong argument for both periodicities having been controlled by similar operating processes. Quaternary and older carbonate cycles also have similar sedimentation rates. For soft Quaternary deep sea sediments this rate is frequently 1 to 3 cm/ka, whereas older, compacted, lithified limestone-marl successions from outer shelf, slope, deep, and epicontinental sea settings mostly show rates between 0.5 and 4 cm/ka. Clustering of the frequencies in the range from about 10 to 50 ka is clearly observable. The 100 ka period does not show up distinctly in the pre-Pleistocene data set because it is usually represented by bundles rather than by bedding couplets. In addition, the 100 ka signal is amplified during the late Pleistocene owing to continental ice buildup and decay (see Fig. 4).
2. The individual periods of precession, obliquity, and eccentricity, however, be hardly recognized, which may be explained by the superposition of bedding related to orbital forcing and background noise. It is assumed that in the investigated data set, this superposition is expressed by a trend showing increasing sedimentation rates with decreasing time periods. Such a pattern would result when bed thickness rather than time period is statistically constant (cf. Fig. 9a and b with c). In other words, thicker sequences with more depositional background noise and diagenetic overprint contain statistically more beds than do time-equivalent thinner sequences, a point recently discussed by Hallam (1986) and Ricken (1986).

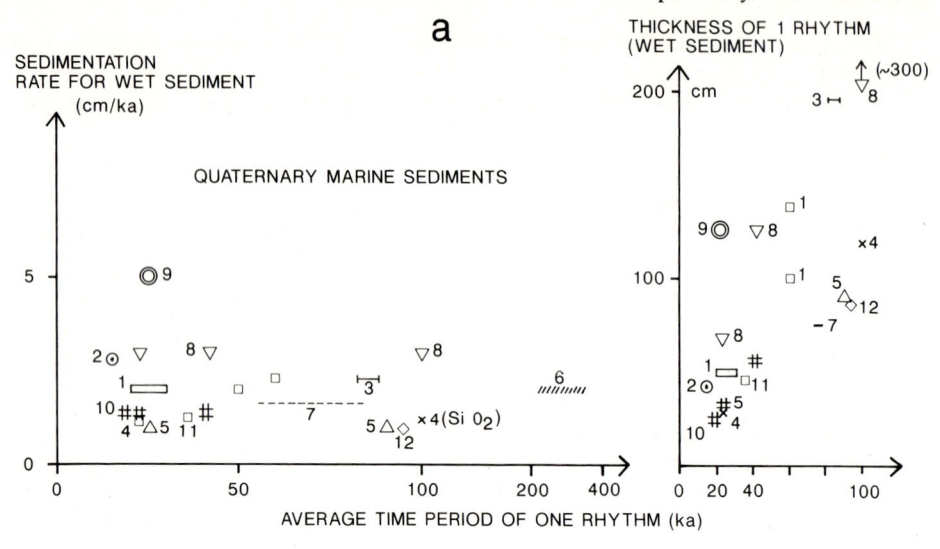

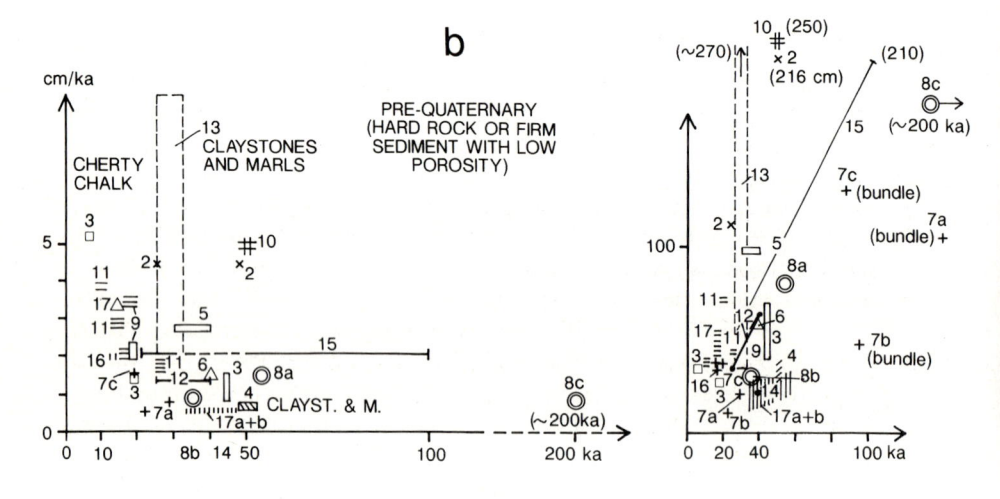

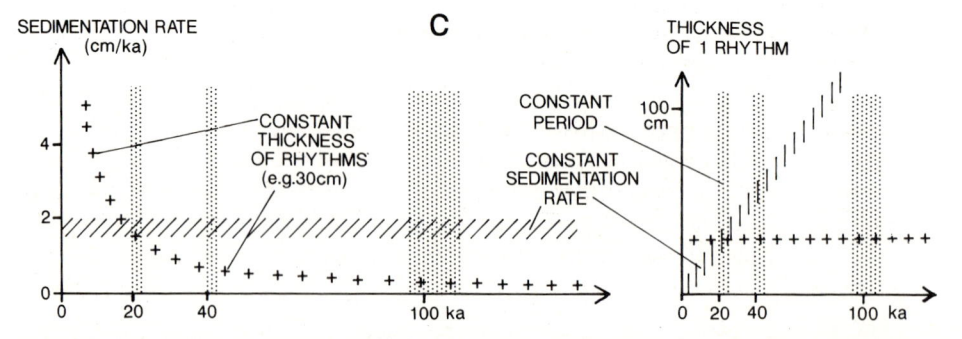

Fig. 9a-c. Relationship between sedimentation rates, average time period of calcareous ooze-marl or limestone-marl rhythms, and mean thickness of one couplet (consisting of a bed and interbed) of **a** Quaternary and **b** pre-Quaternary examples. **c** Schematic diagram indicating three different ways to interpret the data set, assuming either constant time periods, sedimentation rates, or bed thickness. See text for further explanation. References are indicated by *numbers in Table 1*

Table 1. References for the data plotted in Fig. 9

a Quaternary examples (dating radiometric, stable isotopes, partly by magnetostratigraphy; generally mean of several cycles)

1	Colombia Basin, Atlantic	Prell and Hays 1976
2	Western equatorial Atlantic	Bé et al. 1976
3	Northeastern Atlantic	Ruddiman and McIntyre 1976
4	Panama Basin	Pisias 1976
5	Solomon Rise, Pacific	Shackleton and Opdyke 1976
6	Ontong-Juva Plateau, western Pacific	Berger and Mayer 1978
7	Pacific	Volat et al. 1980
8	Indian Ocean	Hays et al. 1976
9	Walvis Ridge, contin. terrace, DSDP Site 532	Diester-Haass et al. 1986
10	N Atlantic, Sierra Leone Rise	Herterich and Sarnthein 1984
11	N Atlantic, 20° N, dissolution cycle, core V26-41	Crowley 1985
12	Eastern Pacific, dissolution cycle	Hays et al. 1969

b Pre-Quaternary examples (biostratigraphic dating, occasionally radiometric using volcanic ash layers generally mean of 50 to several hundred cycles)

1	N Atlantic, Vigo seamount, Pliocene Site 398	Maldonado 1979
2	S Atlantic, Walvis Ridge, Pliocene Site 532	Dean and Gardner 1985
3	Sierra Leone Rise, DSDP Site 366, Eocene-Miocene	Dean et al. 1981
4	Cape Verde Basin, Eocene light-dark clayst. and U. Cret. to Paleocene green.-reddish claystones	Dean et al. 1977
5	NW Germany, Upper Cretaceous	Ernst et al. 1979
6	Western Interior, USA, Upper Cretaceous	Bottjer 1986
7	Umbria, Italy, Middle Cretaceous	Herbert and Fischer 1986
a	Late Cenomanian: single rhythms and bundles,,	
b	Late Albian: single rhythms, bundles, and longer cycle	
c	Barremian: single rhythms and bundles	
8	S Atlantic, Angola Basin, Site 530, M. Cretaceous	Dean et al. 1984a
a	Coniacian-Santonian	
b	Turonian	
c	Albian-Cenomanian	
9	Southern England, Lower Chalk, Cenomanian	Kennedy and Garrison 1975b
10	N Atlantic, Vigo Seamount, Site 398	Aptian/Albian
11	NW Atlantic, Gulf of Mexico, Vocontian Trough in SE France, Lower Cretaceous	Cotillon and Rio 1984
12	NW Atlantic, Blake-Bahama Form., Lower Cret.	Ogg et al. 1987
13	NW Germany, Hauteriv. to Albian clayst. & marls	Schneider 1964
14	Cape Verde Basin, DSDP Site 367, Barr.-Tithonian	Dean et al. 1977
15	N Atlantic, Valanginian, Sites 391 and 534	Robertson 1984
16	Southern Germany, Kimmeridgian	Ziegler 1958
17	Southern Germany, Oxfordian	Seibold 1952
18	Lias and Kimmeridgian England a, Kimmeridgian; b, Lias	House 1985a

7 Type and Thickness of Marl-Limestone Sequences: the Role of Environmental Setting and Subsidence

7.1 Environments Suitable for the Formation of Limestone-Marl Rhythms

The pelagic limestone-marl alternations described herein have been dominant since the Upper Jurassic because they are closely related to times of substantial planktonic carbonate production in the oceans. In earlier periods, particularly in the Paleozoic, fine-grained carbonate deposition was restricted to the upper slopes, shelves, and intra-basinal rises, where condensed sequences with either well-bedded or nodular limestones were formed (Franke and Walliser 1983).

The occurrence of limestone-marl rhythms in field sections requires the maintenance of a certain carbonate to clay ratio on the order of 3 to 4. Since planktonic carbonate production is slow, with only 0.5 to 3 cm/ka, marl-limestone alternations can be easily destroyed by clastic dilution. As a result, all three environments suitable

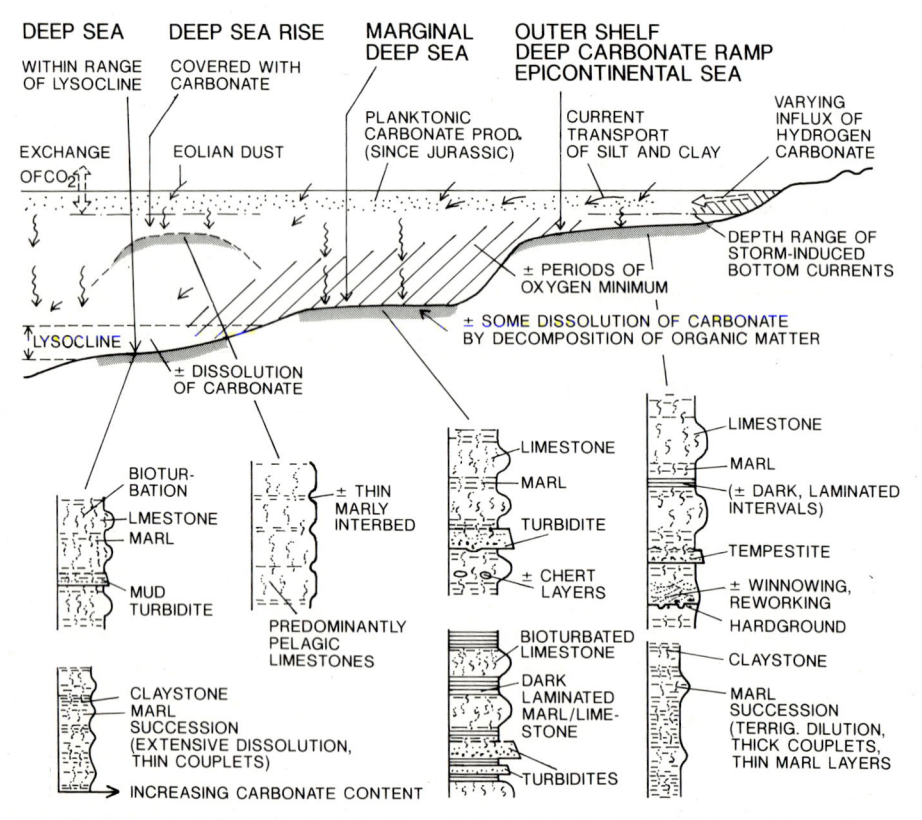

Fig. 10. Summary of depositional environments and processes generating different types of pelagic to hemipelagic limestone-marl, claystone-marl, or calcareous redox rhythms. See text for discussion

for limestone-marl alternations lie in areas of low terrigenous sediment influx and below the range of storm-induced currents (Fig. 10):

1. *Deep outer shelf, deeper part of a carbonate ramp, central epicontinental sea.* Because of the proximity of large land masses, these regions can receive relatively large quantities of terrigenous silt and clay. Therefore, pelagic and shelf-derived fine clastic carbonate tends to be diluted in relation to the climatic conditions on land, and to large-scale sea level variations influencing the carbonate to noncarbonate ratio of the sediment (Eberli Chap. 2.8, this Vol.). All of these successions deposited in relatively shallow water may be intercalated with distal carbonate and siliciclastic tempestites. If the influx of silt and clay becomes too high and reaches a proportion of more than one third to one half of the total sediment volume, limestone-marl successions are replaced by claystone-marl sequences, with thicker couplets. Diagenetic overprint is intensified by the presence of unstable carbonate minerals.
2. *Marginal deep sea, deep sea plateau, and environments around isolated carbonate platforms.* These environments appear to be the most suitable for the development of limestone-marl rhythms. They are formed in water depths above the lysocline, and they receive only small amounts of terrigenous sediment from the continents; thus they are only moderately affected by terrigenous dilution and dissolution. Productivity of planktonic carbonate ooze may change in accordance with oceanic circulation and the nutrient supply; under certain conditions redox cycles with dark laminated interbeds are formed. Below upwelling areas, sequences also show rhythmic accumulations of siliceous ooze (see Decker Chap. 4.1, this Vol.). Under certain circumstances, the carbonate input can be so high that there is not enough clay for the formation of marly interbeds.
3. *Deep sea between the lysocline and the CCD.* Varying carbonate dissolution is the main factor for the development of pelagic limestone-marl rhythms due to considerable oscillations of the lysocline. As a result, sequences with large differences in $CaCO_3$ between succeeding beds are formed (e.g., Farell and Prell 1987; see Fig. 11 in Grötsch et al. Chap. 1.6, this Vol.).

7.2 Factors Controlling the Thickness of Rhythmic Carbonate Sequences

For the development of thick rhythmic sequences, e.g., 100 m, steady conditions over a period on the order of 2.5 to 15 Ma are needed. Consequently, in a comparatively shallow basin, where the sea floor is somewhat deeper than the storm-wave base, subsidence should be higher or equal to sediment accumulation. On the shelves of passive continental margins, which are built up on transitional crust, suitable rates of subsidence are realized between approximately 50 and 120 Ma after rifting (e.g., Steckler et al. 1988). Prior to this time span, rift basins are usually influenced by clastic sedimentation.

In epicontinental seas, the mean rate of subsidence is normally found to be low, but it may reach the order of magnitude necessary for the formation of limestone-marl rhythms, especially if subsidence coincides with periods of global sea level rise (Haq et al. 1987). In several epeiric basins of Europe and North America, limestone-marl rhythms were deposited during short periods of time during the Jurassic to Cretaceous when the sea level was relatively high. However, the development of such sequences was always in danger of being disturbed or modified by (1) terrigenous input too high, with a transition to marls, or too low, with a transition to pure limestones, such as some units in the European chalk which have more than 98% $CaCO_3$; (2) shallowing of the sea with reworking, omission, channelling, interfingering and replacement by reefs, and the transition to carbonate tidal flats or shelf sands (e.g., Seibertz 1979; Einsele 1985; House 1985a; Pratt et al. 1985; Kauffman 1984, 1988a; Strasser 1988). Therefore, complete, pure limestone-marl successions are developed here only for relatively short time intervals on the order of 1 to a few Ma.

In oceanic basins, the occurrence of limestone-marl sequences is thought to be associated with sea level changes and variations in the position of the lysocline on a scale of 10 Ma. During transgressions, a higher production of pelagic carbonate seems to be related to increased water mass mixing and nutrient recycling, while large regressions are periods with higher siliciclastic input (see Vail et al. Chap. 6.1, as well as Eberli Chap. 2.8, both this Vol.). Oscillations of the CCD and lysocline seem to roughly parallel the large-scale sea level variations, because transgressions reduce the size of the continental area, and thus decrease the input of dissolved $CaCO_3$. In addition to this, more carbonate is probably deposited on the increasing shelf areas, depriving the carbonate reservoir of the deep sea (Arthur et al. 1985c).

8 Discussion and Conclusions

1. *Types of limestone-marl alternations and their depositional environments.* Limestone-marl alternations appear in field sections when their mean carbonate content is around 65 to 85%. Below or above this weathering limit, only marl-shale alternations or pure limestones, respectively, are visible. As a result, pelagic limestone-marl alternations occur in environments with carbonate deposition three to four times higher than terrigenous silt and clay input. They are found in areas with generally little clastic dilution from the shelf to the deep sea and in epicontinental basins. Larger cyclic sequences in shelf and epeiric seas need special requirements such as the coincidence of subsidence and sea level rise, which restrict their occurrence to relatively short time spans of 1 to a few Ma. Generally viewed, three basic processes are involved: variations in carbonate production, dissolution, and terrigenous dilution. Each type represents an end member of alternations with simultaneous oscillations of several inputs. Cycles formed by either varying carbonate production or dissolution show rhythms with thick limestone beds and thin marl layers, whereas terrigenous dilution is indicated by the opposite bed thicknesses. The variations in deposition of each of the three parameters must be relatively large, and thus variations in sedimentation rates fairly high in order to obtain the commonly encountered carbonate oscillations. Subsequent to deposition, the carbonate differences between succeeding beds and the bedding rhythm may be considerably enhanced by diagenetic overprint. The depositional nature of the alternating beds can be determined by using various criteria, such as differences in

fauna composition, trace element ratios, stable isotopes, organic carbon-carbonate relationships, and original sedimentary structures.

2. *Marl-limestone rhythms in greenhouse and icehouse climates.* The occurrence of various types of carbonate rhythms controlled by orbital-climatic forcing is a well established fact in the Quaternary (Hays et al. 1976; Berger et al. 1984). In addition to climate, Quaternary marine sediments were also strongly affected by short period and high amplitude sea level changes caused by the buildup and decay of continental ice sheets (Broecker and van Donk 1970). During periods free of large ice accumulations on the continents, the major sea level changes apparently had longer time periods by a factor of 10 to 100 and usually much smaller amplitudes than the Quaternary oscillations (Haq et al. 1987), though high-frequency fluctuations probably also existed (Goodwin and Anderson 1985; Goldhammer et al. 1987). Many examples of rhythmic and cyclic bedding from times with no or minor continental ice are known and show the same characteristics as the Quaternary counterparts. Unlike shallow water carbonate cycles (see Part II, this Vol.), high-frequency sea level oscillations cannot be the primary factor causing the deeper-water pelagic to hemipelagic limestone-marl alternations treated herein. Periodicities of bedding couplets described in the literature are in the Milankovitch frequency band, and cluster between 15 and 40 ka for pre-Quaternary cycles, while they additionally gather 100 ka for Quaternary cycles. The 100-ka rhythm in the Quaternary is explained by augmentation through continental ice buildup and decay.

3. *Climatic-oceanic enhancement of orbital signals by feedback systems.* Since orbital variations result in relatively small changes in seasonality, with a net effect mostly below ± 10%, they must be enhanced by climatic-oceanic feedback systems (Ruddiman and McIntyre 1984; Berger and Tricot 1986). Some of these are: shifting of climatic belts, changes in the global carbon cycle including atmospheric CO_2 and carbonate deposition, changes in oceanic thermohaline circulation, and buildup and decay of continental ice. All of these systems are interrelated, and some of these processes show considerable time lags of several thousand years (Ruddiman and McIntyre 1984). They are further influenced by the changing configuration of the continents and ocean basins during Earth's history. In the following paragraph, changes in the global carbon cycle and their feedbacks with oceanic circulation are considered the most important processes in generating limestone-marl alternations in pelagic environments.

4. *Oceanic circulation and changes in the carbon cycle.* The climate-driven oceanic circulation controls both surface water productivity and carbonate dissolution in the deep sea. In addition, it influences, to some extent, the atmospheric CO_2 content, which in turn affects global climate via the greenhouse and icehouse effects (Barnola et al. 1987; Sundquist and Broecker 1985; Grötsch et al. Chap. 1.6, this Vol.). Bender (1984), for instance, calculated that a 13% increase of atmospheric CO_2 leads to a rise in the CCD of roughly 1 km. During times of reduced temperature differences between low and high latitude regions, the thermohaline circulation, such as known from the present-day oceans, tends to slow down. On the other hand, globally warm,

equable climates may cause an oceanic circulation quite different from that of today (Arthur et al. 1987). Warm, dense saline water, originating from flooded low-latitude shelves and adjacent seas, most likely became the major bottom water source (see Eicher and Diner Chap. 1.4, this Vol.). Since the solubility of oxygen decreases with increasing temperature and salinity, this sinking surface water formed an oxygen-depleted intermediate to deep-water mass. Climatic changes are expressed by varying mineralization of organic matter and recycling of nutrients. Thus, the surface water productivity is affected. High organic matter productivity leads to a rise of the CCD and the lysocline, while low organic matter productivity shows the opposite effect, when other factors influencing this system can be neglected. In addition, the pelagic and shallow water carbonate production and river input of carbonate may vary and influence the depth of the lysocline (Berger and Keir 1984; Broecker 1982). For example, an increase in the carbonate production on the shelves can lead to a rise of the lysocline in the deeper sea.

This balanced oceanographic system is very sensitive to minor changes in the single processes. The system responses, for example, by generating large fluctuations in productivity, in the position of the lysocline or in the vertical and lateral extent of an oxygen minimum zone. Thus, weak orbital-climatic signals are translated into a conspicuous pattern of bedding rhythms. The sensibility of the oceanic circulation system probably was greater during periods of more uniform global climate than exists today (Berger 1979; Arthur et al. 1985c).

5. *Factors controlling redox cycles.* The sensitive oceanic circulation system mentioned above controls the occurrence of calcareous redox cycles. Varying salinity stratification may lead to short-term expanded and reduced oxygen-minimum zones at rather shallow to intermediate water depths (Arthur et al. 1987). They are held to be responsible for widespread anoxic events, which are usually composed of rhythmically bedded sediments indicating minor variations in oxygenation. Climatic periods with better oxygenated waters usually produce lighter-colored bioturbated layers with somewhat higher carbonate contents than the periods characterized by oxygen deficiency and dark, laminated interbeds. The organic matter may be derived from both marine productivity and terrestrial sources. Bed-scale successions of this type occur, for example, in the middle Cretaceous of the Atlantic and Tethyan Oceans (Dean and Arthur 1986; Herbert et al. 1986; Ogg et al. 1987; de Boer Chap. 1.3, this Vol.). High atmospheric CO_2 contents, warm, uniform climates, sluggish oceanic circulation, and seemingly low marine organic productivity most likely prevailed for this time (e.g., Barron and Washington 1985; Bralower and Thierstein 1984, 1987, see Eicher and Diner Chap. 1.4, this Vol.).

6. *Continental processes influencing the global carbon cycle.* The global carbon cycle, and thus oceanic carbonate sedimentation, are also influenced by processes operating on the continents. Global variations in climate depend on and influence the CO_2 content of the atmosphere; they change the regional and total plant cover of the Earth. A dense vegetation cover promotes chemical weathering and thus increases the river input of hydrogen carbonate into the oceans. This process can enhance the production and preservation of biogenic carbonate in shallow and deeper

waters, and cause a depression of the lysocline (Shackleton 1977; Sclater et al. 1977; Einsele 1982a; Bender 1984). At the same time, the widely extended vegetation cover tends to reduce mechanical erosion and transportation of terrestrial silt and clay into the oceans. These effects are reversed in periods of globally diminished vegetation. Particularly in regions of semi-arid climate, relatively small climatic changes may generate substantial fluctuations in terrigenous input through both river and eolian transport, CO_2 released by the decomposition of terrestrial biomass is transferred via the atmosphere into the ocean. There it may lead, in conjunction with lower input of river-transported hydrogen carbonate, to increased dissolution of carbonate. If at the same time the input of terrestrial clay is high, marly interbeds easily form.

Acknowledgments. We would like to acknowledge the fruitful discussions with many friends and colleagues during the CRER conference in 1988 in Perugia and the Tübingen Conference on the subject of this book, held in 1989. André Berger, Wolfgang Berger, Walter Dean, Poppe de Boer, Alfred Fischer, Jim Gardner, Linda Hinnov, Isabella Premoli Silva, and Fred Read provided useful information on various aspects of cyclic sedimentation. Susanne Borchert and Linda Hobert read the text carefully and made numerous suggestions.

1.2 Orbital Cyclicity in Mesozoic Strata

A. G. Fischer

1 Introduction

The world of our experience is governed by an hierarchical cyclicity, the day-month-year pattern driven by the orbits of Earth and Moon. Astronomy shows that more subtle orbital cycles exist at time-scales beyond human experience (Figs. 2 and 3 in Einsele and Ricken Chap. 1.1, this Vol.). Have these affected climate and geological processes? Can their record be read in rocks?

Low-frequency variations in the orbit cause minor changes in the patterns of insolation (Berger 1978) which affect above all the contrasts between summer and winter. They include (Fig. 1):

1. *Earth's axial precession (P)* relative to perihelion, with a variable period, presently showing modes at 19 (P1) and 23 (P2) ka and a mean of 21 ka;
2. the change in *axial obliquity (0),* with a modal period of 41 ka;
3. the degree of *orbital eccentricity (E),* which varies in several cycles. E1 at 98 ka and E2 at 126 ka form the "ca. 100 ka" cycle. E3 has a period of ca. 413 ka, E4 a period of ca. 1300 ka.

The theory that these low-frequency cycles drove the waxing and waning of continental ice sheets was advanced in the last century by Adhemar and by Croll and carried on subsequently by Milankovitch and others. While the mechanisms by which climate is altered remain unclear, the theory is substantiated by the oxygen isotope ratios in the fossils of the deep sea record. Serving as proxies for global ice volume, these record P1 and P2, 0, and E1,2, and 3. The history is charmingly told by Imbrie and Imbrie (1979).

As early as 1895, G.K. Gilbert, convinced of the climatic sensitivity of depositional systems, attributed a cyclic repetition of limestone-marl alternations in the Late Cretaceous of North America's Western Interior Seaway to the precession, and suggested the possibility of a geochronology based on an orbital cyclostratigraphy.

Many stratigraphers have remained skeptical. Duff et al. (1967) for example, questioned the power of orbital forcing to leave a stratigraphic record without the aid of glacial climatic amplification. Others have been concerned with the role of diagenesis, which can certainly enhance or obscure depositional cyclicity (Arthur et al. 1984b; Ricken 1985, 1986, also Chap. 1.8, this Vol.), and might even be able to produce spurious cyclic patterns (Hallam 1986; Bathurst Chap. 3.2, this Vol.).

Other stratigraphers have been too eager to refer any oscillation with apparent timing in the Milankovitch frequency band (20–400 ka) to "Milankovitch cyclicity".

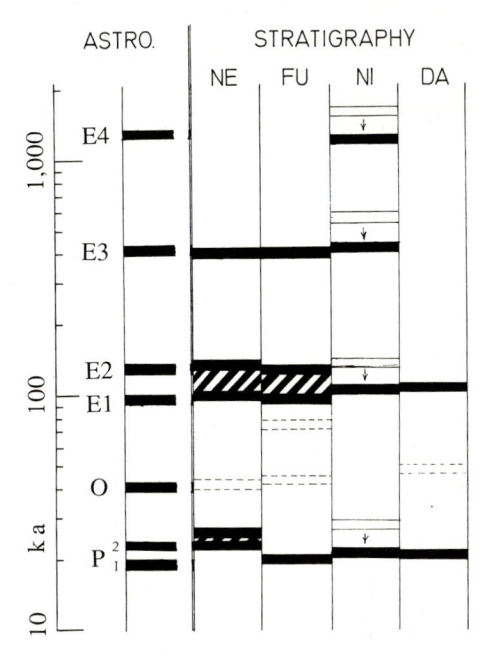

Fig. 1. Periods of orbital variations compared to periods of cycles in four Mesozoic stratigraphic sequences. *Left* main orbital periodicities of Berger's series, plotted logarithmically. The precession *P*, with modes at 19 and at 23 ka; the obliquity cycle, *O*, at 41 ka; the ca. 100 ka eccentricity cycles *E1,2* with modes at 98 and 126 ka; the 413-ka eccentricity cycle E3; and the 1300 ka eccentricity E4. *Right* stratigraphic sequences, timing from mean sedimentation rates per stage after Harland et al. (1982), in NE refined by extrapolated varve counts, in NI modified by a suggested 25% correction. *NE* Newark Supergroup, Late Triassic-Liassis, lacustrine-playa, New Jersey. Recognition of O and of bimodality in P and E1,2 based on spectra. *FU* Scisti a Fucoidi, late Albian, pelagic, Piobbico core, Italy. Recognition of O, a 74-ka peak, and bimodality of E1,2 based on spectra. *NI* Niobrara Formation, Coniacian- early Campanian, hemipelagic, Colorado. *DA* Dachstein Limestone, Norian- Rhaetian, platform, Northern Alps

As pointed out by Algeo and Wilkinson (1988), the average sedimentation rates are such as to place the apparent mean period of most outcrop-scale stratigraphic oscillations into the Milankovitch frequency band, no matter what their origin and whether their apparent period be determined by their initial frequency or by their preservation (the "hiatus factor").

The majority of stratigraphic cycles which have been referred to orbital (or "Milankovitch") forcing remain unidentified with specific orbital cycles, and thus remain useless for cyclostratigraphy at present level of insight. Approaches to such identification are of three sorts: (1), *varve counts,* and their extrapolations; (2), *derivatives of the standard geochronology,* such as estimated durations for stages, biostratigraphic zones or magnetic polarity chrons; (3), *relative timing* within a hierarchy of cycles.

Varve counts are limited to very specific facies, but have proved very useful (Anderson 1982, 1984). The limitations of Mesozoic geochronology are such as to leave most approaches to cycle durations with an uncertainty factor of two or more

– not sufficient to distinguish the various cycles of the orbital hierarchy. But when elements of that hierarchy can be identified, the combination becomes compelling. Schwarzacher (1947) discovered that in Triassic platforms bedding couplet oscillations are subordinated to a cycle of couplet bundles, in ratio 5:1, suggesting the precession-E2 relation. This has been found in other facies, as have additional frequencies of the orbital hierarchy – E3 (1:4 relative to E1,2) and possibly E4.

We shall here consider five hierarchical Mesozoic sequences in which such criteria have provided a strong basis for identifying specific orbital variations. These are the lacustrine Newark Supergroup, the pelagic Scisti a Fucoidi, the hemipelagic Niobrara Chalk, and the Dachstein and Latemar platform limestones. Others are coming to light, but the number of such cases remains small – not, I believe, because there are few, but simply because very few have been analyzed in appropriate manner.

Sequences dominated by the simple obliquity cycle may include the British Jurassic (House 1985a), the British Cenomanian (Hart 1987), the Cenomanian-Turonian Greenhorn sequence of North America (Hattin 1985; Fischer et al. 1985; Kauffman 1988a) and some of the strikingly rhythmic Lower and Middle Cretaceous hemipelagic sequences of southern France (Cotillon and Rio 1984). However, the geochronology of simple cycles is inherently more difficult to establish than is that of hierarchies, and definitive identifications remain to be made.

2 Newark Supergroup: Lacustrine-Playa Facies, Late Triassic-Liassic

Lakes are settling basins likely to accumulate sediment continuously. Climate provides a range from comparatively open flushed systems, in which outflow matches inflow, to closed systems with no outflow. This combination of continuous sedimentation and climatic sensitivity make lakes particularly promising monitors of climatic change.

2.1 Setting

The rifting that split Laurasia left a series of abortive basins along the American Each Coast. These were filled with several km of alluvial and lacustrine sediment as well as basaltic flows and intrusives. The Newark rift, extending 200 km from southern New York through New Jersey into Pennsylvania, is one of the largest, and the site of the studies reported here (Van Houten 1964; Olsen 1986). Pollen and vertebrate remains suggest that the axial lacustrine facies (Lockatong and Passaic Formations) represents a span from Late Carnian to Hettangian – something between 14 and 20 Ma. A thickness of about 4,400 m yields a mean sedimentation rate of 220–310 Bubnoff units (m/Ma or mm/ka).

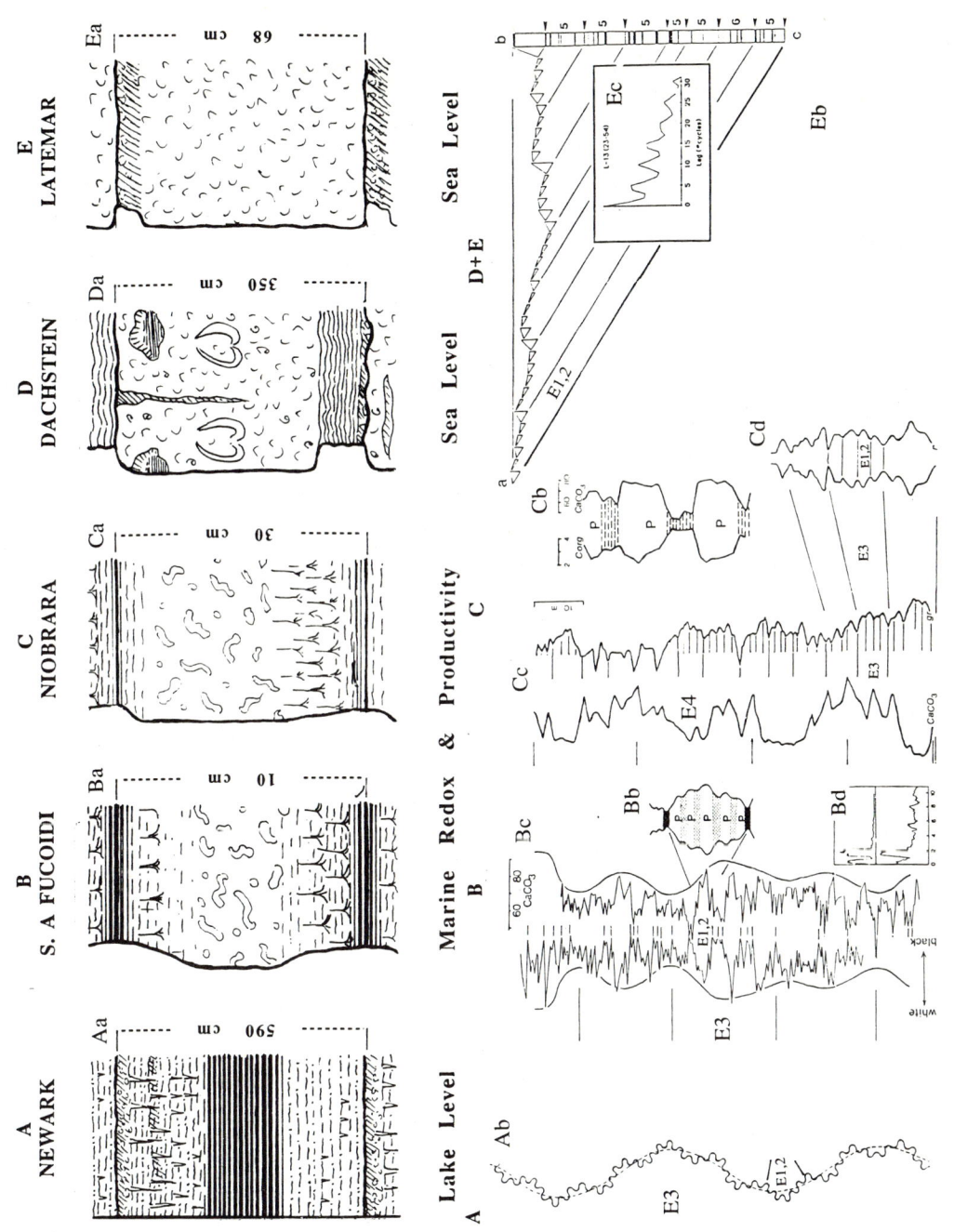

Fig. 2A-E. Five stratigraphic sequences showing hierarchical rhythms identified with orbital cycles. *Upper tier* lithology of precessional cycles. Note variations in scale. *Lower tier* proxy curves showing precessional and eccentricity cycles. **A** Newark Supergroup, Triassic-Jurassic, Eastern North America. Lacustrine-mudflat alternations. **B** Upper Scisti a Fucoidi, Albian, central Italy. Pelagic facies, showing variations in carbonate productivity and bottom redox potential. **C** Niobrara Formation, Coniacian-Campanian, Colorado. Hemipelagic facies showing variations in carbonate productivity, possibly detrital influx, and bottom redox potential. **D** Dachstein Limestone, Late Triassic, northern Alps. Platform facies showing subtidal-intertidal-emergence oscillations. **E** Latemar Limestone, Middle Triassic, southern Alps, showing platform emergence cycles.

2.2 Precessional Signal

A high-frequency cycle, recognizable throughout the sequence, consists of a succession of three rock types (Fig. 2Aa). A basal calcareous claystone or siltstone shows lenticular lamination and burrows. Oolites and stromatolites as well as desiccation cracks may be present, in what is interpreted as a transgressive lake-shore facies.

This is overlain by a microlaminated calcareous shale or limestone, commonly black, generally bearing fish and other body fossils and devoid of mudcracks: a lacustrine facies.

The upper unit is comprised of calcareous claystones or siltstones with abundant desiccation cracks, burrows and root casts, and occasional footprints of dinosaurs and other reptiles. Soil textures are commonly observed. This unit is interpreted as a mudflat or playa facies.

This cycle, with a mean thickness of ca. 6 m, has been termed the Van Houten cycle. The laminae of the black shales average 0.24 mm thick. If varves, they yield a sedimentation rate of 240 Bubnoff units, well within the 220–310 B rate calculated for the sequence as a whole. That rate yields a mean of 24.6 ka for this cycle, close to the timing of the precession.

In view of the paleolatidude, 10–20°N lat, this lake-level fluctuation suggests a latitudinal oscillation in the boundary between the wet tropics and the para-tropical dry belt, just what might be expected from precession-induced variations in the seasonal displacement of the heat equator (Berger 19778; deBoer 1982a).

2.3 Eccentricity 1,2 Signal

In much of the sequence, groups of cycles richer in calcite, dolomite, and analcime are bracketed by one or more cycles depleted in such chemical components. Van Houten reasoned that the "detrital" cycles must represent times when the basin was flushed by discharge of water to the outside, while the "chemical" cycles resulted from times when drainage was internal and soluble salts accumulated in the basin. The mean number of "short" or "Van Houten" cycles in this longer cycle is 5, and that cycle thereby becomes a candidate for the ca. 100 ka eccentricity cycle.

2.4 E3 Signal

In the lower part of the sequence (Lockatong Formation) most of these E1,2 bundles are drab, but reddish ones occur at intervals of four to five such bundles. These record periods of more intense oxidation and their 400–500-ka timing is roughly that of the E3 eccentricity cycle. In the upper portion of the sequence red prevailed throughout.

Thus Van Houten's original analysis (Fig. 2Ab) yielded the major components of the orbital cycles, save the 41 ka obliquity cycle.

2.5 Time-Series Analysis

Olsen (1986) measured additional sections, and designed a scale, from 0 for beds with soil structure and root casts to 7 for varved shale. By this means stratigraphic sequences could be converted in proxy curves of lake level. Spectral analysis of the longest of these, the 400-m Delaware section, shows five major peaks. At a depositional rate of 240 Bubnoff units, the high frequency peaks yield periods of 22.8 and 25.5 ka, suggesting the precession (Fig. 1). A low peak at 41.9 may represent the obliquity cycle. Very strong peaks at 100 and 133 are then good matches for E1 and E2, while a strong peak at 400 ka must represent E3.

3 Scisti a Fucoidi: Albian Pelagic Facies

Pelagic oozes generally contain a mixture of plankton-derived skeletons and largely wind-blown dust. The extreme sensitivity of organisms (plankton blooms) to temperature, chemistry (nutrients etc.), and light penetration renders such sediments sensitive to bottom currents and chemistry. Oxygen supply to the sea floor controls the benthic faunas and the degree of bioturbation; variations in lysocline and CCD will vary the amount of carbonate preserved. Thus the cycles most likely to appear in pelagic sedimentary regimes, given relatively steady sedimentation, are productivity cycles in the plankton; redox cycles of the bottom; and, in deep facies, dissolution cycles (see also Einsele and Ricken Chap. 1.1, this Vol.)

3.1 Setting

The Umbrian facies of central Italy represents pelagic deposition on the southern side of Tethys, between a carbonate platform on the south and the oceanic Ligurian trough to the north (Arthur and Premoli Silva 1982). Deposition occurred below aragonite compensation depth but generally above the lysocline, at depths estimated at 1–2 km. Through the Cretaceous calcareous pelagic oozes (now limestones) prevailed excepting the marly Aptian-Albian Scisti a Fucoidi interlude.

We shall here deal primarily with the upper (latest Albian) portion of the Scisti a Fucoidi (Arthur and Premoli Silva 1982; deBoer 1982a; deBoer and Wonders 1984), and specifically, with studies of an 8-m segment of the Piobbico core (Herbert and Fischer 1986, Herbert et al. 1986). As in the Newark series, our first approach to time is through mean sedimentation rates. At an Albian thickness of 60 m, and a duration of 12 Ma (Harland et al. 1982) the mean sedimentation rate for the Scisti a Fucoidi is 5 B (mm/ka), enormously slower than that of the Newark sediments, but comparable to modern rates of pelagic sediment flux.

3.2 Precessional Signal

The stratification pattern most apparent at first glance is a fluctuation in calcium carbonate content that divides the formation into low-carbonate/high-carbonate couplets with a mean thickness of about 10 cm (Fig. 2Ba). The range is from 50 to 85% CaCO3, i.e, from marl to limestone, but individual couplets do not generally span the full range, going only from shale to marl and back to shale, or from marl to limestone to marl.

In addition to this, there is a cyclic variation in color, body fossils, and ichnofacies (Fig. 2Ba). The shales are generally black (due to some 1–2% of very finely divided C_{org}) and more or less laminated; the marls are drab, and conspicuously patterned by black *Chondrites* burrows, especially near the black shale intervals; and the limestones are whitish, are thoroughly bioturbated, and contain large random burrows (*Planolites*). Planktonic foraminifera occur throughout (though they are notably scarcer in some of the limestones), while the calcium carbonate content reflects coccolith abundance.

At a mean sedimentation rate of 5 B, the mean couplet thickness of 10 cm matches the couplets to the ca. 20 ka precessional cycle. Variations in couplet thickness are proportional to their calcium carbonate content (deBoer 1982a) and the sedimentational fluxes (Herbert et al. 1986) indicate supply rates of detrital matter (wind-borne dust) at 0.25–0.50 g/cm^2/ka and of carbonate at 0.30–3.0 g/cm^2/ka. The couplets thus appear to reflect a cycle in carbonate productivity of the plankton, chiefly the nannoplankton.

This productivity cycle is linked to a redox cycle: lowest productivity occurred at times of bottom anoxia. At such times the wind-blown dust containing soot-like terrigenous organic matter (Pratt and King 1986), not much diluted by nannoplankton, formed the black shales. With increasing oxygen supply, sedimentation passed through a dysaerobic phase (*Chondrites* marl, Bottjer et al. 1986) into the aerobic *Planolites* limestone facies.

3.3 Eccentricity E1,2 Signal

The stratigraphic couplets are conspicuously bundled into sets of ca. 5 by a shift in composition (Fig. 2Bb). A typical bundle begins with black shale-drab marl couplets, evolves to marl-limestone couplets, and ends by reverting to the black shale-marl mode. The bundle structure is accentuated by the thicker nature of the high-carbonate couplets. This structure is not only visually apparent in outcrop but is also demonstrable in plots of calcium carbonate content and in instrumental scans of darkness (Fig. 2Bc). The 5:1 ratio identifies this cycle with the ca. 100 ka E1,2 syndrome.

3.4 E3 Signal

A further modulation is seen in the envelope curves drawn round the carbonate curve and darkness scan shown in Fig. 2Bc. These curves show a grouping of bundles into superbundles, in a ratio of 1:4. In the field these superbundles are visible only on large exposures, where they stand out by virtue of thick central limestone beds. These superbundles thus appear to represent the 400 ka E3 eccentricity.

3.5 Time-Series Analysis

In Fourier spectra in which the stratigraphic dimension is used as the time axis, the precessional couplet does not emerge above background noise: variations in couplet thickness scramble the signal.

In the carbonate and darkness profiles the presence of an obliquity signal is equivocal, but spectra of foraminiferal abundance show a consistent peak at around 43 ka (Tornaghi et al. 1989). An unexplained peak at about 74 ka occurs in various spectra. In all spectra, the ca. 100 ka E1–2 signal forms the dominant peak (Fig. 2Bd). Some spectra show a double peak, and Park and Herbert (1987) believe to have identified the 98 and 126 ka E1 and E2 terms. The spectra do not show the 400 ka E3, but, at the margin of resolution, suggest power at about 800 ka.

3.6 Other Pelagic Cycles

The Barremian (Upper Maiolica) limestones underlying the Scisti a Fucoidi and the Cenomanian Scaglia Bianca Limestone that overlies the Scisti a Fucoidi show rather similar cycle patterns (Schwarzacher and Fischer 1982; Fischer and Schwarzacher 1984). Here the shaly or marly members are generally reduced to bedding planes. Studies of sequential bed thickness show a similar grouping of bedding couplets into bundles. In these limestones, the depositional rate is about double that of the Scisti a Fucoidi. Somewhat similar bedding cycles have been resolved in the red-and-drab marls of the Eocene Scaglia variegata, where their timing matches that of the magnetostratigraphy (Schwarzacher 1987b). But we must also call attention to the fact that the 5:1 bundle pattern is not present or at least not conspicuous in all parts of the Umbrian sequence, including parts of the Scisti a Fucoidi (Tornaghi et al. 1989).

The Lower and Middle Cretaceous sediments of the Western North Atlantic show remarkable parallelism to the cyclic structure of the Scisti a Fucoidi (Arthur and Dean 1986). In the bathyal facies they show a similar alternation between pelagic limestones and black shales (Blake-Bahama Formation), and below carbonate compensation depth the lithologies oscillate between red, green and black shales of the Hatteras Formation (Jansa et al. 1979). While the timing of these cycles remains poorly defined, their resemblance to the Scisti a Fucoidi cycles is so great as to suggest a common cause, and one that affected the whole western Tethys (of which the North

Atlantic was a part). Episodic changes in deep-water circulation due to incursion of dense saline bottom waters (Arthur and Dean 1986) seem a likely cause.

4 Niobrara Formation: Hemipelagic Chalk

Hemipelagic sedimentation is likely to show the same variations that are exhibited in pelagic sedimentation, with an added factor of episodic fluctuations in the detrital supply rate, related to climate, sea-level change or tectonics. Rhythmic patterns of various sorts are widespread in hemipelagic sediments.

4.1 Setting

During Cretaceous time in North America the rising Western Cordillera (Sevier Mountain Belt) was flanked by a foredeep that provided marine connections from the Tethys to the Arctic Ocean during eustatic high stands. In Coniacian-Early Campanian (Niobrara) time the western, tectonically active side of the trough was occupied by deltas fed from the growing highlands, while the eastern side, fronting the low stable craton, accumulated the chalky limestones, chalks, and marls of the Niobrara Formation (Hattin 1982). Exposed and cored sections from Colorado have helped to elucidate the cyclic patterns, but precise dating is inhibited by the endemic nature of the fossils. In the Berthoud cores near Fort Collins, Colo, (Pratt et al. 1990) the formation measures 85 m. An estimated 6 Ma of time yields a mean sedimentation rate of 14 B.

Excepting the basal Fort Hays limestone with abundant inoceramids, depth was probably in the range of 100–200 m, certainly above aragonite compensation depth as revealed by molds of ammonites and other argonitic mollusks.

4.2 Precessional Signal

Bedding couplets recording a low-carbonate/high-carbonate oscillation (Fig. 2Ca,b) average 30 cm thick in the Berthoud core, yielding a period of 28 ka, close to the precessional cycle considering the crudeness of dating. As in the Albian example above, organic carbon content varies inversely with carbonate. On outcrop, Bottjer et al. (1986) found that at least in the Fort Hays Member unburrowed dark shale passes upward into *Chondrites* chalk, and this upward into *Planolites* chalk, again reminiscent of the redox cycle in the Scisti a Fucoidi, though seemingly less symmetrical. Laferriere et al. (1987) have shown that in the Fort Hays Member the precessional couplets tend to become amalgamated eastward, toward the stable shore, while westward they become confused by high-frequency influx of mud.

4.3 Eccentricity E1,2 Signal

The bundling of couplets, into sets of ca. 5, can be observed in the field in the Fort Hays Member at Pueblo (Fischer 1980; Barlow and Kauffman 1985; Laferriere et al. 1987), and in the upper chalks in Kansas (Fischer et al. 1985). While well logs do not generally resolve the couplets, they may show a consistent oscillation in the 1–2-m range, which must represent E1,2 (Fig. 2Cc,d). Pratt et al. (1990) attribute this to variations in mud supply, but a shift in the mean level of carbonate productivity seems a possible alternative.

4.4 Eccentricity E3 and Lower Frequency Signals

A grouping of E1,2 bundles into sets of four is demonstrated by the gamma-ray log of the Berthoud core (Fig. 2Cc, right curve) as well as by the resistivity log of Fig. 2Cd. Again, this appears to be simply a shifting in the base-line of carbonate or mud supply.

A cycle of yet lower frequency is suggested by gross lithology and overall carbonate profiles (Pratt et al. 1990). Over the entire region the formation can be divided into four calcareous units – the Fort Hays Limestone, and the lower, middle, and upper chalks of the Smoky Hill Member, separated by more marly beds. This structure is nicely expressed in the calcium carbonate profile of the Berthoud well (Fig. 2Cc, left curve). The mean period, some 1500–1600 ka, does not correspond to any in Berger's orbital series, but the period derived for the precession is about 25% too high, suggesting that sedimentation was underestimated, in which case this long cycle could correspond to Berger's E4, at 1300 ka, as suggested in Fig 1.

5 Alpine Triassic Carbonate Platforms

Subsiding carbonate platforms grow mainly by accreting the skeletons of invertebrates and by chemical/physiological deposition of carbonate. If net growth exceeds subsidence, the depositional surface rises to the intertidal zone, bringing a change into mud-cracked algal mat sediments (loferites).

Tectonic uplift or a drop in sea level will expose the platform to erosion and the formation of soils with caliche crusts, vadose pisolites and microkarst. Renewed subsidence or a rise in sea level will cause flooding and new colonization by a neritic biota unless so rapid as to drown the platform. While autocyclic processes such as the cutting-away of intertidal-supratidal deposits by tidal channels can induce subtidal-supratidal oscillations (Ginsburg 1971; Zankl 1967, 1971), such processes cannot occur on a regional scale, nor will they develop soils and penetrative dissolution features. Carbonate platforms are thus sensitive recorders of changes in relative sea level, induced by eustatic fluctuations or vertical tectonics (see Strasser Chap. 6.5, Haas Chap. 6.6, this Vol.).

5.1 Settings

Middle and Late Triassic carbonate platforms were widely developed in the parts of Tethys that were to evolve into the Alps, Apennines, and Dinarides (Wilson 1975). These platforms are spectacularly exposed in mountain faces, and the backreef facies generally shows strikingly rhythmic bedding. We shall here review two separate sequences, the Norian-Rhaetian Dachstein Limestone of the Northern Limestone Alps, as exposed in various ranges of the Salzburg region (Austria) and the Middle Triassic (Ladinian) Latemar platform of the Dolomites (northern Italy). The Dachstein cycles of the Northern Alps and Hungary as well as those of Carranante (1971) are mainly upward-deepening, i.e., emergence-peritidal-subtidal, whereas the Latemar and Sicilian cycles (Catalano et al. 1974) are upward-shoaling, with a diagenetic imprint acquired during emergence.

5.2 Dachstein Platform, Northern Limestone Alps (Lofer Cycles)

The Northern Limestone Alps are a tectonic klippe over 1000 km long, consisting of a stack of allochthons. These are largely composed of the disjunct pieces of Triassic carbonate platforms. Despite structural complications, a Late Triassic (Norian-Rhaetian) platform, intervening between the Vindelician shoreline in the north and the deep Hallstatt basin on the south, can be reconstructed (Fischer 1965; for a somewhat different reconstruction assuming simpler structure but more complex facies patterns see Zankl 1967). The southern platform margin, formed by reefs, achieves a thickness of about 2 km. The body of the platform, thinning northward, consists of the back-reef Dachstein limestone and the ultra-back-reef Hauptdolomit. Both of these units show grossly similar cyclicity, involving an alternation between a massive bed and a laminated unit (the loferite), and we shall here deal with the expression of this cycle in the Dachstein limestone. The length of the Norian-Rhaetian interval is slated at 12 Ma by Harland et al. (1982), which yields a sedimentation rate of 166 B for the southern margin of the platform.

5.3 Precessional Signal

Massive m-scale limestones contain a rich and diverse marine fauna, including echinoderms and conspicuous megalodontid clams, and represent subtidal deposition in normal marine waters (Fig. 2D). Beds of this type alternate with thinner (dm-scale) beds of laminated dolomitic limestone whose biota is largely restricted to cyanobacterial mats, a few species of ostracodes and foraminifera, and cerithid snails. Laminar-fenestral ("loferite") fabric is abundant, and prismatic mudcracks occur more rarely. This is a peritidal facies (Fischer 1964).

Locally the succession is interrupted by layers of red or green calcareous mudstone, containing limestone clasts and extending into the underlying limestone in veins. These interbeds are interpreted as paleosols formed during emergence beyond the peritidal zone. The upper parts of the subtidal beds commonly contain cavities of cm to dm dimensions, filled by a combination of geopetal mudstone and sparry calcite linings. Commonly, such cavities are aligned in planes parallel to bedding, and show a correlatable sequence of filling, suggesting development and filling of a microcavern system along a water table. They testify to emergence even when no coherent paleosols remain (Fischer 1964).

It thus appears that the carbonate platform emerged episodically; either only into the intertidal, recorded in algal mat etc. ("loferite") facies, or far enough to undergo weathering, the development of solution cavities, and the accumulation of soils. A 300-m section in the upper part of the formation in the Loferer Steinberge (Schwarzacher 1947, 1954; Schwarzacher and Haas 1986) yielded a mean couplet thickness of 3.5 m. The accumulation rate of 166 B (see above) yields a period of 21.1 ka. Fischer's (1964) study yielded a mean thickness of 5.5 m/cycle, and a correspondingly longer period, but included as single cycles massive beds up to 20 m thick, which now appear much more likely to represent not single cycles but sets of cycles during which emergence was not achieved at the sites measured, or in which such emergences were not recognized.

5.4 E1,2 Eccentricity Signal

These couplets are arranged into sets or "bundles", in which the thickness of couplets decreases upward. Since the massive beds form cliffs while the loferites tend to form shelves, these bundles are apparent in the topography (Schwarzacher 1947, 1954). The bundle period averages 5x that of the precession, or 105.5 ka, and thus matches the E1,2 eccentricity cycle.

Another way to demonstrate these bundles graphically, devised by Fischer (1964), is shown for the Latemar sequence in Fig. 2Eb. These are a form of geohistory plot, in which couplets are spaced equidistantly on a horizontal time axis a-b and at their proper stratigraphic elevation above a line of mean subsidence a-c. If indeed the couplets represent even time intervals and subsidence was linear, then the rise and fall of couplets relative to line ab depicts a longer-term oscillations in sea level.

Such plots are extremely sensitive to the proper recognition of cycles. As noted above, the interpretation (Fischer 1964) of thick (up to 4x normal) couplets as single precessional cycles warped the calculations of mean period; it also vitiated the recognition of the 5.1 precession to E1,2 ratio. I attributed the asymmetrical megacycles thus obtained to variations in subsidence rate. But, given the 5:1 ratio that emerges from Schwarzacher's studies of the Dachstein cycles and in the Latemar platform (see below), and given what we have learned from other sequences, it now seems much more likely that the Dachstein sequences represent a combination of the precessional cycle with the E1,2 eccentricity pattern, albeit imperfectly recorded.

5.5 Time-Series Analysis

A Walsh spectrum of 305 m of section in the Loferer Steinberge (Schwarzacher and Haas 1986) yields a main peak for the bundle at 15.25 m. If this be equated with the ca. 100 ka E1,2, then the next lower peak at 7.25 m = 47.5 ka could represent the obliquity cycle (Fig. 1). Of three somewhat prominent peaks in higher frequencies, one at 3.58 m matches the couplets measured in the field and yields a timing of 23.5 ka.

5.6 Ladinian of Latemar Complex

The carbonate platforms of the Dolomites are autochthonous and relatively undeformed. While most of the region has undergone extensive dolomitization, the Latemar massif has remained largely calcitic. Time control is poor. Harland et al. (1982) arbitrarily assign 6–7 Ma to each of the Jurassic-Triassic stages. The boundaries of the stage are only roughly known in the Dolomites, but in Mt. Latemar the Ladinian appears to be represented by some 400 m of platform limestone. Those figures yield a mean depositional rate (compacted) of ca. 59 B.

5.7 Precessional Signal?

The basic bedding rhythm of the Latemar platform (Hardie et al. 1986; Goldhammer et al. 1987) is the limestone bed, which normally bears a dolomitized cap, with pores enlarged by dissolution and filled with vadose cements of various sorts (Fig. 2Ea). Each such surface represents a horizon of emergence and weathering, and the cycle records a sea level oscillation superimposed on what may have been steady subsidence. If we accept the 7-Ma figure for the duration of the Ladinian and assume that the 615 cycles represent the full Ladinian but nothing more, then the period is one of 11.4 ka. Conversely, if we apply the nominal precessional duration of 20 ka. the Ladinian was 12.3 Ma long. Neither of these possibilities is out of the question – and they illustrate the dilemma of identifying cycles by inferred depositional rates through most of Phanerozoic time. Let us assume for the moment that the cycle is that of the precession.

5.8 E1,2 Eccentricity Signal?

The thickness of these cycles varies in a systematic manner: like the Dachstein cycles described by Schwarzacher, they are grouped into bundles of ca. 5, as demonstrated in a "Fischer plot" (see above), shown in Fig. 2Eb. The submergence-emergence pattern of the precessional cycles rides on a longer eustatic oscillation normally including five emergence cycles, and presumably representing the ca. 100-ka E1,2 eccentricity signal. These cycles in turn ride on a wave of yet lower (but here unresolved) frequency.

5.9 Conclusions for Triassic Platforms

The Triassic carbonate platforms of the Alpine region reveal a complex eustatic oscillation, persisting through Middle and Late Triassic times. Platforms that were subsiding tectonically shoaled or emerged at intervals, and this cycle was modulated by an asymmetrical eustatic cycle having a period five times as long. If the one represents the precessional cycle, the other is that of the E1,2 eccentricity (Fig. 1).

Such cycles are found not only widely distributed through the Northern and Southern Limestone Alps, but also from the Late Triassic of Hungary (Schwarzacher and Haas 1986) and from the Triassic of Italy. That they reflect vertical tectonics, namely stick-slip motion on bounding faults (Cisne 1986), seems unlikely on two counts – their ubiquitous distribution, and their hierarchical character. Alternatingly warmer and colder Triassic poles, small ice caps, and thermal volume effects on the oceans remain tantalizing possibilities.

6 Identification of Orbital Signals in Other Facies and Ages

Anderson (1982) has documented the precessional cycle in a 200 ka varved record of Permian evaporites, and his data suggest the presence of the E1,2 cycle as well. Clearly Mesozoic evaporites pose tempting targets for cyclostratigraphy. Foucault et al. (1987) have made a case for the precessional cycle and the E1,2 eccentricity signal in the Cretaceous Helminthoid Flysch of Liguria. For pelagic limestones of the Southern Alps, troubled by condensed sequences and hiatuses, Weedon (1989 and Chap. 7.4, this Vol.) has made a plausible case for the presence of P,O, and E1,2 by means of Walsh spectra, cleverly filtered and reconstituted to an idealized stratigraphy. Strasser (1988 and Chap. 6.5, this Vol.) has shown cyclic eustasy comparable to that of the Triassic platforms in Neocomian carbonate platforms of the French-Swiss border country. Hart (1987) suggests that bedding in the British Cenomanian Chalk reflects the obliquity cycle, and that the chert bands in the upper chalk of England represent a combination of P and O. Many of the other cyclic patterns recorded from Mesozoic rocks may also be attributable to orbital forcing, but until they are identified with specific cycles, by such analyses as our examples have illustrated, their contribution to cyclostratigraphy lies in the future.

7 Conclusions

All five of the sedimentary sequences considered show a basic bedding oscillation, whose period is referred with varying certainty to that of the 21-ka precessional cycle, (Fig. 1,2). They record this oscillation in different ways, and at different scales related to differences in sedimentation rate. The Newark sequence reveals oscillations in lake level; the Scisti a Fucoidi and the Niobrara Formation depict marine planktonic carbonate production tied to the vigor of water circulation; and the Triassic platforms show a eustatic cycle possibly related to oscillations in polar ice volume.

In each case the precessional oscillation rides on a cycle having a period approximately five times as long. This presumably represents the ca. 100 ka E1,2 eccentricity cycle. In time-series spectra this is invariably the strongest of the signals observed – just as it is in the late Pleistocene. In the Newark Supergroup, this cycle appears to reflect a periodic flushing of the basin by high lake levels that discharged to the outside; in the Scisti a Fucoidi and the Niobrara Chalk it represents a shifting of the productivity base-line, and in the Triassic platforms it involves an asymmetric eustatic cycle.

A third level of cyclicity, apparently the E3 eccentricity cycle of 400 ka, is recorded in the Newark Supergroup by an oscillation of more intense (red) to less intense oxidation of the sediment, and in the Scisti a Fucoidi and Niobrara Formations by a still greater shift toward and away from higher carbonate productivity.

In the Niobrara this shift appears to be repeated at a yet higher level, in the range of 1500 ka or the 1200 ka E4.

In these sequences the obliquity cycle is weak or absent. There are, however, other sequences in which it appears to dominate.

I therefore submit that the case for orbital forcing of stratigraphic sequences has been made in principle. Contrary to Duff et al. (1967) sedimentary sequences of various facies have sufficient sensitivity to record orbitally driven climatic fluctuations in greenhouse as well as icehouse times. Diagenesis enhances cyclicity in some facies, and may obliterate it in others. Whether it can create spurious cyclicity still remains an open question.

But while the demonstration of orbital cycles in a few sequences is heartening, it leaves us a long way from a workable cyclostratigraphy. Cotillon's (1984, 1987) belief that cycle patterns can be correlated in distant parts of the world has yet to be convincingly demonstrated. A beginning has been made in tracing cycle patterns within a basin, from one facies into another (Laferriere et al. 1987), but nowhere have we established a coherent cyclostratigraphy that resolves the various orbital signals throughout a basin, let alone a hemisphere or the globe. Nowhere has the radiometric geochronology of a stage been superceded by orbital timing. The very question of which tectonic settings and facies are amenable to cyclostratigraphy remains to be explored.

Traditional stratigraphic approaches are laborious and lack the consistency and objectivity of instrumental scans. Great promise lies in the physical and chemical scanning of cores and boreholes by high-resolution techniques, tied to studies of paleontology and stratigraphy, and thoroughly integrated by reference of all data to a common stratigraphic computer base such as Stratabase (Ripepe 1988).

The practical applicability of orbital cyclostratigraphy to correlation and geochronology remains to be demonstrated, placing a premium on extending detailed studies through a basin and through adjacent basins. Other questions such as the relationships of cycle patterns to paleogeography, facies, and time call for more far-flung studies, and international cooperation. Application of orbital cyclostratigraphy to problems of paleoclimatology, paleoceanography, and the evolution of orbital characteristics remain distant but exciting goals.

Acknowledgments. I am much indebted to M.A. Arthur, W.E. Dean, L.M. Pratt and P.A. Scholle for permission to use some of their as yet unprinted work on the Niobrara Formation and to the National Science Foundation for its support of my cycle studies.

Chapter 1 Rhythmic Stratification

1.3 Pelagic Black Shale-Carbonate Rhythms: Orbital Forcing and Oceanographic Response

P. L. de Boer

1 Introduction

1.1 Orbital Parameters

The orbital parameters (precession, obliquity, and eccentricity; Berger 1978a; 1988) produce periodic changes of climate – Milankovitch cycles – which, in turn, may influence both terrestrial and marine systems of erosion and sedimentation. These climatic changes, which are due to the changing orientation of the Earth in its varying path around the Sun, involve shifts of the caloric equator between about 10°N and 10°S dragging with it low-latitude climate zones, and causing, at higher latitudes, changes in the intensity of the monsoons and changes of seasonality. At low latitudes the influence of the precessional factor (present-day main periodicities 19 ka and 23 ka) is dominant, towards higher latitudes the changing obliquity of the Earth's axis (present-day main periodicity 41 ka) produces the most significant climatic changes (van Woerkom 1953; Berger 1978a; see Fischer Chap. 1.2, Einsele and Ricken Chap. 1.1, both this Vol.).

Sedimentary deposits record the variation in climatic conditions under which they were formed. On land these climatic conditions concern temperature, rainfall, wind, evaporation, and the formation and action of ice. In the shelf seas and deep oceans climatic changes are noticeable by variation in temperature, current velocity and direction, the distribution and exchange of water masses, the recycling and supply of nutrients, the production of carbonate and organic matter, the supply of oxygen to deep water, carbonate dissolution, etc. Moreover, climate-forced processes acting on the continents influence the amounts of terrigenous sediment and dissolved chemicals which are supplied to the sea, as well as the input rate of fresh water and, in dry areas, of hyper-saline water, which may cause water mass stratifications to varying degrees.

In deep seas, the orbital variations lead to sedimentary rhythms in the 20 ka to 2 Ma frequency range (see de Boer 1983; Fischer Chap. 1.2, this Vol.) by changing the organic productivity in surface waters, the supply of terrigenous matter, the dissolution of carbonate in the deep ocean, and by changing the supply of oxygen to deep water. Examples have been described from pelagic and hemipelagic settings, from platform carbonate successions, from evaporite and lacustrine settings, and from continental paleosols. Especially the deep ocean appears to be sensitive to the effects of orbital forcing. Variations in carbonate- and organic matter content form a part of these sedimentary rhythmicities, and their origin will be discussed in the subsequent pages.

Einsele et al. (Eds.)
Cycles and Events in Stratigraphy
©Springer-Verlag Berlin Heidelberg 1991

1.2 Types of Oxygen-Deficient Facies and Environments

In the present-day seas and oceans, anoxia and the formation of organic carbon-rich fine-grained sediments, commonly referred to as black shales, are restricted to the marginal parts of the system with very high or very low circulation intensities (Fig. 1a,c). In other periods of the Phanerozoic, anoxia of deep waters occurred more frequently and was a more widespread phenomenon, e.g., during the Cretaceous (Schlanger and Jenkyns 1976), and during parts of the Paleozoic (e.g., Wilde and Berry 1982).

In Fig. 1 current models for anoxia in the deep ocean and marginal basins are presented. In these models, circulation velocity is an important factor which influences the supply of nutrients to surface waters and of oxygen to deep water:

1. In stagnant systems, such as the present-day Black Sea (Fig. 1a), the supply of oxygen to deep water is minor or absent, and on the surface organic productivity is low. According to observations of Shimkus and Trimonis (1974) from the present Black Sea, sediments are richest in organic matter in areas with both very slow circulation and very low primary organic production. For polar ice cap-free periods the denser waters from the more saline shelf seas must have been an important mechanism for generating stagnant bottom waters.
2. In systems of coastal upwelling of deep and intermediate waters (Fig. 1c), the amount of nutrients brought to the surface leads to an increased organic production. As a result the sinking of organic matter to deeper water may be in excess of what can be possibly oxidized by the available oxygen supplied by deep currents over the basin floor.
3. A third model, estuarine circulation within semi-enclosed basins (Fig. 1b; cf. Brongersma-Sanders 1971) as often inferred for the Eastern Mediterranean during periods in the Pleistocene (e.g., Thunell and Williams 1989), is comparable to the upwelling model (1c) with a continuous flux of organic matter from the surface to deep water, and a recycling of elementary nutrients which are released during partial oxidation and decomposition of organic matter in deep water. The landward side of the semi-enclosed basin thus acts as a nutrient trap enhancing organic productivity.

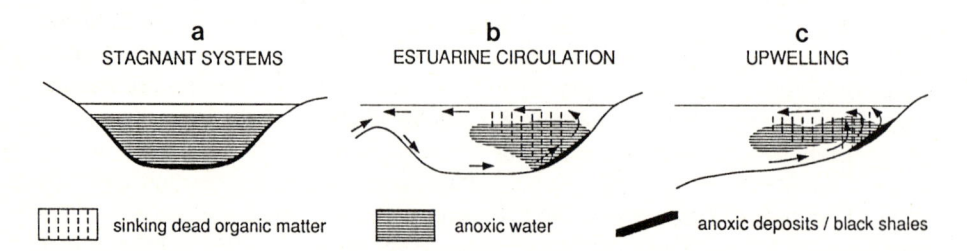

Fig. 1a-c. Models for the production of pelagic black shales. **a** Stagnant basins; **b** estuarine circulation; **c** upwelling. (After Brongersma-Sanders 1971 and Thiede and Van Andel 1977)

In case of deposition under oxygenated conditions, the organic content would be positively correlated with the sedimentation rate if other factors are constant (Müller and Suess 1979; see also Suess et al. 1987). However, obviously the organic content of fine-grained pelagic sediments is often negatively correlated with the rate of sedimentation (Tyson 1987), because indeed "other factors" do vary. In addition, black shale intervals in deep water can be formed by en masse transport of organically rich fine-grained sediment, e.g., by turbidity currents. Stow (1987) discussed the formation of black shales in the South Atlantic Ocean by different resedimentation processes. Degens et al. (1986) even suggested that most of the Mesozoic black shales in the Atlantic might have formed as turbidites. Certainly, this may have been the case for a number of black shale deposits, but for those examples in which the succession of lithologies exhibits Milankovitch cycles, that is, in which chalk (or marl)-black shale couplets represent periods of about 20 ka (precession cycle) or 40 ka (obliquity cycle), a turbiditic origin is unlikely; turbidity currents would produce much more irregular successions.

Ocean circulation largely determines the supply of inorganic nutrients to the photic zone affecting the production and sinking of organic matter. This organic matter, in turn, determines the oxygen demand in the water column. Circulation also determines the supply of oxygen to the various levels of the basin. Generally this oxygen is supplied by dense (cold and/or relatively saline) waters which have descended elsewhere and may have moved laterally over great distances.

Both oxygen supply and production of organic matter depend on circulation intensity. Figure 2a shows the tentative relations between present-day oceanic circulation intensity and the production of organic C and its descent into deep water on the one hand, and the supply of oxygen to deep water on the other hand. Figure 2b tentatively shows the effect of a basin-wide reduced circulation leading more easily to anoxicity. Similarly, a reduced input of O_2 into the system (Fig. 2c) and an increased input of organic C (Fig. 2d) enhance anoxicity, at the stagnant side in the lower left, as well as at the upwelling side in the upper right of the figures.

In situations of a near-equilibrium of the supply of oxygen with that of oxygen-consuming organic matter to the deep ocean, slight changes of external variables (i.e., climate, oceanic circulation intensity) may cause the system to shift to and fro aerobic and anoxic conditions. This may be due to variations in the supply of oxygen, or in the supply of organic matter, or to a combination of both. Regular shifts over the aerobic-anoxic boundary are inferred to have occurred, amongst others, during the Cretaceous, when rhythmic successions of anoxic and oxygenated sediments accumulated (Arthur 1979a; McCave 1979; Dean et al. 1977; de Boer 1983; Arthur and Dean, Chap 1.7, this Vol.). Circulation velocity was low, and warm and saline ocean waters could retain little oxygen in solution. Astronomically induced fluctuations of climate and oceanographic conditions were an obvious factor.

1.2.1 Anoxic Conditions During the Cretaceous

The vast amount of organic matter which was stored within Lower and Middle Cretaceous pelagic sediments (Schlanger and Jenkyns 1976; Arthur 1979a; Tissot

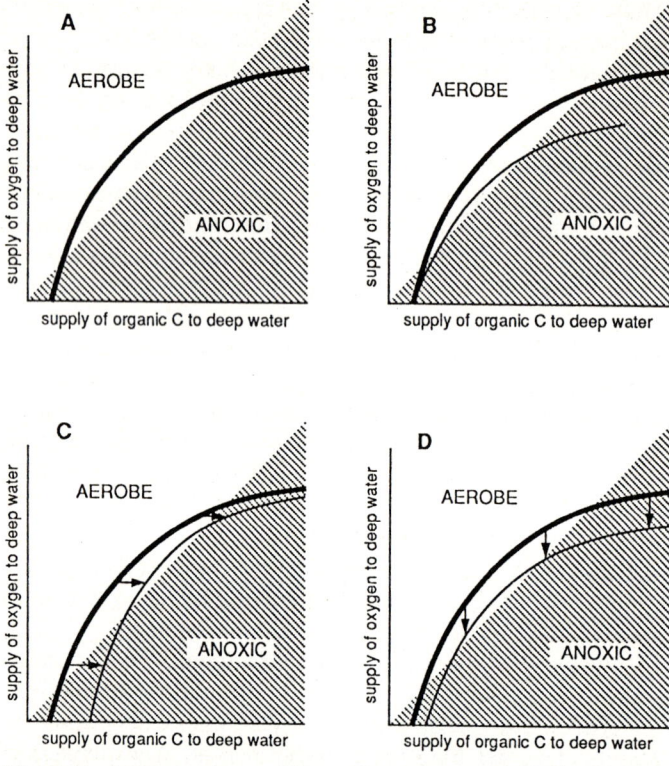

Fig. 2A-D. A Oxygenation conditions of deep ocean water in response to the oxygen supply (vertical axis) and the organic carbon flux (i.e., organic matter supply, horizontal axis) to deep water. Both oxygen supply and production and preservation of organic matter depend on circulation intensity. The *solid line* tentatively delineates the conditions met in present-day oceans: anoxia in deep water at present occurs in very strong and in very slow circulation (in the figure this corresponds with the *upper right* and *lower left* respectively). Under "normal" conditions, with moderate circulation, deep waters are at present sufficiently aerated to oxidize the dead organic matter which descends (*middle part of the curve*). **B** In times of reduced circulation, the production of organic matter and the supply of oxygen to deep water are reduced. Thus, the values along both axes are compressed, making bottom waters depleted in oxygen to expand in volume (*thin line*) and anoxicity to become more obvious. **C** Input of terrestrial organic matter by rivers or wind will shift the curve towards the right into the anoxic part of the field. **D** In times of poor ventilation, due to a decreased O_2-solubility in warm and saline waters (Cretaceous) or to a low oxygen content of the atmosphere [Paleozoic, (Pre)Cambrian], the curve is depressed vertically downward

1979; Scholle and Arthur 1980; de Boer 1986) has given rise to speculation about a period of high organic production (Jenkyns 1980; Habib 1982), and it has been proposed that density-driven circulation caused by salinity differences might have been vigorous (Brass et al. 1982; Southam et al. 1982; Barron 1986). However, it has also been suggested that the climate in the Cretaceous was warm and equable (Barron and Washington 1982). The circulation was slow, due to the (near-)absence of polar ice caps, and (largely) salinity-driven due to evaporation in shallow seas at low latitudes (Brass et al. 1982). The resulting organic production must have been low (Bralower and Thierstein 1984; de Boer 1986; Stein 1986b; Thierstein 1989).

One might wonder if for the great amount of organic matter stored in the Cretaceous oceans, the stagnant basin model (Fig. 1a) would not be more feasible than the upwelling model (Fig. 1c). However, the whole system was sluggish compared to present rates, i.e., not only was the refreshment of deep waters probably slow, but organic production also reached only 10% of present-day values (Bralower and Thierstein 1984). Black shales were probably formed in upwelling as well as in stagnant systems, but the intensity of "upwelling" in the absolute sense was probably at a much reduced level compared to the present day (Fig. 2b,d). The rare reworking of sediment and the scarcity of hiatuses in Cretaceous oceanic sediments further imply a very low-velocity circulation system (Arthur 1979). On the basis of species diversity of calcareous nannofossils, Roth (1987) also concluded that oceans would have been highly stratified and have shown slow vertical mixing and poor ventilation. The widespread storage of organic matter in pelagic sediments thus can be ascribed to a reduction in O_2 supply to deeper water, due to a reduced circulation, and due to the decreased solubility of oxygen in warm saline waters. Douglas and Savin (1975) and Salzman and Barron (1982) estimated that the temperature of bottom waters was about 15 °C higher than today. Under these conditions, upwelling (Fig. 1c), as well as some degree of stagnation (Fig. 1a), led, more easily than in other periods, to anoxicity of bottom waters (cf. Parrish and Curtis 1982; Arthur et al. 1984b). The oxygen-deficient tendency of the Cretaceous oceanic system was strongly intensified during the maximum transgression at the Cenomanian-Turonian boundary (Schlanger et al. 1987; Arthur et al. 1987). The carbon involved in the large-scale storage of organic C during the Cretaceous must have been largely derived from increased volcanism (de Boer 1986). For the organic rich sediments from the early Paleozoic Thickpenny and Leggett (1987) infer similar climatic and oceanographic conditions. However, many features, which eventually may allow a better understanding of the paleocirculation and the paleoproductivity of the oceans of the past, are still poorly understood (see Fleet and Brooks 1987; Hay 1988).

In conclusion. Storage of organic matter and the formation of black shales occurs under the conditions represented in the upper right and the lower left sides in Fig. 2a-d, i.e., under (strongly) upwelling (estuarine circulation) conditions (Fig. 1b,c) and in very slow circulation and (nearly) stagnant bottom waters (Fig. 1a). Moreover abundant organic matter may be stored within sediments deposited at a high rate (e.g., Müller and Suess 1979; Degens et al. 1986).

1.3 Deposition of Pelagic Carbonate

The production of pelagic carbonate (mainly calcareous nannofossils, pelagic foraminifera) depends, amongst other factors, on the supply of nutrients to surface waters, and it is, with some exceptions, positively correlated with the circulation velocity. In the same way the production of organic matter depends on circulation velocity and the consequent supply of nutrients to the photic zone, but the amount of organic matter reaching the seafloor and being incorporated in the sediment depends on the oxygen stock available in the water column. In addition, terrigenous sediment is supplied to the ocean bottom, being transported in suspension by ocean currents or through the air. In the case of a more or less regular supply of non-carbonate sediment, the ratio pelagic carbonate/terrigenous sediment should reflect the rate of carbonate production in the photic zone. Various features, e.g., competition between carbonate-producing and other organisms and dissolution below the aragonite/calcite compensation depth (ACD/CCD), may disturb this picture.

2 Variations of Carbonate and Organic C Content: Character and Origin

2.1 Nature and Origin of Carbonate-Marl Rhythms

Initial variations of caronate content in pelagic sequences may be due to variations of carbonate production in the photic zone, to variations of the supply of other sediment diluting the carbonate, or to dissolution in the water column and at the sediment surface in the case of deposition at or below the lysocline or below the CCD (see Einsele and Ricken Chap. 1.1, this Vol.). For example, variations in carbonate content often concern differences of less than 10–20%, but a small difference of $CaCO_3$ content may result from a large variation in carbonate productivity when the input of non-carbonate matter is essentially constant and when no dissolution occurs. This is exemplified in Fig. 3.

Diagenesis during burial often leads to an increase of initial differences of carbonate content between adjacent beds (Ricken 1986 and Chap. 1.8, this Vol.;

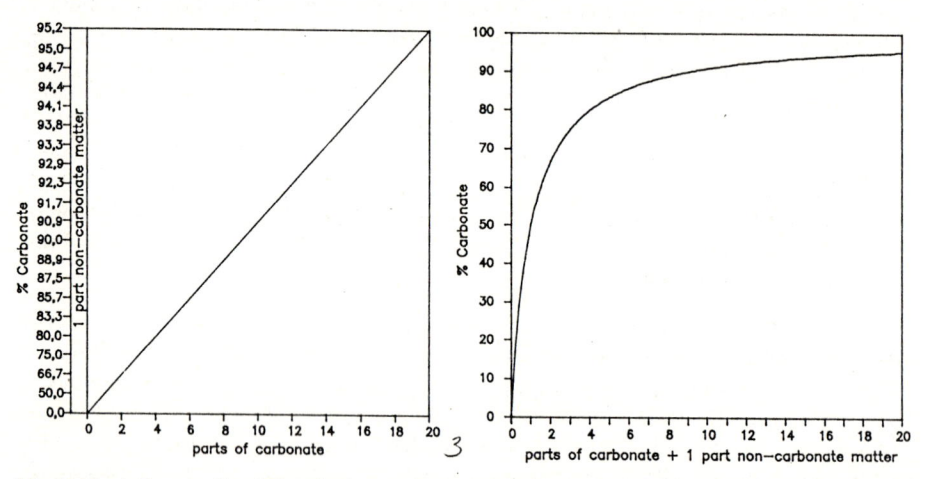

Fig. 3. The influence of variations in the supply of carbonate on the carbonate content of a sediment in case of a constant supply of non-carbonate sediment (see also Arthur and Dean 1986 and Ricken Chap. 1.8, this Vol.)

Bathurst 1987). Moreover, weathering of fossil sequences after uplift to the Earth's surface can accentuate differences of carbonate content, resulting in relatively small differences of the carbonate content between adjacent "carbonate-rich" and "carbonate-poor" intervals leading to an expression as "carbonate" and "marly" beds (de Boer 1983) (see also Einsele and Ricken Chap. 1.1, this Vol.).

Bedding rhythmicity, due to regular variations in carbonate content induced early workers to speculate about "rhythmic unmixing" due to purely diagenetic causes. However, primary differences in sediment composition are a prerequisite for rhythmic bedding (Ricken 1986, 1987, Chap. 1.8, this Vol.; Bathurst 1987). Such differences, when present, are often enhanced by diagenetic redistribution of carbonate, and, if seen in the outcrop, by weathering processes. As pointed out by Ricken (Chap. 1.8, this Vol.), a carbonate content fluctuating around 60% provides the best field exposures of cycles which become recognizable upon weathering.

In addition to variations of carbonate content, the state of oxygenation of the sediment may have varied during deposition. A resulting dark gray or black color due to the presence of organic matter and of pyrite may enhance the rhythmic appearance of successions with varying carbonate content, especially when the organic matter and pyrite are concentrated in layers poorer in carbonate (Fig. 4). This is illustrated by e.g., Dean et al. (1978, their Fig. 5).

2.1.1 Stable Oxygen Isotope Record

Stable oxygen isotope ratios of pelagic carbonate-marl rhythms often show low values for the "carbonate-poor" intervals, e.g., in the Early Cretaceous in the Southern Alps (Weissert et al. 1979), in the Cretaceous of the Umbrian Apennines (de Boer 1983; e.g., see Fig. 4), in the Cretaceous of the Vocontian Trough (Cotillon and Rio 1984), and in the Cenomanian of the Western Interior Basin (Kauffman 1988a).

For an explanation of the low $\delta^{18}O$ values of the carbonate in the marly units as an essential original feature reflecting the environment in the photic zone, two possible mechanisms should be considered:

1. fluctuations of the surface water temperature: the carbonate of the carbonate-rich beds would have been formed in relatively cool surface water (de Boer 1982b, 1983), and
2. fluctuations of the salinity, proposed by Pratt (1984) and by Kauffman (1988a); i.e., the salinity would have been relatively low at the times of formation of the carbonate-poorer beds.

The first mechanism, variations of surface water temperature (on a ka scale), fits in well in a model with variations of circulation and upwelling intensities related to shifts of climatic zones due to changing orbital influences on the Earth's climate (see Fig. 5). In the example shown in Fig. 4, the difference of $\delta^{18}O$ between adjacent carbonate-rich and carbonate-poor beds is some tenths of a per mile. This could be the result of differences in temperature of one to a few degrees Celsius. Of course, the cause of temperature differences of oceanic surface waters should not be sought in changes of atmospheric temperatures and insolation only, but also in changes of

Fig. 4a,b. Pelagic carbonate sequence of alternating gray to black "marly" beds and light-colored "carbonate" beds, Upper Albian, Moria, Apennines, Italy. Scale on the photo is 2 m. Assuming that the input of non-carbonate matter was constant and diagenetic modifications were moderate, each 1% rise of carbonate content in this example implies a rise of input of carbonate of 4–5%. The sedimentation of $CaCO_3$ thus would have varied up to a factor of 2 between the "marly" and the "carbonate" beds (see Fig. 3). The succession was deposited at a water depth of about 1 km (Wonders 1980), i.e., well above the Carbonate Compensation Depth for that period

oceanic currents, upwelling, and supply rate of waters derived from higher and cooler latitudes.

The second mechanism implies for the deposition of "carbonate-poor" intervals a decrease of surface water salinity. The reduced circulation velocity in the Cretaceous oceans might well have led to a greater relative influence of precipitation and/or evaporation, producing variations in salinity. In areas where precipitation exceeded evaporation, a density stratification due to fresh water input would have caused a slowing-down of the already slow circulation and low organic production in the surface waters (Kauffman 1988a) and may have contributed to or caused the observed isotopic effects. Based on physical models of ancient oceans, Barron (1986) stated that (orbitally influenced) surface runoff patterns may have influenced stratification in marginal seas and could have had a global effect. Either or both of the above mechanisms, i.e., (slightly) elevated temperatures and a decrease of surface salinity at the times of deposition of the carbonate-poor beds, may have been active. Further studies are needed to clarify such hypotheses.

Some of the examples of the fossil record, however, show an opposite oxygen isotope pattern with low $\delta^{18}O$ values in the carbonate-rich beds, and are apparently due to fluctuations in the original stable oxygen isotope composition of the seawater. It is to be mentioned, however, that the examples described for the Plio-Pleistocene (Diester-Haass et al. 1986; Diester-Haass Chap. 1.5, this Vol.) are related to an increase in upwelling during (warm) interglacial periods. In the example from the Upper Cretaceous Chalk in Arkansas, USA (Bottjer et al. 1986) the carbonate-*rich* intervals would have formed in relatively warm periods, unless diagenesis has overprinted an original signal, which is unlikely considering the low burial depth. A common feature of these Upper Cretaceous chalk deposits is that they were formed at relatively high latitudes. Bottjer et al. (1986) noted that in the Lower Chalk of SE England the calcispheres are well preserved in the marls but less abundant than in the carbonate-richer intervals, whereas planktonic foraminifera do not show a parallel trend. They suggested that the calcisphere-producing algae may have been less tolerant of turbid water and/or water of less than normal salinity. Considering the high paleolatitude of deposition of many Upper Cretaceous chalks (35°N or more) the temperature of the surface waters might have been a limiting factor for the biogenic production of carbonate in these cases rather than nutrient supply (Ditchfield and Marshall 1989; Leary et al. 1989) which also would explain the observed oxygen isotope pattern (de Boer 1991). The inferred presence of an obliquity signal in the Cenomanian (Hart 1987) and Turonian (Cottle 1988) indeed indicates a high paleolatitude of deposition.

2.1.1.1 Diagenesis?

A diagenetic origin for the difference in oxygen isotope composition between carbonate-rich and carbonate-poor beds in examples with the higher $\delta^{18}O$ values in the carbonate-rich beds (see Fig. 4) is unlikely. Precipitation of carbonate in initially carbonate-richer beds is the common diagenetic feature (see Ricken and Eder Chap. 3.1, this Vol.), and would have an isotopic effect making the bulk carbonate in the "carbonate-rich" beds change to lower values (see Scholle 1977; McKenzie et al. 1978; Jenkyns and Clayton 1986). Considering the pattern observed in the above examples, diagenetic redistribution of carbonate would have lowered the mean $\delta^{18}O$ values of the carbonate-rich beds and thus have diminished the original differences. This suggests a primary origin of the carbonate-marl rhythms. An example of an opposite pattern caused by an inferred diagenetic shift was given by Arthur et al. (1984a). In the Early Cretaceous at Deep Sea Drilling Project (DSDP) site 535, Cotillon and Rio (1984) found systematically lower $\delta^{18}O$ values for the carbonate in the carbonate-rich beds compared to the marly intervals. Whereas throughout that succession $\delta^{18}O$ values are predominantly negative, a diagenetic origin should not be excluded.

2.2 Organic Carbon Cycles

In the case of slow deposition, fluctuations in the content of organic carbon depend on variations in the supply of organic matter and/or variations in the state of oxygenation of the bottom waters. In the case of near-equilibrium between the supply of oxygen and of organic matter to deep water, small shifts of external variables may cause a cyclic pattern of black shales and oxygenated sediments (Figs. 1 and 2).

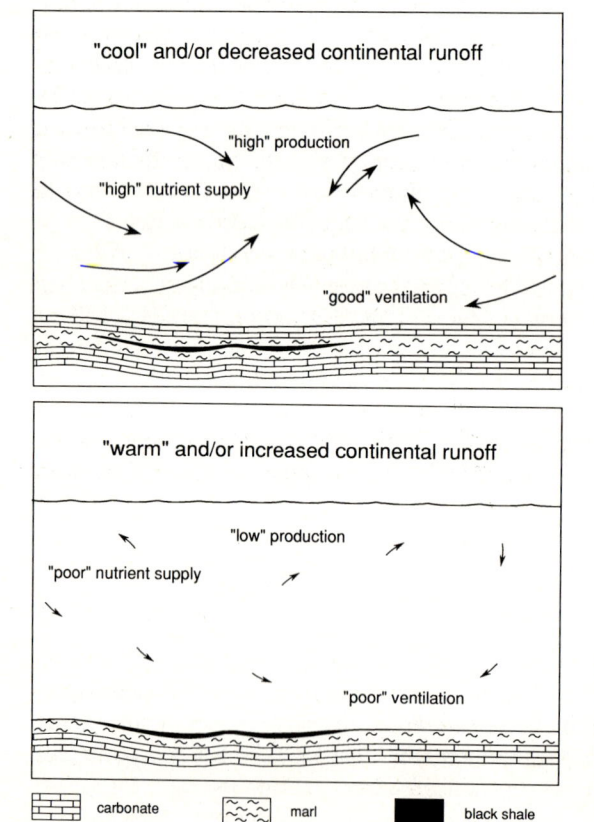

Fig. 5. Model showing the deposition of alternate "carbonate-rich" and "carbonate-poor" intervals in relation to variations of circulation velocity of ocean waters, temperature and salinity. Orbital variations may drive such alternations

3 Astronomical Rhythms

Astronomical influences on climate and sedimentary environment may result in rhythmic successions of different lithologies. For low latitudes, the precession of the Earth's axis is assumed to be the cause of such rhythmicities. At higher latitudes, the changing obliquity of the Earth's axis may dominate the effect of the precession, and be recorded in sedimentary successions.

A particular feature of precession-induced sedimentary rhythms, evolving from field observations and statistical analyses, and predicted also by astronomical theory, is the modulation of the effect of the precession (main cycle periods at present ~19 and 23 ka) by the changing eccentricity (at present ~100 ka). This leads to a bundling of precession-related units in groups of four to five. Often, such bundles are clear in the field (de Boer 1982a, 1983; Schwarzacher and Fischer 1982). Analyses of and conclusions about Milankovitch influences are based on calculations of the mean time-span occupied by lithological alternations (the number of carbonate/marl/black shale couplets per stratigraphic/time unit), on spectral analysis of equally spaced observational and analytical data, or on measurements of thicknesses of beds and/or the content of carbonate, organic matter or other constituents followed by a statistical analysis (see Weedon Chap. 7.4, this Vol.).

One important characteristic of the astronomical variables is that these do not present simple sine wave patterns, since they vary in frequency as well as in amplitude (Fig. 6; Berger 1978b); for example, the precessional period varies between about 14 and 28 ka, and the amplitude of the latitudinal shifts of the caloric equator may be as little as 1° and as much as 10°. Thus, the average period of the astronomical cycles shows some variation, depending on the time-span chosen, but especially if the chosen interval is short. In the example shown in Fig. 7, variations in thickness of the carbonate-marl couplets clearly are largely the result of variations in the amount of carbonate contributed. The pattern is very similar to the effect which is to be expected on theoretical (astronomical-climatological) grounds. If we consider the "irregularity" in the precessional period and the amplitude (see Fig. 6), it is clear that the analysis of a succession sampled at equal space (or equal time) intervals will not reveal ideal frequencies, unless the sampling is done in great detail with complete time control. De Boer (1983) analyzed such a succession with supposedly precessional rhythmicity by measuring bedding thicknesses, and by comparing the amplitude of this signal with the character of the successive amplitudes of the precessional signal, as expressed, for example, in the changing position of the caloric equator.

3.1 Paleolatitude and Sensitivity of Climate and Sedimentary Facies to Orbital Forcing

The orbital parameters have a small seasonal effect (Berger 1978a). Various (feedback) mechanisms, such as continental ice-buildup, flooding of continents, atmospheric CO_2 (greenhouse effect), oceanic circulation and upwelling, etc., can amplify the astronomical signal (see Einsele and Ricken Chap. 1.1, this Vol.).

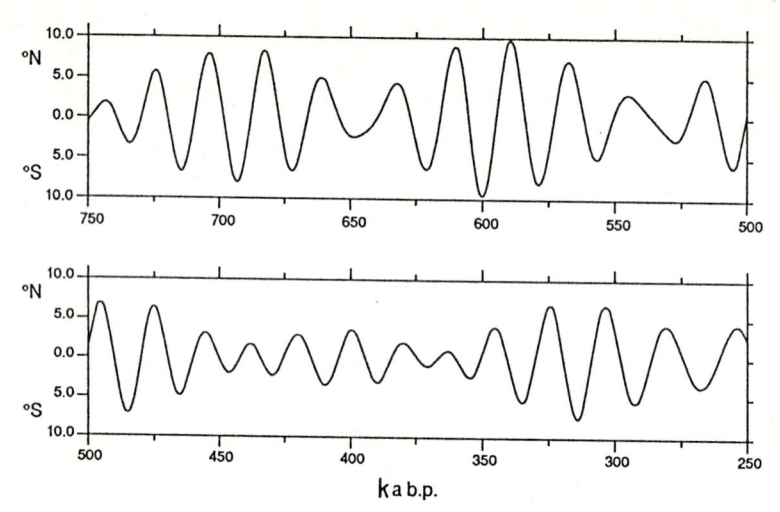

Fig. 6. Representation of the changing position of the caloric equator (Berger 1978b). Note that the frequency as well as the amplitude vary in time

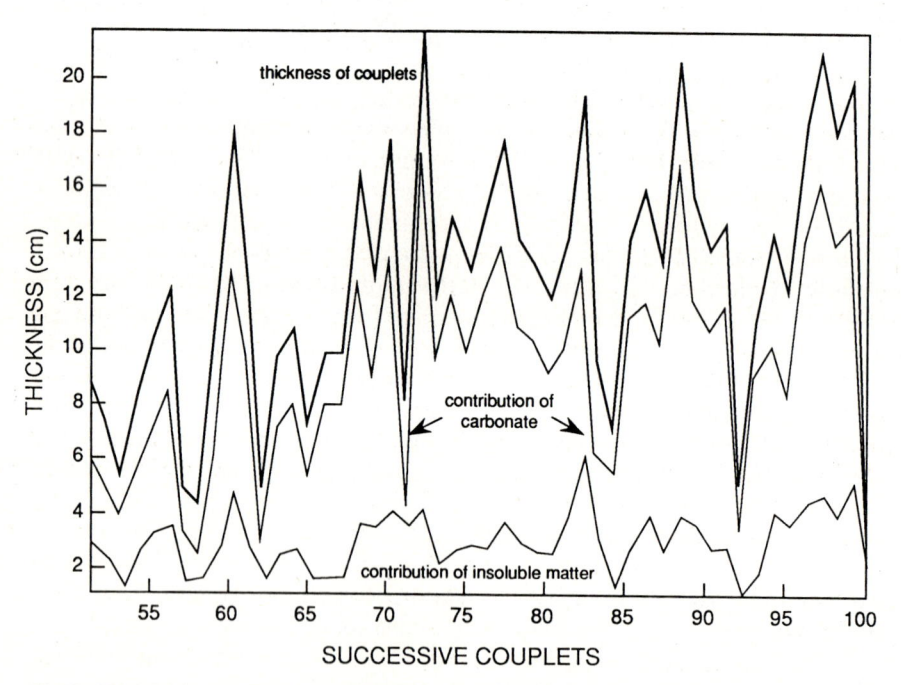

Fig. 7. Thicknesses of a series of carbonate-marl couplets in a rhythmic pelagic sequence with the contributions of carbonate and non-carbonate sediment (inclusive biogenic silica) indicated separately (Albian, Moria, Apennines, Italy; de Boer 1983). Variations of thickness are largely determined by varying amounts of carbonate. The positive correlation of the non-carbonate (HCL-insoluble) matter is (at least partly) due to admixtures of biogenic silica

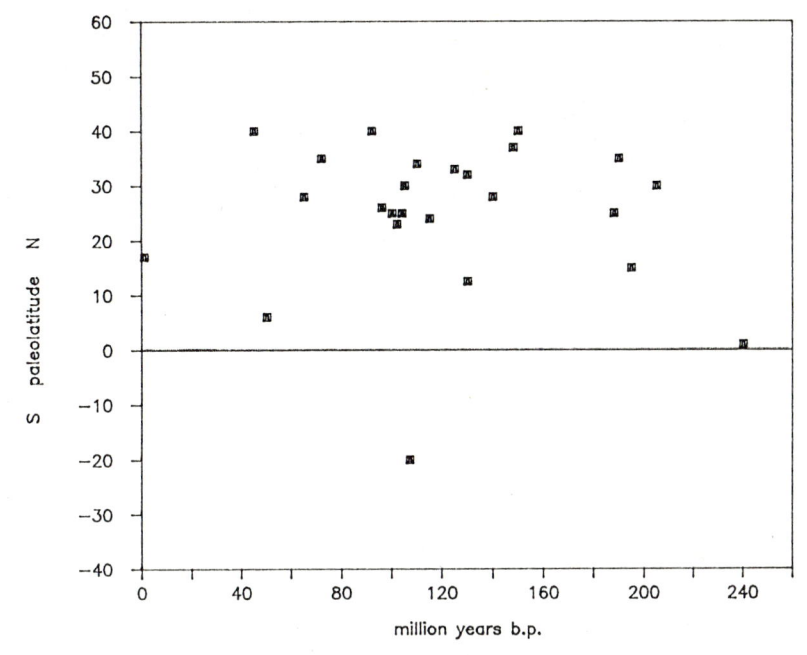

Fig. 8. Age and paleolatitude of sedimentary successions with inferred precession-induced rhythmic character. Sources: Anderson (1984), Bottjer et al. (1986), Bradley (1929), Cotillon and Rio (1984), De Boer (1983), De Boer and Wonders (1984), Dean et al. (1977), Ferry and Schaaf (1981), Fischer (1964, 1980), Foucault and Fang (1987), Foucault and Renard (1987), Foucault et al. (1987), Goldhammer et al. (1987), Hardie et al. (1986), Herbert and Fischer (1986), McCave (1979), Montadert et al. (1979), Olsen (1984, 1986), Rye (1977), Schwarzacher (1954), Strasser (1988), Van Houten (1964), Weedon (1986, 1989)

A compilation of literature examples of precession-induced cyclicities (Fig. 8) shows that examples of astronomical forcing from the sedimentary record are most frequently found at about 30° paleolatitude. In the past there has been no clustering of, presently accessible, sedimentary environments at that paleolatitude nor of DSDP sites. This leads to the conclusion that sedimentary systems are especially sensitive to the (amplified) influences of the precession of the Earth's axis at about 30°. Precession-related shifts of the boundary between the tropical dry and the adjacent moister climate belts obviously explains lacustrine/evaporitic rhythmicities (Castile Fm, Permian, Anderson 1984; Lockatong Fm, Triassic-Jurassic, van Houten 1964; Olsen 1986; Green River Fm, Eocene, Bradley 1929). In the ocean, the zone of subtropical convergence is located at about 30° and, especially in the absence of polar bottom waters (Jurassic, Cretaceous), changes in its exact location must have had a significant influence upon the supply of nutrients to, and organic productivity in surface waters, and supply of oxygen to the deep. Moreover, precession-related variations in precipitation/evaporation, and resulting variations in the production of highly saline, warm and poorly oxygenated waters in shallow shelf seas has probably been an important feature in ice cap-free periods, causing intermittently anoxic conditions in the deep ocean. With respect to the exact latitude of such sensitive zones during geological history it must be kept in mind that, apart from astronomical influences, in the long term various features (ice caps, distribution and relief of landmasses, connections between ocean basins) may have pushed this sensitive zone towards (slightly) lower and higher latitudes.

The data in Fig. 8 also suggest that the northern hemisphere is the more sensitive one, but one should realize that more exposures and DSDP sites are located in the northern hemisphere than in the southern hemisphere, and that the northern hemisphere has been geologically better explored.

4 Cyclic Production of Carbonate and Organic Matter and its Preservation

The supply of carbonate, terrigenous sediment, and organic matter to the deep sea during geological history, as well as the carbonate saturation and the oxygen supply in deep water, are all subject to fluctuations. These fluctuations are ultimately caused by variations in external factors, such as climate and tectonics. In the short term (10-ka to 100-ka range), climate is probably the most important variable producing gradual and regular variations in the sediment composition. Over a longer time period (Ma to 100 Ma), plate movements, the morphology of ocean basins, orogenesis, etc. are important controls. Incidental features such as the closure or opening of seaways, avulsion of large river systems, etc. may result in sudden shifts in sedimentation patterns.

Special attention has been given in the above to organic productivity (of carbonate and organic matter) and the supply of O_2 to deep water. In the Mesozoic examples, the high sea level and the low input of siliciclastics to deep marine settings appear to have especially favored the recording of the relatively weak astronomical signs in pelagic sedimentary successions. The generally slowly circulating oceanic system led to low biogenic productivity in the oceans, and a rather low rate of $CaCO_3$ production. Since the input of siliciclastics (mainly clay minerals) to the open ocean was low as well, the pelagic sedimentary successions are often dominated by carbonate. The low oxygen level of ocean waters meant that, despite the low organic production, anoxia in deep water could occur more easily than in other periods, and that abundant organic matter could be preserved within fossilizing sediments. In this way the excess juvenile carbon which was liberated from the degassing young basalts associated with increased spreading of oceanic plates and volcanism was removed (Budyko and Ronov 1979; Fischer 1981; de Boer 1986). In cases of a near-equilibrium in the supply of oxygen and of oxygen-consuming organic matter to deep water, small shifts in oceanographic conditions could produce successions of alternating oxidized and anoxic sediments. Astronomical influences (precession and changing obliquity of the Earth's axis) may be responsible for such small shifts of oceanographic conditions.

Apart from variations of surface productivity and of oxygen supply to deep water other astronomically driven mechanisms may similarly produce rhythmic successions. For example, where deposition was close to the CCD, intermittent dissolution could influence the character of the succession. Rhythmic successions can develop also with a fluctuating input of wind-blown or water-borne terrigenous matter, irrespective of the processes acting in the ocean. Moreover, shifts of biogenic production between carbonate-producing organisms and organisms producing silica tests or only organic matter could produce rhythmic successions.

For the Cretaceous examples, the often observed negative correlation between the content of organic matter and carbonate in pelagic sequences with black shales (see, e.g., Figs. 2 and 4; Dean et al. 1978, their Fig. 5) suggests, as discussed above, that deposition often occurred in slowly circulating systems, above the CCD, and under the subordinate influence of terrigenous input. In poorly circulating to stagnant systems, an increase of circulation velocity must cause an increase of organic production of carbonate and of organic matter, as well as an increase in the supply of oxygen to deep water in such amounts that much or all of the sinking and depositing organic matter can be oxidized (Fig. 2). This leads to successions of beds with negatively correlated carbonate and organic C contents (see de Boer 1983; Herbert et al. 1986).

The supply of carbonate-poor, organic-rich sediments by turbidites may also produce patterns with a strong negative correlation between carbonate and organic matter (Degens et al. 1986). However, on the scale of the individual turbidite beds a relation to astronomical parameters, such as, e.g., the above bundling of four to five beds, should not be expected. Moreover, sedimentary structures typical of the deposition from turbidity currents will then be present. Nevertheless, on the scale of (long) successions of turbidite beds, astronomically induced climatic and oceanographic changes may affect the character of turbidite deposits by producing gradual changes in the composition of the successive beds (cf. Haak and Schlager 1989).

5 Discussion and Conclusion

Examples of pelagic sediments with rhythmic changes of the content of carbonate and organic matter, produced by orbital climatic forcing, are derived especially from the Mesozoic, when the high sea-level stand led to a large extension of the pelagic and hemipelagic sedimentary domain where sedimentation could continue, little disturbed relatively by fluctuations of sediment input from the land. During periods in the Cretaceous, the oceanic system suffered from anoxia, primarily as the result of increased sea floor spreading in combination with the related large output of juvenile carbon and a warm global climate.

Conditions for the formation of chalk/marl/black shale alternations are met in the deep ocean, above the CCD, in areas with a low and regular input of clastic sediment from the continent. Alternations of carbonate-rich and marly beds depend on variations of the production of biogenic carbonate in the photic zone, relative to the input of non-carbonate sediment. Where a deficiency of oxygen in ocean waters occurs, astronomically induced changes of climate and oceanography may similarly produce rhythmic successions of alternating anoxic and aerobic sediments.

A sensitive sedimentary system is essential for the reflection of orbital variations in the sedimentary succession. In theory such orbital climatic effects should be of influence at all latitudes and in all sedimentary environments, but mostly the effects are overshadowed by other features. The reported occurrences of astronomically induced sedimentary cycles show that the deep sea and lacustrine-evaporitic sedimentary environments must be the most sensitive for producing successions with recognizable cyclic lithological patterns, especially at latitudes of about 30°. Of course, at any given time, different (parts of) sedimentary environments may experience different effects from orbitally driven changes of climate, and they may produce cyclic patterns of differing natures (see Arthur et al. 1986, their Fig. 4; Kauffman 1988a). Further studies of precession-induced climatic changes, their effect upon erosion and transport of clastics, and upon depositional sequences in sedimentary basins as, e.g., performed by Perlmutter and Matthews (1990) may, in the future, well reveal Milankovitch rhythmicities in other sedimentary environments which have not yet been studied from this point of view.

Acknowledgments. I thank George Postma and Eelco Rohling for commenting on earlier versions of the manuscript, and Werner Ricken and Gerhard Einsele for their suggestions and patient assistance in its completion.

Chapter 1 Rhythmic Stratification

1.4 Environmental Factors Controlling Cretaceous Limestone-Marlstone Rhythms

D. L. Eicher and R. Diner

1 Introduction

Limestone-marlstone or chalk-marl sequences are particularly common and widespread in the Cretaceous, although they are known from deposits as old as the Jurassic and as young as the Quaternary. The focus here is on limestone-marlstone rhythms of Cretaceous age and the kinds of paleoenvironments that they must represent. A correct interpretation of the causes of these cyclic sequences will provide a great deal of insight into the behavior of ancient oceans.

In Cretaceous limestone-marlstone sequences, each individual couplet typically represents some tens of thousands of years. The limestone and marlstone beds are composed chiefly of two fractions: (1) pelagic carbonate particles, mostly calcareous nannoplankton but with some planktonic foraminifera and (2) an insoluble (in HCl) fraction, most of which consists of detrital clays, but some of which may be biogenic silica.

In typical Cretaceous limestone-marlstone sequences (Fig. 1), the limestone beds are light-colored, bioturbated, and contain little organic matter, whereas the marlstone beds are typically dark, laminated or slightly burrowed, and relatively rich in organic matter (i.e., greater than 1% organic carbon). The Cretaceous limestone-marlstone cycles thus coincide with inferred redox cycles. The cyclic nature of Cretaceous limestone-marlstone sequences was first suggested by Gilbert (1895). Later, such sequences were described in land sections in many regions (Hattin 1971, 1975; Cobban and Scott 1972; Weissert et al. 1979; Cotillon et al. 1980; Arthur et al. 1984b; Arthur and Premoli Silva 1982; Einsele and Wiedmann 1982; Cotillon and Rio 1984; Herbert et al. 1986; Pratt 1984; Bottjer et al. 1986; Schlanger et al. 1987; ten Kate and Sprenger 1989), as well as at many oceanic sites (Schlanger and Jenkyns 1976; Dean et al. 1977; Ferry and Schaaf 1981; Arthur et al. 1984b; Weissert 1981b; Cotillon and Rio 1984; see also Arthur and Dean Chap. 1.7, as well as Einsele and Ricken Chap. 1.1, both this Vol.).

2 Origin of the Rhythms

Recently the theory that Milankovitch climatic cycles produced the Cretaceous limestone-marlstone sequences has been verified convincingly by identifying, within the same sequence, cycles that are related by ratios identical to those that the various

Einsele et al. (Eds.)
Cycles and Events in Stratigraphy
©Springer-Verlag Berlin Heidelberg 1991

Fig. 1. Limestone-marlstone cycles in the Bridge Creek Member of the Greenhorn Formation (late Cenomanian-Turonian) at Rock Canyon anticline, Colorado. Underlying Hartland Shale Member makes up lower portion of the railroad cut

Milankovitch frequencies have today (Park and Herbert 1987; ten Kate and Sprenger 1989; Fischer Chap. 1.2, this Vol.). These studies have not only verified that Milankovitch climatic cycles were responsible for the limestone-marlstone rhythms, but they have also provided a precise tool for determining rates of sedimentation within the sequences.

The major question remains of how the climatic signals were translated, through the ocean, into limestone-marlstone rhythms. The variations in the clay/carbonate ratio that actually produces the rhythms could be the result either of primary depositional processes or of diagenetic processes (Einsele 1982a; Einsele and Ricken Chap. 1.1, this Vol.).

Substantial debate has centered around whether particular limestone-marlstone cycles are dominantly primary or dominantly diagenetic (Arthur et al. 1984b; Research on Cretaceous Cycles, Group 1986; Bottjer et al. 1986). Limestone-marlstone sequences generally contain evidence for some dissolution and reprecipitation of calcite; typically, coccoliths are partially dissolved, and chambers of foraminifera are filled with sparry calcite. The question is to what extent this solution transfer occurred within beds and to what extent it occurred between beds so as to turn completely homogeneous, massive carbonate-rich clays into pseudo-bedded deposits by the rhythmic unmixing processes hypothesized by Sujkowski (1958) and Hallam (1964,

1986). In most well-studied sequences the marlstone beds, which would have suffered the greatest dissolution effects by diagenetic unmixing, contain well-preserved calcareous microfossils, showing that dissolution of calcite after deposition was limited (Eicher and Worstell 1970; Cotillon and Rio 1984; Herbert et al. 1986; ten Kate and Sprenger 1989). In many cases, individual limestone-marlstone couplets can be traced great distances with little variation in thickness (Hattin 1971, 1986a). Without any primary variation, diagenesis would not be likely to be so consistent over such broad areas where depth of burial and other post-depositional factors differed. In addition, detailed examination of the couplets typically reveals that limestone is piped from above into a marlstone bed beneath by means of burrows, showing that the limestone and marlstone lithologies are a result of primary processes of sedimentation and that later solution transfer of both carbonate and silica occurred mainly at a very local level (Herbert et al. 1986).

The effects of primary dissolution within the water column or on the sea floor can similarly be ruled out as the cause of most Cretaceous limestone-marlstone sequences. Foraminiferal dissolution can be seen in Cretaceous deep sea clays that were deposited near the CCD (Arthur et al. 1984b). However, the resulting deposits are typically varicolored clays with varying carbonate content and not limestone-marlstone rhythms. Typical Cretaceous limestone-marlstone sequences were deposited well above the CCD.

Hence, the cyclicity in most Cretaceous limestone-marlstone sequences reflects variations in the input of one of the principal components. Either detrital material was in fairly steady supply and there was a marked fluctuation in carbonate input (productivity cycles), or carbonate was in fairly steady supply and there was a marked fluctuation in detrital input (dilution cycles). If the couplets represent dilution cycles, then the marlstones were deposited more rapidly than the limestones, but if the couplets represent productivity cycles, then the limestones were deposited more rapidly than the marlstones.

However, this distinction is not easy to make (Arthur et al. 1984b; Research on Cretaceous Cycles, Group 1986) and in some cases the same sequence has been attributed to different mechanisms (Barron et al. 1985; Eicher and Diner 1985). The problem is illustrated in Fig. 2, which shows two ways of graphing the carbonate and insoluble portions of limestone-marlstone rhythms of the Berriasian Miravetes Formation in southeastern Spain (ten Kate and Sprenger 1989). The two curves are based on the same samples, which were selected to reflect the entire range of values of clay/carbonate ratios in the Miravetes. The dilution curve (Fig. 2) assumes constant carbonate input, and it shows that, if the Miravetes rhythms are dilution cycles, the bed with the greatest detrital content would have been deposited about ten times faster than the bed with the lowest detrital content. The productivity curve assumes constant detrital input, and it shows that, if the Miravetes rhythms are productivity cycles, the reverse is true.

3 The Case for Productivity Cycles

In spite of theoretical arguments opposing strong fluctuations in productivity as the cause of limestone-marlstone rhythms (see discussion in Einsele and Ricken Chap. 1.1, this Vol.) field evidence favors the productivity mechanism for several Cretaceous limestone-marlstone sequences that represent settings from epicontinental basins to the deep sea (de Boer 1982a, de Boer and Wonders 1984; Cotillon and Rio 1984; Eicher and Diner 1985, 1989; Bottjer et al. 1986). There are few quantitative indicators of marine productivity (Arthur et al. 1986), but one aspect of some limestone-marlstone rhythmic sequences that points to productivity cycles as their probable cause is the correlation between carbonate content and thickness of individual beds. This relationship has been quantified for the Campanian Annona Formation of Arkansas (Bottjer et al. 1986) and for the Albian-Cenomanian Schisti and Fucoidi a Scaglia Bianca of the Italian Appennines (de Boer and Wonders 1984; Herbert et al. 1986). In these cases, an increased supply of carbonate appears to have caused the beds to become thicker. Had fluctuations in detrital input been the cause of the cycles, the beds with greater detrital content would likely be the thicker, the assumption in both cases being that the two modes were of approximately equal duration, which appears to be reasonable (Arthur et al. 1986).

An additional argument for productivity cycles was made by Bottjer et al. (1986) on the basis of calcispheres, the calcareous remains of marine algae. Their abundance in the limestone beds and paucity in the intervening marlstone beds of the Campanian Arcola Limestone of Alabama is considered as good evidence that the Arcola limestones represent times of high productivity (Boettjer et al. 1986). A brief review

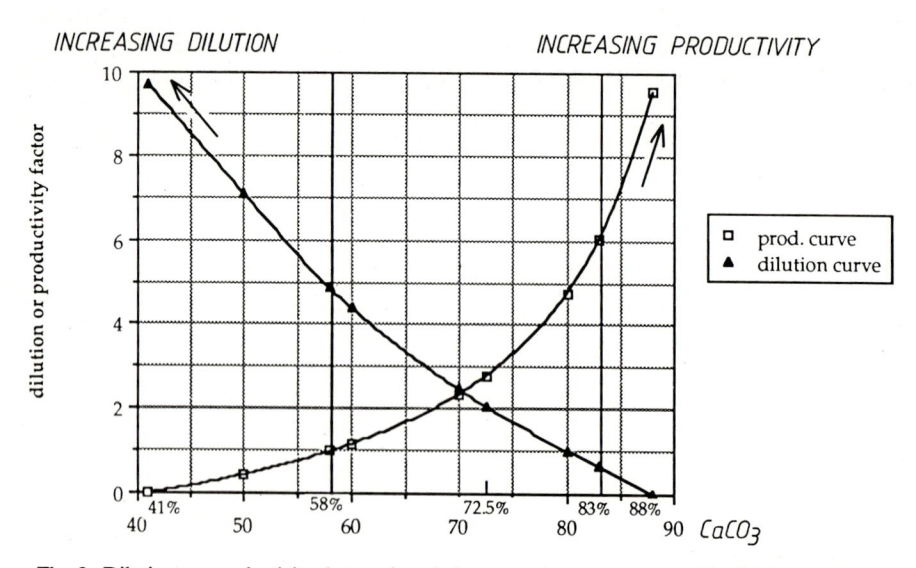

Fig. 2. Dilution or productivity factors in relation to carbonate content of individual limestone and marlstone beds in the Miravetes Formation of Spain. (Productivity and dilution models after ten Kate and Sprenger 1989)

of two well-studied limestone-marlstone sequences, one representing a deep Tethyan basin and the other a comparatively shallow epicontinental sea, will emphasize the emergence of productivity cycles as a percieved common cause of Cretaceous limestone-marlstone sequences. The first is a cored interval of late Albian age in the Schisti a Fucoidi near Piobbico, Italy (Park and Herbert 1987; Herbert et al. 1986), and the second is the Bridge Creek Member of the Greenhorn Formation in the Western Interior of the United States (Eicher and Diner 1985, 1989).

4 The Schisti a Fucoidi

The late Albian portion of the Schisti a Fucoidi consists of typical Cretaceous limestone-marlstone redox cycles, alternating light gray limestones deposited under oxygenated conditions and black calcareous shales deposited under anoxic conditions. It was deposited in a relatively deep water basin in the tropical Tethys (Herbert et al. 1986). In an 8-m interval of core taken near the town of Piobbico, the limestone beds contain much finely divided silica derived from siliceous plankton, which is an excellent indicator of marine productivity (Lisitzin 1972; Berger 1976; Molina-Cruz and Price 1977; Barron and Whitman 1981; Broecker and Peng 1982; Leinen et al. 1986). Most of the silica in the marlstone beds (and a small amount of the silica in the limestone beds) was transported in as detrital clay but this detrital silica is associated with aluminum, and the Al/Si ratio in the detrital fraction is nearly constant (Fig. 3). The silica/carbonate ratio is nearly constant also, and this tight covariance argues against dissolution as an influence on carbonate content, since carbonate and silica have different solubility characteristics. Using the Al/Si ratio, and the quantity of aluminum in each sample, Herbert et al. (1986) determined how much of the silica was detrital and how much was biogenic (Fig. 4). Their ratio in the core is similar to the ratio of biogenic silica and carbonate in the oceans today. The abundance of biogenic silica in the limestone beds is thus excellent evidence that they represent

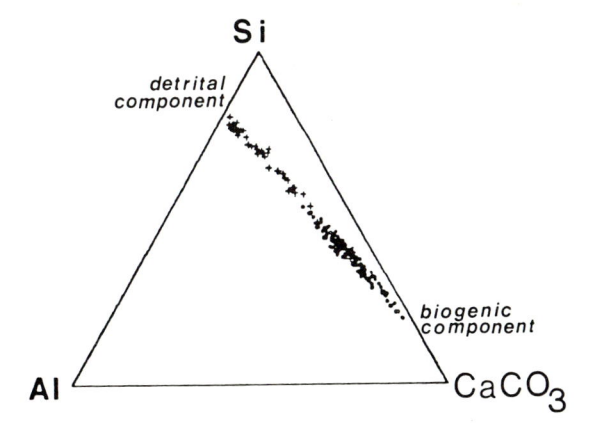

Fig. 3. Mole percent of silica, carbonate, and aluminum in Schisti a Fucoidi strata (normalized to 100%). (Herbert et al. 1986)

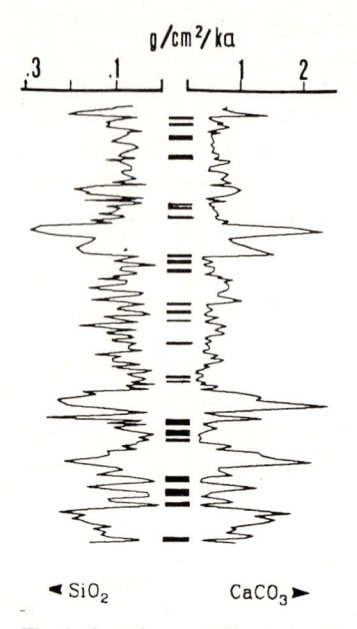

Fig. 4. Covariance of silica and carbonate flux in a 5-m interval of the Schisti a Fucoidi, calculated by normalizing biogenic silica to aluminum. Black shales (*dark bands*) mark times of inferred low productivity. (Herbert et al. 1986)

times of enhanced productivity and that the Schisti a Fucoidi rhythms are productivity cycles (Herbert et al. 1986). Additional analysis of the cycles by Park and Herbert (1987) showed that the limestone beds were deposited more rapidly than the shales. In the 8-m Piobbico core interval in 21 ka precessional cycle frequency, the obliquity cycle with a frequency of 39.2 ka, the eccentricity cycle (strongest signal of all with the two sub-peaks of 97 and 128 ka), and an eccentricity cycle of 413 ka (and see Fischer Chap. 1.2, this Vol.). Using the time framework provided by these repetitive cycles, Park and Herbert (1987) calculated variations in the rate of deposition throughout the core, and they found a linear correlation between content of calcium carbonate and the rate of sedimentation. The limestones were deposited rapidly when the production of calcareous plankton was high and the marlstones were deposited slowly when the production of calcareous plankton was low, indicating that the Schisti a Fucoidi limestone-marlstone rhythms are productivity cycles.

5 Bridge Creek Cycles of the Western Interior

In the Western Interior of the United States, the Bridge Creek Limestone Member of the Greenhorn Formation consists of typical Cretaceous limestone-marlstone rhythms (Figs. 1 and 5), many of which can be traced for hundreds of kilometers (Hattin 1971). The Bridge Creek is typically around 15 m thick and is underlain and overlain by dark calcareous shales. It records maximum transgression of the Cenomanian-Turonian transgressive-regressive cycle (Weimer 1960; Eicher 1967, 1969; Kauffman 1967; Hattin 1975), and its lower half coincides in time with the Cenomanian-Turonian Oceanic Anoxic Event (Pratt 1985), also termed with Cenomanian-Turonian Boundary Event (Thurow and Kuhnt 1986). Bridge Creek deposition occurred in a relatively shallow epicontinental sea, perhaps 500 m or more in maximum depth, some 1000 km north of the main Tethys, and hence it represents a different tectonic setting than the Schisti a Fucoidi. As in the Schisti, the limestone beds of the Bridge Creek contain little organic carbon and are intensively burrowed, whereas the marlstones are carbon-rich and microburrowed or laminated, indicating a redox cyclicity concurrent with the limestone-marlstone cyclicity.

Foraminifera of Tethyan origin are as abundant and well preserved in Bridge Creek marlstone beds as they are in the limestone beds, indicating that dissolution was not a significant factor in producing the limestone-marlstone rhythms. The rhythms resulted primarily from either variations in the supply of pelagic carbonate

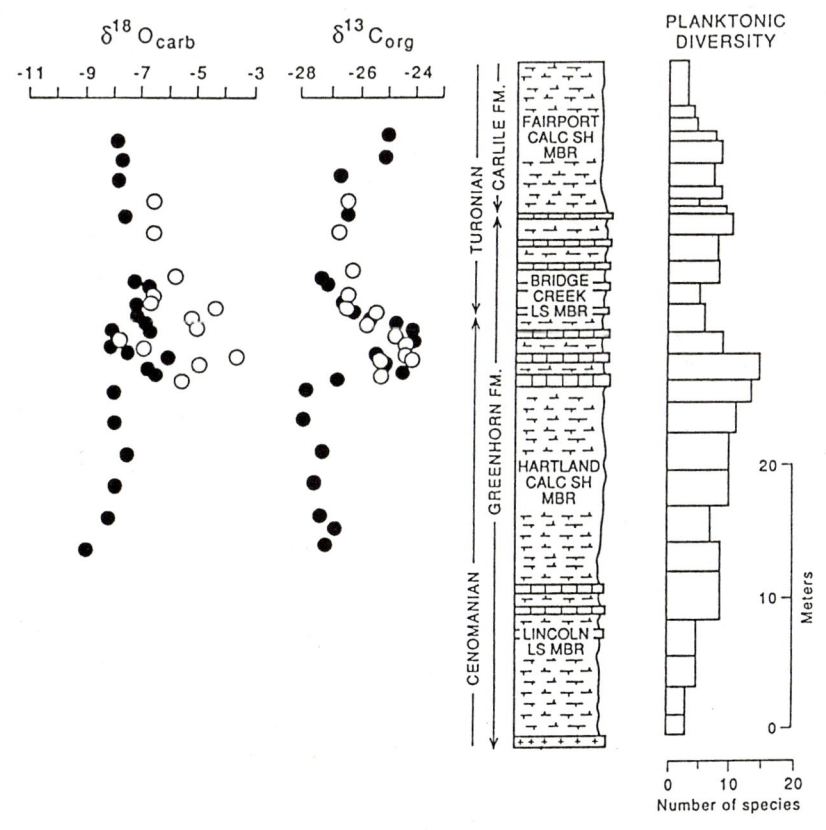

Fig. 5. $\delta^{18}O$ and $\delta^{13}C$ (vs. PDB) values and diversity of planktonic foraminifera for the Bridge Creek and adjacent deposits at Rock Canyon, Colorado. (Data from Pratt 1985; Eicher and Diner 1985).

(productivity cycles) or of terrigenous detritus (dilution cycles). Fischer (1980), Pratt (1984), Arthur et al. (1985c) and Barron et al. (1985) have made a case that the Bridge Creek rhythms represent dilution cycles. Recently, we have argued (Eicher and Diner 1989) that the Bridge Creek limestone-marlstone rhythms more likely represent productivity cycles. The correct interpretation is important because it bears on the accumulation history of marine biogenic materials, the intensity of mid-Cretaceous ocean circulation, and the patterns of climatic response to orbital forcing.

5.1 Depositional Models for the Bridge Creek

In the dilution model of Fischer (1980) and Pratt (1984), the Bridge Creek limestone-marlstone couplets reflect alternating periods of wet and dry climates in the highlands that bordered the seaway on the west. The periods were tens of thousands of years long, corresponding to Milankovitch cyclicity. During wet climate, high volumes of runoff from the western highlands produced a buoyant layer of sediment-laden brackish water across the surface of the Western Interior Seaway. The clays that settled from the brackish surface layer combined with the slow rain of calcareous plankton to produce the marlstone beds. The lid of brackish water effectively stratified the water column, inhibiting vertical mixing and fostering low oxygen conditions on the sea floor. Benthic organisms were rare and considerable organic carbon was buried without being oxidized. As a result the marlstone beds are laminated or only slightly burrowed and rich in organic carbon. During periods of dry climate, runoff from the western highlands was minimal and little terrigenous detritus reached the seaway so that relatively pure pelagic carbonate slowly accumulated. During these dry times, no brackish lid covered the seaway and vertical mixing furnished sufficient oxygen to the sea floor to support benthic organisms. The resulting limestone beds are macroburrowed and nearly devoid of organic carbon.

In the productivity model, limestones were deposited when fertility of the surface waters was high, and marlstones were deposited when fertility was low. These two different modes were not the result of differing environments in the Western Interior as in the dilution model. Instead, as sea level rose, the connection between the Western Interior and the Tethys deepened sufficiently that Tethyan currents began to reach northward and circulate through the southern portion of the Western Interior Seaway making it essentially an arm of the Tethys. The Bridge Creek cycles reflect the alternating stratified and well-mixed modes that were characteristic of the Tethys during the middle Cretaceous (Thierstein and Berger 1978; de Boer 1982a; Cotillon and Rio 1984). In effect, the Cenomanian-Turonian transgression brought Tethyan pelagic conditions into nearby epicontinental seas and thus it expanded temporarily the geographic limits of deposition of the limestone-marlstone rhythms.

5.2 Stable Isotopes

Strata deposited during the Greenhorn transgression, which culminated with the Bridge Creek Member, record progressively heavier whole rock $\delta^{18}O$ values (Fig. 5).

These represent increasing salinity as the Tethys water mass moved more and more freely into the deepening Western Interior Seaway (Pratt 1985), replacing the Boreal water mass. This interpretation is substantiated by the increasing diversity of Tethyan planktonic foraminifera upward in the interval. The Bridge Creek Member represents a peak in $\delta^{18}O$ values, but within the Bridge Creek $\delta^{18}O$ values fluctuate, those from the limestone beds being consistently heavier by 2 to 3‰ than those from adjacent marlstone beds (Fig. 6). Pratt (1985), Barron et al. (1985), and Arthur et al. (1985c) considered the isotopic differences between marlstones and limestones to be primary (but see Arthur and Dean Chap. 1.7, this Vol.).

The consistently heavier whole rock $\delta^{18}O$ values in Bridge Creek limestone beds relative to adjacent marlstone beds suggests that the salinity of the near-surface waters was higher during limestone deposition than during marlstone deposition, and the apparent low salinities during marlstone deposition could be explained by the presence of a brackish surface layer on the seaway. This would support the dilution model in which the brackish lid is inferred to have been in place during marlstone deposition.

The alternation in isotopic values between marlstone beds and limestone beds can be explained another way: perhaps these alternations were not generated in the Western Interior Seaway, but instead reflect the alternating isotopic signatures of the upper part of the water column in the Tethys Ocean to the south. The comparatively low $\delta^{18}O$ values in Bridge Creek marlstones need not represent brackish water. Even the lowest of these values is heavier than the values from the underlying Hartland Shale Member (Fig. 6), and the Hartland contains a planktonic foraminiferal fauna

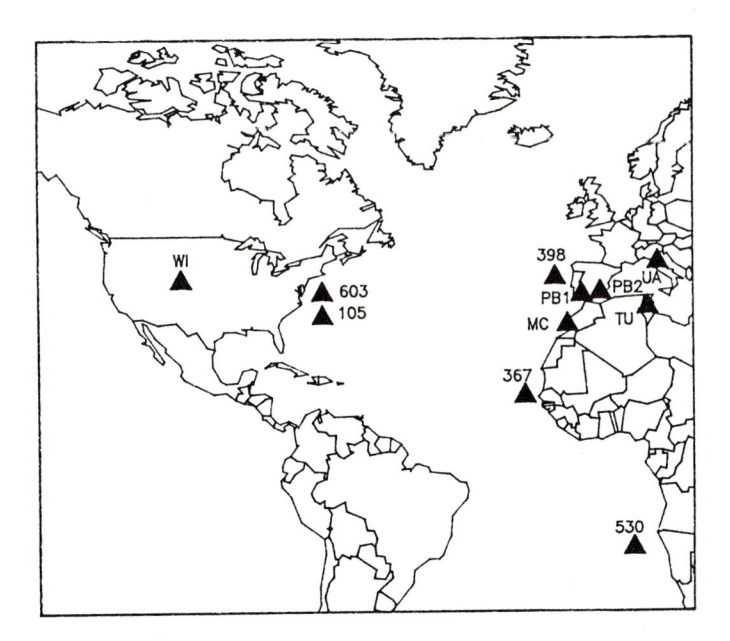

Fig. 6. Map showing localities mentioned in text

that is nearly as diverse as that in the main Tethys, indicating near-normal marine salinities for the upper part of the water column where the planktonic foraminifera lived. Hence, the upper part of the water column during Bridge Creek marlstone deposition is even more likely to have had normal marine salinities.

In the Cretaceous ocean, thermohaline circulation was probably driven chiefly by differences in salinity and not by differences in temperature as it is today (Brass et al. 1982; Hay 1983). That is, bottom water was dense because it was saline and not because it was cold. Carbonate in the warm saline bottom water was enriched in $\delta^{18}O$ as a result of its higher salinity. When rates the vertical mixing in the Tethys Ocean were low, as during deposition of Bridge Creek marlstone beds, the deeper, more saline water with its higher $\delta^{18}O$ value remained relatively isolated from shallow, less saline water with lower $\delta^{18}O$ value so that the difference in $\delta^{18}O$ values between shallow and deep water masses was high. The upper waters of the ocean were not brackish but lay within the range of salinities of Cretaceous oceanic water masses. During times when the rate of vertical mixing in the Tethys Ocean increased, as during deposition of Bridge Creek limestone beds, deeper waters mixed upward and raised the $\delta^{18}O$ values of the upper waters. At the same time, these upwelling waters brought nutrients from below the photic zone that fostered production of calcareous plankton. The Western Interior recorded this mode of circulation with the deposition of the Bridge Creek limestone beds and their attendant higher $\delta^{18}O$ values.

5.3 Planktonic Foraminifera

Perhaps the strongest evidence for the productivity hypothesis comes from the abundant foraminiferal faunas in the limestone-marlstone couplets of the Bridge Creek, which are nearly as diverse as contemporaneous faunas of the open Tethys to the south (Douglas 1972) (Fig. 5). Diverse assemblages of planktonic foraminifera occur only in open marine waters with normal salinities (Bolli et al. 1957; Be 1977). The planktonic faunas from Bridge Creek marlstone beds reflect the same kinds of marine salinities as those of the limestone beds and these faunas are not compatible with the presence of a brackish lid on the seaway during marlstone deposition. One could hypothesize that diverse planktonic foraminifera might coexist with a brackish lid if the lid were thin, and the foraminifera lived in normal marine water below it. However, no modern example is known in which a planktonic foraminiferal fauna lives beneath a low salinity surface layer (Parker 1954), nor has any been demonstrated to have done so in the past.

5.4 Calcispheres

Bridge Creek limestone beds contain an abundance of calcispheres, probably cysts of planktonic algae, which have been considered as an indicator of high productivity (Research on Cretaceous Cycles, R.O.C.C. Group 1986; Bottjer et al. 1986). Calcispheres occur in abundance in Bridge Creek limestone beds in Colorado and Kansas (Hattin 1975; Pratt 1981). Calcispheres have not been found in Bridge Creek marlstones, and hence their abundance in the limestone beds provides an additional line of evidence that the Bridge Creek couplets are productivity cycles.

5.5 Rate of Sedimentation

Estimates for the time represented by the Bridge Creek range from 0.8 Ma (Fischer 1980) to 2.45 Ma (Kauffman 1977a). These estimates translate to sediment accumulation rates of around 1 to 2 cm/ka, which is a typical pelagic sedimentation rate for regions above the CCD. The Bridge Creek appears to have been deposited in a truly pelagic setting, far from shore. A persistent plume of sediment-laden brackish water across the surface of the seaway would have deposited fine-grained detrital sediment at a high rate, much more like that in a nearshore setting than in a pelagic setting. The estimated pelagic rate of sedimentation for the Bridge Creek does not appear to be compatible with the existence of a brackish lid on the Western Interior Seaway during deposition of the marlstone beds. The sedimentation rate thus favors the productivity model for the Bridge Creek rhythms.

6 An Offshore Transect to the Ocean Basins

The Bridge Creek limestone-marlstone rhythms of late Cenomanian-Turonian age, like those of the Schisti a Fucoidi of late Albian age, are interpreted to the productivity cycles. These deposits are believed to be representative of mid-Cretaceous ocean dynamics (Park and Herbert 1987; Eicher and Diner 1985, 1989). This suggests that productivity cycles are more widespread in Cretaceous oceans than has generally been thought. It is possible that fluctuations in productivity were the principal cause of Cretaceous limestone-marlstone rhythms in the Atlantic and Tethys.

The Cenomanian-Turonian Boundary Event (CTBE), which is recorded in the Western Interior by the lower portion of the Bridge Creek, produced organic-rich facies worldwide in a variety of marine depositional environments (Herbin et al. 1986; Thurow and Kuhnt 1986; Kuhnt et al. 1986; Arthur et al. 1985c; Schlanger et al. 1987; Arthur et al. 1987). The CTBE is marked by a positive excursion in $\delta^{13}C$ that has been traced globally (Scholle and Arthur 1980; Pratt and Threlkeld 1984; Arthur et al. 1987), and it coincided with a marked rise in sea level that temporarily brought pelagic limestone-marlstone cyclicity to epicontinental seas and pronounced, abrupt changes in the deposits of marginal basins, pelagic plateaus, and the deep sea. Strata that mark the CTBE represent an unusually extensive record of facies that can be traced from epicontinental seas to the continental margin and to the deep sea with high confidence by means of the partial range zone of the planktonic foraminifer *Whiteinella archeocretacea*, which largely coincides with the positive excursion in $\delta^{13}C$.

Enrichment of $\delta^{13}C$ in deposits of the CTBE must have resulted from the sequestering of ^{12}C in buried organic carbon (Arthur et al. 1987). CTBE organic matter is chiefly of marine origin, and most of the water column, from shelves to the deep oceans, was favorable to its preservation (Herbin et al. 1986). Various oceanographic models have been offered to explain the unusually widespread burial of organic carbon. Basically, these call upon either sluggish deep water circulation (Lancelot et al. 1972) or an increased supply of organic matter (Dean et al. 1977). The first emphasizes preservation of organic matter and the second emphasizes production. Both basic ideas have been expanded extensively, for example by calling upon anoxia resulting from expanding oxygen minimum zones (Schlanger and Jenkyns 1976; Schlanger et al. 1987), stratification of the upper water column (Arthur and Natland 1979; Barron et al. 1985), or upon enhanced productivity over wide areas as a function of vertical mixing (Thurow and Kuhnt 1986; Kuhnt et al. 1986). However, understanding of the CTBE remains elusive. Herbin et al. (1986, p. 416) summarized the problem succinctly: "It is hard to imagine how such vast areas could have been stagnant during the CTBE, but it is also difficult to see how an upwelling system could wind around the different ocean margins – east and west – and even occur simultaneously on the mid oceanic ridge."

During the CTBE, limestone-marlstone cyclicity was widespread in epicontinental seas and other relatively shallow marine environments (Schlanger et al. 1987; Cotillon and Rio 1984). Redox cyclicity accompanied not only the limestone-marlstone cycles, but also the different kinds of cyclic deposits that were produced in bathyal environments and in the deep ocean basins as well (Kuhnt et al. 1986; Thurow and Kuhnt 1986; Hay 1988). Most of the carbon buried during the CTBE accumulated during the anoxic phase of the redox cycles. Early studies of the Cenomanian-Turonian episode of deposition of organic-rich marine sediments and of other "Oceanic Anoxic Events" (Schlanger and Jenkyns 1976) were inclined to view them as measurable in millions of years. Deep-sea drilling (Dean et al. 1977; McCave 1979; Arthur et al. 1987; Cotillon and Rio 1984) and studies of exposed sequences (Arthur and Premoli Silva 1982; de Boer 1982a; Herbert and Fischer 1986; Herbert et al. 1986) showed that in the Atlantic and Tethyan region this anoxia occurred in relatively brief (thousands of years) episodes, and that the redox cyclicity long known from land-based outcrops was also present in the deep oceans in varying lithologies and environmental settings (Fig. 7). Both shallow and deep redox cycles were probably responding to the same fluctuating conditions in the Cretaceous ocean.

Why did Milankovitch cycles leave such a powerful imprint on middle Cretaceous oceans? Apparently they did it by causing factors that are responsible for productivity to fluctuate in an ocean that lay in a delicate balance between two contrasting circulation modes. Precisely what conditions characterized these modes is not understood, but during the CTBE, cyclic deposition dominated almost all environments. The intensified environmental fluctuations that these cycles record appear to have acted as the catalyst for the accelerated rate of carbon burial during the CTBE.

7 Depth Facies During the Cenomanian-Turonian Boundary Event

Arthur et al. (1987) proposed that the flux of Warm Saline Bottom Water (WSBW) increased during the CTBE as a result of widespread flooding of low latitude shelves, and that this resulted in upwelling, particularly along continental margins. Eicher and

Diner (1985) extended this model into the Western Interior Seaway, but they stressed the importance of fluctuations, both in the flux of WSBW and in the corresponding organic productivity. They suggested that the Bridge Creek limestone-marlstone rhythms resulted from this mechanism, the limestones representing increased mixing of the water column and enhanced productivity and the marlstones representing retarded mixing of the water column and depressed productivity. In deeper waters of Tethyan marginal basins and the deep sea the facies are different, but the cyclicity prevails.

Table 1 lists the lithologic changes that accompany the CTBE in major tectonic settings, from shelf seas to the deep ocean basins. Figure 6 is a map that locates the sections used as examples, and Fig. 7 shows the tectonic setting of each and the lithologic makeup of the couplets in each. In all cases, the CTBE generated an abrupt shift in lithologies. The common characteristic of CTBE deposits are (1) an increase in organic-carbon-rich black shales during the CTBE interval and (2) the rhythmicity apparent in all CTBE deposits irrespective of the rock types.

7.1 Epicontinental Seas

Facies of CTBE strata vary in epicontinental seas. In the Western Interior the limestone-marlstone rhythms that mark the CTBE overlie dark calcareous shales. In southeast Morocco (Figs. 6, 7) the relationship is similar except that cherts occur within the limestone-marlstone rhythms (Table 1). In Tunisia, the limestone-

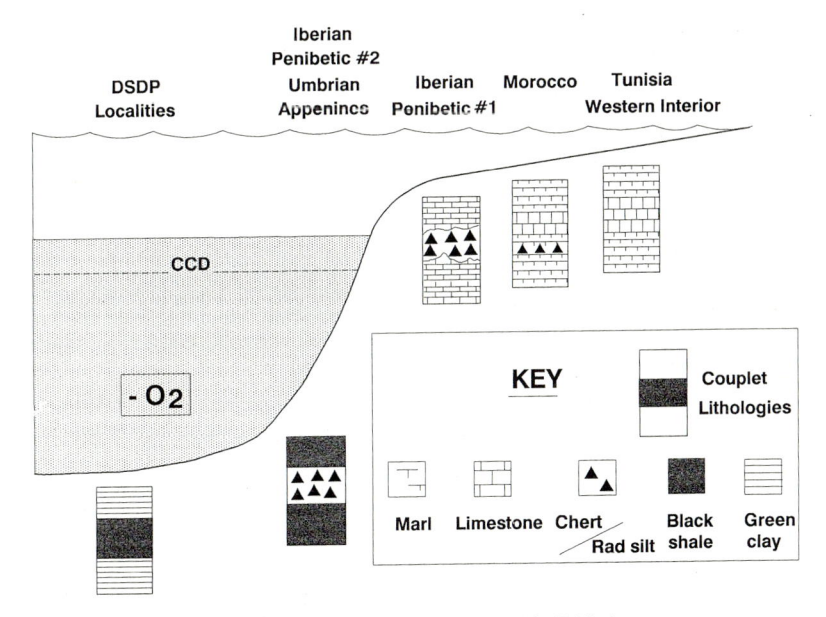

Fig. 7. Couplet lithologies and tectonic settings for several localities listed in Table 1

marlstone rhythms of the CTBE rest uncomfortably on older sandstone. In southern Spain (Penibetic 1 in Table 1 and Figs. 6 and 7), limestones and cherts that mark the CTBE overlie marlstones and cherts. In Europe the CTBE is represented by limestone-marlstone rhythms in some localities, but at others it is represented only by limestone beds or, at still others, by a relatively thin unit of black shale (Schlanger et al. 1987). In all European localities, the CTBE strata overlie Cenomanian limestones. Schlanger et al. (1987) suggested that these facies represent different depths and a different position relative to the inferred thick oxygen minimum zone.

7.2 Bathyal Deposits

In the Italian Apennines, and in part of the Penibetic region of southern Spain (Penibetic 2 in Table 1 and Figs. 6 and 7), massive Pre-CTBE limestones and siliceous limestones are replaced during the CTBE by subregular cyclic alternations of organic carbon-rich black shales and radiolarian sands (the Bonarelli bed in the Apennines) or their diagenetic equivalent, radiolarian cherts. In both regions the CTBE silica-rich black shale deposits are only a meter or so in thickness. Organic carbon content of the shales reaches 23% (Schlanger et al. 1987). These are considered to be deposits of "pelagic plateaus" by Kuhnt et al. (1986). These CTBE deposits contain very little carbonate, and must represent a short-lived rise in CCD as widespread carbonate deposition was shifted to greatly expanded epicontinental platforms during the Cenomanian-Turonian sea level maximum. The alternation of radiolarian sands and carbon-rich shales probably reflects fluctuations in marine productivity.

Table 1. Tabulation of CTBE lithologies within inferred paleoenvironments (data from Cobban and Scott 1972; Schlanger et al. 1987; Thurow and Kuhnt 1986; Arthur and Premoli Silva 1982; Kuhnt et al. 1986; Herbin et al. 1986; Stow and Dean 1984)

Lithology Environment	Pre-CTBE lithology	CTBE lithology	Post-CTBE lithology
Shelf			
Western Interior	Calc shale	Limestone/marl	Limestone/marl
Tunisia	Sandstone	Limestone/blk shale[a]	Limestone/light shale
SW Morocco	Marl	Bitum marl/lmst/chert	Bitum marl/lmst/chert
Iberian Penibetic #1	Limestone/marl	Lmstn/chert (Scaglia)	Limestone/marl
Bathyal			
Umbrian Appenines	Lmstn/chert (Scaglia)	Si-rich black shale	Lmstn/chert (Scaglia)
Iberian Penibetic #2	Lmstn/chert (Scaglia)	Si-rich black shale	Lmstn/chert (Scaglia)
Deep Sea			
DSDP Site 367	Blk shale/grn clay	Black shale	Grn clay/blk shale
398	Grn clay/blk shale	Blk shale/grn clay	Varieg clay
105	Grn clay/blk shale	Blk shale/grn clay	Varieg clay
530	Red/grn clay/blk shale	Blk shale/grn clay	Red/grn clay/blk shale
603	Grn clay/blk shale	Blk shale/grn clay	Grn clay/blk shale

[a]Unconformable with underlying unit.

7.3 Deep Sea

In deep sea environments (Figs. 6, 7, Table 1) below the CCD, the onset of the CTBE is reflected in an increase in thickness and abundance of organic-carbon-rich black shales. At Site 367, the CTBE appears to be represented entirely by black shale, but in most sections (Sites 398, 105, 530, and 603) the black shales are rhythmically interbedded with green shales. Strata above and below the CTBE interval consist of green or green and red shales, generally with some thin interbedded black shale beds. Redox rhythms thus dominated deep sea environments during the CTBE, and are even apparent in the distal facies of turbidite fans that lay marginal to the deep ocean basins (Thurow and Kuhnt 1986).

8 Summary

Within the Cenomanian-Turonian boundary event, CTBE, a depth transect from shallow seas to the deep ocean encounters facies that are distinctive from those above and below the CTBE interval, and that typically include (1) limestone-marlstone cycles, (2) silica-black shale cycles and (3) green shale-black shale cycles. These represent comparatively shallow marine environments, bathyal environments and deep ocean basins. Although CTBE deposits differ in lithology, they share the characteristics of an abrupt appearance of lithologies that differ from those below, an increase in organic-carbon-rich beds, and cyclic sedimentation.

Kuhnt et al. (1986) suggested that the distinctive CTBE facies in the Gibraltar Arch region could be explained by high productivity along the continental margin as a result of upwelling. In their view, productivity was highest in the shallow facies and decreased toward the deep sea. The high productivity at the margins produced an expansion and intensification of the oceanic oxygen-minimum layer, which reached abyssal environments. In the generalized sequences shown in Table 1, one kind of lithology in a given CTBE couplet is similar to the pre-CTBE lithology in the same section. The other lithology in that couplet is similar to the pre-CTBE lithology of the next section seaward (Fig. 7). These facies shifts must reflect in part the important changes in ocean circulation that accompanied the CTBE, and in part the landward shift in carbonate deposition during transgression and the accompanying rise in the CCD.

The evidence suggests that most limestone-marlstone rhythms that characterize epicontinental seas and other comparatively shallow mid-Cretaceous oceans were the result of fluctuations in marine productivity. During the CTBE, these fluctuations intensified and limestone-marlstone rhythms were widely deposited in shallow settings. In addition, contemporaneous cyclic deposits formed widely in the bathyal realm and in the deep sea where black shale-radiolarian sand couplets and black shale-green shale couplets were deposited. These, too, probably resulted directly or indirectly from fluctuations in marine productivity. These fluctuations somehow acted as the catalyst for the extraordinary episode of carbon burial during the CTBE.

Acknowledgments. This work was financed by grants EAR-8319846 and EAR-8618877 from the National Science Foundation. We thank Dr. Werner Ricken for his support.

1.5 Rhythmic Carbonate Content Variations in Neogene Sediments Above the Oceanic Lysocline

L. Diester-Haass

1 Introduction

1.1 Shallow and Deep-Water Carbonate Content Variations

Rhythmic changes in carbonate content (clay-marl, marly clay-marl, marl-ooze) are a widespread phenomenon in Neogene marine sediments which are not markedly affected by diagenesis. They are encountered in all water depths from shallow continental slope areas (Diester-Haass et al. 1973; Diester-Haass et al. 1986) to more than 5000 m depth, below the lysocline (Ruddiman 1971; Gardner 1975; for review see Volat et al 1980). However, the processes producing rhythmic carbonate content changes in water depths below the present-day oceanic lysocline (Berger 1968) are completely different from those in shallower water. The oceanic lysocline is here defined as that level separating well from poorly preserved carbonate fossil assemblages in deep-ocean areas far away from continental margins which create disturbances in the depositional sequences (Berger et al. 1982).

In deep-water areas, cyclic changes in carbonate ion concentration (Broecker and Takahashi 1978) and deep-ocean circulation patterns (Berger 1968; Gardner 1975; Balsam 1983; Peterson and Prell 1985b) produce cyclic changes in carbonate dissolution, which is the primary origin of rhythmic marl-clay-ooze sediments below the present-day or glacial oceanic lysocline (Grötsch et al. Chap. 1.6, this Vol.; Dean and Gardner 1986; Crowley 1985).

In water depths above the presentday oceanic lysocline, three different environmental factors control the carbonate content of the sediments: carbonate production, carbonate dissolution and dilution of carbonate by terrigenous matter or opal (also see Einsele and Ricken Chap. 1.1, this Vol.). Other factors such as calcareous turbidites, slumping or winnowing, volcanic matter input (producing rhythmic carbonate content changes), as well as biogenous opal-carbonate cycles, as observed in ODP Leg 114 in subpolar Antarctic areas (Ciesielski et al. 1988), are not considered here.

This chapter deals with rhythmic carbonate content changes in water depths below the shelf and above the oceanic lysocline. Neogene continental shelves and epicontinental seas are often too shallow and subsidence is too slow to prevent sediments from being reworked and winnowed (Einsele 1982a). Furthermore, knowledge on sediments older than the Late Quaternary from these shallow areas is poor. An example from the Persian Gulf points to the possibility of rhythmic carbonate content changes in arid climates; there Holocene marls are underlain by late glacial aragonitic muds which formed during lowered sea level in shallow embayments (Diester 1972).

1.2 Environmental Factors Controlling the Carbonate Content Variations in Depths Above the Oceanic Lysocline

Variations in carbonate production in hemipelagic sediments are difficult if not impossible to measure. The quantitatively most important carbonate contributors are planktonic foraminifera and coccolithophorids. Numbers of planktonic foraminifera per volume of water increase with growing fertility (Berger 1976). In Holocene sediments off NW Africa which have been influenced by upwelling, this increase in production is reflected by increased planktonic foraminiferal accumulation rates (Diester-Haass 1978). Coccolithophorids, however, increase only slightly with fertility (Berger 1976; Margaleff 1973).

If growing fertility leads to augmented production of tests, dissolution will increase simultaneously, because increasing fertility increases the supply of organic matter and production of metabolic CO_2 (Berger 1970, 1977). On the other hand, increases in sedimentation rate in high productivity areas cause calcareous tests to be more rapidly buried, thereby protecting them from dissolution.

Productivity effects can only be recognized clearly when a group of calcareous organisms completely disappears, although overall carbonate preservation of other groups is good (Diester-Haass and Schnitker 1990).

Carbonate dissolution on continental margins at depths above the oceanic lysocline can only be produced by decomposition of organic matter and consequent production of porewater CO_2 (Berger 1970; Berger et al. 1982; Emerson and Bender 1981). This dissolution is controlled by two processes: (a) surface water productivity and (b) lateral supply of organic matter from the shelf/upper slope (Parker and Berger 1971; Swift and Wenkam 1978). With the migrating glacial-interglacial shoreline, organic matter supply on the continental slope can change: an increased supply during regressions and a reduced supply during transgressions (Thunell 1976; Berger and Vincent 1981; Broecker 1982; Diester-Haass et al. 1986). Furthermore, productivity on continental margins is generally greater in glacial periods as compared to interglacial times. For the eastern Atlantic, this productivity has been calculted to be two to three times greater during the Quaternary glacial periods (Müller et al. 1983). Both the increase in surface water productivity and the increase in lateral supply of organic matter cannot be quantitatively differentiated. Both lead to an increase in net organic matter accumulation rates and thus to enhanced carbonate dissolution.

An increased organic carbon supply cannot be detected in the sediments by means of percent of organic carbon: the sediments might be subject to variations in dilution, produced by carbonate dissolution and terrigenous input. If organic carbon percent values increase with increasing terrigenous supply, organic carbon supply rates may also increase. If, however, the organic carbon supply is constant, the percent values would decrease with increasing terrigenous supply. If the absolute ages of individual high and low organic carbon units are known, their accumulation rates can be calculated. If no ages are known, the dilution factor of organic carbon (due to varying carbonate dissolution) can be eliminated from percent organic carbon values. This can be accomplished by calculating the amount of organic carbon that would exist if carbonate content were constantly high and no dissolution occurs, after the formula of Berger (1971) and Dean et al. (1981):

$L = (1 - Ni/Nf) \times 100/Ci$

L = % loss of $CaCO_3$ by dissolution

Ni = initial organic carbon content in sample with highest $CaCO_3$ content and lowest dissolution

Nf = final organic carbon content which would exist if it were determined by dissolution alone

Ci = initial carbonate fraction.

L, Ci, and Ni are known, while Nf can be calculated. The ratio of measured versus calculated organic carbon content, as used in some cases below, is assumed to indicate real supply changes.

Dissolution of carbonate is measured here by means of the fragmentation of planktonic foraminifera and by benthos/plankton foraminiferal ratios (Berger 1970; Thunell 1976). With increasing dissolution at the sea floor, planktonic foraminifera tests become increasingly damaged, the width of the pores and fissures in the shells becoming larger until finally the shell becomes disaggregated. The stronger the dissolution, the more the numbers of intact tests decrease (Arrhenius 1952). Planktonic foraminifera are more sensitive to dissolution than benthonic ones, which increases the benthos/plankton foraminiferal ratios with increasing dissolution. Without dissolution the ratio of benthonic versus planktonic foraminifera depends only on water depth and nutrient supply (Berger and Diester-Haass 1988). The problem of the degree of coccolith dissolution and its relation to foraminiferal dissolution is not yet solved (see Paull et al. 1988).

With increasing organic matter content, carbonate dissolution generally increases at the sea floor; but there is an upper limit: when organic matter input is so high that the oxygen in bottom waters is totally consumed by bacterial decomposition, an anoxic layer exists at the sediment/water interface which limits bacterial decomposition and thus CO_2 production and carbonate dissolution (Fig. 1). In this case excellent carbonate preservation occurs even with high organic carbon contents, as observed in the Santa Barbara basin off California (Berger and Soutar 1970). These anoxic sediments are laminated, without any bioturbation.

Dilution of carbonate by terrigenous matter is a common phenomenon in continental margin areas and can be produced by eolian, river or glacial supply. Changes in climate, as have occurred in the Quaternary glacial-interglacial cycles, produce strong changes in the amount, composition, and grain size of the terrigenous supply. Furthermore, the glacially lowered sea level tends to enhance terrigenous supply to the continental slopes. In oceanic subpolar and polar areas, far away from the continents, strong variations in the clastic ice-rafted supply occur with glacial-interglacial cycles in the extension of sea ice (Roberts et al. 1984; Ruddiman et al. 1987).

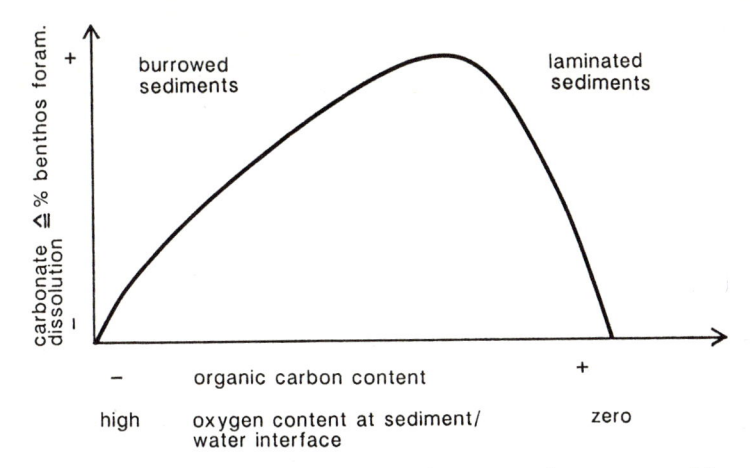

Fig. 1. Relationship between carbonate dissolution (benthonic foraminifera expressed as percentage of the sum of both benthonic + planktonic foraminifera), organic carbon content of the sediments, and oxygen content at the sediment/water interface

For the percentages of terrigenous matter the same is true as for organic matter: it may be subject to varying dilution by CaCO₃ as a consequence of carbonate dissolution. This factor can be eliminated for terrigenous matter in the same way (see formula above) as for organic matter.

In summarizing the effects of dissolution and dilution on the CaCO₃ content of sediments, we conclude that there are three possibilities (Fig. 2):

1. only dilution by terrigenous matter, decreasing the CaCO₃ content without dissolution;
2. dissolution of carbonates, reducing the CaCO₃ content only slightly; and
3. dissolution and dilution, reducing the CaCO₃ content the more, the higher the terrigenous input is.

2 Examples of Neogene Rhythmic Carbonate Content Variations on Continental Slopes

2.1 Walvis Ridge, SE Atlantic Ocean

2.1.1 Neogene of DSDP Sites 532 B and 362

On the eastern Walvis Ridge, at a depth of about 1300 m and about 180 km offshore, DSDP Sites 532 (Hay et al. 1984) and 362 (Bolli et al. 1978) (Fig. 3) contain rhythmic CaCO₃ content changes in the Neogene sediments. The sites are situated in a subtropical area, with no river influence. The oceanographic situation is characterized by coastal upwelling and the SE-NW-flowing Benguela Current, which turns to a western direction at 23–20°S.

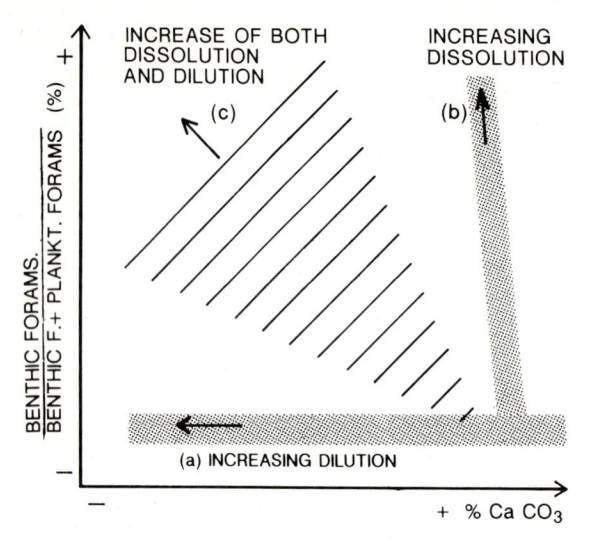

Fig. 2. Relationship between carbonate dissolution (benthonic foraminifera expressed as percentage of the sum of both benthonic + planktonic foraminifera), and % CaCO$_3$ in the sediments, and influence of terrigenous dilution. *a* dilution by terrigenous matter only; *b* dissolution of carbonate only; *c* both dilution and dissolution

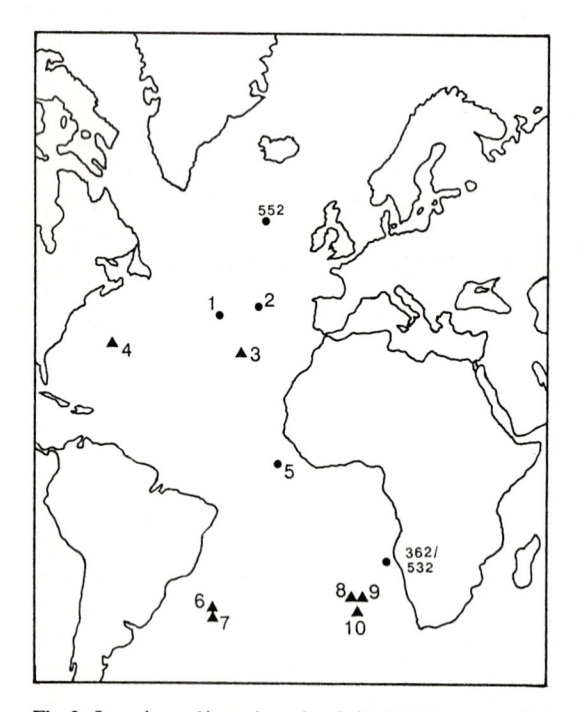

Fig. 3. Locations of investigated and cited DSDP sites and other cores in the Atlantic Ocean. *Circles* sites containing rhythmic carbonate changes in the Neogene. *Triangles* sites containing homogeneous oozes without rhythmic carbonate content changes in the Neogene. *1* DSDP Site 607; *2* DSDP Site 609; *3* DSDP Site 606; *4* Core E2D–79–31; *5* Meteor core 13519; *6* DSDP Site 357; *7* DSDP Site 516; *8* DSDP Site 525; *9* DSDP Sites 528 and 529; *10* DSDP Site 526

The rhythmic carbonate content changes coincide with sediment colour and oxygen isotope data, i.e., climatic changes: high carbonate contents occur in light gray to whitish beds from interglacial periods, while low carbonate contents coincide with darker, olive-green and gray-green interbeds from glacial periods (Gardner et al. 1984; Diester-Haass 1985; Diester-Haass et al. 1986; Diester-Haass and Rothe 1988). The $CaCO_3$ content varies between 20 and 90% in the Neogene sequence; the range of variation within the individual couplets is smaller. In general the difference between maxima and minima within one couplet is about 20–30%. The average thickness of the couplets ranges from 1 to more than 2 m in the Late Quaternary and from less than 1 to 1.5 m in the Early Quaternary and late Pliocene.

The influence of the three possible environmental factors producing carbonate content cycles is the same in all the Neogene sequences. So only one example from the Pliocene/Quaternary boundary, at 70 m below the sea floor, is presented here (Fig. 4). The light-colored, high carbonate content units from interglacial intervals (Fig. 4a and c) have low organic carbon contents (Fig. 4b), abundant planktonic foraminifera (Fig. 4d), weak dissolution (Fig. 4e: low benthos/plankton foraminiferal ratios) and low, >40 µm-sized terrigenous matter input (Fig. 4f). In the greenish glacial sediments the opposite is found. These cycles originate from glacially increased organic matter input during regressions, through near-bottom downslope transport from the shelf/upper slope. The ratio of measured versus calculated organic matter content is >1 in glacial periods, pointing to a net supply increase. Furthermore, these cycles are formed by glacially increased input of terrigenous matter: the ratio of measured versus calculated terrigenous matter >40 µm in size increases strongly in glacial sediments, as a consequence of increased supply rates.

A quantitative evaluation of the effect of dilution and dissolution on $CaCO_3$ content cycles is impossible because no absolute ages are available. But a plot of $CaCO_3$ content versus carbonate dissolution (Fig. 5) allows a semi-quantitative estimate: on the Walvis Ridge strong dissolution decreases $CaCO_3$ content only slightly (see examples of cores 72, 57 and 67 in Fig. 5). Additional terrigenous supply lowers the $CaCO_3$ content much more (cores 23, 26, 56). An intermediate terrigenous supply is found in cores 16 and 17.

2.1.2 Middle Miocene of DSDP Site 362

In the Middle Miocene at Site 362 (core 33), 525 m below the sea floor, carbonate content cycles containing between 35 and 55% $CaCO_3$ have been observed. Although they are related to glacial-interglacial fluctuations as in the Neogene, they are not related to dissolution: benthos/plankton foraminiferal ratios remain constant at about 10%, a ratio normal for a water depth of 1300 m (Figs. 5, 9). Therefore, only terrigenous dilution appears to be responsible for the cycles. This is because during the Middle Miocene no upwelling occurred, and very low productivity led to an organic matter content in the sediments of less than 0.3%. Climatic changes are responsible for varying terrigenous input (Diester-Haass et al. 1990).

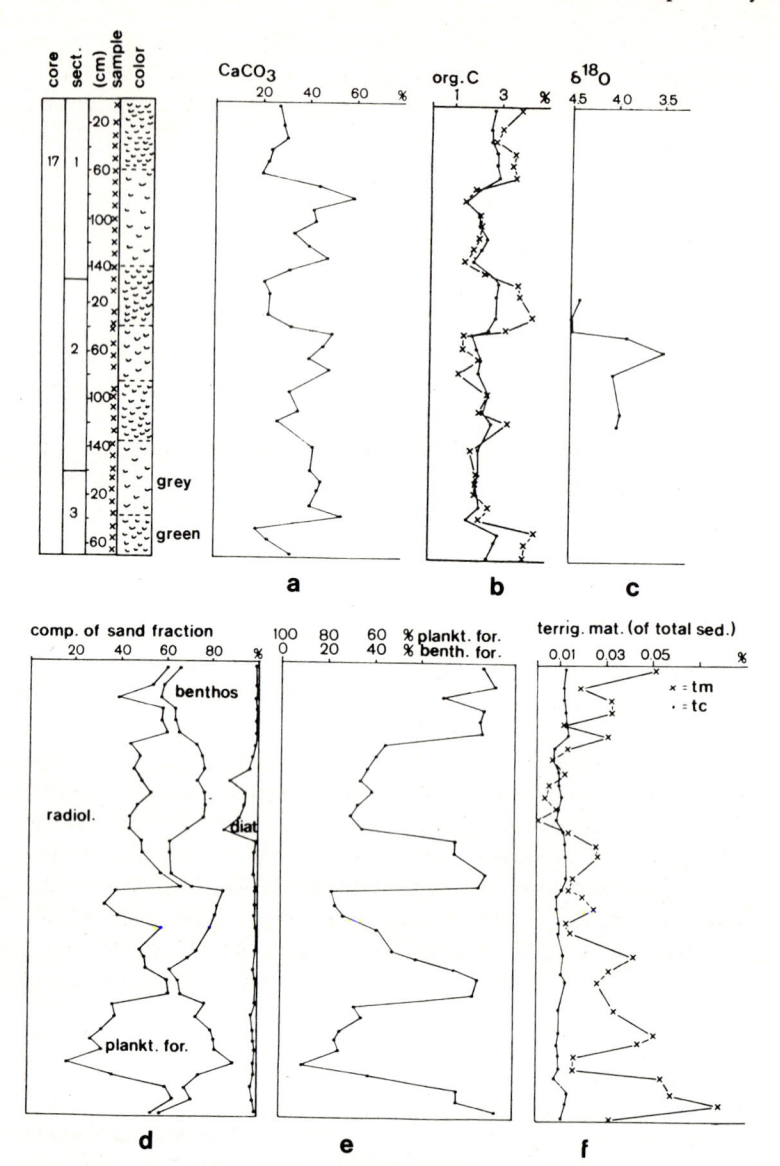

Fig. 4a-f. Site 532 B, core 17, Plio-Pleistocene boundary, 70 m subbottom. **a** CaCO3 content of total sediment. **b** Organic carbon content of total sediment, measured (*crosses*) and calculated (*circles*), see text for explanation. **c** Oxygen isotope curve. **d** Composition of the sand-sized fraction. **e** Dissolution of carbonates, shown by means of the benthonic/planktonic foraminiferal ratios (calculated as % benthonic foraminifera/% benthonic foraminifera + % planktonic foraminifera × 100. **f** Terrigenous matter (>40 μm) % of total sediment. *Crosses* measured; *circles* calculated values. See text for explanation

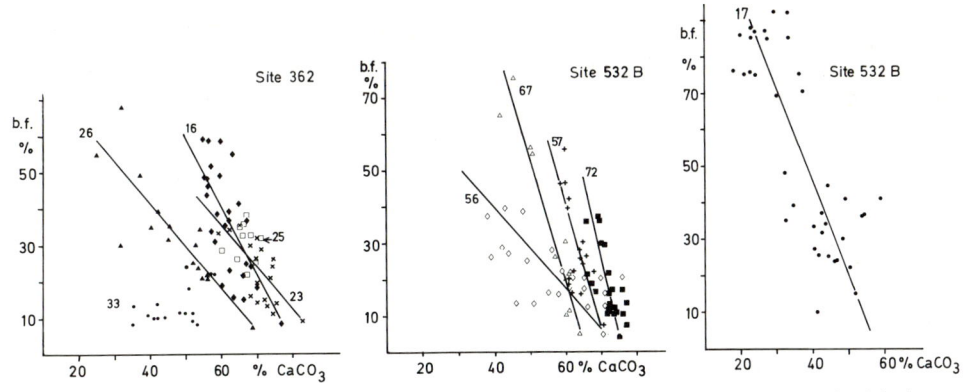

Fig. 5. Sites 532B and 362, plot of carbonate dissolution (shown by means of the benthos/plankton foraminiferal ratios) versus CaCO$_3$ content: for Late Miocene of Site 362 (cores 16–26, about 5–9.5 Ma), for the Middle Miocene of Site 362 (core 33, about 14 Ma), for the Late Miocene of Site 532 B (cores 56–72, about 6.5 Ma), and for the Plio/Pleistocene boundary of Site 532 B (core 17, about 1.8 Ma). % b.f. benthonic foraminifera expressed as the percentage of the sum of benthonic + planktonic foraminifera

2.3 Continental Slope South of New Zealand, DSDP Site 594

2.3.1 Late Quaternary Carbonate Content Cycles

On the continental slope south of New Zealand at a water depth of 1204 m (at 45° 31.41′S, 174° 56.88′E; Kennett et al 1986), cyclic carbonate content changes (5–70%) have been found that are similar to those off SW Africa. Greenish CaCO$_3$-poor interbeds represent glacial periods, while greenish-white, CaCO$_3$-rich beds represent interglacial periods. The accumulation rates at subpolar Site 594 are, however, much higher. The processes for cycle genesis in this case are the same as those on the Walvis Ridge: a markedly increasing glacial organic matter supply (from 0.15 to 0.7%; R. Stein pers. commun.) increases carbonate dissolution and reduces planktonic foraminifera accumulation rates by a factor of 5 to 10. Increased glacial erosion rates on New Zealand produce stronger terrigenous input (accumulation rates increase by a factor of 3 to 5). Glacial carbonate contents are thus reduced by (a) decreasing accumulation rates of CaCO$_3$ as a consequence of dissolution and (b) increasing dilution by terrigenous matter. Terrigenous input has a stronger effect on CaCO$_3$ content variations than dissolution (Figs. 2 and 9).

These results on increased carbonate dissolution in glacial periods reveal, that in the Pacific there are two types of carbonate dissolution stratigraphy: (1) in continental margin areas, far above the CCD, carbonate preservation is good in interglacial periods and bad in glacial periods as a consequence of sea level lowering and enhanced organic matter supply and carbonate dissolution. (Examples: Site 594 and Panama Basin, Swift and Wenkam 1978). (2) In water depths below the oceanic lysocline carbonate dissolution is strongest in interglacial periods and weak in glacial periods (Thompson and Saito 1974; see synthesis in Volat et al. 1980; Crowley 1985), as a consequence of paleochemistry or paleocirculation changes at great depths. In the Atlantic Ocean, however, carbonate dissolution at depths below the oceanic lysocline as well as on the continental slope is strongest in glacial periods. In near-continental areas, carbonate dissolution history is therefore synchronous in the Atlantic and in the Pacific, whereas below the oceanic lysocline the carbonate dissolution history is often antagonistic in both oceans.

2.3.2 Early Pliocene Carbonate Content Variations in DSDP Site 594

In the Early Pliocene sediments of Site 594, cyclic sedimentation processes can be inferred from cyclic color changes (Kennett et al. 1986), which are produced by slightly varying $CaCO_3$ contents (77 to 99%) and related dissolution variations in planktonic foraminifera. Dissolution is produced by processes similar to those in the Quaternary: during cyclic regressions (which are well known also for the Pliocene, Diester-Haass and Rothe 1988), organic carbon supply is increased. Since terrigenous input is very weak in the Early Pliocene, prior to the tectonic uplift of the New Zealand Alps, and since carbonate dissolution lowers $CaCO_3$ content only slightly, the effect on the bulk $CaCO_3$ content is not significant.

Accumulation rates suggest cycles of 21, 25, and 50 Ka in the carbonate content of these Early Pliocene sediments. Similar cycle durations have been found for the same time frame in the North Atlantic (Diester-Haass and Schnitker 1989).

3 Examples of Neogene Rhythmic Carbonate Content Variations on Submarine Elevations at Large Distances from Continents

At submarine elevations which are generally at some distance from the continents, the factors responsible for carbonate content cycles on the continental slopes are no longer effective: the lateral supply of organic matter from the shelf/shelf edge is not possible, and eolian or river loads will not markedly affect sedimentation. Two other factors may produce local carbonate content cycles: cyclic changes in (a) ice-rafted terrigenous matter and (b) the intensity of equatorial upwelling. Submarine elevations that are not influenced by these parameters do not show cycles in carbonate content.

3.1 Rockall Plateau, DSDP Site 552 A

The Rockall Plateau in the North Atlantic (Fig. 3), at Site 552 A (2311 m water depth; Roberts et al. 1984), presents an example of cyclic carbonate content changes. Sediments from the last 2.4 Ma, a period which has been influenced by cyclic ice coverage (Shackleton and Hall 1984), show cyclic changes between white oozes with 80% $CaCO_3$ from interglacial periods and green marls with 20% $CaCO_3$ from glacial periods. Prior to that time, carbonate content was consistently high, greater than 90%.

3.1.1 Carbonate Content Changes from 0–2.4 Ma

During glacial periods (Fig. 6a) carbonate content (Fig. 6b) is drastically lowered because of an increase in the supply of terrigenous matter (Figs. 6c,h). This coarse terrigenous matter is supplied by glacial sea ice, which is furthermore responsible for a reduction in glacial surface water productivity. Accumulation rates of foraminifera are reduced, with those of planktonics more affected than those of benthonics (Fig. 6d,e). So the ratio of accumulation rates of planktonics versus benthonics is low in glacial and high in interglacial periods. This observation shows that benthos/plankton foraminiferal ratios (Fig. 6g) cannot be used in this case as a dissolution index, because they are controlled by surface water productivity.

Fragmentation of planktonic foraminifera, however, is a dissolution indicator and reveals stronger dissolution in interglacials, as a consequence of higher surface water productivity, more organic carbon input, and thus more dissolution at the sediment/water interface. Another factor is species distribution: nearly monospecific (*G. pachyderma*) glacial associations are less sensitive to dissolution than are warmer assemblages. The interaction of foraminiferal production variations and dissolution explains the contradictory behavior of the two "dissolution indicators", fragmentation and benthos/plankton foraminiferal ratios (Fig. 6g). The marls show better preservation, even with aragonitic pteropods (Fig. 6c), than do calcareous oozes.

Ice rafting started suddenly at 2.4 Ma in the area of Site 552 A (Fig. 7). In the first marl-ooze cycles the benthos/plankton foraminiferal ratio and fragmentation of planktonic foraminifera change in parallel (Fig. 7d). The reason for this is that at the beginning of the North Atlantic glaciations the Site was situated at the ice margin and profited from high ice margin fertility, with higher production of planktonic foraminifera and related stronger dissolution during deposition of glacial marls. But dilution by terrigenous matter plays a much greater role in lowering $CaCO_3$ content than does dissolution.

3.1.2 Prior to the First Glaciation in the North Atlantic: No Carbonate Content Variations

In the time period prior to 2.4 Ma, homogeneous oozes were deposited on the Rockall Plateau, containing more than 90% of $CaCO_3$ (Fig. 8b). Oxygen isotope data (Fig. 8a) reveal constant conditions. No terrigenous matter (>40μ) was deposited. The dissolution of carbonates, however, was subject to cyclic changes (Fig. 8c). These have been either too weak to reduce the carbonate content, or the assumption of Peterson and Prell (1985b) has to be considered, namely, that although dissolution may break down foraminifera, the smaller fragments are preserved in smaller grain size fractions and do not completely disappear. Assuming accumulation rates of 2.5 cm/Ka years for core 10 (Roberts et al 1984), these cycles have a periodicity of about 28 Ka years.

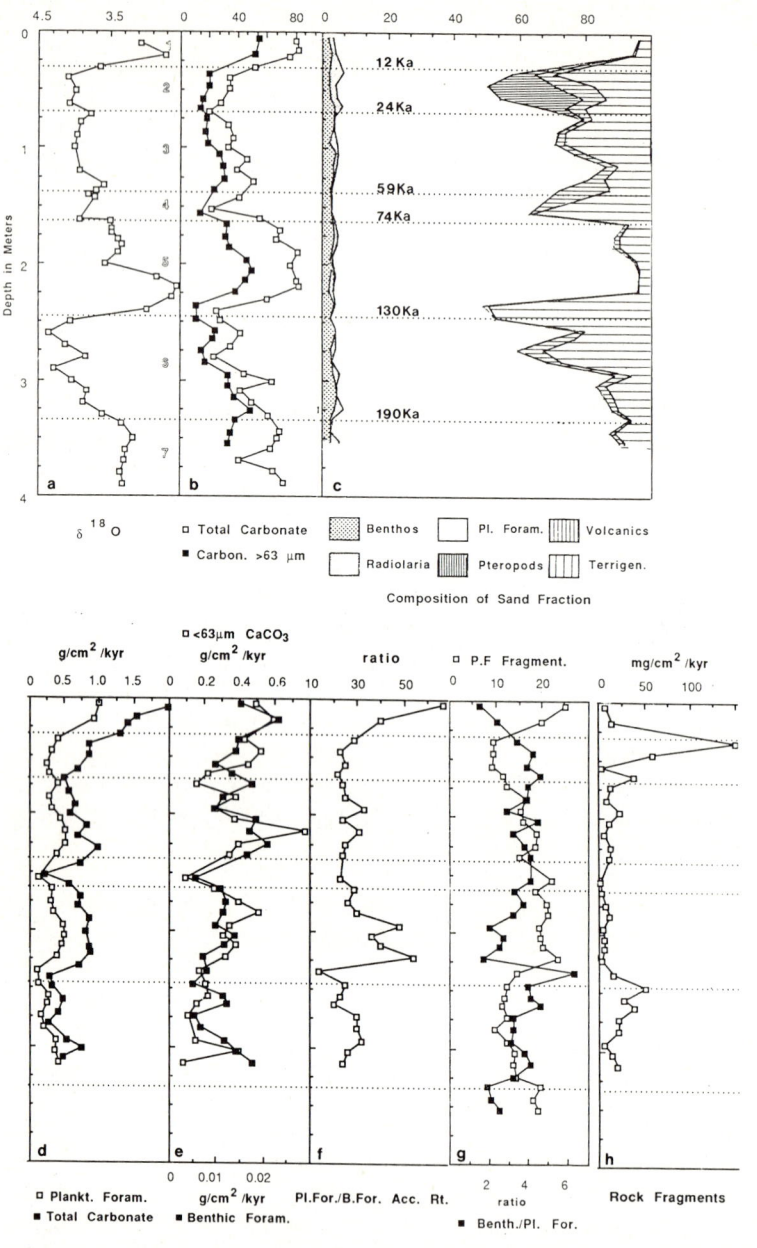

Fig. 6a-h. Site 552 A, core 1, oxygen isotope stages 1–7. **a** Oxygen isotope curve of Shackleton and Hall (1984). Absolute ages for boundaries of stages 1–7 from Martinson et al. (1987). **b** CaCO$_3$ content. **c** Composition of the sand-sized fraction. **d** Accumulation rates of planktonic foraminifera and of total CaCO$_3$. **e** Accumulation rates of benthonic foraminifera. **f** Ratio of accumulation rates of planktonic foraminifera versus accumulation rates of benthonic foraminifera. **g** Benthos/plankton foraminiferal ratios, calculated as (% benthonic foraminifera/% benthonic foraminifera + % planktonic foraminifera × 100) and carbonate dissolution, shown by means of the fragmentation of planktonic foraminifera, calculated as (% fragments/% fragments + % whole tests × 100). **h** Accumulation rates of rock debris >40 μm

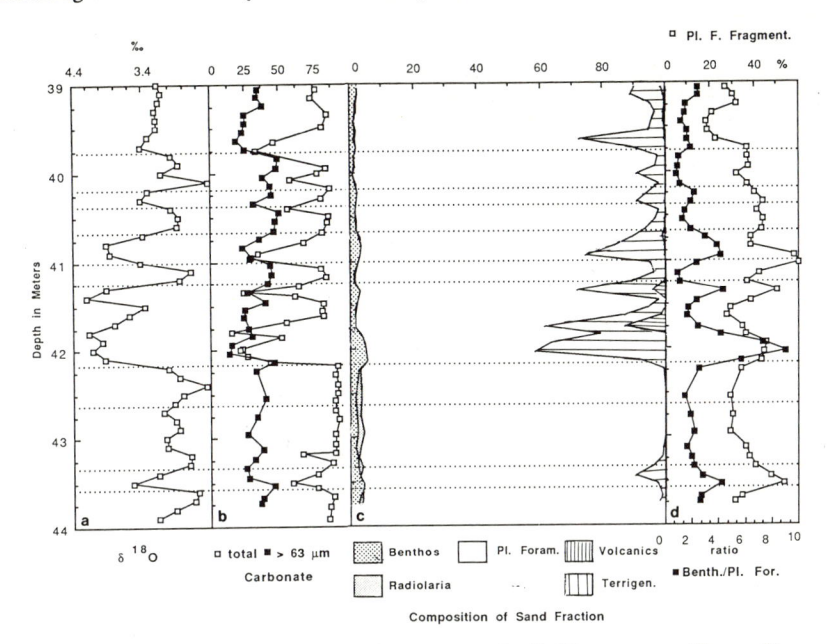

Fig. 7a-d. Site 552 A, core 9, Late Pliocene. **a** Oxygen isotope curve. **b** CaCO$_3$ content. **c** Composition of the sand-sized fraction. **d** Benthos/plankton foraminiferal ratio and fragmentation of planktonic, foraminifera, calculated as in Fig. 6

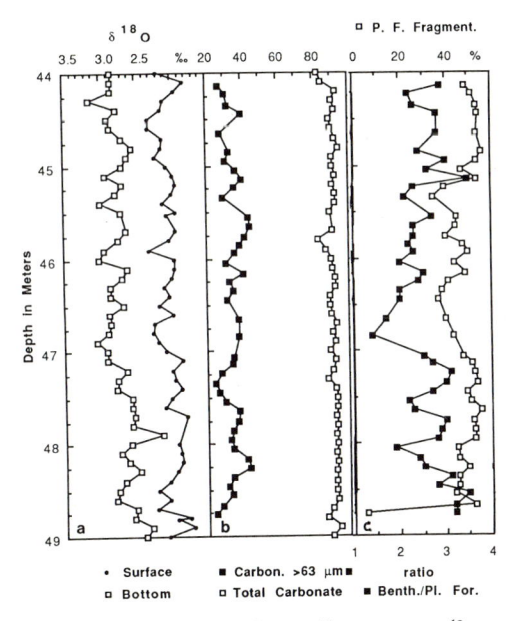

Fig. 8a-c. Site 552 A, core 10, Late Pliocene. **a** Oxygen isotope curve for surface and bottom water (from Shackleton and Hall 1984). **b** CaCO$_3$ content. **c** Fragmentation of planktonic foraminifera and benthos/plankton foraminiferal ratios (calculated as above)

3.2 Testing the Theory: DSDP Sites Above the Lysocline from Other Submarine Elevations

3.2.1 Mid-Atlantic Ridge Sites 607, 609 and 606

Ooze-marl cycles, similar to those described at Site 552 A from the Rockall Plateau, occur at sites on the Mid Atlantic Ridge that have been reached by sea ice during Neogene North Atlantic glaciations, i.e., Sites 607 and 609 at 50 and 42°N (Fig. 3; Ruddiman et al 1987). Site 606, at 37°N, south of the most southern extent of polar ice, however, does not contain cyclic carbonate content changes, but consists of pure ooze (op. cit.). Glacial sea ice and the related dilution of carbonates by ice-rafted detritus is the main cause for rhythmic $CaCO_3$ changes, besides varying carbonate production and dissolution in areas north of the polar front in glacial and interglacial periods.

3.2.2 Walvis Ridge, Rio Grande Rise, and Muir Seamount

Sediments from submarine elevations which have never been reached by sea ice, and which are not influenced by equatorial upwelling, do not contain rhythmic carbonate content changes. Examples are Site 525, 526, 528, and 529 from the Walvis Ridge (Moore et al. 1984), all situated above the oceanic lysocline and all showing constant $CaCO_3$ contents above 90–95% in Neogene sections. Other examples are Site 357 (2086 m water depth; DSDP Leg 39: Supko et al. 1977), and Site 516 (1313 m water depth; DSDP Leg 72; Barker et al. 1983), both from the Rio Grande Rise, where constant, high $CaCO_3$ contents (>80%) and excellent preservation (e.g., pteropods) point to the absence of dissolution and to very weak, constant terrigenous dilution. A third example is from Muir Seamount (Western North Atlantic, see Fig. 3), where in 1780 m of water (Core E2D–79–31), above the presentday lysocline, carbonate content is constant, remaining above 85% during glacial and interglacial periods (Balsam 1983). The great distance of the sites from the continental shelf/upper slope, as well as the existence of a deep-water area between them and the shelf, prevent input of terrigenous matter and organic carbon, which have been found to be responsible for cyclic carbonate content changes.

3.2.3 Sierra Leone Rise

Sediments from the Sierra Leone Rise in the Central Atlantic (Fig. 3) show that cyclic carbonate changes can occur above the oceanic lysocline, without ice-rafted influence (Figs 9, 2b). Meteor core 13519, from 2862 m water depth above the presentday and glacial lysocline (Sarnthein et al 1984), shows cyclic carbonate content changes. The authors attribute them to rhythmic carbonate dissolution changes, which are due to paleoproductivity changes in surface waters. The core has been retrieved in the area of equatorial upwelling, whose intensity changed with time. During glacial periods in the Late Quaternary, stronger equatorial upwelling led to higher organic carbon

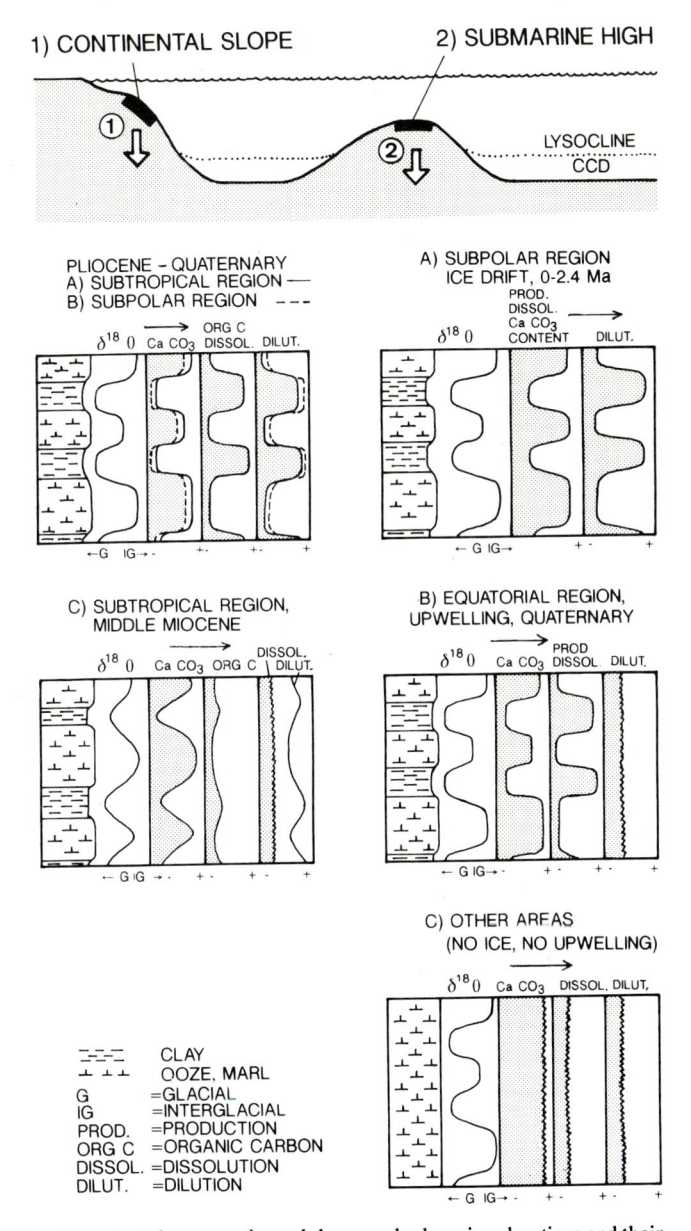

Fig. 9. Schematic diagram of the carbonate cycles on continental slopes and submarine elevations and their relation to $\delta^{18}O$ values (*G* glacial; *IG* interglacial periods), dissolution, productivity in surface water (i.e., organic carbon content from vertical and/or lateral supply of organic matter), and dilution by terrigenous matter

supply to the sea floor, and hence intensified carbonate dissolution, whereas interglacial sediments are well-preserved oozes (op. cit.; Müller et al. 1983).

4 General Conclusions on the Origin of Rhythmic Carbonate Content Changes in Neogene Sediments

1. All Neogene rhythmic carbonate changes observed in the present investigation are related to a glacial-interglacial cyclicity and thus are consequences of orbital parameter variations, i.e., Milankovitch cycles (Imbrie et al. 1984). Related changes in global temperature and atmospheric circulation produced glacial-interglacial fluctuations in sea level, climate, and oceanic circulation. As a consequence of these environmental changes, terrigenous input and the supply of organic matter to the continental slopes also varied, generating fluctuations in dissolution and dilution of pelagic carbonates. Advance and retreat of ice in polar regions controlled carbonate production and clastic input. Strengthened glacial atmospheric circulation intensified coastal and equatorial upwelling and thus productivity and carbonate dissolution. All these factors together controlled the genesis of high (interglacial) and low (glacial) carbonate contents.

2. The effect of varying carbonate production on $CaCO_3$ content can generally not be separated from that of carbonate dissolution: an increase in carbonate production may lead to more dissolution (Berger 1970). Production changes can only be seen when one group of carbonate fossils diminishes or disappears in spite of the excellent preservation of other carbonate fossils.

3. Synsedimentary carbonate dissolution has to be considered not only in great water depths below the oceanic lysocline, but also in shallow continental slope areas. Bacterial decomposition of organic matter is responsible for carbonate dissolution. The organic matter comes to the sea-floor by (a) lateral supply from the shelf/shelf edge during regressions, and (b) vertical supply from surface water production.

4. All continental margin areas are possible sites of rhythmic carbonate changes, at least in the Neogene, because sea level changes produce rhythmic changes in input of organic matter and productivity and thus in carbonate dissolution. Furthermore, rhythmic changes in terrigenous input occur as a consequence of rhythmic changes in climate and erosion on the continents. The carbonate content and its range of variation depend on the climatic, tectonic and morphologic situation: adjacent to arid areas and often on passive continental margins (Diester-Haass 1976 and Diester-Haass et al. 1973), carbonate contents will be higher. Near river mouths (Diester-Haass 1983), and especially on active continental margins (e.g., Peru margin, Oberhänsli et al. 1990) carbonate contents are very low, and variations within cycles are smaller.

5. On submarine elevations at depths above the oceanic lysocline far from the continents generally no rhythmic carbonate content changes will be found, because organic and terrigenous matter supply from continental shelves/slopes is impossible. Sediments from such elevations containing rhythmic carbonate changes are due to either rhythmic influence through ice rafting (i.e., polar/subpolar areas), or to having been situated below equatorial upwelling, with its temporal, interglacial-glacial changes in intensity and thus in carbonate dissolution.

Acknowledgments. The samples for this investigation were provided through the assistance of the United States National Science Foundation, Deep Sea Drilling Project/Ocean Drilling Project.

I thank several colleagues for stimulating discussions on the chalk-marl cycle problem: W. H. Berger, G. Einsele, P. Meyers, P. Rothe, W. Ricken, D. Schnitker, and W. Wetzel. Special thanks are due to G. Einsele for providing the facilities for this investigation.

1.6 Carbonate Cycles in the Pacific: Reconstruction of Saturation Fluctuations

J. Grötsch, G. Wu, and W. H. Berger

1 Introduction

Over much of the deep sea in all ocean basins we find carbonate cycles. They appear as alternations of light and dark sediment layers, in cores of calcareous ooze (Fig. 1).

Chemically, they are expressed as fluctuations in percentage of carbonate in the Pacific and Indian Ocean, as well as in the Atlantic. Their amplitude varies, depending on the location in the ocean itself. It is possible to correlate these cycles over very long distances, even from the Pacific to the Indian Ocean (Fig. 2). The major controlling factors are varying productivity of calcareous and noncalcareous plankton, dilution by terrigenous input, and dissolution of carbonate.

In the Atlantic the carbonate cycles are attributed mainly to fluctuating dilution by lithogenic material, as ice-age aridity leads to deforestation and intensified erosion. A contributing factor, in depths near the lysocline, is increased dissolution during glacial periods which is caused by changes in abyssal circulation, notably a decrease in production of North Atlantic Deep Water (see Droxler et al. 1983).

The Pacific carbonate cycles ((Fig. 2) are generally thought to be dissolution cycles (Berger 1973; Volat et al. 1980; Crowley 1985). If so, these cycles express fluctuations in the degree of saturation of deep waters with respect to calcite. Here we shall attempt to reconstruct and constrain the amplitude of such fluctuations.

2 Carbonate Dissolution Versus Productivity

When carbonate cycles were described for the first time, they were explained as expressions of fluctuating productivity (Arrhenius 1952). This explanation was recently reaffirmed (Arrhenius 1988).

There are good reasons for invoking productivity changes as a cause of carbonate cycles. If we look at a N-S profile of carbonate deposition in the tropical central Pacific, we note a pronounced CCD depression (Fig. 3). This is an effect of enhanced productivity in the equatorial upwelling area. If there are changes in productivity between glacials and interglacials, the depth levels of equal preservation should move up and down accordingly, producing carbonate cycles. Especially depths close to the CCD should be strongly affected. The lysocline, on the other hand, should remain more or less stationary. We know from diatom stratigraphy and other indicators (radiolaria, phosphate, C_{org}) that productivity was higher during glacials, hence the

Fig. 1. Carbonate cycles seen as light and dark layers in core ERDC 132, Ontong Java Plateau, western equatorial Pacific. (Photo S.I.O)

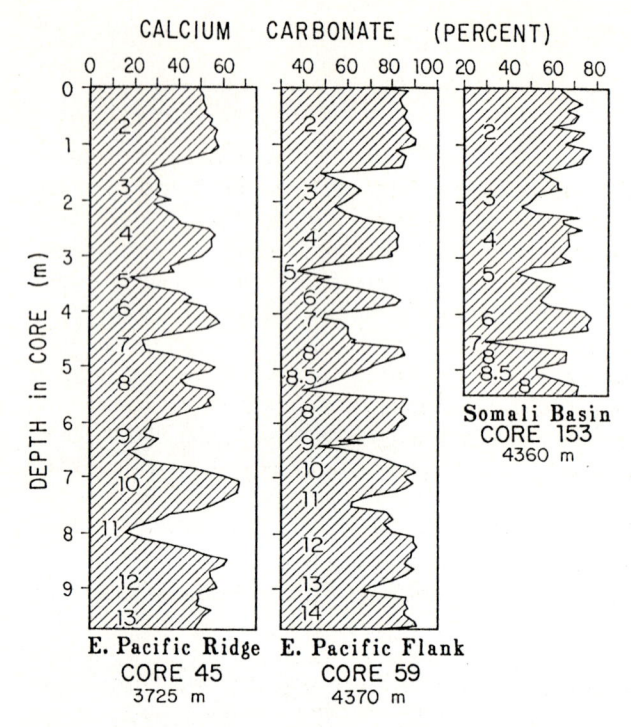

Fig. 2. Carbonate cycles in the Indo-Pacific. Core 45 (7°40′N, 106°21′W), core 59 (3°05′N, 133°06′W) and core 153 (2°18′S, 55°33′E) from the Swedish Deep-Sea Expedition after Arrhenius (1952) and Olausson (1960). (Berger and Vincent 1981)

CCD should be depressed. This effect produces cycles with higher carbonate content in glacial periods. The question is not whether this process occurs, but whether it is as important as the changing intensity of carbonate dissolution.

From preservation stratigraphy, we know that dissolution effects were increased during interglacials (Berger 1973; Thompson 1976). If this change in state of fossil preservation indicates greater rates of dissolution (as seems reasonable), then a portion of the percent fluctuation in carbonate must be assigned to this mechanism. Dissolution rate is driven mainly by saturation state of water in contact with sediment. To some degree it may be increased also by supply of C_{org} and its subsequent oxidation (Emerson and Bender 1981).

As already discussed by Arrhenius (1988), it is of decisive importance to obtain quantitative information on input of calcareous and siliceous plankton, C_{org}, and lithogenic material for short time intervals. This information is not available at present. Therefore the origin of carbonate cycles remains open. In this chapter we discuss the question: how large were the excursions in carbonate saturation, and are these excursions large enough to produce the observed variations in percent carbonate?

3 Reconstructing Saturation History

3.1 Preservation Versus Saturation

Attempts to relate preservation levels to saturation levels were made previously. Berger (1977) showed that the R_0-level (the depth at which dissolution first becomes noticeable in foraminiferal species lists from any one region) closely follows a horizon of equal carbonate saturation in both the eastern and western Atlantic trough. Broecker and Takahashi (1978) related the lysocline to the level of carbonate ion concentration, corresponding to calcite saturation, on a global scale. The saturation index is the ion product of calcium and carbonate measured divided by the same product expected for equilibrium conditions with calcite in the water. Values greater than unity therefore indicate supersaturation, values smaller than 1 undersaturation. As we shall show, saturation equals 1.0 at the lysocline, in normal pelagic sediments.

In order to obtain a proxy for paleosaturation or Apparent Saturation Level (ASL), we applied the following procedure:

1. We obtained today's preservation state of foraminifera tests as a function of depth, using different faunal and sediment parameters to construct a Composite Dissolution Index (CDI).
2. We calibrated the CDI of the surface samples against water column measurements of carbonate saturation in the same region (using GEOSECS data).
3. We computed a transfer function from CDI into ASL.
4. Finally, we obtained CDI's from piston core samples and calculated the absolute values of the ASL. This paleochemical proxy can then be used for comparison between different regions and workers.

3.2 Measuring Preservation State

As indicated in Fig. 3, beginning with the lysocline carbonate dissolution increases rapidly with depth. In the sediment this can be seen in fossil fragmentation, in preservation of individual species and in physical properties, such as grain size. These parameters do not always vary simultaneously, thus there is some advantage in using several variables at once. The first attempt to optimize dissolution indices was by Thunell (1976), who picked several preservation proxies, compared their efficiency in tracking dissolution, and used principal component analysis to produce a "best" index. He found that benthic foraminiferal abundances and foraminiferal test fragmentation are the most reliable indicators. More recently, Peterson and Prell (1985a,b) have pursued this goal, in extracting a Composite Dissolution Index (CDI) from various sediment properties.

The samples which we use to establish the calibration between preservation and saturation are from the surface sediments in box cores from Ontong Java Plateau, western equatorial Pacific. Core names and data are listed in Table 1. The five variables listed (% sand fraction, % *Globigerinoides ruber,* % whole tests of benthic foraminifera, % whole tests of planktonic foraminifera, solution index SI) all track preservation (Thunell 1976; Johnson et al. 1977; Berger et al. 1982).

Table 1. List of box core data from the Ontong-Java Plateau (western equatorial Pacific) with the variables used for calculating a Composite Dissolution Index (CDI) and the Apparent Saturation Level (ASL). The data for the columns % *Globigerinoides ruber*, % whole test of benthonic foraminifera and % whole test of planktonics are from Berger et al. (1982). Using their species counts, the solution index (SI) of Berger (1968) is calculated. Data for the % carbonate sand fraction (% > 63 µm) are based on new observations

Core	Depth (m)	% > 63 µm	% G. ruber (>149 µm)	% wbenth. (>149 µm)	% wplank (>149 µm)	SI (>149 µm)	CDI	ASL
92 Bx	1598	64.0	34.0	0.14	0.84	0.54	1.33	1.19
88 Bx	1924	55.5	28.0	0.18	0.82	0.54	1.14	1.16
112 Bx	2169	58.0	35.0	0.00	0.84	0.52	1.30	1.13
120 Bx	2247	45.0	27.0	0.49	0.83	0.73	0.91	1.12
79 Bx	2767	39.0	28.0	0.57	0.66	0.65	0.78	1.07
123 Bx	2948	35.0	18.0	0.78	0.38	0.69	0.36	1.05
125 Bx	3368	29.5	3.0	1.61	0.17	1.12	−0.30	1.00
135 Bx	3509	24.0	2.0	2.11	0.13	1.35	−0.53	0.98
128 Bx	3732	19.0	0.0	2.50	0.29	1.49	−0.58	0.96
136 Bx	3848	8.5	0.0	6.89	0.26	1.61	−0.96	0.95
129 Bx	4169	9.5	0.2	8.09	0.14	1.65	−1.10	0.91
141 Bx	4324	7.0	0.0	13.77	0.12	1.53	−1.33	0.89
131 Bx	4441	7.0	0.0	9.93	0.38	1.64	−1.04	0.88

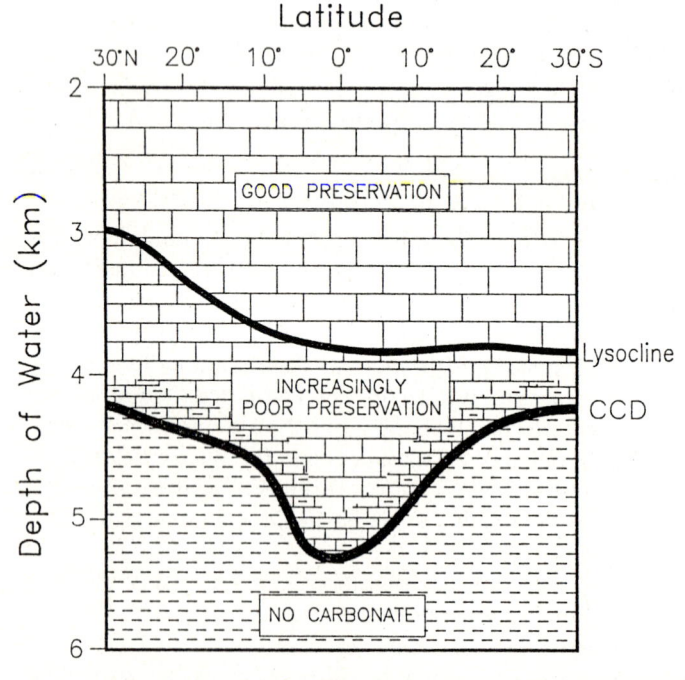

Fig. 3. Depth of today's lysocline and CCD for a N-S profile of the Pacific (extracted from Berger 1981). On an intersecting bottom topography, depending on depth and latitude, the according facies will be deposited. For a detailed definition of CCD and lysocline see Seibold and Berger (1982, p. 190 ff)

We next apply principal component analysis to the various properties listed (% sand, % *G. ruber,* %w benthics, %w planktonics, SI) and use the first factor as a CDI, in the manner described by Peterson and Prell (1985a). This helps in reducing preservational variables from five to only one. The CDI is the factor score of the first principal component of the five parameters, using a correlation matrix. This gives equal weight to each of the variables used. For a detailed description of this procedure see Thunell (1976) and Peterson (1984, pp. 130–134).

3.3 *Calibration of Preservation to Saturation*

Carbonate saturation decreases with depth, and, barring the presence of water layers with unusually low oxygen, the decrease is linear. We first calibrated saturation versus depth for the western equatorial Pacific (Fig. 4). Three GEOSECS stations with the necessary measurements (tCO_2, alkalinity, temperature) are available: 241, 244, and 246. From these data, a saturation index can be calculated using standard equations (see Broecker and Takahashi 1978). We employed a program which originated with Taro Takahashi, and was adapted for interactive use by R. Toggweiler and A. Spitzy.

The plot of depth versus saturation shows some scatter (Fig. 4). However, two of the three stations agree very well, and over the range here considered the relationship is essentially linear, allowing a direct conversion of depth to saturation. The equation we use is

$$\text{Sat} = 1.37 - 0.11 * \text{depth}, \tag{1}$$

where depth is in km. By Eq. (1), the depth at which saturation is unity is 3.364 km, which corresponds to the lysocline in this area (3.4 km; Berger et al. 1982).

A plot of the CDI versus depth (Fig. 4) is, by Eq. (1), also a plot of CDI versus saturation (Fig. 5). To convert the CDI to saturation values, three equations are used, valid over the three different slopes labeled A, B, and C in Fig. 5:

(A) $\text{Sat} = 1.14 + 0.21 * \text{CDI}$ (CDI = –1.4 to –0.98)   (2)

(B) $\text{Sat} = 1.0 + 0.06 * \text{CDI}$ (CDI = –0.98 to 0.6)   (3)

(C) $\text{Sat} = 0.91 + 0.21 * \text{CDI}$ (CDI = 0.6 to 1.4).   (4)

The saturation value 1.0 coincides with the CDI value of 0.0. At that level, which corresponds to the lysocline, the CDI has its maximum sensitivity to a change in depth, that is in saturation. This lends support to the notion that the lysocline marks a mappable boundary between distinct preservation facies.

The relationship described in Eq. (1) indicates that a calibration of preservation indices against depth of deposition is equivalent to a calibration against saturation, for the region considered, and for present conditions.

Western Equatorial Pacific

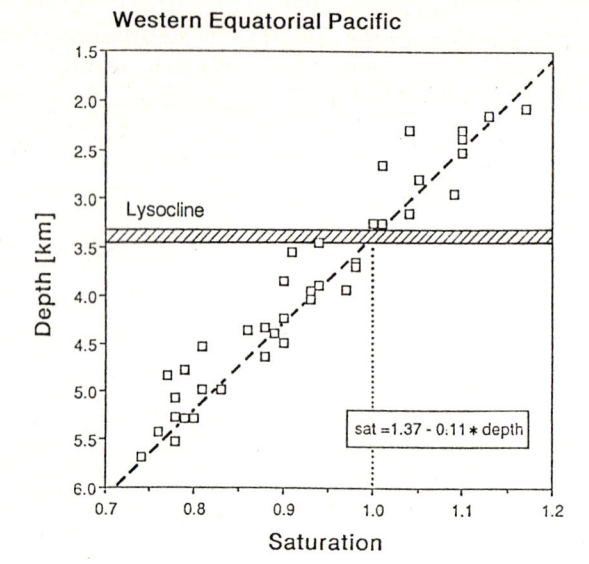

Fig. 4. Combined depth profiles of carbonate saturation for the GEOSECS stations 241, 244 and 246 (Western Equatorial Pacific)

3.4 Emerson-Bender Effect

There is one serious problem with reading preservation records in terms of saturation of deep waters. It is that changing productivity can also change preservation, by changing the flux of organic matter to the sea floor. This provides for corresponding changes in interstitial water saturation, from organic carbon oxidation (Emerson and Bender 1981). We estimate this effect to be on the order of 25% of the preservation signal from comparison of maps of productivity (Berger et al. 1987) with maps of levels of equal preservation (Berger 1978).

In addition, there is another, more subtle effect of productivity. In Eq. (1), the coefficient on depth contains a hidden productivity factor, because the organic flux to the seafloor (J_{sf}) varies as a function of depth (z). The relationship may be expressed as

$$J_{sf} = 0.2 \, PP / z + 0.05 \, PP, \tag{5}$$

where z is measured in units of 100 m (Berger et al. 1987). For the productivity (PP) in the study area (50 gC/m^{2}yr), the ratio of fluxes 250 m above and 250 m below 3000 m is 1.1 [$J_{sf}(2750m)/J_{sf}(3250m) = 1.1$]. That is, the shallower site receives 10% more organic matter, and its preservation index is depressed, therefore, by the equivalent of about 100 m. This decreases the sensitivity of the CDI towards saturation changes, and introduces a source of error for fluctuating productivity. It is clear that additional studies are needed to constrain the Emerson-Bender effect. Here we take it to be 25%. We proceed without considering it further until we interpret the results of the calculations.

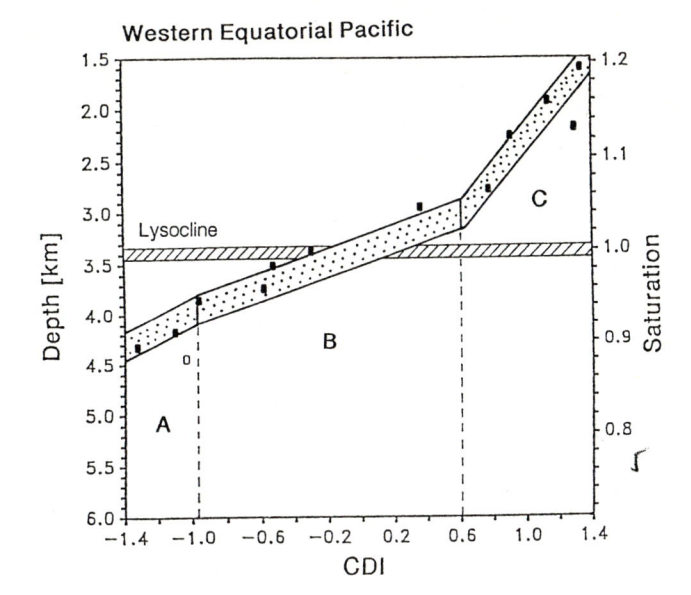

Fig. 5. Composite Dissolution Index (CDI) from ERDC box cores versus depth and corresponding saturation (according to data from Table 1)

3.5 Stratigraphic Application

We next reconstruct the ASL for core V28–238 (1° 1'N, 160° 29' E; depth 3120 m), based on the data of Thompson (1976). To do this, we standardized each variable to a mean of zero and a variance of one. The resulting transformed data points were each multiplied with the appropriate eigenvector from the box core data set, and summed for each core depth. The sum was then divided by the eigenvalue of the first principal component. This yielded the CDI values and hence the ASL, from Eqs. (2) to (4).

By this procedure, then, we gain a quantitative estimate for the conditions of preservation downcore. Figure 6 shows the $\delta^{18}O$ for *G. ruber* and the ASL for the last 960 ka. The ages up to 784 ka were adjusted to the stacked $\delta^{18}O$ curve from Imbrie et al. (1984), and the data in both sets interpolated in 1-ka steps.

There is a general correspondence between glacial conditions and high ASL, and interglacial conditions and low ASL. This is as expected from previous work (Thompson and Saito 1974; Shackleton and Opdyke 1976), including the correlation of high carbonate stages with glacials (Arrhenius 1952). Also, there is an association between deglaciation events and preservation spikes, and between ice-buildup events and dissolution spikes (event lines in Fig. 6, marked D and B, respectively), as noted by Berger and Vincent (1981).

This association may be described, from a purely formal standpoint, as a lag of the carbonate record with respect to the oxygen isotope record (Luz and Shackleton 1975; Shackleton and Opdyke 1976; Moore et al. 1977). Such a lag could be produced, for example, by shifting organic carbon between margin deposits and the deep sea (Keir and Berger 1983) or by analogous shifts in biosphere and soil carbon (Shackleton 1977).

4 Nature of the Saturation Record

4.1 Mechanisms of Change

The carbonate ion concentration in the deep ocean is one of the important "signals" of the workings of the oceanic carbon cycle through geologic time. Other such signals are the $\delta^{13}C$ record (see Berger and Vincent 1986), organic carbon deposition (Müller et al. 1983; Pedersen 1983; Sarntheim et al. 1987) and the CO_2 record in ice cores (Neftel et al. 1982; Barnola et al. 1987). All these proxies are related and trace the dynamics of the same system, in different ways (Keir and Berger 1983; Broecker and Peng 1986; Keir 1988).

Thus, the hypotheses offered here can be tested, in principle, by studying the relationships between these different "C" signals.

The overall correlation of carbonate-rich and carbonate-poor stages with cold and warm periods, respectively, may be ascribed in the main to two mechanisms: (1) exchange of the ocean's carbon reservoir with ocean-external transient reservoirs, and (2) basin-basin fractionation. One important reservoir is carbonate on shelf and platform areas, which competes with the deep sea floor, carbonate being fixed on shelves and platforms during high sea level position, and eroded during periods of low sea level (Berger and Winterer 1974). Similar shifts could be envisaged between high and low latitudes (Luz and Shackleton 1975).

Changes in the rates of supply of organic carbon to the ocean, or uptake from it, likewise affect saturation and atmospheric CO_2 (Berger et al. 1989). The apparent phase shift of the ASL signal with respect to the oxygen isotope record (Fig. 6) probably is the result of an instantaneous response of the carbon system to the rate of change in sea level (Keir and Berger 1983). Certainly erosion and supply of organic matter from shelf and continents is most active during the glacierization phase, adding CO_2 to the ocean, which is then titrated against carbonate on the sea floor. The reverse process, the buildup of margin deposits, forests, and soils during deglaciation, would have the opposite effect, and produce a preservation spike.

In addition to the shifting of carbonate reservoirs, the reorganization of deep circulation during glacierization and during deglaciation has to be considered in the attempt to explain transient saturation events (e.g., Keir 1983; Keir and Berger 1985). Glacial nutrient stripping of upper waters and a corresponding enrichment of deep waters with CO_2, would, during onset of glaciation, produce a dissolution event, while the relaxation of biological pumping in general would produce a preservation spike (Keir 1983; Broecker and Peng 1987).

4.2 Cross-Spectral Analysis

Cross-spectral analysis of the $\delta^{18}O$ and ASL records (Fig. 7) in essence confirms the results of Moore et al. (1977), who previously performed such analysis on the $\delta^{18}O$ and preservation signals of core V28–238 and V28–239. Moore et al. (1977) found that maxima in the oxygen isotope signal precede dissolution minima by about two sample intervals or 10 ka. For the 100-ka period they found a lag for preservation of

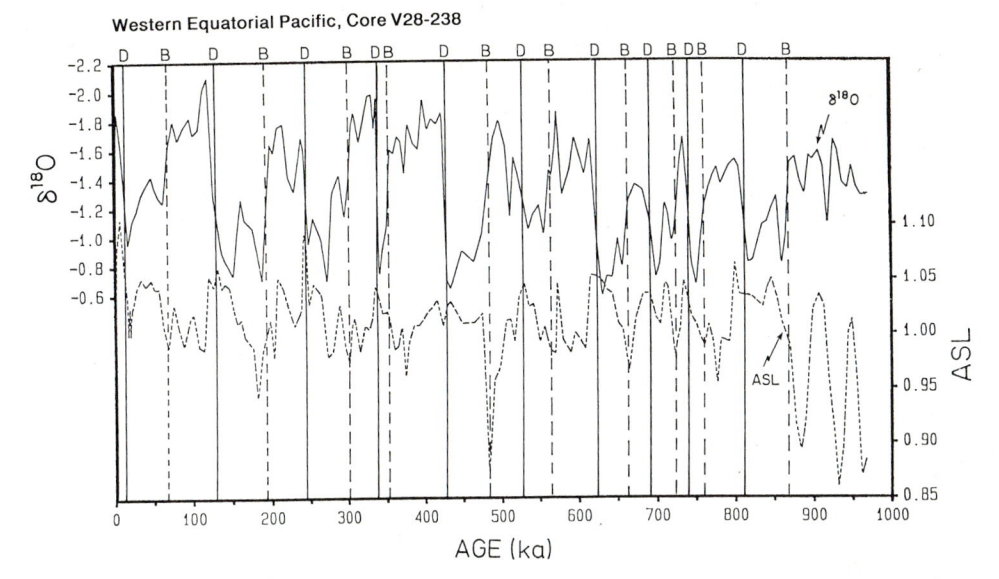

Fig. 6. Comparison of $\delta^{18}O$ record for *Globigerinoides sacculifer*, Core V28–238 (Shackleton and Opdyke 1973), with the ASL record based on data in Thompson (1976). Deglaciation steps marked *D*, ice buildup steps marked *B*

15 ka, for the 47-ka period one of 16 ka and for the 26-ka period one of 10.5 ka. Since the ASL is a linear transform of a preservation index, there is no reason to expect that our results should differ.

The spectrum for the ASL shows a very clear signal for the 100-ka cycle, which corresponds to that in the $\delta^{18}O$ spectrum. The phase lag (bottom of graph) is near 50°, with the carbonate preservation lagging oxygen isotope values about 14 ka. This agrees well with the lag shown in Shackleton and Opdyke (1976, Fig. 1). For the 41-ka period the ASL peak is slightly offset, maximum power appearing near 50 ka. Coherency is well below the 95% level for this part of the spectrum. There is good coherency near the 23-ka period with a lag of preservation of 16.5 ka; however, peak power in the $\delta^{18}O$ record is low. Also, the broad peak near 16 ka in the $\delta^{18}O$ record has no counterpart in the ASL record. On the whole, it appears that the carbonate preservation cycles faithfully reflect the long-term 100-ka and the short-term 23-ka period in the Milankovitch spectrum in this part of the Pleistocene record.

We suggested above that the carbon system may be reacting to rates of change in sea level (Keir and Berger 1983; Peterson and Prell 1985b). A comparison of the derivative of $\delta^{18}O$ with the ASL record (Fig. 8) shows a good correspondence between peaks and troughs. Clearly, however, the ASL record is not detailed enough, or does not preserve enough high frequency structure, to allow comparison on a fine scale. It must be remembered that whenever the ASL drops to below 0.90 we move into the vicinity of the CCD, so that microhiatuses may be present. Such hiatuses – if made by dissolution of part of the stratigraphic sequence – will interfere with spectral analysis. Also, while peaks and troughs agree quite well between the two

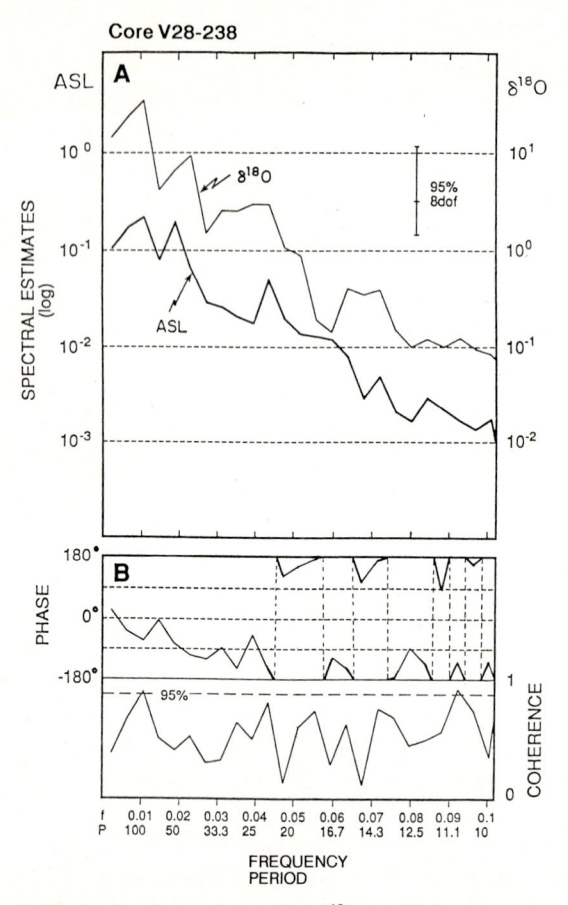

Fig. 7A,B Cross-spectrum for the $\delta^{18}O$ record and the ASL record of Core V28-238 (Fig. 6). **A** Spectral estimates in two different log scales, while phase and coherence are shown in a linear plot. The *vertical bar* gives the 95% confidence interval with 8 degrees of freefom (*8 dof*). The X- axis represents frequency (*f*) in cycles/ka and the corresponding period (*P*). **B** Phase spectrum of $\delta^{18}O$ and ASL. Negative phase indicates a lag of the ASL with respect to the $\delta^{18}O$, that is, a preservation maximum follows the isotopic maximum. The *dashed horizontal line* in the coherence plot indicates the 95% significance level

records, the amplitudes do not. This suggests that some additional factor is at work, perhaps having to do with the size of the carbon reservoirs involved. Cross-spectral analysis of $d(\delta^{18}O)/dt$ and the ASL signal shows that coherency is not affected when using the derivative of the $\delta^{18}O$ record, but that phase shifts are, as postulated, on the whole closer to zero (Fig. 9).

One other factor must be considered when discussing phase shifts between $\delta^{18}O$ and a preservation signal: a change in preservation by itself will affect the $\delta^{18}O$ record in such a fashion as to make it virtually impossible for the preservation signal to lead the isotope signal (see Wu and Berger 1989). The reason is that differential dissolution preferentially removes open-structured shallow-water specimens among the

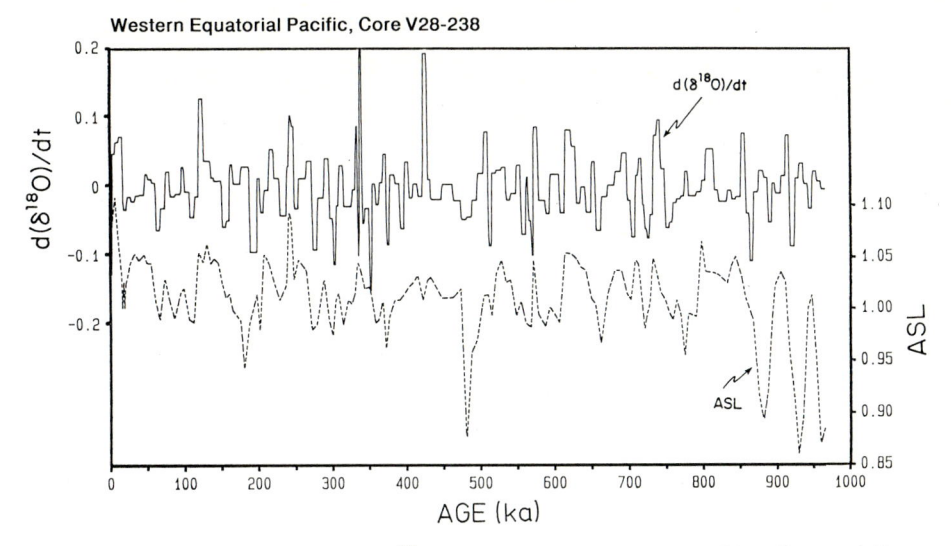

Fig. 8. Record of the first derivative of $\delta^{18}O$ in Core V28–238, compared with ASL record. Data as in Fig. 6

planktonic foraminifera species analyzed, concentrating thick-shelled forms which grew in deeper and therefore colder water. Thus, a dissolution pulse will generate a false signal of "cooling" or "drop in sea level", as seen in the oxygen isotope record.

5 Inter-Regional Comparison

The ASL index allows the direct comparison of preservation records from different regions in the ocean. Peterson and Prell (1985a,b) provide CDI records for the Indian Ocean based on the study of a series of box and piston cores from 6°S at the Ninetyeast Ridge. For comparison with V28–238, we chose a core with high resolution sampling and which was taken close to the lysocline, that is, V34–53 (6°7'S, 89°35'E; depth 3812 m). The method for calculating the ASL is the same as described above. For calibration we use the GEOSECS station 441 which is the closest to V34–53. Lysocline and saturation level 1.0 are at 3800 m. The average sedimentation rate is 0.7 cm/ka, but there is considerable variation downcore. Thus, the resolution is lower in that core than in V28–238, which has a more or less steady accumulation of 1.7 cm/ka.

The ASL curve for V34–53 (Indian Ocean) shows an excellent peak-to-peak correlation with that of V28–238 (Pacific) (Fig. 10). This indicates that at least for the last 800 ka the Indian Ocean and the Pacific have shared a similar response to changing deep circulation, in contrast to the Atlantic (Volat et al. 1980; Crowley 1985). This is also seen when comparing carbonate records (e.g., of Arrhenius 1952; Olausson 1960) or preservation index records (e.g., of Oba 1969; Thompson 1976).

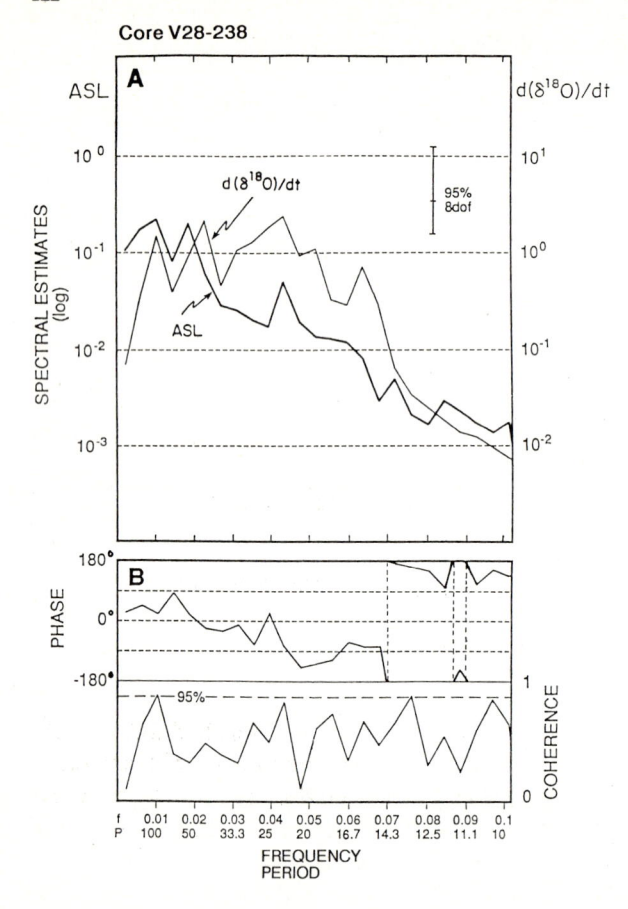

Fig. 9A,B. Cross-spectrum for the first derivative of the $\delta^{18}O$ record and the ASL record of V28-238 (Fig. 8). For further explanation see Fig. 7

Although the correlation is very good, there are distinct differences in the amplitude of the two time series, depending on the time interval. The Apparent Depth Change (ADC) in the lysocline position (scale on right of graph) shows a greater excursion in the Indian Ocean, for the Brunhes dissolution maximum between 400 and 600 ka B.P. This may indicate a greater sensitivity to this disturbance within the deep Indian Ocean, or it may reflect the difference in depth of the two cores. One would perhaps expect a greater amplitude of saturation fluctuations in the Pacific, since the greater age of (interglacial) Pacific deep water would leave more room for fluctuation of age, and hence saturation.

In comparing ASL amplitudes between V28–238 and V34–53, the lower resolution in V34–53 must be kept in mind. If one wishes to emphasize the possible role of sedimentation rate in controlling dissolution rate and preservation state (Arrhenius 1988), one could argue that the reduced tendency in the V34–53 record to display sections of good preservation, compared with Core V28–238, is due to the twice lower rate of accumulation in the Indian Ocean core.

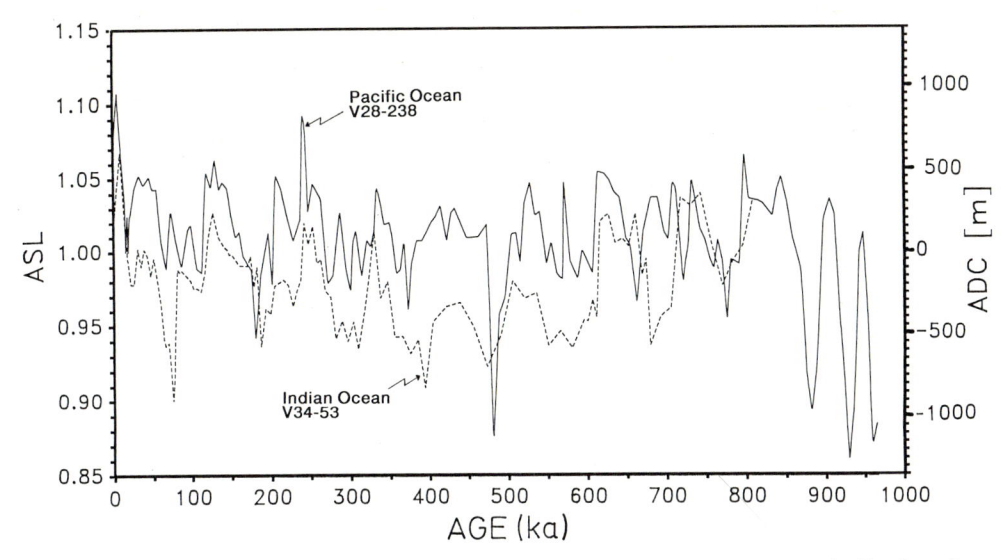

Fig. 10. Comparison of ASL records of core V28–238 (Pacific Ocean) and core V34–53 (Indian Ocean). Scale on the *right* gives an approximation of the Apparent Depth Change (ADC) in saturation level 1.0 (lysocline) in meters

6 Depth-Dependent Fluctuations of ASL

Using the CDI values from the series of piston cores investigated by Peterson and Prell (1985b) and calculating the ASL record for each, we obtain the saturation fluctuations as a function of depth and time (Fig. 11). The present lysocline (ASL = 1.0) is near 3800 m. This may be compared with the lysocline position in the western Pacific, at 3400 m. The Pacific site is "downstream" with regard to deep circulation and water mass aging, hence saturation is decreased for constant depth.

There is some indication that the range of fluctuation of the ASL is greatest in the vicinity of the lysocline, that is, between 3 and 4 km (Fig. 11). Clearly, below 4 km amplitudes are reduced. Assuming that we see both the effects from global ocean saturation changes (e.g., shelf sediment exchange), and from deep circulation changes (i.e., inter-basin fractionation), the sum of these two effects appears greater in deep waters than in bottom waters, in this region. This observation can be rationalized by assigning a North Atlantic Deep Water (NADW) signal to deep waters, and an Antarctic Bottom Water (AABW) signal to the bottom waters. NADW production increases greatly from glacial to interglacial time (Duplessy et al. 1980; Boyle and Keigwin 1982), with the effect of decreasing saturation in the deep waters of Pacific and Indian Ocean. The amplitude of the saturation signal should vary correspondingly.

As before, we add a cautionary note to these tentative interpretations. The ASL values at 0.90 and below are very close to the CCD, where the record must be unreliable to some extent. Preservation events which would increase the amplitude of ASL toward higher values may be eroded during subsequent dissolution events.

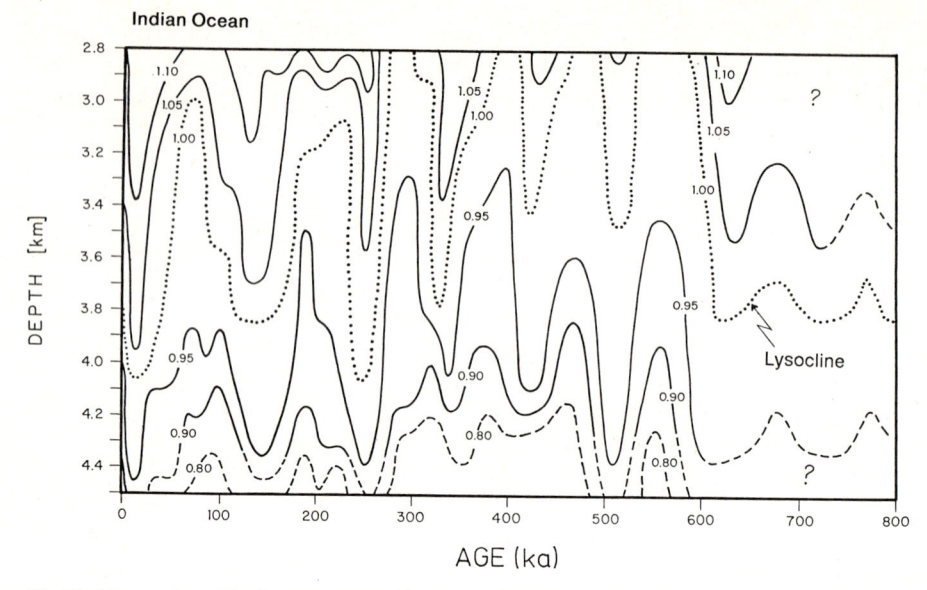

Fig. 11. Fluctuation of the Apparent Saturation Levels (ASL) through the last 800 ka for the Indian Ocean, as a function of water depth. Graph is based on the CDI data of seven piston cores investigated by Peterson and Prell (1985b, Fig. 8) which was converted in ASL values as described in text. *Dashed lines* indicate uncertainties in data

Also, excursions to very low ASL values will simply result in removal, rather than in modification of remaining sediment. Thus, the ranges below 4 km are likely to be truncated, so that the significance of the reduced amplitudes at depth is not clear.

7 Summary and Conclusion

Quantitative descriptors of preservation state can be derived by tying preservation indices to the carbonate saturation of the water in contact with the sea floor. This allows comparison of preservation records between different cores, different workers and different regions. The determination of the Apparent Saturation Level (ASL) rests upon the assumption that interference from varying productivity is negligible or can be constrained. Thus, it is unlikely to work in regions where the supply of organic matter reaching the sea floor is highly variable. We have estimated the effects from the oxidation of organic carbon as 25% of the ASL. We are aware that this estimate is rather crude.

Conversion of preservation proxy data to ASL signals does not affect conclusions regarding periodicities and phase properties of the records. For example, it is readily demonstrated that the ASL is not synchronous with the $\delta^{18}O$ isotopic changes, but has a lag of about 13 to 15 ka. In contrast, the major changes of the $\delta^{18}O$ and the extremes in preservation are in phase. These relationships give rise to explanations employing both inventory models and circulation models.

While there is little effect on the results of spectral analysis, when converting preservation arbitrary indices to saturation levels, the amplitudes of saturation changes or Apparent Depth Changes (ADC) of preservation levels are more meaningful than those of arbitrary indices. Comparison of ASL records from the western Pacific and from the central Indian Ocean suggests that the saturation state of deep waters in the two regions fluctuated together, over a considerable range. These fluctuations show a high coherency within the 100 ka and 23 ka Milankovitch band.

The amplitude of the fluctuations in the saturation index typically reaches 0.1, corresponding to a depth excursion of saturation isolines of about 900 m. This is somewhat larger than the excursion shown by Berger and Keir (1984, Fig. 6), for the last 15,000 years. Allowing for a 25% correction for the Emerson-Bender effect, we retain an amplitude of 700 m due to changing saturation. This amplitude indicates substantial changes in alkalinity in the Pleistocene ocean, which would have markedly affected atmospheric CO_2 levels. Hence, varying carbonate dissolution makes a substantial contribution to the fluctuations of atmospheric pCO_2 on the glacial-interglacial time-scale. Such a contribution from changing carbonate preservation would alleviate problems arising when referring the pCO_2 change observed in ice cores (Neftel et al. 1982; Barnola et al. 1987) to ocean productivity alone (Broecker and Peng 1987).

Finally, we come back to the question posed at the outset – whether the saturation fluctuations are large enough to produce the carbonate cycles seen. The depth difference between lysocline and CCD, in the equatorial Pacific, is typically on the order of 1 km. Near the lysocline, carbonate percentages are near 85%, at the CCD they are zero. Thus, a vertical range of 700 m of saturation fluctuation can readily produce a substantial variation in carbonate content of sediments deposited at a site between lysocline and CCD, in the equatorial Pacific.

We claim that our results provide strong evidence that saturation fluctuations are the main factor responsible for carbonate cycles. We admit that they do not constitute proof. If one wishes to invoke productivity cycles for generating the carbonate cycles one can argue that the concept of tying preservation state to saturation (rather than to productivity) prejudges the issue. In defense of the saturation hypothesis we point to the results of Broecker and Takahashi (1978), who showed that preservation state does correspond to saturation, on a global scale, at least in the vicinity of the lysocline, and away from continental margins.

Acknowledgments. We thank Uwe Send and Sargun Tont, who were very helpful in assisting with software applications and statistical problems. The research was supported by a 1-year fellowship of the Studienstiftung des deutschen Volkes, to Jürgen Grötsch, and by the National Science Foundation (OCE 88–16167).

1.7 A Holistic Geochemical Approach to Cyclomania: Examples from Cretaceous Pelagic Limestone Sequences

M. A. Arthur and W. E. Dean

1 Introduction

Bedding rhythmicity, characterized by distinctive cyclic interbeds of different lithologies, mostly controlled by variation in amounts of carbonate and organic matter is particularly well developed in pelagic carbonate units such as the Greenhorn and Niobrara Formations of the Cretaceous Western Interior Seaway of North America and Neocomian sequences of the North Atlantic and Tethyan realms (e.g., Fischer et al. 1985; Research on Cretaceous Cycles, ROCC Group 1986). All of these carbonate sequences are characterized by cyclically interbedded limestones and shales/marlstones with estimated periodicities in the 20–40 ka range. Despite arguments that attempt to explain limestone/marlstone cyclicity in terms of diagenetic redistribution of carbonate (e.g., Eder 1982; Hallam 1964, 1986; Ricken 1986; Raiswell 1988a), the variations in carbonate, organic carbon, and inorganic geochemical characteristics within these cycles most likely reflect changes in surface-water productivity, bottom-water energy and oxygenation, and detrital flux and composition as the result of periodic climate changes. These inferred periodic climate variations are in the Milankovitch band and may indicate forcing by changes in insolation due to periodic solar-terrestrial orbital variations (e.g., Hays et al. 1976; Imbrie et al. 1984; ROCC Group 1986; Schwarzacher and Fischer 1982; Fischer et al. 1985; Herbert and Fischer 1986).

In this paper we briefly examine the constraints that a "holistic" approach to the geochemistry of Cretaceous pelagic carbonate sequences can provide on interpretation of changes in paleoceanographic and paleoclimatic conditions, primary depositional processes, production and preservation of organic matter and other biogenic components, sources of detritus, and diagenesis. All too often, geochemists obtain their own suite of samples, take them home, and analyze them with their favorite "black boxes", never to relate the data back to the lithologic sequence. Our "holistic" approach involves attention to the details of variation in lithology, texture, and sedimentary structures in a given sequence during sampling, followed by a suite of diverse analyses on splits of each sample; for example, inorganic geochemical, clay mineralogical, selected organic geochemical and stable-isotope data are all routinely collected for each sample where possible. Our preliminary results suggest that the overall geochemistry of each lithology is remarkably similar from site to site in a single ocean basin, but at any one site there is a marked contrast in the geochemistry of interbedded lithologies. These geochemical variations and the implied mineralogi-

Table 1. Information on Deep Sea Drilling Project (DSDP) Sites in the North Atlantic Ocean and wells in Cretaceous strata in the Western Interior Seaway of Colorado for which data are presented in this report

Site or well	Water depth (m)	Depth within interval cored (m)	Age of interval	Formation
DSDP Site 387	5117	631 − 792	Lower Valanginian to lower Barremian	Blake-Bahama
DSDP Site 367	4748	891.5 − 1091	Berriasian to Barremian, Blake-Bahama	
DSPD Site 603	4633	1300.8 − 1511.5	Valanginian to Barremian	Blake-Bahama
Berthoud State #4		868 − 965	Turonian to Campanian	Niobrara
Princeton Univ. Pueblo, outcrop		18 − 25	Cenomanian to Turonian	Greenhorn

cal variations primarily reflect changes in the relative amounts of different components that reach the sea floor, redox variations and authigenic minerals that result from changes in the flux and preservation of organic matter, burial diagenetic recrystallization/cementation of biogenic components, and changes in the composition of detrital minerals delivered to the depositional sites. We illustrate selected examples of cycles from Neocomian sections recovered from Deep Sea Drilling Project (DSDP) Sites 603, 387, and 367 in the North Atlantic Ocean, the Cenomanian/Turonian Bridge Creek Limestone Member of the Greenhorn Formation from a shallow core taken near Pueblo, Colorado, and the upper Turonian to lower Campanian Niobrara Formation from continuous cores in the Berthoud State wells near Fort Collins, Colorado (Fig. 1, Table 1). The geochemical data included in this chapter represent a small part of data being collected for our ongoing studies of Milankovitch cycles in organic-carbon-rich strata. Because of length restrictions, the reader should consult Dean and Arthur (1987) for a description of our sample handling and analytical methods.

2 The Investigated Milankovitch Cycles

2.1 Neocomian Limestones of Sites 367, 603 and 387

Interbeds of laminated dark olive to black marlstone and laminated to bioturbated, white to light gray limestone of Tithonian (Late Jurassic) through Neocomian (Early Cretaceous) age have been recovered, among others, at DSDP Sites 105, 367, 387, 391, and 603 in the eastern and western basins of the North Atlantic (Fig. 1). The oceanic unit comprising these rocks has been called the Blake-Bahama Formation by Jansa et al. (1979), and is equivalent to the late Tithonian to Aptian Maiolica (or Biancone) Formation of Garrison (1967) that crops out in the Tethyan regions of the Mediterranean (Bernoulli 1972; Bernoulli and Jenkyns 1974; Jansa et al. 1979) and

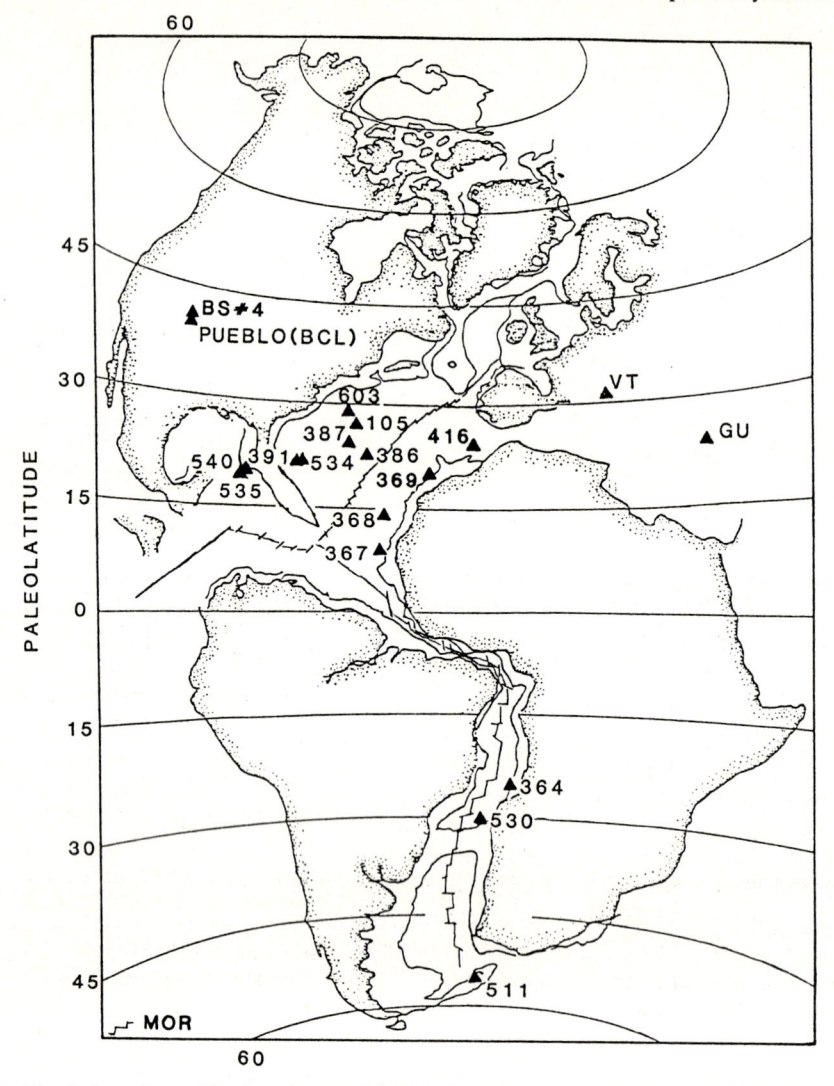

Fig. 1. Locations of land sections and DSDP sites discussed in this report plotted on a continental reconstruction of 100 million years ago. (After Sclater et al. 1977). *VT* Vocontian Trough, southern France; *GU* Gubbio, central Italy; *Pueblo (BCL) Bridge Creek Limestone, Pueblo, Colorado; BS #4* Niobrara Formation, Coquina Oil, Berthoud State #4 well near Ft. Collins, Colorado; *MOR* mid- ocean ridge

to thick limestone sequences of the Vocontian Trough in southern France (Cotillon 1984; Cotillon and Rio 1984; Contillon Chap. 7.3, this Vol.).

The laminated marlstone beds of the Blake-Bahama Formation usually contain relatively high concentrations of organic carbon. For example, at Site 367 in the eastern North Atlantic basin, bioturbated to faintly laminated white limestones are interbedded with dark-colored marlstones (Fig. 2) that are finely laminated and

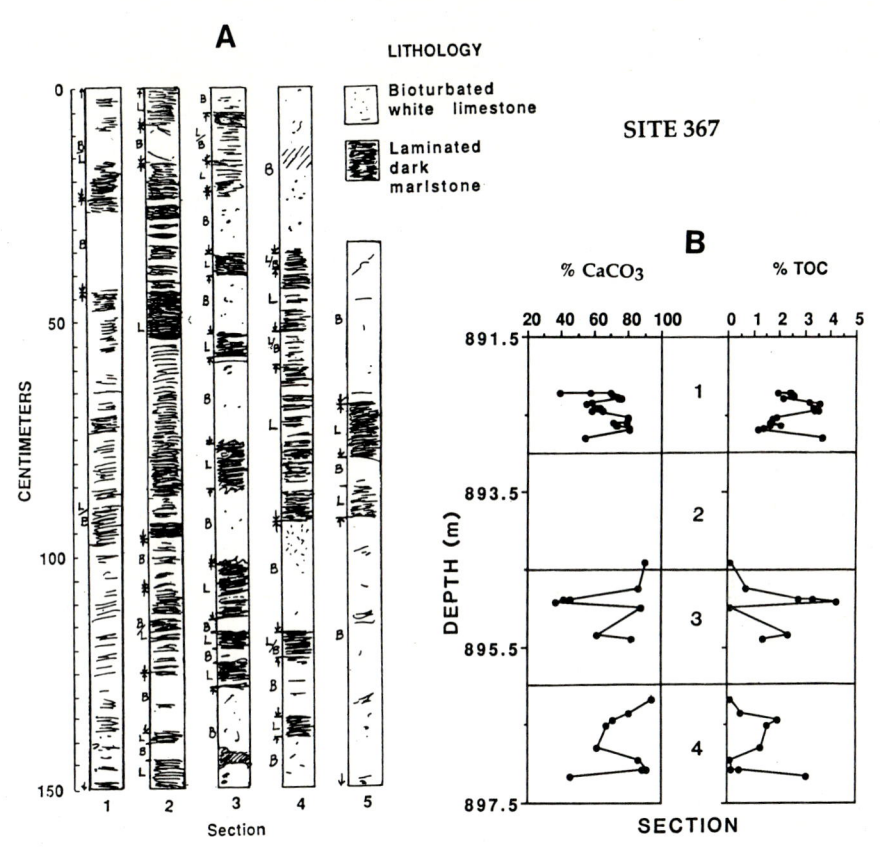

Fig. 2 A,B. Diagrammatic representation of the basic Neocomian lithologic types at **A** DSDP Site 367 as they occur in Core 25. **L** finely laminated intervals; *B* highly bioturbated intervals; *L/B* faintly laminated to sparsely bioturbated intervals. **B** Variations in percent CaCO₃ and total organic carbon (*TOC*) in samples from 891.5 to 898.7 m subbottom in core 25. *Section numbers* are the same as in **A**. Note the overall increase in TOC and decrease in CaCO₃ towards the top of the core in concert with the increasing proportion of laminated interbeds

contain up to 5% organic carbon (C$_{org}$) with relatively high hydrogen indices (HI; around 400 mg hydrocarbons/g C$_{org}$; Herbin and Deroo 1982; Dean et al. 1986), probably indicating enhanced preservation of marine C$_{org}$ under periodic anoxic conditions (Dean et al. 1986; see below). The carbonate contents range from 40 to 98%, with the highest contents in the bioturbated, white limestone beds. The relative thicknesses of the simple alternation of dark- and light-colored units can change considerably through a 9-m core (Fig. 2). The pattern of lamination is somewhat more complex, with some intervals of the core being almost entirely laminated, even though carbonate contents continue to vary cyclicly. The laminae in the Blake-Bahama Formation in the western North Atlantic basin, particularly at Site 603, typically are cryptic, and several lines of evidence suggest a dominantly terrestrial source of organic matter (e.g., Herbin and Deroo 1982; Summerhayes and Masran

1983; Habib 1983; Katz 1983; Dean and Arthur 1987; Meyers 1987). It is therefore not entirely resolved whether these laminae indicate deposition in anoxic bottom waters, as has been commonly assumed (Tucholke and Vogt 1979; Robertson and Bliefnick 1983; Arthur and Dean 1986), or were produced by bottom-current activity (Dean and Arthur 1987; also see Cotillon Chap. 7.3, this Vol.).

The Neocomian sequence at Site 603 provides a good example of the complexity of cycles in some pelagic sequences of the North Atlantic Ocean basin (see Fig. 15A for examples of cycles). Four lithologic types are present in the interbedded Neocomian carbonates: (1) bioturbated white to light gray limestone and marly limestone; $CaCO_3$: 70–92%; C_{org}: 0.07 to 0.28%; (2) medium brown to light gray, finely to coarsely laminated, streaked or microburrowed marlstone and marly limestone; $CaCO_3$: 54–93%; C_{org} 0.43% to 2.05%; (3) dark brown to black homogeneous claystone and calcareous claystone; $CaCO_3$: 8–25%; C_{org}: 1.43 to 2.24%; and (4) gray to tan, graded to massive siltstone and sandstone. The relative proportion of each lithologic type changes with depth: interbedded laminated and bioturbated limestones dominate in the lower part of the sequence (Berriasian-Valanginian) with increasing amounts of terrigenous clastic sediment upsection. Coarsely laminated carbonate beds predominate over bioturbated beds throughout most of the unit.

On the basis of lithologic relationships and carbonate and C_{org} contents, the type 2 laminated beds appear to be intergradational with types 1 and 3, in other words, a mixture of nannofossil carbonate and low-carbonate clay lithologies. Their inorganic geochemical characteristics also are intergradational, as we show below. The black, homogeneous claystones have a narrow range of relatively high C_{org} and low $CaCO_3$ values. The bioturbated limestones also cluster, with low C_{org} and high $CaCO_3$ contents. The laminated (Type 3) carbonates have C_{org} and $CaCO_3$ values intermediate between the other two types, but closer to the bioturbated limestones in $CaCO_3$ content than to the claystones. Cycles in the Blake-Bahama Formation at DSDP Site 387, also in the western North Atlantic Ocean, are more like those at Site 367.

2.2 Bridge Creek Limestone and Niobrara Formation

Two important organic-carbon rich pelagic limestone units were deposited in the North American Cretaceous western interior seaway during two major transgressive episodes (e.g., Kauffman 1977a,b; Arthur et al. 1985b; Kauffman et al. Chap. 7.2, this Vol.). A Cenomanian-Turonian transgression resulted in deposition of the Bridge Creek Limestone Member of the Greenhorn Formation, and a Late Turonian through early Campanian transgression led to sedimentation of the Niobrara Formation. Both

Fig 3A,B. A Concentrations of insoluble residue and estimated outcrop profiles of limestone/marlstone couplets produced by variations in terrigenous dilution (*a*), carbonate dissolution (*b*), and carbonate production (*c*) after Arthur et al. (1984a). **B** Photograph of typical interbedded bioturbated limestones (*light*) and laminated marlstones (*dark*) of the Bridge Creek Limestone in a core taken near Pueblo, Colorado

Arthur and Dean: A Holistic Geochemical Approach to Cyclomania

A

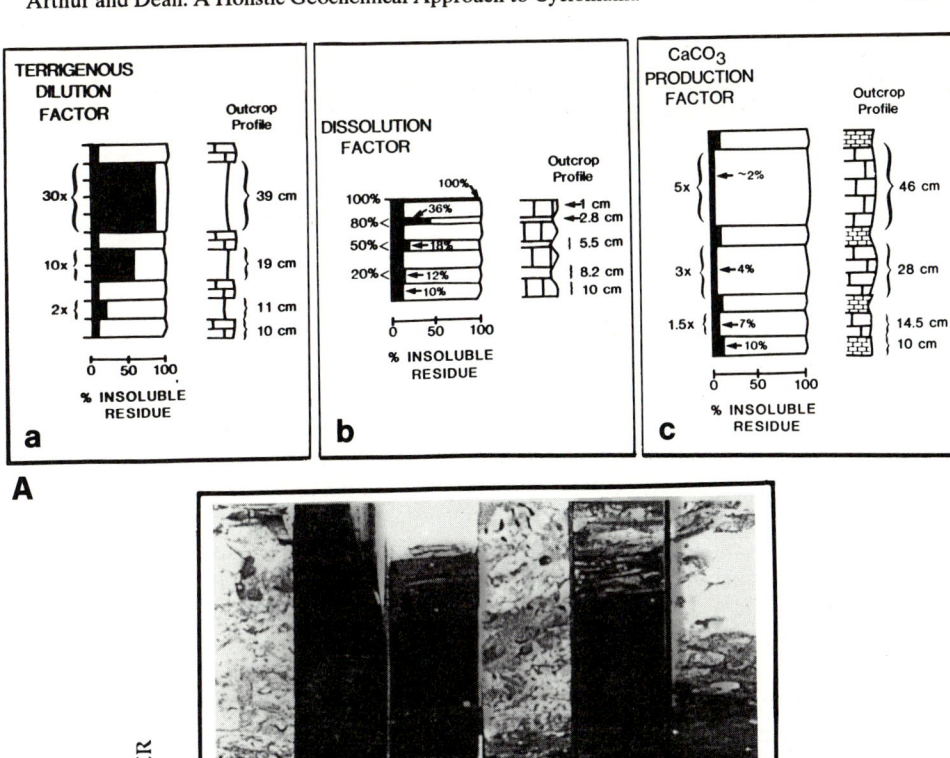

a) TERRIGENOUS DILUTION FACTOR — Outcrop Profile: 39 cm (30x), 19 cm (10x), 11 cm (2x), 10 cm; % INSOLUBLE RESIDUE

b) DISSOLUTION FACTOR — 100%, 36%, 18%, 12%, 10%; Outcrop Profile: 1 cm, 2.8 cm, 5.5 cm, 8.2 cm, 10 cm; % INSOLUBLE RESIDUE

c) CaCO₃ PRODUCTION FACTOR — ~2% (5x), 4% (3x), 7% (1.5x), 10%; Outcrop Profile: 46 cm, 28 cm, 14.5 cm, 10 cm; % INSOLUBLE RESIDUE

B BRIDGE CREEK LIMESTONE MEMBER PUEBLO CORE

0 10 CENTIMETERS

of these pelagic carbonate units were deposited in a rapidly subsiding basin (Bond 1976; Cross and Pilger 1978) that was characterized by substantial clastic sediment input (e.g., Reeside 1944; Ryer 1977) from uplifted tectonic terranes of the Sevier orogenic belt to the west (Armstrong 1968; Weimer 1970) as well as an abundant supply of windblown ash from concomitant volcanic activity in the same region (Kauffman 1977a). A major feature of these pelagic carbonate sequences is their cyclicity (Gilbert 1895; Hattin 1971; Kauffman 1977a; Fischer 1980; Pratt 1981; Arthur et al. 1984a; Barron et al. 1985; Barlow and Kauffman 1985) in which C_{org} and $CaCO_3$ contents vary on scales of 0.2 to 1.5 m or an estimated periodicity of 20–100 ka.

The lower to middle part of the Bridge Creek Limestone (Figs. 3B, 4A) spans the Cenomanian-Turonian boundary. Cores 2–4 (21.8–25.5 m; Fig. 4A) comprise an interval of thin, dark, laminated to somewhat bioturbated marlstone and interbedded light-colored, bioturbated pelagic limestone beds of variable thickness. The relative proportions of each lithology vary, and couplets are 25 to 50 cm thick. Cores 5–7 (17.5–21.4 m; Fig. 4A) are characterized by somewhat thicker interbeds of marlstone and limestone (so-called major limestone beds with bed numbers as shown; e.g., Elder 1985) but lithologically similar to those in cores 2–4. Carbonate contents vary between 42 and 96% overall, but contents are more variable in cores 2–4, and only range between 65–90% in cores 5–7 (Fig. 4A). Organic-carbon contents range from 0.05 to 7.0% overall, but vary widely from carbonate-rich to relatively carbonate-poor beds.

If we assume that each hemicycle in a simple limestone/marlstone alternation represents an equal amount of time, and if we also assume a constant periodicity for the bedding cycles, then the systematic variations in relative thickness of each part of the cycles in the Site 367 and Bridge Creek Limestone sequences must also indicate changing relative fluxes of one or both of biogenic carbonate and detrital material, both across individual cycles and on the longer term. Three simple models to explain the observed variations in carbonate content and relative bed thickness are (Fig. 3A): (A) constant carbonate accumulation rate and systematically changing flux of terrigenous material to the seafloor; (B) constant terrigenous flux and changing carbonate flux as the result of variations in dissolution rate; and (C) constant terrigenous flux and variable carbonate flux resulting from productivity variations (e.g., Dean et al. 1977; Einsele 1982a; Arthur et al. 1984a). A more complex model would have

Fig. 4A,B. A Downhole plots of percent organic carbon and $CaCO_3$, values of $\delta^{13}C$ and $\delta^{18}O$ of whole-rock carbonate, and whole-rock values of Na_2O/K_2O and SiO_2/Al_2O_3 in a core of the Bridge Creek Limestone from near Pueblo, Colorado. Limestone bed numbers are from Cobban and Scott (1972). *Letters* by the intervals between major limestone beds indicate whether the interbedded lithology is bioturbated (*B*), microbioturbated (*M*), or laminated (*L*); all major limestone beds are bioturbated. **B** Downhole plots of % $CaCO_3$ and organic carbon, values of Rock-Eval pyrolysis hydrogen index (in milligrams of hydrocarbons per gram organic carbon), and whole-rock values of Ba/Al, Fe/Mn, Fe/S, S/OC, and Na/K in the Smoky Hill Chalk (878–960 m) and Fort Hays Limestone (960–966 m) Members of the Niobrara Formation in the Berthoud State #4 core

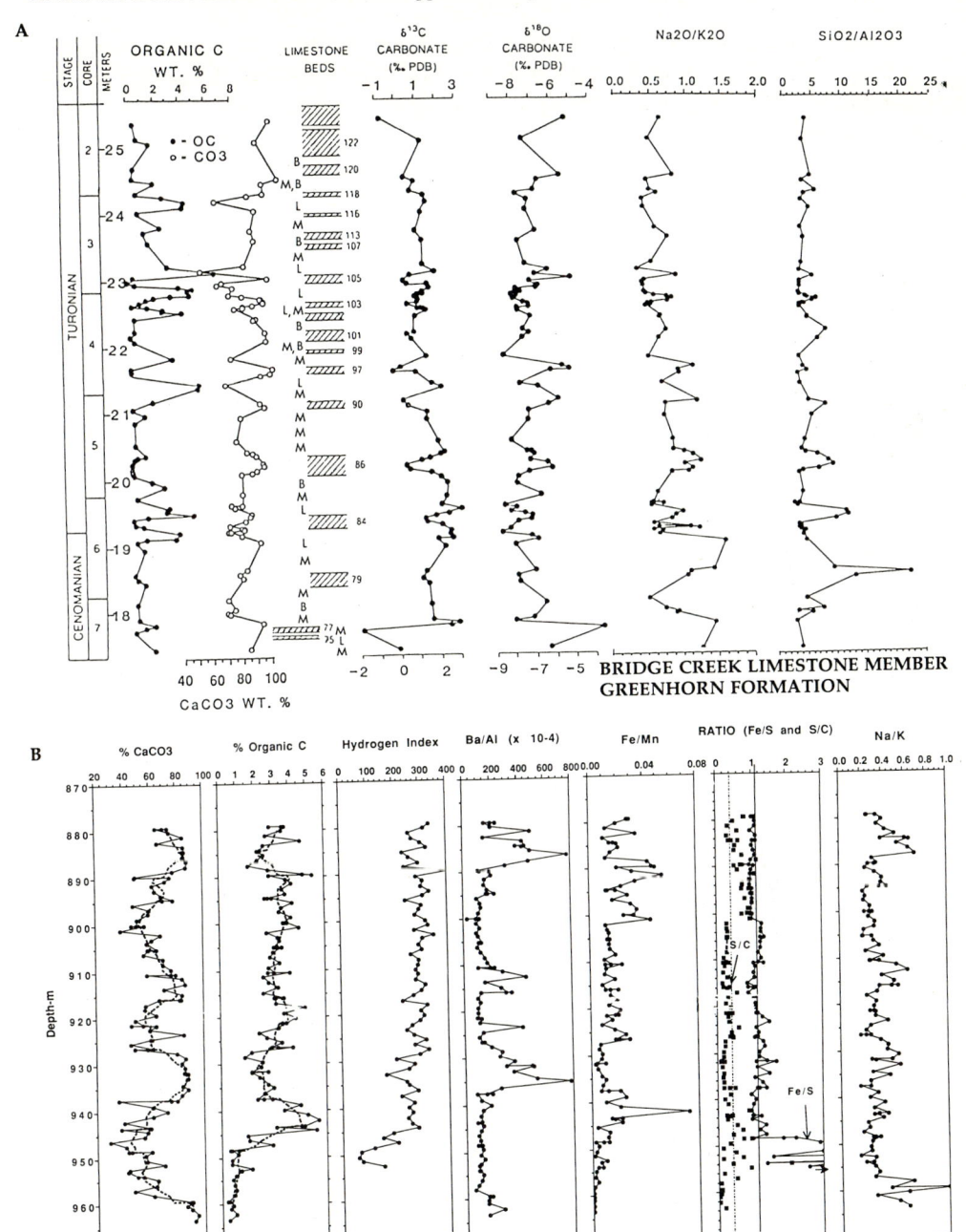

A

BRIDGE CREEK LIMESTONE MEMBER
GREENHORN FORMATION

B

NIOBRARA FORMATION

both carbonate and terrigenous fluxes covarying or varying antithetically. Applying the simple models to the Bridge Creek Limestone sequence, for example, leads us to predict that the bedding cycles are mainly due to variations in terrigenous dilution, perhaps with superimposed slight variations in carbonate productivity, because of the extreme variability of carbonate content, bed thickness, and upsection changes in the relative thickness of carbonate vs. marlstone beds. This is essentially the conclusion reached by Pratt (1984), Barron et al. (1985), and Arthur et al. (1985b). Dissolution is unlikely as a cause of the carbonate variability because of the shallow depth of the interior seaway.

Bedding cycles (20–50-cm scale) in the Niobrara Formation in the Berthoud State #4 well have many of the same characteristics as those in the Bridge Creek Limestone (Fig. 5A,B), and probably had the same overall cause. The variations in $CaCO_3$ as well as other lithologic features (e.g., color, lamination, and bioturbation) have been used to subdivide the Niobrara Formation into two formal members and seven informal chalk/marl units (e.g., Scott and Cobban 1964; Scott 1969; Kauffman 1977b; Hattin 1982). The Fort Hays Limestone Member (below 960 m in Fig. 4B) is generally bioturbated throughout with limestone beds separated by very thin, dark-colored shales, whereas much of the Smoky Hill Chalk Member (above 960 m in Fig. 4B) is dark-colored and finely laminated, or is characterized by cycles of laminated marlstone alternating with bioturbated chalk (Fig. 5A; e.g., see Savrda and Bottjer 1986). Contents of $CaCO_3$ and C_{org} tend to vary antithetically (Figs. 4B and 5B), as in the Bridge Creek Limestone. High $CaCO_3$ and low C_{org} contents are typical of the bioturbated light gray limestones of the Fort Hays Limestone Member at the base of the Niobrara. The Smoky Hill Chalk Member is characterized by high concentrations of hydrogen-rich organic matter (C_{org} >2%; hydrogen indices, HI >200 mgHC/g C_{org}), and variable $CaCO_3$ contents.

3 Diagenetic Geochemical Signals

3.1 Geochemical Indicators of Redox Cycles

The Rock-Eval HI provides a measure of the degree of preservation (and/or source) of sedimentary organic matter (e.g., Espitalié et al. 1977). There is an excellent correlation between percent C_{org} and HI in both the Bridge Creek Limestone and the Smoky Hill Chalk. Data in cross plots of these two variables from the two carbonate units overlay one another almost exactly (Dean et al. 1986), suggesting that they had similar organic sources and depositional conditions. The relationship between HI and C_{org} indicates that increasing C_{org} contents are due to the addition and preservation of hydrogenous marine (sapropelic) organic matter. Because such organic matter is highly "reactive" or easily oxidized, particularly in oxic environments, the enhanced preservation of this organic matter suggests deposition under marginally oxic to anaerobic conditions. The relationship between organic matter degradation and selective loss of hydrogen as a function of degree of oxidation was clearly demonstrated by Pratt (1984) for the interbedded laminated to microburrowed (dysaerobic to anaerobic), C_{org}-rich units versus the bioturbated (oxic) C_{org}-poor units of the Bridge Creek Limestone. The same principle applies to the Niobrara Formation. However, the consistently high HI values (above 200 mgHC/g C_{org}) for the Smoky Hill Chalk (Fig. 4B) indicate the prevalence of oxygen-depleted bottom waters throughout deposition of that unit. Because values of HI do not vary cyclically through C_{org}-rich and -poor units, we conclude that cyclic variations in C_{org} are due mainly to variable C_{org}-flux and/or dilution and not preservation. The HI values probably are somewhat lower than expected for such laminated C_{org}-rich strata because the organic matter is marginally mature as the result of the high paleo-burial depth (>2000 m) of the sequence recovered in the Berthoud State no. 4 well.

Diagenetic Fe-S-C_{org} relationships in the Niobrara Formation and the Bridge Creek Limestone have been discussed by Dean and Arthur (1989). A lower S/C_{org} (sulfide sulfur/organic carbon) weight ratio in the Smoky Hill Chalk (Fig. 4B) relative to that for modern marine sediments (0.4; Goldhaber and Kaplan 1974; Berner and Raiswell 1983), suggests either that seawater in the Cretaceous Western Interior basin contained less sulfate than modern seawater, or that sulfide escaped before being fixed as iron sulfide, probably due to limitation by "reactive" iron. Higher S/C_{org} ratios in bioturbated limestones of the Bridge Creek Limestone (Pratt 1984) probably are due to low initial C_{org} contents and formation of iron sulfide once reducing conditions were established by reduction of available reactive iron and scavenging of upward and downward diffusing sulfide produced mainly in the adjacent C_{org}-rich beds. Most samples of Smoky Hill Chalk have S/Fe ratios close to that of stoichiometric FeS_2 (S/Fe=1.15; Figs. 4B and 6A) suggesting that most, if not all, reactive Fe was consumed, along with organic matter, during bacterial sulfate reduction in the C_{org}-rich units. Data for the Bridge Creek Limestone fall in the same field as those for the Niobrara Formation on an Fe-S-C_{org} ternary plot (Fig. 6A). Iron limitation on pyrite formation in carbonate-rich rocks is common, and explains the typical linear Fe/S relationship in many such rocks.

The Fe-S-C_{org} data for Neocomian carbonate cycles at DSDP Sites 367 and 387 also plot along a line of constant S/Fe ratio, indicating iron limitation on pyrite formation. However, those for Neocomian carbonate cycles at Site 603 (Fig. 6B) exhibit a significantly different pattern, with a tendency towards distribution of points along a constant S/C_{org} weight ratio line of 0.4 similar to data for Holocene C_{org}-rich marine sediments. This pattern has been interpreted by Dean and Arthur (1989) for other strata as indicating a likely limitation of reactive C_{org} on pyrite formation. In this case, there is insufficient bacterially degradable marine organic matter preserved in sediments to allow complete reduction of sulfate and fixing of all available iron as sulfides. This interpretation is consistent with the low HI values and dominance of terrigenous organic matter at Site 603 (e.g., Dean and Arthur 1987; Meyers 1987). Figure 10c shows an example of interbedded carbonate-poor, C_{org}-rich black and C_{org}-poor green claystones in the Albian through Turonian section at DSDP Site 530 at the base of Walvis Ridge, southern Angola Basin. These multicolored claystones have been interpreted as representing either redox cycles resulting from changes in oxygenation in bottom water at the site, or, more likely, resulting from periodic redeposition of marine C_{org}-rich sediment from upslope (Dean et al. 1984a; Stow and Dean 1984). The Fe-S-C_{org} data for black beds, which contain up to 12% C_{org}, exhibit a trend of constant Fe/S ratio indicating iron limitation on pyrite formation in the

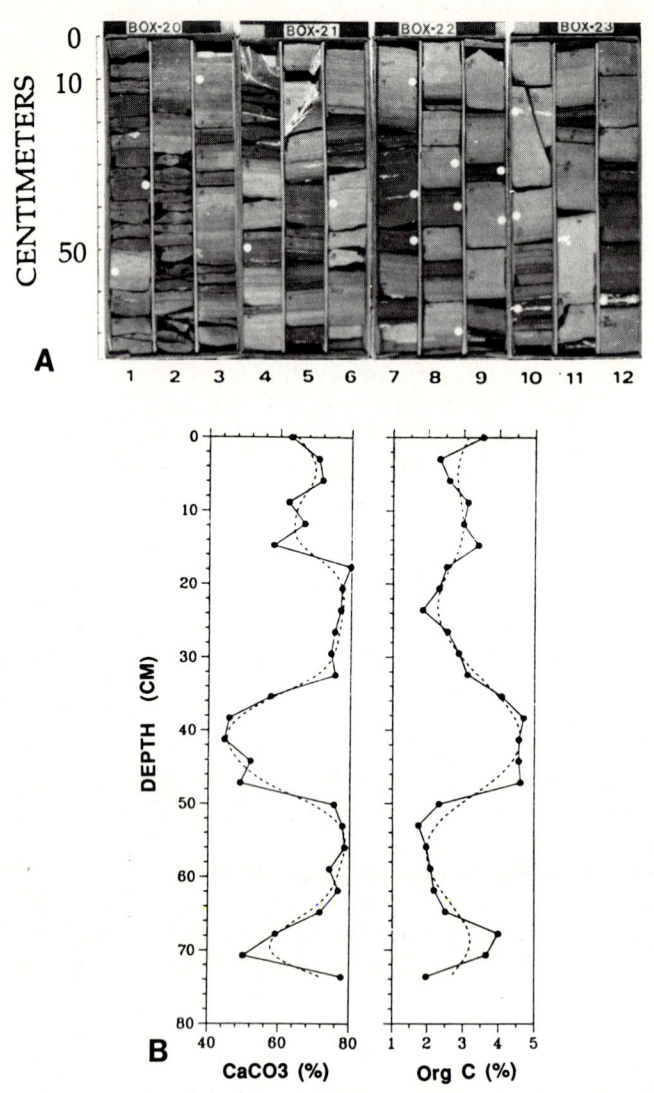

Fig. 5A,B. A Photograph of typical interbedded bioturbated limestones (*light*) of the middle chalk unit (8–12) and laminated to bioturbated marlstones (*dark*) of the middle marl unit (1–7) of the Smoky Hill Chalk Member of the Niobrara Formation in the Berthoud State #3 core. **B** Percentages of CaCO$_3$ and organic carbon through three limestone/marlstone cycles in the Smoky Hill Member of the Niobrara Formation in the Berthoud State #4 core. Beds are from about 940 m depth in the lower marl to lower chalk transition (Fig. 4B)

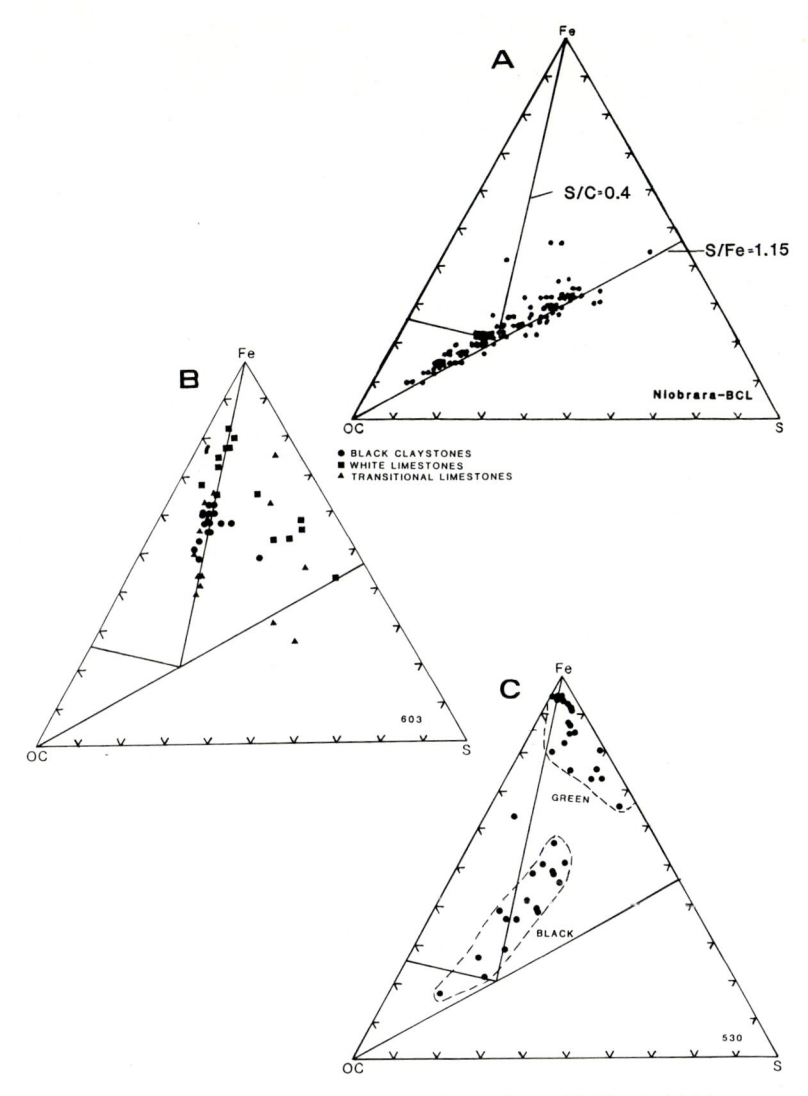

Fig. 6A-C. Ternary plot of Fe-S-OC for samples of the Niobrara Formation and Bridge Creek Limestone Member (*BCL*) of the Greenhorn Formation, Colorado (**A**), Neocomian carbonates from DSDP Site 603, North American basin, North Atlantic Ocean (**B**), and green and black claystones from DSDP Site 530, southern Angola Basin, South Atlantic Ocean (**C**). *Lines* represent constant ratios of S/Fe = 1.15, S/C = 0.4, and OC/Fe = 3. (After Dean and Arthur 1989)

sediment, whereas the green beds show no such trend. Nonetheless, the C_{org}-depleted green beds have had some portion of iron fixed as pyrite, probably by migration of sulfide from Fe-limited black beds to the adjacent C_{org}-limited green beds.

Under variable redox conditions, iron concentrations of original sediment tend to be preserved because, once reduced, iron is fixed in one of several ferrous-iron minerals, such as pyrite. Manganese concentrations, on the other hand, are highly variable and almost entirely dependent upon redox conditions because minerals containing reduced Mn (e.g., rhodochrosite, $MnCO_3$) are rare in marine sediments and rocks. Consequently, concentrations of Mn greater than a few hundred ppm usually indicate more oxidizing conditions and presence of an oxide phase. For example, redox fluctuations are well illustrated in profiles of Mn concentration or Fe/Mn ratio for both the Bridge Creek Limestone (Arthur et al. 1985b) and the Niobrara Formation (Fig. 4B). In general, the rocks at the base of each section were deposited in more oxic environments, as indicated by higher degree of bioturbation and lower C_{org} concentrations (Figs. 4A,B), and consequently the concentrations of Mn are considerably higher (Fe/Mn ratios considerably lower) at the bottom of each section analyzed, such as shown by the very low Fe/Mn ratios in the Fort Hays Member. The low Mn concentrations throughout the Smoky Hill Chalk (high Fe/Mn ratios; Fig. 4B) suggest that these sediments remained reduced throughout most of their deposition. The more extreme variations in Mn content of the Bridge Creek Limestone (Arthur et al. 1985b), on the other hand, suggests that redox conditions in the sediments, and perhaps in the overlying water, were much more variable during the time that they were deposited, in agreement with other lines of evidence discussed above. Concentrations of Mn are highest in the bioturbated limestones of the Bridge Creek Limestone and it is possible that the Mn occurs as a lattice substitution in early diagenetic calcite cements under reducing conditions. Similar patterns of Mn enrichment in bioturbated limestones also occur in the Neocomian cycles at Sites 367, 387, and 603.

3.2 Stable Isotopes and Other Geochemical Indicators of Burial Diagenesis

The light stable-isotope ratios of carbon and oxygen in biogenic components are used in paleoenvironmental studies as indicators of dynamics of carbon cycling, salinity, and temperature. In a similar way, trace element data from biogenic carbonates are commonly interpreted as indicating the effects of seawater chemical composition and temperature. Both techniques offer a tempting possibility in studies of ancient sediments and sedimentary rocks. However, such data commonly are collected and interpreted without regard for potential diagenetic overprints. Such diagenetic problems may typify the pelagic sequences we have analyzed because all are highly lithified. Burial depths of 600 to 1000 m are typical for the sequences examined in Deep Sea Drilling Project sites and as much as 2000 m for the western interior sequences. The initial carbonate mineralogy of these units was most likely relatively stable, low-magnesium calcite. Because of extensive cementation and low porosities (<10%), we would expect, for example, a diagenetic signal characterized by isotopi-

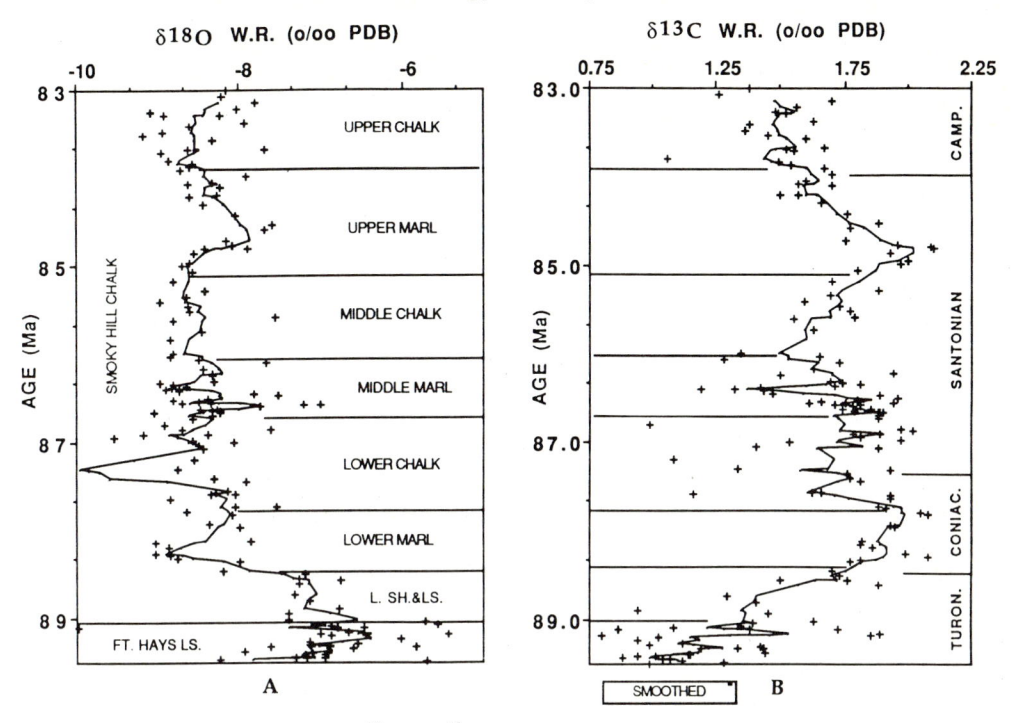

Fig. 7A,B. Whole-rock values of $\delta^{18}O$ and $\delta^{13}C$ in samples of the Niobrara Formation from the Berthoud State #3 core. Individual points (*plus signs*) represent raw data; *line through raw data* is a five-point moving average smoothed curve. Approximate age assignments and informal chalk-marl units are from Scott and Cobban (1964).

cally light $\delta^{18}O$ and relatively low Sr/Ca ratios in the limestones (e.g., Schlanger and Douglas 1974; Scholle 1977).

The $\delta^{18}O$ values for the deeply buried Niobrara Formation (Fig. 7) are more negative than –6‰, which is at lest 3‰ more depleted than expected for carbonate precipitated in isotopic equilibrium with Cretaceous seawater at a temperature of 20 °C. The oxygen-isotope values in the more carbonate-rich intervals are depleted by 1 to 2‰ relative to those from marlstones. Likewise, the $\delta^{18}O$ values for the Bridge Creek Limestone at Pueblo (Fig. 4A) are quite negative. These data suggest that burial diagenetic effects might be important. Pratt et al. (1991) summarize data showing that on average the most deeply buried Niobrara Formation sections are about 3.5‰ depleted in ^{18}O in comparison to shallowest buried sequences. This decrease in $\delta^{18}O$ with increasing burial probably is the result of alteration of volcanic materials and neoformation of clay minerals that decrease the $\delta^{18}O$ of porewaters and, therefore, carbonate cements (e.g., Lawrence et al. 1979). Higher burial temperatures also can produce calcite cements that are relatively depleted in ^{18}O (e.g., Scholle 1977). The Niobrara Formation in the Berthoud State #3 well was not buried to depths as great as sequences farther south and east in the Denver Basin (Pollastro and Martinez 1985) and perhaps has undergone a maximum 3‰ negative $\delta^{18}O$ shift relative to original values (Pratt et al. 1991). Burial diagenesis also has affected the Neocomian carbonate sequences: a typical $\delta^{18}O$ profile across cycles at DSDP Site 387 (Fig. 8) shows that the most negative $\delta^{18}O$ values again are found in the more cemented limestone beds. Although the intervals with lower carbonate contents have better preserved a "primary" oxygen isotope signal, it appears that there is little hope for estimation of absolute or relative temperature changes for each part of a cycle from these data because of the heavy diagenetic overprint.

Sr/Ca and Mg/Ca ratios also provide a measure of the degree of carbonate diagenesis because of the expected loss of Sr and Mg during recrystallization and cementation (e.g., Matter et al. 1975; Schlanger and Douglas 1974; Veizer 1983a). Because the Bridge Creek Limestone and Niobrara Formation consist mostly of "impure" chalks, averaging about 25% insoluble residue in the lower to middle Bridge Creek Limestone and about 32% insoluble residue in the Niobrara Formation, whole rock Mg and Sr contents may be influenced by detrital clastic material and/or volcanic ash. The purer limestone beds (ca. 90% $CaCO_3$) in the Bridge Creek Limestone and Niobrara Formation contain 0.3–0.4% Mg (Mg/Ca about 1×10^{-2}; Arthur et al. 1985b). In both units the Mg/Ca ratio is inversely proportional to the $CaCO_3$ content, and marlstone beds usually contain 0.7–0.8% Mg (Mg/Ca about $3-4 \times 10^{-2}$). All of this variation can be explained by dilution of the carbonate with clastic material and/or volcanic ash. We do not have data on the composition of the insoluble residue from either the Bridge Creek Limestone or the Niobrara Formation, but average marine shale and bentonite from the Pierre Shale, that overlies the Niobrara Formation, have Mg concentrations of 1.5 and 1.8% respectively (Schultz et al. 1980). Assuming dilution of a pure pelagic carbonate containing 0.2% Mg with clay containing 1.5% Mg, a limestone with 90% $CaCO_3$ would contain 0.33% Mg (Mg/Ca

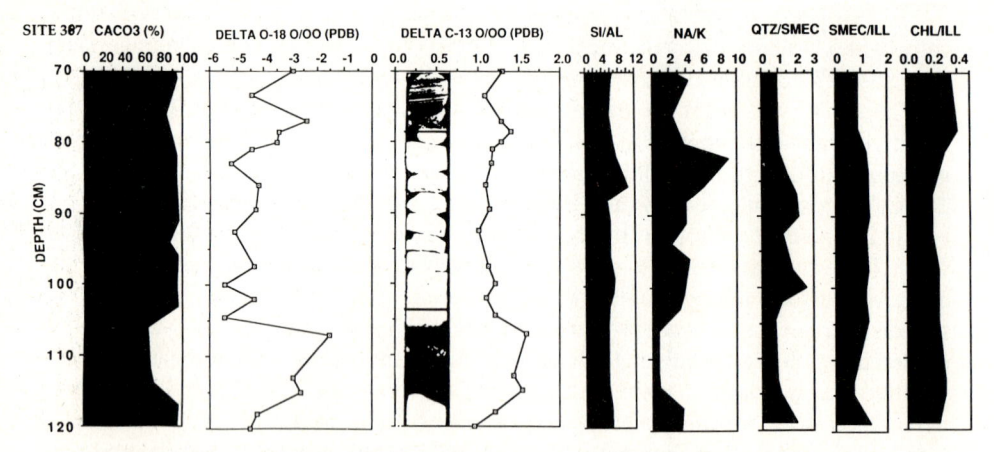

Fig. 8. Downhole plots of percent $CaCO_3$; whole-rock values of $\delta^{18}O$ and $\delta^{13}C$; and ratios of Si:Al, Na:K, quartz:smectite, smectite:illite, and chlorite:illite in samples of limestone/marlstone bedding cycles in Neocomian carbonates from DSDP Site 387, Core 40. Lithology shown pictorially in column for $\delta^{13}C$. Clay mineralogy is from Chamley, Leinen, Arthur and Dean (unpubl.).

= 0.9×10^{-2}), and a marlstone with 60% $CaCO_3$ would contain 0.72% Mg (Mg/Ca = 3×10^{-2}), which are the observed values in both the Bridge Creek Limestone and Niobrara Formation (e.g., Arthur et al. 1985b).

The Sr/Ca ratio also is inversely proportional to the $CaCO_3$ content in the Cretaceous carbonate units (Fig. 9). The Sr/Ca ratios in the Bridge Creek Limestone and Niobrara Formation and in low-carbonate Neocomian samples at Site 603 (Fig. 9) are all higher than we would expect for pure pelagic carbonates that have lost Sr during diagenetic dissolution-reprecipitation reactions (e.g., Scholle 1977; Veizer 1983a). For example, modern pelagic, low-magnesium calcites contain about 1600 ppm Sr (Sr/Ca = 0.40×10^{-4}), whereas relatively pure British chalks (ca. 98% $CaCO_3$) contain an average of about 500 ppm Sr (Sr/Ca = 0.12×10^{-4}) (Scholle 1977). Concentrations of Sr in shale and bentonite in the Pierre Shale are about 250 ppm (Schultz et al. 1980). Assuming that the insoluble residues in the Bridge Creek and Niobrara have similar Sr concentrations, then the observed variations in Sr concentration cannot be explained entirely by clastic dilution, as they were for magnesium and, therefore, must be due, in part, to differences in diagenesis. The same reasoning applies to the results for Neocomian carbonate sequences at DSDP Sites 603, 387, and 367. In the samples from Sites 387 and 367, the Sr/Ca values appear to reflect substantial loss of Sr and there is only a slight difference between the high- and low-carbonate samples in estimated Sr loss (Fig. 9). The samples from Site 603, in contrast, more closely follow the expected trend, with higher Sr concentrations in intervals of lower $CaCO_3$. Although previous studies of pelagic carbonate diagenesis suggest that trends towards decreasing Sr/Ca are related to burial depth, there is no simple relationship between the two in the Neocomian data: the highest Sr/Ca ratios are found in the most deeply buried sequence (Site 603; see Table 1). However, the porosity in the Neocomian limestones at all sites has been reduced to <10%, and the total porosity reduction by compaction-cementation is more important in determining the Sr/Ca ratio than absolute burial depth.

If we assume that the pelagic carbonates of the Bridge Creek Limestone and Niobrara Formation originally had similar Sr/Ca ratios, then the higher Sr/Ca in the Bridge Creek Limestone samples, for a given $CaCO_3$ concentration, suggest that this unit has suffered relatively less diagenetic loss of Sr. An inverse relationship between Sr/Ca ratio and $CaCO_3$ content for all of the examples shown in Fig. 9 suggests that there has been less diagenetic alteration of the carbonate in the marlstone beds (higher Sr/Ca) than in the limestone beds (lower Sr/Ca), as reflected by the divergence of points from the theoretical Sr/Ca mixing curve for unaltered samples (Fig. 9). Limestone beds appear to have been lithified early in the burial history of both the Bridge Creek Limestone and Niobrara Formation, as indicated by the lack of compaction of burrows and fecal pellets, the extensive overgrowths on calcareous nannoplankton, and the precipitation of 4–10 μm rhombohedral calcite crystals in pore spaces. Beds that contain more than about 25% insoluble residue, however, show signs of greater compactional flattening of burrows and fecal pellets and have less pore-filling cements (e.g., Hattin 1971, 1982); these are the units that also contain higher Sr/Ca ratios.

The conclusions concerning relative diagenetic alteration of the Bridge Creek Limestone based on the Sr/Ca ratios seem to contradict the isotopic results because

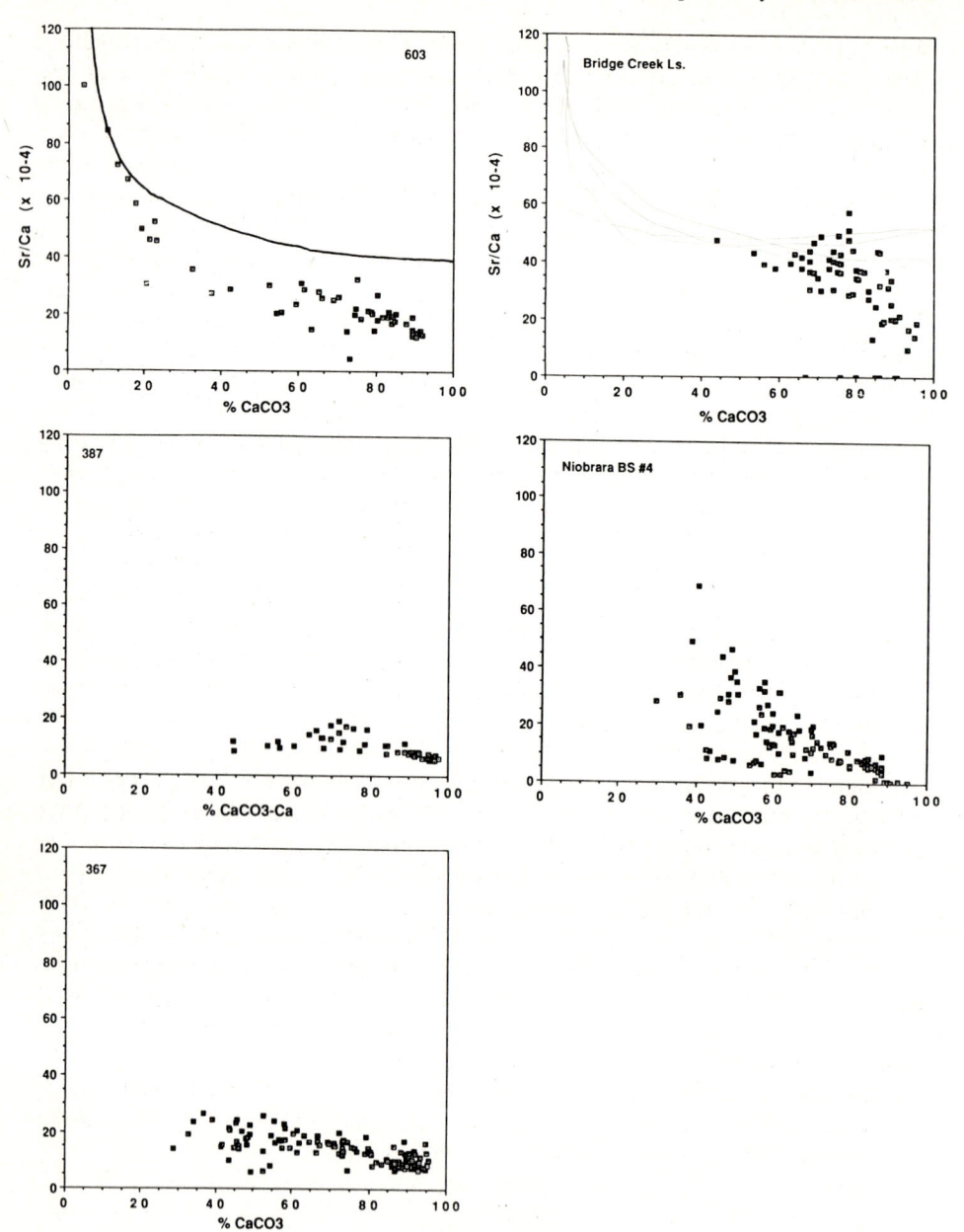

Fig. 9. Scatter plots of percent CaCO3 versus Sr/Ca in samples from limestone-marlstone bedding cycles from DSDP Sites 603, 387, and 367; the Bridge Creek Limestone Member of the Greenhorn Formation; and the Niobrara Formation. The *solid curve* on the plot for DSDP Site 603 represents the Sr/Ca ratio obtained by mixing biogenic marine CaCO3 with a Sr concentration of 0.16% and terrigenous clay with a Sr concentration of 0.023% (the average Sr concentration in black claystones from Site 603)

the relationship between $\delta^{18}O$ and CaCO3 is opposite to that found in other carbonate cycles we have examined. As previously stated, the whole-rock $\delta^{18}O$ values are significantly depleted in $\delta^{18}O$, ranging from –4‰ to –8.5‰ (Fig. 4A). In contrast to the data for the Niobrara Formation (Fig. 7) and Neocomian cycles (Fig. 8), the most negative values in the Bridge Creek generally occur in the low-carbonate, least diagenetically altered parts of the section. If all $\delta^{18}O$ values are shifted 3‰ in the positive direction, to accommodate possible diagenetic effects, the most negative $\delta^{18}O$ values are still about –5.5‰ in the carbonate-poor beds, but values in the carbonate-rich beds all fall within the range –2.5 to –3.5‰, which are reasonable isotopic compositions for marine carbonates formed in a warm, ice-free mid-Cretaceous ocean. The more negative $\delta^{18}O$ values of the dark-colored, carbonate-poor intervals probably represent an effect of lower salinity surface waters rather than higher temperature or differential diagenesis (e.g., Pratt et al. 1991). Pratt (1984), Barron et al. (1985) and Arthur et al. (1985b) suggest that such a salinity effect was responsible for the 2–3‰ difference between carbonate-rich and carbonate-poor parts of each cycle. Variations in surface-water salinity, in combination with development of dysaerobic to anoxic bottom waters, may explain the observed changes in biota (Cobban and Scott 1972; Eicher and Diner 1985; Leckie 1985; Elder 1985), clay mineralogy, sedimentary structures, and Corg content in the Bridge Creek. This is discussed further below in light of the inorganic geochemistry. Similar conclusions can be drawn for the Smoky Hill Chalk Members of the Niobrara Formation (e.g., Arthur et al. 1985b; Pratt et al. 1991), although diagenesis has significantly altered the primary oxygen isotopic values.

Cyclic variations in whole-rock $\delta^{13}C$ values also occur in all rhythmically bedded Cretaceous carbonate units and may provide additional clues to diagenetic history. In general, more negative values of $\delta^{13}C$ correspond to intervals of higher carbonate content (Figs. 4A, 7 and 8). Carbon-isotope values in carbonates can be interpreted in terms of primary processes, such as changes in water-mass stratification and consequent supply of isotopically light CO_2 (e.g., Berger and Vincent 1986). However, d13C values of lithified pelagic carbonates that contain significant amounts of organic matter most likely reflect addition of carbonate cements precipitated in equilibrium with a porewater reservoir of isotopically light dissolved CO_2 derived from the oxidation of isotopically light organic matter (ca. –25‰; Irwin et al. 1977; Scholle and Arthur 1980; also see Ricken and Eder Chap. 3.1, this Vol.). The heavier carbon-isotope values of beds with lower carbonate concentrations probably represent values that are closer to the original primary biogenic components because of less cementation during diagenesis (Arthur et al. 1984a).

4 Geochemical Indicators of Clastic Source Variations

The concentrations of many major and minor elements vary with CaCO3 content in most of the units we have studied, particularly those elements commonly associated with the noncarbonate (detrital-clastic) fraction (e.g., Si, Al, Na, K, and Ti). Because subtle compositional variations often are obscured by fluctuations in CaCO3 that dilute the clastic fraction, we have calculated element/element ratios for some of the more important elements (Figs. 9–14). In this section we will examine trends in ratios of selected elements as they relate to changing sediment sources and depositional environments.

The Bridge Creek Limestone and Niobrara Formation were deposited during overall periods of maximum transgression of the Western Interior Seaway of North America. During these two periods, the seaway was essentially a northwestern arm of the Tethys Sea, and provided a connection between Tethys and the Arctic Ocean. There were three major sediment sources that competed against one another during deposition of the Bridge Creek Limestone and Niobrara Formation: biogenic carbonate, windblown volcanic ash, and terrigenous detritus transported by winds, rivers, and surface currents from the Sevier highland to the west. We cannot assume that any of these inputs were constant, and in fact all probably varied in response to environmental changes, but we suspect that the fluxes of biogenic carbonate and windblown ash were more constant, except of course for discrete volcanic ash layers. The oxygen isotope data and lithologic variations discussed above suggest that there were changes in continental runoff to the basin through time, probably related to changes in sea level. Therefore, the flux of terrigenous detritus probably varied significantly and produced most of the observed variations in carbonate content. The clay mineralogical studies of the Bridge Creek Limestone by Pratt (1984) demonstrated some changes in the relative proportions of detrital clay minerals (illite+chlorite) and quartz superimposed on a background of authigenic mixed-layer illite-smectite, which is predominantly derived from alteration of volcanic ash in the sediment. The quartz could be either detrital or eolian in origin, but the higher percentage of quartz in the bioturbated limestone beds (Fig. 4A), as opposed to adjacent marlstone beds, which contain a higher detrital fraction, suggests that the quartz was eolian. This interpretation is in line with the oxygen isotope evidence suggesting that the limestones represent the more arid portion of each climatic-depositional cycle.

Cyclic variations in the Al_2O_3/TiO_2 and Na_2O/K_2O ratios in the Bridge Creek Limestone probably are caused by changes in the relative importance of discrete illite and other detrital phases versus mixed-layer clays derived from altered volcanic material (Pratt 1984). The Na_2O/K_2O ratio in bentonites is about 3.0 (Schultz et al. 1980), representing the pure mixed-layer-clay (e.g., smectite) end member. Our analyses show that Na_2O/K_2O ratios are highest in bioturbated limestone beds (Fig. 4A), suggesting that the proportion of mixed-layer clays is highest in those beds. Al_2O_3/TiO_2 ratios also generally decrease in limestone beds (Arthur et al. 1985b), perhaps corresponding to decreased detrital terrigenous influx during times of more pure pelagic carbonate deposition. The Al_2O_3/TiO_2 ratio in bentonites of the Western Interior basin is about 30 (Schultz et al. 1980), a value typical of many of the limestone beds, whereas a higher ratio is typical of the marlstone beds.

The same reasoning used in interpreting the geochemistry of the Bridge Creek Limestone applies to the geochemistry of the Niobrara Formation because we would not expect major changes in the chemical composition of the competing sediment sources from the Turonian to the Santonian. Variations in Al_2O_3/TiO_2 and Na_2O/K_2O ratios are caused by variations in influx of volcanic ash and/or detrital clays. The relative proportion of K-rich detrital clays (illite+chlorite) is significantly higher in the Niobrara Formation than in the Bridge Creek Limestone (Arthur et al. 1985b), suggesting greater influence of fluvial sediment source regions to the west during much of Niobrara deposition. The Na$_2$O/K$_2$O ratio is about three times lower in the Niobrara Formation overall than in the Bridge Creek Limestone, in agreement with a higher detrital clay content.

Variations in the Na$_2$O/K$_2$O (Na/K) ratio associated with interbedded limestone/marlstone in the Bridge Creek Limestone and Niobrara Formation (Fig. 4A,B) are minor compared with variations in Na$_2$O/K$_2$O in similar lithologies in Neocomian carbonates of the North Atlantic, particularly at DSDP Sites 367 and 387 (Fig. 10). There are also large variations in the Al$_2$O$_3$/TiO$_2$ ratio in the Neocomian carbonates at all three DSDP sites relative to those in the carbonate units of the Western Interior Seaway (Fig. 11). Based on the data presented in Figs. 10 and 11, we consider Al$_2$O$_3$/TiO$_2$ ratios of 20–25, and Na$_2$O/K$_2$O ratios of 0.3–0.6 to be typical of terrigenous clastics derived from the eastern margin of North America during the Early Cretaceous. The composition of this terrigenous clastic material is represented by the composition of the homogeneous black claystones at DSDP Site 603 (samples with <30% CaCO$_3$ in Figs. 10 and 11) that were deposited by turbidites derived from the continental margin of North America (Dean and Arthur 1987). As discussed above, variations in Sr/Ca (Fig. 9) in the carbonate units can be explained by less diagenetic alteration of calcite in the marlstone beds relative to the limestone beds. However, the much higher Sr/Ca ratios in the claystone turbidite beds at Site 603 may indicate a minor additional Sr source from terrigenous clastic debris. The Na$_2$O/K$_2$O ratios in the white limestones at Sites 367 and 387 are much higher than those in the black marlstones and any of the lithologies at Site 603 (Figs. 14 and 15). One possible explanation for the marked difference in Na$_2$O/K$_2$O ratio is that the Neocomian carbonates at Sites 367 and 387 may have had two clastic sources, a low-sodium, high-potassium source for the black marlstones, and a high-sodium, low-potassium source for the white limestones. This again might be related to volcanic ash windblown to these sites from possible volcanic sources in the White Mountain region of the northeastern U.S. (Foland and Faul 1977).

Alternatively, the compositional differences between lithologies at each DSDP site may be due to fluctuations in climatic conditions, such as alternating drier and wetter conditions on the generally more arid northwest African continent. Similarly, the compositional differences between Site 603 and Sites 367 and 387 may be the result of regional climatic differences on eastern North America (e.g., Parrish et al. 1982; Barron and Washington 1985; Hallam 1985). If the compositional differences are due to differences in climate, then the low-sodium characteristics of the claystones at Site 603 may reflect low clay abundances and weathered feldspars in the humid coastal soils of North America. Intuitively, we expected the Neocomian carbonates at Site 387 to be more similar in composition to those at Site 603 than to those at Site 367. However, it may be that the composition of sediment at Site 603 was dominated by downslope transport of terrigenous clastic material from the continental margin of North America whereas that at Site 387 was dominated by eolian clastic material from North Africa.

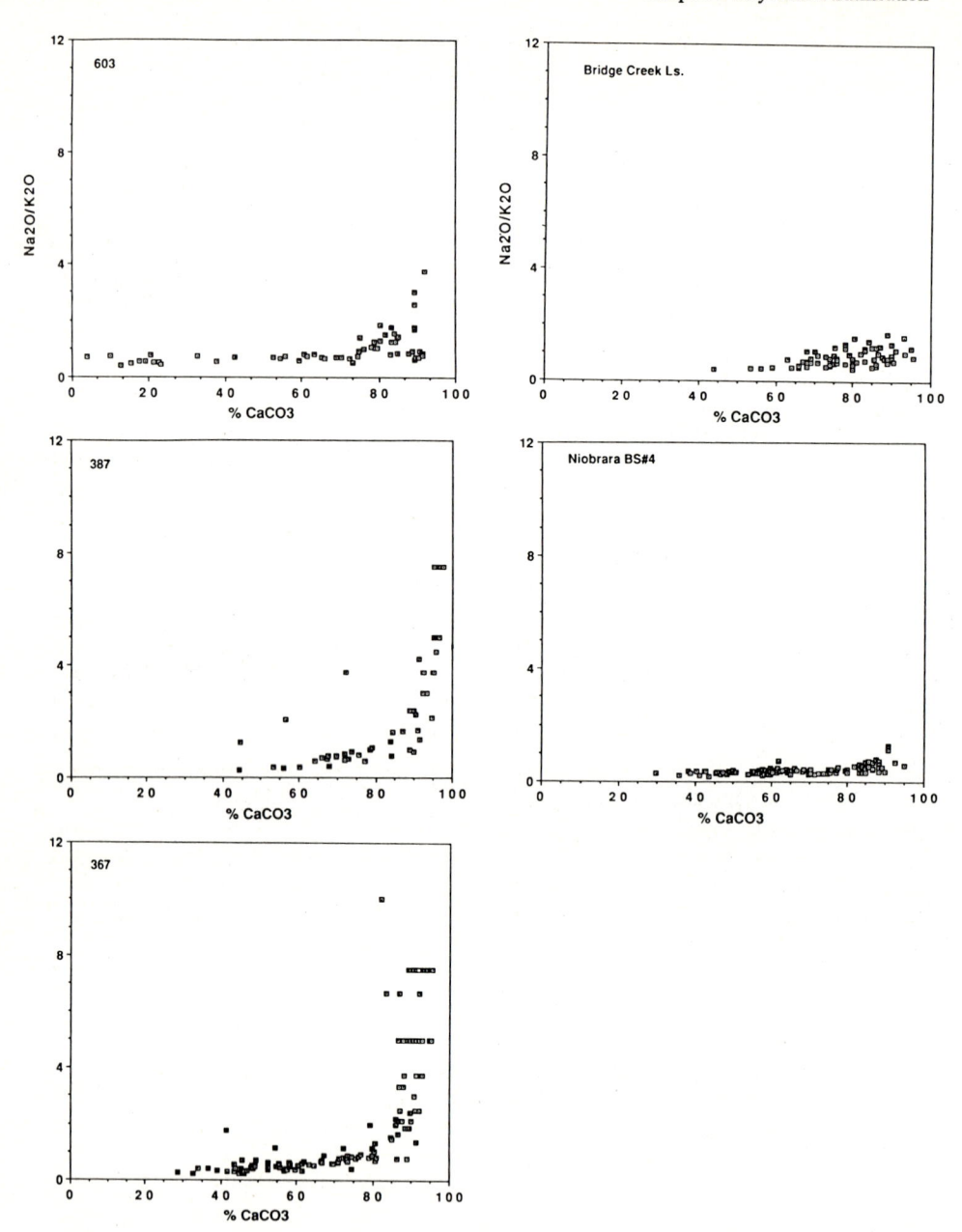

Fig. 10. Scatter plots of percent $CaCO_3$ versus Na_2O/K_2O in samples of limestone-marlstone bedding cycles from DSDP Sites 603, 387, and 367; the Bridge Creek Limestone Member of the Greenhorn Formation; and the Niobrara Formation

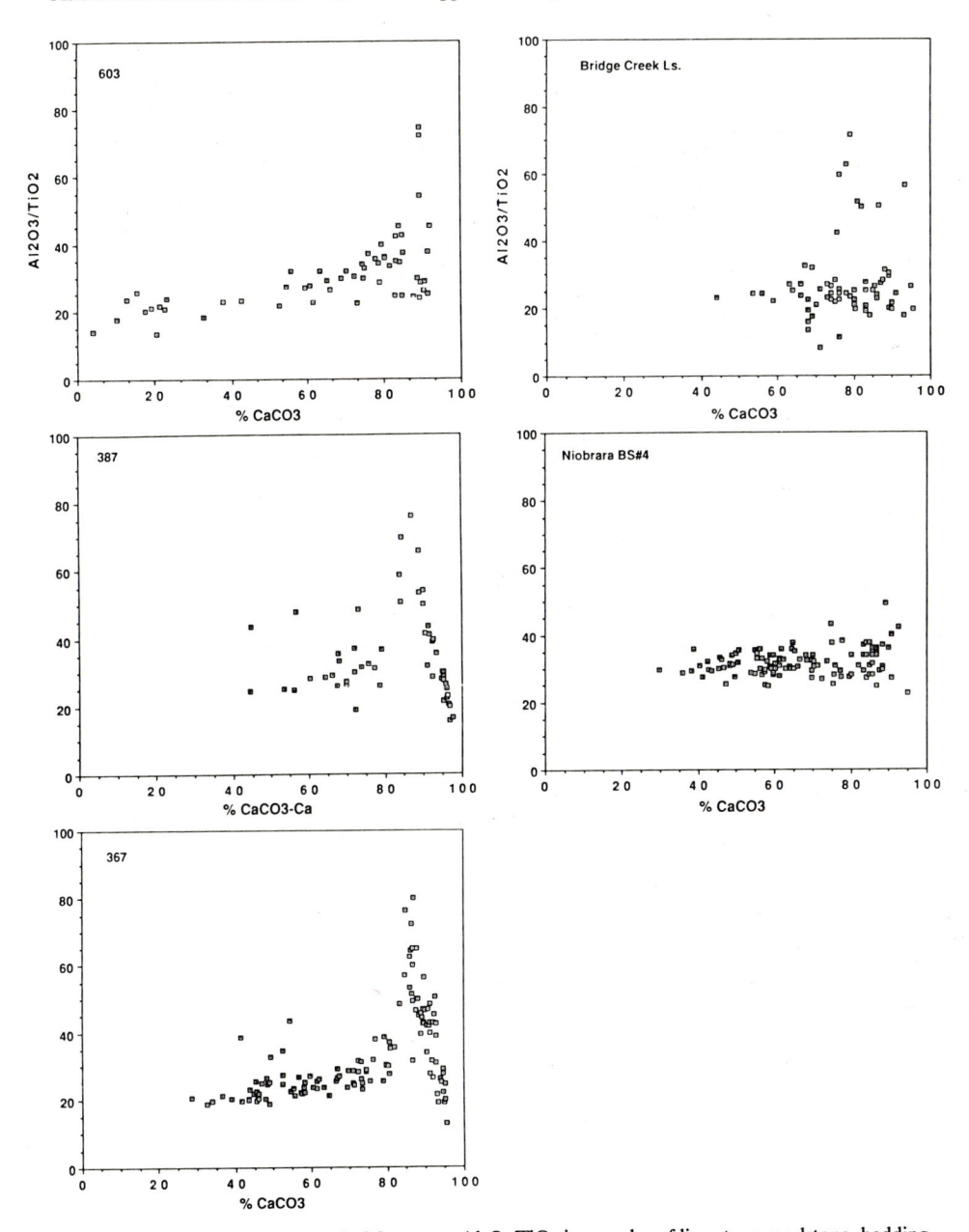

Fig. 11. Scatter plots of percent CaCO3 versus Al2O3/TiO2 in samples of limestone-marlstone bedding cycles from DSDP Sites 603, 387, and 367; the Bridge Creek Limestone Member of the Greenhorn Formation; and the Niobrara Formation.

The limestone beds at all three DSDP sites have highly variable Al_2O_3/TiO_2 ratios. The distributions of Al_2O_3/TiO_2 versus $CaCO_3$ for these sites (Fig. 11) show two very distinct trends, one for beds containing <80% $CaCO_3$ and one for those with >80% $CaCO_3$. Beds with >80% $CaCO_3$ at Sites 367 and 387 are all high in aluminum (high Al_2O_3/TiO_2 ratio), with a high and variable sodium content. The high Al_2O_3/TiO_2 ratio may be the result of greater clay abundances in the arid, less-weathered soils of northwest Africa relative to those of eastern North America. The high-sodium characteristics of the white limestones at Sites 367 and 387 may reflect the eolian supply of less weathered clastic material relatively enriched in labile plagioclase feldspar. By this analogy, the low-sodium black marlstones would represent wetter periods with more decomposition of plagioclase feldspar relative to potassium feldspar and illite (lower Na_2O/K_2O ratio).

We do not have mineralogical data for many samples for the Neocomian carbonates that might help to explain the geochemical data, but reconnaissance mineralogical studies by Chamley and colleagues do provide some information (e.g., Chamley 1979; Chamley and Debrabant 1984; Chamley et al. 1979, 1983; Debrabant and Chamley 1982; Debrabant et al. 1984). These studies have established that there is a predominance of a smectite-rich (>90 wt% of the <2 µm fraction) clay mineral assemblage in the Neocomian marls of the western basin of the North Atlantic and at Sites 367 and 398 in the eastern basin. Chamley and Debrabant (1984) highlight the absence of significant amounts of kaolinite in both eastern and western basins of the North Atlantic. Its absence is unexpected because many continental and shallow-water sediments of the same age are rich in kaolinite. They explain its absence by preferential entrapment of kaolinite in marginal basins. Within this Neocomian smectite abundance, two areas stand out as exceptions: Site 416 on the northeastern African margin and Site 387 which was on the western flank of the mid-Atlantic ridge during the Neocomian (Fig. 1). Both sites have abundant illite (40–50% at Site 416), mixed layer clays (~20% at Site 416; 40–60% illite+chlorite+mixed layer clays at Site 387), and Site 416 also has substantial concentrations of kaolinite (10–20%). The Neocomian sediments at Site 416 are turbidites, but the high concentrations of detrital minerals and low concentrations of smectite suggest that circum-Atlantic continental source areas were varied and that the Atlas range, uplifted during the Jurassic, was, at least periodically, a significant source of detritus to the Neocomian northeast Atlantic.

Paleoceanographic reconstructions (Fig. 1) show that Site 416 was at about 37°N during the Neocomian, within the temperate latitudinal zone characterized today by the production of illite, mixed-layer clays and quartz during weathering processes. The abundance of illite, mixed-layer clays and kaolinite in marlstones from Site 387 are difficult to explain. The Neocomian section has been described as pelagic limestones interbedded with marlstones which represent pelagic turbidites or bottom-current winnowing of rapidly deposited organic-carbon rich detritus (Tucholke and Vogt 1979). Chamley and Debrabant (1984) suggest that the mineral fraction of the marlstones have a western North Atlantic source. They are far richer in primary detrital minerals (illite, mixed-layer clays, and kaolinite), however, than the clay fraction in *any* of the sites on the western margin of the Atlantic. Furthermore, downslope transport processes would tend to increase the smectite concentration relative to illite or mixed-layer clays. The absence of significant concentrations of illite, mixed-layer clays, and kaolinite in any other North Atlantic site suggests that the source of the Site 387 detrital component may have been to the east, in the direction of Site 416 and transported from an arid northwest Africa by prevailing tradewinds. The detrital fractions of these limestones have relatively high quartz contents as indicated by high SiO_2/Al_2O_3 ratios in limestones from Sites 367 and 387 (Fig. 12). This quartz also is probably eolian, although we cannot rule out the possibility that the quartz originally was biogenic silica that was diagenetically converted to quartz.

5 Biogenic Productivity and Cycles

5.1 Elemental Ratio Indicators (Si/Al:P/Al:Ba/Al)

Superimposed on the production, dilution, and dissolution trends (Fig. 3A) is the effect that variation in redox conditions has on sedimentary structures, color, and concentrations of organic matter and trace metals. Knowledge of accumulation rates of components across individual cycles would help to discriminate between the various possibilities, but it is difficult to perform these calculations even with precise age control. We can estimate the average periodicity of the cycles as being between 20–40 ka, but in order to calculate relative changes in element accumulation rates across high- and low-carbonate parts of cycles, we must make assumptions about the amount of time that each lithologic type represents. We suspect that changes in primary productivity (organic-carbon and carbonate fluxes to the sea floor) played a role in the origin of the cyclicity in carbonate and C_{org} contents, but supporting evidence for this hypothesis is not easily obtained.

The interpretation of variations in silica content of Cretaceous strata is difficult because of the complications that result from variations in detrital quartz concentrations and the "no-analog" Cretaceous ocean in which diatoms were not a major component of the phytoplankton, in contrast to modern high-productivity settings. High Si/Al ratios resulting from high concentrations of radiolaria in sediments may indicate initial enhanced preservation of biogenic opal, but do not necessarily indicate high *primary* productivity. The plots of Si/Al for DSDP Sites 387 and 367 and the Bridge Creek Limestone (Fig. 16) all show high values of Si/Al in the high-carbonate beds, which can be considered "excess" silica over the Si/Al ratio in normal background terrigenous material (ca. 5). There is no change in the Si/Al ratio in the Neocomian carbonates at Site 603 (Fig. 12), and in fact, the Si/Al values in samples from Site 603 are lower than those in the background terrigenous material at Sites 387 and 367. The high Si/Al ratios in limestones at Site 387 and 367 could indicate that the accumulation of siliceous plankton on the seafloor was higher during sedimentation of the carbonate-rich parts of the cycles, or simply that fine-grained detrital (eolian?) quartz preferentially accumulated at these times. One might expect to have high concentrations of eolian quartz at Site 367 off northwest Africa, and,

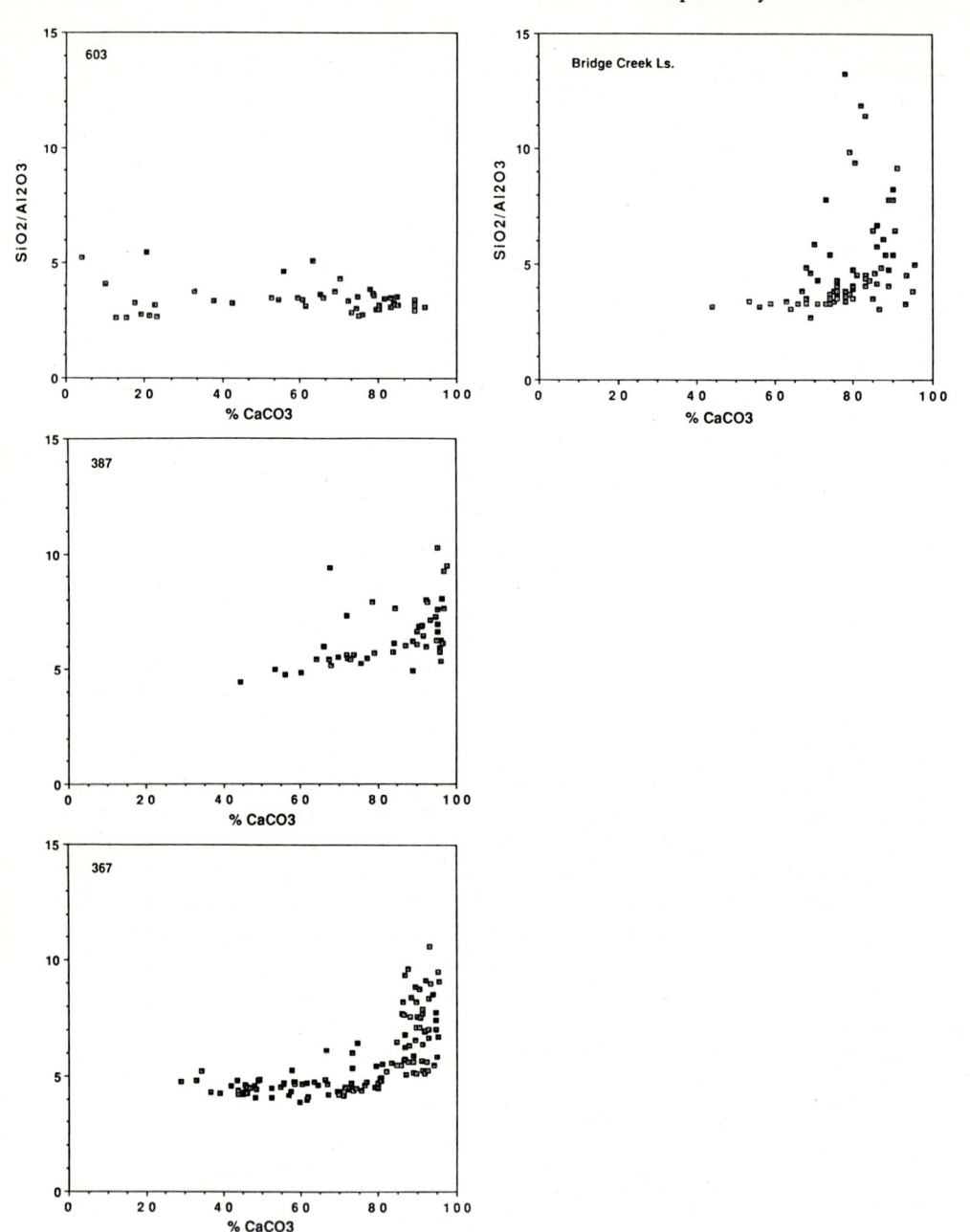

Fig. 12. Scatter plots of percent $CaCO_3$ versus SiO_2/Al_2O_3 in samples of limestone-marlstone bedding cycles from DSDP Sites 603, 387, and 367; and the Bridge Creek Limestone Member of the Greenhorn Formation

because the other geochemical signals (e.g., Na_2O/K_2O and Al_2O_3/TiO_2 ratios) are similar in the limestone beds from both Site 367 and Site 387, eolian quartz might also be present at Site 387.

The P/Al ratio (Fig. 13) could also indicate variations in biologic productivity in that the flux of phosphatic fish debris and/or organic phosphorus to the seafloor might be proportional to surface-water primary production. However, the interpretation of phosphate concentrations in sediments is complicated because of other possible phosphorus sources, such as that associated with, sorption on manganese- and iron-oxyhydroxides (e.g., Froelich et al. 1982) and coatings on planktonic foraminifers and other calcareous components (e.g., Sherwood et al. 1987). The pattern of increasing phosphorus enrichment with increasing $CaCO_3$ is similar for all five sample suites (Fig. 13). The shape of the curves indicates mixing of essentially two end-members, a carbonate-rich component with a high P/Al ratio, and a terrigenous component with a lower P/Al ratio. Because all sample sets exhibit the same relationship between P/Al and $CaCO_3$, we consider it unlikely that the increasing P/Al reflects higher relative productivity in the carbonate-rich beds. The relative phosphorus enrichment in the limestone beds is most likely due to the processes of sorption suggested by Sherwood et al. (1987), especially in light of the fact that the Mn/Al ratio also increases in the limestones.

The relationship between barium concentrations and $CaCO_3$ across cycles (Fig. 14) also may indicate higher Ba fluxes to the sediment during the carbonate-dominated part of the cycle (the Ba/Al ratio increases with increasing carbonate content). Barium, which apparently is formed as barite in some organisms or is associated in some way with organic-matter rich particle aggregates falling through the water column, has been suggested as a proxy indicator for productivity and organic-carbon flux in the modern ocean, with higher Ba fluxes indicating relatively higher primary productivity (e.g., Dehairs et al. 1980; Schmitz 1987; but see Bishop 1987, for a dissenting view). However, Ba commonly is mobile during early diagenesis in organic-carbon rich sequences such as these (e.g., Dean and Parduhn 1984; Brumsack 1986) because of extensive bacterial sulfate reduction and solubilization of sedimentary barite. The higher Ba/Al ratios in some limestone units, particularly in the Bridge Creek and Niobrara (Fig. 14) may indicate diagenetic dissolution of barite in organic-carbon-rich beds and precipitation of barite in organic-carbon-poor, carbonate-rich beds. Variation in the Ba/Al ratio in the Niobrara Formation (Fig. 4B) shows relative enrichment in Ba at the base of each of the major carbonate-rich units in the Smoky Hill Chalk Member, which we infer to represent significant decreases in terrigenous dilution and possibly slower sedimentation related to high stands of sea level. Barite would have a tendency to precipitate at the interface between more rapidly deposited, organic-carbon-rich strata and overlying more slowly deposited organic-carbon-poor beds according to the model of Brumsack (1986). Nonetheless, relative barium enrichment occurs in $CaCO_3$-rich samples in the three sections that offer the best possibilities for high but variable productivity, the Neocomian of Site 367 off northwest Africa (e.g., Roth 1986), and the Bridge Creek Limestone and Niobrara Formations of the Cretaceous Western Interior Seaway of North America (e.g., Parish and Curtis 1982), and not at those DSDP sites where

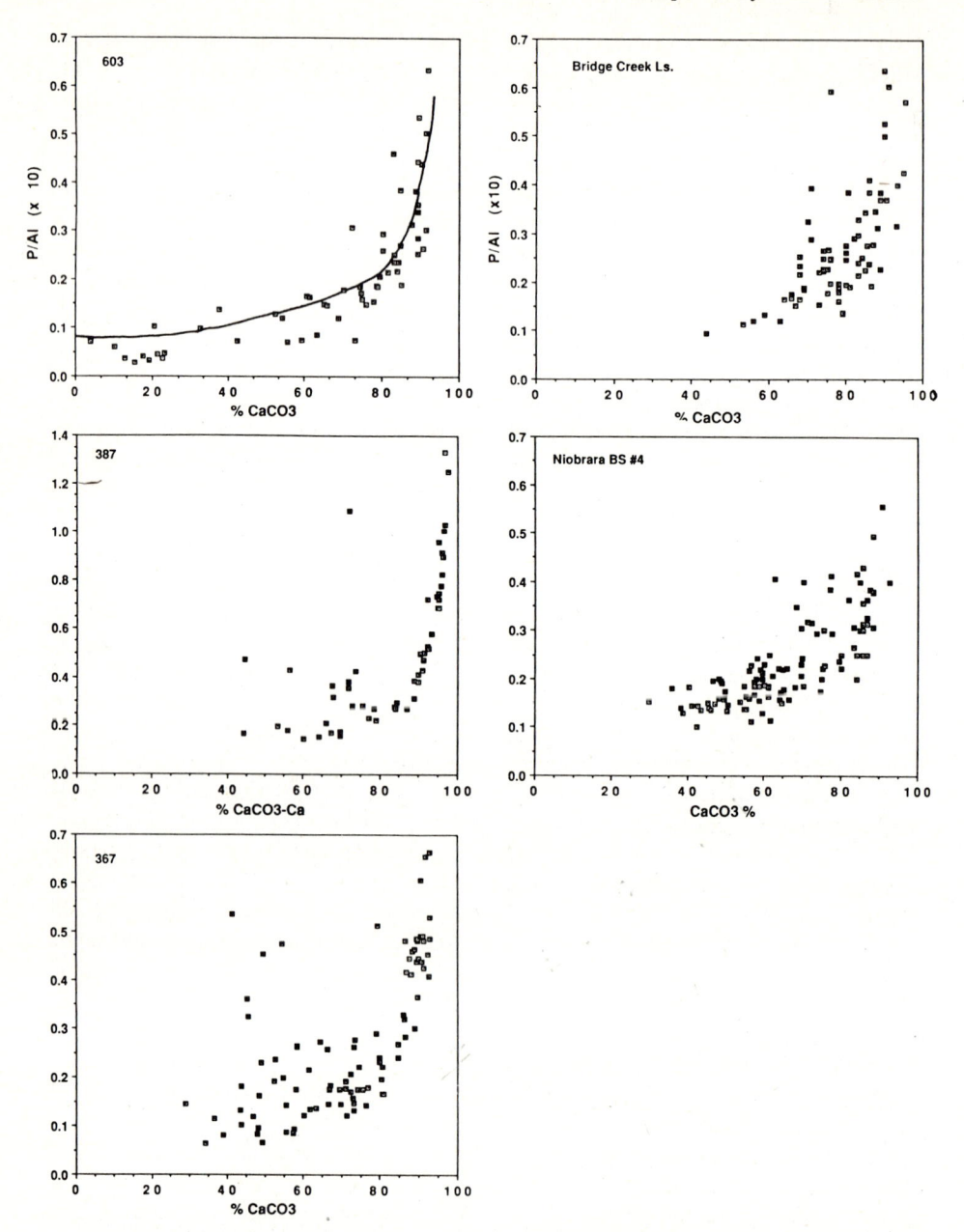

Fig. 13. Scatter plots of percent CaCO3 versus P/Al in samples of limestone-marlstone bedding cycles from DSDP Sites 603, 387, and 367; the Bridge Creek Limestone Member of the Greenhorn Formation; and the Niobrara Formation. The *solid curve* on the plot for DSDP Site 603 represents the P/Al ratio obtained by mixing carbonate with a P concentration of 0.044% and terrigenous clay with a P concentration of 0.065% (the average P concentration in black claystones from Site 603)

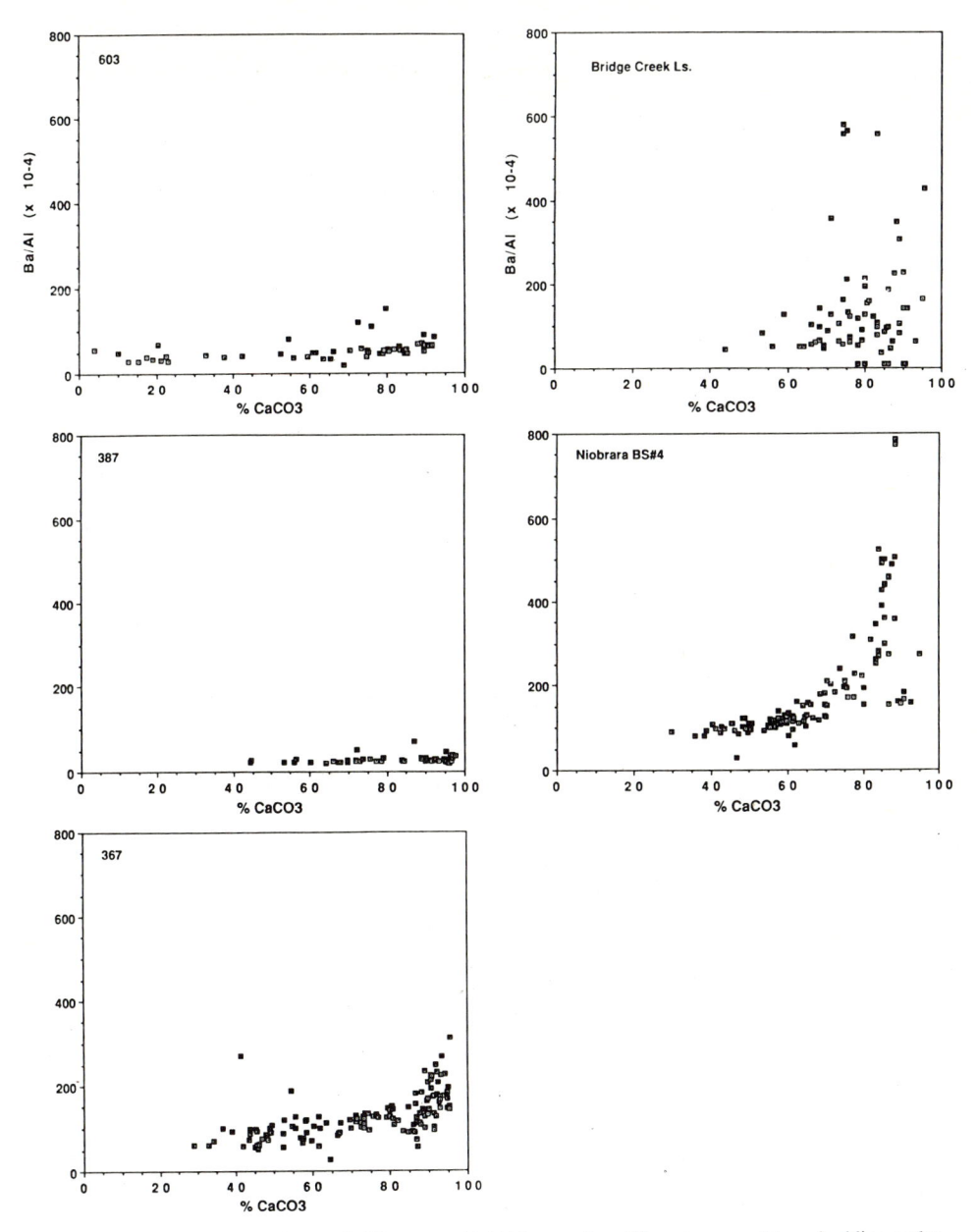

Fig. 14. Scatter plots of percent CaCO₃ versus Ba/Al in samples of limestone-marlstone bedding cycles from DSDP Sites 603, 387, and 367; the Bridge Creek Limestone Member of the Greenhorn Formation; and the Niobrara Formation

productivity probably was lower (Sites 603 and 387). We conclude that there is a primary productivity signal in the Ba/Al ratios, at least in these few sample sets.

It remains possible, therefore, that the barium and silica concentrations are indicators of the role of variable biological productivity in producing limestone/marlstone cycles because of higher Si/Al and Ba/Al ratios in high-carbonate beds at sites where we would except variable organic productivity. What may be surprising is that the highest inferred productivities resulted in the lowest sedimentary organic-carbon contents. Enhanced preservation of organic matter at times of bottom-water oxygen-depletion, therefore, is a more likely explanation of the enrichments in organic matter.

5.2 Organic Matter Indicators

The type and relative preservation of organic matter varies with lithology in all sequences that we have investigated. For example, there were differences in the sources and degree of preservation of organic matter in the Neocomian carbonates in the eastern and western basins of the North Atlantic (Dean and Arthur 1987). Many other studies indicate that, in general, the supply of C_{org} to the western basin of the North Atlantic consisted mainly of terrestrial organic matter and highly degraded autochthonous marine organic matter throughout most of the Cretaceous (e.g., Tissot et al. 1979, 1980; Summerhayes 1981; Herbin and Deroo 1982; de Graciansky et al. 1982; Habib 1983; Summerhayes 1987), and that much of this organic matter may have been redeposited by turbidity currents from shallower environments (e.g., Habib 1979 and 1983; Summerhayes and Masran 1983; Arthur et al. 1984b). Most of the organic matter in Cretaceous C_{org}-rich strata in the eastern North Atlantic appears to have had a similar origin except in the region off central NW Africa, particularly the Cape Verde Rise and Basin Sites 367, 368, and 369, where the organic matter was derived mainly from authochthonous marine sources throughout the Cretaceous. The Rock-Eval HI results shown in Fig. 15B substantiate these previous studies, and we suggest that these two sites (603 and 367) represent end members of two different sources of organic matter to the Cretaceous proto-Atlantic ocean; predominantly terrestrial organic matter (low HI) at Site 603 and predominantly marine organic matter (high HI) at Site 367. Site 387 is characterized by a mixture of the two types of organic matter.

Figure 15B shows that the HI of the Neocomian samples studied at DSDP Site 603 is low, (mostly less than 100 mg HC/g C_{org}); the highest values of HI occur in the relatively C_{org}-rich, laminated limestones, whereas the bioturbated limestones and black claystones contain organic matter with very low values of HI. For comparison, we have also plotted the Rock-Eval pyrolysis results for Neocomian white, bioturbated limestones and black, laminated marlstones from Site 367 in the eastern North Atlantic basin (Dean et al. 1986). Note that the HI values for all Site 367 samples are uniformly higher for both oxidized (white) and reduced (black) lithologies. In most samples we have studied there is a strong positive correlation between percent C_{org} and HI (e.g., Dean et al. 1986). This relationship implies that samples with higher C_{org} concentrations (>1%) have organic matter that is more

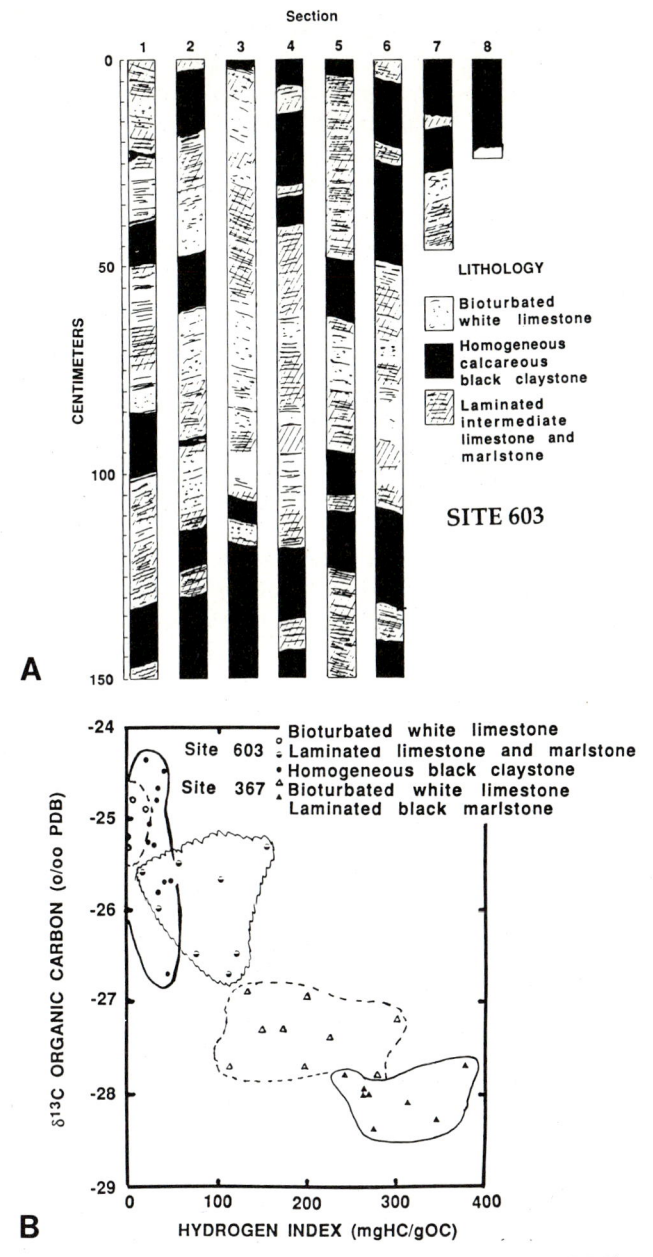

Fig. 15A,B. **A** Diagrammatic representation of the three basic Neocomian lithologic types of DSDP Site 603 as they occur in Core 69, Hole 603B. **B** Scatter plot of Rock-Eval pyrolysis hydrogen index versus $\delta^{13}C$ of organic carbon in samples of Neocomian carbonates from DSDP Site 367 in the eastern North Atlantic basin, and DSDP Site 603 in the western North Atlantic basin. (Data from Dean et al. 1986)

hydrogen-rich. Or, to put it another way, that increases in C_{org} above about 1% are due to the enhanced preservation of hydrogen-rich marine organic matter. Plots of absolute percent amorphous organic matter vs. percent total C_{org} for the western North Atlantic presented by Summerhayes (1981) illustrate this principle: there appears to be a fairly constant background of about 0.5 to 1% terrigenous organic matter and values above 1% are due mainly to addition and/or enhanced preservation of amorphous authochonous marine organic matter. The bioturbated limestones at Site 603 probably contain a mixture of terrestrial and highly degraded marine organic matter, indicated by their low HI values. The only significant amounts of marine organic matter in the Neocomian samples from Site 603 are preserved in the laminated limestones and marlstones, and even in these beds the organic matter is highly degraded and mixed with terrestrial material, as indicated by low HI values relative to other samples rich in marine C_{org}, such as those of equivalent age at Site 367.

The stable carbon-isotopic composition of C_{org} often has been interpreted in terms of the source of the organic matter. The basis for interpreting the isotopic composition of C_{org} as a source signal is the fact that modern terrestrial vegetation generally is depleted in ^{13}C relative to modern marine plankton and therefore has values of $\delta^{13}C$ that are about 5–7‰ lighter (more negative) than those of marine plankton (e.g., Deines 1980). Most organic matter in modern marine sediments, judged by other geochemical criteria to be mainly marine, has a $\delta^{13}C$ value similar to those of marine plankton and particulate organic matter (about –22‰). This appears to be true of organic matter at least as far back as the early Miocene (see Arthur et al. 1985a; Dean et al. 1986). We have found, however, that the $\delta^{13}C$ of Cretaceous organic matter does not appear to follow the same behavior. On the basis of available data (Dean et al. 1986), Cretaceous marine C_{org} is typically isotopically lighter than terrestrial C_{org} by 3‰ or more, with the exception of that in strata of upper Cenomanian to lower Turonian (Arthur et al. 1988) and lower Albian (Pratt and King 1986) ages. Such a relationship is illustrated by our studies of Neocomian carbonate Sites 603 and 367 (Fig. 15B). Here we see that the suspected terrestrial end-member at Site 603 has significantly heavier $\delta^{13}C_{org}$ values (–24 to –25‰) than the relatively well-preserved marine end-member at Site 367 (–27 to –28‰).

Another line of evidence for variations in surface-water primary productivity as a factor in producing sequences of interbedded limestone and marlstone comes from results of nitrogen isotopes (Rau et al. 1987). At two of three locations in the Atlantic Ocean (DSDP Sites 367 and 530) where we have measured nitrogen isotopes, the organic-carbon-rich beds (black marlstones or claystones) were found to have significantly lower $\delta^{15}N$ values (lower $^{15}N/^{14}N$ ratios) than adjacent organic-carbon-poor beds (white limestones or green claystones; Fig. 16). At the third site (Site 603) the black, organic-carbon-rich beds are characterized by somewhat higher $\delta^{15}N$ values than adjacent lighter-colored, bioturbated carbonate beds (Fig. 16). The black beds at Site 603, however, contain predominantly terrestrial organic matter, and, in fact, this terrestrial organic matter has C- and N-isotopic compositions that are very similar to those of modern terrestrial organic matter. In contrast, all of the lithologies at Sites 367 and 530, but especially the black beds, contain predominantly marine organic matter and have $\delta^{15}N$ values that are significantly lower than those found in

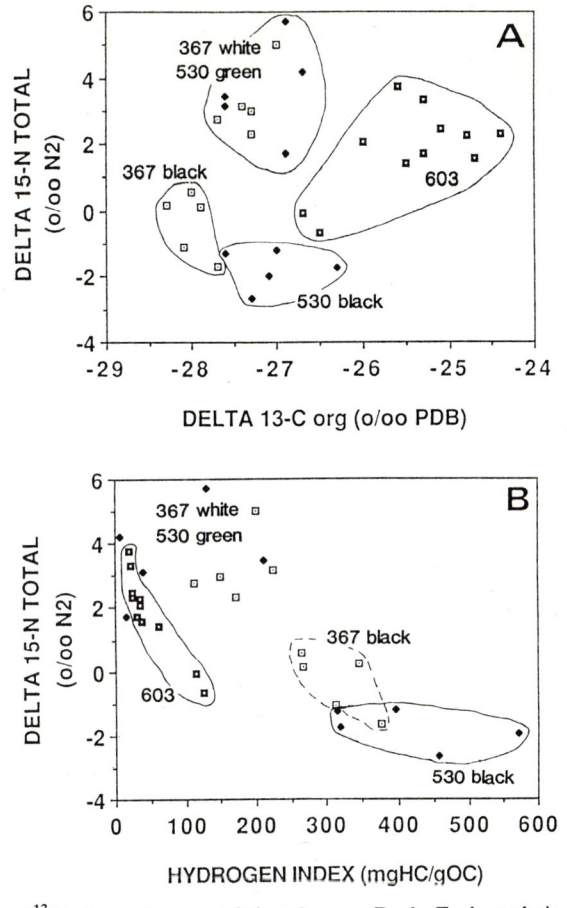

Fig. 16A,B. Scatter plots of $\delta^{15}N$ versus $\delta^{13}C$ of organic matter (**A**) and versus Rock-Eval pyrolysis hydrogen index (**B**) in samples of Neocomian carbonates from DSDP Site 367, eastern North Atlantic basin and DSDP Site 603, western North Atlantic basin, and in samples of middle Cretaceous claystones from DSDP Site 530, southern Angola basin. (Data from Rau et al. 1987)

most previously analyzed contemporary marine organic materials (e.g., Rau et al. 1987).

The isotopic differences between organic-carbon-rich and -poor beds may reflect differences in the diagenesis of nitrogen in adjacent beds having significantly different organic-carbon content. In general, diagenesis results in heavier, not lighter, values of $\delta^{15}N$ (see discussion by Rau et al. 1987). We interpret the data, therefore, as indicating significant differences in the isotopic composition of Cretaceous pelagic marine organic matter relative to that of modern pelagic marine organic matter. The low $\delta^{15}N$ values (^{15}N-depletion) in the black beds at Site 367 and 530, coupled with other geochemical data, suggest that the organic-N was produced in the surface-water column under conditions of lower nitrate availability that possibly resulted in a predominance of N_2-fixing phytoplankton, such as blue-green algae (Rau et al. 1987). The bioturbated white limestone beds at Sites 367 and 603 apparently provide a record of N-biogeochemistry involving normal populations of calcareous phytoplankton during periods of more effective overturn of oceanic deep-water masses. These productivity and redox cycles were possibly climate-induced and reflect changes in wind stress and upwelling rate and/or rates of production of deep water. Periodic slower rates of regeneration of deep water and/or higher surface-water productivity may have produced the anoxic conditions in deeper water that were necessary for preservation of marine organic matter as well as for substantial water-column nitrate reduction and consequent nutrient limitation of phytoplankton production.

6 Insights into the Origin of Cyclicity

The reasons for obtaining a large suite of geochemical data on cyclic C_{org}-rich carbonate sequences are obviously many and varied. However, with regard to understanding the origin of the cycles, we look to the geochemical data for constraints on the main processes that produced the cycles, namely variations in carbonate production or dissolution, dilution by detrital clastic material, and/or C_{org} production or preservation (see also Einsele and Ricken Chap. 1.1, this Vol.). The relative influence of these processes that produce the cyclicity in carbonate and organic carbon can be assessed by constructing another form of ternary plot (Fig. 17) that has Ca, Al, and C_{org} (OC) as end-members representing the relative amounts of biogenic carbonate, aluminosilicate clastic material, and organic matter, respectively, in a sample. To construct this diagram the absolute values of weight percent Ca, Al and C_{org} are added and the sum is set to equal 100% followed by calculation of the relative proportion of each component for each sample, which is then plotted on the ternary diagram. Because the carbonate percentages are relatively high, we have only used the upper part of the ternary diagram (equal to greater than 50% Ca). From such a diagram it is easy to detect changes in ratios of the three components. For example, clustering of data along any one of the chords shown in Fig. 17F represents a case in which the flux of only one component varies causing relative changes in the amount of dilution of the other two components which maintain a constant ratio to one another. An array of data points falling along a line from the Ca apex to a point that bisects the Al-OC axis would represent variable dilution of a constant relative flux of Al and C_{org} by changing carbonate flux. Such a trend could indicate either variable carbonate production or carbonate dissolution. It is also possible, in the above example, to vary the Al and C_{org} fluxes with relatively constant carbonate flux as long as the Al and C_{org} fluxes maintain a constant ratio. In order to provide a unique interpretation of such data, information on thicknesses of the bedding cycles is also necessary, particularly the relative thickness of each of the hemicycles in a simple a-b rhythm, which help to constrain the relative flux of each component.

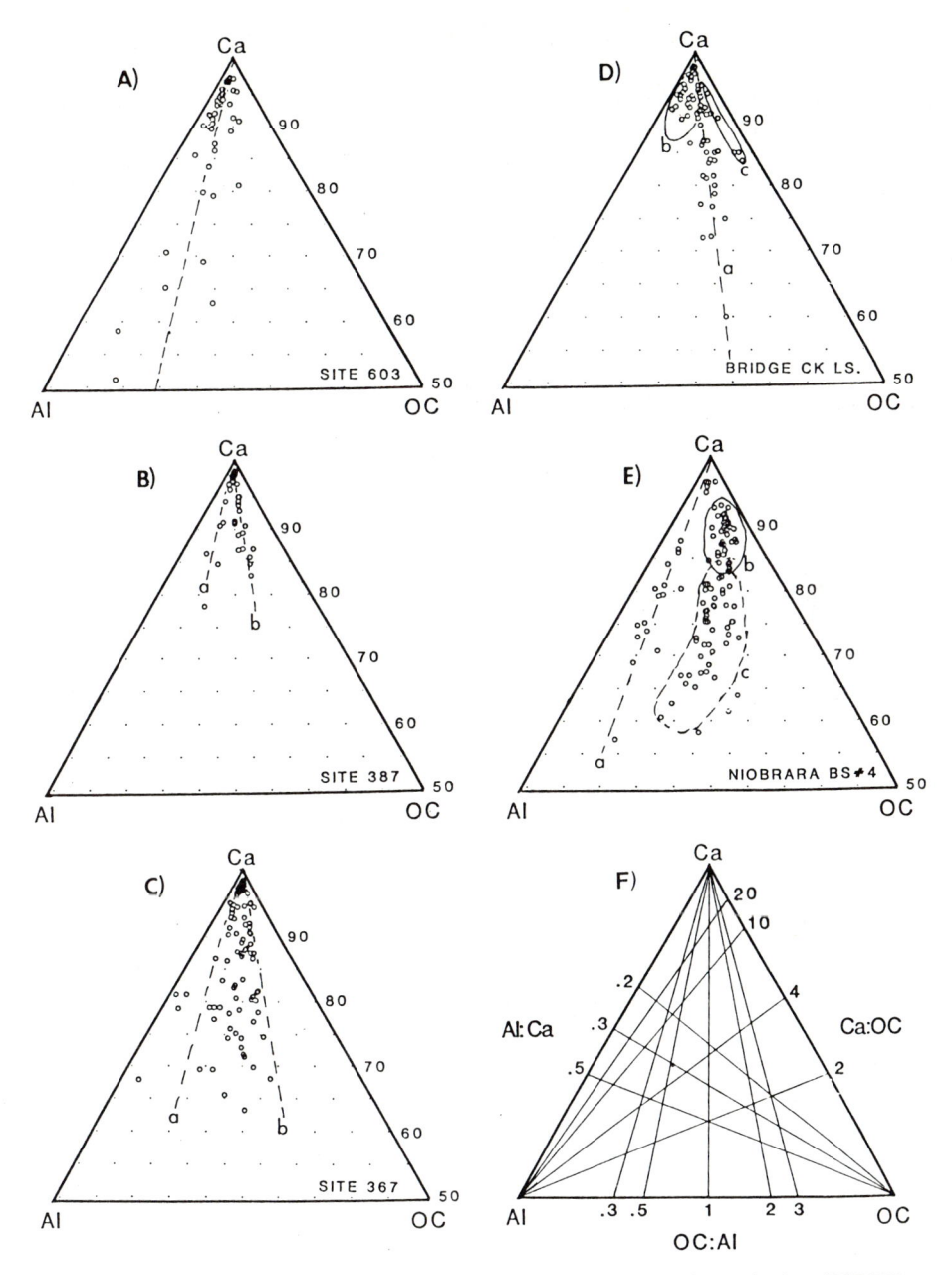

Fig. 17A-F. Ca-Al-OC ternary plots for samples of limestone-marlstone bedding cycles from DSDP Sites 603 (**A**), 387 (**B**), and 367 (**C**); the Bridge Creek Limestone (**D**) Member of the Greenhorn Formation; and the Niobrara Formation (**E**). *Lines* representing trends of constant ratios of the three components relative to one another are shown in (**F**). See text for explanation of lines *a,b,c*, etc

We use a simple computer model to generate trends in the ternary system Ca-Al-OC by varying the fluxes of the three components, and also compute final bed thicknesses and percentages of Ca, Al, and C_{org} for comparison to data sets for actual sequences. The bed thicknesses and carbonate and C_{org} percentages help to constrain the direction taken in computing fluxes (see also Ricken Chap. 1.8, this Vol.). In the above example, we would first establish an average rate of accumulation for a given sequence (knowing the average sedimentation rate and the sediment bulk density) and an average bedding couplet thickness and duration. In order to distinguish between the possibilities of carbonate dilution and carbonate dissolution, we examine the changes in bed thickness that are implied by each model (see, for example, Fig. 3A), assuming equivalent time represented by each hemicycle. Cases 1 and 2 in Table 2 show the results of a simple model for the changes in these parameters as a function of either variable carbonate production or dissolution, maintaining constant Al and C_{org} fluxes. Both cases produce the trend labeled (1) in Fig. 18A, but, as enumerated in the "bed thickness" column of Table 2, they produce very different thicknesses for hemicycles having the same carbonate content. We use the model to examine trends for the Niobrara Formation in the Berthoud State #4 well below (for cycles and data illustrated in Figs. 4 and 5).

Inspection of the ternary Ca-Al-OC diagrams (Fig. 17) indicates that the variability in the Ca/Al and Ca/OC ratios is much greater than that of the OC/Al ratio. Some sequences exhibit relatively straightforward relationships. Data for the Neocomian carbonate sequence at Site 603, for example (Fig. 17a), are arrayed along a line of relatively low OC/Al and a broad range of Al/Ca ratios which is to be expected for a locality characterized by strong terrigenous dilution, generally low inferred primary productivity and variability in preservation of biogenic carbonate. The main variable appears to be dilution of biogenic carbonate by terrigenous material, which has a relatively constant OC/Al ratio; the C_{org} is dominantly of terrestrial origin. These inferences are similar to those derived earlier from other geochemical data. In contrast to Site 603, Site 387 samples (Fig. 17b) have a narrow range of all ratios, even though both sites are in the western basin of the North Atlantic. The Site 387 data exhibit essentially two trends on a Ca-Al-OC diagram that differ only in the OC/Al ratio. These two trends are in stratigraphically different parts of the sequence and indicate that the C_{org} accumulation rate was relatively higher (trend "b") in one part of the sequence characterized by laminated black marlstones. Minor cycles of carbonate dissolution superimposed on the C_{org} preservational variations is the most likely origin of the Site 387 cycles because the C_{org}-rich intervals are generally much thinner than the C_{org}-poor intervals. Data from Site 367 off Northwest Africa (Fig. 17c) are arrayed with widely varying OC/Al ratios (bounded by lines a and b) as well as varying Ca/Al and OC/Ca ratios. Because these samples span a large stratigraphic range, it is likely that a number of factors contributed to the scatter, mainly variable flux of biogenic carbonate and C_{org}, as well as variations in C_{org} preservation as the result of periodic anoxia. These stratigraphic changes in the relative percentages of C_{org} and carbonate at Site 367 can be seen even over a 9-m (about 0.5 Ma) interval (Fig. 2A,B) of carbonate cycles in which the development of more prolonged anoxia is manifested by increasing dominance of darker-colored, laminated marlstones.

Although deposited in somewhat similar depositional environments, the Bridge Creek Limestone Member (Fig. 17d) and the Niobrara Formation (Fig. 17e) display very different patterns on the Ca-OC-Al ternary diagram. Within each unit there is variability in the ratios of the major components that has stratigraphic and environmental significance. For example, a large proportion of the data for the Bridge Creek Limestone Member data fall along a line of constant OC/Al with decreasing Al/Ca (line "a" in Fig. 17d) that corresponds to laminated to microbioturbated marlstone beds. It is difficult to envision an environmental situation in which the relative fluxes of C_{org} and Al are always constant because this would imply that the fluxes of these two components are somehow linearly related. Such a linear relationship seems unlikely, but it is possible that the C_{org} burial flux is proportional to the Al flux because of either (1) a sedimentation-rate control on C_{org} preservation (e.g., Müller and Suess 1979), or (2) that the increase in detrital flux is related to increasing fresh-water runoff to surface waters and development of salinity stratification that results in bottom-water oxygen depletion and consequent enhanced preservation of organic matter. The latter is our preferred hypothesis for the cyclicity in the Niobrara Formation and Bridge Creek Limestone as discussed previously. The points in the area labeled "b" on Fig. 17d correspond to the bioturbated limestones that constitute one type of major hemicycle, so that the trend in "a" actually bends towards lower OC/Al ratios at lower Al/Ca ratios in the bioturbated limestones. This indicates increasing oxidation of C_{org} during limestone deposition as well as decreasing relative flux of terrigenous material. Changes in the carbonate flux are inferred to be minor (see below). The samples represented by trend "c" differ from others in that they come from several laminated beds associated with the "oceanic anoxic event" in the Cenomanian-Turonian transition (see Fig. 4A; 19–20 m), during which the organic matter flux to the sea floor was higher and/or C_{org} was preferentially preserved relative to other parts of the Bridge Creek Limestone.

The data along line "a" in Fig. 17e for the Niobrara Formation are from samples of the C_{org}-poor Fort Hays Limestone Member. The broad range of Al/Ca and OC/Ca ratios and constant OC/Al ratios reflect a major role of carbonate dilution in formation of the cyclicity, while the oxic sea floor that characterized deposition of this highly bioturbated limestone unit resulted in poor C_{org} preservation. The cycles in the Fort Hays Limestone in the Berthoud State wells that we have studied average about 15 cm thick, with a range of 10–25 cm. These individual bedding couplets are thought to represent ca. 20 ka cycles, with thicker limestone portions and relatively thin marlstone partings. Our data suggest a narrow range of $CaCO_3$ variations in the limestone beds of 80–93% and as low as 40% in the thin marlstone partings. The model (not shown) that best explains these data is one that primarily requires changes in $CaCO_3$ flux through carbonate productivity variations. The area marked "b" in Fig. 17e represents limestones and marly limestones of the major chalk or limestone units within the Smoky Hill Chalk (see Fig. 4B). This sample array appears to represent carbonate cycles formed by minor variations in terrigenous dilution and possibly in biogenic carbonate production as well as little variation in C_{org} preservation. The bedding couplets in these intervals are, on average, about 30–35 cm thick and have somewhat subordinate marlstone hemicycles. Model 2 (Table 2, Figs. 18A,B), which varies the $CaCO_3$ and C_{org} fluxes with constant Al flux, is probably most appropriate for the cycles in the chalk units. These cycles have a narrow range of $CaCO_3$ and C_{org} variation, but the darker-colored intervals are relatively enriched in C_{org} and depleted in $CaCO_3$. In the model, average $CaCO_3$ values in light-colored and dark-colored intervals of 92 and 86%, respectively, produce C_{org} contents in adjacent hemicycles of 1.94 and 2.9%, and thicknesses of 17.6 and 15.6 cm, respectively. This is approximately what we observe in the Berthoud State cores. The points in area "c" represent cycles in the marlstone or shale units of the Smoky Hill Chalk that are characterized by strong terrigenous dilution, causing dilution of both the biogenic carbonate and marine organic carbon. Again, these inferences correspond closely to those made on the basis of other data presented previously. For these cycles, we favor a model similar to that of model 4 (Table 2; Figs. 18A,B) which requires that the flux of $CaCO_3$ decreases in some proportion to increases in the Al and C_{org} fluxes. This implies that as runoff to the basin increased, delivering larger fluxes of terrigenous clastic material, the production and/or preservation of C_{org} increased while $CaCO_3$ production decreased. Such a model produces thicker dark marlstone hemicycles and thicker bedding couplets overall (Fig. 18B), as observed for the marlstone units in the Smoky Hill Chalk. More detailed analyses of the Niobrara Formation are in progress.

7 Conclusions

Small-scale and larger-scale cyclic variation in carbonate and organic-carbon concentrations and in degree of bioturbation characterize the cyclic pelagic limestone sequences of the middle to Upper Cretaceous Bridge Creek Limestone and the Niobrara Formation in Colorado and Neocomian sequences at DSDP Sites 603, 387, and 367 in the North Atlantic. The periodicities of the dominant small-scale (bedding) cycles are between 20 and 40 ka/cycle and probably are attributable to Milankovitch forcing. The preliminary results of geochemical studies reported here demonstrate the potential of inorganic and stable data for interpreting the complex paleoenvironmental changes and variable sediment sources that characterized these regions. These changes produced cyclic sedimentary sequences that were the result of (1) cyclic changes in production of biogenic material in surface waters and flux to sediments that resulted from variations in upwelling or salinity, (2) cyclic changes in climate, weathering, runoff and rate of supply of terrigenous detrital material that diluted the fluxes of biogenic components such as pelagic carbonate, and/or (3) fluctuations in water column stability and bottom-water production, circulation, and oxygenation that produced changes in preservation and accumulation of organic matter as well as carbonate.

The Niobrara and Bridge Creek cycles apparently are the product of all three mechanisms, that were interrelated and ultimately controlled by fresh water runoff to the Cretaceous Western Interior Seaway of North America from uplifted highlands to the west. The Neocomian cycles at DSDP Sites 603 and 387 in the western North Atlantic basin appear to be mainly a function of changes in terrigenous dilution with possible superimposed variation in carbonate dissolution. The Neocomian cycles at DSDP Site 367, although simple in appearance, are probably a function of

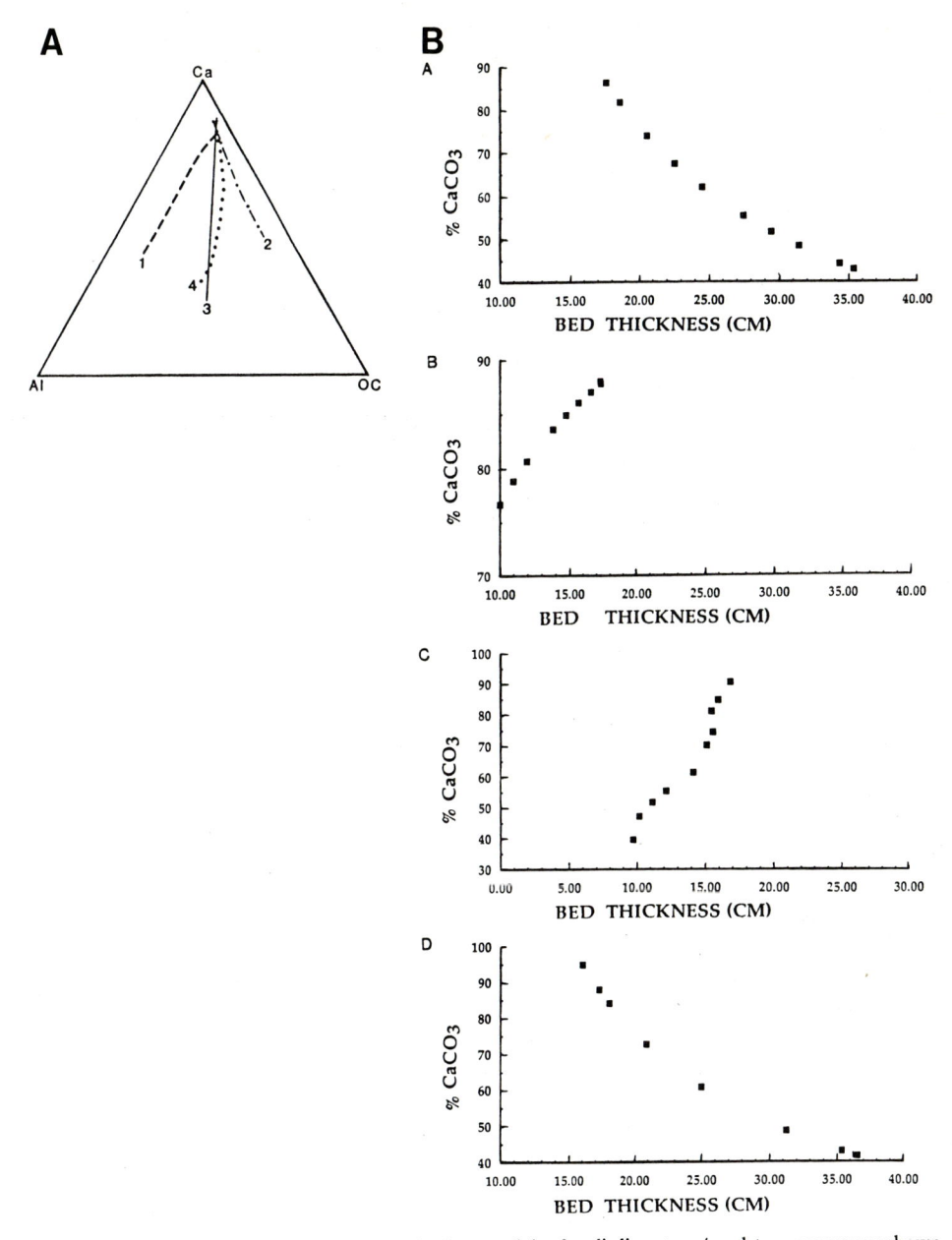

Fig. 18A,B. A Ca-Al-OC ternary plot for the four models of cyclic limestone/marlstone sequences shown in Table 2. Scales as in Fig. 17. **b** Hemicycle thickness (cm) versus CaCO₃ (%) for the models shown in Fig. 18A and Table 2; *A-D* correspond to models 1–4 respectively

Table 2. Results of model simulation for creating limestone/marlstone cycles in the Niobrara Formation by varying the amounts of CaCO$_3$ (expressed as percent Ca), organic carbon (OC), and clastic content (expressed as percent Al). Input to the model consist of mass accumulation rates (MAR) of the three variables expressed in g/cm^2/ka. The models assume an average porosity of 0%, a dry bulk density of 2.7 g/cm^3, an average accumulation rate of 1.6 cm/ka and hemicycle duration of 20 ka. The models predict thickness of hemicycles (single beds), percent CaCO$_3$, and percent OC

Model 1: Constant Ca MAR, constant OC MAR, increasing Al MAR

Ca MAR	OC MAR	Al MAR	Ca/OC	OC/Al	Ca/Al	Ca	OC	Al	Bed thickness (cm)	CaCO$_3$ (%)	OC (%)
1.58	0.120	0.04	13.17	3.00	39.50	0.91	0.07	0.02	17.62	86	2.62
1.58	0.120	0.06	13.17	2.00	26.33	0.90	0.07	0.03	18.61	82	2.48
1.58	0.120	0.1	13.17	1.20	15.80	0.88	0.07	0.06	20.58	74	2.24
1.58	0.120	0.14	13.17	0.86	11.29	0.86	0.07	0.08	22.55	67	2.05
1.58	0.120	0.18	13.17	0.67	8.78	0.84	0.06	0.10	24.52	62	1.88
1.58	0.120	0.24	13.17	0.50	6.58	0.81	0.06	0.12	27.47	55	1.68
1.58	0.120	0.28	13.17	0.43	5.64	0.80	0.06	0.14	29.44	52	1.57
1.58	0.120	0.32	13.17	0.38	4.94	0.78	0.06	0.16	31.41	48	1.47
1.58	0.120	0.38	13.17	0.32	4.16	0.76	0.06	0.18	34.36	44	1.34
1.58	0.120	0.4	13.17	0.30	3.95	0.75	0.06	0.19	35.35	43	1.31

Model 2: Decreasing Ca MAR, increasing OC MAR, constant Al MAR

Ca MAR	OC MAR	Al MAR	Ca/OC	OC/Al	Ca/Al	Ca	OC	Al	Bed thickness (cm)	CaCO$_3$ (%)	OC (%)
1.58	0.090	0.035	17.56	2.57	45.14	0.93	0.05	0.02	17.26	88	2.01
1.58	0.100	0.035	15.80	2.86	45.14	0.92	0.06	0.02	17.30	88	2.22
1.5	0.110	0.035	13.64	3.14	42.86	0.91	0.07	0.02	16.57	87	2.55
1.4	0.120	0.035	11.67	3.43	40.00	0.90	0.08	0.02	15.65	86	2.95
1.3	0.130	0.035	10.00	3.71	37.14	0.89	0.09	0.02	14.72	85	3.40
1.2	0.140	0.035	8.57	4.00	34.29	0.87	0.10	0.03	13.80	84	3.90
1	0.150	0.035	6.67	4.29	28.57	0.84	0.13	0.03	11.92	81	4.84
0.9	0.155	0.035	5.81	4.43	25.71	0.83	0.14	0.03	10.97	79	5.43
0.8	0.160	0.035	5.00	4.57	22.86	0.80	0.16	0.04	10.03	77	6.13
0.65	0.180	0.035	3.61	5.14	18.57	0.75	0.21	0.04	8.67	72	7.99

Model 3: Decreasing Ca MAR, increasing OC MAR, and increasing Al MAR

Ca MAR	OC MAR	Al MAR	Ca/OC	OC/Al	Ca/Al	Ca	OC	Al	Bed thickness (cm)	CaCO$_3$ (%)	OC (%)
1.58	0.095	0.025	16.63	3.80	63.20	0.93	0.06	0.01	16.79	90	2.18
1.4	0.120	0.04	11.67	3.00	35.00	0.90	0.08	0.03	15.89	85	2.90
1.3	0.120	0.05	10.83	2.40	26.00	0.88	0.08	0.03	15.42	81	2.99
1.2	0.140	0.07	8.57	2.00	17.14	0.85	0.10	0.05	15.52	74	3.47
1.1	0.150	0.08	7.33	1.88	13.75	0.83	0.11	0.06	15.09	70	3.82
0.9	0.140	0.1	6.43	1.40	9.00	0.79	0.12	0.09	14.12	61	3.81
0.7	0.130	0.1	5.38	1.30	7.00	0.75	0.14	0.11	12.15	55	4.11
0.6	0.120	0.1	5.00	1.20	6.00	0.73	0.15	0.12	11.15	52	4.14
0.5	0.120	0.1	4.17	1.20	5.00	0.69	0.17	0.14	10.19	47	4.53
0.4	0.120	0.11	3.33	1.09	3.64	0.63	0.19	0.17	9.72	40	4.75

Model 4: Constant Ca MAR, increasing OC MAR, increasing Al MAR

Ca MAR	OC MAR	Al MAR	Ca/OC	OC/Al	Ca/Al	Ca	OC	Al	Bed thickness (cm)	CaCO$_3$ (%)	OC (%)
1.58	0.085	0.01	18.59	8.50	158.00	0.94	0.05	0.01	16.01	95	2.04
1.58	0.095	0.035	16.63	2.71	45.14	0.92	0.06	0.02	17.28	88	2.11
1.58	0.105	0.05	15.05	2.10	31.60	0.91	0.06	0.03	18.06	84	2.24
1.58	0.200	0.1	7.90	2.00	15.80	0.84	0.11	0.05	20.88	73	3.68
1.58	0.240	0.18	6.58	1.33	8.78	0.79	0.12	0.09	24.98	61	3.70
1.58	0.330	0.3	4.79	1.10	5.27	0.71	0.15	0.14	31.23	49	4.06
1.58	0.380	0.38	4.16	1.00	4.16	0.68	0.16	0.16	35.36	43	4.13
1.58	0.390	0.4	4.05	0.98	3.95	0.67	0.16	0.17	36.38	42	4.12
1.58	0.410	0.4	3.85	1.03	3.95	0.66	0.17	0.17	36.46	42	4.32
1.58	0.430	0.4	3.67	1.08	3.95	0.66	0.18	0.17	36.54	42	4.53

coincident changes in biologic productivity in surface waters, in bottom-water oxygenation and in terrigenous dilution. Productivity appears to have been highest during times of better oxygenation of bottom waters and low rate of supply of the detrital fraction. Regardless of the relative importance of changes in terrigenous dilution in the cycles, the geochemical data indicate that there were significant changes in mineralogy of the detrital fraction from one part of a Milankovitch cycle to another at all sites studied. The largest changes occurred within cycles at Site 367 and more subtle changes occurred within cycles in the Niobrara Formation and Bridge Creek Limestone.

Our studies indicate that subtle variations in mineralogy are easily seen in major- and minor-element geochemistry. Stable-isotope and other data on the organic fraction support the notion that changes in biologic productivity through a given cycle are at least partly responsible for the cyclic variations in lithology. However, carbonate stable isotope data from carbonate cycles of varying degree of lithification largely tell a diagenetic story and, except in very unusual cases, do not allow interpretation of primary depositional conditions such as salinity and temperature. The application of geochemistry to understanding complex cyclicity in Cretaceous pelagic sequences is most powerful when a suite of inorganic, organic stable isotopic geochemical analyses in done on splits of the same sample, and when sampling takes into account subtle changes in lithology – the holistic approach to geochemistry.

Acknowledgments. We thank the convenors for their invitation to contribute this chapter to the *Cycles and Event Stratigraphy* monograph. We would also like to express our gratitude to Sheila Hagerty Rieg for her diligent attention to analytical details during different phases of this study and for her help in managing our geochemical data base. We are also indebted to discussions and debates with our fellow cyclomaniacs who have been involved in other aspects of these studies with us – Margaret Leinen, Hervé Chamley, Tom Glancy, Rich Pollastro, Eric Barron, Erle Kauffman, Al Fischer, Lisa Pratt, and Peter Scholle. M. Leinen, H. Chamley, and R. Pollastro provided X-ray diffraction mineralogical data on the clay fraction of some of the samples used in this paper that have enabled us to speculate on the significance of the variations in major and minor element geochemistry. We also greatly appreciate constructive reviews by R.E. Garrison, P. Hansley, and G. Einsele. This research has been supported by NSF OCE 85-16295 (MAA) and the U.S. Geological Survey Climate Program (WED).

1.8 Variation of Sedimentation Rates in Rhythmically Bedded Sediments. Distinction Between Depositional Types

W. Ricken

1 Introduction

Determining the origin of the various types of rhythmic carbonate sediments and distinguishing between them is difficult. Depositional processes generating bedding rhythms are not easy to demonstrate conclusively, and their ultimate determination requires a wide range of paleontological, geochemical, and other methods. Generally, chalk-marl alternations are formed by depositional variation in carbonate, clastic, and organic matter. Rhythms caused by one of these depositions, or a combination of two or all three of them, differ in many aspects including carbonate and organic carbon contents and type of rhythmicity. The aim of this chapter is to show the close interrelationships between depositional inputs, bedding rhythms, carbonate differences between beds, and the related variation in sedimentation rates. These questions are addressed by the investigation of three major subjects. (1) Determination of sediment inputs and relative sedimentation rates based on the evaluation of carbonate and organic carbon data. (2) Introduction of procedures which show idealized rhythms for various types of input. (3) Application of these methods to calcareous bedding rhythms in various environments, especially in deep sea sediments.

2 The Three Major Types of Bedding Rhythms

Rhythmically bedded sediment with oscillating carbonate contents can be explained either by fluctuating deposition from individual sediment fractions or by a combination of several fractions influencing the carbonate content of the sediment (Arthur and Dean Chap. 1.7, Grötsch et al. Chap. 1.6, Diester-Haass Chap. 1.5, Eicher and Diner Chap. 1.4, Einsele and Ricken Chap. 1.1, all this Vol.). Variation in *carbonate deposition* appears to be the most common cause for the formation of bedding rhythms. It can be generated by various factors, either through fluctuations in surface water carbonate productivity and fluctuations in lateral $CaCO_3$ input, or by dissolution near the sediment surface. Variation in *silicate deposition* may form bedding rhythms in deep clastic shelves or in the deep sea below the lysocline. Silicate variation is related to the dilution by terrigenous clay and silt fractions or to variations in deposition of siliceous ooze. Variation in *deposition of organic matter* is related to variations in marine productivity, delivery of terrestrial organic matter, or changes in redox conditions in the bottom waters. As will be shown in the following sections, alternations formed by one of these basic processes can be easily distinguished by the different relationships of their carbonate and organic carbon contents, and by their types of bedding rhythm.

2.1 Types of Depositional Variations as Expressed by Distinctive Carbonate-Organic Carbon Relationships

Variation in the deposition of one fraction (i.e., carbonate, silicate, or organic matter) "dilutes" or "concentrates" the weight percentages of the two remaining fractions. These depositional dilution and concentration processes are distinctly different for each of the three depositional types. A three-component system for *idealized* depositional variation in carbonate, silicate, or organic matter is developed (Fig. 1). In this model, the depositional input of each of these fractions is systematically varied, while the inputs of the two remaining fractions are held constant. The resulting changes in carbonate and organic carbon contents are then expressed in xy diagrams, producing distinct C_{org}-$CaCO_3$ relationships for each of the three types of deposition as described below:

Variation in carbonate deposition (Fig. 1a). Consider a given sediment sample composed of carbonate, silicate, and organic matter (see arrow). With increasing carbonate deposition, the total sediment volume increases, while the absolute amounts of silicate and organic carbon remain unaffected. As a result, the weight percentages of silicate and organic matter decrease (expressed as percentages of the total sample weight). On the other hand, when carbonate deposition is reduced, contents of silicate and organic matter increase. Consequently, in alternations with varying carbonate deposition, the carbonate content tends to be negatively correlated with the percentage of organic carbon (see the C_{org}-$CaCO_3$ graph in Fig. 1a). Such variations in carbonate input also influence the sedimentation rate (s_r). During periods of high carbonate input, the sedimentation rate becomes greater and vice versa.

Variation in silicate deposition (Fig. 1b). When silicate deposition increases (e.g., commonly the clay and silt fraction) while the other inputs remain constant, the weight percentages of carbonate and organic carbon decrease. Since increasing deposition of silicate is accompanied by a decreasing carbonate content, a positive correlation of weight percent of organic carbon and carbonate is obtained. Thus low carbonate contents are related to low organic carbon contents, and vice versa (see the C_{org}-$CaCO_3$ graph in Fig. 1b).

Variation in deposition of organic matter (Fig. 1c). Since organic carbon is a minor constituent of the sediment, carbonate content and sedimentation rate are only slightly affected. In an C_{org}-$CaCO_3$ diagram, variation in organic matter is defined by a regression curve that is nearly parallel to the C_{org} axis, with $CaCO_3$ slightly decreasing with increasing organic carbon content.

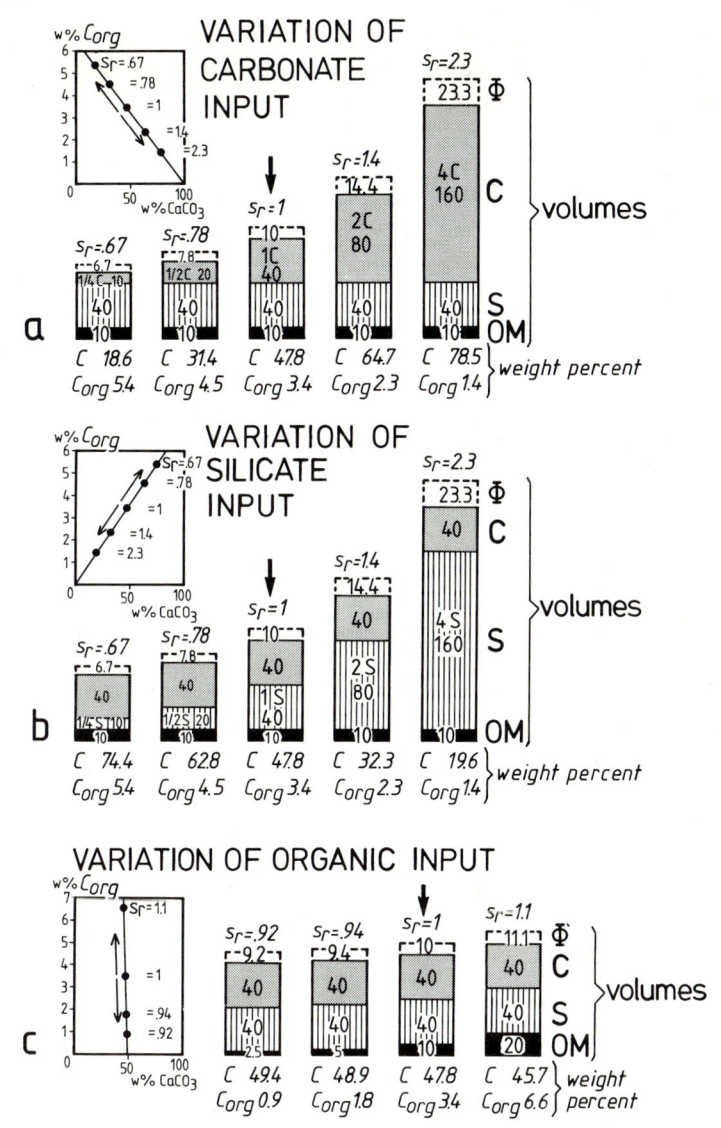

Fig. 1a-c. Type of deposition reflected by the different relationship between weight percent organic carbon (C_{org}) and carbonate content (C). The initial sediment composition (*arrow*) is systematically changed by enhancing or diminishing the input of only one fraction (**a** to **c**). *Left-hand graph* illustrates the consequent C_{org}-$CaCO_3$ relationship (w%). *Boldface symbols on the right* denote sediment volumes, including porosity (Φ), carbonate (C), silicate (S), and organic matter volume (OM). Relative sedimentation rates (s_r) are expressed with respect to the initial sample (*arrow*) which is assigned a relative sedimentation rate of $s_r = 1$.

In addition to depositional dilution and concentration processes, four factors influence the C_{org}-$CaCO_3$ curves. (1) Increasing sedimentation rates may enhance organic carbon preservation because organic carbon is embedded faster, and thus withdrawn from degradation at the sediment surface (e.g., Heath et al. 1977; Müller and Suess 1979). According to Emerson (1985) and Stein (1986b), however, the influence of this factor is questionable. Müller and Suess found increasing rates of sedimentation along with organic carbon accumulation among various marine environments from the deep oceans to shelf seas. This relationship is also positively correlated with increasing productivity and decreasing water depth; both factors additionally augment the organic carbon accumulation rate (e.g., Bralower and Thierstein 1987). Consequently, the role of the sedimentation rate in organic carbon preservation (at constant productivity and water depth) is unclear. Stein (1986) showed that a preservation effect with increasing sedimentation rates is not observed where environments contain oxygen-deficient bottom waters. (2) Bacterial decomposition during shallow burial lowers the organic carbon content initially embedded in the sediment. It is assumed that this decrease equally affects succeeding beds whether rich or poor in carbonate content. Hence, the determination of *relative* sedimentation rates between alternating beds is not significantly influenced. (3) Carbonate cementation and dissolution processes may dilute or concentrate organic carbon contents, respectively (see Ricken and Eder Chap. 3.1, this Vol.). This difficulty is avoided as much as possible in this study, whereby primarily sediments with minor diagenetic overprinting are investigated. Where bedding rhythms were affected by carbonate diagenesis, corrections (i.e., carbonate mass balance calculations) were performed according to methods described by Ricken (1986). (4) Some bedding rhythms are formed by simultaneous variation of several parameters. Such composite rhythms may show a superposition of several individual C_{org}-$CaCO_3$ regression lines.

Thus, for soft sediments with minor diagenetic overprinting, depositional variations in carbonate, silicate, and organic matter can be distinguished by their typical C_{org}-$CaCO_3$ pattern. As will be demonstrated, such patterns occur in various bedding cycles, indicating the origin of bedding rhythms by variation of one or several depositional types.

2.2 Rhythmic Bedding Due to Variation in Carbonate, Silicate, and Organic Matter: Distinctive Features

In addition to the characteristic C_{org}-$CaCO_3$ pattern, bedding rhythms can be distinguished by the shape of their carbonate and organic carbon curves (Fig. 2). Rhythms

Fig. 2a-c. Distinctive types of bedding rhythms formed by varying deposition in either carbonate (**a**), silicate (**b**), or organic matter (**c**). Bedding rhythms show carbonate ($CaCO_3$) and organic carbon (C_{org}) curves (in w%) for alternations with low, medium, and high carbonate content (*A, B,C*). Changing sedimentation rates are expressed by the varying thickness of section intervals representing equal time spans. *Right-hand graphs* represent the carbonate organic carbon relationship (in w%) for each bedding rhythm. Diagrams are based on depositional inputs that vary sinusoidally by a factor of 4

VARIATION OF CARBONATE INPUT

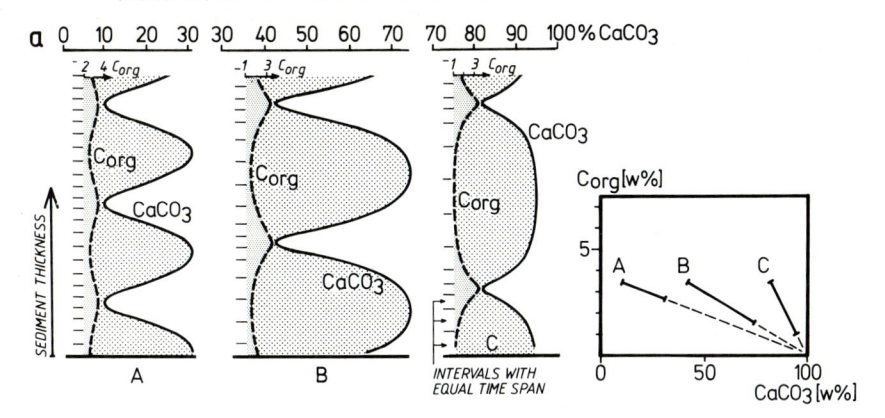

VARIATION OF SILICATE INPUT

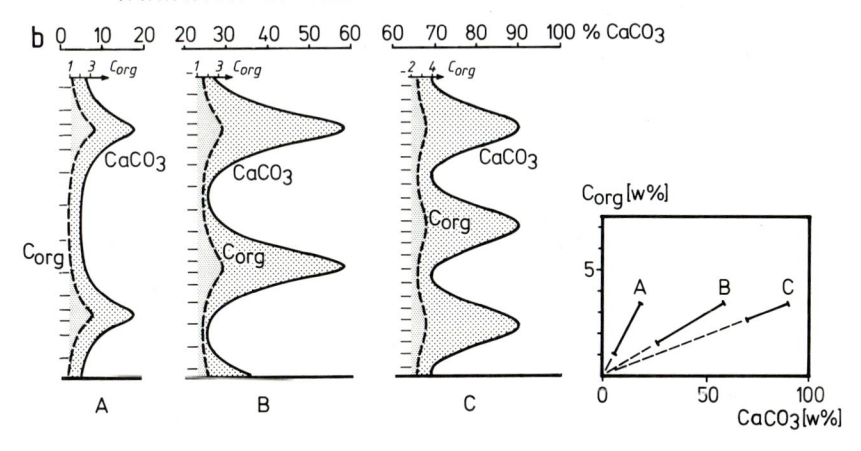

VARIATION OF ORGANIC INPUT

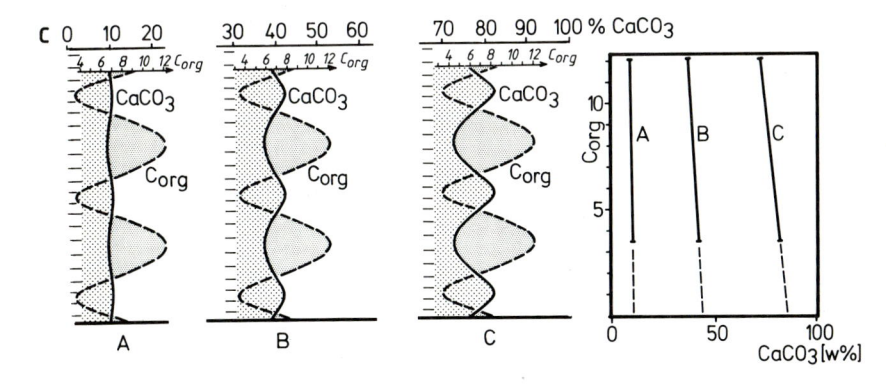

of the three basic types of deposition contain distinct alternating thick and thin beds, representing successive periods with high and low sedimentation rates, respectively. The three basic types of rhythms are described below with their idealized curve forms, assuming that the sediment supply of each of these fractions fluctuated regularly.

Alternations with varying carbonate deposition (Fig. 2a). Ideally, carbonate curves of alternating beds have broad, convex maxima and smaller, sharper minima, because the sedimentation rate and carbonate input are higher in the CaCO$_3$-rich bed. This effect is developed best in alternations with overall high carbonate contents. Since the organic carbon and carbonate contents are inversely correlated, the lowest organic carbon values occur in the middle of the carbonate-rich layers.

Alternations with varying silicate deposition (Fig. 2b). The bedding rhythm essentially shows the opposite curve forms compared to alternations with varying carbonate deposition. Here, carbonate maxima are narrow and distinct, whereas minima are broad and concave. This pattern is most obvious with low carbonate contents and where terrigenous dilution is high, which increases the thickness of the carbonate-poor bed (Fig. 2b). As explained in the previous section, the organic carbon content is positively correlated with the carbonate content. Carbonate-rich layers have high organic carbon contents, while the opposite is true for carbonate-poor beds. The influence on the weight percent of organic carbon is great for alternations with low carbonate contents.

Alternations with varying organic matter deposition (Fig. 2c). Sedimentation rates and carbonate contents are influenced little by variations in the small amount of organic matter deposition. Carbonate contents are significantly affected by variation of organic matter only when carbonate contents are very high (Fig. 2c, curve C).

Similar depositional processes also control how beds are grouped into *bundles* (Fig. 3). Bundles are understood here as being formed by a regular variation of only one type of input (either carbonate or silicate), but with a varying frequency. Bundles related to carbonate variation are composed of several carbonate-rich layers and thin, carbonate-poor beds forming the upper and lower boundaries of the bundle. Such bedding rhythm occurs especially with an overall high carbonate content. The opposite pattern is observed for silicate variations. Here, bundles show relatively thin carbonate-rich beds in the middle which are surrounded by thick carbonate-poor zones.

Hence, carbonate and organic carbon relationships, bedding rhythms, and bundling of beds are all integrated instruments which allow the distinction between the three basic types of depositional inputs. They have in common that regular variation in deposition are expressed in not equally regular, distinctive types of bedding rhythms. The marine environments of such bedding rhythms are presented below with their characteristic organic carbon-carbonate relationships.

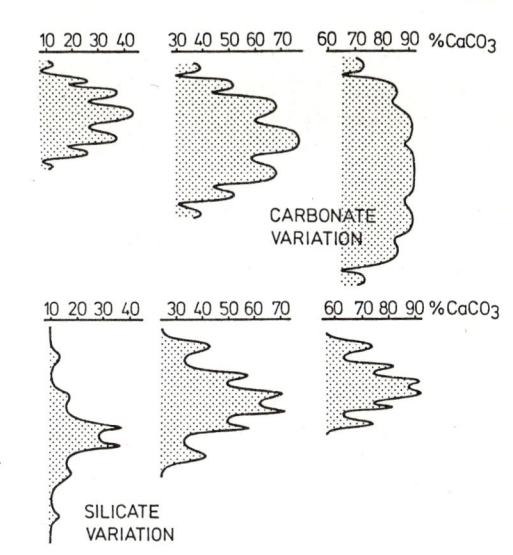

Fig. 3. Type of bundles for carbonate and silicate deposition. The effect of bundling is shown for alternations oscillating at low, medium, and high carbonate contents. Depositional input varies sinusoidally for bundles by a factor of 5 and for smaller oscillations by a factor of 2

2.3 Sediment Input Pattern as Expressed in C_{org}-$CaCO_3$ Data for Various Environments

Marine environments can show nearly any style and combination of depositional inputs that lead to rhythmic bedding. The rhythms are shown in a schematic marine facies transect depicting idealized C_{org}-$CaCO_3$ regression curves (Fig. 4). Two major groups of rhythms can generally be discriminated. These are (1) variation of a single component, and (2) simultaneous variation of several components. The latter group can be distinguished from the first by the development of two or more C_{org}-$CaCO_3$ regression curves, indicating a superposition of several inputs.

2.3.1 Environments with Variation of a Single Component

Evaluation of a large data set of C_{org}-$CaCO_3$ contents from sediments encountered in the deep sea and epicontinental basins suggests that variation in carbonate input is the dominant sedimentation type above the lysocline (Fig. 4). Thus, varying planktonic carbonate production in pelagic environments and epeiric seas of the post-Jurassic seems to be the most important process generating the typical pattern of carbonate variation (i.e., productivity cycles, see Eicher and Diner Chap. 1.4, this Vol.). Seaward of carbonate platforms and calcareous shelves, variation in carbonate productivity is superimposed with the fluctuation in lateral transport of shallow-water, aragonite-rich sediment, forming aragonite-chalk rhythms (e.g., Kier and Pilkey 1971; Droxler et al. 1983; Haak and Schlager 1989). Between the lysocline and the CCD and in alternations above the lysocline containing a greater amount of organic matter, changing intensity of carbonate dissolution is expressed by large $CaCO_3$ differences in succeeding beds (i.e., dissolution cycles, Berger 1973; Emerson

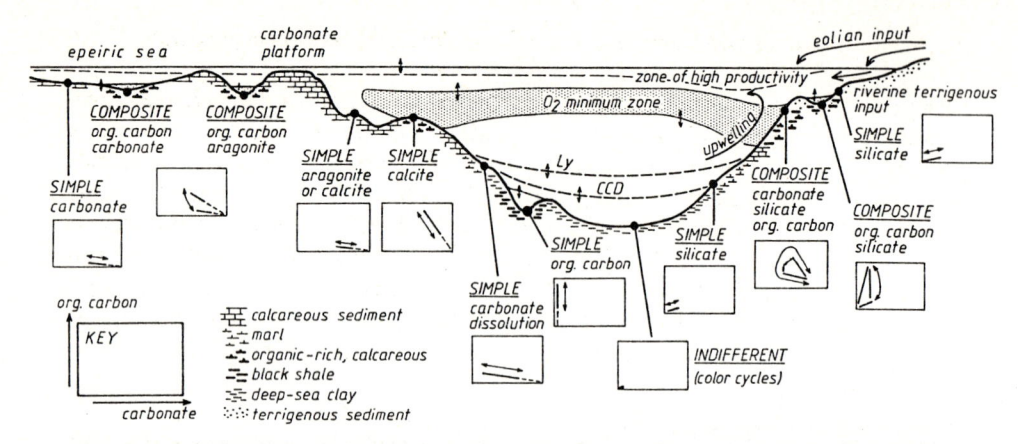

Fig. 4. Types of rhythmic sedimentation expressed in various organic carbon carbonate relationships (w%). Caused by variation in carbonate, silicate, and organic matter deposition, rhythmic sedimentation shows simple and composite types. Simple types of rhythmic bedding are caused by depositional variation of one dominant parameter, while composite cycles show the simultaneous variation of two or three depositional parameters. The figure is based on organic carbon carbonate evaluation using DSDP sites mentioned in Fig. 7

and Bender 1981; Peterson and Prell 1985a; Farell and Prell 1987; Grötsch et al. Chap. 1.6, this Vol.). For all of these environments, the correlation curves between C_{org} and $CaCO_3$ are negative, with various slopes representing the different ratios between organic matter supply and terrigenous sediment. For instance, environments with a high organic matter input, as in the oxygen minimum zone or in other oxygen-deficient bottom waters, show relatively steep slopes of the correlation curves (Fig. 4).

Environments with variations in terrigenous input show a positive correlation between organic carbon and carbonate content (i.e., dilution cycles, Fig. 4). This group includes sites with carbonate admixtures of minor importance, for example some epicontinental seas, deep clastic shelves, and upper continental slopes. Variation in the terrigenous fraction also occurs below the CCD and in largely carbonate-depleted sediment below the lysocline. A positive C_{org}-$CaCO_3$ correlation pattern is assumed for siliceous ooze productivity cycles with constant supply of the calcareous fraction (see Decker Chap. 4.1, this Vol.).

Varying input of only organic matter with the two other fractions deposited at a constant rate, is rarely observed under oceanic and epeiric-sea conditions. Usually, the much greater depositional fluctuations of the carbonate and silicate fractions influence the smaller variation of the organic fraction. In deep waters and far away from terrigenous sources, however, alternations may solely show varying input of organic matter (Fig. 4). It seems that such types were more common in the middle Cretaceous, when sluggish oceanic circulation with restricted marine productivity and a high input of terrestrial organic matter led to a wide distribution of redox cycles (Bralower and Thierstein 1984; Dean and Arthur 1986; de Boer 1986; Stein 1986b).

2.3.2 Environments with Simultaneous Variation of Several Components

Composite rhythms with a simultaneous variation of two or more depositional components are typical of many marine systems. A common situation is carbonate production in the upper water zone underlain by oxygen-deficient bottom waters. They occur either locally or in large areas, e.g., when the oxygen minimum zone touches the sea floor (Fig. 4). In this case, rhythms indicate concurrent carbonate and organic matter deposition. In upwelling areas, the production of carbonate and siliceous ooze and the terrestrial and marine input of organic matter may vary simultaneously with the transport of terrigenous sediment from nearby land sources, leading to a complex organic carbon-carbonate correlation pattern (Fig. 4).

2.4 Carbonate and Organic Carbon Data: Relative Sedimentation Rates and Time Spans Inherent in Beds

Figures 5 and 6 present a few examples of simple and composite bedding types, illustrating the mostly theoretical considerations made so far. These examples support the three basic types of depositional variations. The variations in relative sedimentation rates involved in these rhythms can be determined by employing the characteristic C_{org}-$CaCO_3$ relationships previously described.

2.4.1 Derivation of Relative Sedimentation Rates

Varying deposition of one of the three inputs is reflected by increasing or decreasing sediment volumes depicted in Fig. 1 and, thus, variation in relative sedimentation rates. The relative sedimentation rate is numerically expressed in relation to the actual (but normally unknown) sedimentation rate in the carbonate-rich or carbonate-poor bed of an alternation, which is assigned a value of $s_r = 1$. Consequently, the equations as given below calculate by what factor the relative sedimentation rate of one bed (with $s_r = 1$) must be changed, by employing one of the three basic types of deposition to obtain the $CaCO_3$ and C_{org} contents of a succeeding bed. (For instance, the marl bed is assigned a relative sedimentation rate of $s_r = 1$, while the succeeding limestone bed may have a two times higher sedimentation rate; or vice versa with $s_r = 1$ for the limestone bed and $s_r = 1/2$ for the marl bed.) The relative sedimentation rate (s_r) for the variation in carbonate and silicate deposition is expressed by

$$s_r = \frac{C_{org_1}}{C_{org_2}} \frac{1 + 0.0218\, C_{org_2}}{1 + 0.0218\, C_{org_1}} \frac{100 - \Phi_1}{100 - \Phi_2}, \qquad (1)$$

and for the variation in organic matter deposition by

$$s_r = \frac{C_1}{C_2} \frac{1 + 0.0218\, C_{org_2}}{1 + 0.0218\, C_{org_1}} \frac{100 - \Phi_1}{100 - \Phi_2}. \qquad (2)$$

In these equations, C_{org1} and C_{org2} is the organic carbon content, C_1 and C_2 the carbonate content, and Φ_1 and Φ_2 the porosity of a first and second sample, respectively. Average grain densities were assumed to be 2.7 g/cm^3 for the carbonate and silicate fraction, and 1.01 g/cm^3 for the organic matter. A factor of 1.3 was chosen to convert weight percent organic carbon into the amount of organic matter.

According to Eqs. (1) and (2), variations in relative sedimentation rates are determined between alternating carbonate-poor and carbonate-rich beds. This is performed by comparing the compositions defined by the maximum and minimum values of the plotted C_{org}-$CaCO_3$ data range as indicated in Figs. 5 and 6. In addition, the time span inherent in beds can be estimated. The bed thickness must then be divided by the relative sedimentation rate, assuming that differential compaction and diagenetic carbonate redistribution have not yet occurred.

2.4.2 Rhythms Due to the Depositional Variation of One Dominant Parameter

Fort Hays Limestones, Upper Cretaceous, U.S. Western Interior Seaway (Fig. 5). Here a negative correlation exists between C_{org} and $CaCO_3$, the cyclicity is caused by depositional variation of carbonate input, probably related to changes in productivity (Eicher and Diner 1985; Arthur et al. 1985; Arthur and Dean Chap. 1.7, this Vol.). Sedimentation rates differ by a factor of 7.9 across the entire C_{org}-$CaCO_3$ curve [Eq. (1)]; however, when correcting for carbonate diagenesis by performing mass balance calculations, carbonate-rich layers have a sedimentation rate only 1.9 times higher than the marl beds.

DSDP Site 535, water depth 3450 m, Valanginian to Cenomanian, Straits of Florida (Buffler et al. 1984; Cotillon and Rio 1984; Fig. 5). These middle Cretaceous alternations are described as representing redox rhythms, with transitions from chalk to black shale. The negative C_{org}-$CaCO_3$ correlations indicate that varying carbonate input repeatedly diluted or concentrated the organic carbon content of the sediment, leading to periodically changing oxygenation in the uppermost sediment layer. As a result, changing bioturbation is observed in beds rich in carbonate and poor in organic carbon. The correlation curves show various clustering of the C_{org}-$CaCO_3$ data, with a sedimentation rate 1.8 to 4.8 times higher in the carbonate-rich bed. The Cenomanian alternations have an overall lower C_{org} input, as indicated by a flatly sloping C_{org}-$CaCO_3$ regression line.

Fairport Shale, Cenomanian, U.S. Western Interior Seaway (Fig. 5). This sequence consists of little to intensively bioturbated transgressive shales with minor bedding variations (Glenister and Kauffman 1985). From the positive slope of the C_{org}-$CaCO_3$ data, terrigenous dilution is inferred, with sedimentation rates up to 6.7 times greater in carbonate-poor beds.

DSDP Site 603, water depth 4633 m, Cenomanian, North Atlantic Ocean off New Jersey (van Hinte et al. 1987; Fig. 5). Alternating Upper Cretaceous black shale and claystone show a background variation in terrigenous input superimposed with an oscillating input of organic matter. Although organic matter varies by 10%, the effect on the sedimentation rate is insignificant.

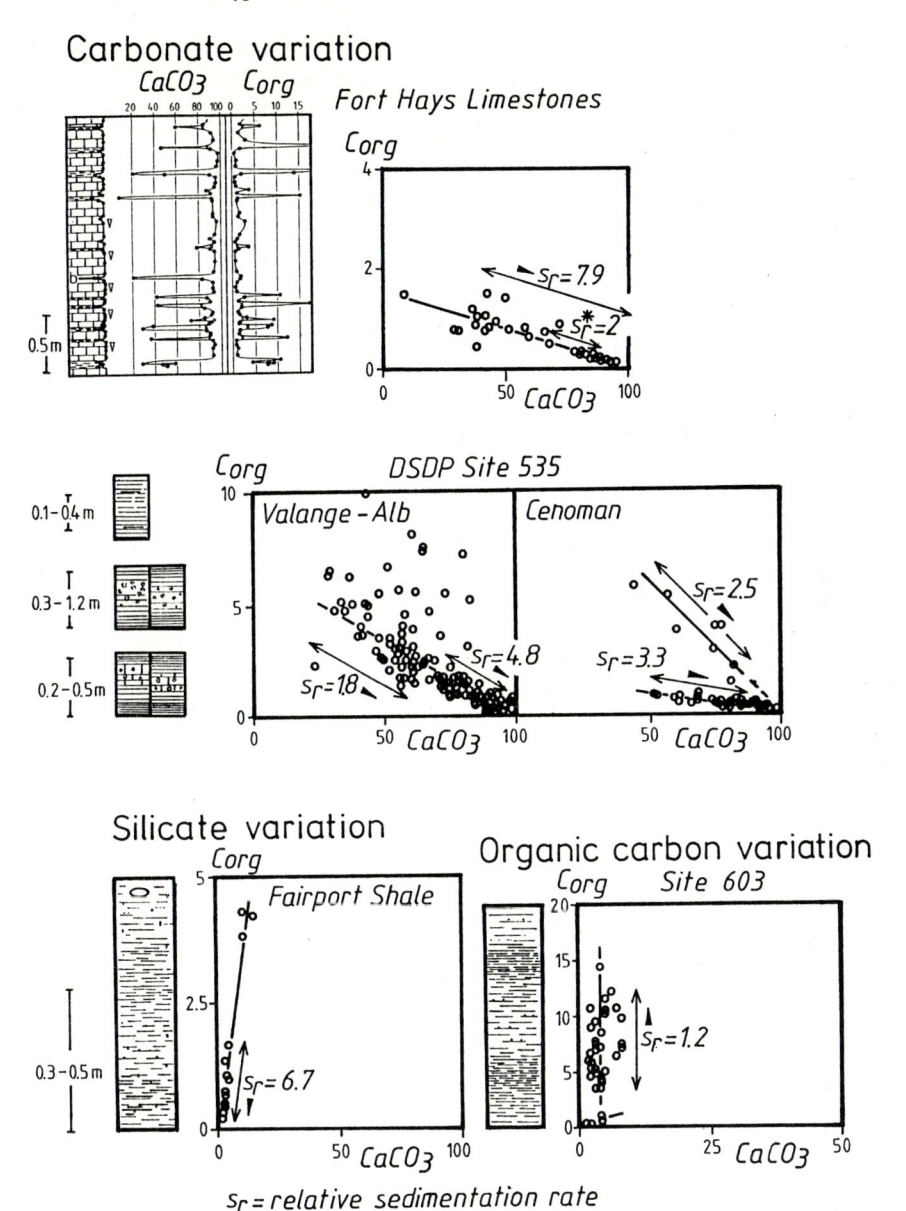

Fig. 5. Formation of bedding rhythms by the variation of one dominant parameter and the associated organic carbon-carbonate relationships (w%). *Arrows* denote cyclic variations in carbonate, clastic, and organic carbon deposition. *Numbers* show factors by which the relative sedimentation rates (s_r) change between succeeding beds; *stars* indicate sedimentation rates after correction for differential carbonate diagenesis; *arrowheads* denote direction of sedimentation rate increase

2.4.3 Rhythms Due to the Simultaneous Variation of Several Depositional Parameters

Rhythms formed by the simultaneous variation in several inputs can be recognized in C_{org}-$CaCO_3$ diagrams by the superposition of different types of organic carbon-carbonate relationships (Fig. 6). This is the case when several regression curves are obviously present, or when one composite regression curve is formed, which does not intersect the $CaCO_3$ axis at 0 or 100%, in other words, at pure silicate or carbonate variations, respectively. The composite regression lines can be treated numerically as the resultant of two (or more) C_{org}-$CaCO_3$ relationships, from which relative sedimentation rates can be derived. Multiplication of these individual rates then gives the relative variation in sedimentation rate for the resultant composite regression curves.

DSDP Site 532, water depth 1331 m, Miocene to Pleistocene, South Atlantic Ocean, Walvis Ridge (Gardner et al. 1984; Dean and Gardner 1986; Fig. 6). As suggested by the negative C_{org}-$CaCO_3$ correlation and the broad, convex carbonate maximum, variation in carbonate is the dominant type of depositional input. A detailed investigation by Gardner et al. (1984) shows that under conditions of coastal upwelling, a complicated interaction between carbonate productivity and terrigenous dilution occurred. Separation of the composite C_{org}-$CaCO_3$ relationship into individual carbonate and silicate variations shows that periods with high carbonate productivity are contemporaneous with increased input of terrigenous sediment. Obviously, periods representing higher current activity and intensive coastal upwelling, higher carbonate production, and larger terrigenous input, alternate with contrasting periods of low current activity. As a result, the carbonate bed depicted in Fig. 6 has a maximum sedimentation rate 3.4 times higher than that of the carbonate-poor bed; this proportion is reflected in the greater thickness of the carbonate-rich bed. Thereore, the thinner carbonate-poor and the thicker carbonate-rich beds probably represent a similar or identical length of time. This result was already obtained for the ideal alternations described in Section 2.2 (see Fig. 1).

ODP Site 722, water depth 2028 m, Miocene, Western Arabian Sea, Owen Ridge (Prell et al. 1989; Fig. 6). Rhythmic bedding is related to changes in monsoonal upwelling intensities. C_{org}-$CaCO_3$ data express a carbonate dilution pattern combined with questionable variation in organic carbon or silicate matter. Differences in relative sedimentation rates are largely through variation in carbonate productivity, with a sedimentation rate in the middle of the carbonate-rich bed being some 2.4 times higher relative to the carbonate-poor bed.

Bridge Creek Limestone, epicontinental chalk, Cenomanian-Turonian boundary, U.S. Western Interior (Pratt 1984; Elder 1987; Arthur et al. 1985c). This alternation consists of bioturbated limestones and laminated marls. The inversely correlated C_{org}-$CaCO_3$ data and the greater thickness of the carbonate-rich peak point to carbonate variation as the main process. After correcting for diagenetic carbonate redistribution, the limestones have a sedimentation rate higher by a factor of 1.7 compared to that of the marl beds. Superposed on this variation is a fluctuating input of organic matter, with virtually no effect on the sedimentation rates.

The previous examples are in close correspondence to the theoretical considerations made so far. The data presented in Figs. 5 and 6 evidently support the following arguments: (1) In principle, the organic carbon and carbonate relationships found in

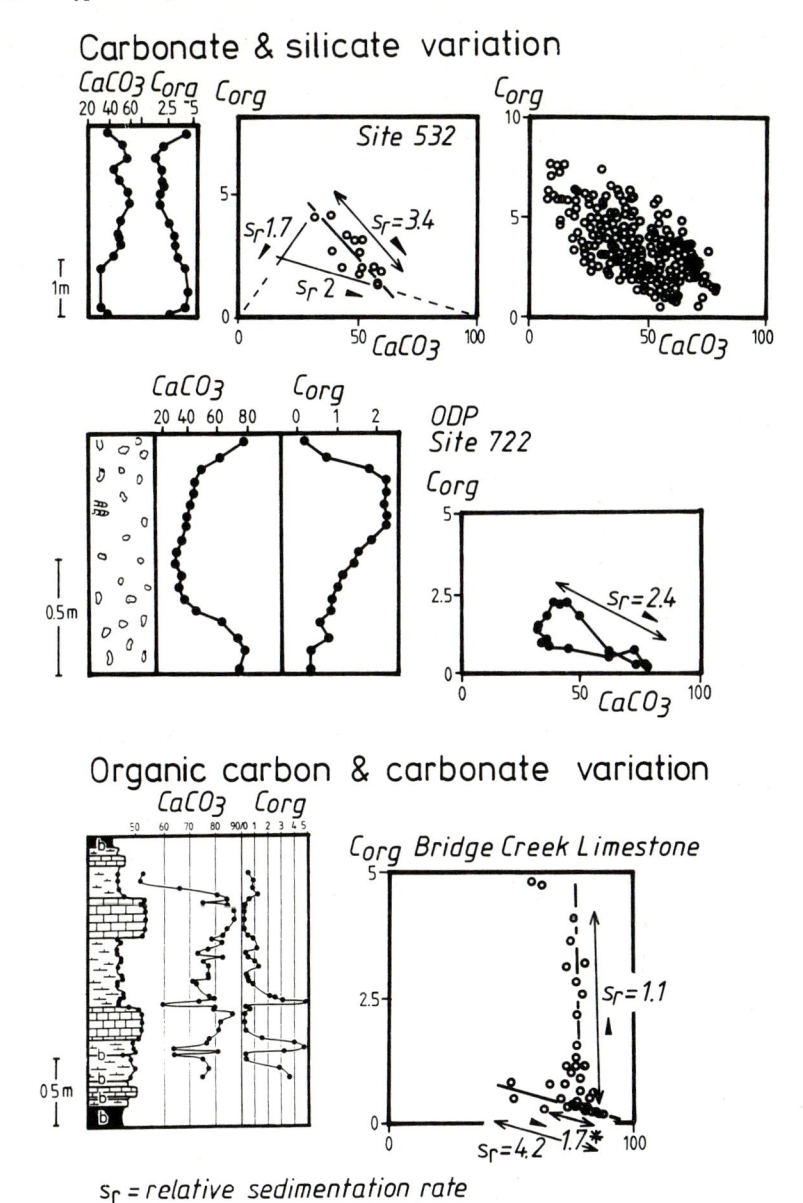

Fig. 6. Formation of bedding rhythms by the variation of several parameters and the associated carbonate-organic carbon relationships (w%). For symbols see Fig. 5

bedding rhythms from various environments show an equivalent pattern as found in the modeled dilution processes in Fig. 1. (2) In addition, the observed bedding rhythms with their specific carbonate and organic carbon curves are similar to the idealized rhythms presented in Fig. 2. (3) The consequent application of input-composition equations show that variation in sedimentation rates associated with rhythmic bedding are relatively large; ranging between factors of 1.7 and 6.7 for variation in carbonate and silicate input. Such values in sedimentation rates are also supported by theoretical considerations performed by de Boer (1980), Dean et al. (1981), Einsele (1982), and Arthur et al. (1984b); see also Arthur and Dean Chap. 1.7, Fischer Chap. 1.2, and Fig. 3 in de Boer Chap. 1.3, all this Vol.

3 Statistical Evaluation of Carbonate Contents in Rhythmically Bedded Sediment

3.1 Carbonate Differences Between Alternating Beds

Deep sea bedding rhythms were evaluated to obtain in relationship between the carbonate difference of alternating beds and related variations in sedimentation rate. Such an evaluation was carried out by using carbonate and organic carbon data described by various authors in the Initial Reports of the DSDP (Fig. 7), where the carbonate content is thought only to be slightly influenced by differential diagenesis. At first, a relationship between the general carbonate content of the environment and the carbonate difference between alternating beds is established.

In Fig. 7 the carbonate content in the middle of each of the alternating beds changes between relatively rich and poor in $CaCO_3$. This is depicted by the ends of a horizontal bar parallel to the x-axis. Such values, representing the general carbonate content of a given alternation (e.g., shale-marl or marl-chalk rhythms), are then plotted versus the absolute $CaCO_3$ span (or difference) between enclosing beds (y-axis). Alternations with small carbonate difference between beds are denoted by small horizontal bars close to the x-axis, whereas interbedded sequences with large carbonate differences have long horizontal bars located in the upper part of the diagram (see key to Fig. 7). By evaluating deep sea bedding rhythms in such a manner, it can be shown, at which overall carbonate content largest carbonate differences between beds occur (i.e., rhythms with shale, marl, or chalk sediment).

The observed carbonate difference between beds in deep sea sediment was characterized as dominantly due to variations in carbonate or silicate deposition (see symbols in Fig. 7). This was performed by using the C_{org}-$CaCO_3$ relationships of the various interbedded sequences or other data reported in the literature.

Three major results were obtained.
1. The carbonate spans between beds in Cretaceous to Quaternary deep sea alternations are found to be much larger in sequences representing environments with generally medium carbonate contents (i.e., marl and marly chalk rhythms) than in those with low or high carbonate contents (i.e., rhythms in shales and chalks, respectively; Fig. 7).

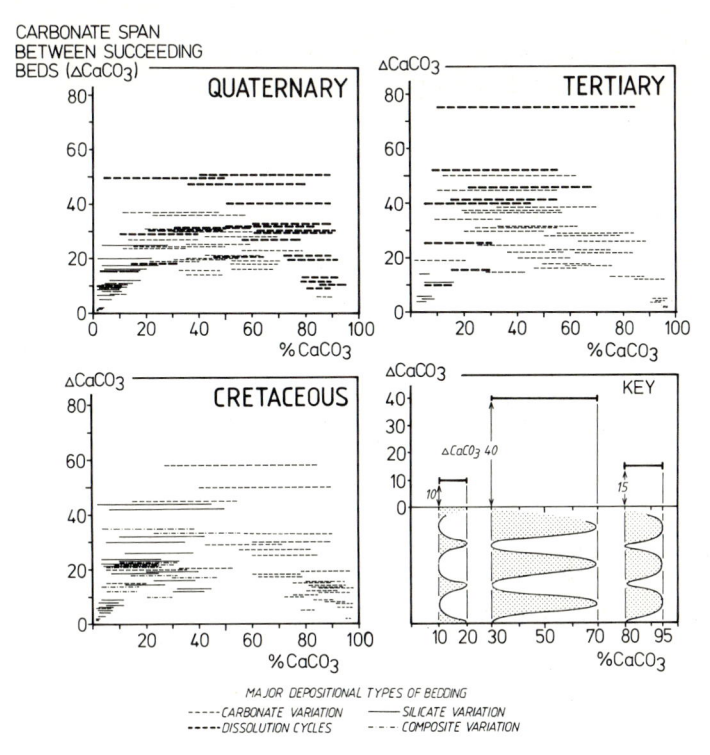

Fig. 7. Carbonate differences between succeeding beds in various deep sea alternations Cretaceous to Quaternary in age. *X*- axis carbonate contents in the middle of succeeding carbonate-rich and carbonate-poor beds, expressed by the ends of a horizontal bar. *Y*- axis absolute carbonate span between succeeding beds (CaCO3). Note that the carbonate difference is largest when cycles have medium carbonate contents. Source of data: Quaternary, DSDP Sites 418, 474, 503, 530, 532, and 593, Berger (1973) and Hays et al. (1969); Tertiary, DSDP Sites 366, 329, 474, 530, 532, 540, and 593; Cretaceous, DSDP Sites 398, 461, 463, 511, 540, and 535

2. Rhythms related to dominant variation in silicate deposition occur more frequently with low carbonate contents (from 0 to 30%), while carbonate deposition is dominantly observed in alternations with a medium to high carbonate content (30 to 100%).
3. Alternations described in the literature as dissolution cycles have larger carbonate variations between succeeding beds than alternations obtained above the lysocline.

To explain the observed carbonate differences between succeeding beds, an equivalent CaCO3 pattern is generated and plotted on the same type of diagram as presented in Fig. 7. Such theoretical carbonate spans are determined by assuming that either the silicate or carbonate flux varies (between beds) by a constant factor (F) for rhythms with various carbonate contents. An example is given in Fig. 8a.

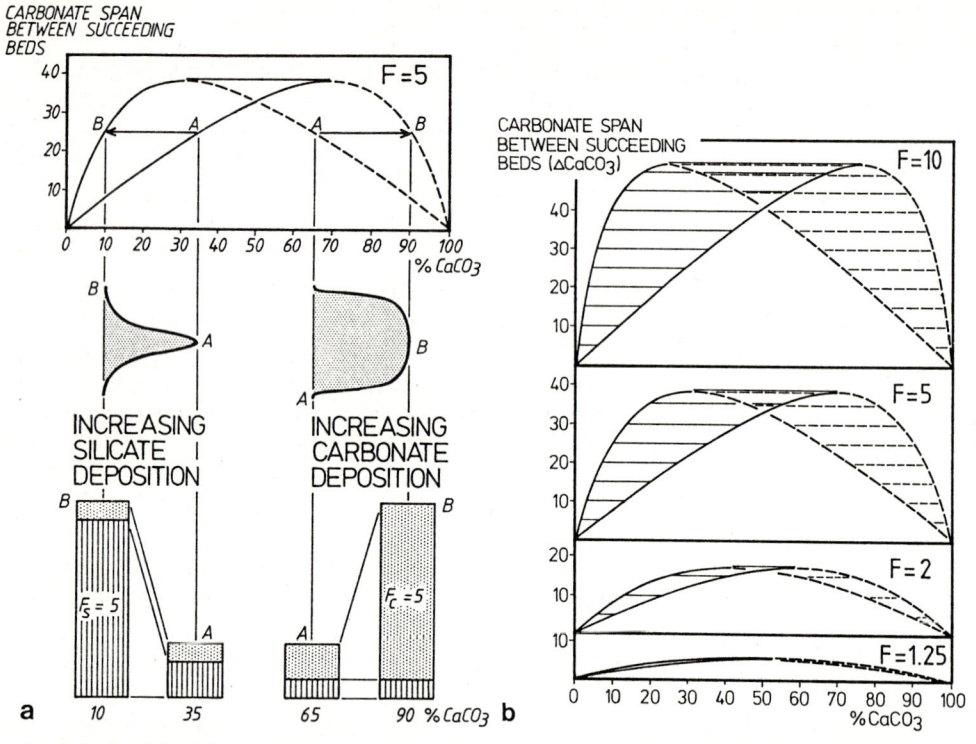

Fig. 8a,b. Model for carbonate differences between succeeding beds. **a** Carbonate differences are generated by varying the silicate or carbonate deposition by a constant factor (F_s, F_c) for alternations at various carbonate contents. **b** Consequent pattern of theoretical carbonate differences between succeeding beds, variation in silicate or carbonate deposition by factors (F_s or F_c) from 1.25 to 10. Carbonate differences are large for alternations with medium carbonate content, whereas low for alternations with low or high carbonate content. Note close correspondence to $CaCO_3$ patterns as depicted in Fig. 7

The consequent pattern in theoretical $CaCO_3$ spans between succeeding beds is represented in Fig. 8b. The carbonate spans generated with equally varying factors of silicate or carbonate deposition are obviously similar to the observed deep sea $CaCO_3$ spans shown in Fig. 7. They are small at both low and high carbonate contents and large at medium carbonate contents. Comparison of the observed and calculated $CaCO_3$ differences demonstrates that carbonate or silicate deposition in interbedded sequences above the lysocline varies by average factors of 3 to 6, and by a factor of approximately 10 in rhythms deposited between the lysocline and the CCD. In the next section, a more comprehensive explanation of the observed pattern in $CaCO_3$ variation is discussed involving changes in sedimentation rates.

3.2 Variation in Sedimentation Rates of Deep Sea Alternations

Considering the fact that the magnitude of carbonate spans between succeeding beds is largest at medium carbonate contents, a pattern in sedimentation rate changes can obviously be determined. One has to keep in mind, however, that sedimentation rates are different for silicate and carbonate input. Let us first consider how variations in carbonate input affect sedimentation rates, for example, assuming that the amount of carbonate deposition between alternating beds varies by a factor of $F_c = 5$ for various alternations with low to high carbonate contents (Fig. 9).

At low carbonate contents the carbonate fraction is so small, that after multiplication by a factor of 5 (to give the carbonate-rich bed), both $CaCO_3$ content and the newly formed sample volume, do not increase significantly. As a result, the change in relative sedimentation rate (s_r) between carbonate-rich and carbonate-poor beds is small. At medium carbonate content, an already large carbonate fraction is increased, resulting in a considerable shift in $CaCO_3$ content. Augmentation of sedimentation rate is moderate, namely by a factor of approximately 2.2. When the carbonate content is very high and comprises most of the sediment volume, multiplication of the carbonate fraction by a factor of 5 results in a large sediment volume, and therefore, in a significant increase of the sedimentation rate (Fig. 9). However, this large increase in carbonate volume, expressed as the percentage of $CaCO_3$, can only have a small effect on shifting the carbonate percentage to a higher value.

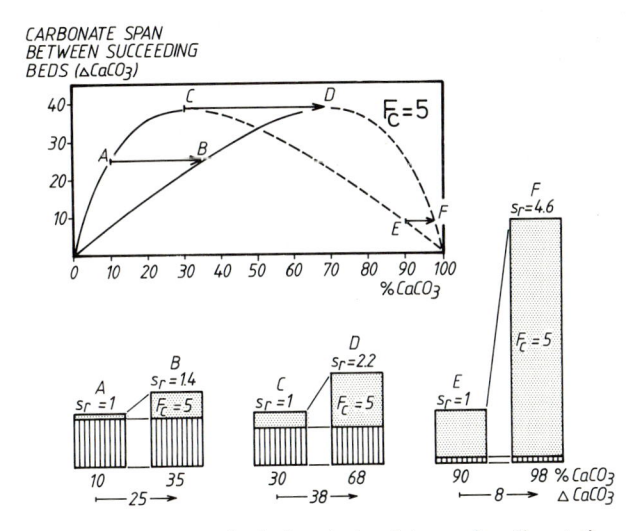

Fig. 9. Contradiction between the carbonate span of succeeding beds and related changes in sedimentation rates. Example shows three marl bed sediments with low, medium, and high carbonate content (*A, C,* and *E*); each of the carbonate fractions is augmented by a factor of $F_c = 5$ to produce the carbonate-rich beds (*B, D,* and *F*). Maximum increase in relative sedimentation rates (or sediment volumes) occurs in bedding rhythms with generally high carbonate content, whereas maximum carbonate difference between beds ($CaCO_3$) exists in rhythms with medium carbonate content. *Upper graph* is equivalent to graphs depicted in Figs. 7 and 8

This example demonstrates a contradiction between carbonate differences (spans) of alternating beds and the related variations in sedimentation rates. Carbonate differences are large in combination with medium carbonate contents, whereas associated variations in sedimentation rates are large with high carbonate contents. The same pattern of carbonate spans between succeeding beds is observed for silicate deposition. However, variation in sedimentation rates shows the opposite values, with small variations at high, and large variations at low carbonate contents.

As a result, carbonate differences between succeeding beds are associated with sedimentation rates in two very different fashions, depending on whether carbonate or silicate deposition is varying. As mentioned earlier, the observed pre-diagenetic $CaCO_3$ differences in deep sea sediment show that, statistically, most alternations below a carbonate content of ca. 30% are dominated by a depositional variation in silicate input, while above this value, alternations are dominated by variation in carbonate input. This means that most alternations with small $CaCO_3$ differences oscillating either at low or high carbonate contents can be related to silicate or carbonate variations, respectively. Both types of alternations are associated with large variations in sedimentation rates, with maximum factors between 3 and 6 for alternations above the lysocline. On the other hand, alternations with medium carbonate content are related to smaller variations in sedimentation rates, with factors between 1.5 and 2.5 (Fig. 10). Consequently, environmental changes must have the largest effect on bedding variations where medium carbonate contents occur. Here, Milankovitch-type, orbital-climatic variations have the potential to be prominently expressed in the bedding rhythms, likewise disturbing variations related to the noncyclic background deposition. On the other hand, depositional variations in environments with low and high carbonate contents will have a much smaller

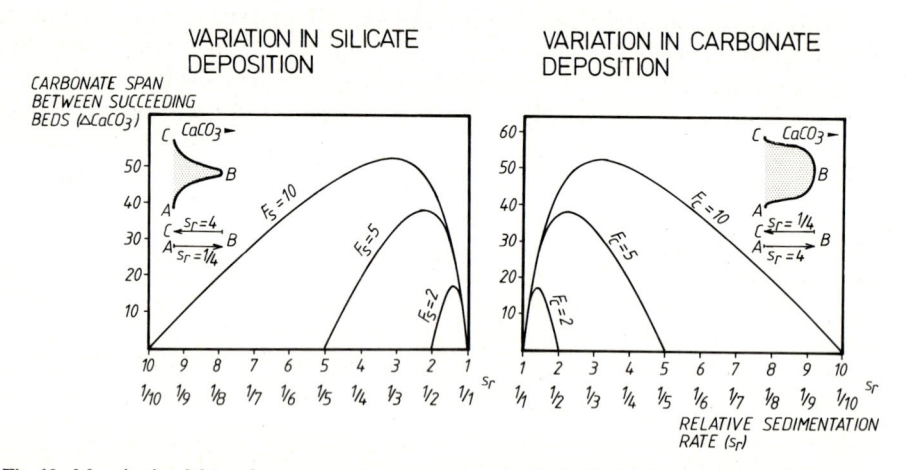

Fig. 10. Magnitude of the carbonate span between succeeding beds ($CaCO_3$) and associated variations in relative sedimentation rates (s_r), depicted for variation in silicate and carbonate deposition by factors (F_s, F_c) from 2 to 10. Note that maximum carbonate differences between beds occur at moderate changes in sedimentation rates. *Inset of each diagram schematically shows changes in relative sedimentation rate between a layer rich in carbonate content and enclosing carbonate-poor beds (arrows)*

expression in bedding variations, although they are associated with higher variation in sedimentation rates.

4 Discussion and Conclusions

This chapter is an attempt to qualitatively treat sediment input variations associated with rhythmic bedding. Based on carbonate and organic carbon data, this study allows both discrimination between different types of depositional variations and derivation of sedimentation rate changes associated with rhythmic bedding.

1. Carbonate-organic carbon relationships and the determination of relative sedimentation rates. Uncemented alternations that are formed by variations in carbonate, silicate, and organic carbon inputs can be related to distinctive C_{org}-$CaCO_3$ correlation patterns reflecting various types of depositional dilution processes. Ideally, variation in (1) carbonate deposition produces a correlation line with a negative slope, intersecting the carbonate axis at 100% $CaCO_3$ (see Fig. 1a); while variation in (2) silicate deposition has a positive correlation line between carbonate and organic carbon, intersecting the carbonate axis at 0% $CaCO_3$ (see Fig. 1b). Variation in (3) organic matter is indicated by organic carbon contents fluctuating largely independent of the carbonate content, resulting in a correlation line which is essentially parallel to the organic carbon axis (see Fig. 1c). Based on such a determination of the depositional type, relative sedimentation rates in bedding rhythms can be derived by using input-composition equations. Thus, carbonate and organic carbon data provide a promising tool for distinguishing different types of depositional inputs in rhythmic bedding and quantifying the magnitude of associated oscillations in sedimentation rates.

2. Distinction of various types of bedding rhythms. Regular variations in deposition do not result in equally regular variations in sedimentary rhythms. Instead, associated variations in sedimentation rates influence the bedding rhythm related to carbonate, silicate, or organic matter deposition in a contrasting and distinctive manner (see Fig. 2): (1) Variation in carbonate input is expressed by alternations with broad carbonate maxima, small minima, and inversely correlated C_{org}-$CaCO_3$ curves. Bundles formed by variation in carbonate deposition have carbonate-rich layers, with similar carbonate contents which are separated from other bundles by small, carbonate-poor zones. (2) Variation in silicate deposition is expressed by alternations with sharp, small carbonate maxima, wide minima, and parallel curves of organic carbon and carbonate contents. Bundles are characterized by sharp, small carbonate peaks separated by relatively thick shale zones. (3) Variation in deposition of organic matter is denoted by alternations with C_{org} variations essentially unrelated to carbonate content.

3. Simple and composite rhythms. Alternations with one varying parameter can be distinguished from those with a simultaneous variation of several parameters by using single and composite C_{org}-$CaCO_3$ correlation patterns, respectively. In marine environments, nearly all types of depositional variations and combinations were found (see Figs. 4 to 6). The most common type of alternations encountered in oceanic and epicontinental

Cretaceous to Quaternary sediments show either pure variations in carbonate input, or composite variations, in which carbonate input is combined with contemporaneous organic or silicate fluctuations. For these variations, the carbonate-rich beds represent higher sedimentation rates than the carbonate-poor beds. Rhythms with generally low carbonate contents may express variation in silicate input or combinations with other simultaneously varying fractions. They are found on deep clastic shelves and in environments deep below the lysocline, with the carbonate-poor layer reflecting the bed with the higher sedimentation rate. Rhythms related to pure redox cycles are seldom encountered, but combinations with other types of input are very common.

4. Deep sea carbonate variations and associated variation in sedimentation rates. Statistically, alternating deep sea sediments of Cretaceous to Quaternary age show the largest carbonate differences between succeeding beds with overall medium carbonate content, whereas the smallest carbonate differences are encountered in alternations with either low or high carbonate contents. This conspicuous pattern is explained by the assumption that the amount of carbonate or silicate deposition fluctuates in various oceanic bedding rhythms by a roughly constant factor. For alternations above the lysocline, carbonate or silicate deposition varies by factors between 3 and 6, while higher factors, around 10, are obtained for alternations situated between the lysocline and the CCD. Such a systematic pattern in carbonate differences between succeeding beds is predictively related to variations in sedimentation rates. In bedding rhythms with low and high carbonate contents, carbonate differences between beds are relatively small, but rhythms are thought to be associated with considerable variations in sedimentation rates. In bedding rhythms with medium carbonate contents, carbonate differences between beds are relatively large, and variations in sedimentation rates range between factors of only 1.5 and 2.5 for both types of inputs.

The major outcome of this investigation is that variations in sedimentation rates associated with rhythmic bedding seem to be much larger than previously thought. Moreover, variations in sedimentation rates seem to be the controlling factor for bedding variations and the formation of the various types of bedding rhythms (expressed in their distinctive $CaCO_3$ and C_{org} curve forms). In addition, rhythmic bedding in the oceanic environment is related to a consistent pattern of carbonate differences between beds. Such results have some serious ramifications regarding the origin and interpretation of cyclic sedimentation.

1. Contradiction of sediment thickness and time span. The interpretation that large variations in sedimentation rates are the rule rather than the exception in cyclic sedimentation shows a general contradiction between bed thickness and the associated amount of time. In interbedded sequences, thin and thick beds commonly represent different sedimentation rates. This problem becomes increasingly obvious in alternations with higher variations in sedimentation rates having a generally low or high carbonate content. The contradiction between sediment thickness and associated time span also influences time series analyses of the bedding rhythm frequencies. The application of time series analyses are based on the assumption that equal section thicknesses also represent equal time intervals. As regular variations in deposition do not result in equally regular bedding rhythms, the time scale becomes considerably distorted (when expressed on a sediment thickness scale). Such effects limit the value of time series analyses, because unrecognized systematic variations in sedimentation rates cause harmonics and other unwanted peaks in the spectras of wave length determination (see Weedon Chap. 7.4, this Vol.).

2. Obliteration versus expression of bedding rhythms. Bedding rhythms with low or high carbonate contents are interpreted as being associated with the largest variations in sedimentation rates. In spite of this, rhythms are usually not detectable at these carbonate contents, as they only give rise to minor shifts in composition, unless redox and color cycles are developed. For example, in some European chalks with approximately 98% $CaCO_3$, bedding cyclicity is not visible although sedimentation rates must have varied considerably in order to generate small variations in the carbonate content. On the other hand, rhythms in environments with overall medium carbonate contents express striking carbonate differences between beds although associated variation in sedimentation rates are much smaller. Consequently, environments with generally medium carbonate contents are certainly sensitive to Milankovitch-like, orbital variations, but also to noncyclic changes of the background deposition.

3. Sedimentary expression of Milankovitch-type oscillations. The most common type of depositional variation observed in oceanic environments above the lysocline and in epeiric seas is denoted by considerable variation in carbonate deposition; thus either high variation in carbonate productivity has to be assumed, or carbonate dissolution is more common than usually supposed (compare Eicher and Diner Chap. 1.4 with Diester-Haass Chap. 1.5, both this Vol.). In addition, carbonate variation by a roughly constant factor may be an argument for carbonate variations being influenced by overall climatic processes, for example orbital, Milankovitch-like variations. If this is true, the weak orbital-climatic changes must be considerably augmented in order to generate the observed large variations in sedimentation rates. As discussed by Einsele and Ricken (Chap. 1.1, this Vol.), this augmentation seems to be related to various feedback mechanisms influencing the global carbon-carbonate system.

Acknowledgments. The author would like to thank G. Einsele, L. Hobert, and S. Borchert for critically reading and improving an earlier version of this manuscript. For discussions on the meaning of weight percent organic carbon, I would like to acknowledge my colleagues F. Niessen, L.M. Pratt, P. Sandberg, R. Stein, and E. Suess. Organic carbon determination of samples from the U.S. Western Interior was performed with the Geological Survey, Denver, Oil and Gas Branch.

Chapter 1 Rhythmic Stratification

1.9 Sedimentary Rhythms in Lake Deposits
C. R. Glenn and K. Kelts

1 Introduction

Rhythmically laminated sequences provide a fascinating stimulation for the human psyche. In addition, laminated sequences hold vital information on recurrence rates of events and the rate of change of natural environments, as well as keys to understanding the importance of forcing functions on our earth system. Lake sediments provide one way to trace a high resolution history of biotic/abiotic interactions, and these interactions are themselves dependent on changing climatic and hydrologic factors (Fig. 1). Relative to marine systems, lakes respond very rapidly to environmental change and a few centimeters of core or outcrop lacustrine sediment may record changes from dilute deep water settings to the development of shallow hypersaline brines. In addition, because of their relatively small size, geochemical signals of such environmental perturbations are greatly amplified in lake basins. Because lakes are often stratified, relatively shallow, have high sedimentation rates, and characteristic annual productivity/eutrophication cycles, lake sediments may provide a chronology of short, long, and rhythmic environmental and climatic change.

The origin of rhythmic laminations is a problem which often confronts the working geologist. Much thought is influenced by de Geer's (1912) concept of "varves" which describes the yearly variation of color and texture in sediments due to the seasonal melting of retreating glaciers in Sweden. Thus, the term varve has become associated with "annual". However, many instances of rhythmic couplets in lakes are due to compositional differences and the yearly designation is often by analogy alone. Laminated sequences are a common feature of lake deposits from a variety of environments because lake systems commonly fulfill the simple requirements for laminations, which are:

1. variable input source mechanisms and
2. a bottom environment with preservation potential.

Many lakes experience a partial or complete anoxia near the sediment/water interface and thus limit bioturbation. In other lakes high sedimentation rates swamp burrowers, and in many lakes with elevated salinities there are few benthic dwellers to churn sediments.

Lake basin sediments are poorly known in the geologic record and studies of modern lake sequences are limited. Few such sequences have yet been analyzed

Einsele et al. (Eds.)
Cycles and Events in Stratigraphy
©Springer-Verlag Berlin Heidelberg 1991

rigorously in terms of their rhythmic signals. Thus, in addition to a brief review of seasonal rhythms and variable input source mechanisms, we focus in this paper on how forcing mechanisms may produce patterns beyond the dominant annual/seasonal variation or the short-term event. It is not the description of varve couplets that interests us here (see O'Sullivan 1983, and Anderson and Dean 1988), but rather the progressions within varve bundles, or the alternations between varved and homogeneous sequences. The aim of this chapter is to stimulate research by reviewing how diverse lake sediment sequences might record cycles or event recurrence over time scales from 10^0 to 10^6 years (a). In contrast to the earlier conclusions of Duff et al. (1967, p. 250), recent literature seems to show abundant evidence of a link between rhythmic patterns in lake deposits and the direct influences of climate. A central question remains: how are signals of climate forcing transferred through various processes into lake sediments?

There are some clear pitfalls in investigations of lacustrine rhythmic sequences. Abrupt sedimentation rate changes of orders of magnitude are common, although directly linked to environmental processes. These may confuse spectral analyses. When matching a population of only a few dozen samples, statistical tests of spectral peak heights cannot be used to discriminate between episodic and noisy periodic models (Lutz 1987). This is especially true if the signal derives from a mixing between a rather slow uniform sediment input (e.g., lacustrine pelagic processes) and abrupt events (multiple episodic flooding on annual or super annual scales). Many simple A:B sets of laminations or beds may simply not have a narrowly definable frequency.

2 Lakes and Their Sediments

Lakes occur in many continental settings. Requirements are a basin and a hydrological balance where inflow exceeds outflow plus evaporation. Classifications are also many, depending on whether geological, chemical, or biological criteria are applied (cf. Hutschinson 1957; Lerman 1978; Dean 1981; Eugster and Kelts 1983 or Kelts 1988). The modern landscape is dominated by lakes related to the Quaternary glacial patterns. The geologic record shows a prevalence of lake deposits from continental sags, tectonic and paralic basins and event basins (crater, dam, glacial). The controls on lake sedimentation are very dependent on the composite type of system (Fig. 1). There is an interplay of autochthonous and allochthonous components. Sediments may be dominated by land-derived clastics and organics, or by authigenic biotic/abiotic processes. The chemistry of the waters is determined by the weathering characteristics of the surrounding drainage basin geology. The residence time of water within the basin determines its geochemical evolution, and, in closed basins, concentrations may range over five orders of magnitude in short time frames concomitant with drastic changes in physical limnology and biota. Lakes may be as small as a pond or over 80000 km^2 as for the Caspian Sea. Depths range from a few centimeters to 1900 m for Lake Baikal. Littoral zones are minor in deep rift lakes and craters, or range to hundreds of kilometers as in the case of saucers such as the Aral sea and Lake Victoria.

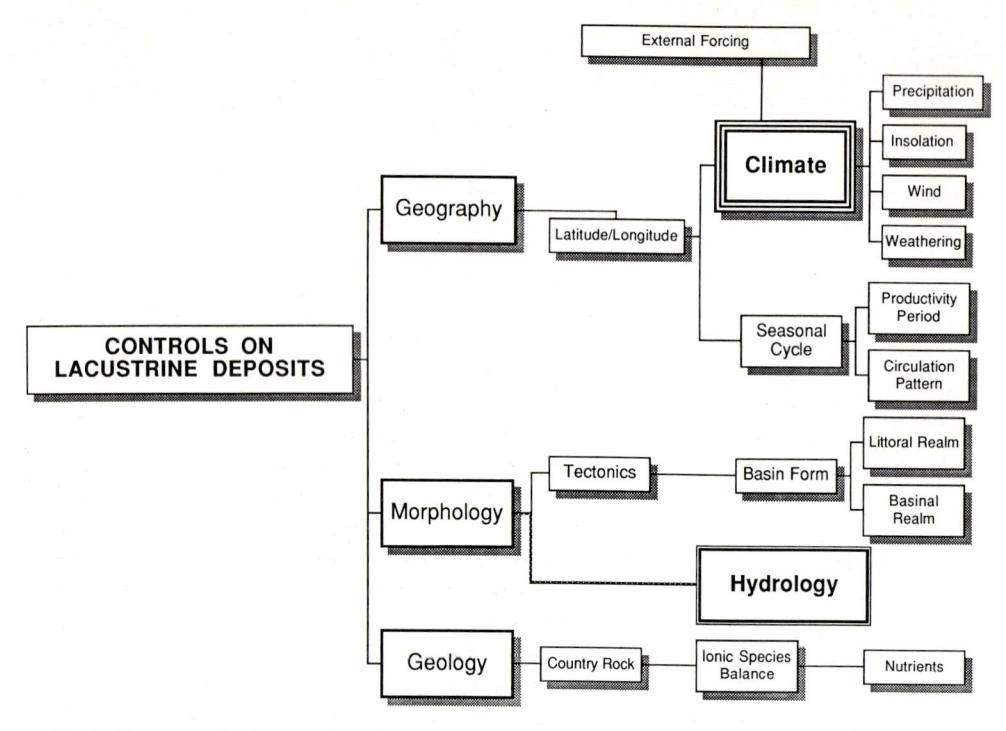

Fig. 1. Summary of main controls on lacustrine sedimentation

Viewed as water bodies, lakes have some features in common and may be variously classified and subdivided accordingly (Fig. 2). Lakes may be well mixed (*poly-* or *oligo-mictic*) to permanently stratified (*meromictic*). They may be nutrient and biota poor (*oligotrophic*) to nutrient and algal rich (*eutrophic*). Anoxic bottom waters may result from thermal or chemical stratification, or from high rates of organic matter input. Seasonal variations in temperature, density, viscosity, visibility, pH, and oxygen profiles show the relative intensity of biological activity, stratification, and mixing. A *thermocline,* defined by the zone of highest thermal gradient, commonly separates a wind-mixed surface water layer, the *epilimnion,* from a more sluggish water mass, the *hypolimnion*. Two principle environments of deposition can be distinguished in open fresh water lakes with a stable shoreline:

1. littoral (near shore, or shallow photic zone) and
2. profundal (pelagic, deep, basinal, redeposits).

In closed basins, shorelines are not as stable, and the littoral zones may be complex deposits controlled by annual flooding, microbial mats, and receding of waters with rapid salinity changes.

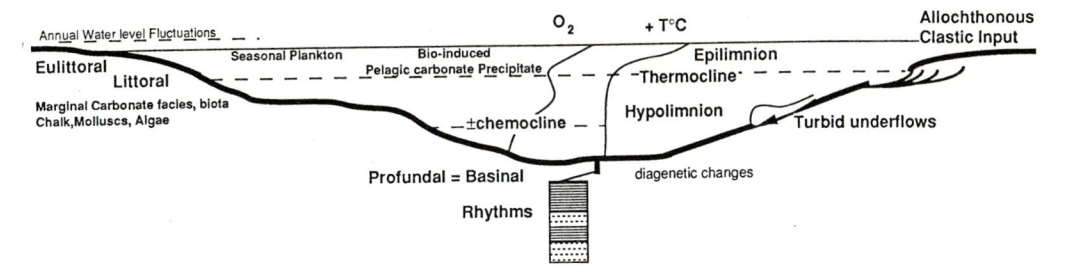

Fig. 2. General elements of lake systems (cf. Hutschinson 1957; Kelts 1988)

3 Seasonal Rhythms in Lakes: Laminations and Varves

The mention of lakes and rhythms immediately conjures up images of varved glacial clays for geologists. Yet sedimentation in most lakes responds extremely sensitively to regular or episodic changes on a regional scale whether these be due to seasonality, wind stress, El Niño, solar variation, or Milankovitch forcing. Annual rhythms will be the dominant cyclic signature in most lake sequences for the simple reason that seasonal forcing may be up to an order of magnitude greater than other forcing mechanisms (Mitchell 1976; cf. Anderson 1986). Thus, because preservation potential is high in many lakes and because tidal and other current influences in lakes are relatively small, annual rhythms often become the dominant cyclic signature in most lake sequences.

3.1 Variable Input Source Mechanisms

Rhythmic couplets may be caused by any mechanism or combination of mechanisms that produces a variable sediment input signal. The mechanisms may include allochthonous variations in the input of clastic suspended loads, a dominant mechanism in glacial lakes, or air-borne dust and fire or volcanic ash. Autochthonous (lake-internal) processes include variations in wind-driven overturn and circulation and the sequential bloom of plankton species. Chemical processes may be biologically induced, such as the precipitation of various carbonate minerals, and there may occur a physical-chemical precipitation controlled by variations in evaporative concentration. Many of these mechanisms follow an annual pattern, but others may reflect periodic, episodic, or even chaotic recurrence of events on widely varying scales. Table 1 presents a summary of some of these aspects from the many lakes on which we have based this work, and Fig. 1 summarizes some of the major mechanisms that interplay in lacustrine systems.

Nonglacial signals may derive from variable biological productivity and biologically induced chemical precipitation. In Cenozoic deposits, the most common form of this are the well-known calcite/organic matter-diatom varve couplets described from many temperate lakes (Fig 3a; also see Kelts and Hsü 1978). In many lakes, even homothermic tropical basins, the organic productivity follows definite seasonal

Table 1. Summary of lacustrine rhythmites, arranged by geologic stage

g = glacial, ng = nonglacial, dg = distal glacial, pg = proglacial, cl = current lamination, t = turbidite, na = nonannual, gr = graded, ngr = nongraded, hw = hard water lake, sl = soda lake, sal = saline lake, skh = karst sinkhole, dim = dimictic (two overturns/a), mon = monomictic (one overturn/a), mer = meromictic (permanently stratified); hol = holomictic (unspecified overturns/a).

Formation/Lake, locality	Age	Origin	Laminae Light	Laminae Dark	Couplet thickness (mm) Range	Couplet thickness (mm) Avg.	Suggested Periodicity (years)	Ref.
Lake Skilak, Alaska	Recent	pg, t dg dg	Silt	Clay	10.0–20.0 1.2–9.6	40	1a, 11a, 22a	(1), (48)
Malaspina Lake, S. Alaska	Recent	pg,gr	Silt	Clay		40	1a	(41)
Hector Lake, Alberta	Recent	1. Proximal pg 2. Intermed. prox. 3. Intermed. dist. 4. Distal pg	Silt Silt	Clay Gr. silt-clay-silt varve triplets Clay Laminated to massive clays	5.0–10.0 2.0–4.0 1.0		— 1a 1a —	(3)
McCay Lake, Canada	Recent	ng	Carbonate	Organic	0.13–0.43	0.028		(4)
Crawford Lake, Southern Ontario	Recent	ng, skh, mer	Calcite	Organic			1a	(5)
Lake Judesjon, Sweden	Recent	ng	Mineral grains	Organic	ca. 0.15–70		1a, 30–40a	(6)
Elk Lake, Recent Minnesota		ng, hw, dim	1. Calcite 2. Clastic 3. Calcite	Org. clay, diatom Organic Diatoms, FeO, MnO		0.8	1a 1a<200a 1a	(46)
Deming Lake, Minnesota							33–36a, 80–90a	(47)
Lake of the Clouds, Minnesota	Recent		Siderite	Organic		0.6		(7)
Cayuga Lake, New York	Recent	ng, t, mon	Org.-poor (light?) (varves laminated)	Org.-rich (dark?)	10–30		1a	(8)

Formation/Lake, locality	Age	Origin	Laminae Light	Laminae Dark	Couplet thickness (mm) Range	Couplet thickness (mm) Avg.	Suggested Periodicity (years)	Ref.
Fayetteville Green Lake, New York	Recent	ng, mer	Calcite + org.	Org. + calcite	<1 – a few		1a	(8)
Third Sister Lake, Michigan	Recent	ng, dim	Clay	Organic			1a	(8)
Stockbridge Bowl, Massachusetts	Recent	ng, dim	Diatom + org.	Org. + diatom			1a	(8)
Great Salt Lake, Utah	Recent	ng, cl, na, sal = ng, na, sal =	1. Arag. pellets 2. Calcitic mud	Arag. mud red algae			2500a	(51)
Lake Urmia	Recent	ng, cl, sal	Arag. pellet	Carb. mud	0.4–1.0	0.5		(18)
Lake Van, Turkey	Recent	ng, sl, mon	Aragonite	Organic, clastic	ca. 0.5–1.8		1a, 10a	(10)
Dead Sea		sal	Aragonite	Clay-size Lmstn	mm's to cm's			(11)
Lake Malawi, Central Africa	Recent	ng, mer	Diatom + clay	Organic + clay	mm's to cm's		1a	(12)
Lake Idi Amin Dada (a.k.a. L. Edward)	Recent	ng	Carbonate laminae					(14)
Lake Kivu, East Africa	Recent	ng, mer	Carbonate laminae					(14)
Lake Turkana (a.k.a. L. Rudolf), Kenya	Recent	ng	Clastic, carb.	Clastic, carb.	0.1-a few		4a 25, 31, 44, 78, 100a, 165a, 200a, 270a	(15), (16), (17)
Lake Tanganyika	Recent	ng, mer	Diatoms	Clay				(14)
Saski Lake, Crimea	Recent	ng	Silt, gypsum	Organic		0.13	50a	(4), (9)
Lake Saki, Crimea	Recent	ng					30a, 70a	(20)
Zurich Lake, Switzerland	Recent	ng, hw	Calcite	Diatom, organic	1.0–5.0	3		(22)
Lake Greifen, Switzerland	Recent	ng, hw	Calcite	Organic	3.0–10.0	8	1a	(49)
Lake Brienz, Switzerland	Recent	ng, hol	Fine silt <10% carb.	Coarse silt 8–36% carb.	<10.0		1a	(50)

Table 1. (*Continued*)

Formation/Lake, locality	Age	Origin	Laminae Light	Laminae Dark	Couplet thickness (mm) Range	Couplet thickness (mm) Avg.	Suggested Periodicity (years)	Ref.
Lake Lugano, Switzerland, Italy	Recent	ng, mon	Calcite	Diatom + organic		0.55	1a	(40)
Lake Soppen, Switzerland	Recent	ng, hol	Calcite	Diatom + organic			1a	(53)
Meerfelder Maar, Germany	Recent	ng	Clay	Diatom/organic	0.5–1.8	0.8		(19)
Russian Lakes (6)	Recent Recent Recent Recent						7.5–9.0 12.0–16.0 21.0–24.0	(20)
Gosciaz Lake, Poland	Recent	ng, mer	Calcite	Organic + clay	0.6		11, 22, 35 200	(44)
Lake Barlow-Ojibway, N. Ontario	U. Pleistocene	pg	Silt	Clay	1.0–400.0		11, (14), 22 200	(2)
Zurich Lake, Switzerland	U. Pleistocene	dg	Coarse silt	Fine Silt	1.0–3.0	1.5	1a	(22)
Lake Lugano, Switzerland, Italy	U. Pleistocene U. Pleistocene	pg pg	Medium silt Silt	Fine silt Clay	24–110 0.2–1.5	34	1a 1a	(23) (40)
Priest Lake, Idaho, Wash.	U. Pleistocene (Wisconsin)	pg	Sandy silt	Silty clay	10.0–40.0		1a, 20–50a[a]	(25), (26)
Lake Hitchcock, Mass.-Connecticut	U. Pleistocene	1. pg, ngr 2. pg, gr 3. pg, gr/ngr	Silt Silt Silt	Clay + org. Clay Clay	2.5–12.5 12.5–760.0	10	1a 1a 1a	(13)
Rita Blanca Lake, Texas	L. Pleistocene	ng, hw, mer	Thick; carbonate, clay	Thin; clay, organic	0.12–3.40	0.93	1a, 10–12?, 22a, 200a	(21), (27) (28), (52)
Lisan Fm., Dead Sea	Pleistocene	ng	Aragonite	Detrit. cal, qtz, dol and clay			1a	(30)
Lake Superior	Pleistocene	g					1a, 6–14a, 20a, 80–90a	(28)

[a] (rhythms produced by Lake Missoula outbursts)

Formation/Lake, locality	Age	Origin	Laminae Light	Laminae Dark	Couplet thickness (mm) Range	Couplet thickness (mm) Avg.	Suggested Periodicity (years)	Ref.
Steep Rock Lake, Ontario	Pleistocene	g					1a, 22a 80–90a	(28)
Lake Biwa, Japan	Pleistocene	ng		none			20ka, 40ka	(31)
Berkshire Hills, Massachusetts	Pleistocene	g					55–77a	(32)
Kashmir Basin, Romushi Section NW Himalayas	Plio-Pleist.	Fluvial-lacustrine					21ka, 25ka, 35ka, 94ka (177–400ka)	(33)
Rubielos de Mora Basin, Spain	Miocene	ng, mer	Calcite	Clay	0.01–1.0			(39)
Florissant Fm., Colo.	Oligocene	ng	1. Diatomite 2. Tuff, diatomite 3. Pumice, diat.	Sapropel Sapropel Sapropel		1 8 16		(4)
Green River Fm., Wyo. and Colorado	Eocene	ng	1. Silt, carb. 2. Carb. 3. Carb. 4. Carb.	Organic Org. (low) Org. (med.) Org. (high)	0.6–9.8 0.14–0.37 0.03–0.114 0.014–0.153	1.16 0.167 0.065 0.037	11a, 50a, 21.6 ka	(34)
Lake Tubutulik, Alaska	Eocene	ng	Siderite	Siderite, clay, and organic	1.0–15.0		1a?	(36)
Newark Canyon Fm., Nevada	Cretaceous	ng	Calcite	Organic	0.002–0.3+	0.2	10–13a	(4)
Todilto Fm., New Mexico	Jurassic	ng ng ng ng ng	1. Clastic 2. Clastic, carb. 3. Clastic, carb, gyp. 4. Clastic gyp. 5. Gyp.	Organic Organic Organic Bitum. Is. Shale	0–6	0.13 6	1a, 10–13a 60a, 85a, 170a, 180a	(29)
Lokatong Fm., NJ, Pennsyl.	Triassic	ng	Dol	Organic	0.5–2		1a, 25 ka, 44ka, 100ka, 133ka, 400ka 2,000ka	(37), (38)

Table 1. (*Continued*)

Formation/Lake, locality	Age	Origin	Laminae Light	Laminae Dark	Couplet thickness (mm) Range	Couplet thickness (mm) Avg.	Suggested Periodicity (years)	Ref.
Lower Beaufort Fm. Milawi	U. Permian	ng	Calcite	Clastic	0.1–0.3		1a	(42)
Dwyka Fm.	L. Permian	g	Silt	Clay	2.0–10.0			(43)
Wellington Fm., Kansas, Oklahoma	Permian	ng	Insect generations		0.5–1		60a	(4)
Kuttung Series, New South Wales	Carboniferous	g	Silt	Shale	2–1000+	20–40		(4)
Middle Old Red Sandstone, Scotland	Devonian	ng, sal-brackish, hol	1. Cal. or dol./org./siliciclastic 2. Siliciclastic 3. Silt	triplets Org. + clay Shale		0.5 <7.0 <10.0	1a, (Milankovitch rhythms,unpubl.)	(24), (35)
Achanarras Ls.	Devonian	ng	Clastic, carb.	Organic	0.5–2			(4)
Wollogorang Fm., N. Australia	Proterozoic	ng, ngr	Chert, dol.	Organic	0.1–0.2?			(45)
Huronian, Ontario	Precambrian	g	Fine sand	Shale		3		(4)

References

(1) Perkins and Sims 1983; (2) Agterberg and Banerjee 1969; (3) Smith 1978; (4) Picard and High 1981; (5) Boyoko-Diakonow 1979; (6) Renberg et al. 1984; (7) Anthony 1977; (8) Ludlam 1979; (9) Perfliev 1927 (cited in Bradley 1931); (10) Kempe and Degens 1979; (11) Garber et al. 1987 (12) Crossley and Owen 1988; (13) Ashley 1975; (14) Stoffers and Hecky 1978; (15) Yuretich 1979; (16) Halfman 1987; (17) Halfman and Johnson 1988; (18) Kelts and Shahrabi 1986; (19)Zolitschka 1990; (20) Anderson 1961; (21) Anderson 1986; (22) Zhao et al. 1984; (23) Lister 1984; (24) Duncan and Hamilton 1988; (25) Walker 1967; (26) Waitt 1984; (27) Anderson and Kirkland 1969a,b; (28) Anderson and Koopmans 1969; (29) Anderson and Kirkland 1960; (30) Katz and Kolodny 1977, Neev and Emery 1967; (31) Kashiwaya et al. 1987; (32) Antevs 1929; (33) Burbank and Grant 1985; (34) Bradley 1929, 1931; (35) Parnell 1988; (36) Dickinson 1988; (37) Van Houten 1962, 1964; (38) Olsen 1984, 1986; (39) Anadón 1988; (40) Niessen and Kelts 1989; (41) Gustavson 1975; (42) Yemane et al. 1989; (43) Smith 1984; (44) Walanus 1989; (45) Jackson 1985; (46) Dean et al. 1984b; (47) T. Johnson, pers. commun.; (48) Sonett and Williams 1985; (49) Hollander 1989; (50) Sturm and Matter 1978; (51) Spencer et al. 1984; (52) Anderson and Dean 1988; (53) Lotter 1989.

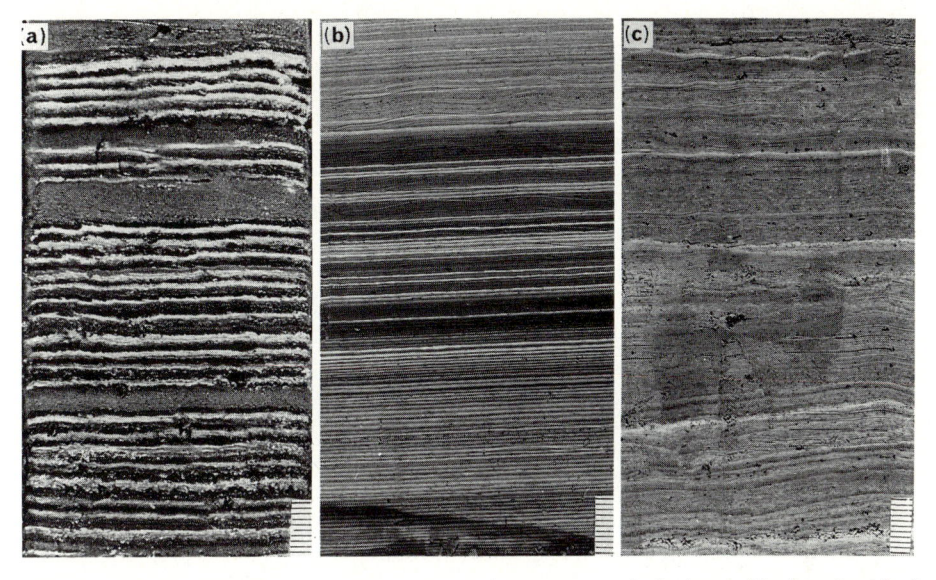

Fig. 3a-c. Examples of patterns recorded in recent carbonate varves, rhythmites and laminae from fresh (**a**), alkaline brackish (**b**) and hypersaline (**c**) lakes. All at same scale (scale bars in millimeters). **a** Section of open, deep (110 m) Lake ZÇrich nonglacial varves with diatom and organic matter (*dark*)/calcite (*light*) couplets from 1900 to about 1923. Examples of recurrence rate information include: (1) thicker (5–7 mm) fine-grained gray mud turbidites related to historical shoreline slump catastrophes; (2) unusually thick dark laminae (2 cm from bottom) that mark exceptional years (1905–1906) of diatom blooms of Melosira helvetica; (3) thicker calcite layers towards the top that record warmer summers. **b** Uniform rhythmic couplets from closed, deep (380 m) and brackish Lake Van representing about 190 years of deposition as based on varve counts. Light layers, aragonite or calcite; dark layers, organic-rich clay. Dark bands are sets of varve couplets in which dark layers predominate. **c** Laminated pelletal aragonitic marl from closed, shallow (12 m) hypersaline Lake Urmia representing about 350 years of deposition as based on sedimentation rate estimates. The precipitation of aragonite is a function of temperature and evaporation, but more importantly in combination with shifts in CO_2 balances due to high seasonal biological activity. Five distinct white laminae are relatively pure pelagic aragonite that separate on average 80-year cycles

patterns of species and productivity. These are linked to annual solar radiation, monsoon, or wind-driven lake upwelling. Even in evaporative settings evaporative rhythms may cause both chemical precipitate and bio-induced precipitate couplets on seasonal to multiseasonal scales (Fig. 3c). Because solar radiation is also a motor for the glacial melting cycles, we may view the dominant varve generating mechanisms as a function of radiation as suggested in Fig. 4.

The variable input mechanisms may be combined in various ways to produce simple to complex couplets that are generally interpreted as varves (Fig. 3 and Table 1). The time element for deposition in a couplet is important for understanding how and what the couplet records from the seasonal cycle. A varve may represent one year, but half a varve rarely represents half a year. Commonly, the quantitatively dominant sedimentation processes are limited to a specific season; for example spring meltwater, or spring and/or fall plankton blooms (Fig. 4).

3.1.1 Glacial Rhythms

In many lacustrine settings, varve thicknesses may reflect annual sediment delivery rates from catchment basins and thus serve as at least a proxy record for precipitation and runoff rates, and such studies should prove important in investigations of the variability of past climates. In areas of prolonged annual freezing, sediment transport is greatly reduced during winter. Beneath a frozen lake surface, quiescent conditions allow even the finest particles to settle out. In glacial areas which lack vegetation and thus humic acids, the clays will not flocculate but rather form dense, oriented, dark laminae that define the annual couplet of clastic glacial varves. Between two such dark clay laminae, glacial varves may display various levels of microlaminations reflecting pulses of supra or subglacial meltwater clouds, wind-driven turnover of homothermic lake waters laden with glacial milk, or inwash from rainfall events (cf. Ashley 1975).

Clastic content of glacial lake sediments record allochthonous input commonly discharged from fluctuating rivers and streams. These may fluctuate either as a function of glacial melting and freezing episodes, or as chaotic recurrences of significant floods in temperate to tropical environments (e.g., Waitt 1984). Their input will nevertheless also reflect long-term variation in paleohydrology. Antevs (1935) suggested that summer weather events, rather than annual mean temperature, are reflected in varve thicknesses. It is thus unclear what parameter glacial varves are actually recording; even more so for nonglacial varves. There has been insufficient calibration of lake systems against modern meteorlogical and instrumental records, and further studies are required.

3.1.2 Allochthonous Events

In contrast to marine settings, sediment suspension underflows and turbidity currents are common features of freshwater lakes because the suspended loads in rivers easily surpass the density contrasts in the lake waters. Storms are therefore one form of sediment forcing that commonly do not follow a uniform annual pattern in the sedimentary record. How can we determine the recurrence rate of such events? Storms produce higher suspended sediment loads in streams entering lakes. In dilute, open-system lakes with stable shorelines, deltas form and suspended sediment commonly flows as density underflows over the basin plain (Lambert et al. 1976). Frequent turbidity currents deposit graded silts that reflect the strength of storms only indirectly. Current meter studies have shown that the prehistory, i.e., the length of time since the last large storm, is an important parameter in the understanding of lacustrine storm surge sedimentation (Lambert et al. 1976). The amount of fine-grained sediment settled out on a river bed before the storm surge determines the suspended load and subsequent deposit thickness. Variable input strengths are thus not proportional to volume or thickness parameters of the deposits. Seasonal storm patterns will also lead to multiple thin layers that may be falsely interpreted as varves (Lambert and Hsü 1979). Lamb (1984) notes how the frequency and strength of storms seems to have increased during the Little Ice Age, related to an enhanced thermal gradient. Although such storms would most likely be reflected in lake laminations as well, no specific studies have yet demonstrated such a link.

Volcanic activity may not only provide an episodic ash fall to nearby lake deposits, but aerosols can affect climates worldwide. One might thus expect to see lamination evidence, for instance, following after the volcanic aerosol "year without a summer", i.e., 1816 AD (Bradley 1985). Major volcanic events, such as the Tambora eruption in 1815, have produced significant flood events that might be recorded in some lake deposits.

3.1.3 Biological Processes

Biological productivity in lakes very often follows a predictable sequence of species growth determined by solar energy (light and temperature) and nutrient supply (as dictated by seasonal overturn). Thus biologic productivity in many lakes is characterized by a definite yearly succession of plankton in open waters in addition to near-shore vegetation processes (Fig. 4). In some instances as many as six or seven populations of plankton may succeed and overlap each other within one annual cycle. Depending on circulation energy and the abundance of predators, these signals will reach and become preserved in the microlaminated sediment record (Dickman 1985). Siliceous diatom frustules are a conspicuous element of the record, but cyanobacteria filaments may be recognized, and siliceous Crysophyte capsules are important seasonal markers (Battarbee 1981; Lotter 1989). Dinoflagellates are important plankton, but seem rarely observed in sediments. Nipkow (1927) was able to determine the early summer from late summer diatom frustule succession in varves from Lake Zürich. Geyh et al. (1971) could even pick out insect wings from specific varve laminae. In addition, laminations may also be produced in situ by microbial mats and preserved as stromatolites (e.g., Casanova 1986). This may occur not only along shorelines, but also on lake bottoms at great depth.

3.1.4 Chemical Processes

Periodic shifts in climatic parameters should affect the residence times of water in lake basins and would thus become recorded by shifts in mineralogy, magnesium content, or in thicknesses of precipitated laminae. Biogenically induced precipitation of calcite occurs in the open waters of Lake Zürich, and many other temperate and tropical lakes, according to a definite annual cycle (Fig. 4; cf. Kelts and Hsü 1978; Dean 1981). Biological productivity removes CO_2, which decreases the total carbonate concentration. Maintaining charge balance requires an increase in alkalinity, which increases pH and abundance of carbonate ions. Calcite supersaturation may occur and a seasonal precipitate results. This seasonal control on productivity and carbonate precipitation typically leads to the basin-wide sedimentation of laminated organic carbon-rich limestones and marls (Fig. 3). If preserved as nonglacial varves, such precipitates contain valuable geochemical and isotopic archives of paleoenvironment. Carbonate content and thickness variations provide clues to the length of

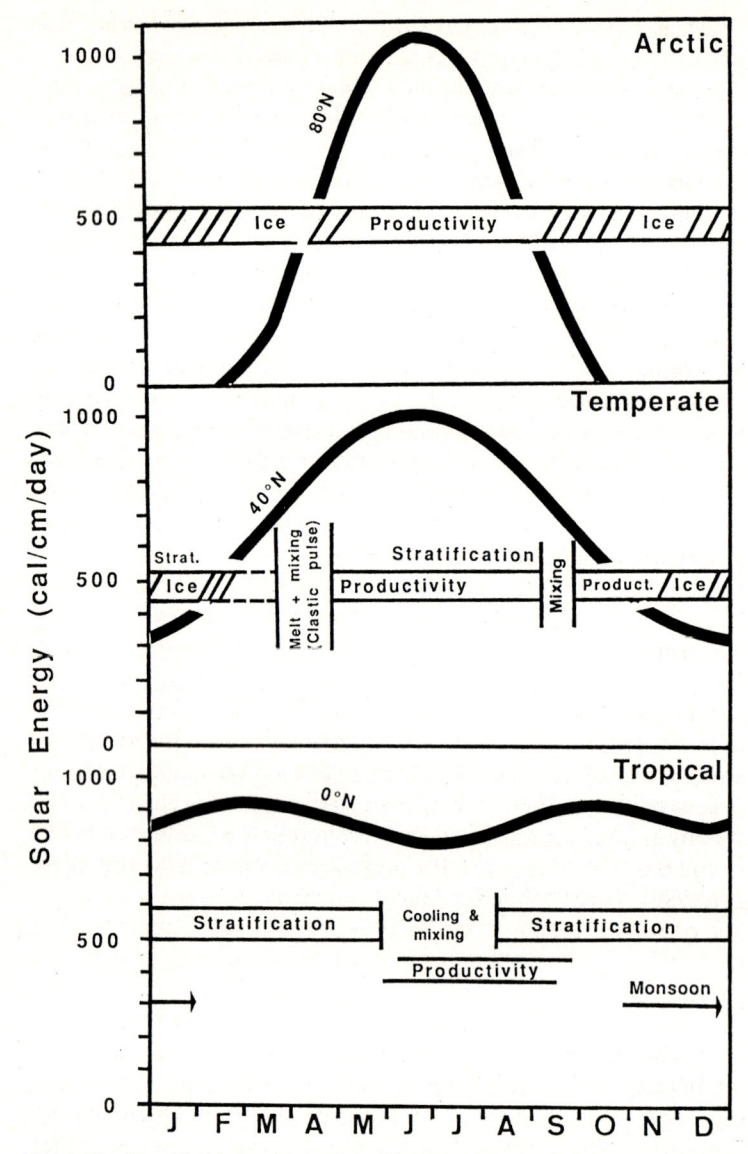

Fig. 4. Northern Hemisphere latitudinal dependence of daily totals of solar radiation (after Taub 1984) as related to various patterns of seasonality-controlled annual sediment couplets. Polar lakes that are ice-free in the summer undergo one period of high productivity during long summer days. Mid-latitude temperate dimictic lakes have two major periods of mixing and algal primary production in spring and fall. In spite of the uniformity of solar input, tropical lakes also have a productivity cycle due to seasonal patterns of wind-mixing and wind-driven evaporative cooling. Many variations can occur at any latitude due to local effects (e.g., wet/dry cycles)

the productivity season, and perhaps the length of the stable stratification season. If the warm season sets in earlier in a given year, the plankton responds earlier and the calcite precipitation cycle begins earlier. Oxygen isotope signals from these carbonates will therefore not be a direct measure of the maximum water temperatures, but tend to record an average temperature band of the upper epilimnion during the productivity cycle (Kelts and Talbot 1990).

The carbonate mineral phase precipitated is related to solute concentration through Mg/Ca ratios (Müller et al. 1972). Most carbonate-saturated temperate lakes have low Mg/Ca and thus precipitate low magnesium calcite. Such systems are however, still sensitive to annual processes. For example, as a productivity season progresses with calcite precipitation, the Ca^{2+} of the surface waters is depleted, increasing the Mg/Ca ratio of the precipitate. This is reflected in the calcite crystals of Lake Greifen, Switzerland, which shift from 0 to 2% magnesium in the calcite lattice over one season's crystal harvest (Weber 1981). Higher magnesium calcite and aragonite are common precipitates which form rhythmic couplets in many lakes with very long residence times where calcium depletion has occurred. This is the case for closed basin salty lakes such as Lake Urmia, Iran (Kelts and Shahrabi 1986) or Lake Van, Turkey (Degens and Kurtman 1978; see Fig.3).

3.1.5 Diagenetic Overprints

Diagenetic processes may also impart laminations or rhythmicities to lake deposits. Siderite and dolomite may form as early diagenetic minerals near the sediment water interface of sediments rich in organic matter and undergoing intensive methanogenesis (Talbot and Kelts 1986). Siderite/organic couplets occur in the Lake of the Clouds (Anthony 1977), numerous crater lakes such as the Eifel Maar-Lakes (Negendank 1989), and Barombi Mbo, Cameron, as well as the Eocene Messel shales of Germany (Bahrig 1989). Whether siderite precipitates in the water column or is a replacement product of primary aragonite has not been ascertained in this cases. The environmental signals held in the early diagenetic phases are, however, still related to biological and aqueous conditions in the basinal areas of a lake. Other important diagenetic facies include chert, analcime, vivianite, and pseudo-hardgrounds.

In contrast, evaporite minerals such as trona may form at various depths in unconsolidated sediment and deform of destroy laminations. Because of the rapid changes in lake levels and concentrations, these minerals may develop within sediments which were deposited in lake conditions radically different from those represented by the overlying hypersaline water body. Interpretations require caution.

4 Environmental Thresholds and Anderson's Concept of the Varve Microcosm

Anderson (1986) suggested that the interaction of the same processes which form varves may ultimately determine certain types of larger-scale cyclic bedding. Fine-scale laminations and varves of cyclically deposited units may thus be microcosms of the rhythmic bedding produced at a larger scale. An example of this comparison of annual and long-term climatic response is given by early Pleistocene Rita Blanca Lake deposits in Texas where the composition of individual yearly varve couplets is repeated in bedding cycles displaying an average period of about 200 years (Anderson 1986). In Anderson's view, the repetition of varve cyclicity in bedding is ultimately the result of the sedimentary averaging of the dominant lithology of varve couplets. Thus, just as long-term climate is an average of the secular trends of seasons, bedding cycles may represent an average of trends in yearly deposition.

Sedimentation thresholds in lakes (i.e., the point of switch-over between differing types of sedimentation) may be controlled by climate (or weather) and may apply to either autochthonous or allochthonous processes (Anderson 1986). Chemical thresholds controlling autochthonous lake sedimentation are perhaps easiest to visualize as an example. An evaporitic lake basin, for example, like a giant beaker, may be perched at the midpoint between the chemical precipitation of two or more compositional end-members such as anhydrite and gypsum (or aragonite). Precipitation of one end-member may dominate annual sedimentation and produce an asymmetric varve couplet (Fig.5). In this example, anhydrite and gypsum (or gypsum and aragonite) may precipitate seasonally to form varves, but the thickness of the gypsum laminae may greatly exceed that of the anhydrite. Similarly, calcite laminae may greatly dominate in thickness over an organic/clay laminae in seasonally-stratified, temperate lake varve couplets. If these conditions persist for hundreds or thousands of years, cumulative beds will be dominated by gypsum (in the first example) or calcite (in the second example). Bioturbation or current reworking may blur or remove the annual input signal of the rock, but the lithologic composition of the beds will reflect an average of its original component rhythms. Thus, in these examples, sedimentation thresholds may be said to be skewed in the directions of gypsum and calcite precipitation, respectively (Fig. 5). The skewing may reverse by rapid transition of sedimentation thresholds which are forced by climatic shifts or other regional changes in a particular drainage basin.

The thickness of one varve half-couplet, relative to another, is usually not a measure of the length of time required for the formation of each. Asymmetric calcite-dominated, calcite/clastic varves such as those of Lake Zug, Switzerland, are produced by a relatively short period of calcite precipitation in association with spring productivity blooms, separated by longer periods with fine-grained pelagic siliciclastic rain. A marl bed, if produced by the sedimentary mixing of such calcite/clastic varves, could be incorrectly interpreted as representing a uniformly long period dominated by calcite deposition. The same reasoning may also apply to clastic-dominated lake systems. Varve couplets in glacial lakes typically consist of a relatively thick clastic layer deposited during thaw conditions, and a thin clay layer that forms from pelagic settling during periods when lake surfaces are frozen. Hence, we might expect permanently frozen lakes or lakes with extended freeze to generally produce couplets with higher clay/silt ratios, and thus ultimately clay dominated beds with thin coarser silt to sand intercalations. The frequency axis in Fig. 5, therefore, should not be interpreted as a measure of time but, rather, as simply a reflection of mass.

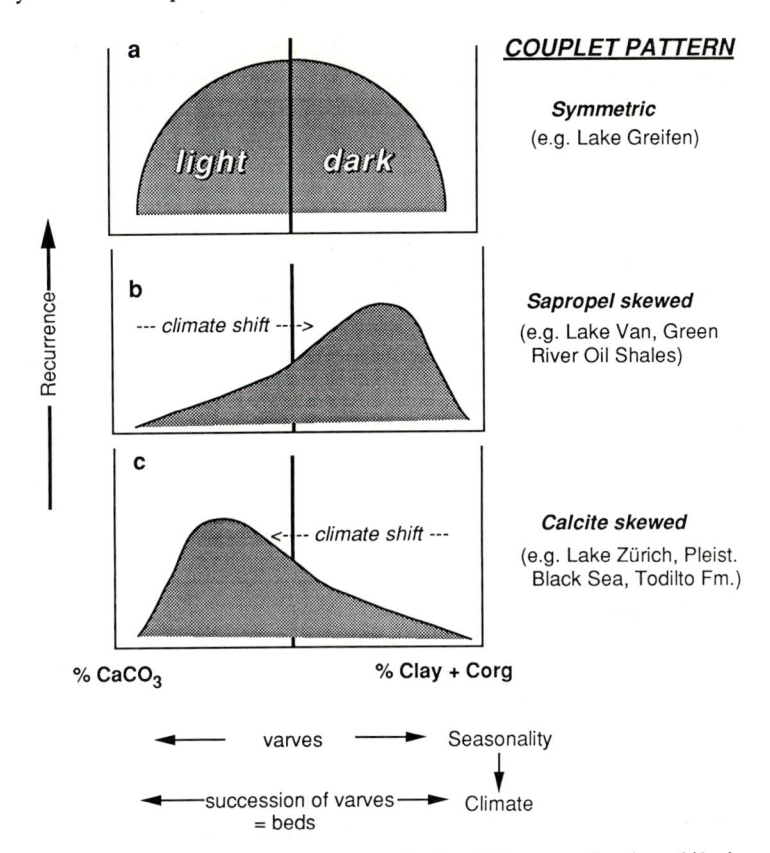

Fig. 5a–c. Model of threshold climatic controls on seasonal couplet thicknesses related to shifts in mechanisms of lamination. Sedimentation in **a** is poised between mechanisms that produce equal half-couplet thicknesses. Skewed couplets (**a** and **b**) are produced by domination of a single sedimentation type

5 Potential Extraterrestrial Forcing Functions

Extraterrestial influences of Earth's climatic variability have been variously attributed to sunspot cycles, lunar cycles, and other planetary and solar orbital parameters. Sunspot modulation is very important to the Earth climate system as it accounts for several tenths of one percent of the Sun's output energy, and thus constitutes the greatest known modulation of the total Earth's receipt of solar energy (Eddy 1988). Yet, it remains a critical question to determine how solar radiation energy output to the Earth varies, how it is modulated, amplified, and how such changes may be reflected in the sedimentary record. Whereas amplitudes of solar variability are small when compared to seasonality, they may still exert a forcing on climate and lake sedimentation.

Empirically determined cycles include the well-known 11-year sunspot (Schwabe), 22-year Hale (solar magnetic), 44-year Double Hale and 80–90-year

Gleissberg (Schwabe modulation) periods which characterize the sunspot index. The 11-year sunspot cycle is a measure of peak to peak sunspot activity; the double Hale cycle is a measure of 22-year bundles of alternate major and minor 11-year maxima of the sunspot cycles, and correlates with the alternate positive and negative polarity of the sun's magnetic field (Fig. 6). Longer-term solar variations have also recently been postulated from times-series analyses of fluctuations of ^{10}Be in ice-cores and ^{14}C in tree rings (cf. Sonett 1984; Beer et al. 1988; Attolini et al. 1988; Stuiver and Braziunas 1989). These studies suggest perhaps quasi-periods of about 50–70, 130, 140, 200–220 and 420 years (Fig. 7). In some way one might expect to find these signals in the high resolution record of varved lake deposits.

Complicating the high frequency solar rhythms are prolonged "minimums" (Fig. 6) of sunspot activity (maximums in ^{14}C production; cf. Stuvier and Braziunas 1988). These appear to correlate with climate cooling episodes on earth (e.g., the Oort Minimum, A.D. 1010–1050; the Wolf Minimum, A.D. 1280–1340; the Spörer minimum, A,D, 1420–1530; the Maunder Minimum, A.D. 1645–1715; the Dalton Minimum, 1795–1830; the Modern Minimum, 1880–1915). The Spörer and Maunder minima, for example, correspond with the two coldest excursions within the Little Ice Age (Siscoe 1978). These minima are neither evenly spaced nor uniformly long, suggesting that they are not a simple, periodic function of solar activity (Stuiver and Grootes 1980; Stuiver and Quay 1980; Eddy 1988). The climate cooling minima are clearly recorded in lake deposits, for example, as an increase in melt water varves in Lake Zug, Switzerland, or as a decrease in calcium carbonate contents in other lakes. Detailed correlations of these solar patterns with sediment records are, however, only just in progress.

Several other planetary-solar relationships also possess periodicities that may force variation in lacustrine sedimentation patterns. These include fortnightly lunar tides, an 18.6-year lunar nodal precession, a 19.9 "Saturn-Jupiter lap" cycle, an approximate 27-year lunar-solar tidal period, a 45.4 "Uranus-Saturn lap" cycle, a 178 "King-Hele" cycle, a ca. 180-year Sun-Jupiter repeat cycle, and a 556-year progression for the lunar Perigee (Fairbridge 1984). On a longer time scale, well-known Milankovitch orbital periodicities of 19 and 23 ka (precession), 41 ka (obliquity), and 100 and 413 ka (eccentricity) must play interrelated roles in lake sedimentation in that these parameters modulate ocean-continent-ice feedbacks.

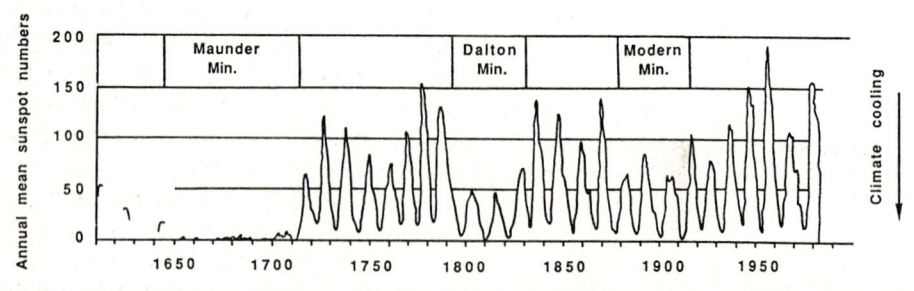

Fig. 6. Annual mean sunspot numbers and prolonged solar minima since the introduction of the telescope in A.D. 1609 as a time series of possible forcing functions for lake deposits. (After Eddy 1988)

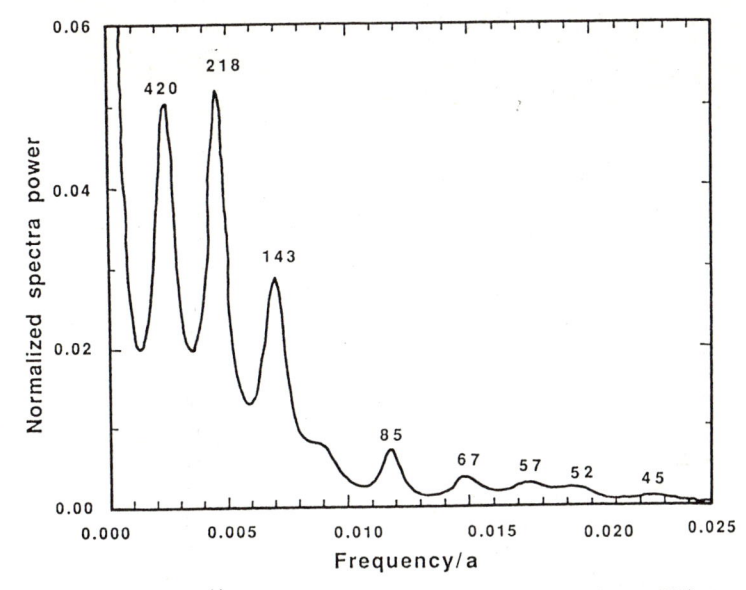

Fig. 7. MEM Power spectra of Holocene ^{14}C production rate (after Stuiver and Braziunas 1989) as a guide for possible terrestrial level climatic signals modulated by solar patterns that might be found in varved lacustrine sequences. The spectral peaks are related as harmonics of a fundamental solar period of 420 years

A summary of these extraterrestrial forcing function periods is presented in Fig. 8, together with examples of periodicities reported from various lake deposits. We examine several of these occurrences in more detail below. There is much evidence for the existence of these effects, but our understanding of the physical linkages involved between such potential forcing functions, climate response, and ultimate impact on the sedimentary record is currently inadequate. Our survey suggests that few studies have had enough rigorous control to test the solar and orbital parameter hypotheses. The response of lake systems to oscillating forcings must have nonlinear components and a great deal of care must be taken in interpreting estimated power spectra from them. The application of chaos analyses techniques (e.g., Gleick 1987; Lutz 1987; May 1989) and the study of aliasing effects (i.e., the production of spurious spectral peaks in time series analyses; e.g., Pisias and Mix 1988) to laminated lake deposits are, indeed, still in their infacy.

6 Decadal Scale Rhythms

6.1 Sunspot Control and 11- and 22-Year Rhythms in Lakes

In recent years, links between the solar sunspot index and global weather effects have been increasingly well documented, although still very poorly understood (cf. Kerr

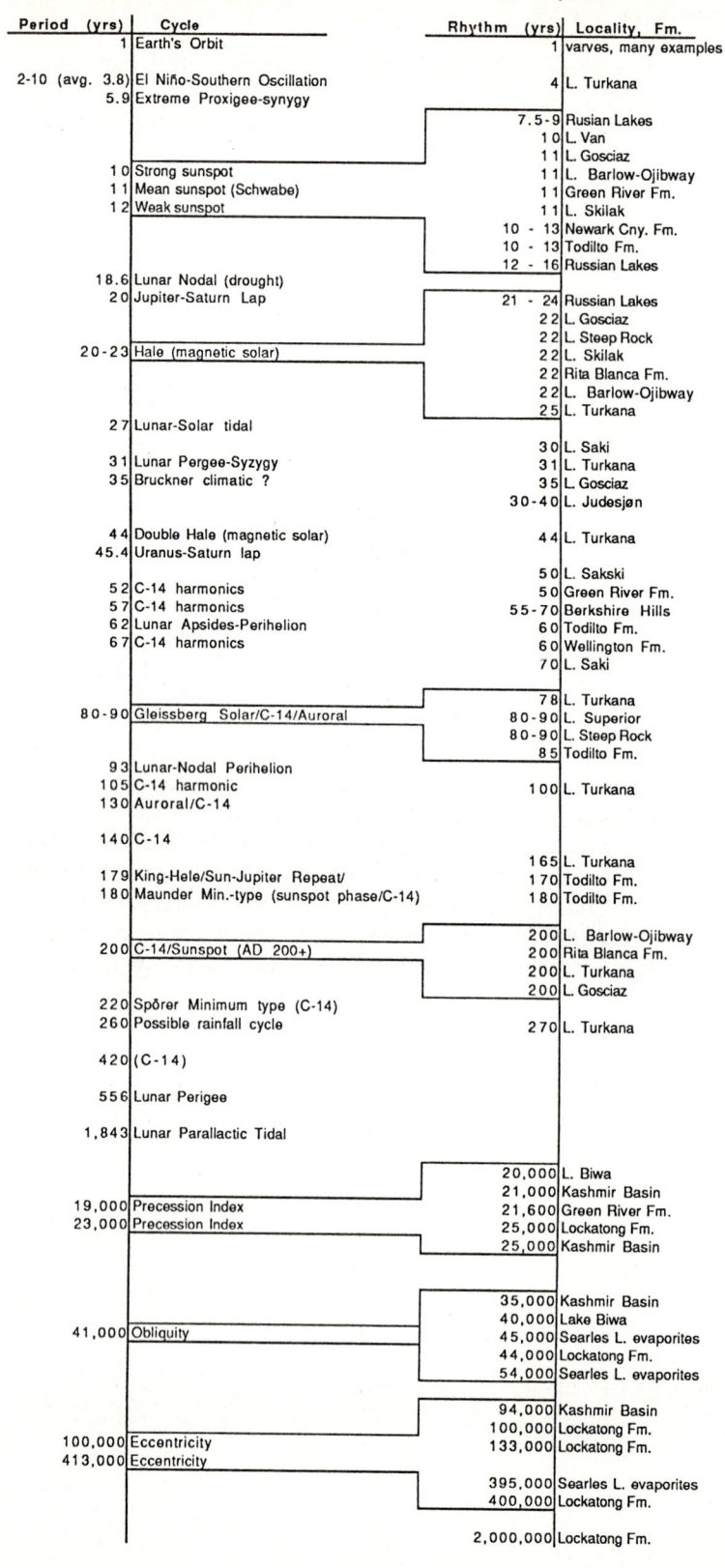

Potential External Forcing Functions		Lake Response	
Period (yrs)	Cycle	Rhythm (yrs)	Locality, Fm.
1	Earth's Orbit	1	varves, many examples
2-10 (avg. 3.8)	El Niño-Southern Oscillation	4	L. Turkana
5.9	Extreme Proxigee-synygy		
		7.5-9	Rusian Lakes
		10	L. Van
		11	L. Gosciaz
10	Strong sunspot	11	L. Barlow-Ojibway
11	Mean sunspot (Schwabe)	11	Green River Fm.
12	Weak sunspot	11	L. Skilak
		10-13	Newark Cny. Fm.
		10-13	Todilto Fm.
		12-16	Russian Lakes
18.6	Lunar Nodal (drought)		
20	Jupiter-Saturn Lap	21-24	Russian Lakes
		22	L. Gosciaz
		22	L. Steep Rock
20-23	Hale (magnetic solar)	22	L. Skilak
		22	Rita Blanca Fm.
		22	L. Barlow-Ojibway
		25	L. Turkana
27	Lunar-Solar tidal		
		30	L. Saki
31	Lunar Pergee-Syzygy	31	L. Turkana
35	Bruckner climatic ?	35	L Gosciaz
		30-40	L. Judesjøn
44	Double Hale (magnetic solar)	44	L. Turkana
45.4	Uranus-Saturn lap		
		50	L. Sakski
52	C-14 harmonics	50	Green River Fm.
57	C-14 harmonics	55-70	Berkshire Hills
62	Lunar Apsides-Perihelion	60	Todilto Fm.
67	C-14 harmonics	60	Wellington Fm.
		70	L. Saki
		78	L. Turkana
80-90	Gleissberg Solar/C-14/Auroral	80-90	L. Superior
		80-90	L. Steep Rock
		85	Todilto Fm.
93	Lunar-Nodal Perihelion		
105	C-14 harmonic	100	L. Turkana
130	Auroral/C-14		
140	C-14		
		165	L. Turkana
179	King-Hele/Sun-Jupiter Repeat/	170	Todilto Fm.
180	Maunder Min.-type (sunspot phase/C-14)	180	Todilto Fm.
		200	L. Barlow-Ojibway
200	C-14/Sunspot (AD 200+)	200	Rita Blanca Fm.
		200	L. Turkana
		200	L. Gosciaz
220	Spörer Minimum type (C-14)		
260	Possible rainfall cycle	270	L. Turkana
420	(C-14)		
556	Lunar Perigee		
1,843	Lunar Parallactic Tidal		
		20,000	L. Biwa
		21,000	Kashmir Basin
19,000	Precession Index	21,600	Green River Fm.
23,000	Precession Index	25,000	Lockatong Fm.
		25,000	Kashmir Basin
		35,000	Kashmir Basin
		40,000	Lake Biwa
41,000	Obliquity	45,000	Searles L. evaporites
		44,000	Lockatong Fm.
		54,000	Searles L. evaporites
		94,000	Kashmir Basin
		100,000	Lockatong Fm.
100,000	Eccentricity	133,000	Lockatong Fm.
413,000	Eccentricity		
		395,000	Searles L. evaporites
		400,000	Lockatong Fm.
		2,000,000	Lockatong Fm.

1988). For example, evidence for the 11- and/or 22-year solar cycle signal in drought-flood proxy records has been reported for western and middle North America (Mitchell et al. 1979; Currie 1984), Argentina (Currie 1983), and northeastern China (Currie and Fairbridge 1985), variations in lake levels and climate in Finland appear to track 22-year Hale periodicities (Keränen 1984), and variations in global air temperatures and pressures now appear to be highly correlative to sunspot numbers (Currie 1987; van Loon and Labitzke 1988). On longer scales, Gleissberg solar activity periods of about 80 to 90 years also appear to be correlated with variations in terrestrial climate. Studies of the past few centuries of temperature for the northern hemisphere show correlations between prolonged periods of minimum sunspot activity with periods of minimum terrestrial temperature, whereas periods of maximum temperature tend to precede periods of maximum sunspot activity (Willett 1987). Evidence for solar cycles in lacustrine deposits has been variously reported (cf. Table 1) but rarely tested rigorously. Rhythmicities in most of these deposits are based on time series analysis of varve thicknesses.

6.1.1 Recent High Latitude Lakes

Because glacial ablation is sensitive to temperature and snowfall, varve thicknesses in glacially fed lakes should be linked to these factors. Increases in temperature promotes glacial ablation, while fresh snow increases the albedo of the glacier surface and protects the sediment-laden ice from the effects of incident solar radiation. Varve thicknesses in Recent glacial Silak Lake in Alaska appear to correlate directly with mean annual temperature and inversely with mean annual snowfall in southern Alaska (Fig. 9a), but not with mean annual cumulative precipitation (rain and snow). Sonett and Williams (1985) found that the varves in this lake display a periodicity in power spectral data of about 11 and 22 years after correcting varve years for hypothesized intra-annual sediment layers, although several other peaks with both longer and shorter periods were also found in the power spectra. In this rather complicated example (see details in Sonett and Williams 1985), solar maxima (as measured by the sunspot index) thus appears positively correlated with maximum varve thickness and annual temperature. A similar relationship between summer temperatures and varve thicknesses is also reported from glacial lakes of the Canadian Rockies (Leonard 1985) and prominent peaks at 21–23 years have been suggested from power spectra data of varve thicknesses from a series of high latitude lakes in Russia (Anderson 1961). Although few, such studies suggest that clastic varve thicknesses from proglacial lakes may vary proportionally with rates of glacial melt-water runoff and thus serve as proxy records of solar maxima and minima. What is greatly needed in such studies are direct correlations relating the magnitude and timing of historic

Fig. 8. Compilation of potential external forcing frequencies and examples of frequencies reported from studies of lake rhythms. Potential forcing functions after Mitchell 1976; Schove 1983; Fairbridge 1984; Fairbridge and Sanders 1987; Quinn et al. 1987; Beer et al. 1988; Stuiver and Braziunas 1989; and others, cf. Table 1

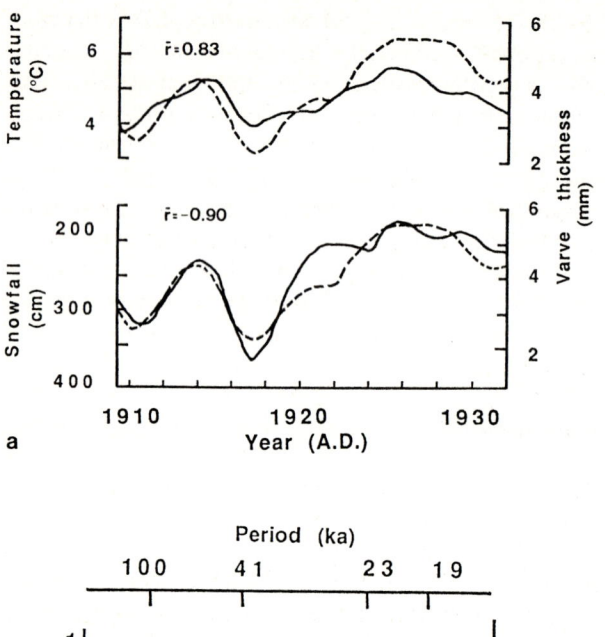

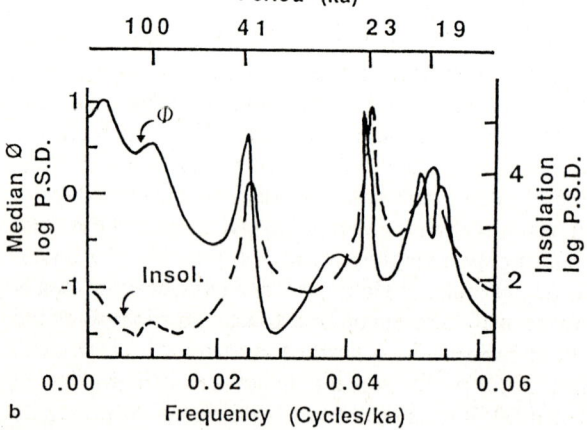

Fig. 9a,b. a Correlation of glacial varve thicknesses (*dashes lines*) and climatological data from Lake Skilak in southern Alaska as based on a weighted five point moving average. (After Perkins and Simms 1983). **b** Comparison of solar insolation curve with log power spectral density of median grain size (*solid line*) from L. Biwa sediments. (After Kashiwaya et al. 1987)

sunspot records to the thickness and chronology (including lag effects, e.g., Agterberg and Banerjee 1969; Sonett and Williams 1985) of varves.

6.1.2 Recent Low Latitude Lakes

Lake Van, Turkey, has no outlet other than evaporation, and is the world's fourth largest closed lake. Kempe and Degens (1979) postulated that the lake level seemed to fluctuate in concert with the sunspot cycle of 11 years, although over only two cycles analyzed. Ten-year periodicites were visually identified from 7-ka aragonite-clastic varve couplets from the Lake Van sediments, the thickness of these couplets being related to annual runoff phases. Approximately 25- and 44-year (Hale and double Hale) power spectra rhythmicities have also been recently suggested from couplet thickness data from Lake Turkana in Kenya (Halfman and Johnson 1988).

These modern examples suggest that solar periods, with feedbacks through weather effects, may regulate input-evaporation cycles in lakes and may thus be recorded in both the chemical and clastic sediment accumulation rates of the lake's sediments. Serious problems persist, however, as both the Lake Van and Lake Turkana sequences lack adequate age control to confirm the suggested timing and frequency of their lamination couplets.

6.1.3 Sunspots in the Ancient Record

Fairly convincing examples of 11- and/or 22-year rhythmicities have also been reported from older portions of the rock record. As part of a very detailed study of Lower Pleistocene Rita Blanca Lake deposits in the Panhandle of Texas, Anderson and Koopmans (1969) examined a 14-ka varve series for variations in time, thickness, and composition. Clay and calcium carbonate are the main couplet components. Results of power spectra and Fourier analyses of thickness suggested a rather weak periodicity for clay contents in the region between 10 and 12 years, but indicated marked 22-year rhythms in both total varve thicknesses and in calcium carbonate contents. These results suggested that varve thickness periodicities in the Rita Blanca deposits are chiefly controlled by rhythms in $CaCO_3$ precipitation, and that these are linked through climate to variations in solar intensity. In this basin, therefore, solar signals in sedimentation appear not so much to stem from changing rates of allochthonous input through runoff (as in the Lake Skilak and Lake Van examples), but rather to autochthonous variations in carbonate production as regulated by water column temperature, evaporation, and primary productivity. As discussed below, longer-term climatic fluctuations also left their impress in these rocks in the form of major lithologic cycles.

From Eocene Green River Formation (Lake Gosiute) organic marlstones, Bradley (1929) identified an average recurrence interval between bundles of thicker varves of a little less than 12 years and attributed these to the sunspot cycle. These meromictic-lake varves (Boyer 1982) are nonglacial and consist of carbonate-rich laminae induced by summer algal blooms with abundant fish debris, alternating with kerogen-rich layers. Bradley (1929) also suggested a 50-year recurrence period for the spacing of calcite-filled salt casts in the oil shales, although the timing of this diagenetic process is unreliable.

Lacustrine varve sets of the Jurassic Todilto Formation, New Mexico, are among the most completely described (Anderson and Kirkland 1960, 1966). From the basal limestone member of this sequence, 1,592 very thin varve couplets consisting of limestone and silty organic layers were counted, measured, and correlated over a distance of several kilometers. Rhythmicity of these varves, as defined by sequential dendrocyclogram analysis of varve couplet thickness (Anderson and Kirkland 1960), showed a 10–13-year period presumably characteristic of the sun spot cycle. The variation of the varve thicknesses was believed to be controlled by seasonal weather changes, principally temperature and rainfall variations, which influenced carbonate precipitation by effecting water temperature, composition, evaporation rates, photosynthesis, and seasonal inflow of water into the basin. Annual cycles where limestone laminae where absent or thin were interpreted as indicating cooler and/or wetter years, these annual variations thus apparently cumulating in cooler/wetter versus warmer/dryer decade-scale variations in climate. In addition to these, four other varve-couplet types (Table 1) were described from this formation, but without time series analyses. Evaporite varves have been well-documented from giant marine-paralic basins such as the European Zechstein (Richter-Bernbug 1955) or Permian Paradox Basin, USA (Anderson and Kirkland 1966).

6.2 ENSO Events

One might expect low latitude lakes to be affected by quasi-periodic El Niño-Southern Oscillation (ENSO) events because recent meteorological studies suggest a global feed-back mechanism between sea surface pressure, temperature, regional circulation, and precipitation anomalies throughout the tropics (cf. Rasmusson and Carpenter 1982; Graham and White 1988). Our survey, however, uncovered only one direct reference to the record of this phenomenon in lake sediments. Using a linear sedimentation rate of 2.7 mm/a and the fact that there are 903 calcite/silt couplets in 935 cm of core, Halfman and Johnson (1988) calculated an average 4-year periodicity in the Recent sediments of Lake Turkana, Kenya. These authors suggested that relatively carbonate-rich, light-colored layers of the couplets correlated with quasi-periodic dry years on the Ethiopian plateau and that these dry years, in turn, correspond to an average 3.8 year periodicity (Quinn et al. 1978) of low-latitude ENSO events. This is an intriguing hypothesis, although there may be complications that make this assertion difficult to prove. ENSO events themselves do not follow a simple pattern but rather appear to reoccur irregularly with varying strengths over a 2–10-year range. A modern calibration is required.

6.3 Other Decadal-Century Scale Rhythms

In addition to the above, numerous other examples of reported sedimentary rhythmicities on century or shorter scales may also be gleaned from inspection of Fig. 8 and Table 1. Results from the Jurassic Todilto Formation and Recent Turkana Lake sediments are particularly interesting in that both display a variety of sedimentary rhythmicities on these time spans. Harmonic analysis of both percent calcium carbonate and couplet thicknesses from the Lake Turkana core revealed spectral peaks corresponding to periodicities of about 100, 200, and 270 years at the 95% confidence level (the harmonics of number of lamination couplets were converted to time intervals by using the mean sedimentation rate and a mean couplet thickness for the core studied), and spectral peaks with shorter periods of about 25, 31, 44, and 78 years in the couplet-thickness data alone (Halfman and Johnson 1988). Rhythmicities in thickness variations of calcite-sapropel varves of the Todilto Fm. apparently occur with periods of 10–13, 60, 85, 170, and 180 years. Some of these periods appear to overlap, but others do not, and several remain very hard to reconcile in terms of the known or suggested potential climatic-forcing frequencies depicted in Fig. 8. This appears especially true for all of the deposits we have outlined with suggested periodicities clustering about 30–40 and 50–70 years (Fig. 8). Although most of the deposits falling into these clusters reflect quite different depositional environments, it appears that 30–40 years clustering of rhythms may in some way relate to the poorly defined Bruckner climatic cycle, which supposedly reflects a quasi 30- to 40-year periodicity in European rainfall patterns (Oliver and Fairbridge 1987, p. 184), and the 50–70 clusterings may relate to the higher-order, yet lower-amplitude solar harmonics expressed in the power spectra of Recent tree-ring data (cf. Fig. 7).

If these rhythmicities are real, the discrepancies between reported sediment and forcing periods suggest either nonlinear responses to climate forcing involving lag times, rhythmicities reflecting local pertubations of the lake systems in question, and/or perhaps even a diagenetic trend such as the case for the 50-year period from the Green River Formation. What may be most important, yet most difficult to ascertain in many of these examples, however, is the compounding (stacking) of several perhaps unrelated rhythmic forcings into the periodic low-frequency oscillations reported. In summary we must conclude that while several of the periodicities suggested from lake deposits are tantalizingly close to those described from modern studies of solar variability, a fully rigorous documentation of such high frequency forcings in lake sediments has yet to be undertaken.

7 Milankovitch Rhythms

As we have emphasized above, lacustrine sedimentation, whether rhythmic or not, is strongly affected by regional to global patterns of temperature, aridity, and rainfall. Over the past 2 million years (cf. Hays et al. 1976), and probably over much of the Phanerozoic record at least (see Fischer Chap. 1.2, this Vol.), the major mechanism for changing global climate on time scales between 10 ka and 500 ka appears to be due to the changing distribution of solar energy by latitude and by season as dictated by three master orbital variables: 41 ka obliquity cycles of changes in the inclination (tilt) of the Earth's rotational axis (between about 22° and 25°), 100 ka and 413 ka eccentricity cycles of the changing shape of the Earth's orbit around the Sun (from circular to more elliptical), and 19 ka and 23 ka precession (index) cycles which are related to changing distance between the Earth and the Sun at any given season and are modulated by the waxing and waning of the cycles of eccentricity. These effects have now been well documented in the marine record (cf. Imbrie et al. 1984).

At low latitudes, rhythmic variations of climate may be more dominated by the eccentricity and precession than by obliquity (axial tilt). This is due to both the diminished effect of obliquity changes and the increased importance of monsoons at low latitudes. Through regulation of the heating intensity of the northern continental masses, orbital precession influences rates of precipitation of 19 ka and 23 ka periods by altering the intensity of low pressure cells that drive monsoonal winds (Rossignol-Strick 1983). This control on low latitude precipitation is modulated by the eccentricity cycle (see Fischer Chap. 1.2, Einsele and Ricken Chap. 1.1, this Vol.) and these effects have been recorded by lake levels and lake deposits. Street-Perrott and Harrison (1984) compiled lake level data for 198 lakes, and modeling experiments (e.g., Kutzbach and Street-Perrott 1985) have shown a strong correlation between low-latitude lake levels, atmospheric circulation, and the Milankovitch regulation of late Quaternary glacial-interglacial change (also see Giraudi 1989). Using a general circulation atmospheric model to simulate climates, Kutzbach and Street-Perrott simulated strengthened monsoonal circulation culminating between 6–9 ka BP with increased precipitation in northern-hemisphere tropics. The computed hydrological budgets show good correlation with the geological evidence for Quaternary lake-level variations. Below, we illustrate some of the evidence which substantiate the orbital theory of climatic change from saline-, clastic-, and mixed clastic/chemical lacustrine deposits.

7.1 Searles Lake: Pleistocene

G.I. Smith (1984) described nine periods of wet, intermediate, or dry paleohydrologic regimes with a mean duration of 396 ka (but with a 200 ka standard deviation) from climatically sensitive evaporitic Pleistocene Searles Lake sediments. These may be related to the 413 ka orbital eccentricity cycle, but the large standard deviation makes the significance of the mean value questionable. While the wettest and driest regimes from the Searles sequence do not appear sensitive to shorter period cycles, sedimentary indicators of intermediate climates show internal rhythmicities about 45 ka and 54 ka that may in part reflect the cycles of precession (if halved) and obliquity if other perturbations are also involved. In attacking the problem from the other direction, however, Smith re-grouped the regimes into periods of 400 ka with higher frequency salinity variations of 100 ka, and arrived at a postulated sedimentation model that in many respect mimicked the observed Searles sequence, the majority of residual discrepancies again being attributable to the inability of this sequence to "see" or record the more extreme wet and dry intervals of climate. In addition, whereas only some of the transitions between the paleohydrologic regimes over the past 1.75 Ma appear to coincide with the transitions in the marine record of ice-volume changes, most appear to approximately coincide in time (but not in direction) with faunal and isotopic sea-surface temperature changes in the tropical Atlantic, and many coincide with North American and European paleoclimatic boundaries. While no single aspect of these correlations are in themselves conclusive of a rhythmic climatic control on Searles Lake sedimentation, the collective evidence suggests that the paleohydrology of this basin was linked in timing with global climatic phenomena, albeit at a tempo that was out of phase with the volumetric wax and wane of polar ice. Much of this may have been due to local variations of glaciation-deglaciation cycles in the nearby Sierra Nevada Mountains which, being fed by orographic precipitation from the Pacific, have themselves yet to be correlated or contrasted with Milankovitch timings of Pacific sea-surface temperature. The evidence needed to resolve the coincidence and contradictions between these many interregional paleoclimatic data sets are few, and this example illustrates that even with long, high-resolution evaporate records much work will be needed before many of the complexities in correlating the spatial and temporal rhythms of climate change can be resolved.

7.2 Lake Biwa: Plio-Pleistocene/Recent

Japan's Lake Biwa, with sedimentation continuous since the Pliocene, is a semi-enclosed mid-latitude basin that receives, with minor exception, fine-grained clastics derived from relatively uniform source rocks of its surrounding catchment. Sedimentary rhythms in Lake Biwa surface sediments are not readily apparent because the monotonous, silty-clay sediments are bioturbated and lack sedimentary structures (Yokoyama and Horie 1974). The uniform grain size of these deposits apparently reflects the relative lack of seasonality and the uniform, maritime climate in this portion of Japan. Surprisingly, however, some promising evidence for a record for recent orbital rhythms of climatic change have been suggested from these rather monotonous deposits. Using an approximated paleotemperature time series derived from radiometric age dating and composite pollen spectra for the past 500 ka, Kanari et al. (1984) found spectral peaks with median values at frequency bands of 104 ka, 44 ka, 25 ka, and 12.7 ka. Except for the latter, these periods in the paleoclimatic data appear close to those predicted from Milankovitch theory. The cyclicity seen in the temperature variations was also later suggested from the Biwa sediments themselves as reflected in variations of organic biomarkers (Ogura 1987), and as 20 ka and 40 ka rhythmicities in several grain-size parameters (e.g., Fig. 9a) measured for grains greater than 4.5 phi for the last 200 ka portion of the core (Kashiwaya et al. 1987). In addition, it appears from these data that increases in grain size correlate with increases in insolation (cf. Fig. 10), these factors presumably relating to either

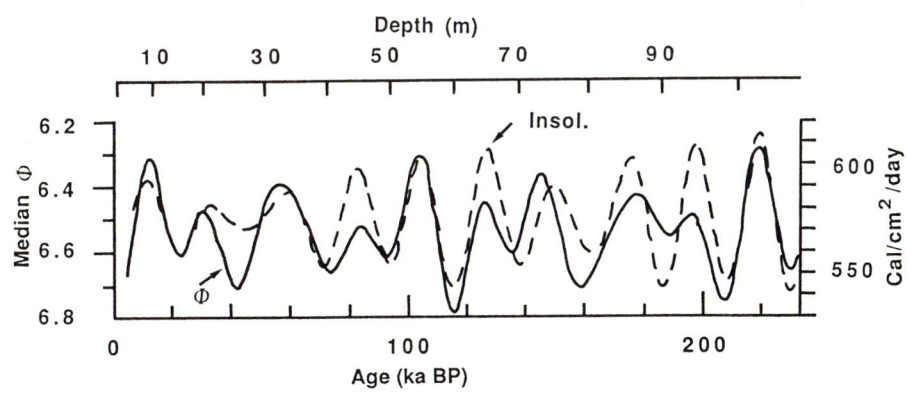

Fig. 10. Harmonic analysis of median grain size (*solid curve*) from L. Biwa sediments compared with 17 ka to 45 ka bandpass-filtered insolation curve. (Kashiwaya et al. 1987)

increased runoff or lake lowerings during climatic warming. Unfortunately, no attempt was made to correlate the paleotemperature pollen records and grain size parameters.

7.3 Newark Rift Basins: Triassic-Jurassic

Triassic-Jurassic rift valley basins of the Newark Supergroup in eastern North America form a linear system of lacustrine deposits 2000 km in length that spectacularly record the rise and fall of lake levels and variations in chemical lacustrine sedimentation in tempo with Milankovitch orbital periodicities. From the Lockatong Formation in the Newark basin, Van Houten (1962, 1964) early recognized short (ca. 5–6 m) chemical and detrital bedding cycles (Van Houten cycles). Detrital cycles are coarsening upwards sequences consisting of: (1) a basal gray mudstone sometimes showing current lamination, stromatolites, desiccation cracks, and oolites and representing lake "transgression" or infilling, (2) a laminated (varved) black organic carbon-rich, fossiliferous, calcareous mudstone (low carbonate-high organic carbon/high carbonate-low organic carbon couplets) representing lake highstand and (3) an upper, gray mudstone with reptile footprints, desiccation cracks, and rootlets and representing a regressive and low-stand deposit (Fig. 11). Chemical cycles consist of alternating gray mudstones and more massive red mudstones rich in analcime, dolomite, and pseudomorphs after evaporite minerals (see Van Houten 1964). The detrital cycles are continuous over the whole of the Newark Supergroup, while the chemical cycles are restricted to the central 60 km of the basin where brines concentrated at low lake levels (Olsen 1984). Van Houten suggested that these short cycles were produced by the periodic wax and wane of rainfall and hence lake levels. Each cycle begins with increased rainfall and inflow, deepens to high stand conditions conducive to the formation of organic carbon-rich shales and marls and then shoals back to shallow water facies as the basin dries up. Salts accumulated and chemical cycles resulted where and when basins were closed and insufficiently filled to reach their outlets. Detrital cycles developed when the basins were fed by through-flowing drainage which exported most dissolved material to the sea. These and other similar cycles are widespread throughout the Newark Supergroup (cf. Gore 1988, 1989).

The short, ca. 5 m Van Houten cycles of the Lockatong and Passaic Formations bundle into alternating 25-m sections (made up of four to six short cycles) each dominated by chemical or detrital facies. Clusters of four of these 25-m cycles make up ~ 100m cycles recognizable at map scales as the main members of the Formations (Fig. 11). Van Houten extrapolated the varve thicknesses to arrive at periods of about 21 ka for the short 5 m cycles, 100 ka for the 25 m cycles, and 500 ka for the 100 m cycles. Using a ranking of sedimentary characteristics, Olsen (1984, 1986) numerically related these cycles to changes in water depth from deeper than 100 m (rank 6) to playas and eventual desiccation (rank 0) (Fig. 11). On the basis of these numeric curves of sedimentary ranking, Olsen refined the bedding periodicity estimates and deduced a 21.8 ka periodicity for lake level oscillations recorded by the 5-cm cycles, and 101.4 ka and 418 ka periodic changes in the magnitude of lake level high stands for the longer 25-m and 100-m-thick compound cycles, respectively. Remarkably,

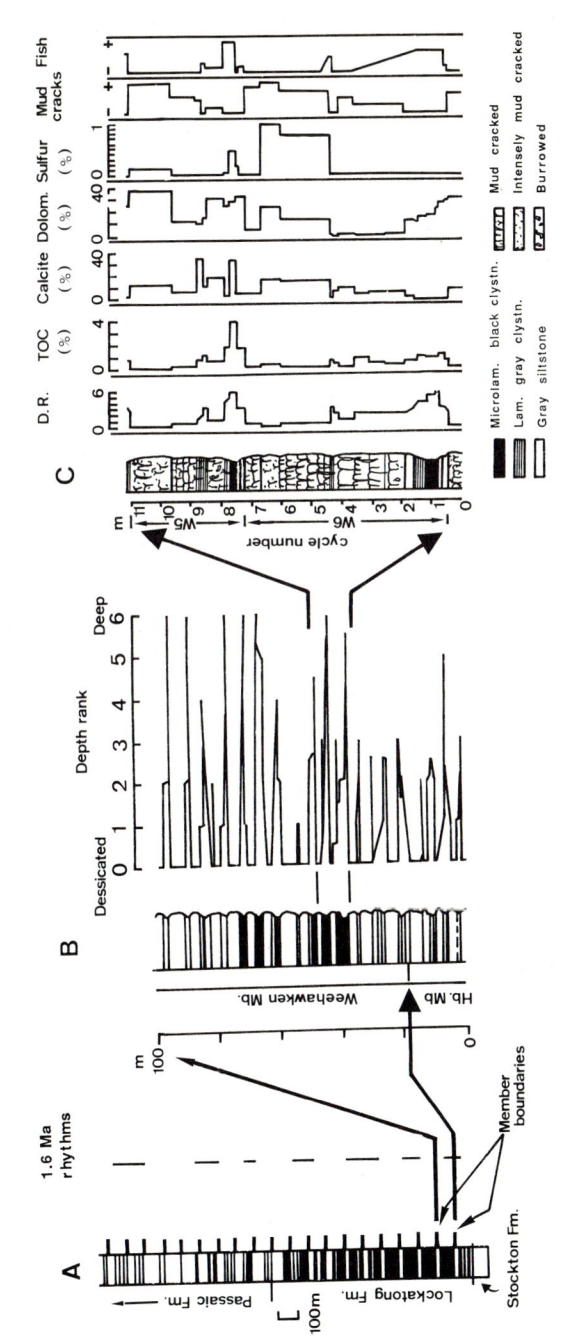

Fig. 11A–C. Comparison of rhythmic bedding at various scales within the Triassic Lockatong Formation in the Newark Supergroup, NE USA. **A** Member-scale (ca. 100 m, 400 ka) cycles. *Horizontal black intervals* represent principally gray to black sediments, *white intervals* are mostly red sediments. The *black vertical bars* indicate intervals that show especially thick and finely laminated unit 2 portions of Van Houten cycles. Approximate 1.6 Ma cycles are indicated by the spacing of these zones. **B** Bundling of short Van Houten lithologic cycles (ca. 25 m, 100 ka) and Olsen's depth-scale ranking as used in Fourier analysis. **C** Depth Rank (*D.R.*), mineralogy, sedimentology and geochemistry of two bed-scale detrital "Van Houten" precessional cycles (ca. 5.9 m, 25 ka). Although thin, division 2 (deep-water microlaminated black claystone) may represent the longest time period of each short cycle. See Fig. 12 and Table 2. (After Olsen 1986)

Table 2. Ratio of the mean period of the precession cycle to other main orbital periods and comparison with the mean thickness of Van Houten cycles to other cycles in the Lockatong Formation. P, precession; O, obliquity; E, eccentricity. (After Olsen 1986)

Item	Orbital periods (ka)				
	P	O	E1	E2	E3
Present day periods	21.7	41.0	95.0	123.0	412.1
Av. thickness of cycles (m)	5.9	10.5	25.2	32.0	96.0
Ratio of modern period of precession to other modern orbital periods	1.0	1.9	4.4	5.8	19.0
Ratio of average periods of Van Houten cycles to averages of other cycles	1.0	1.8	4.3	5.5	16.3

ratios of the average thickness of the Van Houten short cycles to the longer bundles of Lockatong and Passaic formations agree well with the ratios of the periods of precession to the other main Milankovitch orbital parameters (Table 2). Subsequent Fourier analysis of the water depth rankings (Olsen 1986), such as that applied to deep sea oxygen isotope records (cf. Imbrie et al. 1984), further showed bedding thickness periods of 5.9, 10.5, 25.2, and 96 m to correspond to time periodicities in time of roughly 25 ka, 44 ka, 100 ka, 133 ka, and 400 ka. A critical step in the spectral analysis of beds is the assumption of uniform sedimentation rates. Olsen (1986) compared varve thickness counts will overall age data from radiometric methods. The similarities were surprising, giving credence to his approach. The cyclicity of lake

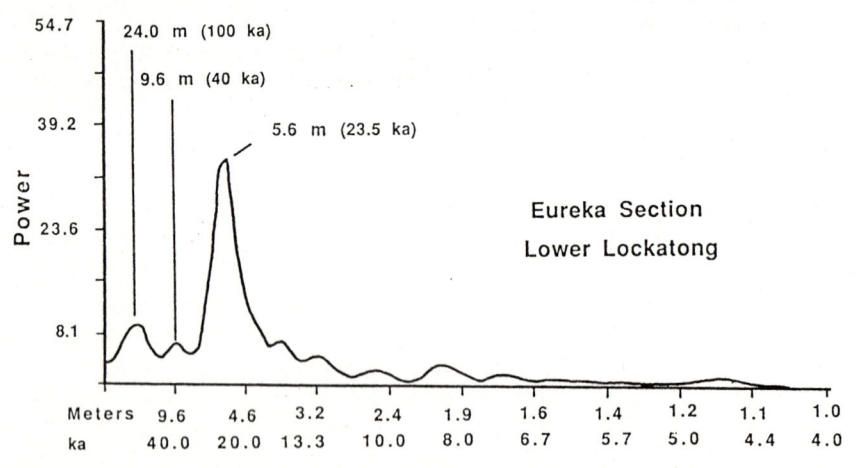

Fig. 12. Example of power spectra from the lower Lockatong Formation suggesting a precessional periodicity at 23.5 ka for short (ca. 5–6 m) Van Houton cycles and a eccentricity period for longer (ca. 25 m) composite cycles. The minor peak at 40 ka was used to suggest that the obliquity term has a relatively minor influence at these paleolatitudes. (After Olsen 1986)

levels recorded by these low latitude deposits may be related to the precessional regulation or regional precipitation by monsoonal winds, as discussed above. The magnitude of the Power Spectra from these deposits (Fig. 12) suggests that the cycles are dominated by precession (Van Houten short cycles) and modulated by eccentricity (long cycles), and that the relative unimportance of the obliquity period (41 ka) in these cycles is in keeping with the Rosignold-Strick and Kutzbach predictions for a monsoon-dominated climate at low latitudes.

8 Lacustrine Life Cycles

In Fig. 13 we schematically portray three hypothetical depositional sequences which may be of use in predicting the rhythmic or cyclic development of lacustrine facies in the geologic record. Each represent progressive facies development in a particular type of lake system, as might be observed in cores from basin centers. They serve as simple guides for vertical successions due to variations in water level, chemistry, and clastic input. Although associations among lithologies should be present, variations and complications exist, and facies shown may be combined to produce new cycles as changing climate dictates. Figure 13 does not include variants on nonglacial predominantly siliciclastic sequences, as these may be dominated by any number of processes. Commonly cited siliciclastic lake models comprise laminated silt or mud coarsening up to sand, leading eventually to deltaic sands and gravels. Such straightforward successions are, however, rarely observed. Lakes seem more likely to simply silt up rather than become infilled by the inward migration of marginal clastic facies. Reviews of various siliciclastic-dominated systems are in Picard and High (1972, 1981), Allen and Collinson (1986) and Cohen (1989).

Closed lake basins (e.g., Fig 13a) occur at all latitudes and are ideal settings for the development of transgressive-regressive cycles because such basins respond very sensitively to variations in evaporation, rainfall, and fluvial input, and their shorelines thus tend to move rapidly and frequently. On average, evaporation exceeds inflow in closed basins and our present climate is generally arid, so most of the world's closed basins are thus now much more desiccated than during preceding stages. Rates of draw-down have commonly been so rapid that a progression from dilute and open to shallow and eventually hypersaline facies may be preserved in a few centimeters of pelagic carbonates. Alternatively, thick carbonate bioherms marking paleoshorelines may represent only a few tens or hundreds of years and evaporite salts on playas developed during final stages of desiccation may typically accumulate at rates of 10 cm/a (e.g., Fig. 13a). Sedimentation cycles produced by repetitive inflow and desiccation in such settings may commonly contain unconformities produced by evaporite dissolution and fluvial erosion following evaporative draw-downs.

Open lakes with permanent outlets tend to have stable shorelines and sedimentation in such lakes may be dominated by clastic input or, if protected from such inputs, chemical and biochemical sedimentation may be important. Cores from temperate perialpine settings provide a good example of both (Fig 13c). As a model, the deep basin Lake Zürich proglacial to present sequence is 35 m thick (Hsü and Kelts 1984). Lower sections are dominated by fine-grained glacial sediments char-

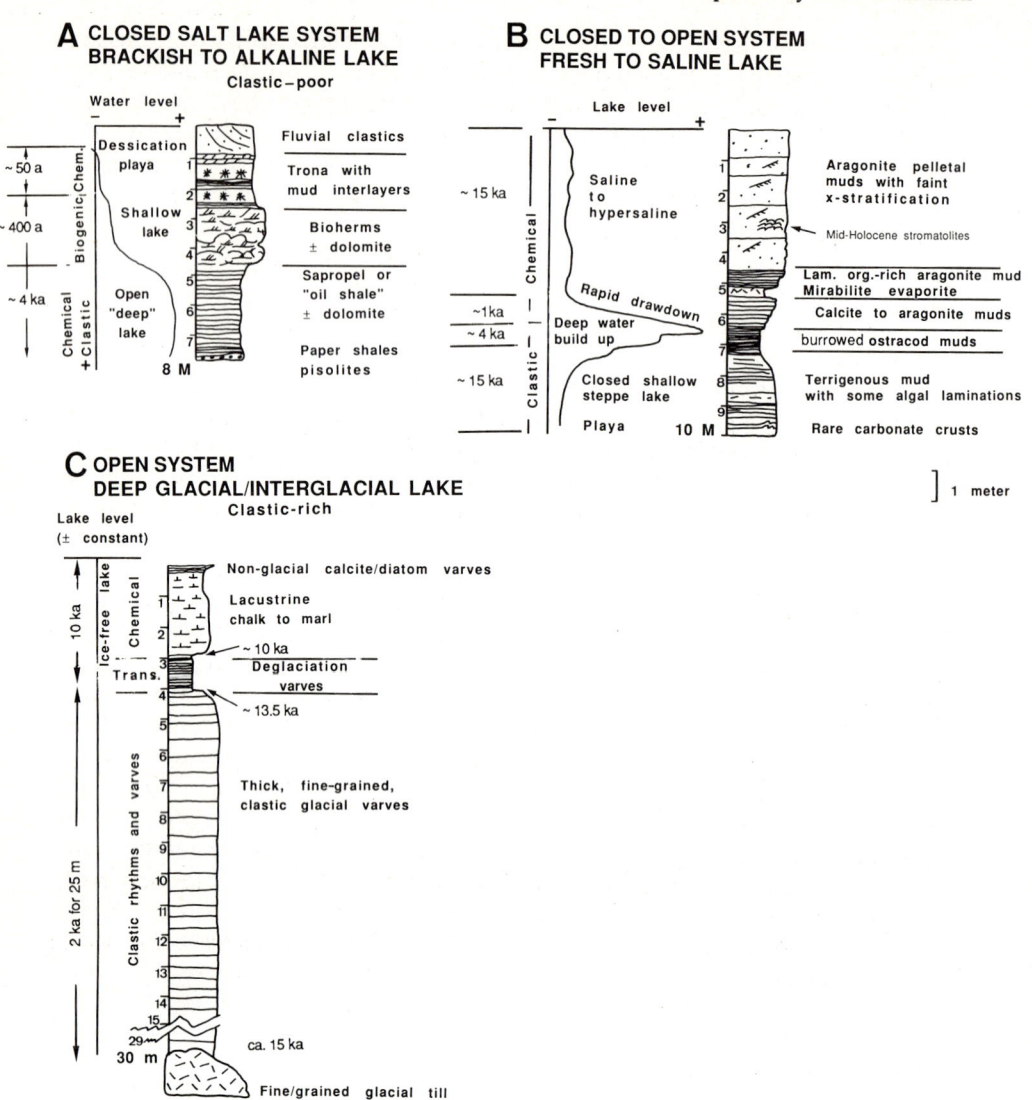

Fig. 13A-C. Comparison of hypothetical sequence development for three very different lacustrine systems driven by climate change emphasizing abrupt changes in lake levels, sedimentation and accumulation rates. **A** Closed, shallow, clastic-poor salt lakes in which a perennial alkaline lake gives way to desiccation (e.g., after Surdam and Stanley 1979; Talbot 1988). **B** Transition from lake filling of closed basin through open-system flushing with abrupt down-draw to a perennial hypersaline lake. Example from Lake Bonneville, Utah (after Spencer et al. 1984; Kelts and Shahrabi 1986). **C** Deglaciation sequence from open system glacial to interglacial lakes. (After Zhao et al. 1984; Hsü and Kelts 1984)

acterized by laminated clay-silt varves deposited during freeze-thaw conditions, respectively. These overlie older glacial tills with dropstones. Sedimentation rates for clastic varves are usually very high, typically being centimeters per year. Following glacial retreat, as meltwater clastic influx subsides with glacier retreat, new vegetation also hinders soil erosion so that chemically precipitated lacustrine chalks and marls become dominant sedimentation mechanisms. These calcareous facies may be laminated or varved and very organic-rich if basinal areas are nearly anoxic. Stacks of such glacial/interglacial lacustrine cycles are relatively rare, however, because older lake deposits are often missing due to cannibalization by subsequent glacial activity. Figure 13b represents a case which combines elements of both closed and open basin settings.

Organic carbon contents provide one sensitive monitor of changing lake conditions. Whereas organic carbon-rich pelagic sediments are a typical feature of many modern dilute lakes due to seasonal productivity blooms and water column stratification (e.g., Fig. 13c, post-glacial chemical facies), algal productivity and organic carbon preservation may reach a maximum in the brackish to saline brines of shallow closed lakes experiencing evaporative drawdown (such as Lake Bonneville transitional sediments, e.g., Fig. 13b). Algal mats bordering ephemeral lakes may also contribute significant organic matter to sediments (Burne and Ferguson 1983).

We emphasize that accumulation rates vary drastically both within and between the different facies illustrated in Fig. 13. In each sequence, varved intervals from chemical sedimentation appear condensed relative to their glacially derived counterparts. Care must be taken when evaluating climatic significance because estimates of the temporal periodicity involve combining different processes into complete cycles. It would be incorrect, for example, to assume that the glacial/interglacial cycle as represented in Fig. 13c represents a long cold period followed by a relatively brief warm period of interglacial chemical sedimentation, as it would likewise be erroneous to assume on the basis of the relative thicknesses in Fig. 13a that deep lake phases uniformly alternate with shallow/playa phases. Errors may result in periodicity studies when sedimentation rates based on varve years are extrapolated over longer cycle lengths.

9 Conclusions

Williams and Sonett (1985) stunned solar scientists with what seemed a patent case of the 11-year sunspot cycle operating since the Precambrian, and the view of stable solar activity has become a welcome standard in astrophysical textbooks. In these deposits, a series of dark bands separated, on average, 11.6 light bands. The analysis was based on the assumption that the deposits were proglacial lacustrine varves. The time-series analyses eventually crumbled, however, as evidence forced a reinterpretation of the 650-Ma-old Elatina Fm. rhythmites as extensive marine tidal-bundle deposits. The varve couplets were only schematically described, and the bundles turned out to comprise considerable sand, a component which is fairly rare in either nonglacial or true glacial varves. The reaction for geoscientists came in 1988: Williams (1988) and Sonett et al. (1988) now regard the Elatina as a marine record of Precambrian lunar tide variation. Unfortunately, the Precambrian "sunspot" dogma will most likely linger for years.

Are our examples of rhythmicities in lakes really trustworthy, and are they really proxies for paleoclimatic variation? To answer these questions, many of the deposits we have reviewed or that are shown in Table 1 are in need of a closer coupling between interpretations of depositional histories as well as rigorous time-series analysis of their varves, bedding cycles, or bundles. We are reasonably convinced by the more recent studies on the periodicities of lake levels and lacustrine bedding cycles such as those within the Milankovitch waveband, yet we would like to caution against the unconstrained use of statistical data in the interpretation of sedimentary rhythmicities. Inspection of Fig. 8 shows a variety of potential external forcing functions which may act on climate and subsequently the sedimentary record. However, Fig. 8 also shows a wide scattering of periodicities recorded in the lacustrine record which show only meager correspondence to the periods of postulated forcing. Several of the reported rhythms deviate by more than 10% from the closest forcing frequencies. Some of these deviations may be due to the inherent uncertainties associated with varve measurements. Varve years for example, may vary in length depending upon freeze-thaw periods and productivity cycles, and there are often long lags in varve years compared to calendar years. Varve thicknesses in similar lakes at similar latitudes may also vary due to local orographic effects or to local seasonal variations in weather (i.e., warmer than average summers, cooler than average winters, etc.) and identification and correction to often subtle allochthonous or intra-annual autochthonous layers is also very important. The addition of a layer due to storm input or similar events, for example, may be misinterpreted as representing 3 years while the varve thicknesses will be approximately halved. The periodicity, if present, is thus scrambled by two effects, one extending the apparent period and the other tending to convert thick varves into thin. Varve identification itself is often a very subjective process and different workers may often arrive at quite different results.

Other discrepancies between lacustrine rhythms and climatic forcings may have to do with the use of extrapolated sedimentation rates over long sections of core or outcrop (e.g., Lakes Turkana and Biwa, Green River Formation, Newark Supergroup), while others may relate to complex overprints of various, quite unrelated cycles. The length of the series being analyzed is also of obvious importance because the length of a cycle cannot be accurately determined if repeated only two or three times (Schwarzacher 1975, see discussions and examples therein). Further complications may relate to not only the inadequacies of our interpretation of the geological record but also to the imprecise resolution of some of the periods of potential climatic forcing.

Despite the above complications, there does appear to be some correlation between several of the rhythmicities reported from lakes and some of the suggested potential climate-forcing functions. These include the 22-year Hale (and to a lesser extent the 11-year sunspot frequency), 80–90-year Gleissberg and 200-year solar cycles well known from growth and isotopic studies of tree-rings, and the master orbital parameters of precession, obliquity, and eccentricity that are now becoming well established in portions of the marine record. Other less well-defined clustering of lacustrine rhythmicities appear to occur with periods of 30–40 years, 50–70 years, and about 180 years. These, too, may relate to the harmonics of solar oscillations but their correspondence to known climatic rhythms is undiscernible. In essence, such findings from different ages and different locals imply a strong potential for detailed correlation between the various rhythms of our solar system, meteorological parameters, and lacustrine sedimentation. However, just how these rhythms were transferred, especially on the regional and local level, is very poorly known and this remains one of the frontiers of paleoclimate research.

All the main Milankovitch periods of insolation changes felt at the top of the atmosphere have apparently been recorded in both mid- and low-latitude lacustrine settings and these have occurred as far back as the Triassic and probably in the Paleozoic as well (Table 1). Of these, the largest variations appear to occur around the obliquity waveband, obliquity insolation cycles being both diminished at lower latitudes and even quasi-periodic at high latitudes as well (Berger 1978a). Of all reports, the Newark Rift Valley sequence seems still the most complete with precessional rhythms bundled into obliquity and both the long and short cycles of eccentricity. To our knowledge, high-latitude basins have not been investigated for any of these patterns. While these studies are important in determining the past variability of solar insolation on the earth, the incomplete record and the degree of uncertainty within the present data set precludes further interpretation with respect to changes in the length of the orbital periods through time.

Paleoclimatic time series such as those from the Searles Lake, Lake Biwa, and the Karewa Formation are encouraging because they highlight the kind of data that might be used to establish a type-terrestrial curve of Pleistocene to Recent Milankovitch variations which can then be compared to that of the as yet better-documented marine record (e.g., Imbrie et al. 1984; Pisias and Imbrie 1986/1987). These and other studies from newer long, high-resolution varved lake sequences will have important implications for future terrestrial-marine correlations and dating, as well as for understanding how Milankovitch orbital parameters are transferred through the Earth climate system into the sedimentary record. In the long term such studies may also aid the "orbital-tuning" of the chronostratigraphic record (cf. Martinson et al. 1987) and become an important step in the development of a new global stratigraphy. There are many rhythmic lacustrine sequences in the geologic record that have yet to be tested.

Acknowledgments. Many colleagues have contributed to our ideas on lacustrine systems. We thank G. Einsele and W. Ricken for suggesting this theme and for their patient editorial comments. CRG acknowledges support and discussions from K.J. Hsü while on a one year fellowship to ETH. We appreciate the help of B. Schwertfeger with the manuscript, L. Glenn with research and K. Yemane and U. Gerber with illustrations. This is contribution 89/4 from the Geology Section, EAWAG and U.H. School of Ocean and Earth Science and Technology Contribution Nr. 2216.

Chapter 2 Event Stratification

2.1 Events and Their Signatures — an Overview

A. Seilacher

1 Events Versus Cycles

Cycles and events are relative terms that defy rigorous definition. Humans are adapted to daily and yearly cycles: gradual and predictable changes that dominate our lives, at least in rural societies. Coastal populations also register the phases of the moon because they control the tides. Events, in contrast, have the notion of being unpredictable, sudden, and therefore often catastrophic: storms, floods, earthquakes, volcanic eruptions. Still, in areas that experience such disturbances more frequently, people have learned to live with them. Lobbyism may also play a role: In early cultures, priests had a professional interest to treat less common cyclic phenomena (eclipse, rainfalls) as unpredictable events. Scientists, in contrast, try to discover periodicities in such disasters in order to make them more predictable. Nowadays, even the most unpredictable events, such as large meteorite impacts, are claimed to occur with a certain periodicity.

This means that for the theme of our book the distinction between cycles and events is disputable. Not only is it a matter of the causative forces, but also of the systems on which they act. Just as human populations are more vulnerable towards physical perturbations in one area than in another, sedimentary systems have different perceptivities. A sea level change that is registered in finely laminated deep sea sediments as a cyclic affair may be recorded in the coastal facies or in condensed slope deposits by a single event bed. This is because different disturbances may interact in the creating of the sedimentary record and it depends on the sedimentation rate, sediment property, water depth, and relative frequencies, as to which kind of perturbation will be preferably recorded in a particular setting.

2 Limits of Presentday Measure Sticks

Cycles are described as sinuous curves with certain periodicities and more or less uniform amplitudes. Thus they can be extrapolated from modern data into the past, unless periods are too long to be actualistically measured. Events, in contrast, are not only aperiodic; they also occur in varying magnitudes, the larger the rarer. Translated to geologic time spans, the open scale of intensities implies an element of nonuniformitarianism, as is well illustrated by the example of the Mt. Helens volcano. Before its dramatic eruption in 1980, information concerning lava flows, ignimbrites, wet mud flows, and airborne ash falls as consequences of volcanic activity were known. A new phenomenon in the Mt. Helens eruption was a debris avalanche on an unprecedented scale. Mobilized by an explosion high up on the flank of the peak, tremendous rock masses could pick up enough gravitational energy to rush at speeds of 350 km/h far over the foreland. These relatively dry masses could carry giant blocks and make them float to the surface, as boulders do in submarine mass flows. Once this mechanism had been observed, then the deposits of a similar, but much larger avalanche of 26 km near Mt. Shasta could be identified and related to an eruption of this volcano 20 000 years ago. In this deposit, the floating blocks reach several hundred meters in diameter – large enough to dominate the hummocky landscape up to the present day. Only one contribution (Schmincke and van den Bogaard Chap. 2.12) in our book deals explicitly with volcanic phenomena, but the general message of Mt. Helens also applies to other events: Disturbances too rare to be sampled in adequate sizes by actualistic observations must be expected in the long-term stratigraphic record.

3 Preservation Potential

The original sedimentary record is subject to secondary distortion. Larger events wipe out the signatures of earlier smaller ones by erosional cannibalism. There is also the bioturbational effect of burrowing organisms that destroys depositional structures and mixes material of originally well-separated layers. This effect has changed through the Phanerozoic, because depth as well as intensity of bioturbation have evolutionarily increased (Sepkoski et al., this Vol.). Deeper below the sediment surface, diagenetic redistribution of carbonate or silica (Ricken Chap. 7.1, Tada Chap. 4.2, both this Vol.) and differential compaction may superimpose their own signatures and introduce a horizontal and vertical rhythmicity that was less pronounced, or absent, in the original state. Since these overprints occur mainly on a minor scale, they all contribute to the overrepresentation of larger cycles and events in the fossil record.

4 The Problem with Beds

The present chapter deals mainly with the identification of the forcing mechanisms on the bed scale. But what, after all, is a bed (or a stratum, if you want to avoid any equivocality)? The miner's definition, namely a resistant layer that can be quarried as a unit, is of little help. In carbonates, diagenesis may have padded the original event layer with a concretionary under- and overbed, or it may have welded several distinct layers into one unit. Laminae in a sandstone, on the other hand, do not qualify as beds either – but do the laminae in a black shale? For the sedimentologists, beds are primarily elements of a facies and should reflect one state of the sedimentary system, for instance event beds alternating with beds of background sediments. This means, however, that lithologic criteria, as used in stratigraphic logs, lose their dominating significance. Maybe the dynamic stratigrapher has to live with the fact that his most basic unit, the bed, cannot be objectively defined. After all, why should he be better off than the paleontologist is with the definition of species?

Nevertheless, we will continue to use the term "bed". In a record full of erosional gaps, individual beds represent the only truly coherent paragraphs though of different lengths. In cyclic sequences each bed may reflect a global signal and a time span in the order of 10^4 years. Event beds, in contrast, form in a matter of hours or days and may be very local affairs. Still the within-bed succession of erosional and depositional features, as well as their relationship to the slower biological and diagenetic processes, provide us with clues about the nature of the particular perturbation, its erosive effect, and the sedimentation rates during the event versus the background situation.

5 Kinds of Event Deposits Neglected in this Book (Fig. 1)

5.1 *Inundites*

In the following section the majority of the papers is concerned with small-scale marl/limestone cycles on one side, and with tempestites and turbidites as the most common (and most familiar) kinds of event deposits on the other. We should not forget, however, that additional kinds of event deposits may become dominant in particular settings. In protected bays, for instance, river floods from adjacent land areas may be more important and effective than storms. Their deposits (inundites) resemble turbidites more than tempestites – particularly if they are associated with a marine fauna rather than with a continental re-bed facies.

5.2 *Mass Flow Deposits*

This important group of event deposits forms, like turbidites, due to downslope gravity transport, but with the particles being supported and tossed by other grains and pressurized pore water rather than being fully suspended. In detail, the interaction of buoyancy, grain-to-grain contact, turbulence, dispersive pressure, and shear is very complex and varies with grain size distribution (Pierson 1981). There may also be a bedload transport with different characteristics below the slurry (Pflueger and Selacher Chap. 2.11, this Vol.).

5.3 *Seismites*

A final class of event deposits that is underrepresented in this book are seismites (Seilacher 1969). The term refers to in-situ shock deformation of sediments rather than to secondary seismic effects, for instance in the form of tsunamis or downslope slides and turbidity currents that lose the signature of their seismic (or impact) origin in the depositional phase. The reaction of sediments to shock depends on their coherence and stiffness. In muds and sands the unconsolidated surface layer will be liquefied; i.e., particles lose their contacts so that original sedimentary structures are

Seilacher: Events and Their Signatures   225

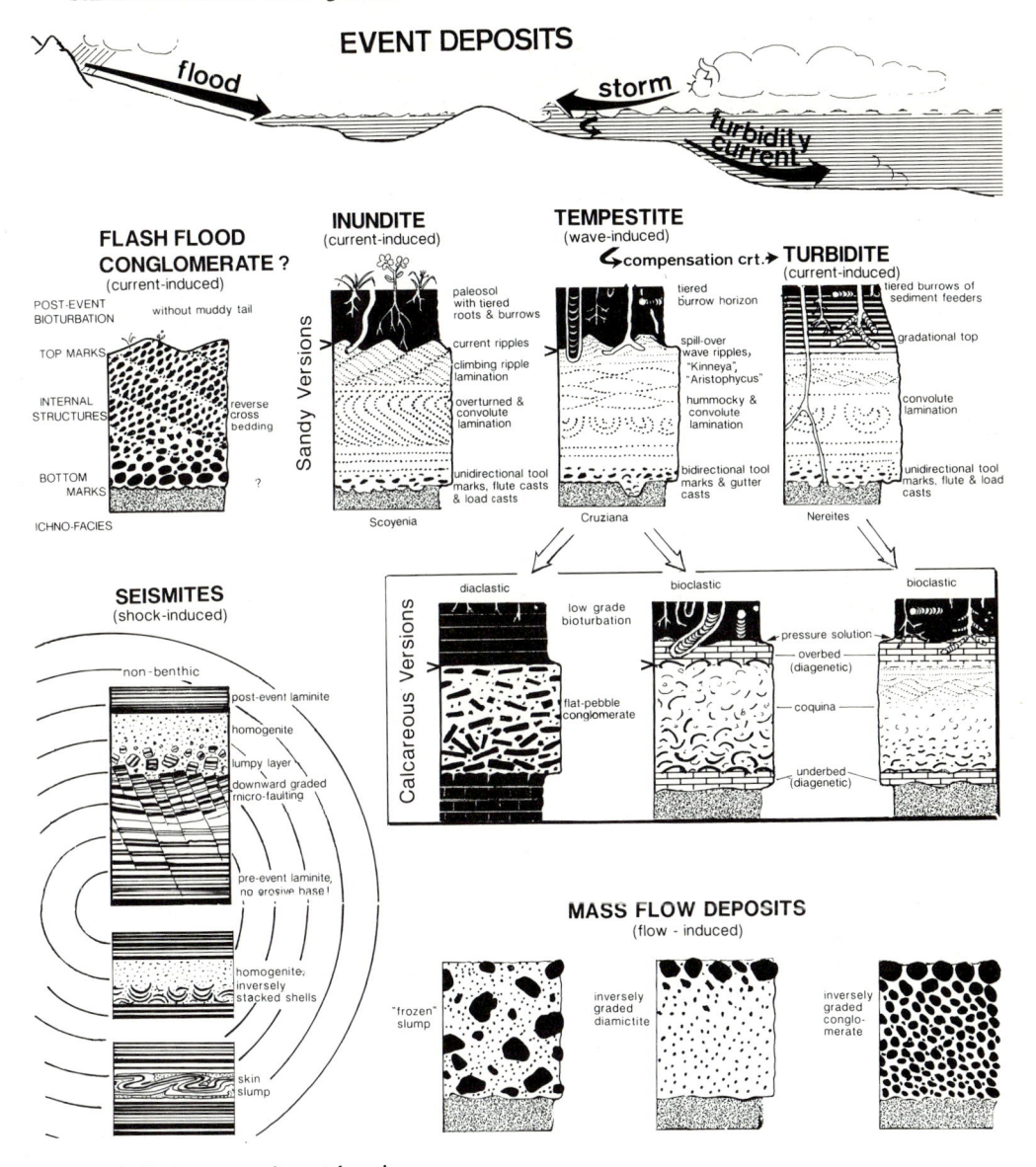

Fig. 1. Basic pattern of event deposits

wiped out, while suspension does not reach the state that allows new ones to develop. The result is a structureless homogenite (Cita et al. 1984; Kastens and Cita 1981), occasionally marked by a basal layer in which outsized shells have been concentrated and stacked during liquefaction in a predominantly concave-up position (Seilacher 1984b). In other settings, more compacted layers underneath the homogenite may react by graded microfaulting, if some instability due to slope, or to a more liquefiable layer underneath exists (Fig. 1).

Compared to the frequency of earthquakes on a geologic time scale, seismites are dramatically underrepresented in the stratigraphic record. Why is this so? Firstly, sedimentologists may not have watched out for them as much as they did for turbidites or storm deposits. Secondly, seismic signatures are, in most settings, likely to be deleted by other events, such as storms, or by bioturbation. Therefore, seismites are most likely to be preserved, and most easily recognized, in the finely laminated muds of quiet anoxic basins. Thirdly, mechanical properties make some sediments more susceptible to shock deformation than others.

The "Wellenkalk" of the Lower Muschelkalk (M. Triassic, Germany) is a good example. It acquired its name from closely set microfaults of probably seismic origin, a feature that is rarely found in the Upper Muschelkalk rocks of the same basin. This does not mean that there were more earthquakes in an early phase, but may reflect more saline and less oxygenated bottom water conditions that reduced the destructive effects of bioturbation. Such conditions would have favored the development of microbial mats that not only counteracted the mixing of sediment and water, but could also have had an osmotic effect if the underlying sediment was a mud turbidite with less saline pore water.

Biomats may also have been involved in the conspicuous wrinkling of individual laminae, or varves, in the otherwise evenly laminated Lissan Marls (Pleistocene) of the Dead Sea Rift valley in Jordan (Fig. 1). In all wrinkled layers the folds are inclined towards the basin center, indicating that a thin skin of mucus-bound sediment began to slide downslope, but could not develop into a major slump or turbidite due to the short duration of the shock.

Problems of dynamic stratigraphy are manifold. Firstly, we must try to remove secondary overprints. Secondly, we should test the sequence for repetitive order in the form of periodic change (cycles), or of events versus quieter and slower background sedimentation. Thirdly comes the task to identify the nature of cycles or events and finally the question how complete the preserved record is at various scales.

In the Dead Sea example, cyclic phenomena (varves) potentially provide the measure stick for charting noncyclic events (earthquakes) over time spans longer than human history. Much more observational and experimental work needs to be done before detailed paleo-chronologies of this kind can be safely established. Nevertheless, this is how cycles and events in stratigraphy should be viewed: not as conflicting antitheses, but as two complementary aspects in the analysis of the rock record that stratigraphy is all about.

Chapter 2 Event Stratification

2.2 Shallow Marine Storm Sedimentation – the Oceanographic Perspective

D. Nummedal

1 Introduction

One of the aims of this book is to provide an actualistic foundation for the interpretation of sedimentological signals in the stratigraphic record. This chapter serves that aim by providing an up-to-date, relatively jargon-free account of the physics of nearshore water movement and sediment transport during storms. This is a necessary foundation for the proper interpretation of shallow marine storm beds (tempestites).

The accompanying chapter by Seilacher and Aigner (Chap. 2.3, this Vol.) provides an in-depth discussion of the sedimentological signatures of ancient storm beds. The two chapters were not designed to provide a series of easy "process-response" explanations for intriguing observations. On the contrary, the evident difference in views between the "oceanographic perspective" of this chapter and the "geological perspective" of the other (terminology from Walker 1984), is intended to highlight our communal ignorance and stimulate intelligent inquiry.

Storms are atmospheric disturbances which affect marine sedimentation only after a complex pathway of energy transfers, first from air to sea and then from water column to sea floor. It is appropriate, therefore, first to consider how this energy transfer occurs. The first step is accomplished through generation of waves and currents by wind shear on the sea surface. The physics of the various energy and momentum transfer mechanisms which operate at this stage is a subject of active research. The waves that result from a given storm can be calculated reasonably well by semi-empirical equations. In contrast, accurate calculation of the nearshore currents still lies in the future. The second step in the energy transfer, from the water column to the sea floor, is even more complex because of wave motions of different frequencies and orientation relative to the shoreline and the interactions between such waves and current motions on different spatial scales. Since these are the very processes that directly drive storm sediment transport and deposition, however, the core of this paper will deal with our present understanding of their nature and effects.

The complexities of shallow marine storm sedimentation are great, and many of the simplified models commonly presented in the geological literature are wrong. For example, the dominant role of long-period infragravity waves and edge waves in surf zone storm sedimentation is rarely considered. The role of the coastal jet in the slightly deeper water of the lower shoreface or inner shelf has generally also gone unrecognized. Finally, the failure to distinguish paleocurrent indicators produced by instantaneous peak shear stresses beneath shoaling waves from those that record

Einsele et al. (Eds.)
Cycles and Events in Stratigraphy
©Springer-Verlag Berlin Heidelberg 1991

longer-term sediment transport easily yields meaningless reconstructions of ancient sediment dispersal paths. This chapter will have achieved its purpose if in future studies of event stratification all geologists will consider the total spectrum of shallow marine storm processes.

2 Storm Energy Transfer

2.1 Waves

The actual mechanisms for generation of waves by wind have been much debated over the years (see review by Stewart 1967) and are still inadequately understood. As a consequence, the available approaches to wave prediction (or hindcasting) have been semi-empirical (see Komar 1976 or CERC 1984 for summaries). It is clear from the many studies that have been conducted that storm wave heights and periods are functions of (1) the wind speed, (2) the fetch length and (3) the duration of the wind. For a given wind speed and fetch there is a maximum sustainable wave height referred to as fully developed sea, a condition characterized by a transfer of energy from the wind to the wave field at the same rate as energy is dissipated internally in the water column and radiated out of the area of the storm.

It is important to recognize that the propagation of deep-water waves is dispersive, i.e., waves move at a speed (celerity) proportional to their period of oscillation. Because all storms at sea generate a spectrum of wave periods (frequencies), the individual frequency bands begin to sort themselves out during propagation away from the center of the storm. The long-period waves (swell) will take the lead. This is why ancient mariners could infer the existence of hurricanes far from their ships by the appearance of very-low-amplitude, long-period "forerunners" of the swell, long before the hurricane winds appeared. In a classical paper, Barber and Ursell (1948) demonstrated dispersion of deep-water waves by measuring, at the shores of England, the long-period, early forerunners of swell generated by a tropical storm off the east coast of the United States.

In some cases the long-period forerunners may be the only evidence of a distant storm. Long period swell arriving at the U.S. west coast in summer often derives from storms in the Gulf of Alaska or the South Pacific. It is noteworthy that wave attenuation during such long-distance travel is relatively minor and occurs within the first 1000 km of travel. Moreover, attenuation is most pronounced in the short-period part of the spectrum (Snodgrass et al. 1966). Consequently, distant storms can propagate great wave energies towards an open-ocean shoreline and, because of their long period, such swell waves affect the seafloor to great depths where they have significant impact on marine sediment transport. Large waves arriving at a shoreline without accompanying local winds are sometimes referred to as wave storms.

2.2 Currents

While some of the energy transferred from the atmosphere is dispersed by storm waves to distant shores, the storm-generated currents operate on a spatial scale comparable to the size of the storm system (Swift and Nummedal 1987). These currents are initiated both by the direct action of the surface wind stress and by barotropic forcing by sea level set-ups or set-downs against the shore. It is well documented that storm-generated currents are widespread in shallow marine environments, and that they easily attain current velocities capable of moving any sediment present on the sea floor. For a review of the general issue of storm-generated currents, the reader is referred to Beardsly and Butman (1974), Csanady (1982), Swift et al. (1986), and Snedden et al. (1988). It is important to stress that storms at sea always generate currents as well as waves. Any wind capable of generating significant waves also generates currents. As a rule of thumb wind-generated sea surface currents are typically about 3–5% of the velocity of the wind.

The physics of shallow marine sediment transport differs depending on water depth. The consequent patterns of water movement must be addressed separately. Four distinctly different dynamic zones are generally recognized (Mooers 1976; Swift et al. 1985; Snedden and Nummedal 1990). These are: (1) the surf zone, (2) the friction-dominated zone, (3) the transition zone, and (4) the geostrophic zone (Fig. 1). Although the boundaries move according to the storm energy these dynamic zones correspond roughly to: (1) the upper shoreface, (2) the remainder of the shoreface, (3) the shoreface-shelf transition zone, and (4) the open continental shelf, respectively.

Recent papers by Swift et al. (1986) and Snedden et al. (1988) present comprehensive discussions of storm sedimentation on open shelves. To allow room for greater depth of discussion, therefore, this paper is focused exclusively on the beach and shoreface.

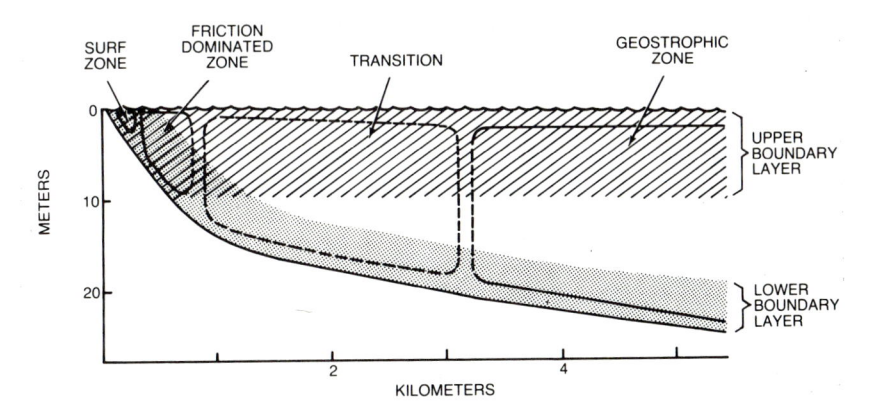

Fig. 1. Coastal ocean dynamic zones. Note the surf zone and friction-dominated zone closest to land where the upper and lower boundary layers overlap. Only in water depths sufficient for separation of these boundary layers is there a core of truly geostrophic flow. (Swift et al. 1985)

3 Surf Zone

Within the surf zone the time-averaged flow is dominated by a longshore current and trans-shore circulation cells driven by momentum transfer from the breaking waves (Putnam et al. 1949; Bowen and Inman 1969; Komar 1983). Superimposed on this system of longshore and rip currents (Fig. 2A) are complex wave motions involving the breaking of the gravity waves, reflection of long-period infragravity waves and the presence of standing or progressive shore-normal edge waves. Although edge waves and infragravity waves are always present, their role in surf zone sedimentation increases during storms (Holman 1981, 1983). This is because gravity waves are limited in height to a magnitude approximately equal to the water depth. As the waves

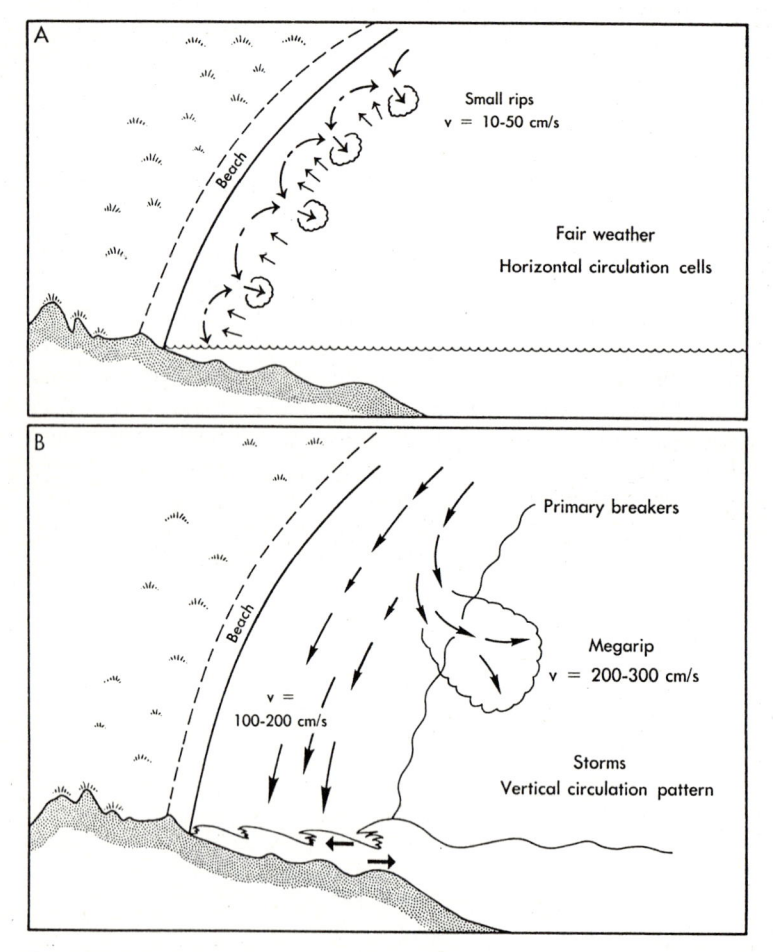

Fig. 2A,B. Generalized surf zone circulation patterns. **A** Fair weather cell circulation characterized by narrow, small rip currents separated by broad zones of weak landward flow. Negligible vertical flow segregation. **B** Storm circulation dominated by strong longshore currents, a few megarips and well-developed vertical flow segregation. The megarips are generally topographically controlled

shoal upon approaching a beach during storms they break once this threshold is exceeded. Long-period (infragravity) waves and edge waves, however, are not depth-limited and carry an increasing percentage of the total wave energy onto the beach as the storm intensifies (Fig. 3). Beach erosion and sediment transport during storms, therefore, is directly related to the action of edge waves and long-period waves (Holman 1981; Wright and Short 1983).

Surf zone sediment transport during storms is directed longshore and offshore. Longshore transport is particularly effective in nearshore bar-troughs where the breaker-induced current generally attains a maximum (Sonnenfeld and Nummedal 1987). The highest transport rates and suspended sediment concentrations generally occur a short distance landward of the zone of primary breakers (Fig. 4B). The current itself generally attains maximum speed a little farther landward and extends some distance seaward of the primary breaker zone because of eddy-viscosity along its seaward margin (Fig. 4A). Offshore-directed sediment transport in the surf zone during storms is driven by rip currents (Shepard and Inman 1950; Cook and Gorsline 1972; Wright and Short 1983; Short 1985), offshore-directed net oscillatory motion beneath long-period (infragravity) waves (Bowen 1980; Holman and Bowen 1982) and mean surf zone bottom currents (Dally and Dean 1984; Fig. 2B). This latter mechanism will induce offshore sand movement along the entire shoreline, not only at laterally restricted rip-current sites.

Offshore movement of a nearshore bar system during storms has been documented by Nummedal et al. (1984), Sallenger et al. (1985), Sallenger and Howd (1989), and others (Fig. 5). These studies yield results consistent with the infragravity model mentioned above. The very precise monitoring of bar movement performed at the Field Research Facility of the Army Corps of Engineers at Duck, North Carolina in October of 1982 by Sallenger et al. (1985) show that the bar moved offshore and increased in height during the rising phase of the storm. Its offshore distance was never more than 10% of the width of the surf zone; therefore, the existence of the bar is clearly unrelated to the position of the primary breakers. Instead, the observations do support the model that the bar is formed due to bottom sediment convergence beneath standing long-period (infragravity) waves in the surf zone. Holman and

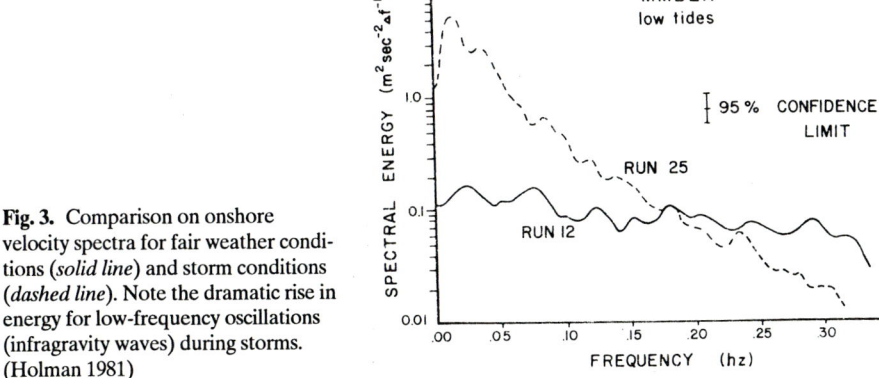

Fig. 3. Comparison on onshore velocity spectra for fair weather conditions (*solid line*) and storm conditions (*dashed line*). Note the dramatic rise in energy for low-frequency oscillations (infragravity waves) during storms. (Holman 1981)

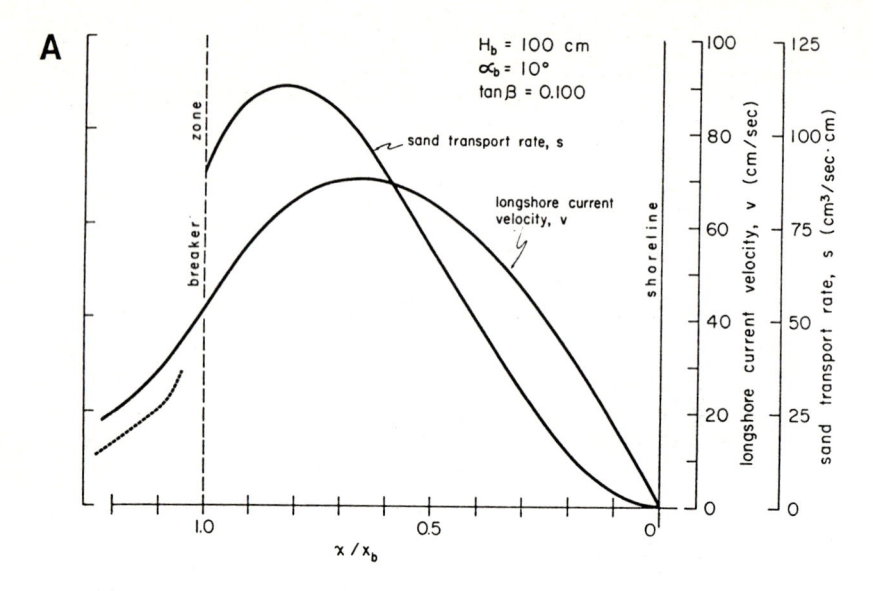

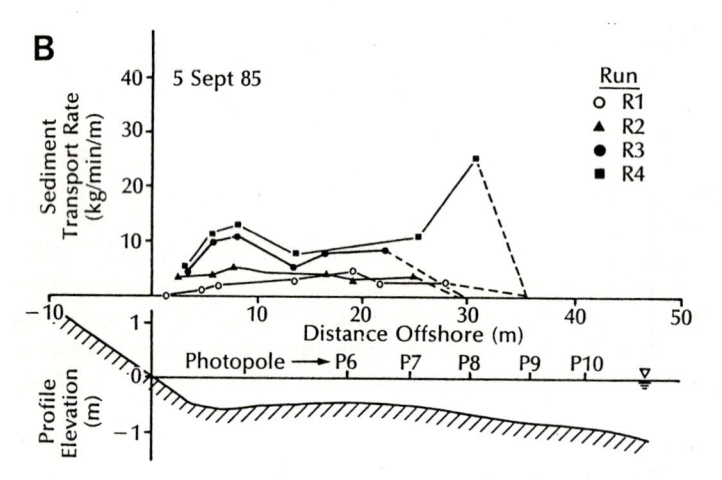

Fig. 4A,B. Theoretical and empirical profiles of longshore current velocity and sediment transport rates across a surf zone. **A** Calculated profiles from Komar (1976). Note that the distance across the surf zone is dimensionless with the primary breaker at position 1.0 on the left. **B** Measured distribution of longshore sediment transport across a 35-m-wide surf zone at Duck, North Carolina. Breaker height on the day of the experiment was 1.2 m. Note that the offshore direction is to the *right* in panel **B**. (Kraus and Dean 1987)

Fig. 5A-C. Observed changes in nearshore bars during storms. **A** Seasonal profile changes at Erie, Pennsylvania (Lake Erie). The outer bar is always best developed and located the farthest lakeward in May of all three years, reflecting early spring storms in the lake (Nummedal et al. 1984). **B** and **C** High-precision profiles obtained with a surf zone "sled" at Duck, North Carolina, documenting offshore movement during a storm (**B**), and landward bar migration during the post-storm recovery period (**C**). The significant wave height during this storm was in excess of 2 m

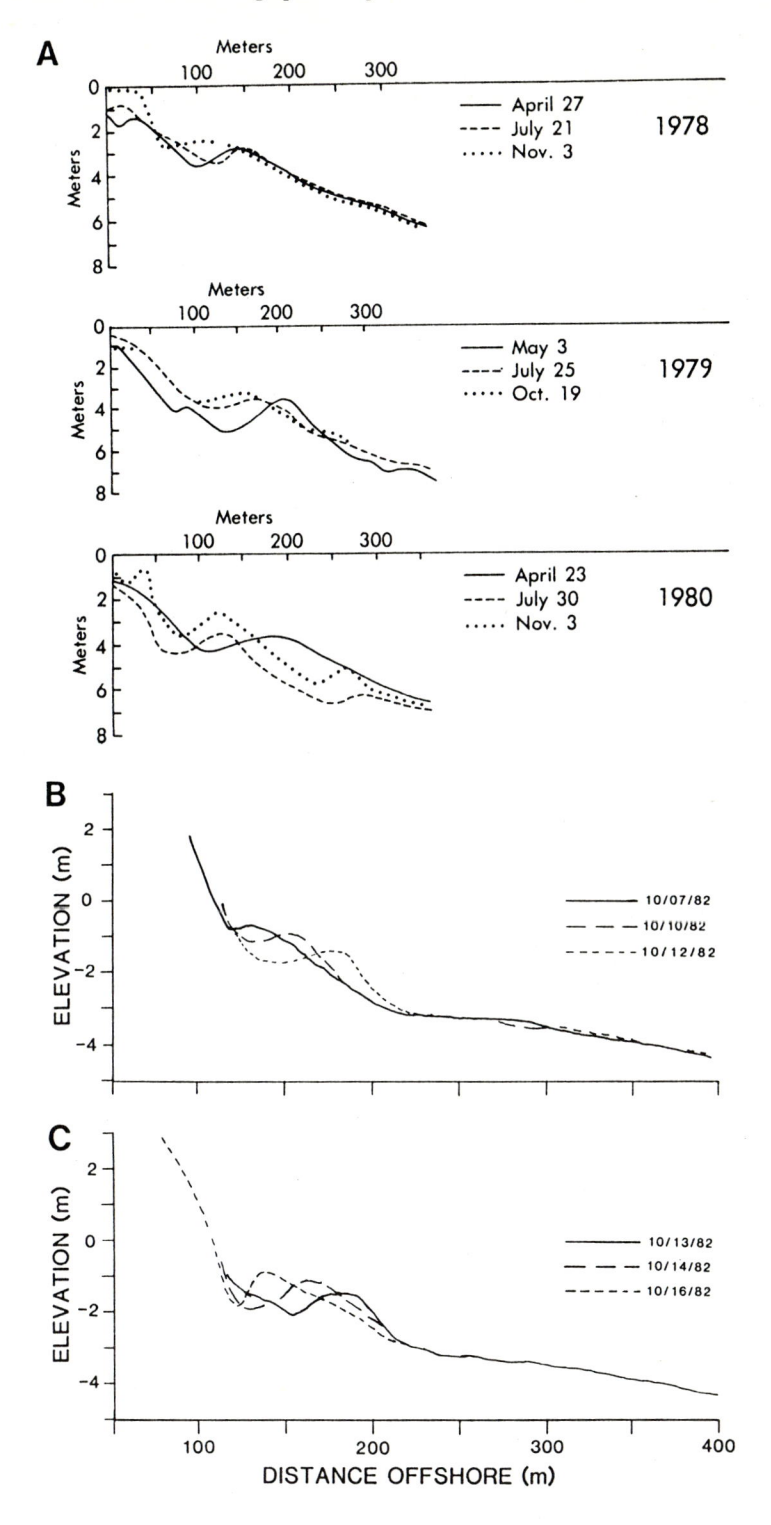

Sallenger (1985) document that the wave energy spectrum did, in fact, contain a significant infragravity component during this storm.

The cause of nearshore bar formation during storms, however, is still controversial. In contrast to the studies just reviewed, investigations by Sunamura and Maruyama (1987), Dally and Dean (1984), and Dally (1987) indicate that bar formation does occur at the location of breaking storm waves. According to their hypothesis, the mean surf zone bottom currents would carry sand to the zone of primary breakers where the bar would form. The uncertainty regarding the origin of longshore bars is reflected in a recent paper by Sallenger and Howd (1989). After evaluating several hypotheses, they concluded that "the offshore migration of near-shore bars is not necessarily associated with break-point processes".

Regardless of the specifics of surf zone dynamics, it is clear that the surf zone circulation pattern is fundamentally different during fair weather and storms. During fair weather the circulation is horizontally segregated into "cells" consisting of narrow, seaward-directed rip currents separated by broad zones of weak landward flow (Fig. 2A). In contrast, during storms the circulation becomes more vertically segregated with offshore flow near the bottom and onshore flow near the surface. This type of circulation is particularly well documented for the high-energy shoreline of south Australia (Fig. 6). Large, topographically controlled mega rips (Short 1985), however, also play a significant role during storms (see below).

Discussions have so far ignored variability in grain size of surf zone sediments. On bimodal or other poorly sorted beaches one may encounter opposite directions of cross-shore transport for coarse and fine grain sizes. For example, Richmond and Sallenger (1984) documented that during the offshore movement of the fine-grained bar at Duck in October 1982 (Fig. 5B) the coarse sand moved onshore to accumulate on the lower foreshore. Bowen (1980) documented theoretically that if a given grain size is in precise equilibrium with the local slope and wave regime, then finer grains will move offshore and the coarser ones onshore.

Sandy surf zones undergo a characteristic morphodynamic succession of erosion, bar formation, and beach accretion in response to storm and post-storm recovery processes (Fig. 7; Wright and Short 1983, Sonnenfeld and Nummedal 1987). The deepest scour in the surf zone occurs within bar-troughs and rip-channels during

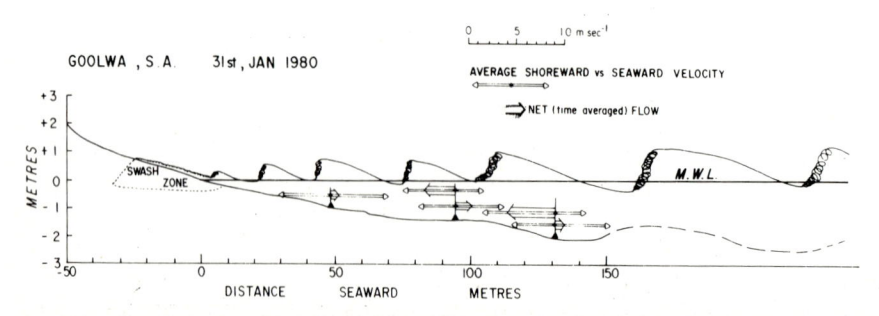

Fig. 6. Documentation of segregated vertical flow across a high-energy, south Australian surf zone. Onshore flow characterizes the near-surface part of the water column while bottom waters move offshore. (Wright et al. 1982)

storms. Bars generally move offshore and are reduced in relief during the peak of the storm; they then reform and migrate landward during the post-storm recovery phase. Bars clearly play a significant role in upper shoreface morphodynamics and in the long-term sediment budget for a beach. Moreover, since bar migration controls the evolution of surf zone channels and troughs, a morphodynamic reconstruction of an ancient bar system should precede the interpretation of surf zone paleocurrents. The bars themselves may rarely be preserved, but the fill of the intervening channels and troughs generally account for most preserved surf zone strata.

4 Shoreface

4.1 Sediment Exchange

The literature on coastal sediment dynamics and engineering has for decades been dominated by studies of surf zone processes. In fact, engineering concepts such as "closure depth" for cyclic changes in the beach profile (Hallermeier 1981), have given the misleading impression that there is little exchange of sand between the surf zone and deeper-water zones. It is now clear, however, that "beaches and surf zones can only be well understood as the uppermost components of a much larger system encompassing the inner shelf and the shoreface" (Wright 1987, p. 26). Sediments are freely exchanged between the beach and deeper-water environments during storms.

In dynamics terms, most of the shoreface corresponds to the friction-dominated zone because the friction-generated bottom boundary layer spans the entire water column. In deeper water, in contrast, the bottom and surface boundary layers are separated by a "core" of interior geostrophic flow (Fig. 1). A summary of processes responsible for sediment transport across the shoreface follows. The review has been arranged to grade from those processes believed to be reasonably well understood to those that are still highly speculative. In that order, the storm-related processes that exchange sand across the shoreface between the beach and shelf are: (1) wind-generated currents, (2) waves, (3) wave-current interactions, (4) tidal currents, (5) rip currents, and (6) turbidity currents. Shoreface sedimentation is also greatly affected by baroclinic currents driven by nearshore water density gradients. Such processes, however, are not storm-related and probably do not produce the event beds which are the main focus of this book.

4.2 Wind-Generated Currents

The friction-dominated zone extends from the edge of the breaker zone, at a water depth of a few meters, to the inner shelf at a water depth between 10 and 20 m, depending on the wave energy. The width of the zone ranges from 100's to 1000's of meters. The most powerful mesoscale patterns of water motion above this shoreface (seaward of the surf zone) are the wind-generated, alongshore coastal jets (Murray 1975; Csanady 1982; Niedoroda et al. 1985; Cochrane and Kelly 1986) and attendant bottom current downwelling or upwelling (Fig. 8). These jets are generated in response to wind shear across the shallow water of the shoreface and attendant

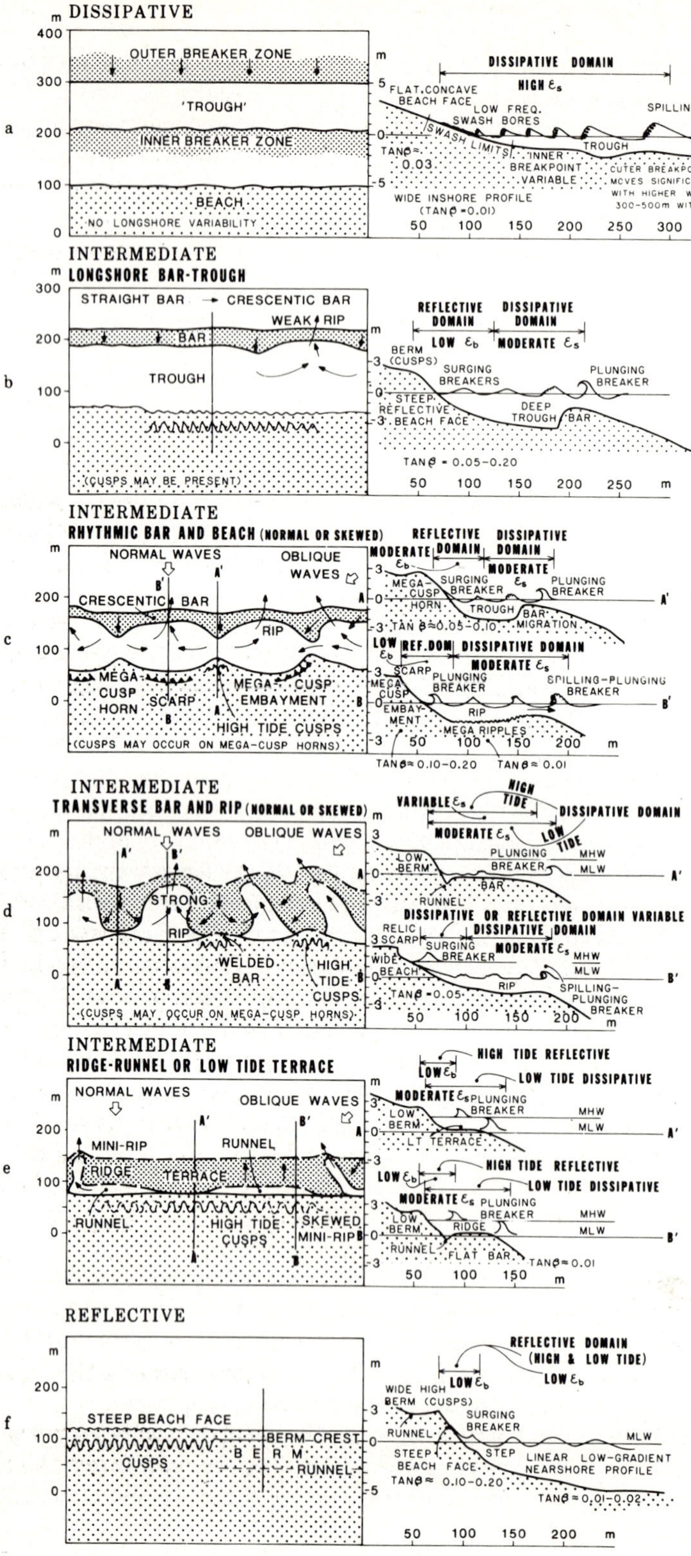

a
b
c
d
e
f
DISSIPATIVE
OUTER BREAKER ZONE
'TROUGH'
INNER BREAKER ZONE
BEACH
NO LONGSHORE VARIABILITY
DISSIPATIVE DOMAIN
HIGH ε_s
FLAT, CONCAVE BEACH FACE
LOW FREQ. SWASH BORES
SWASH LIMITS
SPILLING BREAKERS
MLW
TANβ≈ 0.03
INNER BREAKPOINT VARIABLE
TROUGH
OUTER BREAKPOINT MOVES SIGNIFICANTLY SEAWARD WITH HIGHER WAVES (COMMONLY 300-500m WITH 3m WAVES)
WIDE INSHORE PROFILE (TANβ =0.01)
INTERMEDIATE
LONGSHORE BAR-TROUGH
STRAIGHT BAR → CRESCENTIC BAR
WEAK RIP
BAR
TROUGH
(CUSPS MAY BE PRESENT)
REFLECTIVE DOMAIN
DISSIPATIVE DOMAIN
LOW ε_b
MODERATE ε_s
BERM (CUSPS)
SURGING BREAKERS
PLUNGING BREAKER
STEEP REFLECTIVE BEACH FACE
DEEP TROUGH
BAR
TANβ = 0.05-0.20
INTERMEDIATE
RHYTHMIC BAR AND BEACH (NORMAL OR SKEWED)
NORMAL WAVES
OBLIQUE WAVES
CRESCENTIC BAR
RIP
MEGA CUSP HORN
SCARP
MEGA CUSP EMBAYMENT
HIGH TIDE CUSPS
(CUSPS MAY OCCUR ON MEGA-CUSP HORNS)
MODERATE ε_b
REFLECTIVE DOMAIN
DISSIPATIVE DOMAIN
MODERATE ε_s
MEGA-CUSP HORN
SURGING BREAKER
PLUNGING BREAKER
TROUGH
BAR MIGRATION
TAN β≈0.05-0.10
LOW ε_b
REF. DOM
DISSIPATIVE DOMAIN
MODERATE ε_s
SCARP
MEGA CUSP EMBAYMENT
PLUNGING BREAKER
SPILLING-PLUNGING BREAKER
RIP
MEGA RIPPLES
TANβ≈ 0.10-0.20
TANβ≈ 0.01
INTERMEDIATE
TRANSVERSE BAR AND RIP (NORMAL OR SKEWED)
NORMAL WAVES
OBLIQUE WAVES
STRONG RIP
WELDED BAR
HIGH TIDE CUSPS
(CUSPS MAY OCCUR ON MEGA-CUSP HORNS)
VARIABLE ε_s
HIGH TIDE
DISSIPATIVE DOMAIN
MODERATE ε_s
LOW TIDE
LOW BERM
PLUNGING BREAKER
MHW
MLW
RUNNEL
BAR
DISSIPATIVE OR REFLECTIVE DOMAIN VARIABLE
DISSIPATIVE DOMAIN
RELIC SCARP
SURGING BREAKER
MODERATE ε_s
WIDE BEACH
RIP
SPILLING-PLUNGING BREAKER
TANβ ~0.05
INTERMEDIATE
RIDGE-RUNNEL OR LOW TIDE TERRACE
NORMAL WAVES
OBLIQUE WAVES
MINI-RIP
RUNNEL
RIDGE
TERRACE
RUNNEL
HIGH TIDE CUSPS
SKEWED MINI-RIP
HIGH TIDE REFLECTIVE
LOW ε_b
LOW TIDE DISSIPATIVE
MODERATE ε_s
PLUNGING BREAKER
LOW BERM
LT TERRACE
HIGH TIDE REFLECTIVE
LOW TIDE DISSIPATIVE
RIDGE
FLAT BAR
TANβ≈ 0.01
REFLECTIVE
STEEP BEACH FACE
BERM CREST
CUSPS
BERM
RUNNEL
REFLECTIVE DOMAIN (HIGH & LOW TIDE)
LOW ε_b
WIDE HIGH BERM (CUSPS)
SURGING BREAKER
MLW
STEP
STEEP BEACH FACE
LINEAR LOW-GRADIENT NEARSHORE PROFILE
TANβ≈ 0.10-0.20
TANβ ≈ 0.01-0.02

sea-surface set-ups and set-down. In a theoretical analysis of such jets, Murray (1975) found that the currents should be nearly parallel to the coast and essentially independent of the wind direction. These predictions are supported by results from drogue measurements off the coast of the Florida Gulf coast. While the current direction is controlled by the trend of the shoreline, the current speed is very sensitive to the wind direction with the strongest currents being produced by shore-parallel winds. A two- or three-layered flow field is usually generated in the coastal jet. In the absence of a vertical density stratification in the water column, a coastal set-up would be associated with a slight onshore flow in the upper few meters and offshore flow at depth. Typically, drogue trajectories (which follow water particle paths) would deviate from the shore-parallel direction by less than 10 degrees. In a moderately stratified water column a three-layer flow might develop. In this case, currents near the surface and the bottom are oriented obliquely onshore whereas the mid-depth water masses move offshore (Murray 1975).

In areas with strongly developed density stratification, such as along coasts with great fresh water efflux, the coastal jet may be entirely baroclinic in origin. In this case it becomes a much more permanent coastal feature than the storm-driven barotropic jets (Murray and Young 1985).

Scott and Csanady (1976), Swift et al. (1985), and Niedoroda et al. (1985) have documented in a series of measurements off the south shore of Long Island, New York, that onshore-directed wind stress does indeed generate a system of downwelling circulation and obliquely offshore-directed bottom flows. In addition, the

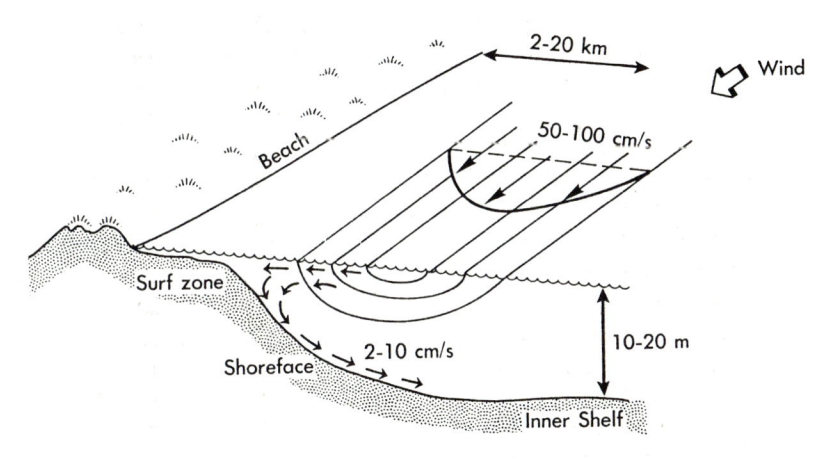

Fig. 8. Generalized diagram of the coastal jet with typical alongshore and cross-shore velocity components measured during moderate storms. Note that the coastal jet is located significantly seaward of the surf zone

Fig. 7a-f. The six stages of beach and surf zone morphology which characterize beaches in Australia, and probably worldwide. The associated circulation patterns and profile configuration are included. During peak storm the beaches are dissipative (first stage). They then gradually recover through four successively lower-energy intermediate stages. The final stage, a reflective beach, is only rarely attained after long periods of fair weather swell. (Diagram from Wright and Short 1983)

alongshore flow trajectories outline the presence of a coastal jet, i.e., a velocity maximum, about 2 to 3 km offshore, essentially above the toe of the shoreface (Fig. 9A). During the relatively mild storm event documented by Niedoroda et al. (1985) the peak surface velocity of the coastal jet was 70 cm/s, and the maximum offshore velocity component (about 2 m above the bed) was 10 cm/s. Niedoroda et al. (1985) also document the presence of a sea surface set-down, an upwelling circulation system, and a weak onshore-directed bottom current during an episode of offshore wind stress (Fig. 9B).

Records obtained at a near-bottom mooring in 8 m of water depth off Duck, North Carolina, during a fall storm in 1985 also document a clear case of shoreface downwelling currents and associated offshore sediment transport (Wright et al. 1986a; Fig. 10). This flow was also associated with a coastal jet flowing to the south seaward of the surf zone. Wright et al. (1986) concurrently obtained high-resolution sonar measurements of sea floor elevation changes. The time-series of the bed response shows initially a slow accretion in response to moderate winds from the south (Fig. 10B). This was followed by a gradual but significant period of scour during peak storm. Sea floor accretion at this 8-m-deep site was initiated during a secondary storm peak but the most rapid net accretion clearly occurred during the waning stages of the storm (Fig. 10B). The scour reached a maximum of –6 cm followed by net accretion to a sea floor elevation of +10 cm, implying a resultant storm-bed thickness of 16 cm. This is a typical thickness for many measured storm beds in ancient shallow marine deposits (Nelson 1982), although some individual shoreface storm beds may exceed a thickness of 1 m. Side-scan sonar images of the North Carolina shoreface obtained in the vicinity of the measurement site document the existence of distinct sediment lobes which had migrated offshore during this storm. The accretion noted at the instrument site probably reflects the migration of one such lobe into the study area.

Coastal jets and their importance to shoreface and shelf sedimentation are well documented along many shorelines. Murray et al. (1980) documented a strong eastward-directed jet off the Nile Delta in the southeastern Mediterranean. Because of the promontory at the mouth of the Damietta River, this jet is deflected offshore to a distance of more than 30 km (Fig. 11) and forms the upwind flank of a complete anticyclonic circulation gyre. The jet is associated with a broad, arcuate zone of shelf sandbodies which extends as a "plume", eastward from the Damietta River mouth (Coleman et al. 1981). Although this jet is a fairly permanent feature, the strongest currents and peak sediment transport episodes are associated with eastern Mediterranean storms.

The nearshore zone affected by high-velocity coastal currents may be very broad. Along shorelines with promontories, such as the Nile delta case discussed above, the

Fig. 9A,B. Current velocity fields across the shoreface off Long Island, New York during **A** a mild downwelling event and **B** a mild upwelling event. In both diagrams the upper part presents the alongshore and the lower part the cross-shore velocity components. *Contours* are in cm/s and *dashed lines* represent either eastward or onshore flow. Case **A** is characterized by onshore flow near the surface and a compensating downwelling flow near the bottom. (All diagrams from Niedoroda et al. 1985)

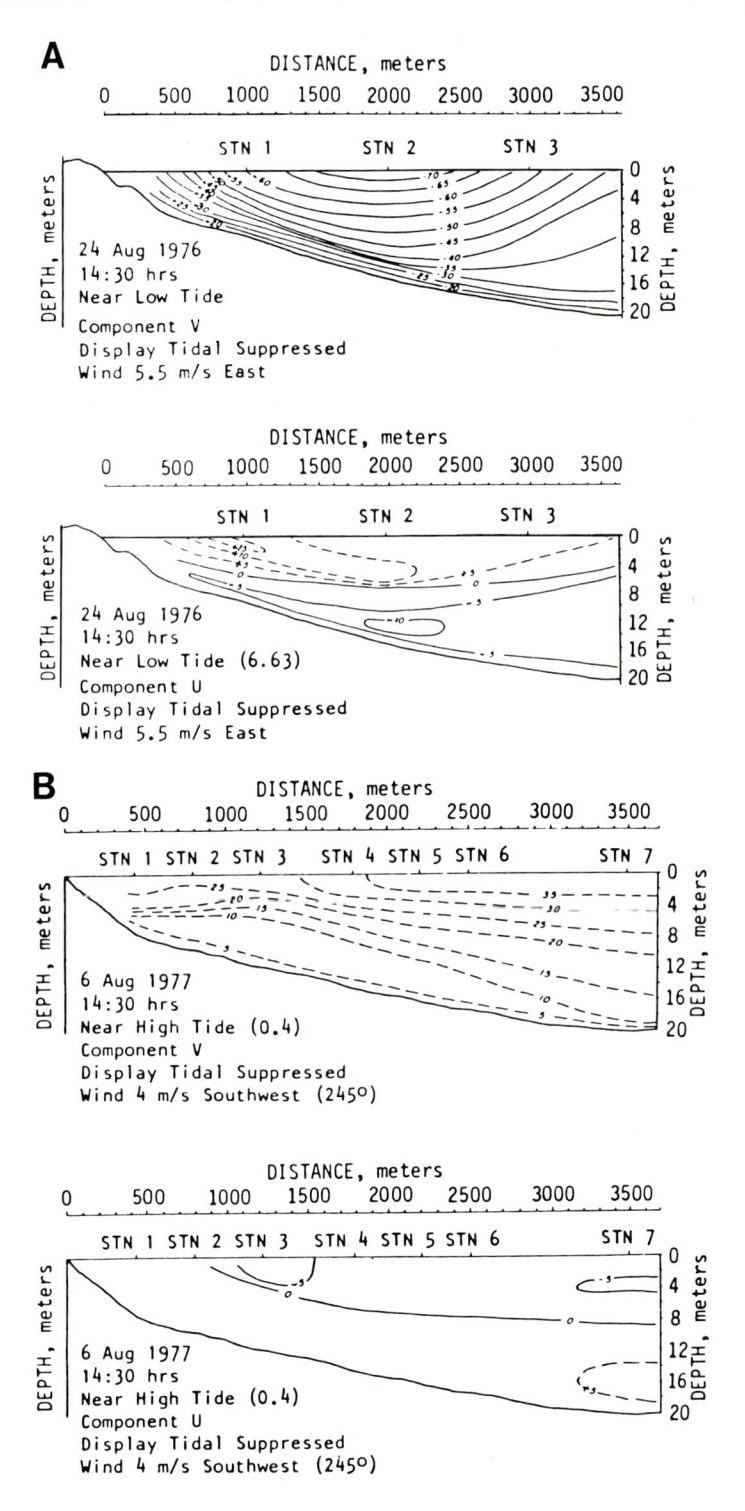

A

DISTANCE, meters

24 Aug 1976
14:30 hrs
Near Low Tide
Component V
Display Tidal Suppressed
Wind 5.5 m/s East

24 Aug 1976
14:30 hrs
Near Low Tide (6.63)
Component U
Display Tidal Suppressed
Wind 5.5 m/s East

B

6 Aug 1977
14:30 hrs
Near High Tide (0.4)
Component V
Display Tidal Suppressed
Wind 4 m/s Southwest (245°)

6 Aug 1977
14:30 hrs
Near High Tide (0.4)
Component U
Display Tidal Suppressed
Wind 4 m/s Southwest (245°)

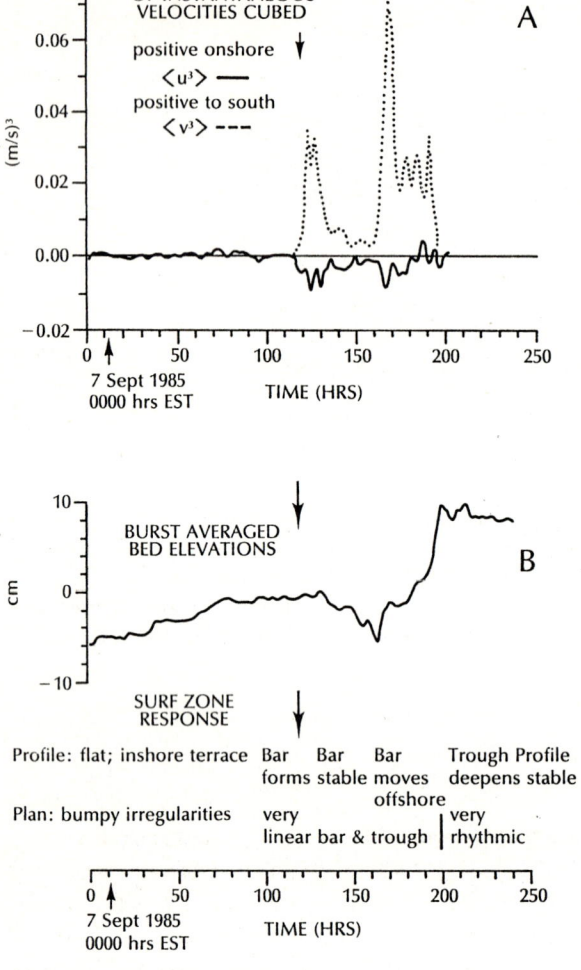

Fig. 10A,B. Sedimentary response at an 8-m-deep shoreface station off Duck, North Carolina, during a moderate fall storm in 1985. **A** Burst-averaged velocities cubed. These values are assumed to be proportional to sediment transport. Note the weak sediment transport before the onset of the northeaster followed by strong transport offshore and to the south during the storm. **B** Response of the bed. Erosion during the early part of the storm was followed by slow, then rapid accretion during the waning stages. (All data from Wright et al. 1986)

seaward extent of the high velocity zone is controlled by the scale of the coastal capes. Along straight coastlines the extent of the zone of jet flow is controlled by the dynamics. During moderate storms along the largely nonstratified Long Island shoreline the peak current is encountered at most a few km offshore (Fig. 9). The jets along the Caribbean coast of Nicaragua (Crout and Murray 1978; Murray and Young 1985) and western Louisiana (Crout 1983), which are largely baroclinic, attain peak current velocities between 10 and 20 km offshore.

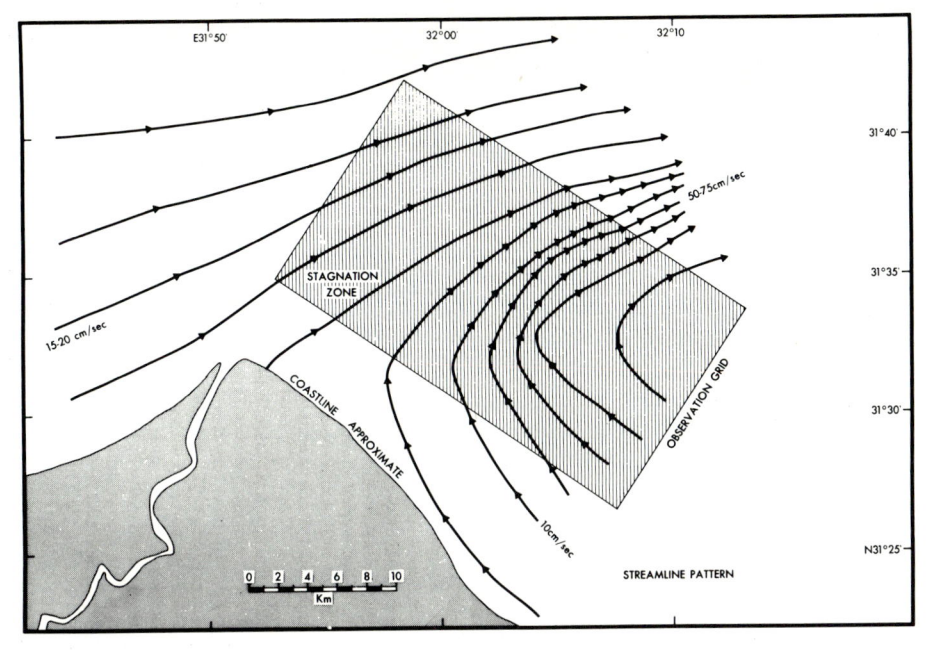

Fig. 11. Inferred streamline pattern off the Damietta River, Egypt. The flow pattern is inferred from interior-layer current observations. Note the coastal jet flow pattern to the east of the Damietta headland. (Data from Murray et al. 1980)

4.3 Waves

From the preceding discussion it is clear that wind-generated currents form a significant part of the shoreface dynamics picture. Traditional models for shoreface sediment transport, however, considered only wave action and were based on the concept of a balance between shoreward movement under shoaling, asymmetric waves and seaward transport down the shoreface by gravity. Using this concept it can be inferred that the observed concave-up profile of the shoreface is a consequence of a constant rate of wave energy dissipation (Dean 1977). It is generally also believed that asymmetric bottom oscillations under shoaling waves are the principal cause of this onshore sediment transport. This simple model of pure wave-driven sediment transport, however, fails to explain the observed seaward sediment transport during storms.

Recent, more detailed studies of wave-induced sediment transport (Bailard and Inman 1981; Dean and Perlin 1986; Nielsen 1979) have made it clear that wave action alone can easily produce net sediment movement in a direction opposite to that of the time-averaged water mass transport, because of the phase lag between maximum sediment concentration above a rippled bed and the peak oscillatory velocity. The groupiness of waves can also produce offshore sediment transport (Shi and Larsen 1984). There clearly are processes that produce net shoreward sand transport; otherwise depositional (progradational) shorefaces and beaches would not exist. It is not at all clear, however, whether onshore transport is primarily a wave phenomenon, a result of upwelling wind-induced currents, or due to nonlinear interactions between current- and wave-induced bed shear stresses.

4.4 Wave-Current Interaction

Sediment movement in all shallow marine environments is a result of currents and waves acting together, the so-called combined flow (Harms et al. 1982). Only in the rare circumstance of the "wave storms" discussed at the beginning of this paper do waves play a significant role on their own. Wave-current interactions affect sediment transport at two levels: sediment suspended by wave action is transported laterally by the superimposed steady current, and (2) nonlinear interactions between shear stresses in the boundary layers of the oscillatory and steady currents produce peak stresses with a magnitude and direction different from the sum of the two components (Grant and Madsen 1979; Grant 1986; Davies et al. 1988). Finally, it is important to recognize that the net, long-term sediment transport may respond to a set of processes which differ from those that control the "instantaneous" peak bed shear stress. The peak stresses, however, may leave distinctive tool marks and other paleoflow indicators which have no relationship to the direction of time-averaged flow of water or sediment. In such cases paleoflow data become irrelevant as indicators of net sediment dispersal paths.

This "paleoflow problem" has recently become a very visible issue. In a series of very precise measurements of different paleoflow indicators in shoreface strata throughout North America, Leckie and Krystinik (1989) documented that sole marks of shoreface and inner shelf storm beds are oriented nearly perpendicular to the local shoreline trend (Fig. 12). Clearly, this orientation is inconsistent with the hypothesis that the sole marks reflect transport by the shoreface currents described above. As proposed by Duke (1990), the explanation for this apparent problem appears to be that the sole marks reflect the instantaneous peak shear stresses exerted on the sea floor by the oscillatory motion of the shoaling waves. Because of refraction, landward-propagating storm waves align themselves nearly parallel to shore, thus the wave orbital plane is nearly shore-perpendicular. The storm bed sole marks are therefore shore-normal, whereas the time-averaged flow across the shoreface probably was oriented nearly parallel to the ancient shorelines, as it is in modern settings (Fig. 8). In a numerical model of bottom shear stresses under combined wave and unidirectional flow, Davies et al. (1988) demonstrated that the direction and magnitude of the wave-induced peak shear stress are little affected by a superimposed current with a velocity typical for the lower shoreface during storms (Fig. 13). In essence, the relationship between the sole marks on storm beds and net sediment transport is the same as the situation on a beach: the parting lineation on the beach face is oriented shore-perpendicular in response to the wave swash. The net sediment transport, however, is invariably parallel to shore; it is driven by the longshore current, not the peak wave stress.

Sediment transport under combined wave and current action on the Long Island shoreface has been modeled by Niedoroda et al. (1985). Their model considered a

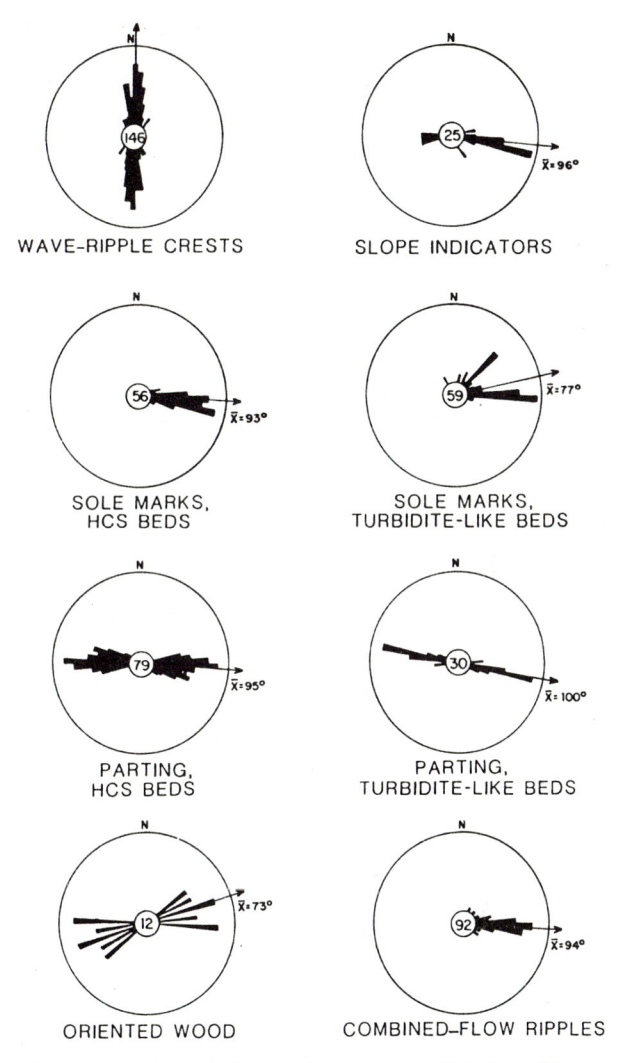

Fig. 12. Summary of all paleocurrent data from a series of shallow marine sandstones. Data from different stratigraphic units were normalized by rotating the mean of the wave-ripple crests to 0 (north). Most of the data were derived form Cretaceous rocks in the Western Interior of North America. If the wave-ripple crests are accepted as being nearly shore parallel then all sole marks and parting lineation have a strong shore-normal orientation. (Leckie and Krystinik 1989)

linear combination of movement in response to shoaling, asymmetric waves and wind-generated downwelling and upwelling. The results suggest that during storms there is longshore sand transport across the entire shoreface. Superimposed on this there is a significant cross-shore transport component. The volume rate of shoreface transport in these models tend to be about two orders of magnitude less than the rate of transport in the surf zone (Niedoroda et al. 1985). The onshore and offshore

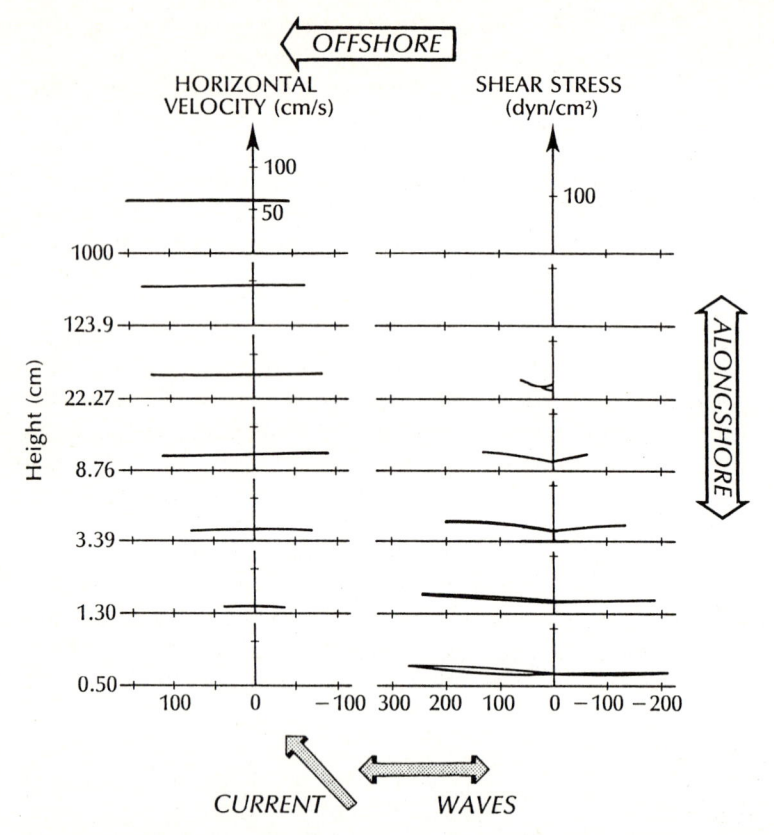

Fig. 13. Results of numerical modelling of velocity and shear stress distribution under combined oscillatory (wave) and unidirectional (current) flow. The diagram shows directions and magnitudes of the horizontal velocities and shear stresses at successively higher elevations above the bed (from 0.5 cm to 10 m). The steady current of 1 m/s had a 45 degree angle relative to the oscillatory plane of the waves. (Davies et al. 1988)

components of shoreface transport, however, are about the same orders of magnitude. In summarizing the long-term transport regime of the Long Island shoreface, Niedoroda et al. (1985) concluded that it is characterized by long periods of slow onshore net sediment transport, punctuated by short episodes (storms) of strong offshore sediment movement. The issue of long-term shoreface accretion versus erosion is therefore more a matter of the nearshore sediment budget than differing shoreface dynamics.

4.5 Tidal Currents

The issue of the role of tidal currents in nearshore sedimentation clearly goes beyond the scope of a review paper on storm sedimentation. It should be recognized, however, that tidal currents play a significant role beyond the surf zone. Moreover, the capacity of tidal currents for sediment transport is commonly greatly enhanced during storms. In an interesting analysis of tidal transport off southern California, Seymour (1980) concluded that symmetrical tidal currents do produce a net alongshore transport seaward of the surf zone. Sediment is entrained by the wave-induced bottom shear stress and advected alongshore by the tides. Since the wave-induced shear stress is greater at low than at high tide, the net transport will, generally, go in the same direction as the tidal current at low tide. It follows from Seymour's (1980) analysis that tidal sediment transport rates are directly related to wave heights, i.e., tidal transport is more effective during storms than fair weather periods.

Although no quantitative measures of tidal sediment transport on shorefaces are available, one would expect greater tidal influence along macro-tidal shores. One field study has been conducted on the shoreface off northwest Australia, where the local spring tide range during the field experiment was 9.5 m (Wright 1981). In this macrotidal setting the tidal currents were comparable in magnitude to the wave orbital currents near the bed. Because of strong asymmetry in the tidal currents, the net bedload transport in this case was found to move in the direction of the current at high tide, but the suspended load, which is directly related to wave-induced sediment transport, moved in the direction of the low tide current (as in the California study). Even in a macrotidal setting the suspended sediment concentration is primarily a function of the bottom wave stress, thus very dependent on the wave height. Therefore, tidal sediment transport is greatly increased in volume during storms, even along macrotidal shores.

4.6 Rip Currents

Rip currents are driven by longshore differences in mean water surface elevation. These variations, in turn, are controlled by differential breaker heights driven by refraction across offshore bathymetry (Shepard and Inman 1950) or interaction of edge waves and incident gravity waves (Bowen and Inman 1969). Semi-permanent, topographically controlled rip currents carry large quantities of suspended sediment offshore beyond the primary breakers (Cook and Gorsline 1972) and even smaller, edge-wave related rips play a significant role. It is unlikely, however, that these latter currents are significant during storms, because the rip-current circulation is then generally overwhelmed by a strong, unidirectional longshore current (Komar 1975).

It is well known that some rip currents extend far out to sea during storms (Cook 1970; Short 1985) perhaps more than a kilometer near major headlands (Fig. 2B). The few available direct observations of their flow, however, indicate that the main body of the current is limited to the upper few meters of the water column (Shepard and Inman 1950; Cook 1970). Therefore, rip currents are thought to carry only suspended sediment beyond the surf zone. Reimnitz et al. (1976) suggested, however, that rip currents off the west coast of Mexico may have been responsible for bottom sediment movement offshore to a water depth of 30 m. The shore-perpendicular orientation of the ripple-floored, shallow channels, and their occurrence in a region where major rip currents were commonly observed were in Reimnitz' opinion strong arguments for formation of these ripple bands by rip currents. Cacchione et al. (1984), in contrast, attribute very similar ripple bands on the shelf off northern California to storm-induced downwelling flow. Until actual current velocity profiles are measured within rip-currents their potential for offshore bedload transport will remain an issue of debate. In my opinion, present evidence does not support the concept of bottom-hugging rip currents beyond the surf zone. As documented by Clifton (1981) and Short (1984), it is clear that rip current transport has left significant records in both ancient and modern upper shoreface sequences.

4.7 Suspension Currents

This last mechanism of shoreface sediment transport to be discussed is clearly the most controversial. Walker (1984) argues for the generation of shoreface suspension currents by a mechanism of liquefaction of shoreface sediments by cyclic wave loading. Nelson (1982) documented graded beds on the shoreface and inner shelf of the Yukon delta in Norton Sound but the specific generative mechanism was not identified. Field et al. (1982) reported slumps and zones of liquefaction on the inner California shelf after an earthquake, but did not find evidence of related suspension currents. Snedden et al. (1988) and Morton (1981) have reinterpreted Hayes' (1967) observation of the graded bed deposited by Hurricane Carla to be a consequence of offshore transport by downwelling wind-induced currents. Thus, there is no documented example of modern shoreface and inner shelf suspension currents. There is some observational support, however, for the possible existence of shallow marine suspension currents. Probably the best-documented case is the turbidity current that moved a series of moored current meters downslope off Oahu, Hawaii, during Hurricane Iwa in 1983 (Dengler et al. 1984). That movement, however, occurred in water depths beyond those generally addressed in this chapter.

If they do exist, shallow marine suspension currents would be extremely difficult to detect because of the overprinting of other shoreface dynamics phenomena during storms. There are, however, valid theoretical arguments for their existence. A true auto-suspending flow can only be maintained if sediments are entrained at a rate sufficient to increase the density and accelerate the flow. From a consideration of the energetics of sediment transport, Bagnold (1962) reasoned that for such an autosuspending flow to occur, the ratio between the fall velocity of the entrained sediment and the mean current speed ("k") must be less than a constant proportional to the slope of the bed. Subsequent studies have supported this analysis, and the "initiation coefficient", k, has been calculated for a series of documented turbidity currents.

Considering the local steep slopes within the upper shoreface it is physically quite plausible to initiate autosuspending currents. Storm waves breaking on the upper shoreface stir up seafloor sediments and produce a vertical concentration gradient. The highest turbulence will be generated at locations where there is a rapid lateral change in wave energy (dE/dx), i.e., at the outer breakpoint bar (Lippmann and Holman 1989). Consequently, the highest suspended sediment concentration during storms is established at the top of the steepest shoreface slope. At the break-point bar on a fine-grained beach during storms it would be quite easy for the ratio between sediment fall velocity and the seaward-directed oscillatory current velocity to satisfy the criterion for initiation of a suspension current, as discussed above. Seymour (1986; 1990) presents both a theoretical model for surf zone autosuspending flows and a review of reported observations that are inferred to be consistent with a surf zone suspension current origin.

Shallow marine suspension currents could also originate as hyperpycnal underflows of riverine effluent in special circumstances. These would be conditions of extremely high suspended sediment concentrations, such as off the Huanghe (Yellow River) delta of China (Wright et al. 1986b), or in restricted embayments where saline waters are entirely replaced by fresh water during floods. Hyperpycnal flow would generally not operate in normal marine water because sea water always has a much higher density than the sediment-laden freshwater effluents.

5 Summary and Conclusions

Atmospheric storm energy is transferred to the sea through the generation of both waves and currents. The currents generally are restricted to the geographic region of the storm. The long-period part of the wave spectrum, however, may be dispersed across wide oceans without great loss of energy. It is well recognized that storms leave distinct imprints in the marine geologic record in their immediate area of occurrence, yet this dispersive property of waves implies that storms also may have geological consequences as far away as the opposite side of oceans.

The surf zone corresponds to that part of the upper shoreface which is subject to wave breaking at any given time. During fair weather and minor storms sediment dispersal in this zone is primarily driven by horizontal circulation patterns consisting of small rip currents separated by broad zones of onshore flow. During storms, in contrast, the circulation pattern appears to become vertical, in part, with onshore flow near the surface and offshore flow at the bed. The wind-generated gravity waves break and sustain strong longshore currents, while at the same time long-period (infragravity) waves play an increasing role in surf zone sedimentation, typically moving nearshore bars to more offshore locations, eroding the beach and deepening the overall surf zone. Some rip currents, mostly those that are topographically controlled, may persist during storms but offshore sediment transport occurs essentially on a broad front along the entire shoreline.

Thick, regressive, shallow marine sandstones most commonly reflect deposition on ancient shorefaces. Some such shorefaces may have formed parts of major delta fronts, others may have formed along shorelines far downdrift of the river mouths. Sediments arrive at any shoreface either from updrift shoreline sources by transport in longshore storm currents and seaward dispersal into deeper water, or from exhumed shelf sources followed by landward dispersal. Shoreface sedimentation, therefore, can only be understood in light of the processes which control sediment exchange between surf zones and continental shelves. This shoreface sediment dispersal is affected by all of the following processes: wind-generated currents, shoaling waves, wave-current interactions, tidal currents, rip currents, suspension currents and, locally, various baroclinic (thermohaline) circulation patterns.

Long-term sediment dispersal is primarily controlled by the wind-generated coastal jets and associated downwelling and upwelling patterns. It is well documented that such flows are present during storms and that storm surge (coastal water level set-up) is associated with downwelling, offshore-directed bottom flow. Offshore net sediment movement during storms, therefore, appears to be the norm. Broad zones of the lower shoreface and inner continental shelf may be influenced by the coastwise jet flow.

Storm waves interact with the currents and affect sediment transport in shallow marine settings in two very different ways: first, they entrain sediment into the water column by exerting high bed shear stress and, second, they generate a combined bed shear stress which often exceeds the sum of the individual wave and current stress components. The high instantaneous wave-induced shear stress may give rise to sole marks oriented perpendicular to shore, while long-term sediment transport is oriented alongshore. Numerical modelling of shoreface sediment transport under this "combined flow" is still in its infancy and needs substantial development before we can make much progress in understanding sediment dispersal pathways in specific shoreface settings.

Tidal currents and rip currents both affect shoreface sedimentation. During storms, the breaking waves will entrain higher concentrations of sediment for alongshore transport by the tides and offshore transport in large, topographically controlled rip currents. The rip currents themselves are generally restricted to only the upper few meters of the water column and therefore carry only suspended sediments offshore.

The highly controversial subject of shallow marine suspension currents can only be addressed on a theoretical basis at present. From principles of basic physics it appears highly probable that such currents do in fact exist during storms.

Acknowledgments. I would like to express my thanks to John Snedden and Don Swift for many stimulating discussions of shallow marine storm sedimentation and to Charlie Adams, Steve Murray, and Bill Wiseman for very helpful reviews of this paper.

2.3 Storm Deposition at the Bed, Facies, and Basin Scale: the Geologic Perspective

A. Seilacher and T. Aigner

1 Introduction

Wind and sea belong together -- not only in sailor's songs. Wind is the motor of most coastal processes and storms dominate and effectively build the coastal shelves. However, it was only in recent years that storm-generated sedimentological signatures, tempestites, have been fully appreciated. In the present chapter we attempt to summarize their characteristics, and the processes involved, without the ambition to provide a complete compendium.

A number of reviews of storm sedimentation have been published in recent years, including Allen 1982, Brenchley 1985, Walker 1984a, Knight and McLean 1986, and Tillmann et al. 1985, among others. We are also fortunate to have this chapter preceded by an excellent review of modern storm dynamics from D. Nummedal, which shows how wind energy is transformed into an irritating variety of wave and current patterns, particularly in the coastal zone. Perspectives, however, change in important ways if we switch from modern processes to their fossil expression. While the interest of coastal engineers is understandably focused on the nearshore zone, geologists are more concerned with storm effects in deeper parts of a basin, where sediments have a higher preservation potential and would therefore be more representative for the fossil record. There is hope that things become less complicated with depth -- for instance by attenuation of minor wave lengths and by merging of coastal current patterns into channelized basinward compensation flows. We must also consider temporal rarefaction and telescoping effects when dealing with the geologic record. In other words, there is perhaps some justification in using simplified models.

Nevertheless, such a review is not an easy task, because storm process and effect change tremendously with the kind of sediments, the physiographic setting, and water depth. There is also a strong connection with constructive and destructive biological agents, so that we must be aware of biohistoric factors. Last but not least, diagenesis not only fixes the final product; it also modifies the record (Ricken 1986) and in some cases interferes with the sedimentational process as an important feedback mechanism. In such a network of interrelationships, logical order is difficult to establish. We shall proceed by first discussing the characteristics of individual tempestites in siliciclastic and calcareous sediments (bed scale), then their accumulation at the facies scale, and finally their long-term paleogeographic behavior at the basin scale.

Einsele et al. (Eds.)
Cycles and Events in Stratigraphy
©Springer-Verlag Berlin Heidelberg 1991

2 Basic Storm Hydraulics

An increasing number of studies in modern environments contribute to a better understanding of storm processes and products (e.g., Aigner and Reineck 1982; Nelson 1982; Swift et al. 1983; Wanless et al. 1988).

Basically, wind energy is transferred to the water mass in three forms:
1. Surface waves. Their lateral propagation towards the shore produces the breakers pounding against cliffs and coastal dams.
2. Ground waves result from the vertical translation of orbital surface motions to the bottom, where they become flattened into oscillating currents. On their way down, the spectrum narrows in favor of larger wave lengths. The waves also lose energy, so that they eventually dissipate. The level at which wave action becomes too weak to move sediment particles is defined as *wave base*. This is a relative term. Wave base is not only subject to seasonal fluctuations (Aigner and Reineck 1983). It depends also on the size of available particles and on their cohesion. Therefore the depth of effective wave base may be considerably reduced by grain-binding biomats, compaction or cementation.
3. Coastal swell. By their tractional force, upland winds raise sea level near the coast (storm flood). The depositional energy of the piled-up water mass is released by a seaward backflow, or compensation current, that is most effective in the waning phase of a storm and may become deflected by the Coriolis force into geostrophic along-shelf flows (e.g., Swift et al. 1983). Loading with suspended mud, however, makes this backflow hug the bottom. Thus it becomes increasingly controlled by the local slope and concentrated into channels, in which it bypasses intervening areas and shoots beyond wave base (Fig. 1 in Brett and Seilacher Chap. 2.5, this Vol.).

At this point we should remember that "storm", as well, is a relative term. In a vernacular as well as a geological sense it implies exceptionality as well as unusual strength. Hydraulic and sedimentological effects of storms also depend on what has happened in the quiescent periods in between; i.e., storm impact has a *historical* component.

Take, for instance, the effects of the coastal swell. At a coast that experiences frequent storms, local grain size will be pretty much in equilibrium with storm level turbulence. Where a storm is a rare event, however, mud influx from adjacent rivers has time to settle in the quiescent years. At the onset of the next storm (Seilacher 1985), this material is quickly resuspended and produces a turbid water mass that will not flow back as a uniform sheet, but tends to concentrate into channelized systems, just as water runoff does on land. Within their submarine channels the turbid streams will also pick up gravitational energy that allows them to proceed beyond wave base.

Fig. 1. Proximity trends in tempestites and their application to facies and basin analysis. *Above* The proximity trends observed in *lateral* shelf sequences can also be recognized in *vertical* progradational cycles, the "parasequence" of sequence stratigraphy. *Below* The packaging of "parasequences" within a basin may monitor the relative sealevel history (Aigner 1985). Note that changes in climate (storminess; humidity) and crustal movements would have similar effects!

facies analysis

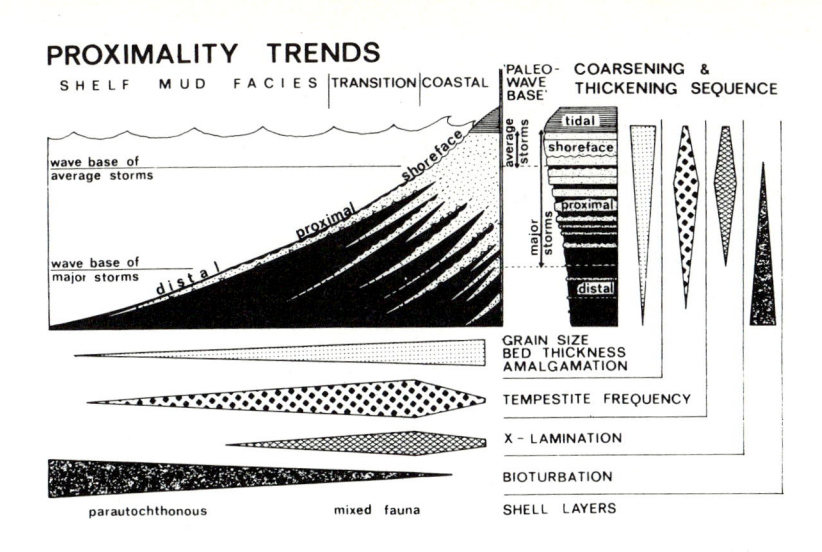

PROXIMALITY TRENDS

SHELF MUD FACIES | TRANSITION | COASTAL | PALEO-WAVE BASE | COARSENING & THICKENING SEQUENCE

- wave base of average storms
- wave base of major storms
- distal / proximal / shoreface
- tidal / shoreface / proximal / distal

GRAIN SIZE / BED THICKNESS / AMALGAMATION
TEMPESTITE FREQUENCY
X-LAMINATION
BIOTURBATION
SHELL LAYERS

parautochthonous — mixed fauna

basin analysis

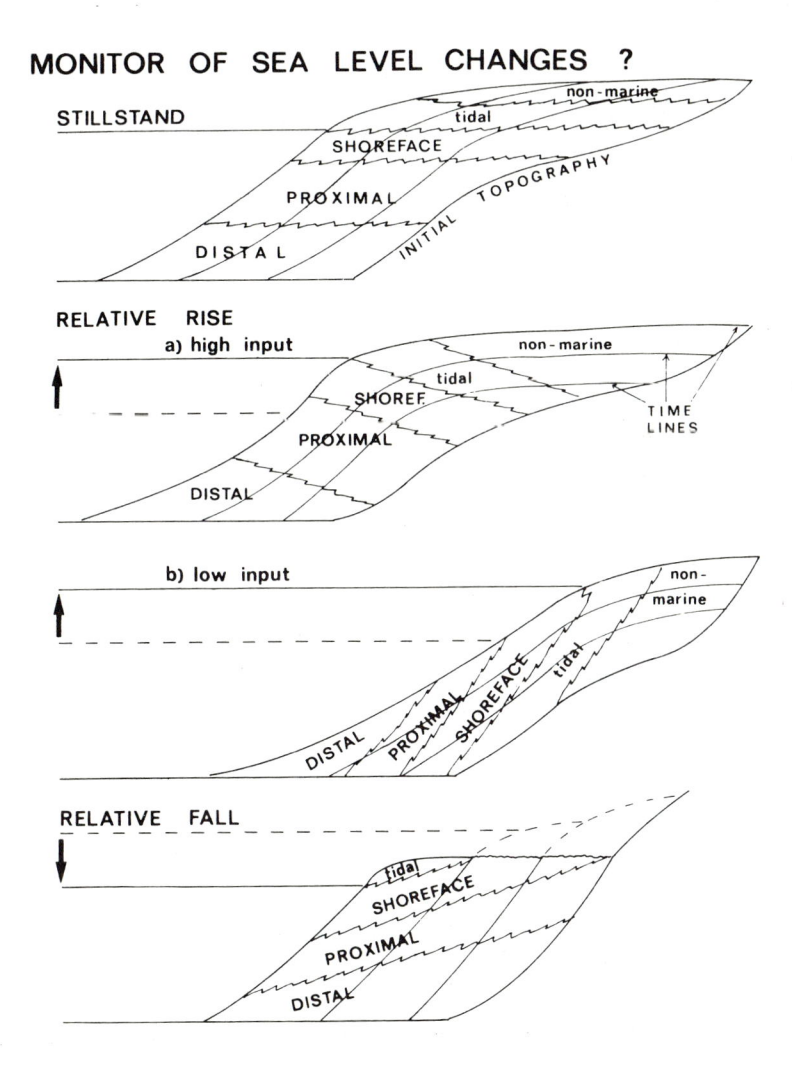

MONITOR OF SEA LEVEL CHANGES ?

STILLSTAND
non-marine / tidal / SHOREFACE / PROXIMAL / DISTAL / INITIAL TOPOGRAPHY

RELATIVE RISE
a) high input
non-marine / tidal / SHOREF / PROXIMAL / DISTAL / TIME LINES

b) low input
non-marine / tidal / SHOREFACE / PROXIMAL / DISTAL

RELATIVE FALL
tidal / SHOREFACE / PROXIMAL / DISTAL

At a continental margin they can thus reach the shelf edge and trigger a wholly new sedimentational system of submarine canyons and turbidite fans. On the gentle slopes of epicontinental basins, however, they will soon lose kinetic energy and drop their silty load, followed by a muddy blanket, in the form of silty or muddy storm turbidites spreading over large fan areas. The Devonian Hunsrück Shales with their smothered fauna, as well as other obrution Lagerstätten (see Brett and Seilacher Chap. 2.5, this Vol.) exemplify this situation. Since storm effects depend on the temporal distribution of the events, and meteorological pattern becomes more accentuated at low latitudes (with a peak in the hurricane zones), there is also a strong *climatic* component to storm sedimentation – the same as it is on land (Pflüger and Seilacher Chap. 2.11, this Vol.).

What has all this to do with the theme of our contribution? Firstly, it means that viewing storm sedimentation only in radial transects is inadequate. Concentration of major erosion to the coastal zone, and of major radial transport to submarine channel systems, leaves out most of the sea floor. It is in the vast level-bottom areas above wave base that storms can produce the most typical sheet tempestites. Secondly, storms, in a geologic sense, should be primarily viewed as relative phenomena, whose effects vary with bathymetric, sedimentological, and climatic parameters as well as with their timing.

3 Bed-Scale Features of Storm Sedimentation

Event stratification builds on the principle that sediments are in equilibrium with ambient turbulence. Fine-grained sediments settle only at low energy levels and become eroded and re-suspended when turbulence rises to levels that are in equilibrium with coarser particles – provided such particles are available. Thus simple stirring in a water jar will bring the upper layers of a mixed sediment into suspension, from where they will settle again on top of an erosional contact, but now in graded succession: coarse first, fine later. In natural environments there will be also some lateral redistribution, with coarser materials being imported from shallower, more turbulent areas and winnowed fines becoming exported into still deeper and quieter depocenters. It should be noted, however, that the depositional behavior of particles depends not only on their sizes, but also on their shapes and densities. This is why shells behave differently from species to species and why amber or originally very porous echinoderm ossicles may be swept towards the shore rather than away from it (Ruhrmann 1971).

Another effect is that, in accordance with the temporal turbulence curve, flow regimes along the bottom do change during an event in a regular sequence. The result is an event bed with an erosive base and with a graded succession of grain sizes and depositional structures (Bouma sequence). Since these basic principles apply to all kinds of turbulence events, individual deposits of floods (*inundites*), of storms (*tempestites*) and of turbidity currents (*turbidites*) share similar characteristics, in spite of differences in turbulence modes and bathymetries (diagram in Seilacher Chap. 2.1, this Vol.). Therefore, we shall try in the first part of this chapter to identify features that allow the distinction of tempestites from other kinds of event deposits.

As may be expected, distinctive features relate mainly to hydrodynamic regimes (current versus wave events), but also to sedimentological and biotic background situations and to the physiography of the depositional environment.

4 Sandy Tempestites

4.1 Sole Face Structures

As outlined before (Seilacher Chap. 2.1, this Vol.), the soles of typical event beds are erosional surfaces and consequently bear the casts of erosional markings. Only some of the markings known from fossil examples have been observed as original impressions on modern mud surfaces or in flume experiments. The reason for this is probably that erosional effects of sand suspensions are different from those of clear water currents and that the markings they produce cannot be observed because they will always be covered by sand right after their formation. Other sole structures may already have originated underneath a thin layer of sand in the form of undermarks (corresponding to undertraces in paleoichnology), as suggested by their invariably perfect and unblurred preservation.

1. Among the two major groups of sole markings, *flute casts* (including their corkscrew and crescent-shaped variations) require unidirectional currents to produce the one-sided excavations. Therefore they are characteristic for current-induced event beds (inundites and turbidites).

    Tempestites, in contrast, are wave-dominated and therefore lack flute casts. Instead they show another distinctive type of erosional features, *pot and gutter casts* (Aigner and Futterer 1978), which occur most commonly in somewhat deeper facies near wave base, where their long axes tend to run parallel to the coast line, rather than radial to it (Aigner 1985). Their orientations and limited sizes distinguish gutter casts clearly from intertidal channels, for which they have often been mistaken. On the other hand, tool marks on the flanks of gutter casts are typically bipolar. At the moment our best guess is that we deal with the effect of combined flows consisting of an unidirectional and an oscillatory component (e.g., Swift et al. 1983). In this process shells and other abrasive particles are probably instrumental – a view that is also supported by the pot hole modification and its similarity to "glacier mills".

2. Erosional structures of the second group, *tool marks*, originate in an instant. Therefore they form not only in unidirectional, but also in the oscillating currents into which waves transform at the bottom. Since impact casts have a broader and kinked end on the down-current side, and the broader prod casts are steeper in the same direction, their statistical evaluation allows the distinction between unidirectional and oscillating flow regimes.

3. *Load casts* of various forms and sizes reflect a secondary, but still syndepositional, deformation. Resulting from instabilities between layers, they tend to form repetitive patterns reminiscent of Bénard cells ("dinosaur skin"). Load casts are common on the soles of turbidites as well as of inundites. Their absence in tempestites is little understood;

probably it has to do with the fact that the erosional and the depositional phase of a storm are locally symmetrical. In current events, however, erosional and depositional peaks shift, so that in distal zones sand may be deposited on unproportionally soft mud that had escaped equivalent erosion.

4. While inorganic markings tell us little about how much sediment has been removed during the erosive phase of the event, such information can be derived from *washed-out burrow casts*. Infaunal communities are typically tiered, with some species penetrating deeper than others below the sediment surface. In oxygenated shelf bottoms since the Mesozoic the lowermost tier is occupied by horizontally branching burrows (*Thalassinoides*) of fairly large ghost shrimps. Their washed-out (exhumed) appearance on tempestite soles indicates erosion of at least several decimeters (Aigner 1985).

4.2 Internal Structures

1. Although being in fact three-dimensional, internal structures of event beds are usually described by their cross-sectional aspects. The *parallel lamination* in the lower part of the Bouma sequence formed while turbulence was still strong enough to move whole sheets of water-saturated sand along the bottom. In this phase linear concentration of mica flakes may also lead to *parting lineation* in turbidite, as well as inundite, versions.

At the lower energy level characterized by *ripple* formation, distinction between regimes becomes easier. Currents produce unidirectional progradation, with climbing ripples in the extreme case (note, however, that sand ripples and boulder megaripples may not prograde in the same sense; Pflüger and Seilacher Chap. 2.11, this Vol.). Wave ripple lamination, in contrast, is less directional and may lead in the accumulative phase to *hummocky cross stratification* as a distinctive tempestite feature (Dott and Bourgeois 1982; Duke 1985; Nottredt and Kreisa 1987; Brenchley 1985; Swift et al. 1983; Walker 1984a; Tillmann et al. 1985).

2. Ripple lamination may occur in a variety of situations. Its *synsedimentary deformation*, however, is a distinctive feature of event beds, because it requires that the sand-package is still water-saturated enough to act as a coherent unit. Most commonly, such deformation occurs in the middle of a turbidite or tempestite bed in the form of *convolute lamination*. In contrast to slump folds, these convolutions are asymmetrical not in a lateral, but in a vertical direction, because differential loading results in broad bag-like structures, between which water escape accentuates narrow tepees. Ten Haff (1956) also showed that – in contrast to load casts – the convolutions did not form at once, but by interaction with the ongoing sedimentation.

In contrast to convolute lamination, *overturned cross lamination* is diagnostic for continental deposits, where it helps to distinguish fluvial, as opposed to marine, parts in nonfossiliferous sandstone series. The structure clearly indicates that the laminae were bent over in current direction, like pages of a telephone book, by the drag of an upper-flow-regime sand sheet. The problem is that this deformation had to happen during the same depositional event, i.e., before the deformed sediment had been dewatered. This appears possible only in braided river systems (and perhaps in deep sea fans) where the current may quickly switch into new stream beds and thereby locally produce multiple energy peaks during a single flood.

4.3 Top Surface Structures

Sandy deep sea turbidites rarely have sharp top boundaries, because on their way down the continental slope, suspension currents pick up such a collection of sediments that their ultimate grain size spectrum is rather continuous. Tempestites, however (and to a certain degree also inundites and the microturbidites of distal storm fans; see Brett and Seilacher Chap. 2.5, this Vol.), deal with material that has, through fair-weather processes, been pre-sorted into a sand and a clay fraction. The gap in the bimodal grain size curve leads to a pause in the depositional process. During this interval the uppermost sand layer can be freely modeled into either asymmetric, and commonly linguoid, *current ripples* (in case of inundites) or into straight or reticulate *oscillation ripples* (in tempestites). The intermission also gives shy animals (such as starfish) a chance to get dug into the sand for protection. This they do in the ripple troughs rather than on the crests, where post-storm undertrack generations (e.g., those made by bivalves) would be preferably preserved. The eventual onset of mud fall from the remaining suspension, however, saves the rippled surface from intensive bioturbation at a later stage. That this mudfall was part of the same event rather than an independent affair, is also shown by other phenomena:

1. The sandy rims around *starfish burrows* in ripple troughs have steeper slopes than could persist at a sand/water interface (Seilacher 1982b, Fig. 5b). This means that the animals left their hiding places only after mud fall had started.
2. *Spill-over ripples* are somewhat flattened at the top and thin sheets of sand drape the upper halves of the flanks like a table cloth. They formed when the ripple troughs were already filling with mud so that the still emergent sand crests could spill over the mud under continuing wave (or current) action (Seilacher 1982, Fig. 3).
3. The dendritic markings of "*Aristophycus*" occur preferably on ripple crests and commonly originate from animal burrows. The inverse relief and a marginal rim suggest that this strange structure is produced by water escaping from the sand and dissipating below the mud cover like a minute distributary river system. Aristophycus markings, however, appear to have formed some time after the event, because they intersect the trace fossil *Gyrochorte*, which has been produced as an undertrace below the overlying mud (Seilacher 1982b, Fig. 4c).
4. The equally rhythmical patterns of "*Kinneya*" markings are also poorly understood. Their common association with ripple marks (Seilacher 1982b, Fig. 4c) tells us that they formed within the sediment, rather than at the sediment/water interface, while their striking similarity with Bénard cells and their association with a magnetite lamina (Bloos 1976) suggest some kind of instability release in connection with sediment loading and dewatering.

5 Calcareous Tempestites

5.1 Diagenetic Maturation

During the depositional phase, calcareous sands behave like siliceous ones. What makes them different in the long run is their susceptibility to secondary diagenetic alterations. These will be most profound if the components have originally been aragonitic, such as ooids and the fragments of mollusc shells or scleractinian corals. Unless the whole bed becomes lithified very early (for instance as a beach rock), aragonitic particles will soon become dissolved and supply the carbonate for initial cementation. Since pore solutions in subaqueous sediments migrate slowly, one might expect cementation to start within the coarse layer itself. If, however, this layer contains little matrix, it might be that the carbonate is first attracted to the finer-grained sediment on top and underneath, or to sites where such sediment is trapped in pressure-resistant vessels, such as ammonite or bivalve shells. At least this is what is indicated by observations in the Upper Muschelkalk (M. Triassic, Germany):

1. In Muschelkalk tempestites the rippled tops and erosive soles are hardly ever exposed, because the calcarenites are "mummified" by micritic *over-* and *under-beds* (Aigner 1985).
2. Ceratite accumulations at the base of such beds have their infills cemented, while the surrounding sediment remained soft. This must have been an early process, because thus *prefossilized* ceratites have commonly become reworked during later events, as indicated by disoriented voids (Seilacher 1971).
3. *Lithoclasts* embedded in the lower part of tempestites are almost exclusively micritic, probably because the shelly beds were not yet hardened in the zone affected by event erosion.

Partly or wholly indurated underbeds become exposed not only during the short interval between the erosive and the depositional phase of a subsequent storm. They can also be more permanently, or intermittently, exhumed to form the sea bed over extended periods of time. This is shown by the association of various trace fossil generations in the same underbed (Aigner 1985). *Teichichnus* and other burrows of sediment feeders represent the soft ground stage. Burrows of suspension feeders with preserved scratches (*Glossifungites* Association) have been made in compactionally stiffened mud. In condensed underbeds they may also be associated with a later generation of borers and oyster-like encrusters (*Trypanites* Association). This indicates that the original firm ground had over time matured into a hardground – probably while being intermittently covered again by loose carbonate sand. Such details tell us not only how the "memory" of previous events can be enhanced by diagenesis. They also give us a feeling for the time hidden in diastems, particularly in situations where background sedimentation was reduced and only few storms (or earthquakes) affected the bottom. In such situations, for instance in red cephalopod limestones, the difference between short- and long-term sedimentation (or resedimentation) rates became very large.

As mentioned before, a cementation front may also advance upwards from the top of the tempestitic calcarenite. The resulting concretionary *overbed* will eventually be as firmly welded to the shelly layer as the underbed. If it comprises only part of the tempestitic top mud, this overbed will not only conceal the rippled intermission surface, but also introduce a bias in the fossil record, because in loose blocks the noncemented top part of the event mud will be lost by weathering. In addition, the already cemented overbed has commonly been reduced by "soft" stylolitization along the shale/micrite interface (Seilacher 1988a). In no case should such diagenetic artifacts be mistaken to indicate erosion at the sea floor or carstification.

5.2 Biostratinomic Features

Biogenic particles concentrated in a shelly tempestite are not just another kind of sand. Already in their original state shells and bones behave unlike quartz grains because of specific properties with regard to size, shape, and density. In the middle of a shelly tempestite, rapid sedimentation and dense packing may reduce the *biostratinomic* consequences of size and shape except for some degree of vertical grading. In the top lag, however, (which was remodeled during the intermission, but is commonly concealed by an overbed) the shells did have a chance to behave individually with all the consequences for stable azimuth orientations, convex-up positions and shape sorting (Seilacher 1984b, Fig. 4).

Microstructural and mineralogic differences between biogenic particles, rather than shape factors, become important in the subsequent phase of *early diagenesis*. Aragonitic shells are more readily dissolved than calcitic ones and may disappear altogether unless their mud fillings had become transformed into pressure shadow concretions (see above) before shell dissolution occurred.

Echinoderm ossicles, although originally consisting of more soluble *high-Mg calcite*, unmix into calcite and submicroscopic dolomite crystals. Thereby they become resistant to dissolution without changing their bulk composition (Richter 1985).

Phosphatic skeletons (vertebrates, inarticulate brachiopods) also suffer dissolution. But under anaerobic conditions they not only persist, but tend to become further enriched in phosphate. This process, probably mediated by bacterial activity, may even transform soft feces and muscles into solid phosphatic bodies.

Under dysaerobic conditions, other, sulfate-reducing, bacteria may similarly strengthen fragile shells or carapaces by providing them with a coating of *iron sulfides* that later turns into durable framboidal pyrite.

Certainly, diagenetic processes require time. But given the right water depth, where sedimentation rates are low and only the most severe storms reach the bottom, diagenetic and sedimentational processes get a chance to interact. Recycled biogenic particles, having gone through the filter of prefossilization, will show not only a drastically distorted taxonomic spectrum. Survivors also come back with an altered sedimentological behavior that favors their abrasion into shell or bone sand, and their sorted accumulation into placer-like *concentration Lagerstaetten*, during subsequent storm events.

Storms, however, do not only degrade the fossil record. They may also bring about the most perfect preservation by killing benthic organisms and smothering and blanketing them with their muddy tail sedimentation. This kind of *conservation Lagerstaetten* is discussed in a separate contribution (Brett and Seilacher Chap. 2.5, this Vol.).

5.3 Biological Feedbacks

Also to be mentioned is the ecologic effect of storm-produced shell layers, particularly of large shells, on post-event encrusting communities (Kidwell Chap. 6.3, this Vol.). Given low interim sedimentation rates, gregareous species can themselves grow beyond mere encrustation to keep pace with sedimentation and form well-sized reeflets (Hagdorn 1982). However, such buildups give us only a minimum value for the actual sedimentation rates, because they may have become intermittently buried and then re-colonized after a subsequent storm had stripped off the mud cover again.

If sedimentation rates are still lower, solitary encrusters, such as corals, rudists, and oysters, survive on mud bottoms by outgrowing their shelly starter substrates to become self-stabilized sediment stickers or recliners. Favoring large and heavy shells, this strategy considerably increases fossilization potential. However, such secondary soft bottom dwellers can become rarely fossilized in life position, because rare storms winnow the mud away from underneath the large shells, make them sink down on a layer of former victims, and eventually blanket them with the mud settling from suspension. Over time, thick layers, mounds, or banks may thus accumulate not only of reclining oysters (Seilacher 1983) and nummulites (Aigner 1982), but also of mud-sticking horn corals or gooseneck barnacles (Seilacher and Seilacher-Drexler 1986).

6 Tempestites at the Facies Level

Facies is defined as the fossil expression of an environmentally uniform and representative slice in time and space, characterized by paleontological, sedimentological, and geochemical criteria. Each event bed, however conspicuous, is only an element of such a larger facies unit, which also comprises the background sediments, as well as other event beds, and their spatial relationships. Typically, comparisons are made between lateral facies, because they are most directly related to paleoenvironmental patterns. This results in an idealized facies transect, which can then be translated into vertical patterns according to Walther's rule of facies succession. Alternatively, analysis can proceed the other way around.

6.1 Proximality Gradients

Right after a storm is over, its tempestite forms a large sheet, but of limited extension. Shoreward this sheet is bounded by the erosional, or winnowed, zone. Alongshore it can reach only as far as the storm was felt and downslope only to the effective wave base. Towards the lateral and distal margins, bed thickness and grain size will also decrease. Locally this sheet may be dissected by the channels through which the turbid compensation currents had flushed sediment from the shore directly to depocenters below the effective wave base. Since limited downslope transport occurred also in the level bottom areas between channels, our model sheet tempestite will tend to be coarser than the host sediment and have a surplus of muddy tail sediment in the more downslope, distal areas (Aigner 1985). This granulometric scheme, however, applies only to terrigenous, and to somewhat transported skeletal detritus. In the deep zones of mud stickers and recliners, locally produced oversized shells may be concentrated at the base of the tempestite without lateral transport, just by winnowing.

The *preservation* of physical event signals is another question. Depending on benthic population densities, bed thickness and the duration of quiescent periods, the original tempestite signature may be wiped out by bioturbation in a matter of years. Bioturbation, however, may also have a conservational effect: large, open-hole burrow systems (such as those made by *Callianassa* shrimps) can selectively trap coarse shell debris reworked during a storm, to form "tubular tempestites" (Wanless et al. 1988). Subsequent storms are another destructive factor that will be discussed in the next chapter.

By and large, proximity gradients can be derived from the general rule that tempestites tend to become finer, thinner, rarer, better preserved and more biogenic in composition towards the deeper parts of an epicontinental basin (Fig. 1). Application of such proximity criteria to bundles of beds may also bypass the problem of bed-by-bed correlation.

6.2 Vertical Gradients and Telescoping

If beds simply accumulated in a layer-cake fashion, there would be no difficulty in interpreting vertical successions. An event bed, however, is inherently erosive at its base and will therefore destroy and incorporate previous background and event deposits. The degree of cannibalism will generally decrease in a distal direction; but it also becomes modified by the erodability of the substrate. Erodability decreases with time by processes such as microbial binding, compaction, and early cementation, as well as by the growth of oversized bioclasts and concretions; but cannibalism also depends on the overall rate of sedimentation and subsidence. In the extreme case, i.e., in zones of starved net sedimentation, repeated events may lead to a single condensation horizon, in which locally produced bio- and diaclasts plus nektonic fallout have been accumulated and winnowed over very long time, while only the ultimate storm had a chance to leave a lasting sedimentological signature (Fürsich 1971).

In still deeper parts of the basin, where cannibalism is reduced and net accumulation of well-bedded sediments persists, the interaction of various bed-scale factors is modulated by long-term changes of turbulence and net sedimentation. Sea level changes (see Sect. 7), can thus be reflected by a vertical cyclic change in bed characteristics, which is the subject of *sequence stratigraphy*. In epicontinental basins, the upward-coarsening and upward-thickening "Klüpfel Cycles" (for a discussion see McGhee et al. Chap. 6.4, this Vol. and Aigner 1985) are the most common type of bed bundles (Fig. 1). They are best developed along "regressive" coastlines, while "transgressive" shelves tend to be characterized by sediment starvation and the occurrence of isolated shelf sand bodies (e.g., Nummedal 1989; numerous examples in Tillmann et al. 1985, Knight and McLean 1986).

6.3 Larger Bodies of Carbonate Sediments

Carbonates are largely produced in situ. Consequently, they react to changing environmental factors not only by modes of sedimentation, but also by biologic and chemical responses. Moreover, carbonate bodies can lithify fast enough to create their own topographies.

The large variety of carbonate systems can be classified according to overall physiography into (1) carbonate ramps, (2) attached platforms, and (3) isolated carbonate platforms. In each of these broad categories storms have different effects.

1. *Carbonate ramps* are buildups of loose particles. Consequently, they have gently inclined depositional surfaces without reef barriers and without sharp breaks in slope and compare well to the "graded shelves" of siliclastic systems, both with respect to hydrodynamics and to facies organisation. Also the style of storm stratification shows the proximality trends described above and it is in carbonate ramps that the most typical tempestites occur.

The coastal zone of carbonate ramps is strongly influenced by the predominant wind/storm direction, which controls the type and geometry of sand bodies. This is well illustrated in the modern Arabian Gulf, where various types of "windward", "oblique" and "leeward" coastlines can be found (Purser 1973; Fig. 2).

2. Attached carbonate shelves are characterized by shelf margins with more or less fixed reefs or banks. Therefore they slope relatively steeply into deeper water areas. The banks and/or reefs are also responsible for various degrees of protection for shallow shelf areas behind them. Consequently, storm effects differ markedly from those in carbonate ramps. Because of the variety of subenvironments, storm beds tend to be different from place to place, and far less continuous than in ramp settings. The modern environments of South Florida provide a good example for the diversity of "tempestites" in an attached carbonate shelf.

3. On *isolated carbonate platforms* of the Bahama type, storm effects are particularly important in controlling the depositional facies along the platform margins. The difference between windward and leeward platform margins, as established in the modern Bahama Banks (Hine et al. 1981a, among others), has served as an actualistic analogue for many ancient platforms (Fig. 3).

Storm-derived base-of-slope deposits on the leeward sides of carbonate platforms are of economic interest, not only by their volume, but also by their high secondary porosity (Fig. 4).

WINDWARD/LEEWARD CARBONATE FACIES
Modern Arabian Gulf

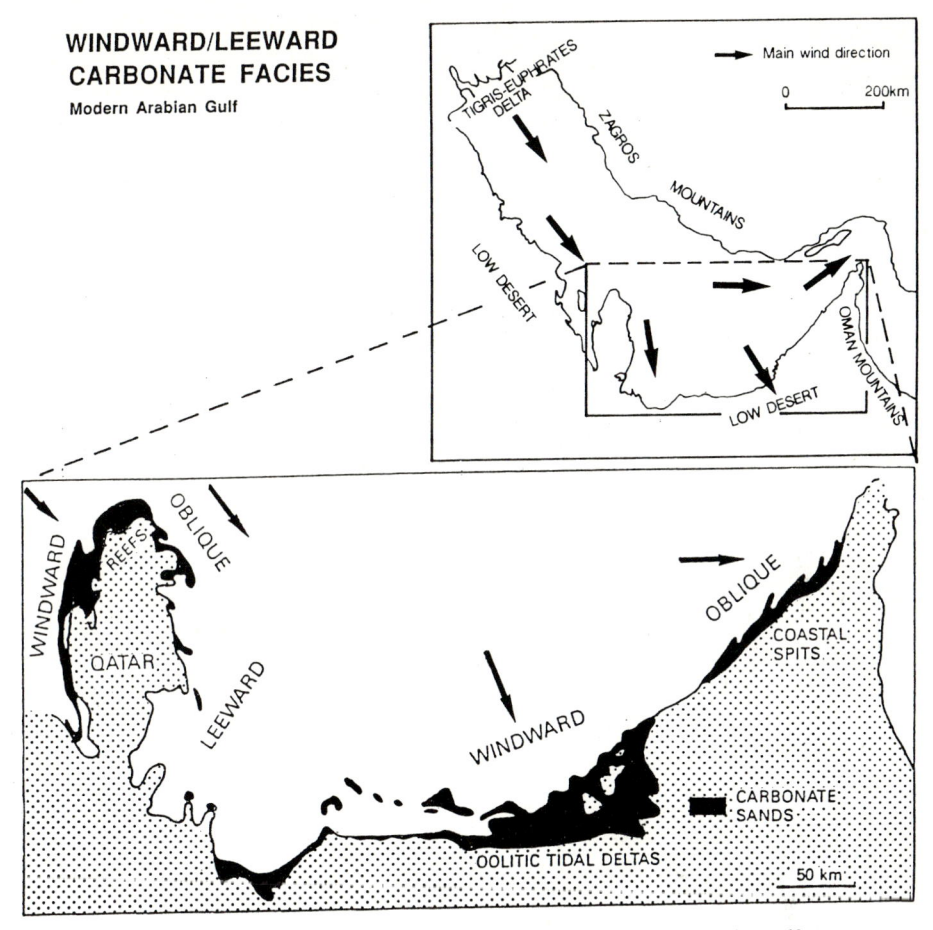

Fig. 2. Coastal carbonate sand bodies in the modern carbonate ramp of the Arabian gulf may serve as actualistic analogs for paleo-windward, leeward, and oblique setup in ancient ramps. (After Purser 1973)

7 Basin Scale Expression

At the facies scale we have been interested mainly in lateral changes of individual storm beds, or bundles of them, in terms of proximality. We did not consider paleogeographic asymmetries, nor temporal changes of boundary conditions such as climate, influx rate, sea level, or vertical movements in the crust (e.g., Swift et al. 1987). A steady-state basin would eventually fill up. The filling, however, would not proceed as in a dish with meniscoid sheets, but (because of the vertical turbulence gradient) by sediment prisms prograding from the margins towards the center of the basin. Within such a prism, each layer will follow a sigmoidal curve, in which a proximal, transitional, and distal facies can be distinguished. Because the whole stack

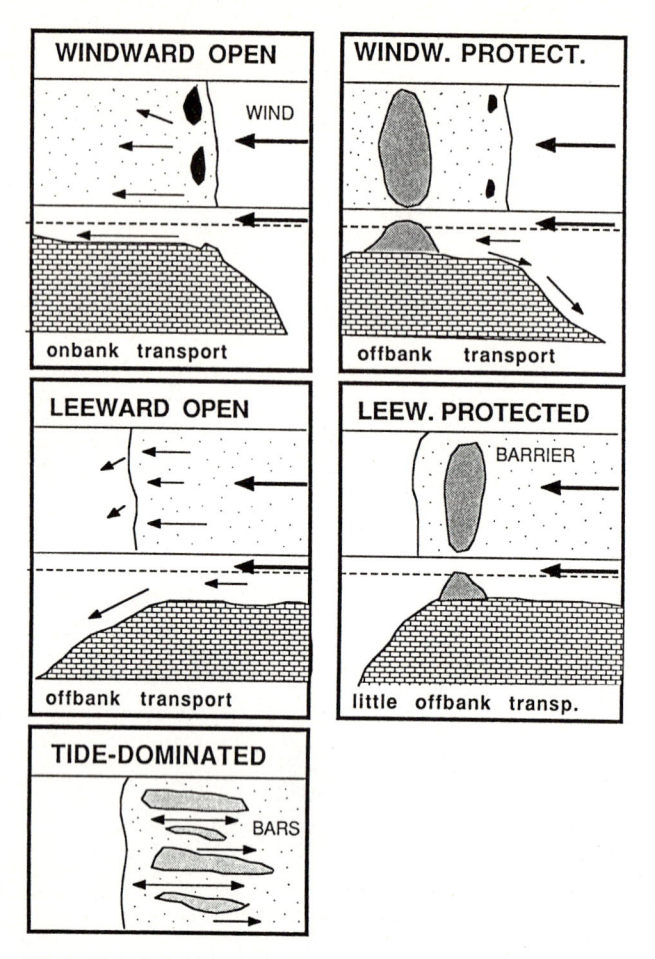

Fig. 3. Variations in windward and leeward margins of the modern isolated carbonate platforms of the Bahama Banks. (After Hine et al. 1981a)

progradess towards the basin center, every vertical core will show, bed by bed, a gradual upward transition from distal to transitional and proximal facies types.

7.1 Sequence Stratigraphy

Sequence stratigraphy (e.g., Posamentier et al. 1988) can be a key into analyzing the infill histories of storm-dominated basins. Commonly, distal-to-proximal, coarsening-upward ("shallowing-upward") Klüpfel cycles (see above) form the basic stratigraphic building blocks. They are called "parasequences" in sequence stratigraphic terminology (see also Fig. 1). Stratal patterns within, and the stacking of, these parasequences reveal larger-scale changes in boundary conditions.

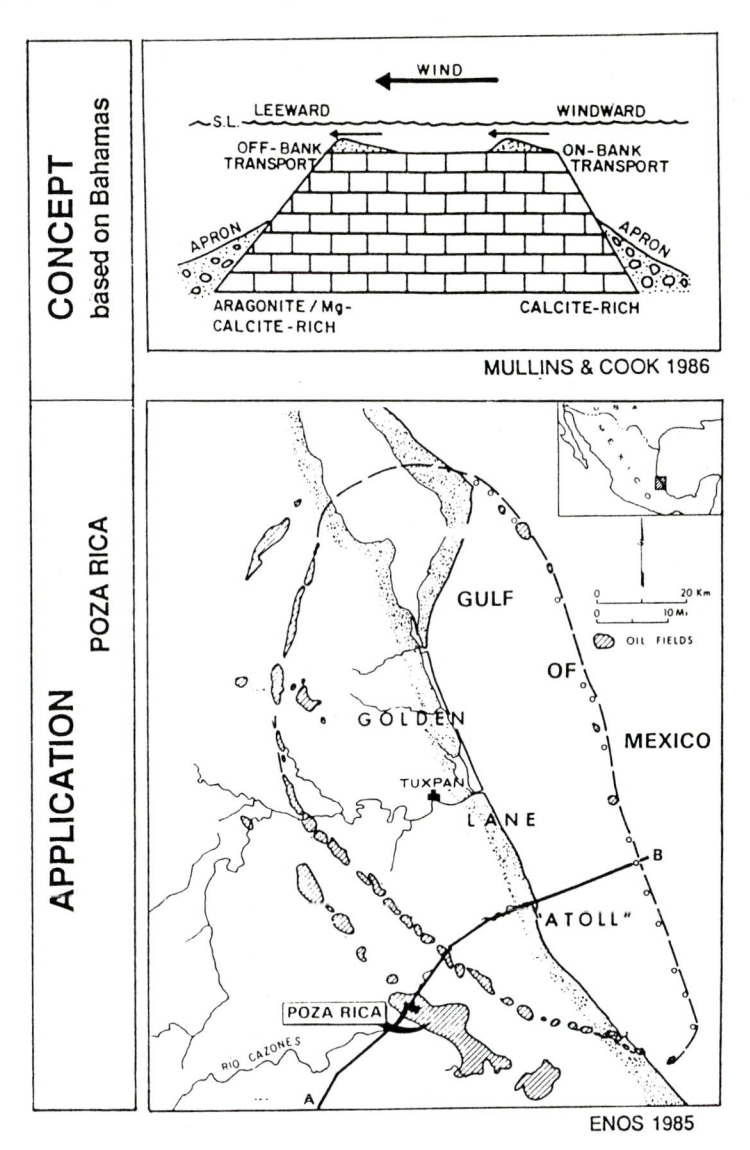

Fig. 4. Application of the windward/leeward concept of the modern Bahamas may explain the preferred occurrence of the Poza Rica giant oil fields in the paleo-leeward base-of-slope deposits of the Cretaceous Golden Lane platform: rudistid skeletal debris was shed preferentially along leeward platform margins. (Mullins and Cook 1986; Enos 1985)

Before we go on to model temporal shifts in boundary conditions, however, a few warnings are in place. (1) When looking at diagrams like the ones in Fig. 1, one should remember that the vertical scale is always grossly exaggerated. Actual submarine slopes (with the exception of erosional and talus slopes) are so gentle that they would otherwise be hardly recognizable. The necessary graphic distortion means also that the interfingering facies boundaries appear much sharper in the diagrams than they are in nature. (2) One should be careful in using the terms "regression" and "transgression". They properly refer to displacements of the water line, which can move in a regressive mode by sediment input and progradation alone, without any change in boundary conditions. (3) What we see in the rocks are changes in turbulence levels (both with respect to the background and the storm situation), for which water depth is only one of the possible controles.

The last statement is important because we all too easily think only about one of the two major forcing mechanisms, climate or water depth. Climate has the attraction that it may be linked to the Milankowitch cycles that were identified as the basic pulses of earth history in the first part of this book (contributions by Fischer Chap. 1.2, and by Einsele and Ricken Chap. 1.1). It has also been suggested that during most of earth history oceans were more sensitive to climatic signals than under present post-glacial conditions (Kauffman 1988a). Water depth, on the other hand, may be linked with global sea level curves; but we should note that the rock bodies recognized in seismic stratigraphy (Vail et al. Chap. 6.1, this Vol.; Posamentier et al. 1989) are commonly of a larger scale than the bed bundles seen in outcrops.

More specifically, climatic forcing may have a variety of causes and effects, e.g., changes of storminess (by relative shifts of the hurricane zone, for instance) or of precipitation patterns. Equally, local water depth can change either by global rises and drops of sea level or by epirogenetic movements of the crust on a regional scale.

Unfortunately, subjective options cannot be easily ruled out here, because sediment prisms react similarly to different forcings. It is easy to see whether the local turbulence level increased or decreased through time, but not whether energy increased by a rise in storminess or humidity, by a rise of the crust or a drop of sealevel. With the exception of dryness (which does not affect turbulence regimes), different forcings produce similar sequences. One of the features in common is the asymmetry of Klüpfel cycles: areas of maximum sedimentation switch shoreward, and more distal zones become sediment-starved whenever the turbulence level drops, be it by climatic change or sea-level rise. This means that the distinction of causes must be based on additional, independent evidence, such as time concordance with Milankovitch rhythms, or geochemical signals (Kauffman 1988a; Röhl 1990), while biological responses may also be ambivalent.

Regardless of interpretation, however, the mere recognition of rhythmicities may allow us to correlate rocks within basins and across facies boundaries at a much higher resolution than biostratigraphy alone would provide (Kauffman 1988a). On the other hand, high resolution stratigraphy is at its best in noncannibalistic facies, i.e., in the deeper parts of a basin or in quiet epeiric seas.

7.2 Paleogeographic Distribution of Sediment Bodies

The other dimension that basin analysis is concerned with, in addition to the sequential spectrum, is the spatial distribution of rock types (particularly of porous reservoir rocks) within ancient sedimentary basins. Real basins are not symmetrical dishes, but have inherent asymmetries, for which local sediment influx from river systems, general outline, topography and paleostructure, but also prevailing paleowind directions are the major causes.

River mouths produce fans that may spread over large parts of a basin. While being primarily unrelated to storm events, this influx provides much of the material from which tempestites will eventually be formed.

Another spatial factor, wind direction, controls the migration of sand bodies relative to larger obstacles such as islands (Fig. 4) and coastal promontories (Fig. 2). Such movements are enhanced as soon as a bank rises above sea level, so that aeolian transport can act in the same direction.

A promising line of research is to apply knowledge about modern storm depositional systems to particular paleogeographic settings. Paleoclimatic modeling and paleowind directions (e.g., Parrish and Curtis 1982; Marsaglia and Klein 1983) help to predict particular depositional facies. Vice versa, paleocurrent measurements from ancient depositional systems constrain numerical simulations. Figure 6 shows an Upper Jurassic paleogeographic and paleowind reconstruction. The predicted paleowind directions appear to agree with depositional facies recorded in three different carbonate realms.

Even though oil geologists like to speak in these cases of "sand bodies", we deal in reality with facies zones that cut diachronously through the stack of layered sediments. But it should also be remembered that such sediment bodies commonly owe their final shaping to storm events rather than to everyday conditions.

8 Conclusion

By integrating actualistic observations and models with field data, research in dynamic stratigraphy proceeds from the bed and facies scale to basin histories and their particular paleogeographic settings (e.g., Swift et al. 1987). In storm-affected basin fills, proximity criteria can be recognized at each of the three hierarchical levels (Fig. 5). Firstly, single *tempestite* beds may show regular changes from proximal to distal. Secondly, lateral facies zones or vertical *facies sequences* may – statistically – be grouped into more proximal or more distal settings. Thirdly, on a *basin scale*, mapping of facies zones may result in the recognition of local sediment influx and of more or less storm-affected (wind-ward versus leeward) realms (Figs. 3–6).

Fig. 5. (p. 266) Storm effects and tempestites at three hierarchical temporal and spatial scales: (1) *individual beds* with characteristic proximity trends, (2) parasequence development on the *facies-scale*, (3) windward versus leeward coastlines on the *basin-scale*. (Partly from Aigner 1985)

Fig. 6. (p. 267) A "vision" for integrated paleogeographic modeling. Model-predicted Jurassic paleowind directions (after Parrish and Curtis 1982) appear to be in agreement with observed, most likely wind-controlled facies patterns in three different carbonate systems: (1) southwesterly winds in the Paris Basin (grainstones most prominent on SW ramp margins, Mégnien 1980); (2) NE winds shedding oolitic base-of-slope deposits from the Friuli Platform in a leeward direction into the Belluno Trough (Bosellini et al. 1981); (3) SE winds on the Arabian Platform accumulating carbonate sand shoals of reservoir quality along coastlines much like in the modern Arabian Gulf. Empirically derived facies data and numerical paleogeographic modeling may thus complement each other

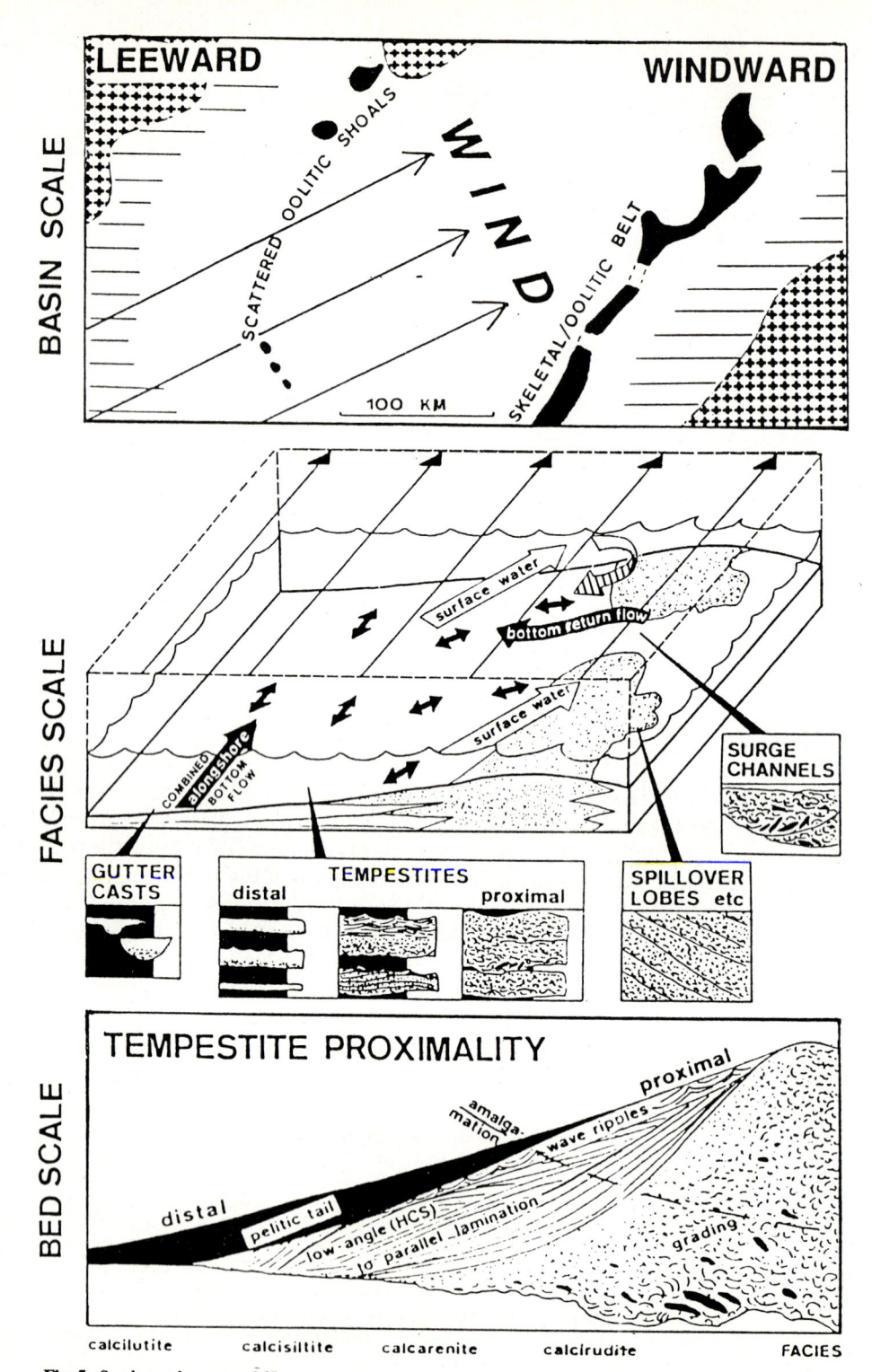

Fig. 5. See legend on page 265

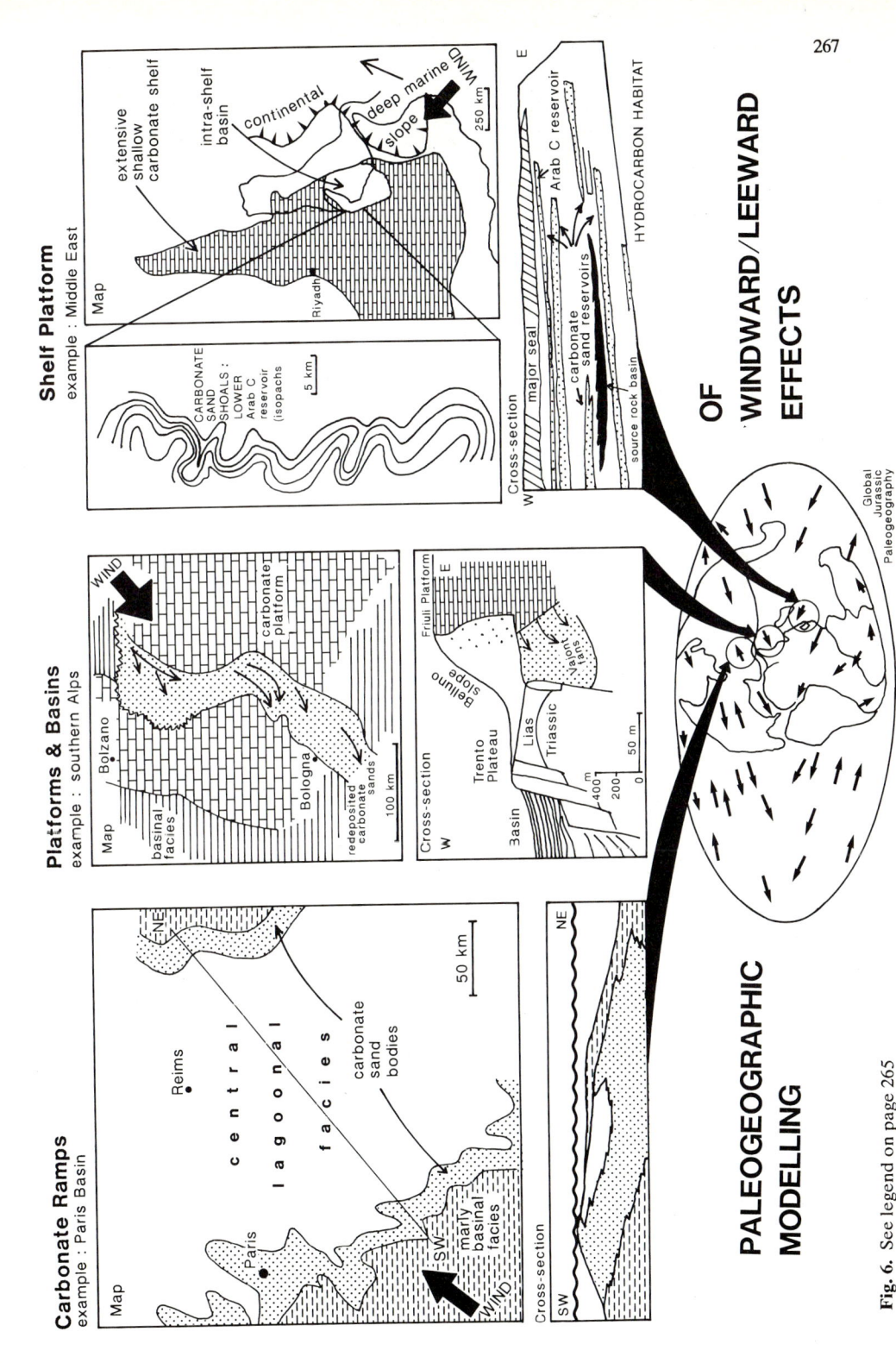

Fig. 6. See legend on page 265.

2.4 Taphonomic Feedback (Live/Dead Interactions) in the Genesis of Bioclastic Beds: Keys to Reconstructing Sedimentary Dynamics

S. M. Kidwell

1 Introduction

Densely fossiliferous beds – known variously as coquinas, lumachelles, and bioclastic limestones – are common features in marine sedimentary records and can have relatively straightforward post-mortem histories. Some of these concentrations form rapidly (e.g., most physical and biogenic event-concentrations; Fig. 1) whereas others accumulate over longer periods (hiatal concentrations; Fig. 5). Concentrated shells may be indigenous or exotic to the accumulation site, and may accumulate on the seafloor, below the seafloor, or undergo multiple episodes of burial and exhumation. Notwithstanding these variations, most skeletal material accumulates in biotic environments where living organisms both influence and can be influenced by the accumulating death assemblage (Fig. 2).

The influence of hardparts upon the ecological success of living benthos was termed *taphonomic feedback* by Kidwell and Jablonski (1983), who focused on live/dead interactions as a driving mechanism for benthic community change. Skeletal material provides islands of hard substrata in otherwise soft-bottom habitats and, where hardparts accumulate in abundance, can transform the seafloor into a coarser, firmer, and topographically more complex benthic habitat. The development of shell-gravel conditions, whether achieved instantaneously or gradually by autogenic or allogenic mechanisms, should facilitate colonization and reproductive success by species that require or prefer these conditions – predominantly epibenthic suspension feeders – and at the same time inhibit earlier species that can tolerate only the initial soft-bottom conditions (Fig. 3). Natural history observations and manipulative experiments provide abundant evidence for the importance of dead shells in structuring Recent benthic communities: dead shells provide domiciles, shelters from physical stress, spatial refuges from predators and competitors, substrata for attachment of eggcases, larvae, and adults, and generally create a heterogeneous habitat that favors higher species diversities of shelled benthos. Where they accumulate in abundance, dead shells also reduce the penetrability of sediments to burrowers and reduce the efficiency of both deposit- and suspension-feeding by infauna, all acting to reduce infaunal survivorship (see reviews by Kidwell and Jablonski 1983; Kidwell 1986b).

Changes in the structure and dynamics of benthic communities, whether driven by taphonomic feedback or by other biotic and abiotic processes, have in turn their own feedback upon the taphonomy of fossil assemblages. New species contribute new kinds or proportions of skeletal elements and also can directly enhance or reduce

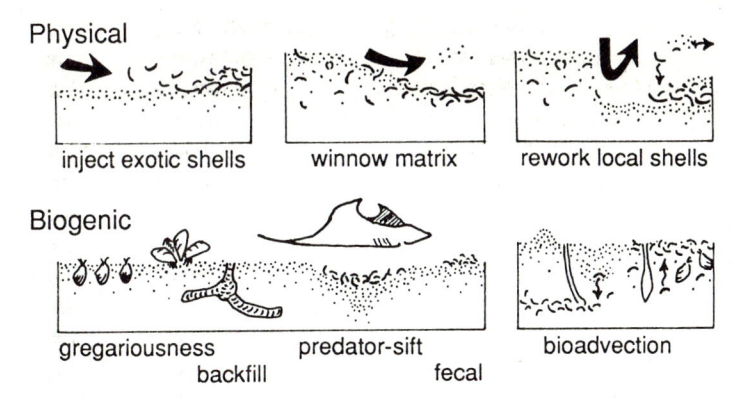

Fig. 1. Event-concentrations of bioclasts form rapidly relative to geological and ecological time-scales. Physical agents of concentration include storm surges, fairweather reworking, and turbidity currents; biological agents include predators, scavengers, bioadvecting and burrow-lining infauna, and gregarious behavior of the bioclast-producing taxa themselves

Live organisms influence
the accumulation of bioclasts

* Bioclasts provide hard substrata for attachment by larvae, adults and eggcases
* Create coarser, firmer and more stable seafloor for colonization
* Create microtopography with spatial refuges
* Alter hydrography of benthic boundary layer
* Decrease penetrability of seafloor by infauna
* Reduce feeding efficiency of both suspension- & deposit-feeding infauna
* Provide domiciles for hermit species

* Contribute bioclasts at particular rate and composition
* Enhance bioclast preservation by encrusting or by inhibiting bioerosion
* Disarticulate and fragment bioclasts by predation, scavenging and bioturbation
* Destroy and weaken bioclasts by boring, rasping and crushing
* Dissolve bioclasts by ingestion and by increased porewater irrigation
* Bury, exhume, concentrate, disperse and transport bioclasts into and out of life habitat

The accumulation of bioclasts
influences live organisms

Fig. 2. Major pathways by which death assemblages are both influenced by and directly influence live benthos

the preservation potential of bioclasts produced by other organisms (Fig. 2). This chapter focuses on the taphonomic aspects of live/dead interactions, and in particular the ways in which paleoecological, taphonomic, and other evidence can be used to reconstruct the short-term dynamics of sediment accumulation in aerobic marine environments. Pathways of taphonomic feedback and their outcome have undoubtedly shifted over the course of Phanerozoic evolution. This history has not been

TAPHONOMIC FEEDBACK

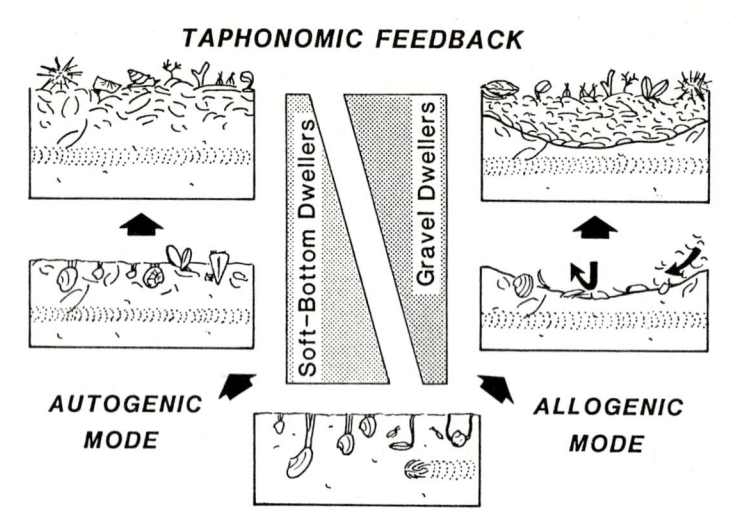

AUTOGENIC MODE — **ALLOGENIC MODE**

Fig. 3. Transformation of an initially soft-bottom substratum into a shell-gravel mediated by live/dead interactions. The accumulation of dead shells – whether supplied by death of in situ organisms or introduced and concentrated by outside agents – changes the structure and dynamics of the benthic community, which shifts toward dominance by taxa that tolerate or prefer shell-gravel conditions (Kidwell 1986b)

investigated systematically but clearly has implications both for paleoecology and for process-level studies of stratification.

2 Taphonomic Consequences of Live/Dead Interactions

Many live/dead interactions reflect active selection of dead shells by organisms for specific ecological ends. Very few of these appear to be obligate relationships driven by co-evolution (but see case of deep-sea limpets that graze exclusively on squid beaks, fish bones, or whale bones; Hickman 1983; Marshall 1987). Instead, shell-utilizing organisms are fairly opportunistic within certain limits, seeking out the most desirable of the readily available death assemblage. For example, hermit crabs (pagurids) and octopods will occupy a range of gastropod shell types, although their populations can be limited by the number of available shells (Mather 1982a,b; McLean 1983; McClintock 1985). The gastropod *Xenophora*, which attaches dead shells onto its own, accepts many kinds of small bivalves for camouflage when young but later prefers elongate shells (high-spired gastropods, scaphopods) presumably as stilts to distribute its weight (Linsley and Yochelson 1973). Many encrusting organisms select substrata by size and/or hydraulic stability and thus do not necessarily prefer shells over lithic gravel, but in other instances ornamented or strongly concave shells are highly advantageous refuges from rasping predators (Bishop 1988; summarized in Kidwell and Jablonski 1983).

Other live/dead interactions are better characterized as unintentional. These include dead shells crushed by bottom-feeding rays, shells buried or exhumed by bioadvection, shell breakage and reorientation below the seafloor by bulldozing organisms, and disarticulation by scavengers. A complete gradation exists between such chance live/dead interactions and those known to be intentional.

All types of live/dead interactions provide information on the behavior and ecological strategies of benthos and all influence patterns of bioclastic accumulation. Some of these interactions such as heavy encrustation (e.g., Balson and Taylor 1982) increase the likelihood that the dead "host" is preserved and thus favor bioclastic accumulation, whereas other interactions are detrimental (e.g., shell attack by bioeroders, delayed burial by pagurids, bulldozing, and exhumation). Still other interactions, such as the preferential colonization of shell gravel by free-living macrobenthos, have few direct consequences for preservation of bioclasts but contribute new shells to the accumulating death assemblage. In this way they can have a strong additive effect on the formation of bioclastic beds.

2.1 Additive Effects

Preferential colonization by shelled benthos of isolated shells on otherwise disadvantageously soft or oxygen-poor seafloors is the first stage in development of many biostromal and biohermal deposits. Initial colonists can be endo- or epi-byssate forms that require only a small shell or shell fragment for attachment when juveniles, but then outgrow the need for such "life preservers" because of snowshoe and various isostatic adaptations when adults (e.g., anomiid bivalves, strophomenid brachiopods; various oysters, see Seilacher et al. 1985a). These colonists in turn provide substrata for colonization by con-specifics and others both while alive and after death. Many shell gravels thus can be "seeded" by death assemblages of deposit-feeders and small infaunal suspension feeders whose dead shells are small and widely dispersed. Lithic pebbles can serve the same function, but these are typically more scarce than indigenous shell debris in soft, level-bottom settings.

Once an initial, local concentration of shell material is formed, live/dead interactions can play a major role in further bioclast accumulation. The epibenthic suspension feeders that require or prefer surficial shell-gravel conditions (i.e., firm to hard substrata) range in size but include larger-bodied species that contribute more skeletal carbonate per individual death. Shell production is further augmented when these shell-utilizing colonists are particularly fecund (as with true oysters) and/or highly gregarious. Sessile and attached epibenthos in particular are highly susceptible to death by burial because they are unable to escape even thin layers of fine sediment (although many have the ability to sweep themselves clean of small amounts of debris, and others can wait out temporary burial by slowing metabolism). Such obrution-related mortality can greatly augment bioclastic accumulation if depositional increments (1) are removed after death to allow for recolonization of the "new" death assemblage and (2) are sufficiently infrequent that the second generation of colonists can grow to a threshold size and thereby survive the post-mortem rigors of the next cycle of burial, exhumation, and recolonization. In this way some sedimentary dynamics should pump up bioclastic production and accumulation on and just below the seafloor.

In addition to facilitating colonization by new, more productive species, local concentrations of shell can trap skeletal debris in motion across the seafloor. Thickets of ramose corals and bryozoan colonies, for example, commonly initiated on dead shell substrata, behave in this way on both Recent and ancient level-bottom seafloors (Nelson et al. 1988; Cuffey 1985). The accumulated allochthonous debris, along with bioclasts produced by the local community, serves to further enlarge the initial concentration by providing an ever-widening halo of appropriate substrata for colonization.

Individually, epibenthic hardparts commonly have higher preservation potential by virtue of their size and (for some groups) calcitic mineralogies, but once accumulated to some threshold abundance they can have further beneficial effects on preservation. Close-spaced bioclasts reduce the erodibility of the seafloor owing to their relatively large size and tendency to interlock, and thus should protect underlying bioclasts from damage they might accrue from further burial-exhumation cycles. Also, by creating a less penetrable substratum for infauna (e.g., Newell and Hidu 1982), bioclasts should reduce sediment irrigation by bioturbators and thus allow favorable alkalinity to build up in porewaters. In situations where the bioclastic sediment is highly porous and contains little fine-grained matrix, bioclastic fabrics can foster free exchange with overlying waters, which are usually saturated or oversaturated with respect to calcium carbonate. Under these conditions of surficial exposure, shells are afforded little protection from boring organisms but should not experience true dissolution.

Live/dead interactions can figure in the genesis of bioclastic beds even when bioclast accumulation occurs primarily beneath the seafloor surface. Unless buried very deeply (i.e., greater than the burrowing depth of common shelled infauna, ~20 cm), subsurface shell concentrations created by storms and bioadvection (Fig. 1) can impinge on benthic ecology by restricting infauna from burrowing to optimal depths. Such individuals are subject to higher metabolic stress from physical environmental fluctuations and/or higher rates of predation and other interference competition, thereby reducing survivorship (Pearce 1965; Haddon et al. 1987; Zwarts and Wanink 1989). Through the in-situ accumulation of infaunal hardparts (assuming that hardpart production exceeds hardpart destruction), the surficial sedimentary layer can become increasing shell-rich, reducing infaunal habitat space. In an idealized sequence (Fig. 3), deep-burrowing infauna and large-bodied mobile infauna become less abundant in favor of shallow-burrowing and smaller-bodied mobile infauna and, when shell-gravel conditions at or just below the seafloor are fully developed, eventually yield to dominance by taxa that tolerate or prefer a nestling or epifaunal habit.

2.2 Subtractive Effects

From the perspective of bioclast accumulation, bioerosion is probably the most detrimental of live/dead interactions in that it both weakens and reduces hardparts to fine sediment (Highsmith 1981). This category includes surface rasping by grazing organisms, micro- and macroscopic boring, pitting and other shell damage by encrusters, and fragmentation by organisms that prey upon post-mortem encrusters and borers. Shell destruction rates can be very high in shallow-water tropical settings [e.g., >8000 $g/m^2/a$ by the sponge *Cliona* (Acker and Risk 1985), 40–168 $g/m^2/a$ for parrotfish (Frydl and Stearn 1978), 80–325 $g/m^2/a$ for echinoids (Russo 1980)]; infestation by borers is high even in temperate and high-latitude shell gravels (Young and Nelson 1988). Shell destruction by bioerosion is primarily a function of shell exposure on the seafloor – burial usually affords good if not complete protection – and is greater in shallow (photic) than in deep waters (Budd and Perkins 1980; Akpan and Farrow 1985). Evidence of boring and encrustation thus generally provides good evidence for delayed burial (and for exhumination of infauna).

Bioturbation has a range of effects: both burial (Meldahl 1987, references cited therein) and exhumation (McCave 1988), concentration (op. cit.) and dispersion have been attributed to mobile infauna, as have disarticulation, reorientation and fragmentation (Brett and Baird 1986b). Any of these processes that reduce the post-mortem survival of individual hardparts or interfere with the formation and survival of hardpart aggregations are detrimental to the development of bioclastic beds. By loosening and irrigating sediment, bioturbation also interferes with colonization by many shelled benthos, particularly suspension-feeding epibenthos, and reduces their preservation potential by maintaining undersaturated porewater conditions in the bioturbated interval. The typically aragonitic compositions and thin-shelled morphologies of species that characterize soft, shell-poor substrata predispose them to early physical and diagenetic destruction.

These kinds of interactions, not all of which are direct or selective interactions between live and dead individuals, act to keep initially shell-poor substrata in a shell-poor condition or to deplete them further. Once a local concentration of shells is formed, however, various kinds of biological, physical, and geochemical feedbacks should tend to enhance the preservation potential and can even favor growth of the bioclastic bed: shell-rich substrata stay rich or become even richer (Kidwell 1986a, 1989). It is unclear what minimum, threshold shell-richness is necessary to insure (or at least significantly improve the probability of) bioclast preservation, but it is clearly an unstable equilibrium point determined by local taphonomic and ecologic conditions. In reef communities, for example, Highsmith (1980) and Hallock (1988) related bioerosion and thus accumulation potential to nutrient levels in the overlying water: as nutrient level increases, coral growth and recruitment decrease, carbonate production decreases, algal growth increases, and bioerosion intensity increases, thereby reducing the likelihood of a good fossil record.

3 Reconstruction of Sedimentary Dynamics

Patterns of small-scale erosion, deposition, and transport on the seafloor can be inferred from the extent and selectivity to which encrusting, boring, and other attaching organisms utilize individual shells. All of these interactions indicate shell exposure at or near the seafloor for some ecologically significant period of time: shell-utilizers must discover, colonize, and grow to a preservable size if exposure is to be taphonomically detectable. Strong correlations between sediment shelliness and taxonomic composition are also consistent with taphonomic feedback on omission surfaces and within slowly aggrading shell gravels (hiatal accumulations). By using these and other ecologic, taphonomic, sedimentologic, and micro-stratigraphic lines of evidence, a high degree of detail regarding the dynamics of stratification can be achieved for bioclastic units.

3.1 Post-Event Colonization (Allogenic Taphonomic Feedback)

Event-concentrations, whether physical or biogenic in origin (Fig. 1), can be composed of (par)autochthonous (= indigenous pre-event community), allochthonous ("syn-event" transported shells), or mixed-origin death assemblages. Any of these types of background assemblage can evoke an ecological response from living benthos if the shells are exposed on the seafloor for an ecologically significant period of time, or if the shells occur in sufficient abundance within the sediment to alter its mass properties. The shelly assemblage produced by the post-event colonizers can be either admixed with or superposed upon shells of the original event-concentration.

The nature and extent of interactions between living benthos and dead shells depend upon burial patterns: post-event burial of the initial shell concentration can be immediate or delayed, permanent or temporary, deep or shallow (=thick vs. thin burial increments) (Fig. 4).

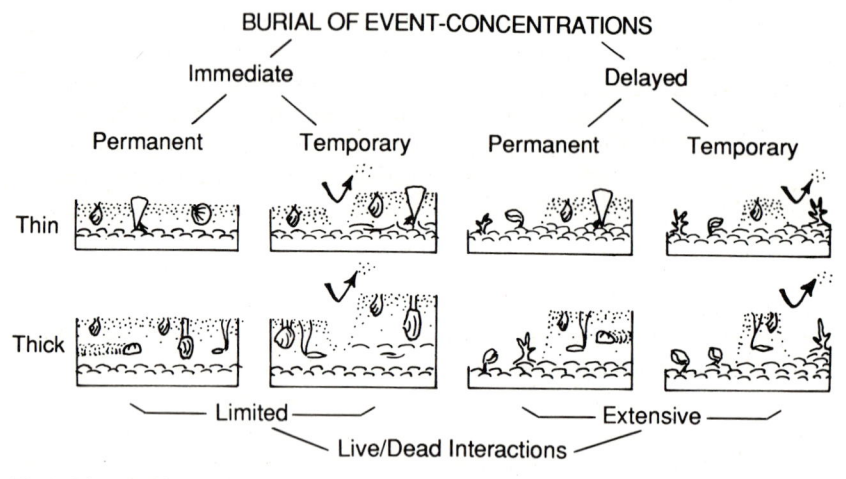

Fig. 4. Live/dead interactions modify the ecologic composition and taphonomic features of simple event-concentrations to varying degrees depending upon the immediacy, permanence, and thickness of post-event burial by sediment. See text for explanation of each contingency

3.1.1 Immediate Burial (Including Obrution)

Immediate and permanent burial is essential for good preservation of articulated and/or lightly skeletized specimens in the pre-event community. The thicker the layer of entombing sediment (= *burial increment*), the less likely that the burial-censused assemblage will be modified taphonomically or that some portion of the living fauna will escape. Bioclastic beds formed in this way offer minimal opportunities for taphonomic feedback because dead shells are sequestered from all but the deepest-burrowing infauna. The faunal assemblages of such concentrations should also have suffered negligible post-mortem damage from seafloor exposure.

When the burial increment is thin relative to typical burrowing depths of infauna and benthic scavengers, and the event-concentration itself is also thin, bioturbators may reorient, disarticulate, and even disperse shells out of the concentration. Alternatively, shells may be concentrated more tightly by "conveyor-belt" biogenic reworking of the burial increment (cf. Meldahl 1987). The fossil assemblage will otherwise resemble that of the deep-burial scenario, since the concentrated assemblage is never exposed at or sufficiently near the seafloor surface for infestation by encrusters and bioeroders. The background assemblage, however, may be augmented by endo-byssate species that can penetrate the burial-increment and use dead shells for attachment (e.g., pinnid and some mytilid bivalves). These post-event fauna in turn can provide substrata for attachment by epifauna (either a live/live or live/dead interaction).

If burial is only temporary, geologically speaking – that is, the entombing sediments are removed by winnowing or reworking – shells produced by colonists of the burial increment can be amalgamated with the original event-concentration of shells. The thickness of sediment available should determine the kind of species that dominate the burial increment – thick increments will be characterized by soft-bottom shallow- and deep-burrowing infauna, whereas thin increments will be characterized by shallow burrowers and endo-byssate forms – so that knowledge of the typical burrowing depths of exhumed species allows the original thickness of the burial increment and depth of its reworking to be estimated (Kidwell and Aigner 1985; Beckvar and Kidwell 1988; Kondo 1989). If the entire burial increment is removed during reworking, the original event-concentration may act as a "reference horizon" (sensu Seilacher 1985) for further bioclastic accumulation. Minor scour surfaces, pods of unreworked sediment from the burial increment, and even laterally continuous remnants of the burial increment can remain to mark this second reworking event. These microstratigraphic features allow the background and post-burial assemblages to be distinguished, as they might otherwise resemble each other in ecology and taphonomy. If the original event-concentration is not resistant to erosional reworking, its assemblage can be completely admixed with its burial assemblage into a single graded bioclastic bed.

Ecologically, immediate burial of event-concentrations maintains continuous soft-bottom, shell-poor conditions on the seafloor. Taphonomic damage from "intentional" live/dead interactions will thus be relatively slight – few shells are available for living organisms to interact with – and will accrue largely from incidental interactions (e.g., damage from bioturbators) and from porewater-related processes (see Sect. 2.2. above). Consequently, taphonomic evidence for most scenarios of immediate burial consists of the conspicuous *absence* of taphonomic feedback: individual shells are clean of post-mortem utilization by benthos even if their orientation and state of articulation indicate exhumation and reworking, and soft-bottom species overwhelmingly dominate the post-event fossil assemblage (Fig. 4).

3.1.2 Delayed Burial

Event-concentrations that have suffered some history of exposure because of delayed burial will show opposite taphonomic features: (1) considerable taphonomic feedback, including both (a) infestation of individuals shells and (b) community-wide response to shell gravel conditions (yielding ecologically mixed assemblages), as well as (2) damage accrued during exposure to physical processes on the seafloor (Fig. 4).

During the post-event period when bioclasts lie on and just below the seafloor, the background assemblage can be augmented by shell-gravel colonists and modified by various shell-utilizing benthos (borers, encrusters, crushers etc.). In addition, carcasses are disarticulated and individual hardparts reoriented and vertically advected both by living organisms and by physical processes. Event-concentrations can undergo a considerable period of colonization and modification by shell gravel taxa before burial, leading to biostromal and biohermal buildups if the delay in burial is sufficiently long (see discussion of hiatal concentrations Sect. 3.3).

Sedimentary burial, when it does eventually occur, terminates this taphonomic/ecologic regime and caps the much-modified bioclastic accumulation with soft sediments. The thickness of the burial increment can be reconstructed using burrowing depths of infauna as discussed in Section 3.1.1.; if sufficiently thick, the last generation of shell-gravel colonists can be preserved articulated and in life position. Johnson (1989) has provided some superb examples of this from the Silurian of Norway, and demonstrated that the age-structure of post-event pentamerid colonists can be used to estimate the length of delay in burial (obrution) of storm shell layers, and thus absolute differences in the frequency of storm reworking in shallow and deep water.

If the burial increment is thin or temporary, post-event colonists have much lower probability of being preserved intact. They will instead be subject to disarticulation, reorientation, fragmentation, and other modification during the next cycle of reworking, during which post-event soft-bottom colonists of the burial increment are admixed or amalgamated with the original event-concentration and its shell-gravel colonists. Once incorporated into the bioclastic bed, the soft-bottom death assemblage can be recolonized and taphonomically modified together with earlier bioclasts until the next temporary (or permanent) burial increment. In this way – by the alternation of shell-gravel and soft-bottom conditions on the seafloor – the final bioclastic bed acquires a taphonomically much-modified fossil assemblage that is ecologically mixed in nature. The mixed fauna reflects both taphonomic feedback (benthic response to the initial event-concentration) and physical amalgamation and mixing of ecologically unrelated faunas (owing to winnowing of death assemblages from temporary burial increments).

3.1.3 Other Situations

In high-stress environments, where salinity or oxygenation extremes restrict benthos, the taphonomy of bioclastic beds is less dependent upon the dynamics of sedimentation. Skeletal concentrations in such settings are typified by quiet-water hiatal accumulations of nekton/plankton, opportunistic benthic colonizations, and rare high-energy event-concentrations that inject allochthonous shells. Skeletal concentrations from the Solnhofen Limestone, Posidonienschiefer, and other fossil-lagerstätten (Seilacher et al. 1985b; Brett and Seilacher Chap. 2.5, this Vol.), for example, differ significantly from bioclastic beds in fully aerobic environments, and this testifies to the significant taphonomic impact that living organisms can have upon the accumulation of dead shells.

Burial of event-concentrations by coarse, relatively porous sediment (e.g., sand, gravel, shell debris) is less effective than fine-grained sediment in smothering fauna and excluding fouling organisms. Many encrusters in fact prefer the cryptic habitats provided by the undersides of shells because of less intense predation, and thus avoid the seafloor surface sensu stricto. Scoffin and Henry (1984) found that encrusting sclerosponges can survive at least several months of burial under 1–2 m of hurricane-deposited rubble on Jamaican reefs, whereas burial under only a few cm of mud (silt-clay mixture) is generally lethal for suspension feeders. The relative sensitivity of bivalves to various thicknesses and grain sizes of anastrophic burial has been examined in considerable detail by Kranz (1974) and others. The minimum effective thickness of burial increments thus varies significantly among sediment types and must be considered in reconstructing the dynamics of interstratified bioclastics.

3.2 Autogenic Taphonomic Feedback

Hypothetically at least, live/dead interactions can figure in the formation of bioclastic beds even without the "seed" of an initial event-concentration and also in the absence of other reworking events. Given early diagenetic regimes that allow for net accumulation of some of the local death assemblage, repeated colonization of a non-aggrading seafloor should gradually transform the initially shell-poor soft-bottom habitat into a more shell-rich and thus coarser, firmer, and eventually topographically complex habitat (Kidwell and Jablonski 1983) (Fig. 3).

This autogenic mode of shell-gravel genesis results in ecologically mixed faunal assemblages because later shell-tolerant taxa occupy the same sedimentary volume as the initial soft-bottom forms. Vertical mixing by bioturbators or by physical reworking would further homogenize the assemblage, obliterating any microstratigraphy produced by the progressively more epifaunal habit of successive colonists. The resulting bioclastic unit will lack the minor scour surfaces and winnowed interbeds that characterize accumulations produced by burial/exhumation cycles (Sect. 3.1. above). Post-mortem infestation will vary: bioclasts produced by the initial soft-bottom community (particularly by the deepest-burrowing taxa) can remain buried for the duration of autogenic feedback and thus can show negligible damage from shell-utilizers, whereas bioclasts produced by later, more surficial shell-gravel dwellers will have suffered proportionately greater taphonomic damage.

The diffuse, interference competition represented by autogenic taphonomic feedback is difficult to demonstrate unambiguously in the fossil record, unlike the direct evidence provided by most live/dead interactions at the individual level (e.g., post-mortem boring, encrusting, and hermiting of shells; Walker 1988). Autogenic feedback is certainly consistent with the well-documented behaviors of Recent benthos in the presence of dead shells (see Sect. 2 and references therein), but such analogies are not necessarily appropriate for much of the fossil record.

If taphonomic feedback plays an important role in shaping benthic communities (and thus bioclastic beds), then strata containing greater densities of shell should also contain assemblages with greater relative abundances of shell-gravel species. This can be tested statistically with the null hypothesis being one of no correlation (Kidwell 1986b), implying that dead hardparts played no role in shaping the living community. Some insight into the life habits and ecological preferences of species is

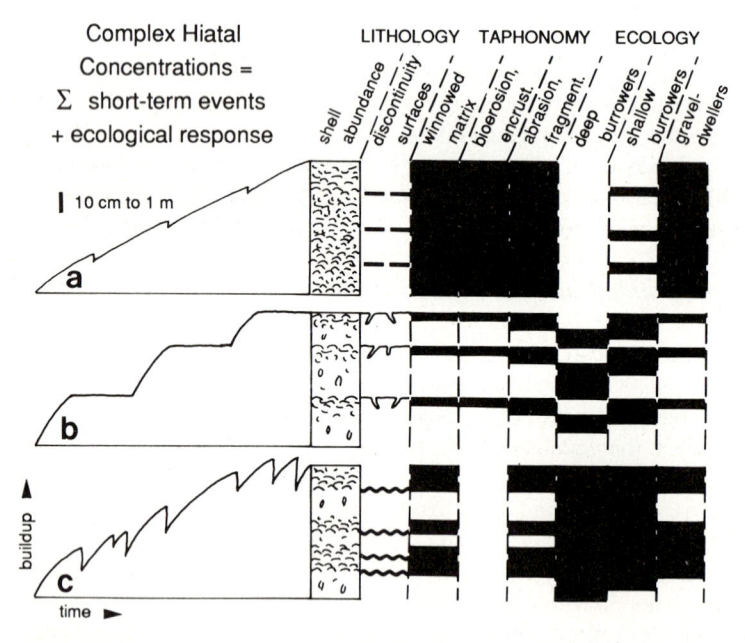

Fig. 5. Detailed, process-level histories can be reconstructed for periods of slow net sedimentation by micro-stratigraphic analysis of the lithology, taphonomy, and ecology of the shell assemblage. Three idealized end-members are described here: sediment starvation or total passing (**a**), episodic deposition and omission (**b**), and episodic deposition and erosion (**c**). (After Kidwell and Aigner 1985 and Kidwell 1989) (See note added in proof.)

essential for such an analysis. In addition, alternative explanations for a correlation between fauna and sediment-shelliness, such as differences in water energy or selective diagenesis, must be accommodated or rejected if the taphonomic feedback explanation is to be accepted.

3.3 Multiple-Event (Hiatal) Accumulations

Many bioclastic accumulations are longer-term buildups that comprise many event-scale concentrations. The subsidiary event-concentrations can be of any type or combination of types described above (Sects. 3.1. and 3.2.). If simply amalgamated, they yield a microstratigraphically complex bioclastic deposit that can include shell-poor intercalations (Fig. 5). Alternatively, hiatal accumulations can comprise little or no internal stratigraphy if (1) physical and biogenic reworking were intense during buildup or (2) shell-gravel conditions went virtually uninterrupted by burial/exhumation cycles. This latter situation results in interlocking bioclastic fabrics and, in some cases, significant boundstone/framestone increments. In all poorly stratified hiatal accumulations, taphonomic and ecological evidence becomes even more valuable for reconstructing event-scale sediment dynamics.

The term *hiatal* refers to (1) the close association of many of these bioclastic accumulations with discontinuity (hiatal) surfaces (see Kidwell Chap. 6.3, this Vol.), (2) their composite, multiple-event nature and inclusion of many minor discontinuities as opposed to simple event concentrations, and (3) their formation during significant slowdowns if not complete hiatuses in the accumulation of nonbioclastic sediment (e.g., siliciclastics, carbonate mud, other allochems) owing to negligible supply, continuous transport, or active removal of such sediment. Many are characterized by some form of faunal condensation sensu Fürsich (1978), and some are stratigraphically condensed sensu stricto, that is, thin relative to coeval strata elsewhere (see Kidwell Chap. 6.3, this Vol., for review and examples). Hiatal accumulations are not necessarily condensed, however, because bioclasts can accumulate in significant thicknesses that equal or even exceed aggradation in adjacent environments. Biohermal and bioclastic shoals are common examples of stratigraphically normal or even expanded shell-rich records. (See note added in proof.)

The three basic pathways of hiatal accumulation characterized in Fig. 5 have been described in greater detail elsewhere (Kidwell and Jablonski 1983; Kidwell and Aigner 1985; Kidwell 1989) and are idealized types: all intergradations are possible (examples in Beckvar and Kidwell 1988). These models were developed for siliciclastic and mixed siliciclastic-carbonate systems, but can be adjusted for purely carbonate systems as well.

3.3.1 Continuous Omission of Nonbioclastic Sediment

Under conditions of sediment starvation or complete bypassing of sediment in suspension, shell gravel conditions can be maintained continuously on the seafloor by local benthic production (± addition of allochthonous shells) (Fig. 5a). Faunal assemblages should be dominated by shell-gravel dwellers, most notably epibenthos (both free-living and attached), nestlers, and shallow-burrowing taxa that are either tolerant of a semi-infaunal habit or capable of exploiting matrix available between bioclasts. Bioclasts should exhibit a relatively high frequency and intensity of infestation by boring and encrusting organisms, a high proportion of disarticulated elements, and a large proportion of fragments (fragmentation can be biogenic or physical and pre- or post-depositional in origin and thus is a poor environmental indicator). Water depths can be inferred from the composition of borers and encrusters and from surface features of individual shells; abraded fragments for example indicate repeated shifting of shells and sediment on the seafloor, consistent with shallow water (unless allochthonous, of course). Age-at-death of autochthonous epibenthos provide a maximum estimate for rates of seafloor aggradation.

3.3.2 Episodic Deposition and Omission

Stepped aggradation of the seafloor repeatedly renews soft-bottom habitats, which through autogenic taphonomic feedback (± allochthonous shell input) become increasingly shell-rich during intervening periods of non-aggradation (Fig. 5b). Shell enrichment by taphonomic feedback will be largely limited to the upper 10–20 cm of any depositional increment, although the seafloor may then aggrade upward by shell accumulation once a predominantly epibenthic shell-gravel community is established. Complex accumulations formed by repeated cycles of deposition/omission thus can include some relatively shell-poor layers if depositional increments are thick. Faunal assemblages will range from soft-bottom dominated (shell-poor layers) to ecologically mixed (shell-rich, upper part of original depositional increments) and shell-gravel dominated (shell-rich buildups above each depositional increment).

Although depositional increments can be intensely bioturbated, at least before shell abundance becomes prohibitive, they are not disturbed by physical reworking; moreover, each increment and its living benthos eventually undergoes permanent (rather than temporary) burial. Consequently, a significant number of specimens (from early and from latest colonists) can be preserved articulated and even in life positions. Shell damage by endo- and epibionts will be variable, with early deep-burrowing colonists bearing lighter infestations than later colonists. Fragmentation can be extensive throughout each increment and will affect all ecological groups, although shell-gravel taxa may suffer disproportionately because of the combined effects of physical and biological processes. The sedimentary matrix in the upper part of each increment may be winnowed.

3.3.3 Episodic Deposition and Erosion ± Omission

Alternating deposition and erosion amalgamates event-concentrations such as described individually in Section 3.1. Once bioclastic material has accumulated to an erosionally resistant threshold thickness, a microstratigraphic record of the discrete event-concentrations can be preserved. The type and extent of taphonomic feedback in these kinds of hiatal accumulations depends upon the length of delay between erosional reworking – which concentrates soft-bottom taxa into a shell-gravel lag – and burial under the next depositional increment. Thus the faunal composition of the complex accumulation can be (1) dominated by well-preserved but reoriented soft-bottom taxa (ecologically insignificant delay in burial) or (2) a mixture of soft-bottom and shell-gravel taxa, both sets having accrued considerable taphonomic damage. All specimens from the reworked increments will be reoriented (± disarticulation); soft-bottom infauna that remain in life position indicate the maximum depth of erosional reworking of the substratum. Infauna preserved in life positions within shell-rich parts of the accumulation should be suspected of being shell-gravel dwellers; such occurrences provide more dependable insights into the ecological tolerances of fossil species than inferences based on the ecology of modern representatives.

Such sawtooth histories of alternating deposition and erosion are most common in shallow marine settings, and are usually characterized by winnowed sedimentary matrices and many minor discontinuity surfaces produced by scour and firmground development. Depending upon the depth of erosional reworking, depositional increments may be only partially truncated, leaving intercalations of less shelly, less winnowed, and less modified substratum within the complex accumulation. These layers and pods provide a useful baseline for background conditions, which have lower preservation potential than the higher energy concentration events.

4 Summary and Conclusions

Live/dead interactions are pervasive in aerobic benthic habitats and can have significant consequences for bioclastic deposits. Organisms with mineralized hardparts contribute bioclasts whose accumulation modifies the physical habitat, facilitating species that tolerate or prefer shell-rich substrata, and inhibiting the success of earlier soft-bottom colonists. New epibenthic colonists commonly produce more skeletal carbonate than precursor benthos, whereas other shell-utilizing colonists destroy or inhibit further accumulation of shell material. The budget of shell production versus destruction by live individuals is not known quantitatively for any environment, but live/dead interactions clearly have both additive and subtractive aspects.

Most bioclastic beds, even many event-concentrations, consequently have complex taphonomic and paleoecologic histories related to the accumulation of shells on or near the seafloor. Together with conventional sedimentologic and microstratigraphic criteria, taphonomic and paleoecologic features can be used to reconstruct detailed, process-level histories of aggradation, erosion, and omission/transport on the seafloor. Post-mortem bioerosion and encrusting indicate exhumation and seafloor exposure of some duration; the burrowing depth of exhumed infauna indicates minimum depths of erosional reworking; age structure of post-event colonists indicates minimum duration of seafloor exposure; distinctive sedimentary fill of skeletal cavities records depositional increments not otherwise preserved; and quality of fossil preservation in the uppermost parts of bioclastic deposits reveals the immediacy, permanence, and thickness of burial deposits.

Given sweeping evolutionary changes in the diversity of species that produce, utilize, and destroy bioclasts, it would be surprising if patterns of bioclastic accumulation did not change over Phanerozoic time. The past 600 million years have seen an increase in the body size and robustness of shelled benthos and progressive infaunalization, both favoring skeletal accumulation. This has been countered by an overall shift from calcitic to less stable aragonitic hardparts, an increase in duraphagous predators and bioeroders, decimation of large-shelled nekton, and an increase in the depth and/or intensity of bioturbation, all of which should reduce the likelihood of individual shells and shell concentrations being preserved (Kidwell 1990). Documentation of how pathways and mechanisms of taphonomic feedback have changed would be of value not only to the paleontologist concerned with fossil behavior and post-mortem bias, but also to the geologist concerned with extracting maximum paleoenvironmental information from sedimentary deposits. Evolutionary changes in the dynamics of bioclastic accumulation would admittedly complicate the straightforward application of actualistic models to reconstructing the past. On the other hand, such evolution might better explain secular trends in bioclastic stratification previously attributed to broad paleogeographic changes or to cumulative diagenesis.

Acknowledgments. Taphonomic research supported by grants from Shell Foundation, Arco Foundation, Arco Oil and Gas Company, Amoco Production Company, and the U.S. National Science Foundation (EAR85–52411-PYI). Many thanks to A. Seilacher and D. Jablonski for fruitful feedback, both paleo and neo.

Note added in proof: See Kidwell (The Stratigraphy of Shell Concentrations, in *Taphonomy: Releasing Information from the Fossil Record*, P.A. Allison and D.E.G. Briggs, eds., 1991) for updated treatment of complex shell concentrations, and specifically the restriction of the term "hiatal concentrations" to stratigraphically condensed shell accumulations. This usage supersedes that of this paper, which was last revised in May 1989.

2.5 Fossil Lagerstätten: a Taphonomic Consequence of Event Sedimentation

C. E. Brett and A. Seilacher

1 Introduction

Physical events are primarily reflected in sedimentary structures, both erosional and depositional; but they also affect fossil preservation by burying the remains of contemporaneous organisms under a protective sediment cover. This improves not only the preservation of single shells that would survive at the surface only for months or a few years (Davies et al. 1989), but much more so of articulated skeletons, whose elements are held together by connective tissue and would normally fall apart in a matter of days. Echinoderm, arthropod, and fish beds connected with storm layers are examples for such whole-body preservation. They have traditionally been classified as "Obrution Lagerstätten" (obruere = to smother), because rapid burial was certainly an important preservational factor. But fossils in such layers not only contain sessile forms, such as crinoids, brachiopods, or oysters, which could not have escaped and therefore became buried alive, they also include vagile organisms like starfishes and trilobites and these occur in densities higher than would normally be expected of carcasses on the sea floor.

Typical obrution deposits thus require coincidence of a number of factors, including the following:
1. Availability, in adequate numbers, of undecayed carcasses at the sea floor. In the case of nonsessile, and particularly of nektonic organisms, this requires mass mortality, which should in some way be related to the sedimentation event itself.
2. Rapid deposition of a sedimentary blanket thick enough to hold the multi-element skeletons together and to protect them against disarticulation by scavengers and bioturbators.
3. Permanent removal of the burial ground from the reach of subsequent event erosion and resedimentation.
4. For softpart preservation: suitable conditions for bacterial mineralization (Allison 1988a,b).

These conditions could be met by any kind of physical event. The volcanic ashes of Vesuvius killed and cast their Pompeian victims at once. In rare instances, turbidite beds may also contain extraordinarily preserved fossils. The majority of known obrution Lagerstätten, however, come from tempestites and their lateral equivalents. Therefore, we first review the processes involved in a storm in order to predict those situations in which obrution would be most likely to occur and its victims be preserved.

On the basis of the above criteria, general models have been developed to interpret and predict patterns of *taphofacies* on a basinal slope combined with lateral gradients in sedimentation rate (Speyer and Brett 1988). In the present context it is particularly the degree of disarticulation that should interest us, since the lack of disarticulation is the most distinctive criterion of obrution deposits.

2 Storm Dynamics and Proximality Gradients

Basically the wind energy of a storm transforms at the water interface into waves, whose orbital movements become flattened into a pattern of oscillating currents towards the sea floor. Therefore it is justifiable to interpret the erosional and depositional features of a typical storm bed in terms of wave action. A second effect is the buildup of a water swell towards the shore and a counteracting gradient current. Where intervals between storms have been long enough for mud to accumulate in the coastal zone, this current will by its suspended load move along the bottom. Therefore in every storm there is also an element of basinward current activity, which will be felt most as the storm wanes. This component, together with the energy decrease in deeper waters, contributes to the proximality gradient in an individual storm bed. At medium depths the current will have little effect because most of the sediment has already been resettled by the time the gradient current reaches its peak and, more importantly, because this current tends to bypass intermediate areas by concentrating into submarine streams (Fig. 1). The perfectly symmetrical oscillation ripples at the tops of many tempestites bear evidence for the near-absence of current action on level bottoms between the channels.

Translating this hydrodynamic system into taphonomic processes (Fig. 1), we expect that in nearshore areas gradient currents will carry away the suspended mud before it can settle again, leaving behind sandy surfaces or exposed shell pavements ready to be colonized by an encrusting community (see preceding contribution by Kidwell Chap. 2.4).

More relevant for the obrution theme is the effect of gradient currents in the distal parts of the storm-affected area. Being released from the channels, the mud-laden compensation currents will travel by their inertia beyond the level of the storm wave base into a zone in which the storm is felt only in its later stage and acquire the

Fig. 1. Model of representative fossil-Lagerstätten in the frame work of storm dynamics. The complicated wave and current patterns in surface waters and coastal areas become simplified with depth: shorter wave lengths attenuate and gradient currents concentrate into downslope turbidity streams that bypass level bottom in between. Shelly *concentration-Lagerstätten* form predominantly in the coastal zone, but may also occur in deeper waters, where secondary soft bottom dwellers produce outsized shells ("storm tells"). Inverse stacking of thin valves in still deeper zones is probably due to seismic fluidization of the host sediment. The articulated skeletons of *conservation-Lagerstätten* typically retain their life positions and lack current alignment if they are associated with tempestites (Muschelkalk sole face with ophiuroid *Aspidura*, Hagdorn collection, Ingelfingen). In low-oxygen bottoms below wave base, storms are felt as turbidity currents that may have become toxic by suspension of fetid mud and silts. Therefore fossils are typically current-aligned and partially disintegrated (Hunsrück shale asteroid and hatchet holothurians, Tübingen collection) or became dorsally flexed (*Leptolepis*, Solnhofen) before they became mud-blanketed

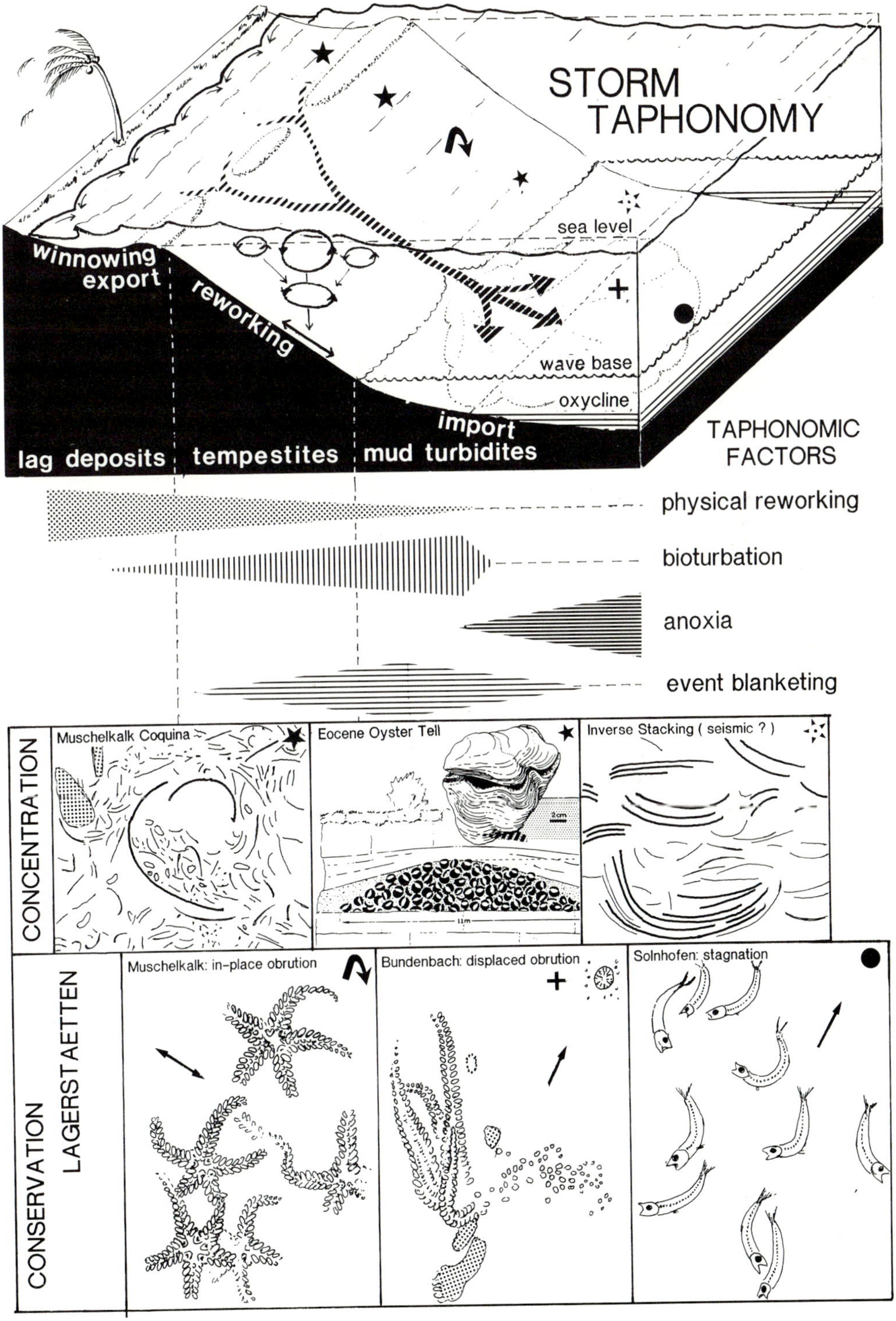
STORM
TAPHONOMY
sea level
winnowing
export
reworking
wave base
oxycline
import
lag deposits
tempestites
mud turbidites
TAPHONOMIC
FACTORS
physical reworking
bioturbation
anoxia
event blanketing
CONCENTRATION
Muschelkalk Coquina
Eocene Oyster Tell
Inverse Stacking (seismic ?)
CONSERVATION
LAGERSTAETTEN
Muschelkalk: in-place obrution
Bundenbach: displaced obrution
Solnhofen: stagnation

characteristics of a turbidity flow. On continental slopes storm-induced currents may pick up enough gravitational energy to develop into full-sized turbidity currents. On the gentle slopes of epicontinental basins, however, carrying capacities will quickly diminish as the rip currents dissipate. In this distal zone, deposits may still reach only the grade of silty micro-turbidites with current-rippled tops, but in their majority consist of mud turbidites without distinctive sedimentary structures.

As far as the burying effect is concerned, obrution could occur anywhere along this proximality gradient. What changes is the probability with which the buried carcasses eventually escape subsequent bioturbation and storm reworking. The killing factor, however, may also be unevenly distributed. In general the preservation potential of obrution layers will increase downslope.

When sea level changes, the proximity gradient can be expressed also in the vertical section. Here the preservation potential will be highest during transgressive phases and at levels in which the particular area fell below the direct reach of storm waves. But it should also be remembered that storm wave base is not a fixed datum but depends on the time interval over which storms are averaged. In the sedimentary record this averaging is a function of sedimentation rate. The more slowly a pile of sediment accumulates, the more likely it will experience a thousand year storm. Conversely, the chance that an obrution layer escapes secondary reworking will be higher as sedimentation rate increases.

To sum up, in a storm-dominated epicontinental basin, obrution deposits are most likely to be found in distal facies just beyond storm wave base, during transgressions, and in periods of increasing mud sedimentation.

Oxygen concentration is another factor that influences preservation potential. At low oxygen levels obrution layers will have a better chance to survive not only because there are fewer scavengers, but also because bioturbation reaches less deeply into the sediment than under fully aerated conditions. Also, bioturbators may be interested only in H_2S-pumping and therefore leave the carcasses untouched (Savrda et al. Chap. 5.2, this Vol.). Because oxygen levels tend to decrease with depth and during transgressive phases, the redox factor works in the same direction as the sedimentological factors just mentioned.

3 Causes of Mass Mortality

If obrution deposits had affected only carcasses that happened to be present at the sea floor at the time of the event, buried objects would be rare and would display various stages of decay. What we characteristically find, however, are large accumulations of fossilized carcasses all in a rather perfect state of preservation. Therefore mass mortality (Brongersma-Sanders 1957) must have been involved in the formation of typical obrution deposits. Given that dead corpses at the sea floor would be torn apart and disintegrate in a matter of hours or days, it is also clear that this mortality could not have had an independent cause but must have been part of the very event that led to the burial.

Unfortunately, observations in modern seas are of little help in this matter. Red tides, in which large numbers of fishes become poisoned in the trophic chain, are irrelevant here: firstly, because they do not involve extraordinary sedimentation; secondly, because they do not affect benthic organisms; thirdly, because the victims drift and end up at the shore (where preservation potential is very low) rather than ending up at the sea floor.

In speculating about other causes of mortality, we must treat different guilds of organisms separately, because vulnerabilities will be different for animals living at the sea floor and those in the water column.

3.1 Infauna

Burrowing organisms are in various degrees adapted to cope with sedimentational emergencies. In the case of erosion they can burrow away from the eroding surface. Conversely, they respond to fast sedimentation by upward burrowing, leaving characteristic escape structures behind. Still they may fail. If erosion is faster than the downward escape reaction, animals become washed out and die at the surface if they have no chance to become re-established. In the case of rapid sedimentation, their upward escape may also be stopped by newly deposited soupy mud, which is unsuitable for their kind of burrowing and respiration. Horizons of clams in life position may be explained in this way (Kranz 1974); but it is not certain whether the failure to escape or a toxic effect caused their final death.

This question becomes critical in the case of rare burrowing clams (*Solemya, Goniomya*; Riegraf 1977) in the bituminous Posidonia Shales. They are always found flat on the bedding planes, with the valves still articulated, but without any sign of erosion in the host sediment. Could it be that they came out of their burrows to escape asphyxiation? Modern *Callianassa* populations were observed (A.S., El Salvador), under the influence of pesticides, to come to the surface – which they never do under normal conditions. Oschmann (Chap. 5.4, this Vol.) records the same phenomenon from polluted modern shelf environments. But evidently we need more extensive experimental evidence before this observation can be applied to fossil examples.

3.2 Immobile Epibenthos

Active escape is impossible in sessile organisms. Hardground encrusters such as barnacles, oysters, bryozoans and edrioasteroids may be killed by burial alone, no matter how slowly it occurs. But in order to remain articulated, oysters and echinoderms must also be effectively removed from the action of scavengers and bioturbators. The same applies to passive recliners and mud stickers. They may, in addition, be concentrated during the erosive phases of storms.

Secondary soft bottom dwellers tend to produce overweight skeletons for stabilization. So storm erosion may not be able to transport the shells, but will sink them to a deeper level (commonly onto a layer in which the victims of former storms have been accumulated) before spreading a new protective mud blanket over them (Fig. 1). Thick beds of oysters, horn corals and the similarly horn-shaped barnacle *Waiparaconus* (Seilacher and Seilacher-Drexler 1987) can be explained in this way.

But although the oysters in such amalgamated shell beds commonly remain articulated, this is not what we mean by typical obrution deposits. In current Fossil-Lagerstaetten classification (Seilacher et al. 1985b), they would fall under concentration, rather than conservation, deposits.

3.3 Flexibly Attached Epibenthos

Another guild consists of elevators. As illustrated by modern species of gooseneck barnacles, tunicates, brachiopods, and crinoids, they are also sessile filter feeders. But in order to reach a higher tier, they have evolved flexible stems or pedicles. In the case of obrution this adaptation has an advantageous side effect: even if the pedicle is buried by a sediment blanket, the animal itself may survive – if there was no associated toxicity. In this way horizons of brachiopods in the Devonian Oriskany Sandstone (Seilacher 1968) were embedded in life position, with the beaks down. Also they maintain a level about 1 cm above the base of the sandstone bed – just corresponding to the length of a fleshy pedicle, with which they may have been attached to the top of the previous bed.

Another case are the crinoids of the Devonian carbonate platform in upstate New York. They occur at the top of a storm-concentrated bed of crinoidal debris, but the crowns again lie a few centimeters above the original substrate. In fact, what appear to be thousands of "decapitated" crinoid crowns exposed on a Helderberg slab in the Peabody Museum, New Haven, may be largely the upper parts of whole animals still attached to the coarse debris layer some centimeters below. But the slab also contains many detached individuals and stems washed together in a poorly oriented fashion (Ch. Wray and R. Vasquez pers. commun.). So there has been more than simple smothering involved.

3.4 Vagile epifauna

Animals crawling on the sea floor could be similarly killed by rapid sedimentation, particularly if they were unable to rapidly swim off. Layers of perfectly preserved starfish, brittle stars, and echinoids would fall in this category (Rosenkranz 1971), but also the famous trilobite clusters in the Devonian of New York (Fig. 4; Speyer and Brett 1985, 1986). These are the cases on which the very concept of obrution deposits was originally based. But it was also recognized that mere smothering cannot have been the only cause, because instead of mixed communities we find clusters of only one species or even age group (trilobites) or of ecologically differing members of one phylum (echinoderms): frozen parties or selective killing? The first explanation appears to fit the trilobite example, the second the echinoderms, if we assume that the shared possession of an ambulacral system communicating with sea water makes them more vulnerable than other groups to mud blanketing (Rosenkranz 1971). In this case death would have almost coincided with the blanketing and would have excluded escape reaction. In a similar way the delicate structure of their exopodial gills could have made some species of trilobites more sensitive than others to mud-clogging.

An argument against simple sediment-trapping, however, is that not all individuals are found in life position as in the ophiuroid examples (Figs. 1 and 2). In trilobite layers the majority of individuals may actually be lying on their backs (Speyer 1987). Another argument for a temporal separation of the killing and blanketing effects comes from signs of incipient decay. Either the tips of crinoid or starfish arms are falling apart (Fig. 1), or the body sack of ophiuroids is smeared out (Fig. 3), or trilobite segments are separated. Certainly hours or days would be sufficient to reach such states of decomposition, but time would be too short for rapid blanketing, to have been the agent of mortality.

3.5 Nektonic Organisms

Typical obrution deposits contain only benthic animals. But the silty turbidite in the Cretaceous of Sendenhorst, N. Germany (Siegfried 1954), which by its sedimentary characteristics suggests rapid deposition, contains a diverse fauna of fishes and shrimps. These nektic or nektobenthic species could not have been killed by blanketing. Also their sheer numbers require some kind of mass mortality.

The same is true for fish horizons in lithographic limestones. Some slabs from Solnhofen, where fishes are otherwise a rarity, contain hundreds of specimens of the

Fig. 2. Obrution deposit of the brittle star *Aspidura* on top of a thin Muschelkalk tempestite (Hagdorn collection, Ingelfingen). Note that 17 out of 21 individuals retain their life position and that they are neither current-aligned nor do they show signs of disintegration. This indicates in-place obrution in a wave regime. Elevated knobs are the result of diagenetic overbed cementation

Fig. 3. Specimen of the Silurian ophiuroid *Protaster* sp. on the base of a thin carbonate siltstone bed displaying evidence of incipient decay and slight reorientation prior to burial by storm silt layer. Note consistent recurvature of tips in four of the rays and extension of fifth ray in presumed current direction (*arrow*); also note smearing out of partially decayed disk in direction of current. This suggests partial decay of ligaments before current reorientation. Rochester Shale, Rochester, New York, x1.4, Rochester Museum and Science Center RMSC 91.481, collected by Peter Debbs

small teleost *Leptolepis* (Fig. 1), as does a certain level in the Liassic Posidonia Shales. In the Solnhofen, beds carcasses show current alignment and the characteristic dorsal bend of the vertebral column. This post-mortem deformation, well known from all *Archaeopteryx* and pterosaur specimens, is usually referred to desiccation; but it may also result from dehydration in a brine. In any case, it probably needed more time than would be available once a mud turbidite had begun to settle.

Similar observations apply to Solnhofen slabs in which dozens of small ammonites are swept together, all with their aptychi (and, hence, their soft parts) in place. Still the shells had time to tilt from their edgewise landing positions into the bedding plane before they were blanketed. We may also extend this argument to black shales, for instance to the shrimp beds of the Carboniferous near Edinburgh, where whole conodont animals also have been found (Briggs et al. 1983). Of course, this brings us into fossil-Lagerstaetten that have been classified as stagnation deposits rather than event beds. But since even on anoxic bottoms multi-element skeletons would eventually fall apart without some degree of blanketing, the presence of so many carcasses at one instance cannot be explained by normal rates of mortality and sedimentation.

We propose that these cases represent poisoning by H_2S that formerly penetrated aerated portions of the water column either by mixing of an oxicline or, more likely, by erosion and suspension of fetid mud in turbid compensation flows during storms that may not have left a recognizable sedimentary signature in these distal mud sediments. Whether or not H_2S poisoning was also involved in benthic obrution remains to be tested from case to case.

3.6 Can Storms Sink Floats from the Surface?

The linkage of storms or other turbulence events with mass mortality becomes even more problematic when we deal with objects that had been floating at or near the surface, such as driftwood with its epiplanktonic passengers, or graptolites supposed to have been suspended like siphonophore colonies from their own balloon floats. Still there are reports that such objects are also found concentrated in certain black shale horizons. Although one feels tempted to speculate about mechanisms of mass-foundering, such a discussion would at the moment be too far-fetched. Instead we prefer to summarize the results with reference to specific geologic examples.

4 Genetic Classification of Obrution Deposits

4.1 Hauptrogenstein Type

In storm-dominated environments, siliciclastic, bioclastic, and oolithic sands lend themselves to extensive winnowing, or redeposition and amalgamation. In the resulting grainstone units, which may reach considerable thickness, individual storms are difficult to distinguish. They may, however, have carried torn-off sessile organisms (such as flexibly anchored brachiopods or crinoids) and buried them in the middle of the coarse bed. There the articulated skeletons may be preserved in their original arrangement if they happened not to be disturbed by bioturbation and not to be exhumed by subsequent storms. Except for the uprooting, no particular killing mechanism is required in this case, although drift may be involved. Preferred sites are near-shore banks, channel fills, and beaches.

Examples:

1. The oolitic mid-Jurassic *Hauptrogenstein* of SW Germany and Switzerland has yielded a wide variety of well preserved echinoderms, but mainly crinoids.

2. The grainstones of the Mississippian *Burlington Formation* in Missouri similarly contain whole crinoids. But these occur, as in the previous case, only in local lenses.

3. The rather coarse-grained Lower Devonian *Oriskany Sandstone* of Maryland contains the molds of large crinoid calyces with complete pinnulation, but no stems. The sandstone beds also contain horizons of bivalved brachiopods in life position with preserved brachidia. The occurrence of the same brachiopods as roll-fragments in other bedding planes suggests abrasion by waves in a near-shore zone, into which the broken-off calyces were imported during storms from deeper environments (Seilacher 1968).

4. Similarly well-preserved echinoderms are known to occur locally in the more fine-grained *Angulatus Sandstone* of the Lower Jurassic of Southern Germany. It was deposited in a more offshore area, but still above storm-wave base (oscillation ripples; gutter casts).

4.2 Gmuend Type

In this case, the well-preserved fossils occur not within the coarse storm deposits, but in the shale immediately above them. Echinoderms are again the dominant group, but besides crinoids members of the asteroids, ophiuroids and echinoids may also be represented. Because of their association with coarse and commonly amalgamated storm beds (which may be largely made up of disarticulated ossicles of the same species) and the lack of current alignment or incipient disintegration, we probably are not dealing with animals swept in from other habitats. Rather the carcasses represent a fauna that became locally established during periods of reduced sedimentation, i.e., during a transgressing phase. Still it was only the last generation buried by the muddy tail of the ultimate storm that became preserved in an articulated fashion. Because the victims were supposedly part of a more diverse local fauna that was selected according to taxonomic rather than ecologic affiliation, a specific killing factor must also have been involved – either mud-clogging, temperature shock, or asphyxiation. Environmentally we are dealing with level offshore areas above storm wave base, where either mechanism could apply (Fig. 1).

Examples:

1. The type example (Schechingen near Gmuend; Rosenkranz 1971; Seilacher et al. 1985b) is at the top of a centimetric conglomerate that drapes the erosive transgression surface of the Jurassic. The conglomerate consists mainly of reworked concretions from the underlying terrestrial Triassic and is overlain by several decimeters of barren black shales. Apart from a few worn oysters, the fauna consists only of well-preserved echinoderms with varying ecologies: crinoids, asteroids, ophiuroids, and echinoids. In spite of the proximity to the transgression surface, we are probably dealing with a fairly deep environment, but above storm wave base.

2. The Upper Muschelkalk (Middle Triassic) of Southern Germany contains at several levels horizons with perfectly preserved crinoids (Linck 1965), asteroids or ophiuroids (Figs. 1 and 2; Linck 1965; Seilacher 1988a). They invariably rest on coarser-grained tempestites that are probably amalgamated, because they consist largely of the dissociated ossicles of former echinoderm generations.

3. Similar echinoderm layers occur at various levels within the Devonian and Carboniferous of the eastern United States (Lane 1973). Large slabs stored in museums of the region show the association with an underlying crinoidal tempestite (Rosenkranz 1971). The faunas consist largely of crinoids, but in some examples encrusting edrioasteroids may also be included – particularly where the underlying bed had already been cemented into a hardground (Koch and Strimple 1968). Geologic considerations suggest depths between normal and storm wave base.

4.3 Hunsrueck Shale Type

Areas below storm wave base are represented by muddy environments, in which bottoms are generally more quiescent and storms have left their record only as silty micro-turbidites deposited from distal gradient currents. Due to reducing pore water conditions (enhanced by sealing biomats) and to relatively shallow bioturbation, the preservation potential for carcasses is high and includes not only benthic organisms, but also nektobenthic fish and arthropods. Echinoderms may still be the dominant group.

Examples:

1. The inclusion of distal gradient currents into the model of storm deposition solves the old problem of the Lower Devonian *Hunsrueck Shales* of W. Germany, namely the association of an autochthonous benthic fauna (including tracks of trilobites and other arthropods!) with soft-part preservation that one would relate to anoxia, but also with many signs of currents (current ripples, tool marks, fossil orientation; Seilacher and Hemleben 1966). In an environment in which a rich soft-bottom epifauna can coexist with anoxic conditions in the sediment immediately below, turbidity events, by erosion of the fetid muds, will automatically acquire an asphyxiating effect. Also the lack of deep bioturbation (except for the H_2S probes of *Chondrites*; see Savrda et al. Chap. 5.2, this Vol.) will leave the buried carcasses undisturbed even if they are covered only by a thin mud blanket.

2. The Upper Ordovician Frankfort Shale near Rome, New York (Beecher 1894; Cisne 1973) appears to represent a similar setting. Although percentages are different (predominantly trilobites and graptolites; very few echinoderms), the faunal spectrum resembles that of the Hunsrueck Shales. More importantly, there is the same contrast between benthic diversity (including burrows, but not of flysch type), soft part preservation (trilobite legs, worm bodies) and the signs of fairly strong currents (microturbidites; alignment of fossils). It is also noteworthy that the trilobites with preserved appendages form a layer a few millimeters *above* the base of a 4 cm turbidite and that about half of them are inverted (Cisne 1973).

4.4 Hamilton Type

The trilobite clusters of the Middle Devonian Hamilton Group (Brett et al. 1988; Speyer and Brett 1986a,b; Speyer 1987) differ from the Hunsrueck example by (1) containing only trilobites, usually of a single species and one size-class; (2) by the absence of preserved appendages; (3) by the lateral consistence of individual horizons; (4) by the absence of strong alignment or other current indicators. This suggests that we are dealing with a yet more distal facies out of the reach of waves and eroding currents, though not necessarily with low oxygen levels. Still, storms could affect the area by importing turbid mud clouds that left enough time between poisoning and burial for extended trilobite carcasses to incipiently decompose or (in

Fig. 4. Slab of mudstone containing a cluster of 22 complete outstretched and partially enrolled *Phacops rana* trilobites. Middle Devonian, Silica Shale, Sylvania, Ohio. Scale in centimeters. Specimen in collections of Rutgers University

specific layers) for the living animals to react by enrollment before they were killed and buried (Fig. 4).

Examples:

1. In the type area of the Hamilton Groups in upstate New York (see references above), careful studies have not only shown the basic characteristics of this kind of obrution deposits, but also characteristic differences between individual layers. These differences may be due to the particular situation of the trilobite communities at the moment of the event, the characteristics of the event, or to both. Since similar studies have not been carried out in other rocks, it is as yet difficult to say how an equivalent taphofacies would look, say, in the Jurassic. As a first attempt to generalize, we tentatively group here two other occurrences.

2. Horizons with well-preserved crinoids in the Silurian Rochester Shales of Ontario (Fig. 3; Brett and Eckert 1982) resemble the Hamilton occurrences in that they are not associated with a coarser bed underneath (as in the Gmuend type) and that they have also yielded articulated trilobites in the inverted, ventral-up position. This may be a link to the famous Crawfordville crinoids from the Carboniferous of Indiana.

Fig. 5. Entire carcasses of isopod crustaceans accumulated in stable convex-up positions; Wealden of North Germany (Tübingen collection). Alignment is induced by a current *from the top* of the picture; but since some specimens point in the opposite direction, this must have been a post-mortem effect. The animals were probably asphyxiated, washed together and blanketed by a muddy storm turbidite

3. More far-fetched may be the comparison with horizons, in Cretaceous mudstones, of current-aligned isopod clusters (Fig. 5). Certainly isopod crustaceans are not trilobites, but similarities in mode of life and body shape may bring them into a similar taphonomic category and open the way to an actualistic test.

4.5 Solnhofen Types

Although lithographic limestones have typically formed in small storm-protected basins (Hemleben and Swinburne Chap. 5.5, this Vol.) and could also be classified as stagnation deposits, they should here be mentioned because of the obrution effects involved in their preservation. In contrast to the examples mentioned above, lithographic limestone faunas are not autochthonous. Like those of bituminous shales they recruit in part from animals that could simply have sunk to the stagnant brine bottom from more oxygenated water layers, or the air, on top. But in contrast to the bituminous shales, we find also a selection of benthic organisms. There appears to a selection of forms that could swim up upon irritation (young limulids, feather stars, flat fish, lobster-like *Mecochirus*), so that turbidity currents could carry them into the abiotic zone. As shown by belly-up landing marks, most victims were already asphyxiated when they reached the bottom at the burial site. Only some hardy crustaceans or limulids could still crawl a few meters before they also died. More surprisingly, nektonic forms (such as ammonites, prawns, and small fishes, Fig. 1) are equally preserved at the bases of, rather than within, the thin lime-mud turbidites.

Their alignment and perfect preservation (in some cases also their accumulation) shows that they had not been lying there for extended periods of time. So they, too, must have been killed by the same event that eventually blanketed them. As already mentioned, the time between landing and blanketing must also have allowed the ammonities to tilt into a horizontal position (even the ones with soft parts holding the aptychus in life position), for a dorsal bend to develop in vertebrates (and a ventral one in prawns) and for limulids to make their final track on the bottom. This would at least be in the order of days – the time that the very fine mud might have needed to settle, particularly in dense brine water.

In contrast to the abundant taphonomic evidence, sedimentological clues to the nature of the individual events are scarce in the Solnhofen limestones. This is largely due to the lack of coarser gradeable sediment particles except for the fossils. Only the jellyfishes with their low density are regularly found inside the limestone beds and only at one locality, while other fossils typically occur at the soles of the beds. This and the grading of calcareous intraclasts in other localities (Nusplingen, Monte Bolca) suggest that individual beds represent turbiditic import from the flanks of the basins. Erosional sole marks are very rare. Where drag marks do occur, they are ruffled (as in the Hunsrueck Shales). This indicates that the muddy substrate was covered by a bacterial scum and therefore soft enough for the formation of arthropod tracks, landing marks, and roll marks, but at the same time rather erosion-resistant (Seilacher et al. 1985b).

Lithographic limestones occur throughout the geologic column and range from fully marine to lacustrine habitats (Karatau). Particular features vary from case to case; but most occurrences share the lack of an autochthonous benthic fauna (including bioturbation), deposition in small basins surrounded by carbonate source areas and the possibility of the bottom waters having been both anoxic and hypersaline. But there are also examples (such as the Eocene Green River Shales; Buchheim and Surdam 1981) that were deposited in large shallow basins, in which the carbonate mud may have been precipitated directly from the water column.

Except for their lithology, lithographic limestones are taphonomically similar to the bituminous black shales of the Posidonia Shale type, whose perfect fossilization (including the preservation of soft parts) has been traditionally attributed to stagnation alone (Seilacher 1982c; Seilacher et al. 1985b). No wonder, because oil shales provide no sedimentary structures other than the fine millimetric lamination and pervasive current alignment of fossils. Still it would be unreasonable to believe that the Posidonia Shale basin remained unaffected by the physical events and their killing and blanketing effect, which controlled whole-body preservation in all the other examples.

5 Concluding Remarks

Taphonomy deals with the post-mortem history of fossils. Its one goal is paleobiological: to remove the overprint that hides, or distorts, original biohistoric signals both at the levels of the individual organism and whole paleocommunities. In the present context we have emphasized the alternative, biogeological perspective of taphonomy.

It views dead organisms and their skeletons as sedimentary particles, but particles that behave and degrade in specific ways. This behavior can be gauged by mapping taphonomic features in different rocks (comparative taphonomy, Brett and Baird 1986b) or by flume and field tests (experimental taphonomy). Thus, paleoenvironmental conclusions can be derived not only from the functional morphologies of organisms, but also from the ways in which they died, decayed and became buried and fossilized.

In the preceding chapter by Kidwell (2.4) the biogeological approach to taphonomy has been applied to the usual situation, where necrolysis is complete and the resulting bioclasts can influence the further course of sedimentational dynamics and benthic community succession ("taphonomic feedback"). Our own contribution has focused on more exceptional cases in which rapid sedimentation and other factors permanently removed the carcasses from continued taphonomic degradation. However, it should be emphasized that obrution of varying degrees, reflecting a predominantly episodic mode of sediment accumulation, is actually far more abundant in stratigraphic sections than has been reported. Obrution deposits deal with mortality factors and rates of decay. They thereby introduce a taphonomic clock into the sedimentary processes acting during stratinomic events and carry the concept of event stratigraphy into facies types, in which the lack of obvious bed units and sedimentary structures might otherwise seduce us to believe in a continuous "rain" of sedimentation.

Whatever approach we choose, however, taphonomic analysis must not end with the detective work of particular case histories. It can be extended beyond the anecdotal level by comparing a broad spectrum of cases and by placing them into the framework of facies patterns, sequences, and basin history. With such perspectives in mind, taphonomy can become an integral part of what this book envisages: dynamic stratigraphy.

Acknowledgments. This paper has benefited from critical reviews by Curt Teichert, David Lehmann, and Gordon Baird. We thank Bill Faggart and Margot Pilopp for typing the manuscript and James Eckert and Werner Wetzel for help with photography. Some of the background research by Carlton Brett was supported by NSF Grant EAR8313103, that of A. Seilacher largely drew from studies in the former SFB 53 funded by the Deutsche Forschungsgemeinschaft.

Chapter 2   Event Stratification

2.6   Secular Changes in Phanerozoic Event Bedding and the Biological Overprint

J. J. Sepkoski Jr, R. K. Bambach, and M. L. Droser

"The importance of animals in forming the final fabric of shallow-water marine sediments can not be overstated. Were it not for burrowing and crawling organisms, the appearance of many ... marine sediments would be entirely different." (Moore and Scrutton 1957, p. 2750)

1   Introduction

Bedding in most sedimentary rocks is the result of three interactive sets of processes: physical deposition, biological modification, and diagenetic alteration. In many shelf environments, storm sedimentation is a dominant process of physical deposition. As summarized by Kreisa (1981) and Aigner (1982), characteristics of storm beds, or "tempestites," include

1. sharp bases, often resting on erosionally scoured surfaces;
2. basal lags of bioclasts, pebbles, and/or coarse sand;
3. flat-laminated, often graded layers of coarse silt, sand, or pelletal carbonate that may be capped by undulatory or ripple cross lamination;
4. overlying clays or lime mudrocks deposited as either "muddy tails" following storms or as hemipelagic sediment during fair-weather times.

Storm beds only rarely escape post-depositional modification in normal marine environments. Infaunal animals rework sediment and distort or obliterate primary bedding features. These activities also influence microbial activity and affect early diagenesis in the sediment. Thus, the bedding that is preserved bears a biological overprint on primary physical stratification.

While physical processes of deposition and thermodynamic controls of diagenesis have remained constant through time, biological activity has not. The evolution of the marine benthos has produced marked changes in the diversity, composition, and behavior of the infauna, and these in turn have changed the biological overprint on bedding (Bambach and Sepkoski 1979; Sepkoski 1982; Larson and Rhoads 1983; Brandt 1986; Droser 1987; Droser and Bottjer 1988). In this chapter, we briefly review some aspects of the biological overprint and how these have changed through the Phanerozoic, and we provide several examples of secular changes in bedding style based on our field observations. (Indeed, it was observation in the field of different kinds and bedding in different parts of the geologic column that led to the idea that evolutionary changes in the marine biota have affected styles of stratification through time.)

Einsele et al. (Eds.)
Cycles and Events in Stratigraphy
©Springer-Verlag Berlin Heidelberg 1991

2 The Biological Overprint on Marine Sediments

The diverse ways in which animals utilize and modify sediment have been the subject of several excellent reviews in McCall and Teversz (1982). We provide here only a cursory summary, emphasizing biotic influence on the production and modification of bedding.

2.1 Modification of Substrate Properties

Activities of marine animals greatly modify the physical and chemical properties of sediment at and just below the sediment-water interface. These activities affect mass hydraulic properties of the sediment and stability of the substrate surface. In turn, these affect how easily sediment can be moved or resuspended by storms or other events and how sediment will be redeposited following such events. Some of these effects, as reviewed by Rhoads and Boyer (1982), are outlined below:

1. Production of fecal pellets converts fine-grained sediment into sand-sized equivalents, increasing the ease with which the substrate can be eroded but also increasing the settling rate of resuspended sediment.
2. Burrowing increases the porosity and thus erodability of sediment whereas secretion of mucus into burrow walls increases stability slightly; this also causes sediment to erode as small clumps rather than individual grains.
3. Production of skeletal material coarsens sediment and provides bioclasts for basal lags of storm beds. Isolated large bioclasts on the substrate surface increase erodability by inducing turbulent vortices on leeward sides, whereas dense skeletal accumulation armor the surface, reducing sediment transport.
4. Grazing of surface microbial mats and disruption of mats by reworking of the upper substrate decrease sediment stability and increase erodability. Conversely, production of holdfast networks or mats of worm tubes decrease erodability.
5. Formation of pits and mounds associated with burrows increases bed roughness and thus also enhances erodability in a manner analogous to isolated bioclasts.

2.2 Modification of Sedimentary Structures

Burrow construction, intrastratal locomotion, and sediment ingestion displace sedimentary grains and thus modify structures created by physical depositional processes. This modification varies from localized distortion of lamination around isolated burrows to complete churning of sediment and obliteration of all bedding features. The degree to which sediment is modified depends on numerous factors, including population densities, animal sizes, kinds and rates of activity, nature of the sediment, frequencies of physical reworking and sedimentation, and rates of diagenesis, especially cementation (see also Savrda et al. Chap. 5.2, and Sagemaw et al. Chap. 5.3, this Vol.).

Sedentary infaunal suspension feeders generally modify sediment the least. Their burrows penetrate and bend lamination and can homogenize sediment over very small scales if water is injected to aid in burrowing. Sediment will become completely bioturbated only if stable substrates are successively colonized over several generations or inhabited by dense populations of sedentary animals.

Mobile deposit feeders and predators, on the other hand, can quickly homogenize sediment, obliterating bedding features and mixing cm-scale beds of coarse and fine sediment. In modern open-marine environments less than 20 m deep, infaunal animals typically homogenize the upper 10 cm or more of substrate within a few months to years (Rhoads and Boyer 1982). In nearshore sands, callianassid shrimp and other infaunal crustaceans can bioturbate sediment to depths in excess of 1 m.

The net results of bioturbation vary with local depositional style: if sediment is deposited in layers greater than 10 cm thick, the infauna may only homogenize the upper portions of beds, leaving the sharp bases and basal lags; thinner beds, however, may be amalgamated with thicker ones, reducing the apparent frequency of depositional events. If sedimentation is generally slow and beds thin, bioturbation can obliterate all bedding and produce massive, poorly sorted units. Relict lag layers of bioclasts or coarse grains may be preserved although matrices may be finer than expected (as a result of downward piping of sediment) and bioclasts may be reoriented from hydrodynamically stable positions.

2.3 Alteration of Chemical Fluxes and Early Diagenesis

Activities of infaunal animals have major influences on chemical reactions in sediment. As reviewed by Aller (1982a), some of the important activities include

1. addition of organic matter into sediment in the form of mucus stabilizing burrow walls and binding fecal pellets;
2. entrainment of surface detritus into sediment by reworking of the substrate surface;
3. enhancement of porosity and permeability of sediment through burrowing;
4. injection of seawater into the substrate for respiration, removal of metabolites, and alteration of local pore water chemistries;
5. cropping of bacteria, thereby increasing growth rates.

These activities promote the productivity of bacteria and thus enhance bacterially mediated reactions such as breakdown of organic molecules and reduction of sulfate, nitrate, and iron.

Pumping of water into burrows typically oxygenates surrounding sediment. This permits growth of aerobic bacteria that oxidize organics and sulfides, producing dissolved CO_2 and SO_2, respectively. These weak acids inhibit early carbonate cementation and can break down and dissolve skeletal material (Aller 1982b). Below the oxidized layer, growth of anaerobic bacteria can have the opposite effect: respiration of bicarbonate buffers bioclasts and can promote early carbonate cementation if there is sufficient concentration of Ca ions in pore waters.

2.4 Restriction of the Infauna

There is a gradient in the degree to which the infauna modifies sediment that is influenced by numerous environmental factors. In general, rates of infaunal activity decline from warm to cold water (Thayer 1983) and from shallow to deep water (partially in response to temperature). Rates also tend to decline from coarse sand to fine mud, in part because lower organic content in sands necessitates more rapid ingestion for equal quantities of food (Thayer 1983).

Adverse environmental conditions variously restrict infaunal activity. Low-oxygen conditions limit the infauna to small, unarmored animals living in the upper few mm's of sediment (Rhoads and Morse 1971). Similarly, abnormally low or high salinities reduce diversities and may limit population densities. Finally, high frequencies of bottom disturbance by storm erosion and/or deposition often locally eradicate the infauna, and several years of recolonization may be necessary to re-establish equilibrium communities (Rhoads and Boyer 1982). If erosion exposes compacted firm grounds or cemented hardgrounds, infaunal activity may be permanently restricted, greatly enhancing the preservational potential of bedding features.

3 Evolution of the Infauna

Just as there is a gradient in infaunal activity with environment, there is a gradient through geologic time. Infaunal animals were absent throughout all but the latest Precambrian and then increased in diversity and behavioral variety through the Phanerozoic (Frey and Seilacher 1980). The precise history of this increase is not well understood, however. This is because many infaunal animals lack skeletons and have low preservation potentials and because there is considerable variation in infaunal communities among environments and paleogeographic settings. Thayer (1983) has provided the most extensive review of the history of the infauna, based largely on times of appearance of modern infaunal animals and an actualistic assumption that their rates and depths of bioturbation have remained constant over geologic time (Fig. 1).

Infaunal animals appeared during the Vendian-Cambrian radiation of metazoans. Vendian ichnofossils, however, principally record surface traces with sediment disturbance only on scales of a few mm's (Fedonkin 1977). In the Cambrian, there was increased sediment reworking by both infaunal animals and surface deposit feeders (e.g., some trilobites). In most platform facies, there was increased pellitization of sediment, but bioturbation was normally limited to scales of 1 cm. Exceptions include nearshore wave-deposited sands where *Skolithos*-producing animals disturbed sediment to depths of 10 cm or more (Miller and Byers 1984; Droser 1987) and shallow-water pelletal carbonates where infaunal animals bioturbated sediments to depths of 6 cm (Droser and Bottjer 1988).

The radiations of the Ordovician led to great expansions of the diversity (Sepkoski and Sheehan 1983) and guild structure (Bambach 1983, 1985) of the marine benthos. Many of the expanding groups included epifaunal taxa with more robust

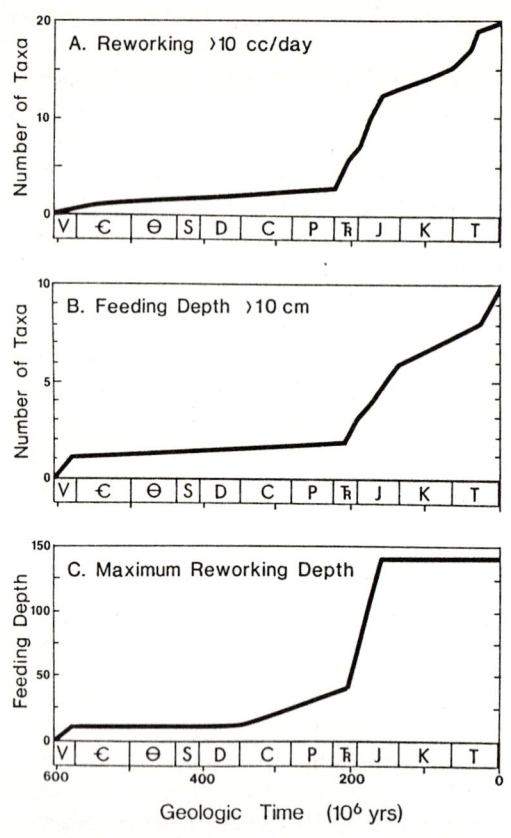

Fig. 1A-C. Estimated bioturbation intensities through the Phanerozoic, modified from Thayer (1983). **A** Number of taxa that rework sediment rapidly (>10^3 cm/day/individual). **B** Number of taxa that rework sediment at substrate depths greater than 10 cm. **C** Maximum average reworking depth of infaunal taxa. All estimates are based on extant taxa (various families, superfamilies, and orders) and assume that members of these taxa had similar behaviors and activity rates throughout their histories

skeletons, contributing coarse bioclasts to basal lags of storm beds. The radiations also involved several shelly groups with infaunal representatives (e.g., free-burrowing bivalves such as nuculoids) as well as some primarily soft-bodied groups with abundant infaunal members (e.g., polychaetes) (see Sepkoski 1981). These groups affected increases in infaunal tiering (Bottjer and Ausich 1986) and in average depth of burrowing (Droser 1987); however, there was only moderate increase in the diversity of actively burrowing animals with mineralized skeletons (Thayer 1983) and of trace fossils (Frey and Seilacher 1980). Expansion of the infauna continued through the remainder of the Paleozoic with some increase in the diversity of taxa capable of rapid reworking and deep burrowing in the Devonian through Permian (Fig. 1); there was also increased movement of some infaunal taxa, particularly bivalves, into offshore environments during this time interval (Sepkoski and Miller 1985).

The Mesozoic was a time of dramatic changes in the diversity, composition, and ecology of the marine benthos (Vermeij 1977; Sepkoski 1981; Bambach 1983, 1985). The radiations affected a major infaunalization of the marine benthos and included expansions of several important taxa with abundant, actively burrowing representatives (e.g., bivalves and decapod crustaceans) and with mobile infaunal carnivores (e.g., neogastropods); these expanded primarily from the Jurassic and Cretaceous onward. Thayer (1983) has estimated that there was a five- to tenfold increase in the number of higher taxa with reworking rates greater than 10 cm^3/day and feeding depths greater than 10 cm (Fig. 1). Expansion of the infauna appears to have continued into the Cenozoic, and bioturbation intensities in modern shelf environments may now be two orders of magnitude greater than those in the early Paleozoic (Thayer 1983).

4 Expected Secular Trends in Bedding Features

The basic expectations from the evolutionary expansion of the infauna are an intensifying biological overprint on sediments, especially in shallow platform facies, and an increasing disruption of physical bedding features through the Phanerozoic. Some of the trends that we expect are

1. decrease in the frequency of very thin storm layers;
2. increase in the degree of biogenic disruption of the upper portions of thicker tempestites;
3. increase in the abundance of poorly sorted sediments as a result of biogenic homogenization;
4. increase in the abundance of massive sedimentary units with only relict bedding;
5. decrease in the frequency of rapidly cemented layers, especially in heterolithic facies, as a result of increased burrow ventilation.

These trends should not necessarily progress slowly and evenly through the Phanerozoic but should follow the general pattern of infaunal evolution. We expect only mild disruption in most (but not all) facies in the Cambrian, moderate increase after the Ordovician radiations and again after the mid-Paleozoic changes, and then considerable increase in the Mesozoic, especially after the Triassic.

All these changes should be statistical in nature. Since there is a gradient in infaunal activity from shallow to deep water, we expect substantial bioturbation to appear in inner shelf facies earlier than in the more offshore. Similarly, since many environmental factors can restrict the infauna, we expect some facies at any time to exhibit reduced bioturbation and to preserve bedding features more characteristic of earlier times.

5 Examples of Changes in Phanerozoic Bedding

Below, we present several examples of the different types of bedding, particularly storm bedding, that are preserved at different times during the Phanerozoic. The examples are neither comprehensive nor definitive but simply illustrate some of the changes in the modal character of bedding caused by the evolving biological overprint in normal, open-marine facies. We emphasize lower Paleozoic sequences in North America where our experience is greatest. We also emphasize heterolithic facies, composed of alternating fine and coarse sediments (i.e., clays and coarse silts, sands, and/or pelletal and bioclastic carbonates).

5.1 Cambrian and Early Ordovician in Western North America

Early Paleozoic platform sediments tend to be characterized by low levels of bioturbation (Droser 1987). In terms of bedding, this is reflected in a prevalence of preserved fine-scale stratification and an abundance of flat-pebble conglomerates. Several of these sedimentary fabrics are common in Precambrian sequences, where bioturbation was absent. We give these sedimentary features emphasis since they can serve as a model for how shallow-water sequences of any age should appear when bioturbation is minimal.

5.1.1 Thin Beds

Excellent examples of Cambrian storm-dominated heterolithic facies are exposed in the inner detrital belt of North America (see Palmer 1971). Sections contain abundant storm beds of coarse siltstone, fine sandstone, and/or pelletal limestone (Sepkoski 1977, 1982). Modal bed thicknesses are typically less than 1 cm, and many beds are less than 0.5 cm. Most beds display planar lamination and some contain basal bioclasts and/or upper undulatory to shallow ripple lamination. Bottoms of beds are sharp and often contain hypichnial casts of epistratal or exhumed intrastratal ichnofossils, attesting to the presence of a normal marine fauna. Few of the tempestites, however, exhibit any significant biogenic disruption. Some do have shallow meandering epistratal trails less than 2 mm deep on upper surfaces or are penetrated by a few scattered vertical burrows, but none is churned or mixed with surrounding sediment.

The tempestites are usually intercalated with green clay shales in layers ranging from mm-thick partings to cm-thick beds (Fig. 2). The muds that compacted to form the shale evidently were bioturbated, as evidenced by the flaky rather than papery nature of the shale chips (Byers 1974) and by the dense hypichnia under some associated tempestites. However, bioturbation did not mix mud with the tempestites, as indicated by the general absence of quartz silt or sand in the shales, probably because the infauna was mostly small and burrowed only to shallow depths (<1 cm).

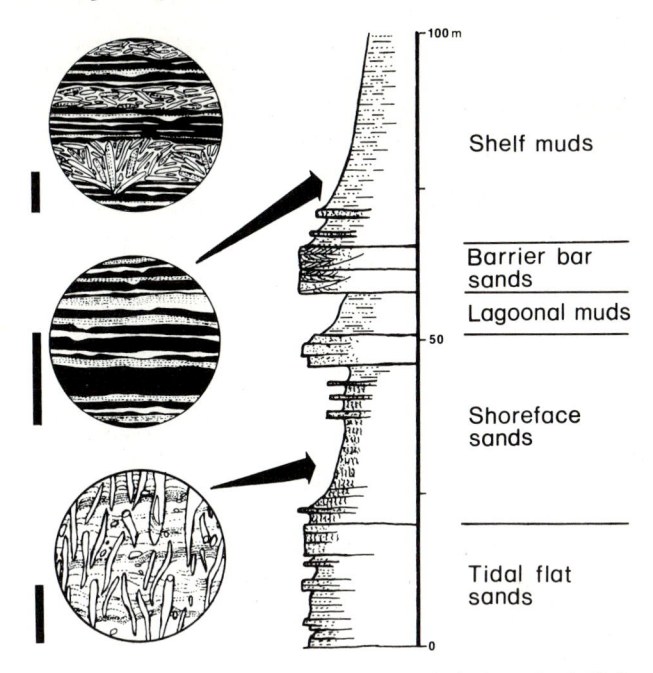

Fig. 2. Examples of bedding character in a Cambrian deepening upward sequence in the inner detrital belt of North America. Basal sandstones vary from unbioturbated in tidal-flat facies to heavily bioturbated by dense *Skolithos* in shoreface facies (*lower circle*). Proximal shelf mud facies exhibit bioturbation only in shale (represented by *black* in the *middle* and *upper circles*); associated very thin storm beds of fine sandstone (*middle circle*) are preserved with little or no burrow disruption. In more offshore sections, where thin tempestites are calcareous, flat-pebble conglomerates (*upper circle*) become abundant, formed by storm reworking of partially lithified, unbioturbated thin carbonate beds. Bedding diagrams are based on outcrop photographs; *scale bars* adjacent to circles represent 10 cm. Stratigraphic section is the Upper Cambrian Ladore Formation at Jones Hole in Dinosaur National Monument near Vernal, Utah

5.1.2 Flat-Pebble Conglomerates

In facies in which thin tempestites are dominantly calcareous, flat-pebble conglomerates are abundant, as described by Sepkoski (1982). These occur in beds from 3 cm to nearly 1 m thick at typical frequencies of one to five beds per meter of section. Intervals with particularly abundant flat-pebble conglomerates can be traced for distances of more than 100 km (McKee 1945; Sepkoski 1977) and are thus useful in stratigraphic correlation. The flat-pebble conglomerates are calcirudite packstones containing rounded, tabular intraclasts, 1 to 30 cm in diameter, with bioclastic or rarely sandy matrices. Intraclast lithologies and thicknesses are identical to the associated thin tempestites and are quite different in size, shape, and lamination from "algal breccias" in coeval stromatolitic facies. Orientations of pebbles range from flat lying in thinner conglomerate beds to edgewise in thicker ones (Fig. 2). The latter have jumbled, swirling, or fan-like fabrics and are not imbricated; this is indicative of reworking by oscillatory waves rather than deposition in unidirectional currents (Futterer 1982). Significantly, even the very thin edgewise pebbles exhibit no plastic deformation; in the absence of any evidence of desiccation or biomats, this is interpreted to reflect early submarine cementation of the carbonate tempestites prior to physical reworking.

Conditions necessary for the genesis of these flat-pebble conglomerates, as posited by Sepkoski (1982), are

1. episoidic deposition of thin, permeable carbonate layers separated by muddy partings (flat-pebble conglomerates decline markedly in abundance in purer subtidal carbonates in the Cambrian; Droser 1987);
2. rapid submarine cementation of undisturbed carbonate layers;
3. erosion and reworking by intense storms to produce the tabular intraclasts (some flat-pebble conglomerates formed in the deeper, outer detrital belt upon slumping of thin carbonate layers).

The first two of these conditions are dependent upon low levels of bioturbation, since

1. the thin carbonate storm layers retained their integrity after deposition and were not biogenically mixed with surrounding mud;
2. in the absence of deep burrow irrigation, the redox boundary should have been close to the sediment/water interface so that bicarbonate produced by anaerobic bacteria could percolate into the permeable carbonate layers, leading to precipitation of carbonate cement.

If bioturbation were intense, pumping of oxygenated seawater into burrows would have promoted growth of aerobic bacteria, leading to production of carbon and sulfur dioxides and inhibition of early cementation, as noted above. The rarity of skeletal fossils in associated Cambrian shales, despite occurrence of mud-filled bioclasts in some storm beds, has been interpreted to reflect dissolution mediated by bioturbation of the muds (Sepkoski 1978).

Flat-pebble conglomerates are abundant in the Cambrian and Lower Ordovician but are not restricted to that interval. They are also common in Precambrian sequences, much more so that realized by Sepkoski (1982). They occur in subtidal to intertidal facies where thin carbonate beds are separated by clayey partings (see Tucker 1982) or by thick stromatolitic biomats. Abundant flat-pebble conglomerates have also been described from scattered post-Cambrian settings (e.g., Kazmierczak and Goldring 1978). We predict that bioturbation was suppressed in these situations either because of rapid deposition (followed by deep erosion) or because of inhospitable environmental conditions (e.g., low or fluctuating oxygen concentrations).

5.1.3 Intense Bioturbation

The bioturbation preserved in Cambrian nearshore siliciclastics and shallow-water carbonates represents a major departure from similar Precambrian facies. In nearshore sandstones of the inner detrital belt, stratification is commonly disrupted or even obliterated by dense *Skolithos* burrows penetrating to depths of 10 cm or more

(Fig. 2). Deep burrowing in other lower Paleozoic quartz arenites has been documented by Miller and Byers (1984).

There is also considerable bioturbation in shallow, inner shelf carbonates of the Cambrian and Lower Ordovician in the middle carbonate belt of western North America. As measured by the ichnofabric indices of Droser and Bottjer (1986), bedding is generally preserved with ichnofabrics in the range of 2 to 4, although nearly complete disruption (index 5) also occurs. In general, bioturbation intensity and obliteration of fine stratification is greater in shallow-water carbonates than in inner-shelf siliciclastics (Droser 1987) and decreases from inner to outer shelf environments (Droser and Bottjer 1988).

In the Cambrian and Lower Ordovician of the Appalachians, stratification is generally similar to that in the inner detrital belt of western North America. Extensive bioturbation is present, however, in some cyclic sequences of both siliciclastics and carbonates. This bioturbation always occurs in the same relative place in the sequence: the deeper-water facies deposited just after rapid sea-level rise (Bova and Read 1987). In deep ramp settings where sedimentation rates were low (just after transgression), shallow horizontal burrowing homogenized sediments as they accumulated. The burrow-mottled facies are superseded by "normal" thin-bedded strata (often with flat-pebble conglomerates) that were deposited under increased sedimentation rates. Thus, homogenization of the deeper-water mudstones was the result of slow sedimentation coupled with shallow burrowing and not deep or rapid bioturbation.

5.2 *Middle Ordovician to Silurian Changes*

In Ordovician and Silurian heterolithic facies of the inner and middle shelf, flaggy sequences of tempestites are still abundant. However, field observations of Ordovician and Silurian sections in eastern North America (e.g., Bambach 1969) suggest that the character of stratification differs from earlier sequences in several significant ways:

1. Very thin tempestites are less abundant: beds of siltstone and sandstone less than 1 cm thick appear to be far less common (Fig. 3) and, when present, are often riddled by burrows and partially churned, disrupting current lamination and leaving the beds laterally discontinuous.
2. Thicker tempestites are more heavily bioturbated: these beds frequently display bioturbation in the upper 1 to 3 cm, and lamination may be obscured near the upper surfaces. Burrow intensity often is "graded," ranging from complete churning near tops to scattered, discrete burrows in lower parts of beds. Many burrows are backfilled with muds piped down from overlying layers.
3. "Gritty" shales and mudrocks are more abundant: these shales, frequently intercalated with flaggy tempestites, reflect mixing of clay mud with variable quantities of silt and sand. While some grittiness may have resulted from wind-borne silt and fine sand, much was probably due to biogenic homogenization of mud with thin tempestites and with tops of thicker tempestites.

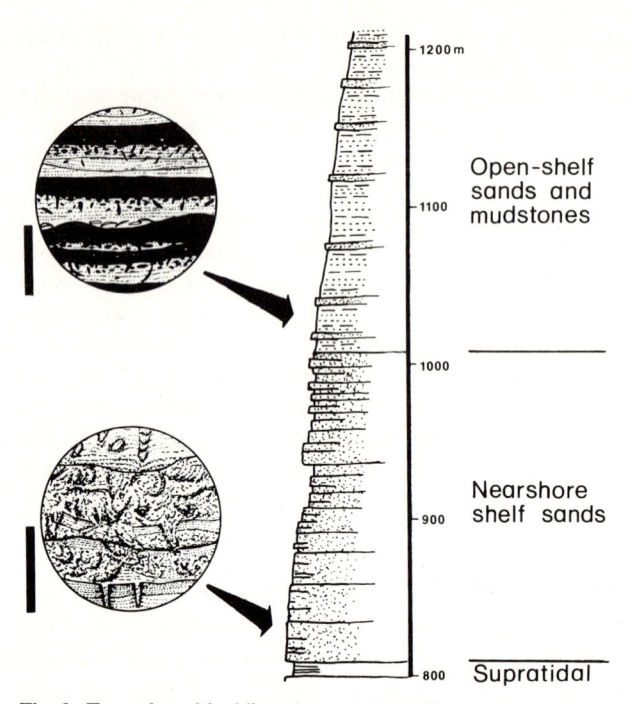

Fig. 3. Examples of bedding character in a Silurian deepening upward sequence in eastern North America. Nearshore sandstones tend to be thoroughly bioturbated, preserving only relict stratification (*lower circle*), as in the Cambrian. Shelf sands with intercalated mudstones (*upper circle*) preserve bedding but are more bioturbated that in the Cambrian: very thin tempestites are absent (biogenically homogenized with mud to form gritty shale) and tops of thicker tempestites are often densely burrowed. Bedding diagrams are based on outcrop photographs; *scale bars* represent 10 cm. Stratigraphic section is the Upper Silurian Stonehouse Formation near Arisaig, New Brunswick (based on Bambach 1969)

4. Flat-pebble conglomerates are much less abundant: intraclast-supported calcirudites are rare in subtidal heterolithic sequences compared to the Cambrian. Tabular intraclasts are common in basal portions of thick storm layers (cf. Kreisa 1981) but are usually sparse and floating in bioclastic matrix. Decline in abundance of flat-pebble conglomerates reflects increased biogenic destruction of thin beds and, perhaps, more limited early cementation resulting from increased bioturbation and expansion of the aerobic boundary layer.
5. Thick tempestites may be absolutely more abundant: it is our impression that tempestites several cm's thick are more abundant than would be expected simply from destruction of very thin beds. There may be several reasons for this. First, because of increased mixing of sand with mud, detrital particles could be resuspended and redeposited as individual grains rather than as lithified clasts; with a larger reservoir of unlithified sediment and fewer cemented layers partially armoring the bottom, even moderate storms could produce sizeable tempestites. Thus, in some senses, the thicker tempestites are the depositional

equivalents of the earlier flat-pebble conglomerates. Second, as a result of the Ordovician radiations of robust benthic taxa (including articulate brachiopods, corals, crinoids, etc.), there were more coarse bioclasts in the environment than typical in the Cambrian; these bioclasts were incorporated into thicker lag layers in Ordovician and Silurian tempestites.

A general increase in bioturbation in purer carbonate facies has also been documented by semi-quantitative investigation of ichnofabrics (Droser and Bottjer 1987). Analyses indicate that by the Late Ordovician, bioturbation in inner shelf carbonates of western North America was nearly an order of magnitude more intense than in comparable facies in the Cambrian, in part because of the presence of three-dimensional *Thalassinoides* (Sheehan and Schiefelbein 1984) which result in average ichnofabric indices of 4 to 5. This bioturbation made bedding less evident than in comparable Cambrian facies. In outer shelf carbonates, bioturbation reached intensities seen previously only on the inner shelf. By the Late Silurian, thin tempestites in outer shelf siliciclastics often became completely bioturbated, leaving only thin relict layers of bioclasts (e.g., Watkins 1979).

These generalizations again do not apply to all environments or sedimentary situations. As in the earlier Paleozoic, very shallow-water open marine facies commonly exhibit intense bioturbation (Fig. 3). Nearshore sandstones, often with preserved infaunal bivalves and linguloids, frequently are reworked to ichnofabric indices 4 to 5, with undisturbed stratification preserved only in storm layers thicker than 10 cm.

There are also many shelf sequences in which bioturbation is not as intense as suggested above. These include rather pure subtidal carbonates that preserve thin, cm-scale bedding, as documented by Larson and Rhoads (1983), and some shaly units that contain intervals with common (although not abundant) flat-pebble conglomerates, such as some Lower Silurian shales in the Appalachians. The sedimentary environments, benthic communities, and relative frequency of these exceptions are not well understood, and more study is needed to document the variation in bioturbation among contemporaneous shelf sequences.

5.3 Mesozoic and Cenozoic Changes

Our data for Mesozoic and Cenozoic strata are more limited than for the Paleozoic. However, we have observed that sediments are generally much more bioturbated and fine bedding is much more limited in platform facies. This observation is supported by the literature review of Brandt (1986), who found a substantial decrease in the reported occurrences of storm beds in the Mesozoic and Cenozoic. She interpreted this to reflect increased post-depositional destruction of tempestites by bioturbation.

Strata in many platform facies of the Mesozoic and Cenozoic in North America tend to resemble the biogenically homogenized strata of the Paleozoic nearshore. Particularly evident characteristics include

1. a much greater frequency of massive fine sandstones and carbonates, often with burrow mottling and bedding preserved only in relict bioclastic layers and in thick (>10 cm) beds deposited by the most intense storms;
2. an increase in abundance of thick units of monotonous gritty shale with only scattered tempestites and bioclastic layers (sometimes preserved in concretion horizons formed after rapid burial); these units represent biogenically homogenized thin tempestites and fair-weather clay muds.

As before, there are many exceptions, as noted below.

6 Anachronistic Facies

The Phanerozoic trend toward increased bioturbation of bedding features is not universal, and any set of environmental factors that restricts infaunal activity will tend to produce a style of bedding that is more characteristic of earlier times. Such "anachronistic facies" are particularly striking in the Mesozoic and Cenozoic when open-marine facies tend to be heavily bioturbated. Several examples we are aware of include the following:

1. The Triassic Muschelkalk of central Europe (Aigner 1982; Hagdorn and Simon 1988), which, in many respects, appears "Ordovician" in the character of carbonate bedding; portions of the lower Muschelkalk even contain flat-pebble conglomerates. Much of this formation, particularly the lower and middle portions, was deposited under conditions of high salinity and, perhaps, low-oxygen concentrations.
2. The Jurassic Sundance Formation of western North America, which contains clay shales and many thin, laminated sandy tempestites that were minimally affected by bioturbation. Through much of its history, the Sundance Sea had only narrow connections to the open ocean.
3. Limited intervals of the Cretaceous Mancos and Cody Shales of western North America that contain thin sandy event beds separated by dark shale. These were deposited at times of reduced oxygen concentrations in the Interior Seaway.

This last example is particularly instructive since the intervals with reduced bioturbation are separated by tens of meters of thoroughly homogenized gritty shale. Where bioturbation is reduced, the preserved event beds have a very "Cambrian" appearance: many are less than 1 cm thick and are separated by partings and thin layers of clay shale. These indicate that physical depositional processes had not changed since the Cambrian; rather, the post-depositional histories of the event beds had been greatly altered by the evolving infauna.

7 Conclusions

The changes in the nature of bedding and storm stratification that we have observed in open-marine shelf sequences roughly parallel the major changes in the diversity

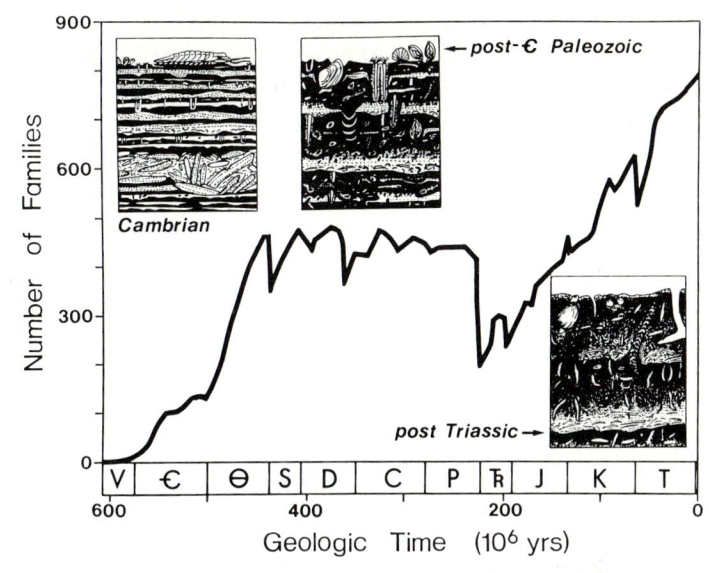

Fig. 4. Representations of changes in storm stratification and bioturbation in shallow shelf settings through the Phanerozoic, set against the record of familial diversity of fossil marine animals. In the Cambrian and Early Ordovician, when diversity was low, bioturbation was minimal so that thin storm beds and fine stratification are normally preserved. In the later Paleozoic, after the Ordovician radiations, bioturbation was more intense, and normally only tempestites thicker than 1 cm are preserved. After the Triassic, during the major Mesozoic-Cenozoic diversification, bioturbation frequently churned all sediment to depths of 10 cm or more, leaving only basal lag layers of tempestites preserved. These diagrams represent the expected norm for open- marine platform facies; abnormal environmental conditions that restrict the infauna will produce "anachronistic facies" with bedding features more characteristic of earlier time intervals. (After Sepkoski 1981, 1982)

and composition of the marine benthos, as depicted in Fig. 4. The Cambrian and Early Ordovician, with relatively low faunal diversity, contain sequences with an abundance of very thin (<1 cm) storm beds and flat-pebble conglomerates. Their production and preservation were dependent on a very shallow infauna that failed to bioturbate sediment below the upper cm of substrate. Exceptions to this generalization largely involve very shallow-water settings where a deeper-burrowing infauna often bioturbated sediments to a much greater extent. In the later Paleozoic, following the Ordovician radiations, very thin tempestites and flat-pebble conglomerates are far less abundant, reflecting pervasive bioturbation to depths of several cm's. Preserved tempestites in proximal to middle shelf settings tend to be thicker, with coarser bioclastic lags and more extensively bioturbated tops; tempestites often are separated by gritty shales, representing biogenically homogenized mud and sand from thin tempestites. After the Triassic, storm stratification seems to be greatly obscured in many shelf settings, reflecting increase in faunal diversity and expansion of the deep infauna. Many shelf sediments resemble certain Cambrian and later nearshore deposits in which deep burrowers thoroughly churned sediments to depths of 10 cm or more.

These generalizations apply to the modal situation in shelf settings, and many exceptions to the trend toward increased bioturbation exist. The predicted secular changes in bioturbation are based largely on patterns of change in the diversity, taxonomic composition, and guild structure of the marine benthos and have been tested in only a few situations (e.g., Sepkoski 1982; Larson and Rhoads 1983; Droser and Bottjer 1988). Considerably more quantitative comparison of bedding characteristics in different time intervals and sedimentary environments is needed to test these predictions and observations and to provide a better account of the kind of depositional and biological information that bedding preserves.

Acknowledgments. Partial funding of research for this chapter was provided by the Petroleum Research Fund of the American Chemical Society to MLD.

2.7 Submarine Mass Flow Deposits and Turbidites

G. Einsele

1 Introduction

The deposits of different types of gravity mass flows belong to the most prominent examples of event stratification, or episodic bedding. In the presentday oceans as well as in many ancient sedimentary basins, they form a great portion of the total sediment body (e.g., Kelts and Arthur 1981; Stow and Piper 1984b; Stanley 1985). For example, thick flysch sequences in the Alps and other orogenic belts consist predominantly of redeposited sediments. Since about 1950 numerous articles and even some special books have been published on turbidites and other gravity mass flow deposits. As far as this author is aware, however, there is no brief summary of the most important features of mass flow deposits, such as siliciclastic and calcareous debris and mud flows, and sandy and muddy turbidites, which includes a discussion of their depositional environment. The need for such a summary is clear, as a uniform, consistent nomenclature for all these phenomena does not yet exist (see, e.g., Stow 1986; Souquet et al. 1987). Therefore, an attempt is made here to describe the most characteristic and frequently related features as well as the environments of these mass flow deposits, using a few, simplified conceptual models. For more detail and examples from the modern and ancient oceans, the reader is referred to Walker (1973, 1978, 1984b), Stanley and Kelling (1978), Saxov and Nieuwenhuis (1982), Schwarz (1982), Prior and Coleman (1984), Mutti and Ricci Lucchi (1978, 1984), Stow and Piper (1984), Mutti and Normark (1987), and others quoted in this article. The flow behavior of the different types of gravity mass movements is treated, for example, in Midddleton and Hampton (1976), Blatt et al. (1980), Stow (1980), Allen (1982), Lowe (1982), Komar (1970, 1985), Postma (1986), and summarized by Stow (1986). The historical progress of our knowledge in this field is reviewed by Görler and Reutter (1968), Walker (1973), Stanley and Kelling (1978), and others.

Gravity mass flows and turbidites are also known from lakes, but large-scale phenomena are generally missing. Nonetheless, most of the general characteristics discussed here for deep sea environments can be applied to lake sediments as well. A special description of lake deposits is not possible here, however.

2 Types of Gravity Mass Movements

The most important gravity mass movements found in both oceans and lakes are summarized in Fig. 1. They can be subdivided into several groups:

a) Mass movements of lithified, jointed rocks: Rockfall along coastal cliffs or steep submarine slopes and fault scarps (Fig. 1a). The transport distance of such fallen rocks is very limited, unless there is a possibility of a composite mass movement as indicated in Fig. 1i.

b) Creep, sliding and slumping of semi-solid to soft sediments (Fig. 1c through e) *on slopes of various angles* (as little as a few degrees): Movement takes place if the shear stress exceeds the shear strength of the sediment at some depth below the sedimentary surface, which is usually tested by stability analysis (Fig. 1b). The shear stress increases with the slope angle and depth below the sea floor. Deep below the surface of a gentle slope, the shear stress can be as high as on a steep slope at shallow depth. For this reason, there is a tendency for thick mass movements to develop on gentle slopes, whereas thin ones are characteristic for steep slopes. The shear stress can, in addition, be significantly enhanced by earthquakes and, in shallow water, by the effect of storm waves on the sea bed. Furthermore, the sudden loading by an approaching slide or slump often generates secondary failure planes and propagation of the mass movement. Similarly, a drop in sea level or tectonic uplift can lead to an additional loading of emerging sediments which when losing buoyancy may trigger slope failure. The shear strength of sediments usually increases with depth below the sea floor, but is reduced by high pore pressure in underconsolidated sediments. Such a situation frequently occurs in areas where fine-grained sediments are deposited rapidly, for example in front of deltas. A further reduction in shear strength is caused by the rather common development of biogenic gas in the uppermost tens of meters in slope sediments rich in organic matter. Finally, the release of methane from crystallized gas hydrates due to increasing water temperature (resulting from climatic changes or shifting of current systems) may locally diminish the shear strength of sediments and thus cause mass movements. In creep and slides, the sediment masses do not change their mechanical state, i.e., they move as a kind of rigid plug without significant internal disturbance. The distance of transport is very short. Slumps show considerable internal disturbance, slump folding, and frequently several slip faces. (Fig. 1e). They often evolve into debris or mud flows.

Fig. 1a-i. Summary of submarine mass movements and gravity mass flows (based on many sources, e.g., Middleton and Hampton 1976, Walker 1978, Moore et al. 1982, Prior and Coleman 1984; Stow 1986, Einsele 1989). Rockfalls (**a**), creep (**c**), slides (**d**), and slumps (**e**) occur if the shear stress exceeds the shear strength of the rock or sediment at some depth (**b**, stability analysis). **f** Grain flow; **g** and **h** debris and mud flows usually originate from slides and slumps by liquefaction of the primary, metastable grain packing (in situ water content, $w \geq$ liquid limit, w_L). Turbidity currents evolve by uptake of additional water. **i** A composite, two-step grain-mud flow mechanism can explain long-distance transport of pebbles and gravel into the deep sea

Einsele: Mass Flows and Turbidites

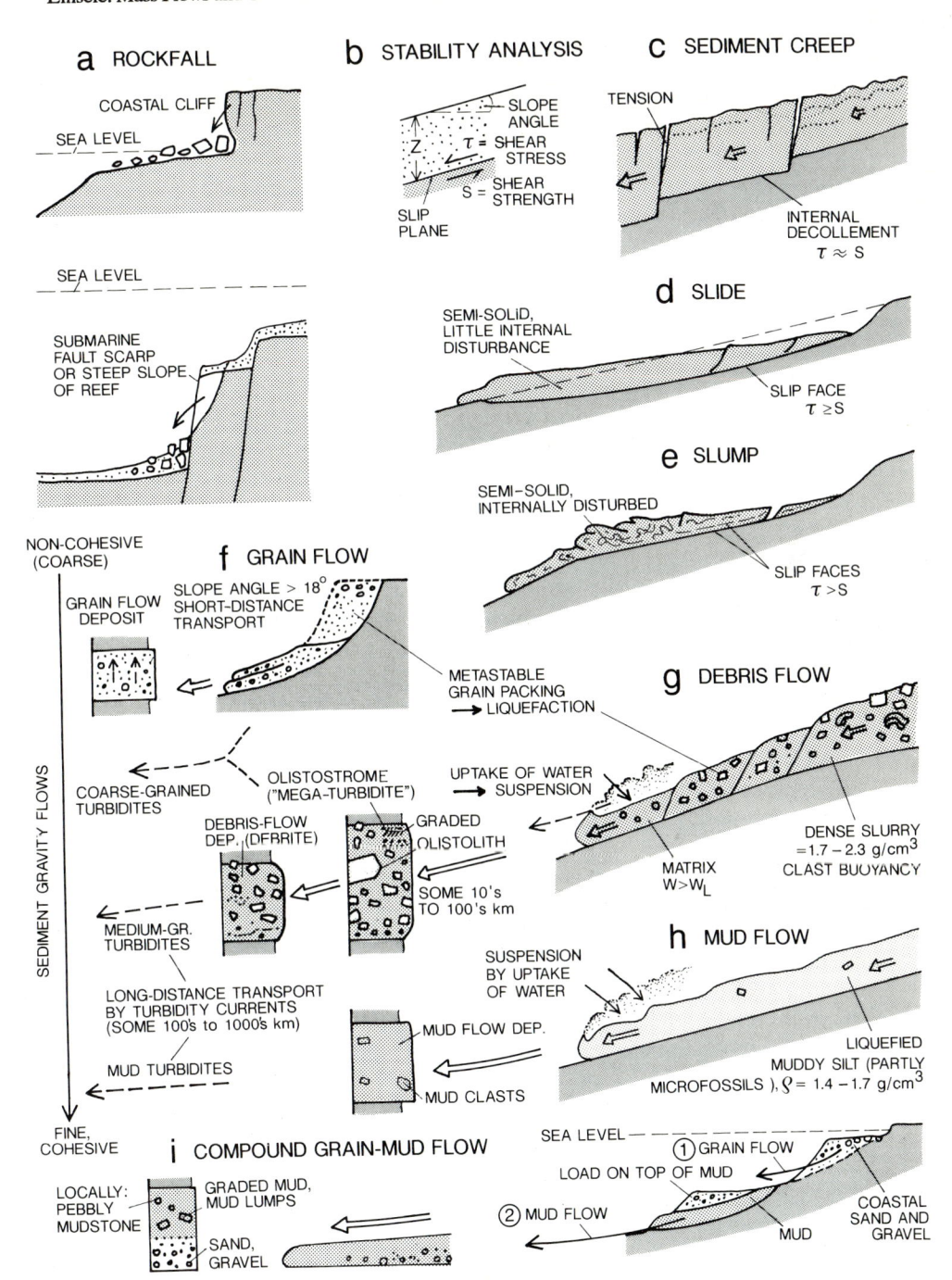

a ROCKFALL
COASTAL CLIFF
SEA LEVEL

SEA LEVEL
SUBMARINE FAULT SCARP OR STEEP SLOPE OF REEF

b STABILITY ANALYSIS
SLOPE ANGLE
τ = SHEAR STRESS
S = SHEAR STRENGTH
SLIP PLANE

c SEDIMENT CREEP
TENSION
INTERNAL DECOLLEMENT
$\tau \approx S$

d SLIDE
SEMI-SOLID, LITTLE INTERNAL DISTURBANCE
SLIP FACE
$\tau \geq S$

e SLUMP
SEMI-SOLID, INTERNALLY DISTURBED
SLIP FACES
$\tau > S$

NON-COHESIVE (COARSE)

f GRAIN FLOW
GRAIN FLOW DEPOSIT
SLOPE ANGLE > 18°
SHORT-DISTANCE TRANSPORT
METASTABLE GRAIN PACKING → LIQUEFACTION

SEDIMENT GRAVITY FLOWS

COARSE-GRAINED TURBIDITES

OLISTOSTROME ("MEGA-TURBIDITE")

g DEBRIS FLOW
UPTAKE OF WATER → SUSPENSION
DENSE SLURRY = 1.7 – 2.3 g/cm³
CLAST BUOYANCY
MATRIX $W > W_L$

DEBRIS-FLOW DEP. (DFRRITE)
GRADED OLISTOLITH
SOME 10's TO 100's km

MEDIUM-GR. TURBIDITES
LONG-DISTANCE TRANSPORT BY TURBIDITY CURRENTS (SOME 100's to 1000's km)

MUD TURBIDITES

h MUD FLOW
SUSPENSION BY UPTAKE OF WATER
MUD FLOW DEP.
MUD CLASTS
LIQUEFIED MUDDY SILT (PARTLY MICROFOSSILS), $\varrho = 1.4 - 1.7$ g/cm³

FINE, COHESIVE

i COMPOUND GRAIN-MUD FLOW
LOCALLY: PEBBLY MUDSTONE
GRADED MUD, MUD LUMPS
SAND, GRAVEL

SEA LEVEL
① GRAIN FLOW
LOAD ON TOP OF MUD
② MUD FLOW
COASTAL SAND AND GRAVEL
MUD

315

*c) **Sediment gravity flows:*** Flow may occur in several different ways:

1. Viscoplastic flows with internal shear planes and virtually no movement at the base of the flow. Such gravity movements usually need a rather high slope angle (5 to 10°) and reach limited thicknesses.
2. Slide-debris flows, or slideflows (e.g., Wächter 1987), move as a more or less rigid plug over a basal shear zone, where the water pressure is in excess of hydrostatic pressure and thus reduces the shear strength of this material, which may even be liquefied. Such flows can reach great thicknesses and occur on very gentle slopes (0.1 to 1°). In addition, they can develop erosional features in the form of broad channels.
3. As a result of their high in-situ water content and meta-stable grain packing, sediment masses on subaqueous slopes can be frequently transformed by earthquake shocks into fluids of high density and viscosity. Partial or entire remolding of the sediment creates a small surplus of pore water which cannot immediately escape. As a result, the shear strength of the material drops drastically and approaches zero without the uptake of additional water (Einsele 1989). The liquefied masses start to flow downslope, even on very gentle slopes ($\leq 0.5°$), and become further remolded and disorganized. Similarly, already moving slump masses may be converted into slow, plastic debris flows or mud flows. Typical examples of this type of liquefaction are noncohesive or low-cohesive sands and silts, but it appears that many sediments rich in diatoms, nannofossils and other micro-organisms are also susceptible to this process.
4. Grain flows consisting of pure sand are characterized by their frictional strength. To overcome friction between the grains, a kind of dispersive pressure must develop. This can only be achieved on fairly steep slopes, such as at the head of submarine canyons and on some prodelta slopes. Therefore, grain flows usually travel only short distances and cover limited areas.

Since many sediment gravity flows evolve from laminar to fully turbulent systems (flow transformation, see also Lowe 1979; Postma 1986), an exact correlation of natural flows to idealized flows (fluidized flow, liquefied flow, grain flow, mud flow, or cohesive debris flow) is often difficult. In the oceans, the most important flow types are mud flows and cohesive debris flows (see below). They contain varying proportions of mud, which provide them with cohesive matrix strength supporting larger particles. When their excess pore water dissipates, the flow masses come to rest.

By uptaking additional water from the overlying water body, individual gravity flows or parts of them can evolve into masses of lower density and viscosity and, if there is a long, sufficiently steep gradient, finally generate turbulent suspension currents of high velocity (turbidity currents).

*d) **Compound mass movements*** result from the sudden loading of slope sediments by other masses, e.g., slides, slumps, or grain flow deposits originating from higher slope areas (Fig. 1i): In this way, the underlying sediment is transformed into a loaded, undrained condition with reduced shear strength. Moore et al. (1982)

reported an interesting case from the southern Gulf of California, where coastal sand and gravel were first transported as a grain flow and debris flow via a steep submarine canyon, down to a mid-fan and lower slope region (around 2500 m deep). There, the superposition of a great load of coarse clastics onto siliceous silty clays (porosity 70 to 80%) triggered a second, considerably larger mass movement on a much gentler slope (about 1.5°), which carried the coarse material down to a 3000-m-deep marginal basin plain (slope angle 0.1°). The mass spread over an area of approximately 300 km^2, forming a sheet several tens to about 100 m thick. Due to the remolding and differential settling of the flow mass particles, the coarse material now forms the base of the mass flow deposit, 70 km away from its source area. From this example one can see that such compound mass movements provide a mechanism for transporting coarse gravel over long distances and on very gentle slopes into the deep sea.

3 Occurrence, Volume, and Transport Distance of Gravity Mass Movements

Figure 2 summarizes the most important tectonic and environmental settings for the occurrence of large gravity mass movements. Most of theses settings provide both high influx of sediment and strong relief. Marine deltas exhibit a variety of mass movements, including mud diapirs and large-scale creep generating growth faults (Fig. 2a). Very common are slides and mud flows on extremely gentle slopes in shallow water as described for example for the Mississippi delta (Prior and Coleman 1984); even more widespread are the huge deep-sea fans in front of the deltas of major rivers. They are fed directly by suspension currents from the river, or by slumps, mud flows, and turbidity currents originating at the prodelta slope. Well-documented examples include the 1929 Grand Banks slump and turbidity current off the Laurentian Channel of North America (Piper and Shor 1988) and the 1979 mass-wasting at the prodelta slope of the Var river in the northwestern Mediterranean (Malinverno et al. 1988). The deep-sea fans of major modern rivers (Bengal, Indus, Amazon, Congo fan) are 250 to 2500 km long and are cut by channels 5 to 25 km wide (Stow 1981; Mutti and Normark 1987). Similarly, submarine canyons collecting shallow-water sediments from the foreshore zone, or directly from river input, can generate large deep-sea fans of many tens to several hundreds of kilometers in length. These fans tend to have high sand contents and even may incorporate some gravel, in contrast to many slope aprons which are predominantly fed by fine-grained, muddy slope sediments. Rapid subsidence and active faulting during rifting and early drifting (right-hand side of Fig. 2a) can create an environment particularly favorable for extensive gravity mass movements along the shelf break of young basins (Bourrouilh 1987; Eberli 1987). Due to "shelf break erosion", the debris flows may carry soft and lithified material of various age.

Subduction-related depositional environments provide several possibilities for different types of gravity mass movements. Deep-sea trenches adjacent to a main continent (not shown in Fig. 2b), such as the modern Peru-Chile trench, frequently receive high quantities of river material, and therefore can be rapidly filled up with both sandy and muddy gravity flow deposits (Thornburg and Kulm 1987). Trenches far away from significant terrestrial sediment sources (Fig. 2b) are fed by slumps and

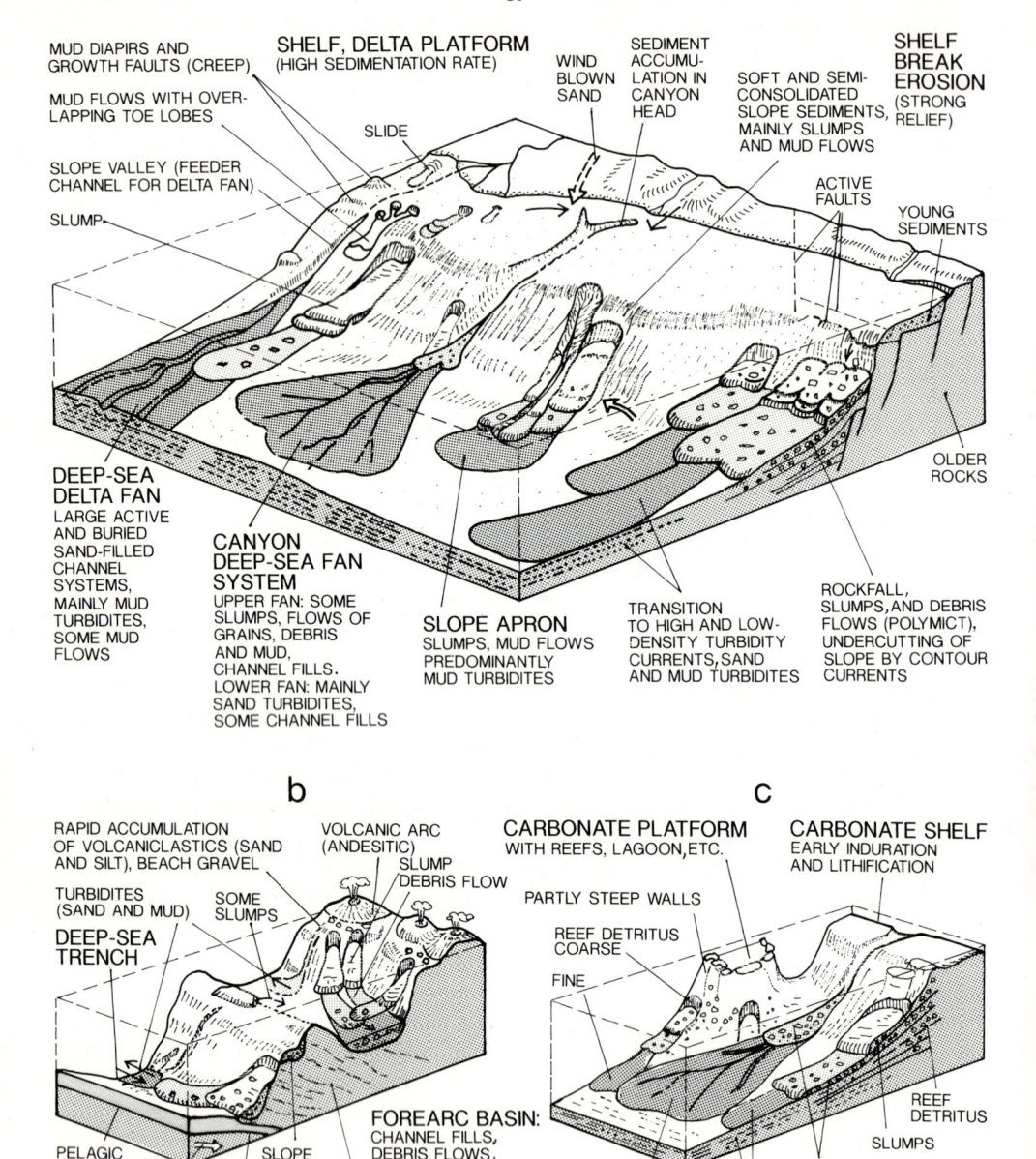

Fig. 2a-c. Summary of the most important depositional environments, where large gravity mass movements and turbidite sedimentation take place. **a** Passive continental margins, including an early rifting-drifting stage with active faults. **b** Convergent margin with forearc basin. **c** Redeposition associated with carbonate shelf and platform. (Based on different sources, e.g., Stow 1986)

debris flows originating from an accretionary wedge as well as young, autochthonous slope sediments. For that reason, they often contain material of varying age and nature (polymict clast composition), sometimes including ophiolites and metamorphic rocks. The sedimentary fill of forearc basins, and to some extent that of backarc basins (not shown in Fig. 2b), is usually characterized by a high proportion of volcaniclastic material, provided by a volcanic arc and transported by gravity mass movements into the basin.

Redeposited gravel-sized rock fragments and skeletal material (rudites), sand-sized shell fragments (arenites), and finer-grained lutites (including pellets) play a great part in deep-sea carbonate depositional environments (Fig. 2c). The main sources of these materials are carbonate shelves and isolated carbonate platforms. Due to early differential lithification and the extremely steep slope of many reef buildups, rock falls including large boulders are fairly common. Oversteepened slopes may also lead to collapse events, generating large-scale slumps and debris flows. Turbidity currents transport coarse and fine-grained shallow-water carbonate farther basinward (see, e.g., Remane 1960; Wilson 1975; Cook and Enos 1977; Scholle et al. 1983b; McIlreath and James 1984; Eberli 1987).

Large mass movements in particular tend, as mentioned above, to evolve from slides or slumps into debris and mud flows which in turn may lose part of their mass to turbidity currents (see below).

In the presentday oceans, a number of extremely large slides have been reported, the volume of which reaches 1000 to 20 000 km^3 (summary in Schwarz 1982); many mass movements range between 0.001 and several 100 km^3. The same orders of magnitude are characteristic of the volume of large mud flows and turbidity currents which evolved from slides and slumps. Of course, there are also numerous smaller displacements of sediments with volumes between 10^3 and 10^6 m^3, which are usually not further described in reports on the modern sea floor. However, this is the category in which we may notice local, large mass movements or thin, more widespread turbidite beds in normal field exposures.

Debris flows and mud flows can travel distances of several 100 up to 1000 km, as observed in the present-day oceans (Akou 1984; Simm and Kidd 1984); turbidite flows may redeposit material as far as several 1000 km away from its primary location. In ancient rocks (Eastern Alps, Apeninnes, Pyrenees) it is possible to trace specific marker beds across 100 to 170 km (e.g., Hesse 1974; Ricci Lucchi and Valmori 1980; Mutti et al. 1984). These observations clearly indicate that gravity flow deposits have a high potential for being widely distributed in relatively large and deep basins.

4 Deposits of Debris Flows and Mud Flows

The principal features of debris flows and mud flows are summarized in Fig. 3. They reflect the flow process prior to their deposition (Lowe 1982), which can be characterized as a more or less laminar, cohesive flow of a comparatively dense, sediment-fluid mixture of plastic behavior. Their sediments are supported by the cohesive matrix and their clasts, at least partially, by matrix buoyancy. Blocks float on the

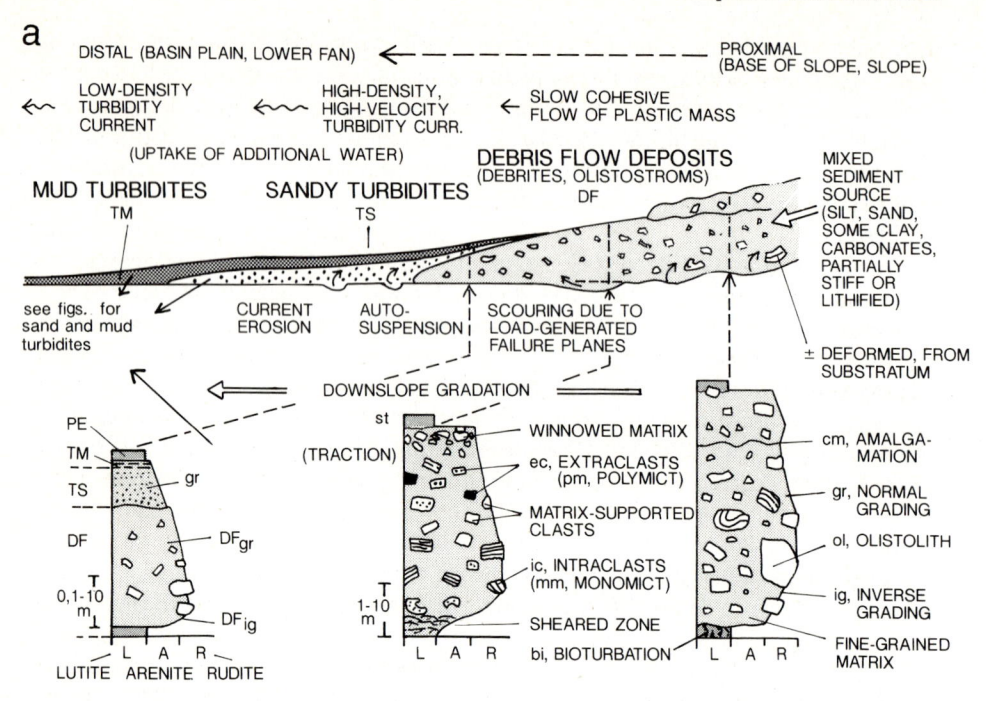

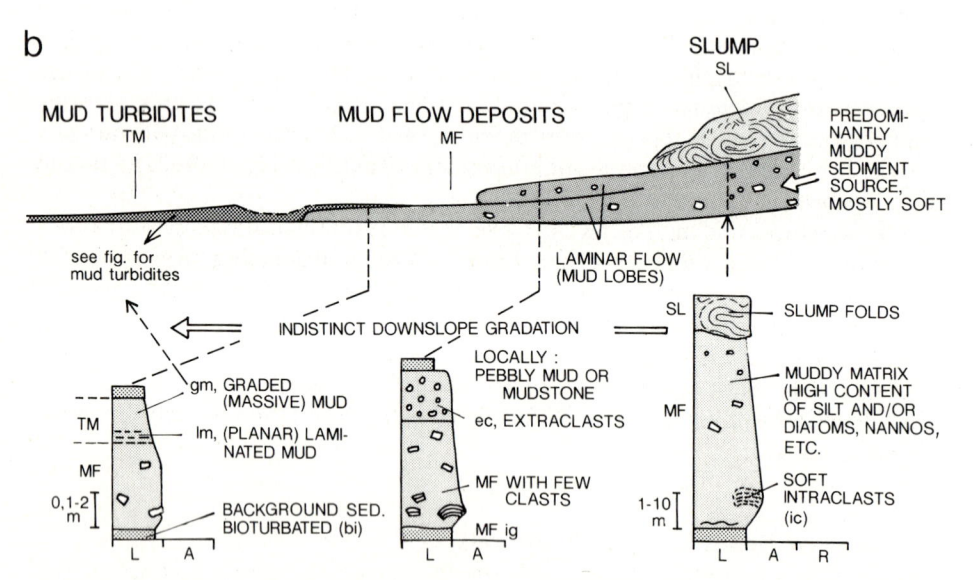

Fig. 3a,b. Conceptual models for proximal and distal debris flows (**a**) and mud flows (**b**), finally evolving into high and low-density turbidity currents. (Based on different sources, e.g., Middleton and Hampton 1976; Lowe 1982; Stow 1986; Bourrouilh 1987; Souquet et al. 1987). For further explanation, see text

debris as a result of small density differences between the blocks and the debris, plus the cohesive strength of the clay-water slurry (Rodine and Johnson 1976). Debris flows come to rest as the applied shear stress drops below the shear strength of the moving material. The flows "freeze", which is accomplished either by cohesive freezing or, in the case of a cohesionless sandy matrix, by frictional freezing; or by both processes.

Debris flow deposits (debrites, DF) and olistostromes (very thick, extensive debrites) consist of a medium to fine-grained matrix and a varying proportion of matrix-supported clasts (Fig. 3a). The typical debrite is rich in clasts of different sizes; the clasts may be derived from older sediments and rocks within the basin (intraclasts), or from sources outside the basin (extraclasts). In olistostromes, single clasts or blocks can reach the size of a house and more. They are called olistoliths, and resemble blocks or intact rock bodies that are frequently observed in tectonic mélange zones. Internally, most debrites lack any bedding phenomena or imbrication of clasts. Except at the top of the debrite, in situ trace fossils are completely missing. In some examples, elongate clasts are aligned horizontally, indicating the direction of flow. The base of debrites may be scoured and show a thin sheared zone. The lowermost part of a debrite is frequently inversely graded, due to prograding frictional freezing. The higher portion may exhibit indistinct normal grading; the top of the bed may be either sharp or grade into an overlying turbidite, thus forming a compound debrite-turbidite couplet (DF-TS/TM, see below). In places, the top of a debrite may be current-winnowed and therefore transformed to a clast-supported layer. However, calcareous debrites resulting from large-scale slope collapse often form sheet-like megabreccia beds, which are primarily clast-supported and contain little fine-grained matrix material (Mullins et al. 1986). There are cases in which a debrite or mud flow deposit is directly overlain by a subsequent debrite or an overlapping lobe of the same mudflow (also see Fig. 2a). The two layers are then combined by amalgamation.

Mud flow deposits have much in common with debrites (Fig. 3b), and there is no sharp boundary between these two end-members of the same group. They have a muddy matrix with a high silt (or micro-fossil) content and contain only a small amount of clasts, mostly intraclasts which are frequently deformed by the preceding slumping and flow process. Locally, the admixture of gravel or other coarse material from submarine canyons leads to pebbly mud or mudstone.

Due to their structureless, homogenized matrix and relation to earthquakes, volcanic eruptions, and tsunamis, the terms "unifites" (Feldhausen et al. 1981), "homogenites" (Cita and Ricci Lucchi 1984) and, for thick beds of frequently compound origin, "megaturbidites" or "seismoturbidites" (Mutti et al. 1984) are used by some workers to characterize gravity flow deposits of exceptionally large volume and areal extent.

Elmore et al. (1979) described a modern megaturbidite, 500 km long, more than 100 km wide, and up to 4 m thick, of upper Pleistocene age from the Hatteras abyssal plain in the western Atlantic. This bed consists predominantly of fluvially-derived sand and shelf mud with a large proportion of mollusk shell fragments. In proximal regions, the poorly sorted lower part of the bed ($\geq 20\%$ mud) may have been deposited as a sandy debris flow, whereas its upper part and more distal portions reflect deposition from a turbidity current. Another compound, but carbonate-bearing, debrite-turbidite was observed in the Exuma Sound, Bahamas (Crevello and Schlager 1980). This bed is 2 to 3 m thick and covers an area of more than 6000 km^2. Hieke (1984) reported a Holocene example from the Ionian abyssal plain in the Mediterranean. Here, a 12-m-thick homogenized mud layer containing around 50% carbonate, partially from intermediate and possibly even from shallow waters, covers an area of 1100 km^2 in about 4000 m of water. Locally, the layer has a sandy base composed of shell fragments.

Couplets of debrites (mud flow deposits) with sandy and muddy turbidites, indicated in Fig. 3 by the symbols DF-TS-TM or MF-TM (see Fig. 4), were also described from several ancient sedimentary sequences (e.g., Stanley 1982; Mutti et al. 1984; Bourrouilh 1987; Souquet et al. 1987).

5 Turbidites

5.1 General Characteristics

Sediments deposited from suspension currents show a variety of distinctive features which vary depending on the magnitude and velocity of turbidity currents, the material in and the distance from the source area, the morphology of the basin, as well as other factors. In fact, turbidity currents were postulated as a mechanism for producing graded, sheet-like sand beds (sandy turbidites) from having studied rhythmic bedding and their internal structures in ancient rock sequences (Kuenen and Migliorini 1950). This fascinating concept can explain all the different observations in the field with one group of simple processes, e.g.:

a) The episodic transport of large volumes of shallow-water sands, including their fauna, into the deep sea and their distribution over wide areas,
b) marked linear erosion or nondeposition in submarine canyons (turbidity current feeder channels) and widely extended erosion of the uppermost centimeters of the pre-existing, soft, deep-sea sediments on deep-sea fans and basin plains,
c) the generation of certain bed types, as well as vertical and lateral successions of sedimentary structures, including traces of bottom-dwelling organisms.

Reports of breakage of submarine telegraph cables on continental slopes, for example as a result of the famous Grand Banks earthquake 1929 off Newfoundland (Piper and Normark 1982; Piper and Shor 1988), or in front of submarine canyons at the mouth of some large rivers, testify to the great power of turbidity currents. However, due to their infrequent occurrence in relation to human life spans, such large-scale turbidity

Fig. 4a-c. Descriptive terms and symbols for the internal structures of redeposited sediments (association of slumps, debris and mud flows, sand and mud turbidites); Bouma divisions in parentheses. Many other characteristics such as sole marks and trace fossils (lebensspuren) are omitted. **b** Minor features and different types of grading. **c** Further subdivisions of the main groups; additional terms and symbols may be introduced as needed

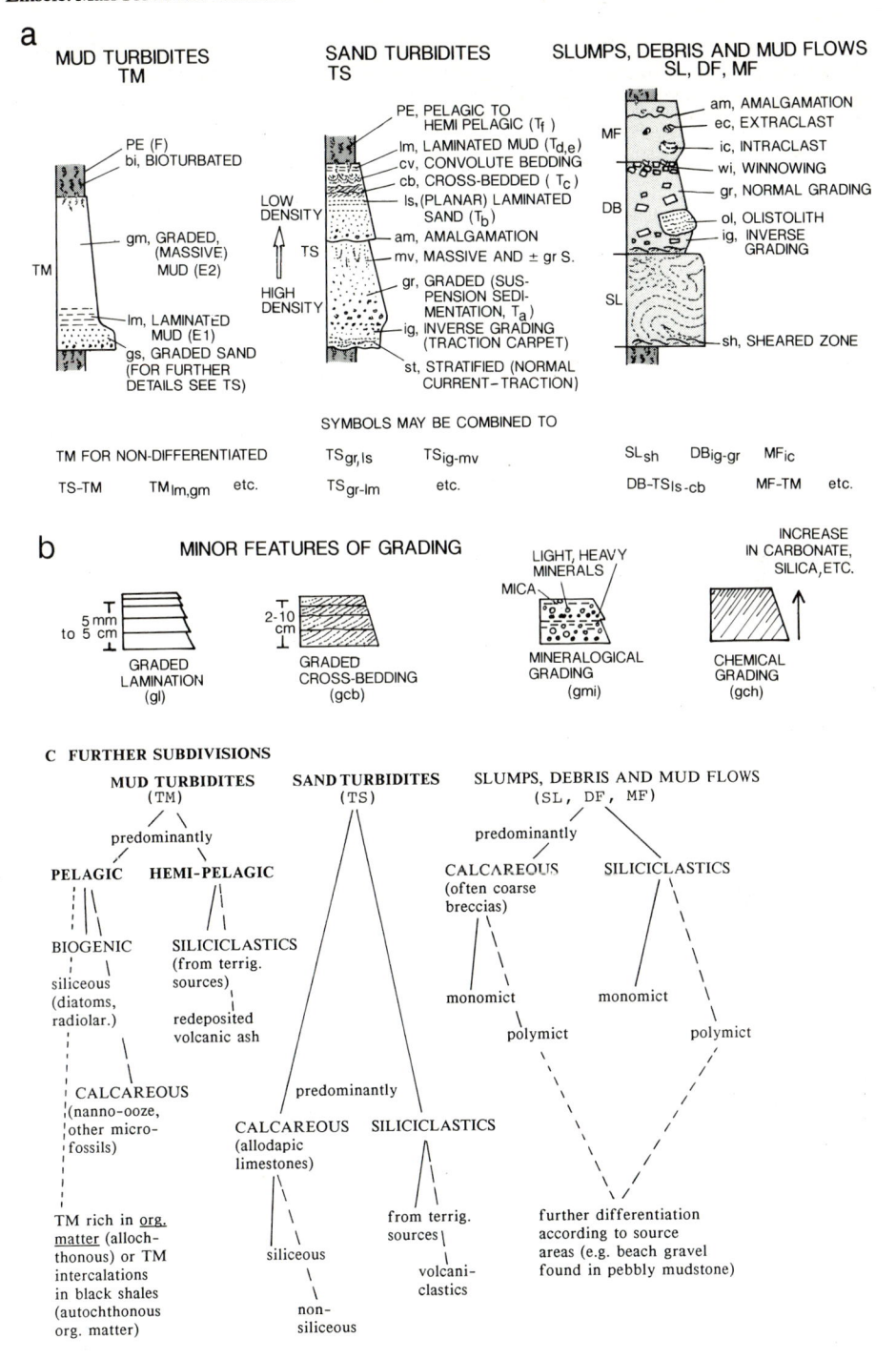

Einsele: Mass Flows and Turbidites

a

MUD TURBIDITES TM

- PE (F)
- bi, BIOTURBATED
- gm, GRADED, (MASSIVE) MUD (E2)
- lm, LAMINATED MUD (E1)
- gs, GRADED SAND (FOR FURTHER DETAILS SEE TS)

SAND TURBIDITES TS

- PE, PELAGIC TO HEMI PELAGIC (T_f)
- lm, LAMINATED MUD ($T_{d,e}$)
- cv, CONVOLUTE BEDDING
- cb, CROSS-BEDDED (T_c)
- ls, (PLANAR) LAMINATED SAND (T_b)
- am, AMALGAMATION
- mv, MASSIVE AND ± gr S.
- gr, GRADED (SUSPENSION SEDIMENTATION, T_a)
- ig, INVERSE GRADING (TRACTION CARPET)
- st, STRATIFIED (NORMAL CURRENT-TRACTION)

LOW DENSITY ↑ TS ↓ HIGH DENSITY

SLUMPS, DEBRIS AND MUD FLOWS SL, DF, MF

- am, AMALGAMATION
- ec, EXTRACLAST
- ic, INTRACLAST
- wi, WINNOWING
- gr, NORMAL GRADING
- ol, OLISTOLITH
- ig, INVERSE GRADING
- sh, SHEARED ZONE

MF / DB / SL

SYMBOLS MAY BE COMBINED TO

TM FOR NON-DIFFERENTIATED	TS$_{gr,ls}$	TS$_{ig-mv}$	SL$_{sh}$	DB$_{ig-gr}$	MF$_{ic}$
TS-TM	TM$_{lm,gm}$ etc.	TS$_{gr-lm}$ etc.	DB-TS$_{ls-cb}$	MF-TM etc.

b MINOR FEATURES OF GRADING

- 5 mm to 5 cm — GRADED LAMINATION (gl)
- 2-10 cm — GRADED CROSS-BEDDING (gcb)
- LIGHT, HEAVY MINERALS, MICA — MINERALOGICAL GRADING (gmi)
- INCREASE IN CARBONATE, SILICA, ETC. — CHEMICAL GRADING (gch)

c FURTHER SUBDIVISIONS

MUD TURBIDITES (TM)
predominantly

PELAGIC — HEMI-PELAGIC

BIOGENIC — SILICICLASTICS (from terrig. sources)
siliceous (diatoms, radiolar.)
redeposited volcanic ash

CALCAREOUS (nanno-ooze, other microfossils)

TM rich in org. matter (allochthonous) or TM intercalations in black shales (autochthonous org. matter)

SAND TURBIDITES (TS)

predominantly

CALCAREOUS (allodapic limestones) — SILICICLASTICS
siliceous
non-siliceous
from terrig. sources
volcaniclastics

SLUMPS, DEBRIS AND MUD FLOWS (SL, DF, MF)

predominantly

CALCAREOUS (often coarse breccias) — SILICICLASTICS

monomict — polymict
monomict — polymict

further differentiation according to source areas (e.g. beach gravel found in pebbly mudstone)

currents could never be directly observed in operation. Direct measurements were performed on low density undercurrents in lakes or water reservoirs, as well as in some small-scale flume experiments.

Suspension or turbidity currents result either directly from suspended sediments delivered by rivers in flood state, or from unstable submarine sediment accumulations which have failed. In the marine realm, the density of suspensions caused by river floods is, however, usually not high enough to produce density currents in the sea (density of sea water 1.027 g/cm^3). Therefore, the failure of large sediment accumulations in shallow waters is probably the most important prerequisite for turbidity currents. Such accumulations are common on prodelta slopes, or they are accomplished by longshore currents carrying material into the head of submarine canyons. Rip currents and storm action generally transport sediment into deeper water and to the shelf edge.

Turbidity currents evolve from high-density gravity mass flows such as slumps, grain flows or mud flows (Fig. 5a) by uptaking additional sea water (or lake water). As a result, the sliding or flowing masses become less dense and therefore turn from a slow cohesive flow into a more rapid, turbulent density flow. The sediment particles lose their mutual contact and go into suspension. They are kept in this state by the upward component of fluid turbulence. The driving force of the turbidity current is primarily a function of the difference in density of the suspension and the overlying water body (in contrast to river flow under air!), the submarine relief, i.e., the angle and length of slope, and the thickness of the suspension current. Thus, high-velocity suspension currents can be achieved only when at least two of these factors become comparatively great. This is usually true for submarine slopes along continental margins and deep, narrow ocean basins. Although these slopes usually have angles of only a few degrees, they are much steeper than the gradients of large, subaerial rivers (0.1 to 0.01°). In addition, large, submarine mass movements also can produce turbidity currents of considerable density (up to 1.10 to 1.17 g/cm^3) and thickness (several hundred meters, see, e.g., Piper and Shor 1988). In combination, these factors can cause high-velocity currents (up to 10 to 20 m/s).

We distinguish between high- and low-density turbidity currents. High-density turbidity currents reach high velocities and can carry relatively coarse-grained sand, pebbles, and intraclasts. Within the confines of submarine channels, large-scale turbidity currents have the competence to transport gravel (up to at least 10 cm in diameter) as bed load, and thus may generate lenses of conglomerate at the foot of prodelta slopes (Komar 1969; Piper and Shor 1988). In addition, they have the

---

Fig. 5a-f. Conceptual model for the generation of and internal structures in proximal and distal sandy turbidites. **a** Transition from slope failures (slumps, debris and mud flows, grain flows) on continental slopes and canyon heads to turbidity currents. **b** "Classic", complete turbidite (more or less proximal) showing the total succession of Bouma divisions (in parentheses) with modified descriptive symbols (Fig. 4a), as well as autosuspension due to turbulence. **c** More distal turbidite, no erosion, basal Bouma divisions missing. **d** Idealized proximal-distal development of turbidite bed; proximal channel fills may show sedimentary structures due to traction by normal currents (*st*), as well as "traction carpets" (*ig*, inversely graded). **e** Different sizes of flute and groove casts in relation to bed thicknesses. **f** Pre-depositional and post-depositional trace fossils. (After Seilacher 1962; Kern 1980).

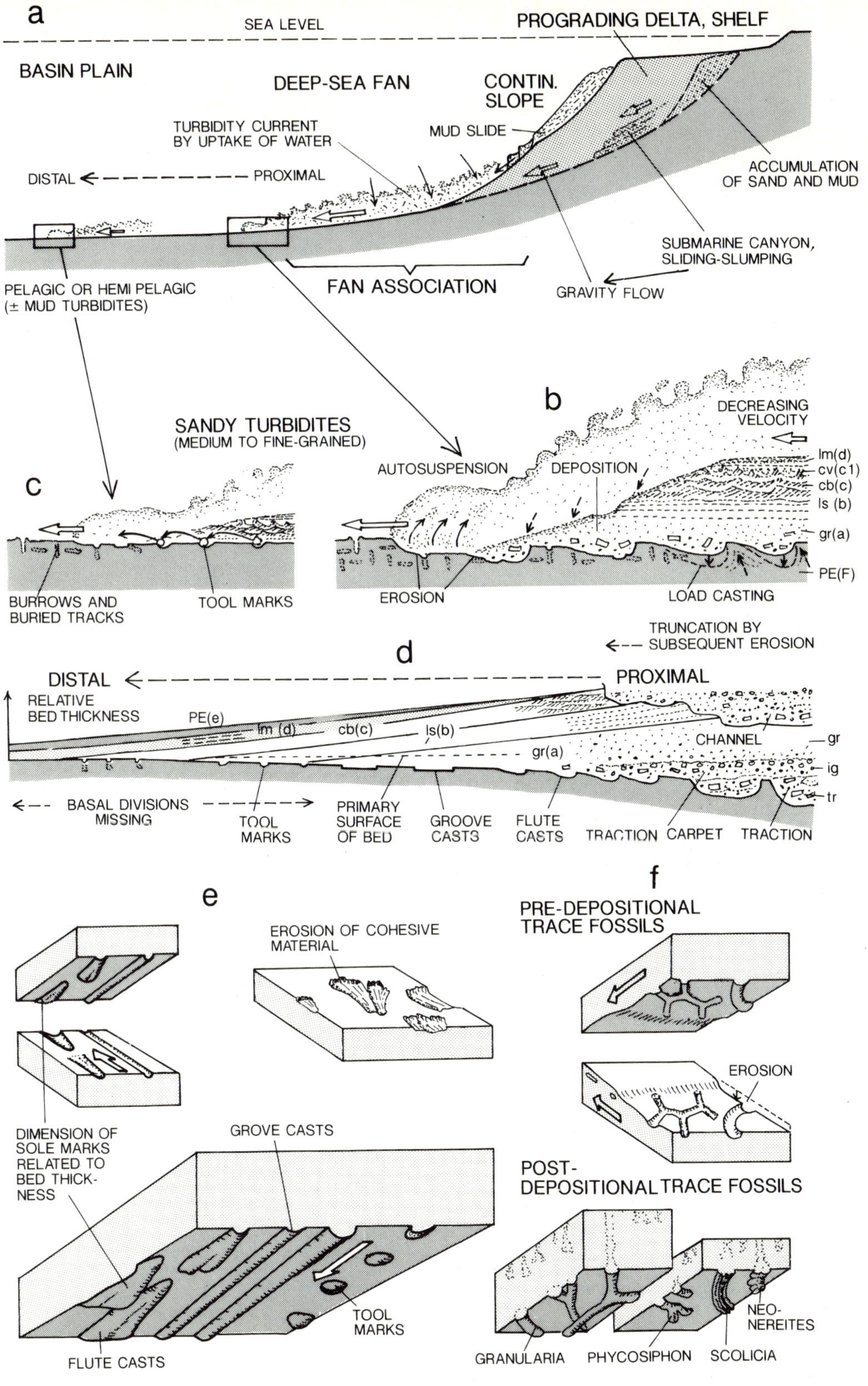
a
SEA LEVEL
PROGRADING DELTA, SHELF
BASIN PLAIN
DEEP-SEA FAN
CONTIN. SLOPE
TURBIDITY CURRENT BY UPTAKE OF WATER
MUD SLIDE
DISTAL
PROXIMAL
ACCUMULATION OF SAND AND MUD
SUBMARINE CANYON, SLIDING-SLUMPING
FAN ASSOCIATION
GRAVITY FLOW
PELAGIC OR HEMI PELAGIC (± MUD TURBIDITES)
b
SANDY TURBIDITES (MEDIUM TO FINE-GRAINED)
DECREASING VELOCITY
lm(d)
cv(c1)
cb(c)
ls (b)
gr(a)
PE(F)
AUTOSUSPENSION
DEPOSITION
c
BURROWS AND BURIED TRACKS
TOOL MARKS
EROSION
LOAD CASTING
TRUNCATION BY SUBSEQUENT EROSION
d
DISTAL
PROXIMAL
RELATIVE BED THICKNESS
PE(e)
lm (d)
cb(c)
ls(b)
gr(a)
CHANNEL
gr
ig
tr
BASAL DIVISIONS MISSING
TOOL MARKS
PRIMARY SURFACE OF BED
GROOVE CASTS
FLUTE CASTS
TRACTION CARPET
TRACTION
e
f
PRE-DEPOSITIONAL TRACE FOSSILS
EROSION OF COHESIVE MATERIAL
EROSION
DIMENSION OF SOLE MARKS RELATED TO BED THICK-NESS
GROVE CASTS
POST-DEPOSITIONAL TRACE FOSSILS
TOOL MARKS
FLUTE CASTS
GRANULARIA
PHYCOSIPHON
SCOLICIA
NEO-NEREITES

capacity to erode cohesive muds on extensive areas of the sea floor. The eroded material feeds the suspension with new sediment (a kind of feedback system), which replaces coarser material settling out of suspension in the slackening body and tail of the current (Fig. 5b). In this way, and by the maintenance of turbulence by gravitational forces (auto-suspension), the current is kept in motion and can travel over long distances. Low-density turbidity currents flow slowly and can therefore keep only silt and clay-sized material in suspension, or larger aggregates consisting of particles of these sizes. Their erosional capacity is very low or nonexistent, but weak turbulence maintains such suspension currents for relatively long periods of time. They can attain considerable thicknesses and distribute their suspended load as a thin bed over wide areas. It can be assumed that low-density, muddy turbidity currents often are the final stage of sand-bearing suspension currents, which have lost their coarser grain size fraction (Chough 1984; Stanley 1985).

With the aid of internal sedimentary structures in the turbidite beds and different sole marks (Fig. 5e,f), one can distinguish between proximal (near-source) and distal turbidites (Fig. 5d). However, this concept should be used with caution, particularly in deep-sea fan associations (see below; further discussion in MacDonald 1986; Stow 1986). Proximal turbidites are often relatively coarse-grained and thick-bedded, and their tops may be truncated by a subsequent turbidity current (amalgamation). Downslope, the turbidites successively tend to lose their basal divisions; they become thinner and finer-grained. In many turbidite sequences, a striking correlation between the dimensions of sole marks and the thicknesses of the corresponding sandy beds can be noticed. For example, large flute casts or groove casts are often associated with particularly thick sandstone layers (Fig. 5e,f). This and the following observations strongly support the turbidity current hypothesis explaining all features of a turbidite by one single sedimentological event.

As Seilacher (1962) has pointed out, we can distinguish between pre-event trace fossil associations living in the pelagic to hemipelagic mud interval, and assemblages which recolonize a freshly deposited turbidite sand layer. The burrows of the first group are exhumed by the erosive force of the turbidity current and filled up with sand (lebensspuren on the sole of the sand bed, Fig. 5g). The recolonizing assemblage has to dig down into the sandy layer from a new, higher level Fig. 5h). If the turbidite is thin, some of the burrowing organisms may reach its base and feed on the background sediments. If the sandy layer is thick, only its top sections can be burrowed, and only those assemblages able to live on sand can persist.

5.2 *Revision of the Bouma Nomenclature*

Bouma (1962) proposed a simple nomenclature to describe the succession of internal sedimentary structures in sandy turbidites (T_a through T_e), which has generally been used for many years. This nomenclature is well suited for medium-grained sandy beds, but it suits neither distinctly proximal types of relatively coarse-grained turbidites nor pure mud turbidites (see below). Therefore, additional terms and symbols were introduced (Lowe 1982; summary in Stow 1986), separate from a classification for larger-scale phenomena such as fining and thickening-upward sequences, channel types, or for deep-sea fan morphology (e.g., Walker 1984b; Mutti and Normark 1987; see Sect. 6). The different nomenclatures describing bed-scale features can be combined into one broad, consistent system with difficulty. For that reason, an attempt is made here (Fig. 4) to revise and simplify the existing proposals and to include debris and mud flows, as well as compound beds deposited from different types of mass flows. Preference is given to purely descriptive terms; in addition, using two characters instead of one (as in the previous systems) could not be avoided. For example, sandy turbidites are discriminated from mud turbidites by using the symbols TS and TM, respectively; ig signifies inversed grading, lm laminated mud, etc. All types of redeposited beds can be described with these symbols. Carbonate turbidites may be referred to as CTS and CTM, and there are of course other possibilities such that one could expand and modify this system.

5.3 Coarse-Grained Turbidites

High-density turbidity currents carrying pebbles and clasts as bed load and finer-grained material in suspension may show an initial stage of traction sedimentation (coarse-grained conglomeratic sand with plane lamination, cross-bedding, and internal scour). This is then overlain by thin, horizontal layers displaying inverse grading. This division represents traction carpet deposits resulting from mixed frictional freezing and suspension sedimentation (Lowe 1982). The uppermost division of such a proximal turbidite is generated by rapid settling from suspension (suspension sedimentation), and may be either structureless or normally graded. Typical features are water-escape structures (pillar and dish structures). This division more or less corresponds with the lowermost division of the "classic" sand turbidite (Bouma division T_a; here referred to as TS_{gr}, see Fig. 4).

5.4 Medium-Grained Sandy Turbidites (Siliciclastics and Carbonate)

The divisions of a "classic", medium-sized sand turbidite deposited from suspension currents of lower density can be interpreted in a similar way. The divisions consisting of plane laminated and cross-bedded sand (Figs. 4 and 5, ls and cb; Bouma division T_b and T_c) reflect traction structures, while the division showing laminated mud, lm (Bouma T_d), may be explained as mixed traction/suspension sedimentation. Finally, the following structureless and indistinctly graded mud interval, gm (Bouma T_e), originates solely from suspension sedimentation. There is also an important group of fine-grained mud turbidites consisting only of the muddy divisions, lm and gm (Bouma divisions T_d and T_e), which are described further below.

Upward-decreasing grain size (normal grading) can be observed not only within division gr (Bouma T_a) and generally from bottom to top in a turbidite bed, but often within the parallel laminated and small-scale cross-bedded divisions, ls to lm (Bouma T_b to T_e, Fig. 4b). In these cases, the thicknesses of individual laminae or cross-bedding sets decrease from bottom to top; this is known as graded lamination, or graded cross-bedding.

It is, however, important to note that the presence and degree of grading in a turbidite bed are also controlled by the availability of a range of grain sizes. Well sorted material in the source area does not allow the formation of distinctly graded beds. An extreme example of this type, indicated in Fig 2a, is nongraded, well-sorted, yellowish sand turbidites in the eastern Atlantic, which are derived from Saharan desert sands. These were blown by offshore winds during glacial low sea level stands onto the shelfbreak, and carried by slumps and turbidity currents into the deep sea (Sarnthein and Diester-Haass 1977).

Vertical sequences in predominantly sandy turbidites do not show only one bed type, for example thick and partially amalgamated, relatively coarse-grained, so-called proximal turbidites, but usually also display thinner and finer-grained beds of a more distal appearance. Nonetheless, a certain trend from sequence to sequence is often apparent (Fig.6a). Such vertical variations in bed types are common in channelized deep-sea fans (see below), where thick-bedded channel fills differ greatly from levee deposits and overbank sediments. If turbidity currents reach abyssal areas below the calcite compensation depth (CCD), they can drop calcareous beds alternating with pelagic clays free of carbonate. This fact has often been used for paleoceanographic reconstructions (e.g., Berger and von Rad 1972; Hesse 1975). Furthermore, sandy and muddy turbidity currents can carry organic matter from terrestrial sources and low-oxygenated slope areas into well-oxygenated deep basins, where it may be preserved as allochthonous organic material (Fig. 6a). Contrarily, autochthonous black shales can alternate with turbidites containing little organic material and displaced fauna characteristic of an oligotrophic environment.

So far we have not considered the composition of the principal material forming sandy turbidites. Essentially there are two groups of sand-sized materials: siliciclastic sands and carbonate sands, which may, of course, occur in all kinds of mixtures. In the case of quartz and feldspar being the main constituents, one observes that grains of equal diameter behave hydraulically in a similar way. If settled from suspension, a poorly sorted sand with a "wide" unimodal grain size distribution will build up a regularly graded bed, without distinct "jumps" in grain sizes, or enrichments of certain minerals in particular divisions of the bed. In contrast, carbonate sands consisting chiefly or partly of skeletal material produced on carbonate platforms may behave quite differently. Therefore, turbidite sands and muds rich in carbonate have often been treated separately from siliciclastic turbidites, and called "allodapic limestones" (Meischner 1964). In this case, grains with diameters much greater than 2 mm may behave hydraulically like quartz sand, or sand-sized microfossil shells may be transported and settle in a manner similar to compact silt grains. As a result, the graded carbonate layer deposited from a turbidity current can show distinct jumps in its vertical internal succession of grain sizes and/or sediment composition. For example, the graded division, gr (Bouma division T_a), can be replaced by a comparatively coarse shell bed, and instead of a carbonate silt division, lm (T_d), a chert layer may be present, which was formed diagenetically from hydraulically equivalent siliceous sponge needles and radiolarians (Fig. 6b). Due to the uptake of eroded

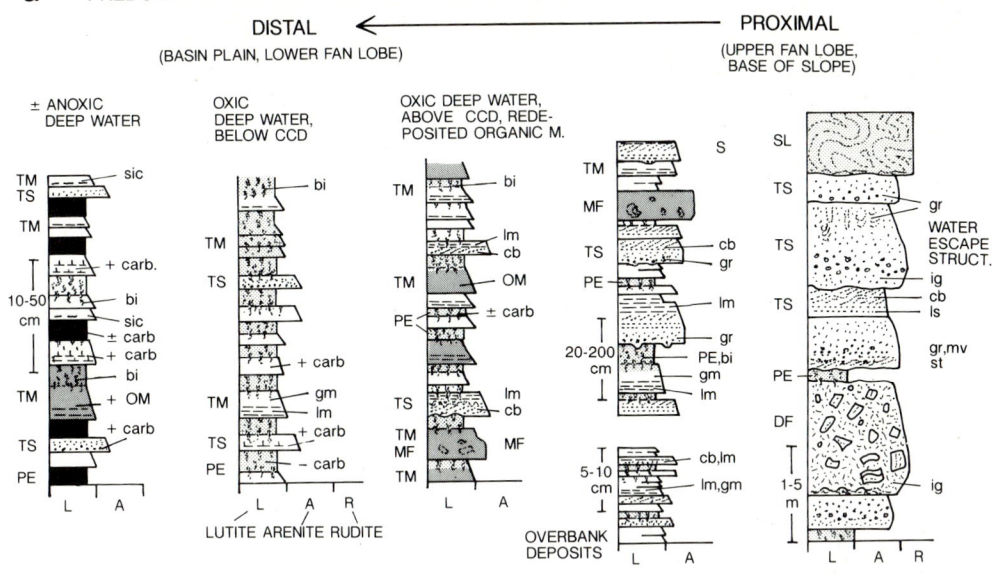

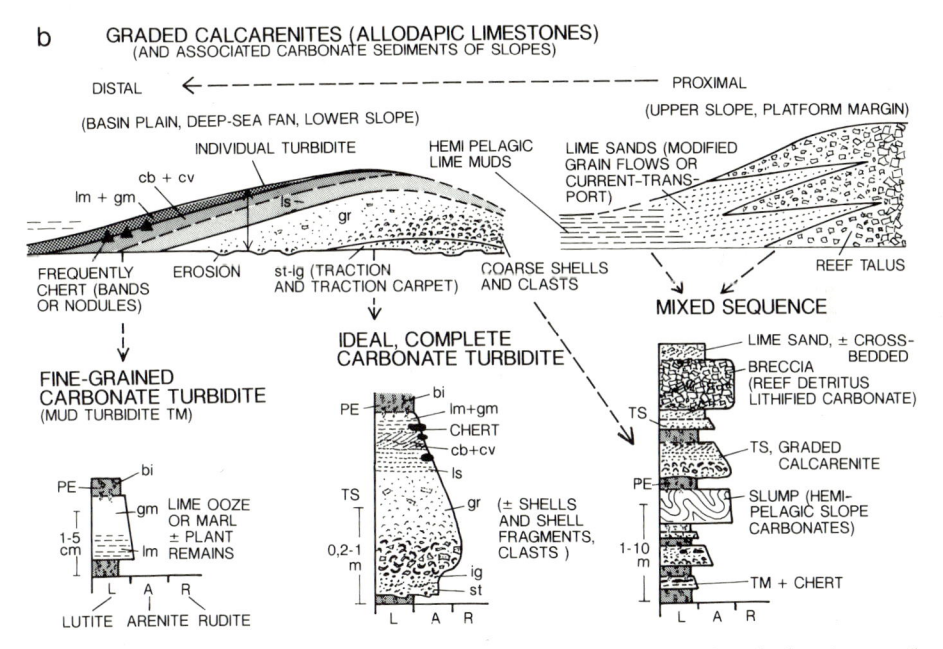

Fig. 6a,b. Vertical successions of pelagic or hemipelagic beds and event deposits (gravity mass flows, sandy and muddy turbidites) in more proximal or distal regions and in different chemical environments (*CCD*, calcite compensation depth). Note the different vertical scales; thin-bedded overbank deposits may occur in mid-fan regions, as well as in more distal areas (cf. Fig. 8). Organic matter (*OM*) is either autochtonous or allochthonous, i.e., laterally transported into the basin by mud turbidites. **b** Proximal and distal carbonate turbidites more or less associated with platform margins and slope sediments; various vertical scales. (After Meischner 1964; Scholle et al. 1983, 1983a; McIlreath and James 1984)

deep-sea mud, the amount of autochthonous fauna (nekton, plankton) often increases toward the top of a turbidite bed.

Redeposited carbonate sediments are often coarser than their siliciclastic counterparts. The primary source of sand-sized and larger carbonate grains are the margins of carbonate shelves and platforms (Figs. 2c and 6b). Abundant shell material, reef detritus, and early lithification of different types of carbonates provide various coarse-grained materials. Hence, allodapic limestones may alternate with carbonate breccias and sands derived directly from platform margins (cf. Sect. 3). The maximum thickness of an individual bed is not commonly attained in the neighborhood of the source area, but at some distance downcurrent (Eder et al. 1983); it then decreases distally. For more details about calcareous sandy and muddy turbidites, see Eberli (Chap. 2.8, this Vol.).

5.5 Mud Turbidites

Overlooked for a long time is the fact that a great part of the fine-grained muddy sediments in a deep basin or submarine fan may also be transported by low-density turbidity currents, instead of being distributed and accumulated by normal, more or less steady pelagic settling. For that reasons mud turbidites are described in a special article by Piper and Stow (Chap. 2.9, this Vol.; also see Stow and Piper 1984b). Here, only a brief summary is presented.

Mud turbidites may be the end-member of gravity mass flows of mixed granulometry, or may be derived from muddy sediment sources such as prodelta slopes and other fine-grained slope sediments. In addition to gravity movements, large river floods or muddy sediments stirred up by storms in shallow seas can contribute to the formation of mud turbidites. The sediments of several modern submarine fans (e.g., Nile cone, Indus fan) on the continental rises of presentday large oceans, and in smaller oceanic basins (e.g., Black Sea, Gulf of California), consist to a large degree of mud turbidites. Since these beds are deposited from low-density suspension currents, they usually are thin; but rather thick, proximal mud turbidites also occur, for example in the Gulf of California (Einsele and Kelts 1982), probably representing a transitional stage from mud flows to muddy suspension currents. For more detail on the hydraulics of muddy suspensions and the process of sedimentation, see Piper and Stow (Chap. 2.9, this Vol.).

Similar to sandy turbidites, mud turbidites can be subdivided into proximal and distal types (Fig. 7):

Proximal mud turbidites may contain a thin sand layer at their base, which displays some of the Bouma divisions for sand turbidites. Distal mud turbidites of the hemipelagic group, and many mud turbidites of the pelagic group often become so thin that they are completely reworked by burrowing organisms. When this happens, their nature as mud turbidites is obscured, and they can only be recognized if their material and/or fauna differ substantially from the pelagic or hemipelagic background sediment (host sediment, PE).

Since biogenic components are often a significant source material for marine muds, many pelagic and hemipelagic mud turbidites also show chemical grading, due

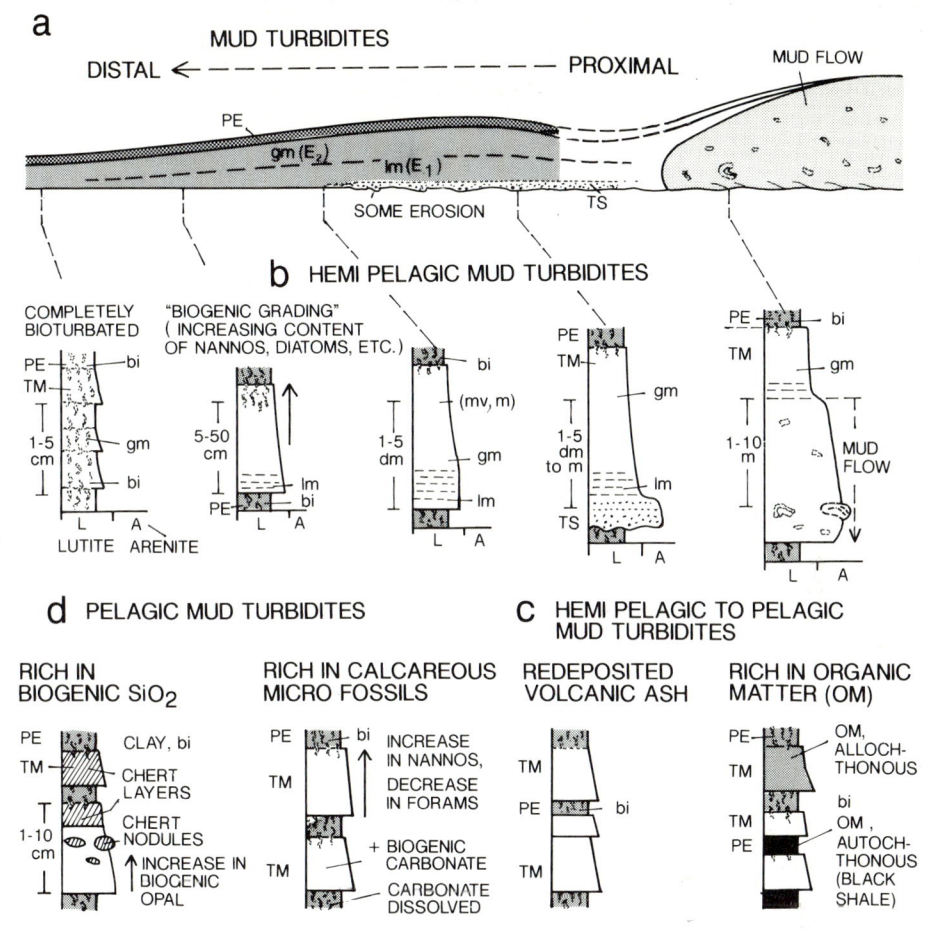

Fig. 7a-d. Different groups of mud turbidites. **a** proximal-distal trend. **b** and **c** Hemipelagic, predominantly siliciclastic (**b**), volcaniclastic, or rich in organic matter (**c**). **d** Pelagic mud turbidites, rich in either carbonate or opaline silica

to the different hydraulic behavior of foraminifera, calcareous nannofossils, diatoms, or radiolarians. Thus, the carbonate content of an individual sandy or muddy turbidite can increase or decrease from bottom to top; likewise, biogenic silica, later forming chert layers, may be concentrated in an upper division of the turbidite.

Frequently, both biogenic carbonate and silica are transported by turbidity currents into deep basins below the calcite compensation depth (CCD), where the background sediments are poor in or free of carbonate or opaline silica. In this way, a succession of layers alternatingly with and without carbonate or biogenic silica ("banded" sequences) can be produced. Similarly, mud turbidites containing high amounts of organic matter from their source area (e.g., slope sediments under regions of upwelling) can alternate with deep sea sediment poor in organic carbon (Fig. 7c).

In this case, the organic matter is rapidly buried, and therefore at least partly preserved from decomposition in oxygenated environments. Examples of this mechanism are known from the presentday oceans, e.g., from the Cape Verde rise or the Biscaya abyssal plain in the eastern Atlantic (Kelts and Arthur 1981; Degens et al. 1986). In contrast, "normal" black shale sediments slowly deposited by vertical settling from the overlying water body in anoxic deep water can be repeatedly interrupted and modified by interbedded, muddy turbidites poorer in organic matter, as in the eastern Mediterranean (Stanley 1986).

6 Deep-Sea Fan Association

The "classical" hypothesis, which explains lateral facies changes within individual turbidite beds in terms of proximal-distal trends, was developed for turbidites which originate from a highly efficient sediment source and are deposited on the plain of elongate basins (type I turbidite deposits, according to Mutti and Normark 1987). This type may include the fringe of the lower, unchannelized part of deep-sea fans or detached fan lobes (Fig. 8D), where turbidity currents are not affected by submarine channels and their levees. This situation is also common at the foot of continental slopes, where relatively low sediment input by slope failures creates a nonchannelized slope apron (Fig. 2a). The predominance of such a "line source" is also characteristic of calcareous deep-sea sands and muds which were produced on carbonate shelves and platforms and later transported into deeper water.

However, as mentioned earlier and as indicated in Fig. 6a, the sedimentary facies of many deep-sea fans fed by sediment input from large rivers, i.e., from an efficient point source, are strongly affected by a channelized sediment distribution system (Fig. 8A through E; type II turbidite deposits of Mutti and Normark 1987). Several thoroughly studied modern examples have shown that submarine canyons do not end at the foot of the slope, but often continue into depositional fans and basin plains (e.g., Mutti and Normark 1987). In addition, the long-term input of large volumes of river sands generates migrating, sinuous channel-levee complexes with crevasse splays similar to those of subaerial meandering rivers (Nelson and Maldonado 1988). The channels of large submarine fans are many kilometers wide and have gradients of the order of 1%. In the upper and middle fan region, their levees may rise above the surrounding sea floor by tens and, in extreme cases, more than 100 m. From the study of both modern and ancient sediments, several models for sediment distribution in such deep-sea fans have been developed (e.g., Mutti 1977; Mutti and Ricci Lucchi 1978; Nelson et al. 1978; Walker 1984b; Shanmugan and Moiola 1985; Mutti and Normark 1987). All of them show a great variety of bed types in the channel system, its levees and overbank regions, and the lobes of the upper to lower fan region.

Moving down the feeder channel, one may encounter more or less channelized deposits of cohesive mass flows, coarse-grained sands and conglomeratic sandstones. The latter two are generated either by normal current traction transport or by mixed "frictional freezing" and suspension sedimentation (traction carpet displaying inverse grading, ig, Fig. 8A). In addition, normally graded beds preferentially showing divisions gr and ls (Bouma $T_{a,b}$), as well as indications of vertical water escape (dish and pipe structures), are common. These channel fills cut into or pass laterally into

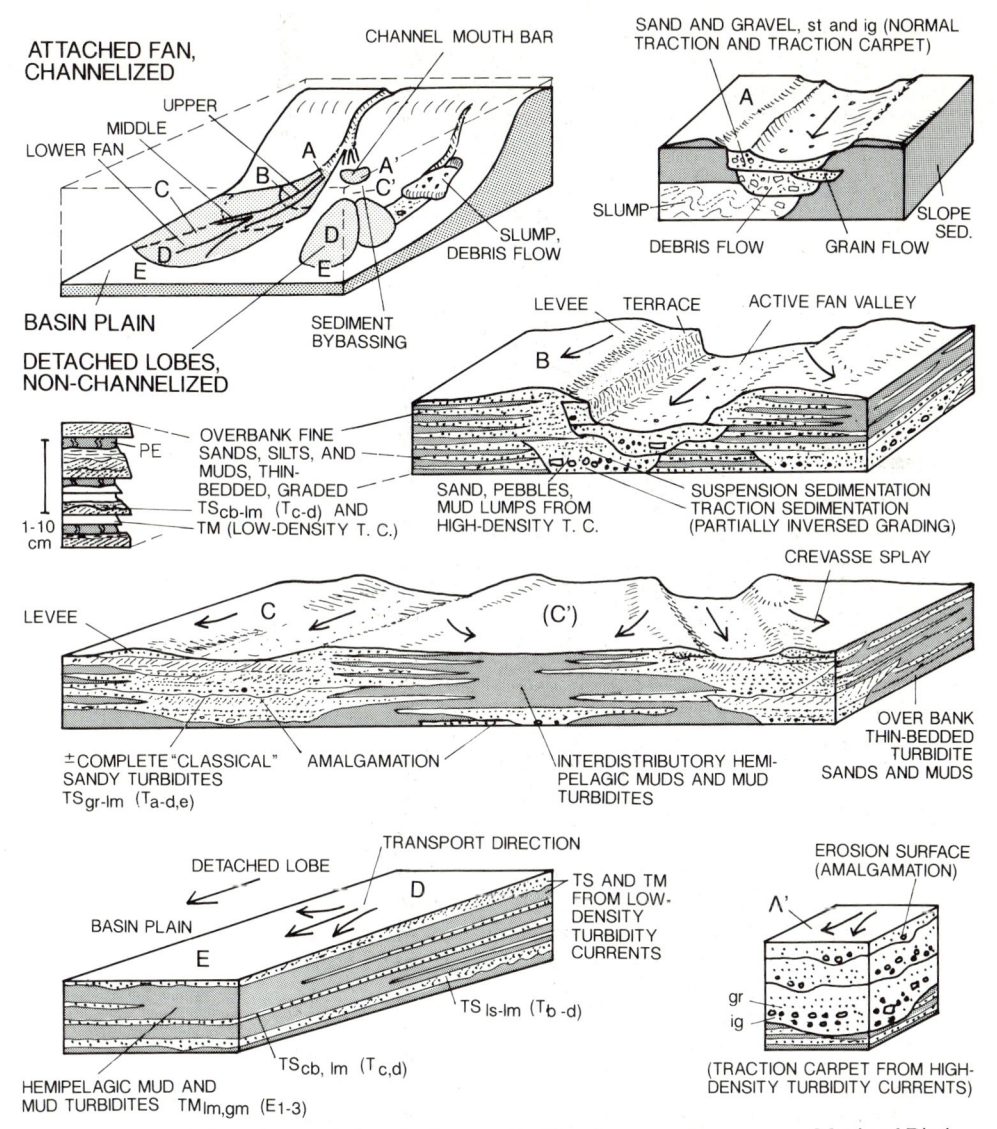

Fig. 8. Model of the facies association of a deep-sea fan (based on several sources, e.g., Mutti and Ricci Lucchi 1978, Nelson et al. 1978, Walker 1978, Shanmugam and Moiola 1985, and author's own observations). Note the difference between channelized, attached fan and detached, nonchannelized fans which show a more regular turbidite sequence. For explanation of symbols see Fig. 4

thick or thin-bedded turbidites of the more classical types on the levees and fan lobes (Fig. 8B,C).

In the transition zone between relatively high-gradient channelized flow and unconfined flow on a gentler gradient at the beginning of a depositional lobe (channel-lobe transition), turbidity currents may suddenly change their flow regime from rapid to more tranquil flow (Mutti and Normark 1987). Such a hydraulic jump is accompanied by increased turbulence and enlargement and dilution of the suspension flow. The sea bed is frequently marked by large-scale scour features, mud clasts, and rapid deposition of sand and coarser material.

Mud-dominated currents tend to deposit the bulk of their suspended load downstream of the hydraulic jump. In the zone of channel-lobe transition, they usually cause less scouring and leave only some cross-stratified sands.

Whereas thin, relatively dense, fast turbidity currents tend to flow basinward within the confines of the channels and their levees, low-velocity, less dense thicker suspension currents build up levees and drop their finegrained load in interchannel areas. Thus, originating in the upper fan area, they form thin-bedded overbank deposits (Fig. 8B,C), showing predominantly the upper divisions of turbidite sands, cb, lm, and gm (Bouma T_{c-e}). Correlation of such beds becomes difficult even over short distances, since levee erosion can take place locally. Fine-grained intercalations between these sand layers for the most part represent redeposited material: mud turbidites. Downslope, on the fringe of the lower fan and on the basin plain, large-scale turbidity currents usually deposit extensive sand and mud layers of the more distal type (Fig. 8D). Only very large, rare debris and mudflows and their subsequent turbulent flows spread their load over large areas of the total fan and basin plain.

According to a comparative study of sand layers in present-day ocean basins (Pilkey et al. 1980), the percentage of sand layers in the total sediment volume decreases distally, as also observed in ancient flysch sequences. The thickest layers were found in basins which have large drainage areas. Single sand beds could be traced over distances as great as 500 km (Hatteras abyssal plain in the western Atlantic). On the levees of mid-ocean channels, far away from any land source, parallel laminated, thin mud turbidites were observed (Hesse and Chough 1980). A more detailed description of the characteristics of deep-sea fans, the adjacent basin plains, and slope aprons is given by Stow (1986) and Mutti and Normark (1987).

As a result of shifting channels and changing sites of mass movement on slopes, vertical sections of deep-sea fan sediments can show a succession of beds of apparently widely differing proximality (Fig. 8B,C, and Fig. 6a). Normal, current-transported material may alternate with debrites and mud flow deposits, sandy and muddy turbidites, and hemi-pelagic or pelagic sediments. Switching of fan lobes in conjunction with migrating channel systems (Fig. 9a) can generate both fining (and/or thinning) upward as well as coarsening (and/or thickening) upward sequences. Under constant sea level and a persisting slope, or a landward migration of the slope, vertical aggradation of fan sediments causes a slight tendency for fining-upward sequences to develop (Fig. 9E).

Type III turbidite deposits, as defined by Mutti and Normark (1987), are composed mainly of fine-grained, thin-bedded levee and overbank deposits. Within these deposits, sands are almost entirely restricted to channel infills.

In response to the amount of sediment input by their feeder system, deep-sea fans undergo phases of inactivity, including some reworking, or periods of upbuilding and progradation. The latter case is often correlated with a lowering of sea level, when the gradients of rivers entering the sea are steepened, and former coastal and shallow water sediments are easily eroded and swept into deeper waters. As a result of slope

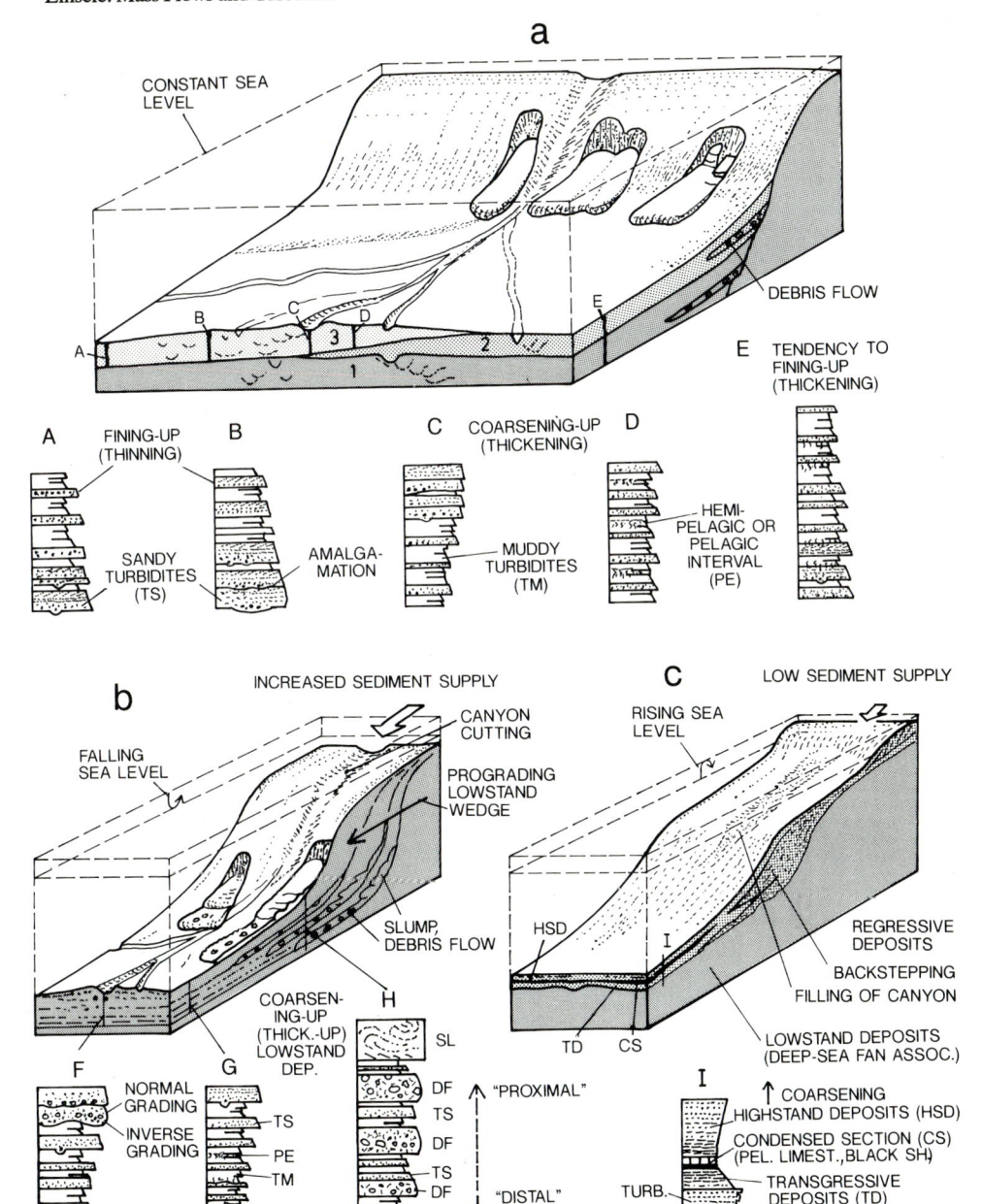

Fig. 9a-c. Facies association and development of deep-sea fan. **a** Constant sea level and steady position of continental slope, permanently high sediment input via submarine canyon; switching fan lobes (*1,2,3*) and migrating fan valleys. Note that both fining (and/or thinning) upward (*A* and *B*) and coarsening (and/or thickening) upward sequences (*C* and *D*) occur; the overall tendency is fining-upward (*E*). **b** Relative sea level fall favors rapid prograding and upbuilding of fan due to high sediment input. Most of the sections tend to coarsen (and/or thicken) upward (*F, G, H*). **c** Relative sea level rise and the resulting strong reduction in sediment supply may terminate fan growth. Channel fills and turbidites are replaced by fine-grained transgressive deposits (*TD*), a thin condensed section (*CS*, black shales, pelagic ooze, etc.), and highstand deposits (*HSD*) of increasing thickness (again, coarsening-upward). (Partially based on Walker 1978; Bally 1987, but significantly modified)

and fan progradation, one may get coarsening (and/or thickening) upward stratigraphic sequences (Fig. 9b, sections F through G). In contrast, sea level rise usually leads to reduced sediment supply from terrestrial sources and may bring about the filling of submarine valleys and canyons. Consequently, gravity mass movements usually come to an end during such periods, and fan deposits may be replaced by normal hemipelagic to pelagic sediments (Fig. 9c, section I). The transition from the transgressive to the highstand phase may be characterized by particularly reduced sediment accumulation. The resulting condensed section is often represented by pelagic oozes or limestones and thin black shales. For further consequences caused by sea level variations see Vail et al., Chap. 6.1, this Vol.).

7 Turbidity Current Directions and Paleocurrent Patterns

Current directions in turbidity currents can easily be derived from sole marks (Fig. 5e), internal structures such as cross-bedding, clast and grain orientation, etc., and current ripples (see, e.g., Collinson and Thompson 1982).

In several early studies on ancient turbidite sequences, one of the most striking phenomena, besides regular rhythmic bedding, was that the directions of turbidity currents were surprisingly constant over large areas. Later it became evident that these examples predominantly represent basin plain and lower fan environments in elongate basins (type 1 deposits), which are fed by one main sediment source. In more proximal fan associations as well as in basins supplied with sediment from varying major sources, the paleocurrent patterns become less regular and sometimes rather complex. In cases where the fan lobes have room to switch (Fig. 8), the sediments are dispersed radially over the course of time.

Small-scale turbidity currents flowing within the levees of a meandering, distributing channel system are forced to change their flow directions significantly. Larger, thicker currents spill over the levees, are deflected from the main current, and may even cause some levee erosion (Piper and Normark 1983). Slackening currents tend to be further deflected by Coriolis forces. Finally, one should take into account that in certain areas the current patterns of turbidites can be overprinted by contour currents (Stow 1986). Slow, low-density turbidity currents may even turn into contour currents (Hill 1984).

Paleoslope orientations can be inferred from slide scars, slump folds, and sometimes from the imbrication of clasts in debrites and mud flows. For reliable measurements, good, large exposures are needed.

8 Average Sedimentation Rates of Deep-Sea Fan Associations and Frequency of Turbidite Events

Deep-sea fans and adjacent basin plains are areas of high sedimentation rate, particularly during relatively low sea level stands or in times of tectonic activity creating increasing relief. In modern fan environments, average sedimentation rates between 100 and 1000 m/Ma are common, but near the sediment source and in over-supplied basins (Mutti et al. 1984), higher values also occur. For example, the giant Bengal fan, Mississippi fan, and some other elongate, presentday deep-sea fans represent wedge-shaped sediment bodies which reach maximum thicknesses on the order of 5 to 10 km, built up in a few Ma to 20 Ma (Curray and Moore 1974; Bouma et al. 1986). The same applies to many thick, ancient flysch sequences.

The frequency of cohesive mass flows and turbidite events is related to the rate of sediment accumulation in their source area, though it also varies greatly among the different fan and basin plain environments (Fig. 10). In addition, frequent earthquakes or volcanic eruptions may cause a relatively short recurrence time for redepositional events. As a result, only thin beds of limited areal extension may be generated. There appears to be an inverse (logarithmic) relation between bed thicknesses and frequency in ancient turbidites (Piper and Normark 1983). In the Sea of Japan, representing a modern backarc basin, one thin-bedded mud turbidite (with an average thickness of 6 mm) has been deposited at time intervals of 50 years (Chough 1984). On the California continental borderland, the recurrence time of gravity mass movements is one event every 200 to 500 years (Malouta et al. 1981). From the Mediterranean, Rupke and Stanley (1974) report one mud turbidite, on the average, every 300 to 400 years for the last 20 000 years. In the central and southern part of the Gulf of California, relatively thick (often several dm) Quaternary mud turbidites have been generated with a frequency between 2000 and 10 000 years (Einsele and Kelts 1982); a similar order of magnitude is typical of many sand turbidites with small to medium thickness in lower fan and basin plan environments. In submarine channels and overbank deposits in middle fan environments, Piper and Normark (1983) assume frequencies of one event per 10 to 100 years and 100 to 1000 years, respectively. This signifies that, according to a rough estimation, one out of ten turbidity currents is thick enough to spill over channel levees and deposit its suspended load. In contrast, most of the smaller turbidite events do not leave much sediment in the higher parts of the channel system, where erosion and amalgamation are common.

The longest recurrence times are to be expected for thick mud flows and turbidites (megaturbidites) which form extensive sheets on submarine fans and basin plains (Fig. 10). In ancient rocks, such key beds, often rich in redeposited carbonate due to high carbonate production in shallow waters, occur once approximately every 50 000 years to 1 Ma, depending on the amount of sediment supply (Mutti et al. 1984). They may have been triggered by extremely strong earthquakes with very long recurrence times. In a deep-sea drillhole at the foot of the slope of Baja California, only one of these event deposits was found in a mud turbidite sequence representing 3 to 4 Ma (Moore et al. 1982). This, and some other even larger mass movements observed in the presentday oceans, occurred during the last glacial sea level lowstand 16 000 to 17 000 years ago. A famous example is the compound slide/mud flow/turbidity current in the Canary basin (Embley, in Saxov and Nieuwenhuis 1982), where a 10- to 20-m-thick mud flow deposit covers an area of 30 000 km^2, and the turbidity current traveled over 1000 km. Although in this and other cases earthquakes are quoted as the triggering mechanism, the recurrence time for such "mega-events" is probably controlled primarily by the period of high amplitude sea level changes, as well as by the availability of large volumes of unconsolidated or only partially lithified sediments at the heads of submarine canyons, along the shelfbreak, and on

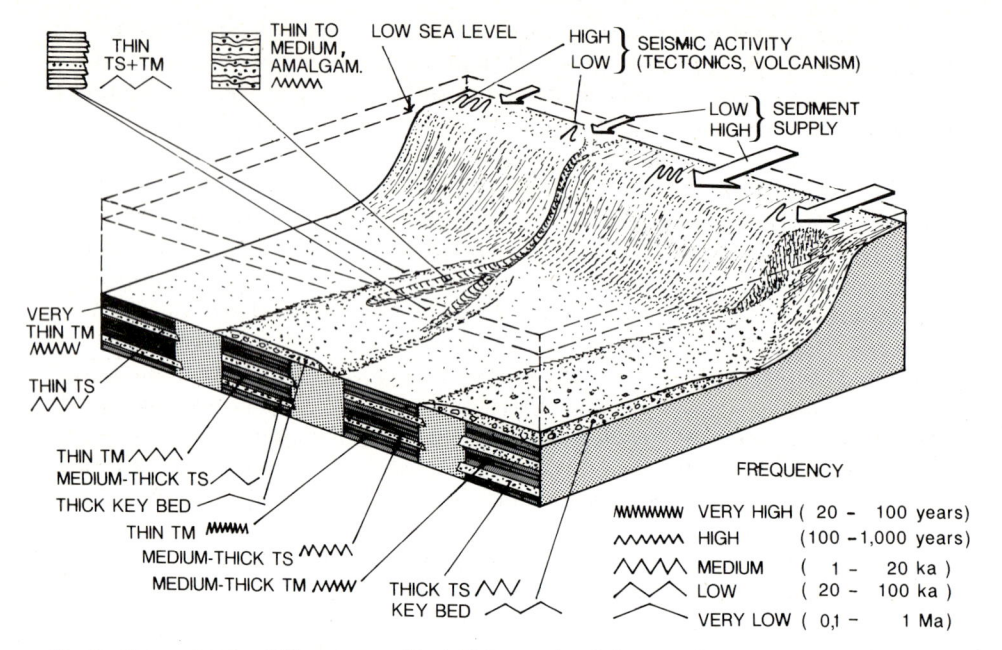

Fig. 10. Frequency of turbidite events and bed thicknesses in relation to sediment supply, seismic activity, and location within a deep-sea fan/basin plain facies association

the upper slope. Tectonic activity alone as the cause of large mass movements appears to be an important factor only under certain conditions, for example during the rifting stage of a rapidly subsiding basin or in subduction-related environments.

9 Summary

In this chapter, all gravity mass movements, particularly debris and mud flows as well as high and low-density turbidity currents, are briefly reviewed, including their occurrence and different bed types in the deep sea. They are treated as a family of related processes and sediments. As demonstrated in some conceptual models, the more distal types frequently evolve from the proximal ones. Sandy and muddy turbidites show a great variety of bedding phenomena. Proximal carbonate turbidites often tend to be coarser-grained than their siliciclastic counterparts; more distal ones may display some special features such as chert nodules. Sandy and muddy turbidites alternate with pelagic or hemipelagic sediments deposited above or below the CCD, in well-oxygenated or euxinic bottom waters. Mud turbidites in particular may contain considerable amounts of allochthonous organic matter as well as datable fauna from their source area. One subgroup of mud turbidites consists predominantly of skeletal carbonate or silica. Textural grading may be complemented or replaced by chemical and mineralogical grading.

The facies patterns of sandy and muddy turbidite beds in deep-sea fan environments, including cohesive mass flows and beds deposited by current traction, differ considerably from the regular bedding and consistent transport directions in lower, detached fan lobes and basin plains. Paleocurrent directions in such systems vary a great deal, and individual beds cannot be traced over long distances. Coarsening and fining-upward sequences may be controlled by several processes.

Submarine fans and basin plains receiving materials from gravity mass movements are regions of relatively high sedimentation rates (100 to 1000 m/Ma and more). The recurrence interval of mass flow and turbidite events varies greatly, from relatively frequent, thin mud turbidites (50 to several 100 years) to thick, extensive key beds ("mega-turbidites", 50 000 to more than 1 Ma). It also varies within a submarine fan association characterized by amalgamated channel-levee complexes and thin-bedded overbank deposits. In addition, the availability of large volumes of unconsolidated sediments and relatively low sea levels are considered a primary control on the abundance and magnitude of gravity mass movements and their deposits. Frequent seismic activity tends to cause more, but smaller mass flows than those found in quieter zones with rare, but larger events.

Lastly, the existing different nomenclatures for mass flow deposits and turbidites are replaced by a uniform descriptive system.

Acknowledgments. I thank Dr. D.J.W. Piper for his critical review of the manuscript and several important suggestions for improvements. Linda Hobert checked the English text.

2.8 Calcareous Turbidites and Their Relationship to Sea-Level Fluctuations and Tectonism

G. P. Eberli

1 Introduction

The pioneering work of Kuenen and Migliorini (1950), and later, Bouma (1962), related specific sedimentary structures to a turbidity current transport mechanism. Since then, such "turbidite" deposits, especially when of siliciclastic composition, have been readily recognized in the rock record, and this interpretative term has become widely accepted as a descriptive term. Carbonate sedimentologists were more apprehensive about calling similarly structured calcareous beds "turbidites", and the term "allodapic limestone" was proposed for limestones deposited from a turbidity current (Meischner 1964). In general, however, the criteria for recognizing a turbidity current deposit are the same for siliciclastic or calcareous sediments, as the two differ "mainly by their composition and to a lesser degree by their texture" (Rusnak and Nesteroff 1964). Both types of turbidites are characterized by graded beds, partial or complete Bouma sequences, lateral continuity and the interbedding of shallow-water biota into deeper-water deposits. The concept that, "each turbidite is the result of a single, short-lived event, and once deposited it is extremely unlikely to be reworked by other currents" (Walker 1984b), applies equally to both siliciclastic and calcareous turbidites.

The turbidity current transport mechanism was a major step towards understanding sediment redeposition processes. During the last 20 years, research efforts have concentrated more on the development of a predictive depositional model rather than on the events producing turbidites and turbidite sequences. The systematic analysis of all facies types in turbidite sequences and their arrangement in facies associations has resulted in depositional models for both siliciclastic and calcareous turbidites, e.g., the fan model and apron model (Mutti and Ricci Lucchi 1975; Walker 1978; Schlager and Chermak 1979; Mullins and Cook 1986).

The purpose of this chapter is not to review these depositional models for calcareous turbidites but rather to discuss the nature of events recorded in individual calcareous turbidite beds as well as to evaluate possible connections between depositional sequences and external cycles, such as sea-level fluctuations and tectonic basin evolution. Because shallow-water carbonate production is only possible when shelves or platforms are flooded, sea-level fluctuations influence the amount of sediment production and turbidite frequency (Mullins 1983; Droxler and Schlager 1985) and, thus, control the basinal depositional pattern (Eberli and Ginsburg 1989). On the other hand, tectonic setting determines the basin configuration and the location of shallow-water and basinal areas. Tectonic activity is considered an important triggering mechanism for tubidites and is directly related to their frequency, while basin morpholgy determines the occurrence and the areal distribution of the turbidite sequences (Hsü 1977; Eberli 1987). Sea-level fluctuations and tectonic activity are not single events, but multi-episodic. This suggests that calcareous turbidite sequences not only record single events, deposited as individual beds, but are also a series of increments recording longer-term cycles of sea-level fluctuations or tectonism.

2 Description

2.1 Composition

The carbonate environment, unlike the siliciclastic one, is productive, and produces debris of organic and inorganic origin. As a result, redeposited carbonates consist of eroded lithoclasts, as well as newly produced skeletal debris and inorganic precipitates. The relative amount of each of these components varies widely and is dependent on the geologic and tectonic setting and the position of sea level (Haak and Schlager 1989).

The amount of biodetrital debris is greatest in turbidites deposited along flooded carbonate platforms. Along the reef-rimmed portions of the platforms, composition is dominated by the debris of reef-building organisms such as corals, bryozoans, molluscs, coralline algae, sponges and associated fauna, including echinoderms, green algae and large benthic foraminifers. Open shelves and leeward portions of platforms usually contain less debris of reef-building organisms but larger amounts of oolites and peloids (Hine et al. 1981b). Monospecific turbidites, with about 90% of the components derived from one faunal group, are not uncommon. For example, nearly pure crinoidal turbidites are often found in Paleozoic sections (Tucker 1969; Davies 1977) or in Jurassic basinal deposits (Eberli 1987). Carbonate lithoclasts are either eroded ancient limestones and dolomites, or reworked penecontemperanous slope and basinal deposits. In many turbidites, mud chips and rounded to sub-rounded micritic clasts indicate partial erosion and subsequent incorporation of semi-indurated background sediment into the turbidity current (Meischner 1964; Weissert 1981; Bosellini et al. 1981). Calcareous turbidites in tectonically active areas have a higher amount of eroded lithoclasts. For example, during the break-up of Tethyan carbonate platforms due to Jurassic rifting, the component assemblage of the turbidites and the associated mass flow deposits is composed of lithoclasts eroded from exhumed parts of the platform (Eberli 1987; Watts 1988).

2.2 Sedimentary Structures

Calcareous turbidites resemble siliciclastic turbidites in that they are recognized by the following sedimentary structures: gradation, partial or complete Bouma sequence, and lateral continuity (Figs. 1 and 2). The carbonate composition, special source area, and post-depositional diagenetic alteration produce, however, some special, distinctive features for calcareous turbidites. These include the following:

1. Sorting. Calcareous turbidites, especially those with medium and coarse-grained components, are poorly sorted compared to their siliciclastic counterparts. In general, sorting is dependent upon pre-sorting, distance of transport, and density and velocity of the transporting current. In calcareous turbidites, poor sorting is the combined result of minimal pre-sorting in the source area plus the divergent shapes of the bioclastic grains, their variable bulk density and the resulting differences in hydraulic behavior for grains of the same diameter (Rusnak and Nesteroff 1964). For example, during a catastrophic storm event, components of all sizes and shapes and with a variable degree of cementation are exported from a reef tract into deeper water. During deposition from the resulting turbidity current, the differential degree of cementation and shape of the clast will allow large, uncemented skeletal pieces to settle out of place within the graded bed (Maiklem 1968).

2. Bimodality. In many calcareous turbidites, there is a bimodal distribution of lithoclasts and biodetrital grains, with the lithoclasts preferentially at the base of the bed, even when they are of smaller grain size than the biodetritus (Fig. 2). This bimodality is caused, similarly to the poor sorting, by the low effective density of the incompletely cemented skeletal debris. As a result, small but denser lithoclasts are deposited first and are overlain by larger, porous skeletal debris, creating an inverse grading. Thus, porosity differences and the resulting density differences are responsible for an inverse grading observed in a number of calcareous turbidites deposited from normal and low-density currents. In contrast, inverse grading in deposits from high-density turbidity currents is achieved by freezing of the traction

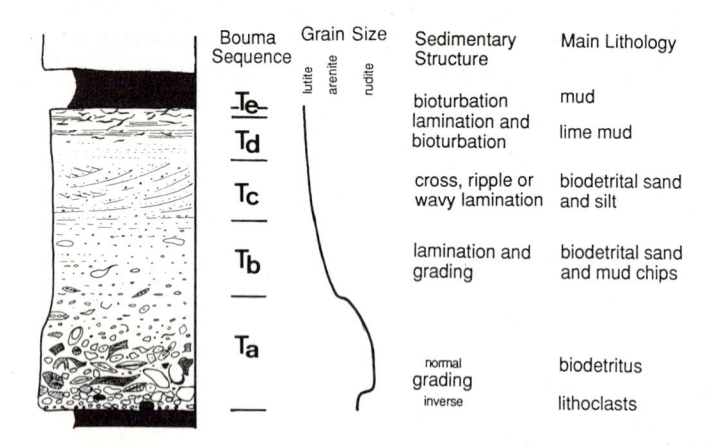

Fig. 1. An ideal calcareous turbidite with grain size range from rudite to lutite, displaying a complete Bouma sequence and, in addition, characteristic features that result from the calcareous composition. These features are poor sorting, inverse grading as a result of different effective density between lithoclasts and biodetritus, and a thin interval T_e of the Bouma sequence

Fig. 2. Calcareous turbidite containing gravel-sized, rounded clasts at its base, grading upward into cross-bedded medium sand and laminated fine-grained limestone. Bioturbation is restricted to the top of the turbidite. Late Albian, Northeast Providence Channel, Bahamas, ODP Leg 101, Site 635B

carpet, a mechanism that is not possible in depositions from low-density currents that produce the Bouma sequence (Lowe 1982). The presence of bimodality can be used as a diagnostic feature in sections where a distinction between fine-grained, uniform turbidites and pelagic beds is difficult. In such series, a thin layer of lithoclasts, such as volcanogenic grains or dolomitic silt, at the base of the beds is often the only indication for turbidites or its base (Tucker 1969; Cook et al. 1976; Davies 1977).

3. Bouma sequence. Using a siliciclastic turbidite with a complete Bouma sequence as a point of reference (see also proposed new nomenclature by Einsele Chap. 2.7, this Vol.), calcareous turbidites often display the following characteristics: (1) a basal inverse grading due to the density differences of the grains, (2) imbrication within the Bouma interval T_a, which is most likely the result of the elongated shape of many clasts (e.g., shell fragments), (3) cross-bedding and convolute lamination are rarely found in ruditic to arenitic calcareous turbidites (Eberli 1987; Price 1977), while they are common in finer-grained, thin-bedded turbidites (Scholle 1971; Hesse 1975), (4) coarse-grained calcareous turbidites often lack the Te interval of the Bouma sequence, possibly because carbonate mud does not develop surface electrostatic charges which would lead to flocculation, as with terrigenous clay minerals. Hence, the muddy part of calcareous turbidites could be easily redispersed into the water column and resettle slowly as background sediment. A second possibility for the lack of the Te interval could be due to flow separation, in which the fine-grained tail of a turbidity current flows along a mid-water pycnocline and subsequently forms a turbidite composed entirely of carbonate lutite (Heath and Mullins 1984). These latter deposits might be difficult to distinguish from background pelagic deposits (Stow et al. 1984b).

4. Bottom marks. Bottom marks are scarce in redeposited carbonates (Meischner 1964; Tucker 1969; Scholle 1971; Eberli 1987). The lack of cohesion in calcareous mud inhibits the preservation of flute and load casts. In addition, diagenesis often welds the turbidite to the underlying limestone bed, obliterating the bottom marks.

2.3 Diagenetic Overprint

Calcareous turbidites, with redeposited shallow-water skeletal debris, have a high amount of metastable carbonates, such as aragonite and high-Mg calcite. Thus, they have a high diagenetic potential and lithify faster than the interbedded calcitic pelagic ooze (Dix and Mullins 1988; Eberli 1988a). The most characteristic diagenetic feature of calcareous turbidites is secondary silicification, which is probably penecontemporaneous with lithification. The silicification does not only occur in layers where opaline silica is primarily enriched according to its hydraulic equivalence during settling but, in addition, proceeds along sedimentary structures with high porosity, where fluid movement allows a diagenetic enrichment of silica. For example, the coarse, graded base and the parallel-laminated top of a calcareous turbidite are often silicified; in such beds, the silicification in the graded base may be the result of diagenetic enrichment. In other turbidites, black chert nodules occur dispersed throughout the bed. In most cases, the source of the silica is siliceous skeletal debris, from which silica is remobilized during diagenesis. If present, secondary silicification along faint laminations helps to identify fine-grained turbidites in pelagic sections.

In coarse-grained turbidites, pressure solution is often manifested by a stylolitic rim around the components, indicating preferential dissolution along the clast boundaries. In extreme cases, redistribution of carbonate by solution/precipitation processes can produce a rim of limestone at the base of the turbidite, mimicking a basal layer (Meischner 1964; Eder 1982; see Ricken and Eder Chap. 3.1, this Vol.).

In deep pelagic realms, the carbonate accumulation rate is below the rate of dissolution, and calcareous oozes are usually not preserved below the calcite compensation depth (CCD). The high sedimentation rate by turbidity currents, however, can inhibit this dissolution process and thus preserve calcareous material or even metastable carbonate phases below the CCD. Examples are reported from the Puerto Rico and the Palau Trenches (Ericson et al. 1952; Yamamoto et al. 1988), the Nauru Basin (Larson et al. 1981), the Blake Bahama Basin (Flood 1978) and from the abyssal plain separating the Madingley Rise from the Seychelles-Saya de Malha banks (Backman et al. 1988). Calcareous turbidites deposited below the CCD are characteristically interbedded with clays or siliceous oozes, and consist of both sand-sized, calcareous turbidites and fine-grained, redeposited carbonate oozes. In many of the fine-grained turbidites of the Nauru Basin, nannofossils are preserved well enough for determination, indicating that dissolution after deposition is minor.

2.4 Variability

Turbidites displaying a classical Bouma sequence (Figs. 1 and 2) form only when the turbidity current has a certain density and the components are of proper grain size distribution (Middleton and Hampton 1976; Lowe 1982). Any variation in the density or grain size distribution results in deviations from the classical sequence (Fig. 3). If the density of the suspension is increased, the components are not only suspended by the turbulence but are hindered from settling, due to bouyant lift provided by the interstitial mixture of water and finer-grained sediment (Lowe 1975). Typical sedimentary structures produced by such coarse-grained, high-density turbidity currents include a basal unit with inverse grading, and/or coarse-tail grading and/or imbrication overlain by a cross-laminated ruditic-arenitic layer (Fig. 4). Similar sedimentary structures are also characteristic of siliciclastic high-density turbidites, suggesting that in high-density turbidity currents the influence of the compositional difference decreases. Only water-escape structures, which are common in siliciclastic counterparts (Lowe 1982), are rare in calcareous turbidites. Beds with poor grading may have been deposited by a mechanism transitional between a high-density

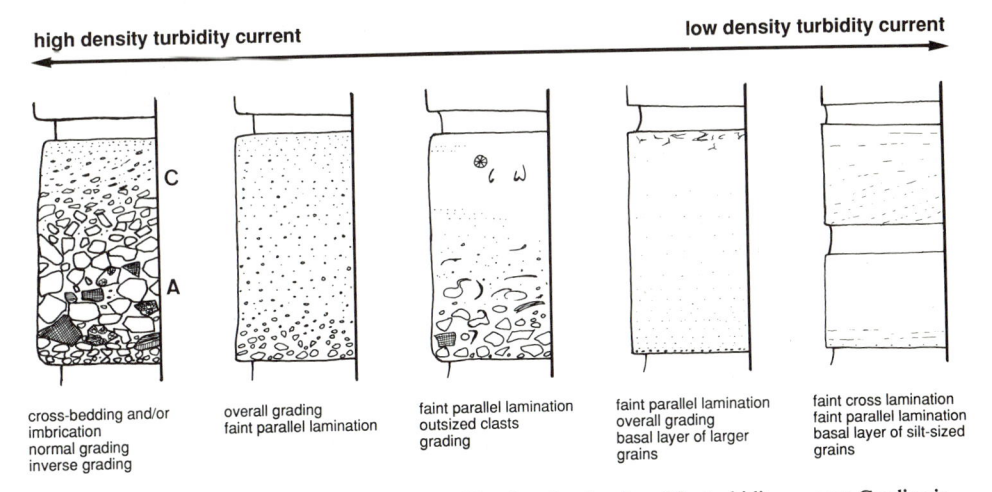

Fig. 3. Variability of calcareous turbidites caused by changing density of the turbidity current. Grading is ubiquitous in all calcareous turbidites except in lime-mud turbidites. The Bouma sequence is never complete and not readily applicable in turbidites, which are not deposited from currents with the proper density. Characteristic sedimentary structures are listed under the respective beds. Note the difficulty of recognizing fine-grained turbidites. For further explanation see text

Fig. 4. High-density calcareous turbidite displaying most of the characteristic sedimentary structures that develop in deposits from high-density turbidity currents. Lithoclasts include both angular older platform deposits (Triassic) and rounded penecontemporaneous basinal marlstone. Taugelboden Formation, Kimmeridgian, Promektal, Austria. (Photo courtesy of Adam Vecsei)

turbidity current and a debris flow with a clast supporting matrix (Crevello and Schlager 1980).

On the other side of the spectrum are fine-grained turbidites deposited from low-density turbidity currents (Fig. 3). They may form as the end member of a long-traveled turbidity current or as a deposit from a large muddy turbidity current generated by slumping of unconsolidated pelagic sediments (Scholle 1971; Hesse 1975; see Einsele Chap. 2.7, this Vol.). During deposition from such currents, only a few faint sedimentary structures are produced, which makes it difficult to distinguish them from pelagic deposits (Hesse 1975; Stow et al. 1984b). The beds are mostly homogeneous or at most faintly laminated. Besides this faint lamination, the lack of bioturbation, an increased organic content in the turbidites, plus noncarbonate grains or a layer of larger foraminifera at the base of the bed, can possibly help to identify such fine-grained calcareous turbidites (Scholle 1971; Hesse 1975; Cook et al. 1976; Crevello et al. 1984; see Piper and Stow Chap.9, this Vol.).

2.5 Associated Mass-Flow Deposits

Turbidite sequences are always associated with other types of mass gravity flow deposits. These associated facies are important because their position within the sequence is indicative of basin morphology and depositional environment (Mutti and Ricci Lucchi 1975; Walker 1978; Mullins and Cook 1986). In calcareous turbidite sequences, the most characteristic facies are sheet-like debris flow deposits; a fact possibly related to carbonate slope morphology. Due to early diagenesis and biological stabilization, carbonate slopes have a higher angle of repose than siliciclastic slopes (Schlager and Camber 1986). These steep slopes are bypassed by turbidites but, nevertheless, have a high sedimentation rate of periplatform ooze; for example, during the Pliocene the accumulation rate is as high as 105 m/ma on the "bypass" slope in Exuma Sound, Bahamas (Austin et al. 1986). Such slopes are prone to oversteepening and, thus, to large-scale slumping and major episodic collapse events (Crevello and Schlager 1980; Mullins et al. 1986). The trigger mechanism is not always unequivocal. In tectonically active regions major earthquakes are the most likely trigger mechanism, while in tectonically quiet areas, overloading of the slope might trigger such large gravity flows. Thick Middle Miocene debris flow units in both the Blake Bahama basin and the Straits of Florida give evidence of tectonic activity on the apparently tectonically quiet margin of eastern North America (Fulthorpe and Melillo 1988).

In the rock record, these collapse events are seen as intraformational truncation surfaces in fine-grained upper slope sequences and as megabreccia beds within the basinal (turbidite) sequences. Such megabreccias are usually spectacular beds with little internal organisation and very coarse (several meters thick) clasts in a lime mud matrix (Fig. 5). Matrix-rich beds are interpreted as deposits from debris flows and

Fig. 5. Megabreccia at the base of a turbidite sequence. The chaotic, approximately 50-m-thick megabreccia is composed of boulders of older shallow-water carbonates, which are eroded along a fault scarp. The megabreccia indicates the beginning of fault activity and is the base of an overall thinning and fining-upward basin fill. Allgäu Formation, Lower Jurassic, Eastern Alps, Switzerland. (Eberli 1987)

are characteristicly widespread, sometimes covering the entire basin floor (Cook et al. 1972; Crevello and Schlager 1980). They can provide distinct marker beds with chronostratigraphic significance (Bernoulli et al. 1981; Labaume et al. 1983), and in tectonically active areas they indicate the beginning of episodes of major tectonic activity (Fig. 5; Eberli 1988).

3 Provenance

In their component assemblage, calcareous turbidites contain information about bathymetry, the geologic setting, and the faunal assemblage in the source area. In addition, along erosional escarpments, the assemblage helps to estimate the amount of erosional down and back-cutting and the onset of renewed fault activity. For example, lithoclasts in rift basin turbidites give direct evidence of the amount of exhumation of older strata along the basin-bounding faults. Component assemblages of calcareous turbidites on the slopes surrounding the Bahamas reflect the depositional setting, for example, between the open-ocean setting north of Little Bahama Bank and the intra-platform setting in Exuma Sound (Kuhn and Meischner 1988). In Exuma Sound, Crevello and Schlager (1980) were able to distinguish between turbidites which were shed from the platform margin, and an underlying debris sheet which originated from the upper slope and was probably triggered by another mechanism.

Faunal assemblages in redeposited carbonates carry a wide variety of information. They can give the age range of the eroded underlying strata (Premoli Silva and Brusa 1981), as well as information on the paleoecology of the source area (Taylor 1976). Cook and Taylor (1977) used trilobite faunal resemblance data from redeposited and in situ slope sediments for plate tectonic reconstructions. In addition, faunal changes in the neritic environment are precisely recorded in calcareous turbidites. Throughout the Phanerozoic, the community of reef-building organisms changed several times (Heckel 1974), each time causing a distinct change in the composition of the basinal turbidites. For example, in the Late Cretaceous when rudists became the major contributors to the reefal buildups, they also became the dominant to nearly exclusive skeletal fragments in the basinal carbonate detritus (Enos 1977). In the Eocene, when nummulites dominated reef margins and fore-reef shoals, nearly pure nummulite turbidites were deposited basinward. The change in faunal composition, sometimes occurring abruptly from one bed to another, is easily detected in the field and indicates a faunal change in the neritic environment.

4 Turbidite Frequency and Sea-Level Variations

The major difference between siliciclastic shelves and carbonate platforms is that carbonate platforms produce sediment in the form of inorganic precipitates and organic skeletal debris. Carbonate production is highest when the platform tops are flooded. During these times, more sediment is commonly produced than can be accommodated on the flat platform top, and the excess sediment is exported into adjacent deep water areas (Supko 1963; Kier and Pilkey 1971; Lynts et al. 1973; Hine et al. 1981b). This "highstand shedding" puts the carbonate environment 180° out of phase with siliciclastic systems, where sediment is stored on the inner shelf during sea-level highstands and exported into basinal areas during low sea level (Droxler et al. 1983; Boardman and Neumann 1984). This different response to sea-level variations is important, because it implies that the two environments create different stratal patterns, which in sequence stratigraphy are used to monitor sea-level fluctuations (Vail et al. 1977a; Vail 1987).

The notion of "highstand shedding" proposed by sedimentologists working in modern carbonate environments has caused some controversy (Mullins 1983). Data sets capable of resolving this controversy are scarce. Within the modern Bahaman environment, however, Droxler and Schlager (1985) and Reymer et al. (1988) were able to demonstrate that turbidites are more frequent during interglacial times when sea level is high and the bank tops are flooded (Fig. 6). A second data set comes from Deep Sea Drilling Project sites along the Line Islands and Marshall Islands atoll provinces in the Pacific (Thiede 1981; Shanmugam and Moiola 1982, 1984; Schlanger and Premoli Silva 1986). There, interbedded within the pelagic sequences, are turbidites with reworked and displaced fossils derived from neritic shallow-water environments (Premoli Silva and Brusa 1981; Premoli Silva 1986). Their occurrence within intervals deposited during times of proposed low eustatic sea level suggests that the shallow-water carbonate debris was likewise shed during sea-level lowstands (Thiede 1981; Schlanger and Premoli Silva 1986). The neritic fossils are, however, never more than 5% of the component assemblage (Thiede 1981). A re-examination of the original core description from Site 462 revealed that most of the turbidites with larger benthic foraminifers are in reality volcaniclastic turbidites, and that clasts from

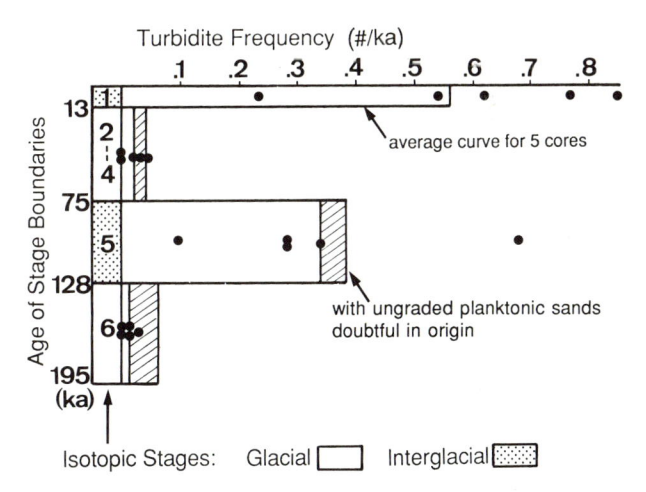

Fig. 6. Turbidite frequency versus glacial/interglacial periods in five cores from Cul de Sac of the Tongue of the Ocean, Bahamas. High turbidite frequency during interglacial times indicates that platforms in the Bahamas shed more turbidites during sea-level highstands. Note that some material is also shed during glacial times, i.e., sea-level lowstands. (Droxler and Schlager 1985)

the neritic and reef environments are "components of the larger-size fraction of the volcaniclastic turbiditic sandstones" (Larson et al. 1981, p. 57). It is therefore not surprising that these mixed turbidites with little carbonate debris respond similarly to sea-level fluctuations as pure siliciclastic turbidites.

Within calcareous turbidites, not all components necessarily follow the same depositional trends. Newly produced sediment is tied to sea-level highstands, whereas lithoclasts are mainly eroded during lowstands. When carbonate and siliciclastic systems are coupled, additional parameters have to be taken into account for evaluating the response to changing sea level. Scenarios for different settings are illustrated in Fig. 7. Two end members, the pure carbonate and the pure siliciclastic environments, and a mixed setting, with a carbonate rim either along a volcanic island or along a continental shelf, are shown. In a pure carbonate environment, newly produced, excess sediment is transported off-bank contemporaneously with production, i.e., during high sea level, resulting in a high turbidity frequency (Fig. 7). When sea level drops, the platform top becomes exposed and carbonate production ceases, while along the margin reef growth might still continue. Erosion and redeposition of older platform deposits and marginal boundstone provide material mainly for talus breccias, whereas the amount of calcareous sand for basinal turbidites decreases significantly (Harris 1988).

In a pure siliciclastic environment during sea-level highstands, sediment is trapped on the inner shelf and turbidite frequency is low. As sea level drops, sediment is brought to the shelf edge, from where turbidity currents carry it to the continental rise and onto the abyssal plain (Fig. 7). In mixed environments, such as atolls and carbonate shelves, the effect of sea-level changes can be either like the carbonate or siliciclastic response, depending on the relative size and contribution of each environment. In the mid-Pacific atoll provinces, such mixed settings are formed by carbonate rims surrounding volcanic islands. When sea level is high, the platform rim is flooded

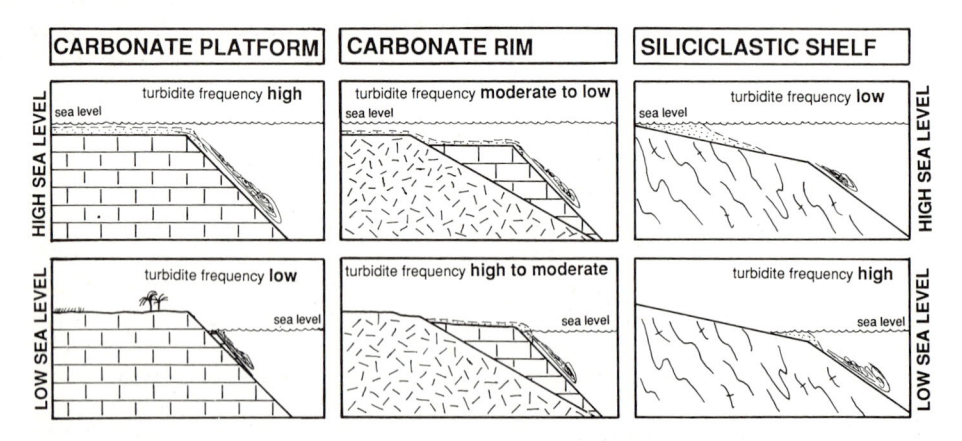

Fig. 7. Turbidite frequency versus relative position of sea level in different settings. There are two end-members, carbonate platform and siliciclastic shelf, which are 180° out of phase in their response to sea-level position, and a mixed environment where the turbidite frequency depends on both sea-level position and respective size and contribution of each environment. For discussion see text

and produces calcareous sediment, while volcaniclastic sand is being stored landward. From the productive rim, newly produced carbonate sediment is transported off-bank, but the limited areal extent of the rim results in a moderate to low turbidite frequency. During low sea level, when the base level is below the rim, volcaniclastic sand bypasses the carbonate rim and is exported into the basinal areas. By passing through the platform rim, volcaniclastic sand erodes older platform material, which is incorporated into the component assemblage (Fig. 7). A narrow rim might have a low storage capacity, so that turbidite frequency during low sea level is high, similar to conditions along pure siliciclastic shelves.

In a carbonate ramp setting with a siliciclastic hinterland, the width and profile of the ramp and the position of the shelf margin relative to sea level are important in determining turbidite frequency in connection with sea-level fluctuations. For example, a distally steepened ramp may indeed overload its shelf margin during sea-level lowstands, when the combination of in situ produced carbonates and siliciclastic sediments transported to the shelf edge causes a high accumulation rate. In such a setting, the frequency of mass gravity flows consisting of mixed carbonate-siliciclastic material is high during sea-level lowstands (Yose and Heller 1989). The basin profile can similarly influence the response to sea-level fluctuations in a pure carbonate environment. A submarine plateau might come into the euphotic zone during a sea-level fall, which results in increased carbonate production and increased turbidite frequency compared to the preceding highstand. Considering all the factors controlling production and accumulation rates in both the carbonate and the siliciclastic environments, the scenarios proposed in Fig. 7 are idealized situations, which in reality are usually modified by the factors discussed above.

5 Turbidite Sequences and Cyclicity

5.1 Sea-Level Fluctuations

Vertical successions of turbidites often display a trend of either increasing or decreasing bed thickness, that usually is coupled with a similar trend in grain size. Such thinning and fining-upward or thickening and coarsening-upward sequences appear on all scales, from a few meters to hundreds or thousands of meters. This vertical trend, in combination with the included facies types, was used to interpret the depositional environment (Mutti and Ricci Lucchi 1975; Mullins and Cook 1986). For example, a thinning and fining-upward sequence was considered as an indication for channel fill (Mutti and Ricci Lucchi 1975) or deposition of a migrating lobe (Walker 1978; see Einsele Chap. 2.7, this Vol.). The thinning-upward trend simply indicates a steady decrease in the volume of the gravity flow for the location studied. This trend can, of course, be achieved during a channel fill or when the channel supplying the sediment migrates laterally away from the investigated locality. But it is also possible that the volume change is controlled by a change in the amount of sediment supply from the source area.

In carbonate turbidite systems, the fan geometry with one main feeder channel, mid- and outer fan, and a characteristic cyclicity in the vertical turbidite succession, as described for many prograding siliciclastic turbidite systems, is very rare. In the carbonate environment, sediment for redeposition is provided along the entire margin. From this line source, it is transported downslope through multiple gullies and deposited in small coalescing fans forming an apron at the slope/basin transition (Schlager and Chermak 1979, Mullins and Cook 1986). In these aprons, channel-levee complexes and migrating lobes are rare and not the likely cause for fining-upward trends. This indicates that in carbonate turbidites, cyclicity in the vertical succession is caused to a lesser extent by migrating depocenters than by variable input from the source area. The volume of the mass gravity flows is a function of relative sea-level position, because each change in sea level translates into a dislocation of the equilibrium point between shelf and basinal sediments, which determines the volume of turbidite and other mass gravity flow sedimentation. Thus, an external cycle of relative sea level change could create vertical stacking patterns in calcareous turbidites. As a result of the superposition of high frequency and low frequency cycles, fluctuations in sea level vary in amplitude and frequency. Medium to small-scale sequences, as observed in outcrop, could easily be produced by small-scale sea-level fluctuations.

Scale is very important in evaluating the cause of trends in the vertical stacking pattern (Fig. 8). A basin filled with turbidites can be divided into several depositional sequences, in the sense of Vail et al. (Chap. 6.1, this Vol.; Fig. 8a). A depositional sequence can be several hundreds of meters thick and, thus, be seismically resolvable. Each of these sequences comprises system tracts, which themselves comprise several depositional units. For example, a lowstand systems tract might have on its base a sheet-like basin fill that is overlain by a base-of-slope apron. Both of these depositional units are characterized by smaller, distinct elements such as coalescing sand lobes (Mutti and Normark 1987; Vail 1978). This hierarchy within a sequence is controlled by the different orders of magnitude of sea-level fluctuation and sediment supply, each one creating distinct sedimentary packages which are not necessarily related to each other, neither geometrically nor genetically (Mutti 1985). This vertical stacking of genetically unrelated sediments also implies that Walther's Law, which postulates a similar horizontal facies distribution as observed in the vertical succession, does not always apply. For example, a thinning and fining-upward sequence of several meters overlain by a coarse talus deposit (Fig. 8b) does not mean that there also exists a lateral transition from talus to turbidite, because the two facies developed at different times; the turbidites are shed during a highstand, whereas the talus breccia is mainly deposited during the following lowstand (Harris 1988).

Carbonate turbidite systems appear to contain less cyclicity than siliciclastic sequences. One explanation for this difference is that the line source for most carbonate mass flows results in an apron-like distribution of redeposited beds and, thus, in the absence of sequences caused by channel filling or lobe migration, as on siliciclastic fans. Basin relief also plays a role in the development of stacking patterns. Low-angle slope aprons display hardly any cyclicity, whereas base-of-slope aprons along steeper slopes show more cyclicity (Mullins and Cook 1986). Smaller-scale cycles, as displayed on Fig. 8b, with thicknesses of several meters, might be caused by either autocyclic depositional variations or small-scale sea level fluctuations. To date, it has not been possible to discriminate between the two causes of these smaller cycles within calcareous turbidites, but the expression of some long-term evolutions

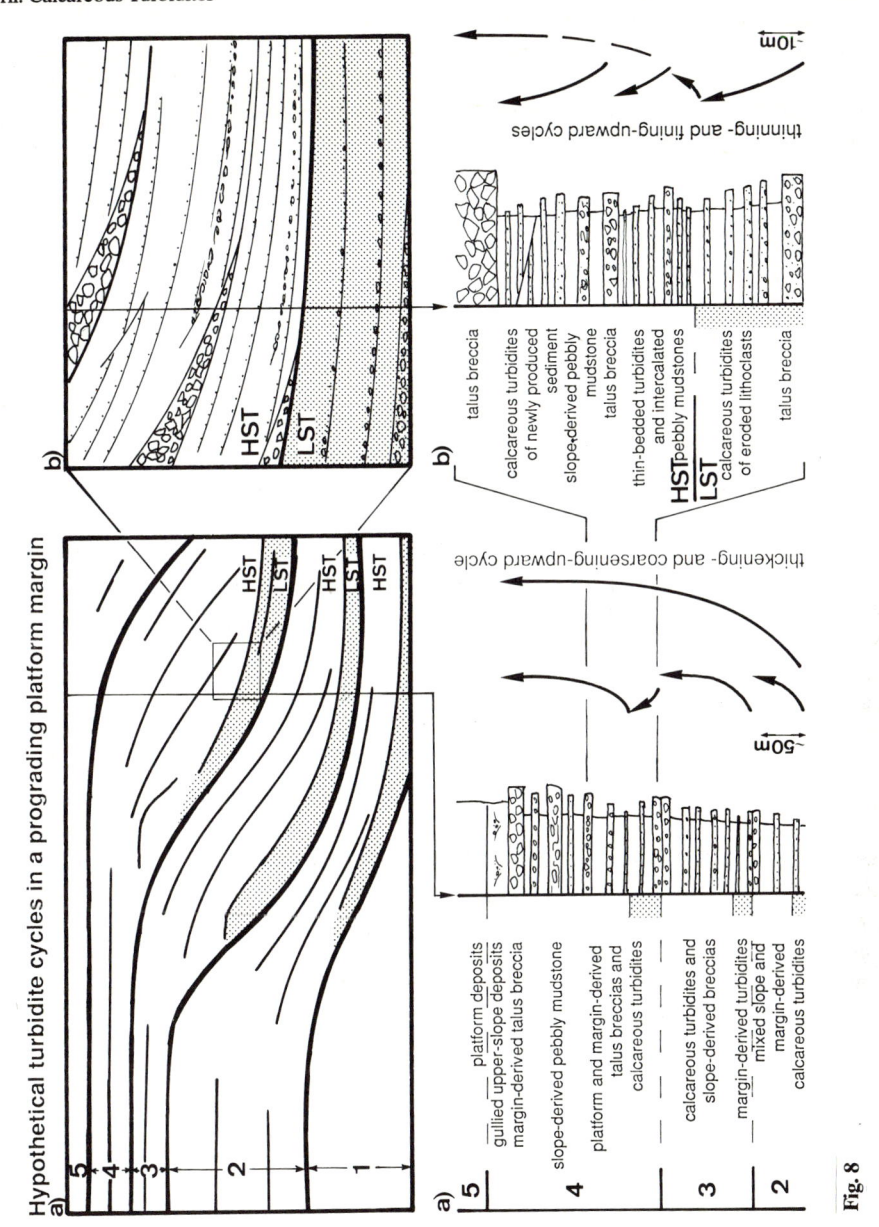

Fig. 8a,b. Cycles in the vertical stacking pattern of calcareous turbidites. **a** A prograding platform margin contains several individual depositional sequences (*1–5*) with different systems tracts (*LST* lowstand systems tract; *HST* highstand systems tract). The overall trend is a thickening and coarsening-upward cycle that is composed of smaller coarsening-upward cycles within the sequences *1–5*. These smaller cycles also show some thinning and fining-upward trends. **b**~ Blow-up of the toe-of-slope of sequence *4*, where a thin lowstand systems tract is overlain by a prograding highstand systems tract. The cycles of this scale do not coincide with the overall trend; the increments of progradation display thinning and fining- upward cycles. Note also the different composition of highstand versus lowstand turbidites. Figure based on following information: Platform progradation geometry (Eberli and Ginsburg 1989), overall vertical trend (Bosselini and Rossi 1974; Mullins and Cook 1986), toe-of-slope blow-up (Harris 1988)

is known. A general thickening and coarsening-upward trend during long-term platform progradation is documented in several studies (Bosselini and Rossi 1974; Cook and Egbert 1981). Long-term progradation is achieved by stacking several prograding sequences (Eberli and Ginsburg 1989); it can be speculated that each sequence might also display a coarsening-upward trend, but, within the individual systems, a variety of stacking patterns can be expected (Fig. 8).

5.2 Tectonic Activity

In basins with shelf sediment supply, turbidite sedimentation is largely controlled by relative sea level, whereby tectonic activity can either enhance or diminish the influence of global sea-level variations (Mutti 1985). In basins, however, with little or no land-derived sediment supply, turbidite sequences potentially can directly record tectonic basin evolution. Such basins, isolated from shelf sediment supply, exist in deep marine areas, e.g., in distal rift and pull-apart basins. In such basins, turbidite sequences are a powerful tool for the reconstruction of basin evolution. In the following example, an application of turbidite sequence analysis is illustrated from the rifted margin of the Jurassic Tethys.

During the Late Triassic, shallow-water conditions were established over a large portion of the Mediterranean-Alpine realm (Bernoulli and Lemoine 1980; Haas Chap. 6.6, this Vol.). In the Early Jurassic as Gondwana broke up, the huge shallow-water platform split and block faulting created a series of asymmetric rift basins (Fig. 9; Winterer and Bosellini 1981; Eberli 1988b). To the east on some high areas, shallow-water conditions were maintained throughout the Jurassic, whereas in the west a topography of submarine highs and basins formed. Basin formation was not coeval along the entire margin but propagated from east to west as rifting proceeded (Eberli 1988b). This rift propagation can be documented with a sedimentological and

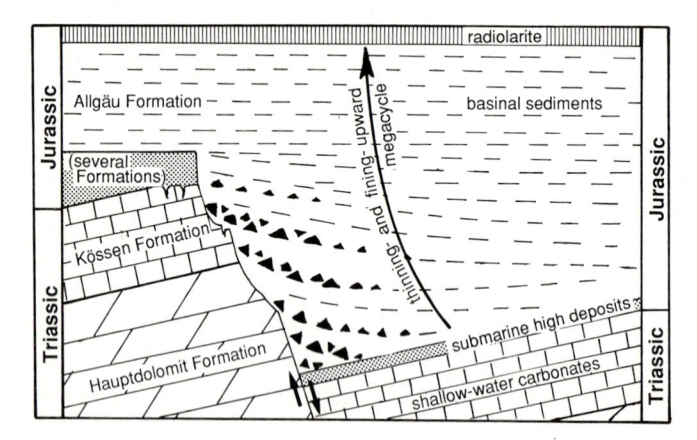

Fig. 9. Schematic diagram of an asymmetric Jurassic rift basin that disintegrated the Upper Triassic carbonate platform in the Tethyan realm and created submarine highs and basins. On the submarine highs, condensed sequences of crinoidal and nodular limestones intercalated with ferro-manganese crusts were deposited. In the adjacent basins carbonate turbidite sequences accumulated that display a thinning and fining-upward trend due to the decrease in sediment supply as the relief became buried

stratigraphic analysis of the basinal turbidite sequences. During rift propagation, each rifting event created a new basin or deepened an existing one and, thus, increased slope instability. As the newly created relief became buried, a characteristic thinning and fining-upward sequence accumulated in the basin (Fig. 10; Eberli 1987). At the base of the prism, huge megabreccias with components up to 20 m in diameter intercalate with finer-grained, calcareous turbidites and limestones (Fig. 10a). This facies association is overlain by conglomerates and thick-bedded turbidites that display smaller-scale thinning and fining-upward sequences of some tens of meters. Turbidites become thinner and finer-grained upsection and, in addition, change composition. The basal parts of the sequence consist mainly of eroded lithoclasts from the underlying Upper Triassic carbonates, whereas the component assemblage

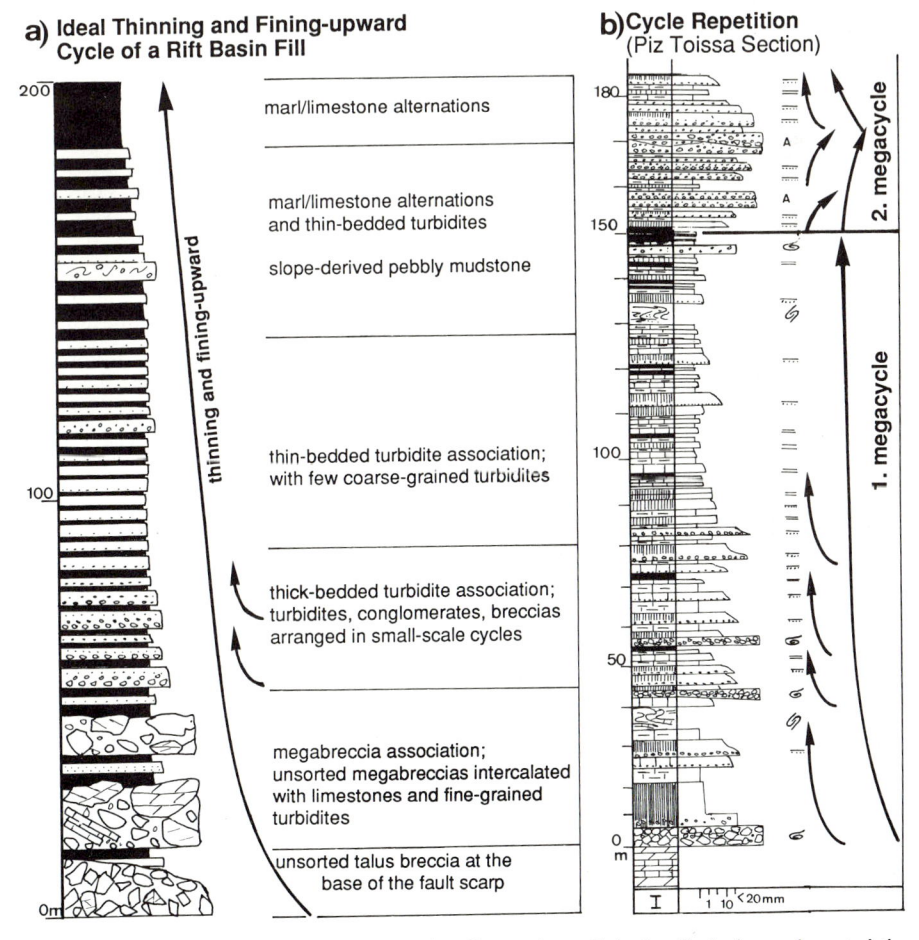

Fig. 10a,b. **a** Idealized megacycle of redeposited sediments in a rift basin, displaying a characteristic thinning and fining-upward trend. **b** Measured section at Piz Toissa, Ela nappe, Switzerland. An incomplete thinning and fining-upward megacycle is overlain by another megacycle. This repetition is indicative of renewed fault activity in the rift basin. *Arrows* indicate fining-upward or coarsening-upward cycles. Location of Piz Toissa (To) is given on Fig. 11

of the thin-bedded turbidites is dominated by skeletal debris of crinoids and sponges, which obviously inhabited the submarine highs. Further upsection, more and more pelagic background sediment is interbedded between the calcareous turbidites, indicating that the basin relief decreased and turbidity currents were less frequent. Finally, the turbidite sequence is capped by marls and limestones (Fig. 10a). In basins with one extensional phase, one thinning and fining-upward megasequence is found. When tectonic activity is renewed, the cycle is repeated. The re-occurrence of megabreccias followed by thick-bedded turbidites records the renewed tectonic activity (Fig. 10b).

In comparing the onset of the turbidite sequences in the different basins, a "time table" of the rifting events along the margin can be established. In the eastern Alps of Switzerland (Fig. 11), ammonite stratigraphy was used to date the onset of the turbidite sequences in the different tectonic units. This stratigraphic analysis revealed a progressive fragmentation from the proximal (east) to the distal (west) part of the margin (Fig. 12), and documented a segmentation pattern that is inconsistent with a simple, symmetric extension of the margin, requiring a mechanism other than linear stretching, i.e., low-angle detachment faults (Eberli 1988b). Because turbidite sequences document lateral fault propagation, it is speculated that, in a basin controlled by growth faulting, repeated vertical fault movement would be recorded in a similar way. Each movement on the growth fault results in an immediate increase in relief that is subsequently buried by a thinning and fining-upward turbidite sequence.

Basins affected by compressional tectonism are usually connected with a shelf area, and the turbidite sequences less clearly reflect tectonic activity. However, thick calcareous turbidites can punctuate these turbidite sequences. For example, in an Eocene flysch in the Spanish Pyrenees, thick mega-beds about 40 m, consisting of a calcarenite/marl couplet, occur at several levels within a 3500–4500 m thick flysch

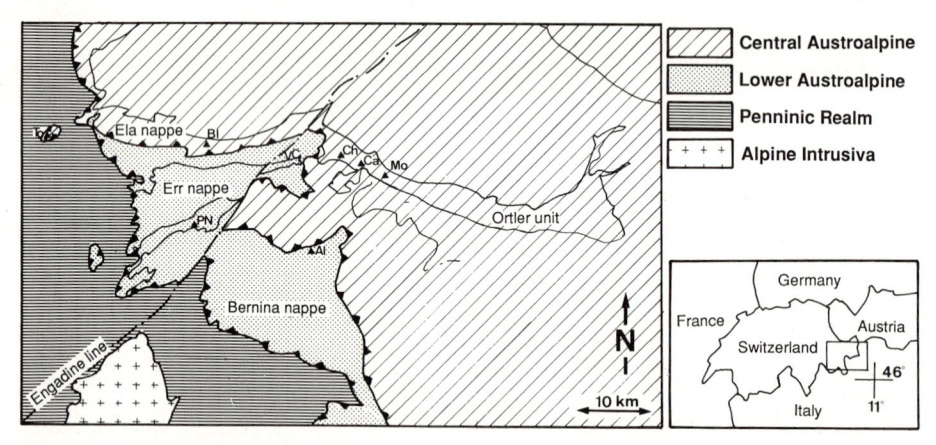

Fig. 11. Tectonic sketch map of part of the Eastern Alps in Graubünden, Switzerland. The Penninic realm (*horizontally shaded*) consists of nappes with oceanic crust and these nappes are considered as relics of the Tethyan Piemont-Ligurian ocean. The Lower and Central Austroalpine tectonic complexes represent paleogeographically the southern margin of the Tethys. The rift sediments were investigated in the labeled tectonic units. Also shown are locations of profiles shown in Fig. 9 and 12; *To* Piz Toissa; *Bl* Piz Blaisun; *Al* Piz Alv; *PN* Piz Nair; *VC* Val Chamuera; *Ch* Chauschaunagrat; *Ca* Casanna; *Mo* Il Motto

Eberli: Calcareous Turbidites

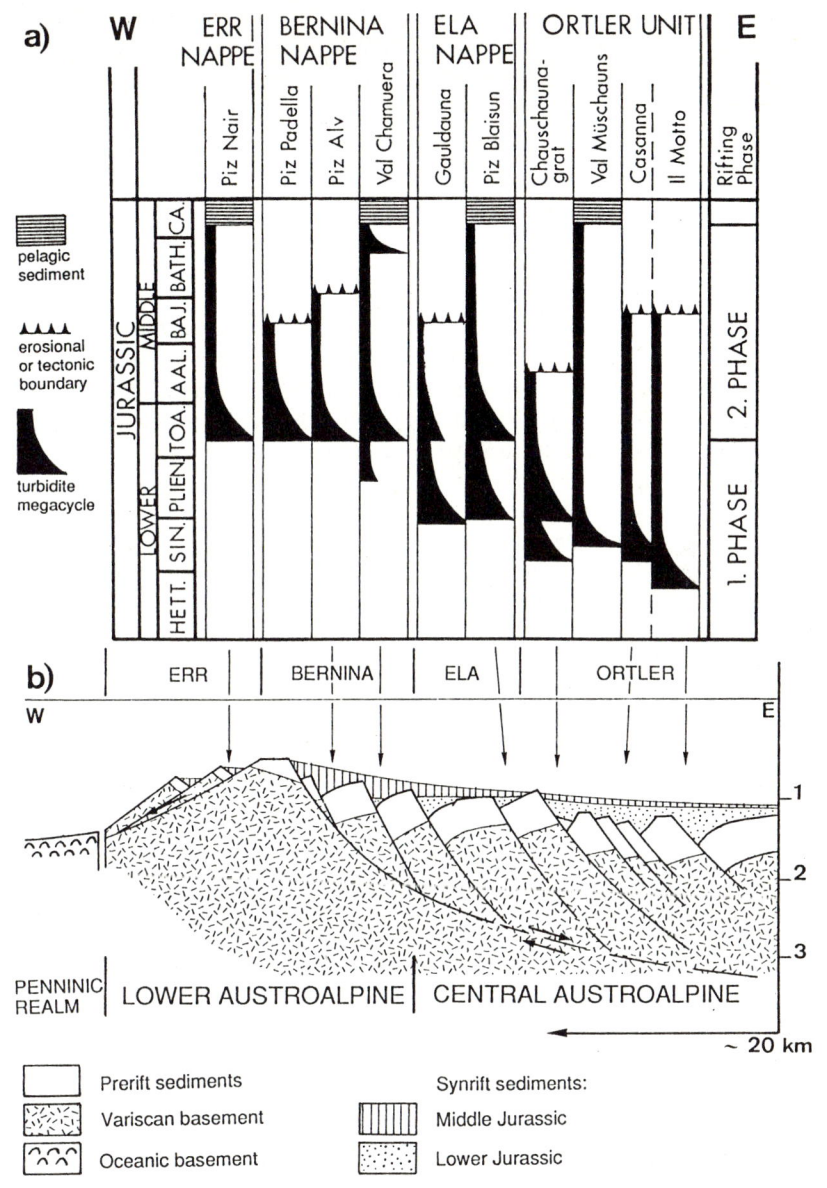

Fig. 12a,b. **a** Onset age of thinning and fining-upward turbidite cycles in rift basins of the Eastern Alps of Switzerland based on ammonite stratigraphy. The oldest turbidite sequence is found in the east (Ortler unit). Note that during 1st phase of rifting the cycles become progressively younger in a westward direction and that the 2nd phase of rifting (Toarcian) mainly affected younger in westward direction and that the 2nd phase of mainly affected the western part of the margin. This stratigraphic succession of rift basin fills indicates an east to west propagation of the rift faults. (After Eberli 1988b). **b** Cross-section of the continental margin in the Eastern Alps of Switzerland based on the stratigraphic succession of the calcareous turbidite sequences and structural analysis of fault orientation (Froitzheim 1988). The progressive fragmentation of the margin from proximal (east) to distal (west) probably is a result of the westward propagation of an underlying, eastward-dipping, low-angle detachment fault, while a younger, westward-dipping, low-angle fault is responsible for extension during the 2nd phase of rifting

succession (Rupke 1976). These very thick turbidites do not occupy a special niche in the facies sequence but rather record unusual seismic episodes in the flysch basin (Rupke 1976). The punctuation of flysch by mega-turbidites was explained in several studies by increased seismicity caused by tectonic basin evolution (Bernoulli et al. 1981; Labaume et al. 1983). The mega-beds represent catastrophic events and are good marker beds with chronostratigraphic significance.

6 Concluding Remarks

Calcareous turbidites are, like their siliciclastic counterparts, records of short-lived events of sediment redistribution. Because the carbonate environment is productive, the component assemblage of calcareous turbidites contains both newly produced sediment and redeposited lithoclasts. This composition does not significantly influence sedimentary structures; grading, a partial or complete Bouma sequence, and lateral continuity remain the principal criteria for the recognition of the turbidity current origin (Figs. 1 and 3). Nevertheless, some special characteristics develop. Among these are: less pronounced Bouma sequences; a bimodal distribution between lithoclasts and biodetritus, with the lithoclasts preferentially at the base of the bed; poor sorting; scarce bottom marks; and a high amount of silicification. Differences between siliciclastic and calcareous turbidites are also seen in the depositional geometry. Calcareous turbidites are deposited in apron-like bodies and not in fans, as are most siliciclastic turbidites (Mullins and Cook 1986).

Calcareous turbidites contain two types of clasts, lithoclasts and newly produced sediment and, thus, carry in their component assemblage a wide range of information about the bathymetry and ecology of the source area. The assemblage also records faunal changes in the shelf area and helps estimate erosional down and back-cutting. The fact that calcareous turbidites are in the simplest case a two-component system (lithoclasts and biodetritus) must be taken into consideration when evaluating the relationship between turbidite frequency and sea-level fluctuations (Fig. 7). Droxler and Schlager (1985) demonstrated that newly produced sediment is exported into deep-water areas mainly during sea-level highstands, when production on the platform is high. Harris (1988) proposed that marginal erosion during low sea level increases the rate of talus deposition, although the frequency of basinal calcareous turbidites decreases. The system becomes further complicated when the carbonate and siliciclastic environments interfinger. A sea level/frequency plot of turbidites from a carbonate platform rim with a siliciclastic hinterland probably gives a mixed signal, with one peak at highstands when carbonate sediment is produced on the rim, and another peak at lowstands when sand bypasses and erodes the carbonate shelf. Therefore, in evaluating the turbidite frequency in relation to sea level, a thorough analysis of the component assemblage is necessary. In addition, the depositional setting, i.e., isolated platform or ramp, and the slope profile have to be determined for such an evaluation.

Turbidite sedimentation is not only caused by sea-level fluctuations but also by tectonic activity; variations in both parameters can produce cyclicity in the stacking pattern. For a prograding platform margin, the progradational stacking pattern is controlled by sea-level fluctuations. In extensional basins, tectonism seems to be the controlling factor during the initial stages of rifting. Steep fault scarps are extremely unstable and tend to release large mass flows. As the relief becomes buried, slope instability decreases, turbidite volume and frequency decrease, and an overall thinning and fining-upward basin-fill sequence is deposited (Evans and Kendall 1977; Price 1977; Eberli 1987). Within these basin-fill megasequences, smaller-scale vertical sequences of both thinning and thickening-upward trends occur; they are either controlled by depositional processes, such as shifting of the depositional center, or by small-scale sea-level fluctuations. Therefore, the application of these small-scale sequences for facies interpretation and long-term evolution is limited.

In tectonically active regions, turbidite sequences have a high potential for monitoring the tectonic basin evolution. Pelagic background sedimentation usually provides the stratigraphy needed for determining the timing of events, while turbidite sequences, with their facies associations, indicate the onset of tectonic activity. In basins with little shelf sediment supply, turbidite frequency and volume are directly related to tectonic activity. In basins connected to a shelf, such as most basins in compressional settings, the distinction between tectonically induced turbidites and those determined by changing sea level is more difficult to establish. There, anomalies in the repetitive stacking pattern of the turbidites, such as megabreccias, potentially indicate episodic tectonic activity.

Acknowledgments. Financial support was provided by the Swiss National Science Foundation (Grant No. 2000 5.091). Discussions with Daniel Benoulli, Emiliano Mutti, Peter Vrolijk, and Adam Vecsei helped to outline the content of the manuscript. I thank André Droxler for reviewing and commenting on an early draft of the manuscript. I am especially thankful to Judith McKenzie for both correcting the English and posing perceptive questions which made me clarify several ideas in the paper. Constructive reviews by Gerhard Einsele and Werner Ricken also improved the manuscript.

2.9 Fine-Grained Turbidites

D. J. W. Piper and D. A. V. Stow

1 Introduction

Fine-grained sediments (terrigenous and bioclastic) of turbidite origin are the most abundant type of deep-water sediment. They form thick sequences on prodelta slopes, deep sea fans, continental rises and abyssal plains. In the ancient geologic record, such sediments are a major component of accretionary wedges and the metamorphic belts of ancient orogens.

When the concept of turbidity currents was first developed, it was thought that mud beds could not be deposited from such currents, unless the mud was transported as aggregates or fecal pellets (Dzulynski et al. 1959). In the late 1960's, the importance of contour-following bottom currents in the modern oceans was recognised and many lithologies now interpreted as fine-grained turbidites were for a time regarded as "contourites" (Bouma and Hollister 1973). Criteria for the recognition of contourites, however, remain a disputed issue (Stow 1979). By the 1970's marine geologists recognised thick mud beds that clearly had a shallower-water source (Rupke and Stanley 1974) and graded fine carbonate beds deposited below the carbonate compensation depth were recognised as of turbidite origin, both in the sea and in ancient rocks (Hesse 1975). Beds with silt laminae, with an upward decrease in lamina thickness and grain size, were recognised as characteristic of mud turbidites (Piper 1972). By the late 1970's, there was widespread appreciation of the volumetric importance of fine-grained turbidites, and distinctive bed sequences were recognised as characteristic of fine-grained turbidites (Piper 1978; Stow and Shanmugam 1980; Stow and Piper 1984a). Fine-grained turbidites were recognised as having a distinctive microstructure (O'Brien et al. 1980). In the last decade, there has been new emphasis on the depositional processes and vertical cycles in fine-grained turbidites.

Because shales are generally extensively weathered in outcrop, most data on fine-grained turbidites is from piston cores of Holocene and late Pleistocene sediments and from DSDP/ODP holes (particularly on the Mississippi and Bengal fans); and to a lesser extent commercial hydrocarbon wells. The best data from ancient rocks is from well-lithified or low grade metamorphic rocks in which mudstones are not preferentially weathered.

This chapter is built on our 1984 paper (Stow and Piper 1984a), in which we provide a more complete bibliography of fine-grained turbidites. In this new review, we deal more fully with the depositional processes, significance of various facies types, and the development of larger-scale cyclicity. Fine-grained turbidites include a wide range of lithologies, just as coarse-grained turbidites are lithologically diverse. For this reason, simple models (such as we present in this review) may be misleading.

2 Deposits of Single Events

2.1 Silt Turbidites

In distal turbidite environments, silt beds (>70% silt-sized particles) are more abundant than sands and commonly occur as thin or medium-bedded turbidites. There is generally a progressive decrease in grain size of sand and silt beds distally in a turbidite system. Silt beds commonly exhibit the same suite of structures as classical sandy turbidites (Fig. 1a). Base-cut-out structural sequences are common (Fig. 2), so that in distal environments, medium and fine-grained silts with fine lamination and common internal load casting may predominate (Piper and Brisco 1975). Thick ungraded massive silts are found less commonly: they appear to be the fine-grained equivalents of AE sand turbidites.

2.2 Mud Turbidites

A mud turbidite may occur overlying a sand or silt bed deposited from the same turbidity current (Fig. 3a), or may occur independantly (Fig. 3d). Overlying any sand or silt bed are three divisions (Piper 1978): mud with silt laminae (which become thinner, finer and less frequent upwards), graded mud and ungraded mud. The latter is generally bioturbated with hemipelagic sediment. Grading is recognised both from grain size and petrography. Colour changes commonly mirror the grain size and textural changes. The interval of mud with silt laminae may show a distinctive hierarchy of structures in the silt (Fig. 1b), systematised by Stow and Shanmugam (1980). The complete set of structures illustrated in Fig. 1b is rarely present in a single bed: both base-cut-out and top-cut-out beds are common (Fig. 2) (e.g. van Weering and van Iperen 1984; Lash 1988). This may lead, for example, to the accumulation of sequences with thin parallel silt laminated mud (Fig. 4d), in which individual turbidite units cannot be readily distinguished: good examples are described by Stow et al. (1982, 1984a) and Walker (1985).

Medium and thick bedded mud turbidites up to many metres thick have been described (Rupke and Stanley 1974; Blanpied and Stanley 1981) in which most of the mud is structureless and not graded (Fig. 1c). Such beds are common in ponded basins, where they probably were deposited from flows with high concentrations of mud resulting from flow expansion and decrease in gradient (McCave and Jones 1988). Flow expansion and gradient reduction are probably more important than ponding, since thick bedded mud turbidites (0.3 to 2 m thick) also occur in situations where ponding is unlikely, such as the distal Bengal Fan (ODP site 717: Stow et al. 1989) (Fig. 3a). The Mississippi mid-fan channel is filled with up to 150 m of

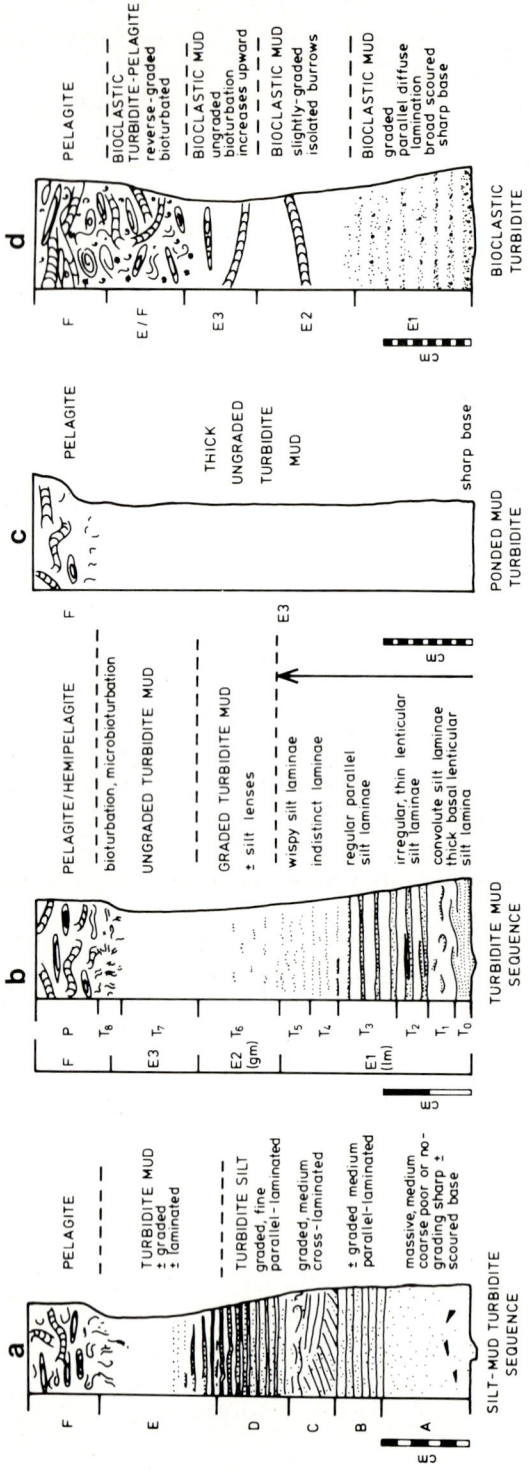

Fig. 1. Idealised vertical sequences for individual turbidite beds showing nomenclature of Stow and Piper (1984a) and Einsele (Chap. 2.7, this Vol.). **a** Silt turbidite (with Bouma *A-E* divisions). **b** Mud turbidite (with Piper *E1-E3*, Stow *T0-T8* and Einsele *m-gm* divisions). **c** Massive and turbidite (with Piper *E3* division). **d** Bioclastic turbidite (with Stow and Piper *E1-F* divisions)

Piper and Stow: Fine-Grained Turbidites

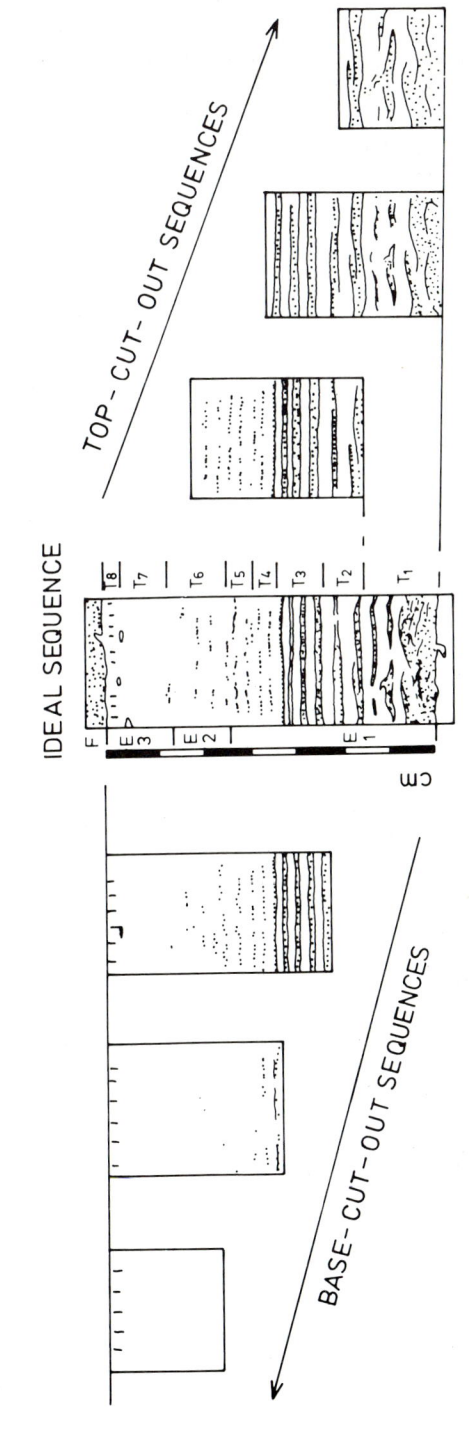

Fig. 2. Modifications to idealised mud turbidite sequence through base- and top-cut-out

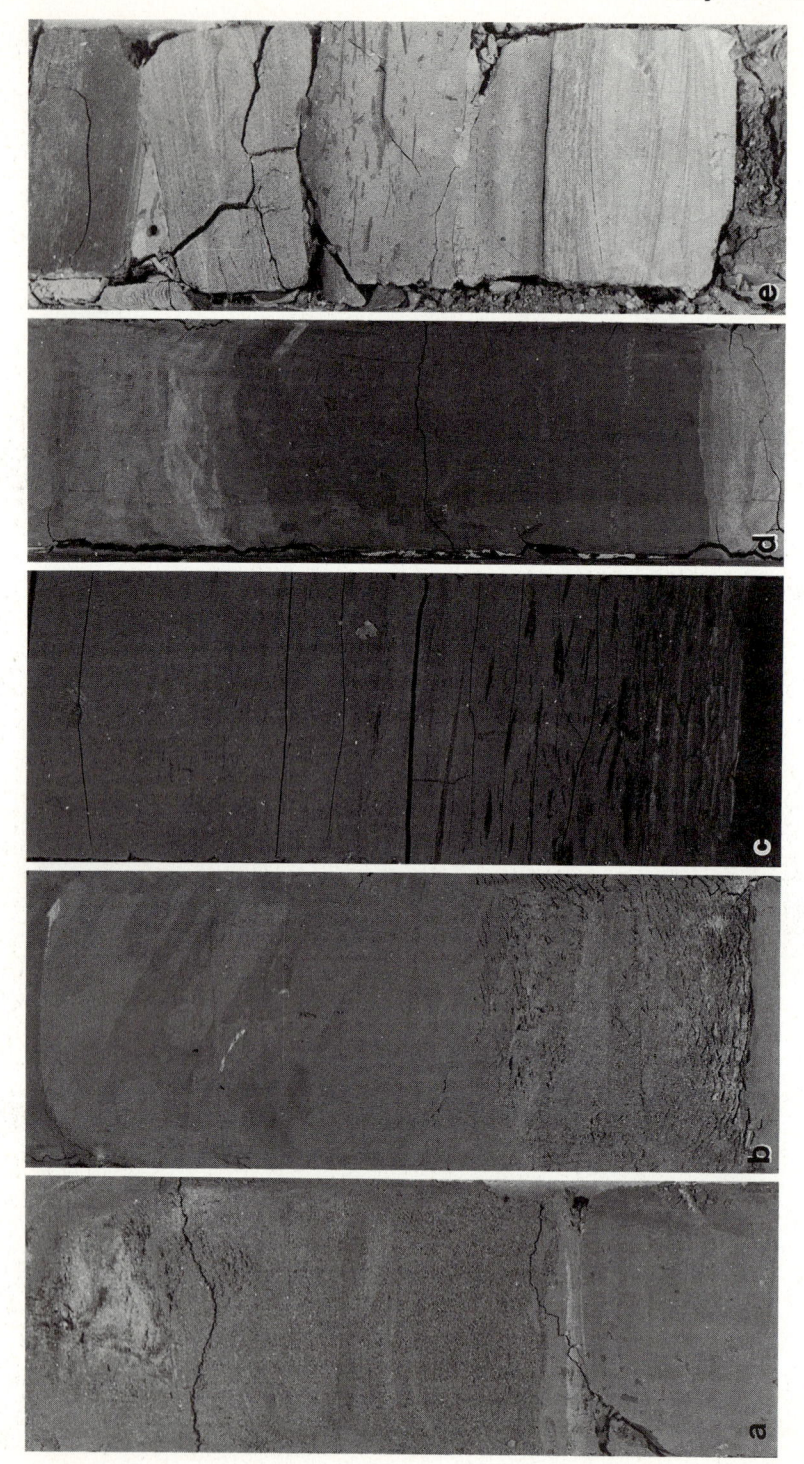

Fig. 3. Photographs of representative fine-grained turbidites. **a** Basal part of 2 m-thick graded silt to mud turbidite (Bengal Fan). **b** Graded, muddy-silt to mud turbidite (15 cm thick) (Bengal Fan). **c** Organic-carbon-rich silt-laminated mud turbidite (Angola Basin) -- upper part bioturbated. **d** Graded mud turbidite (10 cm thick) (Bengal Fan). **e** Two carbonate silt-mud bioclastic turbidites, each approximately 10 cm thick (Bengal Fan) *Scale:* core sections are all approximately 7 cm wide

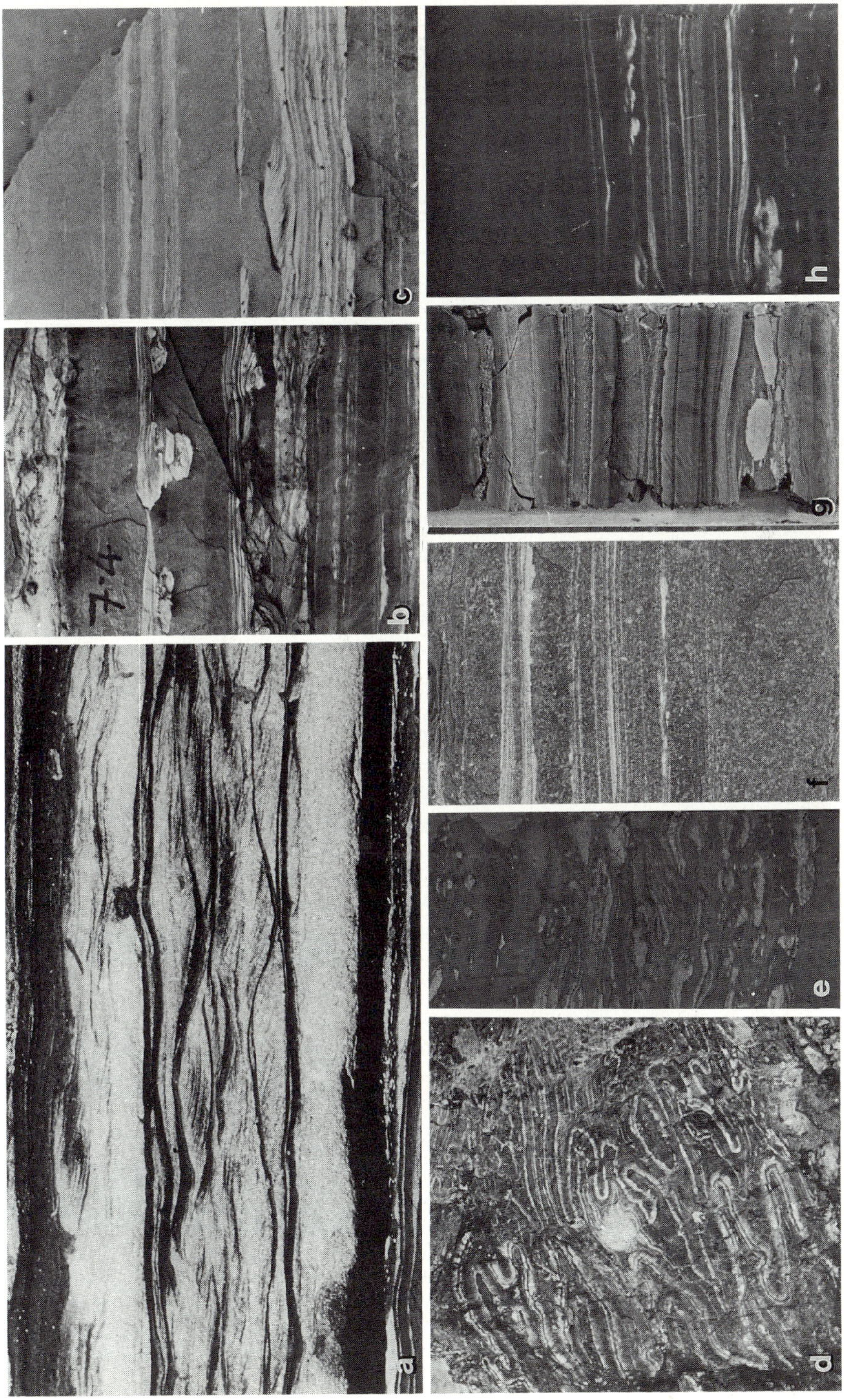

completely structureless and ungraded muds, in which it is difficult to distinguish individual beds, that may have been deposited from high concentration currents developed from muddy debris flows (Stow et al. 1986).

Some thin and medium bedded mud turbidites contain abundant millimetre-sized mudstone clasts, are poorly graded, and lack a well-developed basal division of mud with silt laminae (see Aksu 1984). Such beds appear to have been deposited from muddy turbidity currents derived from debris flows. With further deposition or disaggregation of muddy clasts, such flows deposit the disorganised silty mud turbidites which we have interpreted as the deposits of "immature" turbidity currents (Stow and Piper 1984a).

2.3 Bioclastic Turbidites

Fine-grained bioclastic turbidites consist principally of either fine-grained carbonate (principally "micrite") or of siliceous organisms. They show a sequence of structures (Fig. 1d) similar to that in terrigenous turbidites, except that (1) the graded laminated division contains less distinct silt laminae than in terrigenous turbidites and (2) the transition to hemipelagic or pelagic ooze may be much more gradual and commonly involves an increase in mean grain size. Because individual bioclastic species may consist of individuals of similar size and shape, hydraulic sorting may concentrate particular species at a certain level in a bed (Einsele and Kelts 1982).

3 Criteria for Recognition of Mud Turbidites

The following criteria can be used to identify rocks as fine-grained turbidites. These are shown in photographs of various mud turbidites (Figs. 3 and 4) and schematically in Fig. 5.

1. Distinctive petrography may be diagnostic, particularly in Cenozoic sediments. Many fine-grained turbidites contrast with interbedded hemipelagic sediments in having few pelagic microfossils, different terrigenous components (e.g. clay minerals), different carbonate content (dependant on the position of the carbonate compensation depth), and different organic carbon content (generally

Fig. 4. Photographs showing detailed characteristics of fine-grained turbidites. **a** Lenticular, ripple-laminated silty turbidites – note loads and flame-structures (Cretaceous, California). Width 15 cm. **b** Intense loading of silt laminae into mud turbidites (Cambro-Ordovician, Nova Scotia). Width 10 cm. **c** Parallel lamination, small-scale cross-lamination and fading ripples in fine-grained turbidites (Cambro-Ordovician, Nova Scotia) Width 12 cm. **d** Intense soft-sediment (slump) folding of "zebra-stripe" turbidites (Cretaceous, California), Width 150 cm. **e** Chaotic/lenticular silt-mud turbidite deposited on levee immediately adjacent to channel (Pleistocene, Mid-Mississippi Fan). Width 7 cm. **f** Very fine silt lamination in mud turbidites, with micro-loads and flames, and low-amplitude long-wave length cross lamination (Cambro-Ordovician, Nova Scotia). Width 5 cm. **g** Thin-bedded calciturbites displaying range of typical features (Paleogene, Angola Basin). Width 7 cm. **h** Graded, silt-laminated mud turbidite showing basal and internal loading and low-amplitude rippling (Pleistocene, Mississippi Fan). Width 7 cm

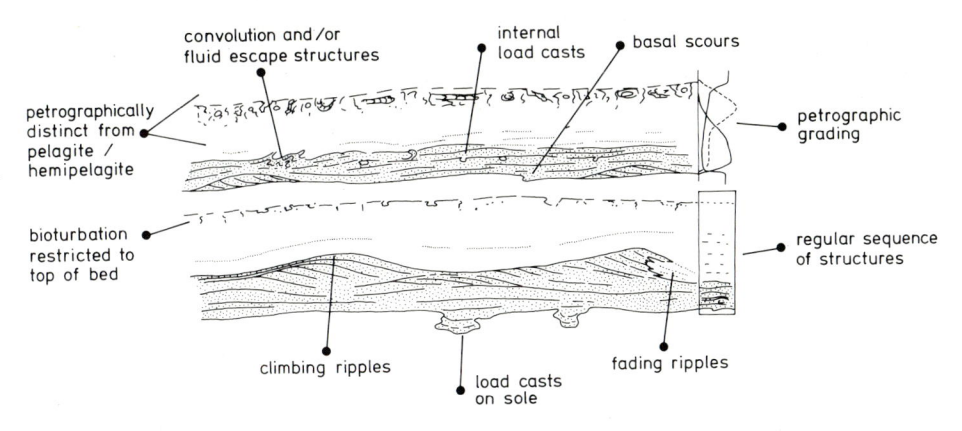

Fig. 5. Diagnostic criteria for the recognition of fine-grained turbidites

higher). Such differences arise from both the source and the high rate of sedimentation, and are therefore difficult to mimic with contour current deposits.

2. Sedimentary structures indicative of rapid deposition from suspension are characteristic of turbidites generally. In fine-grained turbidites, fading ripples (particularly if they are climbing) are particularly diagnostic (Fig. 4c). Sediment instability and fluid escape structures accompany rapid deposition from suspension (Fig. 4d,f). Escape burrows are rarely found, but are very characteristic if present. The restriction of bioturbation to the top of a bed generally indicates rapid deposition in normal marine sequences (Fig. 3c).
3. Regular grading, producing regular sequences of structures (Stow and Shanmugam 1980), becomes an increasingly diagnostic characteristic of fine-grained turbidites as the organisation and complexity of the grading increases.
4. Very thick massive mud beds of uniform petrography, lacking bioturbation and with a mud turbidite petrography are characteristic of fine-grained turbidites deposited as a result of rapid flow expansion. Similar beds may also be deposited from high concentration flows derived from debris flows, but the latter may also contain mud clasts.

We have found paleocurrent measurements to be useful only in particularly cases in distinguishing fine-grained turbidites from contour current deposits. The variability in flow directions both in individual turbidity current and in "abyssal storms" associated with bottom currents makes the interpretation of paleocurrent data difficult.

In the modern ocean, mud turbidites can be readily distinguished from hemipelagic sediments by petrographic criteria such as their paucity of pelagic microfossils: in pre-Mesozoic rocks, such fossil criteria are more difficult to apply and only sedimentological criteria can be applied to recognise fine-grained turbidites.

There are several features which have at times been claimed to be evidence that beds are *not* of turbidite origin. However, since these features commonly occur in beds that are demonstrably of turbidite origin, they are not of diagnostic value. They include:

1. concentrations of microfossils or heavy minerals in silts;
2. starved ripples and silt lenses; and internal erosion above a rippled silt (such features may pass laterally into climbing fading ripples: Stow et al. 1984a);
3. a lack of grading in thick mud beds, or in thick well-laminated silt beds.

4 Turbidite Cycles

4.1 Recognition and Controls

Cyclicity in turbidite successions has been widely recognised and commonly used to infer depositional environments for ancient examples. Earlier work focussed on thinning-upward sequences (channel filling) and thickening-upward sequences (lobe progradation) (Walker and Mutti 1973). More recent work has recognised a greater variety of types and scales of cycles, although the relative importance of allocyclic and autocyclic controls is in many cases unclear (Mutti 1977; Ricci Lucchi 1977). Hiscott (1981) has questioned the validity of "observed" cycles that are not shown to be statistically significant.

Although cyclicity involves both coarser and finer grained turbidites, we use examples from more mud-rich systems. Entirely mud-dominated successions have rarely been subjected to careful sequence analysis. We arbitarily divide cycles on the basis of their thickness into mega- (>100 m thick), meso- (10–150 m) and micro-cycles (<15 m). Some general examples of turbidite cycles are shown in Fig. 6.

4.2 Megacycles

Turbidite cycles commonly change over a few hundreds of metres of vertical section from dominantly coarse-grained to fine-grained or non-turbidite and vice versa. Although this may be at the scale of a single basin-fill episode, repetition of such sequences is observable on seismic reflection profiles of larger turbidite systems (Manley and Flood 1988; Bouma et al. 1986).

Large-scale cyclicity of this type has been generally explained as allocyclic, resulting from sea-level fluctuations (Shanmugam and Moiola 1982) or variation in tectonic activity in the source area (Klein 1985). Possible autocyclic controls include distributary switching (Manley and Flood 1988) and progradation (Lash 1988).

Fig. 6. Characteristic vertical sequences of fine and medium-grained turbidites. **a-c** Halifax Formation, Nova Scotia. **d** Mugu Point, California. **a** Coarsening-upwards mesocycle, with more irregular microcycles near top of section (*left*). **b** No distinct cycles evident within dominantly fine-grained section. **c** Probable symmetric cycle coarsening-up to mid photo then fines upward (to *mid upper left*). **d** Fining-upwards mesocycle (*bottom to top*), composed of coarsening-upwards and symmetric microcycles, and capped by block-like packet of thicker-bedded sandstones. *Scale*: all photos show sections of approximated 20 m stratigraphic thickness

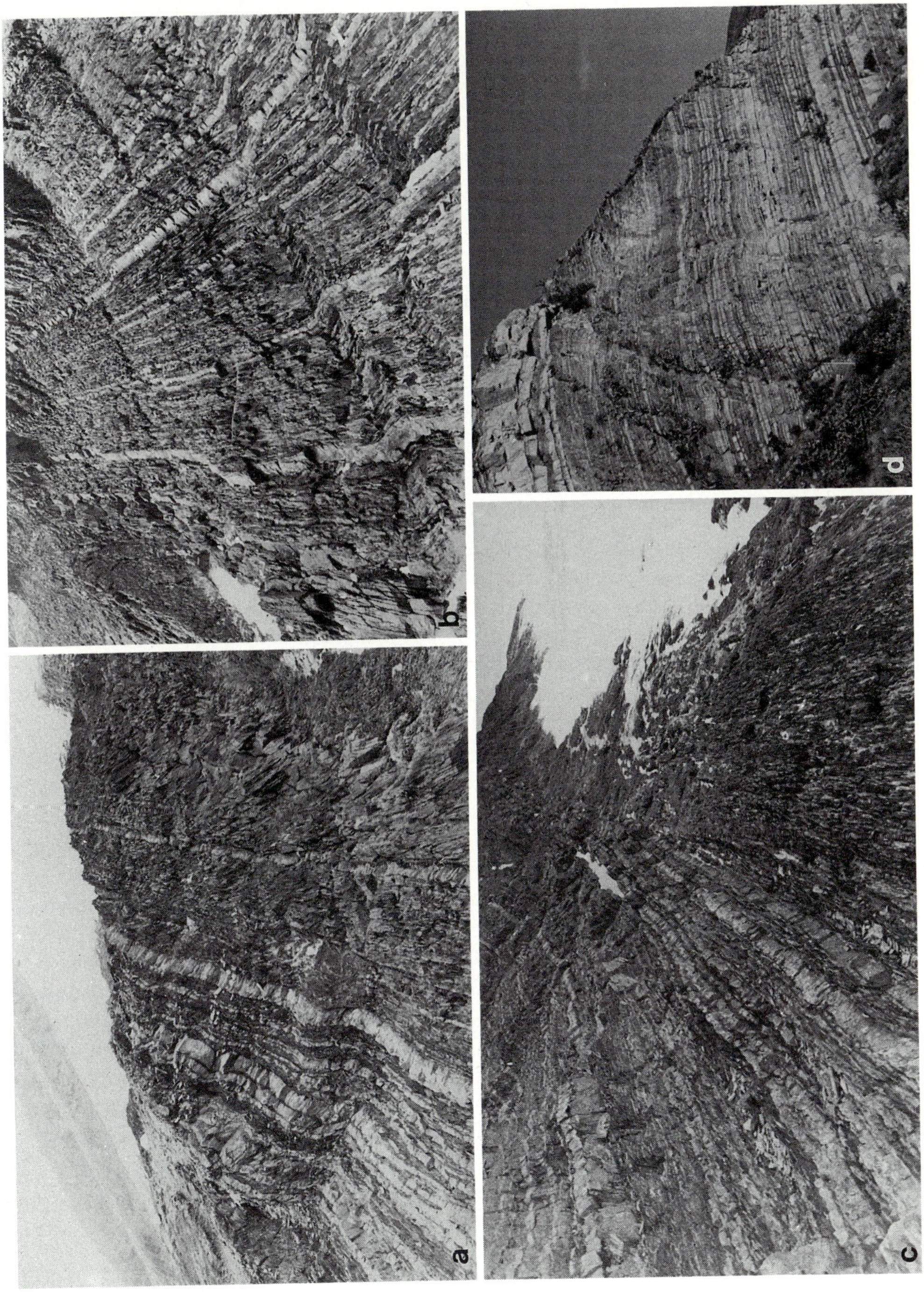
a
b
c
d

4.3 Mesocycles

It is at the scale of good outcrops and within typical cored sections that most debate has centred on the recognition and interpretation of turbidite cycles.

1. Fining-upwards sequences (or thinning-upwards) are most common in ancient coarse turbidite sequences, where they have been interpreted as channel-fill deposits (Mutti and Ricci Lucchi 1975; Stow 1984b), perhaps involving channel migration (Walker 1985). Tectonic control has also been proposed (Stow et al. 1982). Similar sequences are less common in fine-grained sediments (Fig. 6d).

Few modern channel systems have been drilled. DSDP Sites 621 and 622 were cored to about 200 m within a prominent channel on the Mississippi mid fan and recovered predominantly mud. Both showed ill-defined fining-upward sequences, being more silty towards the base (bottoming in gravel at site 621). However, much of the sequence was structureless and unbedded (Pickering et al. 1986). DSDP Site 530 was drilled through a small, mud-dominated fan in the SE Atlantic Ocean. A weak fining-upward trend over the top 80–120 m has been interpreted as resulting from progradation of the inner fan rather than channel fill (Stow 1984b).

2. Coarsening-upwards sequences (or thickening-upwards) are also common in ancient successions and classically interpreted as representing basinward progradation of mid-fan lobes (Walker and Mutti 1973; Mutti and Ricci Lucchi 1975). The facies are typically finer than those in fining-upward sequences and many have been described from mud-dominated turbidite successions (Shanmugam 1980) (Fig. 6a). On the mud-dominated Late Precambrian Kongsfjord fan in northern Norway, Pickering (1981) interpreted coarsening-upwards meso-sequences as the result of catastrophic lobe switching, due to either fan channel avulsion or some source control. Two scales of coarsening-upward meso-sequences described by MacDonald (1986) from a Mesozoic back-arc basin on South Georgia are interpreted as due to autocyclic lobe switching (the thinner sequences) and allocyclic variation in sediment supply from the arc. MacDonald also argued that the lateral shifting of channels and lobes during aggradation is probably equally as common as progradation as a mechanism for producing coarsening-upward (and fining-upward) sequences.

There are few modern examples of mesocycles from fan lobes. DSDP Sites 614 and 615 were drilled up to 520 m through the channel terminal lobe of the Mississippi Fan. Fine sand, silt and mud turbidites were arranged in coarsening-upward, fining-upward, symmetrical and irregular units typically a few tens of metres thick. A similar range of sequence and non-sequence types was observed on mud-dominated channel levees and overbank sites on the Mississippi Fan (Sites 617, 620, 622 and 623: Bouma et al. 1986) and on the distal lobe of the Bengal Fan (ODP Sites 717, 718, 719; Stow et al. 1989).

3. Symmetric packets and bundles. Drilling of modern turbidite sequences (discussed in two preceding Sects.) and statistical analysis of ancient sequences (Hiscott 1981; Walker 1985) has shown that mesoscale cyclicity most commonly involves an approximately symmetrical arrangement of increasing or decreasing bed thickness and grain size. Symmetric packets that show gradational increase and then decrease in coarser-grained, thicker-bedded turbidites are particularly common in mud dominated successions

(Fig. 6c): they occur in a variety of environments and have been interpreted as resulting from both allocyclic and autocyclic processes (Hiscott 1981; Martini et al. 1978).

Blocky packets of coarse-grained thick-bedded turbidites, showing relatively abrupt transitions with the encasing mudstones, are also widely recognised in ancient sequences both in outcrop (Fig. 6d) and from well logs. These have been generally interpreted as channel fill deposits (Surlyk 1987).

4. Random non-cyclic successions. Most turbidite workers would agree that a large proportion of both modern and ancient turbidite successions cannot be assigned to any form of regular cyclicity, symmetric or asymmetric (Nilsen 1980; Melvin 1986) (Fig. 6b). Hiscott (1981) has proposed that many reported cyclical sequences, identified solely on visual criteria, could be explained by chance occurrences within unordered sequences of turbidite beds. Such non-cyclic deposition may be most common in more distal turbidite environments.

4.4 Microcycles

The patterns recognised in mesocycles also appear to be present in microcycles, although grain size is not always closely linked with layer thickness. Mutti (1977) related microcycles (inferred to be autocyclic) from thin-bedded facies in the Eocene Hecho Group in Spain to depositional environments. Thickening-upward and symmetric cycles represent lobe-fringe and fan-fringe areas; bundles of thin-bedded sandy turbidites separated by mudstone units occur in interchannel areas; non-cyclic arrangements of irregularly bedded sandy turbidites were deposited at channel mouths and margins. More regularly bedded but equally non-cyclic sequences characterised the basin plain.

Lash (1988) ascribed fining-upwards microsequences (1–9 m thick) in the Ordovician Martinsburg Formation of the Appalachians to allocyclic controls. His sections also show the presence of symmetrical and a few coarsening-upward cycles, with much of the sequence appearing non-cyclic.

Mutti and Sonnino (1981) describe repeated thickening upward microcycles no more than a few beds thick. They term these compensation cycles and ascribe them to the influence of the slight positive relief of the previous turbidite on the next turbidity current. Other authors have shown more variability in turbidite deposition at this scale in both ancient rocks (Shanmugam 1980; Stow et al. 1984a; Melvin 1986; Kasper et al. 1987) and modern turbidites drilled by ODP (Stow et al. 1986; Stow et al. 1989).

5 Turbidity Current Processes

5.1 Principal Processes Acting in the Deep Sea

There may be a continuum of processes acting on fine-grained sediments in the deep sea (Walker 1978; Stow and Piper 1984a). These include settling of particles through the water column, normal bottom currents and mass gravity resedimentation processes (both debris flows and turbidity currents). The boundaries between these processes are not always clear-cut: for example, settling of silt particles from a deltaic plume may lead to an ignitive turbidity current; a debris flow may evolve into a turbidity current; and the low-density top or tail of a turbidity current may be deflected by contour-following bottom currents. Nevertheless, except adjacent to major sediment sources such as deltas or temperate ice margins, turbidity currents and debris flows are the principal means of depositing large amounts of fine-grained sediment in the deep sea. The deposits of bottom currents and distal pelagic settling tend to accumulate slowly and are thus well bioturbated.

5.2 Dynamics of Mud Transportation and Deposition

The physical processes involved in the erosion, transport and deposition of fine-grained, cohesive marine sediments have been recently reviewed by McCave (1984): we summarise here those aspects of transport and deposition that are most relevant to fine-grained turbidites. This discussion refers principally to terrigenous turbidites: there is insufficient work on bioclastic sediments to know to what extent the same principles apply.

The behaviour of mud suspensions during transport and deposition depends critically on the concentrations present. At concentrations of less than 0.3 kg/m^3, mud suspensions behave as Newtonian fluids; at concentrations of more than 5 kg/m^3 particle-to-particle interactions predominate and flows behave more like a consolidating soil than a turbulent suspension.

Under the conditions found in turbidity currents, mud particles will aggregate or flocculate. Krone (1978) showed that the shear strengths of aggregates from any particular sediment yield a few discrete values rather than a continuum, which he interpreted in terms of distinct orders of aggregation. Thus primary particles flocculate to form zero-order aggregates; several zero-order aggregates yield a less strong first-order aggregate, and these combine to form second-order aggregates of lesser strength. During transport, turbulence will lead to collisions between aggregates, leading to the formation of higher-order aggregates, but these higher-order aggregates will be most susceptible to break-up by turbulent shear.

At concentrations of less than 0.3 kg/m^3, with a bed shear stress of less than about 0.06 Pa, suspended sediment concentrations are observed to decrease logarithmically with time, as a result of entrapment of suspended sediment in the viscous sublayer (McCave and Swift 1976). Such deposition will take place at mean flow velocities of a few centimetres per second. At concentrations of more than 1 kg/m^3, a proportion of any given mud will deposit at shear stresses well above the critical depositional stress for low concentration flows: this appears to take place through the selective deposition of aggregates capable of forming the strongest bonds with the bed (Partheniades 1972): the importance of this effect increases with increasing concentration. For natural flows of concentrations of a few kg/m^3, this adhesive deposition is probably of minor significance and deposition by settling of aggregates trapped in the viscous sublayer results in depositional rates not exceeding a few mm/h.

Fine-grained cohesionless silts behave more like sand than mud flocs. Experiments in a circular flume using mixed grade silts with 30–40% very fine sand and clay at concentrations of 13–44 kg/m^3 (Banerjee 1977) have shown the development of partial Bouma (1962) sequences as a result of flow deceleration. Initial instantaneous deceleration to 0.46 m/s produced normal grading and then further deceleration to about 0.1 m/s over a period of several hours led to progressive development of parallel lamination, ripple lamination, sinusoidal ripple lamination and finally a suspension blanket as the flow ceased. Mantz's (1978) experiments on cohesionless quartz silts with increasing flow velocities again demonstrated that the bedform sequence produced was analogous to that for coarser sand-sized particles. However, for micaceous silts the only observed structure was parting lineation.

5.3 Dynamics of Turbidity Currents and the Deposition of Fine-Grained Turbidites

Because of the practical difficulty of monitoring marine turbidity currents, the physical behaviour of such currents must be inferred from turbidite deposits or from indirect phenomena such as cable breaks (velocity) and erosional structures (thickness). Standard physical properties of flows can be applied to turbidity currents, and in cases where some parameters can be constrained, inferences can be made on the behaviour of specific turbidity currents. Almost all well-known turbidity currents in the deep sea transported principally terrigenous sediment.

In most analyses of turbidity currents, the parameter most difficult to constrain is sediment concentration. Analyses of several currents that have deposited fine-grained turbidites have estimated concentrations in the order of a few kg/m^3 (Stow and Bowen 1980; Bowen et al. 1984). McCave and Jones (1988) have suggested that decelerating flows may reach concentrations of 50–100 kg/m^3, with this high-concentration slurry damping both turbulence and the entrainment of water at the upper interface of the turbidity current. Times available for deposition of fine-grained turbidites have been generally found to be many tens of hours.

Thus deposition through entrapment of aggregates in the viscous sublayer provides an adequate explanation for fine-grained turbidite beds up to a few decimetres in thickness. The structureless mud turbidites in beds metres thick require deposition from high concentration flows either through the processes of adhesion or in ponded settings through dewatering of high concentration slurries, as proposed by McCave and Jones (1988).

The dynamics of lamina deposition in fine-grained turbidites remain uncertain. Several authors have presented hypotheses for the origin of alternating silt and mud laminae, invoking mechanisms within the boundary layer such as turbulent bursts disrupting the boundary layer (Hesse and Chough 1980 – but see Allen 1985), disruption of flocs by shear within the boundary layer (Stow and Bowen 1980) and adhesion of clay onto clay beds (Piper 1978). Although Carey and Roy (1985) have produced such laminae in flume experiments, and invoke aspects of all the three above processes, we know of no specific experimental work directed at further understanding the origin of such lamination. This may be a fruitful area of future research, since variation in flow concentration may have an important influence on the style of mud deposition. Neither do we know of work in which variations in the structure of lamination is specifically linked to variations in mineralogy of fine-grained sediment.

6 Facies Significance of Different Types of Mud Turbidite in the Geologic Record

The variability of mud turbidites with depositional environment was synthesised by Piper (1978, Fig. 12–8), who provided an extensive bibliography. We present a new synthesis (Fig. 7) which draws on the greater understanding of the dynamics of turbidity currents and mud deposition, together with many new descriptions of fine-grained turbidites. There is insufficient data on fine bioclastic turbidites to propose any general synthesis.

Variability in mud turbidite facies results both from variations in initiation processes, which influence the type of sediment and the nature of initial flow in the turbidity current; and variations in transport and depositional processes.

6.1 Turbidite Initiation and Facies Variation

The initiating mechanisms for most turbidity currents are unknown, so that relating facies to initiating mechanism is at present speculative. The influence of initiating

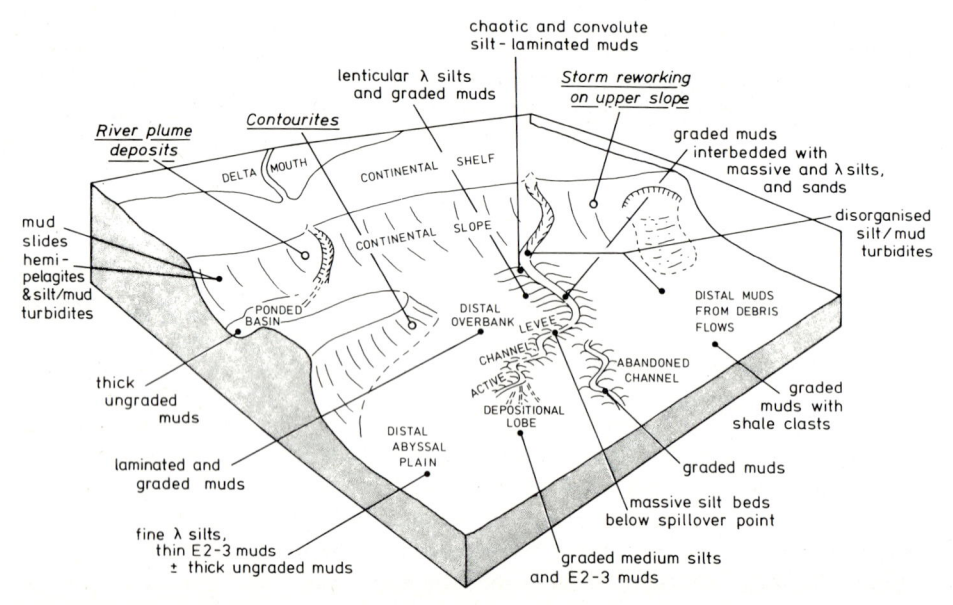

Fig. 7. Schematic facies distributions of different mud turbidites and associated sediments. [Based principally on data in Piper (1978); Stow and Piper (1984a) and references cited in this chapter]

mechanism is greatest on the continental slope and in proximal parts of turbidite systems. Processes such as direct flow of river bedload into prodelta valleys (Prior et al. 1987) and rip current removal of sand from submarine canyon heads (Inman et al. 1976) result in coarse-grained turbidites, except where mud is incorporated by subsequent erosion. The role of large accelerating turbidity currents in eroding older proximal deposits of smaller flows is important in transporting muds to the distal parts of turbidite basins (Piper and Normark 1983).

Processes that put fine-grained sediment into suspension, such as outer shelf storms and river plume discharge, provide sediment that may be initially relatively well sorted. If these processes are effective, they are likely to result in frequent small turbidity currents, and the resulting proximal deposits may consist of thinly laminated well-sorted silts and muds (Hill 1984). The extent to which such small flows may become ignitive (Parker et al. 1986) and erode older proximal deposits is unclear.

The character of muddy turbidity currents derived from slumps is variable. Thick slump masses may break up to produce thick, poorly sorted mud turbidites, which may contain mudstone clasts (Aksu 1984). Mud turbidites derived from slumping of surficial sediment (<2 m burial) may resemble those derived from upper slope suspension.

6.2 Turbidite Flow Processes

The character of more distal fine-grained turbidites depends principally on flow conditions at the time of deposition. Many flows are initially channelized. Flow expansion, which generally results in deposition, may occur from overbank spillover, channel widening or termination, and change in gradient (Bowen et al. 1984). Unchannelised flows (whether distal overbank or lower fan) will experience flow expansion due to change in gradient. In ponded basins both gradient reduction and basin-margin reflection may be important (Hiscott and Pickering 1984). Flow expansion can have a variety of influences on deposition. It will normally result in a rapid decrease in flow velocity and an increase in entrainment of ambient water. Thus while overall sediment concentration will decrease, concentration near the base of the flow may increase temporarily, particularly in coarser silt and mud aggregates. Velocity decrease will result in a decrease in break up of aggregates.

Most source-area muds on outer continental shelves or upper continental slopes have silt contents in excess of 50%. Over flow distances of hundreds of kilometres, the silt content of turbidite muds decreases to less than 40% and there is a concomitant increase in the number of discrete beds of (principally medium or fine) silt in very distal fine-grained turbidite sequences (Piper 1978, Fig. 12–7). The decrease in silt content of muds probably reflects the gradual exclusion of silt from strong floc aggregates through turbulent shear disruption and reforming of aggregates during turbulent flows of long duration. Although this process may increase the abundance of fine silt not bound up in aggregates, source-related processes may also influence the abundance of silt beds. Gradually, as silt is lost almost entirely from the flow, then the number of discrete silt laminae decreases.

6.3 Facies Variation

Deposits on known *levees* and *channel termination areas* consist of muds with well-developed silt laminae, frequently with evidence of sediment starvation such as fading ripples. Silt laminae become thinner, rarer and less rippled away from channels. The features of apparent flow instability on levees and the decrease in granular sediments away from levee crests are features well known from studies of sandy turbidites: in fine-grained turbidites they are represented by irregularly interlaminated silts and muds.

Graded muds, lacking prominent silt laminae, are the most common type of fine-grained turbidite within *channels*. These may result from either deposition from the "tail" of a turbidity current, or from rapid velocity decrease in flows that have been stripped off at low points on levees (Piper and Normark 1983). In addition, thick massive silts appear to be restricted to channels.

Steady *unchannelised flows* appear to deposit poorly graded muds with thin basal silt laminae. In distal environments, these may interbed with thin medium and fine silt beds. *Ponded basins* are characterised by thick ungraded turbidites resulting from rapid decrease in velocity leading to the development of very high concentration flows. Similar turbidites may be deposited in channel termination areas where again there is a rapid velocity decrease.

7 Conclusions

1. Fine-grained turbidites, both terrigenous and bioclastic, are major components of most turbidite systems.
2. Individual beds of silt, mud and bioclastic turbidites show systematic sequences, summarised in Fig. 1.
3. Fine-grained turbidites may be distinguished from other fine-grained facies by distinctive petrography, sedimentary structures indicating rapid deposition from suspension, regular grading and structure sequences, or by the occurrence of thick unbioturbated uniform beds (Fig. 5).
4. Both asymmetric and symmetric depositional cycles are present in fine-grained turbidite sequences. Asymmetric cycles are less common than suggested in the classical models and are not restricted to lobes and channels. It is important that the presence of cycles be demonstrated by objective statistical techniques; there is a particular need for study of microcycles (<10 m thick). There are few studies that convincingly demonstrate the origin of observed cyclicity.
5. Although there is a continuum between turbidity currents and some other depositional processes in the deep sea, in most places debris flows and turbidity currents are the principal means of depositing large amounts of fine-grained sediment in the deep sea. The complex behaviour of flocs has an important influence on the grain size distribution and lamination present in fine-grained turbidites.
6. Facies distribution of fine-grained turbidites is influenced by turbidity current initiation and flow processes. Variation in facies is summarised in Fig. 7.

2.10 Distinction of Tempestites and Turbidites

G. Einsele and A. Seilacher

1 Introduction

Event beds are the result of high energy episodes, in which bottom material (including its fauna) is reworked by current and/or wave action, shock and gravity forces, transported some distance in suspension, and redeposited. Tempestites and turbidites, particularly their distal expressions, have a number of features in common. Therefore, tempestites have often been mistaken for turbidites, because their characteristics were not sufficiently known to many geologists. On the other hand, the correct interpretation of these two bed types is of prime importance for paleogeographic reconstructions and basin analysis. This brief article should provide some general guide lines for this essential distinction.

The general characteristics of sandy and calcareous tempestites are reviewed in articles by Seilacher and Aigner (Chap. 2.3) as well as Nummedal (Chap. 2.2), both this Vol. Similarly, the features of sandy and muddy turbidites are summarized by Stow (1986) and in the review paper by Einsele (Chap. 2.7, this Vol.). Eberli (Chap. 2.8) describes calcareous examples in more detail, and Piper and Stow (Chap. 2.9) particularly deal with mud turbidites (both this Vol.).

2. Distinction Between Siliciclastic Sandy Tempestites and Turbidites

2.1 Sedimentary Structures and Other Sediment Characteristics

The common types of sandy tempestites and turbidites have a number of sedimentary structures in common such as an erosional base with different types of sole marks, graded bedding, small-scale ripple bedding, and amalgamation, but they also differ in several characteristic features (Table 1). Tempestites often display more or less distinct hummocky cross-stratification (Fig. 1) and wave ripples or wave ripple lamination at their tops. The grain size distribution of tempestites tends to be bimodal with a coarser section at the base and, separated by a kind of small hiatus, a finer-grained top section. This phenomenon is less pronounced in turbidites, because suspension currents traveling over long distances are often fed by erosion at their base and in this way take up materials of varying grain size from the sea floor. Proximal turbidites frequently show traction current phenomena and/or inversed grading at their base. Their sole marks show unidirectional current directions which

Table 1. Criteria to distinguish sandy tempestites from turbidites

		Tempestites	Turbidites
Sedimentary structures (from top to bottom of bed)	Wave ripples and wave ripple cross lamination	Common (apart from distal types)	Absent
	Current ripples and current ripple bedding	Less common than in turbidites	Common
	Convolute lamination	Rare	Common
	Hummocky cross-stratification	Common	Absent
	Traction carpet with inverse grading	Absent	Common in proximal types
	Nature of sole marks	Often bipolar, pronounced irregular scouring, gutter casts channeling	Uni-directional
Bio-facies	Benthic background community (in muddy intercalations)	Shallow water fauna, differing with substrate consistency	Deep-water fauna, mainly represented by burrows
	Displaced body fossils within event beds	Shallow water species only	Shallow and deep-water species
	Autochthonous post-event fauna and bioturbation	Fauna similar to pre-event fauna (return to background fauna, if substrate is similar)	Episodic colonization by specific fauna preceding return to background conditions
Stratigraphic context	Amalgamation	Very common and pronounced, including "maturation" of sediment	Less common, no maturation effect
	Continuity of single beds	Mostly limited	Often over wide distances
	Thickness of sequence	Limited associated with shallow-water facies	In general great, associated with deep-water facies

may in many cases, but not generally, indicate a rather constant flow pattern. The sole marks of tempestites, in contrast, often reveal the bipolar direction of wave action superimposed on a seaward directed combined flow. Small-scale channeling and gutter casts are much more common in tempestite environments than in the deep sea. The material of tempestites, which is stirred up by waves and redeposited under combined flow conditions in relatively shallow water, is more or less autochthonous or quasi-autochthonous and therefore does not significantly deviate from shallow water sands. The material of turbidites is allochthonous and usually differs in its composition from the host sediments.

2.2 Body Fossils and Ichnofauna

Tempestites and their interbeds contain only shallow-water fauna. Bioclasts of the event beds are displaced and redeposited according to their hydraulic behavior, but lateral transport is limited and may be absent in oversized skeletons of recliners, which are only winnowed to a lower level. The ichnofauna can be grouped into pre-event and post-event associations. In the case of tempestites, however, the two trace fossil associations contain essentially the same types of dwelling and feeding burrows that characterize the shallow marine *Cruziana* ichnofacies, with local differences due to substrate consistency (soft ground, firm ground, hard ground).

Turbidites may, with respect to body fossils, contain a mixed shallow/deep assemblage, because turbidity currents pick up bioclasts of different bathymetric zones along their way. The autochthonous trace fossils, however, show the character of the deep sea *Nereites* ichnofacies, in which space utilization by systematic patterns (spirals, meanders, nets, dendroids) is a dominating feature. In addition, we observe a sharp divergence, in preservation and trophic strategies, between the infauna before and after the event:

a) The pre-turbidite trace fossils association consists mainly of graphoglyptids. These were originally open burrow systems, whose varied and highly regular patterns suggest that food was extracted indirectly, i.e., by way of farming bacteria or fungi. This association represents the background infauna of hemipelagic muds, whose tiers became blurred by ongoing background sedimentation. Since the hollow tunnels eventually collapse, they require the exhumation and casting mechanism of a turbidity current to leave a lasting record. Thus, what we find on a single sole of a distal turbidite is only horizontal sections of three-dimensional burrow systems that may represent several tiers.

b) The post-turbidite association of trace fossils, in contrast, represents only an episodic colonization of imported muds with an elevated content of detrital food particles. Accordingly, the tiers of this infaunal community are not mixed, but frozen in their original relationship. There is also a marked difference in feeding strategies as compared to the pre-turbidite community. In post-turbidite burrows the dominant mode is direct sediment feeding by sorting and ingestion. Another possible strategy is endosymbiosis with chemoautotroph bacteria fueled by active pumping of H_2S water from deeper zones in the sediment (Savrda et al. Chaps. 5.2, this Vol.). Both strategies require active backfilling of abandoned burrow sections with sorted sediment and fecal material, which allows their fossil preservation without sand casting. In thinner beds the deepest tiers of post-turbidite burrows may penetrate to the sole face of the sandy part of the turbidite, where they are found in association with the exhumed casts of pre-turbidite burrows.

2.3 Proximal Sediment Facies

One of the best means to discriminate tempestites from turbidites is their proximal facies association (Fig. 1). Proximal tempestite beds may alternate with massive sand

TEMPESTITES

DISTAL ← PROXIMAL   PROXIMAL FACIES ASSOCIATION

- COMPLETELY BIOTURBATED
- MINOR DIFFERENCES IN COMPOSITION
- SHALLOW-WATER ICHNOFACIES
- DISPLACED FAUNA (SHALLOW WATER)
- IN-SITU FAUNA (SHALLOW WATER)
- SHARP BOUNDARY

1-5 cm

- WAVE RIPPLES, WAVE R. CROSS-BEDDING
- HUMMOCKY CROSS STRAT.
- BIPOLAR SOLE MARKS
- "JUMP" IN GRAIN SIZE DISTRIBUTION AMALGAMATION
- BIOTURBATION
- GUTTER CASTS, CHANNELING

10-50 cm

- FORESHORE AND BEACH
- CHANNELS WITH DIVERGENT DIRECTIONS
- LARGE-SCALE SWALEY AND HUMMOCKY CROSS STRATIFICATION
- LENSES OF GRAVEL
- MUD

1-3 m

SM = TEMPESTITE MUDS
m = SHELF MUDS
SS = TEMPESTITE SANDS

TURBIDITES

- BIOTURBATION (DEEP-WATER ICHNOFACIES)
- HEMIPELAGIC MUD OR BIOGENIC OOZE (± ORGANIC MATTER, COMPOSITION AND FAUNA DIFFERING FROM PE, PARTLY NODULES)
- ± COMPLETELY BIOTURBATED
- LAMINATED MUD, CROSS-BEDDED SAND
- SOMETIMES BLACK SHALE AND/OR DEVOID OF CARBONATE

1-5 cm

- MASSIVE, GRADED MUD
- LAMINATED MUD
- SANDY BASE
- MUD FLOW
- CURRENT RIPPLES
- CONVOLUTE BEDDING
- POST EVENT TRACE FOSSILS
- ONE-DIRECTIONAL SOLE MARKS (FLUTE CASTS, TOOL MARKS)
- PRE-EVENT TRACE FOSSILS

10-50 cm

SL, TS, gr, WATER ESCAPE STRUCT., ig, cb, ls, gr,mv, st, PE, DF, ig

1-5 m

L A R
LUTITE ARENITE RUDITE

PE = PELAGIC OR HEMIPELAGIC SED.
TM = TURBIDITE MUD, TS = TURBIDITE SAND

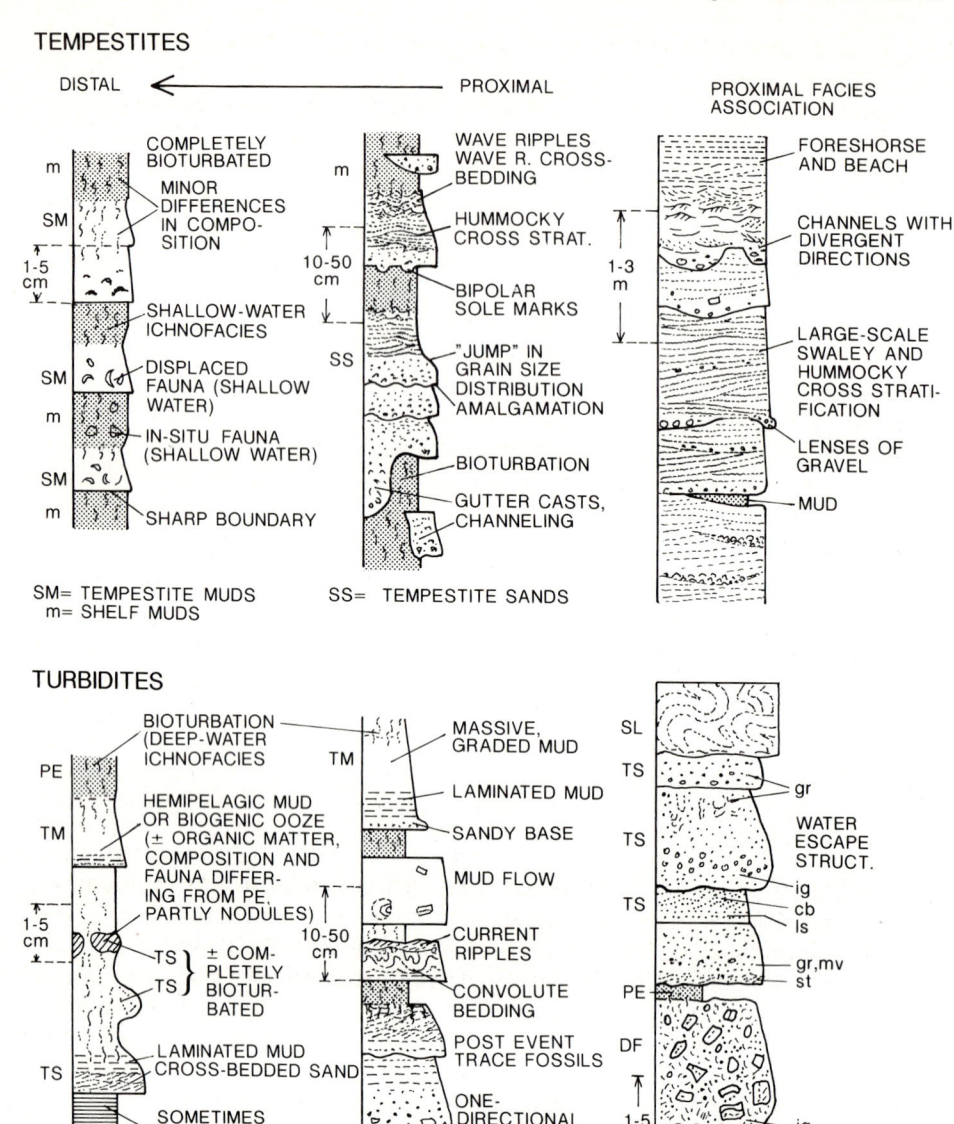

Fig. 1. Characteristic features of proximal to distal, sandy and muddy tempestites and turbidites. Note the particularly great differences in the proximal facies associations. See text and Table 1 for further explanation

beds exhibiting swaley and hummocky cross stratification. They are often associated with shoreface sands, foreshore, or even beach sands and gravel. Normally, the thickness of tempestite sequences is limited, because their shallow depositional environment tends to evolve either into a beach-foreshore-lagoonal complex, or into a deeper basin which is no longer influenced by storms. By contrast, turbidite sequences commonly reach many hundreds and even thousands of meters. Also, proximal turbidite sections frequently contain deposits of mud flows and debris flows, which preferentially accumulate at the foot of submarine slopes.

Large-scale and widely extended deep-sea fan associations are characteristic of turbidite environments. However, they can be studied only in large exposures or in series of smaller outcrops. Individual tempestite beds can rarely be traced as far as many turbidites.

3 Calcareous Arenitic Tempestites and Turbidites

The criteria to distinguish between siliciclastic sandy tempestites and turbidites can also be applied to their calcareous counterparts. In addition, the following observations may be useful:

a) The interbeds of calcareous tempestites also contain biogenic carbonate of shallow-water origin. In the case of turbidites, the host sediment may be devoid of carbonate if the basin floor was situated below the calcite compensation depth.

b) Proximal tempestites tend to incorporate not only rather large and heavy shells, but also a high proportion of broken and abraded skeletal material, because frequent reworking by storms leads to pronounced textural maturation. However, some proximal calcareous turbidites may also contain material which was repeatedly reworked in shallow water, besides particles of finer and less resistant skeletons. The preservation of delicate shell material and articulated skeletons in proximal tempestites is rather unlikely; i.e., we rather observe compositional maturation.

4 Muddy Tempestites and Turbidites of Different Composition

Muddy tempestites in proximal regions are preserved only under exceptional conditions, because they are easily cannibalized by stronger storms. One of the exceptions are storm-induced flat pebble conglomerates composed of mud which had undergone early stabilization and lithification by algal mats. The reworked pebbles alternate with layered beds of the same lithology (Sepkoski et al. Chap. 2.6, this Vol.).

Storms stir up mud in near-shore areas and transport it into deeper water, where it is deposited as top mud of sandy tempestites or as mud tempestites. In contrast, muddy turbidites are also known from proximal regions, particularly from deep basins of limited extent (ponded basins, see Piper and Stow Chap. 2.9, this Vol.). Here, they often represent thick massive beds in association with mud flows.

Mud tempestites alternate with interbeds of similar petrographic composition. They are rarely rich in biogenic silica and organic matter, unless deposited in oxygen-deficient environments. Mud turbidites commonly differ in petrographic composition from their pelagic or hemipelagic interbeds, particularly with respect to carbonate content and clay mineral assemblage. They also tend to contain more tests of microorganisms and organic matter than the host sediments. Below the CCD, only mud turbidites may bear carbonate, which is absent in the interbeds. Mud turbidites frequently alternate with black shales; varve-scale laminations are often interpreted as the result of small low-density suspension currents (see Cotillon Chap. 7.3; Wetzel Chap. 5.1, both this Vol.). As already mentioned, mud turbidites also provide organic-rich layers in well-oxygenated environments. The displaced bioclasts of mud turbidites usually reflect only the source area, because due to the lack of substantial erosion hardly any deep-water organisms are taken up by muddy suspension currents.

A special case is the smothering of complete carcasses (obrution Lagerstätten; see Brett and Seilacher Chap. 2.5, this Vol.). It is mainly to be expected in epicontinental basins below wave base and near an oxycline, where storm-induced compensation currents can kill organisms by H_2S and bury them beyond the reach of scavengers. Significantly, silty layers associated with such mud tempestites show the sedimentological characteristics of thin turbidites (lithographic limestones; Hunsrueck Shales), but lack the ichnological signatures of deep sea biotopes.

Thin mud tempestites are easily destroyed by bioturbation, so there is little chance to recognize them in the fossil record, although a complete mixing of the original beds and interbeds does not take place. In the case of mud turbidites which markedly differ in composition and faunal characteristics from the host sediments, bioturbation cannot entirely obliterate the primary signal.

6 Conclusions

Individual beds of sandy shallow-water tempestites and deep-water turbidites have many features in common, but they can be distinguished by a number of specific criteria (Table 1). Tempestites are characterized by bipolar or multi-directional sole marks, hummocky cross stratification, wave ripples and wave ripple cross lamination, and the lack of deep-water fossils. Turbidites may exhibit traction carpets with inverse grading in proximal types, displaced deep-water body fossils (besides shallow-water fauna), and frequently convolute lamination. In addition, the shallow or deep-water environment of these event beds is indicated by the pre-event benthic fauna in the host sediments, including bioturbation, and by the post-event autochthonous fauna and trace fossil association at the top of the event bed.

Muddy turbidites frequently differ in their composition from the host sediments and may point to source area conditions that were either better or less oxygenated than the environment of redeposition. Thin muddy tempestites are more difficult to identify, particularly after intensive bioturbation.

In proximal regions, tempestites frequently show pronounced amalgamation, but for both tempestites and turbidites, the proximal facies association and thickness of the sequence containing such event beds, are particularly distinctive (Fig. 1).

2.11 Flash Flood Conglomerates

F. Pflüger and A. Seilacher

1 Introduction

Event stratigraphy in its present form refers primarily to marine basins, where turbidity currents, storms, and floods are the dominating turbulence events in a deep-to-shallow transect. However, the term can also be applied to continental deposits. In a lake we expect similar phenomena as in marine basins, though on a smaller scale and with a higher probability of anoxic periods (Glenn and Kelts Chap. 1.9, this Vol.). With respect to fluvial deposits, concepts of event stratigraphy have been less influential, probably because we commonly deal with a complete range of intermediates between ordinary runoff and extreme floods. Nevertheless, the rule that, due to their cannibalistic behavior and higher preservation potential, extreme events are overrepresented in the fossil record must also apply to fluvial deposits.

Flash floods – a term used by Karzc (1972) for episodic fluvial floods in desert wadies – are a different matter. They are separated by very long periods in which the stream beds lie dry and form part of an eolian sedimentary regime, but nevertheless, minimal secondary remodeling by eolian or bioturbational agents during dry periods takes place. Most importantly, however, flash floods are "all or nothing" affairs: when they come they come with all the unpredictability, suddenness and destructive force that we know from historical accounts. In the study area of Petra (Jordan) such a flood killed 28 persons, most of them tourist nuns, in the entrance gorge (El Siq) to the ancient city in 1963. Only after this rare event the authorities decided to restore the safety channel that the Nabataeans had built 2000 years before to prevent exactly such an accident!

2 Reverse Festooning in Aerial Pictures of Wadis

An irritating phenomenon is recognizable, when flying over desert areas or looking at aerial photographs (Fig. 1). The dry flashflood beds (wadis) commonly show dark cross bands that have a distinct seleniform curvature. The bands most probably correspond to gravel bars or megaripples blackened by desert varnish. At first this feature may appear little surprising: ripples should proceed faster in the middle than near the sides of the stream bed. The curvature of the cross bands, however, points in the wrong direction: convex-upstream! Also, the patterns of the main wadis regularly intersect those of minor tributaries. This indicates a sedimentary process

Einsele et al. (Eds.)
Cycles and Events in Stratigraphy
©Springer-Verlag Berlin Heidelberg 1991

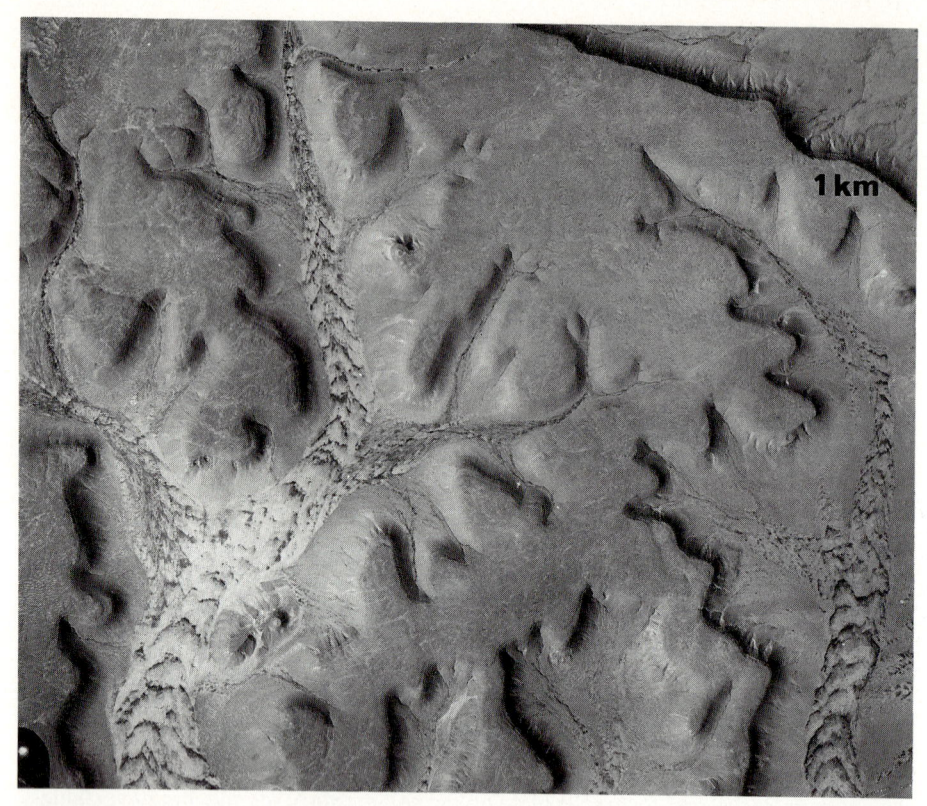

Fig. 1. Seleniform boulder bars in modern wadi beds near Rutba, Iraq. Note that the bars are arcuate, and have a darker edge in the upstream direction. Note also that the patterns of tributaries are intersected by those of the major branches, where runoff lasted longer. Picture about 3 km wide

that makes them prograde in proportion to current velocity, but in an upstream direction. It also suggests that this process stopped earlier in the tributaries than in the main channels. We have not had the chance to check this strange phenomenon in modern wadis. The following fossil example, however, may serve as a provisional substitute.

3 Reverse Cross Bedding in the Saramouj Conglomerate

The Late Precambrian Saramouj Conglomerate of the Petra area in Southern Jordan is a remarkable unit. Locally up to 420 m thick (Bender 1968), it fills the valleys of a considerable basement paleorelief along the northern rim of the Arabo-Nubian shield (Blanckenhorn 1914). Its well-rounded boulders, about 20–30 cm in diameter, are derived from this basement. The Saramouj is overlain by arkoses and marine sandstones whose Lower Cambrian age is established by trilobite trace fossils

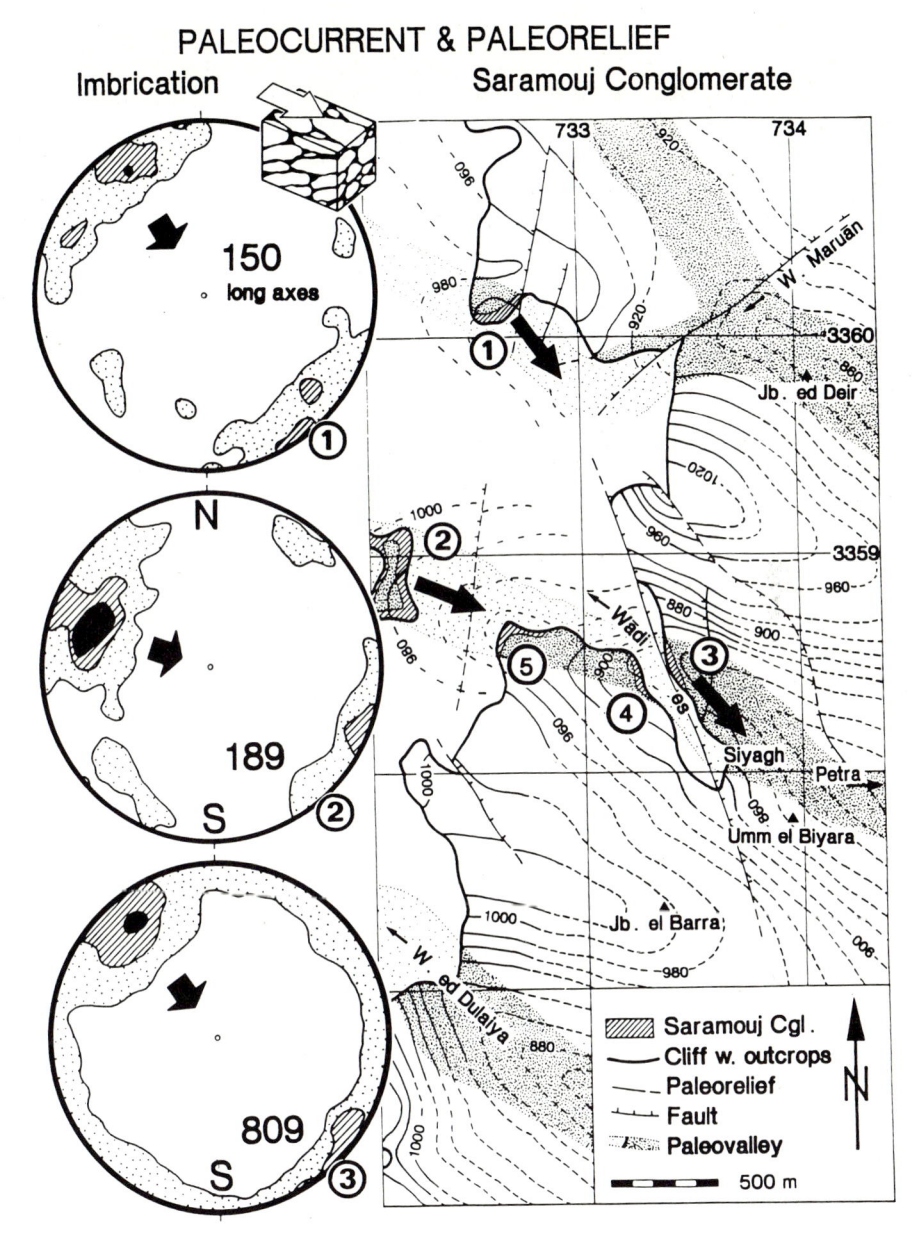

Fig. 2. The paleovalleys in which the Precambrian Saramouj Conglomerate of the Petra region (Jordan) was deposited, as well as boulder orientation imbrication (corrected for tectonic tilt), indicate drainage toward the southeast. This is in marked contrast to the northerly paleocurrent directions in the overlying Cambrian and Ordovician sandstones

(*Cruziana nabataeica*, Seilacher 1990b). It cannot be considered as the basal conglomerate of the Cambrian transgression, because its upper contact is nonconformable. In addition, its SE drainage (Fig. 2) opposes the northerly paleocurrent directions observed in all the Cambrian and Ordovician sandstones on top (Selley 1972; Seilacher 1983). This suggests that a major tectonic reorganization occurred during the time hidden in the upper unconformity.

Within the outcrop area, two lithologic facies of the Saramouj can be distinguished. Facies A of the Petra area is a clast-supported conglomerate of well-rounded cobbles with a prominent imbrication pattern. The coarse-grained arkosic matrix is variably indurated so that in places the boulders weather out readily. In Facies B of the Wadi Araba outcrops (e.g., south of Ghur an-Nmira), the boulders float in an indurated matrix of finer material.

The diamictic aspect of Facies B lead us to initially consider a glacial origin of the Saramouj in connection with the late Precambrian glaciation period. Careful examination of boulder surfaces in Facies A, however, did not reveal any glacial striations. Our present flash flood model does not exclude a glacial climate, but at the same time it does not require ice to explain the diamictite.

Our model is mainly derived from outcrop studies in Facies A along the gorges (Wadi es-Siyagh) and cliffs west of Petra. This implies that, unlike the planar view of wadi air pictures, the rock is seen only in vertical sections. Nevertheless, a similar contradiction between inferred paleocurrent direction and depositional structures is recognizable.

Paleocurrent directions are traditionally derived from bedding phenomena. In coarse, massive conglomerates like the Saramouj, however, bedding is not as obvious as in sandstones, so that paleocurrent direction must be derived from the shingling of blocks. Only from some distance can beds carved by differential weathering be recognized, particularly if the sun is at the right angle. As to be expected in such a coarse sediment, the cryptic bedding planes have the character of large-scale foresets: In dip direction individual sets are up to 80 m long and account for the whole thickness (20.5 m) of the Saramouj as, for example, in outcrop No. 2. Their ESE dip seems to be compatible with the paleocurrent direction derived from paleovalley configuration and pebble imbrication (Fig. 2). There is, however, a severe contradiction in the coincidence of

1. giant foresets (usually created by slipface accretion on transverse bars moving downstream; e.g., Massari 1983; Steel and Thompson 1983), and
2. within these foresets, elongated cobbles that are uniformly a (p) a (i) - imbricated (i.e., the long a-axis is dipping upstream).

The imbrication pattern is so prominent that it implies a smaller-scaled category of bedding dipping upstream at a higher angle. The larger and lower-angle category, on the other hand, corresponds to "false bedding", but nevertheless does dip in the right direction, i.e., ESE (Figs. 3, 4).

From sandy sediments we know that such directional divergence is characteristic for fast clastic sedimentation in unidirectional currents. Ripple foresets dip downcurrent, while a higher, but lower-angle order "false bedding" produced by climbing ripples is inclined in an upcurrent direction. This occurs due to the fact that sand

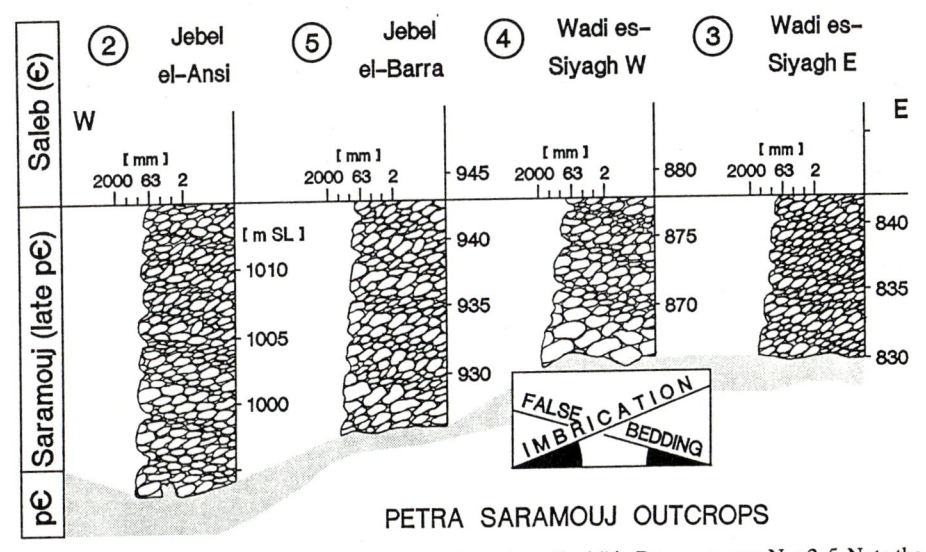

Fig. 3. Comparison of grading cycles and large-scale backset "beds" in Petra outcrops Nos 2–5. Note the paleorelief, the uniform imbrication and the overall fining-upward trend. Section No. 4 comprises in situ-spheroids at the base boulders rounded by in-place weathering

grains in a lower flow regime slide down and come to rest on the lee side of the ripples. The situation may become reversed, however, in a higher flow regime, when vortices become stronger and sand accumulates on the upcurrent stoss side of the ripples (antidunes).

We postulate that the reverse cross bedding of the Saramouj pebbles expresses a mode of bottom transport similar to that of antidunes. Sand grains normally roll up the stoss side of the ripple to the crest. Subsequently, microslides are rhythmically released down the lee slope, with the coarsest grains accumulating near the base. This means that the *remanence time* of sand grains is lower on the stoss than on the lee slope and that starved ripples migrate downcurrent by erosion on the stoss and deposition on the lee sides. With increased sediment supply, stoss erosion will decrease relative to leeside deposition, so that older foresets are preserved, ripples climb downcurrent and false beds dip upcurrent ("ripple-drift cross-lamination", "supercritical cross-lamination", e.g., Allen 1972).

During flash floods, pebbles may have a higher remanence time on the stoss slope of the megaripple where they are trapped in inter-clast gaps. If settled in the hydrodynamically optimal orientation, with the long a-axis dipping upstream, each single clast contributes to the resistance of the stoss-side. In a starved situation, therefore, erosion prevails on the lee slope and the ripple migrates upcurrent, in spite of a downstream transport of its components. Conversely, an increased pebble supply will make the ripples climb in an upcurrent direction (Fig. 4).

The dominant type of imbrication throughout the facies A - Saramouj, with elongated components aligned flow-parallel [a (p) a (i)], is highly diagnostic for vigorous flow and a high concentration of suspended and bed load. This was

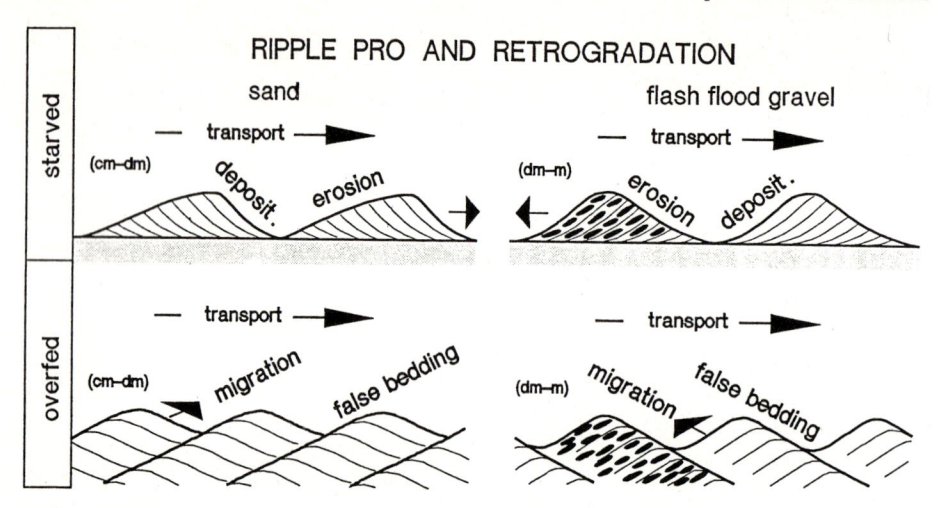

Fig. 4. In flash floods, boulder megaripples prograde in an upcurrent direction. As a result, backsets and imbrication dip upcurrent, while corresponding false bedding dips downcurrent – opposite to what we know from sand ripples. Upcurrent progradation also explains why the boulder ripples curve upcurrent across the stream bed in planar view (Fig. 1)

suggested by Rust (1972), and further corroborated by theoretical and experimental investigation of Johansson (1976), with the same results. The contribution of this type of imbrication to flash flood processes cannot be overemphasized, as it is the reason for the increase of remanence time of cobbles on the stoss relative to the lee side. On the other hand, a (p) a (i) - imbrication in the stoss side beds is, per se, the evidence of a backward accumulation instead of a forward avalanching (e.g., Rust 1984) of the sediment.

False bedding requires extreme currents as well as a sudden and ample supply of imbricatable sediment – a combination that is most likely to be found in the flash flood environment.

The suddenness and intensity of flow corresponding with this special type of imbrication is in excellent agreement with the flow type deduced from the large-scale sedimentary structures (Fig. 3), as well as with the structural pattern observed on aerial photographs (Fig. 1).

The surface ripples themselves are, of course, not preserved in the truncated remnants of the unit in Petra. Their dimensions, however, can be estimated from outcrop parameters. The wavelength l of ripples is calculated from the thickness t of the complete grading cycle (Fig. 5), and the dip angle α of the false bedding surfaces by

$$l = \frac{t}{\sin \alpha} \tag{1}$$

The height h of the ripples can be approximated if we assume the longitudinal section to be a sine curve of wavelength l. This simplification is permissible until the exact shape of backclimbing ripples has been either determined in tank experiments or identified in nature. If we further assume correspondence of the measured angle of imbrication β with the maximum gradient at the flank of the ripple, we find the relation

$$h = \frac{1}{\pi} \tan \beta \qquad (2)$$

Grading cycles of t = 0.8 m, α = 20° and β = 17° (outcr. No. 3) accordingly correspond to ripples of l = 2.3 m and h = 0.23 m, whereas wavelengths of more than 20 m and a height of 3.1 m (t = 3.5 m, α = 10°, β = 26°, outcr. No. 2) represent the upper end of the scale observed in the field. The calculated values of h may be slightly too high, as, for hydrodynamic reasons, the angle of imbrication β should exceed the gradient at the stoss side of the ripple.

Structures similar in texture, bedding type, and magnitude may be generated by floods resulting from sudden drainage of lakes, as thoroughly described by Bretz et al. (1956) and Baker (1973) from the Pleistocene glacial lake Missoula. Blair (1987) reported the buildup of gravelly antidune sediments and well-imbricated "non-cohesive sediment gravity flows" from an alluvial fan that was built after a dam failure. Such giant transverse gravel bars were also discovered in modern low-sinuosity streams (Gustavson 1978). They can be easily distinguished from flash flood boulder ripples

1. by their grain size,
2. by the abundance of internal erosion surfaces (evidencing a frequent shift of the flow direction), and
3. by the lack of typical backsetting features, such as false bedding and imbrication.

Coarse-grained meander lobes, as described by Arche (1983) and Billi et al. (1987) also show giant foresets and imbrication, but imbrication is obliquely oriented to the dip of the lateral accretion. It also seems that those foresets hardly can reach the necessary dimensions. Coarse fillings of cross-cutting channels or abandoned chutes, like those described by Doeglas (1962), are much closer to what we are searching

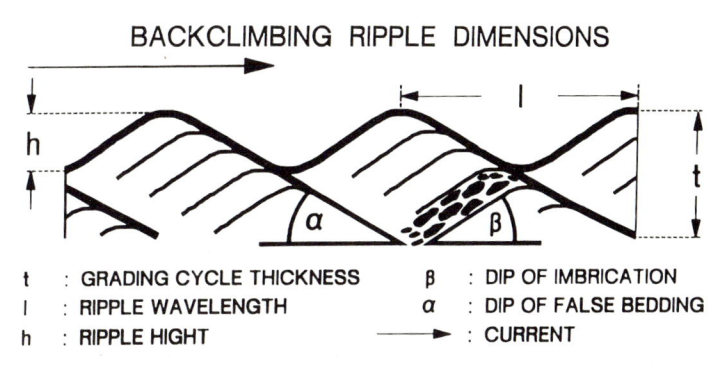

Fig. 5. Measures used for calculation of the original dimensions of boulder ripples in Eqs. (1) and (2)

for. They, too, are related to floods confined to a preexisting negative relief and also display coarse framework foresets dipping downstream. The textural pattern, however, with cobble a-axes transverse to the current, corresponds to a lower current drag (Rust 1972; Johansson 1976, see above), and there is again more than a whole order of magnitude between the depth of the filled-up Durance channels and the thickness of the Saramouj beds. Unchannelized sheetfloods, another desert phenomenon, were reviewed by Hogg (1982).

While it would be difficult to simulate a flash flood in a tank it might be possible to test this model with a sediment of bivalve shells, whose curvatures make them equally more stable on the lee side – provided their rollability allows the formation of megaripples at all.

4 Tentative Model of Flash Flood Sedimentation

Given the scarceness (to our knowledge) of detailed actualistic observations, experiments, and calculations (Pierson 1981; Costa 1983; Jarrett and Costa 1985; Blair 1987), we will now try to intuitively hypothesize the hydraulic and sedimentary properties of flash floods.

Like turbidites on the other end of the clastic facies spectrum, flash flood deposits reflect the temporal alternation of two regimes that differ not only in energy, but also in duration by an order of 10^5. In both cases, long quiescent periods are suddenly interrupted by short sedimentary events of very high energy. In contrast to turbidites, flash flood particles mature in a dry state and become transported in an aqueous medium. Desert soils primarily form by physical weathering that degrades and rounds large bedrock fragments by breaking angular flakes off their surfaces. Additional debris may be apported by wind. The result is a strongly bimodal raw material, in which rounded or wind-faceted boulders float in a matrix of small angular clasts and eolian sand and silt.

In the rare case that a local desert shower is strong enough not to be soaked up, but to produce a runoff, the water will soon transform into a turbid flow by picking up increasingly larger dry grains from the bottom. Eventually, the entire matrix will be transported in suspension, while boulders shuffle, roll, and jump along the bottom. Bottom transport of blocks automatically leads to self-enhancing bar formation; but in contrast to sand ripples, whose foresets form by microslides down the lee slopes, boulders will preferably come to rest on the stoss sides and together act as a resistant imbricated lag, which resists further movement during lower-stage flooding (Bluck 1979). This is what we see in Facies A, where only few, and preferably the heaviest, of the small matrix particles have been preserved. (Therefore it should also be the most promising facies for gold prospectors!)

The rest of the original matrix will move further downstream. However, instead of settling in the graded downstream succession known from normal stream deposits, the mixed suspension of a flash flood will eventually be "frozen" as it was deposited. Within a dry streambed, water will be lost when the current decreases and the sediment underneath has a high porosity. This is the case in the distal fan part of the system, where the suspension first transforms into a viscous slurry that is able to support even large cobbles during transport by buoyancy and dynamic interaction of the grains (dispersive pressure; Pierson 1981). The debris flow finally freezes by interaction of soak drainage and decreasing slope gradient, so that static grain-to-grain contacts and buoyancy enable boulders to maintain their random positions and orientations. Consequently we should expect a diamictic, tillite-like sediment (Facies B). In a way this process partly resembles the formation of gravel lobes in sieve deposits (Hooke 1967), or of swash marks that can be observed on a sandy beach – on a very minute scale, surely, but also under less hazardous conditions.

5 Conclusions

1. Flash floods are a desert phenomenon. Where bedrock is exposed without vegetation, a rare but heavy shower may produce enough instantaneous runoff that is not soaked up by the time it reaches the pediment. Instead, it turns into a turbid flow by picking up the finer particles that have accumulated by physical weathering and eolian sedimentation during extended dry periods.

2. Due to its high speed and density – Blair (1987) calculated a peak-flood density of 1.8, which means that suspended plus bedload sediment accounts for almost half of the flood volume – such a flow is competent to transport roundly weathered boulders of considerable size. Since they move largely as bottom load, boulders accumulate into large ripples. These boulder ripples grow upstream, because the boulders come to rest only on the stoss slopes, in their hydrodynamically most stable positions and with their long axes dipping up-current. This mode of accumulation – reverse to the behavior of ordinary sand ripples – is the reason why boulder ripples festoon and climb upstream, while backset "beds" dip upcurrent.

3. Further downstream, loss of water into underlying dry bed gravels is expected to transform the suspension into a mass flow, in which rounded boulders float in a matrix of finer and more angular debris. Distal flash flood deposits, as represented by Facies B of the Saramouj Conglomerate, can therefore be easily mistaken for glacial tillites.

4. Since our model is relevant not only for facies interpretation, but also for rare metal prospection and hazard protection, it would deserve further testing. Suitable sites are (1) ancient coarse sediments (for instance the Precambrian, gold-bearing Witwatersrand conglomerates of South Africa) and (2) areas which recently suffered flooding due to dam failures or sudden drainage of proglacial lakes and (3), of course, wadi beds.

2.12 Tephra Layers and Tephra Events

H.-U. Schmincke and P. van den Bogaard

1 Introduction

Volcanic eruptions are beautiful and awesome from a distance, disastrous and lethal from close by, and for the scientist in hindsight unique signals of extremely short-lived natural events. Volcanic ash layers deposited from explosive eruptions – if preserved in the geologic record – represent the most precise event signal on a regional, albeit not global scale. Tephra layers are superior to any other geological material for stratigraphic correlation and physical dating purposes and are therefore ideal for establishing high resolution event stratigraphies and event chronologies. Tephra layers are the only sedimentary-type bed than can be traced as recognizable event deposit across marine, terrestrial, and glacial environments. Tephra layers can reflect global climatic and paleoenvironmental changes and geotectonic activities that are the result of – or have caused – major volcanic eruptions. Curiously, study of tephra layers is still in its infancy.

Tephra is a collective term for all types of volcanic fragments regardless of grain size, composition, shape, mode of origin or depositional mechanism. *Pyroclastic fragments or pyroclasts* are particles generated during explosive volcanic eruptions caused or dominated by degassing of magma. *Hydroclasts* are volcanic particles formed during hydroclastic (phreatomagmatic) explosive eruptions, when magma comes into contact with external water and is disrupted by granulation and – at shallow water depth – by steam explosions, commonly but not necessarily accompanied by magmatic vesiculation. Pyroclasts and hydroclasts are called *essential or juvenile*, when generated from the erupting magma, *accessory or cognate* when broken from disrupted older co-magmatic rocks, and *accidental*, when derived from underlying basement rock of any composition. The latter two types of fragments are also collectively called *lithoclasts*. The terms *volcanic dust* (< 63 µm), *ash* (< 2 mm), *lapilli* (2–64 mm) and *blocks* and *bombs* (> 64 mm) merely describe the median grain size of tephra deposits. Lithified ash is called *tuff*. In common usage, lithified tephra layers, especially those interbedded with marine sediments, are loosely called ash layers. *Scoria* (if basaltic) and *pumice* (if evolved) are poorly defined terms for vesicular pyroclasts in the lapilli and bomb size range.

The nature and general significance of tephra have been treated recently in more detail in Fisher and Schmincke (1984, 1990), Schmincke (1988) and Cas and Wright (1987). Selected aspects of marine tephra layers were reviewed by Kennett (1982), Self and Sparks (1981) and Bitschene and Schmincke (1989). Here we will discuss

---

Einsele et al. (Eds.)
Cycles and Events in Stratigraphy
©Springer-Verlag Berlin Heidelberg 1991

tephra fallout and flow beds as documents of short-lived geologic events. We begin with a summary of some of the characteristics of subaerial and marine fallout and flow deposits, emphasizing their stratigraphic relevance and significance for physical dating. This part is followed by a discussion of geological events reflected by instantaneously generated and deposited tephra layers.

2 Tephra Deposits

Primary tephra accumulations comprise fallout, flow, and surge deposits. These differ fundamentally from each other in bed forms, thickness, internal organization, areal extent, and other parameters as a result of differences in mode of eruption, transport, and deposition (Fig. 1).

2.1 Fallout Deposits

Fallout tephra consists of pyroclasts or hydroclasts settled through air and/or water. Tephra layers a few cm to mm thick at medial or distal occurrences may form widespread sheets on land or on the sea floor (Fig. 2). On land, especially close to volcanic source areas, geologically young (Quaternary) fallout tephra deposits may be preserved in loess-paleosol sequences, lacustrine sediments, peat bogs, and glaciers.

2.1.1 Origin and Properties of Widespread Subaerial and Submarine Fallout Ash Layers

Fallout ash layers are produced mainly during pyroclastic eruptions such as Plinian and Ultraplinian but also Hawaiian and Strombolian, and during Vulcanian (phreatomagmatic or hydroclastic) and Phreatoplinian eruptions. Widespread fallout tephra layers form (1) from vertical eruption columns above the vent, and (2) from elutriation clouds rising above pyroclastic flows ("co-ignimbrite ashes", Figs. 1, 2), ash dispersal being governed by (1) voluminous eruptions generating high eruption columns, and (2) high shear wind velocities such as at the tropopause "jet stream" level (troposphere-stratosphere boundary). Plinian eruptions s.l. produce the most widely dispersed volcanic ash layers.

Eruption columns consist of a basal *gas thrust region* – generally several hundred meters in height – in which hot tephra and gas is expelled into the atmosphere, where momentum and acceleration forces dominate and external cold air is entrained at the column/air interface (up to four times the mass of the initial tephra/gas mixture) (Sparks and Wilson 1976) (Fig. 1). Heating and expansion of entrained air generates a tephra-gas-air mixture whose bulk density is less than the surrounding atmosphere. A *convective plume* dominated by turbulent buoyancy develops above the gas thrust and can ascend up to 50 km into the atmosphere until density equilibrium is reached. Above the level of neutral density, the column may

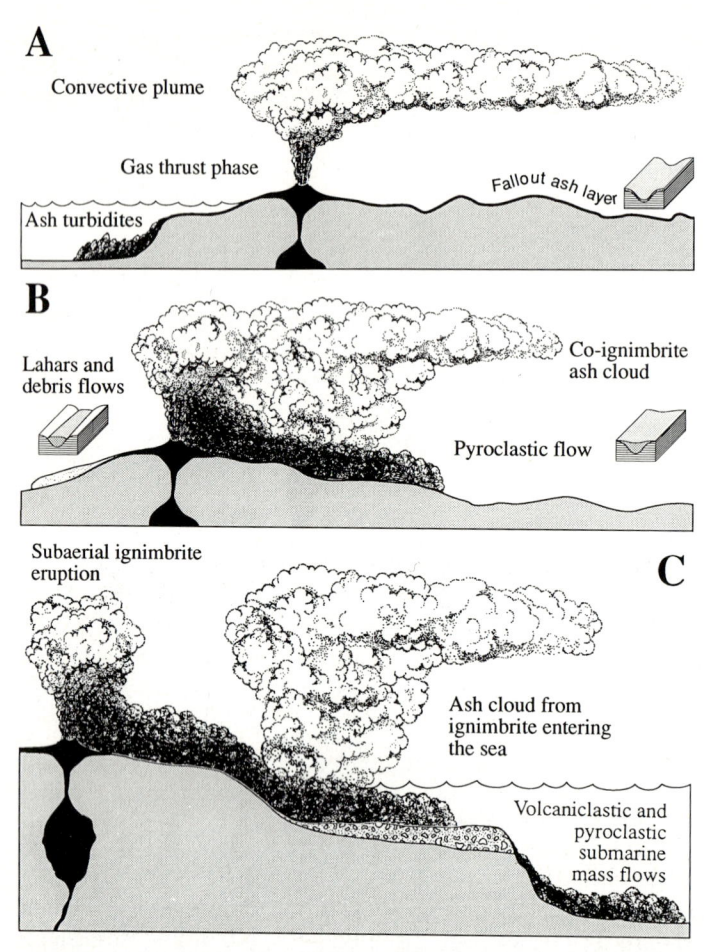

Fig. 1. Cartoon depicting major subaerial and submarine modes of transport of tephra systems: **A** Gas thrust and convective part of a Plinian eruption column generating subaerial fallout tephra layers. Erosion and/or primary pyroclastic flows/lahars and/or ash deposited on the shelf may become destabilized and form submarine ash turbidites. **B** Collapsing eruption column leading to formation of ignimbrites and co-ignimbrite ash clouds rising from pyroclastic flows. Lahars and debris flows may form independently from pyroclastic flows which may also grade laterally into lahars. 3-D diagrams indicate typical morphology of fallout, ignimbrite and lahar deposits. **C** Co-ignimbrite ash clouds generating widespread ash blankets may form when ignimbrites enter the sea. Pyroclastic flows may become submarine ignimbrites or mix with water to become submarine mass flow deposits

continue to ascend due to its momentum. The column then spreads horizontally and radially outwards to form an *umbrella cloud* (Sparks 1986). Radial expansion of the upper regions of large eruption columns significantly influences the fallout and dispersal of ejecta (Carey and Sparks 1986). Eruption column height is governed chiefly by the bulk thermal energy released and hence the magma temperature and magma discharge rate.

Fig. 2. Dark, fine-grained ash layers, some representing overbank deposits of co-genetic ignimbrites, interbedded with pumice lapilli layers. Late Quaternary Laacher See tephra deposits (Germany) at site "Burgerhaus", 8 km east of the vent. Total tephra thickness is ca. 6 m

The areal distribution and shape of fallout tephra blankets reflect the height of an eruption column and the strength and direction of the prevailing wind vectors (Figs. 3 through 5) (Eaton 1963; Walker 1973; Fisher and Schmincke 1984). Single tephra sheets are typically lobate to fan-shaped away from the source (Figs. 3, 5). Isopach patterns of fallout tephra deposits show a logarithmic thickness/distance relationship. The area over which a given size clast will fall is strongly correlated with column height (Wilson et al. 1978), thus column height and wind speed can be inferred from isopleth geometries (Carey and Sparks 1986). Individual beds of fallout tephra are commonly well-sorted and show few internal structures, except for grading (Figs. 2, 22).

Fallout tephra fans systematically thin away from the source, and are best defined by isopach maps constructed from thickness measurements. Source locations can be inferred by extrapolation of fan axes which may change direction (Fig. 5). Classification schemes for different types of fallout-producing eruptions are based on quantitative parameters derived from isopach, isomass, and isograd maps which describe the dispersal and bulk grain size (Walker 1973) or the rates of thinning and maximum clast size of deposits (Pyle 1988), but generally require extensive extrapolation and assumptions on the mathematical nature of thickness/distance and grain size/distance functions. Volumes of fallout tephra sheets can be estimated from integration of

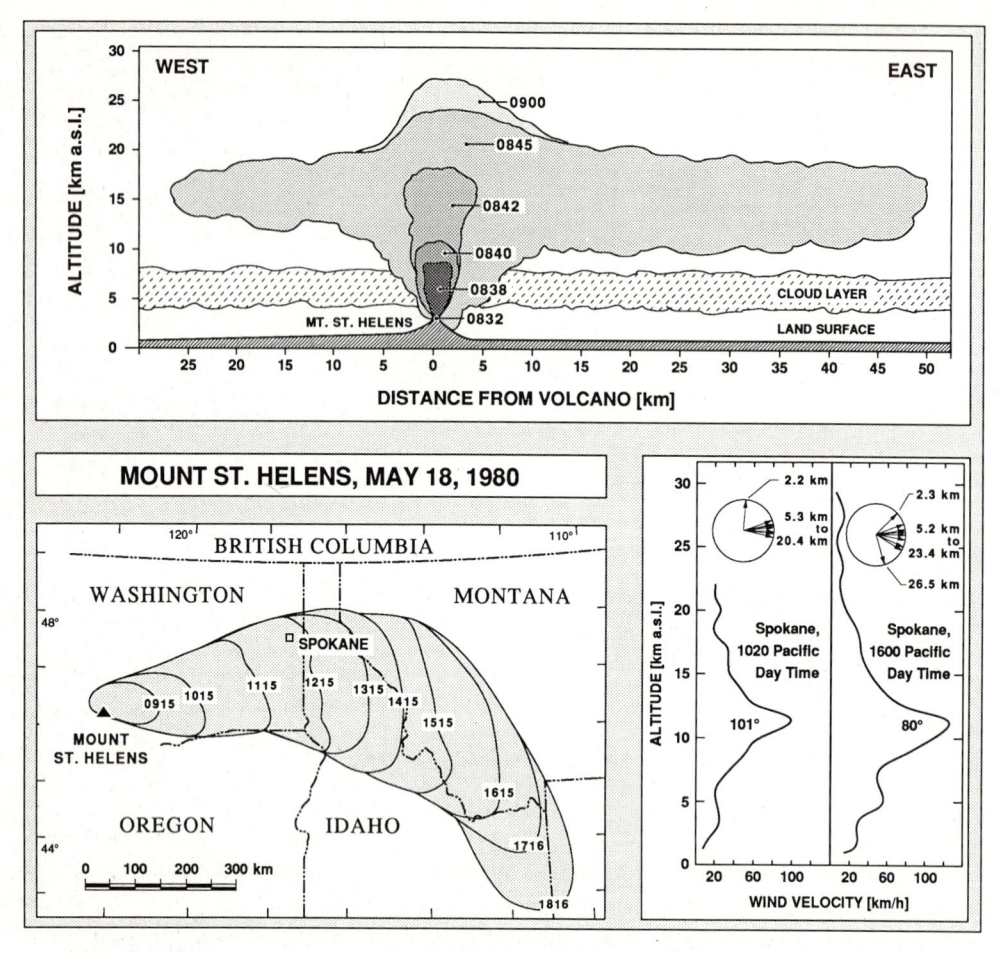

Fig. 3. Diagrammatic east-west profile showing early vertical growth and lateral expansion of plume from the May 18, 1980 eruption of Mt. St. Helens, compiled from visual and satellite observations (*upper*). Wind directions at selected altitudes (2.2 to 26.5 km) and average wind speed profile at Spokane, Washington during the eruption (*lower right*). Isochron map showing maximum downwind extend of ash from airborne ash plume, carried by fastest moving wind layer as observed on satellite photographs (*lower left*). Four digit numbers are Pacific Day Time. (Compiled from Sarna-Wojcicki et al. 1981)

log-isopach-area/log-isopach-thickness functions (Rose et al. 1973) or integration of a simple exponential cone (Pyle 1988).

2.1.2 Marine Ash Layers

Most well-studied marine fallout tephra deposits are generated by eruptions on land and are initially dispersed by wind. They are commonly less than 10 cm thick (rarely

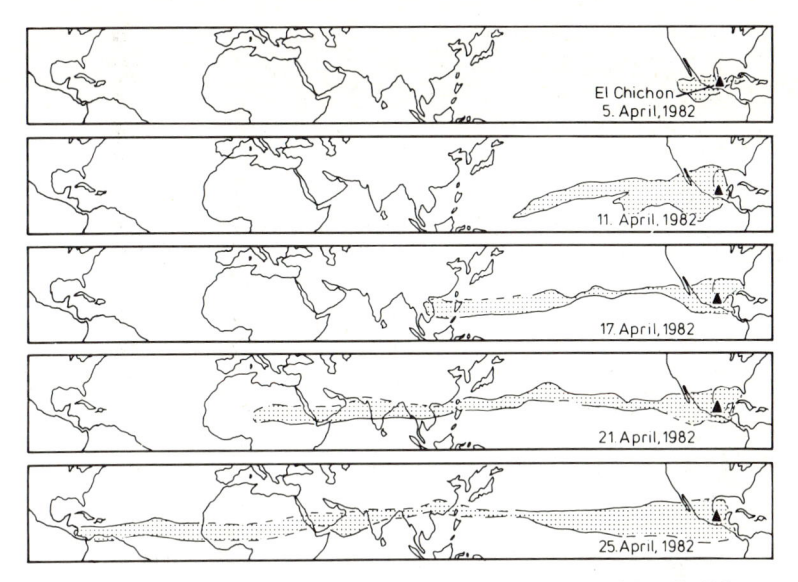

Fig. 4. Migration of stratospheric aerosol cloud of El Chichon eruption, Mexico 1982. (Newell 1985)

> 0.5 m), with a sharp base and a gradational top. Vertical zonation in color and composition may be due to gravity sorting by wind and fall through the water column, may represent a composite layer derived from several eruptions, or may reflect the successive emptying of a single compositionally zoned magma column. More efficient size fractionation of particles sinking in water results in better sorting of marine than subaerial fallout tephra (Ledbetter and Sparks 1979). Many marine ash layers, however, are poorly sorted and show irregular isopach patterns due to (1) local reworking by bottom currents; (2) sediment ponding; (3) bioturbation and compaction; and (4) aggregation with pelagic and hemipelagic materials (Ruddiman and Glover 1972; Watkins et al. 1978, Fisher and Schmincke 1984).

Eruption column height and magnitude and energy of an eruption can be estimated from the downwind grain size patterns of submarine tephra (Shaw et al. 1974, Huang et al. 1975; Ninkovich et al. 1978; Rose and Chesner 1987). The duration of large magnitude explosive eruptions can be inferred from the vertical size grading of nonvesicular ash particles in deep-sea tephra layers (Ledbetter and Sparks 1979). The size grading is a function of the duration of eruption, rate of release, residence time in the atmosphere, settling velocity, and water depth at the site of deposition. The model predicts a zone at the base of a deep-sea tephra layer where the coarsest particles remain constant in size with height above the base. The coarsest particle size decreases with height above this zone, with the break in slope corresponding to the level in the layer where the last large particle, ejected at the end of the eruption, was deposited. The finest particle deposited, together with the last largest particle, was erupted at the beginning of the eruption. Thus, the duration of the eruption is predicted to be the difference in settling time between the largest and smallest

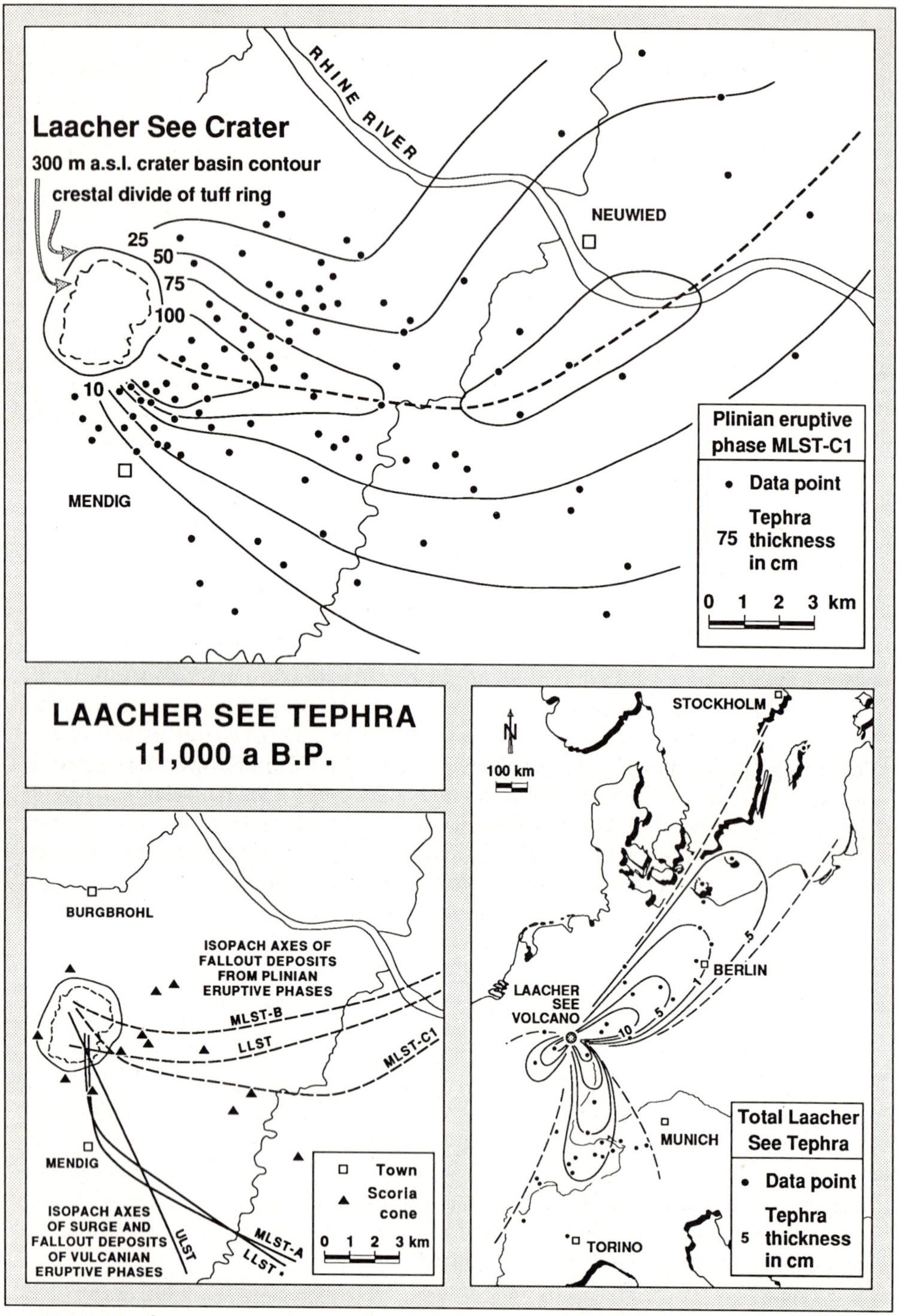
Laacher See Crater
300 m a.s.l. crater basin contour
crestal divide of tuff ring
RHINE RIVER
NEUWIED
MENDIG
25
50
75
100
10
Plinian eruptive phase MLST-C1
Data point
75 Tephra thickness in cm
0 1 2 3 km
LAACHER SEE TEPHRA 11,000 a B.P.
BURGBROHL
ISOPACH AXES OF FALLOUT DEPOSITS FROM PLINIAN ERUPTIVE PHASES
MLST-B
LLST
MLST-C1
MENDIG
ISOPACH AXES OF SURGE AND FALLOUT DEPOSITS OF VULCANIAN ERUPTIVE PHASES
ULST
MLST-A
LLST
Town
Scoria cone
0 1 2 3 km
N
100 km
STOCKHOLM
BERLIN
LAACHER SEE VOLCANO
MUNICH
TORINO
10
5
1
5
Total Laacher See Tephra
Data point
5 Tephra thickness in cm

particles at this level. Specific sources have been determined for a few geologically young deposits.

Shape, vesicularity, color and refractive index of glass shards, characteristic minerals (phenocrysts) and microprobe analysis and major and trace element composition of glass separates and bulk ash samples are the main methods used for characterizing and correlating marine and terrestrial ash layers (e.g., Sarna-Wojcicki et al. 1987; Carey and Sigurdsson 1978, 1980; Watkins et al. 1978).

Most clay rich layers called *bentonite* (or *tonsteins* in coal-bearing strata) are former ash layers, and nearly all bentonites are interlayered with marine sediments. (Fig. 6). Interpretation of their volcanic origin is based on (1) thinness of beds (generally < 10 cm), (2) wide areal distribution, (3) vitroclastic textures, (4) abundance of euhedral (phenocrystic) quartz, sanidine, plagioclase, amphibole, pyroxene, biotite or zircon, (5) expandable sheet silicate minerals, or zeolites like phillipsite, clinoptilolite or analcite, all derived from altered volcanic glass, and (6) geochemical indicators such as high concentrations of immobile incompatible elements such as REE and Zr (felsic, especially alkali-rhyolitic ash) and characteristic element ratios indicative of igneous derivation. Bentonites are the most important stratigraphic marker beds in coal measure-bearing sedimentary sequences.

2.2.2 Terrestrial Fine-Grained Ash Layers

Until about 20 years ago, fine-grained ash layers (modes below 0.15 mm, in very fine-grained layers below 0.07 mm) on land, many interbedded with proximal or medial pumice fallout beds such as those shown in Fig. 2, were almost universally interpreted as dust that had slowly settled during lulls in the eruptive activity following rapid sedimentation of fallout pumice lapilli. Closer inspection following initial studies at Laacher See Volcano and São Miguel (Azores) (Schmincke 1970; Walker and Croasdale 1971), however, showed that many were intimately associated with flow and surge deposits while others were poorly sorted. Although the origin of thin ash layers remains controversial, several different types of deposits formed by – or related to – flow deposits are presently distinguished:

Fig. 5. Near-vent and distal isopach patterns and isopach axes of 11 000 a B.P. Laacher See Tephra (LST) fallout deposits, East Eifel volcanic field, FRG. The multilobate structure of the entire event deposit (*lower right*) reflects the interplay of primary volcanic (eruptive mechanisms, eruption column heights) and environmental controls on the areal distribution of fallout tephra: ash erupted during Plinian stages (high convecting columns) was transported at high altitude to the NE (typical present-day tropopause wind direction); ash erupted during phreatomagmatic stages (low eruption columns) was transported to the S and SW, indicating northerly paleowind directions at lower altitudes. Note that on transsecting the lower atmosphere, Plinian columns were initially shifted to the E-SE as well, with isopach axes bending towards NE only at about 10 km distance from the vent (*upper*; eruptive phase MLST-C1). *Lower left* Near-vent isopach axes of selected LST eruptive phases. Extrapolation of isopach axes of contemporaneous and alternating Plinian and phreatomagmatic eruptive phases suggests a 1.5 km lateral SW-NE-migration of the eruptive vent between the early (*LLST*) and middle (*MLST-B*) Plinian stages. (Compiled from van den Bogaard and Schmincke 1984, 1985)

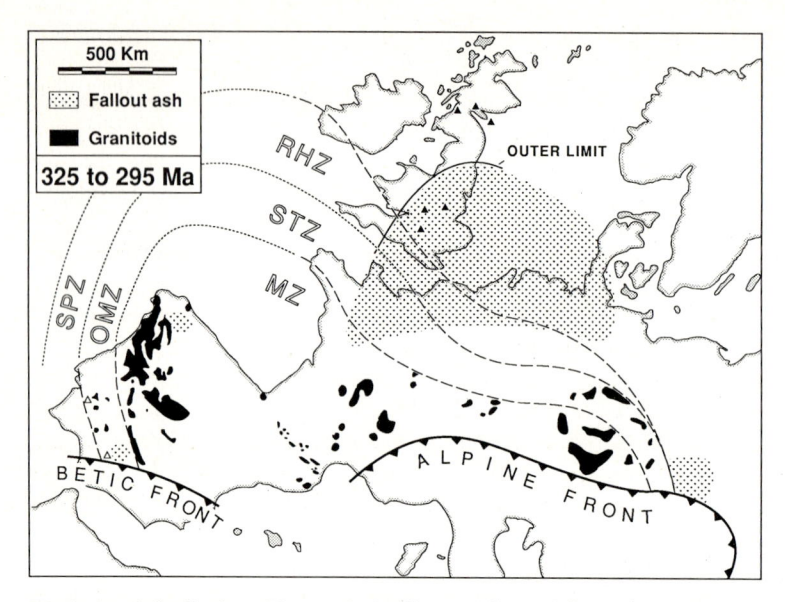

Fig. 6. Areal distribution of Lower Carboniferous fallout ash layers (bentonite or "tonstein" layers) in front of the main areas of granite intrusions in Europe. *Open triangles* (felsic) and *closed triangles* (mafic) indicate occurrences of volcanic rocks. *SPZ* South Portugese zone; *OMZ* Ossa Morena zone; *RHZ* Rhenohercynian zone; *STZ* Saxothuringian zone; *MZ* Moldanubian zone. (After Francis 1988)

Elutriation cloud or co-ignimbrite fallout deposits. Many *ignimbrites* (see Sect. 2.2) are overlain by thin (mm to cm) ash layers which mantle the topography uniformly, some extending laterally for hundreds or thousands of kilometers, well beyond the distribution area of the co-genetic ignimbrites. Following the recognition of enrichment of crystals in the deposits of basal ground hugging glowing avalanche by Hay (1959), some authors hold that up to almost 50% of the total erupted mass of ignimbrites are co-ignimbrite ashes based on the depletion of crystals in the accompanying ashes (deposits of the glowing airborne clouds) (Sparks and Walker 1977). Such co-ignimbrite fallout ash layers thus represent the fine ash elutriated out of the moving avalanche by hot gases rising from the flow, the actual billowing glowing cloud that largely hides – and is often mistaken for – the glowing avalanche itself. Many if not most thin ash layers, both terrestrial and marine, may actually be co-ignimbrite ash layers rather than being derived from high Plinian eruption columns (Sparks and Huang 1980).

Some co-ignimbrite ashes are generated when large pyroclastic flows enter the sea. The resulting secondary explosions – triggered by vaporization of water – are believed to form ash clouds that deposit extremely widespread, thin but voluminous fallout ash layers, some exceeding 5000 km in lateral extent (Walker 1979; Sigurdsson and Carey 1989; Sparks and Walker 1977).

Overbank facies ash layers. In the near-vent area of explosive volcanos, ridges and interfluves next to ignimbrite-filled valleys maybe covered by thin, fine-grained ash

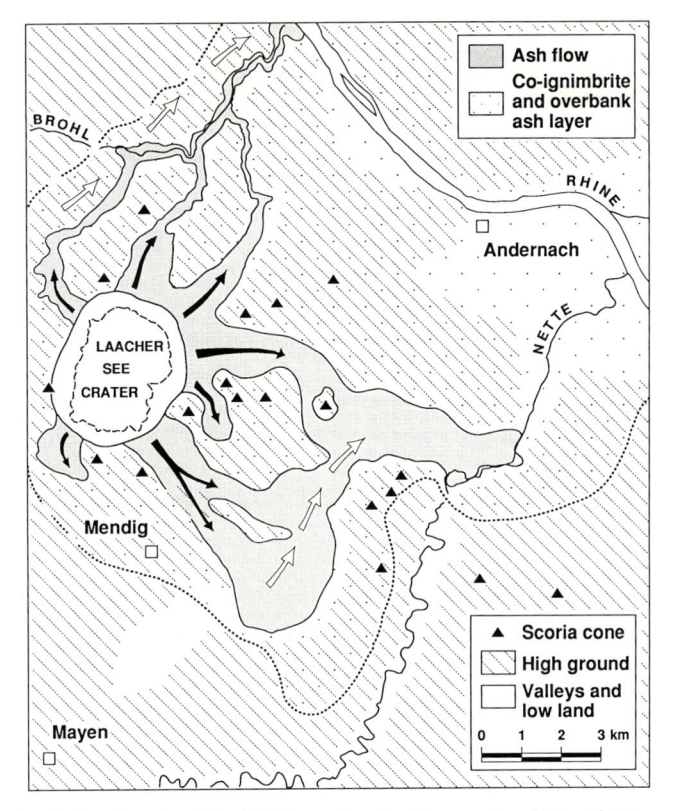

Fig. 7. Areal distribution of main ash flow deposits of the 11 000 a B.P. Laacher See eruption ("Trass"). The ash flows initially descended and spread radially through passes between morphologically prominent older scoria cones (*black arrows*), but flowed tangentially and accumulated to greater thickness (≥ 30 m) in paleovalleys to the NE and SE (*open arrows*). *Solid line* is 1-m contour of valley-filling ignimbrite deposits. *Dotted line* is 5-cm contour of ignimbrite overbank facies and co-ignimbrite fine-grained thin ash layers covering higher ground and paleovalley interfluves. (After van den Bogaard and Schmincke 1984)

layers which are laterally continuous with the main flow bodies such as those at Laacher See Volcano (Schmincke 1970; Schumacher and Schmincke 1989). These thin ash layers show abundant evidence of lateral emplacement from more diluted, finer grained, more turbulent flow lobes overriding higher ground and thus represent the overbank facies of valley-filling ignimbrites (Figs. 7, 10, 11).

At Taupo volcano (New Zealand), thin ash layers on higher ground called *ignimbrite veneer deposit* — related and laterally linked to a valley-filling ignimbrite — have been interpreted to represent material left behind because of ground friction from the basal and trailing parts of a single, highly mobile, high-velocity pyroclastic flow (Walker et al. 1981; Wilson and Walker 1985; C.J.N. Wilson 1985).

Fine-grained fallout ash layers unrelated to ignimbrites. There has been a tendency during the last few years to interpret all fine-grained ash layers on land as having either been deposited by some sort of horizontal transport mechanism or to have been derived from, or be associated with, ignimbrites. At the present state of knowledge, the main criterion for fallout origin of proximal fine-grained ash layers is their field aspect. An independent fallout origin can be assumed if thin proximal fine-grained ash layers maintain an even thickness over irregular terrain – save for thickness decrease away from source – do not show structures indicating lateral transport, and are not overlying or do not grade laterally into surge or flow deposits.

Some thin, fine-grained ash layers, e.g., at Laacher See Volcano interlayered with pumice fall deposits, are interpreted as fallout accumulations from eruption clouds during lulls in the eruption and by premature fallout of wet ash clusters (aggregates) and accretionary lapilli during phreatomagmatic eruptive phases as indicated by grain-size distribution (Schmincke 1990). Flushing out of ash clouds by rain is another mechanism for fallout of fine-grained ash close to source (Walker 1981). If external water gains access to the conduit, fragmentation of magma increases, resulting in higher mass production of fine-grained ash. Many fine-grained ash layers unrelated to ignimbrites indicate phreatic or phreatomagmatic origin.

2.2 Terrestrial Volcanic Mass Flows and Their Deposits

Volcanic mass flow deposits are composed of crystals, glass shards, pumice and lithic fragments, in highly variable proportions. The recognition of the character, widespread occurrence and mode of origin of different types of volcanic mass flow deposits is relatively recent. Each major well-studied eruption of an active volcano greatly expands our understanding, especially of volcanic mass flows and their deposits. At present, most terrestrial volcaniclastic mass flows can be grouped into ignimbrites, surge deposits, block and ash flow deposits, debris avalanche deposits, and lahars.

Pyroclastic flows are volcanically produced hot, gaseous, particulate density currents. Their deposits, called *ignimbrites*, are among the most voluminous and instantaneous volcanic manifestations. Ignimbrites are generally poorly sorted and massive, but may show subtle layering such as graded basal zones, discontinuous trains of large fragments, alternating coarse- to fine-grained layers, crude orientation of elongate or platy particles, and by color or compositional differences.

Ignimbrites range widely in volume. Eruptions producing small-volume pyroclastic flow deposits (0.001 to 1.0 km^3) are from small central vent volcanoes, as for example, 1902 and 1929 Mt. Pelee (Martinique), 1968 Mount Mayon (Philippines), 1976 Augustine volcano (Alaska), 1980 Mount St. Helens (Washington), and 1982 El Chichon (Mexico) (Figs. 7, 8). Intermediate-volume flow deposits (1–100 km^3) originate from small calderas and large strato-volcanoes, some with summit calderas resulting from collapse, such as 1883 Krakatau (Java) and Mount Mazama, Crater Lake (Oregon). Deposit volumes of 100 to > 1000 km^3 are associated with large calderas such as Long Valley, California (Bailey et al. 1989) (Fig. 9). Giant ignimbrite emplacement units may contain > 3000 km^3 tephra and their areal extents can exceed 1×10^5 km^2, rivalled, but not exceeded, only by single flood basalt units.

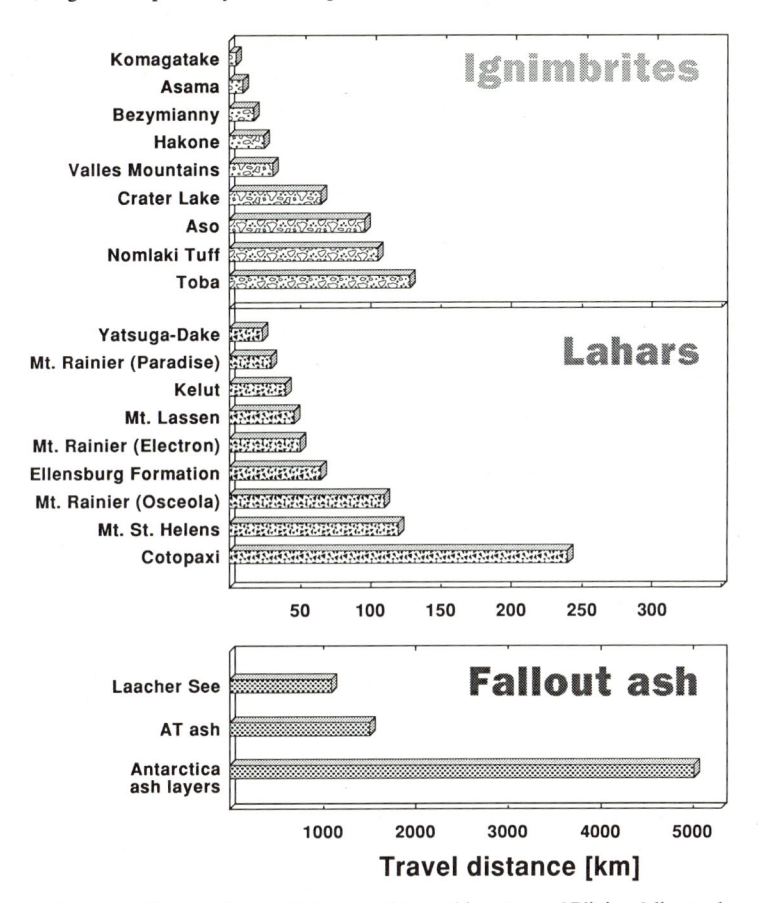

Fig. 8. Typical transport distances of ignimbrites and lahars, and traceable extent of Plinian fallout ash layers. Various sources compiled in Schmincke (1988) and Fisher and Schmincke (1984, 1990). Cotopaxi and St. Helens lahar data may include lahar run-out hyper-concentrated stream deposits as well. (Fallout data from Machida 1981, Kyle and Seward 1984, and van den Bogaard and Schmincke 1985)

Pyroclastic flows originate from collapse of overloaded eruption columns, by ash fountaining from the vent, or as marginal collar of an eruption column (Fisher and Schmincke 1984, 1990). Large volumes of ash with particles of all sizes are transported horizontally as partly fluidized pyroclastic flows. The avalanche part of flows generally concentrates in valleys (Figs. 1, 7). Pyroclastic flows may begin turbulently (and deposit chiefly cross-bedded thin surge layers on the upper slopes), move laminarly in the main canyons and terminate as plug flows (Freundt and Schmincke 1986). The lateral facies of valley-ponded ignimbrites consists of flow units each a few centimeters thick many of which show features of flowage on a minute scale (grading etc.) (Schumacher and Schmincke 1989) (Figs. 10, 11).

The fundamental unit, the *flow unit*, is defined as a volcaniclastic mass flow emplaced within minutes or hours from a single event developed during an eruption from a single pyroclastic flow and/or surge deposited in one lobe (Smith 1960a,b).

Fig. 9. Cooling units of Bandelier outflow ignimbrite (Lower and Upper Bandelier Tuff) at Horn Mesa, ca. 27 km SE of the center of Valles Caldera (New Mexico, U.S.A.). Combined thickness is ca. 100 m

Individual flow units can vary from a few centimeters to many tens of meters in thickness. Structural and textural differences between flow units include changes in grain size and sorting, composition, basal lithic layers, pumice concentration layers and others. Both lahars (volcaniclastic debris flow) as well as pyroclastic flow units commonly show tripartite subdivisions into fine-grained basal, main body and top facies (Schmincke 1967; Sparks et al. 1973) (Fig. 12).

High emplacement temperatures can result in welding, sintering, re-solution of gas phases and loss of pore space, transforming pyroclastic flow deposits into massive lava-like rocks. *Ignimbrite cooling units* are composed of one or more flow units deposited in rapid succession and cooled together from emplacement to ambient temperature (Smith 1960a). Zones in simple cooling units comprise a top and bottom layer of unwelded tephra, and a zone of dense welding in the lower half of the cooling unit, grading upwards into less welded deposits which may be lithified due to crystallization of high-temperature vapor phase minerals.

Surges, turbulent high-velocity, low-density flows, are among the most devastating types of volcanic events. Surge deposits are especially common in phreatomagmatic tephra sequences, where they were first recognized (Moore et al. 1966). Wavy, lenticular, or low-angle cross-bedding, as well as high-velocity antidune and chute-and-pool structures, are most characteristic of base surge deposits, as are discontinuous bedding, U-shaped erosional channels, rounding of pumice lapilli, and

Fig. 10. Lenses of well-rounded pumice lapilli, deposited around lithic block, separated by fine-grained accretionary lapilli-bearing ash layers forming overbank facies deposits of valley-filling ignimbrites. Laacher See Tephra (Germany)

abundance of vesicle tuff layers (Schmincke et al. 1973; Lorenz 1974; Wohletz 1983). Many surge deposits are associated with ignimbrites deposited from pyroclastic flows (Sparks 1976). In contrast to both fallout tephra and ignimbrites, however, surge deposits generally extend only a few kilometers from source and grade into laminated ash cloud and fallout deposits. They are thus not treated further in the context of the present paper.

Volcanic mud and debris flows, generally called *lahars*, are superficially similar in composition and structure to ignimbrites whose distal facies they commonly form. They are distinguished from pyroclastic flow deposits mainly by one or more of the following characteristics: more polymict composition, stronger rounding of clasts, higher proportion of clay-sized matrix, smaller abundance of pumice and general evidence for water and low depositional temperatures.

Lahars can form during the course of a volcanic eruption, but also on older volcanoes. Lahars form a large volume of most composite volcanoes, and are the most abundant type of reworked deposit, forming widespread aprons in the proximal facies surrounding the lower slopes of explosive volcanoes. Lahars form thick sheets and can travel as far as 200 km (Fig. 8). Lahars represent an intermediate facies between

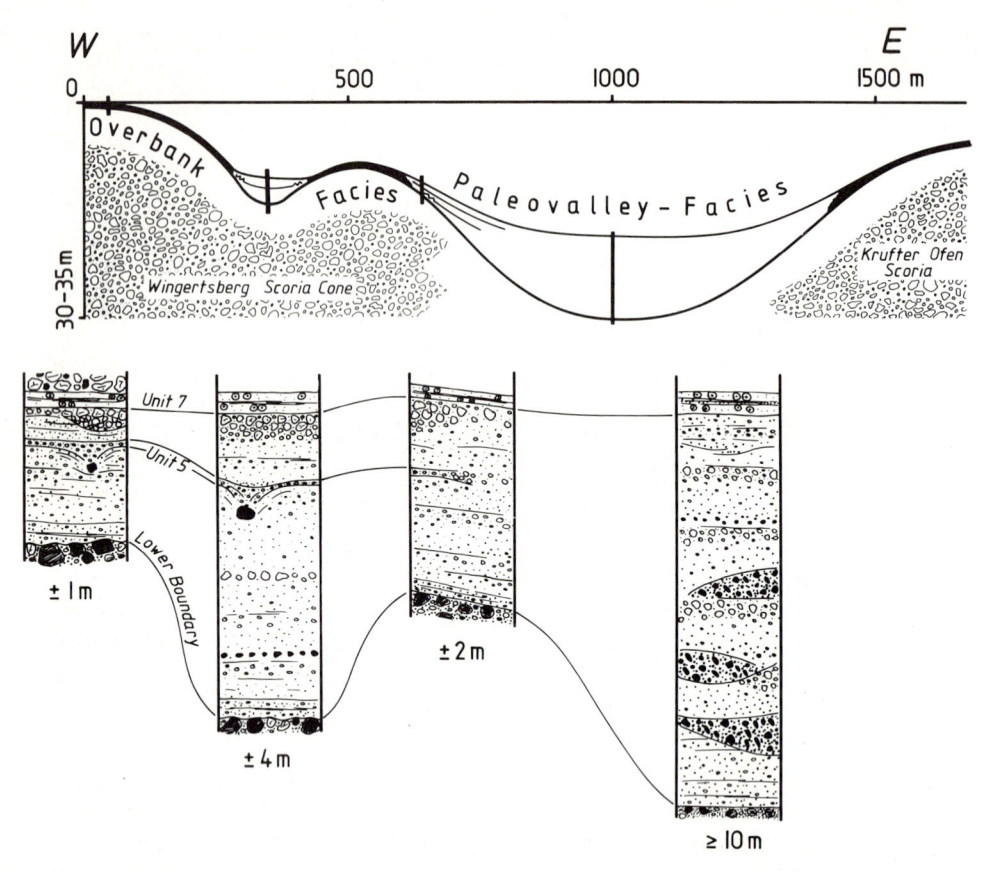

Fig. 11. Schematic east-west section across the Mendig paleovalley (Laacher See area) showing field relations of ignimbrite overbank facies draping the Wingertsberg and Krufter Ofen scoria cone and the paleovalley fill in between. *Vertical sections* illustrate depositional characteristics. (Schumacher and Schmincke 1990)

the primary tephra deposit on the slope of a volcano or, e.g., pyroclastic flows that entered a river, and the hyperconcentrated stream deposits into which lahars commonly grade downstream as the ratio water/clasts increases (lahar run-out; Figs. 13, 14) (Pierson and Scott 1985; Hackett and Houghton 1989). Lahars are a major source of volcanic particles and form an intermediate on-shore depot for volcaniclastic marine debris flows, grain flows, and turbidites.

Rapid growth of volcanoes results in increasingly rapid erosion because slopes become very steep. Reworked primary pyroclastic and epiclastic volcanic debris are carried away by rivers leaving little or no record of sedimentation near the source (Fig. 14) (Smith 1988).

Debris avalanche deposits, a newly recognized but widespread type of volcaniclastic deposit, form extremely coarse-grained, poorly sorted hummocks or small hills surrounded by matrix (Voight et al. 1981). Debris avalanches (and the lahars

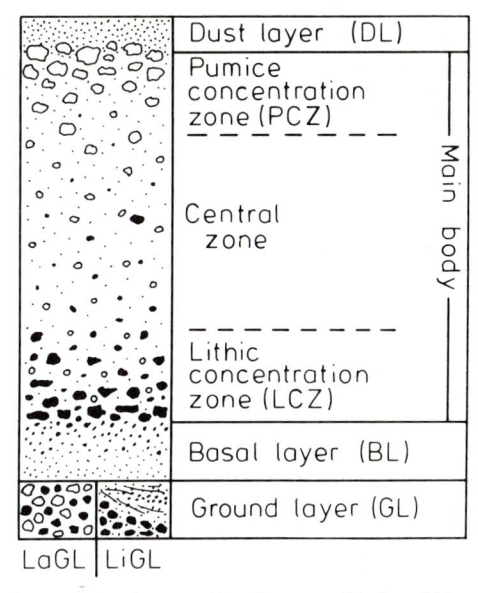

Fig. 12. Schematic representation and designation of structural and compositional zones and layers within, and associated with ignimbrite flow units. (Freundt and Schmincke 1986). *LiGL* lithic-rich ground layer; *LaGL* lapilli-rich ground layer

derived from them) are typically associated with active volcanism and rapid edifice growth and result from failure of oversteepened unstable slopes leading to sector collapse (Ui 1983; Voight et al. 1981; Siebert 1984; Glicken 1986; Fisher et al. 1987). More than 100 occurrences of debris avalanche deposits have now been recognized throughout the world since 1980. Repeated collapse of volcanic edifices obviously is a major process during the growth history of many subaerial and also submarine volcanoes (Lipman et al. 1988).

2.3 Submarine Volcaniclastic Mass Flows and Their Deposits

Volcanic deposits and successions in sedimentary basins may be thick but have often been deposited quickly. Most submarine sediment gravity flow deposits surrounding seamounts and oceanic islands are of basaltic composition (Figs. 15 through 17). A vast amount of volcanogenic debris is also shed into fore-arc and back-arc basins next to island arcs and continental margin volcanic chains, volcanoes being basaltic-andesitic-dacitic-rhyolitic in composition. Single oceanic islands have lifetimes rarely exceeding 20 Ma, but the lifetime of an entire magmatic system and thus the volcaniclastic supply to the basin are much larger in groups of volcanic islands and along volcanic arcs or active continental margins.

In theory, submarine volcaniclastic mass flow deposits, which directly or indirectly reflect tephra events, include deposits from submarine eruptions, deposits

Fig. 13. Photograph of ca. 11 000-a-old reworked facies of Laacher See tephra deposits ca 10 km east of source (locality between Andernach and entry of Nette into Rhine in Fig. 7)

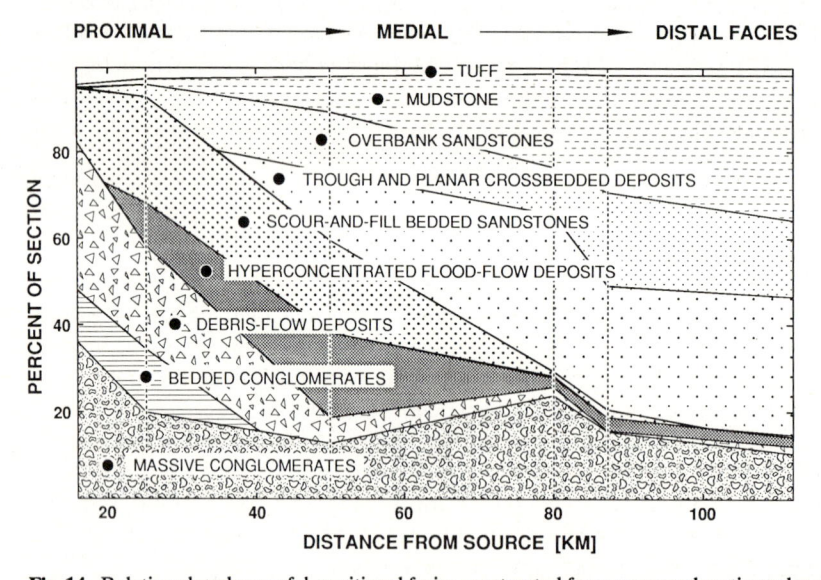

Fig. 14. Relative abundance of depositional facies constructed from measured sections along the depositional axis of upper Ellensburg Formation volcaniclastics in the Naches-Yakima Rivers paleodrainage (Washington, NW U.S.A.). (After Smith 1988)

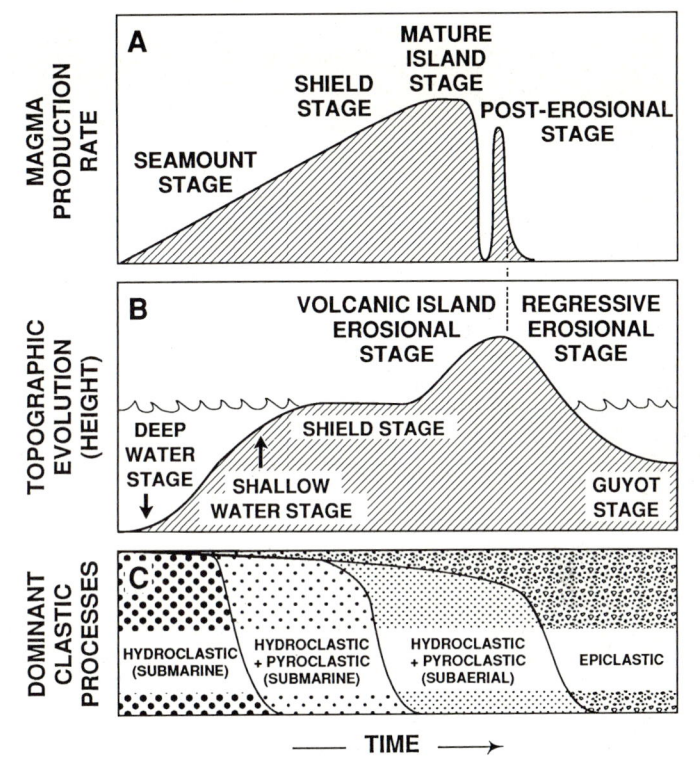

Fig. 15. Change of **A** magma production rate, **B** topographic height and **C** dominant volcanic and clastic processes during the evolution of volcanic ocean islands. The difference in gradient in **A** and **B** mainly results from the abundant production of volcaniclastic material during shallow submarine and initial subaerial shield stages (combination of hydroclastic and pyroclastic processes)

resulting from entry of ignimbrites or lava flows into the sea, redeposition of volcaniclastics formed by shallow water eruptions, redeposition of volcaniclastic material formed by erosion (unrelated to specific volcanic events), and submarine debris avalanche deposits. In practical terms, it is commonly impossible to deduce the primary origin of a marine volcaniclastic mass flow deposit.

2.3.1 Deposits from Submarine Eruptions

Pyroclastic flows may possibly form entirely from underwater eruptions, but we know very little about the nature of volcaniclastic mass flow deposits that are unequivocally the direct result of submarine volcanic eruptions. There is no evidence at present, for example, to suggest that an eruption column entirely beneath water can generate a hot pyroclastic flow, although submarine lava fountains generating spatter deposits appear to be likely, judging from some deposits in ophiolites (Schmincke and Bednarz 1989). Many subaqueous pyroclastic flow deposits resemble turbidites

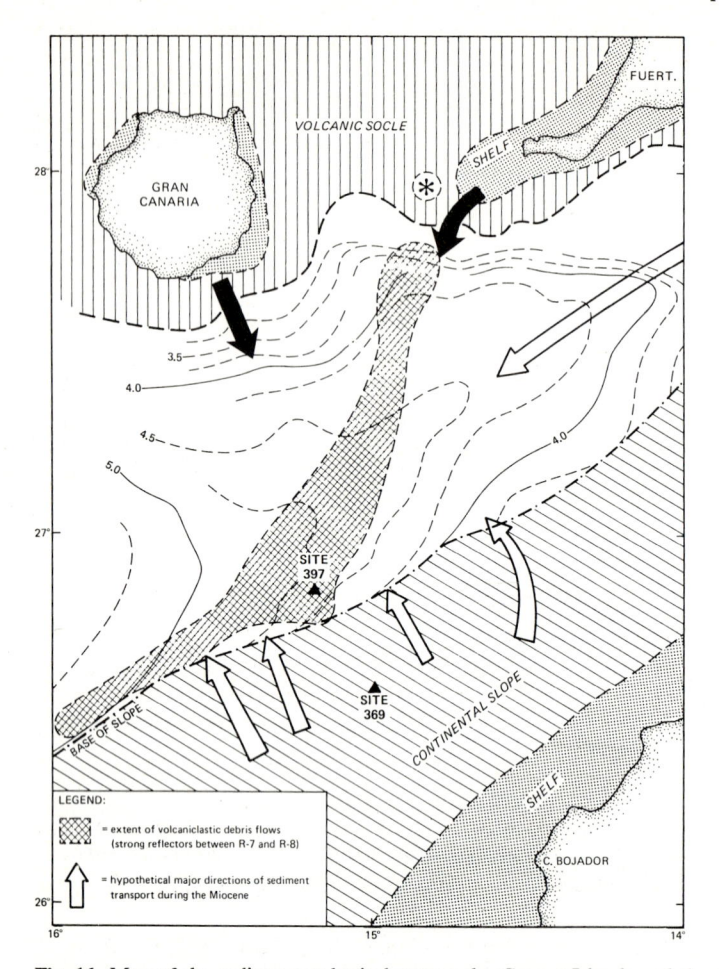

Fig. 16. Map of the sedimentary basin between the Canary Islands and the northwest African coast, showing depth contours (in seconds of two-way travel times) of seismic reflector R7, representing widespread volcaniclastic debris flow deposits probably generated during the basaltic shield building phase of Gran Canaria (V1 and V2 in Fig. 17), overlying middle Miocene volcaniclastic debris flows (V3; from Fuerteventura ?). *Black arrows* indicate island-derived sediment transport directions. *Asterix* indicates basement high. (After Wissmann 1979)

or coarser-grained mass flow deposits and thus most likely represent remobilized pyroclastic debris originally deposited by fallout or other processes along the shoreline or shallow-water flanks of an active volcano. At present, we cannot distinguish flows that originate in this manner from flows that might have been initiated by underwater eruptions. In any case, it is likely that such deposits come to rest at low temperatures. Structures, textures, lateral extent and frequency of occurrence of mass flow deposits resulting directly from submarine eruptions are thus entirely conjectural. Fiske (1963) and Fiske and Matsuda (1964) have discussed deposits believed to be the result of underwater eruptions.

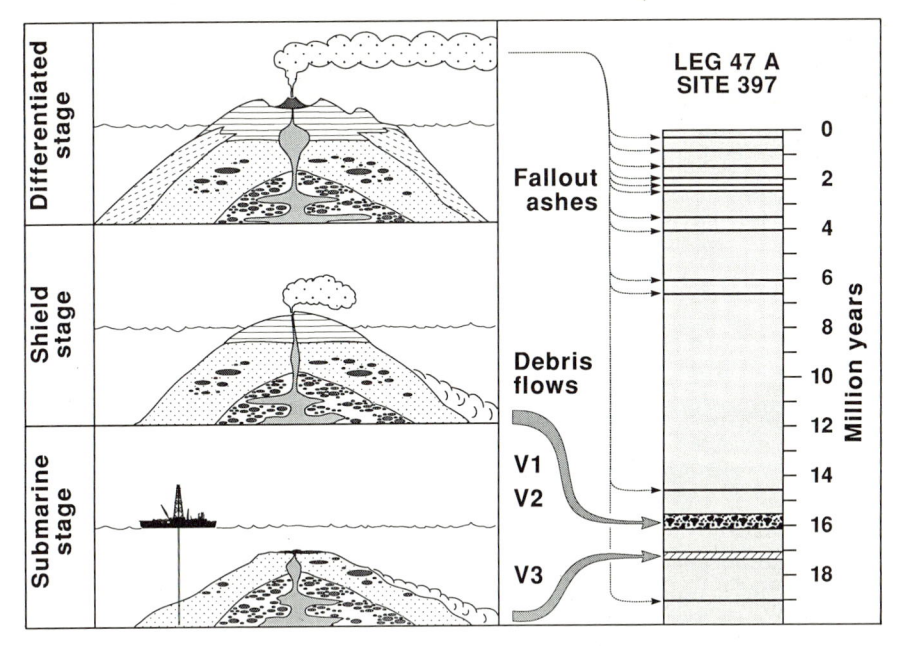

Fig. 17. Three schematic growth stages of an ocean island as reflected in three main types of submarine volcaniclastic rocks in the drill core of DSDP site 397 south of the Canary Islands. Stratigraphic position and age of fallout ashes (*solid lines*), epiclastic volcaniclastic (*V1, V2*) and hyaloclastite debris flows. (After Schmincke 1982)

2.3.2 Deposits Resulting from Entry of Ignimbrites or Lava Flows into the Sea, Directly Reflecting Volcanic Events

Pyroclastic flows are known to have entered the sea following subaerial eruptions and short subaerial transport on land in several cases (e.g., 1902 Mt. Pelee, Lacroix 1904; Augustine Volcano 1976, our observation), but none are definitely known to have continued under water as true pyroclastic flows, either from historic eruptions or from the fossil record. They occur mainly around active island arcs, active continental margins, and intraplate volcanic islands (Fig. 1). On land, eruptions where flows enter water generate complex event stratigraphies, since the land-sea transitions comprise (1) subaerial flows that either pile up along the shore to form complex breccias and/or continue as submarine mass flows; (2) lahars and hyperconcentrated stream deposits, associated with the ignimbrite deposits; (3) tephra fall deposits generated when a flow hit the sea (see above), and (4) submarine grain flow and turbidite deposits.

Hot pyroclastic flows probably mix with water when entering the sea and lose their character as hot pyroclastic flows. They become quenched and piled up and may continue directly or from intermediate depots as mass flows downslope and along the floor of adjacent basins. This is a likely scenario even though there are no well-documented case histories. One example of a pyroclastic flow continuing under water, often quoted, has turned out not to be supported by more detailed evidence: a pyroclastic flow deposit described by Carey and Sigurdsson (1980) from the Grenada Basin was originally interpreted to be a debris flow(s) initiated by the entry of hot subaerial pyroclastic flows into the sea at the mouth of the Roseau Valley, Dominica, as indicated by (1) tracing the subaqueous deposit to a subaerial pyroclastic flow deposit at the mouth of the Roseau Valley, (2) the similarity in composition and grain size characteristics of the subaerial and subaqueous deposits, and (3) the occurrence of charcoal in the subaqueous deposit. These rocks have now been reinterpreted as reworked material of the Roseau ash (Whitham 1988). Other views are discussed in Fisher and Schmincke (1984, 1990).

2.3.3 Redeposition of Volcaniclastics Formed by (a) Shallow Water Eruptions (b) Entry of Lava Flows and Pyroclastic Flows into the Sea, and (c) Erosion Unrelated to Specific Volcanic Events

These scenarios will be discussed jointly since their deposits probably make up the bulk of marine volcaniclastic mass flow deposits. Moreover, the mode of transport, extent and other aspects relevant for this volume are broadly similar, differences lying chiefly in the early history of the transported sediment domains and possibly the triggering mechanisms.

The question of whether a volcaniclastic mass flow deposit originated from a subaerial or subaqueous eruption is difficult to assess. Subaerial volcanic activity is reflected in the following criteria: (1) The presence of tachylitic pyroclasts indicates slower cooling in subaerial compared to subaqueous eruption columns. (2) Iddingsitized olivine is a useful although not unequivocal criterion. (3) High vesicularity may indicate subaerial volcanic degassing. (4) Rounded epiclastic fragments of many types (polymict composition) may indicate subaerial erosion of a multicomponent volcanic terrain (Schmincke and von Rad 1979; Simon and Schmincke 1983; Schmincke 1987; 1988). Reworked volcaniclastic mass flow deposits comprise submarine debris avalanche deposits, lahars, submarine landslides and shield-building lava-derived mass flows. (Figs. 16, 17).

In areas where steep-sided volcanoes have grown near the shore, as in many island arcs, but also around several oceanic islands where pyroclastic flows are much more common than generally believed, there is a common lateral succession from ignimbrites to hot lahars to cold lahars to hyperconcentrated streams to fluvial sediments which are the source for many mass flows transported into deeper water directly or via sediment deposits, depending on the slope. Lahar sediment reservoirs hold vast volumes of volcanic detritus that are easily reworked into adjacent marine and nonmarine sedimentary basins. The sediments stratigraphically record the time and place of volcanism that produced them.

In the deep sea record, medial and distal pumiceous volcaniclastic mass flow deposits are not uncommon. They are most likely the result of either pyroclastic flows or lahars, perhaps derived from pyroclastic flows that mixed with water in the near shore environment and continued downslope as gravity driven mass flows. Thus, while the accumulation of sediment material may reflect one or several tephra events, the formation of the actual deposits is generally a seismic event, which in turn may be triggered by a volcanic event. Quickly accumulated shallow-water tephra can be remobilized and transported into deep water, resulting in further problems of interpretation.

2.3.4 Volcaniclastic Aprons

Volcanic islands and seamounts are surrounded by large *volcaniclastic aprons* (Menard 1956) whose volume of clastic, island-derived material may far exceed that of the volcanoes (subaerial part plus submarine core and flanks). Volcaniclastic aprons appear to be formed chiefly by (1) submarine volcanic activity below the volatile fragmentation level, (2) explosive volcanic activity in shallow water and on land, and (3) lava flows and ash flows entering the sea, and (4) erosional activity, slumping, and mass flows. Widespread conspicuous seismic reflectors near volcanic islands are probably mostly volcaniclastic debris flow sheets (Fig. 16) (Wissmann 1979; Schmincke and von Rad 1979). South of the Canary Islands, for example, such a horizon traced by seismic profiling over an area of 3000 km^2 was correlated with a volcaniclastic grain flow deposit drilled at site 397, its volume amounting to 80 km^3. Seismic reflectors, presumably volcaniclastic layers, extend north of Oahu (Hawaii) for at least 200 km through the moat surrounding the islands (Watts et al. 1985; ten Brink and Brocher 1987).

The clastic aprons harbor a great wealth of material by which the temporal, bathymetric and compositional evolution of the source area and adjacent basins can be reconstructed. They can be divided into several facies domains comprising: (a) the proximal facies, the area of primary deposition of massive clastic extrusives (coarse pillow-fragment breccias, hyaloclastites and various types of volcaniclastics), shallow-water and steep-slope erosion, and the dominant source region for mass flows, (b) the medial facies comprising the steep slope with maximum transport energy and the lower plain as the main area of sedimentation consisting of mass flow deposits and extending to at least 100 km from an island, and (c) the distal facies, consisting of thin distal turbidites and fallout ashes, potentially extending to more than 1000 km.

Deposits from the submarine and shield stages of the Canary Islands have been encountered in DSDP drill holes (e.g., site 397; Schmincke and von Rad 1979), about 100 km SSE of Gran Canaria (Figs. 16, 17). In this hole, hyaloclastite debris flows, 6 m thick (ca. 17 Ma old), are interpreted as representing the submarine stage of an island, transported for more than 100 km to the south and southwest. The hyaloclastite flows contain angular, blocky and only slightly vesiculated sideromelane shards – now replaced by smectite and phillipsite – indicating a submarine growth stage perhaps at water depth below the volatile fragmentation depth (VFD) of ca 500 m. Similar near-source hyaloclastite flow deposits in the uplifted seamount part of La Palma (Staudigel and Schmincke 1984) are overlain by volcaniclastic debris flows containing abundant tachylite and a wide variety of subaerial volcanic and subvolcanic rock fragments, ranging from microgabbro to trachyte but dominated by alkali basalt.

Volcaniclastic debris flows above the hyaloclastite debris flows at site 397 are composed mainly of tachylitic, vesicular basalt, pyroxene, plagioclase and altered olivine crystals, and a variety of epiclastic rock fragments forming excellent seismic reflectors (Wissmann 1979). Their chemical composition indicates these debris flows to reflect the rapid growth of Gran Canaria during the shield stage, a very large amount of clastic debris being generated during the transition period seamount/island by phreatomagmatic and magmatic explosive activity as well as erosion of freshly formed clastic deposits. Younger ash layers of Sites 369 and 397 represent later, differentiated, subaerial ocean island stages (Figs. 15, 17), from Gran Canaria and Tenerife (Schmincke and von Rad 1979).

2.3.5 Submarine Debris Avalanche Deposits

A new type of volcanically induced event deposit has recently been recognized around the Hawaiian islands (Moore and Moore 1984; Lipman et al. 1988). These are giant submarine slide deposits with cumulative volumes up to 2000 km^3. The slumping events are believed to have been triggered by seismic activity during dike injection along the rift zones which characterize the Hawaiian Islands. Accompanying giant waves threw beach gravel >300 m up islands, such as Lanai. Massive slumps, slides, and distal submarine turbidite flows appear to be widespread on the flanks and within the clastic aprons of many oceanic volcanoes. Many beach deposits on oceanic islands, often used to infer isostatic and eustatic vertical movements and sea level changes, may have formed by the above chain of volcanically induced events and probably need to be reexamined.

Although these events have only recently been recognized and although the structures of the deposits are basically unstudied, they are probably quite common. We venture to predict that many studies will be published in the near future identifying and describing such deposits in the fossil record.

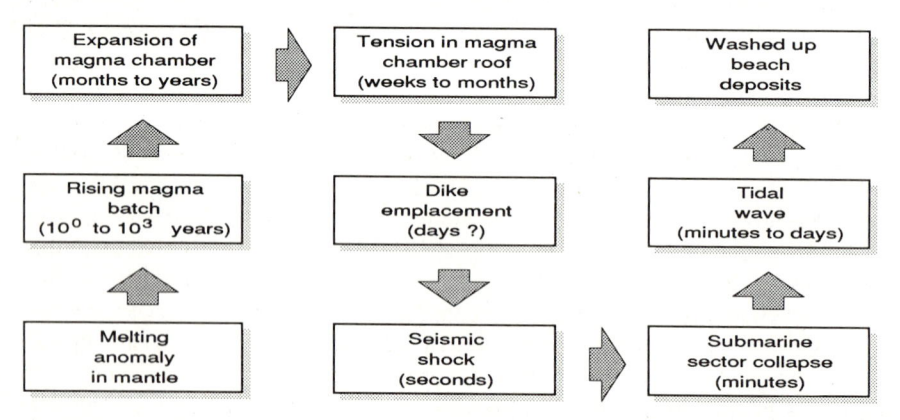

Fig. 18. Flow diagram showing succession of geological events producing seismically induced tidal wave deposits

Submarine debris avalanches are fascinating from a theoretical point of view as they exemplify the great complexities of geological causes and events (Fig. 18): a partial melting event in the mantle leads to the ascent of a magma batch and causes a magma chamber to expand, inducing tension in the roof above the magma reservoir and dike emplacement by magma-fracturing, either in the submarine or subaerial part of the volcanic edifice or both. Seismic shock and/or oversteepening of the volcanoes flanks then causes the collapse of a sector of the submarine cone, the resulting tidal wave substantially affecting beach and other subaerial deposits.

3 Discussion

Many aspects of prehistoric and older volcanic events – as reflected in a single tephra layer or more complex assemblages produced in a geologic instant – can be quantified precisely, including the point source of the event (within a few tens or hundreds of meters), the age (in some cases the season of the year), climatic conditions, mass and heat energy released, wind directions, height, speed, and transport directions, and first and second order mode of transport (fallout, flow, surge; hot or cold emplacement; primary or reworked) or subaerial or subaqueous mode of eruption.

3.1 Volcanic Activity Units

Volcanic phenomena are diverse, ranging from seismic activity, fumarole activity, high rates of heat flow, and from explosive ejection of tephra to quiet effusion of lava. Eruptive events on a volcano can be subdivided into *volcanic activity units* (Fisher and Schmincke 1984) which include: (1) *eruptive pulses* that may last a few seconds to minutes, (2) *eruptive phases* that may last a few hours to days and consist of numerous eruptive pulses, and (3) *single eruptions*, composed of several phases, that may last a few days to months, or in some basaltic volcanoes, a few years (Figs. 19, 20). This formal use of the term eruption differs from the loosely applied term eruption which covers several of the units distinguished here. Eruptive units, as defined by Freundt and Schmincke (1985a,b), comprise one or several eruptive phase(s) that produced compositionally relatively homogeneous sets of successive pyroclastic flow and/or surge deposits. Eruptive units are distinguished from other units by compositional differences clearly visible in the field because of particle size or color changes. At Laacher See, longitudinal variations of individual flow units in lithology, volume, and thickness are interpreted in terms of column collapse evolution. Systematic changes in thickness and inferred temporal succession within sets of flow units are interpreted in terms of vent migration and water access to the vent. Differences between eruptive units, in contrast, are interpreted in terms of compositional changes in the magma and changes in eruptive mechanism.

Some volcanoes (e.g., domes and basaltic scoria cones) may form completely within the time span of an eruption as defined above. Others, such as shield volcanoes, composite volcanoes, or volcanic islands, may show higher-order discontinuities

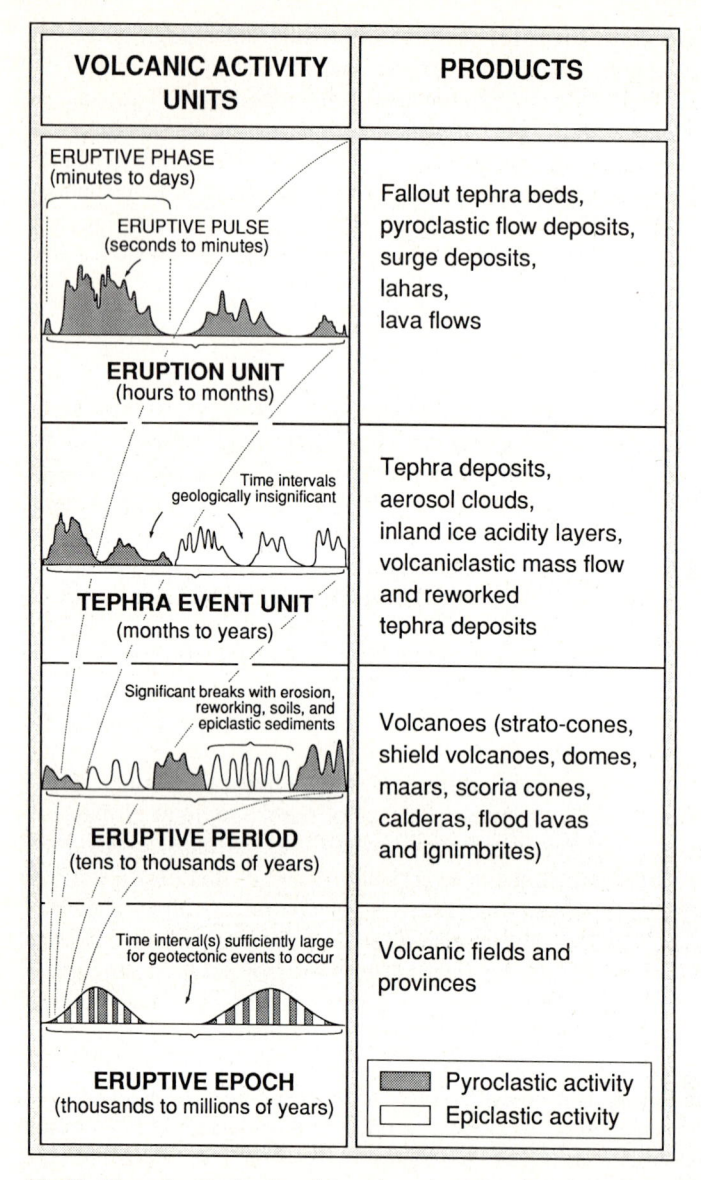

VOLCANIC ACTIVITY UNITS	PRODUCTS
ERUPTIVE PHASE (minutes to days) **ERUPTIVE PULSE** (seconds to minutes) **ERUPTION UNIT** (hours to months)	Fallout tephra beds, pyroclastic flow deposits, surge deposits, lahars, lava flows
Time intervals geologically insignificant **TEPHRA EVENT UNIT** (months to years)	Tephra deposits, aerosol clouds, inland ice acidity layers, volcaniclastic mass flow and reworked tephra deposits
Significant breaks with erosion, reworking, soils, and epiclastic sediments **ERUPTIVE PERIOD** (tens to thousands of years)	Volcanoes (strato-cones, shield volcanoes, domes, maars, scoria cones, calderas, flood lavas and ignimbrites)
Time interval(s) sufficiently large for geotectonic events to occur **ERUPTIVE EPOCH** (thousands to millions of years)	Volcanic fields and provinces ▓ Pyroclastic activity ☐ Epiclastic activity

Fig. 19. Hierarchy of volcanic activity units and associated products. (After Fisher and Schmincke 1984). For discussion see text

Fig. 20. a Well-bedded maar deposits of 1949 eruption on La Palma (Canary Islands), each layer representing a separate eruptive pulse. Trees were debarked but not burned by the relatively cool phreatomagmatic deposits. (Schmincke 1976). **b** Succession of relatively massive, coarse-grained bomb- and scoria breccia at *lower left* (Strombolian eruptive phases), followed upwards by two well-sorted, massive lapilli fallout beds, overlain by well-bedded, xenolith-rich tephra layers (Phreatomagmatic eruptive phases), each layer representing a single eruptive pulse. Quaternary basanitic Eppelsberg volcano, East Eifel Volcanic Field. (Schmincke 1990)

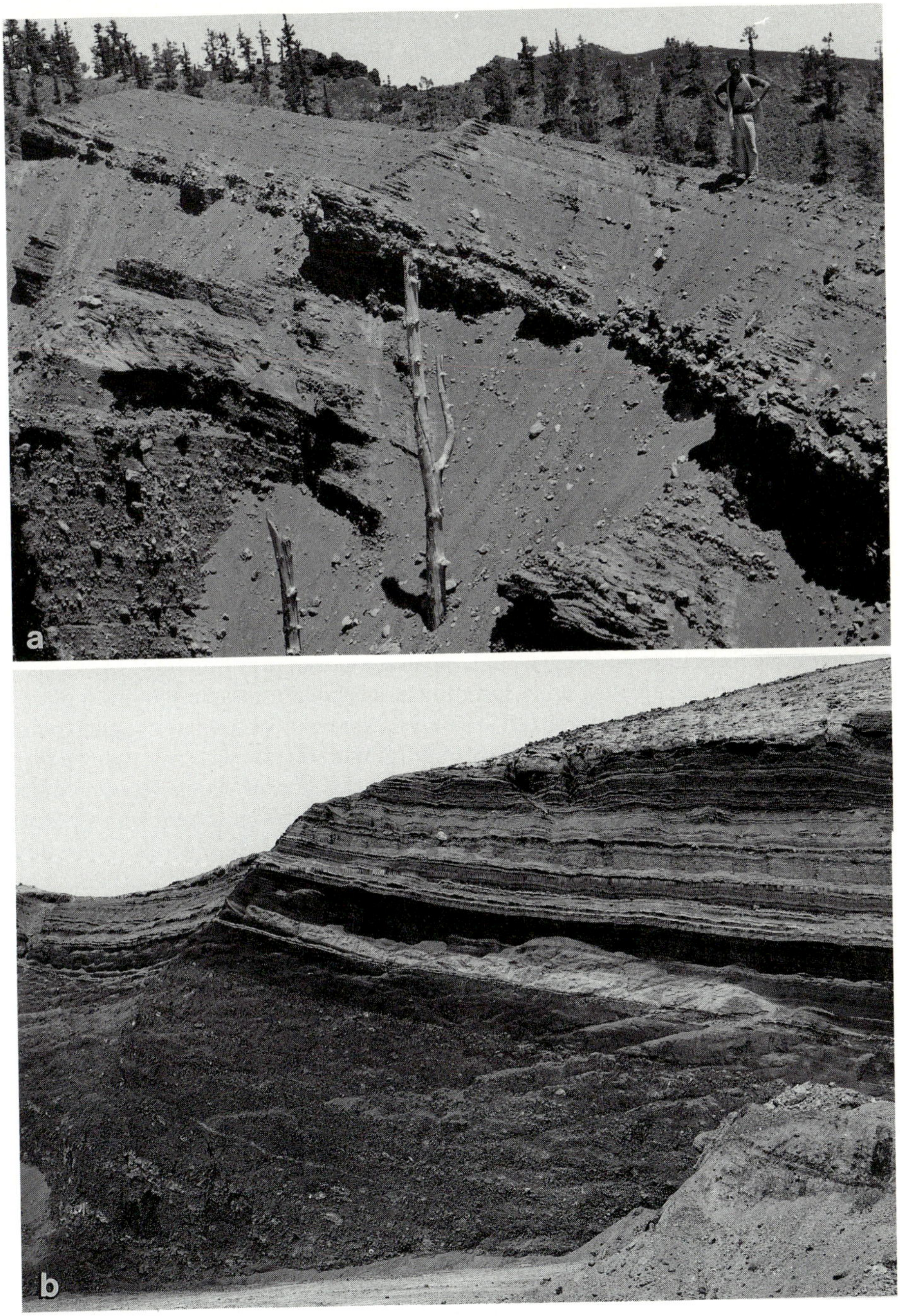
a
b

such as major chemical changes, volcano-tectonic events like caldera collapse, or long erosional intervals, and may last > 10 Ma before volcanism completely dies out. On a still higher order, several volcanoes may form volcanic chains along which volcanic activity may continue for several tens of million of years.

There is a major difference between pyroclastic and hydroclastic volcanic events. Major magmatic events in which magma is fragmented by expansion of magmatic volatiles (pyroclastic processes) last minutes to hours and produce fairly homogeneous deposits save for more subtle grain size and other changes due to pulsating eruption columns. Hydroclastic eruptive pulses, in contrast, typically last only a few seconds, start and terminate quickly and are repeated at intervals of a few seconds to days depending principally on recharge rates of magma and water. The deposits are characteristically well-bedded, each layer representing a single eruptive pulse (Schmincke 1977; Sheridan and Wohletz 1983) (Fig. 20).

3.1.1 Eruption Unit

An eruption unit (Fig. 19) is defined as a thickness of volcanic material deposited from an eruptive pulse, an eruptive phase or an eruption as defined above, conceptually relating rock-stratigraphic and volcanic activity units (Fisher and Schmincke 1984). A mappable accumulation of volcanic material may be defined as a stratigraphic formation and may include several kinds of eruption units (pyroclastic fallout units, pyroclastic flow units, lava flow units, lahar units, etc.). In most places, individual lava flows can be referred to safely as the result of a single eruptive pulse. However, a pyroclastic bed does not necessarily mean that its clasts are derived from a single eruptive pulse, phase, or eruption. Pyroclastic flow sheets (cooling units) commonly consist of several flow units that may be difficult to distinguish. Although beds near the source can be thick and easily recognized, farther away the beds commonly become thin and merge by soil-mixing processes with others. Near eruptive centers, stratigraphic sections are generally too complex and thick to be defined adequately.

3.1.2 Tephra Event Unit

The geological event and the time of transport of tephra to the final site of deposition (disregarding reworking at the sea bottom) generally lasts a few seconds to a few weeks. Well-bedded tephra deposits in the proximal facies (a few hundred meters to a few km) are most commonly due to tens to hundreds of repetitive explosions usually of phreatomagmatic origin, each explosive event generating a single layer (Fig. 21) or a set of layers recording decreasing energy or run-out of a transport system (e.g., Schmincke et al. 1973; Fisher et al. 1983). Where large pumice rafts produce ash by attrition this may be extended to a few years. Ash layers traced sourceward on land or in marine sequences become indistinct a few km around the eruptive center. They merge into, and form part of, a complex medial and proximal facies of many different types of flow and fall tephra deposits that may exceed many tens of meters in

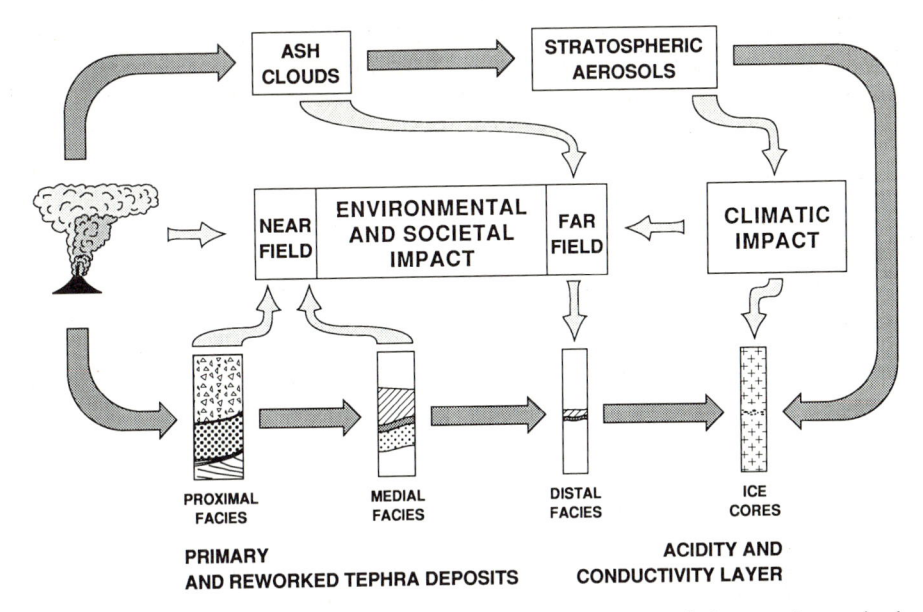

Fig. 21. Scenario of a tephra event unit. A major eruption, lasting << 1 year, results in a complex proximal tephra record, a more homogeneous medial facies, and a single ash layer tens to thousands of km away from the vent. Most glass and mineral particles from the airborne ash cloud are sedimented within days to form tephra fallout layers. Volcanic aerosols (i.e., droplets of sulfuric acid), in contrast, may reside in the stratosphere for several years and eventually form acidity layers in ice sheets which can be detected via conductivity measurements. Environmental and societal (for historic and prehistoric eruptions) impacts include hemispheric temperature decreases due to aerosol emplacement, complete or partial destruction of dwellings and/or inhabitants due to earthquakes, tephra (re-) deposition or gas emission, fertilization of soils, and sealing of the land surface. The global and the near-field direct impacts of explosive eruptions result in environmental changes that may leave longer lasting traces in the geologic record than the tephra deposit itself

thickness – and in the case of Plinian eruptions of evolved magmas – is also erupted and laid down during a few days (Fig. 21). This then is a case of event stratigraphy with single layers in the distant medial and distal facies recording a single event, the same event being represented by many different layers at the source. In geological time terms, this reasoning can commonly be extended to the reworked portion of a tephra event. In ash flow fields, distinct cooling units may be erupted from the same caldera system with time intervals exceeding 0.5 Ma (the large volume systems, generally generating only one or two large ignimbrites) or, in the smaller systems, with time intervals of 0.3–0.5 Ma such as on Gran Canaria where some 15 distinct cooling units were produced over a time span of ca 0.5 Ma (McDougall and Schmincke 1976; van den Bogaard et al. 1988).

We propose a new unit, the *tephra event unit*, encompassing (1) the proximal, medial and distal tephra facies generated by an explosive volcanic eruption during a few days to weeks, (2) the reworked tephra facies formed by rapid reworking commonly within a few weeks to months after a major eruption, (3) syn-eruptive tectonic events and volcanotectonically induced volcaniclastic deposits, (4) the acidity and high-conductivity layer in ice cores formed by precipitates of volcanic aerosols (only preserved with young tephra events), and (5) the environmental impact of the tephra event, which may or may not be detectable in the geological record (Fig. 21).

3.2 Tephrofacies and Tephrostratigraphy

Pyroclastic facies differ from other sedimentary facies in that pyroclastic processes are initially magmatic while clasts are transported and deposited within the atmosphere, on land and under water. Another characteristic especially of near-vent tephra deposits are ubiquitous drastic facies changes within tens to hundreds of meters, the larger eruptive events producing fallout or flow layers that form regionally widespread markers.

A typical example for the complexity of pyroclastic facies is the Laacher See tephra (van den Bogaard and Schmincke 1984, 1985; Fisher et al. 1983; Freundt and Schmincke 1985a,b, 1986; Schmincke et al. 1973; van den Bogaard et al. 1990). With hydroclastic (phreatomagmatic) processes dominating the beginning of the eruption, a central hydroclastic episode, and the end of the eruption, but pyroclastic (Plinian) fallout deposits characterizing the two main phases of the eruption and representing the bulk of the tephra in the medial and distal facies, five main depositional areas (facies) are distinguished from each other:

1. *Fallout lobes* governed by wind vectors and height of eruption column, lower columns resulting from hydroclastic eruptions being transported to the south alternating with high eruption columns and wind transport to the northeast.
2. Radial *proximal flow deposit fans* (five regional fans) governed by low passes between older scoria cones surrounding Laacher See basin. These contain the bulk of the surge deposits and deposits of more viscous pyroclastic flows.
3. *River canyons* contain both proximal and medial facies of more mobile pyroclastic flow deposits extending both radially and at other directions up to 10 km away from the main crater area.
4. Low density ground-hugging ash clouds spread largely independently of local topographic irregularities and form a *veil deposit on higher ground and paleovalley interfluves* throughout the Neuwied Basin east of Laacher See volcano. The distribution pattern of these deposits is intermediate between that of fallout lobes and flow fans and extends up to the eastern boundary of Neuwied tectonic basin, ca. 20 km east of Laacher See Volcano.
5. During the immediate stage of reworking of Laacher See Tephra (prior to soil formation) the lowlands in the Neuwied basin, and flood plains bordering the Rhine River were covered with > 10 m thick reworked fluvial (and lacustrine) volcaniclastic deposits and lahars (Fig. 13).

Near-vent pyroclastic flow deposits of Laacher See volcano show compositional and structural facies variations on four different scales: (1) eruptive units of pyroclastic flows consisting of many flow units, (2) depositional cycles of as many as five flow units, (3) flow units that contain regional intraflow-unit facies, and (4) local intraflow-unit subfacies. The deposits reflect drastic changes in eruptive mechanisms due to access of water to the magma chamber and changes in chemical composition and crystal and gas content as evacuation of the compositionally zoned magma column progressed. In the distal facies, up to > 1000 km distance, the same eruption is represented by a single fallout ash layer that is compositionally zoned with respect to mineral content and glass chemistry where fallout fans from different eruptive units overlap, but compositionally rather uniform in discrete fallout fans.

Tephra deposits represent excellent chronostratigraphic marker horizons because most tephra layers differ in color, thickness, grain size characteristics, sediment structures, mineralogical, chemical and/or isotopic composition from enclosing non-volcanic deposits. Marine ash layers can be used for correlation of piston and drill cores (e.g., Kennett 1982; Kennett and Thunell 1975, 1977; Keller et al. 1978) and identification of marine acoustic reflectors (Ledbetter 1985).

High resolution stratigraphy in sedimentary sequences offers a unique potential, one of the classical cases being the stratigraphy of Wyoming ash layers (Kauffman 1976, 1988a). Bentonites, for example, are extremely valuable tools in intra-basin correlation or correlation of different basins. Ash layers help in stratigraphic correlation when interstratified in marine sedimentary or terrestrial loess and fluvioglacial sequences (e.g., Quaternary of Central Europe; van den Bogaard and Schmincke 1988; van den Bogaard et al. 1989a; Westgate et al. 1985, Sarna-Wojcicki 1985). Ash layers interbedded with loess are useful for correlating magnetostratigraphic records and dating of glacial sediments.

Ash layer correlation is not straightforward, however, because different ash layers may have very similar microscopic and geochemical features, while widespread deposits of a single tephra event may contain a wide range of different whole rock, glass and crystal compositions resulting from the eruption of compositionally zoned magma reservoirs, and grain size and density sorting during transport (Fig. 22). Reliable correlations therefore require a multiple criterion approach (Westgate and Gorton 1981), and tephra identification and equivalence of samples should only be considered firmly established if stratigraphic, paleontological, paleomagnetic, and radiometric age relations are compatible, and if glass and/or phenocryst compositions agree, preferably using grain-discrete methods of analysis.

3.3 Tephrochronology

Tephra layers provide the most precise framework of physical ages presently available, with the exception of impact-generated layers – if these can be identified unequivocally. The numerical age of individual tephra events can be determined by a wide range of physical dating methods. Techniques based on radioactive decay series and isotope analyses include K-Ar, $^{40}Ar/^{39}Ar$, Rb-Sr, U-Pb and U-Th dating (essential lapilli, glass shards, crystals, isotopically equilibrated xenoliths) and radiocarbon dating (i.e., charcoal inclusions in ignimbrites). Another important techniques is fission track dating (uranium-rich phenocrysts, glass); thermoluminescence and obsidian hydration dating are useful but less precise. K/Ar- and $^{40}Ar/^{39}Ar$-

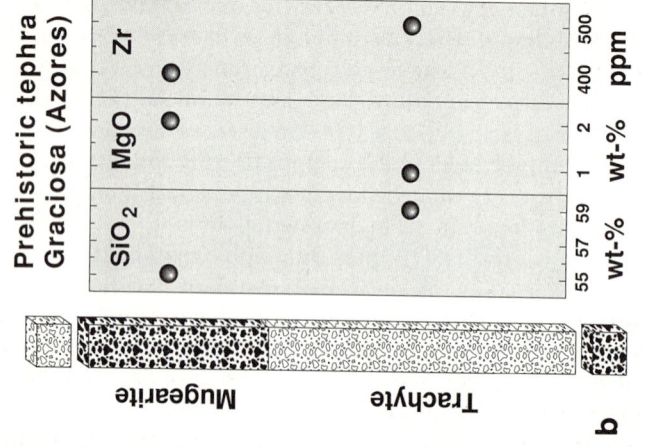

Fig. 22. a Photograph of prehistoric continuous tephra fall deposit from trachyte (*light-colored*) to mugearite (*dark*) reflecting the successive eruption of a compositionally zoned magma reservoir (Graciosa, Azores). Thickness of upper black (mugearite) layer is 1 m. **b** illustrates difference in chemical composition for three selected major and trace elements

dating have become the most widely and successfully applied methods in tephrochronology due to abundant potassium-bearing components in many tephra deposits, straightforward and well-established analytical procedures, and the fact that rocks ranging from several billion to a few thousand years of age can be dated with high precision.

Retentive *essential* mineral and rock components of tephra deposits ideally meet the requirements for K/Ar- or Ar/Ar-dating (McDougall and Harris 1988): moments before the eruption, temperatures are high enough to provide for isotopic equilibrium. During the explosive eruption and/or deposition, the tephra is cooled or quenched to ambient temperatures, transsecting the closure temperature for argon isotopes within a geologically insignificantly short time interval (seconds to minutes in fallout systems; minutes to months in ignimbrites). Tephra deposits, however, generally also contain abundant *accidental* or *cognate* rock fragments and crystals. These can be picked up by the magma prior to eruption, admixed to the tephra during eruption, or added to volcaniclastic sediments by reworking (van den Bogaard et al. 1989a). A major challenge of tephrochronology therefore is the identification of older contaminant xenocrysts which are in isotopic disequilibrium, but may or may not be in apparent chemical equilibrium with the erupted magma.

If the contaminants can be recognized and separated quantitatively, or if the argon from essential and contaminant phases is released at different temperatures, analysis by the $^{40}Ar/^{39}Ar$ incremental heating technique can overcome part of the problem: such analyses yield flat, plateau-shaped age/temperature/volume-^{39}Ar spectra if the sample is homogeneous, but may show significant distortions if contaminants are present (e.g., "saddle-shaped" spectra; Lippolt et al. 1986). In many cases, however, the degree of contamination is difficult or impossible to assess, and only an upper limit of the eruption age can be estimated from poorly defined plateau regions. Bulk K/Ar- or $^{40}Ar/^{39}Ar$-analyses of rock fragments or mineral separates are especially viable to contamination, if a large age contrast exists between the essential and xenolithic components: for example, a mineral separate of 580 000-year-old sanidines with less than 0.1% contaminating Hercynian feldspars (tp = 320 Ma) yielded an apparent total age that was too old by more than 50% (Lo Bello et al. 1987). Only if grain-discrete analytical methods are applied, such as fission track dating (Naeser and Naeser 1988) and U-Pb dating of zircons (Compston et al. 1984; Oberli et al. 1989) and $^{40}Ar/^{39}Ar$ laser dating of potassium-bearing crystals (York et al. 1981; Lo Bello et al. 1987), can the tephra contamination problem be relatively well assessed, the zircon-based methods being especially useful with altered tephra layers (i.e., bentonites).

Single grain $^{40}Ar/^{39}Ar$ laser probe analysis currently represents the most elegant method for high precision dating of ash layers. It is fast and yields the smallest analytical errors and thus highest precision of all methods available: analyses of several sub-mm-size crystals from the same sample are easily done, substantially decreasing the standard error of the mean and yielding age resolutions well below 1% (van den Bogaard et al. 1987; Dalrymple and Duffield 1988). Even the thermal and isotopic history of individual crystals (zoning, relic cores, post-depositional heating events) can be reconstructed, if radiogenic Ar contents (ages) are high enough to permit laser step heating runs (Layer et al. 1987). Laser dating allows high resolution calibration of time scales, which is especially important for understanding the Quaternary geosphere-biosphere system. Precise physical dating of tephra events not only yields "absolute" ages of associated climatic or archeological events, it also enables the intercorrelation of stratigraphic, paleobiological, chemical, geomagnetic, and other isotopic records like marine oxygen isotope stages, thus allowing more precise quantification of rate processes and testing of models such as Milankovich cycles.

For example, in the Middle Rhine area in central Europe, Pleistocene tephra layers from the East Eifel volcanic field (EEVF) are embedded in loess, paleosol, and fluvial sediment sequences (i.e., Rhine and Moselle terraces), many of which contain paleolithic artifact horizons. Based on combined tephrostratigraphic analyses and $^{40}Ar/^{39}Ar$ laser probe dating of feldspar phenocrysts, these tephra layers were correlated with specific tephra event units in the EEVF, and form the age framework for a terrestrial paleoclimate record spanning the last ca. 600 000 years (van den Bogaard et al. 1988; van den Bogaard et al. 1989a). Moreover, high-precision ages were obtained for the appearance of early man in this area (artifact horizons) and for individual paleoclimatic events, as recorded in loess-tephra-paleosol sequences. An illustrative case history is provided by the widespread phonolitic Hüttenberg tephra. It was dated at 215 000 ± 4000 a and occurs precisely at the boundary between a glacial and interglacial stage allowing to verify the main peak in oxygen isotope stage 7 while casting doubt on the beginning of stage 7 at ca. 240 000 years B.P. (Figs. 23) (van den Bogaard and Schmincke 1988; van den Bogaard et al. 1987, 1989b).

3.4 Tephrovolcanology, Tephropetrology, and Tephrogeodynamics

Tephra layers help to answer major volcanological questions as to the age, location, duration and magnitude of explosive eruptions, type of eruptions, fragmentation, transport, depositional processes, including ice rafting. The chemical and physical evolution by cooling, crystallization, and differentiation of crustal magma chambers is obviously an extremely complex process that lasts many thousands to tens of thousands of years. The petrological importance of tephra layers has only been recognized during the last few decades, following the discovery that the vast majority of magma columns is compositionally zoned (Fig. 22). Plutonic rocks, resulting from slow cooling of magma chambers, are the net result of many processes that act after a magma chamber has crystallized beyond the eruptible stage (crystal content < ca. 50 vol. %) (Smith 1979; Marsh 1981). Explosive eruptions, by which 10° to 10^3 km^3 of magma are evacuated instantaneously, provide a snap shot picture of a particular stage of magma chamber evolution. Very distant tephra layers commonly represent only the most explosive phase of an eruption, as reflected in the highest mass eruption rate. They are generally formed during evacuation of the most evolved part of a magma column and are thus not representative of the entire mass erupted.

An entire volcanic complex, whose life-time may exceed 10^6 years, is composed of a series of explosive and hence tephra events, each of which reflects in some way the minimum size of a magma reservoir, compositional zonation of a specific magma column, and many other processes that occur at depth and which are reflected in the

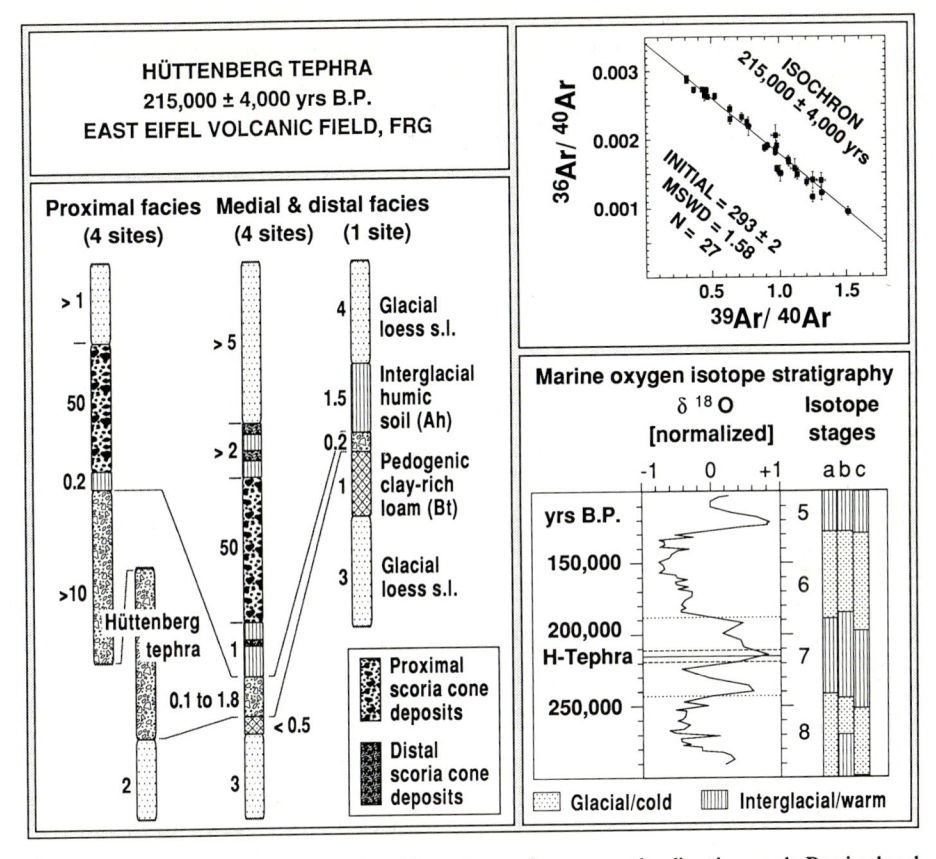

Fig. 23. Tephra events form chronostratigraphic markers and preserve paleoclimatic records. Proximal and distal deposits of the phonolitic-trachytic Hüttenberg Tephra (H-Tephra) in the East Eifel volcanic field (FRG) were emplaced during the brief time interval of transition from a glacial ("cold") to an interglacial ("warm") paleoclimatic episode: the tephra is covered by humic soil layers and/or underlain by loess, or pedogenic loam that formed on a loess substratum (generalized stratigraphic columns to the left; numbers are typical thicknesses in m). $^{40}Ar/^{39}Ar$ laser probe analyses of single, mm-size feldspar phenocrysts of trachytic H-Tephra lapilli consistently yield apparent ages around 200 000 years, the overall isochron age being 215 000 ± 4,000 a B.P. (*upper right* isotope correlation diagram; 1 sigma error bars). The H-Tephra age and stratigraphic setting confirm the existence of an interglacial paleoclimatic episode at about 200 K (stage 7), as deduced from marine oxygen isotope studies (lower right). The tephra data, however, indicate that soil-forming ("present-day") climatic conditions in Central Europe were only established at about 215 000 years B.P., and thus ca. 25 000 years later than indicated by the apparent age and isotope composition of marine sediments. $\sigma^{18}O$ curve from Martinson et al. (1987). Isotope stratigraphies from *a* Imbrie et al. (1984), *b* Herterich and Sarnthein (1984) and *c* Shackleton and Opdyke (1973). *Dashed lines* are 1 sigma error limits of H-Tephra laser age; *dotted lines* indicate marine isotope stage 7 boundaries. (After van den Bogaard et al. 1989)

rapidly erupted and quenched tephra products. Knowledge of the chemical and mineralogical composition of tephra is thus vital to infer and trace the magmatic source region.

Many pyroclastic flows are erupted each year, mostly from central volcanoes but these are nearly all small with volumes << 1 km^3 and confined to river valleys. Larger pyroclastic flows are known from a few historic eruptions: at Tambora volcano, erupted in 1815, pyroclastic flows entered the sea (Sigurdsson and Carey 1989), at Krakatau in 1883 (Self and Rampino 1981) and at Valley of Ten Thousand Smokes in 1912 (Hildreth 1983) ca. 19 km^3. Larger volume ignimbrites are generally associated with calderas from whose ring fractures the ash flows most likely erupted (Smith and Bailey 1968; Lipman 1984). At Valles Caldera in New Mexico, the two main cooling units were erupted at 1 Ma intervals. At Gran Canaria, ash flows of much smaller volume (< 10 km^3) erupted, on average, every 10–30 × 10^4 ka (van den Bogaard et al. 1988).

Ash layer frequency in sediment cores has been used by a number of workers to infer volcanic and magmatic episodicities of source regions, thought to be related to different rates of subduction and sea floor spreading, hot spot activity, mountain uplift, marine transgressions and regressions and other global geologic processes (Vogt 1979).

Ash layers from the North Atlantic around Iceland illustrate the methodical problem of inferring global events from ash layer frequency. Several models were based on ash layers from the Norwegian-Greenland Sea drilled during Leg 38 (Talwani et al. 1976), which suffered from core recovery as low as 4%. Based on these ash layers, Donn and Ninkovich (1980) postulated high explosive activity in the North Atlantic Ocean during the Middle Eocene and Pliocene, while Sigurdsson and Loebner (1981) report apparent peaks in explosive volcanic activity in middle Eocene, middle Oligocene, early to middle Miocene, and Pliocene to Pleistocene using the same data base.

Ash layers in cores from Leg 104 (Eldholm et al. 1987), which had exceptionally good core recovery, provide a nearly complete record of North Atlantic ash layers since the Eocene. No significant high or episodicity in explosive volcanism in the North Atlantic between the Miocene and the Recent can be recognized (Bitschene and Schmincke 1990). Average ash layer frequencies amount to between five and ten layers per Ma during the Neogene.

3.5 Tephropaleoclimatology, Environmental, and Societal Impact

Tephra layers harbor a great wealth of information about events that are rarely reflected in other types of geologic deposits. Large volcanic eruptions can definitely cause rapid, short term temperature changes on a hemispheric scale which are recorded in tree ring widths and frost rings (Rampino et al. 1988; Baillie and Munro 1988; La Marche and Hirschboeck 1984). Tephra layers record paleowind directions and strengths (van den Bogaard and Schmincke 1985; Cornell et al. 1983), and preserve/seal the vegetation and other natural or man-induced environmental features, including dwellings and other records of man's activity. Moreover, tephra deposits enable the most precise dating of climatic changes in an absolute time framework (van den Bogaard and Schmincke 1988).

The potential *climatic impact* of tephra events both at present and in the geological past has been debated for a long time. For decades, the volume and areal distribution of ash layers were believed to be a direct measure of the climatic impact of volcanic eruptions (Lamb 1970). It is now recognized that volcanic aerosols are largely droplets of sulfuric acid formed by oxidation of SO_2. There is good evidence that volcanic aerosols can influence climate significantly (Rampino and Self 1982; Devine et al. 1984; Stothers 1984). Deposition of such volcanic aerosols is reflected in acidity layers of high conductivity in ice cores (Hammer et al. 1980, 1987). The volume and areal extent of tephra layers can thus not be used directly to estimate their potential climatic impact, but must be calibrated for the sulfur content of the magma which can vary drastically (Carey and Sigurdsson 1986).

Mass extinctions such as those at the Cretaceous-Tertiary (KTB) and Permian-Triassic boundary are believed by some as due to cumulative volcanic eruptions. Magmatic volatiles such as SO_2 from powerful basaltic and/or silicic volcanic eruptions could have led to formation of abundant volcanic aerosols and could have caused "volcanic winter" conditions (Stothers et al. 1989) leading to changes in bacterial activity patterns, redox-controlled precipitation and global extinctions of biota (Dia et al. 1989), and possibly even creation of sedimentary Iridium anomalies due to microbial geochemical activity (Betsey et al. 1989). The majority opinion, however, at present holds bolide impacts to be responsible for the catastrophic events at the KTB (Alvarez 1986, 1987) and no widespread tephra layers of undisputed volcanic origin have yet been found at the KTB. Volcanic aerosols believed to have resulted from the vast Deccan flood basalt eruptions and/or preceding silicic volcanism (Courtillot et al. 1986; Javoy and Courtillot 1989) could explain some of the features at KTB, but lead isotope ratios at four famous Ir-rich K-T boundary locations differ from those of carbonaceous chondrites, iron meteorites and Deccan basalts (Dia et al. 1989). Moreover, detailed studies of planktonic foraminifera indicate that species extinctions took place over a prolonged period of time, ranging from about 300,000 a below to about 200 000 to 300 000 a above the KT boundary (and Ir-anomaly), with distinct episodes of accelerated extinctions right below and 50 000 a above the KTB (Keller 1989).

Widespread rhyolitic tuff beds have been detected at the Permian/Triassic boundary, especially in China (Clark et al. 1986; Zhou and Kyte 1988) and may imply that mass extinctions at this boundary may be due to several highly explosive silicic eruptions rather than a bolide impact. The absence of an Ir-anomaly and the trace-element characteristics (Cs, Zr, Hf, Ta, and Th) further suggest that bentonites from this boundary are related to multiple large silicic volcanic eruptions. Tephra layers are quite abundant directly above the KTB in sediments from the Indian Ocean (Dehn, pers. commun.). These are entirely nonfossiliferous, however, and the tephra layer abundance may thus reflect an extreme decline in biological activity rather than an episode of unusually voluminous volcanism, a line of reasoning that may also hold for the Permian/Triassic and other major geologic boundaries.

Fallout tephra layers reflect *paleowind directions* more faithfully than any other type of sedimentary deposit. Interpretations of the shape of fallout lobes, however, are anything but straightforward, except on a global scale (Fig. 4). Multicomponent sheets radiate from a common source or overlap with slightly different distribution trends, depending on shifting wind directions. Contemporaneous sheets can be distributed in opposite directions by contrasting winds at different altitudes. Unusually powerful eruptions give rise to sheets easily recognizable 1000 km or more from their source (Kyle and Seward 1984). The ash from such eruptions may extend far beyond the recognized limits of the dispersal fan as illustrated by the May 18, 1980 eruption of Mount St. Helens, Washington (Lipman and Mullineaux 1981).

Satellite and weather balloon observations of Recent eruptions clearly show that ash clouds (and the deposits they form) respond in a complex way to the distribution of wind direction and strength, both of which may vary significantly with height in the atmosphere (Fig. 3). Wind directions in the troposphere may differ drastically from that of the dominant shear winds in the tropopause (jet stream). Thus, ash clouds and tephra lobes commonly are not straight but curved (Figs. 3, 5). The main fallout lobe of the 11 000-year-old Laacher See Tephra, for example, is directed to the northeast – indicating a general wind profile remarkably similar to present-day conditions (van den Bogaard and Schmincke 1985). Smaller southern and southwestern lobes are thought to be due to lower altitude tropospheric winds. Interestingly, glass shards in the tephra deposits from those smaller lobes are less vesicular, indicating quenching during magma-water contact, and lower eruption column heights due to heat loss (and therefore less heat energy for convective rise) during these eruptions.

During the explosive eruption of Mt. St. Helens on May 18th, 1980, easterly and southerly winds dominated at low altitudes while the main tropopause wind was directed to the east-southeast. The eruption column of Mt. St. Helens in May 1980 was 14 km high for 9 h of consecutive activity (Lipman and Mullineaux 1981). The ash clouds advanced initially with 250 km/h into high velocity wind layers at 10 to 13 km height, which then dispersed volcanic ash at velocities around 100 km/h (Sarna-Wojcicki et al. 1981).

4 Conclusions

Tephra layers record events that range in time from seconds to days or weeks, and in magnitude from small instabilities in magma chambers beneath local eruptive centers to episodic, periodic or long-lasting changes in spreading rates or large seismic events. Their effects span a similarly wide spectrum, ranging from obviously local to global. The larger magnitudes and still speculative global effects are not reached by a single layer or a single eruption, but are cumulative effects from several major tephra events. Geologically speaking, tephra events are instantaneous.

By way of introducing the term "tephra event unit", we want to emphasize that volcanic eruptions may generate superb stratigraphic units that can be lithologically quite diverse. They originate from the explosive discharge of large volumes of magma. Close to the source, they may comprise exceedingly complex assemblages of fall, flow, surge, and reworked tephra deposits, perhaps laid down at a wide range of temperatures. At distances of tens to thousands of km from the source, they may laterally grade into a single ash layer a few mm thick, and eventually become unrecognizable ash particles dispersed in nonvolcanic sediments or acidity layers in ephemeral ice sheets.

The complexity of tephra event units is due to several unique properties: volcanic eruptions produce a wide range of highly contrasting and/or transitional transport systems, many of which overlap and/or grade into each other within minutes. Volcanic eruptions are also hot and rich in volatiles, generating eruption columns that may extend as much as 50 km into the stratosphere. Volatiles thus injected into the stratosphere may form globe-encircling aerosol clouds that can drastically influence climate.

Volcanoes provide shelter, beauty and fertile soils to man. Sudden eruptions can cause death and destruction, but have also preserved the traces/remnants of human settlements, some older than 500 000 years. Thus, being highly destructive while lasting, tephra events also link natural history and human activities over geological time spans, leaving unique markers of time significantly improving our chances to decipher the past.

Acknowledgments. We thank the editors, especially W. Ricken and G. Einsele, for the invitation to contribute to this volume and to the workshop preceding it and particularly for their patience and helpful comments on the manuscript. Our studies on terrestrial and submarine tephra problems are supported by grants Schm 250/37–3 and Schm 250/38–2 from the Deutsche Forschungsgemeinschaft.

Chapter 3 Diagenetic Overprint

3.1 Diagenetic Modification of Calcareous Beds – an Overview

W. Ricken and W. Eder

1 Introduction

Diagenetic overprints in carbonate sediments bring about changes in the original distribution of $CaCO_3$ phases, carbonate and organic carbon contents, sedimentary structures, and bedding rhythms. In the interbedded sequences addressed in this book, in which carbonate-poor beds alternate with carbonate-rich beds, overprinting not only exaggerates the differences in carbonate content but, by carbonate redistribution and differential compaction, may vertically shift the boundaries between beds. Thus original rhythms are both accentuated and changed. At the same time, evidence relating to the original depositional processes becomes more difficult to recognize. Under certain circumstances, a diagenetic stratification may be imposed. Consequently, a knowledge of the diagenetic overprint is an important requirement for understanding initial depositional processes and their bedding rhythms.

Diagenetic overprint results in a change in the relative thicknesses of carbonate-rich and carbonate-poor layers or in a pattern of new layers or nodules. In outcrops, this overprint is also associated with weathering processes. Below and above a carbonate content of 65 to 85%, the so-called weathering boundary, the rock disintegrates either into marl or limestone, respectively. Such weathering processes may affect the number and form of calcareous beds or nodules visible in outcrop (see Fig. 6 in Einsele and Ricken Chap. 1.1, this Vol.).

Only the most common types of diagenetic overprint in carbonates, marls, and calcareous shales are treated here. These are: (1) selective or localized dissolution or cementation of beds changing the carbonate contents and bed thicknesses, (2) patchy localized cementation producing concretions, (3) lateral coalescence of concretions and chert nodules resulting in nodular beds, (4) selective dolomitization accentuating or obliterating stratification, (5) cementation restricted to beds at the sediment-water interface producing hardgrounds, and (6) selective cementation of event beds (Fig. 1). Other topics, such as the large number of diagenetic reactions related to original carbonate phases, organic constituents, interstitial water composition, etc. cannot be addressed in the context of this overview. The reader is referred to several excellent carbonate textbooks, e.g., Lippmann (1973), Bathurst (1975), Flügel (1982), Reeder (1983), Scholle et al. (1983b), Schneidermann and Harris (1985), Schroeder and Purser (1986), Marshall (1987), Scoffin (1987), Füchtbauer and Richter (1989), Moore (1989), and Tucker and Wright (1990). Principally, only three basic processes controlling the most common types of diagenetic overprinting in calcareous rocks

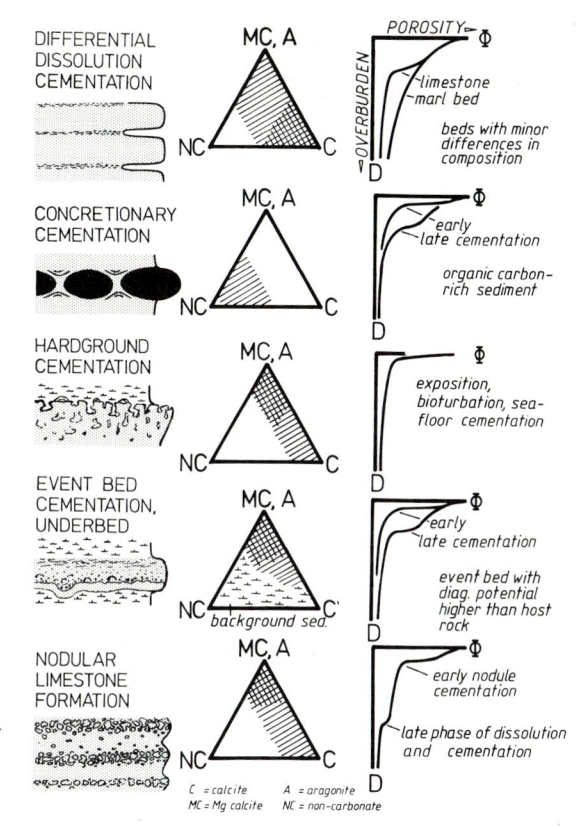

Fig. 1. Overview of diagenetic overprints in calcareous rocks, as related to commonly found original sediment compositions (*shaded portions of triangles*). Symbols indicate instable carbonate phases (*MC* Mg calcite; *A* aragonite), calcite (*C*), and the noncarbonate fraction (*NC*). Diagrams on the *right* indicate general porosity (Φ) versus overburden depth (*D*) curves

exist, such as carbonate cementation, carbonate dissolution, and differential compaction between cemented and dissolution-affected rock portions:

1. Cementation with external source of carbonate. The original carbonate content is enhanced by precipitation of cement derived from a source outside the location of cementation. At the site of cementation, the carbonate content rises until the pore space is filled with cement. This process can be selective: Sites and beds with initially higher contents in carbonate, unstable CaCO3 polymorphs, larger pore space, etc. undergo preferred cementation, while other sites are left unmented or are subject to dissolution. Selective cementation emphasizes certain structures or beds and obliterates others. Thus, it may either reduce or enhance the number of primary beds.
2. Dissolution of carbonate. Dissolution of carbonate is controlled by original carbonate phases, clay and organic matter content, as well as interstitial water composition. Carbonate dissolution can either occur pervasively or locally, in patchy zones or individual beds diminishing the original carbonate content at the site of dissolution. Under high overburden, carbonate dissolution may be related to pressure dissolution or chemical compaction (Bathurst Chap. 3.2, this Vol.).

In calcareous interbedded sequences, original beds affected by chemical compaction suffer a loss in carbonate content and diminish in thickness to form thin, clay-rich interbeds.
3. Differential compaction. Cementation and dissolution may operate simultaneously at different sites in the sediment, giving rise to different degrees in thickness reduction of the original sediment (i.e., differential compaction). Cemented parts are able to withstand compaction and become less reduced in thickness, whereas dissolution-affected parts or beds become more reduced in thickness (i.e., chemical compaction). In interbedded marl-limestone sequences and other types of bedded calcareous rocks, dissolution and cementation processes occur in alternating beds. They are associated with large changes in both the original carbonate content and the degree of compaction, either preserving or reducing original thicknesses in succeeding beds.

2 Diagenetic Overprint Associated with Nonevent Sequences

2.1 Differential Cementation and Dissolution: Contrasting Bedding Enhancement

Differential cementation and dissolution processes between alternating beds (Fig. 2) were first proposed by Wepfer (1926) and were mostly understood as diagenetic modification of original sedimentary "signals", related to variation in carbonate content, $CaCO_3$ polymorphs, or pore space (e.g., Hallam 1964; Eder 1982; Walther 1982; Arthur et al. 1984b; Gluyas 1984; Beiersdorf and Knitter 1986; Ricken and Hemleben 1982; Ricken 1985, 1986; Bathurst 1987; Bathurst Chap. 3.2, this Vol.). However, concepts on complete "diagenetic unmixing" and "self-organization" have also been discussed (Sujkowski 1958; Simpson 1985; Hallam 1986; Dewers and Ortoleva 1990). Besides bedding accentuation in carbonates, differential cementa-

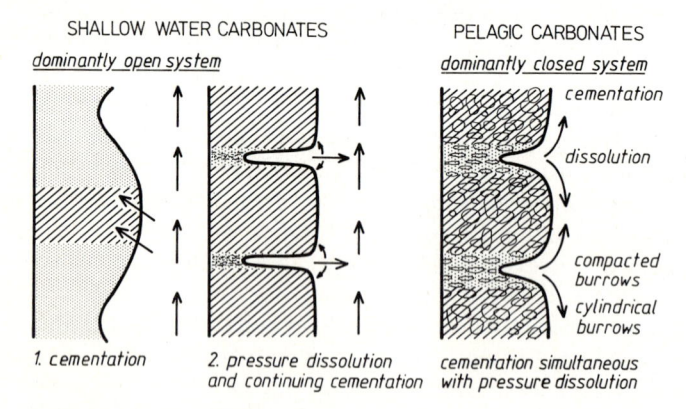

Fig. 2. Differential cementation and dissolution processes in vertical sections of shallow water and pelagic carbonates. *Arrows* denote diffusion and porewater transport of carbonate. Carbonate dissolution is indicated by clay seams and vertical deformation of burrow mottling. *Hatched zones* show cementation

tion and dissolution processes were observed in stratified cherts (see Tada Chap. 4.2, this Vol.)

Compaction studies in coarse and fine-grained carbonates (Bathurst 1987; Ricken 1986) have shown that cementation generally occurs first in the originally carbonate-rich bed. Consequently, the carbonate-poor bed is further affected by mechanical and chemical compaction. High carbonate content may encourage cementation, while clay may inhibit cementation. Pressure dissolution in the marl beds is documented by the products of compaction, such as marl seams, stylolites, and fitted fabrics (Bathurst Chap. 3.2, this Vol.). As a result, the original carbonate content becomes augmented in the limestone layers, but reduced in the marl beds. The marl beds are diminished in thickness, whereas limestone layers are preserved or only slightly reduced. As a result, the bedding rhythm of marl-limestone alternations becomes dominated by the (largely preserved) thicknesses of the limestone layers.

The onset of cementation varies in different limestones. In platform and shallow water carbonates with high proportions of aragonite and Mg calcite, it may begin so early that pressure dissolution postdates limestone cementation (Fig. 2). Here, the diagenetic system is mainly open, and pressure dissolution in the marl beds seems to provide only a small amount of the carbonate cement found in the later limestone layers (see Bathurst 1987, and Chap. 3.2, this Vol.). On the other hand, in sequences with a low potential for the initiation of diagenetic reactions, such as pelagic and hemipelagic foraminiferal and nannofossil (calcite) oozes and marls, mechanical compaction before the onset of cementation is higher, ranging between 10 and 60% (Ricken 1986). This corresponds to overburden thickness between 50 and several 100 m, as documented by the ooze-chalk transition in DSDP cores (e.g., Schlanger and Douglas 1974; Garrison 1981). As carbonate mass balances show, pressure dissolution seems to occur simultaneously with cementation of the limestone layers in calcite oozes, forming a largely closed system for carbonate (Ricken 1986).

2.1.1 Diagenetic Carbonate Contents Resulting from Differential Compaction Between Beds

The magnitude of the diagenetic carbonate changes and its influence on bedding phenomena can be determined using the differences in compaction between succeeding limestone and marl layers (which have low and high degrees in compaction, respectively). The compaction difference between limestone and marl layers can be numerically related to the carbonate content using the "carbonate compaction law" (Ricken 1986, 1987), in which the percentage loss of original thickness by compaction (K) is expressed as

$$K [\%] = 100 - \frac{NC_d}{1-0.01C} \quad . \tag{1}$$

In this equation, where the present porosity is assumed to be low enough to be neglected, C is the carbonate content in percent, and NC_d is the noncarbonate fraction expressed as a percentage of the original or decompacted volume of the sediment.

The application of the compaction law shows that the relationship between carbonate content and compaction is nonlinear (Fig. 3a). Such nonlinear relationships control the various *types of carbonate curves* found in cemented limestone layers (curves record carbonate content along the vertical axis). The carbonate distribution in diagenetically affected limestone layers is expressed in decreasing carbonate contents and increasing compaction values from the middle of the limestone layers

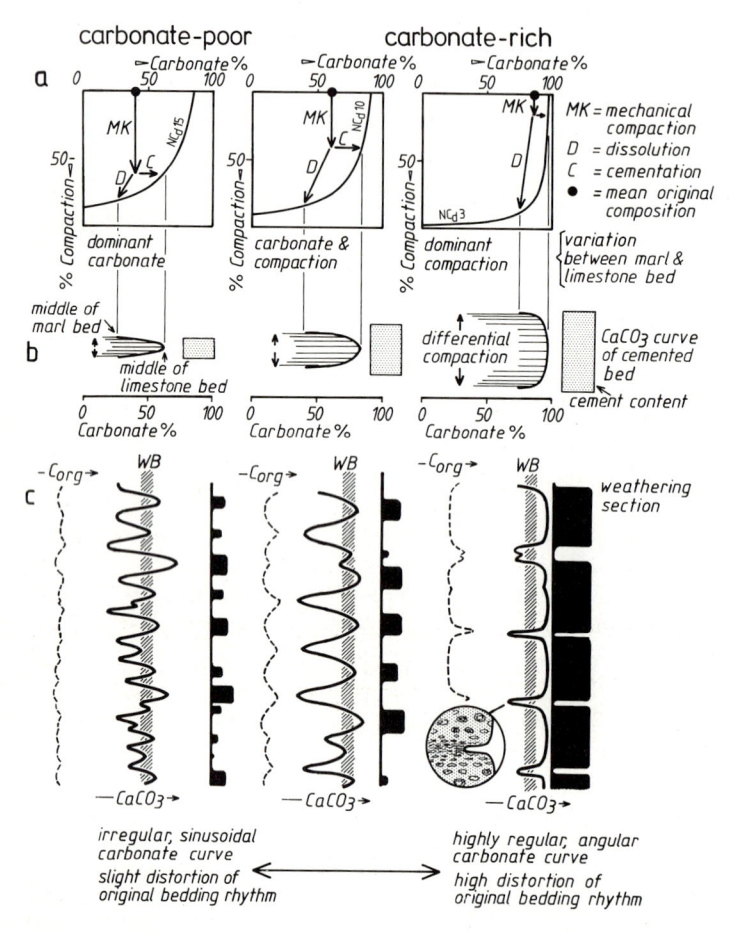

Fig. 3. The effects of diagenetic overprinting on bedded calcareous sediment. **a** Upper part of diagram shows carbonate-compaction relationship [Eq. (1)] for original sediments with low, medium, and high carbonate content. NC_d is the percentage noncarbonate fraction of the bulk original sediment volume. After a phase of mechanical compaction (*MK*), differential cementation (*C*) and dissolution (*D*) between succeeding limestone and marl beds occurs. **b** The resulting carbonate curve of the cemented limestone layer is different for sediments originally low and high in carbonate content. **c** Lower part of diagram depicts diagenetically influenced bedding rhythms with schematic carbonate and organic carbon (C_{org}) curves and weathering sections (*left side* of curves). *WB* is the carbonate content at the weathering boundary. Note contradiction between diagenetic changes of bedding rhythm and changes in carbonate difference between beds

to the enclosing marl beds. Different types of cemented limestone layers are found in carbonate-poor and carbonate-rich sediment. Limestone layers in marly shales have sinusoidal carbonate curves, while those in carbonate-rich sediment are angular (Fig. 3b). In high calcareous alternations, carbonate contents in the middle of several succeeding limestone layers are approximately the same, because of cementation. The original thicknesses of the marl beds are considerably reduced, enhancing the regularity of the bedding rhythm which becomes, during progressive diagenesis, dominated by the thicknesses of the limestone layers.

The *magnitude of diagenetic carbonate changes* can be determined using the degree of differential compaction obtained by measuring the deformation of originally circular bioturbation tubes. Such compaction measurements result in diagenetic carbonate changes spanning approximately 35 to 40% for carbonate-poor sediments, and approximately 20% for carbonate-rich sediments (between original and present values in interbedded alternations). In addition, the organic carbon content is significantly affected; it is enriched in the marl layers because of carbonate dissolution, while it is "diluted" in the limestone layers by precipitation of carbonate cement (Fig. 3c).

2.1.2 Diagenetic Influence on Carbonate Contents and Bedding Rhythm Frequencies: a Paradox

In interbedded limestone-marl sequences, a paradox between diagenetic amplification of carbonate contents and the associated frequency change of the original bedding rhythm is observed (Fig. 3). Such a paradox is obvious when the diagenetic overprints in carbonate-rich and carbonate-poor rhythms are compared. In interbedded *carbonate-poor sequences*, the original bedding rhythm frequencies (i.e., thicknesses of succeeding beds and interbeds) are only slightly changed, although diagenetic carbonate changes from original to present values in marl and limestone layers are large. The opposite is true for interbedded *carbonate-rich sequences*, in which the diagenetic overprint significantly influences the bedding rhythm frequencies (i.e., thicknesses of succeeding beds and interbeds), but creates only slight carbonate changes from original to present values.

Such a seemingly contradictory behavior of diagenetic overprinting is explained with two arguments. (1) Accentuation of the bedding rhythm is related to the magnitude of carbonate masses redistributed between marl and limestone layers. Carbonate redistribution is more intense in carbonate-rich sequences because cementation is relatively early, and the cement content is high. Cementation occurs in a pore space slightly reduced by compaction (Fig. 3). (2) For equal amounts of redistributed carbonate, percentage carbonate can be more easily changed in alternations with generally low and medium carbonate contents than with generally high carbonate contents. This is related to the same numerical requirements as presented in Fig. 9 in Ricken (Chap. 1.8, this Vol.).

When beds in the original stratification are grouped into bundles (see Fischer Chap. 1.2, this Vol.), an additional effect comes into play because bundles become diagenetically augmented or obliterated, depending on the overall $CaCO_3$ content

(Fig. 4). In marl and marly shale sediments, a bundle is accentuated by shifting the carbonate-rich beds to higher carbonate values. Beds contained in the bundle with low carbonate content shift less. In high calcareous sediment, however, beds with relatively poor and rich carbonate contents are transformed to higher but more equal carbonate contents compared to the original composition, diminishing the distinctiveness of the bundle. Calculations depicting this effect in Fig. 4 are based on Eq. (1) and on values for mechanical compaction in micritic marl-limestone alternations as described in Ricken (1986). The following conclusions can be drawn:

1. In interbedded sequences with *low to medium carbonate contents*, the original carbonate differences for beds and bundles are diagenetically enhanced, but the original bedding rhythm frequency remains relatively unchanged (Fig. 4). It is thought that such overprinting is also expressed in time series analyses, in which the significance of frequencies contained in the bedding rhythm are shown in "power spectra" (Weedon Chap. 7.4, Schwarzacher Chap. 7.5, both this Vol.). Commonly, one peak in such power spectra representing a larger frequency is related to bundle thickness, while other peaks with smaller frequencies reflect the thickness of individual beds. It is assumed that power spectra using the weathering section in interbedded sequences with low and medium carbonate contents are dominated by the periodicity of the bundle. This is because the carbonate content of beds situated in the middle of the bundle is diagenetically enhanced, and thus, after weathering, the bundle becomes more visible in outcrop (Fig. 4).

2. Alternations with *high carbonate contents* and angular carbonate curves have fewer carbonate variations compared with the original sediment, because of

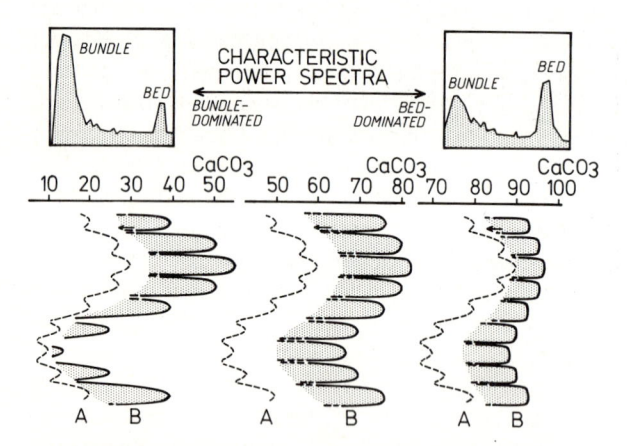

Fig. 4. Diagenetic enhancement versus obliteration of bundles for vertical sequences for generally low and high carbonate contents, respectively. Bedding rhythms show original (*stippled,* A) and diagenetic carbonate curves (*B*). For reasons of simplification, diagrams omit diagenetic carbonate contents in the middle of the marl beds and compaction changes between the curves *A* and *B*. Power spectra of the post-diagenetic bedding rhythm are schematic, recording frequency (thickness) of the bedding rhythm on the horizontal axes. Diagram is based on mechanical compaction values for hemipelagic to pelagic carbonates reported in Ricken (1986)

amalgamation of some cemented beds and the loss of smaller carbonate variations in the marl beds subjected to pressure dissolution. Power spectra are thought to be dominated by diagenetic bed thickness, with slightly lower frequencies than those in the original sequence. The bundle peak in the power spectra is assumed to be relatively insignificant, because bundles with a generally high carbonate content may lose their characteristic distinctiveness (Fig. 4). For the effect of differential compaction on power spectra, see Weedon (Chap. 7.4, this Vol.).

2.2 Carbonate Concretions: Near-Surface Versus Shallow Burial Cementation

Carbonate concretions are most commonly found in shales, black shales, and marls, but they also occur in carbonate-rich sediments. They originate in diagenetic environments ranging from the near-surface to shallow burial and are formed by the decay of organic matter, promoting carbonate precipitation (Berner 1980; Raiswell 1987; Boudreau and Canfield 1988). Compositionally, they are composed of calcite, siderite, ankerite, and dolomite. The early to shallow burial history can be documented using carbonate concretions, that yield information on porosity reduction, compaction, cement content, and pore water composition. Concretions of subsurface and early burial origin can be distinguished by their different shapes. Concretion-bearing horizons may extend over long distances.

2.2.1 Near-Surface and Shallow Burial Concretion Types

Most of the carbonate concretions form within 50 m of overburden, as can be inferred from carbonate data reported in the literature (Lippmann 1955; Seibold 1962; Hoefs 1970; Raiswell 1971; Sass and Kolodny 1972; Hudson 1978; Coleman and Raiswell 1981; Gautier 1982; Dix and Mullins 1986). The depth of concretion formation can be calculated because the carbonate content in the middle of a concretion is related to the porosity of the enclosing sediment during cementation. This porosity can be compared with an average porosity-overburden curve for shales (Baldwin and Butler 1985; Fig. 5a). Where the host rock is essentially carbonate-free, the volume percent carbonate (i.e., cement) in the middle of the concretion corresponds virtually to the former pore space (e.g., Lippmann 1955; Seibold 1962; Raiswell 1971). However, when the host rock contains carbonate (C_h), the former porosity at the onset of cementation (ϕ_0) is:

$$\Phi_0[\text{vol}\%] = Z[\text{vol}\%] = C_c - \frac{(100 - C_c) C_h}{100 - C_h}, \qquad (2)$$

where Z is the cement in vol% and C_c is the carbonate content in the middle of the concretion (in w% if carbonate and noncarbonate fractions have equal grain densities). Equation (2) is valid only for calcite concretions with low present porosity (below approximately 15%), and it is based on the assumption that the carbonate content of the host rock (C_h) is not significantly altered. Application of Eq. (2) shows that, in most concretions reported in the literature, cementation began at porosities

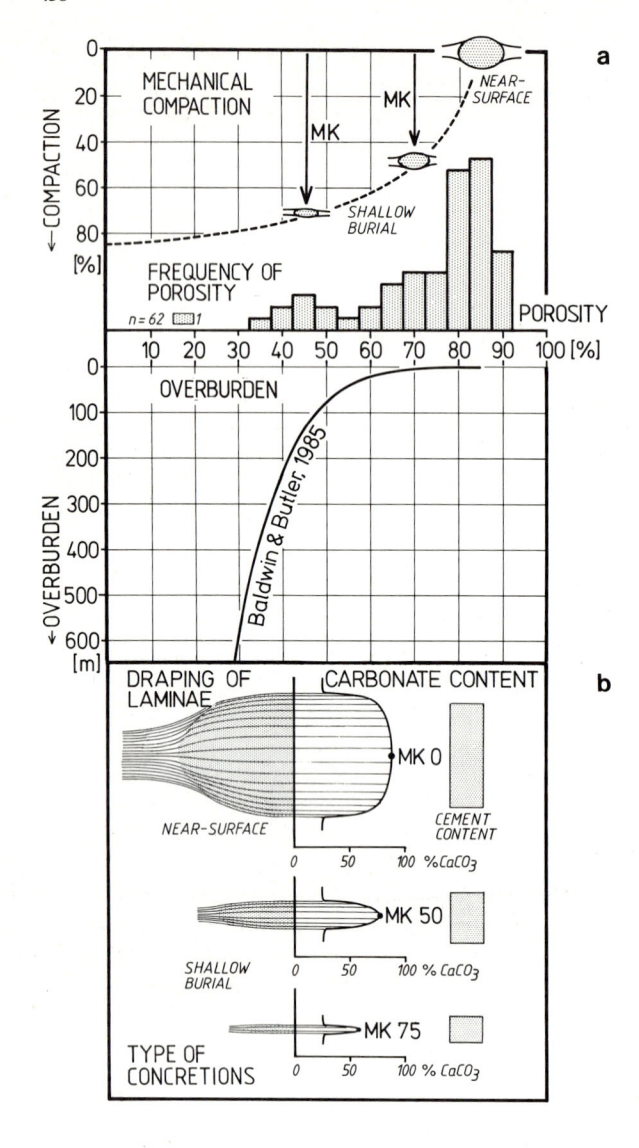

Fig. 5. Near-surface and early burial formation of concretions based on carbonate contents reported in the literature (see text). **a** shows calculated porosity at the onset of cementation [Eq. (2)] and related degrees of over burden and mechanical compaction (MK). **b** depicts shape and laminae draping for near surface and shallow burial concretion types; calculated according to mechanical compaction values (*MK*) as indicated above, and the same increasing compaction from the middle towards the upper and lower concretion margins

between 60 and 90% under an overburden of a few to several 10's of meters (Fig. 5a). On the other hand, a small but important group of concretions began to form at reduced porosities of 30 to 60%, documenting that cement precipitation occurred within an already compacted pore space with roughly a few 10's to 300 m of overburden.

The shapes of near-surface and shallow burial concretions are distinctly different (Fig. 5b). Concretions which originated during near-surface cementation with very low mechanical compaction are relatively sphere-like. They show angular carbonate curves in a vertical cross section, and have a substantial drape of laminae. For shallow burial concretions with more than 50% compaction, however, the concretions are thinner and more elongated, the carbonate curves in a vertical section have sharp maxima, and compactional draping is only insignificantly developed. Small vertical diameters for concretions formed under shallow burial conditions are thought to reflect significant compaction of the original pore space, diminishing the quantities of concretionary cement which can be precipitated. These different shapes of near-surface and shallow burial concretions are essentially those observed by Raiswell (1971).

Near-surface and shallow burial concretions can occur within a single bed (i.e., concretion bed) or several neighboring beds (i.e., concretion horizon) and may be separate or laterally connected, leading to eight basic patterns of concretion occurrence, as indicated in Fig. 6. Laterally connected concretions are essentially bedded concretionary limestones. The positions of swelling and pinching in such concretionary limestones may alternate from one bed to the other, forming a conspicuous diagenetic bedding pattern.

2.2.2 Progressive Cementation and the Isotopic Record

During the progressive cementation of a compacting pore space, concretions "freeze" the chemical and isotopic records of an diagenetic environment which is subsequently buried to increasing depth. From the concretion center to the margin, the $\delta^{18}O$ values of the carbonate change from zero to negative values (Fig. 7). This $\delta^{18}O$ shift can be explained by increasing temperatures, clay mineral dewatering, and neoformation of clay minerals during burial (e.g., Hudson 1977; Scholle 1977; see Arthur and Dean, this vol.); accordingly, Paleozoic concretions with thick overburden depict a pronounced shift to lower $\delta^{18}O$ values.

Carbon isotopes record that most of the concretionary carbonate (CO_3) is derived from bacterial decomposition of organic matter (e.g., Hudson 1977, 1978; Irwin et

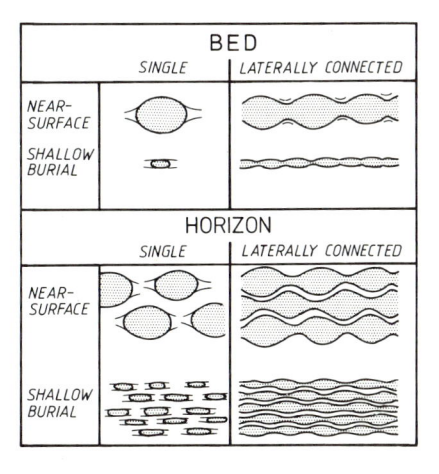

Fig. 6. Most commonly found types of concretions. Occurrence of concretions can be confined to individual beds or to thicker horizons containing several beds with concretions. Concretions can be separated or laterally connected. Near-surface and shallow burial concretions are distinguished. Concretion diameter varies between 0.01 to 1 m

al. 1977; Gautier and Claypool 1984; Raiswell 1987). According to a model proposed by Gautier and Claypool (1984) and Raiswell (1987), growth of concretions can start in the sulfate reduction zone and continues in the underlying methane production zone. During sulfate reduction, bacterial decomposition of organic matter leads to precipitation of δ^{13}-depleted calcite (up to –35‰). With increasing overburden, the concretion passes from the sulfate reduction zone into the methane production zone, where isotopically light bicarbonate is removed for bacteriological methane production. The remaining bicarbonate in the pore water and the precipitated (iron-rich) carbonate therefore continuously shifts from light δ^{13}C values to heavier values, as bacterial decomposition of organic matter increases.

2.2.3 Concretion Horizons

Carbonate concretions commonly occur in distinctive beds or horizons which can by traced laterally for tens of kilometers (Kauffman et al. Chap. 7.2, this Vol.). Lateral

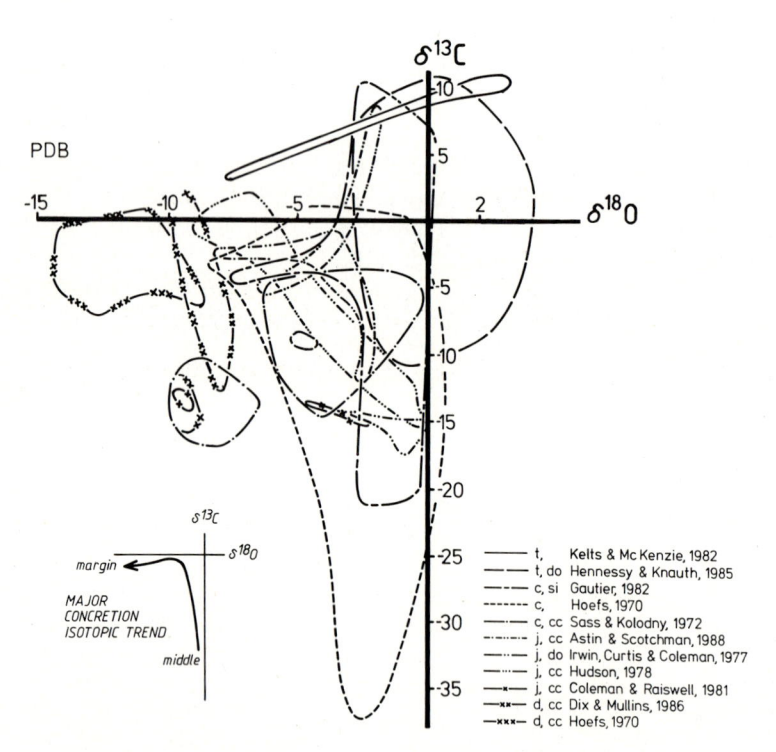

Fig. 7. Diagenetic trends for oxygen and carbon isotopes determined from the carbonate fraction of concretions reported in the literature. Symbols indicate: *t* Tertiary; *c* Cretaceous; *j* Jurassic; *d* Devonian; *cc* calcite; *si* siderite; *do* dolomite. Values expressed in PDB standard

transitions from concretionary limestones (i.e., lateral coalescence of concretions) to horizons containing individual concretions are occasionally observed in which the average carbonate content in the sediment decreases (Pratt et al. 1985). Lateral correlation of concretion horizons is rendered possible basin-wide because the carbonate content of concretions is commonly so high (as a result of cementation) that they are visible by weathering in outcrop.

Stratiform horizons of concretions occur where "diagenetic signals" were emplaced over wide sea floor areas. Breakdown of large quantities of organic matter related to rapid sulfate reduction and succeeding methane production can lead to localized carbonate precipitation in the sediment, because both processes are associated with the release of HCO_3^- to the pore waters (e.g., Claypool and Kaplan 1974; Curtis 1980; Raiswell 1987). High organic matter content is thought to be related to anoxic phases in basin history (see Wetzel Chap. 5.1, this Vol.). The process of sulfate reduction can be especially intense when additional sea water sulfate is allowed to enter the sediment via diffusion. The latter aspect requires very slow or a stillstand of sedimentation as well as relatively high permeabilities, often found in silty sediments (Füchtbauer 1988).

Secondary concretion horizons may form where near-surface concretions underwent erosion and redeposition. Concentration in lag deposits occurred when the enclosing fine-grained sediment was eroded and removed. Such secondary concretion horizons are encountered in European Lower and Middle Jurassic rocks, indicating widespread submarine erosion associated with the transgression-regression history of the basin (see Kidwell Chap. 6.3; Einsele and Bayer Chap. 6.2, both this Vol.). Redeposition of concretions can be distinguished from concretions formed in situ by the presence of breakage, imbrication, borings, organic overgrowth, and association with other lag deposits (Hallam 1964; Voigt 1968).

2.3 Flint Nodules and Chalk Bedding: Development of a Secondary Stratification

In many pure carbonates with more than 95% $CaCO_3$, as in some European chalks, layers of flint nodules form a diagenetic stratification which is very conspicuous. These flints are thought to replace carbonate sediment when pore waters are silica-rich and neutral to slightly acid in pH. They seem to precipitate as opal CT or microquartz (e.g., Williams and Crerar 1984; Hesse 1989; Füchtbauer 1988). Often, flint nodules represent replacements of burrow fills showing virtually no signs of compaction and indicating early formation before the onset of chalk lithification (Kennedy and Garrison 1975b; Chanda et al. 1976; Geeslin and Chafetz 1982; Tada Chap. 4.2, this Vol, see Fig. 10). This assumption of early flint formation is also supported by the dislocation of flints in thin submarine slump sheets (Kennedy and Juignet 1974; Bromley and Ekdale 1987) and rare observations of erosion and redeposition (Carozzi and Gerber 1978; Voigt 1979). Despite early formation of flint, fault-related flint replacements record that a small portion of silica remobilization occurred later in diagenesis. Although the Cretaceous European chalk was heavily bioturbated by various species, concretionary flints predominantly replace arthropod burrows (Felder 1974; Bromley and Ekdale 1984b), forming bands of diagenetic flint nodules every 1 to 3 m. Burrow replacement is presumably associated with horizons or beds with higher silica content and with the repeated occurrence of burrowed omission surfaces and smaller sedimentation rates (allowing intense bioturbation). Individual flint nodule bands can be traced laterally for tens of km (Felder 1974) indicating that flints are largely emplaced parallel to the original stratification. Even so, it is an open question whether the formation of diagenetic flint bands has reduced or augmented the number of original chalk layers (Fig. 8).

2.4 Dolomitization: Accentuation Versus Obliteration of Original Stratification

It is not within the scope of this overview to present a comprehensive treatment on the dolomite problem and on the various models of dolomite formation discussed in the literature (e.g., Shukla and Baker 1988). Only a few comments on the influence of dolomitization on stratification changes are given here. Where dolomitization occurred during or shortly after deposition, dolomitized layers can be parallel or subparallel to original stratification. The crystal size of early formed dolomite is commonly small, so that primary sedimentary and biogenic structures are largely preserved (Füchtbauer and Richter 1988). Early dolomitization is observed in various diagenetic environments, including sabkhas, continental and marginal marine lakes, subsurface water mixing zones, as well as organic-rich settings. Where dolomitization happened relatively late under burial conditions, however, the crystal size of dolomite is relatively large, so that finer sedimentary and biogenic structures are obliterated (such as ripple cross-bedding, bioturbation mottling, and skeletal hard parts). Thus, primary bedding may be difficult to recognize.

Incipient dolomitization can be restricted to either lithological differences or geochemical pore water gradients. When dolomitization is restricted to individual beds, the lithological contrast of the original stratification can be enhanced. Beds

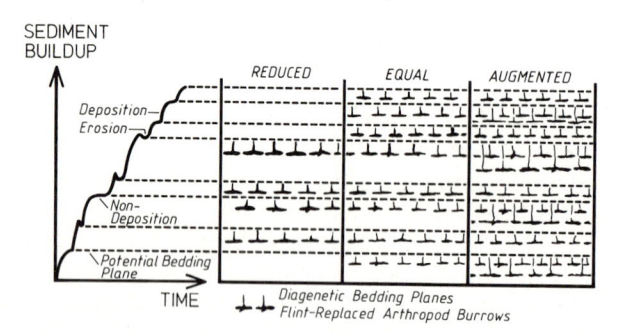

Fig. 8. Chalk stratification controlled by diagenetic flint nodule bands replacing Thalassinoides burrows. Sediment-buildup graph indicates smaller phases of nondeposition or erosion, forming the original stratification largely invisible in poor chalks with intense burrowing by Thalassinoides. A varying portion of burrow fills was later replaced by well-visible flint nodules, reducing or augmenting the original number of beds

originally rich in aragonite and Mg calcite have undergone preferred dolomitization (Zenger and Dunham 1980; Bullen and Sibley 1984). Additionally, marl beds in interbedded marl-limestone sequences may have become selectively dolomitized under burial conditions. Such dolomitization can be associated with processes of pressure dissolution (Wanless 1979) or can be understood as an adjustment of the uncemented marl bed carbonate to burial pore waters. Where dolomitization occurred along geochemical gradients and advection paths of the porewater, irregular dolomite lobes were formed. These cut across the primary stratification and indicate former geochemical reaction fronts.

3 Diagenetic Overprints Associated with Depositional Events

3.1 Hardground Cementation: Diagenetic Accentuation of Erosional Surfaces

Hardgrounds are semi-lithified to lithified erosion surfaces (see Kidwell Chap. 6.3, this Vol.). Borings, encrusting epizoans, and reworking of hardground crusts indicate early cementation (Voigt 1968; Bromley 1975; Kennedy and Garrison 1975b), which is thought to result from biological activity and the unusually long contact time of the sediment with sea water (Jeans 1980; Müller and Fabricius 1974). A modern analogy for hardground formation is *sea floor cementation* known from areas of winnowing, for example, the slope of the Bahama Banks, the Straits of Florida, and the Mediterranean (Schlager and James 1978; Mullins et al. 1980; Müller and Fabricius 1974). Hardground cementation prevents erosion from continuing below the cemented surface; without rapid cementation, erosion would soon destroy the incipient hardground -- a possible reason why hardgrounds seldom develop in marls and marly shales. Composite hardgrounds are due to intermittent deposition and erosion (Bromley 1975), while incipient hardgrounds represent patterns of local cementation which may show transitions to nodular limestones, the so-called honeycomb structures (Garrison and Kennedy 1977).

Some hardgrounds also may be formed by re-exposing previously cemented layers through submarine erosion of soft, overlying sediment (*exhumed cemented beds*). Erosion of soft strata may continue until such a previously cemented layer is reached, which prevents further erosion. Thus, former depositional and diagenetic properties may be inherited ("memory effect", see Seilacher and Aigner Chap. 2.3, this Vol.), so that lithified hardground substratum for boring epizoans and firm-ground dwellers is provided. Such secondary hardgrounds can be distinguished from those formed by sea floor cementation by missing evidence of soft sediment erosion and burrowing representing early stages of normal hardground formation.

Because most of the hardground cementation occurred on the sea floor and thus generally prior to compaction, absolute cement values and carbonate contents in hardgrounds are very high. Consequently, porosities are more reduced than in the surrounding rock (Jeans 1980). Porosity is preserved in some hardgrounds because thin fringes of marine cement on the surfaces of the grains make a rigid framework which resists compaction (see Bathurst Chap. 3.2, this Vol.). Depending on the morphology of the erosional surface, hardground cementation may overprint slightly diachronous sediments. In weathered sections, the bedding pattern of cemented hardgrounds is accentuated, where the original bedding of the intercalated non-hardground sediment is weak or diminished in thickness by compaction. Under such conditions, the stratification becomes dominated by diagenetically enhanced erosional surfaces.

3.2 Event Bed Cementation and Underbeds: Selective Augmentation of Depositional Events

In many calcareous tempestite and turbidite sections, event beds are diagenetically enhanced in carbonate content and relative thickness by early cementation and continuing compaction of the background sediment. After transportation into deeper-water environments rich in clay or calcareous nannoplankton ooze, event beds composed of shallow-water carbonates (containing unstable carbonate phases) may have a higher potential for the initiation of diagenetic reactions than the host sediment. Event bed cementation can occur at the sediment surface, thus may be sea water mediated (Dix and Mullins 1988), or may occur during shallow burial conditions if event beds are embedded rapidly. Where cementation is fast and intense, additional cementation below the event layers may occur, forming the so-called "underbeds" (Meischner 1964, 1967; Eder 1971, 1982; see Seilacher and Aigner Chap. 2.3; Eberli Chap. 2.8, this Vol.). These are small cemented zones a few centimeters thick always lying immediately below the event beds (Fig. 9). They are thought to originate by two mechanisms -- either cementation starts in the event bed and continues below its well-defined base, or two centers of cementation, one in the event bed and one in an underlying layer, amalgamate to form a carbonate curve with two small maxima. In any case, the onset of cementation can occur so early that underbeds may have a sedimentary and biological response providing substrates for firmground dwellers, or it can be relatively late under shallow burial conditions. In Devonian and Carboniferous periplatform turbidites, high original Sr contents are

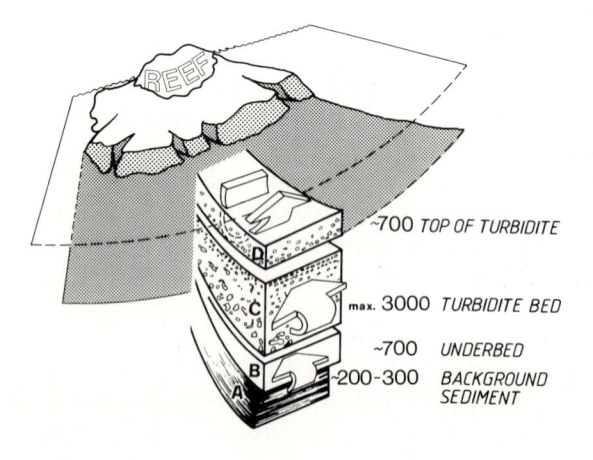

Fig. 9. Cementation and underbed formation of a calcareous turbidite bed (layers *C* and *D*). Preservation of high Sr contents (in ppm) in the turbidite (layer *C*) indicates early cementation of reef detritus. In addition, high Sr contents occur in the cemented zone below the turbidite base (underbed, layer *B*) compared to the nonevent, background sediment (layer *A*). *Arrows* denote carbonate redistribution. Devonian, Germany

preserved in the event beds because of early cementation (Eder 1982; Fig. 9). In this environment and in shallow marine Triassic carbonates of Germany, cemented event and underbeds were exposed indicated by animal borings (Aigner 1985). In contrast, calcareous turbidites of the pelagic Scaglia limestones in Italy still underwent 10 to 20% mechanical compaction before the onset of cementation, which equals roughly 40 to 80 m of overburden (Ricken 1986).

3.3 Nodular Limestones and Related Fabrics: Increase or Decrease of the Number of Original Beds?

The term "nodular limestone" refers to a wide range of nodular fabrics in commonly fine-grained bedded carbonates that are physical, biologic, and diagenetic in origin. Sedimentary structures may include erosional surfaces, slumpings, lag deposits, and bioturbation fabrics, while diagenetic patterns are related to hardground and concretionary cementation, which is followed by burial cementation and pressure dissolution (Jenkyns 1974; Farinacci and Elmi 1981; Wendt and Aigner 1985; Möller and Kvingan 1988). The depositional processes are dominated by repeated phases of stratigraphic condensation and omission, and thus by low rates of sedimentation (as low as 1 m/Ma) in environments ranging from the shelf to the deep sea. According to Wendt and Aigner (1985), Paleozoic nodular limestones were deposited on pelagic carbonate platforms, on slopes, and in sediment starved basins with low terrestrial influx. Bottom waters were well oxygenated. In all environments, pelagic organisms (e.g., cephalopods) predominate over benthic organisms. Various names are used to characterize this sedimentary to diagenetic facies, such as "cephalopod limestone", griotte, and rosso ammonitico (e.g., Tucker 1974).

During early diagenesis, burrows on previous omission surfaces are affected by concretionary cementation (e.g., Gründel and Rösler 1963; Fürsich 1973; Möller and Kvingan 1988; Jenkyns 1974; Kennedy and Garrison 1975b; Farinacci and Elmi 1981; Gluyas 1984). Depending on the bulk carbonate content, types of carbonate phases, and the porewater composition, the onset of nodule cementation may be a few meters to several tens of meters below the sediment-water interface (similar to the formation of concretions), as documented by low compaction values, high carbonate contents, the general absence of borings and epizoans, and nodule displacements in shallow slump sheets (Wendt et al. 1984; Fig. 10). A few examples of nodule redeposition and winnowing are reported in the literature (Hallam 1964; Weber 1969; Kennedy and Garrison 1975b; Elmi 1981; Snavely 1981; Bromley and Ekdale 1987). If, under conditions of increasing burial diagenesis, these early nodules are subjected to pressure dissolution and complete cementation, they may become irregularly shaped, due to the development of wispy dissolution seams and pervasive dissolution in the surrounding matrix. Thus, the lithological contrast between nodules and the marl matrix is enhanced, and the typical features of nodular limestones are generated. The original number of potential beds and the original bedding rhythm may be modified where cementation has occurred selectively; this problem is essentially that of diagenetic bands of flint nodules in chalks (see Fig. 8).

One type of overprint resembles nodular limestones in many ways, but seems to be generated later in diagenesis. It includes flaser chalks and flaser limestones, which

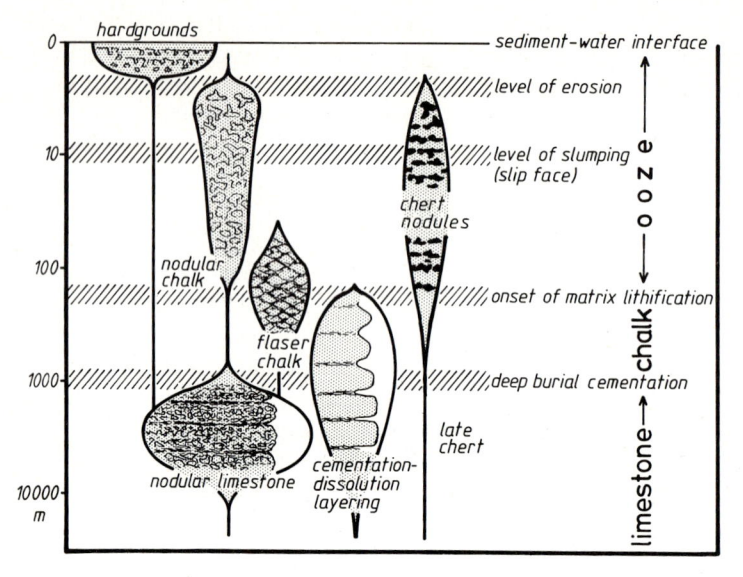

Fig. 10. Schematic development of the most important types of diagenetic overprints in nannoplankton calcareous ooze related to increasing burial depth (in m)

start to form by patchy, incomplete cementation at shallow burial depths, and are modified by deeper burial pressure dissolution-cementation overprinting (Fig. 10). Flaser chalks resemble "augen-structures" in metamorphic rocks; they show small calcareous lenses embedded in a matrix of anastomosing marl seams (e.g., Garrison and Kennedy 1977; Eller 1981).

4 Diagenetic Overprints as Related to Earth History and Changing Carbonate Compositions: Conclusions

The various types of overprint in bedded carbonates, marls, and calcareous shales seem to be associated with $CaCO_3$ compositions and various facies transitions, which changed in certain periods of the Earth's history. Such facies transitions are influenced by the evolution and supply of organic and inorganic calcareous grains, sea-level changes, and climatic variations (Fig. 11). Since the Lower Jurassic, pelagic, calcite (coccolithic) oozes began to intermix with calcareous shallow water sediments, which were dominated by calcite and Mg calcite in the Paleozoic, and evolved to presentday aragonite-dominated carbonates (Fischer 1984; Wilkinson 1979). General maxima in the production of shallow water carbonate sediment are thought to be related to periods with sea-level highstands (Schlager and James 1978; Kier and Pilkey 1971; Droxler and Schlager 1985), such as those in the Middle Paleozoic and the late Mesozoic, when shelves and continents were extensively flooded (see Eberli Chap. 2.8, this Vol.). Additionally, variations in climate and in ocean water chemistry were reflected by the dominant inorganic precipitation of

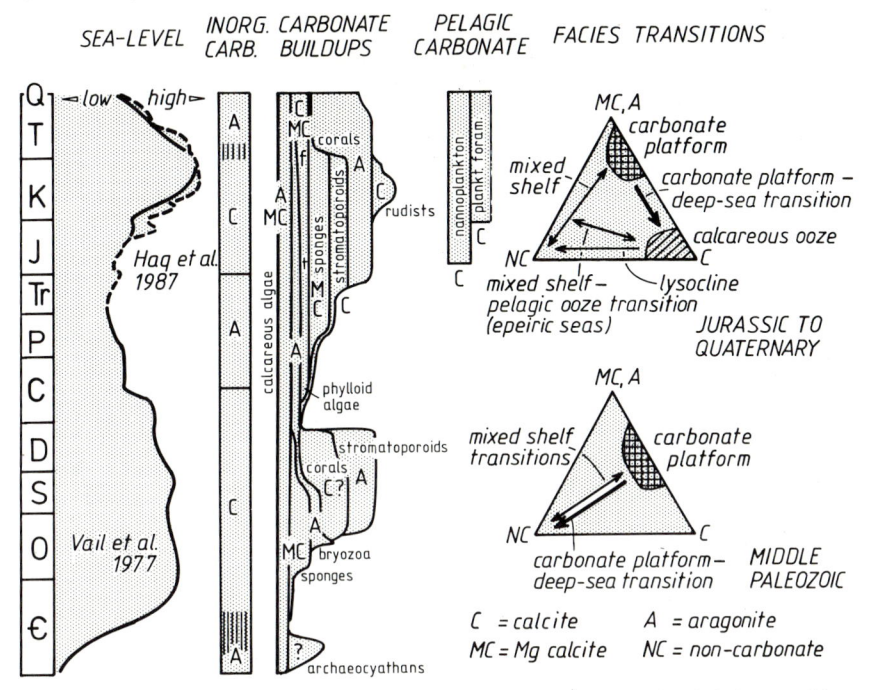

Fig. 11. General trends in Phanerozoic composition of carbonates and associated facies transitions (*triangles*). Sea-level curve from Vail et al. (1977b) and Haq et al. (1987), development of inorganic carbonates after Sandberg (1985), reef evolution after James (1983), and skeletal mineralogy according to Scholle (1978). *Triangles* denote original sediment compositions expressed as mixtures of unstable carbonate content (*MC* Mg calcite; *A* aragonite), calcite (*C*), and the noncarbonate or clastic fraction (*NC*)

aragonite as opposed to calcite (Sandberg 1985; Wilkinson et al. 1985). Aragonite-dominated periods presumably represented cool periods in Earth history with low atmospheric CO_2 and presence of some continental ice, while warm greenhouse periods with relatively high atmospheric CO_2 seem to have favored inorganic calcite precipitation.

1. Diagenetic redistribution of carbonate by *differential dissolution and cementation* between succeeding beds has differing impacts on the original bedding rhythm. If the overall carbonate content is low, the effect of bundling is enhanced, whereas if the content is high, the bedding rhythm is accentuated; this is related to the formation of rhythmic limestones with angular carbonate curves, thin marly interbeds, and only indistinct bundling patterns. In fine-grained, calcite oozes, differential dissolution and cementation processes are assumed to occur in a largely closed carbonate system, whereas the diagenetic system in platform carbonates containing aragonite and Mg calcite is assumed to be mainly open. Under conditions of a closed system, the bedding is more pronounced when the sediment had a high initial porosity, as observed in fine-grained carbonate sediments, and where mechanical compaction was low at the onset of cementation. Consequently, pelagic, low-Mg calcite oozes Cretaceous and

Cenozoic in age, and their transitions into equivalent shelf carbonates become most accentuated in original bedding rhythm. Under conditions of a more open system, the intensity of the diagenetic accentuation of bedding depends on relatively early cementation confining later pressure dissolution to noncemented beds (see Bathurst Chap. 3.2, this Vol.). Early cementation depends on the potential to initiate diagenetic reactions, related to high carbonate content and unstable carbonate phases. Thus, many Phanerozoic shelf and platform limestones are affected by this process.

2. Carbonate concretion horizons are most commonly found in organic-rich marls, clays, and siltstones. A high organic carbon content in the former sediment and intense sulfate reduction seem to be most important for concretion formation. It is assumed that concretion formation was closely tied to Phanerozoic global anoxic events such as the Lower Jurassic, the Middle Cretaceous, and shale-rich periods in the Paleozoic (Jenkyns 1980). Further factors are related to smaller anoxic periods and stillstands in sedimentation resulting from the local basin and sea-level history. Near-surface and shallow burial concretions can be distinguished by their different shapes and the amount of lamina draping. Concretions provide excellent insights into early diagenetic processes and pore water geochemistry.

3. Diagenetic bands of flint nodules are known mainly from pure, pelagic to hemipelagic carbonates which occur since the Jurassic. The reason for this may be threefold: (1) The simultaneous deposition of calcareous and siliceous pelagic sediment began in the Jurassic (Fig. 11). (2) Radiolarians underwent major evolutionary changes at that time, allowing them to form an important minor constituent of carbonate ooze. (3) High carbonate content seems to favor the diagenetic precipitation of amorphous silica, while high clay content seems to inhibit silica precipitation (Baker et al. 1980). Bands of flint nodules in European chalks represent replacements of Thalassinoides burrows and form a very conspicuous, diagenetic stratification.

4. Most hardgrounds are formed by precipitation of sea water-mediated cements in submarine erosion surfaces. In order to accomplish this, large contact times of sea water and supersaturation with respect to aragonite and Mg calcite are required (i.e., warm, tropical, and subtropical seas). In addition, a second type of hardground may be formed by re-exposing previously cemented layers, inheriting diagenetic conditions of the former sediment. Hardgrounds are found in all kinds of calcareous shelf environments and in pure chalks. It is supposed that most hardgrounds contain large quantities of cement, because cementation occurred in a pore space which had been only slightly reduced by compaction. Cemented hardgrounds (i.e., erosional surfaces) tend to stand out in a sequence because they are more resistant to weathering than the enclosing sediments, and may dominate the bedding rhythm.

5. Cementation of calcareous event beds occurred either at the sediment surface or at shallow burial depth when event beds were quickly embedded, and therefore, sealed from sea water influence. Many basinal calcareous event beds consist of allochtonous shallow water carbonates with a higher diagenetic potential than the enclosing sediments. As a result, cementation in these event beds is generally earlier than that in the enclosing, diagenetically more stable basinal sediment. When event

beds contain reactive, instable carbonates, such as reef detritus, cementation can be intense and may continue below the event bed base, forming so-called underbeds.

Largest contrasts in diagenetic potential between shelf (or platform) carbonates and basinal sediments occurred in the Paleozoic, because basins during this time were dominantly free of planktonic carbonates. Thus, shelf-basins facies transitions occurred between reactive shallow-water carbonates and carbonate-poor basinal sediments (Fig. 11). Since the Jurassic, however, shallow-water and platform carbonates, though relatively richer in aragonite, passed into pelagic calcite oozes, reducing the diagenetic potential between allochthonous event beds and the enclosing basinal sediment. Underbeds are, therefore, predominately observed in Paleozoic calcareous turbidite sequences. In addition, underbeds are found in tempestite sequences of shallow water carbonate environments, where cementation may be sea water mediated. Related to early, rapid cementation and underbed formation, event beds become reduced less in thickness than the compacting background sediments, enhancing their stratigraphic significance and visibility in outcrops.

6. Many nodular limestones contain early diagenetic nodules which were later subjected to (deeper) burial pressure dissolution and complete cementation. Sediments with unstable, reactive carbonate grains, patchy distribution of organic matter and porosities seem necessary for the first phase of diagenesis, while carbonates with some clay content are required for the second phase. Nodular limestones are generally associated with periods of stratigraphic condensation. They occur in dominantly fine-grained calcareous sediments at various water depths including shelves, carbonate platforms, and environments in the deep sea. The role of sea-level variations in providing favorable conditions for the formation of nodular limestones is not well understood. Triassic and Jurassic nodular limestones which formed in shallow-water environments may reflect stratigraphic condensation related to global (first order) sea level lowstands (Fig. 11). On the other hand, nodular limestones in the Upper Devonian, Lower Carboniferous and a few occurrences in the Upper Cretaceous are associated with global sea level highstands.

Acknowledgments. The authors would like to thank Robin Bathurst, who carefully read an earlier version of this manuscript and made numerous suggestions and improvements. In addition, we would like to thank Gerhard Einsele, Susanne Borchert, and Linda Hobert for correcting this manuscript, Jochen Hoefs for reviewing the section on concretions, as well as Jobst Wendt for reading the section on nodular limestones. Johannes Weber provided literature on flint formation. Erik Flügel and Philip Sandberg made helpful comments regarding Phanerozoic changes in carbonate mineralogy.

3.2 Pressure-Dissolution and Limestone Bedding: the Influence of Stratified Cementation

R. G. C. Bathurst

1 Introduction

While few geologists would deny that pressure-dissolution has often played a role in the development of bedding in limestones, there seems little certainty about how the process has fitted into the sequence of diagenetic events. The products and process of "pressure solution" (Sorby 1908), or "chemical compaction" (Lloyd 1977), still receive rather scant notice. In most published work on diagenesis, stylolites merit little more than a brief obligatory genuflexion, whilst fitted-fabric and dissolution-seams are ignored. Attempts to account for the distribution of pressure-dissolution caused by overburden in limestones have been dominated by an assumption that clay minerals exert a direct chemical or structural influence on the process. Yet there is growing evidence that suggests that the most powerful control has been the development of an alternation of cemented and uncemented strata early in the history of mechanical compaction during burial. Pressure-dissolution was thereafter concentrated in the less cemented strata. The timing of this stratified cementation varied from syndepositional to some stage during burial. The causes of such a selective inhibition (or inducement) of cementation are only partially understood.

The ideas presented here, on the reasons why pressure-dissolution fabrics in beds are located where they are, were stimulated in the first place by the writings of Chanda et al. (1977) on the mechanical deformation of ooids against early concretions, in a Precambrian limestone, and of Purser (1978, 1984) on the influence of Jurassic hardground cementation. The importance of a resistant body of sediment as a control on the siting of pressure-dissolution was also stressed by Wanless (1979).

Summaries of earlier work can be found in Bathurst (1975, 1980a,b,c, 1984, 1986, 1990), Chanda et al. (1983), Marshak and Engelder (1985) and Halley (1987). Excellent reviews of burial diagenesis in carbonate sediments have been provided by Scholle and Halley (1985) and Choquette and James (1987).

2 Pressure-Dissolution: Terminology and Process

In face of the confusion in the literature, the threefold classification set up by Buxton and Sibley (1981) seems ideally useful (slightly modified with the agreement of the authors) because it explicitly distinguishes between the three quite distinct styles of pressure-dissolution which give rise in limestones to (1) fitted-fabric of interpenetrant or welded grains, (2) dissolution-seams and (3) stylolites.

Fitted-fabric according to Buxton and Sibley is a pervasive microscopic framework of intensely interpenetrant grains (Fig. 4B,C). The authors regarded it as falling within Wanless's *non-sutured seam*. This is understandable because *fitted-fabric* was originally defined by Logan and Semeniuk (1976) as "a fabric where idens are in contact with neighbours along their entire margins", where an *iden* can be any body from a microscopic particle to a whole reef or a geological massif. Buxton and Sibley's more modest usage nevertheless refers to an interpenetration of grains so intense that each grain is entirely in contact with adjacent grains. No allowance is made for grains which are only slightly interpenetrant with some residual intergranular porosity (Fig. 4B,C). Rather than abandon Buxton and Sibley's classification, the definition of fitted-fabric is here extended to include all interpenetration of sedimentary particles (clay-sized to pebbles) as a result of pressure-dissolution. The interpenetrant grains are embayed or truncated by interfaces which may be smooth or sutured (Trurnit 1968).

Dissolution-seam is used here instead of the *solution seam* of Buxton and Sibley so as to keep the word *solution* for the fluid. It is a smooth undulose seam of insoluble (noncarbonate) residue (Fig. 1B) which lacks the sutures of a stylolite. It is equivalent to the *solution seam* of Garrison and Kennedy (1977) and the *nonsutured seam* of Wanless. Unlike a stylolite it lies around grains or particles, in between them, but does not cut through them, though their margins are commonly corroded. Clearly this fabric is an advanced development of fitted-fabric (as discussed later) so that there can be no sharp distinction between the two. It is commonly anastomosing. *Microstylolite* as used by Wanless (1979) and by Marshak and Engelder (1985) is a dissolution-seam. Thus, to avoid confusion, this term is not recommended.

Fig. 1. a Argillaceous wackestones with false bedding planes formed by weathering of concentrations of dissolution-seams. Carboniferous (Dinantian), Clwyd, U.K. *Black* and *white bars* at 10 cm intervals. **b** Dissolution seams in **a**. Slice. *Bar* 500 µm

A stylolite is a serrated interface between two masses of rock: it has a sutured appearance in sections normal to the plane in which it is disposed (Figs. 2B, 3B). Insoluble residue may or may not be present. It is equivalent in part to the *sutured seam* of Wanless. In three dimensions, columns of one rock mass fit into sockets in the opposed mass. A stylolite transects all fabrics irrespective of their composition, cutting *through* both grains and any matrix or cement that may be present (Fig. 3B). It normally has an amplitude greater than the local grain diameter and extends laterally for distances many times that diameter. It is the *stratiform stylolite* of Purser (1984). It seems desirable that, in order to avoid confusion, the term *stylolite* should not be used for individual grain-to-grain contacts, even though they be sutured, but should retain its original meaning as a stratiform feature (Klöden 1828; Marsh 1867; Wagner 1913), with, of course, the inclusion of tectonic stylolites.

The processes by which the fabrics of pressure-dissolution evolve are still a matter of debate. Two theories are currently considered seriously, the *water film diffusion* theory of Weyl (1959), De Boer (1977), Robin (1978), Merino et al. (1983), Guzzetta (1984) and others and the recent *plastic deformation plus free-face pressure dissolution* theory of Tada and Siever (1986) and Tada et al. (1987). Both processes involve dissolution, molecular diffusion and precipitation. Inhibition of any of these activities stops the process.

Fig. 2. Stylolites as bedding planes. **a** Coccolithic chalk, Upper Cretaceous (Upper Chalk), Flamborough Head, Humberside, U.K. *Bar* 0.5 m. **b** Close-up of stylolites from **a**. Scale in cm

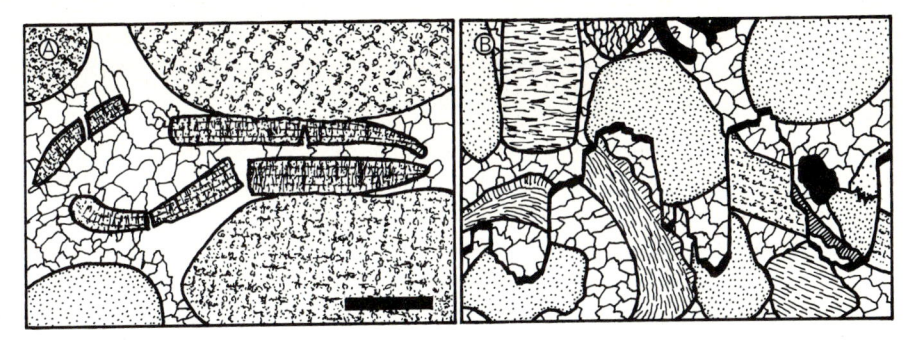

Fig. 3. Patterns caused by compaction: in slices. **A** Mechanical compaction of ooid cortex after leaching of nucleus, Jurassic (Bathonian), Wiltshire, U.K. *Bar* 200 μm. **B** Grainstone with stylolite (imaginary)

3 The Growth of Pressure-Dissolution Fabrics

3.1 The Growth of Fitted-Fabric

The threshold conditions for the start of pressure-dissolution, as overburden increases, are likely to be attained most readily in an unconsolidated sediment, where the stress at grain-to-grain interfaces is concentrated over a minimum total area. In such conditions, also, chemical compaction will have been preceded by mechanical compaction in which particles under load rotate, slide, deform and break (e.g., Fig. 3A).

Reprecipitation of dissolved ions in the pores as cement produces a rigid framework and eventually stops the process. Existing intergranular cement (Figs. 4, 5) must have arrived after the interpenetration was completed (Sorby 1908). In order, therefore, that pressure-dissolution can proceed simultaneously with cementation it is necessary that zones of active pressure-dissolution be separated from zones of cementation in either a patchy or layered pattern (Fig. 5). However, a thin fringe of cement need not impede the continuation of grain interpenetration. Pressure-dissolution of this kind is practically the sole lithifying process in pure coccolith oozes of deep-sea origin and in calcite oolites, owing to the stability of the low-magnesian calcite and the lack of unstable aragonite (Matter 1974; Schlanger and Douglas 1974; Hancock and Scholle 1975; Swirydczuk 1988).

In pure limestones fitted-fabric may develop further into stylolites. In argillaceous limestones it can evolve into dissolution seams.

3.2 The Growth of Dissolution Seams

Where the development of fitted-fabric goes far enough, in *argillaceous* carbonate sediments, the dominant interfaces (normal to the uniaxial compressional stress) tend to coalesce laterally to yield a network of anastomosing dissolution-seams (Fig. 1B). All stages of this development can be seen in some bedded argillaceous limestones

(Bathurst 1987). Dissolution seams generally form in swarms, as in the *flaser structure* of Garrison and Kennedy (1977) or the *horsetails* of Roehl (1967) and Mossop (1972).

3.3 The Growth of Stylolites

Stylolites form only in sediment that has rigidity derived either from a framework of cement, cemented matrix, or from a fitted-fabric or from the growth of a cementing replacement framework, such as dolomite (Schofield 1984). Stylolites seem to be absent in argillaceous limestones with a present composition of more than about 5–10% by weight of clay (Mossop 1972; Wanless 1979; Marshak and Engelder 1985).

4 Factors Controlling the Location of Pressure-Dissolution

4.1 The Influence of Cementation and Initial Mineralogy

The role of cementation in controlling the siting of pressure-dissolution surfaces was emphasized by Purser (1978, 1984), who drew attention to the influence of early cement on the subsequent burial diagenesis of Middle Jurassic grainstones in the Paris Basin. At certain horizons there are beachrocks (hardgrounds) in which syndepositional cement fringes on the grains formed a rigid framework (Fig. 4A), thus inhibiting all compaction. The pores were later filled with sparry calcite cement. Below the beachrock, the carbonate sands were buried without cement fringes and evolved a fitted-fabric (Fig. 4B), perhaps releasing sufficient carbonate to provide the sparry cement for the overlying beachrock. Later the pores in the fitted-fabric were also filled with sparry calcite. Traversing these welded grainstones there are many stylolites (Fig. 4C). The virtual restriction of the stylolites to the grainstones with fitted-fabric is doubtless due to the lateral coalescence of the actively enlarging solution films at the more horizontal grain-to-grain interfaces. A similar contrast in burial history between early (meteoric) cemented and late cemented oolites was made by Hird and Tucker (1988).

Another variation has been reported by James and Bone (1989). Two Cenozoic limestones, one consisting mainly of bryozoan and molluscan debris and the other of bryozoan only, have shared similar histories of vadose diagenesis and burial. Yet the molluscan-rich one, having a source of carbonate for cement, is well lithified and little compacted, while the bryozoan-dominated one, having scarce aragonite, now has a weak fitted-fabric and is soft to friable.

It is important to note that Purser's Jurassic grainstones had a low diagenetic potential because of a low initial content of aragonite. The proportion of grains that could be dissolved and reprecipitated as sparry cement, during meteoric diagenesis, was small. Thus the main supply of dissolved carbonate for cementation was late and derived from pressure-dissolution. By comparison, a sediment rich in aragonite might have suffered an early meteoric dissolution of the aragonite, so permitting a simul-

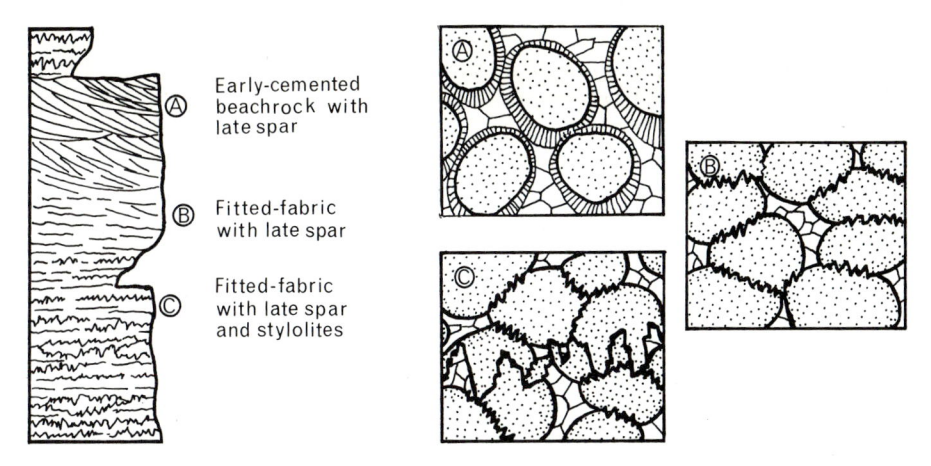

Fig. 4. Shallowing-upward grainstone sequence (diagrammatic). **A** beachrock and pendant cement. **B** fitted-fabric and late spar. **C** Addition of stylolites. Jurassic (Bathonian), Burgundy, France. (After Purser 1984)

taneous cementation with sparry calcite to form a rigid frame. Here a fitted-fabric would probably never have evolved. Initial mineralogy is thus an important control.

The role of original mineralogy in controlling pressure-dissolution is dramatically revealed by the oolites of the Upper Jurassic Smackover Formation, in the subsurface around the Arkansas-Louisiana border in the United States. The more northerly ooids have either been calcitized to a neomorphic fabric or dissolved to yield oomouldic porosity with reprecipitated carbonate as an intergranular cement, so giving the rock rigidity. These ooids were therefore originally aragonite (Sandberg 1984; author's observations, and Swirydczuk 1988). The southern ooids retain their radial-fibrous fabrics and were thus initially calcite. They are strongly pressure-welded and stylolites have begun to grow. It is clear, therefore, that the absence of aragonite saved the southern oolites from cementation and rendered them susceptible to a late pressure-dissolution. So they retained good porosities and permeabilities even at 5000 m subsurface. Similar conditions for the preservation of porosity at depth have been important in many grainstones and oolites, for example in the Jurassic Arab-C in Saudi Arabia, where a fitted-fabric can have a porosity of 20% with a permeability of 921 md. even though it is buried at 2000 m (A.O. Wilson 1985, p. 331).

4.2 The Influence of Clay

Evidence regarding the role of clay in pressure-dissolution is contradictory. While it is true that fitted-fabric and dissolution-seams have developed most readily in the more argillaceous limestones, it is also true that stratiform stylolites have grown predominantly in the purer limestones. Recently Marshak and Engelder (1985) have claimed that clay plays a purely mechanical role, supporting the particle fabric and improving the "interconnectivity" between the sites of dissolution and the open pore system. Yet the experimental work of Baker et al. (1980) indicated that clay in carbonate sediments inhibits pressure-dissolution. It is possible, of course, that simple dilution of carbonate sediment by clay in certain strata inhibits cementation and so encourages compaction. The addition of clay could lead to a reduction in the number of carbonate nuclei available for growth of cement crystals. Adsorbed organics, phosphate or Mg^{2+} may also inhibit nucleation. Finally, many horizons previously regarded, in the field, as shales or marls derived from primary noncarbonate muds, are now known, through petrographic study, to be only concentrations of dissolution-seams in limestones (e.g., Alvarez et al. 1985 and author's observations). The role of clay remains unclear.

The distribution of clay in some argillaceous shelf limestones is rather surprising and significant. Bathurst (1987) examined thin sections from which the carbonate had been leached. A residue of illite was recorded in all microcrystalline fabrics, i.e., in micrite matrix, peloids, ooids, micrite-filled algal bores and in micrite in the pores of echinoderm columnals and plates. Illite is absent in skeletal material and in sparry calcite cement. In the grainstone layers which were largely unaffected by compaction, the clay content ranges from 5–15% by weight, but it is entirely *intragranular*. Any role it might have had in maintaining "interconnectivity" would have been insignificant, but its presence reduced the availability of carbonate mineral nuclei at grain surfaces.

4.3 The Role of Molecular Diffusion Versus Flowing Water

Evidence for the reprecipitation of the pressure-dissolved ions has been adduced from porosity increases, upward and downward, away from stylolites (as in Wong and Oldershaw 1981). There is no doubt about the importance of the diffusion of molecules or ions through the water film (and beyond it through the adjacent porewater) in the removal of dissolved ions from the surfaces undergoing dissolution. Transport of ions over distances of millimetres to centimetres is possible in this way. The symmetrical loss of porosity by cementation above and below stylolites (as in Wong and Oldershaw) suggests the play of diffusion as a dominant control on ion transport, as in the growth of concretions. By comparison, the ability of flowing porewater to carry ions away seems more restricted (discussion of both processes in Berner 1980 and Raiswell 1988b).

Porewater must, of course, flow out of the compacting rock as its volume decreases. During pressure-dissolution the porosity is reduced by a combination of collapse and cementation. In a pervasive fitted-fabric water can flow in any direction, although its passage will become increasingly restricted to a direction normal to the compacting stress as the fitted-fabric matures and dissolution seams develop. Water flowing through the permeable layers of rock that lie between the active dissolution-seams or active stylolites (Fig. 5) must increasingly move in a direction *parallel* to these structures so long as pressure-dissolution is active. It can neither flow across them nor within them. Their final net permeabilities are minimum, as little as 0.05 md, depending on their lateral continuity (Koepnick 1984) so that flow across them would always be greatly inhibited. But in addition, according to the water film diffusion theory, the water dipoles in the films are structured so that the film can transmit a uniaxial stress and does not therefore have the properties of a Newtonian liquid.

The common assumption that upward streaming water, resulting from compaction of *underlying* rocks, can carry dissolved ions away from the dissolution interfaces in the necessary quantities, has been shown to be quantitatively erroneous (Land and Dutton 1978; Bjørlykke 1979; Ricken 1986). The removal of a unit volume of water saturated with dissolved carbonate requires the passage of some tens of thousands of unit volumes of water (Bathurst 1975, p 440). Therefore it is plain that the water dispelled locally, or driven upward from below during compaction, would normally lack the necessary volume by several orders of magnitude. Nevertheless the water from below may be enough to give a distinctive geochemical *signature* to a cement (Land et al. 1987).

It seems, then, that the most important role for flowing water is in flow parallel to bedding, driven by differentials in hydrostatic pressure on a structural or basinal scale. The possibility of flow along or across stylolites or dissolution-seams is not in doubt if they are inactive and have opened as a result of a reduced stress or a change in its orientation.

5 Pressure-Dissolution and Bedding

The distribution in weathered outcrop of fitted-fabric or dissolution seams in argillaceous limestones commonly gives an impression of bedforms that is an illusion (Figs. 1A). In so many clay-rich (and commonly bioturbated) limestones, bedforms (Leeder 1982) are absent, and these late compactional fabrics alone define the bedding. On the other hand, in clay-poor limestones which have not been bioturbated, there is commonly a juxtaposition of fitted-fabric or stylolites with retained bedforms and depositional bedding planes (i.e., two-dimensional interfaces between lithologies). This distribution of pressure-dissolution fabric was emphasized by Buxton and Sibley (1981) in their study of a range of lithologies in Devonian limestones.

5.1 Pressure-Dissolution in Bedded Argillaceous Limestones

The problems have been nicely reviewed by Scoffin (1987a, p 142). Einsele (1982a) summarized current thinking regarding marl-limestone couplets in pelagic-hemipelagic limestones. The precursors to both marl and limestone were bioturbated. Primary differences in carbonate content between them were accentuated by diagenesis, so that the limestone members became more cemented and resistant to compaction. The effects of subsequent compaction, both mechanical and chemical, were thus concentrated in the marls. A similar need for an early stratified lithification (cementation) was implied in the works of Eder (1982) on the Devonian and Carboniferous limestones of the Rheinisches Schiefergebirge in the Federal Republic of

Germany, of Ricken and Hemleben (1982) on the Oxfordian limestones of the southern Federal Republic and of Walther (1982) on Dinantian limestones of Northern Ireland. Alvarez et al. (1985) have argued that the sharply defined couplets of lime mudstones and clay partings (with dissolution seams, Arthur and Fischer 1977) in the Cretaceous-Paleocene at Gubbio, Umbria, are the result of a diagenetic reorganization that took place after the destruction of primary bedding by bioturbation. In all these examples, and in Hattin's (1971) sequences of Cretaceous chalks, it seems probable that the preferential cementation of specific horizons was part of a redistribution of carbonate by dissolution-precipitation over distances of a few centimetres in a closed system. The cementation need only have been partial, just sufficient to impart a resistance to mechanical compaction. Indeed, total cementation would have destroyed the porosity available as an essential sink so that pressure-dissolution would have been impossible.

A study of pressure-dissolution in beds of shallow-water (shelf) argillaceous limestones of various ages (Bathurst 1987) demonstrated the influence of stratified cementation early in compactional history. The bedded appearance of the argillaceous limestones is a result of their division into couplets of *fissile* limestones and *hard* limestones (Fig. 5). The fissile limestones have a pronounced concentration of fitted-fabric and/or dissolution seams and also of clay. The hard limestones are relatively poor in these features. A factor of critical importance in these shelf limestones is that the platy or elongate grains in the fissile limestones have a preferred orientation roughly parallel to bedding, whereas those in the hard limestone are largely random in orientation (Fig. 5). Bathurst argued that, because of the ubiquitous distribution of ichnofabric and of broken shells, all primary depositional fabric (lamination, two-dimensional bedding planes, and grain orientation) must have been modified or destroyed by bioturbation (Sepkoshi et al. Chap. 2.6, this Vol.). This conclusion is sustained by the observed absence of the lamination which normally accompanies depositional bedforms (Leeder 1982) and anoxic shales. Bathurst noted also that pressure-dissolution cannot rotate grains. This is clear in the growth of

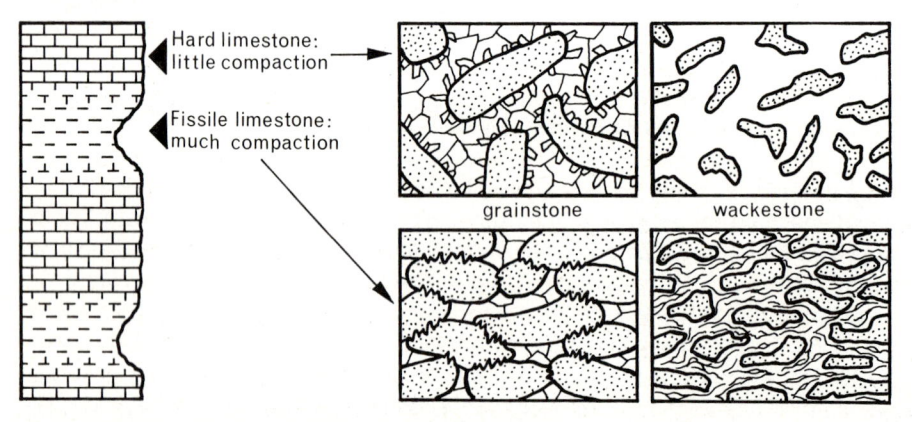

Fig. 5. Diagrammatic sequence of hard and fissile limestones

fitted-fabric which is a development of grain-to-grain adhesion at interfaces that are locking and not subject to shear. It is inferred, therefore, that the greater preferred orientation of particles in the fissile limestones must be a result of mechanical compaction (Shinn and Robbin 1983). Bathurst concluded that the intervening layers (now hard limestone) were made resistant to compaction by a cementation early in the history of mechanical compaction.

It is important to emphasize that this selective cementation was not necessarily early in the post-depositional history of the sediment. It was early only in respect to the history of mechanical compaction. The stratified Holocene cementation discovered by Shinn in the Persian Gulf was syndepositional whereas in the Gubbio section (above) stratified cementation was delayed until the carbonate muds had been reduced to 30% of their original thicknesses (Ricken 1986). This early timing of the cementation with regard to compactional history is also demonstrated by the way in which the preferred orientation of particles has been deflected around nodules (concretions). The junction between hard and fissile limestones is commonly nodular: in extreme cases the hard limestone is composed of a single layer of more or less coalescent concretions (Henningsmoen 1974; Möller and Kvingan 1988).

It follows that the first cementation was stratified, as can be seen in some Holocene shallow carbonate sediments in the Persian Gulf (Shinn 1969). The subsequent positioning of the pressure-dissolution fabrics was preordained by the pattern of this earliest cementation.

Unlike cementation in the deeper-water carbonates, the cementation in shelf carbonates would generally have proceeded in an open meteoric system, as witnessed, for example, by the numerous palaeokarstic surfaces in the Carboniferous sequence of North Wales.

Many Recent equivalents of these various limestones, in DSDP core or Bahamian shelf core, have commonly been heavily bioturbated and show no evidence of bedding planes or lamination, with the obvious exception of beach, bar and turbidite deposits.

Summarizing, there is, therefore, a considerable measure of agreement with the concept that the first rudimentary or incipient cement in argillaceous carbonate sediments was commonly (1) localized in discrete strata and (2) took place early in the history of compaction. Thereafter, mechanical and chemical compaction was restricted largely to the less cemented layers (Figs. 5, 6). In a recent paper, Ricken (1986) has put forward the idea that stratified cementation should be regarded as the norm in pelagic and hemipelagic carbonate sediments.

A nice example of what appears to have been stratified cementation was provided by Byers and Stasko (1978). In a bioturbated lime mudstone of Ordovician age, a vertical sequence containing undeformed *Chondrites* is interrupted at intervals by thin wavy layers of a more shaly composition in which *Chondrites* is flattened.

Controls proposed for this first stratified incipient cementation vary from lithological (e.g., Einsele 1982a; Evans and Ginsburg 1987; Goldsmith and King 1987; Bryant et al. 1988; Schroeder 1988) at one extreme to rhythmic diagenetic reorganization at the other (e.g., Jenkyns 1974; Hallam 1986). Valuable details of near-surface diagenesis are summarized in Berner (1980) and Raiswell (1987). Einsele (1982a) and Bathurst (1987) concluded that an alternating signal (mineralogi-

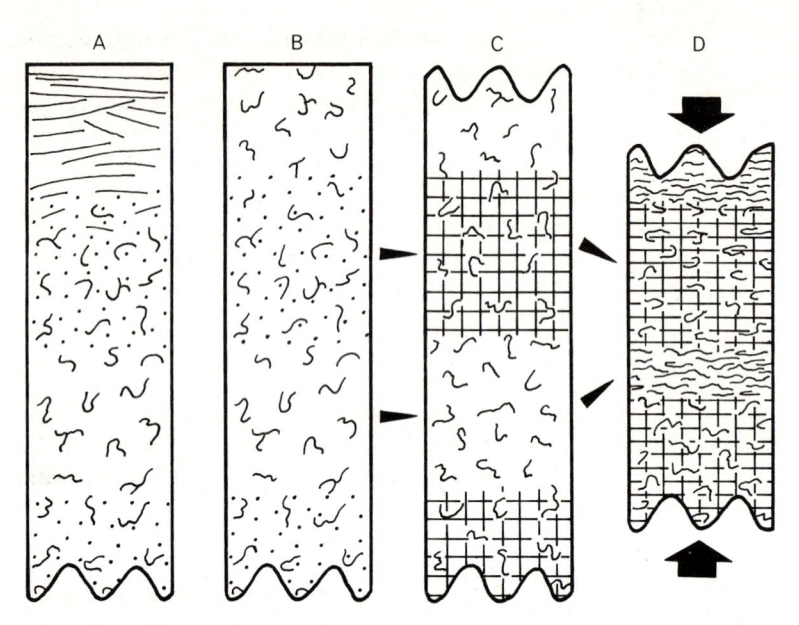

Fig. 6. Depositional and diagenetic history of argillaceous carbonate sediments. **A, B** Accumulation of sediment with or without depositional bedform. Mineralogical/chemical/textural signal implanted at certain horizons (*black dots*). **B** Replacement of any bedforms by ichnofabric with slight upward diffusion of signal. **C** Horizon with signal is selectively cemented (*squares*). **D** Overall compaction but concentrated in the less cemented layers

cal, chemical or textural) was printed on the sediment at the time of its deposition. This would have been related to the positions of successive sea floors (Schwarzacher and Fischer 1982). Bioturbation would have diffused it but could not have destroyed it altogether (discussion Berner 1980). Or, indeed, bioturbation may itself have imposed a textural signal (Bathurst 1987) through a process akin to the tiering of burrows (Bromley and Ekdale 1986).

Other limestones, though lacking evidence of pressure-dissolution, show similar constraints on mechanical compaction following stratified cementation. The Liassic limestone-shale couplets of southern Wales and southern England reveal an early stratified cementation followed by mechanical compaction in the intervening layers. This cementation was so early that lithified nodules were at times eroded and reworked (Hallam 1964). Reports on the Liassic Bridport Sands of the south of England by Bryant et al. (1988) and Kantorowicz et al. (1987) refer to an alternation of well-cemented bioclast-rich layers with other layers that are poorly cemented, clay-rich and bioclast-poor. Mechanical compaction was restricted to these clay-rich layers.

The couplets of marl (shale) and limestone commonly display a nodular structure which Wanless (1979) regarded as a product of pressure-dissolution. Nodules are commonly embedded in wavy anastomosing partings of dissolution-seams. Recent studies, however, have shown that nodular (concretionary) structure has commonly been an early diagenetic development, often related intimately to ichnofabric (Mullins et al. 1980; Raiswell 1987; Möller and Kvingan 1988; other references in Bathurst 1987, pp. 768, 773).

It must be admitted that these early cements have not been isolated with the help of the microscope. Crystal fabrics tend to be fine-grained, comparison of fabrics between layers is hampered by an overprint of dissolution seams and, above all, the quantity of early cement needed to bring about some rigidity and local inhibition of compaction was probably small and is now masked by later more general cementation of the whole sequence.

5.2 Pressure-Dissolution in Bedded Pure Limestones

A rather loose juxtaposition of stylolites and of fitted-fabric with depositional bedding planes in pure limestones seems in places reasonably clear. Nevertheless, in many sequences of so-called bedding-parallel stylolites, the stylolites are the only visible planar structures (Simpson 1985): no depositional layering is discernable (Fig. 2). The "residual seams" of Barrett (1964) in the Te Kuiti Group in New Zealand (which are spaced undulose sheets of fitted-fabric: unpublished observations of Barrett and the author) are a case in point.

The stylolitic grainstones of Dinantian age in part of a quarry at Horton-in-Ribblesdale, North Yorkshire, and in another at Hillhead, Derbyshire, U.K., show no laterally continuous bedforms. Instead, the rock has a streaky appearance, with lithologies of different composition and coarsenesses occurring as thin, laterally discontinuous laminae, a few millimetres to centimetres thick. The stylolites, singly or in groups, are repeated at vertical intervals of tens of centimetres. They appear to be sited at the sharper lithological transitions. Schwarzacher (1958) followed ten or so of the more prominent stylolitic surfaces over distances of some 30 km.

This general association of stylolites with the sharper lithological transitions is well known. Textural misfit is a preferred site for a water film. Extreme examples are contacts between limestone and, say, dolomite or chert. Buxton and Sibley (1981) observed that, in the Devonian Alpena Limestone of Michigan, some 75–80% of stylolites in the grainstones and packstones are at lithological transitions. In some limestones, stylolites coincide precisely with depositional bedding planes, as in well-sorted, cross-bedded grainstones of Silurian (Ludlow) age at Snögrinde, near Klinte, on the Swedish island of Gotland. Stylolites also occur at sharp lithological breaks, at the upper surfaces of marine hardgrounds, at Sandvik on the west coast of the Swedish island of Öland in Ordovician (Lower Arenig) lime mudstones (author's observations).

In some sequences, on the other hand, the stylolites seem, at first sight, to be unrelated to any discernable lithological transition. Simpson (1985), in his study of parallel seams (concentrations) of stylolites in a Dinantian shelf limestone in South Wales, noted that the stylolites bear no direct relation to the small amount of primary depositional bedding. Another striking example is the hard coccolithic Cretaceous

Chalk of Flamborough Head, northeast coast of England, where the bedding planes are no more than wavy dissolution seams and stylolites (Fig. 2). The same tendency is apparent (author's observation) in the massive, oncolitic, lime mudstones of the Jurassic Comblanchien Formation in the Paris Basin (Purser 1972).

It is clear that the control of the location of stylolites is not as obvious as it is for fitted-fabric and dissolution-seams. Nevertheless, if, as seems possible, some stylolites begin as zones of fitted-fabric in which appropriately oriented grain-to-grain interfaces coalescence laterally, then they should be subject to controls similar to those operating in argillaceous carbonate sediments. Moreover, dissolution, like precipitation, depends on transport of dissolved ions and should therefore be related to the distribution of permeability (Goldsmith and King 1987). In their study of the Alpena Limestone, Buxton and Sibley (1981) noted that grainstones composed of crinoids with cement overgrowths are largely uncompacted whereas the intervening, less crinoid-rich, grainstones lack cement but display fitted-fabric. These authors made the point specifically that "cementation controls pressure-dissolution". The influence of aragonite as a source of cement was noted earlier.

6 Conclusions

Evidence gathered over the last few decades indicates that the distribution and growth of dissolution seams and fitted-fabric in bedded *argillaceous* limestones was influenced by an earlier stratified cementation which gave rise to an alternation of more cemented and less cemented strata (Fig. 6). The more cemented strata resisted further compaction, which was restricted to the less cemented strata. The degree of cementation was such that enough porosity remained to act as an essential sink for the reprecipitation of ions released by pressure-dissolution. This stratified cementation took place early in burial history but not necessarily early in diagenetic history if burial had been delayed. It commonly occurred in sequences where original bedforms (if any) had been destroyed during bioturbation (Fig. 6). It acted variously in closed and open systems. The signal that stimulated this localized cementation has rarely been specifically identified but an enhanced content of carbonate nuclei at selected levels seems a reasonable supposition. This condition might have arisen as a result, for example, of a reduction in clay or an increase in echinodermal fragments (large single nuclei). As a consequence of the resultant differential compaction, the textural and compositional differences between strata were thus progressively exaggerated so that a paired layered structure of couplets became increasingly obvious.

In sequences of stylolites and fitted-fabric in low-clay limestones, the influence of such a constrained cementation early in burial history is less clearly demonstrated but was probably significant. Any such cementation would have been related to lithological transitions, especially in permeability, and of course to the availability of aragonite, if any, as a source of carbonate for cement.

The process of pressure-dissolution requires the coexistence of contiguous layers or patches where dissolution and collapse in one is accompanied by cementation in the other (e.g., Merino et al. 1983).

Controls that have been proposed for the stratified cementation vary among authors and from sequence to sequence. They include vertical changes in primary lithology/mineralogy and rhythmic diagenetic changes related to the rate of deposition. An alternating signal was certainly imposed on the sediment as it accumulated and the signal was doubtless diffused but escaped total destruction during any subsequent bioturbation.

Finally, one result of stratified cementation is worthy of special notice. This is the decrease in permeability normal to bedding that has taken place early in the burial history of many carbonate sediments. Fluid flow tended, therefore, to be constrained parallel to bedding at that stage. The implications of this development are of obvious significance in the consideration of fluid migration in compacting sedimentary basins.

Acknowledgments. The author owes a profound debt of gratitude to four colleagues who generously gave of their time and energy to make extremely helpful reviews of an earlier draft, namely to Ian J. Fairchild, Peter Gutteridege, Jim D. Marshall, and Maurice E. Tucker. He is grateful also to Walther Schwarzacher for striving to keep him on the path of logical argument.

Chapter 4 Cherts and Phosphorites

4.1 Rhythmic Bedding in Siliceous Sediments – an Overview

K. Decker

1 Introduction

Rhythmic bedding of alternating nonsiliceous and siliceous layers is one of the most prominent features of biogenic siliceous sediments. Examples of such rhythmites are Mesozoic ribbon radiolarites (e.g., Barrett 1982; Jenkyns and Winterer 1982; Imoto 1983; Vecsei et al. 1988); Tertiary diatomites, displaying rhythmical bedding on both varve (lamina) and bed scale (e.g., Pisciotto and Garrison 1981; Curray, Moore et al. 1982; Donegan and Schrader 1982; Pokras and Winterer 1987); and limestone-chert sequences of the equatorial high productivity belt (e.g., Cook 1972; Theyer et al. 1985).

Five major factors govern accumulation and bedding of biogenic siliceous sediments: production of opaline skeletons by planktonic organisms in marine surface water, rate of dissolution of skeletons both within the water column and in the surface sediments, dilution of opaline matter by intra- and extrabasinal nonsiliceous material, redeposition by turbidity and bottom currents, and diagenetic overprints (Tada Chap. 4.2, this Vol.). The main goal of this chapter is to review mechanisms producing (1) cyclic and (2) discyclic (event related) interbedded chert sequences.

1. Cyclic sequences result from periodic variations of siliceous plankton productivity and dilution by terrigenous sediment or carbonate ooze. Various combinations of these mechanisms are related to distinctive types of bedding rhythms. Variations of plankton productivity responding to Milankovitch climatic cycles are the most important process generating pelagic limestone-marl alternations (e.g., Einsele 1982; Fischer and Schwarzacher 1984; Cotillon 1987; Einsele and Ricken Chap. 1.1, this Vol.). Very likely the productivity of siliceous organisms also responds to climatic changes (Garrison and Fischer 1969; Jenkyns and Winterer 1982; De Wever 1987). Aspects of the response of siliceous productivity to changing silica supply, ocean dynamics, and climate are discussed.
2. Bedding rhythms and sedimentary structures resulting from discyclic redeposition of radiolarian oozes by turbidity and bottom currents are common in radiolarian cherts (e.g., Nisbet and Price 1974; Barrett 1982; Imoto 1983; Iijima et al. 1985; Vecsei et al. 1988; Decker 1990). Structures are in many respects similar to fine-grained siliciclastic turbidites (Piper and Stow Chap. 2.9, this Vol.) and siliciclastic contourites (Stow 1979; Stow and Piper 1984a). Hydrodynamic properties of biogenous and clastic grains are, however, different and may lead to distinctive sedimentary structures (see Sect. 4).

Einsele et al. (Eds.)
Cycles and Events in Stratigraphy
©Springer-Verlag Berlin Heidelberg 1991

1.1 Planktonic Opal Production

Throughout the geological record radiolarians and diatoms have been the most important opal producing organisms. Diatoms are photosynthesizing unicellular algae (Bacilarophyta) that depend on the availability of anorganic nutrients (i.e., Si, P, N) and light. In contrast, radiolarians are heterotrophous protozoans (Actinopoda), depending on the same anorganic nutrients, but also on organic matter as prey and, in some cases, on symbionts (i.e., dinoflagellates; Anderson 1983). The productivity of radiolarians and diatoms closely responds to changing advection of anorganic nutrients. Of these, available dissolved silica is the most limiting factor (Berger and Roth 1975; Anderson 1983). High productivities in modern oceans coincide with areas of high anorganic nutrient supply by upwelling along continental margins, north-equatorial and polar divergences (Fig. 1; e.g., Casey et al. 1982; Lisitzin 1985). Furthermore, there is a good correlation of surface water siliceous productivity and abundance of opal in sediments below (Berger 1976; Lisitzin 1985). Siliceous plankton productivity therefore seems to control the rate of opal accumulation. Similar controls can also be assumed for the geological past. Other paleo-ecologic parameters, such as the availability of prey for radiolarians, are difficult to evaluate.

1.2 Pre-Diagenetic Opal Dissolution

Opaline skeletons dissolve both during settlement through the water column (Berger 1976; Nelson and Gordon 1982) and in the uppermost sediment layers. Less than 10% of skeletons enter sedimentary record. The rate of dissolution depends on thermodynamic factors, skeleton morphology (i.e., size and specific surface area), chemical composition of skeletons, and the presence of protective organic or authigenic mineral coatings (Hurd 1972, 1973; Schrader 1972; Kamatani et al. 1988; Walsh et al. 1988; Van Bennekom et al. 1989).

The rate of opal preservation on the ocean floor is positively correlated to both the amount of opal supply to the sea floor and the amount of dilution by other sediments (Fig. 2; Goll and Bjorklund 1971; Berger 1976; Broecker and Peng 1982). These correlations are of particular importance for the formation of rhythmic bedding in opaline sediments, as variations in opal supply and changing dilution also influence the rates of opal preservation (Fig. 2). Consequences for productivity and dilution cycles are discussed below.

2 Cyclic Bedding in Chert Sequences

Cyclic bedding of chert sequences forms by periodic changes of sediment supply. Bedding is more diverse than in marl-limestone alternations (see Part I.1, this Vol.), as supply rates of three sediment fractions can oscillate. These are variations in opal deposition and variations of the nonopal fraction which is composed of calcareous ooze and terrigenous sediment. Six major types of primary variations result in distinctly different bedding rhythms (Figs. 3 to 5). Periodic alternations formed by rhythmic variations of opal supply are refered to as opal productivity cycles, whereas

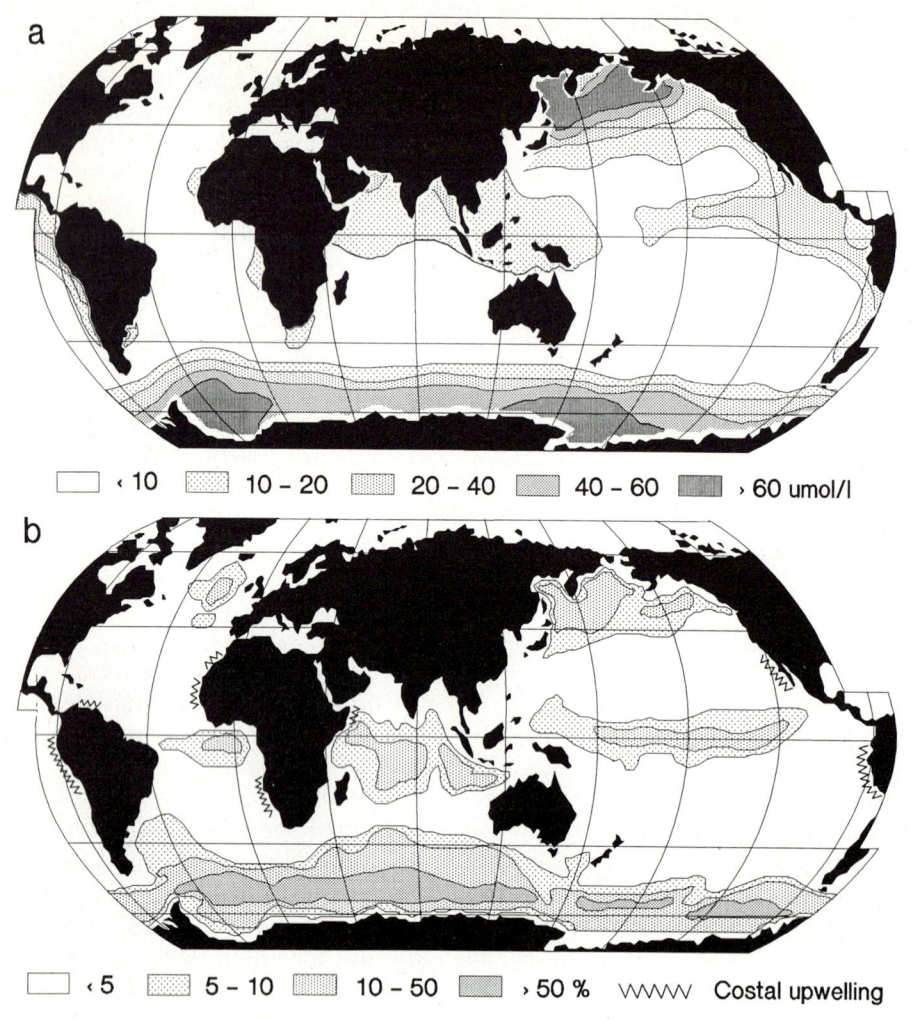

Fig. 1. a Distribution of dissolved Si in present-day surface oceans. **b** Percentage of biogenous opal in pelagic sediments. Distribution of opal-rich sediments reflects high silica content in surface water and related high plankton productivity. (After Lisitzin 1985)

those produced by oscillating opal dilution with biogenic carbonate and terrigenous clay are termed dilution cycles.

Repeated changes in the intensity of pre-diagenetic opal dissolution overprinting sequences of previously homogenous composition are insufficient for the generation of siliceous rhythmites. Dissolution rates can only vary within the range of 10% according to changes of silica saturation, temperature and pressure (Hurd 1972; Walsh et al. 1988). Productivity and dilution cycles are, however, substantially modified by pre-diagenetic opal dissolution. The correlation of the opal preservation

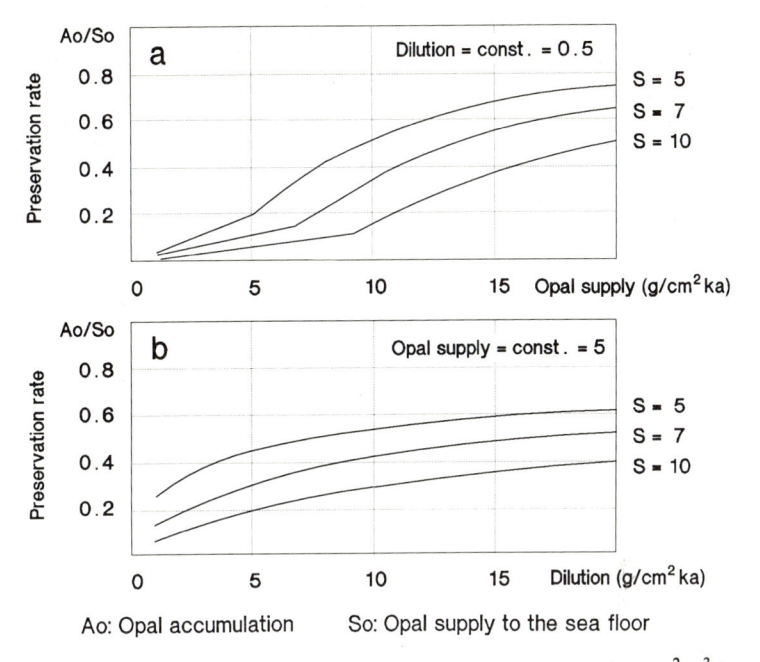

Fig. 2. Opal preservation in the upper sediment layers related to various opal solution rates (S, in g/cm^2 10^{3}a). The opal preservation rate (Ao/Ro) correlates positively to both opal supply (**a**) and dilution (**b**). S depends on thermodynamic factors and on properties of the sediment forming opaline skeletons. (After Broecker and Peng 1982)

rate to opal supply and dilution (Fig. 2) has different consequences for the two types of cycles: in opal productivity cycles, the positive correlation of opal preservation rate and opal supply results in good preservation in periods of high opal supply, whereas stronger dissolution occurs during periods of low supply (Fig. 2). Thus, the resulting variations of opal contents are enhanced (Fig. 3a). In contrast, variations of opal contents formed by dilution cycles decrease through sea-floor opal dissolution (Fig. 3b). Opal-rich layers coinciding with low dilution are subjected to stronger opal dissolution than opal-poor layers produced during high dilution. Consequently, productivity cycles are more effective than dilution cycles in generating rhythmically interbedded siliceous sediments.

2.1 Types of Bedding Rhythms

Periodic sequences of opal productivity and dilution cycles differ significantly. In opal productivity cycles, oscillating opal supply during constant clay or carbonate flux results in variations of total sedimentation rates with highest rates during opal deposition. Therefore, opal-rich beds are thicker than clay or carbonate-rich beds deposited during the same interval of time, and thicker chert beds dominate over

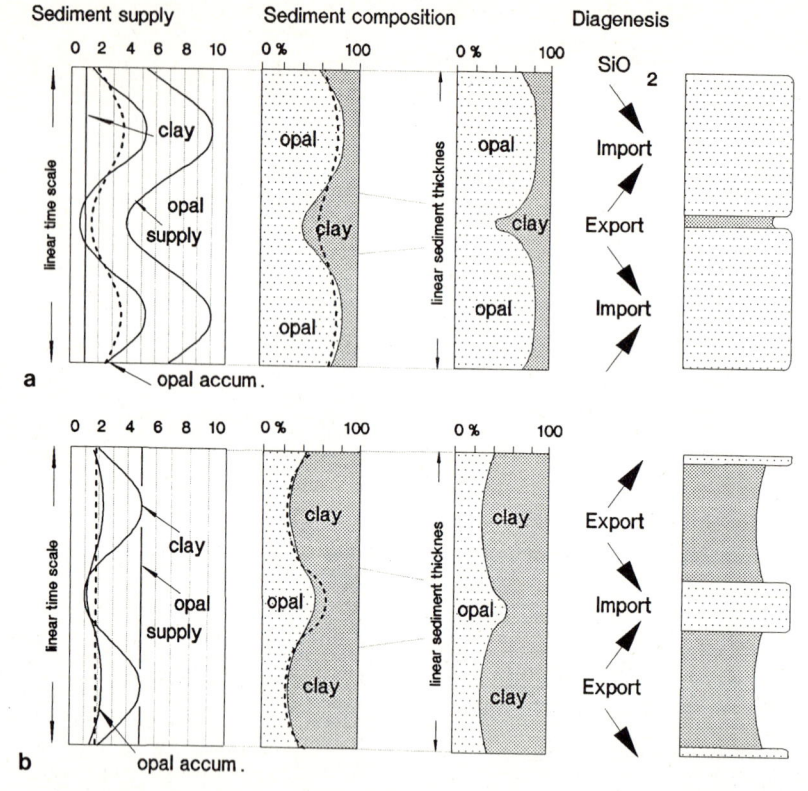

Fig. 3. a Model for productivity cycles. Opal supply (in g/cm² 10³a) varies by a factor of 2.5, clay accumulation is held constant at a rate comparable to red deep-sea clays. Symmetrical oscillations produce a sequence of thick siliceous beds and thin clay-rich layers. Variations in opal accumulation rates, opal contents, and bed thicknesses are amplified by the positive correlation of opal supply and opal preservation rate (Fig. 2); the *hatched lines* illustrate the hypothetical effect of a constant preservation rate independent from opal supply. **b** Model for terrigenous dilution cycles. The sedimentation rate of clay varies by a factor of 2.5, opal supply is constant (values in g/cm² 10³a). Symmetric oscillations form thick clay-rich beds and thin opal-rich layers. Opal dissolution is higher in periods of low dilution (Fig. 2). Variations in opal contents therefore are diminished. *Hatched lines* refer to a hypothetical constant dissolution rate

thinner shaly or calcareous interbeds (Fig. 4a). Increasing but oscillating rates of opal supply result in increasing opal/clay (or opal/carbonate) ratios. Thicker chert beds with only slightly higher SiO_2 contents are generated (Fig. 4a). In contrast, terrigenous dilution cycles and periodic variations of carbonate flux produce interbedded sequences with thick clay (or carbonate) beds and thin opal-rich layers. Opal/clay or opal/carbonate ratios are lower than in opal productivity cycles (compare Fig. 4a with 4b). Higher dilution rates result in thicker clay or carbonate beds and in decreasing opal/clay (opal/carbonate) ratios.

The generation of calcareous-siliceous alternations is complex, as productivities of both siliceous and calcareous plankton vary in response to environmental changes (e.g., Berger 1982). In addition, the highly variable intensities of carbonate dissolu-

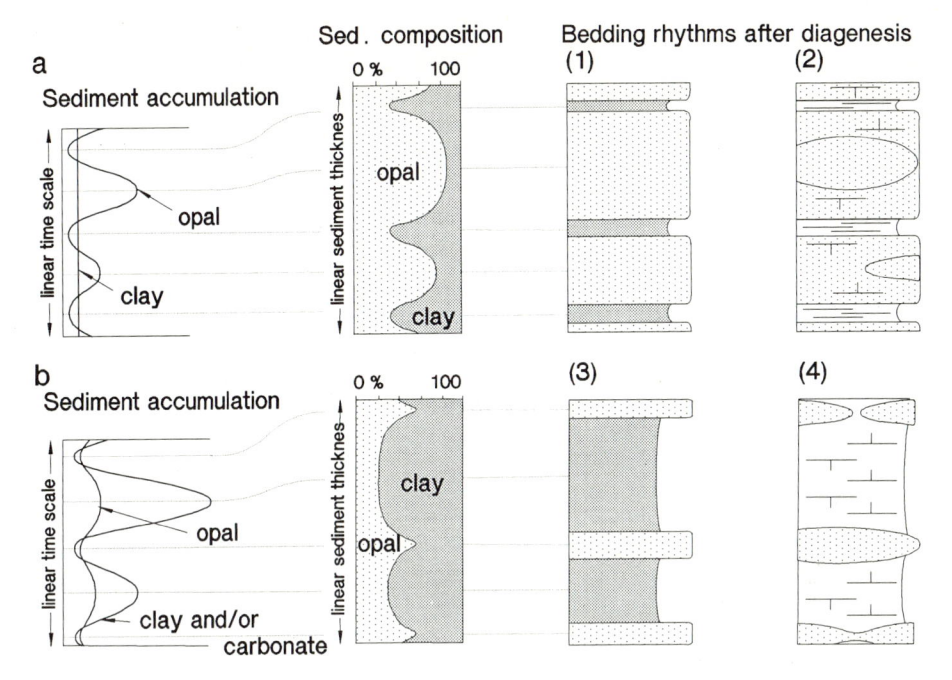

Fig. 4. a Opal productivity cycles: Varying opal flux during constant clay (carbonate) sedimentation forms thick opal-rich and thin clay rich-layers (*1*). Increased opal supply produces thicker chert beds and higher opal/clay ratios. Addition of a constant carbonate supply results in bedding rhythms that display intercalations of calcareous cherts with pure chert nodules and marls (*2*). **b** Dilution cycles: Oscillating clay dilution produces thick clay beds with thin chert layers. Higher dilution rates increase the thicknesses of clay beds and lower the opal/clay ratios (*3*). Cycles with varying carbonate flux (*4*) form limestones with chert nodules due to diagenetic redistribution of SiO_2 and carbonate

tion and lysocline fluctuations contribute to variations in carbonate accumulation (Einsele et al. Introduction, Grötsch et al. Chap. 1.6, both this Vol.). In-phase oscillations of carbonate and opal supply during constant clay deposition (Fig. 5a) result in large variations in sedimentation rates. Accordingly, sequences form with thick beds containing both opal and carbonate, and thin clay-rich interbeds (Fig. 5a). Out-of-phase oscillations of productivities are characterized by more uniform total sedimentation rates that prevent the formation of proper shale beds (Fig. 5b). The lithology of the resulting chert-limestone sequence depends on the actual ratios of clay to biogenous supply.

2.2 Diagenetic Enhancement of Cyclic Bedding

The primary differences in soft-sediment composition of productivity and dilution cycles are substantially enhanced during burial diagenesis (Fig. 3; Tada Chap. 4.2, this Vol.). In opal-clay rhythms, redistribution of SiO_2 from opal-poor to opal-rich layers also decreases the thickness of clay beds. The diagenesis of opal-carbonate sediments differs from that of opal-clay mixtures (Lancelot 1973; Kastner et al. 1977; Maliva and Siever 1989; Tada Chap. 4.2, this Vol.). Redistribution of SiO_2 and carbonate result in limestones with chert nodules and stringers which preferentially form in originally opal-rich layers (Fig. 5; Ricken and Eder Chap. 3.1, this Vol.).

3 Oceanographic and Climatic Controls on Interbedded Chert Formation

3.1 Climatic Controls of Productivity

Climatic influences on plankton productivity and deposition of siliceous sediments include (1) modifications of ocean dynamics, especially of intensity and geographic distribution of upwelling, and (2) variations of silica input into the oceans through continental weathering and river run-off.

1. One of the driving mechanisms of ocean currents and hence upwelling are climate-related thermal gradients. These operate on regional or global scale that influence wind and surface water current patterns. Steep latitudinal thermal gradients during cooler climatic periods result in a more vigorous atmospheric and oceanic circulation (Janecek and Rea 1984; Rea et al. 1985). In cooler periods of the Cenozoic, stronger trade-winds lead to higher upwelling-related silica productivities in the equatorial belt and to an expansion of this productive zone (Fig. 1; Parkin 1974; Stabell 1986a). Interglacial to glacial transitions shift the northern and southern high-latitude silica deposition belts towards lower latitudes (Fig. 1; Moorley and Hays 1979; Stabell 1986b; Ciesielsky et al. 1988). As a consequence, glacial cycles form rhythmic alternations of siliceous and

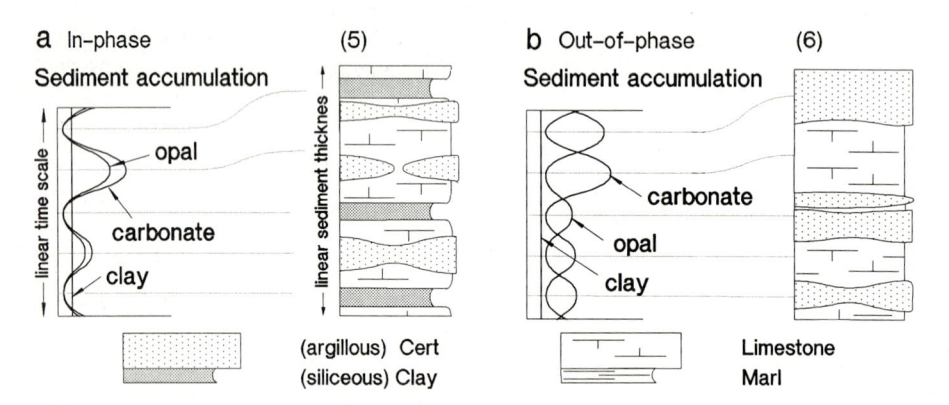

Fig. 5. Opal-carbonate productivity cycles. **a** In-phase oscillations of opal and carbonate accumulation with constant clay supply produce repeated intercalations of thin shale beds and thick layers of opal-carbonate oozes. The latter alter diagenetically to limestones with chert nodules (5). **b** Out-of-phase opal-carbonate productivity cycles form limestone-chert sequences with varying clay contents but without proper clay beds (6)

nonsiliceous sediments along the margins of these belts (Tucholke et al. 1976; Lisitzin 1985).
2. Chemical weathering on the continents and the supply of dissolved silica to the oceans depend on morphology and size of the exposed land area, climate, and the amount of river discharge. High precipitation on continents within the tropical zone combined with large solute transport of rivers provide higher supplies of dissolved silica. The hydrologic cycle closely responds to changing thermal gradients and orbital rhythms. Periods of steep thermal gradients correspond to periods of high precipitation and run-off (Barron et al. 1989) and thus probably to times of higher Si supply, which may result in higher siliceous plankton productivity. On the other hand, periods of high run-off can be associated with higher siliciclastic deposition which may dilute productivity cycles.

3.2 Opal Productivity Cycles

Siliceous productivity mainly corresponds to the availability of dissolved silica in the oceanic surface water which is advected by river run-off and upwelling. Effects of variations in silica supply on opal sedimentation, e.g., through climatic changes, can

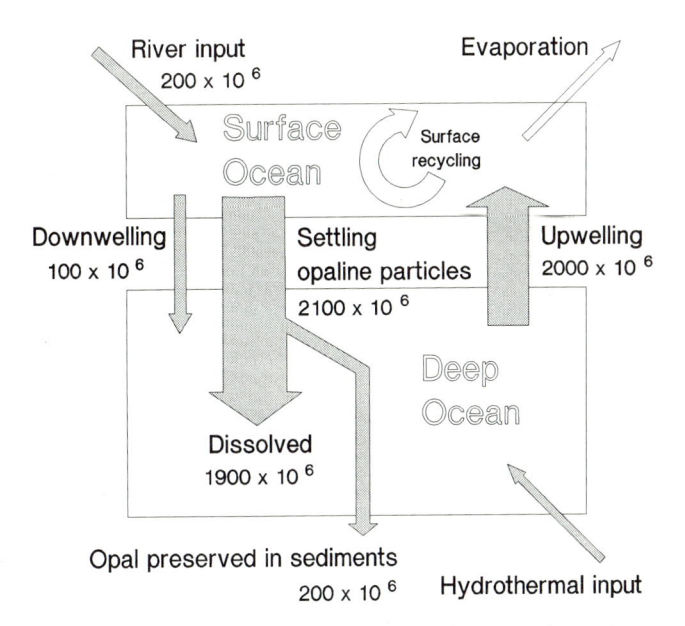

Fig. 6. Simplified model of the global oceanic silica cycle after Broecker (1971). In a steady state, two major silica fluxes are balanced: (*1*) silica inputs to the whole oceanic system (river and hydrothermal input) equal the loss through opal sedimentation, and (*2*) silica inputs to the surface ocean (river input plus upwelling) balance the removal to the deep ocean in form of settling opaline skeletons and downwelling. *Numbers* refer to silica fluxes in t/a. (After Broecker and Peng 1982; Lisitzin 1985)

be described with Broecker's model of the global oceanic silica cycle (Fig. 6; Broecker 1971; Broecker and Peng 1982). During steady-state conditions, two balanced major silica fluxes occur in this cycle (Fig. 6): taking the entire ocean into account, the gain of dissolved silica is balanced by the loss of SiO_2 through sedimentation of opaline skeletons (Berger 1976). When considering only the surface ocean, which is separated from the deep ocean by the main thermocline, the sum of silica input by river run-off and upwelling equals the rain of opaline skeletons to the deep ocean plus the small amount of silica lost by oceanic downwelling. The concentration of dissolved silica in the surface ocean is kept very low by extensive biogenous opal extraction. The deep ocean is relatively rich in dissolved silica because most of the settling skeletons dissolve here. The effects of stepwise changing upwelling and river input on opal deposition are described below.

In the first step, a sudden increase of the global upwelling rate is assumed. Consequently, surface production and the flux of opal skeletons to the deep ocean increase (Fig. 7a). Assuming a constant rate of preservation of the skeletons raining to the sea floor, the total amount of SiO_2 preserved in sediment also rises. This extracts more and more silica from the oceanic cycle, causing a silica deficiency, as the gain of silica (i.e., by river input) remains unchanged. As a result, silica concentration in the deep ocean drops and the amount of Si advected to the surface by upwelling diminishes. In turn, biogenic production and SiO_2 deposition decrease (Fig. 7a). Finally, the cycle reaches a new equilibrium with a SiO_2 sedimentation rate equal to the initial one, balancing riverine silica input. A similar feed-back loop develops when global upwelling is reduced to the initial rate: reduced silica advection to the surface causes a drop of productivity and of opal deposition. Riverine silica supply exceeds the amount of opal deposition and silica concentrations in the deep-ocean increase until the initial opal deposition rate is obtained again. Re-establishing of steady state conditions requires time spans in the order of 10^4 years (Fig. 7a). The results of this assumed two-step change of upwelling are positive and negative spikes of global opal deposition.

In contrast to changing upwelling rates, higher riverine silica input leads to a comparably slow asymptotic increase of opal deposition. Opal sedimentation is again slowly reduced after cutting down the supply rate to its initial value (Fig. 7b). Feed-back mechanisms significantly retard the imprints on opal deposition.

Consequences of this conceptual model for the generation of rhythmical interbedded cherts are:

1. Variations of upwelling induced by climatic changes are more effective in producing short-term (10^4 years) spikes in world-wide SiO_2 deposition than changes of riverine silica input. Positive and negative spikes caused by varying upwelling sum up to significant short-term variations in opal sedimentation rates. Such variations, enhanced through both ocean floor opal dissolution and diagenetic SiO_2 redistribution (Tada Chap. 4.2, this Vol.), may be expressed as discrete siliceous beds within sedimentary sequences and as Milankovitch-type rhythms.
2. Silica spikes only form when upwelling rates change rapidly, e.g., during glacial terminations. In the case of slower gradual changes, feed-back mechanisms are able to keep the system in balance and the silica sedimentation rate stays nearly

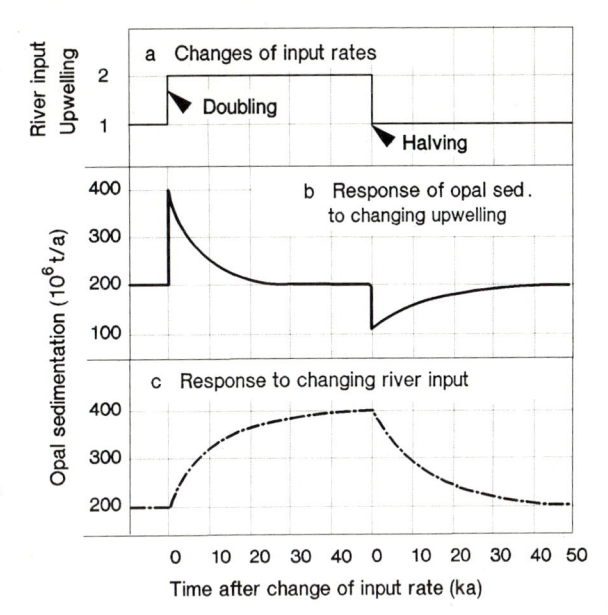

Fig. 7. a Assumed step-like changes of global upwelling/river input rates. **b** Response of the sedimentation rate of biogenic opal to changing upwelling. **c** Response of opal sedimentation to changing river input. The models start from a steady-state oceanic silica cycle with input and upwelling rates of present oceans (Fig. 6)

constant. Response times of silica deposition depicted in Fig. 7 may vary due to the intensities of upwelling and changing recycling rates in the cycle. However, they lie within the range of 10^4 to 10^5 years, i.e., within the periodicity of orbital rhythms that govern climatic changes. As a consequence, Milankovitch-type climatic cycles that are associated with changing upwelling intensities (Sarnthein et al. 1987, 1988) may be reflected by rhythms with varying opal contents.

3.3 Dilution Cycles

Periodic dilution of opaline oozes by terrigenous sediments is important for the formation of siliceous rhythms in land-near basins and on continental margins. Terrigenous sedimentation varies with climate as a function of fluvial or aeolian input or as a consequence of sea level fluctuations. For open-ocean settings, variations in terrigenous sediment supply are insufficient for the formation of dilution cycles.

For oceanic environments, cyclic variations of biogenous carbonate flux diluting siliceous sediments are of major importance for the generation of rhythmites. Intercalations of calcareous and siliceous sediments are widespread especially in the equatorial high productivity belt (e.g., Tertiary sequences of the equatorial Pacific: Von der Borch et al. 1971; Cook 1972; Theyer et al. 1985).

3.4 Varve-Scale Variations

In settings close to continents with very high sedimentation rates, annual fluctuations of terrigenous input and seasonally changing diatom productivities may form varved diatomites. The genesis of varves combines both effects of productivity and dilution cycles. Sediment fluxes vary with seasonal changes of river discharge, wind pattern, and variations of upwelling intensity (Fig. 8). The resulting small-scale dilution and productivity cycles in such varves may oscillate in or out of phase (Pisciotto and Garrison 1981; Donegan and Schrader 1982; Pokras and Winter 1987). Varves are preserved only in the oxygen minimum zone where homogenisation of sediment by burrowers is prevented (Wetzel Chap. 5.1, this Vol.). Depth fluctuations of the oxygen minimum zone are recorded as second order cycles, i.e., alternations of laminated and homogenous diatomaceous sediments (Govean and Garrison 1981; Pisciotto and Garrison 1981).

4 Discyclic Redepositional Processes

Rhythmic bedding in cherts may also be related to short-term event deposition. Bedding features of such events are observed on both bed- and lamina scale, indicating redeposition of siliceous sediments by bottom and turbidity currents. These structures are very abundant in ribbon radiolarites, but they are only seldom described from diatomaceous sediments (e.g., Iijima et al. 1985). This chapter therefore focuses on redeposited radiolarian sediments.

Repeated redeposition of argillaceous sediments or siliceous oozes by turbidity currents can produce intercalations of chert and shale beds similar to those found in productivity or dilution cycles. In event-related interbedded radiolarites, both shale or chert beds have been interpreted as turbidites intercalated with hemipelagic sediments (e.g., Nisbet and Price 1974; Iijima et al. 1985). The presence of sedimentary structures of fine-grained turbidites serves to distinguish these event-related types from cyclic interbedded chert sequences.

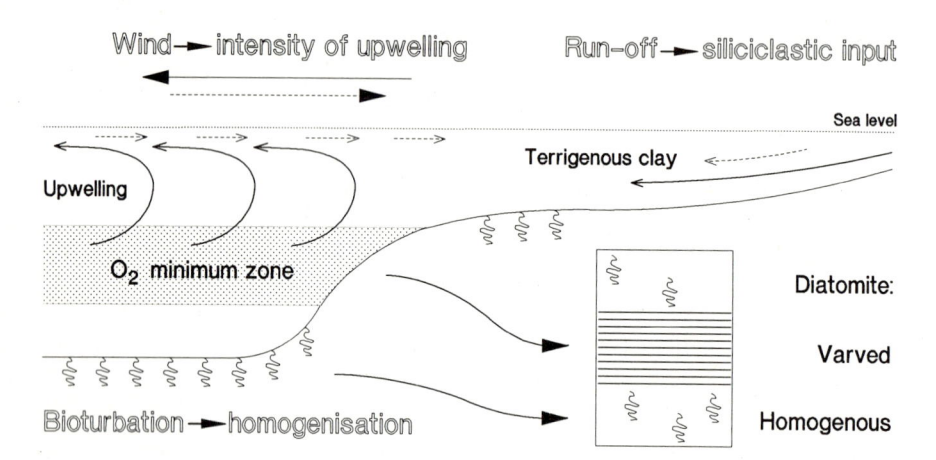

Fig. 8. Generation of varved diatomites by seasonal changes of upwelling intensity and siliciclastic input. Varves only are preserved in the O_2 minimum zone where bioturbation is prevented. Fluctuations of the O_2 minimum zone may result in alternations of homogeneous and varved diatomites

4.1 Hydrodynamic Properties of Radiolarian Skeletons

The hollow shape of radiolarian skeletons and the low specific weight of biogenous opal (2.00 to 2.07 g/cm^3 in contrast to 2.65 g/cm^3 of quartz; Calvert 1974) result in very low effective densities (1.01 to 1.5 g/cm^3) and significantly lower settling velocities as compared to clastic grains of equal size (i.e., 0.015 to 1.0 cm/s versus 1.0 to 10 cm/s; Barrett 1982). As a consequence, radiolarian skeletons are 1.1 to 2.5 times larger than hydrodynamically equivalent quartz grains (Fig. 9a; Decker 1987). The wide range of these values is caused by highly variable skeleton morphologies,

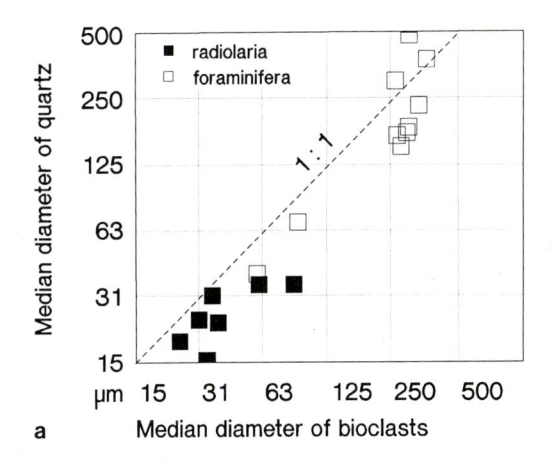

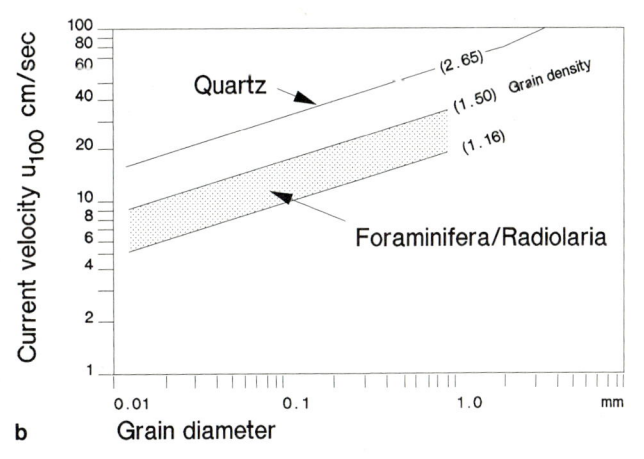

Fig. 9. a Median diameters of radiolarians and planktonic foraminifera plotted against median diameters of hydrodynamically equivalent quartz grains. Measurements of bioclastic and quartz grain diameters were carried out in same turbidite intervals. Bioclasts are 1.1 to 2.5 times larger than hydrodynamically equivalent quartz grains. (After Decker 1987; Sarnthein and Bartolini 1973). **b** Threshold velocities for the erosion of unconsolidated cohesionless foraminifera and radiolarians are only about half of those for quartz grains of equal size. u_{100} = current velocity 100 cm above sediment surface. (After Miller and Komar 1977)

wall thicknesses, and internal sediment fills. Radiolarian size grading during deposition from suspension clouds therefore is poor.

Hydrodynamic properties of radiolarians are comparable to those of planktonic foraminifera (Berger and Piper 1972; Sarnthein and Bartolini 1973). For a first approximation, the threshold current velocities for the erosion of planktonic foraminifera (Southard et al. 1971; Miller and Komar 1977) are applied to radiolarian-rich sediments. The erosion of skeletons occurs at significantly lower current velocities than erosion of equally sized clastic grains (Fig. 9b). Very likely, threshold velocities of silt-size radiolarians are even lower than those for the erosion of cohesive clay aggregates. Winnowing therefore rather removes silt-size radiolarians than the clay matrix from radiolarian oozes (Weissert 1979; Baumgartner 1987). Consequently, sedimentary structures may differ from siliciclastic sediments.

4.2 Bedding Features of Radiolarian Turbidites

Sedimentary structures of radiolarian turbidites are comparable to those of fine-grained siliciclastic turbidites (Piper and Stow Chap. 2.9, this Vol.), except for poor size-grading. Deposition from low-density, low-velocity turbidity currents forms sets that may display sharp and erosive base, flute and load casts, upward-decreasing radiolarian abundance, plane lamination, micro-crossbedding, and fading ripples (Fig. 10; Nisbet and Price 1974; Barrett 1982; Imoto 1983). In redeposited radiolarian-coccolith oozes, carbonate contents may increase from base to top (Decker 1987). A distinctive feature of radiolarian turbidites are concentrations of lithoclasts (e.g., volcanic detritus) and intraclasts at the base of sets (Fig. 10; Barrett 1982). Interbedded calcareous radiolarian turbidites in carbonate-free sediments may indicate down-slope transport below the CCD (Decker 1990). Bioturbated and oxidized red tops of green turbidites intercalated with red hemipelagites also can be distinctive (Fig. 10).

4.3 Bedding Features of Bottom Current Deposits

Bottom currents with fluctuating velocities form laminae that record alternations of sedimentation, winnowing and erosion. Such laminae may display indistinct or scoured bases, normal or inverse grading, plane lamination and micro-crossbedding (Stow 1979; Stow and Piper 1984a).

Figure 11 summarizes effects of increasing and decreasing current velocities. (1) Gradually increasing velocities progressively prevent settling of fine-grained particles and the size interval of settled grains becomes smaller. The resulting lamina shows an upward increasing median grain size and better sorting from bottom to top. For the same reason the sedimentation rate decreases. Traction and bed-load transport of silt-size radiolarians form plane or cross lamination. Above a given threshold velocity, sedimentation is impossible and erosion occurs. In radiolarian oozes, radiolarians are very likely to be removed first, while cohesive clay remains in place and forms a kind of lag sediment. (2) During a decrease of current velocity, a lamina

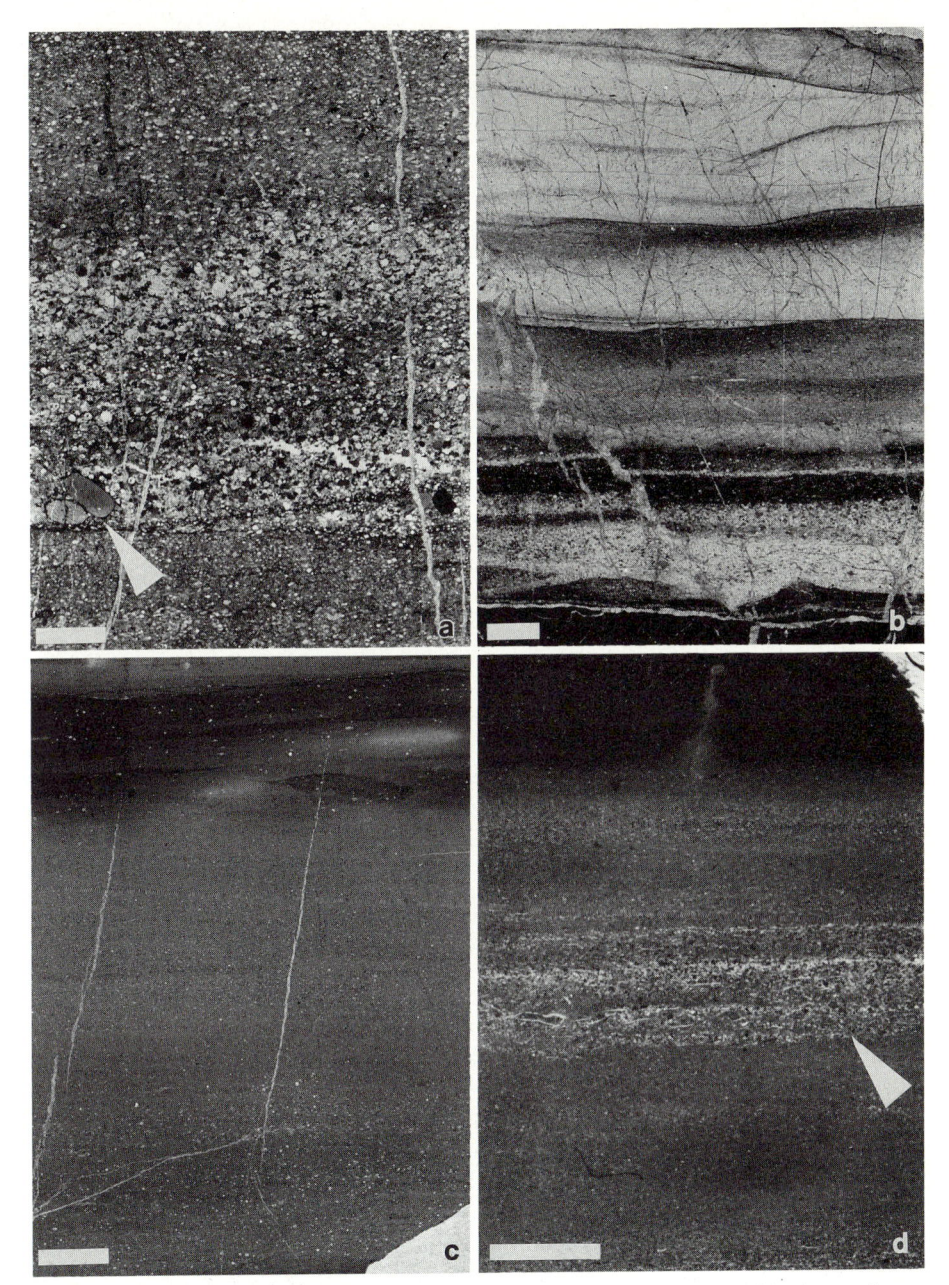

Fig. 10. Sedimentary structures of radiolarian turbidites. **a** Set with large imbricated lithoclasts on base. Radiolarians are not well graded, except for the top of the set. **b** Radiolarian silt laminae with sharp bases grading into siliceous mud. Note that very similar structures may be generated by oscillating bottom currents (see Fig. 11). **c** Green turbidite set (*light*) with red, oxidized and bioturbated top (*dark*) grading into a red hemipelagic interval. Radiolarian abundance is graded within the set. **d** Turbidite set displaying sharp erosive base and alternations of radiolarian silt with siliceous mud laminae. The thickness of silt laminae decreases upwards

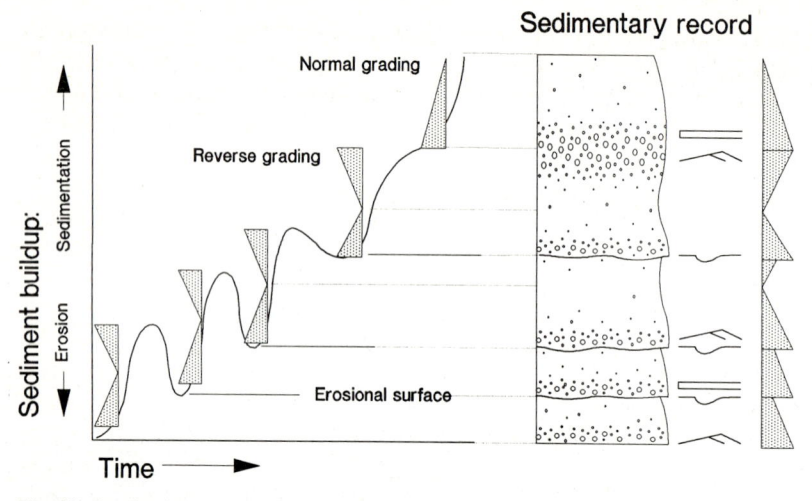

Fig. 11. Lamination produced by changing bottom current velocities through constant sediment supply from the overlying water column. Fluctuations result in reverse or normal grading and varying sedimentation rates. Alternating deposition and erosion can produce successions of graded laminae similar to radiolarian turbidites

forms displaying normal grading of radiolarian size and abundance. Under certain conditions, alternating erosion and deposition during current deceleration can produce successions of graded laminae similar to turbidite sequences (Fig. 11). Interbedded laminae with indistinct, nonerosive bases and inverse or inverse-to-normal graded laminae may be used to distinguish such sequences from radiolarian turbidites.

5 Conclusions

Rhythmic bedding in biogenic siliceous sediments is generated by both cyclic and discyclic (event-related) depositional processes.

Cyclic interbedded sequences form by (1) periodic changes of opal supply caused by varying siliceous plankton productivities (opal productivity cycles) and (2) oscillating supply of the nonsiliceous fraction that is composed of terrigenous sediment and biogenic carbonate (dilution cycles). Six major combinations of cyclic variations are introduced, each generating a distinctive cyclic interbedded chert sequence as depicted in Figs. 4 and 5. Productivity, dilution, and combined opal-carbonate productivity cycles differ in their chert/shale and chert/carbonate ratios. The analysis of bedding rhythms in chert sequences, therefore, is an important tool for determining primary variations of particle fluxes in basins related to oceanographic and climatic changes.

Cyclic variations in sediment flux may be related to Milankovitch-type climatic cycles including glacial-interglacial changes, or to seasonal variations. Among the discussed mechanisms, productivity cycles caused by climate-induced global changes of upwelling intensities are particularly effective in generating siliceous rhythmites, as feed-back mechanisms within the oceanic silica cycle amplify variations of opal sedimentation rates. Primary variations in opal contents generated by productivity and dilution cycles are strongly modified by ocean floor opal dissolution. Opal preservation rates correlate positively to opal supply and dilution. These relations increase the compositional differences in productivity cycles, whereas opal variations in dilution cycles are reduced. Variations in sediment composition are further enhanced by diagenetic redistribution of SiO_2 (Tada Chap. 4.2, this Vol.). Calcareous-siliceous cycles are transformed into bedded limestones containing chert nodules and stringers.

Rhythmic bedding generated by redepositional events is particularly well-known in radiolarian sediments. Hydrodynamic sorting of radiolarian ooze by turbidity and bottom currents produces rhythms on lamina and bed scale. Radiolarian turbidites may form rhythmically bedded radiolarian cherts with chert beds that correspond to single depositional events intercalated by hemipelagic shale. In contrast, chert beds that are composed of a couple of laminae indicate that several depositional events contributed to the formation of one bed. Radiolarian-shale sequences containing such composite chert beds may be interpreted as cyclically interbedded sequences superimposed by small-scale events of bottom current or turbidite deposition.

4.2 Compaction and Cementation in Siliceous Rocks and Their Possible Effect on Bedding Enhancement

R. Tada

1 Introduction

Biogenic siliceous rocks, such as radiolarian, spicular, and diatomaceous cherts, porcelanites, and siliceous mudstones, are some of the main constituents of pelagic and hemipelagic deposits and often show rhythmic bedding. The origin of rhythmic bedding has been previously studied, and various mechanisms have been proposed (Iijima et al. 1985 and references therein). However, before deducing the sedimentation mechanisms, the effect of diagenesis should be evaluated and subtracted so as to obtain primary signals (Ricken 1986; Bathurst Chap. 3.2, Ricken and Eder Chap. 3.1, both this Vol.).

In siliceous rocks, the diagenetic transformations of silica phases seem to play a crucial role in diagenetic enhancement of bedding. For this reason, a brief overview on silica phase transformations in siliceous rocks is provided before evidence of diagenetic enhancement of bedding is presented and a possible mechanism proposed. Finally, the effect of pressure (dis-)solution on bedding enhancement during the late stages of diagenesis is discussed.

2 Silica Phase Transformations During Diagenesis

2.1 Silica Phase Transformations

It has been well established that silica phases are transformed from opal-A (biogenic opal) through opal-CT to quartz with an increase in burial depth (Bramlette 1946; Heath and Moberly 1971; Murata and Nakata 1974; Pisciotto 1981a, and others). Experimental work also supports this transformation sequence (Ernst and Calvert 1969; Mizutani 1970). Although direct transformation from opal-A to quartz has also been suggested as a possible pathway (Kastner et al. 1977), an intermediate opal-CT phase has been recognized in most diagenetic sequences in siliceous rocks (Murata and Nakata 1974; Keene 1975; Mitsui and Taguchi 1977; Hein et al. 1978; Tada and Iijima 1983). The transformations both from opal-A to opal-CT and from opal-CT to quartz have been interpreted as solution reprecipitation processes based on both oxygen isotope and petrographical evidence (Stein and Kirkpatrick 1976; Murata et al. 1977; Tada and Iijima 1983).

In the opal-A zone, there is a slow but progressive dissolution and weakening of the frustules, although many of the more robust species appear well preserved. The specific surface area of these siliceous fossils decreases from several hundred m^2/g to a few m^2/g, with an increase in age from Recent to Eocene (Moore 1969; Kastner et al. 1977; Hurd et al. 1979). This aging process gradually decreases the solubility of opal-A due to a decrease in excess surface energy, finally causing opal-CT precipitation (Williams et al. 1985).

During the opal-A to opal-CT transformation, opal-A siliceous fossils are extensively dissolved and reprecipitated as tiny opal-CT particles within the matrix (Tada and Iijima 1983). When opal-CT precipitates in relatively large open spaces (>10 μm), it forms spherical aggregates known as "lepispheres" (Wise and Kelts 1972). Only the most robust radiolarian skeletons and sponge spicules survive dissolution (Tada and Iijima 1983). When opal-CT is first precipitated, it is poorly ordered, but its ordering is gradually improved with increasing burial depth (Murata and Nakata 1974; Mitsui and Taguchi 1977; Pisciotto 1981a; Iijima and Tada 1981). This gradual ordering of opal-CT was previously considered to be a solid state reaction based on oxygen isotope data (Murata et al. 1977). However, close examination by high magnification TEM (transmission electron microscope) revealed a gradual growth of opal-CT crystallites from 30 A to 300 A in diameter with increasing burial depth (Hurd pers. commun.). The specific surface area of newly precipitated opal-CT is as high as 170 m^2/g (Kastner and Gieskes 1983), which may decrease to several tens of m^2/g, a typical value for opal-CT porcelanites, with the progress of burial diagenesis. The solubility of opal-CT decreases with decreasing specific surface area, eventually leading to quartz precipitation (Williams et al. 1985).

Extensive dissolution of opal-CT and reprecipitation as quartz occurs during the opal-CT to quartz transformation, which further dissolves siliceous fossils. In Neogene siliceous rocks of northern Japan, quartz reprecipitated as particles of ca. 1 μm in diameter within the matrix of porcelanites (Tada and Iijima 1983). The crystallinity and crystal size of quartz gradually increases as diagenesis proceeds (Murata and Norman 1976; Hein et al. 1981).

2.2 Controls on Silica Phase Transformations

Silica phase transformations are affected by such factors as time, temperature, pressure, porewater chemistry, and host rock lithology, with time and temperature as the principal controls (Mizutani 1970; Kastner et al. 1977; Hein et al. 1978; Pisciotto 1981b; Iijima and Tada 1981). Both the opal-A to opal-CT and opal-CT to quartz transformations occur at higher temperatures in younger sediments, whereas more time is required for both transformations at lower temperatures (Fig. 1). The time required for the completion of each of the transformations within a single section with minor lithologic variation is generally of the order of 0.1 to 1 Ma (e.g., Murata and Larson 1975; Murata et al. 1977; Thein and von Rad 1987).

Lithology is another important factor which controls silica phase transformations. Lithology probably affects the rate and mode of the transformations through

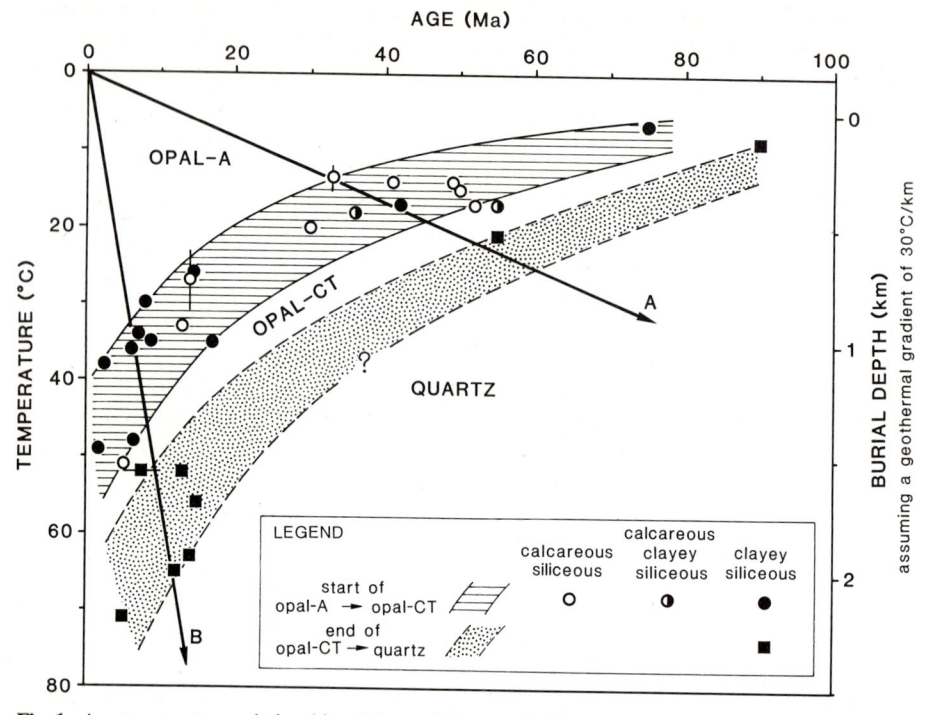

Fig. 1. Age-temperature relationship of the opal-A to opal-CT and opal-CT to quartz transformations within sediments of various lithologies from MITI test wells in northern Japan and selected DSDP sites which have good stratigraphic and temperature control. Line A represents an age-temperature path for a low sedimentation rate and/or a low geothermal gradient, whereas line B denotes a higher sedimentation and/or a higher geothermal gradient. Original database is available upon request to the author

its influence on porewater chemistry, porosity, permeability, and content and specific surface area of the silica phase(s). It is widely observed that silica phase transformations are accelerated in calcareous siliceous sediments (Heath and Moberly 1971; Lancelot 1973; Keene 1975; Riech and von Rad 1979; Thein and von Rad 1987). Experiments by Kastner and coworkers confirm that the presence of carbonate minerals accelerates the opal-A to opal-CT transformation through enhancement of opal-CT nucleation by the presence of magnesium hydroxide as a nucleus (Kastner et al. 1977; Kastner and Gieskes 1983). The higher porosity and permeability of calcareous siliceous sediments compared with clayey siliceous sediments also seem to be favorable for opal-CT precipitation (Lancelot 1973). On the other hand, the presence of clay tends to retard the opal-A to opal-CT transformation (Isaacs 1981), probably through consuming dissolved magnesium (Kastner et al. 1977) and/or reducing the effective surface area by adsorbing cations or by covering the surfaces of siliceous particles. The type of siliceous fossil may also affect the timing of the opal-A to opal-CT transformation through the difference in their specific surface area (Williams et al. 1985). As can be seen in Fig. 1, however, no obvious difference in the timing of the first opal-CT precipitation can be observed between calcareous siliceous and clayey siliceous sediments subjected to the same temperatures. This is partly because the uncertainty in the temperature estimation obscures the effect of lithology. The scatter in the data plots in Fig. 1 gives the upper limit for the time lag of initial opal-CT precipitation among different lithologies, which is on the order of 10^6 to 10^7 years.

The opal-CT to quartz transformation tends to start earlier in chert nodules compared to surrounding siliceous chalks or calcareous porcelanites (Keene 1975; Hein et al. 1981). The earlier formation of quartz within chert nodules is possibly inherited from the earlier formation of opal-CT in these nodules (Kastner 1981). Kastner et al. (1977) speculated that clay retards the opal-CT to quartz transformation in a similar way as it does the opal-A to opal-CT transformation, as discussed above. However, Isaacs (1981) found that the opal-CT to quartz transformation occurred earlier in clay-rich siliceous rocks compared to clay-poor siliceous rocks of the Miocene Monterey Formation. The time lag for completion of the opal-CT to quartz transformation between clay-rich and clay-poor lithologies is less than 1 Ma in the case of Neogene siliceous rocks (Tada unpubl. data).

3 Compaction of Siliceous Sediments and Rocks

3.1 Porosity Reduction Related to Phase Transformations

The compaction of siliceous rocks is closely related to silica phase transformations. Dissolution-reprecipitation processes during the transformations drastically change the pore structure and cause an abrupt decrease in porosity (Isaacs 1981; Tada and Iijima 1983).

Opal-A diatomites and radiolarites originally have an extremely high porosity due to abundant intraskeletal pores, which gradually decreases with an increase in burial depth (Fig. 2). This porosity reduction is caused mostly by mechanical compaction, since no significant dissolution of diatom frustules is observed, and no authigenic silica phase is found [except opal-A' inferred by Hein et al. (1978)], whereas a slight breakage of diatom frustules is observed throughout the opal-A zone (Hein et al. 1978). The rate of porosity reduction in diatomites with depth is small compared to that of mudstone, suggesting that they are resistant to compaction (Hamilton 1976; Isaacs 1981). Diatomites, for example, still retain porosites as high as 75% after a burial of 500 m (Hamilton 1976). During the opal-A to opal-CT transformation, an abrupt decrease in porosity of as much as 30% occurs in diatomaceous rocks, which is caused by nearly complete dissolution of diatom frustules and consequent destruction of intraskeletal porosity (Isaacs 1981; Tada and Iijima 1983). Most of the porosity in opal-CT porcelanites exists as ultramicropores, a few hundred angstroms in diameter, which probably occur as interparticle pores among the opal-CT crystallites within the matrix (Tada and Iijima 1983; Hurd pers. commun.). Besides the ultramicropores, small amounts of micro- (2 ~ 10 µm) and macropores (>10 µm) are present, mostly as molds of siliceous fossils (Tada and Iijima 1983). The amount of an abrupt reduction in porosity during the opal-A to opal-CT transformation decreases with an increase in clay content (Isaacs 1981; Tada

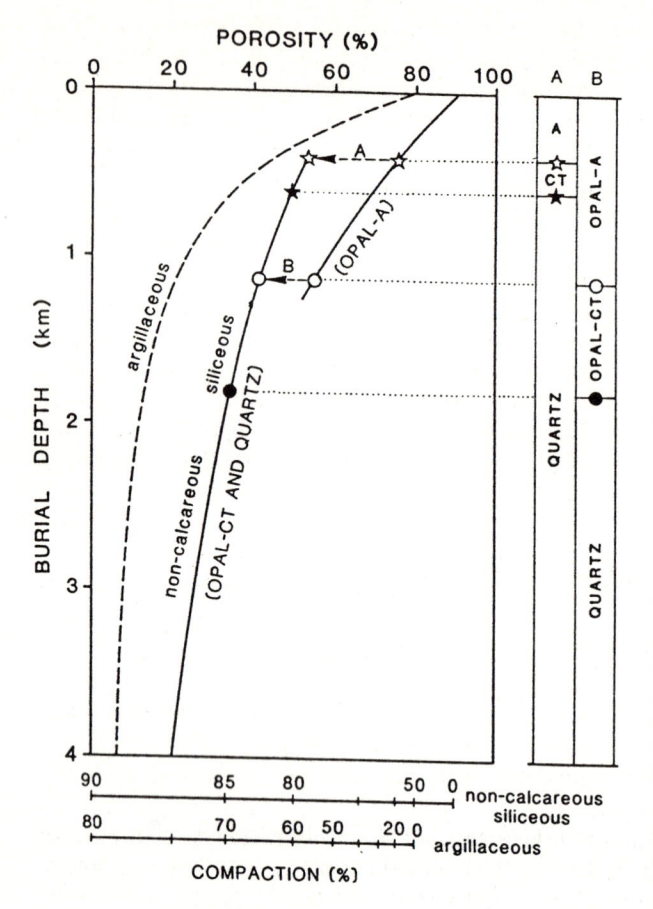

Fig. 2. Porosity-depth relationship for noncalcareous siliceous sediments (*solid lines*) and for argillaceous sediments (*broken line*). Note an abrupt porosity decrease during the opal-A to opal-CT transformation which takes place at a shallower depth in case *A* than in case *B*. Cases *A* and *B* correspond to lines A and B in Fig. 1. (Data from Tada and Iijima 1983 and Hamilton 1976)

and Iijima 1983). It is not certain whether such an abrupt reduction in porosity also occurs in radiolarites, because radiolarian skeletons are more resistant to dissolution than diatom frustules, and more intraskeletal pores may have survived after the opal-A to opal-CT transformation. Opal-CT porcelanite generally has a porosity ranging from 30 to 50%, which decreases slightly with increasing burial depth (Isaacs 1981; Tada and Iijima 1983). This decrease in porosity is insignificant compared to that of mudstones because of the rigid framework in porcelanites. With an increase in clay content, the porosity of opal-CT porcelanites decreases (Tada and Iijima 1983).

No abrupt decrease in porosity has been observed during the opal-CT to quartz transformation in Neogene noncalcareous siliceous rocks of northern Japan. There, quartzose porcelanites have porosites as high as 35% and no difference in porosity has been found between opal-CT and quartzose porcelanites of the same burial depth (Tada and Iijima 1983). A quartzose porcelanite obtained from a deep well in northern Japan retains a porosity of 20%, even after burial of over 4000 m (Tada and Iijima 1983). Most of the porosity in quartzose porcelanites exists as micropores among quartz particles a few microns in diameter (Tada and Iijima 1983). On the other hand, Isaacs (1981) found an abrupt porosity decrease of as much as 20% in calcareous siliceous rocks of the Monterey Formation, California, which she believed was caused by compaction, though it may also be possible that it was caused by cementation. The porosity of quartzose porcelanite tends to decrease with increasing clay content (Tada and Iijima 1983).

3.2 Differential Compaction Around Nodules

The compactional draping around dolomite nodules (differential compaction) provides further evidence for compaction as a principal mechanism of porosity reduction. Dolomite nodules of a few tens of centimeter to a meter in diameter are common in Neogene diatomaceous siliceous rocks of northern Japan, the coastal areas of California, and in other organic-rich siliceous sediments (Matsumoto and Matsuda 1987; Garrison et al. 1984). These dolomite nodules started to form during the very early stage of diagenesis at a depth of several meters to a hundred meters (Pisciotto and Mahoney 1981; Kelts and Mackenzie 1982, 1984; Matsumoto and Matsuda 1987), when the porosity of surrounding diatomaceous sediments was 80 to 90% (Fig. 2). They continued to grow until a depth of several hundred meters (Pisciotto and Mahoney 1981), where the porosity of surrounding diatomaceous rocks decreases to 60 to 70% (Fig. 2).

As is shown in Fig. 3, the amount of compaction for surrounding siliceous rocks with respect to the dolomite nodules averages 30% for diatomites in the lower part of opal-A zone, 40 to 60% for opal-CT porcelanites and cherty porcelanites, and 50 to 65% for quartzose porcelanites and cherty porcelanites, respectively. Assuming an average porosity of 75% for surrounding siliceous rocks during the growth of dolomite nodules and using the amount of compaction with respect to the whole nodules, the porosity is reduced to 64% in diatomites in the lower part of the opal-A zone, to 38 to 58% in opal-CT porcelanites, and to 29 to 50% in quartzose porcelanites. Similar porosity values are obtained when the amount of compaction with respect to the core of nodules is used and a porosity of 85% is assumed for surrounding siliceous sediments at the onset of dolomite nodule formation. These values are in good agreement with the measured porosities for diatomites in the lower part of the opal-A zone (60 to 70%), opal-CT porcelanites (30 to 50%), and quartzose porcelanites (20 to 35%) (Fig. 2). Thus it is concluded that the porosity reduction in diatomaceous siliceous rocks was basically caused by compaction, whereas additional cementation from external sources played a minor role.

Nodules and nodular beds of opal-CT and quartzose chert are commonly found in calcareous siliceous sediments (Lancelot 1973; Keene 1975; and others). They also show differential compaction, which amounts to 15 to 55% for opal-CT porcelanites and 40 to 75% for quartzose porcelanites (Fig. 3). These values are slightly smaller than those for dolomite nodules, suggesting that chert nodules were formed at slightly deeper depths than dolomite nodules. The nodular chert beds in Neogene siliceous rocks generally show pinch and swell beds, with the pinch to swell ratio of 0.4 to 0.8, a value comparable to the ratio for siliceous rocks with respect to chert nodules.

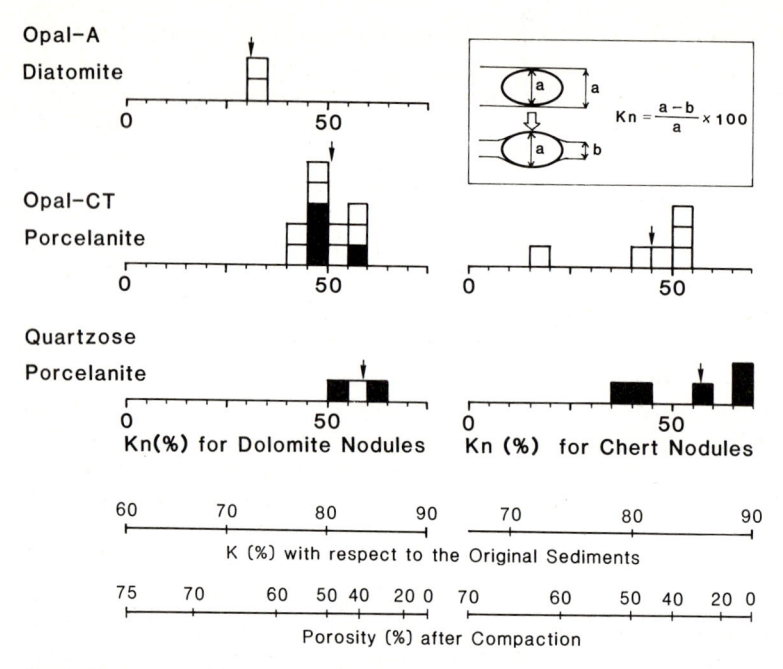

Fig. 3. Compaction with respect to diagenetic nodules of dolomite (*left*) and chert (*right*). Diagrams show compaction (*Kn*) in diatomites of the lower part of the opal-A zone (*upper diagram*), in opal-CT porcelanites (*middle diagram*), and in quartzose porcelanites (*lower diagram*) from Neogene siliceous sequences of northern Japan (*solid square*) and coastal areas of California (*open square*). Average values are indicated by *arrows*. *Lowermost axes* give compaction and porosity relative to the primary sediment

Some observations indicate that radiolarian cherts have also lost their porosity mainly by compaction. Webb (1986) showed that originally cylindrical burrow tubes in Ordovician bedded cherts of eastern Victoria, Australia, were compacted by approximately 75%. Kakuwa (pers. commun.) found extremely pure chert nodules about 1 cm in diameter with differential compaction around them within Permo-Triassic chert beds of southwestern Japan. The amount of compaction for host cherts with respect to the nodules is calculated as 45 to 70%. Since chert nodules were generally formed after a burial of a few hundred meters, the amount of compaction with respect to the original sediments should be larger. Iijima et al. (1988) observed differential compaction around an undeformed log of silicified wood in the uppermost Triassic bedded cherts of central Japan. They calculated the minimum amount of compaction as 60% relative to the log. If these examples are representative, the original thickness of radiolarian oozes is reduced by at least 60% and more likely by 75%, which thereby reduces their porosity to 20 to 60% assuming the initial porosity of 80 to 90%.

4 Silica Redistribution During Phase Transformations

The differential compaction and mineralogy of chert nodules and nodular beds as mentioned above suggest that they were formed by additional silica cementation while surrounding siliceous rocks were still in the opal-A stage. The additional silica presumably was supplied from surrounding rocks by diffusion, unless forced fluid flow is invoked. The driving force for diffusion is the concentration gradient of dissolved silica, which stems from a solubility difference between opal-A and opal-CT. As has been described before, slight differences in primary lithology (e.g., difference in clay content) may affect the timing of the silica phase transformations. When the timing of initial opal-CT precipitation differs between two succeeding beds with contrasting lithologies, the concentration gradient of silica is expected. The bed in which opal-CT is first precipitated tends to import dissolved silica, whereas silica tends to be exported from the neighboring bed in which opal-CT precipitation is retarded or inhibited (Fig. 4). A similar mechanism may operate when quartz precipitates in opal-A or opal-CT host rocks. This redistribution process may be important for the diagenetic enhancement of bedding in siliceous rocks. To evaluate the significance of this mechanism, it is worthwhile to estimate the duration necessary to form a chert bed.

Let us consider the formation of a 10-cm-thick chert bed with a final porosity of 10%, within diatomite with 60% porosity at a temperature of 30 °C. Let us assume that the average distance from source to sink is 10 cm, and that diffusion is the rate-controlling process. Using a grain density of 2.0 g/cm^3 for opal-CT, 5 g of silica should be imported per unit area (1 cm^2) of each side of a bed to form a 10 cm thick chert bed. According to Fick's first law, the diffusive flux (J) through a unit area per unit time (1 s) is given by the following equation:

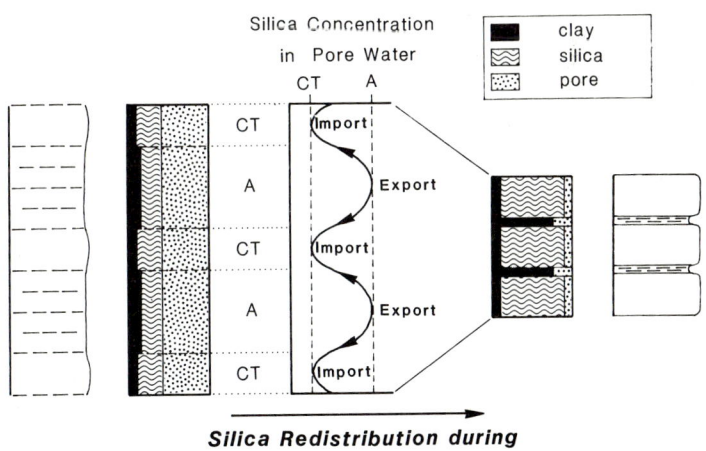

Fig. 4. A possible mechanism of silica redistribution and related enhancement of bedding in siliceous rocks during the opal-A to opal-CT transformation

$$J = -\varphi \frac{D}{\theta^2}\left(\frac{dc}{dx}\right),$$

where φ is the porosity of the surrounding rocks ($0 < \varphi < 1$), D is the diffusion coefficient for dissolved material, θ^2 is the tortuosity of the host sediments, and dC/dx is the concentration gradient of the dissolved material perpendicular to the bedding.

In our case, φ is 0.6, D for dissolved silica at 30 °C is approximately 10^{-5} cm^2/s (Wollast and Garrels 1971), and θ^2 for the diatomite is approximately 3 (Berner 1980; Boyce 1981). The concentration gradient amounts to approximately 6×10^{-6} g/cm^4, based on the solubility difference between opal-A and poorly ordered opal-CT (using the value for beta-cristobalite) at 30 °C (Kastner and Gieskes 1983; Walther and Helgeson 1977). Using these values, J is calculated as 1.2×10^{-11} g/cm^2/s [$= 4 \times 10^{-4}$ g/cm^2/a]. Thus, about 10,000 years are required to form a chert bed 10 cm thick. The solubility difference may be overestimated, since the concentration of dissolved silica in opal-A siliceous sediments is generally lower than the solubility of inorganic opal-A (Gieskes 1981), but this will not affect the result by more than factor of 2. The time interval on the order of 10^4 years is shorter than the maximum time lag of 10^6 to 10^7 for the initiation of the opal-A to opal-CT transformation among different lithologies. It is even shorter than the interval of 10^5 to 10^6 years necessary for the completion of the opal-A to opal-CT transformation within a single lithology. Consequently, significant redistribution of silica within a scale of 10 cm to 1 m is possible between siliceous beds with contrasting lithologies during the overall transformation from opal-A to opal-CT. In noncalcareous siliceous sequences with slight variations in primary clay content, this mechanism may enhance the variations in composition, making silica-rich layers more siliceous and silica-poor layers more clayey (Fig. 4).

It is not certain whether significant redistribution of silica occurs during the opal-CT to quartz transformation. However, such a phenomenon is not necessarily rare or absent, but may have been simply overlooked due to its less obvious effect on bedding enhancement. Let us again consider the formation of a 10 cm thick chert bed at a temperature of 60 °C and a average transport distance of 10 cm. This time, the host rock is opal-CT porcelanite with a porosity of only 30%. Assuming the final porosity after additional quartz cementation is 5% and grain density is 2.6 g/cm^3, 3.3 g of silica should be imported per unit area on each side of a bed. Using D of 2×10^{-5} cm^2/s for dissolved silica at 60 °C, θ^2 of 9 for opal-CT porcelanite (Boyce 1981), and dC/dx of 2×10^{-6} g/cm^4, based on the solubility difference of well-ordered opal-CT (using the value for alpha-cristobalite) and quartz (using the value for chalcedony) at 60 °C (Walther and Helgeson 1977), a flux of 1.3×10^{-12} g/cm^2/s (or 4×10^{-5} g/cm^2/a) is obtained. Consequently, a duration of about 80 000 years is necessary to form a 10-cm-thick chert bed by additional cementation with quartz. It is possible that the diffusion coefficient within opal-CT porcelanite is lower by as much as two orders of magnitude because of electroviscous effect due to the extremely small pore diameters (Tada et al. 1987). Thus, the duration can be as high as 10^7 years, which is comparable to the maximum time lag for the opal-CT to quartz transformation among different lithologies. It is safe to say that significant redistribution of silica is possible only within a short distance of the order of 1 to 10 cm. Besides, if the observation by

Isaacs (1981) that the opal-CT to quartz transformation occurs earlier in clay-rich siliceous rocks is correct, then the silica redistribution between clay-rich and clay-poor siliceous rocks during the transformation would reduce, not enhance, the contrast in composition.

From the calculations described above, it becomes clear that significant local redistribution of silica is possible during the opal-A to opal-CT transformation, whereas the redistribution seems more restricted during the opal-CT to quartz transformation. The time lag for the transformations between contrasting primary lithologies is the probable cause of silica redistribution and enhancement of bedding.

5 Silica Redistribution by Pressure Solution

Porosity data for quartzose porcelanites show that a porosity of as much as 35% can be preserved after the two silica phase transformations (Tada and Iijima 1983). If these quartzose porcelanites can be regarded as precursors of cherts, the porosity should be further reduced either by compaction or cementation. Mechanical compaction seems less important, considering the well-indurated framework of quartzose porcelanite, and the fact that as much as 20% porosity is preserved even after a burial of over 4 km. Slow chemical compaction is possible, for example by intergranular pressure solution (IPS). Both petrological and theoretical studies suggest that the rate of IPS in quartzose sandstone increases with a decrease in grain size, and that chert grains are more susceptible to IPS compared to single crystal quartz grains (Tada et al. 1987; Tada and Siever 1989). Since quartzose porcelanites are composed of quartz particles of a few μm in diameter, IPS can be significant. However, it is difficult to identify IPS contacts in quartzose porcelanites, especially because cathodoluminescence techniques are difficult for such small grain sizes. The effectiveness of IPS in quartzose porcelanites is a problem to be solved in the future. A compaction of approximately 75% estimated for some bedded cherts as described above could not eliminate all the porosity in radiolarian ooze. The remaining porosity of 20 to 60% after compaction must be reduced by additional silica cementation in order to form chert with zero porosity. If compaction by 75% is typical for cherts, we have to consider sources of the additional silica cementation. Calculation of diffusional transport suggests that silica sources should have been nearby, within a distance of a few centimeters to a meter. In the case of bedded cherts, the only possible sources of silica are either shale partings or the chert beds themselves.

There are several possible mechanisms for supplying silica to chert beds, as listed below:

1. Silica may have been released from clay minerals within shale partings during smectite to iilite and/or chlorite transformations (Boles and Franks 1979; Lahann 1980). The thickness ratio of a chert bed to a shale parting is generally less than 10:1 (Iijima et al. 1978, 1985), and the content of clay minerals within the shale is no more than 50%, judging from their Al_2O_3 content (Iijima et al. 1981; Kakuwa 1988). Based on the calculation by Boles and Franks (1979), the maximum amount of silica released from a shale parting of 1 cm^3 is 0.33 g, which

is equivalent to 1.3% by volume of a 10-cm-thick chert bed. Thus, clay mineral transformation is insignificant as a source of silica cement.

2. Silica may have been released by IPS within shale partings. It is known that the presence of clay in small amounts tends to enhance IPS in sandstone through inhibiting cementation and/or providing effective diffusion paths, although too much clay may inhibit IPS through a "cushioning" effect (Tada and Siever 1989). However, generally well-preserved radiolarian skeletons without any evidence of IPS argues against significant IPS within the shale partings (Kakuwa pers. commun.).

3. Silica may have been released by stylolitization within chert beds, shale partings, and/or along the chert/shale contacts. Stylolites are commonly found within cherts and chert/shale contacts (Iijima et al. 1985), and to a lesser extent within shale partings. A loss of several millimeters to a centimeter in thickness along a stylolite is commonly observed. Cox and Whitford-Stark (1987) estimated a stratigraphic thinning of 3.5 to 5.6% in the Caballos Novaculite, based on the stylolite column amplitude. Stylolite amplitudes give only a minimum estimate, which can be as low as 1/10 of the actual value (Heald 1955). Obviously, stylolites within chert beds, shale partings, and along chert/shale contacts are capable of supplying enough silica and are the most promising source for silica cement during late stage of diagenesis (Fig. 5).

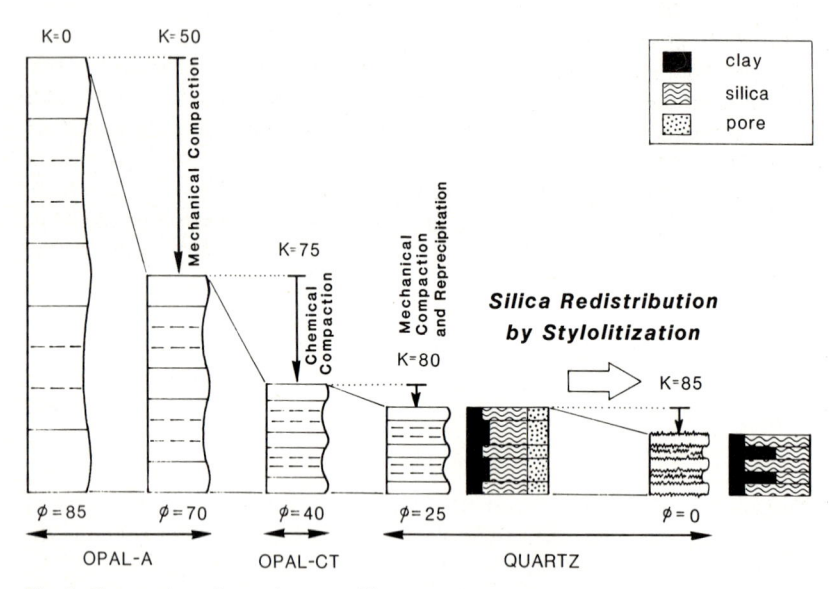

Fig. 5. Compaction of noncalcareous siliceous sediments during the opal-A to opal-CT and opal-CT to quartz transformations, and silica redistribution by stylolitization during the final stage of compaction

6 Conclusion

The silica phase in most siliceous rocks is transformed from opal-A through opal-CT to quartz during progressive diagenesis. The trasformations are principally controlled by temperature and time, but host rock lithology, together with the porewater chemistry, also plays an important role. Petrographic and experimental studies suggest that the presence of carbonate tends to enhance the opal-A to opal-CT transformation, whereas the presence of clay tends to retard it.

For a primary alternation with slight lithologic variations, an onset of the opal-A to opal-CT transformation can be different, depending on the lithologies. The beds with an early onset of this transformation import silica from the beds in which the transformation is retarded or inhibited. Calculation of diffusional transport suggests that considerable local redistribution of silica between the contrasting lithologies is possible during the opal-A to opal-CT transformation, whereas the redistribution seems more restricted during the opal-CT to quartz transformation. Field evidence also suggests the formation of chert nodules and pinch and swell beds by additional silica cementation within the opal-A zone, especially within calcareous siliceous rocks. Silica redistribution during the opal-A to opal-CT transformation can enhance the variations in composition between originally clay-rich and clay-poor layers (Fig. 4), although the majority of noncalcareous siliceous rocks lose their porosity mostly by mechanical compaction, and by chemical compaction and reprecipitation associated with the opal-A to opal-CT transformation.

It is not clear however, whether compaction alone can eliminate all the porosity to form chert. Some field evidence suggests that a compaction of approximately 75% has occurred in bedded cherts, but a porosity of 20 to 60% might still remain after the compaction. Pressure solution, especially stylolitization within chert beds, shale partings, and at chert/shale contacts, is the probable mechanism for differential compaction and cementation, and thus bedding enhancement at the final stage of chert formation (Fig. 5).

Acknowledgments. I express my sincere thanks to Drs. C.M. Isaacs and J.S. Compton who introduced me to many good exposures of the Monterey Formation, and to Dr. D.C. Hurd and Mr. Y. Kakuwa who kindly allowed me to use their unpublished data. Thanks also go to Drs. A Iijima, R. Siever, D.C. Hurd, and W. Ricken for their reading the manuscript.

4.3 Stratification in Phosphatic Sediments: Illustrations from the Neogene of California

K. B. Föllmi, R. E. Garrison, and K. A. Grimm

1 Phosphatic Sediments

Marine phosphogenesis occurs under exceptional and still not well-understood (bio-)chemic and hydrodynamic paleoceanographic conditions (Froelich et al. 1988; Bentor 1980). Modern phosphatic sediments are being formed in two contrasting environments: (1) along west-coast margins, where dynamic upwelling of nutrient-rich waters enhances primary productivity, and the presence of a well-defined oxygen minimum zone favors the accumulation of organic-rich sediments and consequent phosphogenesis (offshore Baja California, Peru, Chile, and Namibia; Froelich et al. 1988; Jahnke et al. 1983; Baturin 1982; Burnett 1977), and (2) along current-dominated margins, where low bulk sediment accumulation rates permit intense biogenic mixing of surficial sediments, and favor physico- and biochemically induced super-saturation of porewaters with respect to apatite (offshore southeast Australia; O'Brien and Heggie 1988).

Commonly, ancient phosphatic sediments have been attributed to either one of these sedimentary regimes (e.g., Föllmi 1989; Garrison et al. 1987; Riggs 1984). In particular, maxima in phosphogenesis have been correlated with the establishment or intensification of such regimes, i.e., to intervals of ice-volume growth and glacio-eustatic sealevel fall, where increased latitudinal thermal gradients induced vigorous oceanic circulation and resulted in strong and persistent coastal upwelling (e.g., Garrison et al. 1987; Sheldon 1980), and to intervals of warm and humid global climate and eustatic sea level rise, where well-stratified oceanic waters with expanded oxygen minimum zones developed and nutrient-rich currents shifted onto the shelves, contributing to low bulk sediment accumulation rates and phosphogenesis (Föllmi 1989; Arthur and Jenkyns 1981). Some ancient deposits, however, have been related to environments for which actualistic counterparts are not known; e.g., to lagoonal (Soudry and Lewy 1988), peritidal, or semi-emerged environments (Soudry and Southgate 1989; Southgate 1986).

Currently, the process of phosphogenesis, i.e., of precipitation of apatite (generally $Ca_5\{PO_4, CO_3\}_3 \{F, OH, CO_3\} [PO_4>>CO_3, F>>OH, CO_3]$; carbonate fluorapatite; Nathan 1984) is the focus of intense research, and a broad range of viewpoints exists, including (Fig. 1): (1) low net sediment accumulation rates (< 10 m/Ma) may favor phosphogenesis, probably because they allow porewaters to become highly concentrated in fluoride and phosphate; (2) fluoride may function as a limiting factor and may confine phosphogenesis to the upper tens of centimeters (Froelich et al. 1983; Jahnke et al. 1983); (3) porewater concentration of dissolved

Einsele et al. (Eds.)
Cycles and Events in Stratigraphy
©Springer-Verlag Berlin Heidelberg 1991

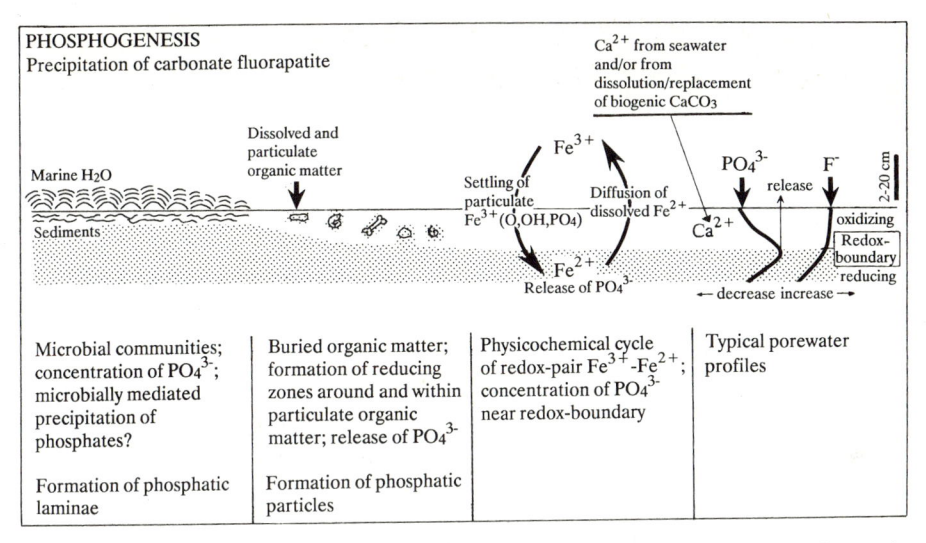

Fig. 1. Overview of biological, biochemical, and physicochemical processes during phosphogenesis; *stippled pattern* refers to anoxic chemical environments

phosphate may be enhanced by microbial activity (e.g., Reimers et al. 1990; Soudry 1987; Lucas and Prévôt 1985), by physicochemical cycling of the redox-pair Fe^{3+}–Fe^{2+} (Fig. 1; e.g., Froelich et al. 1988; O'Brien and Heggie 1988; compare Shaffer 1986), and by the decay of buried organic matter (e.g., Suess 1981; Burnett 1977); furthermore (4) apatite precipitation may be mediated by microbial activity (e.g., Reimers et al. 1990).

Once formed, sedimentary phosphates occur either as singular phosphatized particles, or as continuous or discontinuous laminae (Figs. 1 and 3). Due to the physical and/or biological dynamics of the above discussed paleoceanographic environments, phosphatic sediments commonly experience phases of hydraulic reworking and reburial, during which additional phases of phosphogenesis may occur ("Baturin" cycling; e.g., Föllmi 1989; Mullins and Rasch 1985; Kennedy and Garrison 1975a; Baturin 1971). Baturin cycling embodies an important factor in the genesis of economically viable phosphatic deposits.

The diversity of paleoceanographic environmental conditions at the location of phosphogenesis, the different bio- and/or physicochemical conditions that lead to apatite precipitation, and the commonly complex history of burial, early diagenesis, and hydraulic reworking results in a broad array of stratification patterns, which range from regularly stratified phosphatic sediments to highly complex, thin and condensed, multi-event winnowed phosphatic beds.

This chapter reviews the spectrum of stratification patterns in phosphatic sediments. For this purpose, we choose the Neogene phosphatic sediments of California as an illustrative data-base, including typical examples of each class within the herein proposed classification (Garrison et al. 1987, 1990; Reimers et al. 1990; references in these articles).

2 Important Variables in Stratification of Phosphate Shelf Sediments

Stratification in phosphatic sediments is the result of interacting physical, chemical, and biological processes, which occur before, during, and after phosphogenesis. Since hydraulic energy commonly dominates the interplay of these variables, one may view the stratification of some phosphatic sediments as a residual product, especially in the case of condensed multi-event winnowed phosphatic beds (Baturin cycling; Sect. 3.2).

2.1 Hydraulic Energy

Hydraulic energy, varying through time, leaves a remarkable imprint on the stratification of phosphatic shelf sediments, because

1. it actively lowers sediment accumulation rates, and hence favors phosphogenesis,
2. it determines modes of phosphate deposition (pristine vs. condensed vs. allochthonous; compare Sect. 3), and
3. it drives concentration processes during the development of phosphatic deposits (single-event vs. multi-event winnowing; Sect. 3.2).

Hydraulic energy is controlled by oceanographic and climatic dynamics, and ranges from quasi-periodic (waves, currents) to aperiodic (storms), and from slowly varying in intensity (e.g., geostrophic currents) to rapidly changing (e.g., storms, tsunamis). Tectonic energy, on the other hand, is provided by earthquakes and volcanic eruptions, and represents stochastic, aperiodic high-energy events (Fig. 2).

2.2 Physical Sediment Properties

Physical sediment properties such as bulk density, grain size and shape, and substrate cohesiveness constitute important variables in the process of phosphate stratification, especially in environments of hydraulic activity. The efficiency of winnowing depends on these properties and is enhanced by the presence of poorly sorted grains (small particles will be entrained, whereas larger particles will remain in situ), of high density contrasts (phosphatized particles and heavy minerals will remain in situ, whereas lighter sediment particles will be transported), and of low substrate cohesiveness.

2.3 Microbial Communities

Prominent microbial mats, which develop below oxygen-deficient bottom waters, act as sediment stabilizers by colonization of sediment surfaces (e.g., communities of Beggiatoaceae and/or filamentous Cyanobacteria; e.g., Reimers et al. 1990; Gerdes and Krumbein 1987). Consequently, higher initial entrainment velocities are required and winnowing becomes less effective. The presence of prominent microbial mats in

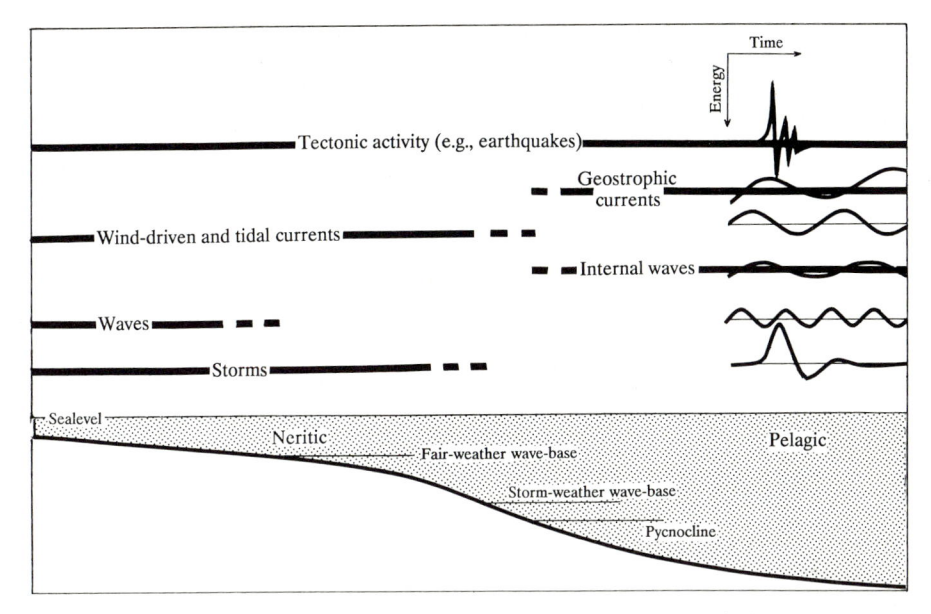

Fig. 2. Overview of important energy carriers in marine environments and their typical frequency spectra

the zone of phosphogenesis results in the formation of phosphatic laminae, rather than particles (Fig. 5; Sect. 3.1).

3 Stratification Types in Phosphatic Shelf Sediments

The proposed classification of stratification types in phosphatic sediments is based upon differences in the morphology of phosphates (particulate vs. laminated phosphates), and upon differences in the mode of deposition (pristine, i.e., nonreworked, vs. condensed, i.e., reworked vs. allochthonous; Fig. 3).

3.1 Pristine Phosphatic Sediments

The term *pristine* is applied to phosphates which lack signs of any reworking. Pristine phosphatic particles and laminae appear as discrete layers in, or scattered throughout their host sediments (Fig. 3).

Pristine phosphatized particles consist of buried and phosphatized organisms, and organic debris such as microbial assemblages, foraminifera, bivalves, faecal pellets, fish-debris and vertebrate bone fragments (Figs. 1 and 4A). Phosphatization may incorporate preserved organic hard-parts, particularly calcareous or calcified tests, may include internal sediments in organism cavities, or may extend into or be limited to a peripheral rim, particularly around detrital particles ("coated" particles; Fig. 4B). Scanning electron microscope surveys of coated grains commonly reveal elongate phosphatized structures within the phosphatized coatings; these are interpreted as

Stratification in phosphatic sediments as a function of time and energy

Stratification

Phosphogenesis

In-situ phosphatic diaclasts

In-situ phosphatic lamina

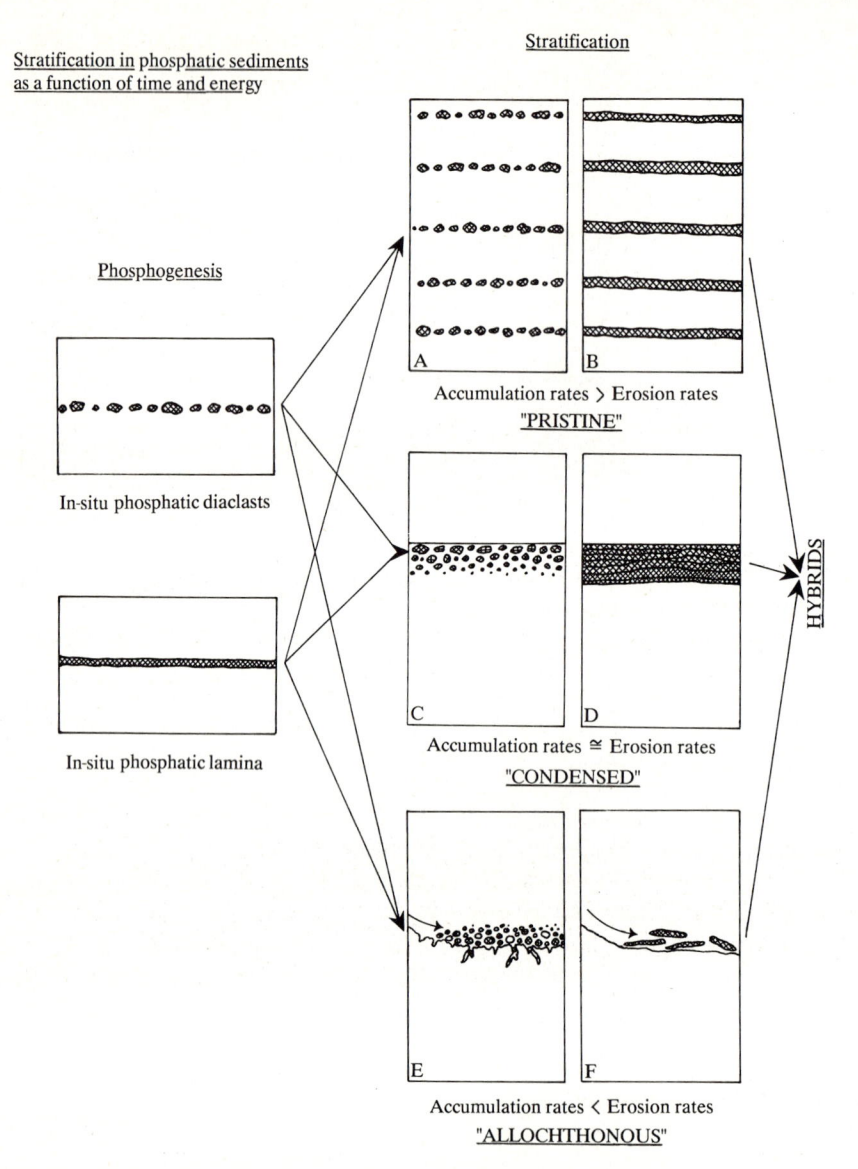

A | B
Accumulation rates > Erosion rates
"PRISTINE"

C | D
Accumulation rates ≅ Erosion rates
"CONDENSED"

E | F
Accumulation rates < Erosion rates
"ALLOCHTHONOUS"

HYBRIDS

Fig. 3. Genetic classification of stratification types in phosphatic sediments

preserved bacterial or cyanobacterial corpuscules, which suggests microbial participation in the build up of these rims (Fig. 5A; e.g., Garrison et al. 1987; Dahanayake and Krumbein 1985; Soudry and Champetier 1983).

Pristine phosphatic laminae appear in fine-grained, thinly laminated and organic-rich sediments, where they form mm- to several cm-thick phosphatic layers and lenses, commonly rhythmically interbedded with the host sediments (Fig. 6). Pristine

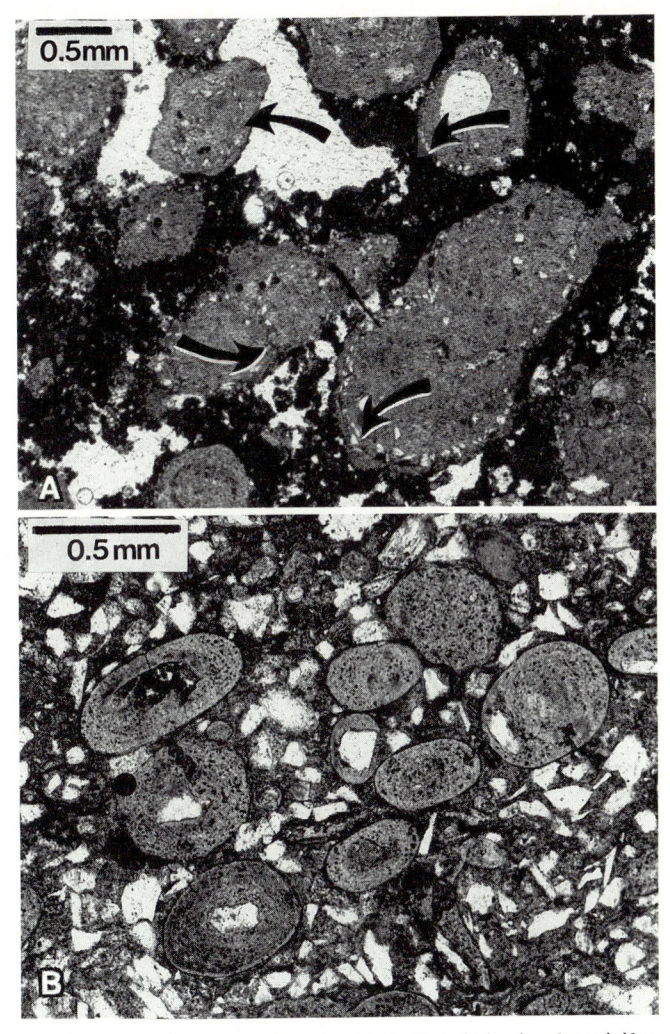

Fig. 4. Photomicrographs of **A** pristine phosphatic particles: phosphatized agglutinating foraminifera (*arrows*), embedded in a phosphatized microbial community. Middle Miocene, Monterey Formation, Naples Beach section, Santa Barbara Basin. **B** Coated phosphatic particles, embedded in a siliciclastic turbidite (Fig. 13). Nuclei consist of quartz grains and foraminifera. Middle Miocene, Monterey Formation, Carmel Valley Road, Salinas Basin

phosphatic laminae occur as well as thin, more unevenly spaced filaments in coarser sediments (e.g., siliciclastic sandstones).

The majority of pristine and condensed phosphatized laminae include fossilized microbial components of different affinities. e.g., the middle Miocene pristine phosphatic laminae of the Californian Monterey Formation, rhythmically interstratified with organic-rich muds and marls, contain phosphatized microbial filaments, which are generally attributed to Beggiatoaceae (Reimers et al. 1990; Williams and Reimers 1983); upper Miocene multi-event condensed, laminated phosphatic sediments, present in the form of reworked boulders in the Pliocene Sisquoc Formation of California, include rod-like corpuscules, very similar to cyanobacteria (Sect. 3.2; Figs. 5B and 11). Condensed phosphatic crusts from Albian strata of the northern Tethys margin display well-preserved columnar "stromatolites" (Föllmi 1989). We conclude, therefore, that microbial mats commonly serve(d) as templates in the genesis of phosphatic laminae.

3.2 Condensed Phosphatic Sediments

Condensed phosphatic sediments include phosphatic particles and laminae, which have been concentrated during one or more intervals of reworking. Sediment reworking may be caused by bioturbation, and, in particular, by episodic winnowing and/or erosion in an environment of variable hydraulic energy.

Phosphatic particles and laminae are recycled during dynamic sediment reworking and may therefore experience several phosphatization events, during which

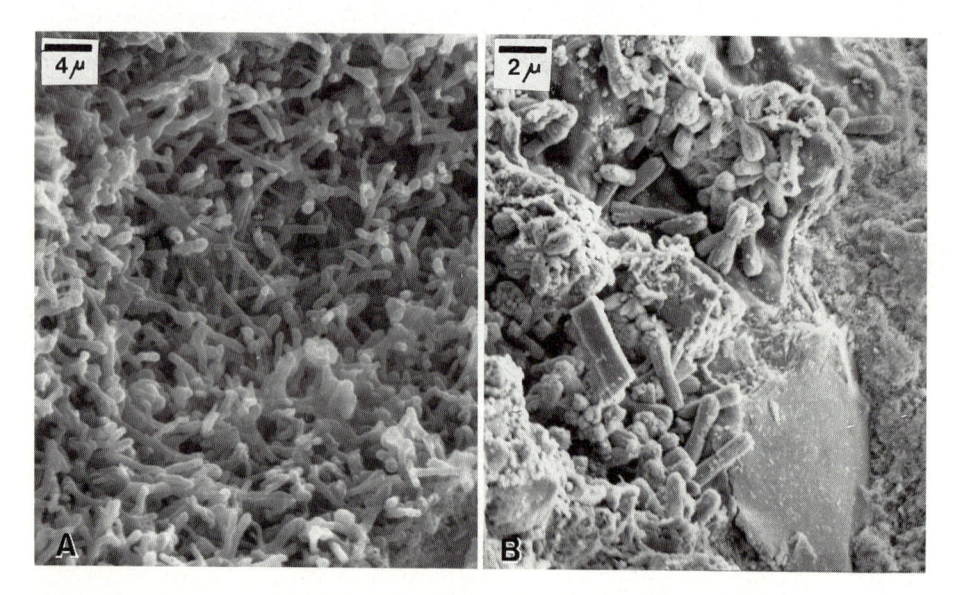

Fig. 5. Scanning-electron micrographs of phosphatic coatings. **A** Phosphatized coating of a particle, shown in Fig. 4B. The coating surface consists of a loosely packed aggregation of elongated apatite crystals, which resemble corpuscular microbial structures. Middle Miocene, Monterey Formation, Carmel Valley Road, Salinas Basin. (Garrison et al. 1987). **B** Phosphatized lamina within a multi-event condensed laminated phosphatic bed (shown in Fig. 11). The phosphatic coating includes elongated, ellipsoidal rod-like apatite crystals of probable cyanobacterial origin. Upper Miocene float boulder from the lower Pliocene Sisquoc Formation, Naples Beach, Santa Barbara Basin

Fig. 6. Light-colored pristine phosphatic laminae and lenses, interbedded with dark, organic-rich muds. The dark, partly phosphatized muds include scattered phosphatic peloids and white phosphatic nodules. Middle Miocene, Monterey Formation, Naples Beach, Santa Barbara Basin

additional phosphate is accreted (Baturin cycling; Figs. 8, 9, 10, 11, and 12C). The different, superimposed phosphate generations may be distinguished by differences in grain sizes of included detrital components, and/or by pronounced generation interfaces, which may be impregnated by iron oxyhydroxides, and/or colonized by microbial mats or encrusting epifauna (e.g., sessile foraminifera, bryozoa, serpulids).

Condensed particulate phosphatic sediments which result from a single sediment reworking event may form a deposit of uniform phosphate particles (Fig. 7A; but compare also Fig. 7B). When subjected to a suite of reworking events (Baturin cycling), a heterogeneous, strongly condensed lag deposit is built up, in which the phosphatized particles commonly display several phosphate generations, and fossils of different biostratigraphic zones and/or ecologic habitats may occur in an intimate mixture (Fig. 8). In special cases, the "event energy" increases with each reworking episode, until sediment accumulation rates approach zero or even negative values (= erosion). These circumstances may lead to the formation of a condensed, possibly coarsening upward bed, which is topped by a hiatus (Figs. 8 and 9).

Condensed laminated phosphatic sediments consist of two or more directly superimposed, phosphatized laminae. Superposition may be regular and conformable (nonerosive), or irregular and truncating, depending on the strength of intervening erosional events relative to the substrate cohesiveness (Figs. 10, 11). In some detri-

Chapter 4 Cherts and Phosphorites

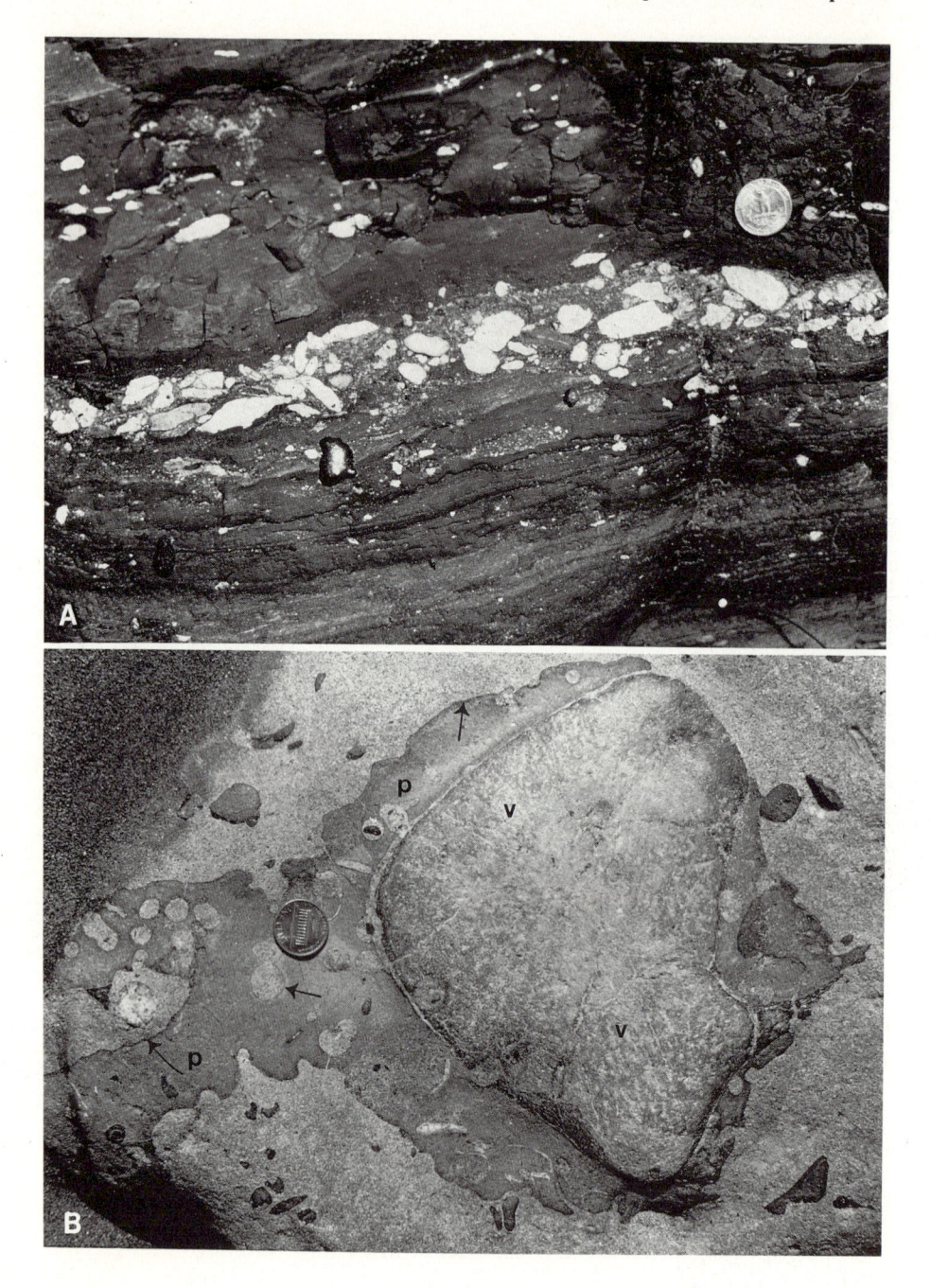

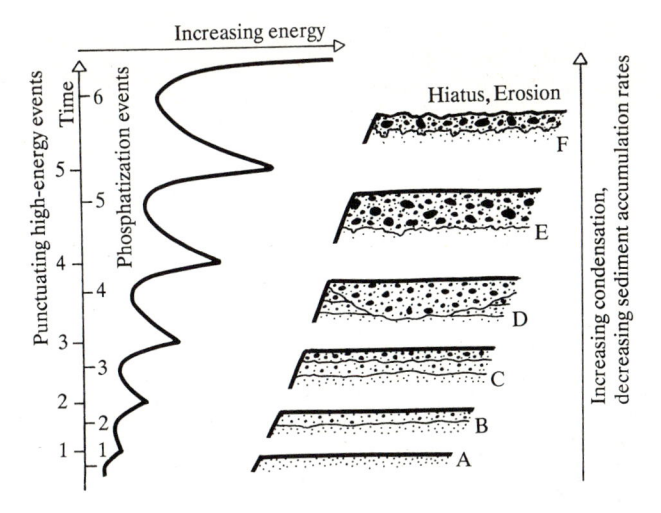

Fig. 8. Formation of a condensed and multi-event winnowed particulate phosphatic bed, in a suite of progressively stronger high-energy events and intervening phosphatization events (Baturin cycling). As long as the winnowing events do not perturbate the boundaries to earlier winnowed layers, a coarsening upward sequence may develop (*A-C*). Coarsening upward may be obliterated when winnowing events involve all sediments of former winnowing events (*D-E*). Strong high-energy events, portrayed at the *top* of this diagram, prevent sediment accumulation, forming a hiatus at the top of the condensed phosphatic bed (*F*)

tal-rich sediments, the laminae may build up dome-like "stromatoloid" structures, suggestive of microbial mats which pursued a strategy of avoiding detrital contamination (Fig. 11).

The phosphatic laminae are interpreted as fossilized remains of microbial mats, which stabilized their sediment substrate (Sect. 2.1; Fig. 5B). Due to the sediment-binding properties of microbial mats, threshold entrainment energies are increased, and winnowing becomes less efficient. Once the hydrodynamic energy is high enough to break up the microbial bonds within the mat, it scours the microbial mat and sediments, and entrains larger particles as well.

Fig. 7. A A condensed, single-event winnowed particulate phosphatic bed. The phosphatized particles are relatively unaltered and are nearly identical to sub- and superjacent nodules within the dark muds (ø coin = 2.4 cm). Middle Miocene, Monterey Formation, Naples Beach, Santa Barbara Basin. **B** Phosphatic nodule (*p*) accreted around a whale vertebra (*v*) within a condensed, single-event winnowed particulate phosphatic bed. Note the numerous Pholad-type borings, which indicate a re-exposure of the nodule on the sea bottom after early diagenetic formation. The periphery of the nodule and of the borings display a thin black phosphatic rim (*arrows*), indicating a second phase of phosphatization, probably after burial in the surrounding coarse sandstone. Smaller allochthonous phosphatic particles, floating in the surrounding sandstone, were partly derived from these larger phosphatic nodules (ø coin = 1.9 cm). Lower Pliocene, Purisima Formation, New Brighton Beach, La Honda Basin

Fig. 9. Condensed and multi-event winnowed particulate phosphatic bed. The phosphatic particles display a distinct coarsening and an increasing degree of heterogeneity towards the top. This bed may have been generated during a series of progressively stronger high-energy events (ca. 10), which is indicated by the increasing irregularity of the boundaries towards the top and by the presence of an unconformity at the top of this bed (not visible on photo; Fig. 8). Each winnowing event was followed by a phosphatization phase, in which the underlying layer experienced phosphatization (including phosphatic cementation). The upper layers include transported, nonphosphatized particles (*arrow* points to a volcanic clast). Middle Miocene, Monterey Formation, Naples Beach, Santa Barbara Basin

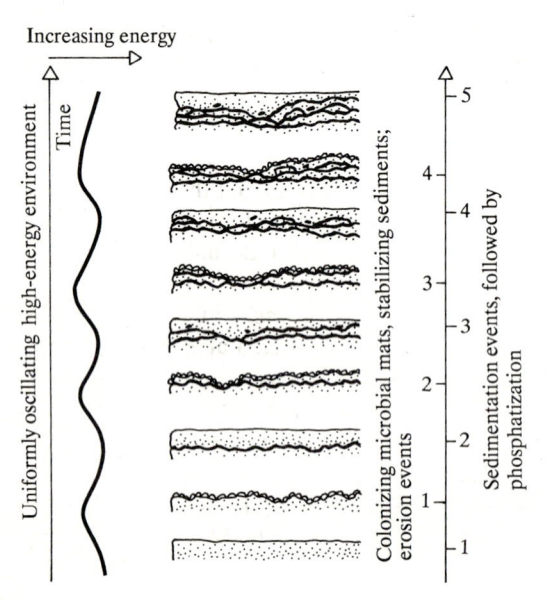

Fig. 10. Formation of a multi-event condensed laminated phosphatic bed, resulting from the influx of siliciclastic sediments and subsequent phosphogenesis, intermitted by phases of nondeposition, erosion, and overgrowth by substrate stabilizing microbial mats (Fig. 11)

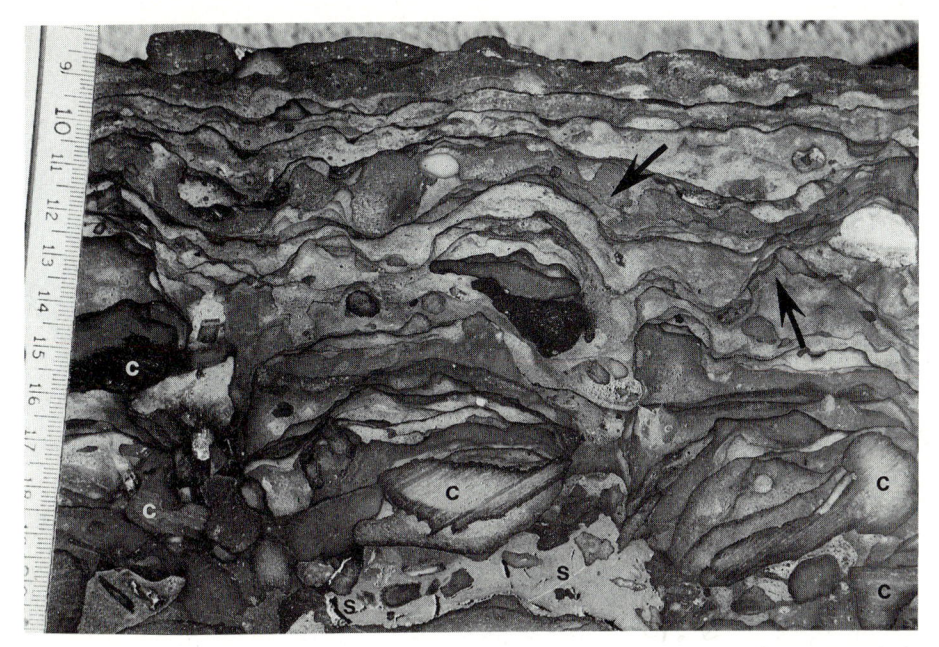

Fig. 11. Multi-event condensed laminated phosphatic bed. This bed consists of a multitude of phosphatic laminae, forming stromatoloid buildups (*arrows*). Each lamina is topped by a thin, dark, phosphatized coating, which includes cyanobacterial corpuscules, and which is interpreted as the fossilized remain of a microbial mat (Fig. 5). Note the exotic clasts (*c*), and the burrow-like infills with younger, nonphosphatized sediment (*s*). Upper Miocene float boulder from the lower Pliocene Sisquoc Formation, Naples Beach, Santa Barbara Basin

3.3 Allochthonous Phosphatic Sediments

Allochthonous phosphatic sediments result from erosional events, in which phosphatic beds are eroded and transported by gravity-driven flows. Phosphatic event beds consist of rounded and/or abraded phosphatic diaclasts, which range from silt- to boulder-size and are commonly mixed with non-phosphatic lithoclasts (Figs. 12 and 13). An array of distinct sedimentary structures such as a sharp and erosive base, and internal cross-bedding, lamination and/or size-sorting, underline the event character of such beds.

Internal structures and bulk composition of the diaclasts, included in allochthonous phosphatic sediments, may give valuable information about the character of the original source beds, otherwise commonly not available. For example, some uppermost Miocene allochthonous phosphatic conglomerates in California contain phosphatic pebbles, which are transported fragments of an otherwise unknown Upper Miocene condensed phosphatic bed (Fig. 12C).

Allochthonous phosphates originate during high-energy pulses, attributable to high current velocities, strong wave agitation, slope failure and seismic activity.

504　　　　　　　　　　　　　　　　　　　　　　　　　　　　Chapter 4 Cherts and Phosphorites

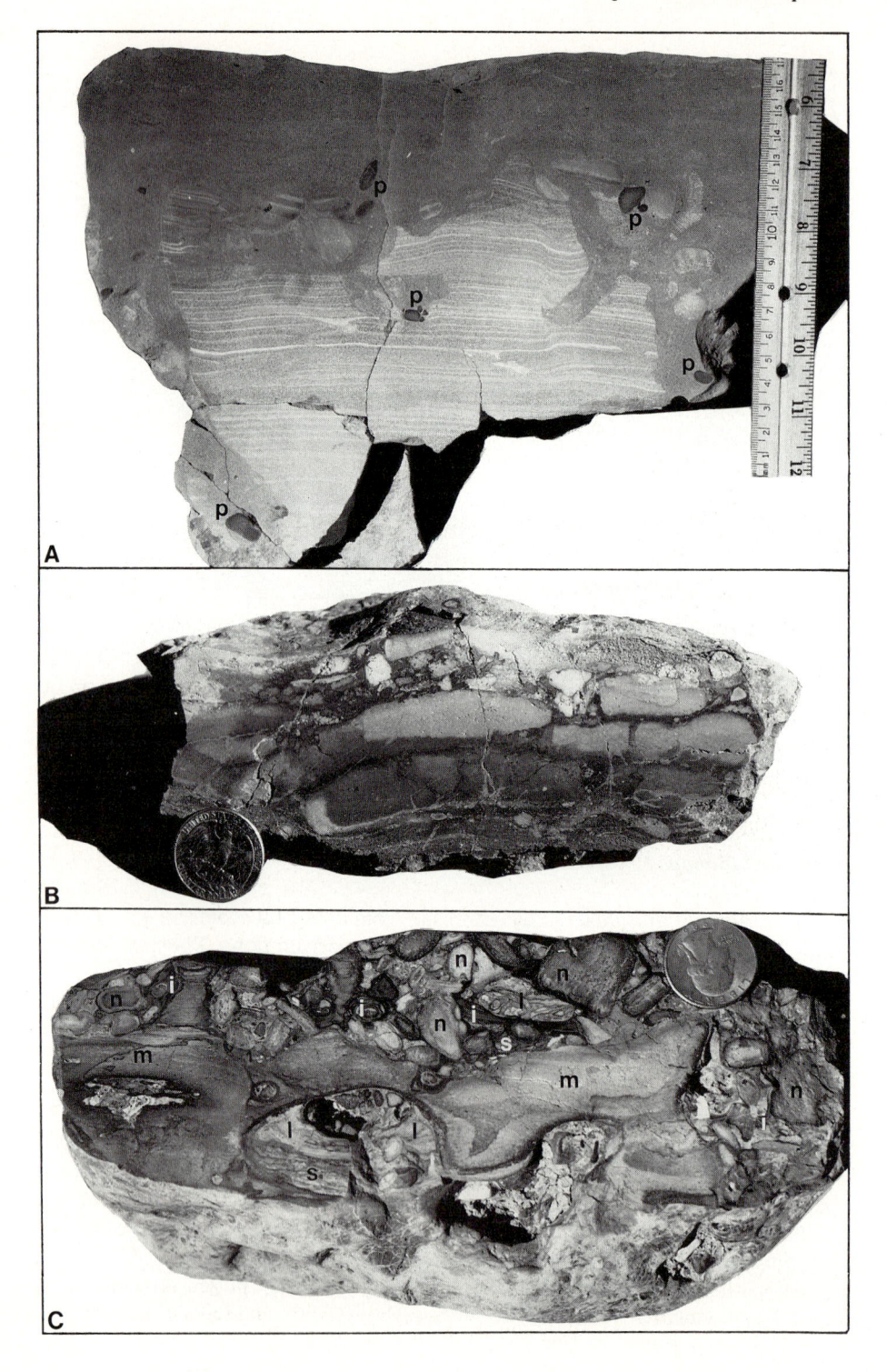

3.4 Hybrids

Hybrids of the above discussed stratification types are very common in phosphatic sediments. Different, superimposed stratification types may occur within a single phosphatic sequence, or even in one bed. Hybridization is observed, for instance, in

1. condensed phosphatic laminae, which may include pristine phosphatic particles and coated grains (Fig. 4A);
2. in pristine laminated phosphatic sediments, where the pristine laminae are interstratified with event-sediments, containing phosphatic diaclasts (Fig. 6);
3. in condensed particulate and/or laminated phosphatic sediments, which include a finer-grained, allochthonous fraction (Fig. 7B);
4. in beds of allochthonous phosphatic sediments, which experienced a subsequent phase of phosphogenesis (e.g., phosphatic cementation; Figs. 11 and 13).

4 Paleoceanographic Implications

Interpreted within their sedimentary context, phosphatic sediments may be useful indicators of paleoceanographic environments, particularly of the competence and persistence of hydraulic energy levels.

Energy intensity is roughly encoded in the stratification type:
1. pristine phosphates form under intermediate energy levels, where background sedimentation rates are probably lowered, but where phosphatized components do not experience reworking;
2. grain-size spectra of particles in winnowed phosphatic strata may serve as an energy indicator (plotted in velocity-grainsize diagrams; e.g., Hjulström diagram);
3. single-event versus multi-event condensed strata point to the character and regularity of the energy environment. Single-event winnowed beds are generally observed in proximal shelf areas, where sediment accumulation rates are high, and high-energy events have a more episodic character (e.g., storms; Fig. 2); multi-event winnowed beds preferentially occur in more distal shelf settings, where sediment accumulation rates are generally low, and energy conditions oscillate in a more steady rhythm (e.g., geostrophic currents; Fig. 2);

Fig. 12. A Gravity-flow deposit, including allochthonous phosphatized particles (p). Note the irregular sharp base and the *Thalassinoides* burrows, piping down into underlying laminated diatomaceous mudstones. Upper Miocene, Monterey Formation, Mussel Rock, Santa Maria Basin. **B** Disrupted and crudely imbricated phosphatic laminae, consisting most probably of phosphatized microbial mats, overlain by an allochthonous nodular phosphatic bed (ø coin = 2.4 cm). Middle Miocene, Monterey Formation, Naples Beach, Santa Barbara Basin. **C** Composite phosphatic boulder from an allochthonous particulate phosphatic bed. It consists of a phosphatized laminated diatomaceous mudrock (m), which is overlain by a layer containing allochthonous phosphatized particles (n). The space between the nodules and depressions or cavities within the laminated mudrock is infilled with three different types of winnowed laminated phosphate matrix: a first silt-rich phase (s), a second iron-rich phase (i), and a third light-colored phase (l) (ø coin = 2.4 cm). Miocene-Pliocene boundary, Sisquoc Formation, Mussel Rock, Santa Maria Basin (phosphatic sediments in this clast yield a late Miocene diatom fauna)

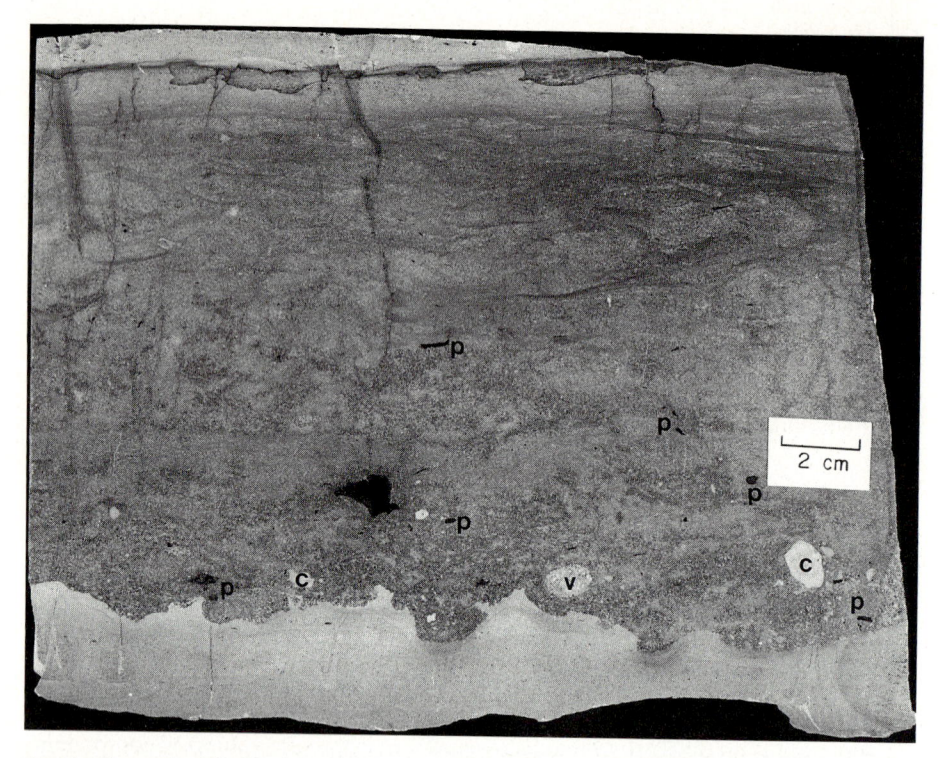

Fig. 13. Fine-grained siliciclastic turbidite, containing abundant coated phosphatic particles (Figs. 4B, 5A). This bed is intercalated in laminated opal-CT porcelanite. Note sharp, irregular base, normal grading within the turbidite, light-colored rip-up clasts (*c*), dark phosphatic particles (*p*), and volcanic clast (*v*). The bed itself shows a phosphatized matrix, due to a subsequent phosphatization event. Middle Miocene, Monterey Formation, Carmel Valley Road, Salinas Basin. (Garrison et al. 1987)

4. allochthonous phosphatic beds are commonly due to single high-energy events such as tectonic activity, tsunamis or high-energy current pulses.

Upwelling-related phosphates may be recognized by the siliceous and organic-rich sediments with which they are commonly associated. In current-dominated phosphatic provinces, upwelling-related sediments do not necessarily occur; instead, nonupwelling calcareous sediments may be included.

5 Conclusions

The classification presented here of stratification types within phosphatic sediments is based on two parameters:

1. on morphometry; i.e., on the distinction between particulate phosphates, consisting of phosphatized organic remains such as fossil debris and fecal pellets, or of phosphatized grains and lithoclasts, coated or interspersed with microbial

colonies, and laminated phosphates, probably representing phosphatized microbial mats; and
2. on the presence or absence of phosphate reworking. Pristine phosphatic sediments include non-reworked phosphatic particles and laminae; condensed phosphatic sediments are the result of single- or multi-event reworking; allochthonous phosphatic sediments are commonly found in gravity-flow deposits.

This classification masks the fact that a continuum exists among all three classes. Transitions are known both between pristine and condensed phosphatic sediments (condensed phosphatic laminae may include pristine phosphatic particles; Fig. 4A), and between condensed and allochthonous phosphatic sediments (condensed, winnowed particulate phosphates may include a finer-grained, allochthonous fraction; Fig. 7B).

Phosphate stratification is the result of physical, (bio)chemical and biological processes before, during and after phosphogenesis. Three factors appear to be of special importance:

1. energy, in form of hydraulic energy (internal and surficial waves; wind-driven, tidal and geostrophic currents) and tectonic energy (seismic activity). The ambient energy level and frequency influence sediment accumulation rates and the mode and degree of sediment-reworking. Hydraulic energy is more randomly distributed in inner-shelf and nearshore environments, and more uniformly in outer-shelf environments (Fig. 2);
2. sediment physical properties such as density, grain size, cohesiveness and particle-shape; and
3. the presence of prominent microbial mats, commonly within oxygen-depleted environments, which act as sediment stabilizers and counteract the efficiency of winnowing by increasing the initial entrainment velocities of waves and currents.

The specific location on the shelf and the oxygen content of bottom waters are less important to phosphogenesis. In Californian Neogene sediments, for instance, we observe phosphates which formed in a suite of different environments, ranging from well-bioturbated and detrital-rich inner-shelf areas to nonbioturbated and detrital-poor pelagic areas (Garrison et al. 1987).

Phosphates formed and form preferentially in areas of coastal upwelling, during periods of increasing ice volume, glacio-eustatic sea level fall, and intensified oceanic circulation, and in current-dominated areas, during periods of warm and humid climate, eustatic sea level rise, and well-stratified oceans. Stratification types within phosphatic sediments, interpreted in context of their associated sediments, are useful indicators, which help to identify general hydraulic and environmental conditions, as components of the depositional paleoceanographic setting during phosphogenesis and phosphate accumulation.

Acknowledgments. We are grateful to the U.S. National Science Foundation for financial support of this study (EAR85-19113) and to the Swiss National Science Foundation for support of a postdoctoral tenure in Santa Cruz (for K. Föllmi). We thank John Barron, Clare Reimers, and Miriam Kastner for discussions and the exchange of preprints, and Eugenio Gonzales for skillful preparation of thin sections.

5.1 Stratification in Black Shales: Depositional Models and Timing – an Overview

A. Wetzel

1 Introduction

"Black shale" is a comprehensive term subsuming fine-grained, dark-colored sediments that show distinct lamination and contain a considerable amount of organic matter. Black shales can form in a number of different environments – in lakes, swamps, and in the ocean from shallow water to the deep sea (e.g., Demaison and Moore 1980). Accordingly, the black sediments vary widely in their principal composition. There are black shales free of carbonate, black marls or limestones. Many of them contain a substantial amount of other constituents, such as phosphatic minerals or biogenic silica.

In the following text, the term "black shale" is used for sediments characterized largely by two typical lithologic features:

1. Dark color, which is due to the presence of finely dispersed iron sulfides and/or an organic matter content normally exceeding 0.2–0.5%.
2. Lamination, which in black shales is related mainly to the formation and preservation of primary small-scale sedimentary changes. (a) Laminae are formed by the cyclic input of different components or organic compounds, and clay minerals are aligned parallel to each other and to the depositional interface (e.g., Moon and Hurst 1984; Gerdes et al. Chap. 5.6, this Vol.). (b) Laminae are preserved due to low oxygen levels at the depositional interface, preventing bioturbation. Normally, below a level of 0.2 ml O_2/l, benthic organisms are absent (e.g., Wetzel 1983; Savrda et al. 1984).

Cycles, rhythms, and events within the various black shale depositional environments are well recorded. In this paper, the various time spans of black shale formation are discussed. It has been discovered that the time encompassed by black shale deposition is often associated with specific environmental conditions.

2 Formation of Black Shales

Black shales accumulate when the bottom water in the depositional area is poor or devoid of oxygen. The oxygen depletion results from oxidation of organic matter which sinks through the water column or is transported laterally, to the extent that oxygen consumption exceeds oxygen supply. In other words, oxygenation at the sea

---

Einsele et al. (Eds.)
Cycles and Events in Stratigraphy
©Springer-Verlag Berlin Heidelberg 1991

floor is controlled by the organic matter input (e.g., Calvert 1987), oxygen content of the bottom water, and circulation intensity. Principally, two scenarios can be distinguished (Fig. 1):

1. oxygen-poor bottom water is formed when circulation is very slow or non-existent. This may result from a pronounced thermohaline stratification of the water body in lakes or silled marine basins, for instance, in Lake Tanganyika or the Black Sea;
2. an oxygen-depleted zone may develop where high productivity in the surface water leads to significant oxygen deficiency in the underlying water, as found in upwelling areas, for instance off Peru. Between these two end members various transitions can occur (Fig. 1), for example, slow circulation and intermediate to low organic productivity, leading to an oxygen minimum layer as is the case in the modern Indian Ocean (for examples, see Brooks and Fleet 1987).

3 Timing of Black Shale Formation

To analyze the timing of black shale formation, it is necessary to distinguish between periods of permanent anaerobic conditions and durations during which numerous beds of black shales and intercalated oxygenated sediments may accumulate. The

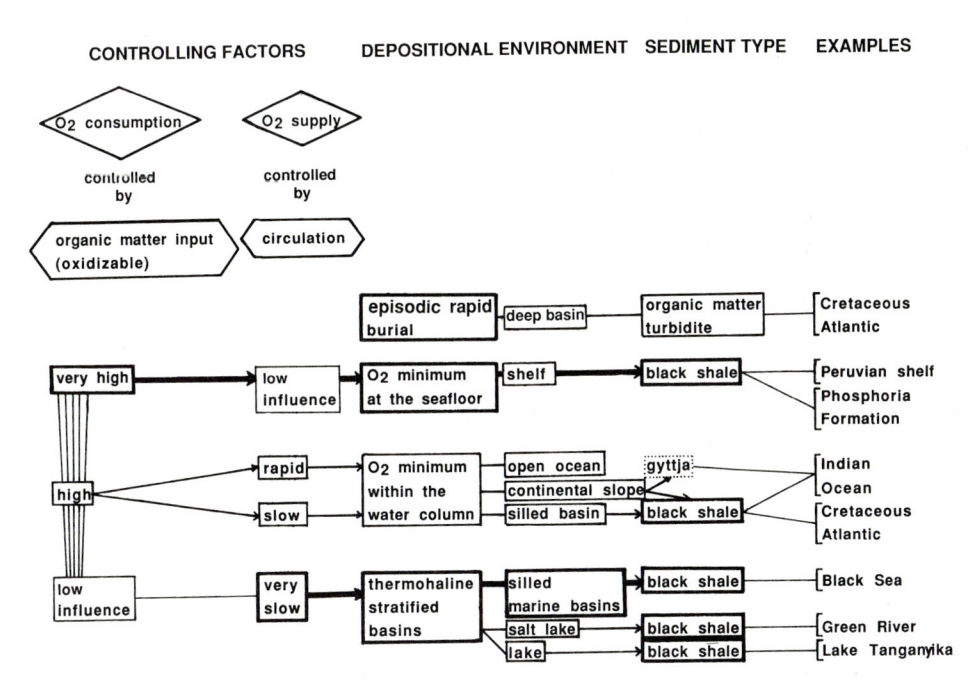

Fig. 1. Generalized scheme for "black shale" formation, with examples of the main types of anaerobic environments. (After Demaison and Moore 1980)

latter case represents a time envelope comprising periods of both aerobic and anaerobic conditions. The recurrence time of black shales is the time span from one black shale unit to the next, including intercalated deposits.

Such time spans can be of several different orders of magnitude (Fig. 2). The scheme used to classify black shales is based on an evaluation of some 400 occurrences reported in the literature. The existence of four distinct frequency maxima was found.

1. Mega-scale periods are time envelopes covering times favorable for black shale formation, their duration being in the range of several 100 Ma.
2. Sequence-scale periods represent times that last 1 to several 10's of Ma. They represent periods of repeatedly occurring black shales (time envelopes) as well as continuous black shale deposition.
3. Bed-scale variations are in the range of 10 to 100 ka and are typically represented by Milankovitch-type cycles 400, 100, 40, and 20 ka in duration.
4. Varve-scale variations are documented by single laminae or bundles of laminae, clearly representing less than 1 ka.

In this paper, only the sequence-scale, bed-scale and varve-scale variations are discussed; mega-scale processes are beyond the scope of the book.

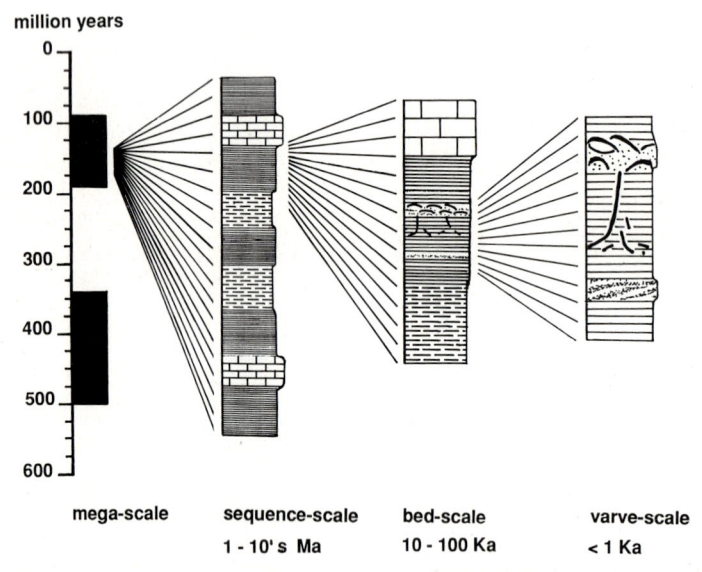

Fig. 2. Different scales of variations documented in black shales. *Mega-scale* variations correspond to long periods in Earth's history, with recurring black shales. *Sequence-scale* variations form lithologic units representing 1 to 10 Ma of permanent anaerobic conditions as well as time envelopes. *Bed-scale* variations are normally shorter than 1 Ma and often document Milankovitch-type cycles. *Varve-scale* variations document short-term variations within a black shale environment

4 Sequence-Scale Variations (1 to several 10's of Ma)

The duration of oxygen-depleted conditions is controlled by various processes. Essentially four environmental situations can be distinguished which control oxygen depleted situations: thermohaline stratification, upwelling, sea level fluctuations, and lake stagnation, are related to the major environmental settings of black shales (Fig. 1).

4.1 Thermohaline Stratification

In a given basin, denser water can form a more or less stable bottom water layer that can become increasingly oxygen-deficient over time. In such a case, organic productivity may vary from either high to low. For black shale formed under such conditions, Ryan and Cita (1977) and Bralower and Thierstein (1984) assumed low to medium surface productivity, e.g., the Cretaceous black shales deposited in the North Atlantic Ocean. An anoxic period is commonly terminated by increasing water circulation. This may be due to tectonic movements changing the configuration of the basin or opening sea ways, or climatic changes. The deposition of such sequences takes 1 to 30 Ma.

4.2 High Supply of Organic Matter (Upwelling)

Black shale units resulting from an increase in organic matter supply are commonly found in upwelling regions. They are often situated on the eastern margins of large basins (e.g., Thiede and Suess 1983). The occurrences of such deposits were extensively modeled by Parrish and Curtis (1982) and Parrish et al. (1983).

Upwelling provides high surface productivity that may lead to the development of an oxygen minimum in the underlying water (Wyrtki 1962). This can cause the formation of black shales. However, upwelling does not necessarily generate anaerobic conditions at the sea floor, because the oxygen supplied by circulation may compensate for excessive oxygen consumption through the oxidation of organic matter (Demaison and Moore 1980).

Changes in water circulation due mainly to climatic changes, relative changes in sea level, and changes in the basin configuration may terminate an anoxic period caused by upwelling. Sequences formed under these conditions correspond to a period of 1 to 15 Ma.

4.3 Periods of High Sea Level

High sea level favors increasing surface productivity in coastal waters, and hence anaerobic conditions in the underlying water and on the sea floor. Therefore, black shales are often formed during transgressions. The addition of comparatively rapid subsidence may lead to the repeated formation of black shales, e.g., in marginal marine environments or in restricted marine basins.

In shelf environments, sequences containing black shales related to rising sea level demonstrate a relatively low proportion of black shales, because of the frequent intercalations of terrestrial or shoreline-derived sediment. The formation of such sequences spans between 1 and 20 Ma, with a modal value of about 1 to 4 Ma.

4.4 Sediment-Starved Lakes

The most famous black shales deposited in a continental environment are those from playa lake complexes in semi-arid regions, e.g., from the Eocene Green River Shale (Eugster and Hardie 1978) or the Mesozoic sediments offshore of Gabon (Brink 1974). These black shales were deposited in a slowly subsiding area. The formation of black shales depends on (a) climate and geography providing shallow water conditions necessary for algal growth on the lake floor, (b) low input of terrigenous material, (c) low input of sulfate, which prevents oxidation of organic matter by sulfate reduction, and (d) the salinity not exceeding the limit tolerated by algae. The formation of these units is usually terminated by infilling of the basin or rapid subsidence. Such black shale units represent time spans of about 1 to 2 Ma.

5 Bed-Scale Variations

The formation of units consisting of black shale beds and other sedimentary rocks is chiefly controlled by (1) fluctuation of oxygen content in the bottom or porewater, (2) changes in the supply of terrigenous or calcareous sediment, and (3) varying organic matter input (e.g., de Gracianski et al. 1987). Bed-forming processes often appear to be cyclic within the Milankovitch frequency band, whereas sedimentological events occur over shorter time intervals. However, bed successions may also be formed by diagenesis (Raiswell 1988a) or by diagenetic enhancement of primarily small differences in the sediment composition, for instance, limestones in black shales. The geometry and/or preservational features of fossils are criteria for distinguishing primary and diagenetically formed strata (for details, see Seilacher et al. 1976).

Principally, there are two types of lithologic successions, formed in different environments (Fig. 3):

1. pelagic sequences commonly consist of only two alternating members; and
2. with decreasing distance from the shoreline, the sections tend to consist of an increasing number of lithologic members. Nevertheless, in stable shallow water environments uniform sequences may also be found (Fig. 3). Event and cyclic deposits have been separated in the classification scheme presented here (Fig. 3), although both types may be found together in the same section.

The various types of successions containing black shales are grouped with respect to the depositional processes which formed them, but some of these processes are

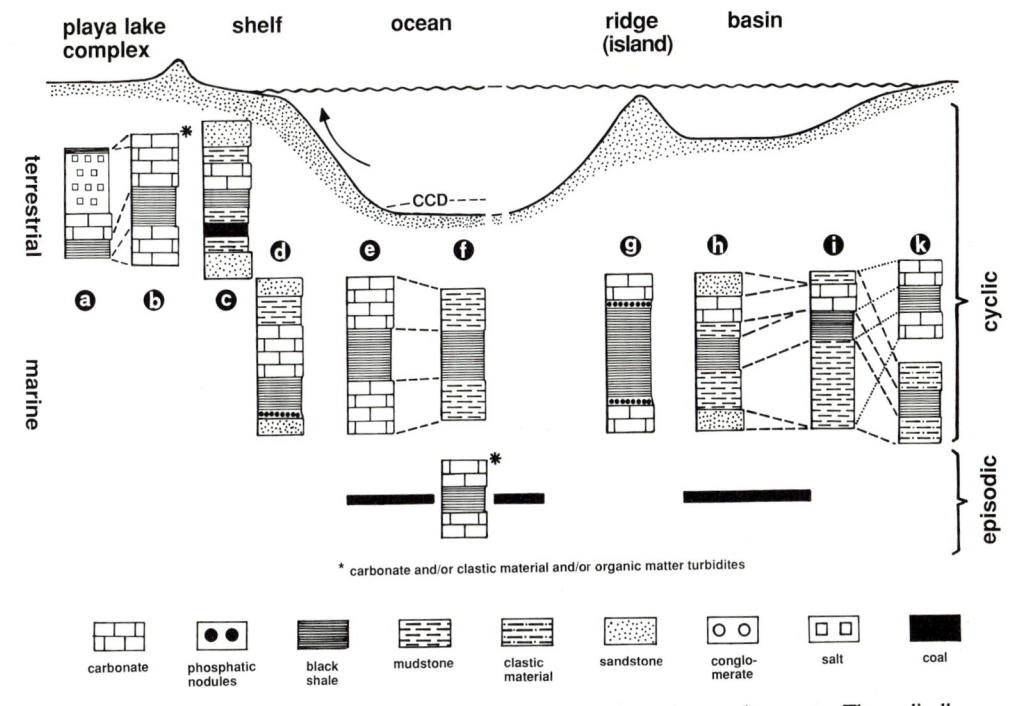

Fig. 3. Schematic successions representing black shales formed in various environments. The cyclically formed sections are deposited as *a* chemical cycle; *b* detrital cycle; *c* coal cyclothem; *d* coastal upwelling cycle; *e* + *f* productivity or oxygenation cycle (*e* above and *f* below the CCD); *g* upwelling cycle in an open ocean close to islands or ridges; *h* + *i* deep-basin cycle (*h* Mediterranean type, *i* Black Sea type); *k* shallow basin type cycle. - - - - Possible lateral correlation. ----- Possible range of episodic beds intercalated in cyclic deposits

closely interrelated, e.g., oxygenation and supply of organic matter, while other processes can occur subsequently within one depositional area (e.g., Jacquin and de Gracianski 1988). Factors creating these successions include:

1. Relative changes in sea level are the predominant factor in shallow water environments, because sea level fluctuations affect the supply of sediment and organic matter, and hence, the position and extent of the oxygen minimum zone.
2. Variations in the oxygenation of bottom water are due to changing circulation patterns, while the input of organic matter remains more or less constant.
3. Organic matter is continuously supplied into a basin with slow circulation.
4. The discontinuous input of organic matter due to event sedimentation or changing productivity may form bed successions consisting of black shales interbedded with other sediments.
5. The supply of biogenic sediment components, such as biogenic silica or carbonate, can change due to varying primary productivity, CCD fluctuations, or terrigenous input.

5.1 Relative Changes of Sea Level

Sea level changes influence the input of particulate as well as dissolved matter from the land and, furthermore, the circulation in shallow water areas, on both wide or narrow shelves. Additionally, rising sea level may favor the influx of high salinity water into adjacent basins separated from the ocean during low sea level stands.

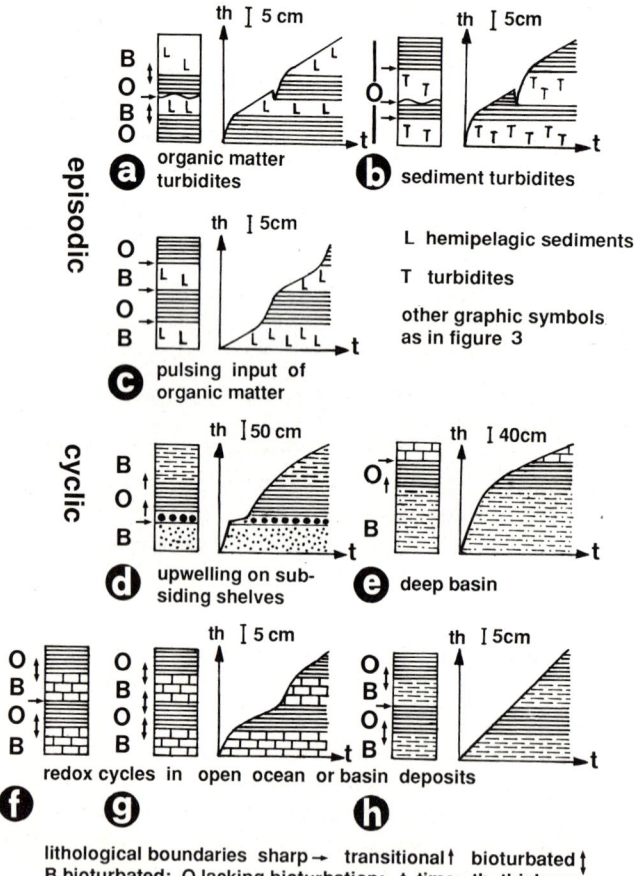

Fig. 4. Time-buildup diagrams for some sections containing bed-scale black shales after descriptions of the following authors. Episodic deposits: **a** de Graciansky et al. (1979); **b** Berger and von Rad (1972); **c** Dean et al. (1978). Cyclic deposits; **d** McKelvy et al. (1959); **e** Ross and Degens (1974); **f** McCave (1979); **g** Weissert et al. (1979), Dean et al. (1978); **h** McCave (1979)

5.1.1 Littoral and Sublittoral Environment

In shallow water, black shales are usually deposited during a transgression, often close to the upper boundary of the oxygen-depleted zone. They occur in numerous regions and consist mainly of marine deposits (in contrast to Sect. 5.1.2) but also contain some terrigenous organic matter. Shelf regions without upwelling are often lithologically different from those *strongly* influenced by upwelling.

1. Shelf without upwelling. A cycle starts with deepening and deposition of (calcareous) mudstones, which are overlain by black shales formed during the transgression. An oxygen-depleted zone is caused by sluggish circulation below a relatively stable thermocline (Heckel 1986). These black shales accumulate at depths between 30 and 200 m. Later on, limestones, marls, and sandstones are deposited in a coarsening upward sequence during slow regression. The lithologic boundaries are usually transitional and bioturbation is common. Depending on the regional setting, the calcareous sediments may be replaced by silty or sandy material (de Klein and Willard 1989). The thickness of the black shales varies considerably, depending on how long sea level remains high. Normally, a cycle is several meters thick, and the black shale strata represent part of it. On the average, one cycle represents several 10's to a few 100's of ka.
2. Coastal upwelling. Black shales alternate with sediments rich in biogenic carbonate and/or silica and/or terrigenous sediments. Furthermore, diagenetically formed layers, or nodules of phosphatic minerals or chert, may indicate minor cyclic variations in environmental conditions (Figs. 3d, 4d, 5). Sediments rich in phosphates indicate slow deposition close to the upper or lower boundary of the oxygen minimum (Figs. 4d and 6; Burnett et al. 1983, see also Föllmi et al. Chap. 4.3, this Vol.). Therefore, minor fluctuations in the circulation pattern may produce cyclic layers rich in phosphates due to an episodic lowering of the sedimentation rate (Notholt 1980), or increased water action winnowing other material away. Intercalations of chert or limestone may have formed similarly, but more commonly they are of diagenetic origin or accentuated by diagenesis.

Major fluctuations, i.e., larger variations in sea level, may lead to fluctuations in input of terrigenous or carbonate material. Fining- or coarsening-upward cycles may indicate deepening or shallowing conditions, respectively. The cycles are generally episodic with durations of 10 ka to 1 Ma. They are between 1 and 10 m in thickness.

5.1.2 Deltaic Environment

Typical sediments of this environment are deltaic lagoonal cycles consisting of various lithologic members. These cycles also contain sediments formed in a non-marine environment (in contrast to Sect. 5.1.1). Within a cycle black shales can be formed mainly in two situations:

1. Black shales accumulate prior to the first marine transgression in a swamp environment with stratified water conditions. It has been proposed that black

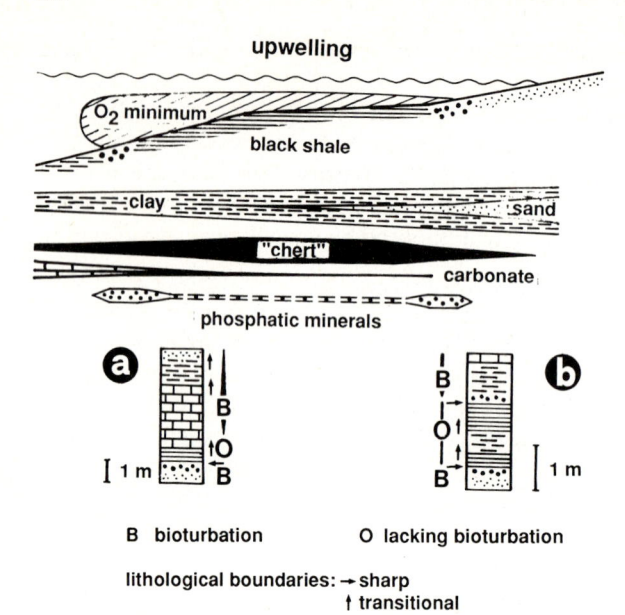

Fig. 5a,b. Bed-scale black shale formation in coastal upwelling regions (for details, see text)

shale organic matter in this case is the product of floating algal/bacterial mats that reduce the oxygen supply to the bottom water (Zangerl and Richardson 1963). The duration of these cycles may vary between 10 ka and 1 Ma, but periodicity has rarely been demonstrated.

2. Black shales may also form in marine environments during transgressions or sea level highstands very similar to those described in the previous sections (5.1.1) under point (2). However, the cycles contain a considerable amount of terrestrial sediments.

5.1.3 Silled Basin Environment

Sea level changes and/or tectonic movements may form seaways, and thereby favor the influx of sea water into adjacent basins, leading to stable water stratification. Well-known examples are documented from the Black Sea (Ross and Degens 1974; Degens and Stoffers 1980). Two types of depositional processes can be observed.

1. A chemically controlled depositional system with the end members, fresh water carbonate (seekreide) and sapropel, usually develops in shallow water (Fig. 3k). In this system, carbonate is produced in an oxygenated water layer, whereas the black shale facies develops under anaerobic conditions. The vertical fluctuation of the pycnocline due to climatic changes is responsible for the alternation of shale and limestone beds; long-term cycles (0.5 to 1 Ma) as well as short-term cycles (several ka; "megavarves") can be distinguished (Degens et al. 1978).

2. A detritus-dominated system with organic-rich terrigenous muds (sapropel) and coccolith limestones usually developes in deeper portions of the basin (Fig. 3i). Terrigenous muds accumulate during the fresh/brackish water oxygenated stages, while the influx of sea water establishes anaerobic bottom water. Further inflow of saline water makes the aerobic/anaerobic interface rise. Simultaneously, sapropelic muds are deposited. After some time the aerobic/anaerobic interface stabilizes at a certain depth, with diffusion taking place and coccolith limestone/terrigenous mud successions being deposited. In the Black Sea, this situation corresponds to the Quaternary interglacial periods with high sea levels. The sedimentary cycles are a few meters thick, with a black shale unit about 50 cm thick. Individual cycles span 10 to 120 ka (Fig. 3e).

5.2 Oxygen Supply

Bed-scale redox cycles are caused by variations in the intensity of circulation, assuming an essentially constant supply of organic matter. Such variations in the oxygen content of the bottom water are typically reflected in cyclic sediments, whereas short-term oxygenation provides only thin layers of oxygenated deposits (see Sect. 6.1.2).

5.2.1 Discontinuous Circulation Due to Salinity Changes Caused by Fluctuating Evaporation

In observing the early stages of the Atlantic Ocean, Arthur and Natland (1979) suggested that more saline water may be produced on the shelves due to increased rates of evaporation. This water would eventually sink to the bottom of deeper seas. Therefore, they assumed a climatically induced circulation pattern which corresponds to the intensity of evaporation.

Depending on the depth of the CCD, gray carbonate/black marl or green/black shale successions might be deposited. The lithologic boundaries of the black sediments are sharp to bioturbated (Fig. 4f,g,h), the individual cycles lasting about 10 to 100 ka with thicknesses of several decimeters.

5.2.2 Discontinuous Circulation Due to Temperature Changes

Two situations are discussed in the literature.

1. A rapid change from a warm, dry climate to a cooler humid one was found to be documented by Tethyan black shales by Weissert et al. (1979). Barremian black shales, which are intercalated with carbonates deposited in the Tethyan Ocean, may have been produced during cooling periods. In this case, one can observe 16 to 67 black shale episodes (0.2–0.5 m thick) within about 6 Ma (6–12 m

carbonate). The lower and upper boundaries of the black shales are usually bioturbated (Fig. 4g).
2. A complex history of a black shale section has been described by Cita and Grignani (1982) from the eastern Mediterranean Sea (Fig. 3g). Here, black shales formed during Quaternary periods of increasing temperature, that provided a strong fresh-water influx from the Black Sea and lead to a stable thermohaline stratification. Due to sea level rise turbidity currents occurred at an increased frequency and an enlarged amount of organic matter was transported from the Nile delta into the deep sea. When climatic conditions stabilized, anoxic conditions were terminated with the development of the modern-day Mediterranean circulation pattern.

5.3 Input of Organic Matter

An increasing input of organic matter may result in black shale formation. One can distinguish: Cyclic processes with periodically increasing and decreasing surface productivity, and turbidites rich in organic matter or finely dispersed organic material transported by currents.

5.3.1 Variations in Surface Productivity

Surface productivity can basically vary superimposed on a low or a high background level. It is probable that productivity fluctuations are climatically controlled (e.g., Sarnthein and Fenner 1988). Two cases need to be distinguished.

1. Black shale formation due to an increase from low to medium surface productivity has been described by McCave (1979) from the Cretaceous Atlantic Ocean. It has been called a "long bloom cycle", because the black shale strata frequently contain layers rich in radiolarian skeletons. Depending on the depth of the CCD, black/green shale or black marl/gray limestone sequences are formed. The lithologic boundaries are gradual at the bottom ("slow blackening") and sharp at the top ("rapid return"; Fig. 5f, h). A sedimentary cycle may have an average thickness of about 25 cm and represents about 20 ka. Because of the recurrence time of these cycles, McCave (1979) assumed that the described changes might have been climatically controlled.
2. Upwelling (see also Sects. 4 and 5.1.4). Several observations have been made of upwelling conditions in open ocean environments due to diverging surface currents. The resulting high surface productivity causes an oxygen minimum. Therefore, on slopes of islands or ridges, black shales or hemipelagic, gray sequences can be episodically deposited. Usually, this type of black shale (Fig. 4g) is not very thick (10's of cm) and represents time spans of several ka (Thiede et al. 1982a).

5.3.2 Turbidites Rich in Organic Matter

In an oxygen-depleted environment where the sea floor is still settled by benthic organisms (characterized by green to gray, bioturbated sediments, and the aerobic/anaerobic interface a few centimeters below the benthic boundary layer), black shales may form if additional organic matter of either marine or terrigenous origin is brought in by turbidites. Due to an increased oxygen demand by the "surplus" organic matter, the aerobic/anaerobic interface moves upward. A sharp contact between black shales and underlying sediments is formed (Fig. 4a). Gray limestone or green mudstone/black shale alternations may also occur in this situation. The carbonate content of the host sediment is related to biogenic carbonate production and the depth of the CCD.

De Graciansky et al. (1979) described repeatedly occurring organic-rich turbidites, and calculated an average recurrence time of about 20 to 40 ka for two different units containing such layers. They assume a climatic role in the formation of these organic-rich slope sediments, for instance, the input of organic matter by rivers during wet periods. This organic matter originate mainly from terrestrial sources, as deduced from kerogen analysis. A black shale bed has been described by de Graciansky et al. (1979) to be between 20 and 50 cm thick, sometimes with intercalations of terrigenous layers (see Sect. 5.4).

Lithologically similar successions have been investigated by Dean et al. (1978, 1984). They interpret these phenomena as "pulsating inputs" of organic matter. The black shales contain plant fragments and show sharp lithologic boundaries at the top and bottom; erosional phenomena have not been described (Fig. 4c). Dean et al. (1978, 1984a) have also calculated average recurrence times of about 40 to 50 ka. Because of this duration they assume "cyclically" occurring "pulsating events" induced by climatic, e.g., wet-dry, changes on the neighboring land mass.

5.4 Input of Other Sediment Constituents

Sequences containing black shale beds which result from the input of allochthonous siliciclastic or bioclastic sedimentary constituents are formed episodically; these particles are imported into the black shale environment by turbidity currents. The lower contact of each turbidite is sharp, while the upper contact may be sharp, gradational or on occasion bioturbated (Fig. 4b).

Turbidity currents can supplement other bed-forming processes, such as the cyclic or episodic input of organic matter (e.g., Arthur et al. 1984a). In these cases the intercalated turbidite layers are several centimeters or decimeters in thickness.

5.5 Climatic Control on the Land

The economically most important black shales formed on land masses are known from ancient semi-arid regions (see Sect. 4.4); in lakes organic matter is mainly produced by algae. Depending on the ratio between precipitation and evaporation in a given drainage area, cycles mainly contain terrigenous material indicative of wet conditions (Fig. 3b) or cycles consist of chemical sediments indicative of dry conditions (Fig. 3a; van Houten 1964; Dean et al. 1984). A bed-scale couplet or rhythm represents about 20 ka, and the black shale stratum several ka.

5.6 Timing of Bed-Scale Processes

In general, cyclic processes predominate in the formation of black shale beds. Relative changes in sea level or changes in the oceanic circulation pattern may be due to a host of factors, such as global climatic changes (e.g., ice ages), global tectonics (e.g., sea floor spreading), and regional tectonic movements. If several of these variations are superimposed on each other, no regular recurrence time can be expected.

Climatic fluctuations of short duration (< 100 ka) are commonly well preserved, especially, in regions with a warm climate. Arid/humid cycles are particularly well documented, because both flora (protecting soil against erosion) on the neighboring land mass and hydrography are directly influenced by temperature, the water budget, and wind.

Cycles formed in semi-arid regions on land masses are very sensitive to these changes. In Mesozoic rocks in the Mediterranean region, Ripepe and Fischer (1989) and de Boer (Chap. 1.3, this Vol.) found a coincidence of black shales with times of maximum eccentricity and minimum precession. In contrast, cool/warm climatic cycles at higher latitudes appear to be formed less regularly.

In summary, the duration of bed-scale cyclic fluctuations varies between 10 ka and about 1 Ma. The majority have periods on the order of 10, 20, 40 to 50, and about 100 ka, mostly corresponding to periodic climatic changes. The periodicity is closely related to variations in the orbital elements of the earth (see Einsele and Ricken Chap. 1.1, Fischer Chap. 1.2, both this Vol.). The black shale within such cycles generally represents 1/2 to 1/4 of a cycle's duration.

6 Varve-Scale Variations

Varve-scale lithologic variations represent short time spans and are documented as thin layers or laminae. Lamination is due to fluctuating deposition of at least one of the various sediment constituents. Varves form due to seasonal fluctuations, or they are generated episodically as varve-type laminae during longer time spans.

It is obvious that laminae must have a minimum thickness to be observed at all. Data from the literature show that such a threshold thickness is in the range of 0.05 to 0.1 mm (independent of the state of compaction), otherwise annual fluctuations are not documented. Consequently, annually formed laminae are only found in rapidly accumulated sediments (> 5–10 cm/ka), whereas in more slowly formed deposits laminae normally represent episodic sedimentation.

6.1 Cyclic Processes

Seasonal varves can be formed by various processes. The most important are:

1. Continuous supply of detrital material plus seasonal input of organic matter, occasionally mixed with other biogenic constituents (e.g., Calvert 1964), or the seasonal formation of algal/bacterial mats (see Gerdes et al. Chap. 5.6, this Vol.).
2. Continuous supply of organic matter superimposed by a seasonal input of detrital material (e.g., Eugster and Hardie 1978).
3. Seasonally alternating input of both organic matter and other particles (e.g., von Stackelberg 1972).
4. Seasonal fluctuations of the aerobic/anaerobic interface, while organic matter and other sediment particles are supplied continuously. This type of lamination is known from subtropical swamp areas. During rainy periods, the water level is higher, favoring less oxygen-deficient conditions, while during dry periods highly oxygen-deficient sediments are deposited (Zangerl and Richardson 1963).

6.2 Episodic Processes

The most important episodic processes are sediment input, turbid layer sedimentation, bottom current activity, and oxygenation (Fig. 6).

Sediment input of organic matter and other sediment constituents, when episodic, causes the formation of laminae. Examples include plankton blooms that provide organic matter and skeletons, or the rapid input of terrigenous material, superim-

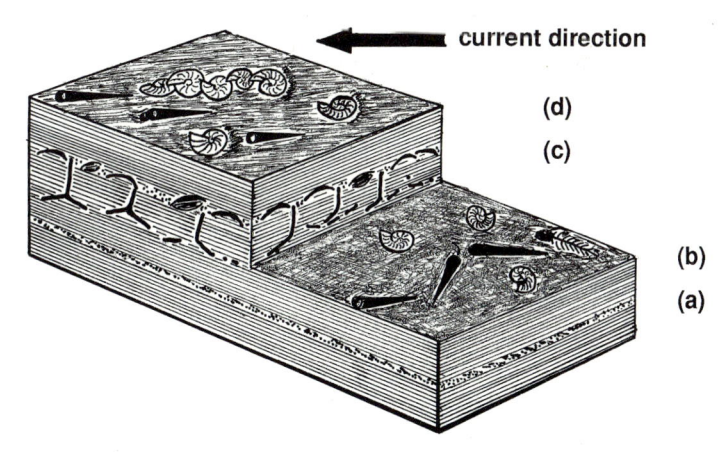

Fig. 6. Varve-scale variations documented within black shale beds. *a* Circulation event forming winnowed deposits; *b* layer of fossils due to fluctuating aerobic/anaerobic boundary; *c* circulation event combined with short-term oxygenation of the sea floor and subsequent mass mortality of benthic organisms; *d* circulation event leading to shell banding, oriented fossils, and small-scale pot and gutter casts

posed on more or less continuous background sedimentation (e.g., Robertson 1984). However, these laminae are difficult to distinguish from those formed by cyclic processes, except by less regular thickness, or their number in relation to the time span represented by the whole section.

Turbid layer sedimentation provides fine-grained particles which have been transported by low density, low velocity flows called turbid layers (Moore 1969). This sediment type often shows lamination as well as graded bedding. The laminae are laterally continuous over several decimeters only. The lower contact of the turbid layer is commonly sharp, but erosional structures can be rarely found (e.g., Stow and Piper 1984a).

Bottom current activity results in characteristic fabrics preserved in lag deposits, oriented fossils, shell stringers, banding consisting of shells, shell pavements, pot and gutter casts, ripples of coarse-grained particles, small channels and lenticular bedding, or uneven laminae sometimes associated with scour marks which result from the winnowing of fine-grained particles (e.g., Wignall 1989a). Besides sedimentary structures, bottom currents also may lead to oxygenation episodes.

Oxygenation represents fluctuations of the aerobic/anaerobic interface that may result from changes in the ratio of organic matter input to the oxygen supply. Furthermore, short-term fluctuations may be caused by bottom currents.

Layers with predominantly planktonic organisms may document a rising aerobic/anaerobic interface within the water column. When parts of the euphotic zone become anaerobic, mass mortality of the organisms can result (Berry and Wilde 1978).

In oxygen-depleted environments, a certain succession of epifaunal and infaunal organisms can be observed in relation to the oxygenation level (Savrda and Bottjer 1988 and Savrda et al. Chap. 5.2, this Vol.). Because of the "opportunistic character" of these animals, often faunas with high population density, but low diversity are found. With increasing oxygen content, a more diverse fauna colonizes the sea floor. The phenomena related to oxygenation episodes are (1) algal/bacterial mats stabilizing the sea floor thereby allowing epifaunal colonization and protecting against erosion, (2) single phase bioturbation (often dominated by Chondrites; Bromley and Ekdale 1984a), and (3) macrofossil layers. Oxygenated intervals are normally overlain by some well preserved epifaunal remains, reflecting mass mortality due to a rise of the aerobic/anaerobic interface.

7 Conclusions

Environment changes documented by black shales may be grouped with respect to their duration. Generally, four categories can be distinguished.

Mega-scale, or first order, periods of black shale formation are time envelopes which are about several 100 Ma in duration. They coincide with times of first order sea level highstands, which reflect increased global or regional tectonic activity that leads to conditions favorable for black shale deposition. Increased spreading and subduction result in the formation of additional basins; in addition, associated greater atmospheric CO_2 content from increased volcanic activity favors warming of the climate, with its consequent effect on oceanic and atmospheric circulation.

Sequence-scale black shale formation is about 1 to 10 Ma in duration. Sequence-scale black shale show a frequency distribution similar to that of the corresponding sea level fluctuations. However, sea level stand is not the primary cause of black shale formation. Instead, neighboring basins may be coupled or de-coupled and hence influenced in their circulation pattern and in the exchange of water masses of possibly differing CO_2 content or density. In this way can basins become (at least partially) anoxic for such time spans.

Bed-scale black shale occurrences last 10 to 100 ka, and they may in part represent Milankovitch-type cycles which are driven by climatic variations. The time span of black shale formation within these successions is often less than half of the cycle length, indicating that conditions favorable for black shale formation occur only during a relatively short time span during a climatic cycle.

Varve-scale laminations in black shales represent time spans shorter than 1 ka, and they are caused by various processes. Seasonal varves can be observed only in rapidly accumulated deposits (> 5 cm/ka). Consequently, most of the laminae found in black shales represent small-scale sedimentological events.

5.2 Redox-Related Benthic Events

C. E. Savrda, D. J. Bottjer, and A. Seilacher

1 Introduction

Analyses of process in event stratigraphy typically include storms, floods, gravity flows, seismic shocks and meteoritic impacts (e.g., see Seilacher and Aigner Chap. 2.3; Sepkoski et al. Chap. 2.6; Brett and Seilacher Chap. 2.5; Eberli Chap. 2.8, this Vol.). Since such physical events leave prominent and fairly distinctive sedimentological signatures, in addition to their ecological (or even evolutionary) consequences, this emphasis is well justified.

In a broader view, however, sedimentation and ecology are also controlled by factors other than turbulence, for instance water temperature, salinity and oxygenation. Like turbulence, these parameters may directly or indirectly depend on water depth, but their change in time is likely to appear gradual rather than in the form of discrete events – all the more since there will be no erosional accentuation in the sedimentary protocol.

Changes in turbulence are reflected by changing grain sizes and sedimentary structures. When it comes to oxygen deficiency, however, turbulence levels are generally so low that the sediment fails to respond in the same manner. Rather, sedimentation will be slow and vertical and the only sedimentary structures to be expected in black shales are laminae recording seasonal fluctuations in plankton production or terrigenous sediment input. Just as the origin of lamination is largely controlled by biological activity, so is its bioturbational destruction in such settings. The organisms responsible for this destruction are infaunal sediment feeders that rework the sediment thoroughly for its content in detrital food.

Since in this dual system only the bioturbators will be affected by the establishment of anoxic conditions in the lower part of the water column, the sequential transition from bioturbationally homogenized to finely laminated layers may be taken to indicate the onset of anoxia. Similarly, the redox facies cycle may be extended on the oxic side to include layers in which homogenizing bioturbation is associated with a diverse, shelly epifauna. With this basic scheme it is possible to map the sequential and lateral transition between anaerobic-dysaerobic-aerobic biofacies in the field (e.g., Rhoads and Morse 1971; Byers 1977).

This tripartite oxygen-related biofacies approach has been effective for basin analysis and recognition of long-term fluctuations in redox conditions. However, recent advances have been made that allow still higher resolution of redox histories. These include:

Einsele et al. (Eds.)
Cycles and Events in Stratigraphy
©Springer-Verlag Berlin Heidelberg 1991

1. trace-fossil models that permit the recognition and evaluation of minor cycles within the broad dysaerobic range, and
2. the recognition of special kinds of short-term benthic episodes that punctuate the generalized redox sequence and, in some cases, appear to defy traditional biofacies models.

In the present contribution, we discuss these advances, in part by review and in part through presentation of new concepts.

2 Tracking Redox Events and Cycles with Trace Fossils

Bioturbated horizons of variable thickness commonly occur within sequences of laminated, organic-rich strata. The anatomy of these horizons is highly variable and this variability dictates the specific approach with which trace fossils are employed, as well as the level of detail at which redox histories can be reconstructed. In order to better understand this phenomenon, we shall first review burrow stratigraphy in modern fine-grained sediments and summarize general ichnologic criteria.

2.1 Burrow Stratigraphy in Fine-Grained Substrates

Bioturbation in modern fine-grained substrates can be divided into three levels:

1. a surface mixed layer,
2. a transition layer, and
3. a historical layer (Ekdale et al. 1984b) (Fig. 1).

The surface mixed layer represents an interval of rapid and complete biogenic mixing and homogenization. Under static depositional conditions, the biogenic structures produced in this zone (including surface traces) are not normally preserved due to low sediment shear strengths, continual rapid biogenic stirring, and subsequent overprinting by transition-layer burrows with ongoing sediment accretion (Berger et al. 1979).

The transition layer, a zone of heterogenous mixing, is characterized by burrows produced by organisms that live or feed at greater depths in the sediment column, below the mixed layer. With sediment accretion and the associated upward migration of the actively bioturbated zone, these burrows become a part of the historical layer, a zone within which no new disruptive burrowing takes place. As a consequence, transition-layer burrows are more likely to be preserved in the stratigraphic record. Owing to vertical partitioning of organisms below the sediment-water interface, biogenic structures produced within the transition layer are typically tiered. In the rock record, the tiering structure can be inferred from cross-cutting relationships between burrows in historical-layer sediments (Wetzel 1983; Bromley and Ekdale 1986; Savrda and Bottjer 1986, 1987a).

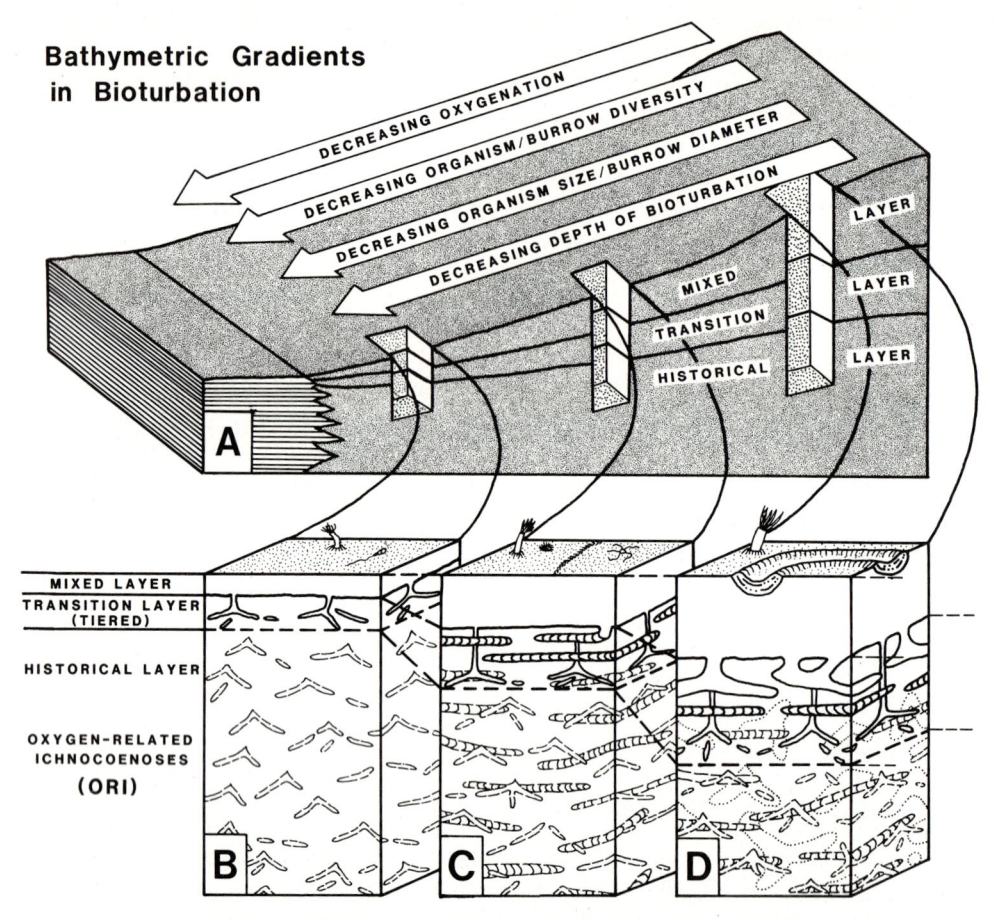

Fig. 1. A General burrow stratigraphy within fine-grained pelagic substrates and response of biogenic sedimentary structures to declining bottom-water oxygenation. Oxygen-related ichnocoenoses (**B-D**), defined on the basis of diversity, diameter, and depth of penetration of transition-layer burrows preserved in the historical layer, provide the basis for evaluating paleo-oxygenation histories for continuous record beds. (After Savrda and Bottjer 1987a)

2.2 Trace-Fossil Response to Changing Oxygenation

Although general burrow stratigraphy remains unchanged, the character of the faunas that occupy the substrate change predictably as bottom-water oxygen levels fluctuate. As bottom-water redox conditions deteriorate, infauna becomes less diverse through the progressive loss of larger organisms, which typically have higher oxygen demands. An associated rise of critical redox boundaries in the substrate also results in progressive reduction in depths of occupation and feeding of most infauna (Savrda and Bottjer 1986, 1987a).

The response of bioturbating infauna is preserved and reflected by the ichnologic characteristics. Summarily, as bottom-water oxygenation decreases, either along a lateral seafloor redox gradient or through time, the thickness of the surface mixed layer and the diversity, diameter and depth of penetration of transition-layer burrows generally decrease (Fig. 1). These trends, synthesized within a trace-fossil tiering framework, form the basis for evaluating the magnitude and relative duration of oxygenation events (Savrda and Bottjer 1986, 1987a).

The level of retrievable information regarding the redox conditions responsible for bioturbated beds varies with event-bed anatomy and degree of preservation of biogenic structures.

2.3 Anatomy of Bioturbated Beds in Black Shales

Bioturbated redox event beds contained within black shale sequences can be divided into two components; a primary stratum and a piped zone (Savrda and Bottjer 1989a). Primary strata are thoroughly bioturbated portions of individual event beds and represent sediments that, at some point in time during the associated redox episode, have passed through the surface mixed layer. Piped zones occur at the interfaces between primary strata and subjacent darker laminated strata, and are characterized by exquisitely preserved, discrete trace fossils. Burrows of the piped zones represent the emplacement of transition-layer burrows, during the earliest phases of the redox event, into dark unoxidized sediments that have *not* passed through the surface mixed layer.

On the basis of preservation, two general types of bioturbated event beds are recognized; *continuous record* and *discontinuous record* beds (Savrda and Bottjer 1989a). Both are typically characterized by well-developed piped zones. However, these bed types differ significantly in terms of degree of preservation and discreteness of transition-layer burrows in primary strata. Primary strata of continuous record beds are characterized by discrete and identifiable biogenic structures throughout, whereas those of discontinuous record beds are generally homogeneous, exhibit vague burrow mottling, and lack discrete biogenic structures. The reasons for the comparatively poor preservation of discrete trace fossils in primary strata of discontinuous record beds are presently unclear. Regardless of the causes, these differences impose different limitations and dictate the specific approach taken towards the evaluation of redox conditions.

2.4 Continuous Record Beds

In continuous record beds, relative degree of oxygenation can be evaluated on the basis of the diversity, burrow diameters, and vertical extent of discrete burrows within primary strata (Savrda and Bottjer 1986, 1987a). Taken together, these criteria permit the recognition and delineation of oxygen-related ichnocoenosis (ORI) units (Fig. 1B-D). The ORI unit concept provides the basis for comparing relative magnitudes of the oxygenation events responsible for discrete bioturbated beds within laminated anaerobic intervals. Moreover, taking the tiering structure of these trace-fossil assemblages into account, vertical changes in trace-fossil assemblages can be translated to interpreted paleo-oxygenation curves that reflect not only relative degree of paleo-oxygenation, but also the rates and magnitudes of temporal changes in oxygenation during longer redox events (Fig. 2). As such, this approach is important for demonstrating redox cyclicity, particularly for situations where oxygenation fluctuates but never moves outside of the dysaerobic range.

This approach is illustrated by an example from the Upper Cretaceous Niobrara Formation of the U.S. Western Interior (Savrda and Bottjer 1989b; in press). In addition to laminated unbioturbated strata, this unit contains bioturbated intervals characterized by four trace-fossil assemblages, whose characteristics and interrelationships indicate that they represent ORI units (Fig. 3). With the tiering structure of these assemblages taken into account, application of the trace-fossil model in cm-scale vertical sequence analysis has been employed in the construction of detailed paleo-oxygenation curves (Fig. 3).

An inverse relationship between interpreted oxygenation and organic-carbon content supports the validity of the model, but also reflects the utility of the trace-fossil approach in the evaluation of petroleum source bed potential of mudrocks (Fig. 3). Curves generated for both the Fort Hays and Smoky Hill Members of the Niobrara suggest a redox cycle periodicity of approximately 20 ka, which closely matches the precession cycle of Milankovitch (Savrda and Bottjer 1989b). The ability to evaluate magnitude and rates of temporal change in redox conditions within the dysaerobic range provides an important tool for future studies that attempt to evaluate more precisely the mechanisms responsible for rhythmically bedded sequences such as these (see Einsele and Ricken Chap. 1.1, Fischer Chap. 1.2, this Vol.).

2.5 Discontinuous Record Beds

Owing to the dearth of well-preserved, discrete trace fossils in the primary strata of discontinuous record beds, a complete record of oxygenation for redox events cannot be generated. However, some information on the duration and magnitude of these events can be derived on the basis of:

1. the thickness of primary strata, and
2. the composition, size characteristics (i.e., burrow diameter and depth of penetration), and interrelationships (vertical partitioning and cross-cutting) of ichnofossil assemblages in the piped zones.

Two general types of events can be recognized in terms of event duration (Savrda and Bottjer 1989a). *Short-term events* are oxygenation episodes that are essentially instantaneous (Fig. 4). Owing to the lack of vertical accretion of the surface mixed layer over such a short time span, primary strata are generally thin. Considering a maximum of 15 cm for the surface mixed layer in well-oxygenated uncompacted modern pelagic muds, 6–8 cm is probably a conservative estimate for maximum thickness of primary strata associated with short-term events in the rock record. However, this maximum thickness would progressively diminish for events of lower

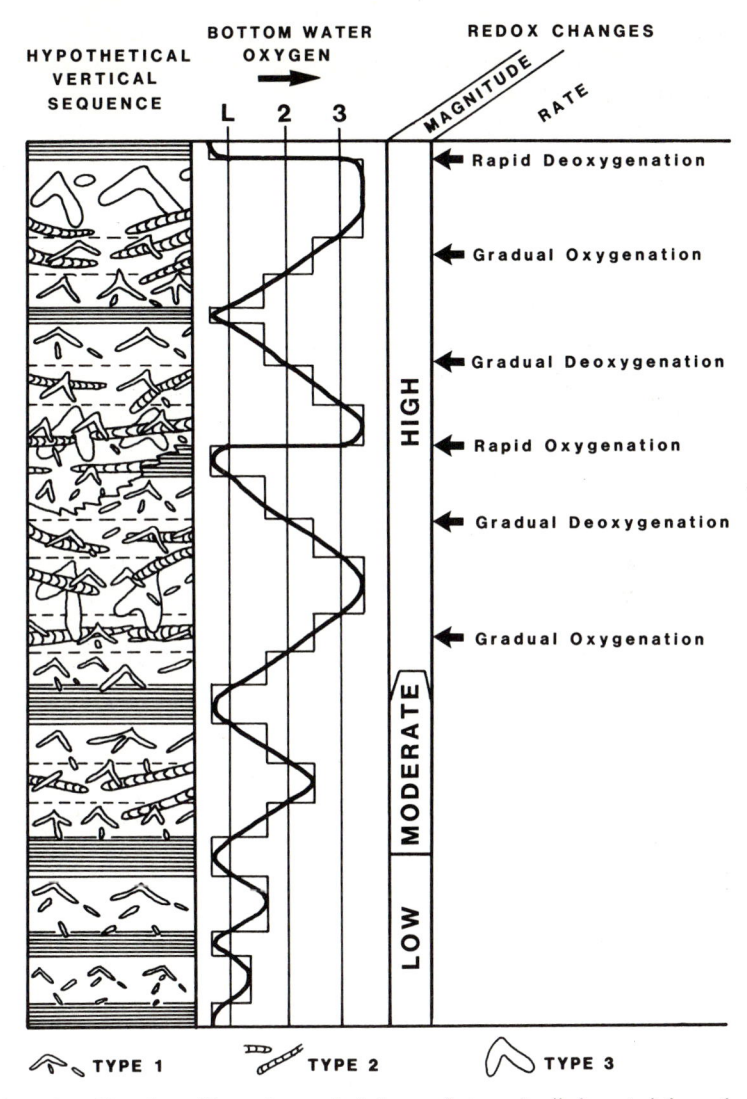

Fig. 2. Hypothetical stratigraphic column illustrating vertical changes in trace-fossil characteristics and their translation into paleo-oxygenation curves that reflect both rates and magnitudes of redox events. In this example, a transition-layer tiering structure similar to that illustrated in Fig. 1 is assumed (i.e., larger burrows occupied shallower tiers). *L* line represents oxygen levels below which lamination is preserved and above which producers of trace type 1 can survive. Reference lines *2* and *3* represent threshold levels of oxygen required for the survival of the progressively larger producers of trace types 2 and 3, respectively. Levels of oxygenation between reference lines can be inferred on the basis of maximum burrow diameter. Details of curve construction are outlined in Savrda and Bottjer (1986, 1987a)

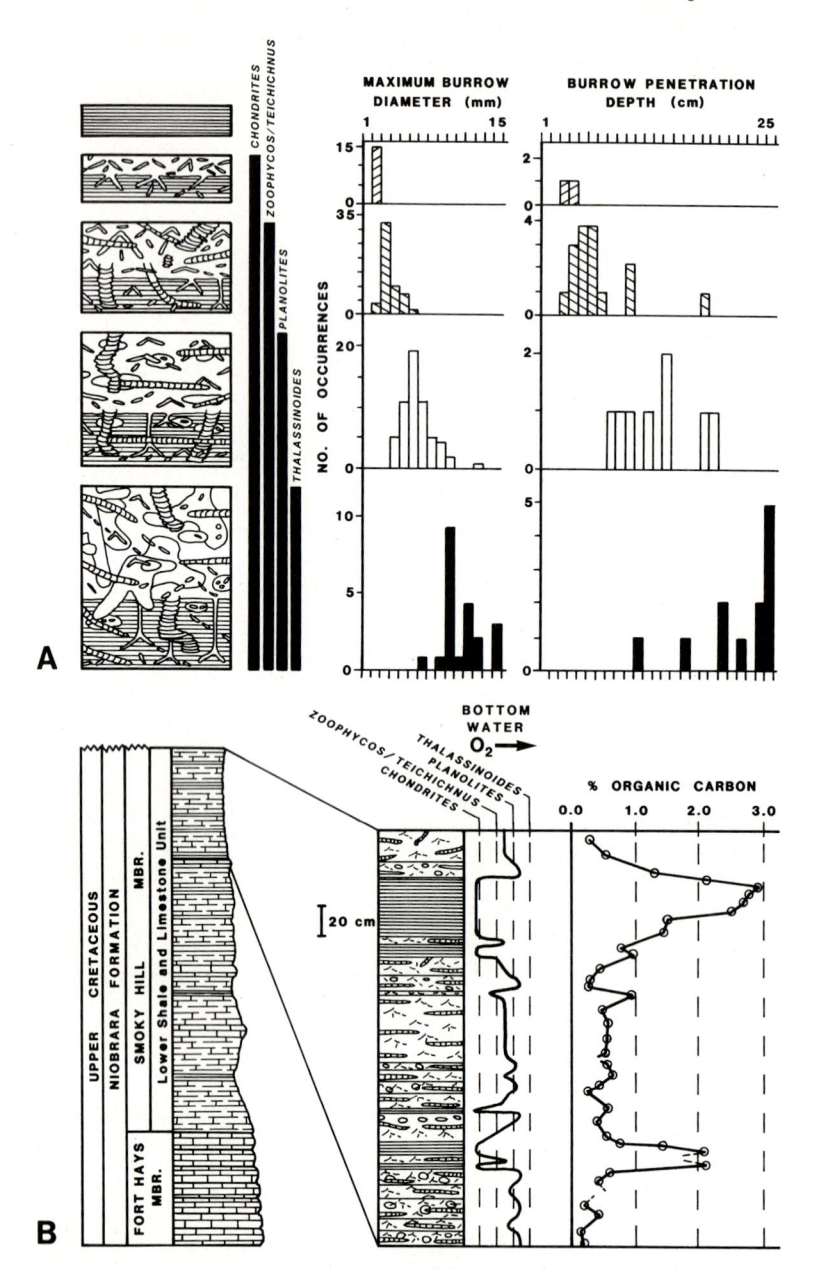

Fig. 3. A Trace fossil assemblages in the basal portion of the Upper Cretaceous Niobrara Formation (Colorado, U.S.A.). On the basis of systematic changes in burrow diversity, diameter, and depth of penetration, assemblages are regarded as oxygen-related ichnocoenoses. (After Savrda and Bottjer 1989b, in press). **B** Interpreted oxygenation curve for the upper 2.5 m of the "lower shale and limestone unit" of the Smoky Hill Member mirrored by organic-carbon trends. (After Savrda and Bottjer 1989b)

magnitude. Owing to the lack of vertical migration of a tiered transition-layer community through time, the piped zones associated with short-term events are characterized by vertical segregation of ichnogenera and little or no cross-cutting between biogenic structure types.

Extended events (Fig. 4) represent oxygenation episodes of longer duration and are reflected by generally thicker primary strata, owing to the temporal accretion of the mixed layer in pace with sedimentation. In addition, extended events are indicated by extensive cross-cutting of ichnofossils in the piped zones as a consequence of the upward migration of the tiered infauna through time.

Relative degree of oxygenation can be interpreted on the basis of the diversity, size, and depth of penetration of ichnogenera in the piped zones (Fig. 4). Progressive increases in these parameters generally reflect increasing magnitude of the associated redox event. This determination is relatively straightforward for short-term events. For extended events, these criteria are useful only for earlier phases of the oxygenation episode that are represented by transition-layer burrows within the piped zones. Changes in magnitude during these earlier phases may complicate the cross-cutting relationships of piped-zone burrows owing to associated changes in ichnofossil diversity and depth of burrow emplacement (Savrda and Bottjer 1989a).

The evaluation of redox events responsible for discontinuous record beds is illustrated by bioturbated event beds in the Posidonia Shales of Germany (Fig. 5). Primary strata in these beds differ only in thickness and are uniformly characterized by vaguely burrow-mottled, greenish-gray claystones. On the basis of the thickness of primary strata and the character of piped-zone ichnofossils, these beds reflect a spectrum in terms of duration and magnitude of associated oxygenation episodes (Savrda and Bottjer 1989a).

The majority of redox event beds in the Posidonia Shale represent short-term, low-magnitude events (Fig. 5). These are characterized by very thin primary strata (0.5–1.0 cm). Corresponding piped zones are also thin (3–4 cm) and contain a single ichnogenus, *Chondrites*. Two forms of *Chondrites*, distinguished on the basis of burrow diameter, are vertically segregated with the larger form (2–3 mm) restricted to the upper 1 to 2 cm of the piped zones and the smaller form (<2 mm) observed at greater depths (3–4 cm).

Other bioturbated beds in the Posidonia Shale, particularly the Seegrasschiefer near the base of the unit, exemplify the high end of the spectrum. These beds are characterized by thicker primary strata (16–35 cm) and piped zones (4–8 cm). The latter contain trace-fossil assemblages that are more diverse, contain larger burrows, and exhibit systematic cross-cutting relationships (Fig. 5). Primary stratum thickness and cross-cutting of piped-zone burrows demonstrate that these are extended events. Higher diversity, larger burrow diameters, and greater depths of burrow emplacement are indicative of higher event magnitudes. The cross-cutting relationships of piped-zone burrows provide a means of evaluating the relative timing of burrow emplacement, which, in turn, provides the basis for interpretation of redox dynamics during the earliest phases of the redox events (Fig. 5).

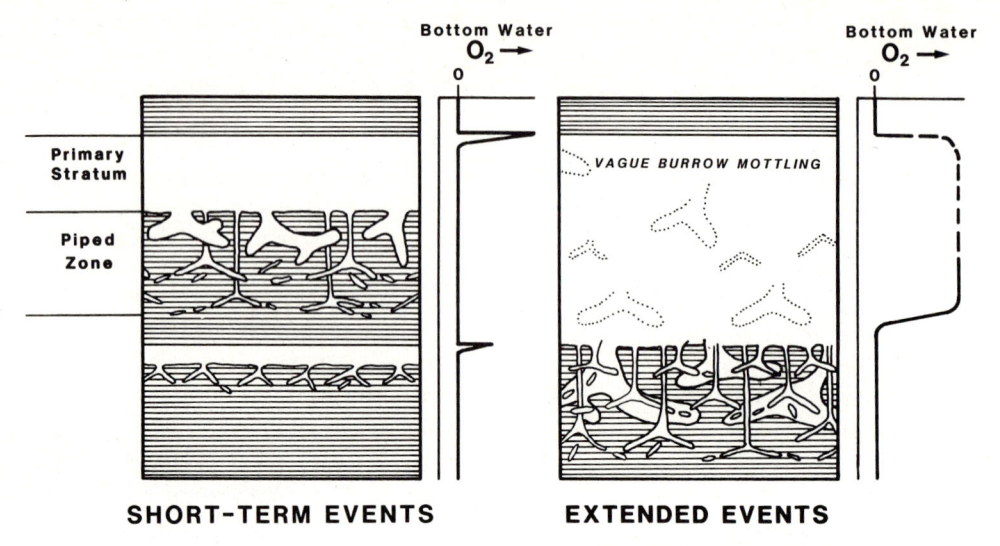

Fig. 4. Characteristics of discontinuous record beds generated by short-term and extended oxygenation events. Short-term event beds have thin primary strata and lack cross-cutting of piped-zone burrows. Extended-event beds have thicker primary strata and exhibit extensive cross-cutting of piped-zone burrows. As illustrated by the two short-term events, relative magnitude of oxygenation episodes can be inferred from size, diversity, and depth of penetration of piped-zone burrows. (After Savrda and Bottjer 1990a)

3 Short-Term Bioevents in Black Shales

The trace-fossil models outlined above are useful for evaluating redox events and cyclicity in most fine-grained strata deposited above the anaerobic threshold. It is necessary, however, to also consider two particular types of short-term bioevent horizons that commonly punctuate predominantly laminated, anaerobic sequences, and which, in some cases, appear to defy the general biofacies scheme:

1. Monospecific horizons of probing burrowers; and
2. Monospecific layers of epibenthic bivalves, brachiopods and echinoderms.

In scrutinizing these irregularities, the Jurassic Posidonia Shales will be used as a standard example and functional morphology will serve as the principle approach.

3.1 Prober Horizons

Sediment feeders of various affiliations, but mostly worm-like creatures, have evolved specific behaviors that allow them to efficiently mine the nutritious substrate. These behaviors are reflected in the patterns of the burrows left behind and are the basis for distinguishing a great variety of trace fossils, the producers of which remain largely unknown. Characteristics of feeding burrows include the tendency:

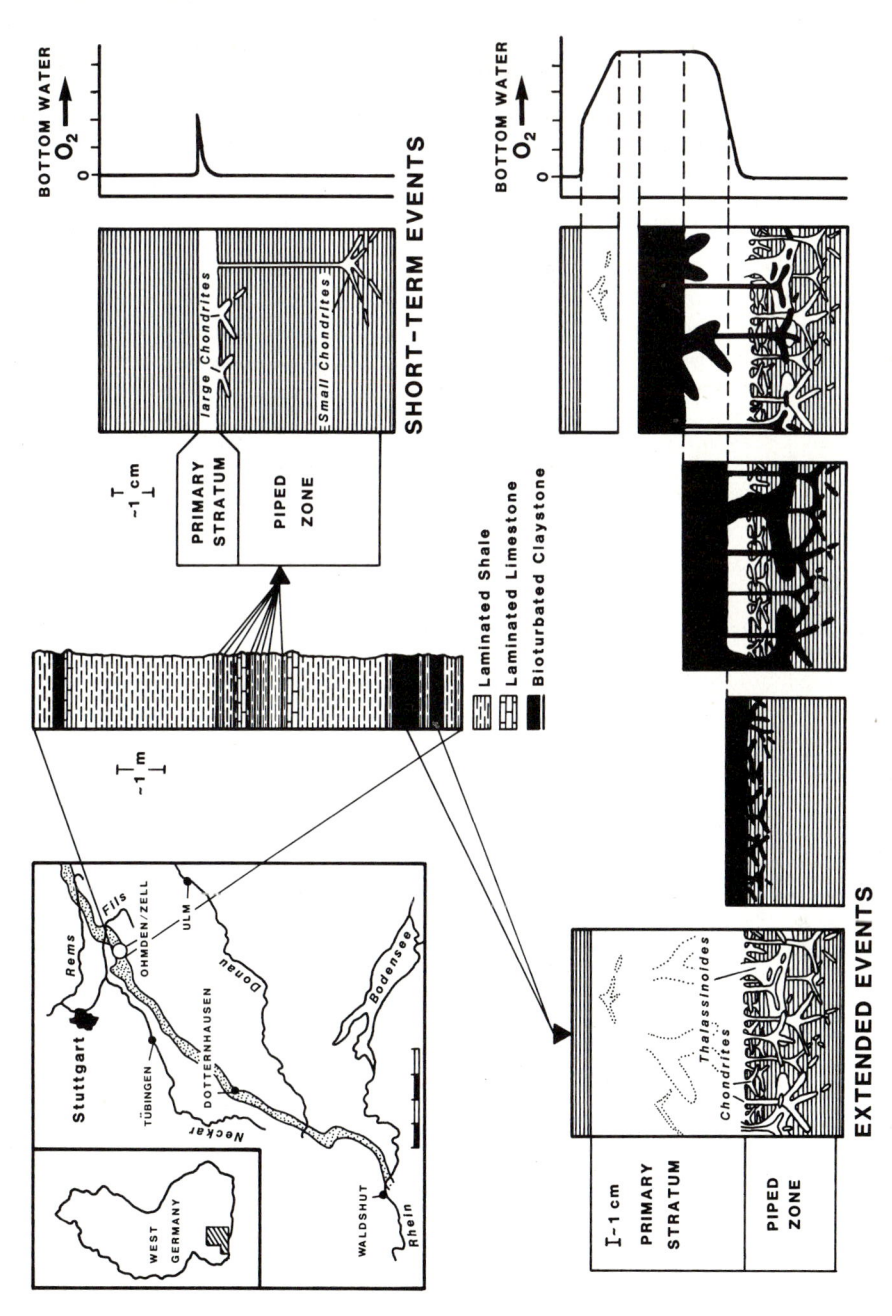

Fig. 5. Short-term and extended oxygenation event beds from the Posidonia Shale in the Ohmden/Zell area of southern Germany. Interpreted sequence of piped-zone burrow emplacement for bioturbated beds of the Seegrasschiefer near base of section exemplifies the evaluation of oxygenation history for the earliest phases of extended events responsible. (After Savrda and Bottjer 1989a)

1. to utilize a given bedding surface or sediment volume with a minimum of space left unused,
2. to do this with a minimum of double coverage, and
3. to regularly backfill the reworked sediment in accordance with the progress of the mining operation.

Burrow systems of *Nereites, Zoophycos,* or *Lophoctenium* may serve as examples.

Chondrites, the chief component of black shale burrow horizons and common in shallow marine as well as deep-sea environments, shows some of these characteristics. Its branching burrow systems cover a given surface very evenly and avoid collision with other branches of its own kind (but not of crustacean-made *Thalassinoides*) by phobotaxis. However, *Chondrites* fails to conform with the paradigm of a feeding burrow on two important counts:

1. Branches are too widely spaced to exploit a given space efficiently. This is the same argument that led to the recognition of graphoglyptid burrows as "mushroom gardens" (agrichnia).
2. In contrast to graphoglyptid tunnels, *Chondrites* burrows are actively backfilled, but in a way that would not allow feeding and backfilling to go on at the same time. Also, the backfill may contain sediment from the surface, which led Osgood (1970) to believe that the burrows were left open and became filled, after having been abandoned, by sediment passively falling in from above.

Further analysis reveals the probability that *Chondrites* is probably a heterogeneous group. In the Posidonia Shales, for example, typical *Chondrites* is represented by the smaller ichnospecies *Ch. bollensis*, characterized by smooth walls and regularly spaced branches. Also, the phobotactic interaction reveals that at all levels the axial part was made first, after which side branches were added in a retrograde succession. The larger species *"Ch." granulatus* shows a similar branching pattern, but the walls are granulated by ovoid fecal pellets. These pellets have their axes oriented transverse to the burrows and probably constitute the whole backfill. In addition, branches commonly show a palmate arrangement, with the basal parts intersecting almost as in a "spreite" structure. These features suggest that *"Ch." granulatus* is the work of a true sediment feeder and should properly be placed in a different ichnogenus.

In addition to the functional discrepancies, there is also the strange preference of *Chondrites* for low-oxygen conditions. Not only does it reach deeper into the reducing sediment than associated burrowers of its size, but it is also the last of the recognizable trace fossils to disappear when conditions become oxygen deficient. Thus, it is well justified to call *Chondrites* a low-oxygen indicator (Bromley and Ekdale 1984a), but it remains an open question as to why this preference should exist. The second kind of bioevent commonly observed in black shale sequences provides some important clues towards a solution.

3.2 Epibenthic Horizons

The small bivalves *Bositra* and *Steinmannia* (formerly called "Posidonia") have for a long time been a paleoecological riddle, because they occur in parts of the Posidonia

Shale section where perfect lamination would indicate anaerobic bottom conditions. To escape this dilemma, it has been suggested that these and other exotic bivalves actually lived in the oxygenated upper part of the water column and, like ammonites, reached the bottom only after death. This required that they were either permanently swimming (Jefferies and Minton 1965; a life form unknown from adult bivalves today) or lived pseudoplanktonically attached to floating objects. The issue was further confused by the fact that *Bositra* and *Steinmannia* are commonly found together with *Inoceramus dubius*, for which an epipelagic life style is documented by its regular association with driftwood, and in rare cases also with ammonites (Seilacher 1982a,b). Accordingly, *I. dubius* could die and drop to the bottom irrespective of the redox conditions in the lower part of the water column, while its wooden float would most likely end up at a beach and be deprived of fossil preservation.

Bositra and *Steinmannia*, in contrast, were true epibenthos. Not only are they never found attached to driftwood or ammonite shells; they are also restricted to distinct laminae, on which they appear as a nonpatchy monospecific pavement except for the occasional admixture of "airborne" *Inoceramus* and ammonite shells. It is also noteworthy that the majority of the *Bositra* and *Steinmannia* shells are still articulated in an open butterfly position, which excludes accumulation by bottom transport or local winnowing.

3.3 Biomats

In modern anoxic basins, the final living things to survive under decreasing oxygen concentrations are bacteria that tend to form a consistent scum on top of the sediment (see Gerdes et al. Chap. 5.6, this Vol.). In black shales their presence is difficult to prove because cellular structures are not preserved. Nor do laminae split readily enough to show the wrinkle patterns known from the Solnhofen Limestones (Seilacher et al. 1985b). However, there is indirect evidence in the observation that individual stacks of laminae can be laterally correlated like tree rings in dendrochronology over many miles. This would hardly be possible without a microbial film that protected the sediment against erosion during the occasional currents recorded by shell alignment (Brenner 1976). Therefore, the lamination is assumed to be biologically enhanced.

3.4 Chemosymbiosis

Having established the bacterial connection, another microbiological issue, namely that of chemosymbiosis, can be approached. The phenomenon was first observed in deep-sea vent habitats, where various and unusually large tube worms and bivalves gain their livelihood from an association with autolithotrophic bacteria that can oxidize the H_2S from the vents (e.g., Felbeck 1983; Hand and Somero 1983; Jannasch 1984a). Subsequently, it was recognized that bacterial chemosymbiosis is found also in other habitats where H_2S occurs in close proximity to sufficiently oxygenated water, such as in organic-rich sea floors near the outlets of paper mills (Reid and Brand 1986) or in eelgrass beds (Fisher and Hand 1984), whose root systems provide a seal on the sediments similar to that of bacterial mats in more stagnant environments. Given the incomplete state of reconnaissance, chemosymbiosis can be assumed to be a widespread trophic strategy that has been independently adopted by many groups of animals throughout earth history. Since this study is based mainly on post-Paleozoic examples, brachiopods and other groups that occupied this niche in earlier times are omitted from further analysis, while concentration is centered on bivalves and chondritid probers.

3.4.1 Bivalves

All present-day bivalves known to be chemosymbiotic house the bacteria within their gills, where they can be flushed with O_2 and H_2S in the traditional bivalve fashion, although the two substances should be spatially or temporally separated to avoid direct interaction (Reid and Brand 1986). In most chemosymbiotic bivalves, shell morphology remains rather unaffected. Not even the gigantism of some vent bivalves

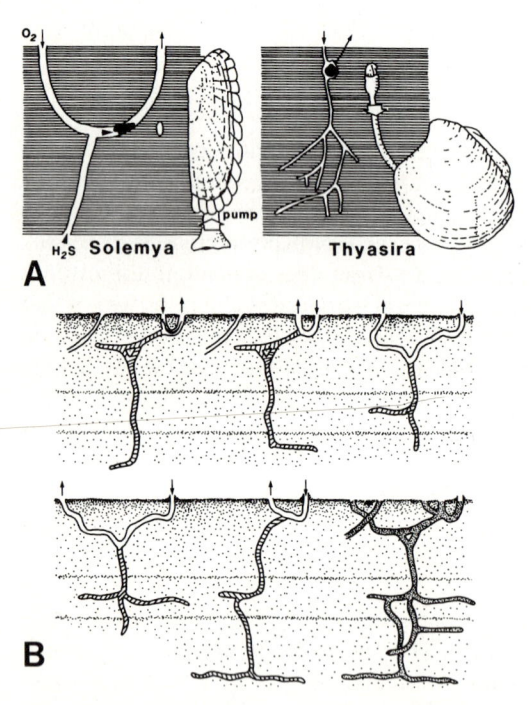

Fig. 6. A Characteristics of chemosymbiotic bivalves *Solemya* and *Thyasira* and their burrows. Extremities of burrows penetrating below the organisms are believed to represent sulfide well systems. Morphology of *Thyasira* burrow networks is reminiscent of *Chondrites*. (Seilacher 1990a). **B** Similar deviated well system observed in the annelid *Nereis diversicolor*. (Seilacher 1953). Burrow branches changed and disappeared during several days of observation; but the worm spent most of the time in U-shaped upper tier of its burrow, probably using its mucus net to sieve from the respiration current

is observed in other H_2S habitats. There are, however, in some fossil genera a number of deviations from standard bivalve shell morphology that would best be explained by the "strange attractor" or bacterial endosymbiosis (Fig. 6).

Unfortunately, none of the Jurassic "Posidonias" and few of the monotids that occur in similar situations in the Triassic show such distinctive features. Their generalized shapes are not even indicative of a particular life position with regard to the sediment. Nevertheless, we may, from biostratinomic evidence alone, place them in this guild.

3.4.2 Burrow Systems

More telling than shell morphologies are the burrows that some chemosymbiotic bivalves produce. *Solemya*, for instance, may have a simple U-shaped tube with a characteristic ovoid cross section. In some modern and fossil species, however, they regularly have an additional tube (Fig. 6; Stanley 1970, Plate 3) that reaches from the bottom of the U steeply into deeper, more reduced sediment layers. This extension is probably used as an H_2S well, with the expanded foot acting as a piston.

The foot of the lucinid *Thyasira* has a similarly expanded tip, but as a whole it has become worm-like and may extend for many times the shell diameter (Fig. 6). This foot is used to clean and mucus-line the inhaling tube (because lucinids significantly have an exhaling but no inhaling siphon), but more important for H_2S production are the deep wells that it makes in the sediment below the animal. This well system resembles the trace fossil *Chondrites* both by penetrating well below the local level of bioturbation and by its root-like mode of branching (Fig. 6). These branches probably do not act at the same time, because this would dissipate the pumping force of the foot. Rather, they most likely are used one at a time.

While it remains to be tested whether the *Thyasira* well system is made with the same retrogressive program as in *Chondrites* and whether the abandoned branches are also inactivated by backfilling, the functional model is intriguing. It explains why *Chondrites* branches are more widely spaced than is desirable for a sediment feeder. It also resolves the dilemmas of backfill volume versus stomach capacity and the introduction of surface sediment that have been problematic for the sediment feeder model.

Another analog of *Chondrites* may be seen in *Nereis diversicolor*, a common worm in the dark muds of the German Wadden Sea. Its burrows appear to be rather irregular. Observation through several days, however, revealed a consistent behavior pattern (Fig. 6). The animal spent most of the time in the upper few centimeters of the sediment within an irregular U-shaped burrow where it uses a mucus net (like *Chaetopterus* and *Urechis*) to strain food from the respiration current. But in addition, it regularly produced deep probes that later disappeared – possibly by active backfilling – and in the protocol combined into a root-like system. Pending the test for bacteria, we may tentatively assume that this species also has a subsidiary chemosymbiotic mode of nutrition.

This means that fossil *Chondrites* may be the product of very different infaunal animals that convergently used a "deviated well" system to extract the fuel for their endosymbiotic bacteria (H$_2$S or methane) from an organic-rich reduced sediment. Whether this organism is a bivalve or a worm – the rules of the game are the same. Because these rules are similar for cases in which bacteria are farmed within the tunnel system, but outside the body (graphoglytids and various corkscrew burrows), it would also be convenient to group sulfide wells in the ecologic category of farming burrows (agrichnia).

4 The Exaerobic Window and the Revised Redox Facies Sequence

4.1 The Exaerobic Zone and Biofacies

Monospecific layers of still articulated benthic shells, particularly brachiopods and bivalves, such as the "Posidonias" described herein, occur in black shales throughout the Phanerozoic. Comparable layers of *Thyasira* shells have also been found in the organic-rich muds of modern fjords (Dando et al. 1985). Although the fine lamination of these sediments would indicate anaerobic conditions, the presence of what are believed to be in situ epibenthic macroinvertebrates defies the traditional view of the anaerobic biofacies. For this reason the exaerobic biofacies was proposed to encompass these intervals or layers (Savrda and Bottjer 1987b). The exaerobic zone and resulting biofacies were originally thought to develop at a critically low bottom-water oxygen level along continuous oxygen gradients laterally across the seafloor (i.e., between dysaerobic and anaerobic zones). However, there may be instances where *vertical* redox gradients across the sediment-water interface are periodically steep enough so that the critical redox boundary is perched directly at the sediment-water interface in a manner that precludes the establishment of burrowing infauna, regardless of bottom-water oxygen levels (see also Sageman et al. Chap. 5.3, this Vol.). In either case, exaerobic faunas reflect either windows along redox gradients or relatively higher-oxygen pulses in otherwise stagnant basins, and may be indicative of chemosymbiosis or of bacteria grazing.

4.2 Refined Biofacies Sequence

Incorporation of the exaerobic biofacies concept and trace-fossil trends with traditional models provides an expanded and refined picture of the oxygen-related biofacies sequence, which, in turn, can facilitate the recognition and evaluation of fluctuations in paleo-oxygenation expressed either laterally or vertically in stratigraphic sections (Fig. 7).

4.2.1 Aerobic Phase

Fully aerated mud bottoms typically show an upper layer inhabited by infaunal suspension feeders (such as burrowing clams), but also by a variety of sediment feeders (largely worms) that exploit the substrate in a tiered fashion. As sedimenta-

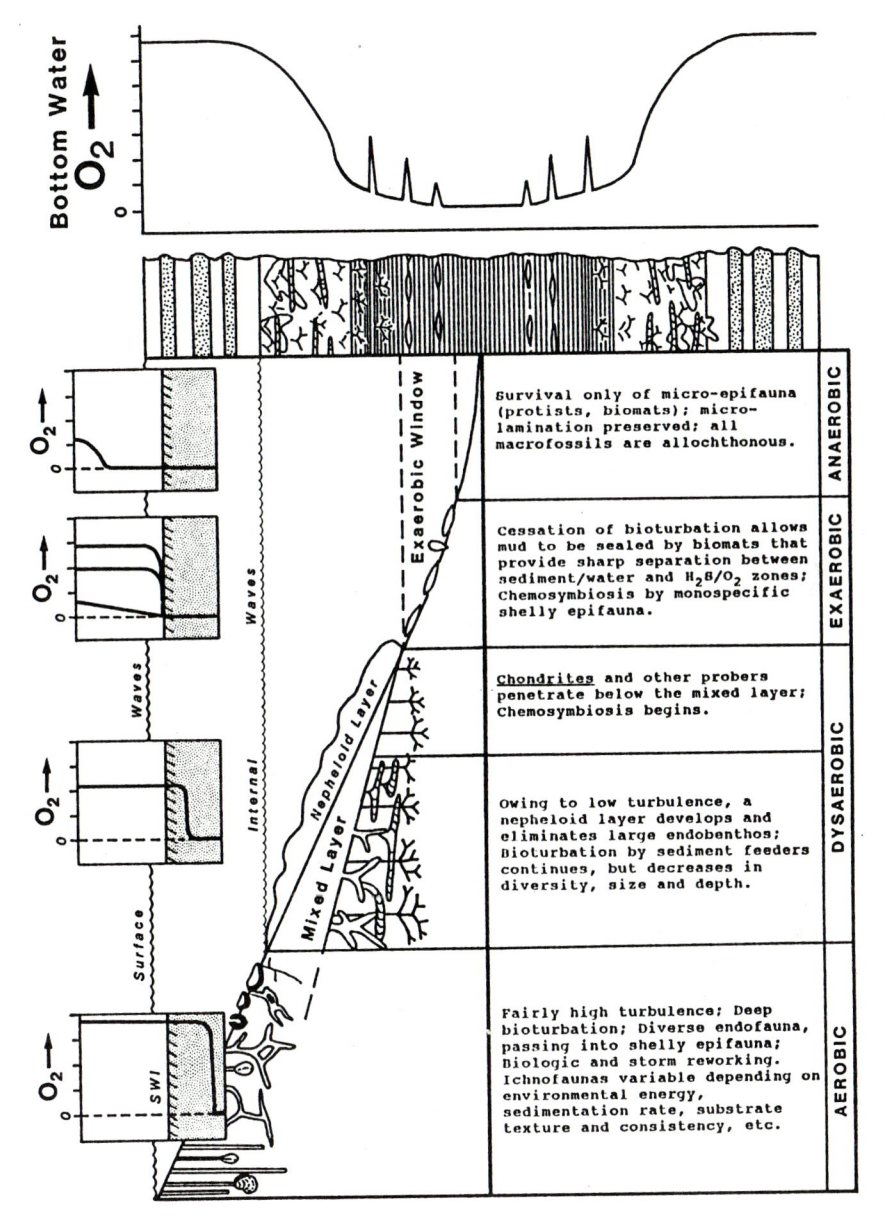

Fig. 7. Refined oxygen-related biofacies scheme summarizing criteria employed for evaluation of long- and short-term redox events in vertical sequence analyses. Model considers dynamic vertical oxygen gradients across the sediment-water interface (SWI), as well as lateral oxygen gradients along the sea floor

tion continues, deeper tiers of burrows will mix with the upper ones, as revealed by cross-cutting relationships. Only in post-event situations (turbidites, tempestites) may the tiering be preserved, because sedimentation was too rapid during the event for bioturbation to develop, with the exception of sparse escape burrows.

4.2.2 Dysaerobic Phase

In quieter water, the activities of the infauna will produce a fluffy layer of fecal material on top of the sediment and eventually a nepheloid layer in the water above. This eventually leads to the disappearance of the shelly fauna (particularly the reclining suspension feeders), so that bioturbation by soft-bodied sediment feeders remains as the only benthic record. Along this gradient, the mixed and transition layers will become thinner as bioturbation structures decrease in size and penetrate less deeply into the sediment.

It is in the lower, less oxygenated part of the dysaerobic zone that this trend may become reversed, in part, by the appearance of chemosymbiotic interactions. As exemplified by *Chondrites, Solemya,* and possibly some crustacean burrows, sulfide pumpers may penetrate deeper. As oxygen levels decrease still further, this other type of bioturbation will also decrease in size and penetration depth before it finally disappears.

Since deep sulfide probing depends not only on the oxygen level in the water column, but also on the availability of an organic-rich substratum, we might also expect some asymmetry in downward versus upward gradients. Probing horizons should be most pronounced in parts of sections indicating deposition that was moving away from anoxia, because in this case probers could exploit the organic-rich deposits of previous stages. *Chondrites* horizons in the Posidonia Shales conform to this expectation.

4.2.3 Exaerobic Phase

As the critical redox boundary rises to the sediment-water interface and bioturbation ceases, biomats can develop on the mud surface. This new biological regime has a double effect. Firstly, the biomat sealing inhibits the development of the nepheloid layer. Secondly, it may allow the H_2S zone to reach right to the top of the sediment. Both effects, in turn, open the window for the members of the exaerobic zone, which may recruit either from epifaunal or endobyssate bivalves and brachiopods (i.e., from suspension feeders that developed bacterial chemosymbiosis without the need to probe deeply for their sulfide fuel). This setting may also include specialized small grazers, such as ostracodes, foraminifera and small echinoids, that are able to stand low oxygen levels and feed directly on the microbial film.

4.2.4 Anaerobic Phase

The exaerobic window closes when anoxia exceeds tolerance levels, which may be different for different species. Therefore, it is not surprising to find benthic microorganisms such as foraminifera in laminated strata, beyond the limits of the exaerobic macrofauna. For practical purposes, the term "exaerobic zone" is limited to the macrofaunal event, although similar interactions may actually extend into the lower dysaerobic and upper anaerobic zones.

5 Conclusions

The approaches and concepts reviewed herein provide new ways of viewing sequences of mudrocks that were deposited under variable conditions of oxygenation. There is, however, a need for continued development and testing of these models. Applications of the trace-fossil approach to paleo-redox reconstructions have been successful, but have been restricted thus far to a limited number of stratigraphic packages that are Jurassic or younger (e.g., Miocene Monterey Formation, Cretaceous Niobrara Formation, and Jurassic Posidonia Shale) in age. Each of these units, deposited in distinctly different oxygen-deficient settings, is characterized by its own set of ichnofossil associations, which respond differently to changes in paleo-oxygenation. This variability provides the incentive for a more global survey of bioturbated event beds contained within stratigraphic sequences deposited in a greater variety of oxygen-deficient settings of a more complete age range. This global approach will provide the larger data base required to fully establish the utility and limitations of trace fossils as tools for assessing oxygenation events and cycles for the whole of the Phanerozoic.

The models for short-term bioevents and chemosymbiosis suggested here are still speculative and need to be tested in modern environments as well as in microstratigraphic sequences of different ages. Such a survey would tell us what kinds of animals have occupied the exaerobic niche through earth history and what were the adaptive strategies employed.

In addition to its biohistoric and evolutionary interest, continued research will sharpen a new tool in the differentiation of redox facies and of basins that have become the sites of hydrocarbon source rocks.

5.3 Biofacies Models for Oxygen-Deficient Facies in Epicontinental Seas: Tool for Paleoenvironmental Analysis

B. B. Sageman, P. B. Wignall, and E. G. Kauffman

1 Introduction

Oxygen-deficient facies in epicontinental seas comprise a spectrum of normally dark-colored, organic carbon- and pyrite-enriched rocks which range from extremely well-laminated oil shales and bituminous shales to various types of marls, mudstones, and limestones (Morris 1980; Potter et al. 1980; Hallam 1987). These facies dominate the sedimentary record of many epicontinental basins formed during periods of relative sea level rise and highstand (Kauffman 1984). Most organic-rich sequences in epicontinental seas have been interpreted as being representative of long-term stagnation with anoxic conditions predominant in all but the uppermost wave-mixed layers of the water column – the stagnant basin model (e.g., Seilacher 1982c). This view of black shales has traditionally discouraged extensive research in these monotonous, apparently fossil-poor sequences. However, in recent years increased interest in the study of organic-rich facies has produced diverse evidence of short-term benthic colonization events, and, in some cases, longer-term resident communities, suggesting the need for a more dynamic view of oxygen-deficient environments (Aigner 1980; Kauffman 1981; Savrda and Bottjer 1986; Wignall and Myers 1988; Brett et al. 1989; Sageman 1989; Wignall 1989b; see also Savrda et al. Chap. 5.2, this Vol.). Black shales are, in fact, highly dominated by evidence of short-term oxygenation and sedimentation events which constitute recognizable physical, chemical, and biological event units. Consequently, these facies offer a promising objective for studies in event stratigraphy.

    The focus of many paleoecological studies in organic-rich facies has been on the development of black shale biofacies models that describe characteristic associations of organisms, sediment fabric, and projected benthic oxygen levels (e.g., Rhoads and Morse 1971). In this chapter we discuss the development of, and revisions to, the original Rhoads and Morse model (Fig. 1), and propose a new, more comprehensive biofacies model based on a broader consideration of data from modern and ancient oxygen-deficient environments. Such biofacies models are powerful tools of paleoenvironmental analysis because biologic communities and their component species are the most sensitive indicators of environmental parameters in modern and ancient environments. Ultimately our goals in the development of a more comprehensive biofacies model are to: (1) reconstruct black shale paleoenvironments in terms of diverse physical, chemical and biological characteristics of the benthic zone; (2) describe the nature of benthic communities and their ecologic strategies that allow

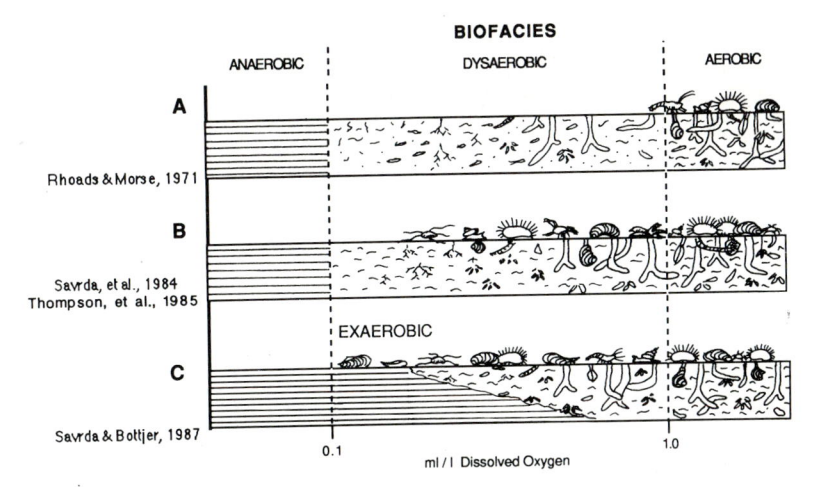

Fig. 1. A The Rhoads and Morse (1971) biofacies model for oxygen-deficient environments includes anaerobic, dysaerobic, and aerobic biofacies. Characteristic sediment fabric and fauna are illustrated within each division. Revisions of the Rhoads and Morse (1971) model include: **B** the models of Savrda et al. (1984) and Thompson et al. (1985) in which new dysaerobic faunas are recognized; and **C** the model of Savrda and Bottjer (1987a) in which a new subunit of dysaerobic biofacies, the "exaerobic zone," is proposed. Anaerobic and aerobic biofacies remain as in the Rhoads and Morse (1971) model

survival in these relatively inhospitable environments; and (3) develop a working methodology for environmental analysis of organic-rich facies that is sensitive to the dynamic short-term oxygenation and sedimentation events which characterized ancient epicontinental oxygen-deficient environments.

1.1 Black Shale Biofacies Models

Black shale biofacies models (Fig. 1) typically follow the Rhoads and Morse (1971) scheme describing organism-sediment relationships according to specific levels of benthic oxygen (as ml/l oxygen in sea water). This scheme was developed from analysis of biotic distributions in certain modern stable-stratified basins (e.g., the Black Sea) and has had a strong influence on the interpretation of ancient oxygen-deficient environments (e.g., the stagnant basin model). Three distinct biofacies were originally characterized by Rhoads and Morse (1971) as *anaerobic* with laminated sediments and no macrofauna (oxygen <0.1 ml/l, or effectively anoxic), *dysaerobic* with slightly burrowed sediments and a sparse soft-bodied infauna (oxygen 0.1 to 1.0 ml/l, or dysoxic), and *aerobic* with bioturbated sediments and abundant and diverse faunas (oxygen >1.0 ml/l, or oxic) (Fig. 1A). Similar or related biofacies have been described from black shale deposits by Byers (1977), Morris (1980), Savrda et al. (1984), Thompson et al. (1985), Savrda and Bottjer (1986, 1987b), Hallam (1987), Wignall and Myers (1988), Sageman (1989) and Savrda et al. (Chap. 5.2, this Vol.).

Most revisions of the original Rhoads and Morse (1971) model have focused on the character and biota of dysaerobic biofacies (Fig. 1B,C). Current research on modern oxygen-deficient environments has shown that abundant shelly and soft-bodied infauna and epifauna can survive at oxygen levels as low as 0.3 ml/l (Savrda et al. 1984; Thompson et al. 1985), significantly altering the biotic and sedimentologic character described by Rhoads and Morse (1971) for the dysaerobic zone (Fig. 1B). Thus, dysaerobic biofacies, under normal circumstances, would be characterized by bioturbated sediments, making the distinction between aerobic and dysaerobic biofacies based on sediment fabric quite difficult in the rock record (Savrda and Bottjer 1986).

The use of body fossils to reconstruct oxygen-deficient paleoenvironments, representing an alternate approach to this problem, has focused the interest of black shale workers on an interesting debate concerning dysaerobic biofacies. The occurrence of numerous extremely well-laminated black shales containing fossil assemblages of in situ epifaunal organisms (Cluff 1980; Kauffman 1981; Boardman et al. 1984; Kammer et al. 1986; Brett et al. 1989; Sageman 1989) has suggested the need for further revision of accepted models. Typically, such fossil assemblages have been interpreted as allochthonous pseudoplankton or otherwise transported debris, since benthic colonization would be unlikely in a stagnant basin scenario (Seilacher 1982c; Boardman et al. 1984). Yet increasing evidence that the bulk of such fossils in Phanerozoic black shales are actually in situ suggests the need for re-evaluation of the pseudoplanktonic interpretation (Wignall and Simms 1989). Arguments against a pseudoplanktonic mode of life include:

1. the great abundance of fossils compared with relative scarcity of suspected pseudoplanktonic floats (e.g., logs);
2. restriction of supposed pseudoplankton to certain black shale intervals, and absence of these same taxa in other time-equivalent or successive facies; a relatively even stratigraphic distribution of pseudoplankton would be expected in most oxygen-deficient epicontinental basins;
3. size-frequency data suggesting benthic colonization events and subsequent mass mortality of juveniles belonging to a single generation whereas a more normal population distribution would be expected among pseudoplankton;
4. taphonomic evidence of in situ occurrence;
5. association of normally benthic epifaunal species (suspected pseudoplankton) with free-living epifaunal and infaunal taxa which are functionally incapable of attachment to floating sites;
6. alternation of beds containing predominantly infauna or predominantly epifauna, suggesting ecologic exclusion among benthic taxa (trophic amensalism: Rhoads and Young 1970); and
7. relatively rare pseudoplanktonic encrustation of potential hosts in modern and ancient marine environments (including floating logs, seaweed, cephalopods, etc.) (Kauffman 1975, 1981; Boston et al. 1988; Brett et al. 1989; Sageman 1989; Wignall and Simms 1989).

In an effort to account for the in situ occurrence of epifaunal body fossils in well-laminated organic-rich facies (otherwise considered to be anaerobic biofacies on the basis of sediment fabric), a new subzone of the basic Rhoads and Morse (1971) model was introduced by Savrda and Bottjer (1987a). The *exaerobic zone* represents a narrow "window" in which shelly epifauna colonize substrate surfaces during transitions between anoxic and dysoxic oxygen levels (Fig. 1C). Savrda and Bottjer (1987; see also Savrda et al. Chap. 5.2, this Vol.) suggested that exaerobic faunas were able to survive at low levels of oxygen (normally associated with high H_2S) due to a symbiotic relationship with chemosynthetic sulfide-oxidizing bacteria, a strategy recently discovered in molluscs and tube worms of deep sea hydrothermal vent communities (e.g., Cavanaugh 1985). Thus, the exaerobic concept has focused attention on new possibilities in the ecology of ancient oxygen-deficient faunas (i.e., chemosymbiosis). However, the fact that most modern chemosymbiont-bearing organisms are equilibrium or K-selected species with large lecithotrophic larvae suggests they are best adapted to stable environmental conditions (i.e., a relatively constant supply of H_2S). Yet ancient shallow epicontinental basins were probably more dynamic, with frequent perturbations (such as storms) causing breakdown and reestablishment of oxygen-deficient water layers; r-selected opportunists would be more successful under such conditions (Wignall 1989a). Thus, chemosymbiosis, though possibly a facultative strategy in oxygen-deficient environments, is not likely to have been the dominant adaptation of all shelly taxa found in ancient well-laminated sediments.

1.2 Development of a New Biofacies Model

In order to develop a comprehensive black shale biofacies model for epicontinental environments, several factors must be addressed:

1. A more complete review of organism-sediment relationships in organic-rich facies is needed that includes the total range of modern and ancient oxygen-deficient environments. In the following discussion (Sect. 2.1), a literature review of modern oxygen-deficient environments encompasses data representing open ocean oxygen minimum zones, silled basins, fjords, nearshore reducing sediments, stratified lakes, and deep sea hydrothermal vents and cold seeps. These data are compared with evidence from numerous high-resolution studies of black shale sequences ranging from Devonian to Cretaceous in age (Sect. 2.2). The recognition of common trends in modern and ancient examples forms the basis for development of a biofacies model that is more finely tuned to the conditions of shallow epicontinental seas.
2. In response to recent discoveries that the benthic environment may not vary uniformly above and below the sediment-water interface with respect to bottom water oxygenation (e.g., Jorgensen and Revsbech 1985), the first part of our proposed biofacies model (Sect. 3.1) describes variations in the nature and position of the redox boundary and predicts the biotas and sediment fabrics associated with different boundary positions. Two principal types of oxygenation gradients related to the different boundary positions are identified in the review of data from modern and ancient oxygen-deficient environments.
3. Typically, biofacies models for oxygen-deficient environments describe the distribution of benthic communities according to the chemical limitations of low

oxygen and H₂S. However, the biologic characteristics of the organisms and benthic communities that manage to survive in these environments are of equal or greater importance to an understanding of benthic ecology in low-oxygen settings. These include special anatomical or physiological adaptations of living taxa to low-oxygen conditions, the ability of opportunistic species to colonize rapidly on or in substrates during brief oxygenation events, substrate preferences and adaptations among low-oxygen adapted taxa, and the trophic strategies which largely control the distribution of benthic communities based on nutrient availability. These factors shape the ecologic structure of oxygen-deficient communities and are discussed in the development of the new biofacies model in Section 3.2.

4. With the increasing recognition that epicontinental oxygen-deficient environments may be highly dynamic on ecologically significant time scales (<1 to 10 years) has come the need for a biofacies model that accounts for short-term changes in benthic communities. In Section 4 a methodology for the application of the new biofacies model to black shale sequences is given which allows for the recognition and interpretation of both short-term and long-term events in ancient oxygen-deficient environments.

2 Faunal Trends in Modern and Ancient Oxygen-Deficient Environments

2.1 Modern Oxygen-Deficient Environments

Modern oxygen-deficient environments can be classified into two general categories (Type 1 and 2) based on the nature and position of the redox boundary and its affect on benthic communities. In *Type 1 Environments* (Fig. 2), the uppermost sediment layer (approx. 0.5 m) and much of the overlying water column are dysoxic, and can become anoxic, due to the consumption of oxygen through bacterial decay of descending organic matter, and to decreased oxygen supply through stratification of water masses (Demaison and Moore 1980). The depth of water mass mixing and intensity of oxygen depletion depends on basin geometry and the climate of the surrounding land masses (Demaison and Moore 1980). Examples of Type 1 environments include silled or blind basins (Black Sea, Baltic Sea, and boreal fjords), continental borderland basins (e.g., California, Venezuela), areas where oceanic oxygen minimum zones impinge on the continental slope or shelf (California, Peru), and upwelling zones (e.g., Peru, West Africa).

In *Type 2 Environments* (Fig. 2), the anoxic zone is sharply delineated at or near the sediment-water interface; higher oxygen levels characterize the overlying water masses, which are commonly dysoxic but can be well oxygenated, creating a sharp redox discontinuity at the sediment-water interface. In some cases this sharp boundary results from the rapid uptake of oxygen by bacterial decay at the surface of organic-rich substrates and is accompanied by the presence of a diffusive boundary layer which may or may not be associated with a biological boundary (e.g., microbial mats; Jorgensen and Revsbech 1985). Such conditions occur in lacustrine, intertidal, and subtidal environments where organic-rich, anoxic, H₂S-saturated muds or pol-

	FAUNA	DIVERSITY	TROPHIC DOMINANCE
TYPE 1 *SUBSTRATE AND BOTTOM WATER OXYGENATED*	Sulphur Bacteria and bacterial mats Benthic Foraminifera *Nematodes *Polychaetes & Oligochaetes *Infaunal Bivalves Crustaceans Echinoderms Other Molluscs Other Annelids Other Arthropods Brachiopods Coelenterates Bryozoa	Moderate to Very High (10 to >300 species)	*Deposit Feeders Scavengers Suspension Feeders
TYPE 2 *SUBSTRATE ANAEROBIC, BOTTOM WATER OXYGENATED*	Sulphur Bacteria and Bacterial mats Nematodes *Bivalves *Pogonophorans Oligochaetes Crustaceans Other Arthropods Polychaetes Gastropods	Low (1 to 10 species)	*Chemosymbiosis Bacterial "Farmers" Suspension Feeders Scavengers

Fig. 2. Modern oxygen-deficient environments are summarized in terms of faunal content, diversity, and trophic strategies represented in benthic communities. Two general types of oxygen-deficient environment are recognized: *Type 1* environments are characterized by oxic to dysoxic substrates and bottom waters whereas *Type 2* environments have anoxic substrates and oxic to dysoxic bottom waters (see oxygen gradient models, Fig. 4). Faunal groups and trophic strategies that are marked with a *star* represent the dominant (most abundant or conspicuous) types represented in benthic communities. Diversity was estimated from published species lists of modern oxygen-deficient environments (see text for references)

luted sediments are found (Jorgensen and Revsbech 1985; Thompson and Ferris 1988; Rosenberg 1977). A related example comes from the deep sea where anoxic, H_2S-enriched waters issue from hot or warm vents and cold seeps which support specially adapted communities of chemosymbiont-bearing organisms (Paull et al. 1984; Cavanaugh 1985).

2.1.1 Faunal Assemblages of Type 1 Environments

Data on the faunal characteristics of Type 1 oxygen-deficient environments reveal a surprisingly high level of benthic community diversity and biomass at bottom water oxygen levels between 0.1 and 1.0 ml/l (Fig. 2). The predominant faunal and microbial elements include anaerobic and low-oxygen adapted microbes (e.g., the sulfur bacteria *Oscillatoria* and *Thioploca*), benthic foraminifera, nematode and polychaete worms, and infaunal protobranch bivalves (Hartman and Barnard 1959; Hulsemann and Emery 1961; Bacescu 1963; Calvert 1964; Leppakoski 1969; Theede 1973; Nichols 1976; Jumars 1976; Gallardo 1977; Smith and Hamilton 1983; Savrda et al. 1984; Thompson et al. 1985). In addition, a wide range of taxa comprise less abundant elements in these communities, including arthropods, echinoderms, coelenterates, and others (Fig. 2). Total diversity can reach ca. 300 species in Type 1 environments (Jumars 1976).

The highest species diversity and biomass are among polychaete worms (up to 150 species in California borderland basins: Hartman and Barnard 1958), which bioturbate the sediment at oxygen levels as low as 0.3 ml/l (Thompson et al. 1985). In many cases, nuculoid and other infaunal bivalves are also important in bioturbating the sediment. In Type 1 environments infaunal deposit feeding is the dominant trophic strategy. It is significant that, with the exception of scattered occurrences of tiny, thin-shelled Pectinidae like *Cyclopecten* and *Delectopecten* (Hartman and Barnard 1958; Calvert 1964; Smith and Hamilton 1983), epifaunal suspension feeding organisms are relatively rare in Type 1 environments. This reflects ecological exclusion (trophic amensalism) resulting from alteration of the substrate to a soupy, thixotropic condition and the production of turbid bottom waters which clog suspension-feeding apparati (Rhoads and Young 1970).

2.1.2 Faunal Assemblages of Type 2 Environments

Type 2 assemblages (Fig. 2) are characteristic of environments in which physical, chemical, or biological conditions create a sharp boundary at or near the sediment-water interface between anoxic sediments and a dysoxic to oxic water column. Infaunal organisms are unable to colonize the sediment either because it is impenetrable (relatively rigid or overgrown), or chemically hostile (low oxygen, high H_2S). Sulfur bacterial mats are common and thrive at the anoxic/dysoxic boundary (Spies and Davis 1979; Jannasch 1984; Tuttle 1985); in some cases nematodes have been found within the bacterial mats (Spies and Davis 1979). The most conspicuous Type 2 macrofaunas, which occur with normally lethal quantities of H_2S, consist of specially adapted Bivalvia (five families: Lucinidae, Thyasiridae, Solemyidae, Vesicomyidae, and Mytilidae) and annelids (Pogonophora, and possibly some Oligochaeta) (Cavanaugh 1985). All of these organisms contain sulfur-oxidizing bacteria in the gills, mantle, and other tissues; some lack a well-developed digestive system and may derive nutrition wholly from endobacteria (Cavanaugh 1985; Dando and Southward 1986). In toxic mud substrates the chemosymbiont bivalves are derived from infaunal suspension feeding groups (lucinoids, thyasirids, and solemyids). Deep sea hydrothermal vent and cold seep substrates are typically hard and inhabited by bivalves from both epifaunal (e.g., Mytilidae) and normally infaunal suspension feeding groups which have adapted to an epifaunal habit (e.g., Vesicomyidae). Survival in this habitat may be due more to the action of H_2S-binding enzymes in the host organisms than to the unique mode of nutrition (Somero 1984; Cavanaugh 1985). Chemosymbiosis, in which organic compounds for host nutrition are produced by the endobacteria via the oxidation of sulfide, may be the dominant trophic strategy. However, some organisms may also feed directly on the bacteria by "farming" gill or mantle tissue for symbionts, or by ingesting bacteria-coated organic fragments and digesting the bacteria, then excreting the organic matrix back into or onto the sediment where it will once again collect a bacterial population (e.g., Lucinidae, Thyasiridae).

2.2 Ancient Epicontinental Oxygen-Deficient Environments

Fossil assemblages in epicontinental black shales suggest that epeiric seas were typically characterized by warm water temperatures, potentially variable salinities, frequent stratification of water masses, low levels of water circulation, and reduced oxygen levels (most were dysaerobic seas). Commonly these conditions resulted in exceptional fossil preservation, including articulated skeletons and soft parts. Excellent examples of epicontinental black shales are known from the Paleozoic Appalachian Basin and midcontinent region of North America (Devonian, Mississippian, Pennsylvanian), as well as from the Mesozoic basins of northern Europe (Jurassic, Cretaceous) and from the Western Interior Basin of North America (Cretaceous). Two types of fossil assemblages are identifiable in these black shales, following the pattern recognized among modern oxygen-deficient communities (Fig. 3). These assemblages are commonly associated with two types of facies: "dark gray mudstones" and "laminated black shales" (Boardman et al. 1984; Brett et al. 1989). Dark gray mudstone facies are thought to indicate dysoxic conditions in and above the substrate, resembling Type 1 environments, whereas laminated black shales are typically interpreted as representative of anoxic conditions at, and sometimes above, the sediment-water interface, similar to Type 2 environments (Byers 1977; Boardman et al. 1984).

2.2.1 Fossil Assemblages of Type 1 Paleoenvironments

Fossil assemblages from Paleozoic dark-gray mudstones strongly resemble the Type 1 assemblages of modern oxygen-deficient environments (Fig. 3). Infaunal soft-bodied deposit feeders are represented by distinct ichnotaxa, such as *Chondrites* (Ch), *Planolites* (Pl), *Zoophycus* (Zo), and small nonspecific horizontal feeding burrows; Byers 1977; Cluff 1980). Body fossils include small bivalves (nuculoids and other protobranchs), ammonoids, archeogastropods, articulate and inarticulate brachiopods (especially linguloids and orbiculoids), and a few other taxa representing most major phyla (Boardman et al. 1984; Kammer et al. 1986). Faunal diversity is moderate to high in these assemblages and deposit feeders numerically dominate benthic assemblages (Fig. 3).

A similar pattern is found in many Mesozoic black shales of Jurassic and Cretaceous age (Fig. 3). Agglutinated and calcareous benthic foraminifera are common constituents of Type 1 facies (Hart and Bigg 1981; Eicher and Diner 1985; Riegraf 1985). Common trace fossils are *Planolites* (Pl), *Chondrites* (Ch), *Teichichnus* (Te), *Trichichnus* (Tr), *Thalassinoides* (Th), and *Rhizocorallium* (Rh) (Rhoads et al. 1972; Byers 1979; Morris 1980; Hart and Bigg 1981; Savrda and Bottjer 1986; Pratt et al. 1985). Typical body fossils include infaunal filter-feeding and detritus-feeding bivalves (e.g., Nuculidae, Nuculanidae, Lucinidae, Thyasiridae, Tellinidae,

		FAUNA	DIVERSITY	TROPHIC DOMINANCE
M E S O Z O I C	1	*Benthic Foraminifera Ostracods *Burrows (Pl, Ch, Te, Tr, Th, Rh) *Infaunal Bivalves Nuculoids Lucinoids Linguloid Brachiopods Gastropods Other Molluscs	Moderate to High (10 to >50 species)	*Deposit Feeders Suspension Feeders
	2	Bacterial Mats *Bivalves "Flat Clams" Pteriids, Ostreids, and Pectinids Brachiopods Other Molluscs Arthropods Echinoderms "Other Infauna"	Low (1 to 10 species)	*Suspension Feeders Scavengers Deposit Feeders
P A L E O Z O I C	1	*Burrows (Pl,Ch,Zo) *Small Molluscs Infaunal Bivalves Gastropods Ammonoids *Brachiopods Arthropods Other Molluscs Porifera Coelenterata Echinodermata	Moderate to High (10 to >50 species)	*Deposit Feeders Scavengers Suspension Feeders
	2	*Bivalves "Flat Clams" Pectinids *Brachiopods Other Molluscs	Low (1 to 10 species)	Suspension Feeders

Fig. 3. Ancient oxygen-deficient environments are summarized as in Fig. 2. Compilations of data from Mesozoic and Paleozoic organic-rich strata are presented separately. *Type 1* and *Type 2* environments (representing oxic to dysoxic substrate and bottom water, and anoxic substrate with oxic to dysoxic bottom water, respectively) are identified for each era on the basis of similarities of fossil assemblages to modern Type 1 and Type 2 faunas (see Fig. 2). Faunal groups and trophic strategies marked with a *star* represent the dominant types represented in fossil assemblages. Diversity was estimated from published descriptions of black shale fossil assemblages (see text for references and abbreviations of burrow types)

Corbulidae, etc.), linguloid brachiopods, aphorraid gastropods, pelagic and epibenthic ammonites, and other molluscs (Waage 1964; Duff 1975; Rhoads et al. 1972; Morris 1980; Kauffman 1984; Wignall and Myers 1988). Diversity is moderate to high in most cases and deposit feeding is the dominant trophic strategy.

2.2.2 Fossil Assemblages of Type 2 Paleoenvironments

In Devonian and Carboniferous black shales, in situ benthic fossil assemblages in well-laminated sediments suggest the presence of Type 2 environments. These fossil assemblages are typically dominated by small, thin-shelled epifaunal bivalves or brachiopods (Fig. 3). The bivalves represent groups of "flat clams" (Kauffman

1988b) including *Dunbarella* and *Canyella* (Boardman et al. 1984; Kammer et al. 1986) that may have evolved special adaptations for survival in low-oxygen environments, e.g., thin, broad shells for habitation on soft substrates, and expanded gill and mantle tissues for greater oxygen absorption and/or housing of chemosymbiont bacteria. Brachiopods in well-laminated strata include small, thin-shelled *Leiorhyncus* (Thompson and Newton 1987). Small pectinid bivalves and a few other molluscs comprise the remainder of these benthic associations; ammonoids and fish scales or bones are the most conspicuous residue of nektonic communities. Diversity is generally low in these Paleozoic Type 2 assemblages and suspension-feeding is the dominant trophic strategy represented (Fig. 3).

In Mesozoic black shales, evidence for Type 2 environments is even more developed (Fig. 3). Maceral studies of laminated Jurassic and Cretaceous black shales have revealed fossil remains of bacteria resembling modern *Oscillatoria* (Glikson and Taylor 1985; Loh et al. 1986), a filamentous sulfur bacterium which forms microbial mats in many modern Type 2 environments. Macrofossil assemblages are characterized by abundant epifaunal bivalves and rare inarticulate brachiopods. Flat clams, including the Inoceramidae, Monotidae, and Posidoniidae, are numerically dominant in facies that are otherwise commonly barren. They typically occur in background assemblages as well as discrete event communities, and form shell islands for colonization by other taxa unable to survive on the toxic sediment-water interface (Kauffman 1981). Shell island epibionts include pteriid, pectinid, gervilliid, plicatulid, and ostreid bivalves, orbiculoid brachiopods, serpulid worms, gooseneck barnacles, and other increasingly diverse taxa as bottom water oxygen levels increase (Kauffman 1981; Sageman 1989). Small "paper pectens" are also common and may occur in great abundance as event assemblages on single bedding planes, suggesting brief opportunistic colonization events in response to changes in bottom water oxygen or the nature of the substrate (Sageman 1989). Other fossils include various molluscs, arthropods, linguloid brachiopods, echinoderms, and scattered infaunal elements which appear to increase in abundance as benthic conditions improve (Fig. 3). Overall diversity is typically low in Mesozoic Type 2 intervals, and suspension feeding is the dominant trophic strategy.

3 A Comprehensive Black Shale Biofacies Model

3.1 Summary of Biofacies Characteristics from Modern and Ancient Oxygen-Deficient Environments

Fossil assemblages from ancient organic-rich facies parallel the trend observed in modern oxygen-deficient environments. Poorly laminated dark gray mudrocks in black shale sequences contain assemblages dominated by infaunal deposit feeders with relatively high diversity at the species, genus and family level, representing a "Type 1 biofacies." In contrast, proven resident taxa of well-laminated black shales are dominated by assemblages of epifaunal suspension feeding organisms with relatively low diversity. Yet many fossils in these "Type 2 biofacies" occur in large populations on single bedding planes, suggesting brief events of benthic colonization by opportunistic species. Although these two biofacies types may appear to represent disparate environments by analogy with modern examples, they may occur interbedded within closely spaced stratigraphic intervals. Further, faunal elements of Type 2 biofacies are commonly found in low abundance in Type 1 intervals, and vice versa, suggesting that these biofacies actually represent end members in a spectrum of oxygen-deficient paleocommunities.

Modern studies illustrate that neither infaunal nor epifaunal taxa are ecologically limited by dysaerobic oxygen levels (e.g., Savrda et al. 1984; Thompson et al. 1985), suggesting that different biofacies may not simply reflect varying bottom water oxygen content but also the ability of taxa to colonize (in or on) the substrate. The presence or absence of a boundary condition at or near the sediment-water interface (due to physical, chemical, or biological factors) would strongly influence the nature of the oxygenation gradient, and thus the predominance of one or the other of the biofacies types described above (Benthic Boundary Biofacies concept: Sageman 1989).

3.2 Benthic Boundaries: Oxygenation Gradients at the Sediment-Water Interface

Based on Type 1 and Type 2 faunal assemblages, two main types of oxygenation gradients can be identified as end members for the biofacies model (Fig. 4). The biota and sediment fabric of Type 1 environments indicate that oxic to dysoxic conditions

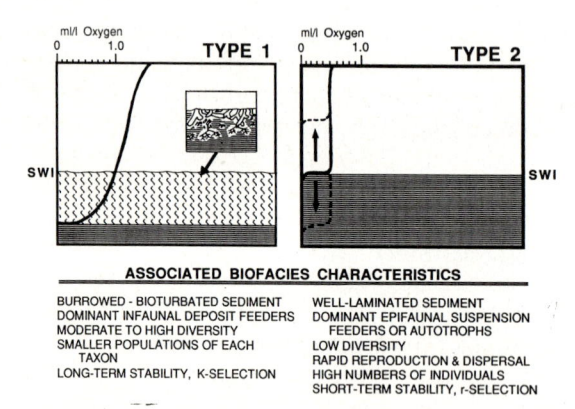

Fig. 4. Models for benthic oxygenation gradients illustrate the difference between Type 1 and Type 2 environments. The *Type 1* model shows a gradual decline in oxygen levels with the redox boundary located at some depth within the substrate (boundary depth varies with nature and extent of burrowing). The zone above the boundary (including sediment and bottom water) may be characterized by oxic to dysoxic oxygen levels. Yet, as illustrated by the *inset panel*, the actual nature of the redox boundary at any given time may be quite irregular as a result of burrowing patterns and aeration halos around burrows. In the *Type 2* model a sharp redox boundary occurs at the sediment-water interface (*SWI*) separating an anoxic substrate from oxic to dysoxic bottom waters. This boundary may episodically fluctuate upward, resulting in anoxic bottom waters, or downward, resulting in Type 1 conditions. Predictable biofacies characteristics associated with each oxygen gradient are also listed

characterize the bottom water and uppermost sediment layers. Sediments are typically clay-sized and quite soft due to high water content, and are extensively colonized by deposit-feeding, sediment-dwelling infauna (Fig. 4). Due to substrate habitability, sediments can be partially to completely burrowed, resulting in effective oxygenation of the burrowed zone. Below this zone (which varies in thickness depending on bottom water oxygen levels: see Savrda and Bottjer 1986; Savrda et al. Chap. 5.2, this Vol.), oxygen disappears rapidly and H_2S is present at the upper boundary of the sulfate reduction zone, marking the position of the redox boundary.

Taxa and sediment fabric in Type 2 environments reflect oxic to dysoxic bottom waters with oxygen levels declining rapidly to zero in the vicinity of the sediment-water interface (Fig. 4). The upper limit of the sulfate reducing zone may be very close to, or coincident with this interface. Thus, the sediment is anoxic, rich in H_2S, and normally uninhabitable except for anaerobic bacteria. The lack of infaunal colonization results in good preservation of primary sedimentary lamination and prevents downward mixing of oxygen. Specially adapted, low-oxygen-tolerant epifauna characterize the sediment-water interface of most Type 2 environments. Although some modern chemosymbiotic bivalves build discrete dwelling burrows in organic-rich anoxic muds (Dando and Southward 1986; Seilacher 1989), this strategy is not well documented in the fossil record. The sharp oxygen gradient associated with Type 2 environments may result from the presence of relatively firm, nonbioturbated substrates, or the establishment of biologically or chemically regulated boundary conditions (diffusive boundary layers and/or microbial mats) at the surface of organic-rich sediments. At times, anoxic conditions may extend above the sediment-water interface, causing mass mortality of epifaunal communities. The sharp redox boundary may also fluctuate below the interface, allowing brief episodes of shallow infaunal colonization.

The biological and sedimentological factors that characterize these two types of oxygenation gradients allow the recognition of patterns applicable to the development of a general biofacies model (Fig. 4). In Type 1 environments the substrate is habitable. Associated biofacies are characterized by: burrowed to bioturbated sediment fabric; numerical dominance of infaunal deposit feeders; small species populations, high levels of interspecific diversity, and extensive niche partitioning (e.g., burrow tiering: Savrda and Bottjer 1986; see also Savrda et al. Chap. 5.2, this Vol.) suggesting long-term environmental stability (10's to 100's of years). This favors K-selected or equilibrium species such as nuculoid bivalves and polychaetes. In contrast, Type 2 environments are characterized by substrates uninhabitable for infauna reflecting an abrupt chemical boundary at the sediment-water interface. Associated biofacies are characterized by: a predominance of epifaunal suspension feeding taxa (including possible chemosymbionts) occurring on

1. single bedding planes in well-laminated sediments; or
2. distributed throughout well-laminated shales in mono- to paucispecific assemblages with large species populations. Evidence for repeated episodes of colonization followed by high juvenile mass mortality suggests frequent environmental fluctuations (e.g., migrations of the redox boundary above the sediment-water interface). This favors r-selected opportunistic species with rapid reproduction and dispersal mechanisms.

3.3 Ecological Aspects of the Biofacies Model

3.3.1 Adaptive Strategies and the Tolerance Gradient

Specific low-oxygen adaptations that have been documented among modern eurytopic benthic marine taxa include:

1. shell closure and metabolic retardation which allows some bivalves and polychaetes to survive for up to a month under anoxic, H_2S-enriched conditions;
2. chemosymbiosis as a feeding and metabolic strategy and the development of sulfide-binding enzymes in H_2S-enriched environments; and
3. physiological or morphological modifications to enhance oxygen absorption (Theede 1973; Rosenberg 1977; Somero 1984; Cavanaugh 1985). Most taxa in modern oxygen-deficient environments are small in size and live for only a few years, yet have high stress tolerances and good reproductive potential, making them well adapted for survival under physically and chemically stressful conditions (Hartman and Barnard 1958; Rosenberg 1977). In fact, modern descriptions of short-term changes in benthic oxygen levels show a pattern of mass mortality and rapid recolonization (within 1 year) following oxygen depletion events (Leppakoski 1969; Rosenberg 1977).

The ecological framework of the biofacies model (Fig. 5) is based on a categorization of organisms according to their adaptive strategies for survival under conditions of low oxygen and high H_2S. These categories (arrows, Fig. 5) reflect ranges of habitation organized in a stepwise manner, along a gradient of tolerance levels inferred from modern and ancient faunas. An idealized sequencing of organisms through the transition from the anoxic to oxic zones is represented by the letters A through F (Fig. 5). This ecological sequence can be recognized in the case of either Type 1 or Type 2 oxygenation gradients. However, different taxa sets occupy the respective adaptive categories in Type 1 vs. Type 2 biofacies as a result of different substrate preferences, trophic strategies, life habits, and tolerance levels. The following ecologic groups are recognized (Fig. 5).

Group A. Microbial elements include sulfur- and methane-metabolizing anaerobic bacteria, as well as algae, fungi, benthic foraminifera, ostracodes, and all other microbial residents of the oxygen-deficient benthic zone. Survival at anoxic to lowermost dysoxic oxygen levels is a function of chemoautolithotrophic metabolism in certain groups (Jannasch 1984a). The low physiological requirements associated with small size might play a role in tolerance to dysoxic conditions among other types of microbiota (e.g., foraminifera). In Type 1 biofacies, sulfur bacteria are generally restricted to the area below the bioturbated zone, whereas they may form an important part of Type 2 communities as substrate mats. In contrast, benthic foraminifera and ostracodes are rare in Type 2 settings unless oxygen levels increase or substrate conditions change, resulting in a breakdown of the sharp redox boundary. However, foraminifera and ostracodes may occur commonly in lower dysaerobic biofacies of Type 1 settings (Hulsemann and Emergy 1961).

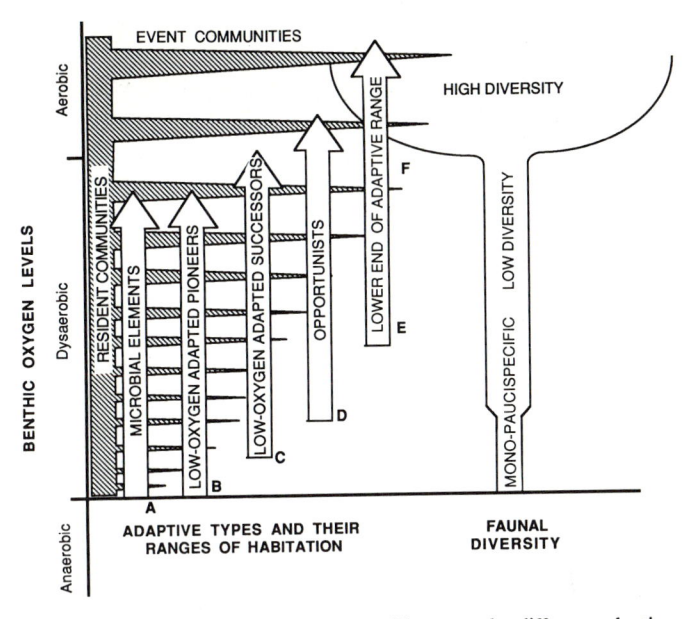

Fig. 5. An ecological classification of oxygen-deficient benthic faunas illustrates the different adaptive types that inhabit oxygen-depleted environments, and their ranges of habitation relative to established biofacies zones. Associated faunal diversity of benthic communities is also illustrated. Biofacies levels, defined by an ascending sequence of the lower end points of habitation ranges, and a corresponding sequence of increasing diversity in benthic communities, are represented by the letters *A-F*. The shaded area indicates that organisms of each adaptive category may occur in black shale biofacies as resident communities or event communities, depending on the nature and timing of changes in the benthic environment. Event communities increase in diversity and duration of habitation as oxygen levels increase

Group B. Low-oxygen-adapted pioneers are soft-bodied metazoans or shelly organisms that are specially adapted to conditions of low oxygen and increased H_2S. Nematodes and larger soft-bodied burrowers (represented by common trace fossils in black shales; e.g., *Planolites, Chondrites, Teichichnus, Trichichnus*, etc.) typify the group in the Type 1 model, whereas epifaunal suspension feeding or potential chemosymbiotic bivalves (especially flat clams) numerically dominate Type 2 benthic boundary biofacies. Special adaptations include morphological, anatomical, and/or physiological improvement of oxygen absorption, enzymatic protection from H_2S poisoning, chemosymbiosis as a feeding strategy, and opportunistic or r-selected reproductive strategies.

Group C. Low-oxygen-adapted successors are organisms whose presence depends on initial colonization of hostile substrates by the above mentioned pioneers. Infaunal deposit-feeding bivalves appear to be less tolerant than polychaetes in Type 1 environments and may colonize following polychaete burrowing and partial aeration of the sediment. In Type 2 benthic boundary settings, pteriid, inoceramid, monotid, and ostreid bivalves, serpulid worms, and rarely other taxa colonize whole or large fragments of dead shells such as ammonite conchs, or larger whole living shells of pioneers such as inoceramid bivalves to establish "shell island" communities above the toxic sediment-water interface (Kauffman 1981). Type 2 low-oxygen adapted successors are inhibited by substrate characteristics or chemistry of the sediment-water interface, but survive when slightly elevated into more oxygenated or less toxic bottom waters, or if given a firm surface for colonization. Their presence suggests steep environmental gradients at or just above the sediment-water interface.

Group D. Low-oxygen-adapted opportunists are species which occur in low numbers in communities representing normally oxygenated environments, owing to competitive restriction. However, such taxa take advantage of highly stressed, sparsely inhabited ecospace in dysoxic environments and produce large populations during brief improvements in benthic conditions. They differ from pioneer species which are specially adapted residents of dysoxic environments. Although many taxa in Type 2 oxygen-deficient environments are opportunistic species in the sense of Levinton (1970), the opportunist category as used here refers to a specific ecologic group in the tolerance gradient (Fig. 5). Certain deposit-feeding taxa which mine specific horizons would qualify in the Type 1 model, whereas tiny, thin-shelled pectinid or inoceramid bivalves and brachiopods comprise Type 2 opportunists.

Groups E and F. Lower end of adaptive range taxa are those found in limited (E) to moderate (F) diversity near the margin of their normal tolerance limits. Although optimum living conditions are at oxic oxygen levels, these organisms (especially adults) can survive for extended periods in dysoxic environments where their occurrence may be characterized by low numbers or high variability in form. Many normal marine benthic molluscs, echinoderms, arthropods, and annelids may occur under such conditions, especially in upper dysoxic benthic environments.

Overall diversity trends shown in Fig. 5 correspond to the tolerance gradient represented by biofacies levels A through F. In addition, elements of the different ecologic groups add cumulatively to benthic resident and event assemblages as oxygen levels increase. Thus levels A and B are typically mono- to paucispecific, levels C through E are of low diversity (3–10 species), and level F marks the beginning of normal high diversity benthic assemblages (>10 species).

3.3.2 The Biofacies Model

By integrating characteristics of dominant benthic oxygenation gradients with ecologic observations from a comprehensive range of oxygen deficient environments, described above, a new biofacies model for black shales is developed (Fig. 6). The two-part division of the model reflects the end member oxygenation gradients associated with contrasting substrate conditions (Type 1 and Type 2). Biofacies levels are defined by discrete biotas and associated sediment fabrics corresponding to the ecologic categories in Fig. 5. Biofacies levels A through F represent an idealized

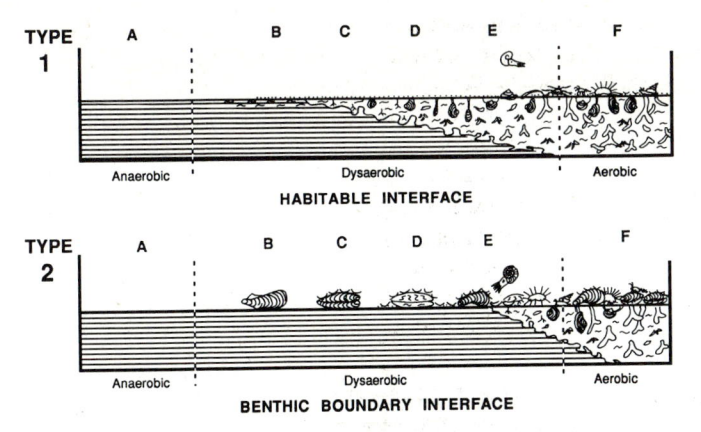

Fig. 6. A comprehensive biofacies model for black shales integrates the Type 1 and Type 2 redox gradients with the ecological classification defined in Fig. 5. As a result, two general biofacies categories, *Type 1* and *Type 2*, are recognized with correlative ecologic sequences of adaptive types in each. Different taxa occupy the respective adaptive groupings according to their different substrate, oxygen, and trophic preferences. The letters *A-F* thus represent biofacies levels (defined in text) that correspond to a specific level of bottom water oxygenation and a specific substrate type: habitable substrates in Type 1 biofacies and benthic boundary substrates in Type 2 biofacies

linear sequence from anoxic to oxic bottom water oxygen levels. Biofacies are commonly cumulative in their faunal content (i.e., biofacies level D may contain elements of biofacies levels A-C). Following the model for benthic oxygen gradients (Fig. 4), oxygen content does not vary uniformly across the sediment-water interface as bottom water oxygen levels vary. The slanting irregular contact between laminated and nonlaminated sediment fabric reflects the increasing depth of the redox boundary as bottom water, interstitial oxygen levels, and associated burrowing activity increase (Fig. 6).

Level A biofacies are characterized by well-laminated sediment fabric and contain no macrofauna. They represent anoxic oxygen levels in the sediment and lower water column in both Type 1 and Type 2 substrates.

Level B biofacies are mono- to paucispecific assemblages composed of low-oxygen-adapted pioneers. Sediment fabric in Type 1 biofacies is slightly disturbed by tiny horizontal or vertical burrows (cf. *Trichichnus*, or tiny planolitid burrows), and benthic foraminifera or ostracodes may occur. In Type 2 biofacies level B is characterized by well-laminated black shales with flat clam pioneers (e.g., *Inoceramus, Bositra, Daonella,* and *Dunbarella*; Kauffman 1988b; Wignall and Simms 1989). These taxa commonly occur in great abundance. Maceral studies of associated organic matter have revealed evidence of microbial mats and possible symbiont bacteria (e.g., in the Posidonienschiefer; Riegel et al. 1986).

Level C biofacies consist of paucispecific associations of low-oxygen pioneers and successors. Type 1 biofacies at this level may contain nuculoid bivalves and discrete trace fossils or moderately burrowed sediment fabric. In contrast, Type 2 biofacies contain flat clam-based assemblages with scattered shell island epibionts (e.g., the bivalves *Phelopteria* and *Ostrea*; Sageman 1989) in well-laminated strata.

Level D biofacies do not change significantly from level C in Type 1 environments as infaunal opportunists are difficult to detect unless they occur as a discrete event horizon in an otherwise barren interval. Thus, levels D through F in the Type 1 model are distinguishable mainly on the basis of increasing density and diversity of trace fossils and other infauna (e.g., Savrda and Bottjer 1986; Savrda et al. Chap. 5.2, this Vol.). Level D in the Type 2 model, on the other hand, marks a significant change in benthic boundary biofacies representing the first evidence of substrate surface habitability by organisms other than specially adapted pioneers. Opportunists of Type 2 biofacies at level D include tin-shelled "paper pectens" (e.g., *Entolium*, *Syncyclonema*) and small brachiopods (e.g., *Leiorhyncus*) that dominate event assemblages on single bedding planes of well-laminated black shales (Thompson and Newton 1987; Sageman 1989; Brett et al. 1989), but may also occur as secondary elements of resident low-oxygen communities.

Biofacies levels E and F reflect changes in bottom water oxygen and the redox gradient such that the benthic boundary condition is effectively diminished. As a result, levels E and F in Type 2 biofacies can resemble Type 1 biofacies due to a somewhat burrowed sediment fabric. Normally the two community types will remain distinguishable on the basis of trophic dominance. Type 2 biofacies are commonly dominated by suspension feeding taxa due to inhibition of infaunal colonization whereas suspension feeders may be ecologically restricted in all Type 1 settings (trophic amensalism; Rhoads and Young 1970). Although the model accounts for cases where the benthic boundary remains intact up to oxic bottom water oxygen levels, in which case high diversity epifaunal assemblages are found in well-laminated sediments, this kind of biofacies is relatively rare in the rock record. Under conditions of sustained increase in bottom water oxygen levels, however, the Type 2 community may shift entirely to a Type 1 condition, and Type 1 biofacies will then dominate. Thus Type 1 and 2 biofacies tend to merge in character at the upper end of the dysoxic zone (levels D to F).

4 Methodology for Application of the Biofacies Model to Black Shale Sequences

The proposed biofacies model provides a framework for the recognition of relationships between substrate conditions, paleo-oxygen levels, and benthic communities in organic-rich strata. Yet in order to produce detailed paleoenvironmental reconstructions, the nature and timing of benthic events must be assessed. To achieve this, a complete set of physical, chemical, and biological data are collected from black shale sequences using the methods of High Resolution Event Stratigraphy (Kauffman 1986, 1988a; see also Kauffman et al. Chap. 7.2, this Vol.). From these data taphonomic and preservational aspects of fossil occurrences are determined and incorporated with biofacies designations derived from fossil diversity and abundance data to produce a high-resolution record of paleoenvironmental history.

Three modes of occurrence characterize fossil assemblages in organic-rich facies:

1. *Resident or background assemblages* consist of taxa that were adapted to low-oxygen and chemically toxic environments and which are distributed relatively continuously through black shale sequences, reflecting almost constant habitation over long time periods (100's to 1000's of years);

2. *Event assemblages* represent populations of normally resident or opportunistic taxa that occupied the substrate for short time spans (<1 to several years) following abrupt changes in benthic environments favoring habitation, for example: brief increase in bottom water oxygen; changes in substrate type and organic input; depositional events related to storms or other gravity flow deposits; and sediment bypass and condensation events that stabilize or partially cement the sediment surface. After colonization, these populations commonly experienced mass mortality when benthic conditions returned to background levels, resulting in dominance of a single (commonly juvenile) size class in fossil event assemblages. Event assemblages typically occupy single bedding planes as shell plasters;

3. *Allochthonous* deposits are composed of fossils transported from a different environment into the oxygen-deficient setting (i.e., a thanatocoenosis), either at a predictable background rate (e.g., planktonic rain) or abruptly as a stochastic "bioevent" associated with tempestites or mass flow deposits (see Brett and Seilacher Chap. 2.5, this Vol.). These deposits are not reflective of environments at the site of deposition. If no resident biotas are found with allochthonous assemblages, it may indicate anoxic conditions in the lower water column. Detailed morphologic, taphonomic, and sedimentologic analyses are required to identify the correct biostratinomic sequence and to differentiate allochthonous deposits from resident and event assemblages.

A hypothetical example of the biofacies model applied to a black shale sequence is shown in Fig. 7. The curve in Fig. 7 graphically illustrates trends in environmental variability in the benthic zone: changes in biofacies levels indicate the extent of environmental change whereas the mode of fossil occurrence reflects the timing of events. Short-term event assemblages are assigned a biofacies level (lower case a-f) according to their faunal content and sediment fabric. Resident fossil assemblages that represent longer-term background conditions are similarly assigned a biofacies level (upper case A-F). The biofacies levels correspond to those in Fig. 6. When points for the different biofacies are plotted (Fig. 7), the resulting curve represents a detailed record of relative changes in benthic conditions.

4.1 Example of Model Application: Middle Cretaceous Greenhorn Limestone, Western Interior, United States

The Late Cenomanian-Early Turonian sequence of the Greenhorn Formation in the Western Interior Basin records both a regional oxygen-depletion event (the Hartland Shale Member Event; Frush and Eicher 1975; Sageman 1985, 1989) and the Cenomanian-Turonian boundary global Oceanic Anoxic Event (OAE) in the Bridge Creek Limestone Member (Elder 1985; Kauffman 1988a). Both stratigraphic units are thought to represent long-term stagnation intervals resulting from expansion and

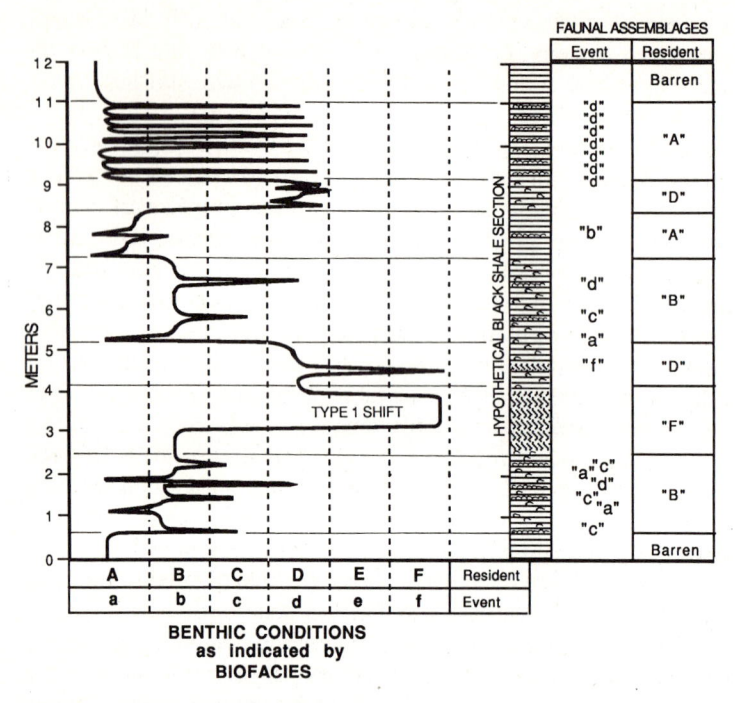

Fig. 7. Application of the biofacies model to black shale stratigraphic sequences is illustrated for a hypothetical black shale section. Taphonomic and preservational aspects of fossil occurrence allow recognition of two types of faunal assemblage relevant to the nature and timing of benthic paleocommunities: these include long-term resident assemblages and short-term event assemblages. Based on faunal content and sediment fabric, each assemblage can be assigned a biofacies level (*A-F* for resident assemblages, *a-f* for event assemblages) following the biofacies model in Fig. 6. The *curve* that is drawn according to these biofacies levels represents a high-resolution reconstruction of changes in benthic paleoenvironmental conditions. In this example, laminated sediment fabric and abundance of suspension-feeding trophic groups indicate dominance of a Type 2 environment. Note, however, that between meters 2 and 4 there is a shift to Type 1 biofacies, reflecting a change in sedimentation or a long-term increase in bottom water oxygen levels. Commonly when this occurs burrowing will extend into the sediment column beyond the point at which conditions improved, giving a false appearance to the timing of these events. As a result of this effect, the record of short-term benthic oxygen fluctuations is rarely well preserved in Type 1 biofacies. In Type 2 settings, however, a boundary maintained at the sediment-water interface insures the record of short-term fluctuations in benthic conditions, as illustrated by the numerous *spikes* in the curve

incursion of the oceanic (Tethyan) oxygen minimum zone into the basin (Frush and Eicher 1975; Eicher and Diner 1985). HIRES analysis of this sequence at the Cretaceous reference section, near Pueblo, Colorado has produced a detailed data base (Fig. 8) which provides an excellent opportunity for testing of the new biofacies model (see also Pratt 1984; Elder 1985; and Sageman 1985, 1989). The interval was dominated by Type 2 biofacies based on the following criteria:

1. predominantly high organic carbon levels (2–5%);
2. generally well-laminated sediment fabric (note bioturbation levels); and

3. dominance of suspension-feeding taxa. Resident and event biofacies were identified for each sample interval and plotted with vertical bars and horizontal spikes, respectively (Fig. 8).

The Hartland Shale Member is predominantly characterized by resident biofacies of low-oxygen adapted pioneers, "flat clams" of the family Inoceramidae. Event biofacies of low-oxygen adapted successor epibionts, such as pteriid and ostreid bivalves, are associated with inoceramids in many samples, and opportunists, such as tiny paper pectens (*Entolium*), comprise event communities on hundreds of bedding planes in the lower part of the member (Sageman 1989). The Bridge Creek Limestone, in contrast, contains resident biofacies that reflect improved benthic conditions, despite the presumed influence of a global OAE. These include diverse communities of predominantly bivalves, gastropods, and epibenthic ammonites, as well as complex trace fossil associations reflecting significant infaunal colonization during limestone deposition. It is hypothesized that the effect of 21–100 Ka Milankovitch climatic cycles on benthic oxygenation and sedimentation/substrate patterns, represented by the successive bioturbated limestone-laminated marlstone rhythmic bedding couplets of the Bridge Creek Member (Kauffman 1984, 1988a), worked to periodically break down benthic boundary redox gradients allowing development of Type 1 biofacies. Although the difference between the two intervals (regional event vs. global event) is made clear by application of the biofacies model, the most striking feature of the analysis is the high incidence of benthic events in both intervals. A number of studies in organic-rich facies have linked benthic event assemblages to circulation, oxygenation, or sedimentation events related to storms (Stel 1975; Aigner 1980; Wignall and Myers 1988; Wignall 1989b) and it is thought that storm events may be partially or wholly responsible for most benthic events in these oxygen-deficient epicontinental deposits. Due to the widespread effect and synchronous occurrence of many benthic redox events related to such oceanographic phenomena, they provide useful markers for high-resolution event stratigraphy and correlation.

5 Summary and Conclusions

Biofacies models offer a powerful paleoecologic tool for the reconstruction and interpretation of benthic oxygen levels, substrate conditions, and paleocommunities in ancient epicontinental black shale environments. The basic pattern established by the Rhoads and Morse (1971) model has provided the foundation for biofacies interpretations in a wide range of organic-rich facies (e.g., Byers 1977; Morris 1980; Hallam 1987; Savrda and Bottjer 1987b). In addition, recent discoveries of diverse shelly and burrowing taxa in modern and ancient oxygen-deficient environments (e.g., Savrda et al. 1984; Thompson et al. 1985; Jorgensen and Revsbech 1985) have resulted in refinement of the model (especially dysaerobic biofacies), and a major change in our views of the oceanography of epicontinental oxygen-deficient basins. The stagnant basin model has been refined to a more dynamic scheme in which benthic events are the predictable consequence of normal climatic-oceanographic processes (such as storms) in shallow epicontinental seas.

Chapter 5 Laminated Sediments

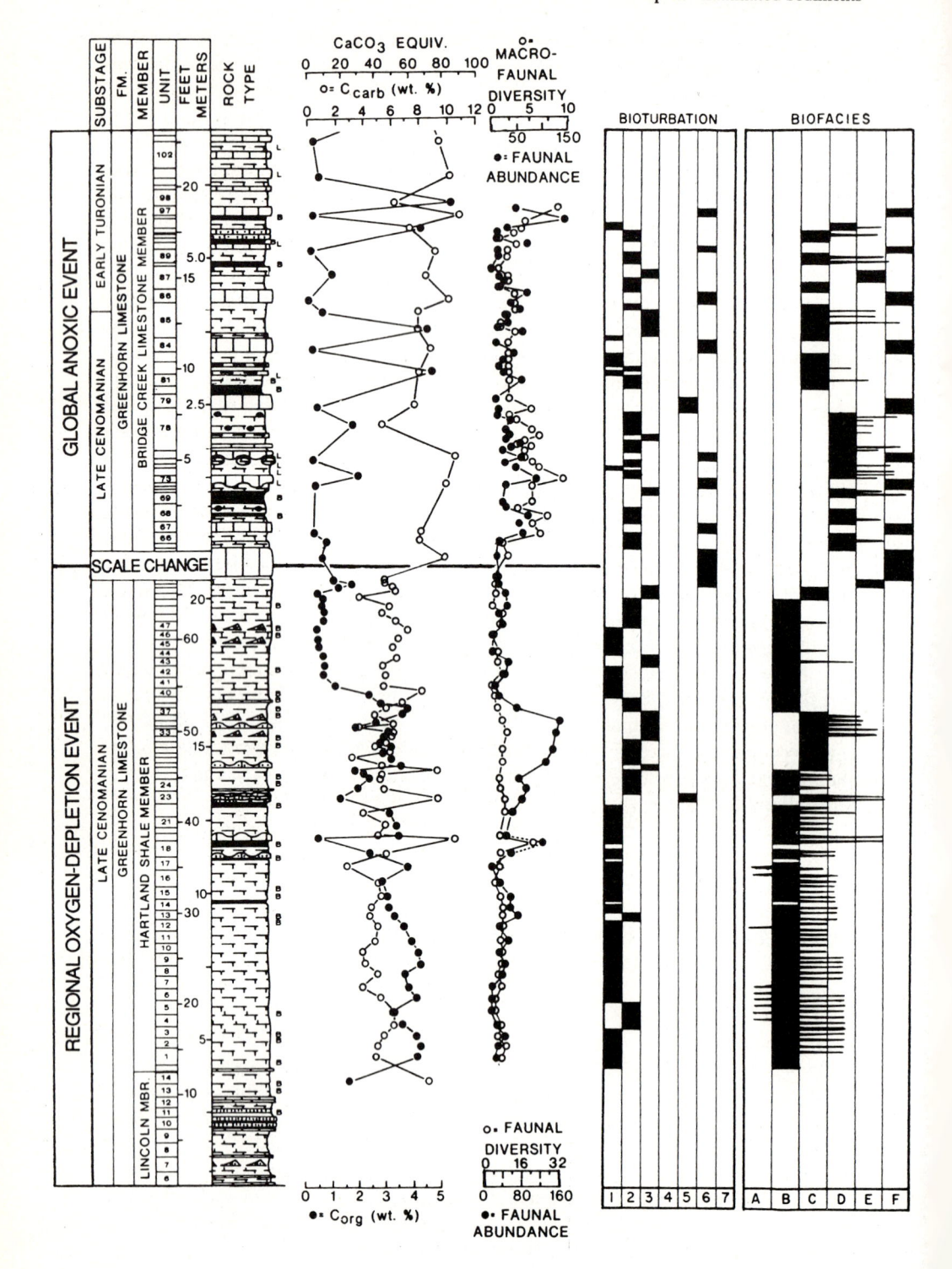

The comprehensive biofacies model proposed herein synthesizes current perceptions of oxygen-deficient environments (as summarized in Sect. 2) into a unifying model. The model is based on a new understanding of redox gradients at and around the sediment-water interface (e.g., Jorgensen and Revsbech 1985) and thus begins with a description of "benthic boundaries" as a possible explanation for the shelly laminite facies of Hallam (1987) or the exaerobic biofacies of Savrda and Bottjer (1987a). Two general biofacies types, Type 1 and Type 2, are identified as end members in a continuum of redox gradients which produce habitable to uninhabitable substrates, respectively. Substrate habitability is a primary controlling factor in the ecology of benthic communities (influencing trophic dominance patterns, epifaunal vs. infaunal life habits, etc.) and thus plays a primary role in the biofacies model. Since tolerance of low bottom water oxygen levels and H_2S are equally important in determining the nature of black shale paleocommunities, an ecological classification of adaptive types representing a gradient from anoxic to oxic oxygen levels constitutes the next part of the model. Biofacies levels corresponding to an idealized sequence of adaptive types (A through F) are defined according to life habits, trophic strategies, tolerance limits (primarily to low oxygen and H_2S), and diversity trends of component taxa. The comprehensive biofacies model illustrated in Fig. 6 combines redox gradient and sediment fabric characteristics with corresponding benthic paleocommunity data representing ascending biofacies levels A through F to produce a two part biofacies model. Biofacies levels are comparable between the two end member redox gradient types resulting in a series of Type 1 and Type 2 biofacies which hypothetically represent equivalent bottom water oxygen levels (in practice Type 1 biofacies are commonly reported to occur in the shallower and presumably better-oxygenated portions of basins whereas Type 2 biofacies typically coincide with the most oxygen-depleted central areas; e.g., Boardman et al. 1984; Brett et al. 1989).

Application of the comprehensive biofacies model to black shale sequences is facilitated by a methodology of high-resolution stratigraphic and sedimentologic documentation and fossil collection (HIRES). Using taphonomic and preservational data, the model allows detailed interpretations of the nature and timing of benthic events in oxygen-deficient epicontinental basins. Testing of the model in a number of cases has supported the view that oxygen-deficient epicontinental basins are not time-stable, stagnant, and generally azoic environments, but instead are dynamic systems sensitive to the perturbations of storms and other environmental disturbances

Fig. 8. The black shale biofacies model is applied to a section of the Middle Cretaceous Greenhorn Formation of the Western Interior Basin in North America. The section includes a regional oxygen-depletion event and the global Cenomanian-Turonian boundary event. Data shown include the lithologic column, curves for organic carbon and carbonate content, total faunal diversity and abundance, and relative levels of bioturbation (*1* = no bioturbation to *7* = totally homogenized; levels *2* through *6* correspond to suites of trace fossil taxa of increasing diversity including *Planolites, Chondrites, Thalassinoides,* and various others). Biofacies defined according to the model in Fig. 6 are shown at the far right. Resident assemblages are shown as *vertical blocks,* event assemblages as *horizontal spikes*

typical of normal shallow marine basins throughout Earth history. In situ resident and event communities were common in these basins, and were regulated by dynamic changes of oxygen, H_2S, and substrate character at the sediment-water interface rather than simply the oxygen content of the overlying water column. Recognition of these communities provides important markers for event stratigraphy and local to regional correlation in ancient organic-rich facies.

Acknowledgments. We wish to thank the organizers of the symposium and the students and staff of the Tübingen Institut für Geologie und Paläontologie for an excellent meeting. Early drafts of the manuscript were reviewed by P. Harries, and C. Johnson and W. Ricken somehow found the time to work editorial wonders on a later version. We are grateful for their efforts.

5.4 Anaerobic — Poikiloaerobic — Aerobic: a New Facies Zonation for Modern and Ancient Neritic Redox Facies

W. Oschmann

1 Introduction

Most examples of modern and ancient oxygen-controlled environments refer to slope and deep sea environments (e.g., Rhoads and Morse 1971; Byers 1977; Arthur et al. 1984a; Savrda et al. 1984; Thompson et al. 1985; Kammer et al. 1986; Savrda and Bottjer 1987a). The standard zonation of these environments is basically depth related, relatively stable and defined as:

1. Anaerobic environments (<0.1 ml/l O_2) which lack benthic fauna (Figs. 1, 3).
2. Dysaerobic environments (0.1 to 1 ml/l O_2) with a soft-bodied and shelly benthic fauna of low diversity (Figs. 1, 3).
3. Aerobic environments (>1 ml/l O_2) with a diverse shelly and soft-bodied benthic fauna (Figs. 1, 3).

An additional zone, the exaerobic environment, defined by Savrda and Bottjer (1987), is characterized by the symbiotic association of sulfur-oxydizing bacteria with certain bivalves at the anaerobic-dysaerobic transition. Bivalves known to contain chemoautotrophic bacteria in part have reduced or lost their digestive system (e.g., Solemyidae) and occur in deep sea hydrothermal vent communities, but also in other environments where hydrocarbons are available such as stagnant basins or mangrove swamps (Cary et al. 1988; Seilacher 1990a). This chemosymbiosis requires long-term stable conditions with a balanced juxtaposition of oxygen and H_2S.

Throughout earth history oxygen-controlled environments are also known from open marine shelf and epeiric seas (e.g., the Liassic Jet-Rock, Posidonia Shale, Kimmeridge Clay). There, fluctuations of the O_2/H_2S-interface in and out of the substrate are assumed, partly due to algal-fungal mats, partly due to a stratified water column (Kauffman 1978, 1981; Morris 1979, 1980; Tyson et al. 1979; Wilson 1980; Seilacher 1982c; Oschmann 1985, 1988a,b). This temporal pattern produced a cyclic stratification in the order of several cm to m (Kauffman 1978, 1981; Morris 1979, 1980; Seilacher 1982c; Oschmann 1985, 1988b), corresponding to several thousand years (Dunn 1974; Tyson et al. 1979; Wilson 1980; Oschmann 1985, 1988b). However, this cyclicity represents only a long-term fluctuation. The real temporal dynamics was probably similar to what we observe in modern open marine shelf and epeiric sea anoxia.

Einsele et al. (Eds.)
Cycles and Events in Stratigraphy
©Springer-Verlag Berlin Heidelberg 1991

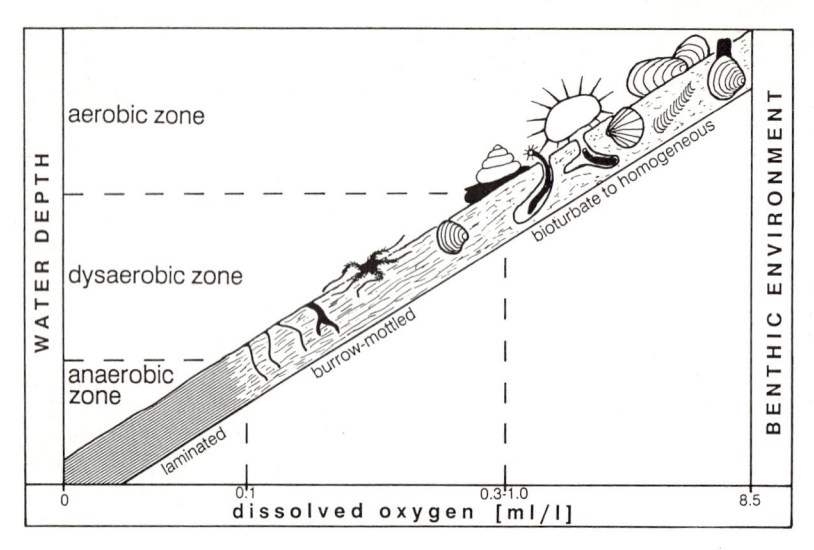

Fig. 1. Schematic facies model for oxygen-deficient slope and basin environments. The zonation in general is a depth-related, permanent feature. (After Rhoads and Morse 1971; Thompson et al. 1985; and Kammer et al. 1986)

2 Modern Shelf Anoxia

Modern shelf anoxia, other than caused by upwelling (e.g., Peru; Namibia) or land-locked basins (e.g., Baltic Sea), is rare, due to reduced area of present shelves and epeiric seas (low sea level) and due to enhanced water circulation (strong climatic gradient). Nevertheless, during the last 20 years certain shelf areas have become increasingly eutrophic due to the influx of anthropogenic phosphates and nitrates (e.g., the North Sea and Mid-Atlantic Bight off New York). These areas are now annually affected by more or less severe phytoplankton blooms and subsequent oxygen consumption in the bottom water (Steimle and Sindermann 1978; Swanson et al. 1979; Rachor 1980, 1982; Rachor and Albrecht 1983; Dethlefsen and Westernhagen 1983; Westernhagen and Dethlefsen 1983; Kempe et al. 1988; Lohse et al. 1989). Phytoplankton blooms begin in late spring or summer, when temperature stratification within the water column is becoming established, and last several weeks depending on warm weather periods. In the North Sea areas affected by oxygen deficiency reached more than 15 000 km^2 and result, if recurring annually, in a significant reduction of faunal diversity (Rachor 1980, 1982; Rachor and Albrecht 1983; Dethlefsen and Westernhagen 1983; Westernhagen and Dethlefsen 1983; Kempe et al. 1988; Lohse et al. 1989). Oxygen deficiency occurs also in parts of the inner neritic zone (e.g., German Bight; Rachor 1980, 1982; Rachor and Albrecht 1983), at a water depth of less than 30 m (Dethlefsen and Westernhagen 1983; Westernhagen and Dethlefsen 1983), and even in a silty to fine sandy facies.

Plankton blooms are commonly associated with mass mortality of the benthic fauna (Steimle and Sindermann 1978; Swanson et al. 1979; Dethlefsen and

Westernhagen 1983; Westernhagen and Dethlefsen 1983; Taylor et al. 1985) caused by oxygen consumption in the substrate and bottom waters. The O_2/H_2S interface, which in fine-grained substrates is normally situated only a few cm below the sediment surface (e.g., Fenchel and Riedl 1970), migrates from the substrate into the bottom water and possibly farther up into the water column and eventually kills the benthic fauna. Before dying, the mobile infauna such as benthic molluscs migrates out of the substrate in order to follow the rising O_2/H_2S interface. Thus, hardparts of the dead infauna accumulate on the sea floor (Swanson et al. 1979; Dethlefsen and Westernhagen 1983; Westernhagen and Dethlefsen 1983; Taylor et al. 1985).

Astonishingly, water stratification is a relatively stable feature. After a storm one would expect mixing of water and, as a consequence, higher oxygen values in the bottom waters. However, stable stratification of water with anoxic bottom conditions was re-established only a few days after a hurricane had passed through the area (Steimle and Sindermann 1978). A single storm can even worsen the situation: Reworking of anoxic sediment enriches H_2S in the bottom water, making it more toxic than it was before (Rachor 1982).

Autumn and winter cooling and seasonal series of storm events do remove the temperature stratification in the water column of shelfs and epeiric seas so that the oxygen conditions in bottom waters and in the uppermost parts of the substrate change to normal; but, within a few months environmental conditions in the substrate and bottom water change back to anoxic. Environments with such short-cycle fluctuations cannot be described with existing models. Consequently, we propose a modified classification scheme for shelf and epicontinental sea anoxia, in which the biologically important temporal aspect is also represented.

3 Oxygen Zonation in Modern Shelf Seas

Presentday eutrophic shelf areas, e.g., the North Sea and the New York Bight, are situated off highly productive agricultural and industrial zones. They are normally affected by strong winds and currents (westerlies and west wind drift) that force mixing of the water column. In more quiet environments prolonged eutrophism results in extended plankton blooms and in anoxic conditions within the substrate and in the bottom water for the major part, or even the whole year. Oxygenated periods, if at all present, are too short to support a macroscopic benthic community. In traditional terms, this facies would correspond to the anaerobic facies (e.g., Rhoads and Morse 1971; Byers 1977), while the other end member (year-round oxygenated conditions in substrate and benthic environment) is the aerobic facies (e.g., Rhoads and Morse 1971; Byers 1977). Since the definition of the dysaerobic facies does not apply to the described seasonal fluctuations, the term *poikiloaerobic zone* (def.: poikilos, greek, variable) is proposed (Figs. 2. 3).

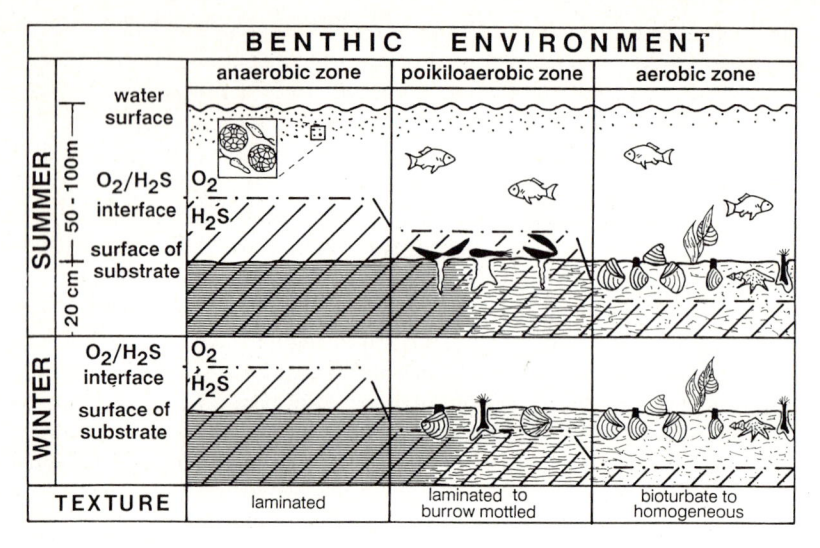

Fig. 2. Schematic facies model for oxygen-deficient shelf and epeiric basin environments. Depending on seasons the amount of oxygen in the poikiloaerobic zone is fluctuating from well-oxygenated to anaerobic

4 The Kimmeridge Clay — an Example of Ancient Shelf Anoxia

A well-studied example of an ancient oxygen-controlled epeiric basin is the Kimmeridge Clay in NW Europe (Dunn 1974; Tyson et al. 1979; Aigner 1980; Wilson 1980; Oschmann 1985, 1988a,b; Wignall and Myers 1988; Wignall 1989b). Parts of the sections are microlaminated. Dark clay-rich lamina alter with coccolith-rich ones in intervals less than 1 mm, indicating very regular plankton blooms. Occasionally distal storm events disturbed the microlaminations and produced reworked sediments with compacted and graded rip-up clasts of the coccolith-rich lamina (Oschmann 1985, 1988b; Wignall 1989). There is also a very low diversity but occasionally very dense benthic macrofauna of small infaunal suspension-feeding bivalves. Apparently the amount of oxygen in the substrate and bottom water was reduced and fluctuating. In the dysaerobic zone of modern and ancient environments nonskeletal or weakly calcified organisms (e.g., burrowing polychaetes, irregular echinoids, and epibenthic ophiuroids) are more tolerant to low-oxygen conditions and thus more common than more heavily calcified taxa such as molluscs (Rhoads and Morse 1971; Byers 1977; Arthur et al. 1984a; Savrda et al. 1984; Kammer et al. 1986). This is different from our example, in which bioturbation is insignificant, and the dominant taxa are bivalves.

slope and basin environments		
anaerobic: <0.1ml O_2/l H_2O	dysaerobic: 0.1-1ml O_2/l H_2O	aerobic: >1ml O_2/l H_2O

benthic macrofauna	none	rare soft-bodied and shelly fauna	diverse soft-bodied and shelly fauna
life habit and trophic groups	----	dominantly infauna; suspension- and deposit-feeders	in- and epifauna; suspension- and deposit-feeders
texture/ bioturbation	laminated	laminated to burrow-mottled	bioturbate to homogeneous

shelf and epeiric basin environments		
anaerobic: 3 to 12 months lack of O_2 per year	poikiloaerobic: max. 3 months lack of O_2 per year	aerobic: year-round aerobic to very short lack of O_2

benthic macrofauna	none	rare soft-bodied and shelly fauna	diverse soft-bodied and shelly fauna
life habit and trophic groups	----	dominantly infaunal susp.-feed. bivalves; lack of bivalves with lecithotrophic larvae;	in- and epifauna; suspension- and deposit-feeders
texture/ bioturbation	laminated	laminated to burrow-mottled	bioturbated to homogeneous;
bivalve stratinomy	----	migration of infaunal bivalves onto surface of substrate, and arrangement of paired valves on bedding plane when dying due to lack of O_2	occasionally in life position; normally reworked and disarticulated

Fig. 3. Comparison of the oxygen-deficient facies and their zonation from slope and basin environments and from shelf and epeiric basin environments

5 The Poikiloaerobic Facies in the Kimmeridge Clay

Surprisingly, none of the infaunal bivalves within the microlaminated sediments was found in life position, although depositional conditions must have been very quiet. All shells are plastered along bedding planes, but commonly remain paired and lack abrasion. This recalls the behavior of infaunal organisms in the modern shelf environments. As long as the O_2/H_2S interface was situated below the sediment surface colonization by deep burrowers and deposit-feeders was inhibited (e.g., Bader 1954). Suspension-feeding bivalves, however, could drain nutrients and oxygen from the as yet unpoisoned water above and survive as long as this condition persisted. During plankton blooms (coccolith-rich lamina), the anoxic boundary rose slowly from the substrate into the bottom water, or possibly even further up into the water column. In response, the mobile infauna tried to follow the rising O$_2$/H$_2$S interface, and migrated out of the substrate and eventually died.

The general lack of infaunal bivalves preserved in life position in a physically unperturbed environment clearly shows that the fauna died regularly from oxygen depletion. On the other hand, the continuous presence of benthic fauna demonstrates that life at the sea floor was not permanently excluded for longer periods (years, tens of years, or longer). The fluctuations of the O$_2$/H$_2$S interface between positions below, at, or above the sediment/water interface must have occurred at periods shorter than the life spans of small infaunal suspension-feeding bivalves, which last between one and a few years (Thorson 1957; Thompson et al. 1980). Most probably, this was an annually recurring event. One more sign for a very short period of recurrence is the absence of bivalves with lecithotrophic larval development (e.g., nuculoids). By their relatively large yolk mass such larvae are able to remain on the sea floor without a planktonic stage (Sastry 1979; Rachor 1982). Therefore, eggs and larvae would be killed along with the adults during anoxic events.

6 Strategy of Survival in Poikiloaerobic Environments

Resistance of modern bivalves to anaerobic conditions depends largely on water temperature. At 10 to 15 °C, corresponding to the North Sea and Mid-Atlantic Bight bottom water temperatures in summer (Dethlefsen and Westernhagen 1983; Steimle and Sindermann 1978), many specimens of well adapted species die within 1 to 2 months (Theede et al. 1969; Oertsen and Schlungbaum 1972; Dries and Theede 1974). On the other hand, many short-lived modern bivalves require only about 8 to 9 months to reach their reproductive age (Thorson 1957; Sastry 1979; Thompson et al. 1980). Their planktotrophic larval development lasts between 10 and 50 days (Sastry 1979). Therefore, spawning with the onset of anoxic conditions enables the planktic larvae to persist in well-oxygenated surface waters unaffected by the temporarily lethal bottom water conditions.

Oxygen isotope data from the Kimmeridge Clay (Irwin et al. 1977; Salinas 1984) suggest bottom water temperatures of 15 to 20 °C, at which anoxic conditions of at least two months would be sufficient to entirely kill the benthic fauna (Theede et al. 1969; Oertsen and Schlungbaum 1972; Dries and Theede 1974). Bivalve prodissoconchs, which occur sometimes in abundance on bedding planes (Fig. 4), show that larval shools did try to resettle, but failed due to extended periods of anoxia.

In view of these limits the poikiloaerobic environment of the Kimmeridge Clay was probably oxygenated for 9 to 11 months, but anoxic for 1 to 3 months. This allowed only a few well-adapted bivalves (r-strategists or opportunists) to survive. In order to persist in this environment, an early reproductive age and an extended planktonic larval stage seem to have been more important than low oxygen tolerance.

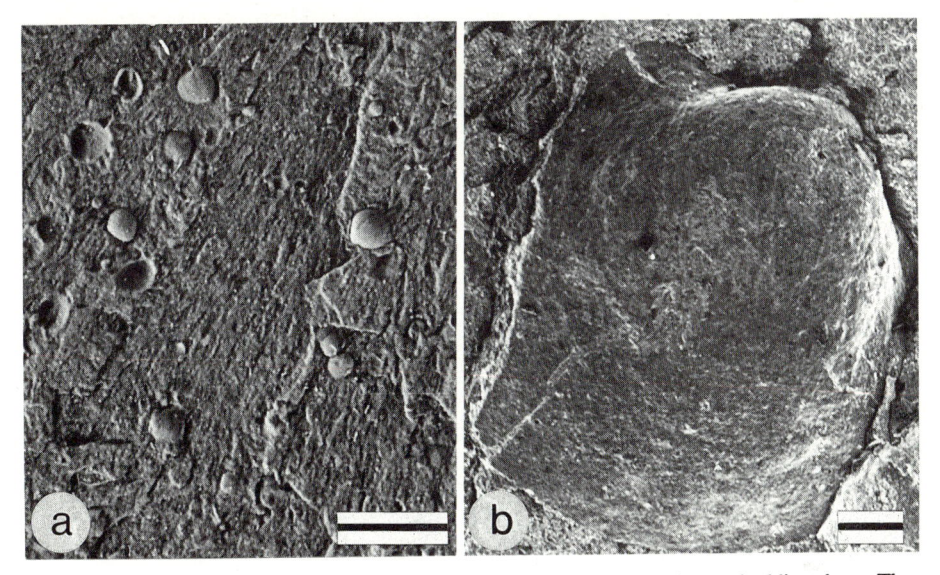

Fig. 4. Bivalve prodissoconchs and small dissoconchs occur sometimes abundant on bedding planes. They confirm that survival of larvae settling on seafloor sometimes failed due to extended periods of anoxia (**a** *bar* 2 mm ; **b** *bar* 100 μm)

7 Conclusions

Classical models of marine redox facies defining an aerobic, dysaerobic, and anaerobic facies zone fit deep (slope and basin) environments, in which oxygen values at a certain depth remain rather constant. Shelf and epeiric basin environments, in contrast, are subject to remarkable annual oxygen fluctuations. While both end members, the aerobic and the anaerobic environments do occur, the shortness of the cycles prevents the establishment of a dysaerobic zone. Instead there is a seasonal low-oxygen to anoxic environment colonized by only a few opportunistic species. The characteristics of this zone, for which the name *poikiloaerobic zone* is proposed, can be identified in modern and ancient shelf anoxia.

Acknowledgments. The author is grateful to M. Aberhan, F.T. Fürsich (both Würzburg), S. Jacobson (Chevron; La Habra), and A. Seilacher (Tübingen/Yale) for stimulating discussions and for reviewing the manuscript.

5.5 Cyclical Deposition of the Plattenkalk Facies

Ch. Hemleben and N.H.M. Swinburne

1 Introduction

The facies we know as plattenkalk represents an extreme of limestone sedimentation. It is an unusually regular sediment in the sense of very flat, tabular beds, evenly spaced at intervals of centimeters or millimeters, a product of cyclic sedimentation. On the flat bedding planes fossils may be exceptionally preserved inside fine sediment and the deposit be called a fossillagerstätte. Plattenkalk, with its composition of almost pure carbonate, is at one extreme of a lithological continuum; there exist all intermediates between this and black shales (see Wetzel, Chap. 5.1, this Vol.).

In this chapter we use the term plattenkalk to mean, at its narrowest, a pure, micritic limestone with regular, parallel, and thin bedding on a centimeter to millimeter scale. The classic example of this is the type locality of the Solnhofen Plattenkalk, where limestone beds suitable for lithography may be up to 30 cm thick without internal partings. However, in practice the plattenkalks we will describe deviate from this narrow definition. Some plattenkalks do not cleave at distinct levels to form bedding planes until they are weathered (Haqel and Hjoula plattenkalks). Then they may split into such thin laminae that they are known alternatively as paper shales. At the other extreme, some beds in a plattenkalk sequence may be unusually thick as a consequence of algal growth or redeposition of sediments. Some plattenkalks do not have the exceptionally flat bedding planes but slightly irregular, undulose planes (e.g., Hvar facies) due to stromatolitic lamination. Although predominantly made of micrite or lime mud, there may be some laminae with larger particles, particularly peloids, or lumps of micrite, set in a micrite or even a sparite matrix (e.g., Bear Gulch). Larger particles, such as coarse grains of fossil debris also occur at the base of graded beds (e.g., Bolca). The original limestone may be secondarily altered either by recrystallization or by dolomitization (e.g., Montral). Other plattenkalks, although pure in the sense of >95% calcium carbonate may still contain a relatively high content of bitumen (>0.5% is quite common) which will give the beds a brown-black color on fresh surfaces (e.g., Haqel and Hjoula). Finally, although the facies may be known as plattenkalk, there may be interbeds with a constitution other than pure limestone. Most common are interbeds of argillaceous limestone (e.g., Solnhofen, Bear Gulch, etc.) or even shale (e.g., Cerin).

In addition, the beds are frequently laminated on a millimetric scale, the preservation of which must be due to an absence of bioturbation at the site of deposition. This in turn is attributable to anomalous conditions of oxicity and/or high salinity in the bottom waters of the plattenkalk basins. In an absence of benthic scavengers,

---

Einsele et al. (Eds.)
Cycles and Events in Stratigraphy
©Springer-Verlag Berlin Heidelberg 1991

organisms which fell onto the plattenkalk basin were not dismembered, and, with a slow rate of decay articulated specimens were often preserved, sometimes with details of the soft parts. The composition of the plattenkalk fauna and flora also tends to be abnormal because the unusual bottom water conditions excluded all but the most tolerant species. Many of the fossils may therefore be representative of quite different conditions prevailing in the surface waters, or even allochthonous and derived from outside the basin.

Most plattenkalks require a protected environment for their formation, such as a basin with barriers isolating it from the ocean, or else a depression in the shelf sheltered from turbulent waters above. Plattenkalks of this kind show features of paleoslope such as slump structures and graded bedding. Some plattenkalks which are algal laminates do not seem to require a depression, but the binding activity of the microorganisms suffices to produce the planar surfaces. Secondly a source of fine-grained carbonate sediment is required. In many cases the micrite is periodically delivered, giving rise to a cyclic interbedding of limestones and thinner argillaceous limestones.

The different plattenkalks vary according to the geographic position of the depositional site, the type of sediment deposited and the mechanism of its delivery, as well as by the conditions in the overlying water column such as oxicity, salinity, and organic productivity. Plattenkalks may be grouped using a combination of all these criteria, but we have decided to use the paleogeographic position because the others are mainly dependent variables. On the basis of geographic position we can separate plattenkalks into four categories (lake, lagoonal, inner and outer shelf) and we will present examples of each which differ in sediment type and in oxicity of the overlying water.

We will describe this facies as plattenkalk. However, the same type of limestone has been reported in the literature under the names of platy limestone or lithographic limestone. Platy limestone is the English translation of the German plattenkalk and records that the limestone is thinly bedded, but does not convey the connotation of flat, tabular bedding implicit in 'plattenkalk'. Lithographic limestone is also a poor term, because it is not directly descriptive, and these days few people know what kind of limestone is suitable for lithographic printing.

2 Case Studies

2.1 Lake Environment

In general, marine environments which favor the deposition of plattenkalks are more common than their lacustrine equivalents. Intramontane basins, although widespread, and frequently including a freshwater phase in their development, are only rarely the sites of plattenkalk deposition as fossil lagerstätten. This may be due to relatively stable conditions of the environment, a comparable small bioproduction, and a rather high clastic sedimentation rate.

Plattenkalks are more common in playa lakes, where much of the calcium carbonate may be of evaporitic origin and the plattenkalk is associated with other salts in a cyclic sequence. In more permanent, perhaps deeper lakes, the calcium carbonate may be derived from planktonic or benthic production in the lake. Thus lake plattenkalks include suspension deposits, turbidites, and algal/microbial laminates.

2.1.1 Green River Formation (USA), Lake Gosiute

Plattenkalk is an important lithology of the Green River Formation as it outcrops in Wyoming, Colorado, and Utah in the western United States. The Green River rocks were deposited in a large inland lake, Lake Gosiute, caught between uplifted areas during the Laramide Orogeny (Eocene). The carbonate units of the formation are exceedingly organic-rich and are frequently referred to as oil shale, a commodity for which they have been commercially exploited. The Green River plattenkalk is also well known to the paleontological community for its fossils, and particularly by its fish.

Our description, mainly based on the work of Surdam and Stanley (1979) and Boyer (1982), concentrates on the Wyoming basins of the former lake, and on the Laney Member, the uppermost of three members of the Green River Formation. This consists of up to 630 m of chemical, organic and fine-grained terrigenous sediment, divided into four lithofacies. One of these, the laminated carbonate lithofacies, reaches around 80 m in thickness and contains plattenkalk as one of its lithologies.

The Laney Member covers the largest area of the Green River Formation and contains the deposits which represent a relatively high stand of the lake. The other two members are typical nearshore, possibly playa lake deposits, but the Laney Member, with its organic-rich and fish-bearing carbonates (Grande 1980), was probably deposited under a somewhat more permanent water body. The change from one lithofacies to another in the Laney Member is a basin-wide event of probable tectonic origin and has resulted in the large scale, layer-cake stratigraphy. However, on the smaller scale, minor transgressions of the lake shoreline, perhaps the result of climatic variation, have effected the lateral facies transitions.

The laminated carbonate lithofacies is sandwiched between layers of evaporites, which must have been deposited at the lowest water stands when the water was most saline. The carbonates and evaporites pass laterally into the sandstones and mudstones of the Wasatch Formation which was deposited by marginal alluvial fans and deltas. These effectively trapped all the clastic sediment which poured off the surrounding uplands, leaving the basin center mud free. These laminated carbonates are overlain by the molluscan-ostracodal calcareous mudstone lithofacies and then by the sandstone and mudstones deposited during the freshwater phase.

The laminated carbonates contain clearly recognizable cycles which vary in thickness from a few centimeters to several meters. Typical cycles are shown in Fig. 1. Most beds also show a stromatolitic lamination on a submillimetric scale with alternating dark organic-rich and light organic-poor layers. However, at other localities beds show laminae with greater separation which grade upwards from an organic-rich base to organic-poor micrite and these resemble varves. Fossils are found on bedding planes in the organic-rich portions of the plattenkalk. Fish are most abundant, many exceptionally preserved. Of particular significance are the catfish;

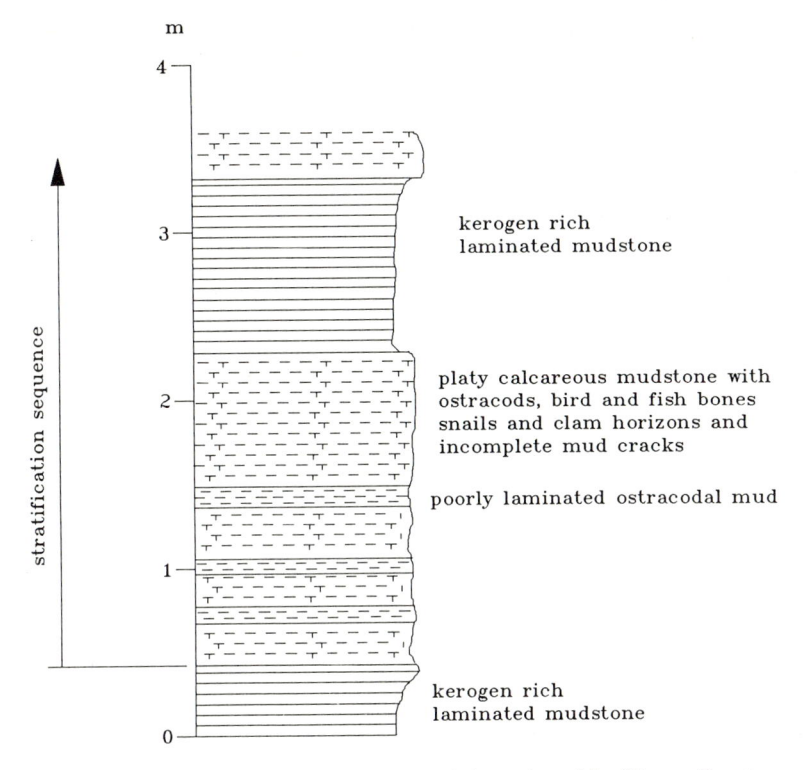

Fig. 1. Lake Gosiute – stratification sequence of alternation of fossiliferous limestone and plattenkalk characterizing the high stands of the lake (kerogen-rich, laminated) and low stands indicated by the kerogen-poor, fossiliferous limestone. (After Surdam and Stanley 1979)

today's related forms live on the bottom of sluggish streams and in lakes and so the Green River catfish were probably also benthic. Other fish occur and many of these are exceptionally preserved. The deposit has also produced birds, bats, flying insects, and terrestrial plants.

Depositional environment. Lake Gosiute seems to have had a gentle underlying topography judging from the transgression-regression cycles of the laminated carbonates. The lake itself probably developed from a saline, alkaline lake to a freshwater lake after a more open-basin hydrologic regime was established (Surdam and Stanley 1979). The water depth in the basin is the subject of much controversy. Classically (cit. in Buchheim and Surdam 1977) the depositional environment was envisaged as a moderately deep, perennially stratified (meromictic) lake (cit. in Boyer 1982). The surface waters were productive and sediment and fauna preserved in the underlying hypersaline and anoxic waters.

These views were challenged by a playa-lake model already applied to other members of the Green River Formation (Surdam and Wolfbauer 1975). The playa-lake model explains features in the carbonate sequence such as stromatolites, oolites, ripples, and flat-pebble conglomerates as well as the increasing Mg/Ca ratio towards

ECTOGENIC – MEROMICTIC LAKE STAGE

Outflow ←

Increased Freshwater Inflow

less saline O_2
H_2S more saline no O_2

Mixolimnion
Chemocline
Monimolimnion

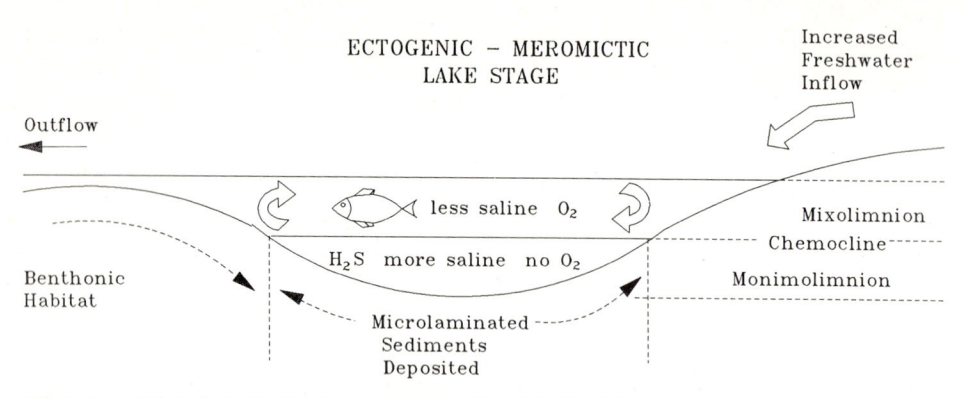

Benthonic Habitat

Microlaminated Sediments Deposited

Fig. 2. A model of a Lake Gosiute during the stage of stratification, which depends greatly on the seasons. The outflow shown in meromictic stage is not strictly necessary, but would help maintain freshness of mixolimnion and facilitate colonization by fish. With continued inflow, the meromictic (only seasonally stratified) lake could become holomictic (unstratified); with reduced inflow it could revert to playa lake stage. (After Boyer 1982)

the basin margins. The carbonates must have been deposited under a depth of water that varied from nothing at the start of a cycle to perhaps a few meters during deposition of the plattenkalk.

More recently, Boyer (1982) tried to combine the two models suggesting a previously saline lake was refilled by fresh water and thus stratified. He interpreted "much of the lamination as a series of varves produced by regular seasonal variations in the physical-chemical or biogenic carbonate precipitation". In addition, he argued, that in a shallow basin there would be a broad marginal area where catfish could have lived on the bottom, in regions where they would not necessarily be preserved. After death they were transported into the conserving milieu of an anoxic, hypersaline bottom. This is summarized in Fig. 2.

2.1.2 Gürün (Turkey), Lacustrine Sedimentation in a Semi-Arid Area

Another example of a limnic plattenkalk comes from the Gürün Series of central Turkey. The deposition of marine sediments over the central Anatolian plate came almost to an end during the Late Eocene although in the Oligocene-Miocene there were some minor marine transgressions. Underlying the Gürün series there are mostly shelf carbonates of Late Cretaceous and Late Eocene. The Oligocene-Miocene sequence contains numerous hiatuses and shows a change in facies from fully marine to an extreme biotope with a very low diversity but high abundance of almost one species (*Rotalia* sp.). Subsequently, intramontane basins of erosional as well as tectonic origin developed and became filled with terrestrial and lacustrine sediments.

The lower Gürün series (Late Miocene) consisting of molasse-like sediments such as conglomerates and sandstones in the basal portion, which are derived from rocks exposed exposed to the north of the basin. Fanglomerates as well as mud and debris flows are quite common. With the beginning of the Upper Gürün Series (Late Pliocene) the basin was no longer actively subsiding and began to be filled with lake deposits which typify the former lake Gürün (Hemleben unpubl. data).

Sediments of the Upper Gürün Series deposited around the lake margin are predominantly charophytic limestones and reed beds, with subsiduary gastropodal, oncolitic, stromatolitic, and pelletal limestones, sometimes intercalated by *Botryococcus* marl. There is also some coal and plant debris. A major river entered from the north bringing eroded Cretaceous sediments of planktonic foraminifers and coccoliths in suspension and built fan-like structures out into the lake. This marginal facies grades into the central lake facies which consists of plattenkalk with flat varves of light and dark laminae. The dark clay-rich laminae can be correlated with a winter/spring season and contain diatoms and reworked Eocene and Cretaceous coccoliths. The light laminae consist either of aragonitic needles sometimes recrystallized to calcite, original calcite, protodolomite, or rarely gypsum which were deposited in summer. Especially during the highly productive spring times, ostracods produced fecal pellet limestones and most layers are highly bioturbated. To this general picture can be added the coarser clastics which flooded the central lake during periods of heavy rainfall. In contrast, during a series of very warm summers the lake evaporated producing a sequence with fresh water sediments as indicated by ostracods and gastropods, followed by brackish ostracods and finished by a bivalve horizon. These cycles (Fig. 3) are in the order of mm-cm thick.

Summarizing our observations from Gürün lake we can conclude that the sedimentological features demonstrate seasonally determined cyclicities. Considering the microfossils, it is indeed a "fossil lagerstätte". However, vertebrates are not known from Gürün and this can be explained by the seasonal fluctuations in salinity of the lake water and by the relatively low bioproduction of this semi-arid area.

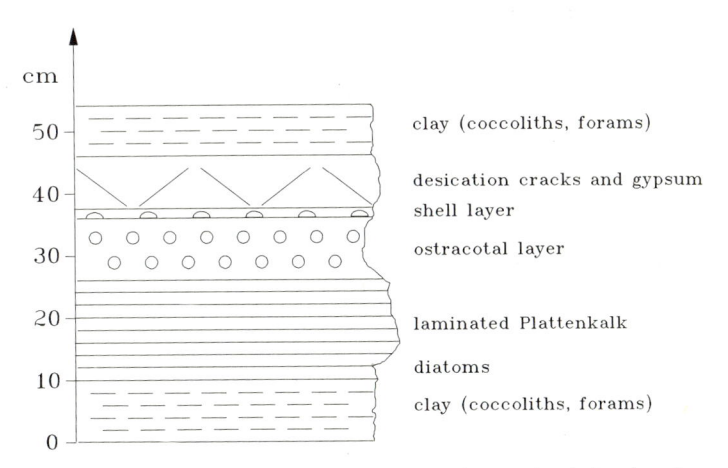

Fig. 3. Lake Gürün sediment sequence of a seasonally determined stratification starting during the rainy erosional season forming the clay followed by laminated plattenkalk a playa lake situation

In contrast to the Gürün plattenkalk the Jurassic (Late Lias-Tithonian) plattenkalk from Karatau in southern Kaszhakstan (USSR) is very fossiliferous (Hecker 1948). This fresh water lake deposit exemplifies the typical facies sequence beginning with coarse clastics and culminating in carbonates. The uppermost carbonate is a finely, fissile argillaceous limestone (papierschiefer) and contains well preserved fossils of almost all taxonomic classes representing fresh and brackish water environments. The deposit is best known for its insect fauna from which more than 18 000 specimens have been recorded. The carbonate lamination is caused by stromatolitic growth and by a somewhat coarser episodic turbiditic sedimentation. Most of the fossils are embedded in the stromatolitic layers similar to Hvar (see below).

2.2 Lagoonal Environment

Under this title we include all those coastal plattenkalks in the supra-, inter- or subtidal zones. Salinity encompass the full range of brackish to normal marine to hypersaline. The sediment is deposited directly from suspension, or redeposited as turbidites, or else produced in situ by microorganisms producing a stromatolitic lamination.

2.2.1 Montsech (Spain), a Supratidal Coastal Lake

One of the better known plattenkalk fossil lagerstätten occurs in the eastern part of Sierra de Montsech in the Pyrenean foothills of northern Spain (Sierras Marginalas Catalanes). It is famous for articulated skeletons of vertebrates and insects. On account of the microfauna and flora it is regarded as Lower Cretaceous (Berriasian – Valanginian, Peybernes and Oertli 1972). There are several deposits of plattenkalk, none exceeding 50 m in thickness, found at different stratigraphic levels, inside a 100 m thick sequence of charophytic limestones. Of the several quarries in plattenkalk, that of La predrera de Meia yielded stones for lithography in the early part of this century and this quarry also produced most of the fossils.

The Montsech plattenkalk is an intercalation of limestone and argillaceous limestone. There are few sedimentary structures indicative of water movement, yet there are indications of palaeoslope in the way of small scale deformational features. These were formed at various stages of sediment lithification, and range from small faults to wrinkled surfaces due to sliding of layers, to internally folded or slumped beds. Each plattenkalk bed shows a cyclic lamination bounded by erosional surfaces. There is a basal layer of biomicrite with ostracod debris, followed by a middle layer of light-colored micrite, which is topped by a laminated clay-rich unit a rich in organic matter (algal/microbial mat). Microfossils, especially some agglutinated foraminifera (figured by Schairer and Janicke 1970) are typical of brackish or hypersaline environments (Groiß 1967), characterizing probably the basin center, whilst the allochthonous charophytes and ostracods indicate fresh to brackish water representing conditions of the surrounding area.

The marine macrofossils are generally found on the bedding planes and split evenly between top and bottom slabs. Terrestrial organisms such as plants and insects predominate, and of the insects, most are either freshwater nymphs, or adults derived from fresh water larvae. Frogs, very rare reptiles and bird remains have also been recovered. There is a complete lack of the exclusively marine groups such as cephalopods and echinoderms. Evidence of autochthonous macrobenthos is scarce apart from the few crustacean tracks and some bioturbation. Nekton is well represented by fish and their abundant coprolites. The fish are not concentrated on particular bedding planes and therefore did not die in distinct mass mortality events as in certain other fossil lagerstätten. However, some of the fish may have died quite suddenly as there are examples of prey still in their stomachs.

Depositional environment. The Montsech plattenkalk is thought to have been deposited during the Early Cretaceous in a coastal lake at the northern edge of the continent. The reconstruction of Barale et al. 1984 (see also Viohl 1989), shows the Montsech area enclosed by the continent to the south and by a barrier to the north (Fig. 4).

The plattenkalk was deposited at the stagnant bottom of a seasonally stratified coastal lake. Whether the stratification is due to evaporation and fresh water inflow or results from the inflow of brackish water from the seaward side has not yet been discussed in detail. However, in an absence of bioturbation, the fine lamination was preserved, possibly because of a seasonally biased sedimentation. Sediment was mainly produced on an intertidal flat and travelled through channels to be disseminated into the lake. It was deposited either directly out of suspension of the water, or

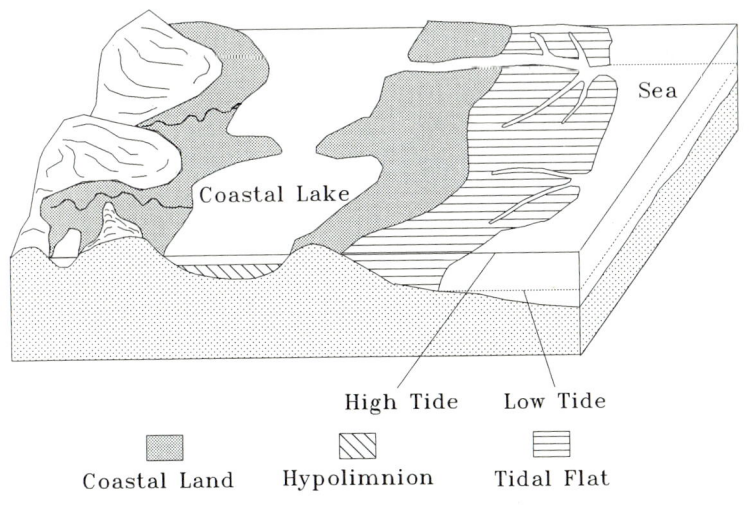

Fig. 4. Coastal lake Montsech – paleogeographical reconstruction of the Montsech area. The coastal lake represents the depositional environment of the plattenkalk. The stratification lead to oxygen poor conditions during times of isolation and normal situation during periods of marine floodings. (After Barale et al. 1984)

redeposited during storm events. The clay-rich layer which tops each cycle is rich in organic matter, which could be of terrestrial origin, or from plankton production in the lake. In summary, a detailed investigation on the origin of the Montsech plattenkalk is lacking. However, the lamination and thus the preservation may be due to sedimentary events as well as to a stratification of the water body in the course of the year.

2.2.2 Hvar (Yugoslavia), Lagoon and Coastal Flats, Supra-, Inter- and Subtidal

The Hvar plattenkalk facies outcrops along the Dalmatian coast of Yugoslavia south of the Istrian peninsular, both on the mainland near Primosten and on the island of Hvar. From the second half of the nineteenth century until around 1920 the numerous quarries at Hvar (e.g., Starigrad and Vrboska) were mined for plattenkalk, which was used locally as a building material. There are several plattenkalk horizons ranging in age from Late Cenomanian to Early Turonian, which are well known for their well preserved fossil fish (Hemleben and Freels 1977a).

During the early Late Cretaceous the Dalmatian carbonate platform had a gentle topography of highs and lows resulting from different tectonic settings. Thus plattenkalk grades laterally and vertically into other shallow water facies, such as rudist mounds, biomicrites with rudist debris, and shallow-water benthic foraminiferal limestone, with pellets, intraclasts and algal mats. Slumping and channels are rare. The plattenkalk deposits result either from cyclic algal/microbial mat growth or other intertidal sedimentary processes.

Lithology. The Hvar plattenkalk can be divided into three different facies varying in the amount of algal/microbial influence. The first facies consists of irregular bedding planes representing hemispherical stromatolites. Algal layers and sedimentary layers are interbedded on a centimeter scale. The sedimentary layers are either finely laminated, including fine debris of bivalves and ostracods or pass abruptly into lenses of miliolid foraminiferal limestone. The stromatolitic layers are up to 2 cm thick and also contain bioclastic laminae or sparite-filled cavities which include birds-eyes. Facies one is limited in lateral extent and most beds wedge out after a few meters. The sedimentary features suggest a deposition in a supra- to intertidal area.

The second facies has fewer of the thick dark layers and is laminated on a millimetric scale with light and dark laminae. The light-colored laminae are graded and consist of silt-sized carbonate whereas the dark laminae, when not recrystallized, are made of micrite and penetrated by long sparite-filled cavities. There is often a sharp boundary between the base of a light coloured layer and the underlying darker layer, which would suggest that the light-coloured sediment evened out an irregular underlying topography. There are 20-50 of these couplets until the next thick sedimentary layer which is strongly graded from micrite to midium-sized sand. There are isolated examples of small erosional unconformities, of convoluted bedding, and shrinkage pores but no obvious indications of subaerial exposure or of bioturbation. Therefore this stromatolitic facies has been interpreted as intertidal to subtidal in origin.

Facies three is entirely sedimentary without algal layers and is finely and flatly laminated. The beds vary in thickness from 0.1 to 15 mm and are graded from a base containing shallow water foraminifers and other bioclastic detritus in a micrite matrix. Between these graded units, the bedding plane has been enhanced by pressure solution so that it parts easily along these surfaces. The sedimentary facies tends to have a lower component of organic carbon (0.73 wt %) compared to the first two facies which average 1.23% (Hemleben and Freels 1977a).

Well-preserved fish and rarer reptiles are found in all three facies, although the best come from the sedimentary facies (facies 3). In the stromatolitic facies (facies 1 and 2), the preservation deteriorates as the lamination becomes less pronounced and the stromatolitic and sedimentary layers become more closely interwoven. The fossils usually occur on the upper surface of a bed, although when the fish lies directly on the stromatolitic layer then it may remain with the upper slab when the rock cleaves.

Depositional environment. The three facies which differ in the amount of stromatolitic fabrics were all deposited around the shoreline of the shallow sea which swept the carbonate platform.

The first facies, with the most developed stromatolitic lamination, was deposited in the shallowest water, in ponds above the supratidal level, and hence contains many horizons indicative of subaerial exposure. The second facies, with its flatter bedding and alternation of stromatolitic and sedimentary layers is typical for inter- and shallow subtidal areas. Its cyclic character could represent a day-night growth pattern of an algal/microbial mat, overprinted by an ebb/flow sedimentary cycle, moderated by seasonal weather patterns (Fig. 5). Similar cyclicities have been shown in the loferite (alpine Triassic) cycles of Fischer (1964) and in Precambrian sedimentary sequences demonstrating tidal cycles and paleorotation of the earth (Elder and Smith 1988; Williams 1989).

The third facies with bedding which is exclusively sedimentary in origin was deposited in the subtidal domain in of small depressions laterally situated compared with the stromatolitic facies. Each bed represents a pulse of sediment delivered by a tidal current and was laid down in a relatively short space of time.

Cerin (France). The French fossil lagerstätte of Cerin is similar in lithology to the stromatolitic facies of Hvar. It was also deposited under very shallow water and stirred by tidal currents. Current orientation of fossils, burrowing, stromatolites and a wide variety of shallow water features can be observed (Fabre 1981; Barale et al. 1985; Bernier 1984). Although the geographic setting – a lagoonal environment – and the lithographic limestones are similar to Solnhofen, the preservation of fossils was due to the cyclicity of algal mats growth and lime detrital mud (Gall et al. 1985).

2.2.3 *Solnhofen (Germany), Late Jurassic Lagoons*

The Solnhofen plattenkalk (Early Tithonian) is perhaps the most celebrated plattenkalk fossil lagerstätte. It has yielded the world's oldest known bird, *Archaeopteryx*, as well as exquisitely preserved flying reptiles and numerous fish and crustaceans. So many Solnhofen fossils are known because the rock has been

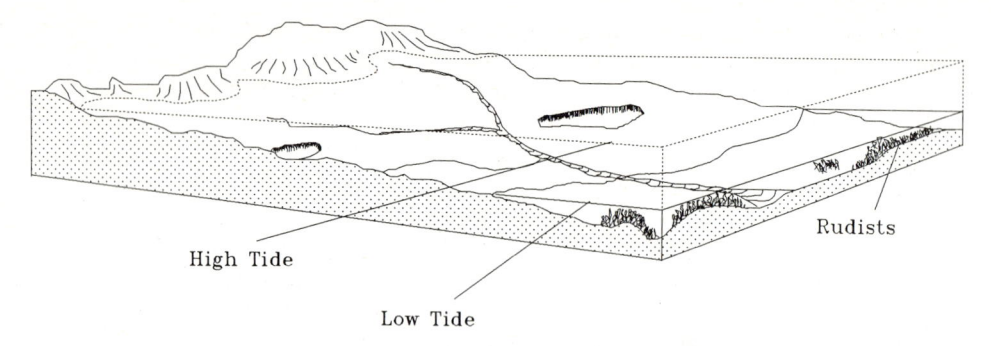

Fig. 5. Depositional model of the Hvar area with rudist biostromes, tidal chanels deepenings acting as sediment traps and intertidal flats the area of algal mats. (After Hemleben and Freels 1977a)

extensively quarried, firstly as a building material, and secondly as a lithographic printing stone. However, exceptionally preserved fossils are quite rare.

The Solnhofen plattenkalk is part of a sequence of shallow water carbonates deposited on the northern margin of the Tethys. The Solnhofen waters were shielded from major currents by sponge-algal mounds, which on the seaward side were overgrown by corals to form a chain of fringing reefs (Fig.6). The plattenkalk is restricted in outcrop to small basins separated and underlain by sponge-algal mounds. The variable underlying topography has caused local differences in thickness as well as in lithology. Thus, in the Solnhofen area the plattenkalk sequence reaches a maximum thickness of 90 m and produces some of the best lithographic stones, 20–30 cm thick, whereas in the Eichstätt area, where only 30 m of plattenkalk is developed, the bed thickness rarely exceeds 5 cm. The plattenkalk in the Solnhofen-Eichstätt area was perhaps deposited in the most protected waters. Beds are very continuous and may be traced over distances of kilometers in the same depositional basin, differing only in thickness as they follow the gentle underlying contours. Slumps are localized into two composite beds, each some tens of meters thick, one dividing the upper and lower Solnhofen plattenkalk and the other marking the end of plattenkalk deposition.

Most localities show a repetition of two lithologies: the pure micritic limestone, locally termed flinz, and thinner interbeds of argillaceous limestone, called fäule. In the Solnhofen basin, argillaceous limestone interbeds range from marly partings to beds 2–3 cm thick which occur about every 50 cm. On the edge of the Eichstätt basin, where beds are only about 1 cm thick, no true fäule is developed and only marly partings separate the flinz beds. In thin section the plattenkalk beds are flatly laminated and this shows up either as weak color bands, or, when aided by pressure solution, as distinct white lines. Occasionally the thicker Solnhofen beds show small scale cross-lamination, and, under suitable illumination, bioturbation. The Solnhofen plattenkalk is not organic-rich in comparison to some other plattenkalks and does not contain noticeable pyrite. It has values of $\delta^{13}C$ of around +2 for the flinz and slightly higher at +2.5 for the fäule (Barthel et al. 1990), which is commensurate with a carbonate of marine origin unaffected during diagenesis by any organic carbon which

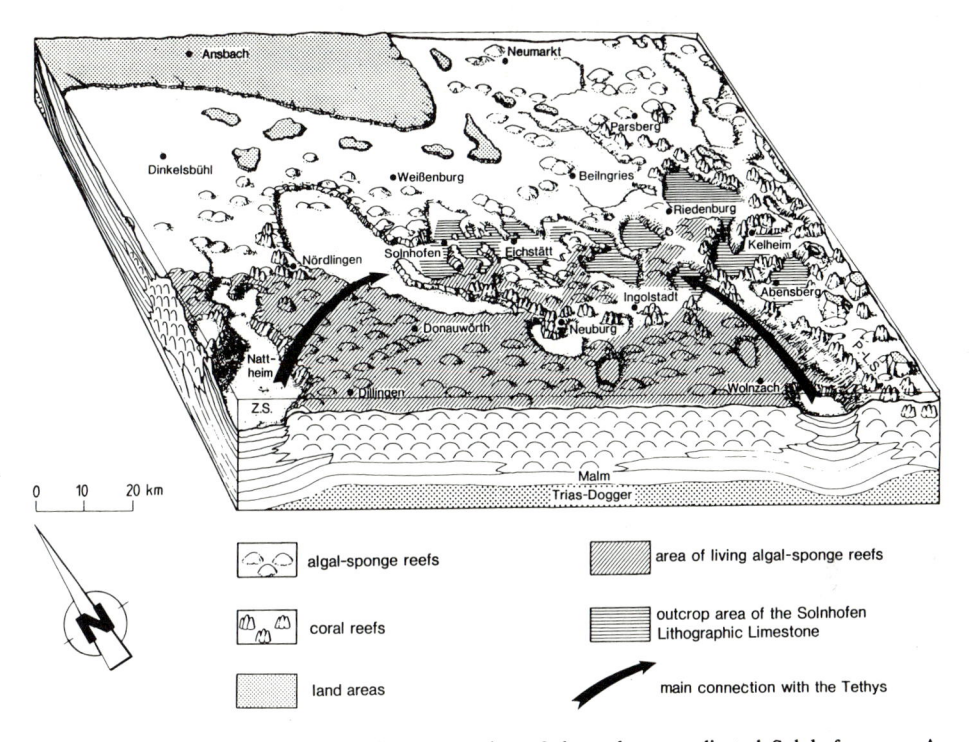

Fig. 6. Solnhofen – paleogeographical reconstruction of the rather complicated Solnhofen area. A composition of geographical position of deepenings between reefs, storm floods, erosional events, dry seasons and nutrient supply results in the deposition of the Solnhofen plattenkalk. (Courtesy of Dr. Viohl)

may have been present in the sediment. This indirect information is helpful because the carbonate of the flinz beds is in general too diagenetically altered for the nature of the particles to discerned with the microscope.

Yet much information had been gained from microscopic study of the fäule and marly intercalations. There are well preserved coccoliths, some in coccospheres frequently incorporated in coprolites as well as calcispheres, coccoid cyanobacteria (Keupp 1977), foraminifera (Groiß 1967), and ostracods (Gocht 1973). Macrofossils, which are rare constituents, are always located on the underside of flinz slabs. Small nektonic marine animals, mainly fish, crustacea and ammonites, together with their coprolite *Lumbricaria* are most common, followed by terrestrial insects and plants. There are several examples of mass mortality of small fish, crinoids, jellyfish, or the ophiuroids, which suggest that the lagoonal waters were at times hazardous. The famous spiral death tracks of the limulids also suggest hostile conditions in the bottom waters.

Depositional environment. Although a large body of literature exists, the origin of the sediments in the Solnhofen-Eichstätt area is still controversial. The Solnhofen basins must have lain in the littoral zone, for the coral reefs to have flourished further south and so that cyanobacterial mats could grow over the hypersaline, stagnant basin floor (Keupp 1977). Mixing events exchanged the lagoonal water creating a normal marine situation which may be represented by the fäule. This changing environment would account for many taphonomic features of the fossils (Viohl 1983, 1985). Whether or not the stagnating brine received sufficient organic input to achieve anoxia is open to question. Most of the sediment was derived from the carbonate ooze which lay around the coral reefs. This was thrown into suspension during bouts of bad weather, crossed the coral reefs, and came into the lagoon (Barthel 1970, 1972, 1978; Hemleben 1977a) forming the allochthonous flinz. Only the finest sediment was transported as far as Solnhofen, making a smooth cover over the irregular floor. In the intervening periods, as the lagoon stagnated, sediment was produced at a slow rate by the microbial mat and some lagoonal microorganisms like coccolithophorids to which was added a continual flux of clay (autochthonous fäule). The two mechanisms of sedimentation alternated regularly and gave rise to the repetitive sequence of flinz and fäule. However, there is no direct evidence for the case of the cyclicity though it could well be due Milankovitch cycles, the regular variations in the earth's orbit which induce a change in the weather pattern.

Very closely related to this type of laminated limestone is the Eocene fossil lagerstätte of *Monte Bolca*, near Verona in Italy. Stagnant bottom conditions and red tide-like conditions in surface waters may have produced the laminated plattenkalks which are enormously rich in fish and deposited in a small basin sheltered by reefs and detrital barriers (Massari and Sorbini 1975; Viohl 1983).

Another deposit similar to Solnhofen in terms of depositional environment is the fossiliferous, dolomitized limestone from the Middle Triassic (Upper Muschelkalk) near *Montral*, Prov. Tarragona, Spain. The so-called plattenkalk is part of a sequence of shallow water facies deposited on the inner shelf of the epicontinental Tethys in rather small basins surrounded by dasycladaceen reefs. It is made of a cyclic sequence of dolomite with marl interbeds containing a uniquely preserved fauna, demonstrating the original sedimentary sequence of limestone and marl, similar to the Solnhofen-Eichstätt outcrops (Hemleben and Freels 1977b; Via Boada et al. 1977).

2.3 Depressions in the Inner Shelf

This section is devoted to those shallow water deposits not associated with patch reefs or mud mounds. The basins are thought to be depressions in the sea floor mostly of tectonic origin, but others may be due to factors such as salt migration, erosional scour, underlying karst etc. If the basins were situated near land, there may be some fresh water influence, otherwise they will be of normal marine character. The micrite sediment is the product of marine plankton, benthic algae or microorganisms and is deposited from suspension as turbidites, or bound by algal/microbial mats.

2.3.1 Bear Gulch (USA)

Introduction. The Bear Gulch plattenkalk of Montana, USA (latest Early Carboniferous) was first recognized in 1968, when the first fossil fish was found and since then it has yielded one of the most diverse and exceptionally preserved collections of Carboniferous fish ever known. The Bear Gulch plattenkalk was laid down in the Big Snowy Trough, a narrow embayment which ran from the Williston Basin in the east and the Cordilleran "Miogeosyncline" in the west. The plattenkalk deposit was formed in an isolated basin which was probably one of many which ran parallel to the structural grain of the trough (Williams 1983).

The trough is filled with a broadly transgressive sequence of fine to medium clastics and evaporites deposited in continental to subtidal environments. At the top of the sequence the Bear Gulch Member consists of black shale enclosing three limestone lenses not more than 30 m in total thickness. These lenses show lateral differentiation into a marginal, slope, and basin facies and a gradual facies transition to the typical basinal plattenkalk. Both basin and slope facies show very regular alternations of nonfissile limestone (< 30 cm thick) with thinner interbeds of argillaceous limestone (< 3 cm thick) whilst such rhythmicity is not present in the marginal facies. The limestone beds of these facies are finely laminated by micritic and bioclastic layers. There are occasional rippled laminae, dewatering faults, rare slump structures (but these are developed to a much greater degree in the slope facies).

Though the basin plattenkalk facies is not abnormally organic-rich (TOC 0.52 wt %), the organic content increases markedly towards the marginal facies (TOC av. 1.3, max. 2.2 wt %) which yields kerogen with an almost exclusively algal signature (Williams 1983). It also contains microscopic organic fibers which are believed to be of cyanobacterial origin (Williams and Reimers 1983).

The basinal facies has the most diverse collection of fossils of the three facies and there is a wide variety of fish, some cephalopods, conularids covered with epibionts, and various forms of macroscopic algae. However, there is a complete absence of those stenohaline groups such as echinoderms, bryozoa and corals, perhaps because of turbid bottom water conditions (Williams 1983). The fossils are not concentrated in particular horizons and, at least for the fish, a variety of different growth stages are known. Of the few benthic fossils these decrease in number towards the center of the basin.

Depositional environment. The Bear Gulch basin shows gradual facies transitions and slump structures which indicate only gently dipping sides. It was probably also reasonably shallow, as the rocks which cover the marginal, slope, and basin facies are all of *supratidal* facies. Allowing for compaction, the water depth would have been of the order of 40 m during deposition. The basin was most likely situated nearshore because of the common occurrence of nonmarine facies in the Bear Gulch sequence. However, the fauna preserved in Bear Gulch plattenkalks represents a normal marine environment. Williams (1983) favored a model where the more hypersaline and anoxic bottom water was separated by a chemocline from the overlying brackish water (Williams 1983). Westerly trade winds promoted water mixing (Fig. 7).

The source of the carbonate must have been local as the Big Snowy Trough was not a site of major carbonate deposition. The skeletal carbonate particles found inside some plattenkalk laminae probably came from organisms which lived in the marginal and upper slope areas. A source for the micrite is more obscure as there were no reefs in the neighborhood and coccoliths had not evolved by the Carboniferous. Williams

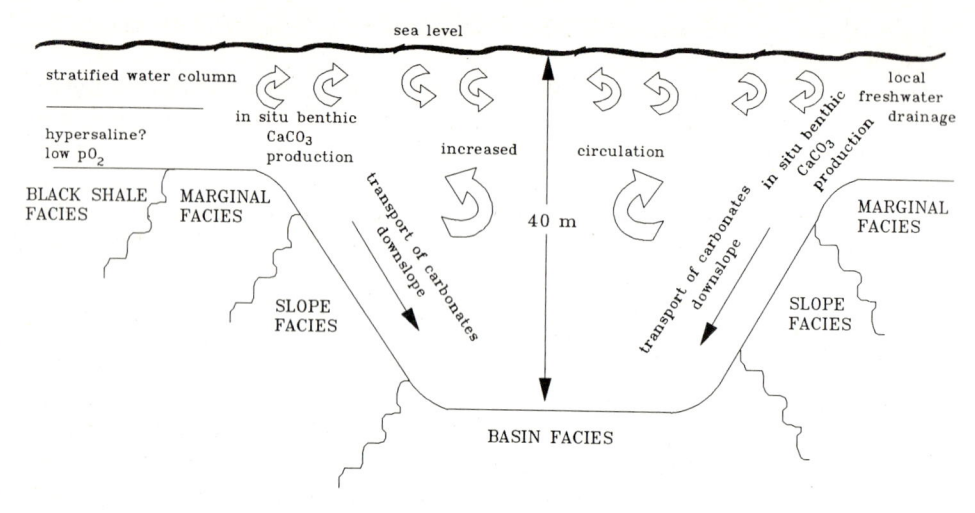

Fig. 7. Depositional model for the Bear Gulch Basin. A variety of factors such as upwelling, wind direction, carbonate precipitation, and stratification of the water column caused the plattenkalk deposition

(1983) suggested either whitings or a biochemical precipitation of carbonate by noncalcareous algae and microorganisms (e.g., cyanobacteria) which lived around the basin margin. The sediment was carried downslope, stronger currents bringing the skeletal grains and micrite settling out in the quieter intervals, so producing the couplet. As suggested for the Solnhofen limestone, the purer limestone beds were deposited in times of greater water movement, coarser sediment transport, and its deposition. The more clay-rich beds were deposited in quieter intervals thus containing a higher proportion of suspension load.

2.3.2 Kozji Rat (Yugoslavia) Tectonic Basin, Not Associated with Reefs

The Kozji Rat plattenkalk (Hemleben and Freels 1977a) was deposited on the same carbonate platform as Hvar, but is slightly younger in age (Campanian-Maastrichtian). The outcrop is only 200 m in diameter and contains a facies of plattenkalk which is almost identical to the nonstromatolitic sedimentary facies of the Hvar plattenkalk except that it contains more organic carbon. The fine lamination is completely planar and parallel. The sediment was derived locally from the shallow water sediments containing a rich assemblage of miliolid foraminifera as well as a minor pelagic component. In this organic-rich facies the fossils are even better preserved than in the Hvar plattenkalk.

The basin was more steep-sided and further from land than is was the case for the Hvar plattenkalk-basin and it acted as a sediment trap. This is shown by the slump and breccia deposits at the edges of the basin. The steep sides prevented circulation in the basin water which stagnated and became anoxic, aiding preservation of the bitumen and fossils.

2.3.3 Nusplingen (Germany) Basin Between Sponge Mounds off the Coast

The plattenkalk of Nusplingen (Temmler 1966) has many similarities to that of Solnhofen. The two deposits are only some 300 km apart and very similar in age (Kimmeridgian). They were deposited in similar geographical settings with an underlying relief composed of sponge mounds, except that the Nusplingen area was further from the shore and the water perhaps deeper. The Nusplingen basin was also steeper sided than the Solnhofen basins, and coarse breccia beds, representing material eroded from the sides of the sponge mounds, interfinger with the basinal plattenkalk. Both turbiditic and laminated facies occur. Generally there is more coarse detritus, sponge spicules, and a color banding due to varying concentrations of organic matter. There is no equivalent of the limestone/argillaceous limestone interbedding which is present in the Solnhofen sequence, and all the beds are equally impure with around 10% HCl-insoluble residue. Many of the fossils are as well preserved as at Solnhofen and the assemblage is similar in content in so far as there are the bodies of both marine and terrestrial organisms with marine nekton and plankton predominating.

2.4 Depressions in the Outer Shelf

This division differs from the previous category in the greater water depth and further distance from shore. These plattenkalks are fully marine and the sediment is partly pelagic and either deposited directly from suspension or, most commonly, as turbidites.

2.4.1 Haqel and Hjoula (Lebanon)

The small plattenkalk outcrops of Haqel and Hjoula are Cenomanian fish lagerstätten recognized for around 100 years, but mapped in detail by Hückel (1970). The fish beds are part of small basins each a few 100 m across and now filled with 270 m of plattenkalk which is interbedded with coarse breccias. The basins lay on the outer part of the Lebanese carbonate platform, mostly occupied by rudist, or oyster mounds, and patch reefs. The relief was only gentle as semi-pelagic sediment covered both swells and basins. In contrast, the plattenkalk basins were steep-sided and were undoubtedly tectonic in origin. Hückel (1970) envisaged a system of strike slip faults which intermittently opened the basins. Periodic deepening of the basins accounts for the different thicknesses of the basin units compared with the surrounding facies. All units of the surrounding facies are represented as blocks in the slide deposits. As the blocks slid into the basins, parts of the basin walls were also eroded, thus some plattenkalk clasts are found in overlying breccia beds.

The Haqel plattenkalk horizon is around 35 m in thickness, whilst the Hjoula outcrop is much thinner, so that the underlying breccias sometimes protrude through. The Haqel and Hjoula plattenkalks are blue-gray in color but weather white in an outer zone which may be up to 1 m in depth. Weathered material cleaves easily into beds less than a few centimeters thick, whilst there are no distinct cleave planes in unweathered material. The Haqel plattenkalk contains both laminated and graded beds. The laminated beds show a TOC content of 0.2–1.3% and the HCl-insoluble residues vary between <0.1 and 5%. The clearly graded beds range in thickness from millimeters to up to 12 cm in thickness, are extremely pure $CaCO_3$ (99%) and, if they have been leached, have virtually no organic matter (Hückel 1974). These graded units are more common around the basin margins and their tops or the overlying units may be bioturbated. There is also some silica, usually in the form of thin, flat chert beds rather than nodules. Allochthonous shallow water benthic foraminifera accompanied by molluscan and echinoderm debris occur in the graded beds, particularly at their base. The laminated beds contain an autochthonous planktonic assemblage predominantly of pithonellids and a small proportion of planktonic foraminifers.

The Hjoula plattenkalk is similar in most respects to the Haqel unit. However, at Hjoula the graded beds are not known (but exposure is limited) and all beds have a slightly higher component of clay (6%) some of which is distributed as distinct marly partings. There is also a higher component of iron which occurs in the form of small pyrite spheres and a very high concentration of bitumen (TOC up to 2.36%) and the HCl residue reaches up to 3 and 18%, respectively.

In both outcrops macrofossils are distributed unevenly through the sequence and are found usually in the laminated units. Some horizons are very rich whilst intervening beds may contain none. In the fossil assemblage fish are most common followed by echinoderms (holothurian sclerites, crinoids and ophiuroids) and then by crustaceans. In some horizons in the Haqel plattenkalk terrestrial plant fossils may be quite common.

Depositional environment. The Haqel and Hjoula basins were sharp fault-bounded depressions (Fig. 8). After fault movement the basins began to fill with olistostrome deposits. Quiet water conditions then prevailed and the water column became stratified. Under these stagnant conditions the plattenkalk beds were deposited. The anoxic bottom water ensured the preservation of the organic matter which accumulated there. Fish which lived partly or wholly in the basin were exceptionally preserved.

Hemleben (1977b) suggested that these fish beds were produced by planktonic blooms (red tides) which were induced by the local upwelling and formed in a relatively short space of time. Toxins produced by the microorganisms poisoned the water and caused the death of the fish (see Brongersma-Sanders 1957). The turbidites representing slides of shallow water sediment laid down in oxic environments, down the unstable sides of the basin, into the anoxic bottom. The turbidites may have oxigenated the bottom waters thus allowing benthos to invade the basin bottom and bioturbate some of the directly overlying beds. Thus, the observed cyclicity may be due to the more or less annual red tides and superimposed by weather events causing a higher sediment input, e.g., by turbidites.

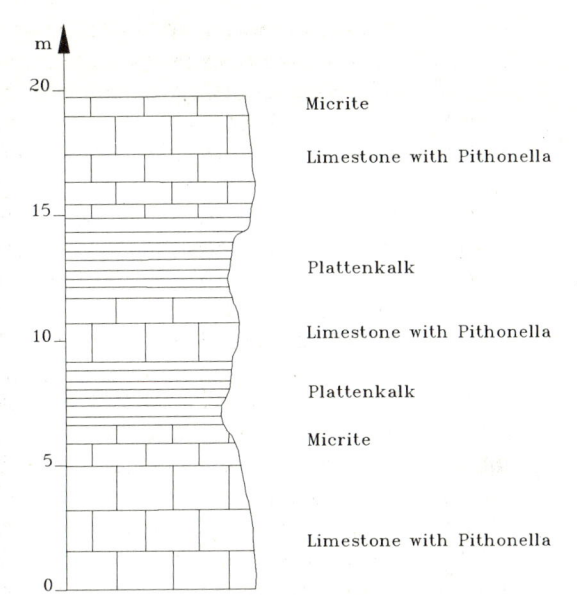

Fig. 8. Lebanon – alternations of fossiliferous (*Pithonella*) limestone and plattenkalk which characterize the time of stratification due to high phytoplankton production (red tides). (Hemleben 1977b)

3 Discussion and Conclusions

To establish rhythms in a calcareous shallow water environment needs a combination of various parameters such as a special geographic setting, unique physico-chemical properties of the water column, and several biotic influences. Each of these factors is again linked either to intrinsic processes or is determined by other extrinsic and higher ranking factors, such as ebb/flow tidal cycles, seasonal cycles, storm, events, etc.

In reconstructing the geographic setting, the sedimentary environment gives information about the site of deposition and its distance from land. The water properties (e.g., temperature, salinity, and oxygen content) may be deduced both from the sediments and from comparisons between the fossil and the living fauna and flora. The biotic factors are directly observable as long as they are preserved in the sediment. Allochthonous and autochthonous components of the fauna and flora must be considered separately and the mixing mechanism may be deduced again by analyzing the sedimentary environment. In the case of hostile conditions on the bottom of a depression, most bottom living creatures, as well as those which are washed in, will not survive; they will be preserved as the enzymatic and bacterial activity comes to a minimum.

Factors determining the physicochemical conditions of the bottom waters have been the subject of greatest debate in discussion of plattenkalk fossil lagerstätten. Two questions dominate the discussion: what kind of factors (e.g., extreme temperatures, salinities, and/or oxygen content) cause the preservation and what kind of mechanisms (e.g., seasonal changes, ebb/flow rhythms, storms, red tides) trigger the factors finally resulting in a cyclic determined facies. Yet the explanations in literature are rife with speculation and show a tendency to be copied from one example to another, despite the obvious differences in depositional environments.

Several reasons may be responsible for the speculative interpretation: lack of basic knowledge (scientific gaps) or an inability to use information gained from other disciplines. For instances when we discuss "high salinity environments", do we mean salinities of 40, 60 or 100‰; do we know the upper and lower limitations of the faunas and floras under extreme conditions? How does a normal marine environment becomes a hypersaline environment? Evaporation and a closing barrier system may play an important role. Thus, a chemocline will be established and a stagnant bottom water body may be developed. What are the decay rates under extreme conditions? How does the preservation process develop which finally results in an excellent fish conservation? However, what are the limiting data for a normal or excellent preservation; especially if one considers that fish do not remain on the bottom after death if the water temperature is above 16 °C (Elder and Smith 1988). There are many other questions concerning the precipitation of carbonate under hypersaline and/or anoxic conditions in respect to other factors, e.g., changing temperatures or biotic influences, e.g., bacteria.

These and many more questions remain to be solved in order to explain the origin of cyclicities which occur in connection with plattenkalk fossil lagerstätten. However, we can summarize the data (Fig. 9) by stating that the plattenkalk facies is either linked to small basins or depressions in which the water body develops a chemocline, a varve-like sedimentation, or is influenced by stromatolitic growth. The observed auto- and allocyclic mechanisms in sedimentation include the daily growth of algal mats, the yearly buildup of varves, changing watermass properties according to the seasons and finally events caused by red tides and storms or tectonic movements (tempestites and turbidites). An anoxic and/or hypersaline calcareous environment or stromatolites inhibits the destruction of animal and plants remains, resulting mostly in an excellent preservation of these fossils. After all, it is the exceptional preservation of many of the fossils which has prompted most explanations of the abnormal bottom water or sediment environment. A more detailed study of all plattenkalk deposits available may lead to more information concerning the depositional environment, but, without further research in the field of Recent environments a general breakthrough will not result. In addition, any cyclicity which possibly or definitely exists should be investigated under the light of Milankovitch cycles. Both rhythms or cyclicity caused by irregularly occurring events or planetary changes can be deciphered only by combining the observations obtained from the Recent and the Past.

Acknowledgments. We gratefully acknowledge the assistance received from I. Breitinger, P. Halla, and R. Ott. We gratefully acknowledge the discussion and supply of Fig. 6 by Dr. G. Viohl, Eichstätt. This study was supported by the Sonderforschungsbereich (SFB 53) Palökologie at various stages of research.

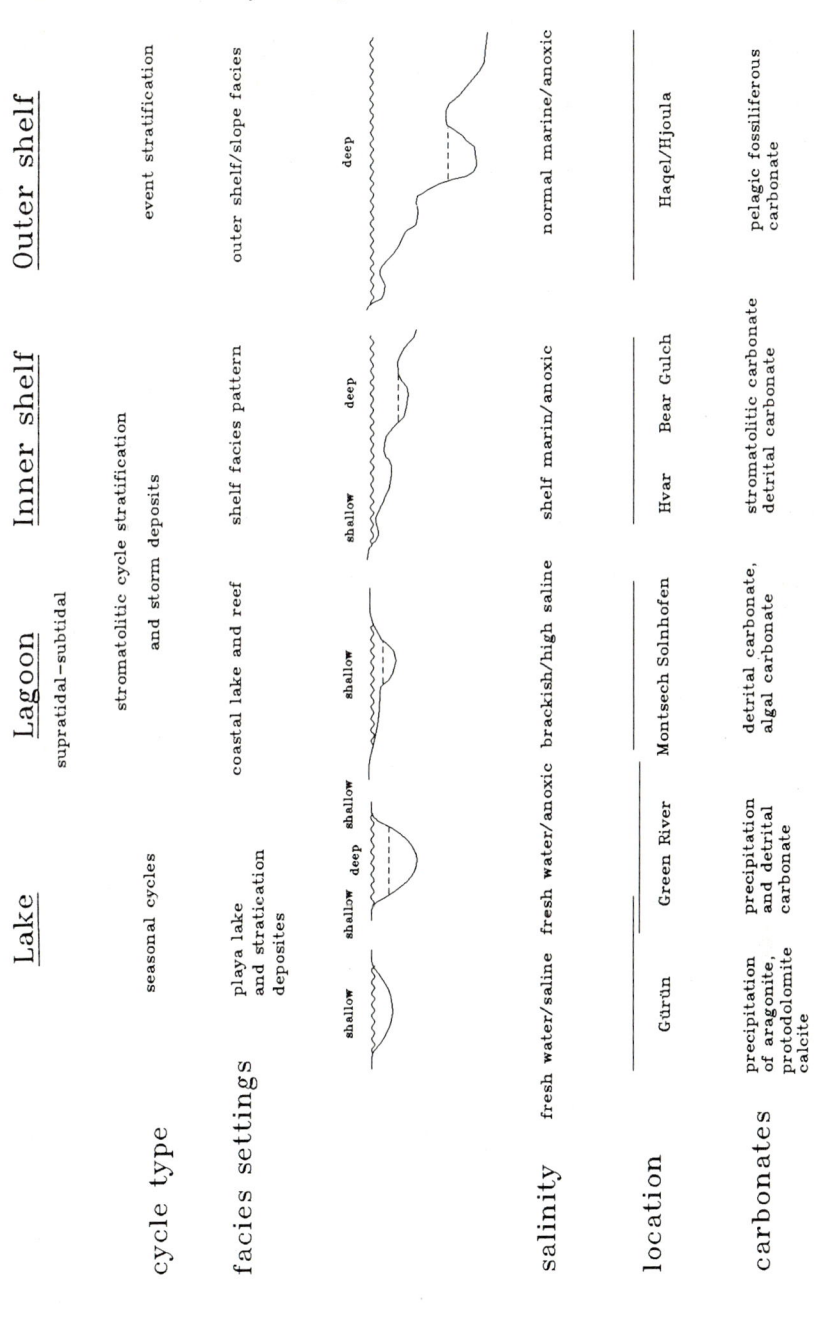

	Lake	Lagoon supratidal–subtidal	Inner shelf	Outer shelf
cycle type	seasonal cycles	stromatolitic cycle stratification and storm deposits	shelf facies pattern	event stratification
facies settings	playa lake and stratication deposites	coastal lake and reef		outer shelf/slope facies
salinity	fresh water/saline · fresh water/anoxic	brackish/high saline	shelf marin/anoxic	normal marine/anoxic
location	Gürün · Green River	Montsech Solnhofen	Hvar · Bear Gulch	Haqel/Hjoula
carbonates	precipitation of aragonite, protodolomite calcite · precipitation and detrital carbonate	detrital carbonate, algal carbonate	stromatolitic carbonate detrital carbonate	pelagic fossiliferous carbonate

Fig. 9. Summarizing graph showing the various geographic settings, cycle types, facies, salinity ranges, and types of carbonate

5.6 Biolaminations – Ecological Versus Depositional Dynamics

G. Gerdes, W. E. Krumbein, and H.-E. Reineck

1 Introduction

Biolaminations are products of benthic microbial communities. Unlike the more familiar hemispherical forms (Logan et al. 1964), they represent the flat laminated type of stromatolites. Usually, laminae of different composition and structure alternate, at least one being biogenically formed in situ by microbes that spread microfibrous, slime-interwoven coatings over sediments and rocks. Formerly known as algal mats, they are now called microbial mats (Brock 1976; Krumbein et al. 1979; Krumbein 1983; Cohen et al. 1984). The study of microbial mats has increased tremendously over the last few years, largely because of their seeming consistency through time (Awramik 1981; Knoll 1985).

The terms stromatolite (Kalkowsky 1908) and biolamination refer to laminated bodies, while individual laminae, ranging usually from micrometers to millimeters in thickness, may be termed stromatoids (Kalkowsky 1908) or biolaminites. The present paper emphasizes (1) the biolaminite level because of its role in the lamination process: Since the organic matter in biolaminations indicates the in situ production of biomass, each biolaminite is seen to represent a biological event that corresponds to a period of nonburial. The recurrence of these processes is evident in biolaminations. Emphasis will be further given (2) to two basic pathways of the biolamination process: burial by sedimentation versus self-burial by temporarily competing microbial populations, and to the timing of the respective pathways. (3) Strata corresponding to local scale dynamics (Fig. 1) may span hundreds to thousands of years. Within these time dimensions, biolaminations usually represent smaller episodes which may be rhythmic or event-related. The environmental significance of these biolaminated episodes will be considered. (4) In outcrops and cores, the recurrence of strata containing biolaminations may also indicate processes that made ecospace repeatedly available for mat-producing microbes. A well-documented example of cyclically recurring biolaminations in the Permian San Andres dolomite in west Texas (Shinn 1983, Fig. 46) was used in Fig. 1 to demonstrate possible frames of reference of which the local environmental situations are emphasized in the following text.

Einsele et al. (Eds.)
Cycles and Events in Stratigraphy
©Springer-Verlag Berlin Heidelberg 1991

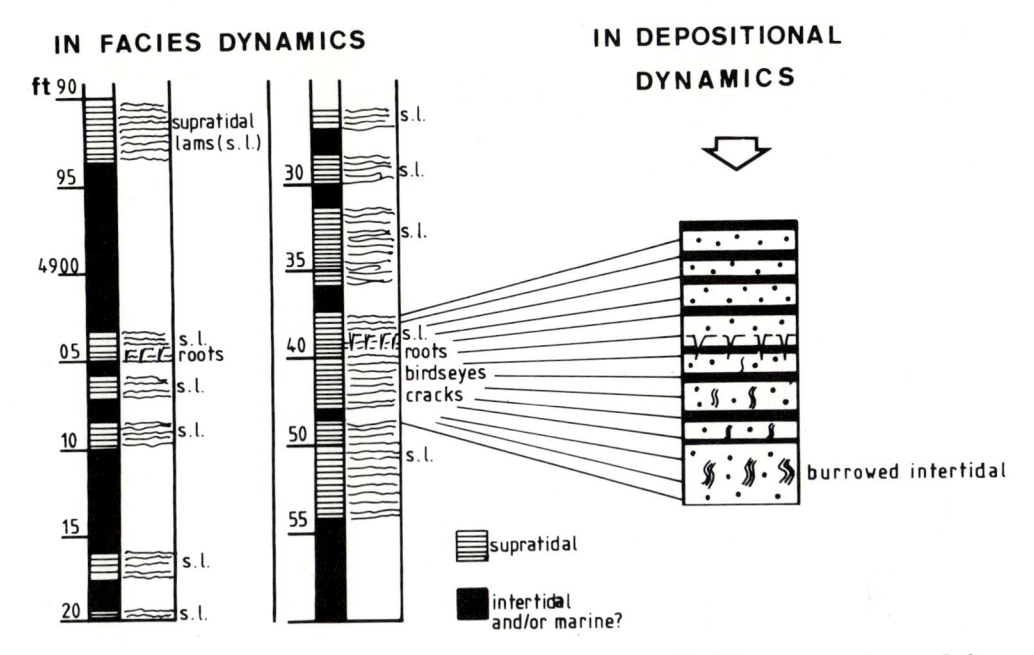

Fig. 1. Two scales at which recurrence of biolaminites can be analyzed in outcrops and cores. *Left* Macroscale variations interrupt repeatedly microbial mat records (length of core sections in feet; after Shinn 1983 Fig. 46). *Right* Microstratinomic record of microbial response to burial. The present chapter is concerned mainly with the scale on the *right*

2 The Biolaminite Level: in Situ Synthesis of Organic Matter

Microbial mats (Fig. 2) are the actively accreting units in biolaminations. In terms of ecology, they deal with succession in which formative and consuming stages are performed by microbes.

Formative stages. The organic matter usually forms via microbial primary production, defined as the process by which cells produce reduced organic carbon compounds from inorganic radicals using energy input from either solar energy or chemical bond energy of inorganic compounds (Krumbein and Swart 1983). Obviously, the two main pathways of primary production, photosynthesis and chemosynthesis, are used in the mat-forming tradition. Photosynthesis is a shallow marine phenomenon, while chemosynthesis is not depth-dependent (Jannasch 1984b). The greatest number of reports, particularly on stromatolites (Knoll and Awramik 1983), relates to the photosynthetic activity (Fig. 2). In deeper shelves and basins, mats are preferentially formed by chemosynthetic bacteria, although their sedimentological record is less conspicuous. Another reason may be the purely accidental lack of data due to the difficulty to sample undisturbed core material.

A third but rarer type of biolaminations is produced by fungal mycelia. Fungal mats have rarely been explored in modern environments; most reports are related to the fossil record (Bubenicek 1971; Gygi 1981; Dahanayake et al. 1985). Fungi may

Fig. 2. Microbial mats are microfibrous, slime-interwoven, coherent coatings of sediments and rocks. **a** Thick mat in Canary Islands salterns. **b** Mat from siliciclastic tidal flats, Mellum Island. (SEM microphotography, Scale: 300 μm)

flourish where large amounts of organic matter are available and salinity is reduced, as in tropical black water lagoons.

Consuming stages. The initial producer community prepares the way for a diverse anaerobic community with heterotrophic, and finally anoxygenic phototrophic bacteria (Fig. 3). These late members of the ecological succession drastically increase the chemical activity of the benthic community. A redistribution of chemicals in turn facilitates mineralization (Berner 1980; Berthelin 1983). Heterotrophic microbes are known to induce the transformation of various mineral elements to silicates, oxides, phosphates, carbonates, and sulfides. Such minerals cement the originally soft biolaminated substrate and increase preservation potential.

In conclusion, biolaminites indicate in situ production of organic matter. Rapid and frequent sedimentation impedes primary production, and thus indirectly also the course of microbial mat succession into mineralogically important anaerobic stages. These can develop also post burial, after a foundation of microbial biomass has been laid at the surface. Thus in situ synthesis of organic matter corresponds to periods of sediment starvation. This principal relationship will be stressed in the next section.

3 Laminated Growth as a Record of Time

3.1 Basic Processes: Upward Escape and Re-Establishment of Mats

The most important agents of biolaminated structures are filamentous bacteria (Fig. 4). The long tradition of these organisms is documented by stromatolites of Proterozoic (Knoll 1985) and Phanerozoic age (Monty 1981). Their significance is based on the following facts:

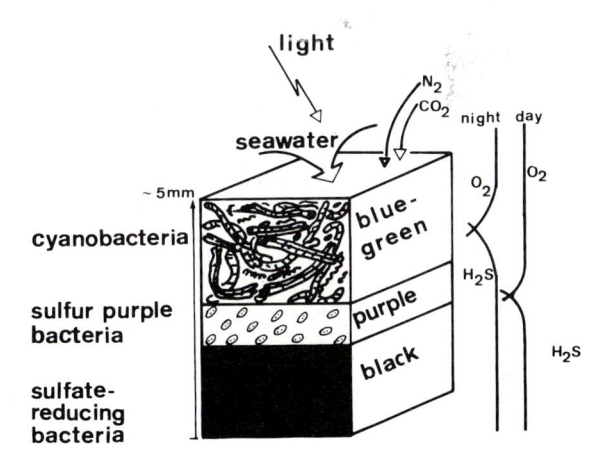

Fig. 3. Tiering of bacteria within a microbial mat community. (After Stal et al. 1985). Top layer (blue-green): photosynthetic cyanobacteria (earliest stage in succession and initial accumulation of biomass), bottom tier: heterotrophic sulfate-reducing bacterial (decay zone, iron sulfide enrichment), purple zone: sulfur purple bacteria (anoxygenic photosynthesis in the presence of free sulfide, latest stage in succession)

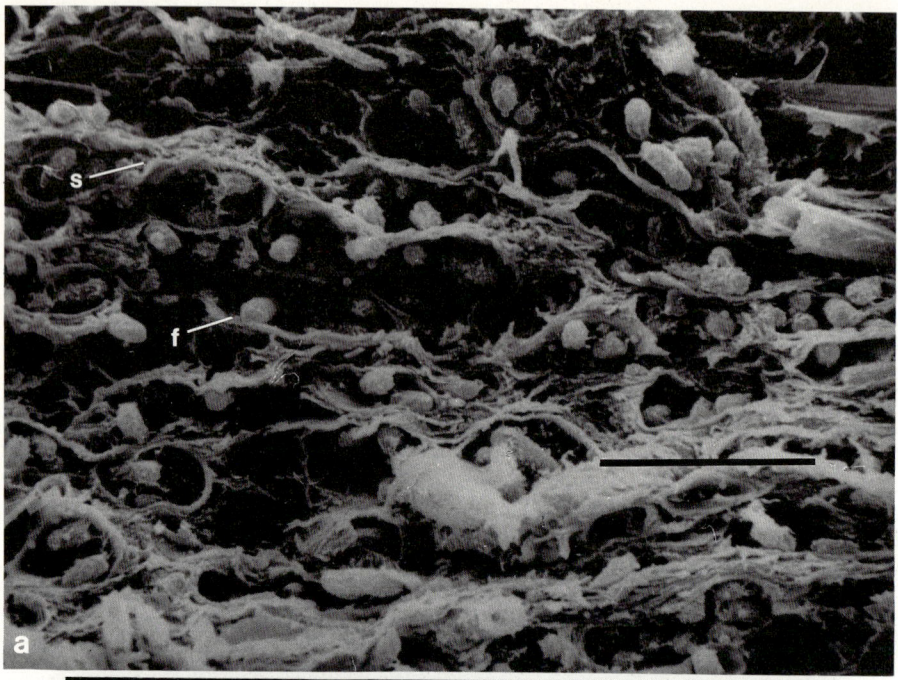

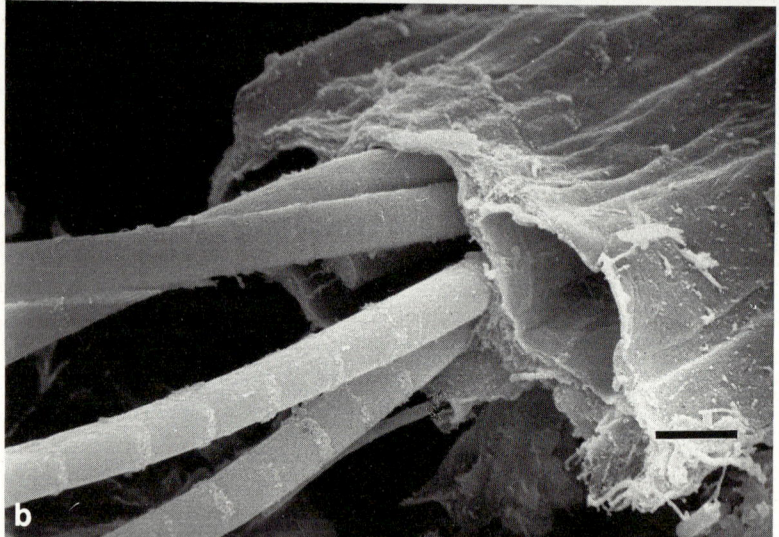

Fig. 4. Filament mats. **a** Cross-section of a mat containing various decay-resistant sheaths of filamentous cyanobacteria (*arrows: s* sheaths; *f* filaments); SEM-microphotography. Scale: 30 µm; total mat thickness is about 100 µm. **b** Filaments surrounded by thick polymer material. Scale: 3 µm

1. Filamentous bacteria are able of coat sediments and rocks with a characteristic jungle of filaments and polysaccharid sheaths (Fig. 2, 4). In modern coastal lagoons, such coatings are known to cover hundreds of square kilometres (Horodyski 1977).
2. A rapid gliding mobility (up to 3 cm/h; Castenholz 1969) allows them to escape from burial (Fig. 5), probably via gel excretion. Shading (Holtkamp 1985) as induced, e.g., by burial, serves as a trigger mechanism. After burial, loose filamentous aggregates appear at the surface after only a few hours, while a dense microfibrous meshwork (Fig. 5) is estimated to need several days of almost zero deposition, and a diverse anaerobic community (Fig. 3) needs several weeks to develop.
3. Filamentous microbes produce biopolymer-enriched extracellular sheaths (Fig. 4) which appear to resist microbial attack and remain after the protoplast has disappeared (Boon et al. 1985). Sheaths, slimes, immotile and dead cells are left behind when filaments escape from burial. The organisms re-establish the sheath

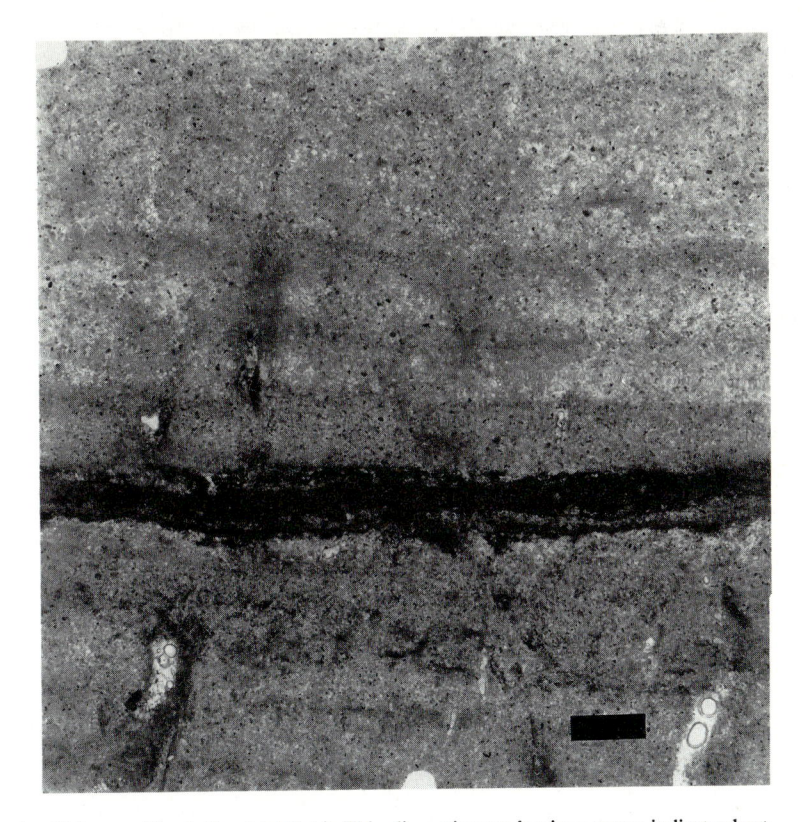

Fig. 5. Buried microbial mats (North Sea tidal flats). Thin discontinuous laminae at *top* indicate short productivity rates between sedimentation events. *Below* longer intervals between sedimentation events have allowed for increased biomass production so that sharp bedding planes and distinct lamination could develop (Scale: 1 cm)

material at the new surface level, provided almost near-zero sedimentation gives them the necessary time.

By gliding, overriding sediments and escaping from burial microbes can multiply into several new mats atop new sediment layers (Gerdes et al. 1985a; Gerdes and Krumbein 1987). Regeneration of buried mats is reflected by a characteristic rhythmic alternation of thin and relatively regular laminae (Fig. 6).

The two chief modes of burial will now be discussed:

1. burial by sediments,
2. self-burial by temporarily competing microbial populations without sedimentation (biogenic varvites).

3.2 Buildup Driven by Sedimentation

The sedimentation-driven buildup is characteristic for environments in which periods of nondeposition alternate with millimetric sedimentation events (Fig. 7a). Under such conditions, sediment blanketing serves as a trigger that forces the buried filamentous cyanobacteria to move upwards and override the new sediment surface. To underline the importance of episodic sedimentation for the formation of laminations, we may envisage continuous fall-out of sedimentary particles in pace with

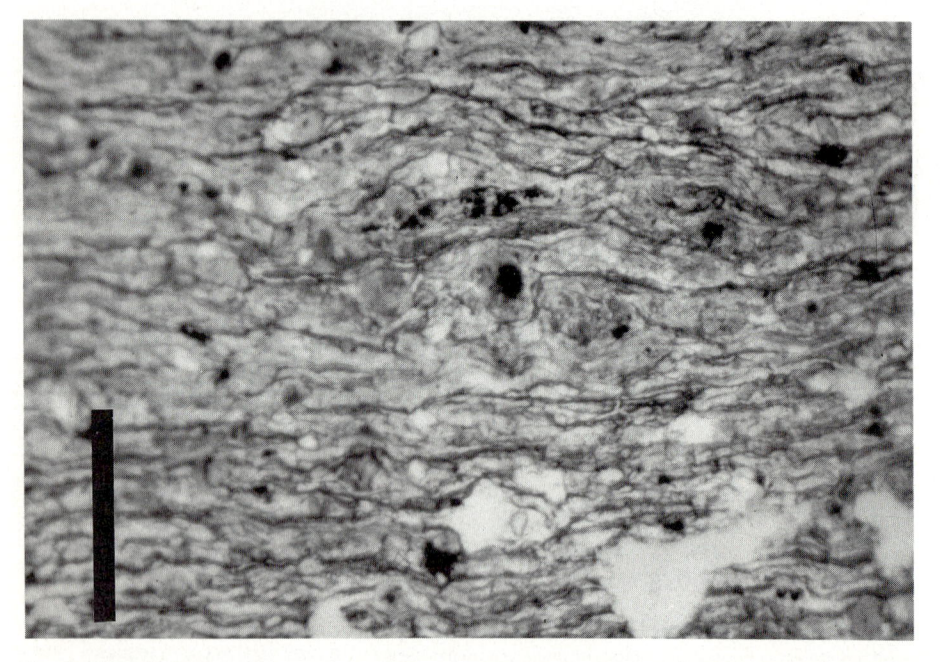

Fig. 6. Biogenic varvites (thin section made of soft Lanzarote material. One cm of sediment (scale) corresponds to about ten couplets of dark L_h- and light L_v-type lamina)

microbial migration. In this case, the bacterial front keeps continually abreast the sedimentation front, and the outcome is a nonlaminated sediment (Fig. 7b). Thus, lamination results from the interplay between physical processes and biological response in a regular succession:

1. Biological response to burial by low-rate sedimentation: organisms move from buried mats upwards.
2. Biological response to zero sedimentation: primary production, condensation of biomass.
3. In the next burial event, dead biomass is left behind as an organic-rich lamina (Fig. 7a).

The ultimate *product* of this mechanism is a biolamination with regular alternating couplets of sediments and microbial biomass.

Timing. The minimal time frame for filamentous mats to cope with sedimentation lies in the range of several days to several weeks. Assuming that a burial event takes a few hours and the re-establishment of a mat at almost zero sedimentation requires a minimum of 4 days, a biolamination of 3 cm including 16 lamina couplets of 2 mm reflects a time span of at least 2 months (Fig. 7a). Longer phases zero sedimentation will proportionally increase the thickness of the biomass increments as well as their preservation potential. Timings in Fig. 7a and b are thus minimum estimations.

3.3 Seasonal Zero Sedimentation (Biogenic Varvites)

Studies in natural environments have documented another pure type of biolaminated buildup termed "biogenic varvite". It presupposes lack of sedimentation over extended periods of time (Fig.7c).

Underlying mechanism. In this case, burial proceeds via mat-by-mat overriding and is triggered by temporal changes in ecologic conditions (temperature, light intensity, salinity). In response to such fluctuations, microbes override other topmats in order to gain the most favorable place within the environmental gradient. Recurrence is most likely in a seasonal pattern (Figs. 6, 7c).

Sedimentological products. Biogenic varvites are uniformly interlaminated sequences (Fig. 7c), consisting solely of microbial biomass. Material and texture of laminar couplets alternate due to a summer dominance of coccoid cyanobacteria and a winter dominance of filamentous cyanobacteria.

Summer layers. Coccoid cyanobacteria produce large amounts of extracellular gel. In the fresh state, the water-saturated viscous gel contributes to the vertical extension of biogenic varvites (Figs. 6,7c). Such gel-enriched interlaminae are known from various marginal marine environments. Krumbein et al. (1979) pointed out that sun light is channeled through the viscous gel and allows for photosynthesis even in deeper subsurface laminae. Being equipped with pigments and tremendous gel

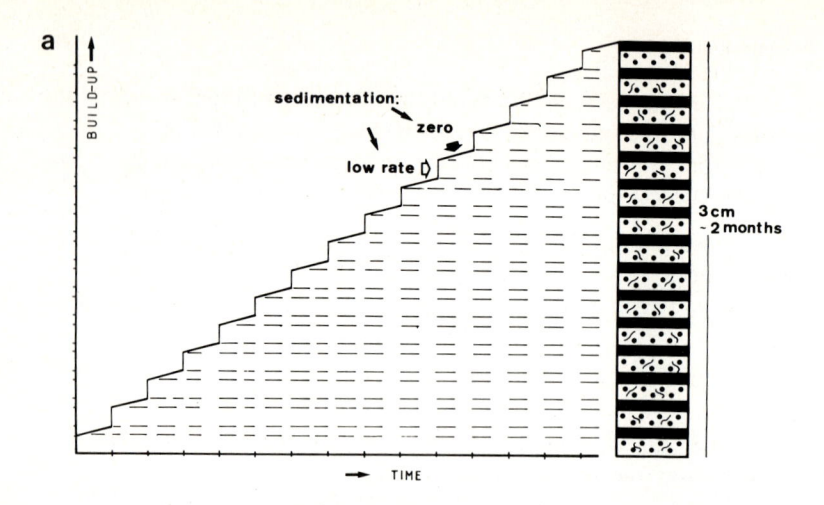

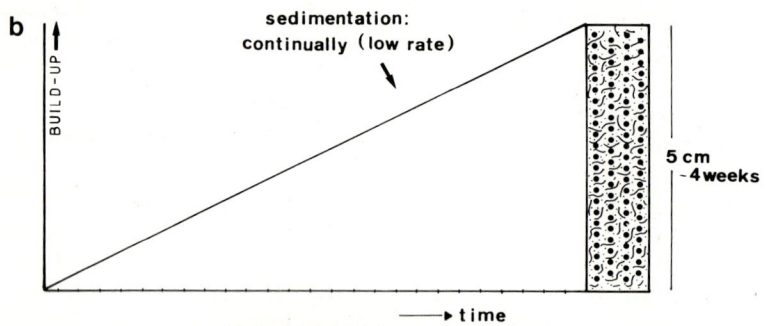

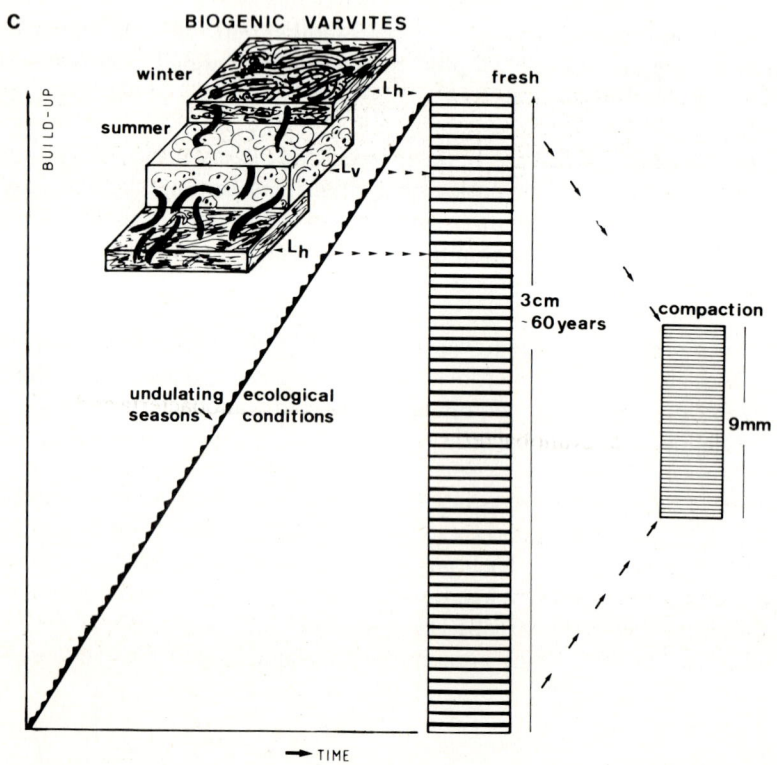

a

BUILD-UP ↑

sedimentation:
zero
low rate

→ TIME

3 cm
~2 months

b

BUILD-UP ↑

sedimentation:
continually (low rate)

→ time

5 cm
~4 weeks

c BIOGENIC VARVITES

winter
summer

L_h
L_v
L_h

fresh

BUILD-UP

undulating ecological
seasons conditions

→ TIME

3 cm
~60 years

compaction

9 mm

production, coccoids can tolerate solar irradiation better than filamentous cyanobacteria, which accounts for their dominance in summer.

Winter layers. During the summer, filamentous cyanobacteria thrive usually below a coccoid blanket that protects them against phototoxic effects. As soon as the intensity of solar irradiation decreases, however, they tend to override the hydroplastic surface layer formed by coccoids.

Alternating summer and winter dominance results in couplets of L_h-(laminae *h*orizontally oriented) and L_v-(laminae *v*ertically extended; Dahanayake et al. 1985). L_h-laminae are produced by filamentous microbes and appear denser than the L_v-laminae of coccoid origin which are swollen by water-saturated viscous gel. Accordingly, laminae dominated by filamentous cyanobacteria tend to be thinner and darker than the coccoidal L_v-laminae in the fresh state (Fig. 7c, see also Fig. 6).

Since coccoid cyanobacteria are less motile than filamentous species and need far longer periods of nonsedimentation to coat sedimentary surfaces, a viable record of alternating L_h-L_v-biolamination is possible only at sites where periods of nonsedimentation persist over a minimum of two seasons. This is not realistic in environments of open marine contact. Biogenic varvites have a better chance to develop in completely restricted lagoons. These interrelationships will be discussed in the next section.

Time frame. As calculated from fresh material from the Solar Lake (Krumbein et al. 1977), the overall growth of biogenic varvites is in the order of 0.5 mm per year corresponding to one couplet of light L_v- and dark L_h-laminae. In this case, a stack of 3 cm (which contains only 16 couplets in the sedimentation-driven case; Fig. 7a) would include 60 lamina couplets, each 0.5 mm thickness, and correspond to a time span of 60 years (Fig. 7c), which in this case is an exact, not a minimal, estimate.

Figure 7c also considers the effects of compaction and dehydration upon deeper burial which may reduce the fresh material to about 30% of its original thickness (Park 1976). In this process, the hydroplastic gel layers of coccoidal origin become disproportionally reduced. These calculations were made at Solar Lake, where biogenic varvites reach more than 1 m in thickness (Cohen et al. 1977a,b; Krumbein et al. 1977; Gerdes and Krumbein 1987).

Fig. 7. a Buildup of biolaminations, driven by sedimentation. The regular laminations result from the biological response to episodic low-rate sedimentation and intervals of zero deposition. Response to low-rate sedimentation: migration from buried mats below to new surface. Response to zero sedimentation: production and condensation of biomass (re-establishment of mats takes days to weeks). **b** Hypothetical model with a continuous rain of sediment particles atop a microbial mat, slow enough for microbial migration to cope with. The bacterial front follows the sedimentation front without producing laminae. **c** "Biogenic varvite", made up solely of microbial biomass. Depositional dynamics: zero over many seasons. Lamination process: mat-by-mat overriding (see also *left*). Summer populations of coccoid cyanobacteria (laminae swollen by gel = L_v-laminae) rhythmically alternate with winter populations of filamentous cyanobacteria (condensed filament layers = L_h-laminae). *Right* Effects of compaction and dehydration

In conclusion, the timing of biolaminated episodes must consider the two different modes of growth. In both, the underlying process is a biological response to environmental stimuli, but biogenic varvites and sedimentation-driven buildups differ with respect to both material and time signature. Biogenic varvites need much more time than the fastest sedimentation-driven buildups, but they are also more compactable. This difference is emphasized in the buildup-time diagrams, Fig.7.

3.4 "False Sediments" in Biogenic Varvites

In biogenic varvites of lower latitude environments, authigenic carbonates in particles develop, which could suggest sediment deposited by mechanical means, although they obviously generate within the mats (Friedman et al. 1973; Gerdes and Krumbein 1987). The formation of these particles is obviously climate-controlled. There is no in situ production of carbonates in biolaminations of the temperate humid climate zone, even not in biogenic varvites of salterns which grow under similar hypersaline conditions as in lower latitudes (see for example Fig. 8a: carbonates do not occur in microbial mats of salterns of Bretagne/France at 47°N.L. but within adequate mats of salterns of Canary Islands at 28°N.L). The grains grow predominantly within the hydroplastic, gel-supported L_v-laminae (Fig. 8b). Bacterial cells embedded in the gel act as biochemically active nuclei which catalyze or limit diffusional processes. This process may account for the particle forms of the carbonates. To differentiate between in situ-formed and sedimentated particles is still a problem. Useful criteria may be a common multitude of grain types within one and the same biolaminite (ooids, oncoids, peloids), furthermore grape-shaped arrangements resembling Kalkowsky's "ooid bags", finally characteristic sedimentary "eye structures" (Dahanayake et al. 1985), which evolve from the pressure of the growing particle on the surrounding soft organic matrix (Fig. 8c).

4 Biolaminations Stacked into Larger Sequences (Case Histories)

The history of a depositional environment (as indicated in Fig. 1) may span hundreds and thousands of years. Within the sedimentological record, biolaminations usually

---

Fig. 8. a Coastal areas of the microbial community type as discussed in this Chapter. *1* Sedimentation-aided growth of biolaminations: North Sea tidal flats. *2-6* growth without the aid of sedimentation (biogenic varvites): Salterns of France and Canary Islands (*2-4*) and closed hypersaline lagoons (*5, 6*). Biogenic varvites containing authigenic carbonate particles occur only in lower latitudes (*3- 6*). **b** Carbonate production in microbial mats (Canary islands salterns, thin section): dark L_h-laminae and light L_v-laminae, the latter containing carbonate particles of various size and shape (peloids, ooids, oncoids, grapes). Grain morphologies are induced by bacterial cells, cell colonies, gas and liquid bubbles, intra- and extraclasts serving as nucleation centers (Scale: 200 µm). **c** In situ formation of a peloid containing diatom tests in a matrix of L_h-laminae. Since the structure that evolves from the pressure of the growing particle on the soft matrix resembles augen gneiss, it is termed sedimentary augen structure (Dahanayake et al. 1985). SEM-microphotography. (Scale: 10 µm)

MICROBIAL MATS

1. Supratidal flats, North Sea
2. Salina, France
3. Salina, Lanzarote
4. Salina, Teneriffa
5. Gavish Sabkha, Sinai
6. Solar Lake, Sinai

○ carbonate formation

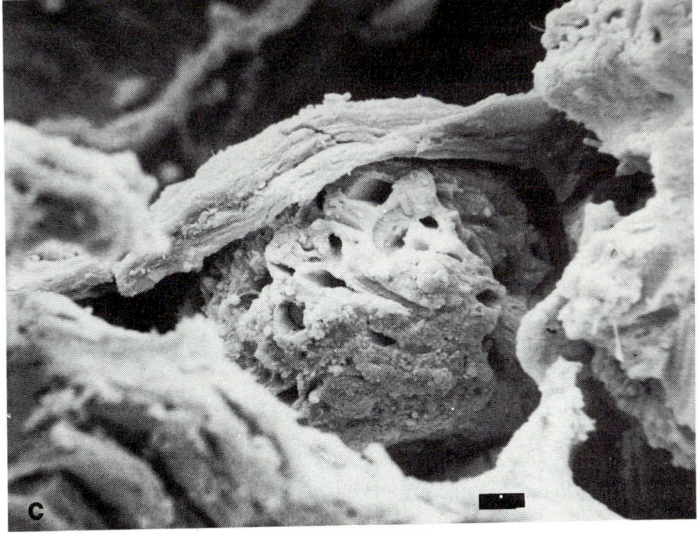

represent recurrent episodes. Figure 9 presents some case histories on the local environmental scale. Examples are based predominantly on own studies in modern environments and reflect marginal marine settings. The core sections compared are from

1. open high tide flats of the North Sea (Fig. 8a setting No. 1) and
2. peritidal lagoons that only have a subterranean connection with the open sea (Fig. 8a settings No. 5 and 6).

a) North Sea tidal flats. Microbial mats are exposed to marine as well as atmospheric dynamics. Cores (Fig. 9a) show biolaminated episodes of the sedimentation-driven type, buildup mainly by wind drift. Interspersed are clastic deposits of spring tides and storms. Storms are able to increase the tidal ranges and may occur throughout the year, as typical for temperate humid climates. The intercalated episodes of regular biolaminations indicate periods of quietness allowing for the

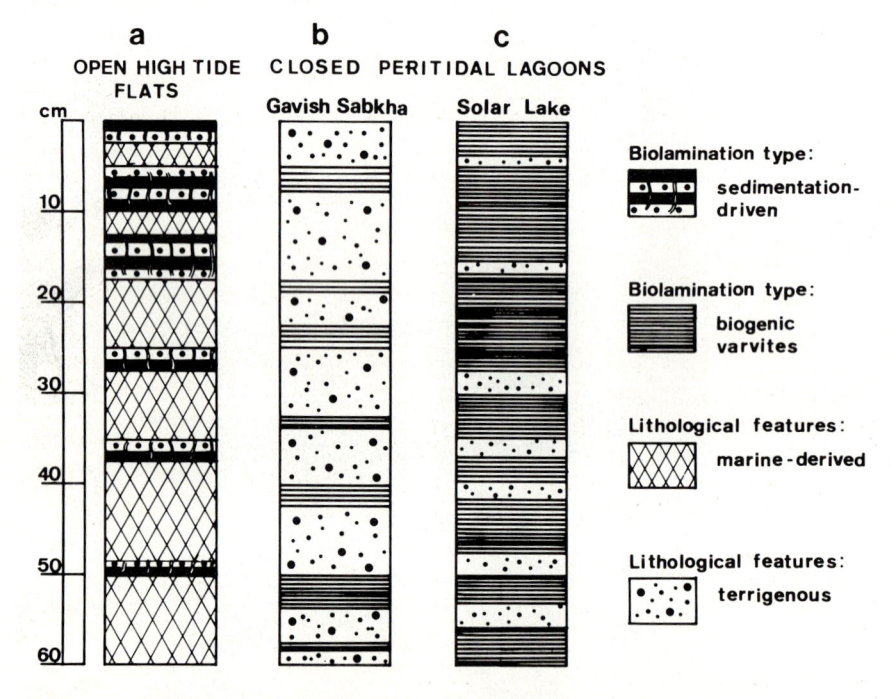

Fig. 9a-c. Schematic vertical sections derived from core studies of modern coastal environments. The stratinomic record shows environmental consequences. In open supratidal flats (**a**), marine contact leads to more massive beds (marine lithological characteristics) alternating with biolaminations of the sedimentation-driven type (situation as also indicated in Fig. 1: intertidal/supratidal boundary zone). In closed peritidal lagoons (**b, c**), biolaminations of the varvite type (long-term patterns) alternate with terrigenous sediments (e.g., from sheet floods). Grades of protection towards land are expressed in the two cores. **b** Gavish Sabkha, open towards land; biogenic varvites embedded between massive terrigenous sediments, **c** Solar Lakes, protected towards land by mountain ridges; biogenic varvites of extraordinary thickness alternating with thin sheet flood (pure biolaminations approx. 5 cm/100 years)

step-by-step raise of the bacterial front towards newly sedimentated bedding planes (sedimentation-driven buildup). Typical is also the upwards flattening of the buildup (left core, Fig. 9a), which reflects the gradual raise of the terrane from the intertidal into the supratidal zone corresponding with the decreasing tidal regime. The more microbial mats are protected against rapidly succeeding burial, the more the concentration of biomass increases. Sharp bedding planes and distinct laminations between organic-rich and inorganic material (see also Fig. 5) thus emphasize periods of quietness (starvation of sedimentation). Around coasts of open marine contact, this is in the supratidal rather than in the intertidal zone (cf. also Fig. 1).

b) Peritidal lagoons (cores Fig. 9b,c) usually have a complex burial/nonburial history. Being completely tide-protected, they may experience sedimentation events derived either from washovers (Hayes 1979) or when situated further inland, from sheet floods. Thus, biolaminated buildups represent in most cases only smaller episodes within a nonlaminated sequence. The comparison of two sequences (Fig. 9b,c), both from closed peritidal lagoons, illustrates also the importance of protection from terrigenous dynamics. In the core from Gavish Sabkha, sheet flood deposits dominate over biogenic varvites, as typical for settings open towards the land. In the core from Solar Lake, biogenic varvites dominate over thin sheet flood deposits. This lagoon is protected on the landward side by a mountain range. The biogenic varvites in both systems indicate sediment starvation over several years. This means they are controlled by terrestrial conditions rather than by the nearby sea.

In conclusion, biolaminites in cores and outcrops may aid in the recognition of paleoenvironmental conditions. Sedimentation may, but must not, be involved in growth dynamics. Both types (sedimentation-driven buildup versus biogenic varvites) indicate continued sediment starvation, due to a topographic protection against tides and other hydrologic events. Since biogenic varvites can form only when near-zero sedimentation persists over several seasons, their occurrence may suggest environments completely closed against normal marine dynamics (fail of transport by tide prism). Biogenic varvites buried by massive layers indicate storm events (washovers further seaward), or sheet floods (lagoons further inland). In a stratinomic sequence containing biolaminations, ecologic time scales may be gauged against depositional events (Fig. 9). A biolaminated episode may span over years, even if its record is only a few millimeters thick. Sedimentary blankets, in contrast, may have been deposited in a matter of hours, even if they are many times thicker than the biogenic varvites. Thus, rather different rates of deposition may be preserved in strata containing biolaminations.

5 Discussion and Conclusion

Biolaminites are flatly laminated accretional structures of microbial origin. Physical (erosion) and biological agents counteract the formation of regular biolaminites. Physical agents are wave action and currents. On the other hand, mat-producing biological systems are by no means only passive recorders of depositional dynamics, but serve as stabilizators, barriers, traps, and filters that produce and structure the inorganic substrate (Krumbein 1987). Flume experiments (Führböter and Manzenrieder 1987) have shown that microbially bound sands effectively protect the sediment against erosion, raising the critical point of grain movement five to eight times higher on microbially coated versus uncoated sand flats. Also biohermal stromatolites signal increased energy, modern examples being reported from current-dominated subtidal channels and wave-dominated lower intertidal areas (Dill et al. 1986; Pratt 1979).

Biological agents counteracting the formation of regular biolaminites include grazing and bioturbation (Garrett 1970) as well as epibenthic space competition (Pratt 1982). Stromatolites suffered a major crisis when metazoans radiated near the Proterozoic/Phanerozoic transition (Awramik 1981). Probably as a consequence, modern microbial mats are typically found in hypersaline or "schizohaline" environments (strongly fluctuating but never normal marine conditions, Folk and Siedlecka 1974). Here, their distribution is world-wide, for example at carbonate-evaporite (Shinn 1983; Stolz 1984) as well as quartzose shorelines (Gall and Larsonneur 1972; Reineck 1979; Gerdes et al. 1985a,b; Stal et al. 1985).

Besides limited erosion and benthic competition, one should also consider the sedimentational conditions that favor the production of biomats, namely sedimentation rates in terms of microbiological time. Where burial events follow in too rapid succession, microbes have no time to form recognizable mats even in azoic environments. Therefore the presence of biolaminites also indicates reduced sedimentation. Biogenic varvites require sedimentary starvation over years. This restricts the types of environments in which biolaminites can occur.

Protected lagoons and supratidal flats have yielded various model examples. However, the geological record yields also examples of subtidal and even deep marine stromatolites (Schieber 1989; Southgate 1989). Modern deep sea mats of several centimeters thickness, made of chemosynthetic bacteria, were found near hydrothermal vents at depth of 2,500 m (Jannasch 1984b). These mats are typical oxic/anoxic interface phenomena which may develop were the chemocline shifts around the sediment-water interface. Even benthic cyanobacteria are well equipped for thriving in deeper water (Monty 1970). Cells growing in dim light and anaerobic conditions develop especially high amounts of photopigment-containing vesicles. As recorded in the laterally persistent stromatolitic laminae of the Solnhofen platy limestone (Keupp 1977), depth-adapted microbial mats may have counteracted burial by calcareous planktonic nanno-organisms. Thus, although in open sea environments wave action has been proved to go as deep as about 200 m, microbial mats can develop below wave base since the mat-forming microbes are able to form under sulfide as well as under reduced light.

In conclusion, the following background conditions control the formation and preservation of biolaminites. Involved microbial taxa have the tendency to coat sedimentary surfaces with a slimy film and to react to burial by extraneous sediment or by other members of the same community by overriding. While the motile cells thus escape burial, sheaths, dead cells, immotile cells, and gels remain behind as an organic-rich interlayer. The amount of the buried organic material depends on the time span in which biomass production could go on undisturbed at the former surface.

Along with the bacterial conversion of biomass, biomineralization becomes catalyzed. The biochemically precipitated minerals forming in and around microbial cells, sheaths and capsules cement the soft substrate and increase the preservation potential of the biolaminated structures.

As seasonal climatic factors take over controls from sea dynamics, the buildup of biomass may go on for longer periods of zero sedimentation. Buildups produced under such conditions are composed solely of biomass and are structured by laminae different in material and composition: condensed meshworks of filaments and sheaths (L_h-type laminae) interlayered with gel-supported laminae of coccoids (L_v-type laminae). Low rates and frequencies of sedimentation and high rates of productivity and self-burial improve the chance laminations to leave a sedimentary record.

In a stratinomic sequence containing biolaminations, ecologic time scales contrast with the pulses of depositional events. A biolaminated buildup only a few millimeters thick may correspond to a time span of many years, while the much thicker clastic layers may have been deposited in a few hours. By filling the gaps between event deposits with a precise calendar, biolaminites may play a unique role in event stratigraphy.

Part II
Larger Cycles and Sequences

Introductory Remarks

Chapter 6   Sequences: Hierarchies, Causes, and Environmental Expression

Chapter 7   Timing and Correlation

Part II Larger Cycles and Sequences

Introductory Remarks

G. Einsele and W. Ricken

1 Introduction

Sequential ordering of beds is found in nearly all stratigraphic successions on various scales. It is thought to be the result from the combination of regional to global causal factors and modifying local environmental processes. The *causal factors* include tectonic, climatic, and eustatic changes (allocyclic processes), expressed as cyclic variations or single and multiple short-term events. However, all of these causal mechanisms are subject to *modifying factors* in the specific depositional environment. These factors include local differences in subsidence, supply, and composition of sediments, various hydrographic regimes, and autocyclic processes. In this way, a great variety of depositional sequences can be originated.

2 Transgressive-Regressive Cycle Hierarchy

Transgressive-regressive processes occur on a wide range of scales. On the one side of the spectrum, eustatic cycles occur with very low frequencies such as the first and second order cycles, while on the other side, high-frequency eustatic cycles exist (Vail et al. Chap. 6.1, this Vol.). Altogether, cycle frequencies are described in the literature to range from 50 Ma to 10 ka. These cycles reflect processes connected to global tectonics, ocean volume, and climatic changes. When various orders of cycles are superimposed, a very complex cycle pattern is generated.

The evolution of an individual sedimentary basin also reflects regional tectonic movements. It may comprise several tectonically controlled phases, for example the transition from a compressional to an extensional regime (Vail et al. Chap. 6.1, this Vol.). As a result, the subsidence rates change in the course of time and the subsidence-time curve shows a transition from a convex to a concave-upward form (Fig. 1a). This leads to a regional relative rise and fall of sea level (Fig. 1b) which is associated with a major transgressive-regressive facies cycle. Such a second order, tectonically controlled cycle typically lasts several tens of Ma.

Transgressive-regressive cycles commonly observed in the field range from the third order to the high-frequency eustatic cycles (Fig. 1c). The sedimentological expression of a third order cycle is the typical succession of depositional systems as defined in sequence stratigraphy, while the higher-frequency cycles are reflected by the smaller parasequences and beds. The time period of the *third order cycle* is usually

Einsele et al. (Eds.)
Cycles and Events in Stratigraphy
©Springer-Verlag Berlin Heidelberg 1991

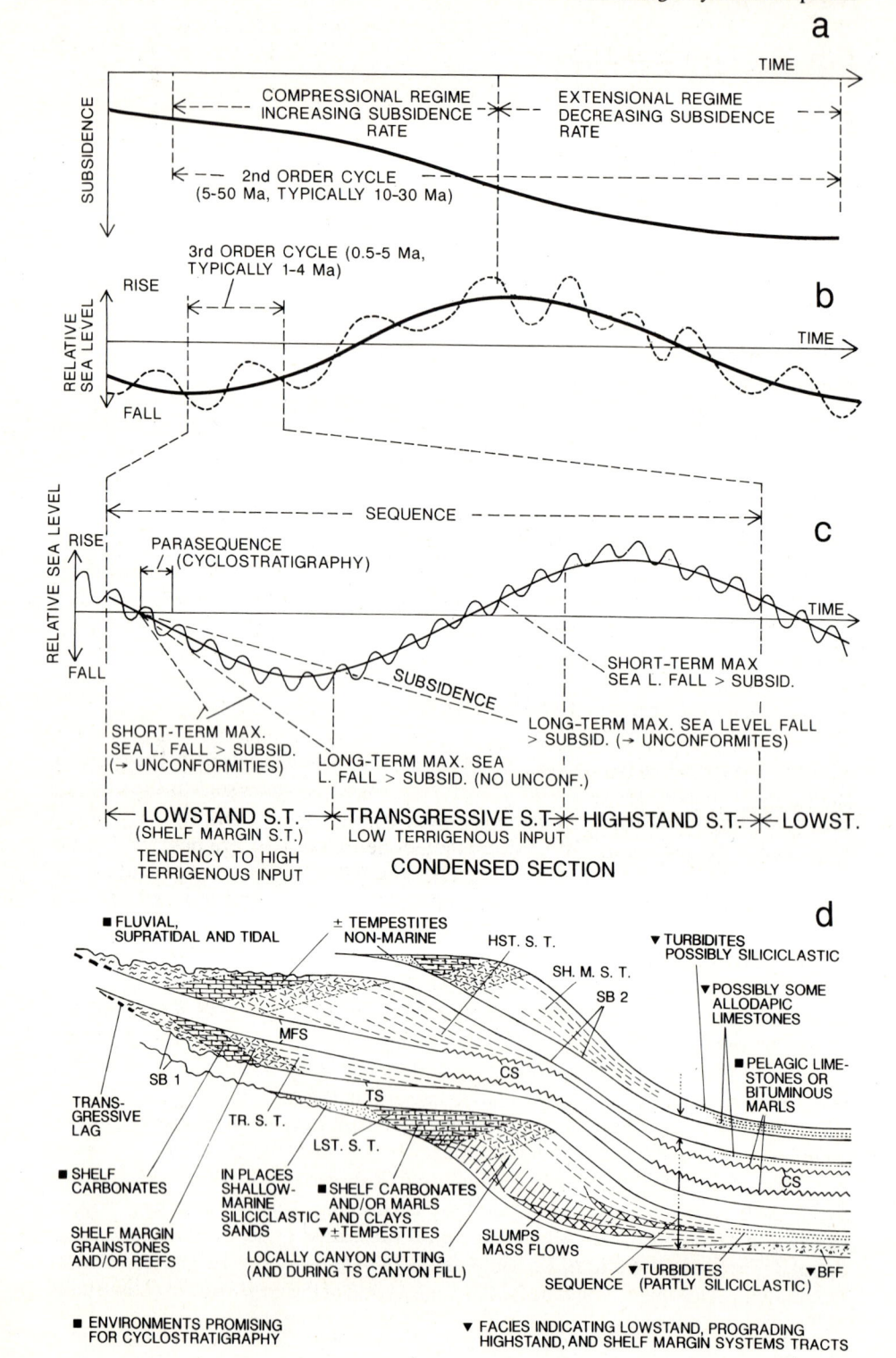

a

SUBSIDENCE / TIME

COMPRESSIONAL REGIME
INCREASING SUBSIDENCE RATE

EXTENSIONAL REGIME
DECREASING SUBSIDENCE RATE

2nd ORDER CYCLE
(5-50 Ma, TYPICALLY 10-30 Ma)

b

3rd ORDER CYCLE (0.5-5 Ma, TYPICALLY 1-4 Ma)

RELATIVE SEA LEVEL — RISE / FALL — TIME

c

RELATIVE SEA LEVEL — RISE / FALL — TIME

SEQUENCE

PARASEQUENCE (CYCLOSTRATIGRAPHY)

SUBSIDENCE

SHORT-TERM MAX. SEA L. FALL > SUBSID. (→ UNCONFORMITIES)

LONG-TERM MAX. SEA L. FALL > SUBSID. (NO UNCONF.)

SHORT-TERM MAX SEA L. FALL > SUBSID.

LONG-TERM MAX. SEA LEVEL FALL > SUBSID. (→ UNCONFORMITES)

LOWSTAND S.T. (SHELF MARGIN S.T.) TENDENCY TO HIGH TERRIGENOUS INPUT

TRANSGRESSIVE S.T. LOW TERRIGENOUS INPUT — CONDENSED SECTION

HIGHSTAND S.T.

LOWST.

d

- FLUVIAL, SUPRATIDAL AND TIDAL
- ± TEMPESTITES NON-MARINE
- HST. S.T.
- SH. M. S. T.
- SB 2
- MFS
- SB 1
- TRANSGRESSIVE LAG
- TR. S. T.
- TS
- LST. S. T.
- CS
- SHELF CARBONATES
- SHELF MARGIN GRAINSTONES AND/OR REEFS
- IN PLACES SHALLOW-MARINE SILICICLASTIC SANDS
- SHELF CARBONATES AND/OR MARLS AND CLAYS
- ± TEMPESTITES
- LOCALLY CANYON CUTTING (AND DURING TS CANYON FILL)
- SLUMPS MASS FLOWS
- TURBIDITES POSSIBLY SILICICLASTIC
- POSSIBLY SOME ALLODAPIC LIMESTONES
- PELAGIC LIMESTONES OR BITUMINOUS MARLS
- CS
- TURBIDITES (PARTLY SILICICLASTIC)
- SEQUENCE
- BFF

■ ENVIRONMENTS PROMISING FOR CYCLOSTRATIGRAPHY

▼ FACIES INDICATING LOWSTAND, PROGRADING HIGHSTAND, AND SHELF MARGIN SYSTEMS TRACTS

between 1 and 5 Ma, and the amplitude of sea level change during this time spans some tens of meters (Van Wagoner et al. 1988; Vail et al. Chap. 6.1, this Vol.). Third order cycles are subdivided from top to bottom into (Fig. 1d):

1. Highstand systems tract.
2. Transgressive systems tract.
3. Lowstand systems tract or shelf margin systems tract,
4. Type 1 or type 2 sequence boundary, respectively.

Each of the systems tracts has a defined seismic signature and is expressed by detailed sedimentological attributes (Vail et al. Chap. 6.1, Kidwell Chap. 6.3, both this Vol.). The development of a lowstand or shelf margin systems tract with related type 1 or type 2 sequence boundary depends on the relationship between sea level fall, subsidence, and sediment supply.

3 Eustatic Cycles in the Milankovitch Frequency Band

The Milankovitch climatic cycles discussed in Part I of this book have time periods which are 10 to 100 times shorter than the third order eustatic cycles. They represent the "parasequences" or "simple sequences" in the Vail nomenclature. If the orbital signals of the Milankovitch cycles are transformed into climatic variations causing minor oscillations of sea level, then the third order variations in sea level are superimposed by a great number of high-frequency but (in nonglaciated times) low-amplitude eustatic oscillations (Fig. 1c). Increasing evidence indicates that such *high-frequency sea level variations* occurred throughout the Phanerozoic: $\delta^{18}O$ variations with periods equivalent to those of the Quaternary (Shackleton and Opdyke 1976) are describes in deep-sea sediments since the Neogene (e.g., Keigwin 1986); shallow water carbonate and evaporite cycles are known from many examples since the Precambrian (e.g., Grotzinger 1986b; Goldhammer et al. 1987; Koerschner and Read 1989; Strasser Chap. 6.5, Haas Chap. 6.6, both this Vol.). Such small sea-level variations (amplitude 1 to 10 m) seem to be related to the storage and release of continental water, affecting ground water reservoirs and lakes, vegetation, as well as

---

Fig. 1. Some general principles of the interrelationship between tectonically controlled regional subsidence and eustatic sea level variations of different time period and amplitude. **a** Regional subsidence-time curve caused by transition from compressional to extensional tectonic movements (based on Vail et al. Chap. 6.1, this Vol.). **b** Second order transgressive-regressive cycle caused by **a**, superimposed by third order eustatic cycles. **c** Third order sea level variation and associated sequence and systems tracts (based on van Wagoner et al. 1987), superimposed by short-period Milankovitch-type oscillations (fourth to fifth order cycles, Vail et al. Chap. 6.1, this Vol.). Note that the short-period oscillations display higher rates of maximum sea level fall and, therefore, may be better recorded in coastal environments than the long-period variations. **d** Systems tracts of a third order sequence composed of mixed carbonate and siliciclastic sediments, passive continental margin, *SB* Sequence boundary; *LST S.T.* lowstand systems tract; *TR S.T.* transgressive systems tract; *HST S.T.* highstand systems tract *SH.M.S.T.* shelf margin systems tract; *TS* transgressive surface; *MFS* maximum flooding surface; *CS* condensed section. (After Posamentier and Vail 1988; Sarg 1988)

some possible mountain and polar ice. Bundling of some of these cycles into roughly 5:1 units favors the interpretation of orbital control. Yet, most shallow water carbonate cycles also seem delicately balanced between high-frequency eustatic variations and autocyclic tidal flat migration processes (Ginsburg 1971; Strasser Chap. 6.5, this Vol.).

4 Interrelationship Between Sea Level Fall, Subsidence, and Sequence Type

Transgressive-regressive cycles on all scales are controlled by *eustatic variations*, the degree of *sea floor subsidence*, and the rate of deposition. Erosion of previously deposited sediments can only occur when the rate of sea level fall is greater or equal to the rate of sea-floor subsidence (Einsele and Bayer Chap. 6.2, Vail et al. Chap. 6.1, both this Vol.). In this case, shallow-marine sediments may undergo subaerial erosion, or storm-induced currents can remove older marine deposits and produce type 1 sequence boundaries. When, however, the rate of maximum sea level fall is lower than the rate of subsidence, sediment accumulation at a specific site is continuous, and no distinct sequence boundary is generated.

It is interesting to note that the maximum rate of sea level fall is much greater for higher frequency oscillations than for the lower ones (Fig. 1c). Consequently, high-frequency eustatic variations have a greater potential to cause erosion and, thus, sequence boundaries. This holds true even for phases of long-term sea level rise (Fig. 1c). Such an erosion potential is, however, restricted to shallow-water environments. Particularly peritidal carbonates which undergo early cementation and lithification, have been found to record and preserve such high-frequency eustatic variations (Haas Chap. 6.6, Strasser Chap. 6.5, both this Vol.).

Whether peritidal carbonate cycles are shallowing-upward (e.g., Strasser Chap. 6.5, this Vol.) or deepening-upward (i.e., Lofer Cycles, Haas Chap. 6.6, Fischer Chap. 1.2, both this Vol.) depends on the differing ratio between sedimentation and subsidence rates. The so-called punctuated aggradational cycles (PACs) of Goodwin and Anderson (1985), therefore, describe only one end member (i.e., the unconformity-bounded shallowing-upward cycle) of the various types of shallow water cycles related to high-frequency eustatic variations.

5 The Position and Attributes of High-Frequency Eustatic Variations Within the Third Order Cycles

High-frequency eustatic variations are expressed differently within the third order transgressive-regressive cycle. Koerschner and Read (1989) describe high-frequency sea level variations with low amplitudes during both falling and rising phases of third order sea level changes for Cambrian shelf carbonates (Fig. 2). During the slowly rising sea level (i.e., transgressive and partly highstand systems tracts), the superimposed high-frequency oscillations permit the deposition of thicker shallow-water carbonate cycles with higher proportions of subtidal sediments than during slowly falling sea level. If the rate of the longer-term sea level fall is equal to subsidence,

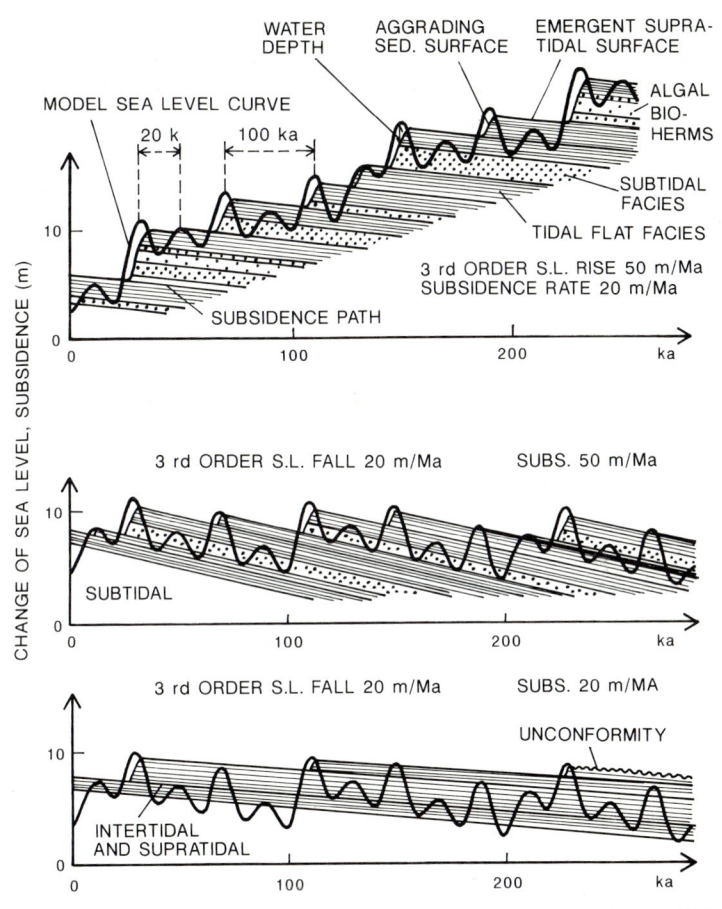

Fig. 2. Simulation of the effects of short-period Milankovitch-type sea level oscillations on carbonate tidal flats. The small oscillations are superimposed on longer-term (third order) rise or fall of sea level and are associated with a changing subsidence rate. Subtidal sediments are partially represented by reefs, supratidal sediments by algal mats. Note the differences in facies and thickness of individual cycles. (After Koerschner and Read 1989)

only intertidal and supratidal sediments can accumulate, interrupted by unconformities.

In other environments, the sedimentological expression of the short-term Milankovitch-type sea level oscillations and their relation to the systems tracts of the third order sequences are less known. One of the reasons for this uncertainty is the fact that most experts interested in short-period cyclostratigraphy did not consider the larger-scale phenomena that partly govern the lithological expression of small-scale cycles. In siliciclastic shallow water environments, the potential to generate and preserve high-frequency cycles is low compared with peritidal carbonate cycles which undergo early lithification. Some of the coal cycles described by Riegel (Chap. 6.7, this Vol.) may represent siliciclastic counterparts of the shallow-water carbonate cycles. Whether it is possible to obtain a more precise time-span estimation for the third order transgression-regression cycles by utilizing high-frequency eustatic sea level variations as the basic, quasi-periodic units is an open question. Peritidal carbonates with moderate subsidence rates might be suitable for such an attempt; various timing methods may be used as presented in Chapter 7.

Marl-limestone sequences, event beds, and black shales occur during different phases of a large-scale transgressive-regressive cycle. Periodic limestone-marl successions and black shale-limestone alternations occurring on the outer shelf or in the deeper sea, preferentially form during the transition from transgressive to highstand deposits (condensed section, Fig. 1c, d). Episodic event beds such as mass flows, tempestites, and turbidites, however, seem to be associated with the prograding phases of the lowstand and the highstand systems tracts (Fig. 1d).

6 Additional Problems

Although eustatic sea-level variations have become the most comprehensive concept in sequence stratigraphy, three modifying factors are frequently neglected:

1. The overall tectonic subsidence of a region and additional subsidence due to sediment and water loadings (e.g., Steckler et al. 1988),
2. The ability of a certain depositional environment and sediment facies to record smaller and larger sea level variations, even in the deep sea (Henrich Chap. 6.8, this Vol.), and
3. The location of the sequence in question relative to the margin or center of a basin (Vail 1987; Vail et al. Chap. 6.1, Einsele and Bayer Chap. 6.2, both this Vol.).

Such parameters control the thicknesses of individual sedimentary cycles, the sequence type (i.e., coarsening or fining upward, etc.), the occurrence and duration of unconformities at the bases and tops of such sequences (Kidwell Chap. 6.3, this Vol.), and, to some extent, the faunal evolution (McGhee et al. Chap. 6.4, this Vol.).

Chapter 6 Sequences

6.1 The Stratigraphic Signatures of Tectonics, Eustasy and Sedimentology – an Overview

P. R. Vail, F. Audemard, S. A. Bowman, P. N. Eisner, and C. Perez-Cruz

1 Introduction

Stratigraphic signatures result from the interaction of tectonic, eustatic, sedimentation, and climatic processes. Tectonic and eustatic processes combine to cause relative changes of sea level which control the space available for sediments (accommodation space). Tectonism and climate control the amount and types of sediments deposited. The resulting sediment supply determines how much of the accommodation space is filled. Sedimentation processes in fluvial and marine environments respond to currents which interact with topography and bathymetry to determine the facies of the sediments that fill the space. Sequence stratigraphic analysis coupled with subsidence and tectono-stratigraphic analysis can be applied to understand the stratigraphic signature of each variable within the stratigraphic record and quantify them in terms of time and space. From this analysis, each stratigraphic signature can be related to a particular cause.

An important requirement for understanding processes in the rock record is to link the appropriate process with an observation and correlate similar observations from different regions which occurred within an equivalent time period. Therefore, the geologist must define time-equivalent rock packages that are genetic intervals. A genetic interval is bounded by correlative physical surfaces and may contain rocks with differing facies, but it is deposited contemporaneously and lacks significant hiatuses.

The two most important criteria necessary for distinguishing tectonic, eustatic and sedimentologic signatures are the distribution in time and space of their effects on accommodation space (Fig. 1). Each signature can be linked to a process when it is understood how each process affects accommodation space. These effects can be differentiated by determining:

1. Their rates;
2. The duration over which the rates are active;
3. Their rates of change or pattern of change, such as periodic cycles, random or step functions; and
4. Their distribution in space, by documenting their local, regional, or global extent.

For purposes of discussion, the signatures are grouped in terms of orders of duration (Table 1). These groupings define one or more of the following:

Einsele et al. (Eds.)
Cycles and Events in Stratigraphy
©Springer-Verlag Berlin Heidelberg 1991

Signatures	Tectonics			Eustasy			Sedimentation
	Sedimentary basin	Major Transgressive/ Regressive facies cycle	Folding Faulting Magmatism and Diapirism	Major continental flooding cycle	Major Transgressive/ Regressive facies cycle	Sequence cycle Systems tracts Periodic parasequence	Depositional system Lithofacies tracts Episodic parasequence Marker beds Beds Laminae
Distribution in space	Regional	Regional	Local	Global	Global	Global	Local
Distribution in time	1st Order Episodic Event	2nd Order Non-periodic	3rd Order Episodic Event	1st Order Cycle	2nd Order Cycle	3th-6th Order Cycles	Episodic Event
Causes	Crustal extension Thermal cooling Flexure loading	Changes in rate of 1) Tectonic subsidence 2) Sediment supply	Local and regional stress release	Changes in ocean basin volume	Changes in ocean basin volume	Changes in climate, water volume	Local sedimentary processes

Fig. 1. Signatures in the stratigraphic record. This diagram presents the distribution in time and space of the six stratigraphic signatures in the rock record

Table 1. Duration of each order

Order	Duration
1	50+ Ma
2	3–50 Ma
3	0.5–3 Ma
4	0.08–0.5 Ma
5	0.03–0.08 Ma
6	0.01–0.03 Ma

1. The time period over which the process is active,
2. The time period over which the process will not change appreciably in character, or
3. The period of a cycle or duration of an event.

Integrating the results of sequence stratigraphic, subsidence, and tectono-stratigraphic analyses yields a geologic history interpretation with improved temporal resolution and greater accuracy in predicting lithology. This permits the unraveling of the stratigraphic signatures due to tectonic, eustatic, and sedimentation processes. Sequence stratigraphic analysis divides the stratigraphic record into physical chronostratigraphic units in which the lithofacies are genetically related. Subsidence analysis determines the total and tectonic subsidence history of a locality. Tectono-stratigraphic analysis determines the kinematics of faulting, folding, and igneous activity and their relationship to plate-tectonic events. The interpretations gained from these approaches can be further complimented with basin fill simulations to reconcile the geologic interpretations with a geologic database (Lawrence et al. 1990).

The stratigraphic signature of tectonism results from a wide range of processes and has the most profound effect on accommodation. Its imprint on the sedimentary record can be divided into three hierarchical groups (Fig. 1), including:

1. Uplift and basin evolution;
2. Changes in subsidence rates; and
3. Folding, faulting, magmatism, and diapirism.

The evolution of a sedimentary basin is interpreted as a first-order tectonic event. A basin is its stratigraphic signature. Second-order tectonic events are indicated by changes in the rate of tectonic subsidence in the basin or by changes in the rate of uplift in the sediment source terrain. Major transgressive-regressive facies cycles are their stratigraphic signature. Folding, faulting, diapirism, and magmatism make up the third-order tectonic events. Folding, faulting, and diapirism tend to make sequence boundaries easier to identify in areas of structural activity and more difficult to identify during periods of rapid subsidence. However, even in these tectonically active areas, the ages of sequence boundaries when dated at the minimum hiatus at their correlative conformities match the age of the global eustatic falls and not the plate tectonic event causing the tectonism. Therefore, tectonism may enhance or subdue sequence and systems tract boundaries, but does not create them. Third-order tectonic events may trigger the deposition of stratigraphic marker beds, such as slumps and megaturbidites. Magmatism will produce bentonite marker beds and datable igneous sills and dikes.

Eustatic changes in sea level produce systematic cycles within the accommodation space created by tectonism and the rising and falling of eustasy. Eustasy profoundly influences the distribution of sedimentation and erosion. Rapid rates of eustatic change are the major controlling factor on the timing of stratigraphic discontinuities. These discontinuities create both the boundaries between sequences and between systems tracts within a sequence. Eustatic effects are divided into two stratigraphic signatures: major continental flooding cycles and depositional sequence cycles. Major continental flooding cycles are believed to be caused by tectono-eustasy (changes in ocean basin volume). Depositional sequence cycles are thought to be caused by glacio-eustasy (changes in water volume).

Sedimentation effects are episodic in nature and local in distribution. Their stratigraphic signatures are depositional systems. Depositional systems are comprised of laminae, lamina sets, beds and bedsets, and episodic parasequences (Campbell 1967). These units contain the synsedimentary structures that reflect the environment of deposition and associated processes. Particular depositional systems and lithofacies tracts predominate during the formations of particular types of systems tracts.

The association of certain types of depositional systems with specific systems tracts provides a unique approach for more accurate mapping of paleogeography. This enables more accurate interpretation of geologic history and predictions of rock type ahead of the drill for resource evaluation.

1.1 Methodology

The physical chronostratigraphic units defined by sequence stratigraphic criteria are important for making more accurate basin analyses, paleogeographic reconstructions, geologic history interpretations, resource evaluations of sedimentary basins, and global stratigraphic correlations. Traditional approaches to stratigraphic analysis have different shortcomings in interpreting paleogeography and geologic history. Physical lithostratigraphic methods use boundaries which may be time-transgressive. Chronostratigraphic methods use boundaries that are not physical surfaces. Marker-bed correlations are physical surfaces which may be chronostratigraphic, but their selection may be arbitrary and differ from one basin to another, or from one worker to another. Sequence stratigraphy is an interpretative approach that can integrate outcrop, well-log, and seismic data. This approach provides correlation tools with a conceptual model for the geologic response of depositional and erosional processes to cyclic base-level changes that identifies and defines the genetic character of the different types of physical surfaces and stratigraphic intervals within the rock record.

Stratigraphy is the discipline that studies the vertical and lateral relationships of sedimentary rocks. These relationships are defined on the basis of physical and chemical rock properties, paleontologic characteristics, age relationships, and more recently, geophysical properties. Prior to the 1970's, stratigraphy was mainly concerned with the classical concepts of lithostratigraphic, biostratigraphic and chonostratigraphic successions, and their lateral correlations. Codification of stratigraphic procedures like those presented by Hedberg (1976) dominated much of the discussions concerning stratigraphic principles.

Lithostratigraphy deals with rock packages, their facies relationships and their organization into mappable units (formations, groups, etc.). Biostratigraphy is the study of correlative units (zones) based on their fossil content. Chronostratigraphy deals with rocks deposited within a particular interval of time; the classical chronostratigraphic units are called systems, series, and stages. Magnetostratigraphy is a branch of stratigraphy which deals with the signature of the magnetic polarity reversals in sedimentary successions and layered volcanic rocks. Magnetostratigraphy can supplement paleontologic data and help refine their temporal equivalency. These approaches in stratigraphic analysis have been used to subdivide and map most of the outcropping sedimentary rocks in the world.

Transgressive-regressive facies cycles are observed in basins throughout the world. They are characterized by a landward to basinward oscillation of lithofacies in association with the shoreline through the geologic development of each basin. In the early part of this century, Suess (1906) introduced the term eustasy and defined it as global sea-level change. He believed eustasy caused these transgressions and regressions of the shoreline. The eustatic mechanism became unpopular as studies on other continents showed that transgressions and regressions did not correlate globally. The poor correlation of transgressive-regressive facies cycles around the North American continent has been documented by Hubbard (1988). He shows how these cycles are related to local uplift and subsidence events.

In studies of the Pennsylvanian and Permian of West Texas, Brown (1969) developed the concept of depositional systems. Depositional systems are three-dimensional assemblages of lithofacies that were deposited under similar depositional conditions. Depositional systems have consistent occurrences within the various systems tracts produced during a eustatic cycle. The depositional systems observed in sequence stratigraphic studies include submarine fans, contourites, levee channel turbidites and attached lobes, soils, incised valley fills, deltas, slumps, offshore shelf deposits, coastal belt sands, ravinements, estuaries, marshes, lakes, dunes, fluvial and coastal plain deposits in clastic settings, as well as reefs, lagoons, tidal flats, talus slopes, and evaporites in carbonate settings. These depositional systems are based upon models established in the literature (Reading 1982). Depositional systems are very useful for relating sedimentologic criteria to depositional environments.

Correlation of marker bed patterns utilizing wireline well logs has been in use for several decades to map geologic time lines using physical criteria. This technique involves the matching of a series of unique log patterns between wells. It works well in marine areas where there are small contrasts in bathymetric relief. Depositional settings such as epicontinental seas or deep marine basins commonly have time-stratigraphic marker beds which can be traced for hundreds of kilometers (Hattin 1971). However, in nonmarine and marginally marine areas and where there are large differences in bathymetric relief, as occurs along a basin margin, correlations are difficult at best.

Physical units such as formations, transgressive-regressive facies cycles, and depositional systems are not reliable for making intrabasinal correlations because the boundaries of these units may be time-transgressive. A time-transgressive boundary is one in which the units above the boundary may be the same age as units below the boundary elsewhere. A time-transgressive boundary often does not form a continuous physical boundary. Instead, the boundary consists of a series of different physical boundaries that step up or down in terms of geologic time. The unwitting use of time-transgressive boundaries for mapping can result in misleading interpretations of structure, distribution of lithologies, and paleogeography. Likewise, chronostratigraphic units defined by paleontology or magnetostratigraphy have problems because their boundaries cannot be defined by physical criteria and cannot be traced in the field or subsurface. Intervals bounded by marker beds are physically defined chronostratigraphic intervals. However, marker beds are not always present, and when they are, the criteria used for their selection is often inconsistent. It must also be kept in mind that intervals which are defined by marker beds may contain significant hiatuses and thus may not be genetic.

Sequences, which are unconformity bound units, were proposed (Sloss et al. 1949; Sloss 1963) as an alternative way to subdivide and map major rock units. Although considered as lithostratigraphic units by Sloss, they can be used as chronostratigraphic units if the bounding unconformities are traced to the minimal hiatus at their conformable position and age dated with biostratigraphy. In this way a sequence represents all the rocks deposited in the interval of time between the age of the two conformities. The rocks above the sequence boundary are always younger than the rocks below.

Vail et al. (1977) demonstrated that primary seismic reflections follow geologic time lines and thus seismic sections show chronostratigraphy at the resolution of the seismic data plus or minus the breadth of one half of a reflector. Based on this observation and the concept of sequences of Sloss et al. (1949, 1963), a seismic stratigraphic interpretation procedure was developed utilizing discontinuities as the basis for subdividing sedimentary rocks (Vail et al. 1977). These studies resulted in a set of cycle charts for the Mesozoic and Cenozoic showing the distribution of discontinuities on a global scale (Haq et al. 1987, 1988). The development of the understanding that these discontinuities were controlled by relative changes of sea level, and that relative changes of sea level could also be recognized on well logs and outcrops as well as on seismic sections led to the concept of sequence stratigraphy.

2 Sequence Stratigraphic Analysis

Sequence stratigraphy, the new globally valid system of stratigraphy, is a precise methodology to subdivide, correlate and map sedimentary rocks. Traditional stratigraphy defines stratigraphic units in terms of lithostratigraphy within imprecise biostratigraphic boundaries. Sequence stratigraphy defines units that evolve in response to changes in shelfal accommodation. These units are bounded by stratal discontinuity surfaces on seismic profiles and geologic cross-sections, and are specific surfaces between vertical changes in facies stacking patterns on well logs and outcrops. Where the sediments are thick enough, these surfaces can be identified on seismic data and outcrop sections and well logs, and can be dated biostratigraphically. Thus sequence stratigraphic units are time-equivalent rock packages (Wheeler 1958) caused by lithologic transitions created by abrupt changes in the sediment supply delivered to each locality. The basic units of sequence stratigraphy are depositional sequences, systems tracts, and periodic parasequences or simple sequences. They develop on time scales of 0.5–5 Ma, 0.2–1.0 Ma, and 0.01–0.5 Ma, respectively.

A sequence stratigraphic analysis develops a chronostratigraphic framework of cyclic, genetically related strata. Sequences are bounded by surfaces of stratal discontinuity or their correlative conformities. These surfaces are created by erosion or nondeposition on the shelf and an associated increase in both supply and grain size of the sediment deposited in the basin. Within this framework, the distribution of depositional environments and lithofacies tracts is defined. These facies units may be confined to synchronous intervals that are bounded by stratal surfaces, or to diachronous intervals that step across stratal surfaces. The packages so defined allow the geologist to make precise correlations and understand the tectonic and sedimentary processes across the shelf-to-basin-floor transition, and to accurately interpret the geologic history for improved structural, paleogeographic, and resource evaluation.

Sequence and systems tract boundaries are always present in the rock record, although at times in certain situations they may be subdued or one boundary may be a composite of many. In general, sequence boundaries are regional onlap surfaces (Figs. 2, 3, 4, and 5). In deep-water basins they are characterized by onlap of turbidites and debris flows, or apparent onlap of prograding deltas. In shallow water or nonmarine settings they are characterized by onlap of strata deposited in deltaic, coastal, or fluvial environments. Subaerial and submarine erosional truncation is commonly present below a sequence boundary. Toplap (indicative of sediment bypassing) is a common pattern found below sequence boundaries in areas of rapid progradation (Fig. 3).

Each depositional sequence is composed of systems tracts (Posamentier et al. 1988; Posamentier and Vail 1988; Jervey 1988). A systems tract is a set of linked contemporaneous depositional systems (Brown 1969). Each systems tract is bounded by a physical surface that is, in part, a discontinuity marking the boundary of a similar set of accommodation patterns such as sigmoidal to oblique, oblique to aggradational, or retrogradational. Depositional systems within each systems tract are linked by changes in sedimentary facies (Fig. 5). There are four systems tracts: lowstand, transgressive (retrograding highstand), highstand (prograding highstand), and shelf-margin.

At the base of the lowstand systems tract is a Type 1 sequence boundary. A Type 1 sequence boundary is characterized by subaerial erosion with valley incision on the shelf, and turbidite fan complexes and a lowstand prograding complex in the basin. The lowstand prograding complex pinches out by onlap in the vicinity of the fair weather wavebase of the previous highstand. This position is called the offlap break (Fig. 3). Previous publications have called this point the shoreline break or the

624   Chapter 6 Sequences

SURFACES

SB   Sequence boundaries
       SB1 = type 1
       SB 2 = type2
DLS Downlap surfaces
       mfs = maximum flooding surface
       tsfs = top slope fan surface
       tbfs = top basin floor fan surface
TS   Transgressive surface
       First flooding surface above maximum progradation.

SYSTEMS TRACTS

HST   Highstand Systems Tract
TST   Transgressive Systems Tracts
       ivf = incised valley fill
LST   Lowstand Systems Tract
       ivf = incised valley fill
       lsw = lowstand prograding wedge
       sf = lowstand slope fan
       bf = lowstand basin floor fan
SMST Shelf Margin Systems Tract

SEQUENCE STRATIGRAPHY DEPOSITIONAL MODEL
SHOWING SYSTEMS TRACTS

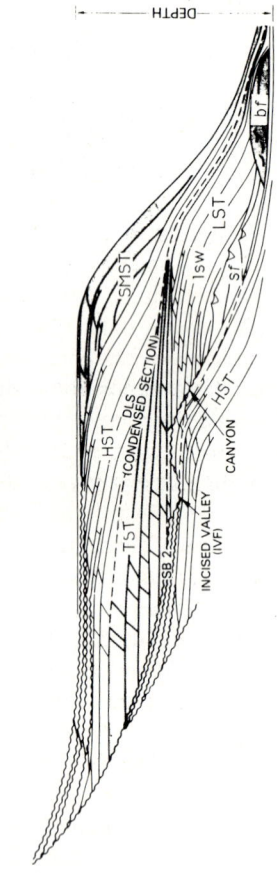

A) IN DEPTH

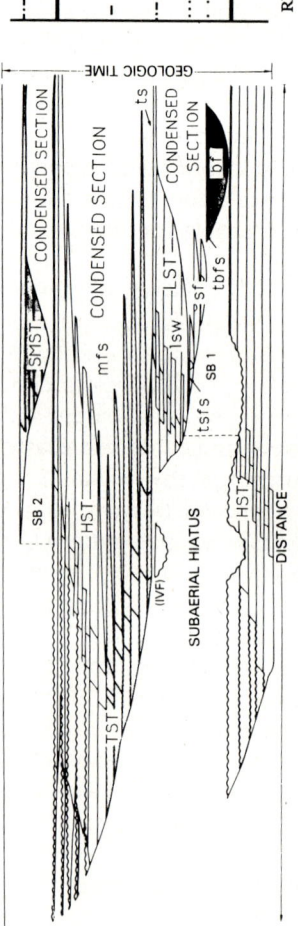

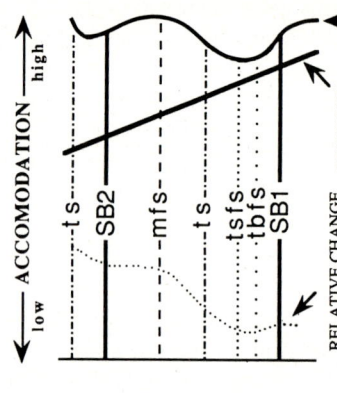

B) IN GEOLOGIC TIME

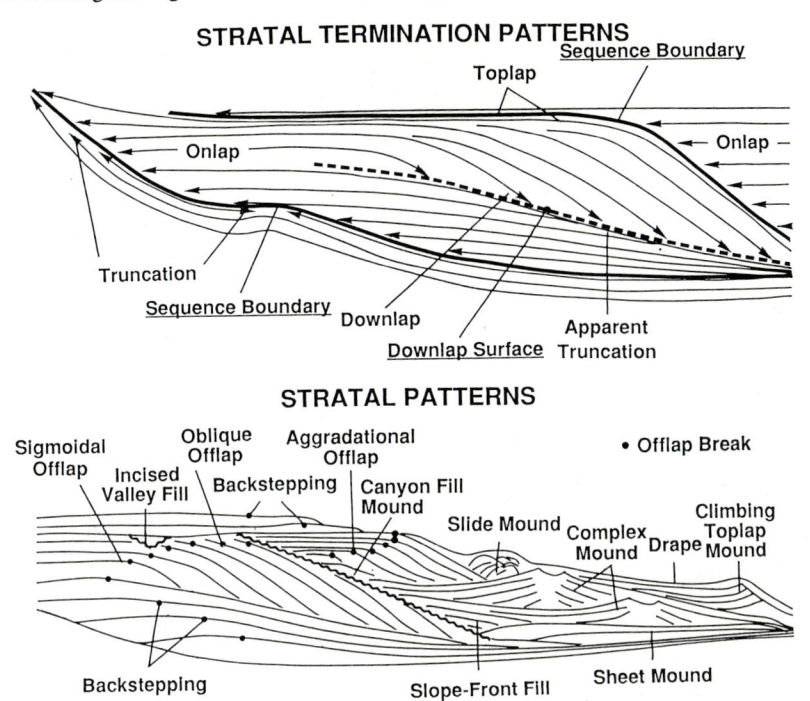

Fig. 3. Diagrammatic sketches showing stratal termination patterns, stratal patterns, and stratal discontinuities

depositional shelf edge. These names have been dropped because of the confusion that resulted from their use.

The physical boundary between the lowstand and transgressive systems tract is defined by the first flooding surface. In previous publications it was called the transgressive surface. Again this name has proven to be confusing and has been dropped in favor of top lowstand surface. The top lowstand surface merges with the basal unconformity landward of the point where the lowstand or shelf margin systems tracts pinch out. The top lowstand surface marks the change from lowstand progradation to retrogradation (Fig. 2).

The physical boundary between the transgressive and highstand systems tract is called the maximum flooding surface. It is a submarine condensed section characterized by downlap above and apparent truncation below.

The boundary between the highstand and shelf-margin systems tract is a Type 2 sequence boundary. A Type 2 sequence boundary is characterized by subaerial exposure on the shelf, minor erosion, and a shelf-margin prograding complex that pinches out by onlap landward of the offlap break of the previous highstand. The top

Fig. 2. Diagrammatic sketch of sequences and systems tracts in depth and geologic time in relation to tectonic subsidence, sea level, and relative changes of sea level

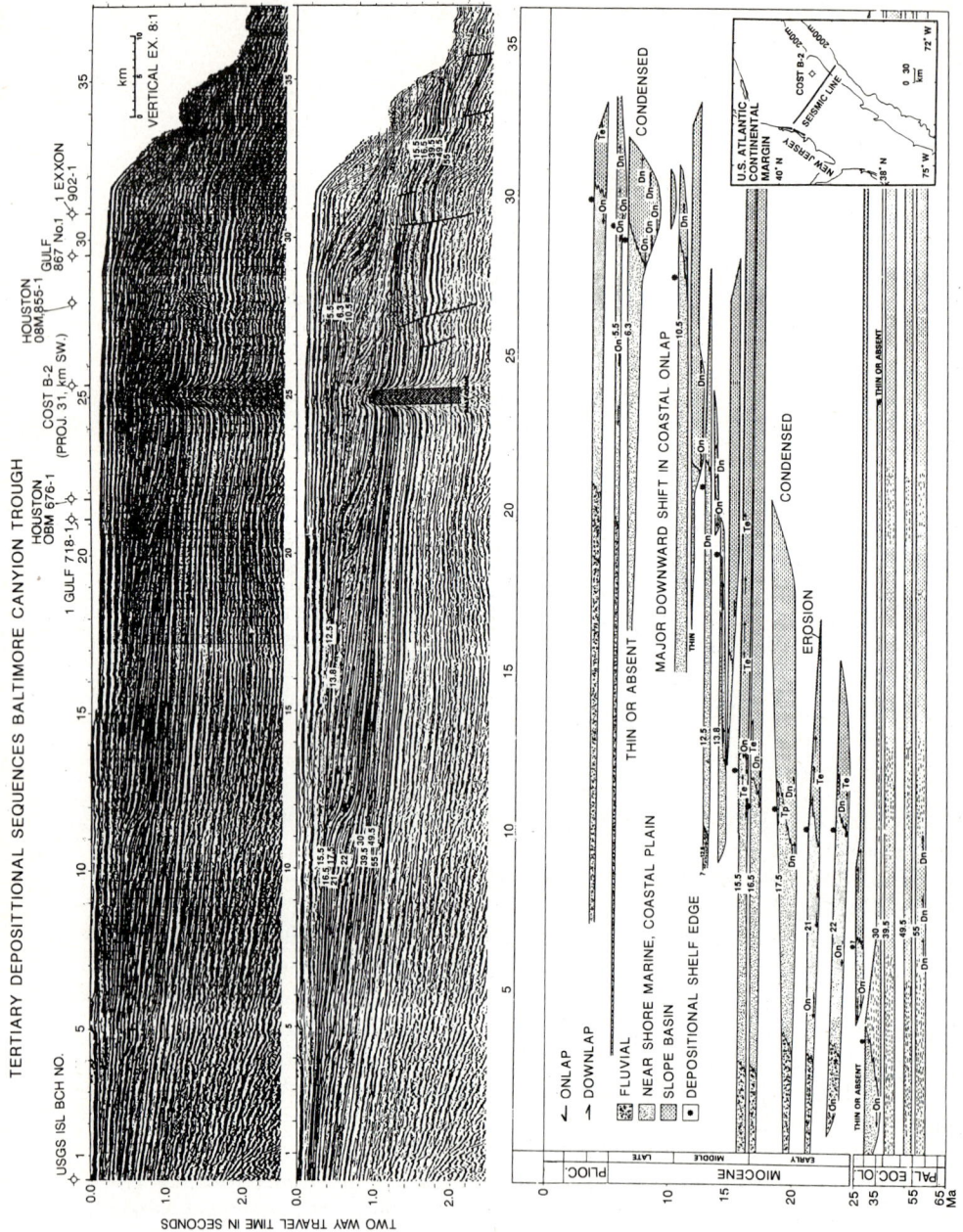

Fig. 4. Interpreted and uninterpreted seismic profiles and chronostratigraphic chart from Baltimore Canyon Trough, Western Atlantic Margin of the United States. (Greenlee et al. 1988)

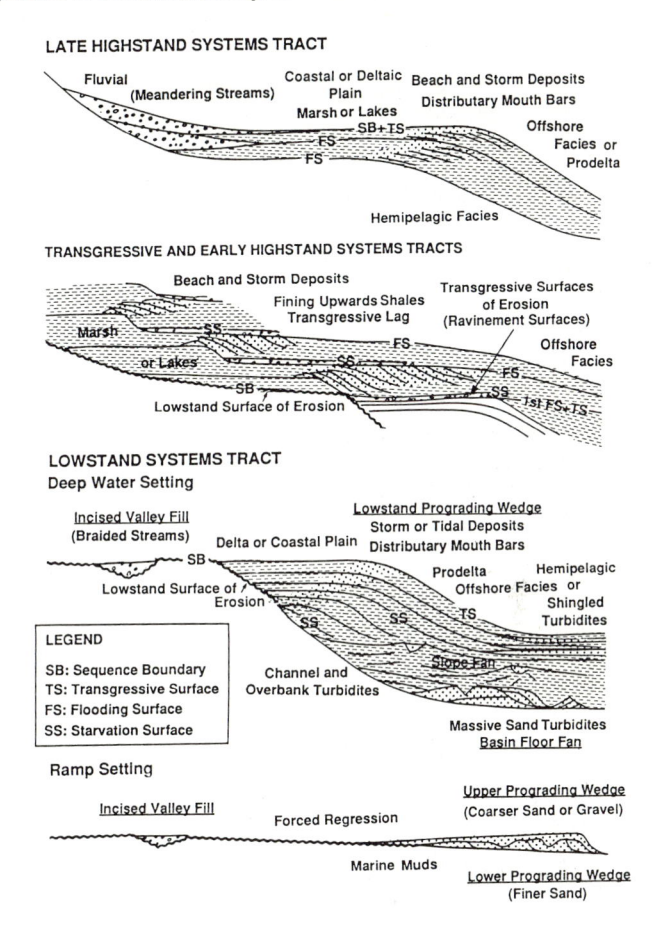

Fig. 5. Diagrammatic examples of systems tracts, parasequences, and depositional systems

of the shelf-margin systems tract was called a transgressive surface in previous publications. This name has been dropped in favor of top shelf-margin surface.

There are four types of systems tracts, but only three will be present in any one sequence. A Type 1 sequence has a Type 1 sequence boundary at its base and is composed of a lowstand, transgressive and highstand systems tract. A Type 2 sequence has a Type 2 sequence boundary at its base and is composed of a shelf-margin, transgressive, and highstand systems tract (Fig. 2).

Depositional sequences and systems tracts are chronostratigraphic intervals because they represent all the rocks deposited within a particular interval of time. Each boundary is age-dated at the minimum hiatus where the strata are conformable. The interval of time between the lower and upper conformities defines the time period over which the rocks were deposited (Fig. 2). Sequence stratigraphic studies from basins around the world indicate that chronostratigraphic intervals defined in this way are global (Vail et al. 1977, 1984; Vail and Hardenbol 1979; Bartek et al. 1990). These intervals are defined on the basis of physical stratigraphy from seismic, wells or outcrop data, and dated by biostratigraphy to identify the age of the interval. The ages of the global Cenozoic sequences and systems tracts are shown in Fig. 6.

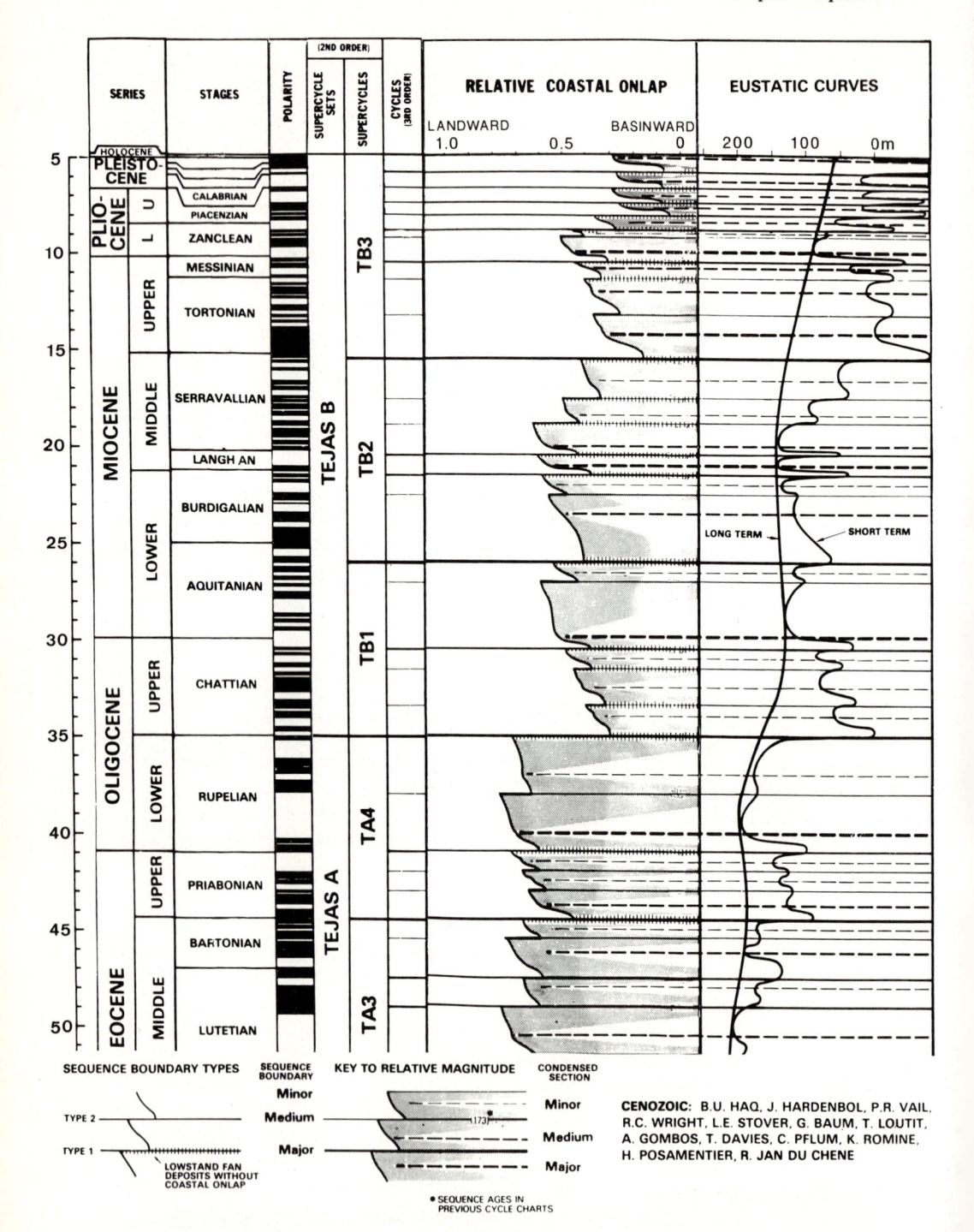

Fig. 6. Simplified global sequence cycle chart for the Tertiary (Haq et al. 1987). Diagram shows approximately two-thirds of the Tejas megacycle

Depositional sequences and systems tracts are believed to be caused by short-term (maximum rate >2.5 cm/ka, maximum acceleration >0.01 cm/ka^2) changes in rates of relative sea level (Fig. 2). In deep water basins with a shelf/slope break, a fall of relative sea level below the offlap break causes a Type 1 sequence boundary and deposition of basin-floor fans. As a relative fall reaches a minimum, lowstand progradation starts. Sediment provided from the collapse of oversteepened slopes of the previously eroded highstand, and slumping of the newly developed prograding complex produce an increase in the sediment supply, which is deposited as levee channel turbidites, debris flows and slump deposits. A continued slow relative rise results in greater slope stability and the domination of the prograding complex over slumping and turbidity flows. The lowstand prograding complex is accompanied by increased accommodation and filling of incised valleys by braided streams on the shelf. A rapid relative rise causes base level to rise above the offlap break of the previous highstand, and initiates flooding of the exposed shelf. The combination of new space available for sediments on the shelf and the increasing rate of relative rise of sea-level causes retrogradation and deposition of the transgressive systems tract. A slowing of the rate of relative rise causes highstand progradation and the basinward progression of the late highstand fluvial wedge. The fluvial wedge builds above sea-level, enabling the streams to maintain their equilibrium profile. An increasing relative rise without a relative fall in the vicinity of the offlap break of the previous highstand causes a Type 2 downward shift of coastal onlap and the development of a shelf-margin systems tract.

The idealized relation between sea level and subsidence that causes the observed relative changes of a sea-level is shown on Fig. 2. In general, the amount of subsidence is high in magnitude, but changes slowly with time. Thus subsidence creates most of the space (accommodation) for the sediments and, along with climate, controls the type and amount of sediment carried to a locality. Sea level causes the short-term changes of slope of the relative change of sea-level curve. These short-term changes create the stratal discontinuities and systems tracts. Evidence for a eustatic control on sequences and systems tracts is the repetitious character of cycles shown in the cyclic relation of highstands and lowstands, and their global correlation in widely separated basins of different tectonic settings (Vail et al. 1977, 1984; Vail and Hardenbol 1979; Bartek et al. 1990).

Relative falls of sea-level occur when eustasy falls at a greater rate than subsidence or when uplift exceeds the rate of eustatic rise (Posamentier et al. 1988). In the case of uplift the time-equivalent sequence boundary will be tectonically enhanced. Lowstand progradation reaches a maximum at a low relative stillstand when eustasy and subsidence are falling at similar rates. Typically lowstand progradation will change upward from oblique to aggradational, indicating an increasing relative rise due to an increasing rate of relative sea-level rise. Retrogradation is indicative of a rapid relative rise caused by rapid rise of relative sea level. Highstand progradation is caused by a decreasing rate of relative rise due to the slowing down of the eustatic rise and the initiation of a slow rate of sea-level fall that is less than the rate of subsidence. Highstand progradation also reaches a maximum at a relative stillstand during the oblique offlap stage. Highstand progradation typically varies from sigmoidal to oblique, indicating the upward decrease in shelf accommodation.

In sequence stratigraphic analysis, it is common practice to display the distance-time geometry of systems tracts and sequences on chronostratigraphic diagrams in which stratigraphy and lithofacies are plotted against geologic time. These chonostratigraphic diagrams (Figs. 2 and 5), are essential to proper understanding of sequences and systems tracts and their relation with geologic time.

The stratigraphic units deposited within fourth- and fifth-order cycles are called *parasequences* or simple sequences. All packages of sediment, except perhaps annual varves and tidal cycles, below this time scale are considered to be episodic and to lack potential for interbasin correlation. Parasequences and simple sequences are the building blocks of systems tracts (Van Wagoner et al. 1990). A parasequence is a chronostratigraphic unit composed of a largely conformable succession of genetically related beds or bed sets bounded by flooding surfaces. A simple sequence has the stratal and lithologic characteristics of a sequence, but its duration is that of a parasequence. Simple sequence and parasequence boundaries do not coincide, but they represent a similar cyclicity. Simple sequences are picked where possible, but in most cases simple sequence boundaries are difficult to identify, therefore the more readily recognizable parasequence boundaries are used.

Parasequences and simple sequences have characteristic *stacking patterns* within each systems tract, as shown in Fig. 5 (Van Wagoner et al. 1990 in press). Parasequences are best developed within the transgressive and early highstand systems tracts. In the transgressive systems tract they retrograde. The first parasequence flooding surface over the lowstand prograding complex and over shoreface and nonmarine rocks within the transgressive systems tract are ravinement surfaces. Ravinement surfaces are transgressive surfaces of erosion (Weimer 1984) which step up in time with retrograding parasequences. They are the most prominent surfaces in the stratigraphic section, but they are time-transgressive because they can correspond to different parasequences. They should not be confused with a sequence boundary, which is in part a lowstand surface of erosion (Weimer 1984). Basinward of the ravinement surface, parasequences are characterized by shallowing-upward cycles that become finer and thinner upward. Landward of the ravinement surfaces, parasequences are difficult to identify in the nonmarine marsh or lake deposits.

The early highstand systems tract is similar to the upper portion of the transgressive systems tract, except that shallowing-upward parasequences in the highstand systems tract prograde. In a marine setting highstand parasequences tend to be thin during the early highstand, thicken to a maximum just below the offlap break, and finally thin abruptly above the offlap break. In the late highstand less space per unit time is available to accommodate coastal sediments; consequently, parasequences become more progradational. Simple sequences commonly develop in the more landward nonmarine areas during the late highstand.

The lowstand systems tract has a variety of parasequence packaging patterns. Parasequences within the lowstand prograding complex are similar to the highstand parasequences, except their shallow water and nonmarine facies tend to be relatively thicker due to the aggradational progradation. Simple sequences commonly develop in the early part of the lowstand prograding complex. Slope and basin floor fan complexes do not have conventional parasequences. The equivalent to a parasequence in the slope fan is a channel overbank levee lobe, and a sheet mound lobe in the basin floor fan.

3 Subsidence Analysis

In subsidence analysis of the stratigraphic record one estimates the total and tectonically controlled subsidence that a basin has experienced. Burial history curves (Van Hinte 1978) are constructed by plotting the depth of each horizon, after a correction for the compaction state at their respective depths, at the times of formation of each horizon. If the datum is adjusted to the long-term sea-level curve and paleobathymetry is included for each age-depth pair, the plot shows total subsidence for the bottom of the stratigraphic column. When the total subsidence curve is corrected for isostatic compensation of the sediment load (assuming the crust has no strength), the result is a tectonic subsidence curve (Steckler and Watts 1978). This curve shows the water-loaded space that tectonics would create if no sediments were deposited. This is the curve to use for calculating a first approximation of rates and magnitudes of tectonic subsidence and isostatic loading. It is assumed that compaction occurs instantaneously and does not affect the subsidence of the surface of deposition. However, subsidence due to flexure loading by sediment is a major contributor to the subsidence that the surface of deposition responds to and may be an important factor when calculating relative subsidence.

To separate tectonic from eustatic effects, the subsidence curve can be compared with theoretically calculated subsidence curves for various amounts of crustal stretching. A first-order eustatic cycle can be estimated from the difference between the load-corrected subsidence curve and the interpreted tectonic subsidence curve as discussed by Hardenbol et al. (1981). The tectonic subsidence curve may also be influenced by flexural loading due to the distribution of adjacent sedimentary units.

4 Tectono-Stratigraphic Analysis

Sequence stratigraphic analysis of seismic, well, and outcrop data can help resolve the interaction between plate tectonics, tectonically enhanced unconformities, magmatism, and local faulting and folding and their effect on sedimentation and their consequent imprint on the stratigraphic record. The purpose of this analysis is to document the age, distribution, and cause of the major transgressive-regressive facies cycles. The boundaries of these cycles are associated with tectonically enhanced unconformities which form during a period of uplift or slow subsidence that precedes an increase in subsidence rates. The maximum surface of transgression of a transgressive-regressive cycle is marked by an enhanced flooding surface which can be regionally extensive. The maximum surface of transgression and tectonically enhanced unconformities define ideal tectono-stratigraphic units for analyzing the geologic history of a region.

Tectono-stratigraphic analysis can be applied by executing the following nine analysis steps:

1. Calculate a tectonic subsidence profile.
2. Determine the age and shape of each second order tectonic event on the tectonic subsidence curve and relate them to major transgressive-regressive facies cycles.
3. Determine the relationship of each of these second order events to plate tectonic events.

4. Assign a cause to each tectonically enhanced unconformity.
5. Relate the age of magmatism to age on the tectonic subsidence curve.
6. Map the tectono-stratigraphic intervals and calculate sediment volumes as a function of time.
7. Determine the structural style and the distribution and orientation of structures associated with each tectono-stratigraphic interval and determine their paleostress conditions.
8. Simulate by computer modeling the infill history to reconcile the interpretation with the geologic database.
9. Analyze the gross distribution of lithofacies within the tectono-stratigraphic intervals and relate to tectonics and model simulation.

The second step requires defining the age, shape and character of each tectonic event on regional tectonic subsidence curves. As discussed later (Fig. 7), tectonic sub-

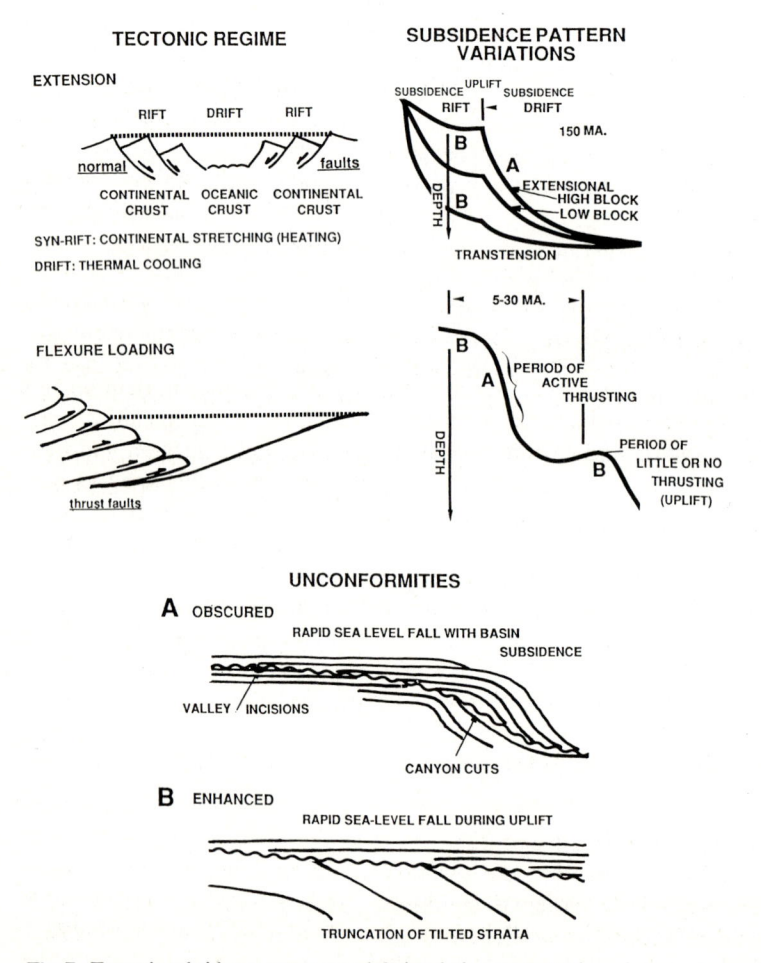

Fig. 7. Tectonic subsidence patterns and their relations to tectonic regimes and unconformities

sidence curves show two basic patterns; a concave-upward curve caused by thermal cooling and a convex-upward curve caused by flexure loading. Tectonic events start when the rate of tectonic subsidence increases. Tectonic events are terminated at the end of periods of slow subsidence or uplift.

In extensional regimes the initial subsidence event is caused by crustal extension and the structural style is rotated fault blocks. The early phases of extension are characterized by active deformation mechanisms which rotate fault blocks in a direction parallel to the separation direction and the least principle stress. The later phases of extension become strong influenced by flexure loading of subsiding crust and therefore may change rotation angles to parallel the basin margin and influence subsidence patterns much more broadly.

Compressional settings will show maximum shortening rates during periods of maximum subsidence within the basin. The shortening direction will be parallel to the plate margin or may be oblique for basement-involved thrusting within a plate. The crest of an active compressional structure will show uplift and enhanced sequence boundaries until the regional tectonic subsidence rates begin to wane.

The third step relates the subsidence patterns to plate tectonic events in the region. In general, the start of each event can be related to a particular plate tectonic change. Examples of such events include initiation of rifting, termination of crustal extension and transition to thermal cooling (rift to drift), volcanic intrusions (hot spot), initiation or ending of thrusting or transpression, initiation or increase of rate of thrusting (which may relate to collisional events). Some events may be related to increases in the sediment supply delivered to the region (due to uplift elsewhere) which cause a change in rate of uplift of salt and shale diapirism, or a change in rate of movement on listric normal faults. Any of these tectonic events which are associated with basement involved tectonics are generally correlatable with a major transgressive-regressive facies cycle. These cycles will be apparent on the subsidence history curve with or without any correction for sea-level changes because the accumulative magnitude of the tectonic adjustments are commonly much greater than any sea-level fluctuations.

The fourth step assigns the cause of the tectonically enhanced unconformities which provides further information about the amount of section missing as well as its regional significance. Some of the different unconformity types include:

1. Break up unconformity,
2. Thermal uplift unconformity,
3. Basal foredeep unconformity,
4. Internal foredeep unconformity due to slowing of tectonic subsidence,
5. Internal foredeep unconformity due to structural inversion,
6. Diapiric uplift unconformity,
7. Listric fault rollover unconformity.

The break up unconformity is related to the termination of crustal extension and transition to thermal cooling (rift to drift) (Falvey and Middletow 1981). Major erosion on high blocks is common, yet basins may contain a relatively continuous section. Unconformities related to thermal uplifts are wide-spread and can record major amounts of erosion. Basal foredeep unconformities (Bally 1989) are due to change in subsidence distribution due to a change to flexure loading within the region.

This causes early sediment starvation in some regions before progradation or onlap onto a deflecting crust, and usually does not record major amounts of erosion. Internal foredeep unconformities are due to both slowing of tectonic subsidence and structural inversion. Internal foredeep unconformities caused by slowing of subsidence record periods of sediment bypass and may not record much uplift. In contrast, internal foredeep unconformities related to structural inversion may locally record major erosion. Diapiric uplift unconformities mark the initiation of diapirism. The tops of salt and shale diapirs tend to maintain a stable position within the density-stratified stratigraphic column and therefore do not record major erosion, but local truncation and sediment bypass are typical. Listric fault rollover unconformities document the initiation of fault movement and record minor erosion of the high regions and a relatively continuous section in low rapidly subsiding regions.

The fifth step requires relating the time of magmatism to the tectonic subsidence history.

The sixth step requires defining the tectono-stratigraphic units bounded by tectonically-enhanced unconformities and the surfaces of peak transgression. Tectonically enhanced unconformities form during periods of slow subsidence or stability. A surface of peak transgression is an enhanced maximum flooding surface that occurs during the period of maximum subsidence rates and minimal sediment supply. The intervals between these surfaces form ideal packages for regional mapping and isopaching. Isopach maps of these tectono-stratigraphic units document the regional variation of subsidence and uplift during discrete tectonic phases. It is during periods of stability that basins may fill completely and hence record a surface which was approximately horizontal at the time of deposition. The variations in thickness reflect the regional variance of subsidence and uplift. The estimation of sediment volumes will document changes in sediment supply which will influence subsidence and the distribution of lithofacies.

The seventh step is to determine the structural style and the distribution and orientation of structures during each tectonic event (Bally 1987). In general, each tectonic event on the subsidence curve will be associated with a particular structural style and one or more tectonically enhanced unconformities.

Orientation of the principle stress axes and kinematics of deformation, as well as rates of deformation can be related to plate-tectonic settings and associated deformation mechanisms. The results demonstrate that during certain periods, active deformation structures are forming in response to plate boundary interactions. This may express itself in one basin as a compressional setting with rapid uplift on tops of folds, while the surrounding region is subsiding rapidly, and another basin as an extensional setting with active rifting.

The eighth step involves running a computer simulation of the geologic cross-section through the region (Lawrence et al. 1990; Bowman et al. 1990; Strobel et al. 1990). This requires defining all the initial observations such as subsidence history, sea level, and sediment supply history and type for a simulator and reproduce the observed section. The models will produce stratigraphic concepts for testing within the cross-section as well as out of the region of immediate interest. A detailed history of temperatures can also be integrated with other analyses for thermal maturation.

The ninth step involves defining the distribution of lithofacies and comparing with the geologic data and results of the simulation.

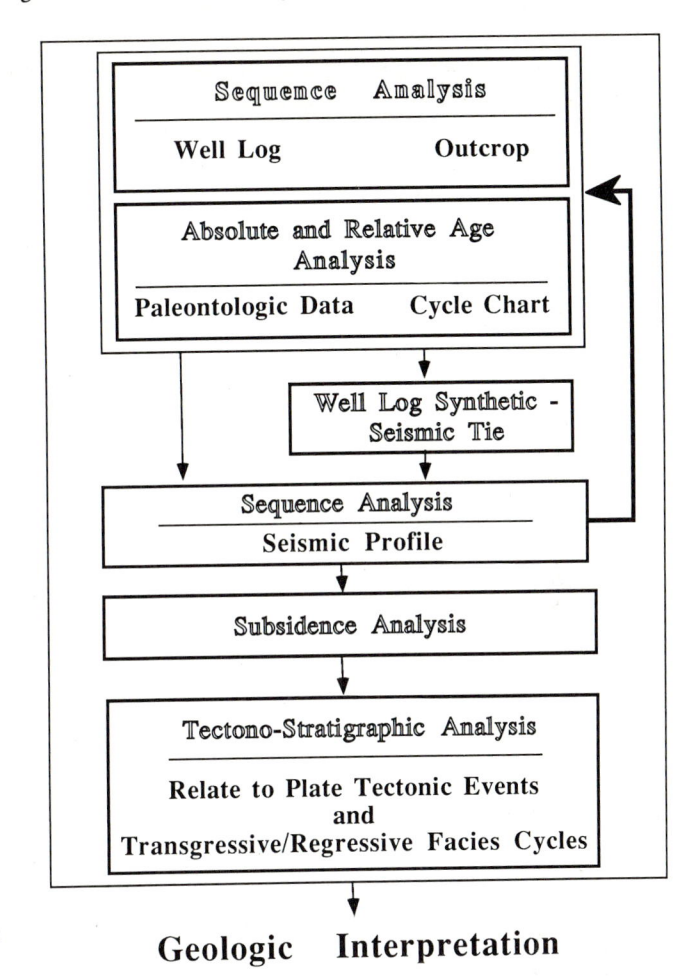

Fig. 8. Sequence stratigraphic interpretation procedure

Geologic Interpretation

4.1 Integrated Interpretation Procedure

The following sequence of procedures for interpretation of paleogeography, geologic history, stratigraphic signatures, and resource evaluation is recommended (Fig. 8; Vail 1987a,b).

1. Determine physical chronostratigraphic framework by interpreting sequences, system tracts and parasequences and/or simple sequences on outcrops, well logs and seismic data and age data with high resolution biostratigraphy.
2. Construct geohistory, total subsidence, and tectonic subsidence curves based on sequence boundary ages.
3. Complete a tectono-stratigraphic analysis.
4. Define depositional systems and lithofacies tracts within systems tracts and parasequences or simple sequences.

5. Interpret paleogeography, geologic history and stratigraphic signatures from resulting cross-sections, maps and chronostratigraphic charts.
6. Locate potential reservoirs and source rocks for possible sites of exploration.

5 Stratigraphic Signatures

The signature of tectonism, eustasy, and sedimentation can be distinguished by understanding the distribution of each of these variables in terms of their effects on the stratigraphic record in time and space.

5.1 Tectonism

Tectonism has the greatest influence on increasing or reducing accommodation space. Also, when coupled with climate, it controls the type and amount of sediments filling that space. The corresponding signatures of each tectonic process can be distinguished on the basis of rates and duration in time and regional distribution. The stratigraphic signature of tectonism results from a wide range of processes and has the most profound effect on accommodation. Its imprint on the sedimentary record can be divided into three hierarchical groups (Fig. 1), including: (1) uplift and basin evolution; (2) changes in subsidence rates, and (3) folding, faulting, magmatism, and diapirism.

Tectonism is characterized by active processes related to plate interactions, and equilibration responses to conditions created by the active processes. Active deformation, or deformation which can be directly linked to fault movement is characterized by high strain rates, offset strata, rotation, and folding. These events create third order episodic events. Faulting is a manifestation of volumetric accommodation driven by strike-slip, collisional or extensional crustal plate boundaries, or density contrasts in the lithologic column. The high rates of strain are caused by relative plate velocities which can exceed 10 cm/a. Equilibration responses to active processes include subsidence due to thermal cooling and subsidence or uplift due to flexural loading of thrusts, erosion of exposed strata, and subsequent deposition of the sediment in subsiding regions. These equilibration responses are responsible for the development of major transgressive-regressive facies cycles. Equilibration responses which adjust to changes at the base of the lithosphere include hotspot activity, which may be related to slow relative plate motions with respect to the mantle, and uplift due to delamination of the lower lithosphere. Equilibration responses are characterized by slow rates of strain and longer durations, and regionally distributed bathymetric or altitude adjustments.

First-order tectonic events result from thermodynamic processes in the asthenosphere, which drive the plates and deform the earth's crust and upper mantle. Crustal extension and transtension, crustal shortening and transpression, and associated thermal heating and cooling are long-term processes that cause sedimentary basins to form and disappear. Their stratigraphic signatures are uplifts and sedimentary basins.

Second-order tectonic events are characterized by changes in subsidence rates during the evolution of sedimentary basins. They may result from reorganization of the plate-tectonic regime, or from local thermodynamic perturbations. These changes in subsidence rates are apparent on tectonic subsidence curves as either concave-upward, or convex-upward (Fig. 7). The stratigraphic signatures of both curves are major transgressive-regressive facies cycles.

Third-order tectonic events are folding, faulting, diapirism, and magmatic activity. The stratigraphic signature of these events is tilted and ruptured strata. They are associated with penecontemporaneous events such as slides, slumps, megaturbidites, bentonite beds, datable extrusive flows, and intrusive sills and dikes (Fig. 1).

The initial stage of an extensional basin typically shows an inflection when passing from the syn-rift (crustal extension) to the drift (thermal cooling) phase. High blocks show low rates of subsidence in the crustal extension phase and higher rates during the thermal cooling phase (Fig. 7). Transtensional basins tend to show high subsidence rates in the crustal extensional phase and slower rates in the thermal phase. Pure extension and transtension create the two end-members of a complete set of subsidence curves possible for extensional basins.

Tectonic subsidence curves of basins that have evolved during extension are characterized by a concave-upward pattern. In general, extensional basins show at least two concave-upward subsidence patterns, one during the syn-rift crustal extension phase and another during the drift thermal cooling phase. In many extensional basins there are additional concave-upward patterns due to other crustal extension episodes, as well as thermal perturbations due to volcanic hot spots.

Tectonic subsidence curves of basins that evolved under compression are characterized by a convex-upward pattern which is characteristic of flexure loading processes. The period of maximum shortening is associated with maximum sub-

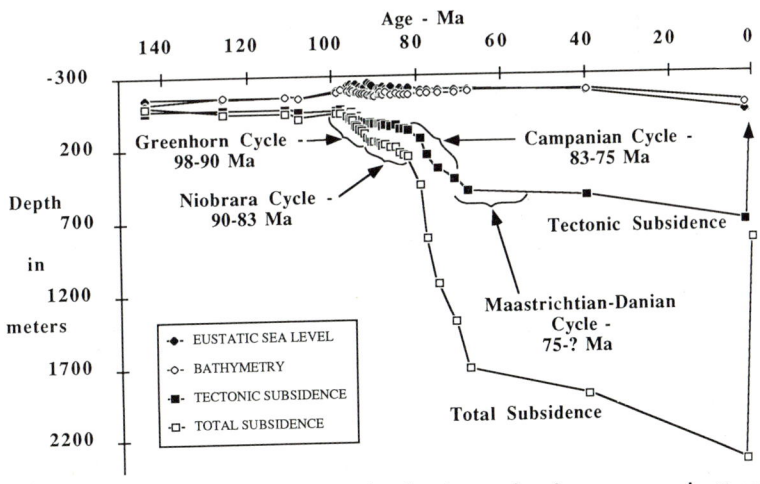

Fig. 9. Tectonic subsidence history in a flexure loading setting showing a series of convex-upward patterns related to tectonic events in the foreland basin of the Western Interior of North America

sidence rates, because this is the time when thrust sheets are stacking at their maximum rate on the border of the foredeep basin. Several convex-upward subsidence patterns are commonly present within compressional basins, indicating changes in the rate of thrust movements. As an example, the subsidence history from the Western Interior Seaway of North America illustrates this pattern (Fig. 9). The subsidence history consists of two weak flexural events (beginning about 98 and 90 Ma, respectively) followed by an intense event (beginning about 83 Ma). Three transgressive-regressive facies cycles relate to these tectonic events, including the Greenhorn cycle from 98–90 Ma, the Niobrara cycle from 90–83 Ma and the Campanian cycle from 83–75 Ma. There often is a period of stability or uplift between convex-upward subsidence curves. The uplift may be due to reduction of the flexural load by erosion of raised topography or the heating of the depressed crust. Transpressional basins show similar convex-upward patterns indicating loading of a region adjacent to the basin. Pure compression and transpression create similar patterns on the tectonic subsidence curve.

Some basins were initiated during an extensional regime and subsequently deformed under a compressional regime. This change will be reflected in the basin subsidence curve, which will show a concave-upward pattern in the early history followed by a convex-upward pattern.

Each concave- or convex-upward pattern on the tectonic subsidence curve is associated with a major transgressive-regressive facies cycle. The transgressive phase of the cycle occurs when subsidence rates increase. The regressive phase of the cycle occurs when subsidence rates decrease. In a flexure-loaded system, there may also be an associated variation in the sediment supply. As the rate of thrusting and associated crustal thickening increase, uplifts are producing and storing potential sediment supply. When thrusting stops, both the storage capacity of the sediment supply and its delivery rate is at a maximum, which produces a regression. A regression produced by an increase in sediment supply can be recorded far from the source of the sediments (100 to greater than 2000 km), as long as the fluvial system is connected.

Tectonically induced folding and faulting occur during particular periods of the tectonic subsidence curve, depending on whether the tectonic setting is compressional or extensional. In extensional settings, faulting is most active during rifting. In compressional settings, faulting is most active in the maximum subsidence phase. A tectonic subsidence curve influenced by third-order tectonic events may show a deviation from the regional subsidence pattern. For example, a local tectonic subsidence curve from a basin formed under a compressional regime will show uplift or slower subsidence rates corresponding to the development of the structure causing the flexural loading. This anomaly will be superimposed on the regional curve that will show maximum subsidence at the corresponding time.

Structure tends to enhance or subdue eustatically caused sequence and systems tract boundaries, but does not affect the age of the boundaries when dated at the minimum hiatus at their conformable position. To illustrate this point, Fig. 7a shows typical stratal patterns associated with a major sequence boundary during times of rapid basin subsidence. When rapid subsidence occurs, sequence boundaries are subdued. They are characterized by minimized headward erosion and valley incision.

Type 1 sequence boundaries generally do not change to Type 2 sequence boundaries in regions with high subsidence rates. Stratal patterns along subdued sequence boundaries tend to be parallel or subparallel and show subtle onlap. During times of tectonic stability or uplift, sequence boundaries tend to be enhanced above eroded contacts of strata which were deformed during the active phase of the uplift (Fig. 7b).

The relations discussed above are illustrated with an example from a well (COST B-2) located along the western Atlantic continental margin, in the Baltimore Canyon trough, offshore New Jersey (Fig. 10). This figure shows a back stripped total subsidence curve of the early Jurassic (188 Ma). These were the oldest sediments drilled in the COST B-2 well. The position of this basal unit through time was calculated following the procedures described by Van Hinte (1978), and Steckler and Watts (1978). The construction and interpretation of the subsidence curves for this well are discussed in detail by Greenlee et al. (1988).

The tectonic subsidence curve (Fig. 10) shows a relatively high rate of subsidence during the rift phase, which is followed by a normal thermal cooling phase after the opening of the Atlantic (157 Ma). During the Aptian (116–109 Ma) a thermal perturbation due to a hotspot and/or rifting in the North and South Atlantic with associated igneous intrusions resulted in slight uplift. Thermal cooling resumed after the uplift.

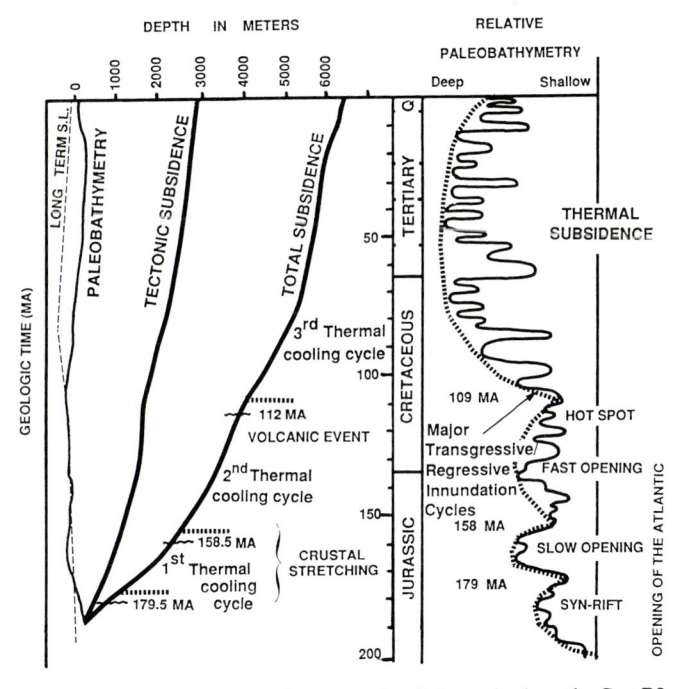

Fig. 10. Example of tectonic subsidence curve and transgression-regression facies cycles from the Cost B2 well, Baltimore Canyon Trough, Western Atlantic. (After Greenlee et al. 1988 and Poag and Valentine 1988)

The tectonic subsidence curve shows three second-order tectonic events which start with increased rates of subsidence. The first one extends from late Triassic to late middle Jurassic (230?–157 Ma), the second from late middle Jurassic to late Aptian (157–109 Ma) and the third from late Aptian to Present (109–0 Ma). Figure 10 also relates the tectonic subsidence curve to variations in the paleobathymetry as determined by Poag and Valentine (1988). Notice that four major transgressive-regressive facies cycles are present. However, only three second-order tectonic events are visible on the tectonic subsidence curve. This is probably due to lack of stratigraphic resolution near the bottom of the well. The Albian to present and the Callovian to end Aptian transgressive-regressive cycles correspond to the thermal cooling events following the Aptian hot spot and fast opening of the Atlantic. The two pre-Callovian transgressive-regressive cycles are thought to coincide with the crustal extension phases that precede the slow opening (end lower Jurassic) and fast opening (end of Bathonian) of the Atlantic. Angular unconformities indicating tectonically enhanced erosional surfaces (Fig. 10) precede the boundaries of the cycles.

Folding and faulting within the New Jersey marginal basin occurred during the four transgressive-regressive facies cycles. Normal faulting and other structures due to rifting were associated with the syntectonic phases of rifting, which preceded both the slow and rapid opening of the Atlantic. During the early Aptian magmatic activity induced doming. Listric growth faults developed later as younger sediments prograded into deep water.

5.2 Eustasy

Five orders of eustatic cycles have been identified in the geologic record including continental flooding cycles (first order) and four orders of sequence cycles with periods ranging from 5 Ma to 10 ka (second to fifth order, Table 1).

5.2.1 Continental Flooding Cycles

Continental flooding cycles are defined on the basis of major times of encroachment and restriction of sediments on the cratons. They represent first-order eustatic cycles and their stratigraphic signature is the megasequence (Fig. 6). There are two Phanerozoic continental flooding cycles (Hallam 1977; Vail et al. 1977; Fischer 1981, 1982). The first begins in the uppermost Proterozoic and ends in the latest Permian. The younger starts at the base of the Triassic and extends to the Present.

The Triassic represents a time of gradual encroachment of sediments onto the craton. Great thicknesses of nonmarine sediments were deposited in grabens and bordering marine basins. This general pattern is believed to be caused by a slow relative rise of sea-level due to a slow long-term rise in eustasy. Nonmarine rocks were widespread because base-level was low and the sediment supply exceeded the space being created by the slow relative rise, resulting in major regression.

The Jurassic and early Cretaceous represent times of extensive encroachment of sediments onto the continental margins. These sediments are predominantly marine. During this time long-term sea level rose rapidly. Marine rocks are dominant because the sediment supply did not keep up with the new space being created by the rapid relative rise, resulting in overall transgression. The lower Turonian is believed to have been the time of the maximum eustatic high (Kauffman 1983; Haq et al. 1987; Sahagian 1987).

The upper Cretaceous and Cenozoic are characterized by an overall gradual restriction of sediments to the continental margins and basinal areas. This pattern is believed to be caused by a gradual long-term fall of eustasy, causing a regression or a relative fall and exposure.

The older first-order eustatic cycle starts in the uppermost Proterozoic and extends to the end of the Permian. The latest Proterozoic represents the time of slow encroachment with regression, the Cambrian represents the time of extensive encroachment with transgression, the Ordovician represents the eustatic high, and the Silurian to Permian represents the time of gradual restriction.

The two continental flooding cycles are recognizable on all continents and are believed to be global. The cause of these cycles is believed to be tectono-eustasy (change in ocean basin volume). Many variables cause changes in ocean basin volume, but the most significant is changing rates of sea-floor spreading (Rona 1973; Pitman 1978). Rapid spreading rates produce broad, high mid-ocean ridges. Slow rates produce narrow, lower mid-ocean ridges. Thus during times of rapid sea-floor spreading the average depth of the ocean basins decreases and causes sea-level to rise onto the continents. During times of slower spreading, the average depth of the ocean basins increases, causing sea level to fall, and sedimentation is restricted to the ocean basins and areas of rapid tectonic subsidence. Other factors that contribute to changes in ocean basin volume are continental collisions, subduction of trenches, submarine magmatism, and sediment infill. Pitman (1978) estimated that the maximum rate of tectono-eustatic change caused by a combination of all these variables would be 1.2–1.5 cm per thousand years.

The movement of the major continental plates through the Phanerozoic is illustrated in Fig. 11 (Scotese and Denham 1987). Each line represents the relative distance of one plate from another through time. Upward-branching lines represent break-up and spreading. It is during this time that sea-floor spreading rates are most rapid and mid-ocean ridges are most extensive. When the lines converge toward the center of the figure, the plates are converging and colliding. Plate convergence and collision causes slowing and reduction in the total length of mid-ocean ridges. Comparison of the first-order eustatic curve to the plate spreading patterns (Fig. 11) shows a close relationship. Rising eustasy coincides with times of continental break-up while falling eustasy coincides with times of continental aggradation. First-order eustatic lows correspond to the times when supercontinents exist, and first-order eustatic highs correspond to times of maximum continental break-up.

The Permo-Triassic low relates to the convergence of plates and stabilization of the Pangea supercontinent. The late Proterozoic low relates to the convergence of plates and stabilization of the Pan-African supercontinent which may have broken up

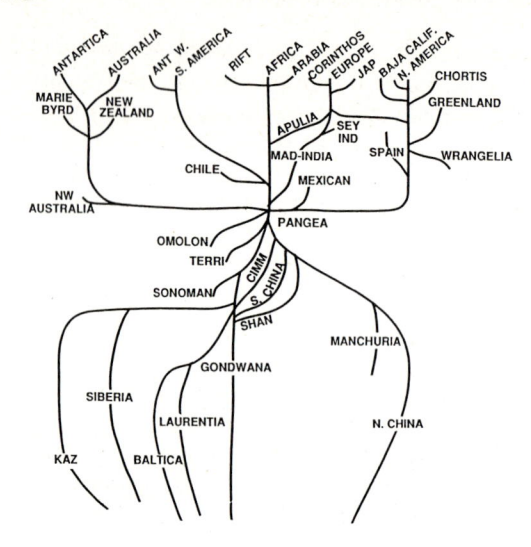

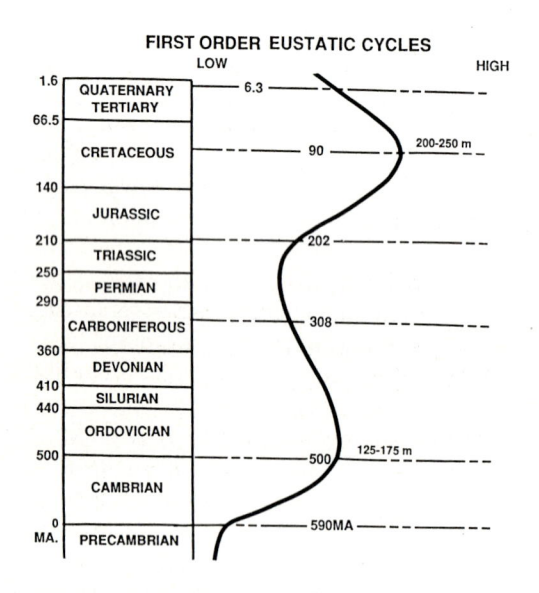

Fig. 11. Comparison of the first-order eustatic cycles with plate motions for the Phanerozoic. (Plate motions from Scotese and Denham 1987)

625 to 555 Ma (Bond et al. 1984). Precambrian radiometric ages indicate there may have been a third supercontinent forming around 1.8 Ga which underwent rifting around 1.2 Ga (Hoffman 1989). Thus a third first-order eustatic cycle may have been present during the Proterozoic. If this is true, the first-order eustatic cycles would have had durations of 295 Ma, 350 Ma, and 600 Ma each.

Estimation of the maximum height of the first-order eustatic cycles above presentday sea level has been the subject of considerable research. Most workers estimate that the eustatic high of the youngest cycle occurred during the early Turonian and had a magnitude of 200–250 m above the present day (Pitman 1978; Kominz 1984; Sahagian 1987). The maximum height of the Paleozoic cycle occurred in the Ordovician. Its absolute value above present day is difficult to determine, but Bond et al. (1988) estimate the maximum to be between 100 and 150 m above presentday sea-level. An example of the first-order eustatic curve (long-term eustatic curve) for the Tertiary is shown in Fig. 6. Its maximum rate of change is less than 1.5 cm/ka. This rate is not rapid enough to cause the observed sequences and systems tracts.

5.2.2 Sequence Cycles

Second- to fifth-order eustatic cycles are recorded by sequence cycles, systems tracts, and periodic parasequences or simple sequences. They are believed to be glacio-eustatic cycles (Vail et al. 1977; Bartek et al. 1990) with a smaller magnitude, but higher frequency than tectonically induced transgressive-regressive facies cycles. Glacio-eustatic variations produce high frequency variations similar to those on the sea-level curve of Haq et al. (1987). The second and third-order cycles for the Cenozoic are shown in Fig. 6.

Second-order eustatic cycles consist of sets of third-order cycles (Haq et al. 1987, 1988) bounded by major eustatic falls. In general, a set of five to seven third-order cycles form a second-order cycle with a duration averaging 9–10 Ma. The boundaries of second-order eustatic cycles are characterized by especially large eustatic falls (>50 m). The stratigraphic signature of a second-order eustatic cycle is a supersequence. The second-order supersequences can be grouped into sets of either three or four. These are called supersequence sets and result in an apparent cyclicity of 27–30 or 36–40 Ma.

Second-order cycle sets with a periodicity ± 30 Ma (three second-order cycles) may have a characteristic reflection pattern on regional seismic sections. This will occur if the cycle sets are deposited during a period of slowly changing subsidence rates and there is enough sediment to prograde into deep water. The global cycle chart of Haq et al. (1987) is based on the recognition and biostratigraphic age dating of the discontinuities of similar stratal patterns in reflection profiles from basins on different continents. An understanding of the control of relative sea level on the stratal patterns and the ability to separate the tectonic and eustatic components led to the global cycle chart of Haq et al. (1987). Continued research has demonstrated that there are more

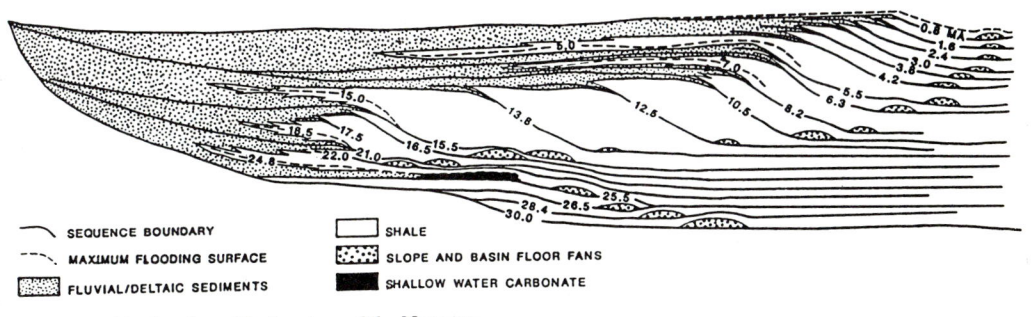

Fig. 12. Stratigraphic signature of the Neogene

sequences than recorded on the Haq et al. (1987) chart, especially in the older rocks, but the basic patterns of the global synchroneity remains valid.

An example of a second-order cycle is the characteristic stratigraphic signature of the Neogene (Vail et al. 1989). Figure 12 shows the stratigraphic signature of the Neogene. It is characterized by the following pattern:

1. Lower Oligocene landward thickening.
2. Upper Oligocene wedge, which laps out at or near the shelf margin and thickens basinward.
3. Basal Lower Miocene flooding.
4. Lower Miocene (Aquitanian) aggradation, commonly ending with lowstand deposits.
5. Lower Miocene (Burdigalian) aggradation, commonly ending with major lowstand deposits.
6. Middle Miocene (Langhian and lowermost Serravallian) flooding.
7. Middle Miocene (Serravallian) major progradation.
8. End of middle Miocene major downward shift of onlap and lowstand deposits.
9. Upper Miocene aggradation commonly ending with lowstand deposits.
10. Lowermost Pliocene flooding.
11. Pliocene- Lower Pleistocene aggradation with multiple lowstand deposits.
12. Upper Pleistocene high-frequency sequences.

The Baltimore Canyon is an example of this signature where this system is well developed. This signature can be related to glacio-eustatic changes recorded in the Neogene of the Ross Sea of Antarctica (Bartek et al. 1990).

The stratigraphic signature of the Neogene is easier to identify than the older second-order cycle sets, but each ±30 Ma period has a recognizable reflection pattern on seismic sections given the right tectonic conditions.

The ability to recognize the stratigraphic signature of the second-order cycle sets provides a means to predict the age of sedimentary deposits ahead of the drill or biostratigraphic evaluation.

Third-order sequence cycles are the fundamental units of sequence stratigraphy with a cyclicity varying between 0.5–5.0 Ma. They are discussed in depth in the preceding section on stratigraphic analysis methodology.

Parasequences and simple sequences are fourth- and fifth-order cycles. They may be episodic or periodic. Episodic parasequences are caused, for example, by delta lobe shifts. They are limited in distribution and of very short duration (usually less than 10 000 years). Periodic parasequences are characterized by regional continuity and by systematic changes in thicknesses between high frequency cycles within a stratigraphic section. If the environment is shallow marine they often correlate for hundreds of miles across different depositional settings. These periodic parasequences are believed to be caused by climatic fluctuations associated with Milankovitch scale orbital cycles (less than 500 ka). The Milankovitch orbital cycles have dominant periods of approximately 20, 41, and 100 ka. These orbital cycles influence the amount of solar energy received on the earth's surface and thus affect climate. It is believed that these climatic variations induce changes in continental ice volume, which cause eustatic changes and consequently small relative changes of sea level.

The stratigraphic signature of sequences and systems tracts is illustrated with an example from the Baltimore Canyon Trough (Greenlee et al. 1988). This section is an interpretation of depositional sequences which can be correlated throughout a seismic grid. The depositional environments associated with these sequences are interpreted using seismic facies analysis techniques and synthetic seismograms to tie the lithologic and biostratigraphic data from wells to the seismic sections. Figure 4 shows a sequence stratigraphic interpretation of the key seismic section in the Baltimore Canyon Trough. Beneath it is a diagram showing its location, and a chronostratigraphic chart with the sequences and systems tracts that were interpreted on the seismic line (Greenlee et al. 1988).

Sedimentation patterns noted on seismic reflection profiles from the Tertiary of the Baltimore Canyon Trough are controlled by long-term tectonic subsidence, lithospheric flexure due to sedimentary loading, long- and short-term changes of sea-level, and changes in the supply of sediments. Superimposed on the long-term tectonic events are much shorter periodic cycles manifested on seismic profiles as cyclic landward and basinward shifts of coastal onlap (Fig. 4).

Three orders of sea-level changes are recognized in this region. A long-term first-order sea-level fall during the Tertiary was in part responsible for continued seaward movement of the shelf edges. Cyclicity with a period of 10 Ma (second order) appears to have controlled the deposition of supersequences or larger packages of third-order cycles. Depositional sequences having a duration of approximately 1.5 Ma further subdivide the stratigraphic section and represent the shortest interval of cyclicity that can be recognized on the seismic reflection data of this particular area. The depositional sequences identified in this study, as well as the ages of the boundaries coincide with the eustatic sea-level changes observed by Haq et al. (1988) (Fig. 6).

5.3 Sedimentation

Sediments fill the space created by a relative rise of sea level. Sediments are deposited episodically and are local in distribution. The stratigraphic signature of the sedimentologic effect is the depositional systems. Depositional systems are composed of laminae, laminae sets, beds and bed sets, and episodic parasequences (Campbell 1967). These units contain sedimentary structures related to currents or settling processes within each depositional setting which are caused by wind, waves, mass flow, tides, floods, marine and fluvial currents, precipitation, and flocculation. These episodic packages accumulate to form depositional systems composed of lithofacies tracts. They may be within a sequence or include several sequences. Stratigraphic marker beds may be created by certain unique depositional events. Parasequences, bounded by flooding surfaces, package the beds and bed set into characteristic upward-shallowing cycles that stack together to form the systems tracts and depositional sequences.

The objective of sequence stratigraphy is to identify and correlate the genetic chronostratigraphic sequences, systems tracts, and parasequences, and then relate them to the depositional systems and lithofacies tracts. Application of this procedure to many different types of basins has shown a close relationship between the type of systems tract and the dominant depositional systems. This relationship is discussed below by systems tracts for siliciclastic and carbonate rock types.

The four systems tracts, lowstand, transgressive, highstand and shelf margin (Fig. 3) are discussed in order for both carbonate and siliciclastic depositional settings.

5.3.1 Lowstand Systems Tract – Siliciclastics

Lowstand systems tracts composed of siliciclastic sediments have been studied in three settings. One is associated with a deep-water shelf-margin setting, another with a growth-fault setting, and the third with a ramp setting (Fig. 13).

Deep water shelf margin setting. In the deep-water shelf-margin setting, lowstand systems tracts may have four parts: basin floor fan complex, slope fan complex, lowstand prograding complex and incised valley fill (Posamentier and Vail 1988; Sangree and Widmier 1977).

Siliciclastic basin floor fan complexes (Fig. 14) are largely composed of sheet mounds made up of massive sands deposited as lobes or sheets in a deep marine setting (Damuth 1980; Damuth et al. 1983; Kolla and Coumes 1987). Basin floor fan complexes may be preserved in several forms, including the original unamalgamated or winnowed amalgamated depositional lobes, submarine eroded remnant mounds, redeposited massive sands characterized by an offlap pattern bordering a remnant lobe, or massive sand mounds redeposited by deep marine bottom currents (Mutti 1985, 1989; Mutti and Normark 1987).

Slope fan complexes (Fig. 15) are made up of levee channel complexes with attached lobes and associated suspension deposits and chaotic slumps (Damuth and Embley 1981). A typical siliciclastic levee channel with attached lobes consists of five parts (Sangree and Widmier 1977) including:

5. The uppermost portion consists of an abandonment facies which grades upward from thin-bedded turbidites to hemipelagic shales.
4. Channel fill deposits may consist of massive amalgamated sands with very thin mudstone partings, multi-story upward-fining sands, or silts or mudstones. The channels have erosional bases and may contain internal erosional surfaces.
3. An upper section of thin-bedded laminated turbidite overbank deposits that build up as levees and bound the channel. Thicker crevasse splay sands may be present on the lower flanks of the levee.
2. A thickening upward section of attached channel lobes with turbidite sands, commonly ranging up to 3–5 m.
1. A basal fine-grained turbidite mudstone apron.

Minor fauna abundance peaks are commonly present at starved horizons between channel levee lobes. Suspension and chaotic slump deposits are common within the slope fan. The proximal portion of the slope fan complex may consist of massive sands, conglomerates, debris flow or slump deposits. The distal portions of the slope fan complex consist of fine-grained laminated turbidites. A condensed section with

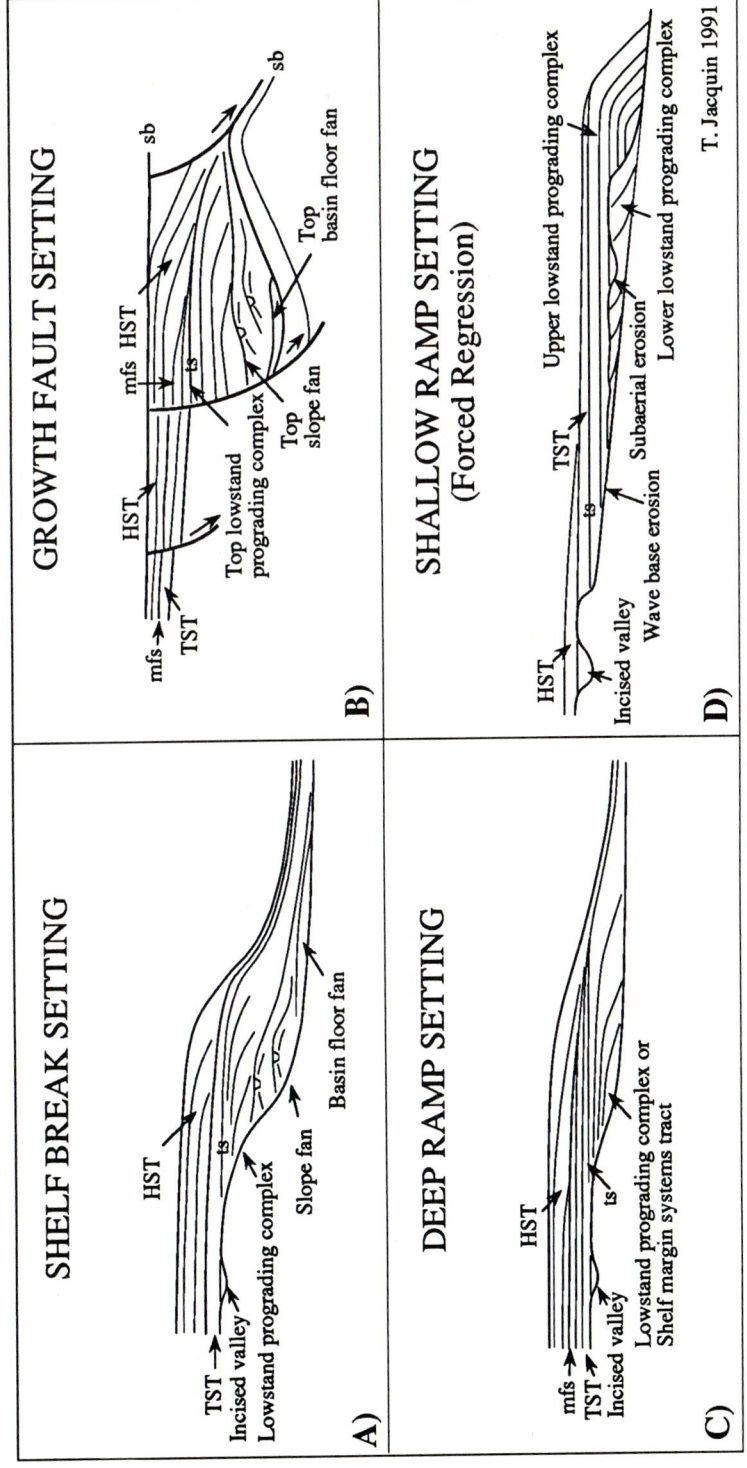

Fig. 13. Siliciclastic systems tracts in four different tectonic settings. A) The shelf break setting showing the development of the basin floor fan (BF), slope fan (SF), lowstand prograding complex (lpc), incised valley fill (ivf), transgressive systems tract (TST), and highstand systems tract (HST). Variations in this pattern are shown for (B) Growth fault setting, (C) Deep shelf setting, (D) Shallow shelf setting.

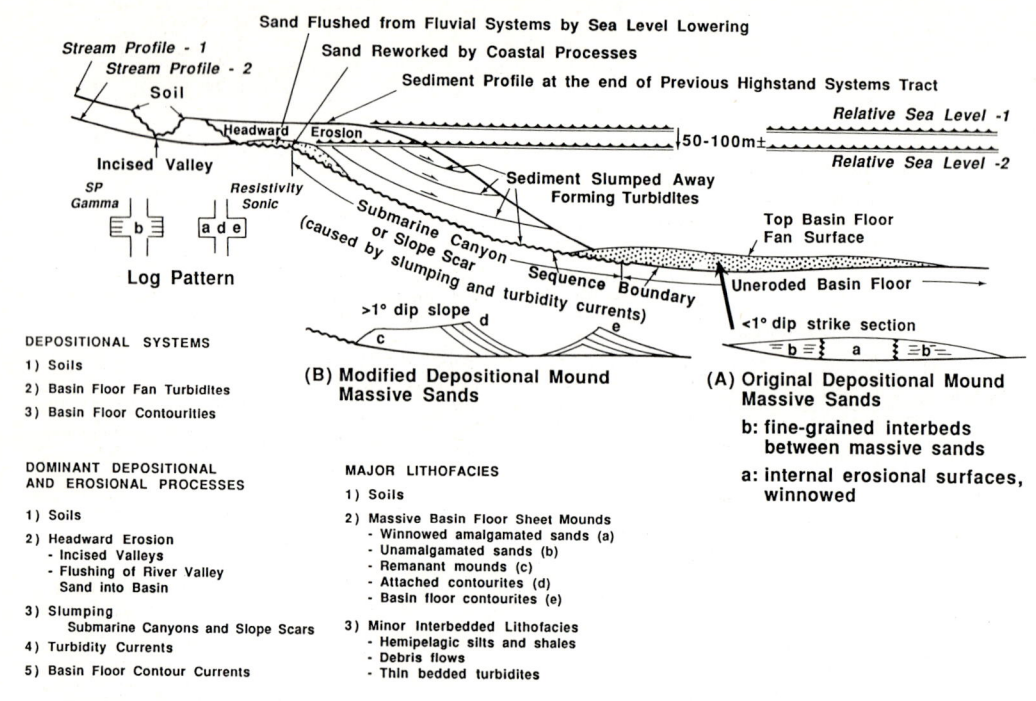

Fig. 14. Lowstand systems tract – basin floor fan complex, siliciclastics (explained in text)

a faunal abundance peak is commonly present at the base of the slope fan complex over the basin floor fan complex.

Lowstand prograding complexes composed of siliciclastic sediments (Fig. 16 and 17) are largely composed of shallowing-upward lowstand deltas or shorelines that prograde basinward and onlap landward with a pinchout in the vicinity of the offlap break of the preceding highstand. They are characterized by aggradational-offlap stratal patterns (Fig. 3). Shingled turbidites may interfinger with the toes of the lowstand prograding complex in certain areas with very high depositional rates. These shingled turbidites may be attached to or detached from the clinoform toes of the lowstand prograding complex. They are attached where the toes of the clinoforms reach the basin floor. They are detached where the clinoforms are prograding into very deep water. In this setting the toes of the clinoforms often pinch out on the slope. In this case, the shingled turbidites collect at the base of slope and are separated from the prograding clinoforms by a bypass zone. Shingled turbidites characterized by basal turbidite sands and pelagic caps may develop very thick basin filling successions that onlap onto the basin slopes. Such successions are usually associated with small tectonically active basins.

Slump blocks and olistostromes are characteristic of the lowstand prograding complex. In general, they are associated with the end of the lowstand prograding complex and early phase of the transgressive systems tract.

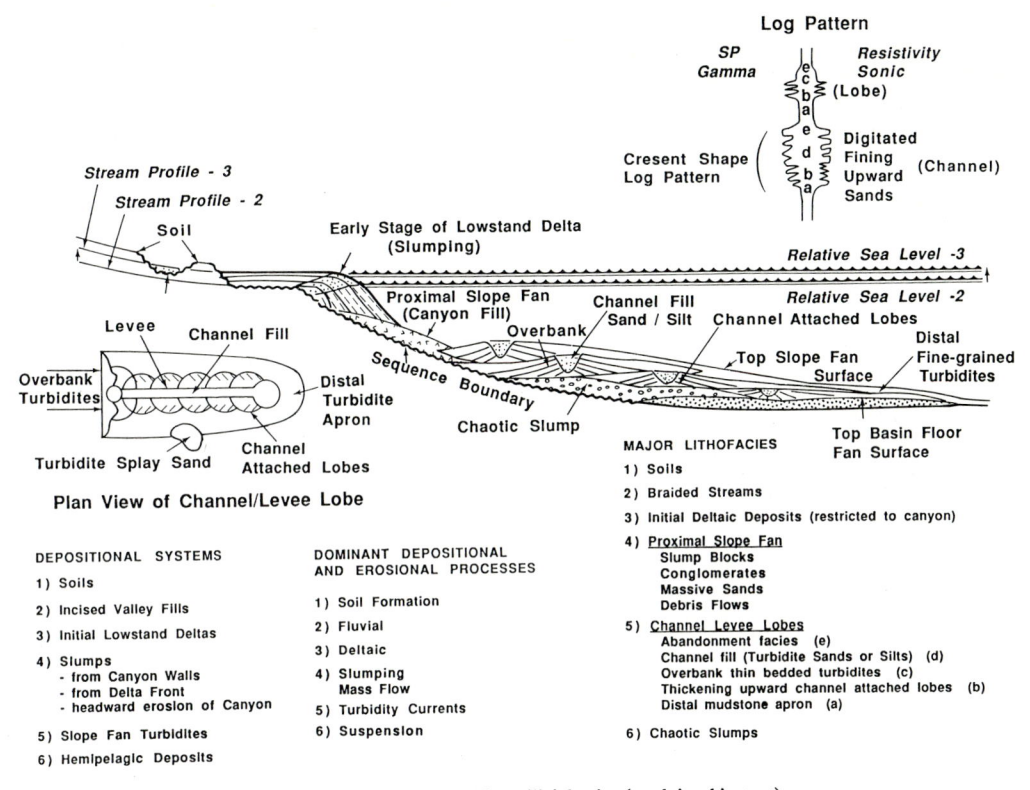

Fig. 15. Lowstand systems tract – slope fan complex, siliciclastics (explained in text)

Temporally associated incised valley fill is composed of braided stream sediments that fill previously cut valleys. Incised valley fill may be contemporaneous with the lowstand prograding complex or may be deposited later during the time of the transgressive systems tract. Incised valley fill deposited during the transgressive systems tract is commonly estuarine sediments.

Listric normal fault setting. Depositional settings in the presence of listric normal faults and/or mobile salt and shale structures typically have well-developed lowstand, transgressive and highstand systems tracts (Fig. 13c). The principle difference is that onlap and offlap patterns tend to be very subtle and often difficult to identify. Turbidite-dominated, slope and basin-floor fan complexes are common in the deep marine settings. Regional studies indicate that these fans originally developed extensive landward onlapping surfaces, but these surfaces typically are involved in the listric normal faulting and become subdued. The offlap pattern tends to be subtle because the shallow-water sediments are carried downward into the basin by subsidence along the listric faults, and not by progradation into deep water. Thus the systems tracts have a different appearance from the more typical shelf-margin setting;

650   Chapter 6 Sequences

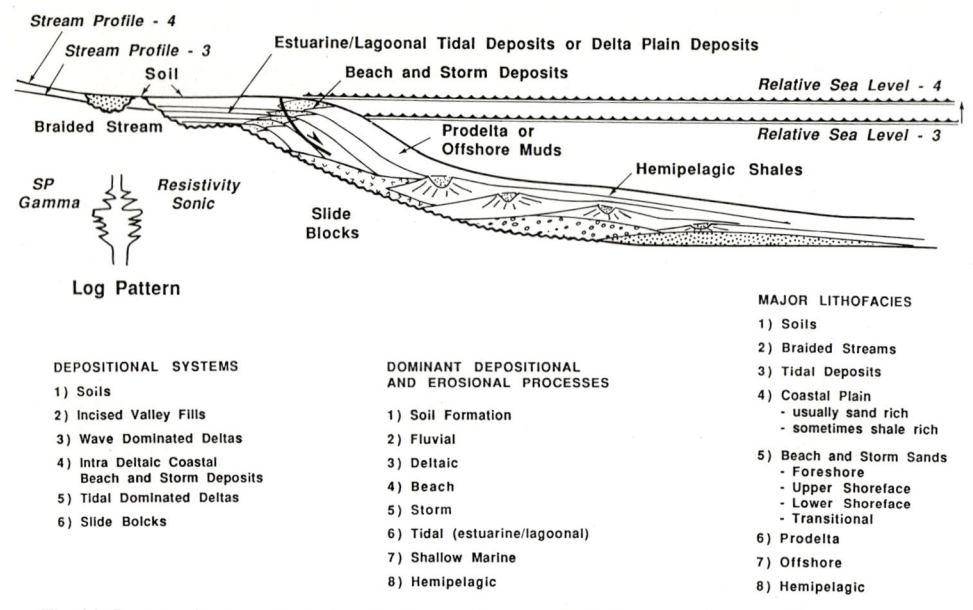

Fig. 16. Lowstand systems tract – lowstand prograding complex in low to moderate sedimentation rates

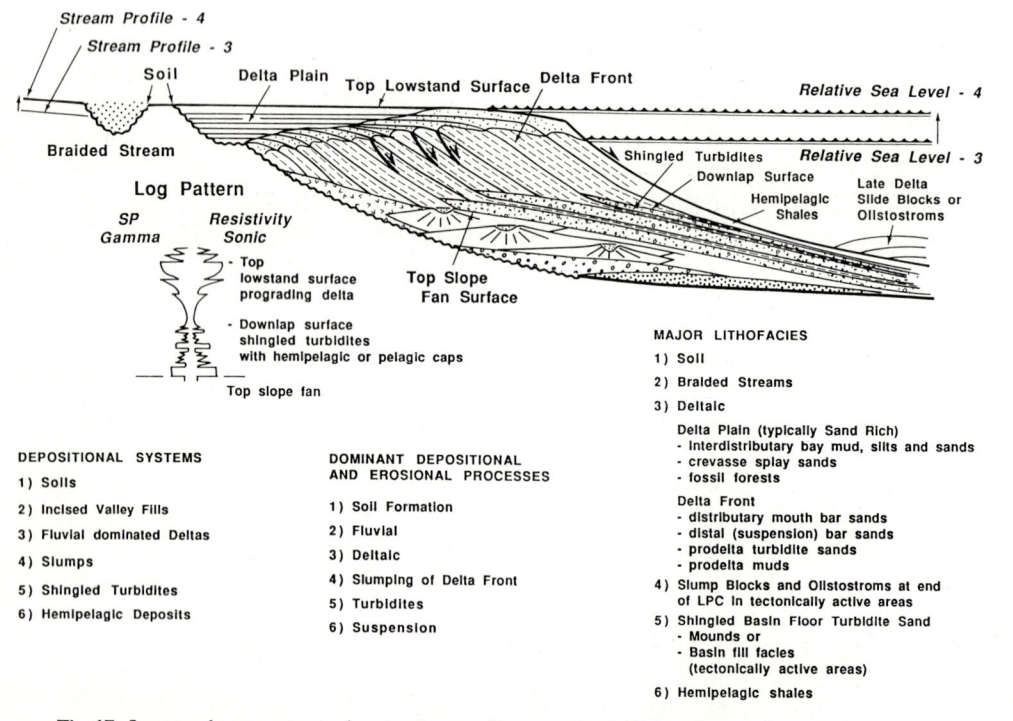

DEPOSITIONAL SYSTEMS
1) Soils
2) Incised Valley Fills
3) Wave Dominated Deltas
4) Intra Deltaic Coastal Beach and Storm Deposits
5) Tidal Dominated Deltas
6) Slide Bolcks

DOMINANT DEPOSITIONAL AND EROSIONAL PROCESSES
1) Soil Formation
2) Fluvial
3) Deltaic
4) Beach
5) Storm
6) Tidal (estuarine/lagoonal)
7) Shallow Marine
8) Hemipelagic

MAJOR LITHOFACIES
1) Soils
2) Braided Streams
3) Tidal Deposits
4) Coastal Plain
   - usually sand rich
   - sometimes shale rich
5) Beach and Storm Sands
   - Foreshore
   - Upper Shoreface
   - Lower Shoreface
   - Transitional
6) Prodelta
7) Offshore
8) Hemipelagic

Fig. 17. Lowstand systems tract – lowstand prograding complex in high sedimentation rates

DEPOSITIONAL SYSTEMS
1) Soils
2) Incised Valley Fills
3) Fluvial dominated Deltas
4) Slumps
5) Shingled Turbidites
6) Hemipelagic Deposits

DOMINANT DEPOSITIONAL AND EROSIONAL PROCESSES
1) Soil Formation
2) Fluvial
3) Deltaic
4) Slumping of Delta Front
5) Turbidites
6) Suspension

MAJOR LITHOFACIES
1) Soil
2) Braided Streams
3) Deltaic
   Delta Plain (typically Sand Rich)
   - Interdistributary bay mud, silts and sands
   - crevasse splay sands
   - fossil forests
   Delta Front
   - distributary mouth bar sands
   - distal (suspension) bar sands
   - prodelta turbidite sands
   - prodelta muds
4) Slump Blocks and Olistostroms at end of LPC in tectonically active areas
5) Shingled Basin Floor Turbidite Sand
   - Mounds or
   - Basin fill facies
     (tectonically active areas)
6) Hemipelagic shales

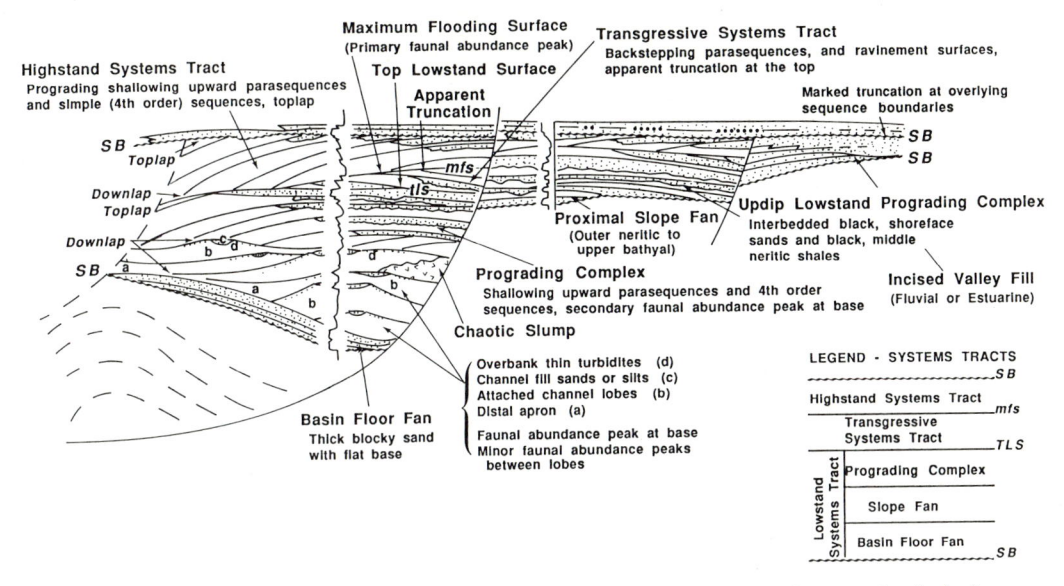

Fig. 18. Characteristics of sequences in a growth fault setting systems tracts in expansion faults in a Cenozoic composite sequence, Gulf of Mexico

however, each systems tract is typically well developed and identifiable when the proper criteria are used (Fig. 18).

Ramp setting. In the ramp setting, lowstand systems tracts have three parts: lower prograding complex, upper prograding complex, and incised valley fill (Fig. 5 and 13) (Posamentier et al. 1988a; Posamentier and Vail 1988). The lower prograding complex is characterized by offlapping stratal patterns with downlap at the base and erosional truncation at the top. The sediments are commonly deposited in a lower shoreface or offshore environment and are finer grained than the overlying upper prograding complex. The lower prograding complex is deposited during the fall of relative sea level (at the same time as the basin floor fans of deeper-water basins).

The upper lowstand prograding complex has an erosional base that in places may cut through the lower prograding complex to form incised valleys. When this happens they are usually filled with coarser-grained sediments which vary from marine conglomerates to tidal sands. The upper lowstand prograding complex is formed during the slow relative rise of sea-level, following the relative fall (lowstand prograding complex time). The incised valleys are filled either by braided stream deposits if they were filled during the lowstand, or by estuarine sands if they were filled during the transgressive systems tract time.

5.3.2 Transgressive Systems Tract – Siliciclastics

The transgressive systems tract (Fig. 19) is made up of a set of retrograding parasequences that thicken landward until they thin by onlap at their base (Fig. 3). In general

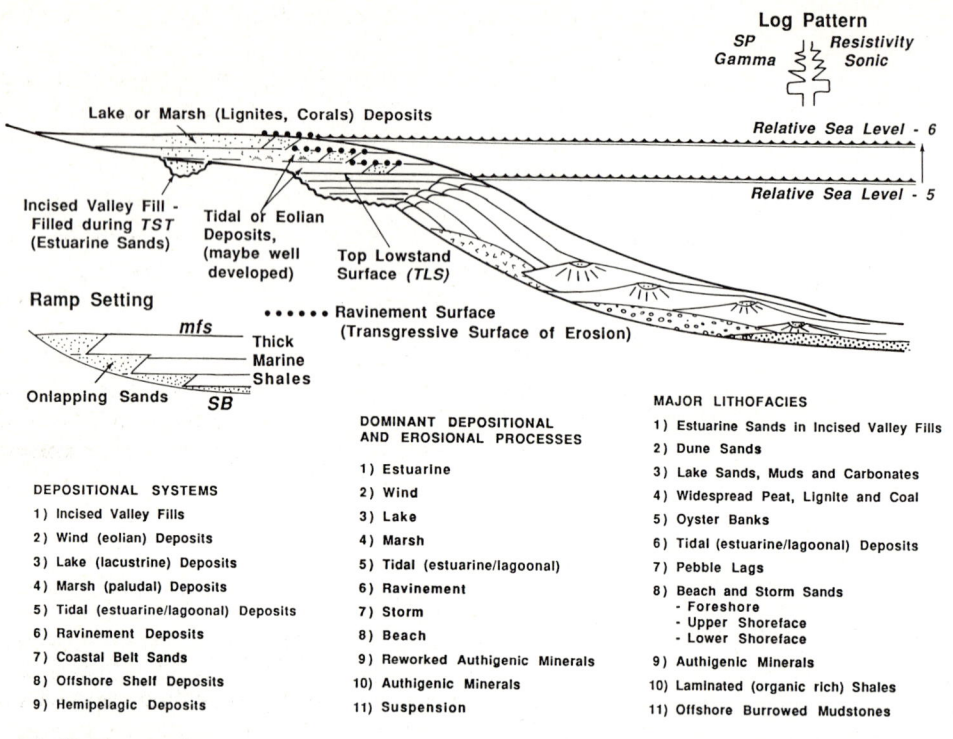

Fig. 19. Transgressive systems tract

Log Pattern
SP   Resistivity
Gamma   Sonic

Lake or Marsh (Lignites, Corals) Deposits

Relative Sea Level - 6

Relative Sea Level - 5

Incised Valley Fill - Filled during *TST* (Estuarine Sands)

Tidal or Eolian Deposits, (maybe well developed)

Top Lowstand Surface *(TLS)*

Ramp Setting

••••• Ravinement Surface (Transgressive Surface of Erosion)

mfs
Thick Marine Shales
Onlapping Sands   SB

DEPOSITIONAL SYSTEMS
1) Incised Valley Fills
2) Wind (eolian) Deposits
3) Lake (lacustrine) Deposits
4) Marsh (paludal) Deposits
5) Tidal (estuarine/lagoonal) Deposits
6) Ravinement Deposits
7) Coastal Belt Sands
8) Offshore Shelf Deposits
9) Hemipelagic Deposits

DOMINANT DEPOSITIONAL AND EROSIONAL PROCESSES
1) Estuarine
2) Wind
3) Lake
4) Marsh
5) Tidal (estuarine/lagoonal)
6) Ravinement
7) Storm
8) Beach
9) Reworked Authigenic Minerals
10) Authigenic Minerals
11) Suspension

MAJOR LITHOFACIES
1) Estuarine Sands in Incised Valley Fills
2) Dune Sands
3) Lake Sands, Muds and Carbonates
4) Widespread Peat, Lignite and Coal
5) Oyster Banks
6) Tidal (estuarine/lagoonal) Deposits
7) Pebble Lags
8) Beach and Storm Sands
   - Foreshore
   - Upper Shoreface
   - Lower Shoreface
9) Authigenic Minerals
10) Laminated (organic rich) Shales
11) Offshore Burrowed Mudstones

the younger marine parasequences in the set are progressively thinner due to sediment starvation. The transgressive systems tract thins basinward and upward, forming a condensed section at the top, seaward of the inner shelf.

The boundary between the marine and nonmarine sediments is commonly a set of retrograding ravinement surfaces (transgressive surfaces of erosion). Well-developed beach and storm sands are often associated with this transition.

The nonmarine sediments are typically marsh or lake deposits that onlap the underlying sequence boundary or incised valley fill. Thick, widespread coals and extensive lacustrine deposits are characteristic of the transgressive and early highstand systems tract. In regions where sediment supply is low, the incised valley may be filled with sediments deposited during the time corresponding to the transgressive systems tract. These sediments are commonly estuarine or coastal plain.

The marine shales basinward of the ravinement surfaces tend to be organic rich and contain authigenic minerals. Pyrite, glauconite and authigenic dolomite are most common. Glauconite and dolomite occur in detrital sand beds. Phosphate nodules and siderite are found along the parasequence flooding surfaces as well as in the condensed section associated with the maximum flooding surface.

5.3.3 Condensed Section at the Maximum Flooding Surface

The transgressive and highstand systems tracts are bounded by the maximum flooding surface. This surface commonly occurs within the top of or at the base of a condensed section caused by very low sedimentation rates (Fig. 20) (Loutit et al. 1988). Condensed sections usually coincide with zones of maximum diversity and abundance of fossils. However, in deeper marine areas the fossil tests may be destroyed or dissolved, and a submarine biostratigraphic hiatus may form with associated concentrations of authigenic minerals such as phosphate, glauconite, siderite, pyrite, and dolomite, as well as airborne particles such as volcanic ash and iridium. Undisturbed authigenic glauconite is often found in the condensed section.

In certain settings, where ocean currents impinge upon the slope and outer shelf, during times of rising sea level, substantial amounts of winnowing or erosion may occur (Popenoe 1985; Snyder et al. 1990). In oceanic settings below the calcite compensation depth (CCD), maximum flooding surfaces coincide with dissolution surfaces (Berger and Winterer 1974; Haq 1990).

5.3.4 Highstand Systems Tract – Siliciclastics

Highstand systems tracts (Fig. 21) are made up of three parts: early highstand, late highstand prograding complex, and late highstand subaerial complex (Fig. 2). The early highstand is characterized by upward- and outward-building sigmoidal prograding stratal patterns (Fig. 3). The late highstand prograding complex is characterized by outward-building oblique prograding stratal patterns. The late highstand subaerial complex is characterized by fluvial sediments deposited during a relative sea-level stillstand.

The late highstand prograding complex and the late highstand subaerial complex are deposited contemporaneously. The boundary at the base of the highstand systems

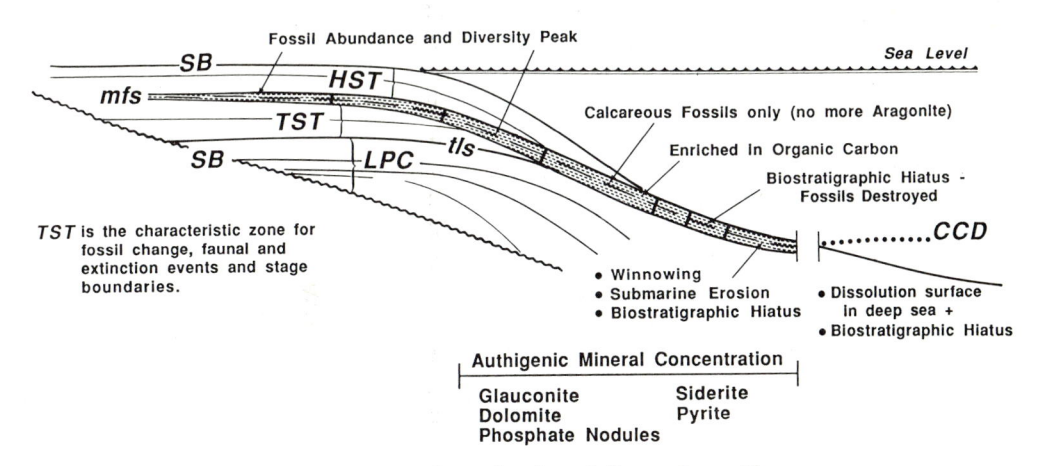

Fig. 20. Character of the maximum flooding surface from shallow to deep settings

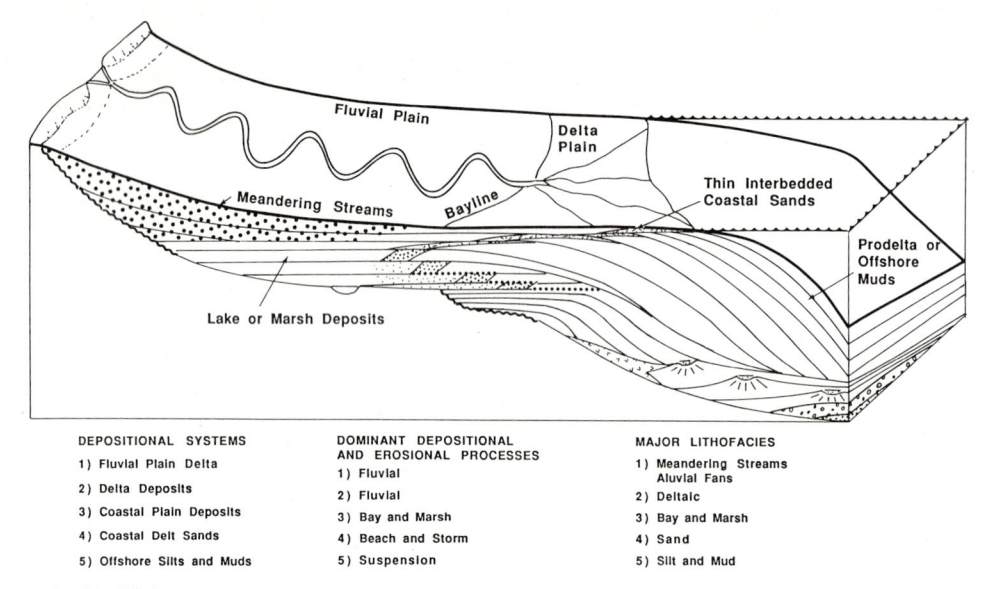

Fig. 21. Highstand systems tract

tract is a downlap surface (maximum flooding surface) that is associated with a condensed section (Fig. 2). It becomes conformable on the inner shelf and is difficult to detect within the coastal plain sediments. However, it is sometimes identifiable within lacustrine sediments. The early highstand is very similar to the late transgressive systems tract. The most important difference is that the prograding parasequences retrograde during the transgressive systems tract and prograde during the highstand.

The late highstand prograding complex is typically made up of deltaic and interdeltaic or beach and storm deposits. Because of the decreasing relative rise of the late highstand, the coastal and delta plain sediments are thin-bedded. The fluvial subaerial complex tends to trap the coarser fraction and cause the highstand deltas to be finer-grained than the lowstand deltas in the same region. The late highstand subaerial complex builds above sea-level, enabling the fluvial systems to maintain their optimum equilibrium gradient as the highstand systems tract progrades seaward (Posamentier and Vail 1988). Typically, meandering stream deposits tend to coalesce and become fine-grained and more widespread upward. In areas of high topographic relief, alluvial fans may develop.

5.3.5 Shelf-Margin Systems Tract – Siliciclastics

The shelf-margin systems tract (Fig. 22) is a wedge that overlies a Type 2 sequence boundary and laps out on the shelf landward of the underlying offlap break. This systems tract is characterized by both aggradation and progradation (Fig. 3). Its lower boundary is a conformable sequence boundary and its upper boundary is a top lowstand surface. An unconformity exists landward of where it pinches out. This is where the top shelf-margin and the unconformity merge. The landward portion of the shelf-margin systems tract is commonly a nonmarine wedge that thickens seaward, while the marine portion is similar to the lowstand prograding complex.

5.3.6 Depositional Sequences – Carbonates

Carbonate depositional systems appear to respond to relative changes of sea level much the same as siliciclastic depositional systems (Bosellini 1984; Sarg 1988; Jacquin et al. 1990; Glaser et al. 1990). Lowstand and shelf-margin systems tracts are well developed in carbonate depositional basins. Retrograding and/or aggrading

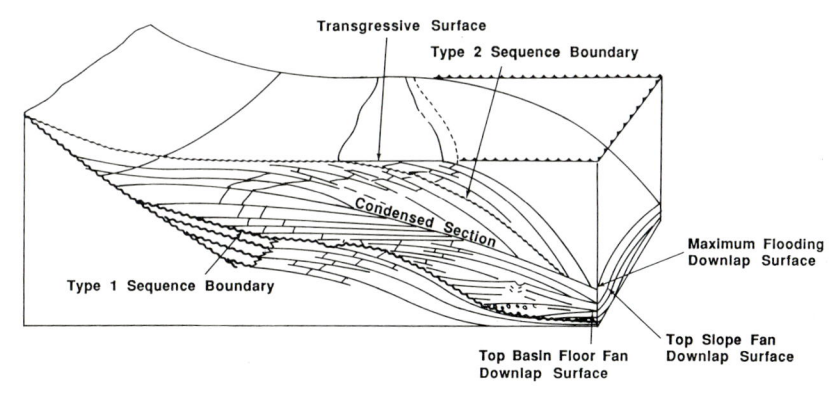

Fig. 22. Shelf-margin systems tract

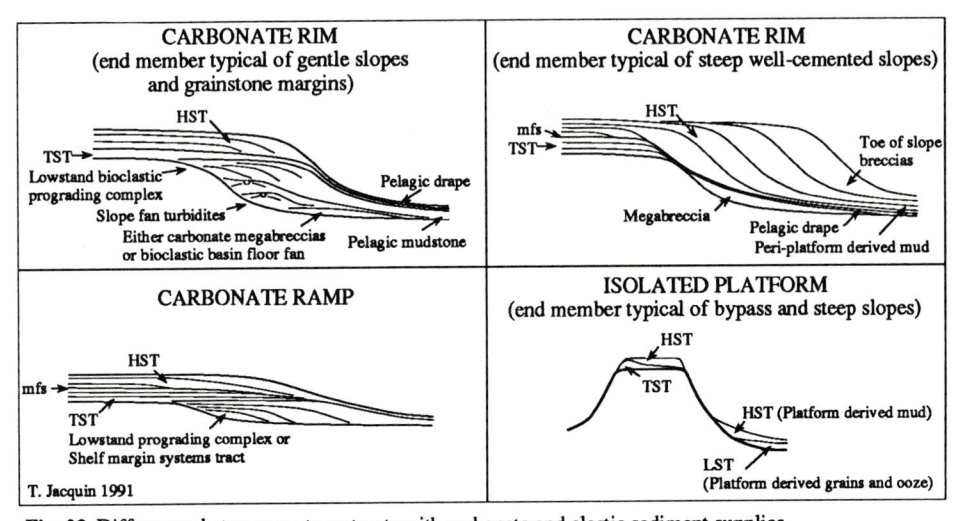

Fig. 23. Differences between systems tracts with carbonate and clastic sediment supplies.

transgressive systems tracts may build thick carbonate shelves and banks on the platform, but tend to be very thin and starved in the basin. Prograding highstand systems tracts may fill in the shallow water depths of the platform and prograde into deep water as clinoforms, but they do not appear to form massive basin fill deposits. A dominance of basin-fill carbonates during the highstand, as observed by Droxler and Schlager (1985), in modern sediments in the Bahamas, is not common in ancient rocks. A possible reason for this is that the slopes in ancient rocks tend to be low and less well cemented in comparison to modern carbonate banks and the carbonate sediments are deposited as clinoforms rather than basin floor turbidites.

A major difference between carbonate and siliciclastic prograding complexes is the steeper depositional slopes associated with carbonates and the fact that grainstones onlapping the platform may be marine. In siliciclastic settings, the shoreline tends to be coupled with the progradational pattern and shows regression. In contrast for carbonate settings, the lowstand and shelf-margin complexes may be transgressive because onlapping marine carbonates can be deposited as the lowstand or shelf-margin systems tract prograde. Also because of the capacity of carbonates to generate massive amounts of sediments, part of what would be a transgressive systems tract in a siliciclastic setting may develop into a shelf-margin systems tract in a carbonate setting.

Despite the general similarities between carbonate and siliciclastics, there are some important differences within each systems tract (Fig. 23). These differences are discussed below.

Basin-floor fan complexes composed of carbonate sediments are common in deep water basins with well-developed platform margins. However, these basin-floor fans are typically composed of massive carbonate debris. These sediments are believed to be derived from the collapse of the platform margin caused by the relative fall of sea-level at the start of the lowstand. Major debris flows are rare in transgressive and highstand systems tracts (Sarg 1988; Jacquin et al. 1990; Glaser et al. 1990). Bioclastic basin-floor fans do exist, but they are not common (Jacquin et al. 1990). These fans are probably caused by platform margin collapse during the initial stages of relative sea-level fall before the platform is exposed. Generally, carbonate sediments are relatively better ("well") cemented compared to siliciclastic sediments and are less prone to collapse. Therefore, there are fewer carbonate basin-floor fan complexes.

Carbonate slope fan complexes have a stratal geometry pattern that is similar to siliciclastic slope fans (Jacquin et al. 1990). Carbonate slope fans also consist largely of levee channels and attached lobes. In a stratigraphic section, the beds of each lobe characteristically first thicken upward and then thin upward. Channel fills are thick-bedded lenses of wackestone and packstone. The channel fills tend to grade to grainstones toward the source and to mudstone in the distal direction. The abandonment facies consists of thinning-upward mudstones. The overbank deposits are primarily mudstones. The lime mud appears to be platform derived and is occasionally interbedded with thin organic-rich laminae when depositional rates are low. The proximal portions of slope fans are often characterized by large erosional channels and collapse features, coarse carbonate grainstones, and carbonate debris flows (Jacquin et al. 1990).

Lowstand prograding complexes and shelf-margin systems tracts composed of carbonate sediments are often the major prograding units that build the platform outward. Lowstand prograding complexes onlap in a landward direction in the vicinity of the platform margin (offlap break) of the previous highstand. Shelf-margin systems tracts onlap onto the margin of the platform, landward of the offlap break. Seaward they prograde into the basin. The shallow-water portion is massive grainstone either with or without reef material. The grainstone facies changes to packstone, wackestone and ultimately mudstone within the prograding beds as the water depth increases. As in their siliciclastic counterparts, shingled bioclastic turbidites or debris flows are common at the base of the prograding clinoforms. They are believed to be formed from slumping of the upper slope. Massive slide slumps are very common within the mudstone facies and olistostromes are common in areas of deep water and contemporaneous tectonic activity, especially at the end of the lowstand prograding complex.

Of special interest is the fact that planktonic organisms tend to form thick, cyclic deposits during the lowstand prograding or shelf-margin systems tract time in the basin and become thin and starved during the deposition of the transgressive systems tract. Highstand systems tracts are starved in the basin in the lower part and gradually accumulate thicker carbonate beds upward as deposition during the highstand fills in the platform. The carbonate-starved transgressive and early highstand systems tracts in the deep basin are rich in a diverse assemblage of fossils but they do not commonly form massive limestone beds. This indicates that the massive blooms of planktonic organisms that characterize carbonate lowstand systems tracts do not occur. This may be caused by the $CaCO_3$ budget being depleted by the development of platform carbonate during the transgressive and early highstand systems tracts (Berger and Winterer 1974; Haq 1990).

During the rapid relative rise associated with the retrograding transgressive systems tract, carbonates may build up on the platform as shelves or banks. The carbonates close to the platform margin are typically drowned, or "give up", near the top of the transgressive systems tract. More landward they maintain a shallow-water facies or "keep up". During this time, the basin is starved of sediments. A thin, often fossil-rich, condensed section records the basin equivalent of the transgressive and early highstand systems tract.

During the decreasing relative rise of sea-level, the highstand systems tract fills in the flooded platform with carbonate sediments that prograde basinward. Sometimes the progradation reaches the platform margin and may extend into deep water. In that case, deep-water highstand deposits are prograding clinoforms with or without a time-transgressive deposit of carbonate debris at its base (Bosselini 1984; Sarg 1988). Thick, flat-lying carbonate basin fill deposits are uncommon during the deposition of this systems tract. In many cases the highstand systems tract never completely progrades across the flooded platform, and the highstand basin deposits are thin basinal muds.

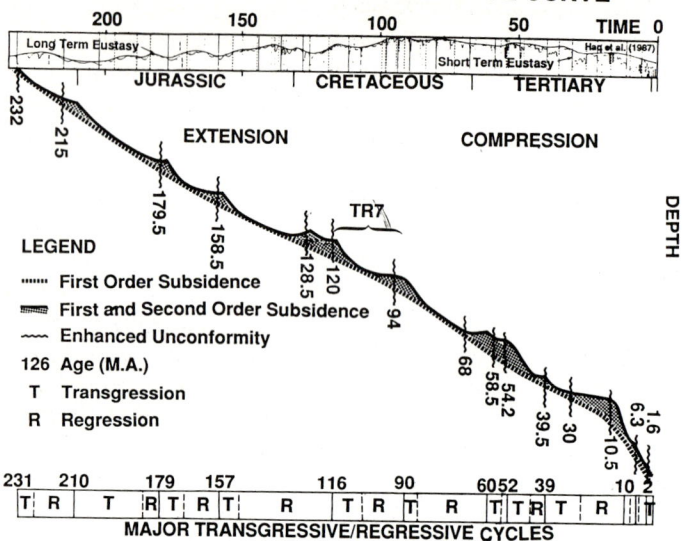

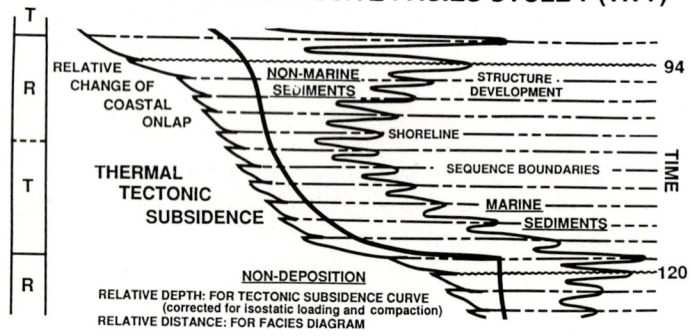

Fig. 24. A schematic integration of transgressive-regressive cycles and the subsidence history of a region

6 Conclusions

The stratigraphic signature of the many processes that control the development of the stratigraphic record can be identified with an integration of sequence stratigraphic, subsidence and tectono-stratigraphic analyses. Figure 24 summarizes the recommended stratigraphic analysis of an area. Major transgressive-regressive facies cycles are identified and plotted along the horizontal axis. Sequences and systems tracts are identified within the transgressive-regressive cycles (shown in Fig. 24). Correlation with the global cycle chart documents any association with global (eustatic) sequences and systems tracts. A subsidence history documents the changes in subsidence patterns, which can be related to plate tectonic readjustments and faulting events.

Faults and folds form at predictable times within the second-order tectonic events, depending on the existing tectonic regime. Tectonic subsidence curves document the relation of the transgressive-regressive cycles to changes in rates of subsidence.

Transgressive-regressive facies cycles provide a method of analyzing gross facies distribution and their relation to structure. In hydrocarbon exploration transgressive-regressive facies cycles provide a framework for stratigraphic and structural play analysis and prediction.

Sequence cycles provide the means to subdivide sedimentary strata into genetic chronostratigraphic intervals that have predictable relationships to depositional systems. Correlation of parasequences provides a powerful method for the analysis of time and rock relationships in sedimentary strata for detailed reservoir analysis. Sequence, systems tract, and parasequence surfaces provide a framework for correlating and mapping. Interpretation of systems tracts provides a framework to predict facies relationships within the sequence. Parasequences further subdivide the sequences and systems tracts into the smallest genetic units for detailed prediction, correlation and mapping of depositional environments.

The following sequence of procedures for an integrated interpretation of paleogeography, geologic history, stratigraphic signatures, and resource evaluation are recommended (Fig. 8):

1. Determine physical chronostratigraphic framework by interpreting sequences, system tracts and parasequences and/or simple sequences on outcrops, well logs and seismic data and age date with high resolution biostratigraphy.
2. Construct geohistory, total subsidence and tectonic subsidence curves based on sequence boundary ages.
3. Complete a tectono-stratigraphic analysis including
   - Relate major transgressive-regressive facies cycles to tectonic events.
   - Relate changes in rates on tectonic subsidence curves to plate-tectonic events.
   - Assign a cause to tectonically enhanced unconformities.
   - Relate magmatism to the tectonic subsidence curve.
   - Map tectono-stratigraphic units.
   - Determine style and orientation of structures within tectono-stratigraphic units.
   - Simulate geologic history.
4. Define depositional systems and lithofacies tracts within systems tracts and parasequences or simple sequences.
5. Interpret paleogeography, geologic history, and stratigraphic signatures from resulting cross-sections, maps and chronostratigraphic charts.
6. Locate potential reservoirs and source rocks for possible sites of exploration.

Acknowledgments. This paper represents the efforts of many Exxon and Rice University scientists who contributed to the development of the concepts presented here. We are especially indebted to J.B. Sangree, R.M. Mitchum, J. Hardenbol, T.S. Loutit, H. Posamentier, and J.F. Sarg, who contributed significantly to concepts developed in the report, and to A.W. Balley, B.U. Haq, and D.S. Sawyer, whose critical review of the manuscript provided a great many important improvements. E. Mutti, F. Francesco, V. Kolla, and C. Ravenne provided insight and helpful suggestions for the lowstand systems tracts. We also thank Malcolm Ross for his work in the initial development of the stratigraphic signature analysis chart, Fig. 1. We thank Maria Bowman for her help in typing the manuscript.

6.2 Asymmetry in Transgressive-Regressive Cycles in Shallow Seas and Passive Continental Margin Settings

G. Einsele and U. Bayer

1 Introduction

The shape and amplitude of pre-Pleistocene sea level oscillations and their effects on sedimentary cycles are only poorly known and, therefore, topics which need further investigation. From information about the last Pleistocene sea level change, we assume that older sea level curves had more or less the shape of a sinusoidal function with a periodic rise and fall around a mean datum line. The true sea level curves, however, can be derived only from vertical sedimentary sections which often appear to be asymmetric and usually are not dated with sufficient resolution (Bayer et al. 1985b; Bayer 1987). The purpose of this chapter is to show that ideal sinus-shaped sea level oscillations commonly generate asymmetric sedimentary cycles, but the type and intensity of asymmetry vary between the margin and deeper parts of a sedimentary basin.

The general interest is to infer the ancient state from the remaining record. For a long-term trend in subsidence superimposed by short-term sea level fluctuations, McGhee and Bayer (1985) proposed a rather simple model involving only elementary calculus. The following chapter describes a graphical method which was already used in a similar form to demonstrate effects of sea level changes (e.g., Baum and Vail 1987; Loutit et al. 1988). We shall provide a set of simple models, which take into account the most important factors contributing to the formation of the sedimentary record. It is, however, beyond the scope of this paper to integrate the basic concepts into a general, three-dimensional model taking into consideration all the possible interrelationships between these and additional factors.

In the context of this chapter it is not crucial to know whether the sea level changes are global phenomena or only regional relative rises and falls of the sea level.

2 Some Basic Considerations

The sediment accumulation in a marine basin primarily results from the interactions between sediment supply, subsidence of the basin floor, and the hydrodynamic regime controlling the water depth at which permanent deposition can take place. A "transgression" is the landward displacement of the shoreline indicated by landward migration of the littoral facies, and a "regression" is the opposite seaward displacement of the shoreline (Vail et al. 1984).

Einsele et al. (Eds.)
Cycles and Events in Stratigraphy
©Springer-Verlag Berlin Heidelberg 1991

In this chapter, we focus on the processes taking place at water depths below the lowstands of sea level, i.e., in areas which do not emerge and therefore possess a high potential of having preserved their sediments in the ancient sedimentary record (Fig. 1). If the amplitude of the sea level oscillations is limited (up to some tens of meters), both the inner and outer shelf and, of course, deeper regions of a depositional basin meet this requirement. Thus, we avoid the complications of an irregular coastal zone. The surface topography of the deeper basin, however, can also significantly affect the sedimentary record related to transgressions and regressions (Sawyer 1986). Our two-dimensional models cannot directly solve these problems.

On the other hand, the response of the shoreline and coastal sediments to sea level oscillations was frequently described, either in front of or at some distance away from a river delta (Fig. 1). Particularly the consequences of the sea level lowstand during the last glacial maximum and the subsequent Holocene transgression were studied in many regions (e.g., Einsele et al. 1977; Morton and Price 1987; Nummedal and Swift 1987).

Subsidence of a basin and sediment supply contrast in their functions with regard to the vertical dislocation of the sediment surface. The term "subsidence" as used here may be defined as "effective subsidence". It is the vertical dislocation of the sea floor resulting from the additive effects of basement subsidence, sedimentation and erosion, and sediment compaction. However, sediment compaction is considered a minor factor, which is neglected here.

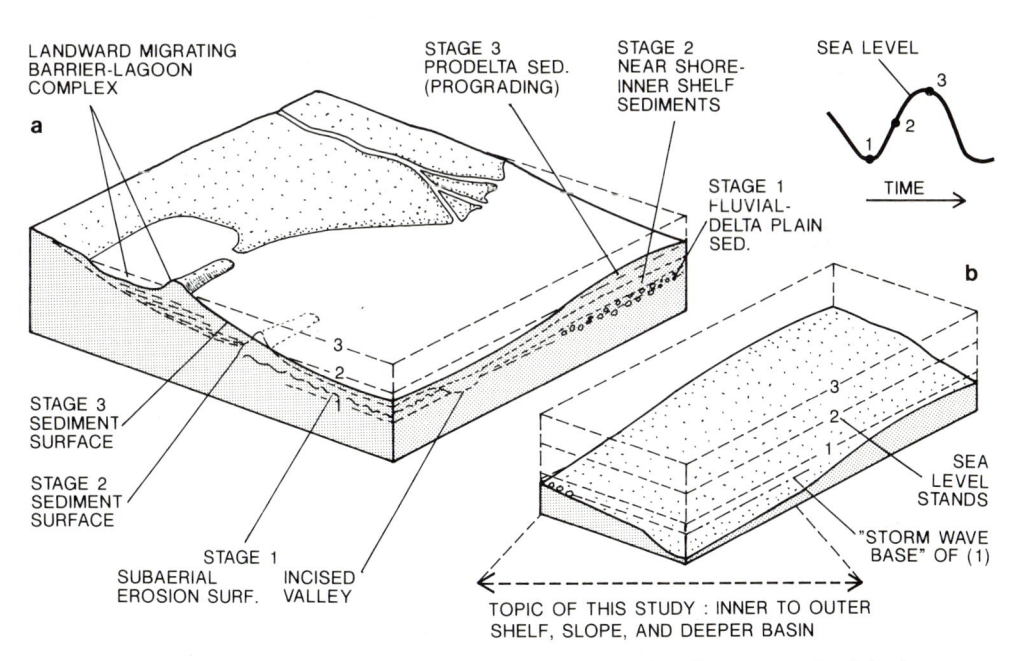

Fig. 1. a Simplified model showing the response of coastal and deltaic sediments to sea level rise (stages *1, 2, 3;* based, e.g., on Nummedal and Swift 1987; Posamentier and Vail 1988). b Topic of this chapter: marine sediments preserved at or below the "storm wave base" of sea level lowstands

The velocities of subsidence and average sediment buildup may be considered constant for a time interval considerably longer than the period of a transgressive-regressive (T-R) cycle as discussed here. Consequently, the subsidence of the sediment surface is given by the time-integral over the difference between the two processes. This leads to an overall deepening sequence if subsidence dominates, or to a shallowing sequence should sediment supply overwhelm subsidence. For our models we mostly assume that subsidence and average sediment-buildup are approximately equal, i.e., there are no long-term significant changes in water depth and extension of the basin.

3 Conceptual Models of Sediment Deposition

3.1 Interrelationship Between Sea Level Fluctuations, Sedimentation Rate, and Subsidence

The models of Figs. 2 and 3 illustrate sediment accumulation and submarine erosion at certain spatial locations (I through VIII, Fig. 2a) within a shallow marine basin. All of these locations are situated at or below the current action of storm waves (for simplicity here referred to as "storm wave base") at sea level lowstand. Sedimentary processes along the shoreline and shoreface are not considered (Sect. 2). The models are based on the following assumptions:

1. The sea level oscillation is sinusoidal, ideally a sine function. The amplitude and period of this function remains constant through time. The storm wave base varies parallel to the sea level.
2. The rate of subsidence of the basin floor is constant throughout the basin.

Constant subsidence and an oscillating wave base generate a variable rate of sediment accommodation, i.e., space available for sedimentation (Vail et al. 1984; Baum and Vail 1987). During a period of rising sea level, the space for sediment accumulation increases, but sediment supply is often not sufficient to fill the gap created by the divergence of subsidence and sea level rise. In contrast, with the drop of the sea level the gap closes and a percentage of the sediment supply cannot be accommodated. As a result, it is swept into deeper parts of the basin in a process referred to as "bypassing". In addition, older sediments may be eroded (see below) and transported to deeper locations (Fig. 2b).

Fig. 2. Conceptual model for the interrelationship between periodic change of sea level and its storm wave base, constant subsidence (rapid or moderate), different (but constant with time) rates of sedimentation (or sediment supply, *I* through *IV*), and resulting chronostratigraphic sequences. **a** Cross-section of uniformly subsiding shallow basin (cf. Fig. 1b) with locations of vertical sequences I through IV and V through VIII (demonstrated in **b** and in Figs. 3 and 4b); chronostratigraphic sequences *I* through *III* (IV shown only in Fig. 4). See text for further explanation

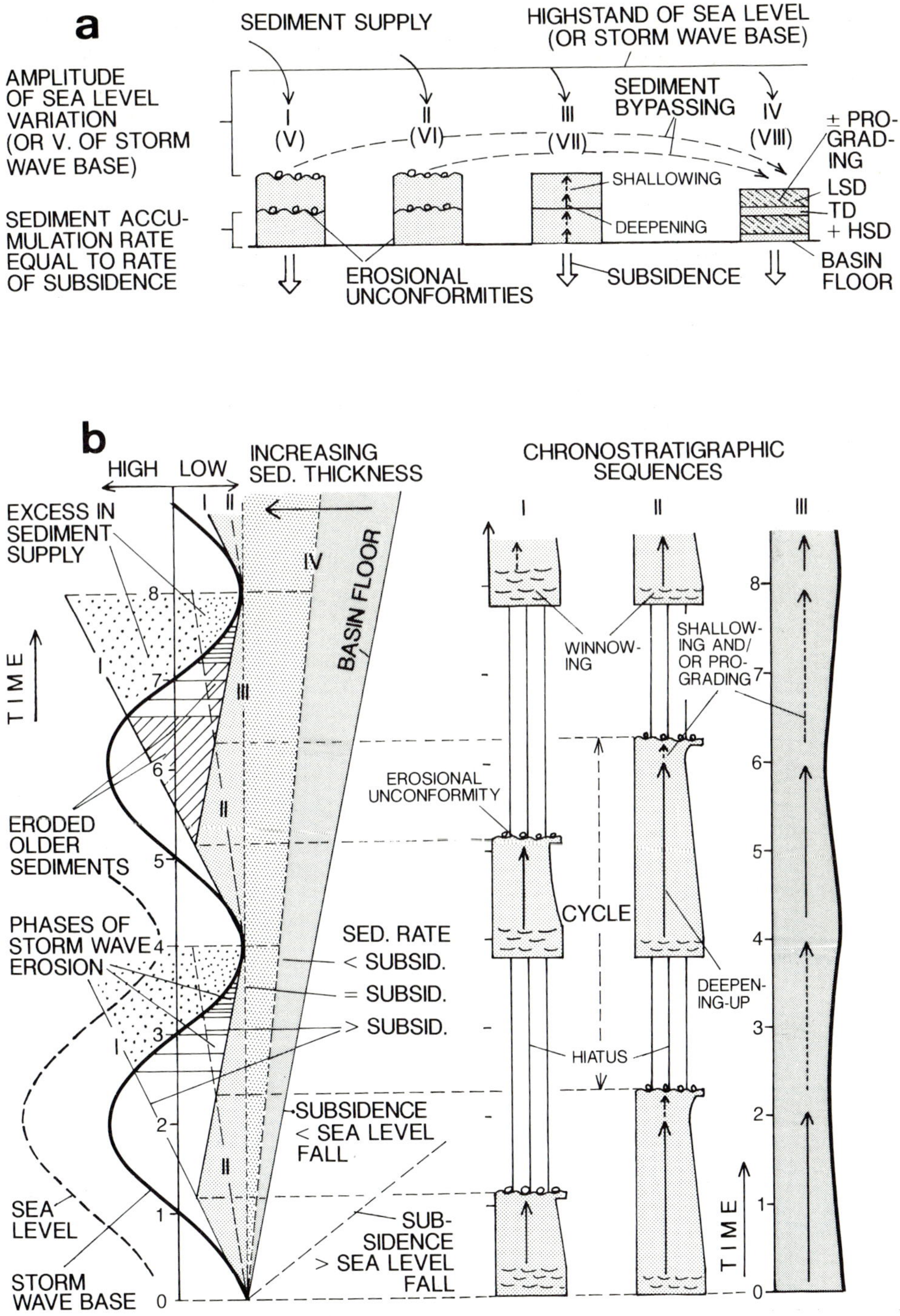
a
SEDIMENT SUPPLY
HIGHSTAND OF SEA LEVEL (OR STORM WAVE BASE)
AMPLITUDE OF SEA LEVEL VARIATION (OR V. OF STORM WAVE BASE)
SEDIMENT ACCUMULATION RATE EQUAL TO RATE OF SUBSIDENCE
I (V)
II (VI)
III (VII)
IV (VIII)
SEDIMENT BYPASSING
± PROGRADING
LSD
TD
+ HSD
BASIN FLOOR
SHALLOWING
DEEPENING
EROSIONAL UNCONFORMITIES
SUBSIDENCE
b
HIGH
LOW
INCREASING SED. THICKNESS
EXCESS IN SEDIMENT SUPPLY
TIME
BASIN FLOOR
ERODED OLDER SEDIMENTS
PHASES OF STORM WAVE EROSION
SED. RATE
< SUBSID.
= SUBSID.
> SUBSID.
SUBSIDENCE < SEA LEVEL FALL
SUBSIDENCE > SEA LEVEL FALL
SEA LEVEL
STORM WAVE BASE
CHRONOSTRATIGRAPHIC SEQUENCES
I
II
III
IV
WINNOWING
SHALLOWING AND/ OR PROGRADING
EROSIONAL UNCONFORMITY
CYCLE
DEEPENING-UP
HIATUS
0
1
2
3
4
5
6
7
8

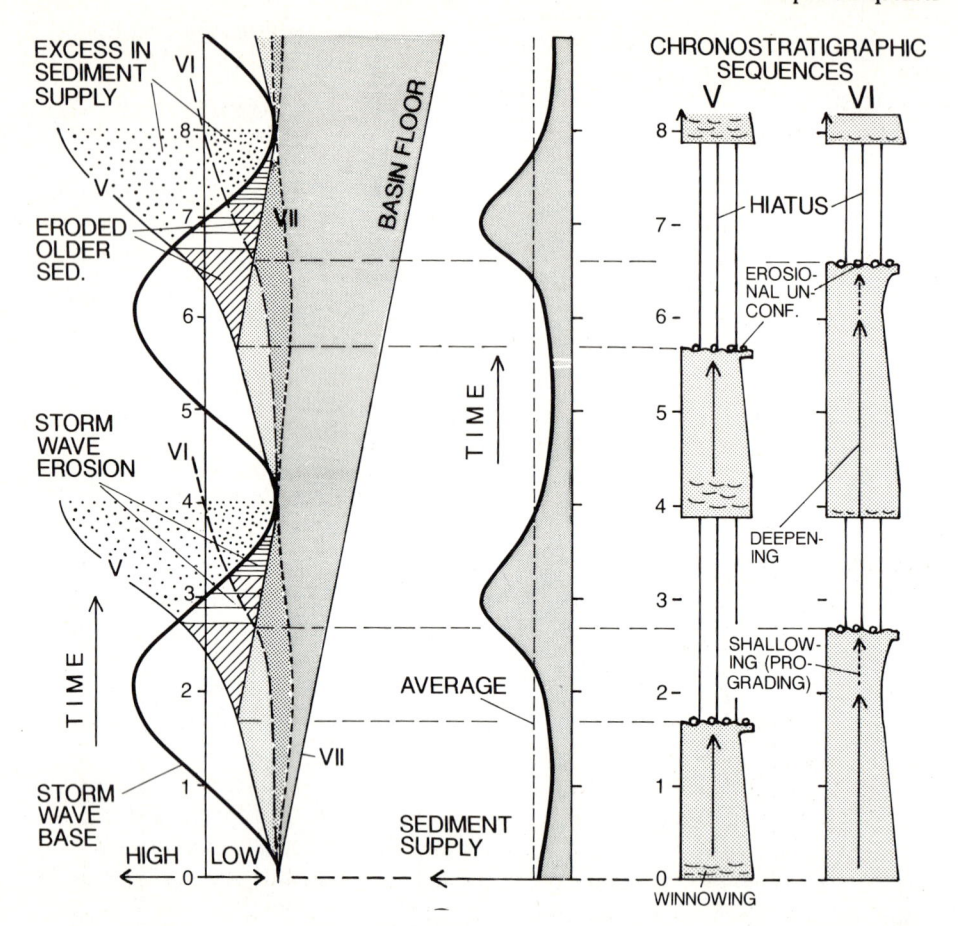

Fig. 3. As Fig. 2b, with rate of subsidence lower than maximum sea level fall, but varying sediment supply (locations V, VI, VII) versus time. Resulting chronostratigraphic sequences of locations V and VI (see Fig. 2a; other sequences are displayed in Fig. 4)

If the rate of subsidence is constantly greater than the maximum rate of sea level fall, then there is no possibility that previously deposited sediments are eroded. Sediment accumulation, however, may be hampered or prevented during a sea level fall.

Erosion of older sediments can take place if the rate of subsidence drops below the maximum rate of sea level fall (cf. Pitman and Golovchenko 1983), provided that sediment accumulation compensates for subsidence in the long run. A frequent consequence is the formation of lag deposits and the incision of channels with the eroded sediments deposited as lowstand fans in deeper parts of the basin.

The complicated interrelationship between the three factors: subsidence, constant sediment supply, and oscillating storm wave base is shown in Fig. 2b. Thereby subsidence is considered constant for all locations and the sediment supply varies only spatially, but is held constant at a specific space coordinate. Taking into consideration a cross-section from the shoreline to deeper parts of the basin, the sediment supply usually decreases from the locations I to IV as indicated in Fig. 2a. The situations at these locations can be characterized as follows:

Locations I and II. The sedimentation rate exceeds subsidence. Sediments deposited during sea level highstands are partially eroded during the regressive phase. The resulting erosional unconformities are marked by lag deposits and incised channels. The chronostratigraphic sections display stratigraphic gaps (hiatuses) of varying duration (Fig. 2b, sediment columns). With increasing sedimentation rates at locations I and II, the duration of the hiatuses increases. The sedimentary section of location I shows repeated deepening-upward sequences (reflecting times of rising sea level), which are truncated by erosional unconformities. The sequence at location II is characterized by deepening-upward followed by a short section of shallowing-upward below the unconformity. With a decreasing sedimentation rate, the proportion of this shallowing-upward section increases as part of the total cyclic sequence.

Location III. The sedimentation rate equals the subsidence rate and thus enables a continuous vertical sediment aggradation. There is no long-term trend in sediment characteristics, but cyclic patterns due to sea level oscillations are recorded in the form of deepening and shallowing-upward sections.

Location IV. The normal sedimentation rate is lower than subsidence, thus causing the water depth to increase with time. Two characteristic sequences arise: (1) If no additional sediment is supplied by bypassing, the resulting sedimentary sequence will indicate an overall deepening trend superimposed by indistinct periods of shallowing-upward. (2) Sediments which bypass locations I and II as well as eroded older sediments accumulate in deeper parts of the basin, generally around location IV (below location III). As will be demonstrated later, such sedimentary bodies tend to shift laterally during a regressive-transgressive cycle and ultimately lead to prograding sand bodies (Figs. 6 through 8). These bodies may be topped by winnowed residual layers in the vicinity of location III. Consequently, enlarged shallowing-upward sections in the deeper parts of the basin (between locations III and IV) are frequently observed. These exhibit coarsening upward close to location III, while they develop into thickening upward sequences at location IV.

3.2 The Role of Temporarily Varying Sediment Supply

Until now, we assumed constant sediment supply at a certain location and during a given time interval which exceeds the period of a transgressive-regressive cycle. In nature, this assumption often will not hold. The supply of allochthonous sediments, for example terrigenous material from the continents or from coastal areas, to a location below the storm wave base will usually decrease during times of transgression. In contrast, sediments stored in coastal areas during a highstand are remobilized in periods of falling sea level and transported basinward. Such variations in sediment supply are considered in Fig. 3. Locations V-VIII now replace locations I to IV at the respective spatial position (cf. Fig. 2a). The peak of sediment supply is reached at the maximum rate of sea level fall, i.e., at the inflection point of the sea level curve.

The resulting chronostratigraphic sequences in relatively shallow water (locations V and VI) do not differ much from those derived for constant sediment supply (Fig. 2, locations I and II). The main difference is somewhat shorter hiatuses in cases with variable sediment supply. However, the differences between the two models are more pronounced in terms of lithological sections and "field water depth curves", particularly in the more distal areas of the basin (see below).

3.3 Asymmetry and "Field Water-Depth Curves" of the Model Sequences

The asymmetry of transgressive-regressive sequences as shown in the chronostratigraphic sections (Figs. 2b and 3) is also obvious in standard sedimentary columns (Fig. 4). Furthermore, relative water depths can be inferred from these columns using sedimentological and faunistic reasoning. The resulting paleowater-depth curves (Fig. 4) are here referred to as "field water-depth curves", because they are based on direct observations in the field. In the models, the sea level fluctuations were given by a sine function. The resulting sequences and their derived field water-depth curves, however, display the following characteristics:

Locations I and V. Deep-reaching submarine erosion during the regressive phase generates pronounced erosional unconformities and hiatuses representing relatively long time intervals. Preserved sediment sections at the top of the unconformities begin with thin, shallow-water deposits indicating frequent winnowing. The subsequent thick deepening-upward section accumulates during the first half or two thirds of the transgressive phase and is truncated by the next unconformity. Therefore, highstand deposits (HDS) are poorly developed, and lowstand deposits (LSD) are also very thin. The sedimentary cycles and the field water-depth curves are extremely asymmetric, reflecting an apparently slow rise and sudden fall of sea level.

Locations II and VI. Submarine erosion during the transgressive phase leads to erosional unconformities and hiatuses. II: Well-developed asymmetric sedimentary cycles and field water-depth curves with relatively thick deepening-upward and thin shallowing-upward sections and apparently slow rise and rapid fall of sea level. VI: Similar to II, but asymmetry less pronounced.

Fig. 4. Vertical sequences and "field water-depth curves" for locations I through VIII (Fig. 2a) of the models shown in Figs. 2 and 3. *I-IV* Different, but time-constant rates of sediment supply; *V- VIII* sediment supply varying with time. Sediment thicknesses plotted on linear scale, time scale given in numbers (in general nonlinear scale). Note the changing contributions of deepening-up or shallowing-up sections to a total sedimentary cycle, as well as the marked change of the "field water-depth curves" based on sedimentary thicknesses and paleobathymetric interpretations. *LSD* lowstand deposits; *TD* transgressive deposits; *HSD* highstand deposits; *CS* condensed section

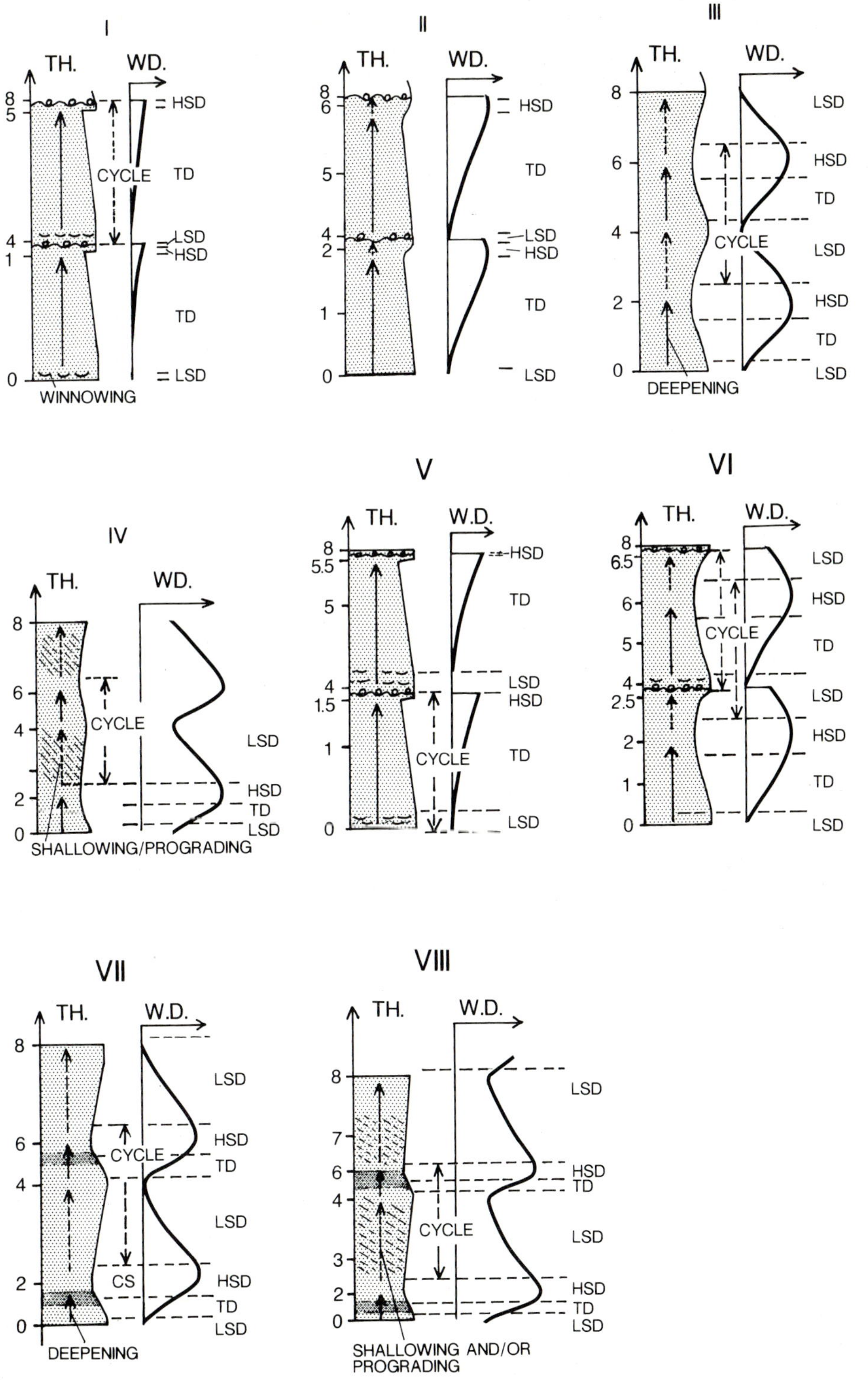
I
TH.
WD.
HSD
CYCLE
TD
LSD
HSD
TD
LSD
WINNOWING
II
HSD
TD
LSD
HSD
TD
LSD
III
LSD
HSD
TD
CYCLE
LSD
HSD
TD
LSD
DEEPENING
IV
CYCLE
LSD
HSD
TD
LSD
SHALLOWING/PROGRADING
V
W.D.
HSD
TD
LSD
HSD
CYCLE
TD
LSD
VI
LSD
HSD
CYCLE
TD
LSD
HSD
TD
LSD
VII
LSD
HSD
CYCLE
TD
LSD
CS
HSD
TD
LSD
DEEPENING
VIII
LSD
HSD
TD
CYCLE
LSD
HSD
TD
LSD
SHALLOWING AND/OR
PROGRADING

Locations III and VII. These locations do not receive additional sediment supply from bypassing sediments or sediments eroded at locations I and II or V and VI, respectively. III: Symmetrical sedimentary cycles consisting of alternating sections of deepening-upward and shallowing-upward sediments; the field water depth curve is symmetric. VII: Moderately asymmetric sedimentary cycles and field water-depth curve, implying faster rise than fall of sea level. Relatively thin transgressive sediments may contain a condensed section (CS, see, e.g., Lutit et al. 1988) of pelagic oozes, marls, black shales, glauconitic sands, etc., dependent on the paleogeographic and environmental setting of the basin.

Locations IV and VIII. During the regressive phase, additional sediment is supplied from bypassing and submarine erosion in shallower water. IV: Moderately asymmetric cycles with field-water depth curves similar to VII; no distinct condensed section. Progradation of shallow water sediments may occur. VIII: Enhanced asymmetry of sedimentary cycles and field water-depth curve implying apparently rapid rise and slow fall of sea level. Possibility of a pronounced condensed section during the transgressive phase and prograding shallow-water sediments during regression.

Summarizing the results, we found that in a single basin and under the same ideal symmetric sinusoidal sea level fluctuation, various types and grades of asymmetric sedimentary cycles may develop. The asymmetric cycles are solely a function of the interplay between subsidence, sediment supply, submarine erosion, and sediment bypassing at different locations within the same basin. Symmetric cycles develop only under very special conditions, one of which is constant sediment supply. In shallow-water areas, where sediment supply exceeds subsidence, erosional unconformities develop frequently and the field water-depth curves imply a slow rise and rapid fall of the sea level. In somewhat deeper water, where the total sediment supply approximately compensates for subsidence, the field water-depth curves show an apparently short rise and long decline of the sea level.

In sequences with erosional unconformities (model sequences and field examples, Fig. 7, 8 and 9), the latter are conventionally used as boundaries of sedimentary cycles (type 1 sequence boundary, Mitchum et al. 1977; van Wagoner et al. 1988; Vail et al. Chap. 6.1, this Vol.). Where such unconformities are missing, the base of the lowstand deposits is proposed as a sequence boundary (type 2 sequence boundary). This agrees with traditional lithostratigraphic practice, in which such coarser-grained beds have long been used as marker beds mappable in the field. As can be seen from Figs. 2 through 4, the base of the sequence boundaries usually does not coincide with the lowest stand of sea level, a fact which also is pointed out by Vail et al. (1984).

3.4 Amalgamated Sedimentary Cycles

A succession of incomplete, asymmetric sedimentary cycles as indicated in the model sequences (Fig. 4, particularly sequences I and V) as well as in field examples (Figs. 9b) is referred to as stacked cycles or amalgamated sedimentary cycles. Such cycles occur frequently in the geologic record, and are always associated with shallow-marine environments. In these cases, the rate of maximum sea level fall exceeds subsidence and thus enables erosion. A field example of this type of stacked cycles and a discussion of some further problems involved are given in Section 5.

4 Differential Subsidence and Variations in Period and Amplitude of Sea Level Changes

A major restriction of the models discussed so far is the assumption of constant subsidence throughout the basin. Pitman (1978) and Watts (1982) showed that the onlap patterns along passive continental margins, which previously had been attributed to eustatic sea level changes (Vail et al. 1977a), may be derived from temporally and spatially varying subsidence. Frequently, subsidence increases from a coastal hinge line toward the center of the basin and, therefore, modifies the interplay between subsidence and sediment accumulation.

Besides regionally varying subsidence, the situation may be complicated by variations in period and amplitude of sea level changes. Such a scenario, also taking into account the factors discussed earlier, was used to model the sequences shown in Fig. 5. Figure 5a displays a cross-section through a comparatively shallow marine basin. If the period of sea level changes is sufficiently long (on the order of 1 Ma in pre-Pleistocene times, Haq et al. 1987), the hydrodynamic regime of a wave-dominated water body, as described, e.g., by Johnson and Baldwin (1986), will cause a kind of "equilibrium topography" of the sea floor (Swift 1978). Such an equilibrium profile is based on the assumptions that sediment buildup is limited both by storm wave-induced bottom currents and, near the coast line, by the transformation of wave energy into surface currents. In response to sea level oscillations, both the storm wave base and the equilibrium profile are shifted in space parallel to the transgressive-regressive (T-R) dislocation of the coast line. Similar to annual coastal periods, sediment will accumulate at a certain depth, and will be transported to deeper areas during the level fall, where sandwaves and bars may be preserved and covered later by mud. Such a situation is probably also responsible for the accumulation of carbonate and iron oolites indicated in Figs. 5b and 8 and discussed by Bayer (1989).

In order to derive a sufficiently simple model for the situation given in Fig. 5a, we introduce the following further assumptions:

1. Sediment is predominantly supplied from neighboring land areas. The sediment influx into the area of investigation is higher during the regressive phases than during transgressions (cf. Fig. 3).
2. At water depths below the storm wave base, the total sediment supply approximately compensates for subsidence. Landward from the zone in which the storm wave base hits the sea floor, sediment supply exceeds subsidence.

The resulting chronostratigraphic (Fig. 5b) and normal stratigraphic cross section (displaying sediment thicknesses, Fig. 5c) are typical for many ancient epicontinental basins (Einsele 1985) as well as for shelf seas on continental margins (also see field

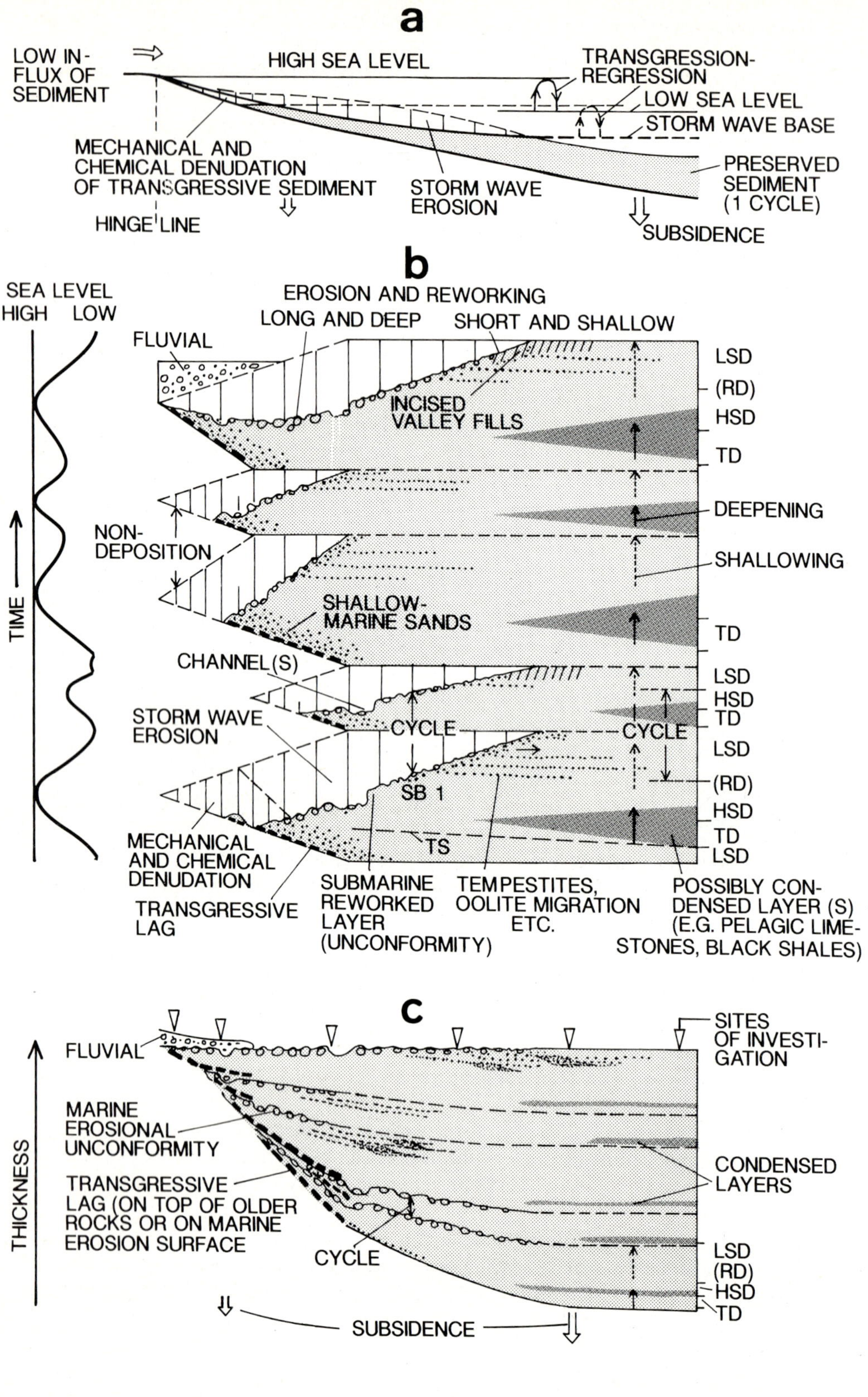
a
LOW IN-
FLUX OF
SEDIMENT
HIGH SEA LEVEL
TRANSGRESSION-
REGRESSION
LOW SEA LEVEL
STORM WAVE BASE
MECHANICAL AND
CHEMICAL DENUDATION
OF TRANSGRESSIVE SEDIMENT
STORM WAVE
EROSION
PRESERVED
SEDIMENT
(1 CYCLE)
HINGE LINE
SUBSIDENCE
b
SEA LEVEL
HIGH LOW
EROSION AND REWORKING
LONG AND DEEP
SHORT AND SHALLOW
FLUVIAL
INCISED
VALLEY FILLS
LSD
(RD)
HSD
TD
DEEPENING
NON-
DEPOSITION
SHALLOWING
TIME
SHALLOW-
MARINE SANDS
TD
CHANNEL(S)
LSD
HSD
TD
STORM WAVE
EROSION
CYCLE
CYCLE
LSD
(RD)
SB 1
HSD
TS
TD
LSD
MECHANICAL
AND CHEMICAL
DENUDATION
TRANSGRESSIVE
LAG
SUBMARINE
REWORKED
LAYER
(UNCONFORMITY)
TEMPESTITES,
OOLITE MIGRATION
ETC.
POSSIBLY CON-
DENSED LAYER (S)
(E.G. PELAGIC LIME-
STONES, BLACK SHALES)
c
SITES
OF INVESTI-
GATION
FLUVIAL
MARINE
EROSIONAL
UNCONFORMITY
CONDENSED
LAYERS
THICKNESS
TRANSGRESSIVE
LAG (ON TOP OF OLDER
ROCKS OR ON MARINE
EROSION SURFACE
CYCLE
LSD
(RD)
HSD
TD
SUBSIDENCE

example below, Figs. 7 and 8). The following points are specified for a better understanding of these cross-sections:

1. Especially if the period of a sea level oscillation is rather long, a considerable part of the transgressive and highstand deposits accumulated near the coast may be removed by chemical and mechanical denudation (Walling and Webb 1986; Gerrard 1988), particularly in regions of wet and tropical climate. Consequently, coastal sediments of limited thickness, which either represent the peak of transgressions of T-R cycles of a duration of more than 1 Ma or which later emerged for a considerable time period, are rarely preserved.
2. Transgressive lag deposits of a particular T-R cycle and unconformities resulting from storm wave erosion during the regressive phase are diachronous marker horizons. While transgressive lags become younger in a landward direction, storm-induced unconformities become younger and less pronounced basinward. The latter may truncate either transgressive, highstand, or lowstand deposits. Exhumed lithified and repeatedly reworked concretions indicate erosion. Particularly the lowstand deposits may be accompanied by tempestites and seaward prograding sands (including ooids formed previously during highstand in more coastal waters, also see Fig. 8).
3. Seaward of the zone of storm wave action, the sedimentation rate may decline to very low values during times of transgression and approach to sea level highstand. The resulting condensed sections are often represented by thin pelagic limestones, marlstones or black shales. Under sediment-starved conditions, condensation may be represented by hardgrounds and/or very thin layers of autochthonous minerals such as glauconite or phosphorite (cf. Fig. 9a; Loutit et al. 1988; Vail et al. Chap. 6.1, this Vol.).
4. Changes in the period and the amplitude of sea level variations primarily control the thickness of the sedimentary cycles, as well as the landward and basinward extension of the submarine erosional unconformities as previously discussed.

In Fig. 5c, which illustrates the sediment filling of the basin, the top of the sequence is taken as the datum line. Due to differential subsidence, the thickness of the sediment filling increases seaward until a certain maximum is reached. Toward the coast, the erosional unconformities of the different T-R cycles merge. It is important to note that the unconformities caused by storm wave erosion during the regressive

◄───

Fig. 5 Cross-sections through shallow basin subjected to irregular variations in sea level and storm wave base, subsidence increasing basinward, sediment supply approximately sufficient to compensate for subsidence. **a** Basin morphology and erosion during regressive phase. **b** Chronostratigraphic chart resulting from irregular sea level changes in **a**. Basinward, certain sections of the sequence may become condensed. *TD* transgressive deposits; *HSD* highstand deposits; *LSD* lowstand deposits; *RD* regressive deposits (nomenclature from Baum and Vail 1987). **c** Normal stratigraphic cross-section through model basin **b**, displaying sediment thicknesses of cyclic sequences. Transgressive lags, in the shallowest part of the basin, may rest directly on erosional unconformities generated during falling sea level. Note the difference in type and possibly also in number of cycles to be observed at locations along the cross section. (After Einsele 1985)

phase can be overlain directly by or mixed with a lag deposit of the subsequent transgression. If sediment columns at different sites would be investigated, then the vertical facies successions of the sedimentary cycles should vary considerably. If the condensed layers are missing, the cyclic nature of the entire sequence may be more difficult to recognize in somewhat deeper parts of the basin. Towards the coast, the number of erosional unconformities differs from one location to the other. The cycles are amalgamated and some may be missing as previously discussed. Therefore, it often appears difficult or even impossible to determine the real number of larger and smaller sea level oscillations for a certain time interval.

The various asymmetries of the sedimentary cycles and the field water-depth curves along a cross section through a shelf/slope/deep-sea setting are summarized in Fig. 6. This basin undergoes differential subsidence and receives increased sediment supply during times of regression. As already mentioned above, the sequences affected by submarine erosion (locations I and II) show an apparently long transgressive and rapid regressive phase, although the triggering signal is considered symmetrical. Areas which receive additional sediment from erosion and bypassing, in contrast, show the opposite development (location III). A virtual slow fall and rapid rise of sea level appears again at the foot of the continental slope (location V). Within the range of this model, approximately symmetric cycles occur either on the outer shelf (location IV), where additional sediment supply from reworking and bypassing can be neglected, or on the deep-sea plain (location VI) beyond the zone significantly affected by deep-sea fans and mass flows. The small degree of asymmetry reflects the slower sedimentation rate during transgressions and highstands.

The asymmetry of cycles is not only indicated by the field water-depth curves in Fig. 6b, but also by the time lines (isochrones 0 through 8). In times of slow sedimentation, the distance to neighboring isochrones is small and sea level change appears to be rapid (provided the true sea level curve follows a regular sine function). Towards the shoreline, several isochrones merge into an erosional unconformity and thus indicate an apparently instantaneous fall of sea level.

The landward merging isochrones also clearly show the increasing time span (hiatus) represented by the unconformities, which are thus difficult to date precisely. Reworked lithified beds or concretions often contain a mixed fauna of differing ages, especially if they are overlain by the lag deposit of a subsequent transgressive phase. Fossils found below the unconformity may be much older than the time of the erosional event, and fossils from the transgressive lag cannot be used to date the regressive phase either.

As indicated in the following section, our models satisfactorily describe many ancient sedimentary sequences. However, further complications arise if subsidence varies temporarily and spatially, or if other factors not discussed here play an important part. The topography of the basin may be altered under slowly changing equilibrium conditions and thus affect the type of T-R sequences at different locations in a less predictable manner than in our two-dimensional, simplified models.

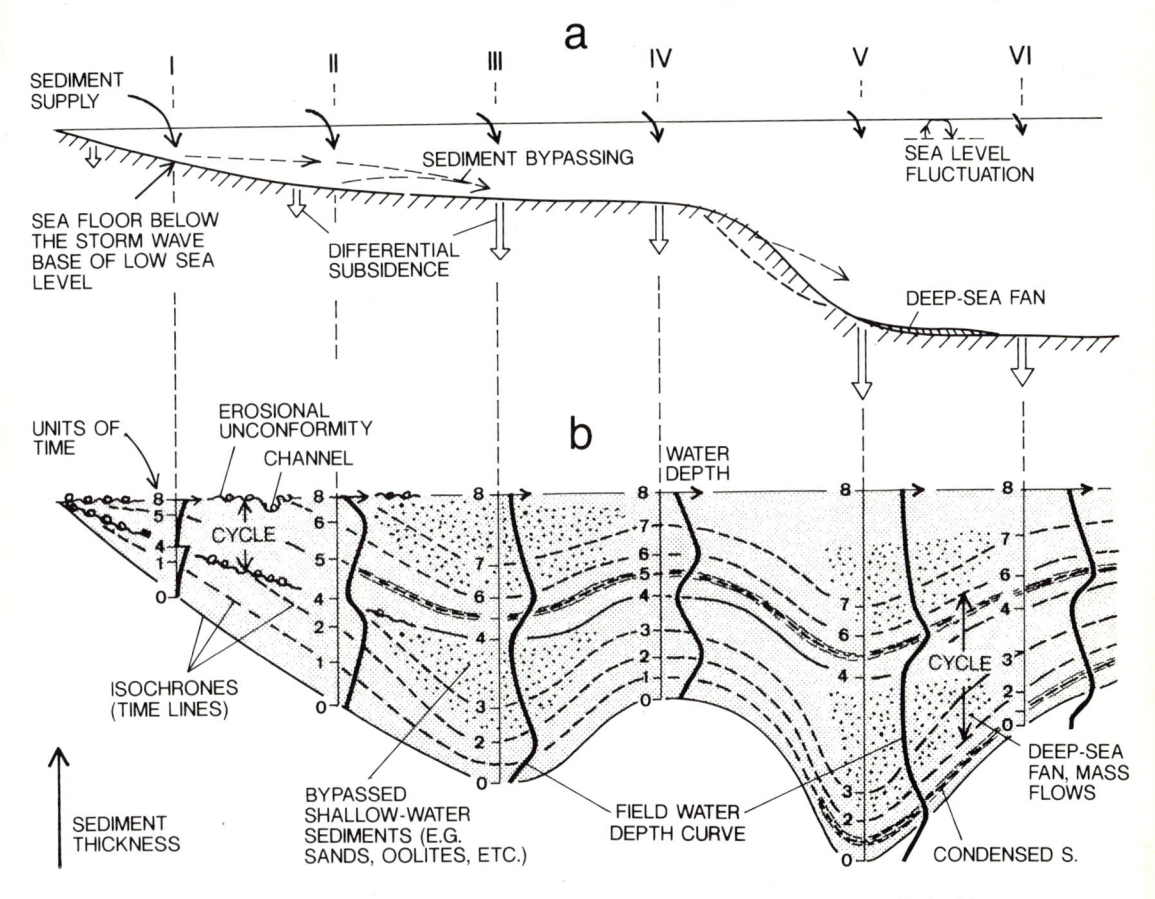

Fig. 6 a Cross-section through shallow basin with transition to deep basin, subjected to variations in sea level and sediment supply (higher during lowstands), differential subsidence, and basinward decreasing rate of sedimentation as long as there is no reworking and bypassing of sediments in shallower water. **b** Resulting sediment thicknesses, erosional unconformities, redeposited sands etc., and asymmetric "field water-depth curves" of two transgressive-regressive cycles (comprising 8 units of time). Note the landward truncation of isochrones. For further explanations see text

5 Field Examples

Due to a limitation on space, we can briefly describe only a few field examples to support the general rules derived from the model basins. One of the prerequisites for such examples is a relatively high stratigraphic resolution. One has to take into account, however, that former coastal and nearshore sediments are often incompletely preserved. Therefore, our examples begin with a site already some distance away from the paleocoastline, after which we proceed to locations farther basinward.

Example 1. Lowermost Jurassic of Southern Germany:
shallow-water cycles and merging unconformities

In the lower Liassic (Hettangian and lower Sinemurian) epicontinental sea of South Germany, Bloos (1976) demonstrated that the number of submarine unconformities, represented by single beds with iron ooides and reworked calcareous concretions, decreases toward the paleo-coast (Fig. 7). The reworked horizons partially merge as shown in the model basin (Figs. 5c and 6b). These unconformities near the coast mostly rest on shales and thus truncate a transgressive sequence (locations I and III in Fig. 7). The field-water depth curves for this part of the basinal cross-section therefore resemble the extreme asymmetric type with a slow rise and sharp drop of the water depth at the unconformity (cf. Fig. 6b, location I). Farther basinward (location III), the unconformities often cap shallowmarine sandstone, which were deposited during the first half of the regressive phase before submarine erosion terminated further deposition. Here the cyclic sequences render field-water depth curves with apparently rapid deepening and slow shallowing (cf. Fig. 6b, location III). The total thickness of this sequence only reaches 30 m and represents a time span of 2 to 3 Ma.

Example 2. Lower Dogger, Southern Germany:
stacked cycles and seaward migrating sand bodies

Figure 8 illustrates the facies distribution and sedimentary sequences in the South German Aalenian. During regressive phases, the sand bodies prograde toward the basin center (Weber 1967), indicating a general shallowing trend. Well-developed coarsening upward cycles occur in parts of the outcrop area (Fig. 8a). The location of these cycles shifts through time toward the basin center as indicated in the cross-section (Fig. 8b), parallel to the shift of sand bodies and the general shallowing trend. Sections taken along a datum line show a typical gradient: Proximal sections at location I are condensed and show frequent erosional unconformities. The few preserved minor cycles are typically fining upward, especially in the lowermost Bajocian (Oechsle 1958), and are similar to model sections I and V in Fig. 4, Basinward, at location II in Fig. 8b, these beds evolve into coarsening and thickening upward cycles with dark shales in the lower part and an increasing number of intercalated sand layers toward the top. These cycles are frequently terminated by an erosional unconformity, forming the base of oolitic limestones, ironstones, and shell beds. Siliciclastics are absent, due to the fact that this material is swept into deeper water. The oolitic beds are usually terminated by an "omission" horizon or a sediment-starved layer which exhibits nondeposition and winnowing. The corresponding asymmetric relative field water-depth curve and more or less symmetric sea level curve are shown in Fig. 9a. Because of several erosional surfaces (indicators of amalgamation), the position of the sequence boundary at the base of the lowstand deposits cannot clearly be defined. The starved horizon probably falls into the transgressive phase of the sea level oscillation and therefore causes an apparently rapid increase in the field water-depth curve. Between locations I and II, the sedimentary sequences are largely amalgamated and thus more or less incomplete. In the deeper parts of the basin, the cyclic pattern vanishes and is replaced by thick shale sequences frequently interrupted by tempestites and winnowed layers, for example at location III in Fig. 8b.

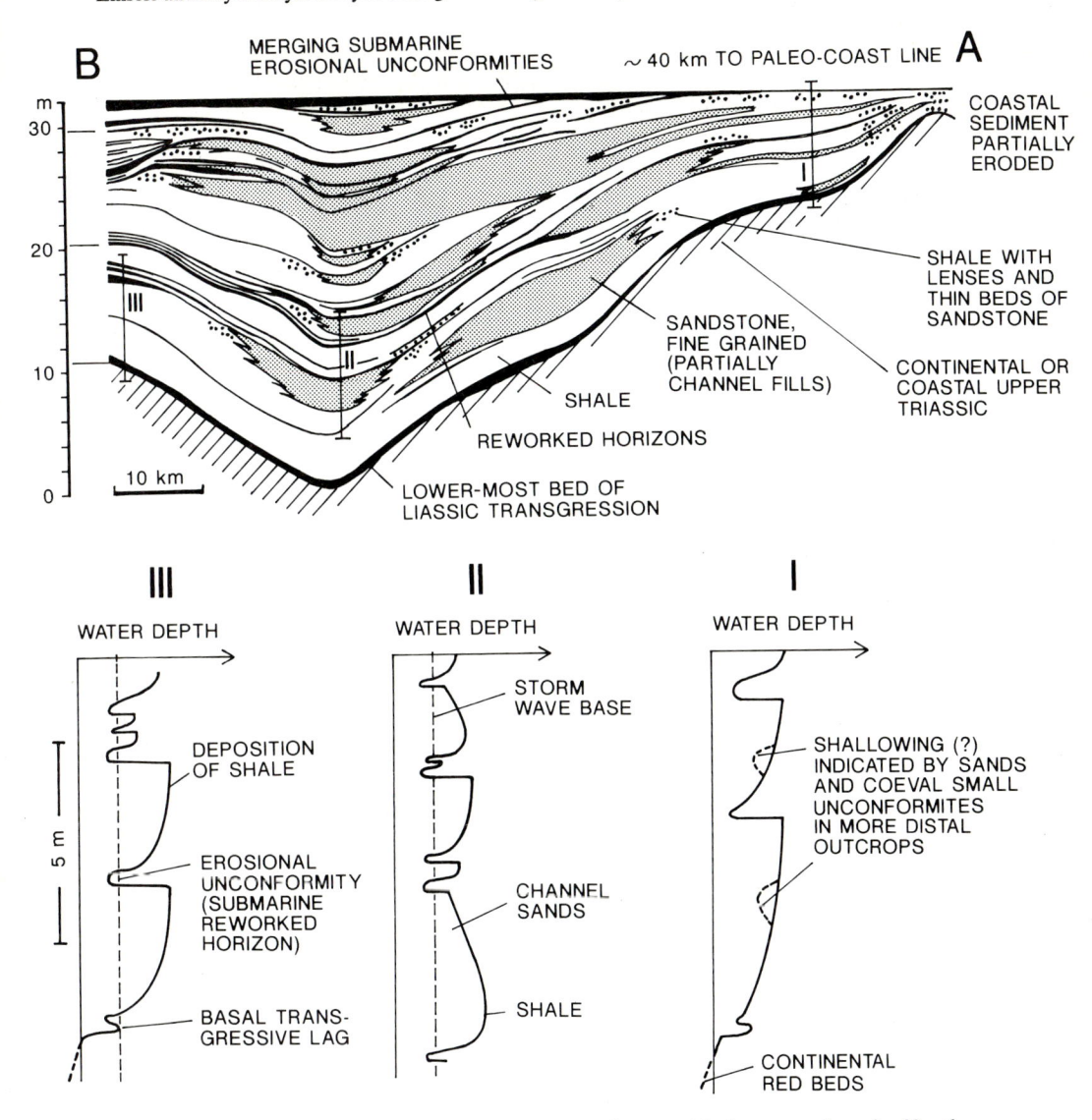

Fig. 7 Generalized cross-section through shallow-marine sediments of the lowermost Jurassic of Southern Germany (for location see Fig. 8a, line *A-B*; based on Bloos 1976, from Einsele 1985), with field water-depth curves for locations *I*, *II*, and *III*. Note the prevailing deepening-upward trends of *I* and *III*, truncated by submarine erosional unconformities, in contrast to *II*, where basinward migrating sands generate shallowing (coarsening)-upward sequences in somewhat deeper water

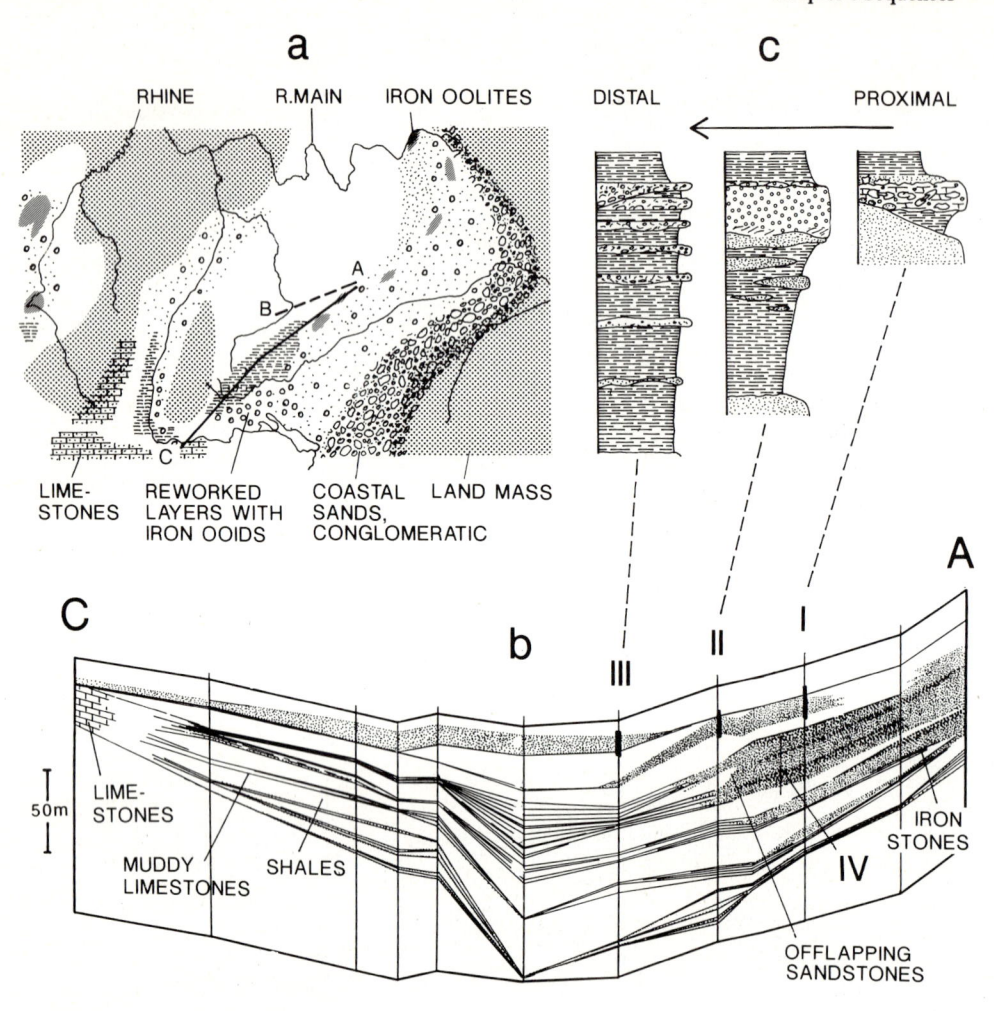

Fig. 8. a Generalized facies distribution within the upper Aalenian of Southern Germany and parts of Eastern France. Iron-oolitic offshore bars run parallel to the paleo-coastline. **b** Cross-section *A-C* of upper Aalenian to lower Bajocian sediments along the present outcrop line. The offlap of sandstone bodies indicates a major regressive trend which is subdivided into minor asymmetric coarsening-upward cycles. **c** Examples of asymmetric cycles from proximal (*I*) to more distal areas (*II*, *III*, and *IV*) within the basin. (After Bayer and McGhee 1985)

Fig. 9. a Example of asymmetric coarsening-upward cycle from location II (Fig. 8). Normal lithologic section (thickness on linear scale) is transformed to chronostratigraphic section (linear time scale) applying ammonite subzones. Note difference between sea level curve derived from chronostratigraphic section and field water-depth curve derived from lithologic section. Upper part of regressive phase (lowstand deposits) is truncated, and laterally transported iron ooids and shell fragments rest on sequence boundary which cannot be clearly defined. **b** Series of minor asymmetric, coarsening (shallowing)-upward cycles from location III in Fig. 8, superimposed on longer trends which also show coarsening-upward. On a linear time scale based on ammonite subzones, the asymmetric cycles may yield sinus-type sea level fluctuations. The lowstands of this curve probably do not exactly coincide with the tops of the asymmetric sedimentary cycles

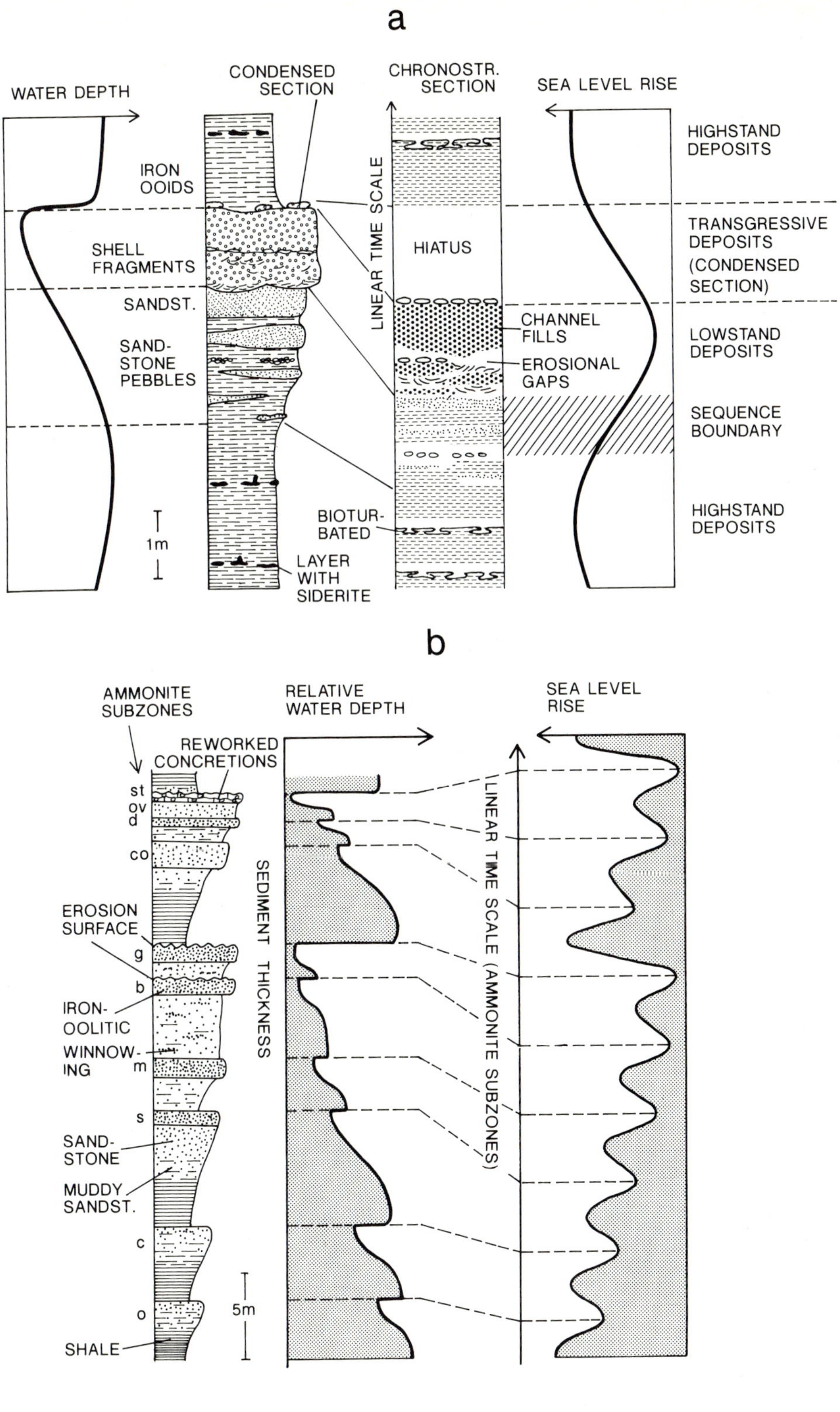
a
WATER DEPTH
CONDENSED SECTION
CHRONOSTR. SECTION
SEA LEVEL RISE
IRON OOIDS
SHELL FRAGMENTS
SANDST.
SAND-STONE PEBBLES
1m
LINEAR TIME SCALE
HIATUS
CHANNEL FILLS
EROSIONAL GAPS
BIOTUR-BATED
LAYER WITH SIDERITE
HIGHSTAND DEPOSITS
TRANSGRESSIVE DEPOSITS (CONDENSED SECTION)
LOWSTAND DEPOSITS
SEQUENCE BOUNDARY
HIGHSTAND DEPOSITS
b
AMMONITE SUBZONES
REWORKED CONCRETIONS
RELATIVE WATER DEPTH
SEA LEVEL RISE
st
ov
d
co
EROSION SURFACE
g
b
IRON-OOLITIC
WINNOW-ING
m
s
SAND-STONE
MUDDY SANDST.
c
o
SHALE
5m
SEDIMENT THICKNESS
LINEAR TIME SCALE (AMMONITE SUBZONES)

Figure 9b demonstrates a series of stacked cycles from location IV in Fig. 8b. Their field water-depth curves exhibit a sharp increase of water depth on top of the coarse-grained beds and slower shallowing-upward conditions for each individual cycle. In addition, the longer-term shallowing-upward trend mentioned above leads to a prograding degree of submarine erosion at the tops of the minor cycles, until a new and relatively large transgression initiates a further long-term trend. Such a pattern is explained by the superposition of minor oscillations onto a larger-scale level change (e.g., Hallam 1981a; Vail et al. 1984; McGhee and Bayer 1985). Such an interaction between a low- and high-frequency signal can essentially contribute to the asymmetry of the sedimentary record, an aspect which is not considered in the above models. In the example of Fig. 9b, the shallowing (coarsening)-upward of the regressive phase of the minor cycles is enhanced and, conversely, the transgressive phase shortened by the long-term shallowing trend. If the ammonite subzones are taken as time equivalents, the stacked and partially amalgamated sedimentary cycles of Fig. 9b may be explained as the result of minor, more or less sinuoidal symmetrical sea level oscillations with a rather constant time period.

Example 3. South German Jurassic:
variations in sedimentation rate

Sea level fluctuations are accompanied by temporarily variable sedimentation rates which have to be measured in terms of absolute time. Figure 10 illustrates the strong variation in sediment accumulation of the South German Jurassic and relates these changes to coarsening-upward and fining-upward cycles. During highstands, the incoming sediments are trapped near the coast. Therefore, most of the thick mudstone sequences of the Jurassic are deposited during the regressive phase around the maximal fall of sea level in a rather short time interval. Several of the pronounced shale intervals with thicknesses up to 100 m represent only a single ammonite zone or even less (e.g., the obtusum Zone and opalinus Zone as the most striking cases, Fig. 10). In contrast, the associated more condensed sequences (black shales, carbonates, oolites, phosphorite horizons) are usually of much longer duration in terms of biostratigraphy.

There are numerous additional examples of asymmetric sedimentary cycles in the literature which may result from more or less symmetric sea level oscillations. the depositional models of passive continental margins based on seismic stratigraphy (Vail et al. 1984; Sarg 1988; Loutit et al. 1988; Vail et al. Chap. 6.1, this Vol.) also display different types of asymmetric sequences. On the middle and outer shelf, where lowstand deposits and transgressive deposits are often thin and incomplete, the cyclic sequences tend to show a rapid rise and slow fall of the field water-depth curve (cf. Fig. 6b, location III). This results from relatively thick prograding highstand deposits. Such a curve type is more pronounced if a marine delta progrades over the shelf (Loutit et al. 1988). Upper Cretaceous stacked deltaic sequences, displaying a shallowing-upward with coal seams at the top, are explained by Cross (1988) as the result of sinusoidal sea level oscillations. Here, the shallowest facies at the top of the cycle is capped abruptly by the deepest facies of the subsequent sedimentary cycle. By contrast, the seimic models of passive continental margins

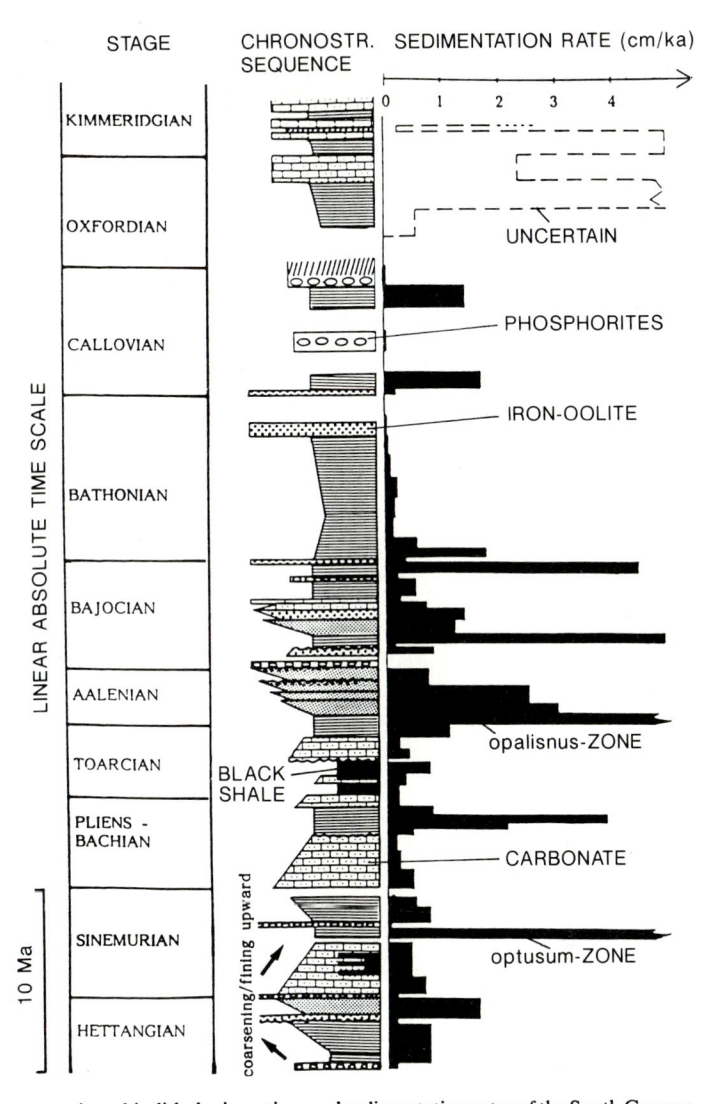

Fig. 10. Generalized chronostratigraphic, lithologic section, and sedimentation rates of the South German Jurassic; subdivision of linear absolute time scale based on ammonite zones. (Bayer and McGhee 1986). Note that high sedimentation rates (*long horizontal black bars*) signify thick subsections. Many thick shales were deposited rapidly during regressive shallowing-upward phases

describe a thick lowstand wedge at the foot of the continental slope overlain by a condensed section and thin highstand deposits. The limited thickness of the latter sediments is a result of the landward shifted coast line. Under these conditions, the field water-depth curve may display an apparently slow fall and rapid rise in water depth. Olsson (1988) applied paleo-depth information from foraminiferal assemblages of slopes to interpret asymmetric Upper Cretaceous depositional sequences. He explained the observed coastal onlap in terms of symmetric sea level changes which were characterized by differing periods and amplitudes.

An example of a passive continental margin setting with mixed carbonate and siliciclastic sediments is given by Sarg (1988). During sea level lowstands, carbonate production on the shelf is strongly reduced and terrigenous input enhanced. This results in either accumulation of largely siliciclastic, carbonate-poor sediments at the foot of the slope and on the adjacent basin plain, or redeposition of older shelf and upper slope carbonates in deeper water (see Eberli Chap. 2.8, this Vol.). If the shelf is flooded again and siliciclastic input hampered, the transgressive and highstand deposits on the shelf are characterized by various types of shallow-water carbonates, while pelagic carbonates dominate in deeper water. Thus, an asymmetric sequence may also be characterized by a marked change in composition.

6 Discussion and Conclusions

Simple two-dimensional sedimentation models show that sinus-shape sea level oscillations mostly generate asymmetric sedimentary cycles and asymmetric "field water-depth curves" at water depths around or below the storm wave base of the sea level lowstands. The models take into account (1) constant, but spatially differential subsidence, (2) varying terrigenous sediment supply, (3) erosion and bypassing of sediment in the zone of storm wave action and, in part, (4) changes in the period and amplitude of the sea level oscillations.

Inner shelf sequences affected by significant submarine erosion during the regressive phases commonly show either amalgamated and relatively thin, stacked fining-upward sedimentary cycles, or thicker, apparently long transgressive fining (deepening)-upward sections (Fig. 6b, locations I and II). The sedimentary record of the regressive phases is thin and incomplete or missing. Areas in somewhat deeper water, which receive additional sediment from erosion and bypassing, display the opposite development (Fig. 6b, location III; Fig. 9b). A relatively slowly rising and more rapidly falling field water-depth curve appears again at the foot of the continental slope (Fig. 6b, location V), where thick reworked lowstand deposits are common. Within the range of this model, approximately symmetric cycles occur either on the outer shelf (Fig. 6b, location IV), where additional sediment supply from reworking and bypassing can be neglected, or on the deep-sea plain (Fig. 6b, location VI), beyond the zone of deep-sea fans and mass flows.

Coarsening or fining-upward of cyclic sequences are intensified, if short-period minor cycles are superimposed on longer-term trends of the same tendency (Fig. 9b), or viceversa.

Transgressive lags and submarine erosional unconformities are diachronous and exhibit landward or basinward trends in their age and facies. Transgressive lags may mix with underlying storm-reworked layers of the earlier regressive phase (Fig. 5b and c).

The asymmetry of sedimentary cycles is also indicated by the time lines (isochrones 0 through 8 in Fig. 6b). In times of slow sedimentation, the distance of subsequent isochrones is small and hence the sea level change appears to be rapid.

Towards the shoreline, several isochrones merge into an erosional unconformity and thus indicate an apparently instantaneous fall of sea level.

Both the landward merging submarine erosional unconformities (in places including subsequent transgressive lags) and the modelled isochrones (Fig. 6b) clearly show the increasing stratigraphic gaps (hiatuses) in marginal parts of the basin. Reworked and mixed faunas hamper an exact dating of both the unconformities and the lower sections of asymmetric sequences characterized by an apparently slowly rising field water-depth. The most reliable dating of cyclic sequences is accomplished with the aid of condensed horizons, which are free of redeposited sediments and represent the transitional phase from transgressive to highstand deposits (also see Vail et al. Chap. 6.1, this Vol.).

If period and amplitude of sea level oscillations vary irregularly with time and condensed sections are partially missing in the investigated part of the basin, then it becomes increasingly difficult to determine the real number of sea level changes.

The quantification and interpretation of varying sedimentation rates within cyclic sequences (Figs. 6b and 10) should, of course, be regarded with caution. The absolute time scale is still uncertain, in many cases (Bayer and McGhee 1986). Furthermore, changes in climate possibly paralleling the sea level fluctuations may cause variations in the primary sediment supply from the continent. Tectonic processes in the hinterland, on the other hand, probably do not lead to short-period fluctuations in sediment supply which are needed to explain the marked variations observed in the sedimentary record.

Some field examples demonstrate that our models can satisfactorily describe ancient sedimentary sequences. Since certain types of asymmetry are associated with a specific environmental situation, the models also provide a tool in basin analysis.

However, not all repetitious depositional sequences originate from sea level oscillations. An alternative to a set of sea level signals is the subsidence rate, which may be a variable of time at a certain fixed location within the basin. It could cause specific sequences and affect the shape of derived water-depth curves. A further discussion of this problem is beyond the context of this article (see, e.g., Watts 1982; Watts and Thorne 1984; Whittaker 1985; Cloetingh 1986).

6.3 Condensed Deposits in Siliciclastic Sequences: Expected and Observed Features

S. M. Kidwell

1 Introduction

In the last decade, seismic analysis has yielded invaluable three-dimensional pictures of stratigraphic sequences (sensu Vail 1987) and, following the conceptual lead of the Exxon Group, has revolutionized the study of sedimentary basins. At present, models relating sequences to relative sealevel are highly stylized, at least in published form (e.g., Vail 1987; Jervey 1988; Posamentier et al. 1988) and are controversial (e.g., responses to Haq et al. 1987 in Science 241:596–602), but the basic anatomy of siliciclastic sequences is little contested. Stratal patterns clearly reveal landward then basinward shifting of depocenters within each unconformity-bound sequence and reciprocal switching of nondepositional (i.e., persistent starvation, transport, or erosion) and depositional regimes. This cyclicity is evident regardless of its precise relation to sealevel fluctuations.

The stratigraphic element that figures prominently in the most recent generation of sequence models are deep-water ("toe-of-slope/basin"; Vail 1987), hemipelagic to pelagic condensed sections (and see Loutit et al. 1988). These lie along surfaces of stratigraphic downlap and are interpreted as products of sediment starvation during rapid sealevel rise and maximum transgression (Vail et al. 1984 and Vail 1987; Loutit et al. 1988; Van Wagoner et al. 1987). Seismic-based models recognize only this one kind of condensed deposit, resulting in a narrow redefinition of the term. Outcrop studies over the past 50 years have shown, however, that condensed deposits form in a variety of shallow- and deep-water settings by sediment bypassing as well as by starvation and that they can coincide with regression as well as with transgression (review by Jenkyns 1971; papers in Hsü and Jenkyns 1974, Einsele and Seilacher 1982, Bayer and Seilacher 1985). The actual distribution of condensed deposits within siliciclastic sequences is probably more complex than depicted by recent models.

This chapter follows a mainstream European definition for condensation: a condensed section (syn. condensed interval, condensed sequence, condensed deposit, condensed facies) is a local record that is complete but thin relative to the same age strata elsewhere and reflects a period of slow *net* sediment accumulation (paraphrased from Jenkyns 1971). Thin but relatively complete records can result from episodic or continuous starvation of siliciclastics at a site, with stratigraphic accumulation determined by the rate at which autochthonous sediment builds up (benthic and pelagic skeletal material, authigenic minerals). The composition, taphonomy, and early diagenesis of these locally produced sediments reflect conditions during the interruption in siliciclastic supply. Thin, stratigraphically condensed deposits can also accumulate at sites of persistent sediment transport both by dynamic bypassing of bedload (episodic or continuous small-scale erosion winnows or largely truncates ongoing siliciclastic deposition) and by total passing of suspended load (sediment maintained in transport, without even temporary deposition). Under these conditions, condensed deposits tend to be coarser grained and better sorted; autochthonous sediments are admixed with variable amounts of coarse, resedimented grains concentrated by reworking of siliciclastic deposits (lag material).

The 3 m-thick Bajocian record (Inferior Oolite) of Dorset, southern England, provides a classic and much-cited example of a biostratigraphically highly condensed section (Arkell 1933:189 ff; Wilson et al. 1958:68 ff). Guide taxa from all zones and most subzones are represented in correct microstratigraphic order; quartz-rich interbeds and conglomerates, fossils in various states of preservation, circumrotary algal structures ("snuffboxes"), complex diagenesis, and small-scale discontinuity surfaces indicate intermittent shallow-water erosion and winnowing during slow net accumulation. Tectonic rather than eustatic movements are usually invoked to explain the persistent, local condensation (op cit.; Sellwood and Jenkyns 1975); this interpretation of course does not diminish the observed degree of condensation.

Regardless of the details of their origin – and regardless of terms applied to them – all condensed deposits present special challenges to paleontological analysis owing to post-mortem mixing and overprinting of faunas (e.g., Kidwell 1986a). They also represent special opportunities to reconstruct the paleoenvironment and short-term dynamics of regimes of sedimentary (near-)omission (e.g., Kidwell and Aigner 1985 on fossil-rich condensed deposits). Because condensed deposits commonly mantle or are capped by significant erosional and omissional discontinuities (i.e., unconformities of various sorts and non-sequences, where geologic time is recorded by zero thickness; Fig. 1), information on the nature and distribution of condensed deposits should provide insights into the origin of these hiatuses, which are analytically less tractable and represent a large portion of many basin histories.

This chapter explores the expected and observed features of condensed deposits in siliciclastic sequences. Hypothetically, condensed deposits can occur anywhere that stratal surface (= time-lines) converge, that is along surfaces of onlap, "backlap", downlap, and toplap (Fig. 2a). In many instances, as described below, condensed deposits formed during onlap become mixed with material condensed during preceding toplap because of erosional transgression; similarly, downlap deposits can be amalgamated or mixed with those of "backlap". It is, however, realistically possible to have any of the four types preserved in simple form, and so they are treated here as end-members. Because they typically accumulate in different bathymetric environments and under different sedimentary regimes, the condensed records that mark these four phases of sequence formation should differ systematically in lithologic, paleontologic, and diagenetic features, in chronostratigraphic attributes, and in likelihood of preservation (Table 1). More than one can occur in any given sequence (Fig. 2b). Analogous deposits are expected in marine carbonate and nonmarine sequences. (Lap-out terminology follows van Wagoner et al. 1987.)

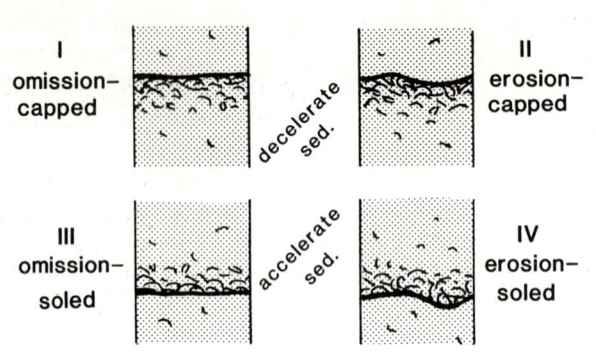

Fig. 1. Condensed deposits, whether shell-rich as illustrated here or not, can be classed according to bed contacts: condensed beds can rest on or be capped by discontinuity surfaces (condensation following or preceding a complete hiatus). The bed contact criterion is more appropriate than lap-out terms in early stages of sequence analysis when three-dimensional relations are not yet well known. (After Kidwell 1986a)

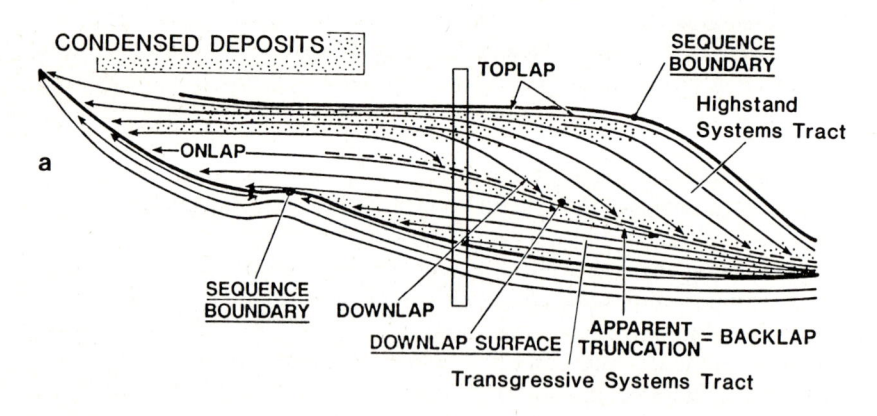

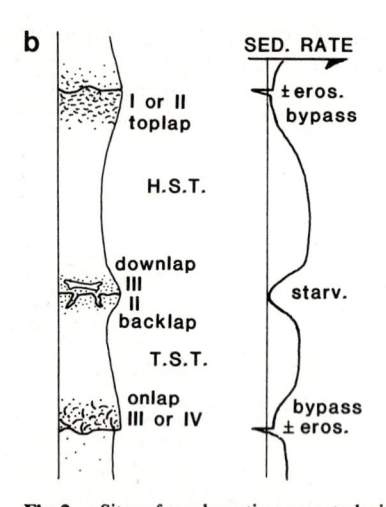

I omission-capped — decelerate sed.
II erosion-capped
III omission-soled — accelerate sed.
IV erosion-soled

Fig. 2. **a** Sites of condensation expected within an idealized siliciclastic depositional sequence. (Cross-section after Vail 1987). **b** Expected occurrence of several condensed intervals in vertical section of a single sequence (a) Condensed deposits are characterized according to their physical stratigraphic relation to discontinuity surfaces (following Fig. 1) and by lap-out directions; sedimentation rate curve is schematic only. Condensed marine onlap facies can rest on erosional unconformities or on transgressive ravinement surfaces and, like condensed toplap facies, are typically coarser-grained than condensed downlap facies

Table 1. Summary of expected differences among condensed deposits as a function of position within stratigraphic sequence. (After Kidwell and Jablonski 1983 and Kidwell 1983)

CONDENSED FACIES TYPES	SEDIMENTARY DYNAMICS	BATHYMETRY SETTING	BATHYMETRY HISTORY	LITHOLOGY	PALEONTOLOGY ECOLOGY	PALEONTOLOGY TAPHONOMY[a]	CHRONOSTRATIGRAPHY ISOCHRONEITY	CHRONOSTRATIGRAPHY DURATION	PRESERVATION POTENTIAL
Toplap	bypassing ± starvation	shallow water	shallowing-upward	benthic skeletal residua, winnowed sand	low diversity, mixed benthos	concentration of durable elements, K Type I or II	diachronous (HIATUS)	similar over extent	poor
Downlap	starvation	range shallow to deep water	shallowing-upward	concentration of authigenic minerals & benthic + planktic fossils	high diversity, mixed benthos & plank/nekton	variable, poor to excellent preserv. K Type III	isochron. base	increases basinward	excellent
Backlap	starvation	range shallow to deep water	deepening-upward	ditto	ditto	ditto K Type I	isochron. top (CONDENSED)	ditto	good
Onlap	bypassing ± starvation	shallow water	deepening-upward	benthic skeletal lags in well-sorted sand, ± authigenic minerals	high diversity, mixed benthos	mix of new & reworked fossils, K Type III or IV	diachronous ("NORMAL" SEDIMENT)	similar over extent	good

[a] K Type I-IV are taphonomic models developed in Kidwell (1986a) (Fig. 1)

2 Expected Features of Condensed Deposits

2.1 Condensation in the Context of Marine Onlap

Onlap is the depositional lap-out or termination of beds against an initially sloping surface, such that younger beds extend further upslope than older beds. Condensation should occur where stratal surfaces asymptotically terminate against the surface of onlap (Fig. 2a). Where stratal terminations are abrupt, no attenuation – and thus no condensation – of beds exists: beds step against the onlap surface in buttress relation.

Condensed facies formed during transgressive onlap of marine beds will rest on an erosional surface (subaerial unconformity or submarine ravinement), grade upward into less condensed marine facies (type IV in Fig. 1), and show evidence of deepening water in very shallow, paralic to inner shelf environments. The paleontologic assemblage can be diverse owing to the species richness of shallow marine habitats; taphonomic feedback during shell accumulation on the seafloor can augment the assemblage (Kidwell and Aigner 1985; Einsele et al. Introduction, this Vol.). Because of taphonomic feedback and changing water depth, plus the effects of continuous shallow-water erosional reworking, assemblages are likely to be ecologically mixed in composition and have taphonomically complex histories. The abundance of glauconite and phosphate is determined by oxygenation and organic carbon content as well as by the duration of condensation.

Developed along the margins of basins and lying at the base of depositional sequences (at least where transgressive facies tracts rest directly on subaerial unconformities), condensed onlap facies record a combination of sediment bypassing and starvation. Locally derived siliciclastics are passed across the site of onlap either in continuous suspension or as bedload that temporarily comes to rest. Any trapping of sediment along the coast during this period starves the shelf accordingly and contributes to condensation of the transgressive record.

Depending on the depth of erosional truncation during initial marine transgression and the durability of skeletal material so released, fossils from older depositional sequences can be incorporated into condensed onlap deposits, particularly along their base. Where this reworked material (remanié) constitutes a significant part of the total assemblage, the deposit is typically called a transgressive lag. A gradation thus exists between (1) transgressive lags formed by erosion and dominated by "old" shells that pre-date transgressive erosion, and (2) condensed onlap deposits formed by relatively passive accumulation of "new" shells that post-date transgressive erosion. Both transgressive lags and condensed onlap deposits (and their combinations) occupy the same position within depositional sequences; the development of one rather than the other reflects such factors as depth of erosion, nature of eroded material, and autochthonous production and preservation potential during the period of slow net sedimentation following erosion.

From a regional perspective, condensed onlap deposits are thin, diachronous facies that reflect the landward migration of an environment of slow net accumulation; in detail, they consist of isochronous condensed horizons arranged in a landward-shingled series (Table 1). These isochronous components can be preserved as discrete beds (e.g., associated with a series of flooding surfaces), or can be physically amalgamated into a single, laterally continuous and microstratigraphically complex unit along the base of the depositional sequence. Regardless of internal anatomy, condensed onlap deposits should have a high likelihood of preservation because of burial under a significant thickness of later transgressive and regressive marine sediments.

2.2 Condensation in the Context of "Backlap" and Downlap

"Backlap" refers to the termination of beds at the distal, basinward edge of a backstepping (= retrogradational) body: younger beds extend less far out into the basin than older ones, that is, they retreat toward the basin margin (Table 1). Seismic interpreters refer to these relations as "apparent truncation" (e.g., Vail 1987; Fig. 2a). Downlap denotes the termination of beds at the distal, basinward edge of a prograding body: younger beds extend further out into the basin than older ones, that is, they advance toward the basin axis. In both situations, condensation should occur where stratal surfaces converge asymptotically.

Backlapped condensed facies grade upward *from* normal depositional facies and typically will be capped by an omission surface (type I accumulation in Fig. 1); in marine settings they should reflect deepening water. Downlapped condensed facies grade upward *into* normal depositional facies, rest upon an omission (or erosion) surface (type III or IV in Fig. 1) and, in marine settings, record shallowing water. Backlap and downlap condensed sequences are thus mirror images of each other in many respects (Fig. 2b, Table 1). Condensed sections depicted in seismic-based models (e.g., Vail 1987; Fig. 2a) are compounds of backlap and downlap and lie mid-position within the sequence between transgressive and highstand facies tracts.

Both backlap and downlap condensed facies indicate local starvation of siliciclastic supply; more precisely, they record transitions from deposition to starvation (backlap) and vice versa (downlap) due to the retreat and advance of siliciclastic depocenters. During the period of effective starvation, only sediments from local benthic production, authigenesis, and pelagic rain accumulate. The proportion of pelagic material, the taxonomic composition, the state of preservation and diagenesis will vary with the environment of starvation. Starved condensed facies in shallow subtidal settings can resemble onlap facies formed by sediment bypassing because of high water energy, firmer initial substrata (sand, muddy sand), and diverse shelled benthos. Starved facies in deeper and more distal sites on the other hand will contain a higher ratio of pelagic to benthic material and are more likely to show evidence of low-oxygenation and mineralization, including seafloor cementation. The downlapped condensed sections referred to in seismic-based models are of the latter type.

In general, starvation can be more widespread at any time and more persistent than sediment bypassing. Starvation requires only that siliciclastic supply be cut off (curtailed production in source area, interrupted transport or deflected transport routes). In contrast, bypassing reflects a sensitive balance between the volume and grain size of sediment supply and hydraulics at the seafloor (erosion must match or nearly match deposition, but never exceed it); such regimes are bathymetrically more restricted and shorter-lived. Condensed backlap and downlap deposits thus can be widespread and can vary laterally in features more so than condensed onlap (and toplap) deposits.

In contrast to onlap and toplap facies, condensed backlap and downlap facies (and their compounds) will in most instances bracket a time-plane and thus can serve as isochronous marker beds (Table 1). This relationship has been verified biostratigraphically by industry workers in the last several years (Haq et al. 1987 and references therein). The length of time recorded by any backlap or downlap facies is expected to vary from a minimum near the basin margin to a maximum towards the basin axis, which is the first area to experience starvation and the last to see sedimentation resume. Backlap and downlap facies should have high preservation potential because they will typically be mantled by regressive marine deposits.

2.3 Condensation in the Context of Toplap

Toplap is the asymptotic termination of beds at the updip edge of a prograding body (e.g., within topset beds of clinoform bodies). Younger beds extend less far toward the basin margin than older beds, that is, the updip limits of beds retreat basinward (Fig. 2a).

Toplap condensed facies grade upward from normal depositional facies, culminate in an omission or erosion surface (type I or II in Fig. 1), and record progressive bypassing of sediment during shallowing-upward conditions. Bypassing in very shallow marine to marginal marine environments can be typified by low-diversity fossil assemblages, owing to low diversity benthic communities and strong taphonomic culling in physically disturbed habitats. Meteoric diagenesis and weathering during late regression will further reduce the diversity of autochthonous grains that are preserved. Size sorting and bias toward high-energy event beds may be more pronounced than in any other condensed facies type.

Like onlap condensed facies, toplap facies will be diachronous when examined regionally; in detail, they will consist of a series of isochronous condensed horizons amalgamated laterally or in a basinward shingle (Table 1). Toplap deposits have low preservation potential owing to likelihood of destruction by

1. subaerial exposure,
2. submarine erosion and transport downslope, where distinctive grains are dispersed among basinal facies, and
3. complete or partial reworking into transgressive lags, thus forming compound condensed sections.

3 Examples from Outcropping Late Cenozoic Strata in North America

3.1 Miocene of Maryland, U.S. Atlantic Coastal Plain

Unconsolidated Miocene strata (Burdigalian-Tortonian) exposed along shorelines of the modern Chesapeake estuary yield variants of all types of condensed facies in a

~70-m-thick record of marine and marginal marine environments (intertidal to mid-shelf depths) (Fig. 3). The section consists of five depositional sequences of ~1 Ma each that are predominantly marine in facies composition, and four much thinner "sequences" that are predominantly marginal marine (Kidwell 1988a); the overall trend is regressive.

3.1.1 Condensed Onlap Deposits

Stratigraphically condensed, densely fossiliferous coquinas mark the base of each depositional sequence in the marine part of the section; each shell deposit onlaps an erosional (probably subaerial) disconformity and can be traced over hundreds to thousands of square kilometers in outcrop (Kidwell 1989). These contain both fragmental and whole specimens in ecologically mixed but little-transported (parautochthonous) benthic assemblages that are dominated by diverse open-marine molluscs and subsidiary echinoids, bryozoans, barnacles, corals, fish, and marine mammals. Negligible material appears to have been reworked from older strata. Within each of

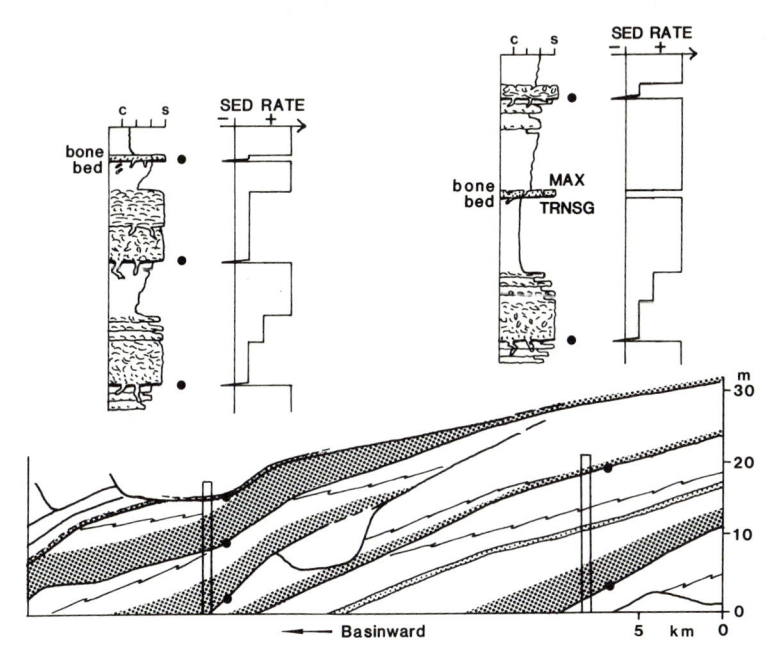

Fig. 3. Sequence boundaries (*solid circles*) in the lower, open marine part of the Maryland Miocene section are onlapped by stratigraphically condensed skeletal sands (30–70% shell carbonate by volume). Thin bone-rich sands mark (1) a burrowed surface coinciding with maximum transgression ("downlap" surface) and (2) a sequence boundary (*solid circle*) at the base of the predominantly paralic part of the local Miocene record. This second, "end-cycle" bone sand is coarser, patchy, and contains finely comminuted rather than whole vertebrate specimens and is thought to represent a transgressively reworked toplap concentration. (Cross-section from the Calvert Cliffs, datum is modern mean sealevel; after Kidwell 1989)

these shell deposits, faunal composition changed up-section in response to increasing water depth and varying degrees of taphonomic feedback (Kidwell and Jablonski 1983). Pods of undisturbed silty sand in the otherwise well-sorted sand matrix, the mixture of soft- and shelly-bottom fauna, variable fossil preservation, and minor discontinuity surfaces all indicate condensation by dynamic bypassing in shallow marine environments below fairweather wavebase but within reach of storms. In outcrops nearest the original basin margin, the condensed shell deposits span the entire thickness of the transgressive systems tract, whereas in more basinward outcrops the shell deposits comprise only its lowermost portion.

Technically shell gravels containing up to 70% coarse shell carbonate by volume, these condensed onlap deposits are the coarsest units in the Miocene succession. All four examples have complex internal microstratigraphies, but differ considerably in detailed anatomy. They can exhibit both simple onlap patterns (Fig. 4a) and internal pinchout and thinning (Fig. 4b) over paleohighs with simple expansion of beds into paleolows on the erosional disconformity. Others consist of an internally complex, condensed blanket that cuts across paleolow deposits of different nature (i.e., expanded sections of intertidal or shallow subtidal facies; Fig. 4c). Although physical stratigraphic relations indicate onlap, none of the deposits is demonstrably diachronous within the outcrop belt. Each is estimated to record 10's to 100's of thousands of years (Kidwell 1989), beyond the resolving power of available biostratigraphy. In lithology they resemble temperate carbonate and sand ridge facies such as characterize modern open shelves that are still starved by Holocene sealevel rise.

3.1.2 Condensed Backlap/Downlap Compound Deposits

The Maryland Miocene provides a single example of this type: a thin (<1 m), laterally extensive (2000 km^2), bone-rich glauconitic sand that is bracketed by the deepest water facies in the section (Kidwell 1989) (Fig. 3). The bone bed contains a rich, ecologically mixed and time-averaged assemblage of marine mammals and fish plus terrestrial and estuarine vertebrates. Although most fossil occurrences consist of disarticulated elements (some broken, abraded, bored, or encrusted), a large number of articulated specimens have been quarried. Benthic invertebrates, primarily barnacles and marine bivalves, are sparse and poorly preserved. Planktonic microfauna by contrast are maximally abundant and diverse. Glauconite also reaches an abundance maximum in this bed, which appears to reflect a prolonged period of terrigenous starvation on the distal open shelf.

Although bioturbation has obscured stratal surfaces that might have been used to physically demonstrate backlap or downlap of surrounding sandy silts and clays, the coincidence of the bone accumulation with maximum water depth (maximum transgression) is most consistent with that kind of relationship: the bone bed lies at the junction of transgressive and regressive facies tracts. Also, when traced updip toward the paleoshoreline, the bone bed splits into several more diffuse bone-bearing tongues that are lost within sandier facies of the basin margin; this too is most consistent with the bone bed marking distal starvation. Contrary to expectations (Sect. 2.2. and Fig. 2b), this "mid-cycle" condensed deposit does not have a major

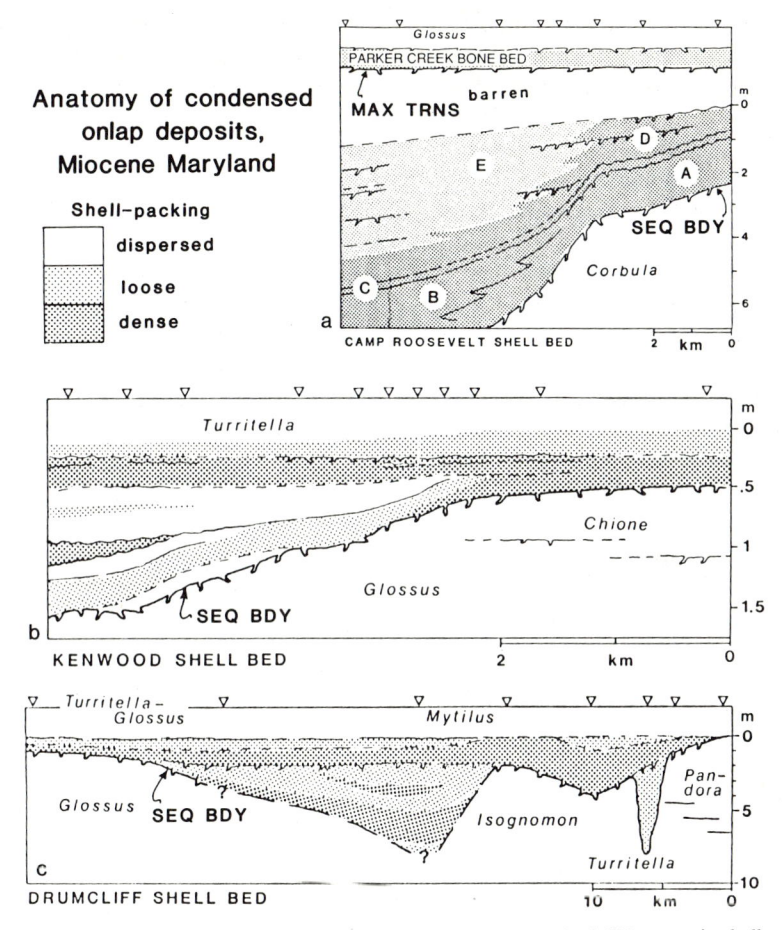

Fig. 4 a-c. Condensed onlap deposits exhibit considerable within- and between-bed differences in shell-packing and thickness, accompanied by changes in taxonomic composition, taphonomic condition, and siliciclastic matrix. *Triangular symbols* along top edge of each cross-section indicate spacing of measured sections. (After Kidwell 1989)

discontinuity surface at its waist, but instead is bound both above and below by burrowed discontinuities (many minor discontinuities also subdivide the bed). This suggests that the changeover from deposition to starvation during backlap was more abrupt than gradual, and that the resumption of deposition during downlap was similarly abrupt and final.

3.1.3 Condensed Toplap Deposits

A very patchy bone- and tooth-rich sand, having a maximum thickness of only 8 cm, is the only indication of toplap-type deposits in the Maryland Miocene (Kidwell 1989) (Fig. 3). Unlike the taphonomically mixed nature of the deep-water mid-cycle bone bed, this "end-cycle" bone bed contains only highly comminuted and polished bone debris of coarse sand and granule size; this is densely packed in the layer with coarse quartz sand, phosphatic steinkerns of the small shallow-subtidal bivalve *Caryocorbula*, phosphatic pellets of indeterminate origin, local concentrations of glauconite, and minor clay and silt infiltrated from above. The bed is associated with an erosional disconformity that lies at the top of the open marine part of the Miocene record: facies in stratigraphically higher sequences are predominantly paralic.

The taxonomic composition, sedimentary matrix, taphonomic features, and facies context of the bone sand are all consistent with its formation in very shallow waters during the latest phase of marine regression. This bone material was then reworked into a transgressive lag, as it now lies concentrated on top of rather than underneath a basin-margin unconformity (i.e., it is a type IV rather than the expected type I or II kind of deposit; Fig. 1) and is conformably overlain by upward-deepening strata (Kidwell 1988a). Because none of the transgressive deposits in the stratigraphically lower, open marine part of the Miocene section are characterized by similarly comminuted vertebrate debris, this highly distinctive bone sand is attributed to initial accumulation under toplap-type conditions, whose signature it still bears despite transgressive reworking.

3.2 Neogene Imperial Formation of Southeastern California

This thick (~5 km) succession records earliest marine transgression and deltaic infill of a basin in the northern part of the nascent Gulf of California, an oblique rift zone at the boundary of the North American and Pacific plates. Depositional sequences sensu strictu are not obvious, but a series of smaller depositional cycles (probably parasequences) can be recognized. These include densely fossiliferous downlap, toplap, and erosional lag deposits that are analogs of condensed facies but are not all demonstrably condensed themselves.

3.2.1 Onlap/Downlap Compound Deposits

Impure skeletal limestones of shallow subtidal origin (water depths of a few m or 10's m) occur at the base (bottomset position, type III in Fig. 1) of progradational coastal alluvial fan sequences along the margin of this tectonically active basin (Kidwell 1988b). The limestone bodies interfinger with proximal conglomerates and coarse arkosic sands, and provide a microstratigraphic record of fan abandonment and reactivation (Fig. 5). Skeletal carbonate is relatively well preserved and taxonomically diverse and typically includes coral ecomorphs that are sensitive to water turbidity. Despite proximity to point sources of siliciclastics, the shallow nearshore was apparently starved of both sand and finer sediment (or bypassed by fines). Starvation is indicated by the low proportion of terrigenous matrix in the marine facies and by the assemblage of predominantly large-bodied, long-lived sessile and cemented epifauna, suggesting slow net aggradation of the seafloor and persistent shell gravel conditions.

Fig. 5. Minor toplap, downlap, and compound onlap-downlap shell concentrations from the Imperial Formation. Fan progradational cycles are 6–10 m thick each; distance from shoreline to site of starved, skeletal limestone formation is several hundred meters. (After Kidwell 1988b)

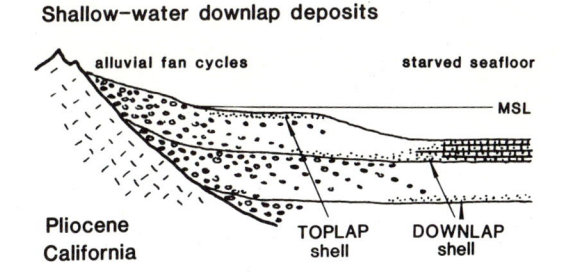

3.2.2 Toplap Deposits

A few thin (<1 m), laterally restricted "toplap" shell concentrations (type I in Fig. 1) are present in coastal alluvial fan cycles (Fig. 5). None of these shows evidence of the lateral amalgamation or seaward shingling expected in condensed toplap deposits (Fig. 2a, Table 1).

Better examples of toplap skeletal concentrations cap progradational, delta-front cycles higher in the Imperial Formation (Kidwell in prep.). Shell material in these oyster coquinas (~1 to 6 m thick) is very densely packed, highly fragmental, low in diversity, and physically stratified (large-scale epsilon cross-sets). The predominantly calcitic, aragonite-poor composition of these deposits probably reflects original ecology rather than selective taphonomic and diagenetic removal: autochthonous biohermal buildups of oysters and anomiid bivalves in other delta-front facies suggest that the source habitats of allochthonous shell material were low in species diversity themselves and biased toward calcitic forms. The allochthonous nature of the concentrated shell material and the laterally amalgamated shingling of these cross-bedded channel fills are all consistent with expected features of toplap condensed deposits, although condensation in this case cannot be demonstrated independently.

3.2.3 Transgressive Lags

One example of this type is known from the Imperial Formation: a laterally extensive (several km^2 minimum) oyster coquina from the delta-front facies. Transgressive reworking of precursor toplap oyster deposits (and limited marine recolonization) is indicated by

1. the tabular rather than lenticular form,
2. absence of cross-stratification or amalgamated cosets,
3. inclusion of a sparse but fairly diverse marine benthos and shark teeth, and
4. truncation of the underlying progradational cycle.

4 Conclusions and Outlook

A small sampling of marine siliciclastic sequences provides ready examples of a variety of end-member and compound condensed deposits.

a) The starved, laterally extensive deep-water condensed sections that figure prominently in seismic-based models can be recognized even in the attenuated, basin-margin reaches of marine sequences where outcrops are available. The lithologic and paleontologic expression of these deposits differs significantly from the pelagic/hemipelagic features typical of basin axes. Compound backlap/downlap deposits can be characterized by time-averaged but rich accumulations of well-preserved vertebrates; in addition to the Miocene of Maryland, examples are known from the Pliocene Purisima and other Tertiary formations of California (e.g., Norris 1986).

b) Condensed onlap deposits and transgressive lags are common in basin-margin sections and can be characterized by abundant shell carbonate. These deposits can be laterally extensive and in my experience are very heterogeneous, both within a single deposit and from deposit to deposit in a single basin (e.g., Fig. 4). Onlap deposits can be compounded by downlap deposits, at least in thin sequences and in parasequences when transgressive deposits are thin or lacking. This situation is illustrated by alluvial fan-related deposits in the Pliocene of California (and in the Pleistocene of Sonora; see Beckvar and Kidwell 1988) but is also ongoing on many modern shelves that are starved across their width. In the Miocene of Maryland, the nearshore area of condensed onlap was bypassed by fine siliciclastics that accumulated offshore; only distal of this depositional area did the "mid-cycle" starved bone sand accumulate (Kidwell 1989). The existence of such a mosaic of sedimentary regimes results in a sequence with multiple condensed intervals, specifically a shell-rich interval at the base of the transgressive systems tract and a second, bone-rich interval marking the top of the tract.

c) Condensed toplap deposits are relatively rare, at least in part owing to low preservation potential. All "end-cycle", type I or II deposits that I have encountered in siliciclastic records have contained abundant shell material that is nonetheless much lower in diversity and much less well preserved than that of condensed onlap deposits. The implication is that regressive facies tracts are qualitatively distinct from transgressive tracts – one is not the mirrored image of the other – and this results in paleoecologically and taphonomically distinct fossil assemblages as well as in distinctive stratigraphic anatomies. Much of this distinctive taphonomic signature is retained when toplap deposits are reworked into transgressive lags.

d) Empirically, the hypothetical array of condensed deposit types (Fig. 2, Table 1) could be modified into several different classes: (1) onlap deposits composed of "new" particles (type III or IV arrangement, Fig. 1); these intergrade with (2) transgressive lags composed of "old" particles (type IV); (3) downlap deposits compounded either with backlap or directly with onlap deposits (condensed deposit may span or be delimited by discontinuity surface(s)); and (4) comparatively rare toplap deposits (type I or II in their larger context, but commonly type IV channel fills in detail).

The modified Exxon diagram in Fig. 2, with its several types rather than single kind of condensed interval, may provide a more realistic conceptual model of siliciclastic depositional sequences and might facilitate comparison and integration of outcrop and subsurface observations, which typically sample different parts of depositional sequences (basin-margin vs. -axial portions). Condensed deposits of all types are important but complex repositories of biostratigraphic and paleoecological data. They are also the preserved trace of regimes of slow net accumulation, whatever their cause, and can provide unique insights into the short-term dynamics of these regimes (Kidwell and Aigner 1985; Kidwell Chap. 2.4, this Vol.). Because skeletal material can retain the imprint of previous taphonomic conditions, fossil-rich condensed deposits can provide unique information on basin history otherwise lost to erosion.

Acknowledgments. Research supported by NSF grant EAR85–52411-PYI and funds from Shell Foundation, Amoco Production Company, Arco Oil and Gas Company, and Arco Foundation. I am grateful to Steven Holland, David Jablonski, and the volume editors for their helpful reviews.

6.4 Biological and Evolutionary Responses to Transgressive-Regressive Cycles

G. R. McGhee Jr, U. Bayer, and A. Seilacher

1 Introduction

Transgressive-regressive cycles involve environmental change. Therefore, biological response is expected under the predictions of the theory of natural selection. The nature, magnitude, and instrumentation of that response is, however, less well understood and difficult to predict. Generally, the magnitude of biological response would be expected to be a direct function of the magnitude of the environmental perturbation during the cycle. However, this may not be the case. Moreover, environmental perturbation itself may not be a direct function of the magnitude, or range, of sea-level rise or fall. The relationships between organisms, environment, and transgressive-regressive cycles will here be examined in a series of empirical case studies, augmented by discussions of the possible forcing mechanisms behind observed faunal changes.

2 Morphological Responses to Transgressive-Regressive Cycles at the Species Level

2.1 An Empirical Example: Ammonite Species Sequences in the German Lower and Middle Jurassic

A well-studied example of biological change in response to transgressive-regressive cycles are the iterative evolutionary trends in ammonite faunas of the south German basin (Bayer and McGhee 1984, 1985). During the Aalenian and Bajocian numerous coarsening upward sedimentary sequences ("Klüpfel cycles") can be recognized. They correspond to the regressive phases of repeated transgressive-regressive events, in which the progressive shallowing of the basin is reflected in a coarsening-upward succession of clays to sands to oolitic sands to hardgrounds (Fig.1).

The striking asymmetry of "Klüpfel cycles" is probably a sedimentational artifact. As we can see in modern, post-glacial shelf settings, coarse sediment is held back in river systems and nearshore areas when sea level rises. As a consequence, deeper zones become sediment-starved, or at least receive less and finer-grained sediments than during equivalent bathymetric conditions in the regressive part of the cycle. In other words, the sedimentary record of the transgressive phase will be underemphasized relative to the regressive one.

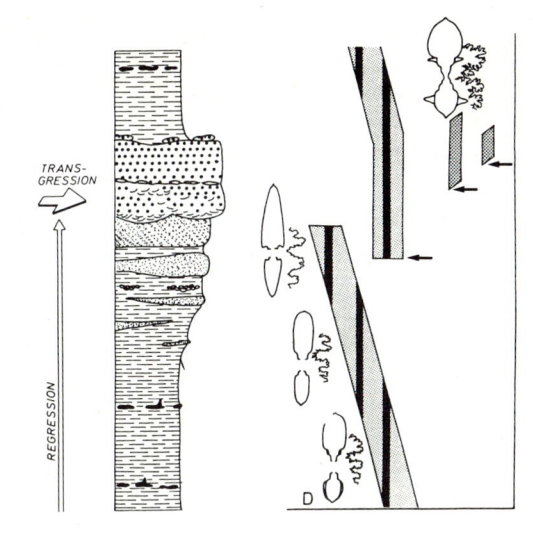

Fig. 1. Observed patterns of morphological change within successive ammonite species with respect to transgressive-regressive cycles. Illustrated is an idealized Külpfel regressive-transgressive facies sequence (lithological column) with respect to the distribution of ammonite morphologies in the same sequence (shown on the *right*)

However, a reduced rate of background sedimentation may also have another, biologically mediated, effect.

1. It favors the colonization of the mud bottoms by an epifaunal community (secondary soft bottom dwellers; Seilacher 1984) which tends to have calcitic and relatively thick shells. Thus the production of durable bioclasts will be increased.
2. In addition, muddy bottoms, low sedimentation rates and slightly reduced oxygen levels (which also tend to be associated with transgression) favor the development of microbial communities that are largely responsible for the production of "oolitic" grains and other early diagenetic "diaclasts".

Both kinds of coarse particles require not only slow sedimentation, but also sufficiently long quiet intervals to grow; i.e., they will not form in very shallow water. But as long as storms (however rare in terms of biological life spans) reach such bottoms, they will winnow away the muddy matrix, leaving behind a new layer of parautochthonous bio- and diaclasts – even if this layer is blanketed by a "muddy trail" as the storm wanes.

In summary, grain size should not be blindly taken as a measure for turbulence levels. In contrast to the terrigenous sands, bioclastic and diaclastic grainstones are largely autochthonous and should be attributed to the transgressive, rather than the regressive, part of the cycle of which they appear to form the top.

Three of the Aalenian/Bajocian cycles are of particular interest in that they are accompanied by morphological changes in the associated ammonite fauna. Within each coarsening-upward sequence we find that ammonites with inflated, evolute, and ornamented shells and with complex suture lines are progressively replaced by species that are more discoidal, involute, and smooth-shelled, and have simpler suture lines (Figs. 1 and 2).

At this point it is important to remember the functional significance of the changing characters. In one way or the other, involute coiling, inflated cross-section and corrugation of the shell contribute to strengthen the shell against ambient water pressure. Similarly, frilling of the suture line, while not significantly contributing to septal corrugation, allowed the ammonite soft body to "crawl" to the new septal position without completely detaching the septal diaphragm from the shell wall (Seilacher 1988b). Simplification of the suture ("ceratitization") means a reduction in the number of tie-points, which may have been admissible if the range of diving was reduced by diminishing water depth. Jurassic ammonites did not reach the extremes known from ceratites and pseudoceratites of Triassic and Cretaceous epicontinental seas; yet their sutural simplification points in the same direction. In summary, the evolutionary response of Jurassic ammonites to cyclic sea level changes corresponds to trends observed in other shallow epicontinental basins as inhabitants became isolated from their oceanic stocks.

In the Jurassic examples it is of particular interest that the observed trends appear to have resulted from one of two totally different processes:

1. in situ speciation of ammonite clades within the basin, and
2. selective immigration of ammonite species from the Tethyan open ocean realm. In the *Staufenia* lineage (Fig. 2) the morphological trend is produced by speciation of endemic forms, while in other regressive phases a corresponding morphological trend is produced by the successive immigration of ammonite species from extra-basinal sources (Bayer and McGhee 1984, 1985). How such a phenomenon may be produced is the subject of the following section.

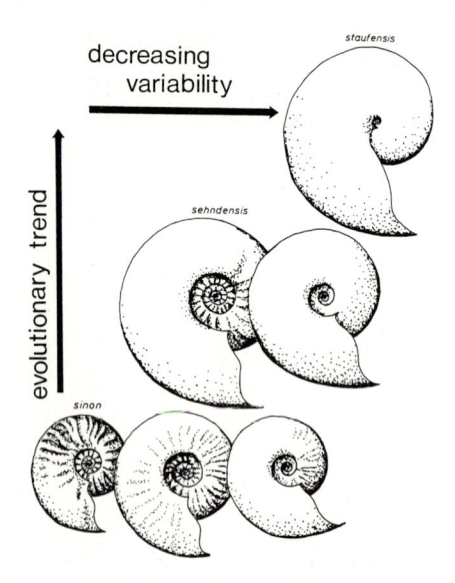

Fig. 2. Example of a single morphological trend in the *Staufenia* lineage, produced by in situ speciation of endemic forms

2.2 Theoretical Models of the Observed Morphological Sequences

2.2.1 Closed Ecological Systems

Starting with the simplest case, we consider a restricted basin as a closed ecological system that had minimal biological contact with external environments (Fig. 3A). Isolation mechanisms other than geographic ones are conceivable, but less easy to visualize. To further simplify things, we shall consider only two species (a and b) that belong to the same higher taxon but prefer different habitats (basin center and basin margin, respectively).

During regression the environment of the basin center species (a) decreases in area and increases in fragmentation and patchiness. Environmental patchiness and geographic restriction leads to fragmented and increasingly isolated populations. It is also likely that the environment experienced differs between populations. The net result will be a series of allopatric speciations (a' to a") from the ancestral central basin species as the basin shallows. Increasing habitat isolation, decreasing areal extent of the patches with consequent reduction in population sizes, and increasing

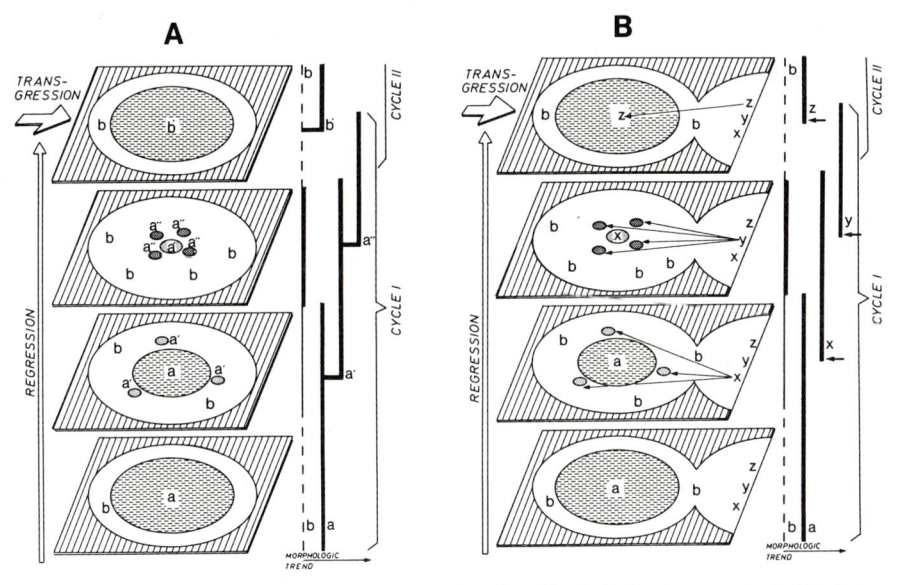

Fig. 3 A,B. Evolution of a theoretical ammonite fauna in a closed basin (**A**) and an open basin (**B**) during a regressive-transgressive cycle. During each regressive phase, the central basin habitat space is progressively reduced and fragmented (*a* to *a'* to *a"*). As the environmental trend is directional and the same in both basins, a similar morphological trend is expected. In the closed basin (**A**) the morphological trend is produced by in situ speciation of the central basin ancestral species (*a* to *a'* to *a"*); in the open basin (**B**) the morphological trend is produced by a series of ecological species substitutions (*x* to *y* to *z*) of previously adapted species which sequentially immigrate into the basin. In both basins the shallow water marginal species (*b*) waxes and wanes in abundance in response to the expansion and contraction of its habitat space. The central basin stock of the new cycle may be produced either by speciation from the marginal stock (**A**) or by introduction of an external species (**B**)

geographical/ecological distance between the habitat patches are the driving forces for intrabasinal allopatric speciation as seen in the species of *Staufenia* (Bayer and McGhee 1984, 1985).

Simultaneously with the habitat changes of the central basin species (a), the basin margin species (b) experiences habitat expansion as the basin shallows. Populations of this species are predicted to increase in size and geographic dispersion. In the simplest case this process would produce a stratigraphic record in which the basin margin species appears and fluctuates in abundance during low sea-level stand (cf. Figs. 3A and 1). As shallow water basin margin environments tend to become eroded, it is perhaps only at low sea-level stand that the basin margin species may appear in the fossil record. Such a pattern is exhibited by repeated appearances of sonniniid and stephanoceratid faunas in the central clay facies of the North German Bajocian (Bayer and McGhee 1985).

In the event that the regression is severe enough to totally eliminate the acceptable environments of the central basin species/clade (and hence terminates the lineage), an "empty environment" is produced when renewed transgression restores the original basin's environmental configuration. In this case we would expect subpopulations of the basin margin species (b) to adapt to and invade this now vacant environment (producing species b', Fig. 3A). If morphology is linked to environment (as hypothesized), then we would expect the origination of a new species (b') which is morphologically convergent on the previous species inhabitant (a), but originated from a different ancestral lineage (b). This convergence in morphology can be expected if the constructional/historical constraints of the basin margin species (b) allow it to parallel the basin center species (a). These conditions are met in the case that the two separate lineages belong to the same higher taxon (subfamily or family). Hence the mysterious "orthogenetic factors" invoked by previous workers to explain such morphologically convergent traits (see Bayer and McGhee 1984, 1985) may simply result from adaptation to similar environmental conditions by lineages having similar constructional/historical constraints.

2.2.2 Open Ecological Systems

Similar morphological trends in response to regressive-transgressive cycles can be expected in open ecological systems, although they are produced by a totally different process. In this case we assume that the basin is not biologically isolated, but open to an external pool of various species (x, y, and z; see Fig. 3B) of the same higher taxon. As before, the initial basin is assumed to be occupied by two species: a basin margin species (b) and a basin center species (a). The expected response of the basin margin species (b) during the regression is the same as in the previous model (Fig. 3B); i.e., it appears in increasing numbers and expands throughout the basin during the same time intervals (cf. Fig. 3A and B).

The response for the basin center species, however, will be different. As regression goes on, the progressive habitat reduction and fragmentation will not induce speciation of local populations of species (a), but rather offers the opportunity for successful invasion of the basin by a previously adapted species, which already existed but was so far extrabasinal in distribution (Fig. 3B). Again, under the assumption of linkage between environment and morphology in species belonging to the same higher taxon, a similar morphological trend is produced in response to the same environmental alterations (cf. Fig. 3A and B), but a trend which is now produced by a series of ecological species substitutions (x to y to z) rather than speciations from the ancestral basin center species (a to a' to a"). Such a pattern of species substitutions is seen among species belonging to the hammatoceratids and sonniniids in the south German basin in response to regressive-transgressive cycles (Bayer and McGhee 1984, 1985).

3 Responses to Transgressive-Regressive Cycles in Multispecies Communities

3.1 An Empirical Example: Benthic Marine Communities in the New York Upper Devonian

A classic area of Devonian paleontological studies are the well exposed stratigraphic sequences of the "Catskill Delta" facies in New York State (U.S.A.). These strata preserve the overall regression of epicontinental seas out of the Appalachian Basin during the Late Devonian. In this case, however, regression was not caused by falling sea level, but by infilling of the basin by clastic sediments eroded from the Acadian Mountains on the eastern margin (Fig. 4A). This overall regression was punctuated by a series of smaller scale and seemingly abrupt (see above) transgressive events that temporarily halted the progradation of shallow water facies (Fig. 4A).

The distribution of benthic marine communities along the bathymetric slope of the "Catskill Delta" during this interval of time corresponds to the facies pattern (Fig. 4B). During the Frasnian, each transgressive pulse produced a noticeable, though minor, reorganization of benthic marine communities (Sutton and McGhee 1985). Most noticeable is the fluctuation in relative population sizes of species following each transgressive pulse; and even here, subdominant species fluctuated more in population density than dominant species (as reflected in the community nomenclatural designations of Fig. 3B, where dominant species are listed first and subdominants second).

In general, larger transgressive events induced larger reorganizations in benthic species assemblages (cf. community dominant/subdominant species names and community geographic distributions pre- and post-event T3 versus pre- and post-event T4 or T5). On the other hand, however, shallow water nearshore communities changed less in ecological structure than deeper water offshore communities did during the same time (Sutton and McGhee 1985).

The most significant reorganization that occurred during the Frasnian in Appalachian marine ecosystems was the introduction of new species migrants from the Old World Province of Europe and western North America (McGhee 1981; Sutton and McGhee 1985); the immigrants (particularly species of the brachiopod genus *Cyrtospirifer*) replacing their ecological equivalents among the endemic Appalachian fauna. Migration of new species into the Appalachian Basin itself was made possible by global sea-level high-stands during the middle and late phases of the Frasnian

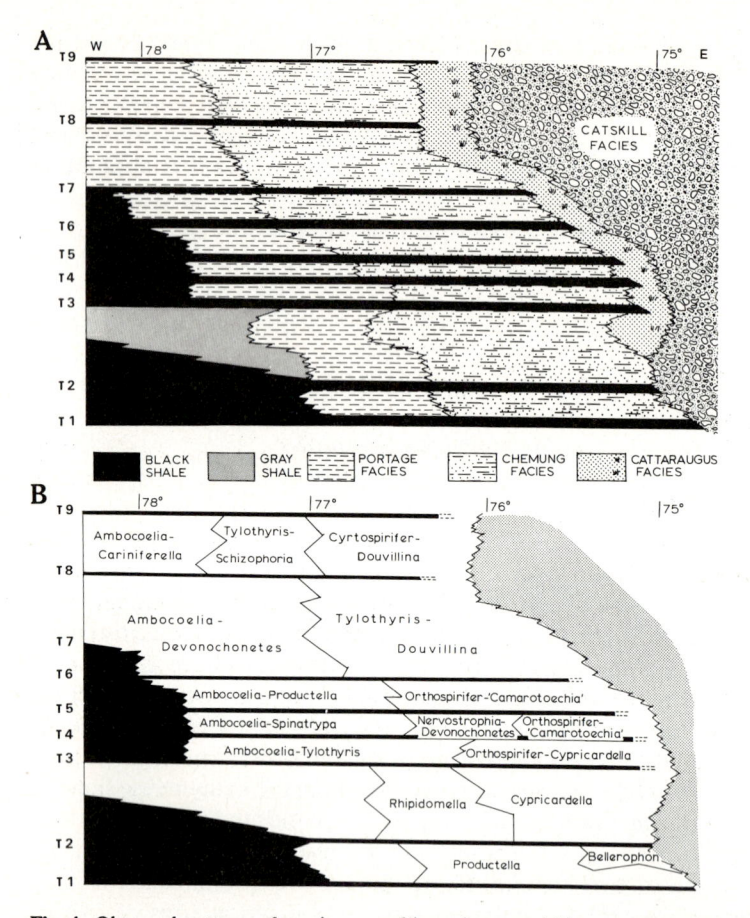

Fig. 4. Observed patterns of species assemblage changes within successive communities with respect to transgressive-regressive cycles. **A** illustrates the progradation of sedimentary facies into the Appalachian Basin (New York State, U.S.A.) during the Frasnian (early Late Devonian). Black shale facies represent deepest water conditions, gray shale, Portage, and Chemung facies represent progressively shallower water conditions, Cattaraugus facies are shore sediments, and Catskill facies are nonmarine terrestrial. The continuous progradation of facies to the west (marine regression) is punctuated by several transgressive events (marked *T* in the *left margin*). **B** gives the geographic distribution of benthic marine communities in New York State within the same interval of time. Each transgressive pulse produced noticeable reorganizations of benthic species assemblages

(McGhee and Bayer 1985; Johnson et al. 1985), which opened migration routes for western North American faunal elements by breaching the mid-continental terrestrial lowland barrier.

3.2 Discussion

Multi-clade species assemblage ("community") responses to transgressive-regressive cycles have been much more extensively studied than morphological responses within species clades. Within the earth's biosphere, shallow water benthic marine ecosystems are generally considered to be the ones most directly affected by a transgressive-regressive event. These ecosystems immediately experience either a shallowing or deepening of average water depths with changes in sea-level, and further experience all of the related physical environmental changes which occur as a result of a change of the bathymetric gradient (depth of wave base, extent of light penetration, etc.).

As multi-clade species assemblages are much more ecologically complex than are ecologically similar and related species within a given taxonomic clade, their behavior is less predictable. There are many parameters whose change can be registered: the diversity (number) of species within the community, their ecological diversity (number of guilds), their dominance (rank structure), the complexity of trophic interrelationships between species, etc.

In the example given above, at least, it would seem that species substitutions and frequency shifts in population densities among ecologically equivalent species are the first responses of benthic marine communities to minor environmental perturbations precipitated by a change in sea-level. Major reorganization of benthic marine communities is induced only by major environmental change; thus multispecies communities may seem to respond directly to the magnitude of environmental alterations. This generalization does not hold, however. Shallow water nearshore communities (which experience major environmental alteration) appear to be more immune to perturbation than the deeper water offshore communities (which experience less environmental alteration) in transgressive-regressive events. Actually (as already suggested by Bretsky 1969), the frequency of environmental alterations experienced by marine organisms may be more influential than their magnitude.

4 Evolutionary Diversity Fluctuations in Response to Transgressive-Regressive Cycles

4.1 An Empirical Example: Biological Diversity Fluctuations in the Devonian of the World

In the previous section we have discussed small scale transgressive-regressive events within a limited span (the approximately 9 Ma span of the Frasnian; Bayer and McGhee 1986). Here we will expand our scale to discuss the entire Devonian history of the Appalachian Basin of eastern North America (a span of some 53 Ma; Bayer and McGhee 1986).

At this scale much larger cyclic patterns of change in environments and biota begin to emerge, patterns which moreover appear to be global in extent (House 1985b). The analysis of sediment accumulation rates in the Appalachian Basin during the Devonian reveals repeated pulsations in sediment influx (Fig. 5). Further, the observed local pulsations appear to correlate global transgressive-regressive events (House 1985b) and with inflections in stratigraphic onlap-offlap curves (Johnson et al. 1985) on a global scale.

The biota within the Appalachian Basin also exhibit evolutionary fluctuations by speciation and extinction, which appear to be in phase with these large-scale sedimen-

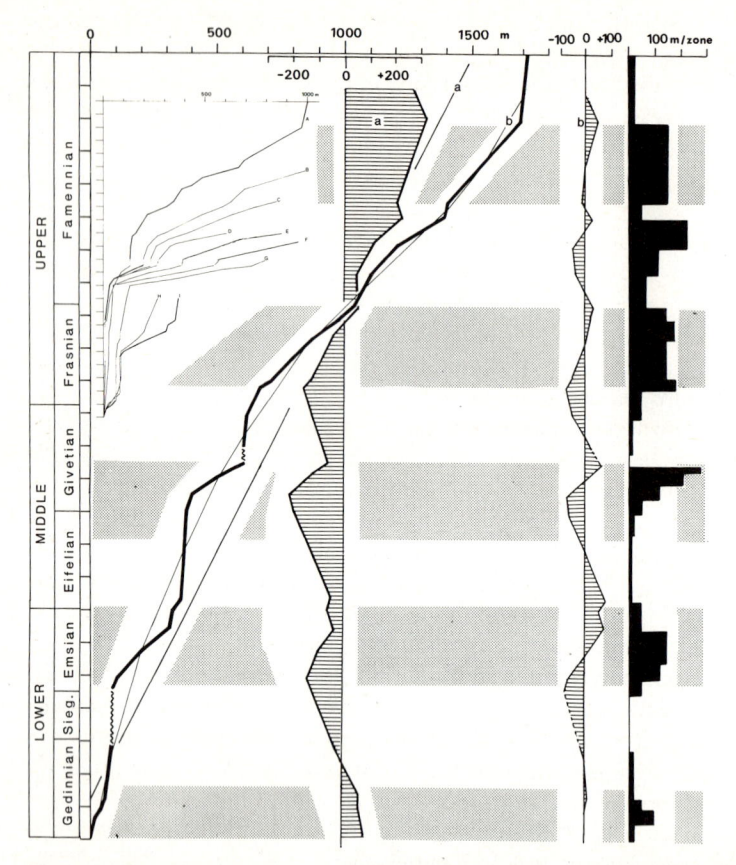

Fig. 5. Analysis of sediment accumulation rates for the New York Devonian. The *heavy line* extending from the lower left to upper right in the figure gives the composite rate curve for the entire basin (computed from local rate curves given in the *inset* in the upper left of the figure, see Bayer and McGhee 1986). The empirical rate curve is then compared with two model curves of basin infilling: constant rate of sediment input (linear function; *thin line a*) and increasing rate of sediment input (nonlinear function; *thin line b*). Both model functions are best fit to the empirical curve, and deviations (−/+ 200 m) of the empirical curve from the linear function are given in the vertical graph in the *center* of the figure (labeled *a*), while deviations (−/+ 100 m) of the empirical curve from the nonlinear function are given in the vertical graph at the *right* (labeled *b*). At the extreme right is given a histogram of sediment accumulation rates per conodont zone. Correspondences in maximum deviation (linear function), deviation reversals (nonlinear function), and pulses in sediment influx at the zonal level are indicated by the *shaded bands* which extend *horizontally* across the figure

tary perturbations. While similar patterns can be seen in several taxonomic groups, the ammonoids are here singled out for discussion as they allow a comparison of local ecological events with global evolutionary ones.

Figure 6 shows the total diversity of ammonoid species as preserved in New York strata during the Devonian. This curve shows periods of expansion and contraction that are approximately in phase with the observed large scale pulsations in sediment

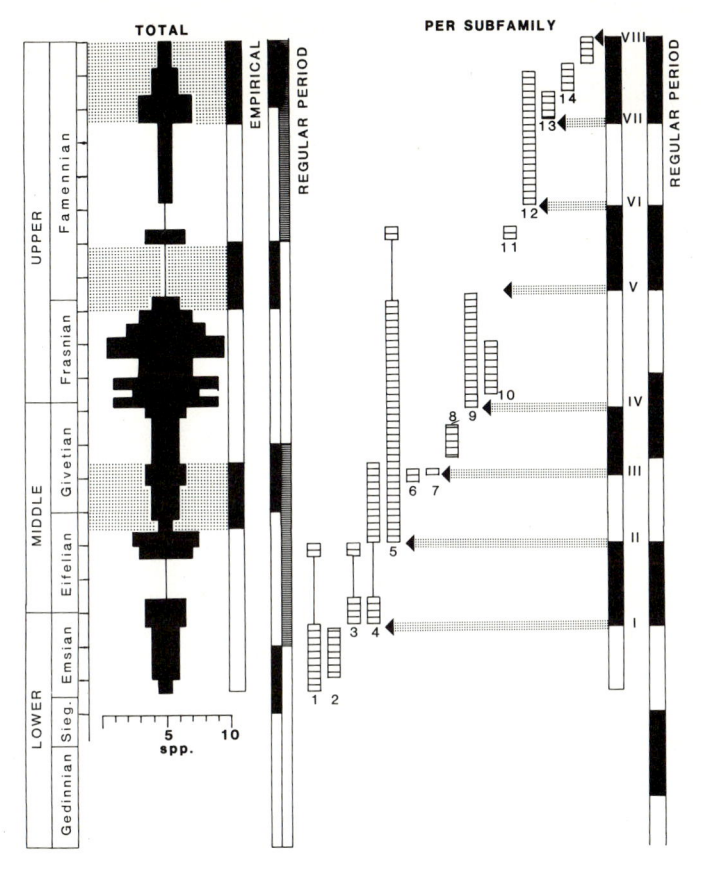

Fig. 6. Observed pattern of ammonoid diversity fluctuations and subfamily substitutions in the Appalachian Basin (New York State, U.S.A.) with respect to the eight global transgressive regressive cycles of House (1985b). On the *left* is given the total ammonoid diversity within the basin with respect to observed ("empirical") and theoretical ("regular period") episodes of diversity expansion and contraction (see Bayer and McGhee 1986), on the *right* is given the pattern of ammonoid subfamily substitutions with respect to the eight global transgressive-regressive events of House (1985b): *I* Daleje event; *II* Kacak event; *III* Taghanic event; *IV* Frasnes event; *V* Kellwasser event; *VI* Enkeberg event; *VII* Annulata event; *VIII* Hangenberg event

influx (Bayer and McGhee 1986). Each pulse in diversity expansion/contraction generally corresponds to the interval spanned by each pulse in high/low sediment accumulation rates (cf. the temporal position and duration of each shaded/unshaded banded couplet in Fig. 5 and the pattern of diversity expansion/contraction in Fig. 6).

The observed sediment perturbations and diversity fluctuations are not simply a local phenomenon, confined to the Appalachian Basin. House (1985b) documents similar physical and biological phenomena on a global scale for the same time interval. Nor is the biological response confined to the ammonoids, in that similar diversity pulses are observed in both benthic brachiopods and nektic conodonts

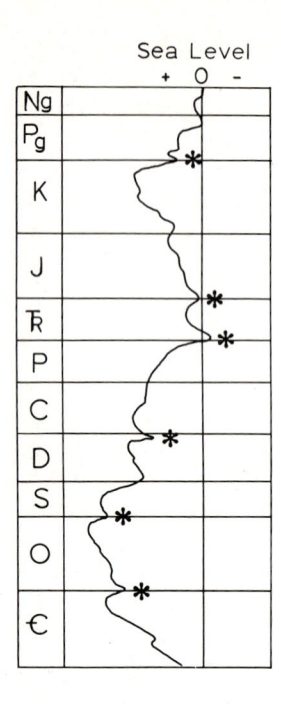

Fig. 7. Correlation between regressive events and intervals of mass extinction in geological time. Excursions of the curve *to the right* indicate regressive phases, excursions *to the left* indicate transgressive phases. Intervals of mass extinction in the geological record are marked with an *. (Hallam 1984; Flessa et al. 1986)

(Bayer and McGhee 1986). Also, the ammonoid diversity pulses are not purely ecological phenomena, but in part reflect evolutionary origination of entirely new subfamilies and the extinction of previous ones. Figure 5 (right side) also shows the temporal duration of the 15 ammonoid subfamilies whose species constitute the total diversity seen within the Appalachian Basin. Their originations and extinctions clearly correspond with the eight "evolutionary events" of House (1985b). As can also be seen, these eight events in ammonoid evolution (which are global phenomena) are correlated with an identical number of local alternations of high/low sedimentation in the Appalachian Basin during the same interval of time (cf. the bar graph in Fig. 6 with the shaded/unshaded bands in Fig. 5).

The sedimentary perturbations in the Appalachian Basin and across the Devonian world are thus correlated with global eustatic pulses (House 1985; Johnson et al. 1985). These global transgressive-regressive pulses produced environmental alterations which in turn triggered the observed evolutionary responses seen in many inhabitants of the marine ecosystem. In general, House (1985) has noted that phases of biological low diversity and/or extinction are often correlated with the spread of euxinic conditions, which is associated with global transgressive phases.

4.2 The Largest Biological Diversity Fluctuations: Mass Extinction Episodes

Mass extinctions are the largest perturbations seen in the geologic record. One consistent feature of mass extinctions is that they affect many environments and are global in scale. Marine as well as terrestrial organisms suffer extinction; within the oceans, both benthic and pelagic ecosystems are affected, but notably the infauna less so than the epibenthos. In sum, mass extinctions are felt in the entire biosphere, not just in the waters of continental margins and in epicontinental seas.

Clearly, a globally pervasive factor must be responsible for mass extinction. One such factor is temperature. Global refrigeration (usually accompanied with glaciation) has recently been proposed as the primary forcing mechanism of mass extinction (Stanley 1987).

Another globally pervasive factor is relative sea-level (Hallam 1981b). Many mass extinctions coincide with periods of world-wide regression (Fig. 7). Since marine regression is, among other causes, related to the formation of polar ice caps during global cooling periods, it may be problematic to deal with temperature decline and regression as separate factors. Still, there are periods in which regression occurred without demonstrable glaciation, so that the causal link between lowered sea-levels and elevated extinction levels is debatable. Lastly, many major regressions had no obvious biological effect. This last observation may be expected, however, as discussed below.

A claimed link between regression and extinction is the species-area effect. Given the hypsographic profile of ocean margins, marine regression would clearly result in the reduction of habitable shelf area. Following predictions of the equilibrium theory of island biogeography, this would reduce population sizes and eventually lead to the extinction of many species. The most serious crisis in the history of life on earth, the Late Permian event, may have been solely due to the habitat reduction effect, since massive marine regressions and low sea-level stands characterize that interval of earth history. Regressions also occurred during the Late Ordovician, Late Triassic, and Late Cretaceous mass extinctions, and perhaps the Late Devonian, though regressions appear to have been minor in scale at the time of the late Frasnian mass extinction.

There are, however, many problems with the species-area scenario. Firstly, regression produces habitat reduction and destruction in the marine realm, but increases the areal extent and habitat space of the terrestrial realm. Yet, in most mass extinctions, terrestrial ecosystems are as severely affected as the marine ones. Secondly, within the marine realm itself, regression produces habitat expansion around oceanic islands, that may harbor high species diversities in modern examples (Stanley 1987). Lastly, even given the highly provincial nature of marine biota today, Jablonski (1986) argues that only 13% of recent families would become extinct if all of the modern shelf biota were eliminated.

Further complications arise concerning the extent of continental inundation and regression. A sea-level drop of given magnitude would have had larger species-area effects during times when the continents were covered with large epeiric seas than during intervals when absolute sea-level stand was low and continental shelf zones were narrow. Thus loss of species diversity due to the species-area effect is not a simple function of regression magnitude, but depends also on the pre-regression condition of the marine biosphere.

A more promising causal link between regression and extinction may exist in climatic effects (Jablonski 1986), though again the pre-regression configuration of land and sea is of major importance. Global climatic extremes may be buffered during intervals when the continents are covered with extensive shallow seas, which may simultaneously warm oceanic waters and dampen seasonal temperature fluctuations on land. Regression would not only destroy this ameliorating effect and accentuate climatic gradients between land and sea. Also, oceanic waters would become colder in the absence of warm water influx from epicontinental seas, and seasonal temperature fluctuations on land would become extreme in the absence of the dampening effect of large shallow water masses. In addition, the larger exposed land area would heighten the albedo of the earth, and would reduce the carbon dioxide content of the atmosphere, due to increasing weathering of exposed land masses. Both effects would contribute to global cooling. In fact, temperature decline is a consistent ecological signal for times of biotic crisis (Stanley 1987; McGhee 1989).

5 Conclusions

Transgressive-regressive cycles involve environmental change. Such changes may be minor in scope and local in distribution, or amount to major alterations in the physical environment of the entire globe. The primary environment to be affected by any transgressive-regressive event is the marine realm, and organisms that feel it most immediately are those of the shallow water epibenthos. As a result, biological response is predicted under the theory of natural selection, and is indeed observed, as discussed in this chapter. Even when sedimentological and biological responses to eustatic changes may at first appear independent, they nevertheless enhance each other in the actual stratigraphic record.

The nature of biological responses to transgressive-regressive cycles is myriad. Examples discussed at the local geographic level have illustrated shifts in morphotype frequencies within species clades of ammonites, and species substitutions and population density fluctuations within multispecies benthic assemblages or communities. At both levels, changes may be due to immigration or truly evolutionary origination/extinction events. These local responses have been triggered by environmental alterations in connection with relative changes of water depth. For such local effects it makes no difference whether we deal with local transgressive-regressive events, or with global eustatic events of minor magnitude. It also makes no difference whether the critical factor was water depth (as in ammonites) or substrate (as in benthic organisms).

Larger-scale eustatic events produce similar environmental alterations but of higher magnitudes and larger geographic ranges. Many of the global "bioevents" (Walliser 1986) correspond to such larger-scale transgressive-regressive cycles, but again it is the marine ecosystem along continental margins and in epicontinental seas that are primarily and most strongly affected. Disruption of the entire global biosphere in periods of "mass extinction" may also correlate with transgressive-regressive events, but in this case the precise forcing mechanism remains problematic.

6.5 Lagoonal-Peritidal Sequences in Carbonate Environments:
Autocyclic and Allocyclic Processes

A. Strasser

1 Introduction

Peritidal sediments form in the intertidal and in the adjacent supratidal and shallow subtidal zones. Much of the carbonate deposited on beaches, tidal flats, and sabkhas, however, is produced in the marine-lagoonal environment, where organic and inorganic carbonate production abounds, and is then transported landward during storms or spring tides (Hardie and Ginsburg 1977). Sediments of shallow lagoonal and peritidal origin are commonly well stratified and display recurring depositional sequences. Vertical stacking of such sequences reflects cyclic processes which controlled formation, transport, final deposition, and early diagenesis of the sediment.

The aim of this chapter is to propose models which explain the observed cyclic sedimentary record, and to present criteria which help to discriminate autocyclic from allocyclic factors (see Einsele et al. Introduction, this Vol.). Processes of sediment production and diagenesis differ fundamentally from siliciclastic to carbonate environments, so that many principles cannot be generally applied. In this chapter, only shallow lagoonal and peritidal carbonate systems are discussed.

2 The Stratigraphic Record

The superposition of different facies in the stratigraphic record reflects environmental changes through time. Organic and inorganic carbonate production is controlled mostly by the physical and chemical properties of the water and by water depth; it is water depth and water energy which define the accomodation potential of the produced sediment. Accordingly, the facies evolution and thickness of a sequence depend mainly on the three parameters which determine water depth: sediment accumulation rate, subsidence rate, and the rate of eustatic sea level change. These parameters are partly interrelated and, directly or indirectly, controlled by climatic and tectonic factors (Fig. 1).

Small-scale depositional sequences in shallow water and peritidal carbonates commonly show a dominant shallowing-upward (regressive) trend (James 1984; Wright 1984). This implies that, at least temporarily, the rate of sediment accumulation outpaces the combined rates of subsidence and sea level change (Table 1; Schlager 1981; Kendall and Schlager 1981). Sediments build up to the water level,

Einsele et al. (Eds.)
Cycles and Events in Stratigraphy
©Springer-Verlag Berlin Heidelberg 1991

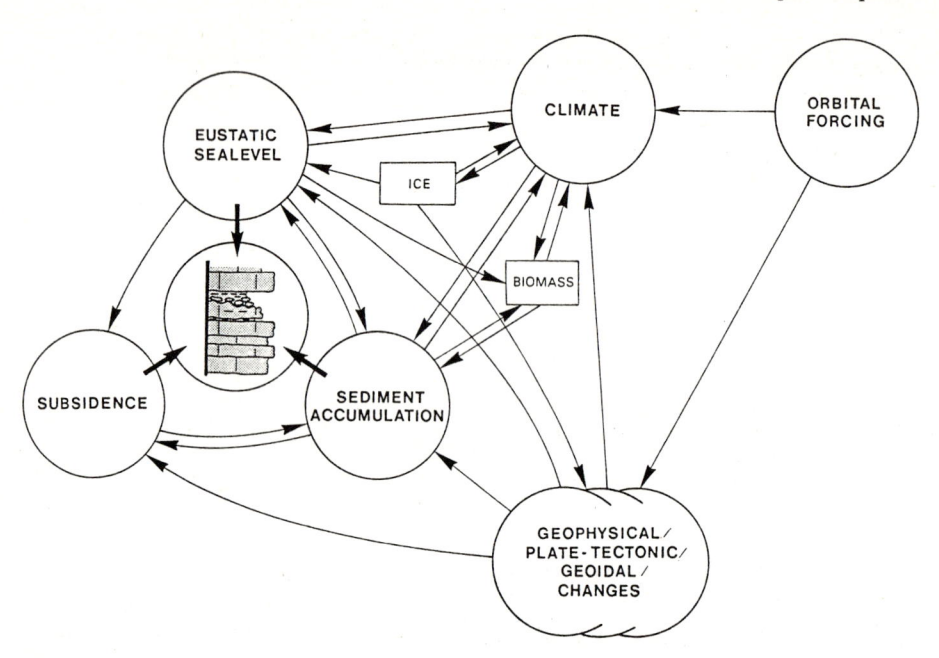

Fig. 1. Interrelationships between parameters controlling eustatic sea level, sediment accumulation and subsidence, which in turn determine the final sedimentary record. Periodicities vary greatly in this very schematic diagram. Feedback effects may be rapid or slow, and their relative importance varies through geographic position and geologic time

passing from subtidal to intertidal facies. If the sediment supply continues, progradation will set in. High-energy events such as storms or spring tides accumulate the sediment above the fair weather high tide level, where supratidal and terrestrial conditions predominate. Wind may redistribute originally marine sediment to form eolian dunes (Strasser and Davaud 1986). Complete sequences contain a thin transgressive bed at their base, which forms when water depth increases but carbonate production is reduced (see below).

The vertical facies distribution in peritidal shallowing-upward sequences is dictated by coastal morphology, water depth, wave and current energy, abundance and type of biota, and climate (Table 2; James 1984; Wright 1984). Facies evolution may go from deeper to shallower lagoonal facies and end with a tidal flat facies (e.g., Shinn 1983), with evaporites and sabkha-type sediments (e.g., McKenzie et al. 1980), with tepee facies (e.g., Kendall and Warren 1987), or with pedogenic features (e.g., Esteban and Klappa 1983). In other cases, birdseyes or keystone vugs overprint facies of shallow subtidal composition and indicate intertidal exposure on beaches or partly emergent bars. High-energy events accumulate sediment in the supratidal zone, where it may be locally submitted to pedogenesis. Some peritidal sequences are capped by green marls (e.g., Fischer 1964; Bechstädt 1979), which formed through repeated wetting by marine waters and subsequent drying (Deconinck and Strasser 1987). A shallowing-upward tendency can equally be expressed by intertidal, supratidal or terrestrial overprinting of deeper subtidal facies, or by subaerial erosion

Table 1. Comparison of some rates of carbonate production, subsidence, and sea level rise. Carbonate production usually outpaces subsidence, but may be slower than short-term sea level rises

Holocene Carbonate Production
(vertical accumulation, no export or import, no compaction)

Environment	Rate (mm/a)	Reference
Coral reef	7–9	Kinsey (1985)
Halimeda bioherm	1–2	Marshall and Davies (1988)
Carbonate mound	2	Bosence et al. (1985)
Ooid shoal	1	Harris (1979)
Sabkha	0.5–1	McKenzie et al. (1980)
Tidal flat	0.3–3	Hardie and Ginsburg (1977)
Algal mud	0.1–0.3	Neumann and Land (1975)
Nontropical shelf	0.05	Collins (1988)
Average for shallow carbonate platforms	1	Schlager (1981)

Subsidence

	Rate (mm/a)	Reference
Mature passive margins	0.05–0.1	Grotzinger (1986b)
Cratons	0.01–0.05	Grotzinger (1986b)

Sea level rise

	Rate (mm/a)	Reference
Long-term	0.01	Schlager (1981)
Short-term (late Holocene)	0.5–3	Schlager (1981)
Short-term (early Holocene, leading to drowning of platforms)	6–10	Schlager (1981)
Short-term (estimated from 1 m ancient tidal flat sequence, decompacted by 50%, deposited in 10 000 years)	0.2	(this chapter)

of the top of the sequences (Fig. 2). Such features indicate a drop in relative sea level, which exposes the subtidally deposited sediments.

In lagoonal settings, the facies of a depositional sequence may remain subtidal, whereas in peritidal environments the sequence can be composed entirely of intertidal to supratidal facies. In some cases, no facies change is visible throughout the sequence, but hardgrounds or a higher degree of bioturbation may indicate a decrease in sedimentation rate at the top of the sequence. Migrating tidal channels, deviated tidal currents, and storms can locally erode the tops of previously deposited sequences.

In the sedimentary record, the above described sequence types commonly form laterally persistent beds and represent the elementary stratigraphic units. As such they are comparable to the "punctuated aggradational cycles" (PACs) of Goodwin and Anderson (1985). Thicknesses vary between a few tens of centimeters and a few meters. These elementary sequences are vertically stacked and may compose larger

Table 2. Typical facies of shallow lagoonal and peritidal carbonate environments which may form shallowing-upward sequences

SUPRATIDAL	eolian deposits soils and calcretes storm and washover deposits levees and beach ridges algal marshes freshwater to hypersaline ponds sabkhas
INTERTIDAL	tidal flats beaches tidal channels
SUBTIDAL	carbonate muds and sands shoals stromatolites reefs

(shallowing-upward sequence)

sequences with deepening or shallowing facies evolution trends (Goodwin and Anderson 1985).

3 Autocyclic Processes

Autocyclic processes leading to repetitive deposition of small, shallowing-upward sequences occur independently of cyclic changes in eustatic sea level and subsidence, or of climatically controlled cycles of carbonate productivity. A classical autocyclic model has been proposed by Ginsburg (1971), based on observations in Florida Bay and on the tidal flats of the Bahamas and the Persian Gulf: sediment produced in open marine environments moves shoreward (e.g., during storms) and accumulates on the tidal flats which begin to prograde seaward. Through this progradation, the open marine source area decreases in size, sediment production slows down, and sediment accumulation is outpaced by continued subsidence. Water depth increases, and a new source area is created. After a certain lag time which allows carbonate-producing

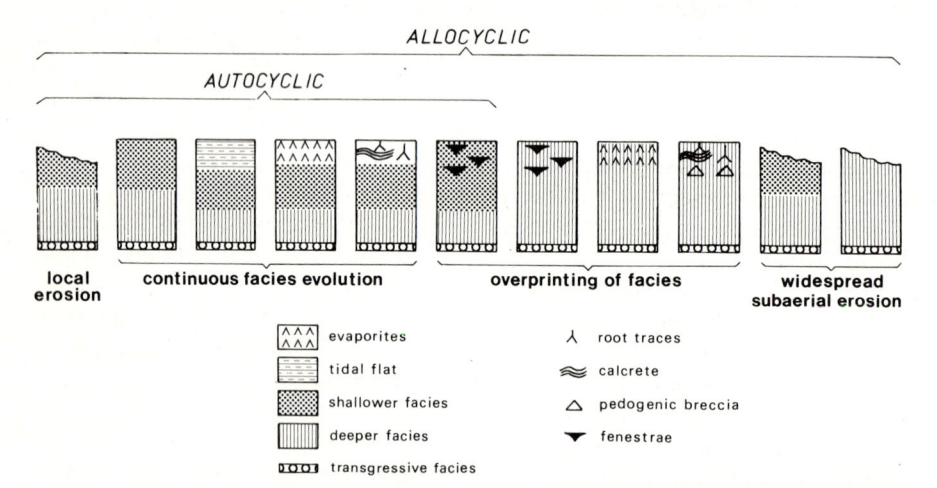

Fig. 2. Possible types of sequences expressing a dominant shallowing-upward tendency in shallow lagoonal to peritidal environments, and their inferred autocyclic or allocyclic genesis

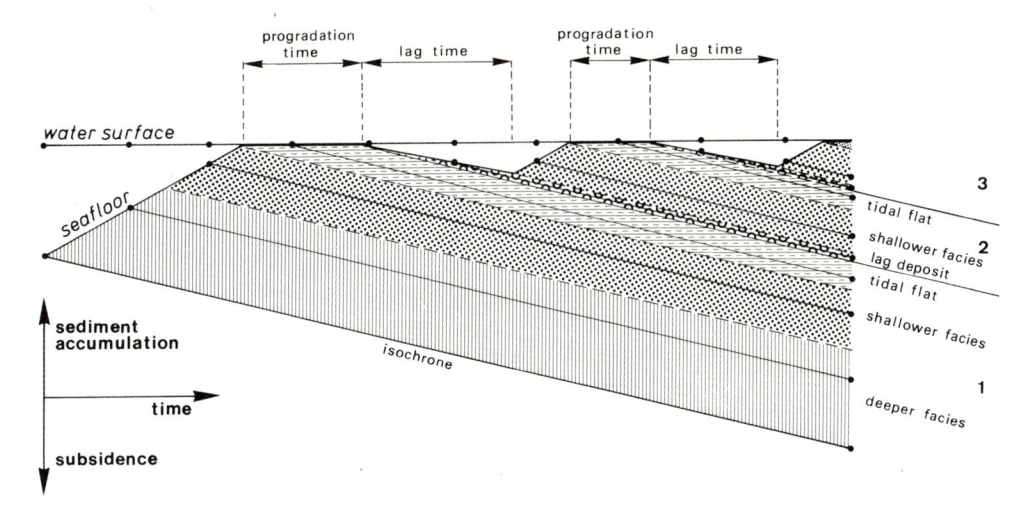

Fig. 3. Diagram illustrating an autocyclic model. Subsidence rate and sediment accumulation rate are assumed to be constant, and accumulation outpaces subsidence. Sea level does not vary. The stratigraphic record shows three shallowing-upward sequences. *Points* on water surface, sea floor, and stratigraphic record mark equal time increments

benthic organisms to colonize the drowned tidal flat, sediment production is renewed, and a new cycle commences. This concept is summarized in Fig. 3. Similar models have been proposed by Bosellini and Hardie (1973), Wong and Oldershaw (1980), and Selg (1988).

Pratt and James (1986) developed a model of local, vertical accretion of islands where tidal flats form and prograde laterally. If the focus of sedimentation shifts due to hydrographic changes, the tidal flat subsides. The focus of sedimentation may then move back over the drowned tidal flat, and a new shallowing-upward sequence is created.

A different autocyclic mechanism has been suggested by Matti and McKee (1976) and by Mossop (1979). Sediment is produced in a shallow water environment, builds up and progrades seaward. As the carbonate-producing area moves in front of the prograding sediment wedge towards deeper water, conditions become less favorable, and carbonate production slows down. Continued subsidence then leads to transgression, until, after a lag time, sediment production starts again in shallow water.

Autocyclic processes are inherent to a given lagoonal-peritidal system. The resulting sequences can be laterally consistent over the length of the system, if seaward progradation is regular. In many cases, however, correlation of sequences even over short distances is not possible (Selg 1988). Areas of aggradation and progradation can migrate laterally and through time (Pratt and James 1986), or sediment can be eroded at one site and accumulate at the same time further along the shore (Shinn et al. 1969; Strasser and Davaud 1986).

4 Allocyclic Processes

Allocyclic processes are controlled by factors external to the sedimentary system (Einsele et al. Introduction, this Vol.). Fluctuations of the eustatic sea level with periods of one to several million years strongly influence shallow and deep water sedimentation and determine the general trend of facies evolution (see Vail et al. Chap. 6.1, this Vol.). The periodicity of laterally persistent, small-scale depositional sequences lies between 10 000 and 400 000 years and is basically related to climatic cycles (Fischer 1986).

4.1 Climatic Cycles

Insolation rate changes with the irregularities of the Earth's orbit. The precession cycle of the equinoxes has a periodicity of about 21 000 years (frequency peaks at 19 000 and 23 000), the obliquity cycle 41 000 years, and the two eccentricity cycles 100 000 and 400 000 years, respectively (Milankovitch 1941; A.L. Berger 1978a). Shorter climatic cycles have been documented by Fairbridge (1976) and Pestiaux et al. (1988). The total amount of insolation and the seasonal insolation changes together with long-term variables such as ocean-land distribution and land relief, determine the climate which influences a given sedimentary system. Carbonate production is favored in warm, shallow, normal marine waters, but is slowed down if colder, brackish, or hypersaline conditions prevail. Climatically induced cycles of rainfall, desertification, and vegetation cover modulate terrigenous run-off which, in turn, may inhibit carbonate production.

Insolation rate equally influences eustatic sea level. Glacial eustacy is an important factor during periods when polar ice is present (Pisias and Imbrie 1986), but in addition, water bound and released by alpine glaciers (Fairbridge 1976), thermal expansion of the upper layers of the ocean (Gornitz et al. 1982), or desiccation of confined basins (Donovan and Jones 1979) can cause climatically driven sea level fluctuations. Ice-rafted blocks at high latitudes indicate that even during the warm Cretaceous, polar ice was present (Frakes and Francis 1988). Furthermore, changes in the mass distribution of ice and water, together with the Earth's orbital geometry, lead to cycles of geoidal eustacy (Mörner 1984).

Cyclic climatic changes control, directly or indirectly through complex feedback paths (Fig. 1), carbonate production and eustatic sea level. Figure 4 presents a greatly simplified model which assumes constant sediment accumulation and constant subsidence. Facies changes occur when a critical water depth is reached. In deeper water, even at low eustatic sea level, the facies stays the same, and the sea level cycle is not directly recorded (Fig. 4a). Lower sea level may, of course, be implied by higher terrigenous input, or by climatically induced changes in carbonate production. In the example presented in Fig. 4b, sediment builds up to low sea level, where a tidal flat develops. Carbonate production slows down, as in the autocyclic model, and eustatic sea level begins to rise. After a certain lag time, carbonate production starts again. If the initial water depth is shallow, or if the rate of sediment accumulation is close to the highest rate of sea level rise, eustatic sea level drops below the level up to which the sediment has been deposited. Consequently, emersion, vadose diagenesis and

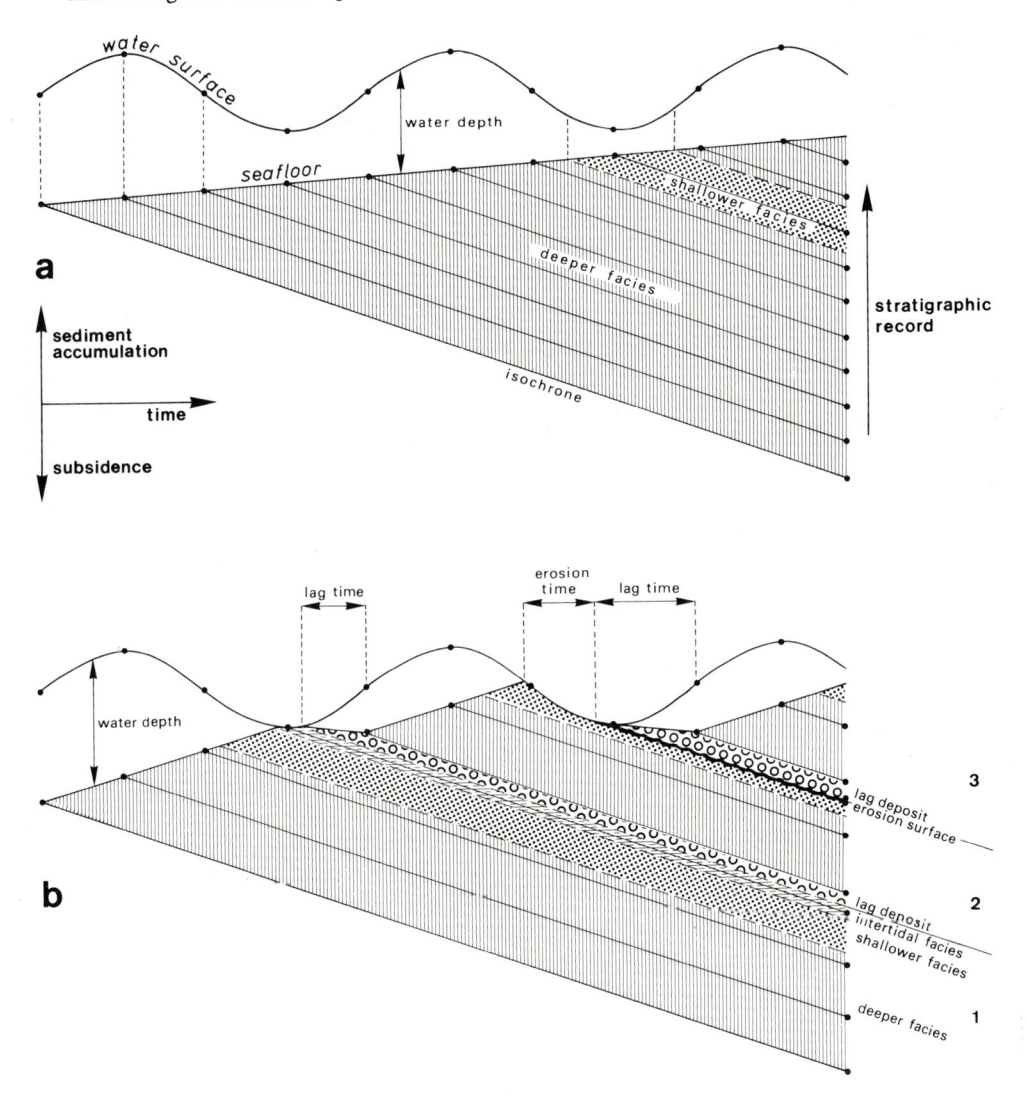

Fig. 4. Diagram illustrating an allocyclic model, assuming regular fluctuations of eustatic sea level, constant subsidence rate, and constant sediment accumulation rate which outpaces subsidence. *Points* on water surface, sea floor and stratigraphic record mark equal time increments. **a** Shallower facies develops if a critical water depth (such as the wave base) is reached. In deeper water sea level cycles are not, or only indirectly, recorded. **b** The sea floor reaches water level where intertidal facies develop or erosion takes place. Three sequences are recorded

erosion occur (top of sequence 2 in Figure 4b). In this case, the lag consists of erosion products and a thin transgressive deposit.

As in the autocyclic model, carbonate production is fully developed at high sea level. Sediment builds up to the water surface and begins prograding. With current and wave action, facies belts migrate laterally (Fig. 5a). Autocyclic processes may

716   Chapter 6 Sequences

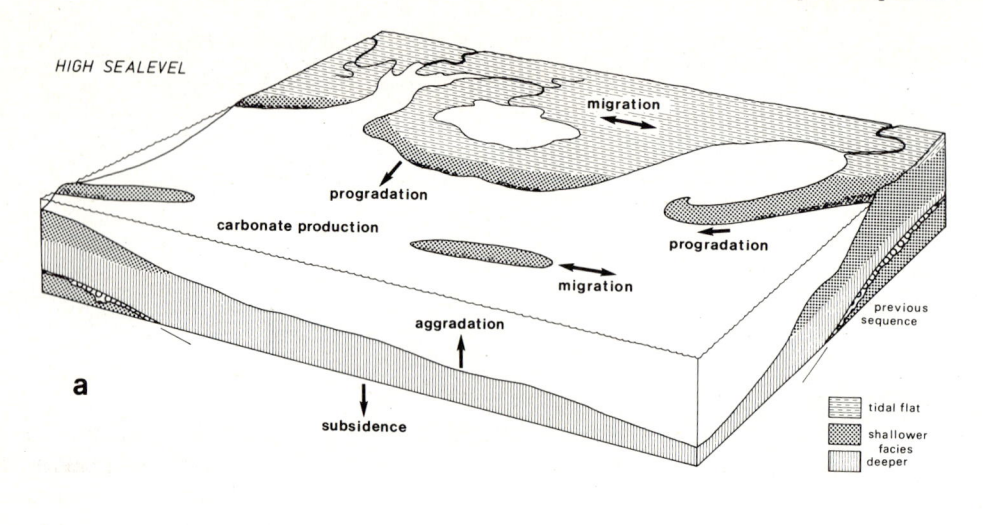

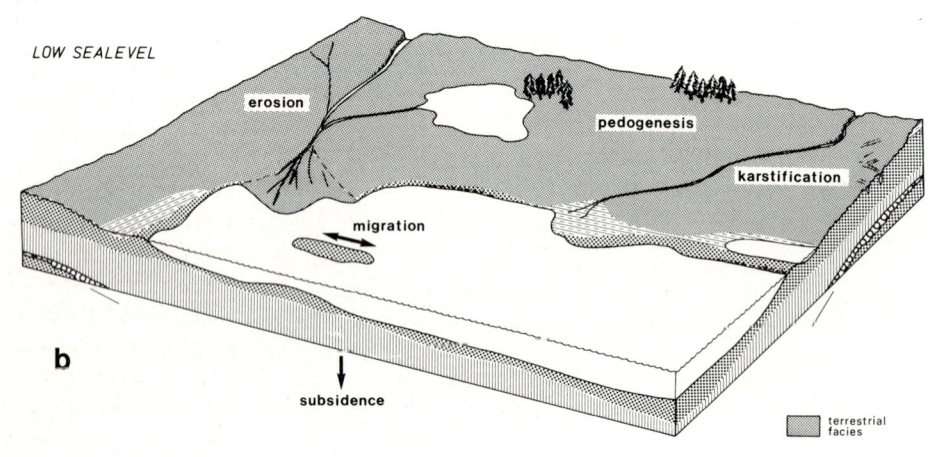

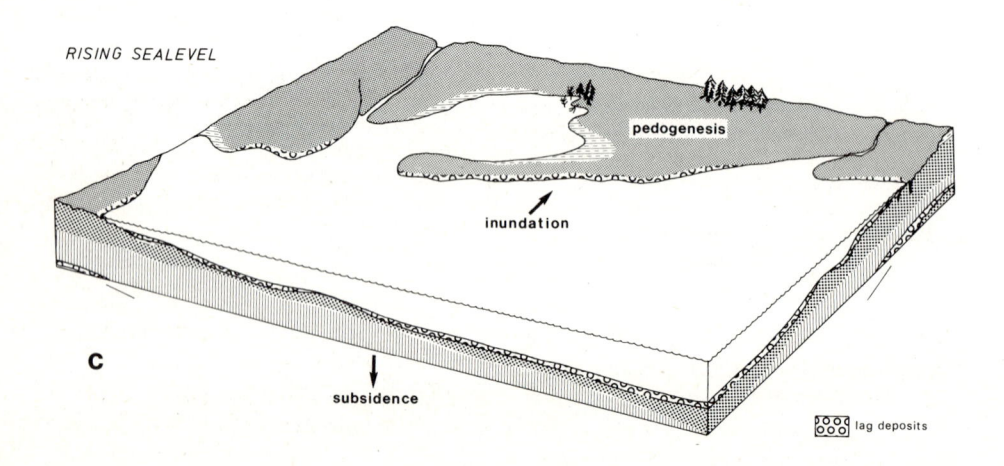

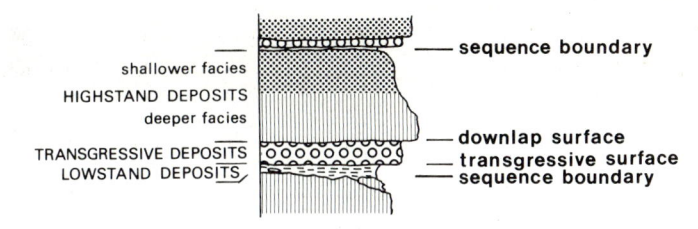

Fig. 6. Small-scale depositional sequence with a dominant shallowing-upward facies evolution. The bulk of the sequence consists of regressive highstand deposits. (Terminology adapted from Vail et al. 1984)

occur locally. Falling sea level exposes parts of the sedimentary system to erosion, karstification and pedogenesis (Fig. 5b). Products of carbonate pedogenesis, such as clay minerals, are washed into the low-lying areas. Terrestrial input may occur from the hinterland and become interbedded with the carbonates (Kendall and Schlager 1981; Mack and James 1986). Nutrient excess may favor plankton growth, which reduces water transparency and thus the growth potential of carbonate-producing organisms such as corals and calcareous algae (Hallock and Schlager 1986). Subsequent rising sea level inundates the previously exposed land. Sediment production is much reduced, and pre-existing material is reworked, leading to lags and thin transgressive deposits (Fig. 5c).

The stratigraphic record resulting from such a sea level cycle is an asymmetric sequence with a thin transgressive and a thick regressive component (see also Einsele and Bayer Chap. 6.2, this Vol.) Glacio-eustatic cycles are characterized by rapid sea level rises and slower sea level falls (Fairbridge 1976; Pisias and Imbrie 1986), which further accentuate the regressive, shallowing-upward trend. Using the definitions of Vail et al. (1984) and Vail et al. (Chap. 6.1, this Vol.), the sequence boundary corresponds to the erosion surface created by sea level drop, or to the base of the lowstand deposits overlying deeper facies (Fig. 6). Commonly, lowstand sediments are missing altogether, and transgressive deposits mark the beginning of a new sequence.

The small-scale shallowing-upward sequences, as elementary units of the stratigraphic record, correspond in many cases to the climatic cycle induced by the precession of the equinoxes (Goodwin and Anderson 1985; Strasser 1988). Sediment production, deposition, and diagenesis required for the making of one sequence are usually much faster than the 20 000 years available. A considerable amount of time is taken up by reworking, nondeposition, and erosion (Fig. 4b). Internal erosion surfaces occur, but are of limited extent (Fig. 7). They are probably related to autocyclic processes such as channel migration.

Fig. 5. Sketches of facies distribution and sedimentary processes in shallow lagoonal to peritidal carbonate environments during high, low and rising sea levels. Most sediment is produced and deposited during sea level highstands (**a**). Terrestrial facies develop generally during lowstands (**b**). Transgressive lag deposits indicate rising sea level, before abundant carbonate production sets in (**c**)

Fig. 7. Example of small-scale sequence displaying a local erosion surface (channel floor ?) and laterally persistent sequence boundaries (Purbeckian of Mount Salève, French Jura)

Due to the complex interrelationship between parameters which control the sedimentary system (Fig. 1), it is difficult to estimate the duration of a lag. Without counting the erosion time, it can probably take from a few hundred to several thousand years. It is equally difficult to propose absolute values for the amplitude of eustatic sea level fluctuations. Depth of the vadose zone marks the lowest point of sea level drop, but nothing is known about the amount of erosion at the top of the sequence when sea level started to fall. However, it can be assumed that sea level amplitudes were in the same range as the decompacted thicknesses of the small-scale peritidal sequences, i.e., around a few tens of centimeters to a few meters.

The 20 000-year cycles are modulated by larger cycles in the Milankovitch frequency band (A.L. Berger 1978a), and by cycles of the million-year scale (Haq et al. 1987). The resulting sedimentary record is a complex vertical stacking, where in some cases the hierarchy of the orbital cycles can be recognized (Fig. 8; Strasser 1988). The composite larger sequences (100 000, 400 000 and million-year scales) commonly are asymmetric and display a dominant regressive trend of general facies evolution. Diminishing water depth during a composite cycle may be expressed by thinning-upward composite sequences (Fig. 8). In many sections, the 100 000-year eccentricity cycle seems to be particularly well developed (Fig. 9), as is also sug-

Fig. 8. Example of highly structured sedimentary record displaying composite peritidal sequences corresponding to the Milankovitch eccentricity cycles. Note thinning-upward tendency of the sequences. The base of Pierre-Chatel Formation is a transgressive surface related to a million-year cycle (Strasser 1988). Many 100 000-year sequence boundaries can be correlated over 200 km (Goldberg Formation, Purbeckian, of Mount Salève, French Jura)

Fig. 9. Well-developed erosion surfaces with calcrete and conglomerates delineating a 100 000-year sequence. Limits of probably 20 000-year sequences are faintly visible (Purbeckian of Col du Banchet, French Jura)

gested from Pleistocene climatic changes (Pisias and Imbrie 1986). In low paleolatitudes, records of the obliquity cycle are often missing (Anderson 1984).

Small-scale shallow lagoonal to peritidal sequences related to cyclic sea level fluctuations have been documented from many geologic epochs and by many authors (e.g., Fischer 1964; Cotillon 1974; Bechstädt 1979; Grotzinger 1986a,b; Hardie et al. 1986; Wright 1986; Goldhammer et al. 1987; Strasser 1988). A study of the Lofer cycles is presented by Haas (Chap. 6.6, this Vol.) Anderson (1984) gives an example of evaporite sedimentation controlled by Milankovitch cycles. Grotzinger (1986a,b), Read et al. (1986), Goldhammer et al. (1987), Read and Goldhammer (1988) and Koerschner and Read (1989) have modeled peritidal carbonate cycles. Statistical analyses of peritidal cycles have been performed by Schwarzacher and Haas (1986).

4.2 Tectonic Cycles

An episodically changing subsidence rate due to large-scale or local tectonic activity is common and can strongly influence the sedimentation pattern, especially if block-faulting and tectonic uplift are involved, and if the rate of subsidence exceeds the rate of sea level change.

Regular, cyclic tectonic processes with a periodicity comparable to that of Milankovitch cycles are more difficult to evoke. Cisne (1986) proposed a model of repeated stick-slip faulting. Carbonate builds up on a platform margin bounded by a fault, strain accumulates until the fault slips, the platform edge rapidly subsides, and the cycle recommences. The resulting sedimentary sequences will be well developed close to the fault zone, but will die out away from the platform margin.

Subsidence rates change due to isostatic response of the crust to eustatic sea level fluctuations (Guidish et al. 1984). Water depth will thus be increased following a sea level rise, and the rebound effect after a regression may accentuate emersion and erosion of the deposited sequence. However, here again it is difficult to estimate amplitudes and delay times.

5 Preservation Potential

In shallow water environments, sedimentary sequences will be optimally preserved if the subsidence rate and drop of eustatic sea level are so adjusted that minimal erosion takes place. Drastic sea level fall or tectonic uplift may cut away several previously deposited sequences. On the other hand, with very low subsidence rates and low-amplitude sea level changes, the accomodation potential may be too small to allow deposition. In such cases, a condensed deposit comprising several sequences will develop.

Freshwater lenses advancing with a prograding sediment body, or forming when a sea level drop has exposed the sediment, may induce rapid cementation, especially in permeable carbonate sands (Halley and Harris 1979; Heckel 1983; Strasser and Davaud 1986). Calcrete caps formed during emersion can further stabilize the sediment. Cemented sequences resist grain-by-grain reworking during transgression and regression, and relic pebbles may point to once deposited, but now eroded sequences. Muddy sediment is less permeable, and cementation is slower. Algal binding and cohesion, however, have a stabilizing effect. Erosion leads to mudclasts and flat pebbles, which can be found reworked in lag deposits.

6 Conclusions

Autocyclic sequences display a shallowing-upward tendency up to intertidal and supratidal facies (Figs. 2, 3). Erosion may occur locally. Widespread erosion, on the other hand, and intertidal, supratidal or terrestrial overprinting of subtidal facies indicate a drop of sea level (Fig. 4b; Grotzinger 1986b). Long-time emersion leads to well-developed sequence boundaries and to pervasive cementation and diagenesis.

It is well established from the Quaternary record that orbitally driven climatic cycles controlled sea level fluctuations during that time (Hays et al. 1976; Pisias and Imbrie 1986). There is no reason to believe that such cycles did not exist in older geologic periods. The amplitude of sea level changes during times when no polar ice was present was probably much reduced, as compared to glacial eustacy, which can attain more than 100 m. On very flat lagoonal-peritidal settings, however, even a small sea level change can have significant effects. Most small-scale shallowing-upward sequences probably formed during sea level oscillations of less than 10 m (Grotzinger 1986a,b).

Sea level fluctuations and climatically controlled cycles of carbonate production and terrigenous input superimpose local or regional episodic subsidence and local, autocyclic processes. The amplitude of the respective processes and their interference define the final stratigraphic record, and in many cases it is difficult to determine the relative contribution of each mechanism (Burton et al. 1987). However, if sequences and erosion surfaces can be correlated over large distances and from one sedimentary system to another, and if the sequences display a vertical stacking which reflects the hierarchy of Milankovitch cycles, a predominantly climatic-eustatic control can be assumed. Calibration of the sequences by biostratigraphy or magnetostratigraphy is, of course, indispensable.

Shallow lagoonal to peritidal depositional environments with all their associated facies are well suited for precise recording of even small changes in the parameters controlling sedimentation. Consolidation of the deposited sequences is rapid and permits a relatively good preservation potential, although in some cases sequences may be condensed, and facies can be determined only from reworked pebbles. Distinguishing autocyclic from allocyclic sequences may often prove difficult if criteria indicating widespread overprinting or erosion are absent (Fig. 2). Detailed analysis of the facies and sedimentary structures in each depositional sequence, good biostratigraphic or magnetostratigraphic time control, and basinwide bed-by-bed correlation are needed to determine to what extent autocyclic and allocyclic processes contributed to the cyclic stratigraphic record.

Acknowledgments. I thank Alex Waehry and Eric Davaud for the long and fruitful discussions in the field and behind the desk, and for critically reading a first version of the manuscript. I am grateful to Werner Ricken for his valuable comments and for editorial help. Research leading to the concepts presented in this paper has been supported by the Swiss National Science Foundation (Project No. 2.897.083).

6.6 A Basic Model for Lofer Cycles

J. Haas

1 Introduction

Sander (1936) was the first to recognize small-scale cycles in the Upper Triassic Dachstein Limestone (the Austrian Loferer Steinberge), and described a regular alternation of laminated dolomitic limestones with thick bedded limestones, the Lofer facies. He explained this cyclicity as having been caused by sea level changes, and

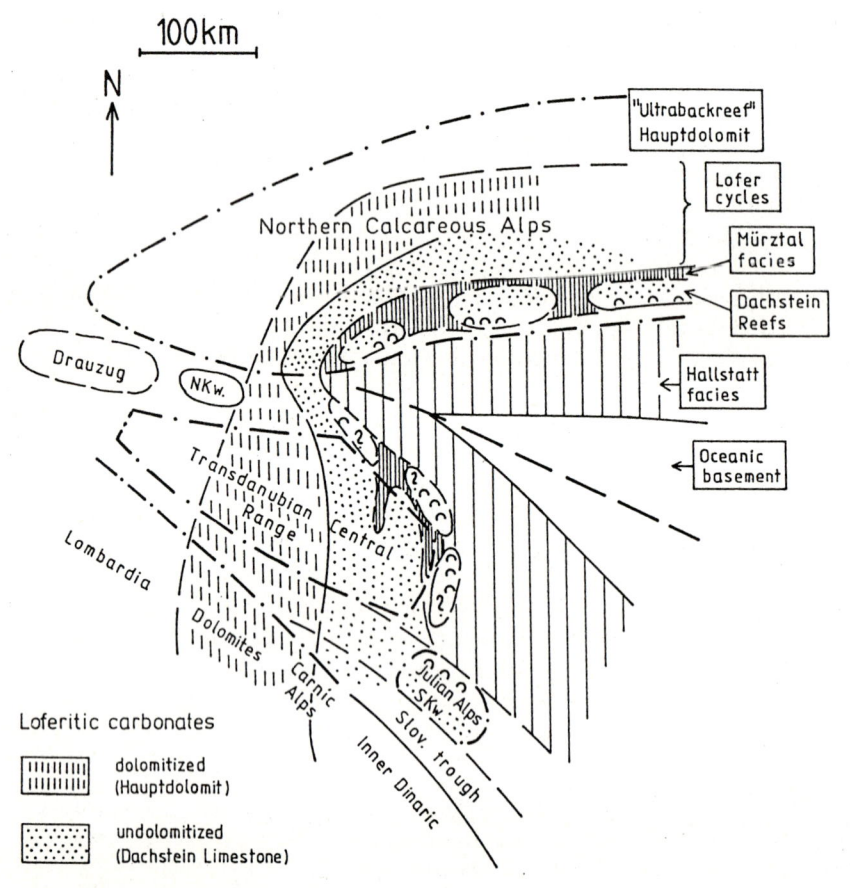

Fig. 1. Reconstruction of the facies zones on the Tethys shelf in the Lower to Middle Norian

Einsele et al. (Eds.)
Cycles and Events in Stratigraphy
©Springer-Verlag Berlin Heidelberg 1991

assumed that the dolomitic laminites were formed in deeper water under the wave base. Schwarzacher (1947, 1954) recognized an overall pattern in the smaller-scale sequences. The exact definition, characterization, and modern genetic interpretation of the Lofer cycles were performed by Fischer (1964), through his study of the Dachstein Limestone sections in the Northern Calcareous Alps of Austria.

The characterization of the Lofer cycles performed in the present investigation is based mainly on studies carried out in the Transdanubian Central Range in Hungary; however, when similar observations from the Northern Calcareous Alps, Southern Alps, Inner Carpathians, and Hellenids are taken into account, a basic model for the explanation of the Lofer cycles can be applied to most of the Upper Triassic carbonate platforms on the western Tethys margin (Fig. 1).

2 Main Lithologic Features of the Lofer Cycles

As Fischer observed (1964), an ideal cycle is composed of the following sequence (Fig. 2):

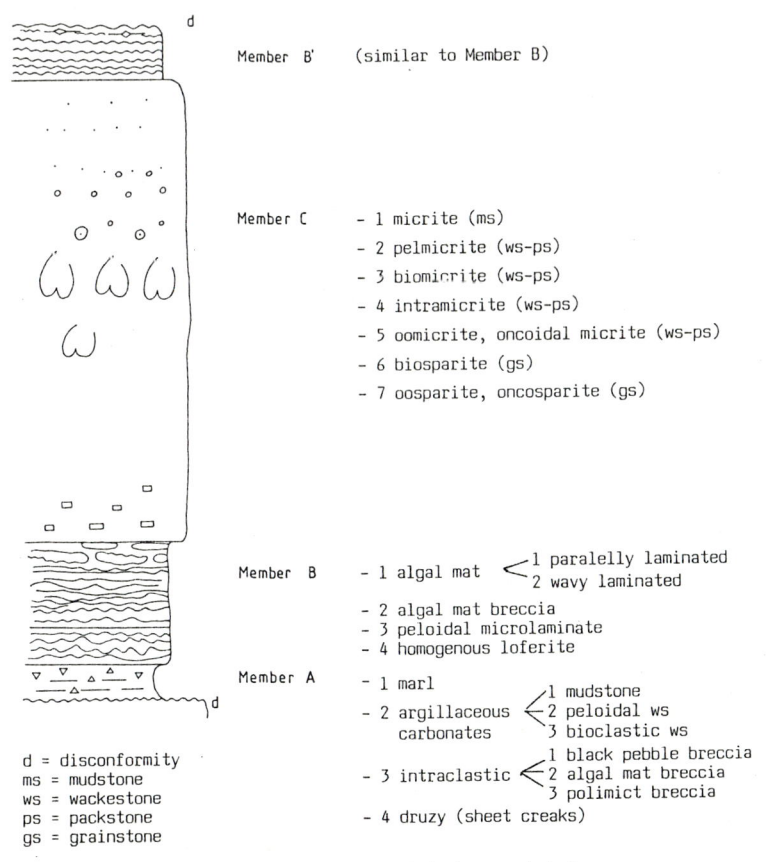

Member B'   (similar to Member B)

Member C
- 1 micrite (ms)
- 2 pelmicrite (ws-ps)
- 3 biomicrite (ws-ps)
- 4 intramicrite (ws-ps)
- 5 oomicrite, oncoidal micrite (ws-ps)
- 6 biosparite (gs)
- 7 oosparite, oncosparite (gs)

Member B
- 1 algal mat <1 paralelly laminated
               <2 wavy laminated
- 2 algal mat breccia
- 3 peloidal microlaminate
- 4 homogenous loferite

Member A
- 1 marl
                           <1 mudstone
- 2 argillaceous   <2 peloidal ws
     carbonates   <3 bioclastic ws
                           <1 black pebble breccia
- 3 intraclastic   <2 algal mat breccia
                           <3 polimict breccia
- 4 druzy (sheet creaks)

d = disconformity
ms = mudstone
ws = wackestone
ps = packstone
gs = grainstone

Fig. 2. Members of the ideal Lofer cycle and their characteristic features

1. Member C: massive calcarenite with rich marine biota – subtidal facies.
2. Member B: laminated carbonates containing abundant fenestral pores – intertidal facies.
3. Member A: red to green argillaceous intraclastic mudstone (frequently present as cavity infillings) – supratidal facies.
4. Disconformity at the base (d).

Investigation of cores from deep boreholes in the Bakony Mountains (Hungary) has revealed that, in many cases, cycles contain an additional Member B above Member C and below the disconformity at the base of the overlying cycle. This was interpreted as representing the regressive portion of the cycle (Member B'; Haas 1982).

Some characteristic subfacies within each member have already been described by Fischer (1964). A summary of these subfacies, based on Fischer as well as on observations by the present author, is presented in Fig. 2. A paleo-environmental interpretation of these subfacies can be seen in Fig. 3.

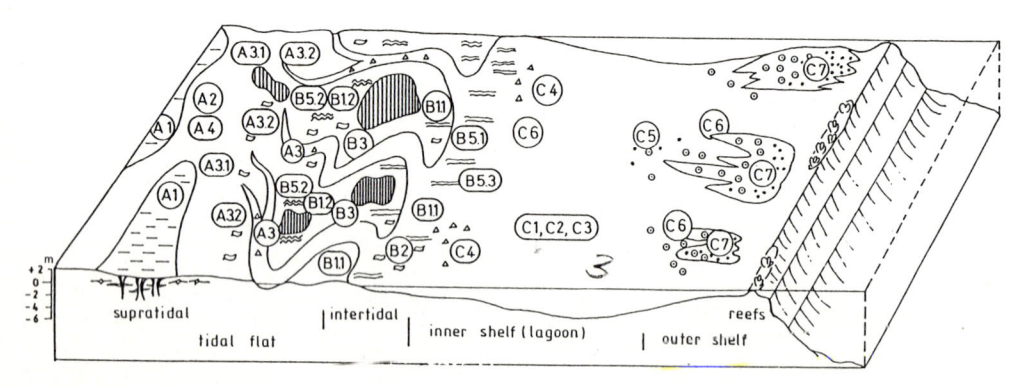

Fig. 3. Paleo-environmental interpretation of facies types in the Lofer cycle members. (For description of symbols, see Fig. 2)

3 Sequential Types of Lofer Cycles

Systematic study of Lofer cycles has revealed that in addition to the ideal cycle composed of the layers d-A-B-C-B', alternative sequences may appear as well. Observations indicated that the sequence in certain formations is predictable and can be characterized by dominant types of cycles. To confirm these observations, approximately 500 cycles were studied.

Types of sequences were identified (Fig. 4), and their frequency within the studied formations was determined (Fig. 5).

The results can be summarized as follows:

1. The Carnian-Lower Norian Hauptdolomit Formation is characterized by the predominance of d-B-C sequences. The appearance of Member A is rare, while Member B is generally thick (although in some cases it is difficult to distinguish between B and B'.

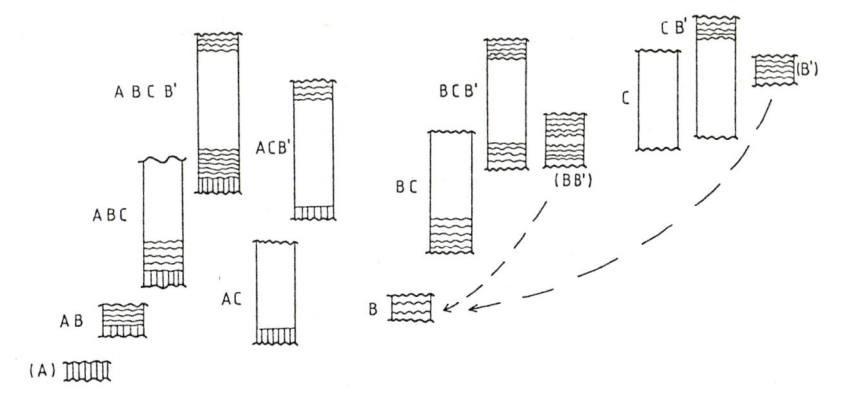

Fig. 4. Theoretical variations of the Lofer cycles. (The difficult to distinguish variations are in parenthesis)

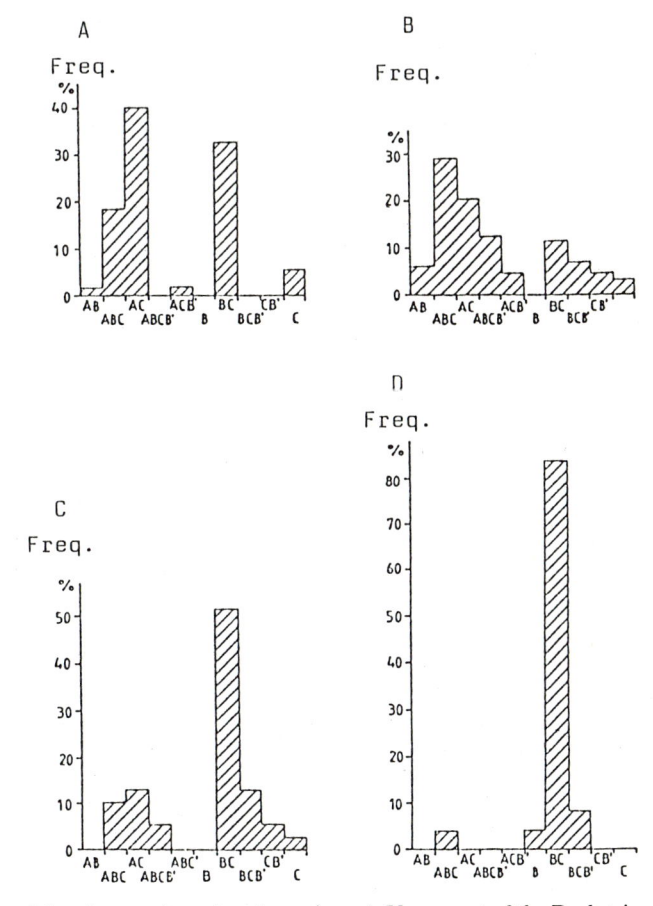

Fig. 5. Frequency of the sequential cycle types in various formations. **A** Upper part of the Dachstein Limestone. **B** Lower part of the Dachstein Limestone. **C** Hauptdolomit-Dachstein Limestone transitional unit. **D** Hauptdolomit

2. A transitional unit several hundred meters thick between the Hauptdolomit and the Dachstein Limestone (Norian) shows transitional features in its sequence. While d-B-C sequences prevail, sequences of d-A-C and d-B-C-B' may also be present.
3. In the lower part of the Dachstein Limestone Formation (Norian), Member A is present at the base of the cycles as a rule, and Member B beds are frequently missing. Therefore, the d-A-C sequence type is dominant, while d-A-B-C and d-A-B-C-B' variations are subdominant. However, almost every combination can be found.
4. In the upper part of the Dachstein Formation (Rhaetian), d-A-C sequences are most common; nonetheless, sequences composed of d-B-C and d-A-B-C are moderately frequent. However, the thicknesses of A and B beds are not more than 10 to 20 cm.

4 Problems of Dolomitization

A significant part of the formations showing Lofer cycles is dolomitized. It is evident that this dolomitization was synsedimentary or occurred during early diagenesis.

Intertidal Member B is generally dolomitic. Even the otherwise undolomitized Dachstein Limestone contains anywhere between 5 and 80% of dolomite in the Member B beds. This synsedimentary selective dolomitization can be related to the activity of blue-green algae (Shinn 1983).

During the final regressive phase of the cycle, selective or total dolomitization of Member C and further dolomitization of Member B (or B') can take place under suitable conditions. This process may be explained by the sabkha model (McKenzie et al. 1980). In the case of the Dachstein Limestone, however, the process of sabkha dolomitization could not occur due to frequent flooding of the supratidal zone. In the transitional unit between the Hauptdolomit and the Dachstein Limestone, dolomitization generally began but then was interrupted. Complete dolomitization of the Hauptdolomit can be explained by intense postsedimentary, early diagenetic dolomitization in the upper supratidal zone which flooded only rarely.

5 Facies Model for Lofer Cycles

The formation of the Lofer cycles is the result of facies migration on a carbonate platform approximately 200 km wide. A general facies model is depicted on Fig. 6, which is based on stratigraphic correlation (Fig. 7), sequence identification, microfacies studies, and observations on the distribution and intensity of dolomitization.

A wide peritidal flat was present, with a subtidal inner shelf or lagoon separated from the open sea by organic reefs or calcarenite mounds (see Fig. 1). Within this scenario, Lofer cycles formed as a result of facies migration due to relative sea level fluctuations (see Strasser Chap. 6.5, this Vol.).

According to this model, divergent development of the lithologic units was caused by variation in facies migration.

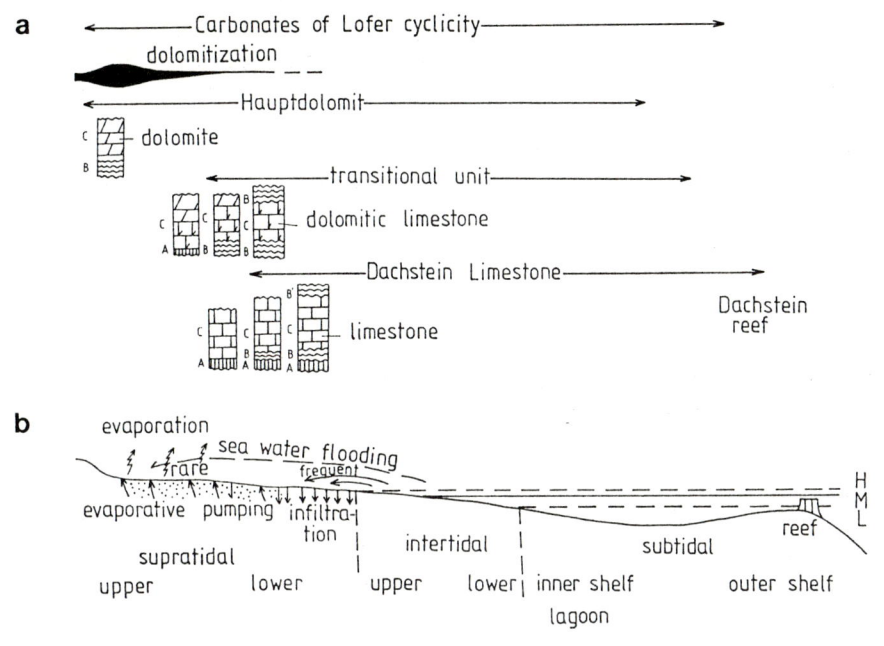

Fig. 6. General genetic model of the Lofer cycle carbonates

Sedimentological features of the lithologic units can be interpreted as follows:

1. Hauptdolomit Formation. During periods of high sea level, sedimentation took place on the inner shelf. During regressions, the upper supratidal zone migrated into this area (1–3 m above the mean sea level). This zone was rarely flooded by sea water, only during strong storms, and consequently the process of sabkha dolomitization was pervasive. Frequent truncation of the cycles and the absence of Member A can be explained by erosion of the semilithified carbonate in the upper supratidal zone.

2. Dachstein Limestone. Sedimentation extended from the reef front to the lower part of the supratidal flat. During periods of low sea level, the lower part of the supratidal zone was frequently flooded by sea water, which hampered early diagenetic dolomitization. Analysis of the cycles reveals that complete cycles are most completely preserved in the Dachstein Formation, particularly in the lower part of Dachstein Limestone. This indicates that during regressions, sedimentation occurred primarily in the intertidal zone (Member B'), and that the length of the regressions might have been limited. The frequent absence of Member B (algal lamination) may indicate rapid transgression, or bioturbation.

3. In the transitional unit between the Hauptdolomit and Dachstein Limestone, some cycles show complete dolomitization, while in others dolomitization is incomplete and selective. This reflects an intermediate state in the sedimentary environment.

4. The general model outlined above explains conditions under which the Dachstein Limestone, the transitional unit, and the Hauptdolomit were time-equivalent facies types. This situation was realized mainly in the Early Norian. A slightly different model

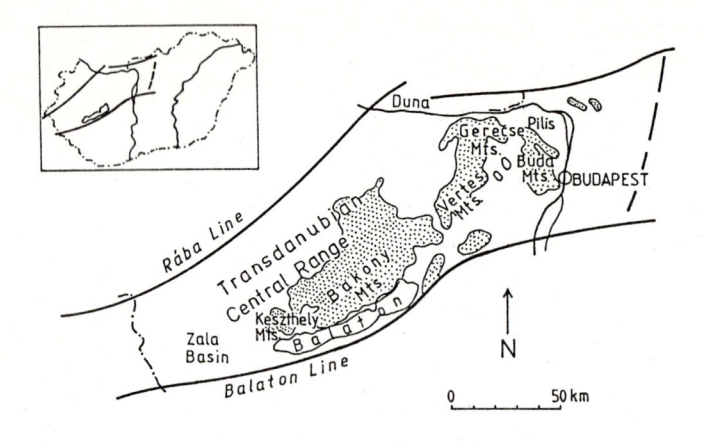

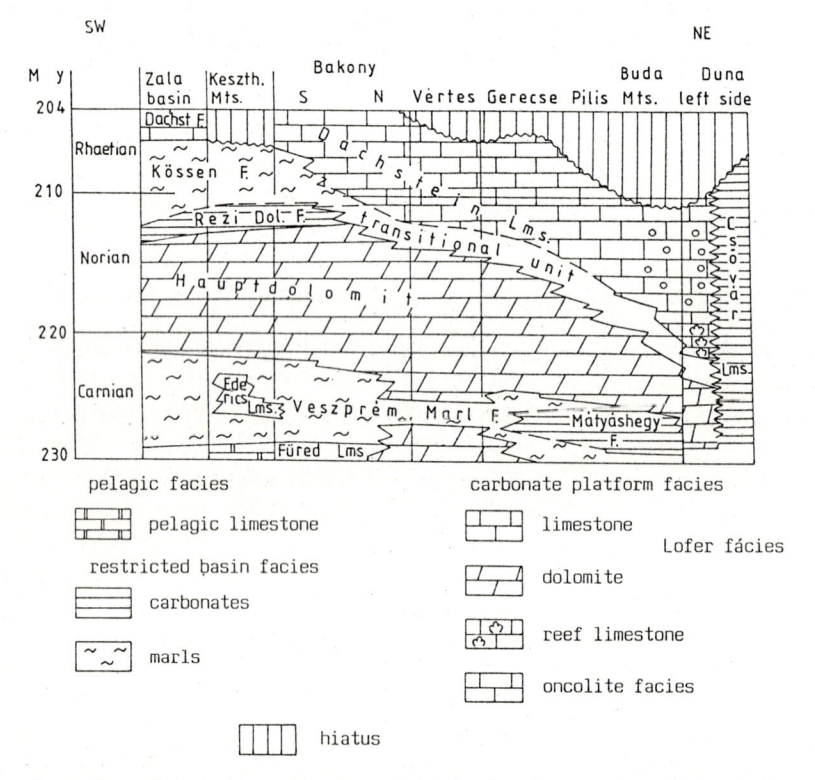

Fig. 7. Upper Triassic lithostratigraphic units of the Transdanubian Central Range. (a) Location of the Transdanubian Central Range, Hungary. (b) Connection of the main lithostratigraphic units along a profile approximately perpendicular to the facies zones

can be outlined for the period when Lofer dolomites were formed over a large part of the shelf ("Hauptdolomit phase" – Late Carnian) and for the period when the Dachstein Limestone became ubiquitous (Middle Norian).

In the "Hauptdolomit phase", during periods of elevated sea level, most of the shelf was covered by a shallow lagoon, separated from the basin by oolitic mounds. During sea level regressions, the peritidal zone prograded into what was previously the lagoon area; at sea level minima, sabkhas formed on the surface of the lagoonal sediments (Fig. 6).

The "Dachstein facies model" is characterized by a reef-fringed platform, with oolitic to oncolitic sand bodies and patch reefs in the outer shelf zone, and muddy, peloidal, bioclastic, green algal facies in the inner shelf area. It is assumed that during regressions, progradation of the peritidal zone occurred to the inner shelf, while a significant part of the outer shelf remained in the subtidal zone. Those Dachstein Limestone successions which do not show Lofer cycles accumulated here.

6 Origin of the Lofer Cycles

Facies analysis has clearly shown that cyclic sedimentation of lagoonal-peritidal carbonates may reflect relative sea level changes. At the same time, there was intensive subsidence over the area. However, carbonate accumulation was able to keep pace, and so an extremely thick peritidal-shallow marine sequence formed. There are two basic models (autocyclic and allocyclic) for cyclic sedimentation (see Einsele et al. Introduction, and Strasser Chap. 6.5, both this Vol.). It is possible that due to the fairly regular nature of the Lofer cycles, allocyclic sedimentation was of prime importance.

However, some elements of Ginsburg's (1971) autocyclic progradational model can be considered. During periods of low sea level, a decrease in the areal extent of the lagoon results in a decrease in carbonate production and consequently in the sedimentation rate (see Strasser Chap. 6.5, this Vol.). This leads to a new transgression, which is followed by further autocyclic progradation. This process may have contributed to cyclic facies changes in the Lofer sediments, but it could not have been the regulating mechanism. For example, in some cases lateral correlation of the cycles is possible for a distance of nearly 10 km (Fig. 8), an additional argument for the mainly allocyclic nature of the Lofer cycles.

Statistical analysis of Lofer cycles from the Transdanubian and Alpine sections (Schwarzacher and Haas 1986) revealed an average thickness of 3.1 m for the Lofer cycles in the Dachstein Limestone (standard deviation 2.1 m) in the Transdanubian Central Range, and 4.0 m (standard deviation 2.8 m) in the Eastern Alps. In a sequence representing 16 Ma (± 6 Ma), or 2000 m (±500 m), of Norian and Rhaetian deposition in the Transdanubian Range, a mean duration of 24.7 ka/cycle was calculated. It must be noted, however, that the majority of the cycles are eroded, which means that their original thicknesses were in fact greater. Thus, a time span for the Lofer cycles of 20 to 50 thousand years seems an appropriate estimate, representing the frequency band of Milankovitch cycles. Thus, the 21 000 year precession period, or 41 000 year obliquity cycle, may be associated with the basic cyclicity of the Lofer facies. As

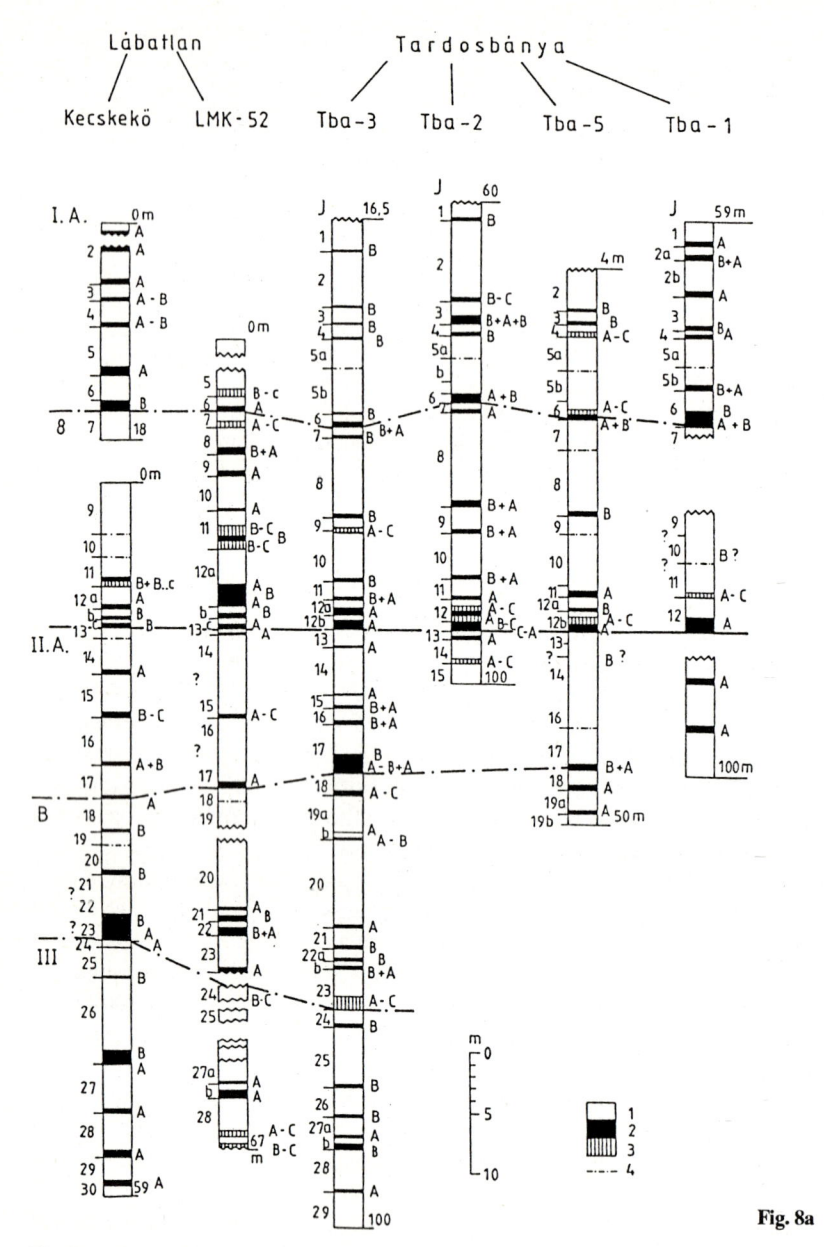

Fig. 8a. An example of correlative cycles in the Gerecse Mountains. Borehole locations are depicted on the inset map. *1* Member C. *2* Members A and B (A+B). *3* Transitional features between A and/or B and C (A-C; B-C). *4* Boundaries between larger cycles

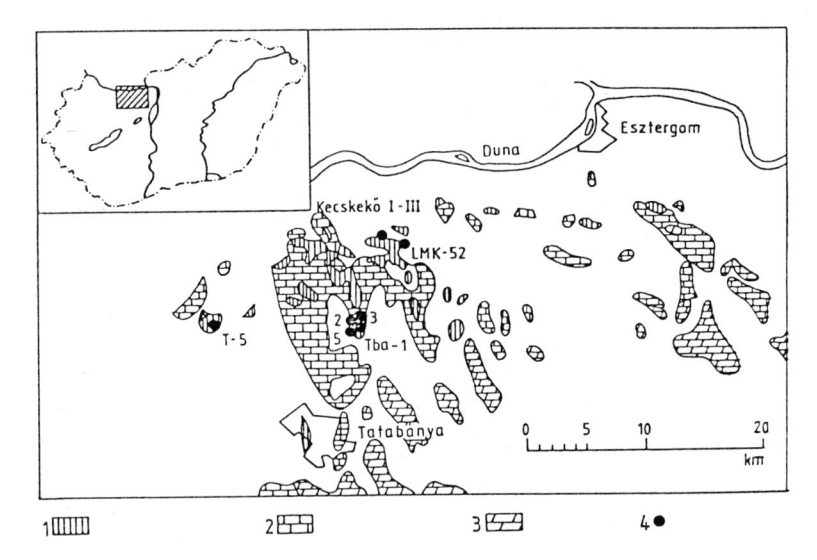

Fig. 8b

discussed by Strasser (Chap. 6.5, this Vol.), Milankovitch cycles influence global climate and small-scale sea level variations.

Larger, superimposed cycles have been described by Schwarzacher in the Alps (1954) and can be observed mainly in the upper part of the Dachstein Formation in the Transdanubian Central Range. Larger cycles are separated from each other by extremely pronounced disconformities and thick, basal Member A beds (Fig. 8). They consist of five to seven basic Lofer cycles, and might be correlated with the Milankovitch eccentricity periods related to significant climatic anomalies.

7 Conclusions

Lofer cycles are lagoonal-peritidal cycles that are a highly common feature in extremely thick, widespread carbonate formations on the western margins of the Triassic Tethys. They are several meters thick, and consist of asymmetric to symmetric facies sequences. These sequences are related to relatively small-scale sea level variations, which resulted in considerable lateral facies migration on the wide (10's to 100's of km), shallow carbonate shelf. The average rate of carbonate sedimentation was able to keep pace with a high rate of subsidence. The ideal, transgressive cycles overlying an erosional unconformity consist of a supratidal argillaceous basal layer, which is followed by intertidal algae laminites, followed in turn by a calcarenitic subtidal bed.

The more symmetric transgression-regression cycles end with a second intertidal unit, before being truncated by the unconformity. Cycles preserving only the transgressive units are thought to occur in the landward zone of the carbonate shelf, as found in the Hauptdolomit. The more symmetric cycles, however, are typical of the shelf lagoon, as in the Dachstein Limestone. From these ideal types of cycles, many derivations and combinations of the subunits exist.

Lofer cycles are organized in bundles of five to seven individual cycles that are separated by relatively thick supratidal beds. Individual Lofer cycles can be correlated over distances of more than 10 km. Periodicities are estimated to range between 20 and 50 ka, suggesting orbital control, though autocyclic processes are believed to have contributed to the cyclicity.

6.7 Coal Cyclothems and Some Models for Their Origin

W. Riegel

1 Introduction

Coal-bearing sequences were probably the first in which the phenomenon of cyclicity was recognized. This early recognition of cyclicity was certainly inspired by the sharp contrast in color and composition of lithic units. In addition, fossil biota in coal cyclothems often show the most rapid changes between marine and nonmarine environments. Thus it is not surprising that these sequences attracted much thought and speculation concerning the causes of their origin soon after their constant appearance over wide areas and their potential for correlation in the Carboniferous were realized. Originally, the desire to find an instrument for local and interregional correlation may have been an incentive to find external, possibly even global factors for their origin. Later, however, with the rapid development of sedimentology, the idea of sedimentary control of coal formation became so prevalent that the term "cyclothem" was almost forgotten. Rahmani and Flores (1984) even divided their recent historical review of the "sedimentology of coal and coal-bearing sequences" into a "cyclothem era" and "post-cyclothem innovations". However, with the advent of cyclic and event stratigraphy as an accepted modern discipline, the term cyclothem and the discussion of causes of their origin have been revived.

2 Review of the Cyclothem Concept

Udden (1912, p. 27) is commonly cited as having first recognized the occurrence and significance of cyclic deposits in coal-bearing sequences in his study of the Pennsylvanian system of the North American midcontinent. However, Westoll (1968) points out that Forster (1809) already clearly implied the recognition of regular repetition in coal-bearing sequences of the North Pennines (England), probably relying on centuries of previous mining experience.

Nevertheless, Udden (1912) was clearly the first to suggest that transgressive-regressive marine cycles were responsible for the repetition in coal-bearing sequences of the Upper Carboniferous. Working in western Illinois, Weller (1930) and Wanless and Weller (1932) expanded on the role of these cycles, stressing their great lateral continuity and practical use in correlation, and introduced the hypothesis of diastrophic control. Later, Wanless (Wanless and Weller 1932) formally introduced the term "cyclothem", which quickly found acceptance for cyclic sequences of Upper Carboniferous coal-bearing sediments.

The term cyclothem was proposed by Weller (Wanless and Weller 1932) "to designate a series of beds deposited during a single sedimentary cycle of the type that prevailed during the Pennsylvanian period. A cyclothem ranks as a formation in the scale of stratigraphic nomenclature". The term was introduced to replace the term "formation", since "formation" is usually applied to beds of uniform character in important respects.

Appropriately, the region in which the cyclothem concept was first formulated has been the basis for continued fundamental discussion and further development of the cyclothem concept. Regional differences in the coal-bearing Pennsylvanian of North America led to the distinction of different types of cyclothems. Cyclothems in Kansas, first described as such by Moore (1936), included a much greater proportion of limestones with marine fossils in addition to marine shales, and suggested a much more lasting influence of marine conditions than was indicated by the Illinois cyclothems described by Weller (Fig. 1). Early attempts at defining a prototype of coal cyclothems led Weller (1930, 1931) and Wanless and Weller (1932) to propose the "typical Illinois cyclothem", while Moore (1936) suggested the Kansas cyclothem, as represented by the Wabaunsee Group, as an "ideal cyclothem" from which all Midcontinent cyclothems can be derived by intermittent addition or loss of individual units (Fig. 1). These types are made up of nine or ten units. The main difference between the two is in the greater prominence of channel sandstones at the base of the typical Illinois cyclothem and the occurrence of three to five marine limestones above the coal in the ideal Kansas cyclothem. The environmental interpretation attached to the lithic units in Fig. 1 and the sea level curve derived from them indicate a difference in marine influence between Illinois and Kansas.

The concept of an ideal cyclothem clearly implies the assumption of control by external forces effecting an orderly succession of events as well as corresponding beds. All variations have been thought to be caused by local topography and have been related to this succession of events, which are connected to marine pulsations (Moore 1936).

Since the occurrence of marine limestones, especially those containing the typical offshore marine fusulinids, has been considered to represent the peak of transgression and thus the turning point of a cyclic event, Moore (1936) suggested that each of the limestone beds in Kansas may represent such a transgressive culmination and, together with adjacent thin shale beds, is equivalent to a complete typical Illinois cyclothem. The original Kansas cyclothem containing four to five marine limestones and being bounded by thicker plant-bearing sandy shales and sometimes coals would then represent a sequence of smaller individual cyclothems. For this Moore (1936) introduced the term "megacyclothem", which was widely adopted by other authors and applied to other regions, such as the Ruhr area in Germany (Jessen 1956a,b). The distinction between different categories of superimposed cycles certainly stimulated further discussions and hypotheses regarding global control of cyclothem formation, though Moore and other early workers on Pennsylvanian cyclothems (Wanless and Weller 1932; Weller 1930; Udden 1912) restricted their genetic considerations to sedimentary and especially tectonic controls.

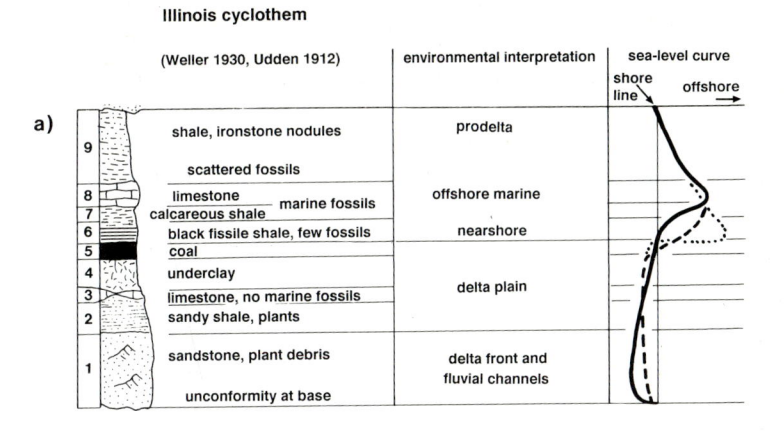

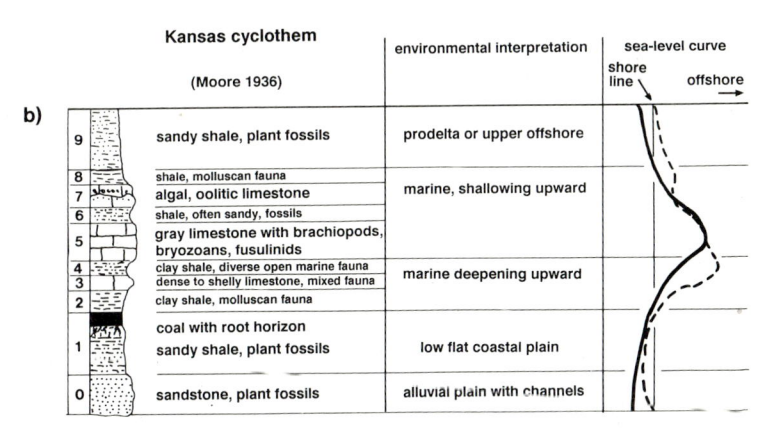

Fig. 1. Examples of "typical" respectively "ideal" cyclothems as inferred originally from the Pennsylvanian of the North American midcontinent (After Heckel 1984). Facies description abbreviated, environmental interpretation this chapter. Sea level curve: *solid line* original interpretation; *dotted line* alternative proposed by Heckel (1984); *dashed line* alternative proposed here. **a** "Typical" Illinois cyclothem of Weller (1930). Note that according to the interpretation as a deltaic sequence this cyclothem should begin with the marine transgression rather than a minor local unconformity at the base of a fluvial channel. **b** Kansas cyclothem proposed as "ideal cyclothem" by Moore (1936)

A major innovation in the cyclothem concept was introduced by the assumption of orbital influences and controls such as the Milankovich periodicity in the earth's rotation and resulting solar energy influx (Van Leckwijck 1948; Wanless and Patterson 1952, see Einsele and Ricken Chap. 1.1, Fischer Chap. 1.1, both this Vol.). This was carried to an extreme by some authors in Europe (Jessen 1956a), as synchroneity of cyclothems was implied.

For some time in the 1940's and 1950's, the discussion of orbital control was overemphasized, which made it difficult for many coal geologists to recognize cyclic deposits in modern analogs. In a somewhat different context, however, this discussion has been renewed in the light of the recent recognition of Milankovich cycles. Eustatic glacial control of cyclothem formation likewise implies global synchroneity of cyclothems and of the onset of regressive-transgressive events underlying them. However, in contrast to orbital processes, the effects of glacial eustatic control on sedimentation of coal-bearing sequences can be demonstrated by modern examples (Spackman et al. 1969), and possibly provide a basic for correlation in the Pennsylvanian and early Permian.

The application of the results of the Mississippi Delta investigations by Fisk (1960) and his school to the coal-bearing sequences in the Pennsylvanian of the Appalachian region (Ferm 1970) focused the interpretation of Pennsylvanian and younger coal-bearing sequences on sedimentary processes, facies shifts, and autocyclic effects. The term "cyclothem", previously burdened by the discussion of orbital as well as glacial eustatic influences, apparently came into disrepute and has rarely been applied since to designate cyclic coal-bearing sequences. It has additionally been recognized, based on increasing sedimentological and environmental proof, that considerable variation of quite unrelated situations could be involved in producing cyclic sequences and that there is no unifying principle justifying the term and concept of a cyclothem.

Perhaps it is time for the term cyclothem and its conception to regain their original meaning, stripping them of later metaphysical preoccupations, and demonstrating their usefulness in designating a multi-elemental sequence of beds that is repeated many times and is commonly associated with certain paleogeographic settings and crustal conditions.

3 Types of Cyclothems

3.1 Cyclothems from the Pennsylvanian of North America

Continuing studies of cyclic deposits in various areas and time intervals have revealed a broad spectrum which apparently involved different sedimentary, paleogeographic, and tectonic settings. Even within the Pennsylvanian of the Illinois Basin it could be demonstrated by Wanless et al. (1969) that deviations from the standard cyclothem are not merely due to lateral facies variation in progradational coastal sedimentary wedges.

The distribution of the first coal seam in the Pennsylvanian of the Illinois Basin, the Rock Island coal, is clearly controlled by paleotopographic features (Wanless et al. 1969). The coal was formed from peat swamps developing in ancient valleys during a relative rise in the water table in the adjacent sea. Subsequent transgression led to the superposition of the coal by marine limestone (Fig. 2, Sect. F). This shows considerable similarities with the recent situation in the Florida Everglades, and contrasts sharply with the fluvial control and deltaic progradation apparent in the later and more common Illinois cyclothems. As in Illinois, paleotopographic control also appears to be more pronounced in the Early Pennsylvanian of Western Pennsylvania (Williams and Bragonier 1974), while fluvial-deltaic sedimentation is more effective

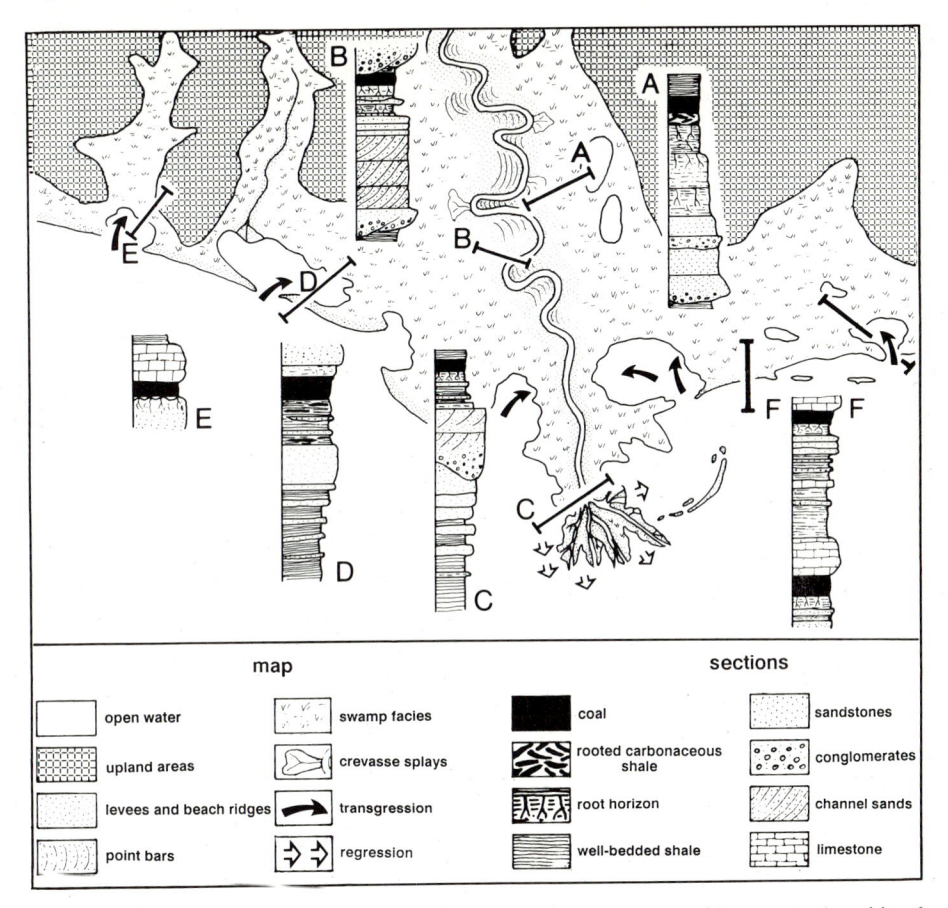

Fig. 2. Schematic representation of various types of cyclothems associated with different depositional environments in the fluvial-deltaic coastal plain complex. *A* Alluvial plain succession with overbank deposits (crevasse splays, levees), rooted flood plain clays, back swamp deposits (carbonaceous shales and coals) and well-bedded lacustrine shales. *B* Fluvial channel facies with lag deposits, point bars, overbank deposits and thin coals. *C* Lower delta plain succession with prodelta clays, coarsening upward sequence of delta front sands, channel fill and delta plain sediments including thin coals. *D* Prograding shoreline succession with basal marine shale unit including lower shoreface sands, littoral unit, lagoonal unit and coal. *E* Succession developed by marine transgression over topographic relief with laterally continuous disconformity on top of underlying unit, superceded by coal and marine carbonate respectively shale member. *F* Carbonate shoreline succession with open marine limestone immediately overlying thick coal and shallowing upward sequence above

higher up in the succession. Cyclic development in paleo-relief-controlled coal-bearing sequences is limited, as the effect of the paleo-relief is reduced by continued sedimentary build-up.

Other coal seams and underlying facies described by Wanless et al. (1969) show little or no relation to fluvial or deltaic control; for example, the Colchester coal, which is very widespread from Pennsylvania to Kansas and uniform in lateral extent and thickness. It probably resulted from the gradual drowning of a very broad coastal plain on which the topographic relief was completely obliterated by sedimentation. Many of the Kansas-type coal cycles fall into this category.

In the Appalachian region (i.e., especially in Pennsylvania, West Virginia, and Kentucky), due to the greater mobility of both the rising orogenic belt and the subsiding foredeep, cyclothem development is more varied and shows a greater dominance of coarse clastics and fluvial control. The extensive and classical sedimentological work carried out during the 1960's and summarized, e.g., by Donaldson (1974) and Ferm (1974), concentrated on the demonstration of lateral facies gradation rather than vertical cyclic repetition. Thus, variability was more emphasized than regularity. Nevertheless, a general trend in the make-up of cyclothems from the proximal alluvial facies exposed in small outliers within the Appalachian valley and ridge province to the lower delta plain facies in western Pennsylvania and eastern Ohio (Ferm and Cavaroc 1968) has been confirmed for the Allegheny Group (Middle Pennsylvanian).

Coal cycles are dominated by gravelly sands fining upward and followed by rooted seat earth, very thin coals, and some silty shales in the most proximal portions of the alluvial valley. Sequences attributed to alluvial plain deposition are most widespread in the Allegheny Group of central West Virginia. Here, channel fill and back swamp deposits form two types of succession that replace one another laterally. The former is made up of fining-upward sandstones showing a sharp erosional base. The basal massive beds are succeeded by festoon cross-beds and rippled beds. Backswamp deposits are characteristically represented by a sequence of seat earth, coal, shale with abundant plant fossils, siltstone, and sandstone, the latter resulting from overbank flood deposition (compare sections A and B, Fig. 2).

Sequences attributed to upper delta plain deposition occur northwestward in a zone reaching from southwestern Pennsylvania to northeastern Kentucky. The main difference to the alluvial plain sequences is the greater thickness of shales and coals in the back-swamp facies and the occasional occurrence of marine and brackish beds. Sandstones are often flat-bottomed and show less scouring at their bases. They have been variously interpreted as distributary mouth bar sands, as well as levee and crevasse splay deposits.

In sequences arising from lower delta plain deposition in Ohio and northwestern Pennsylvania silty and clayey interdistributary bay deposits often including marine limestones of considerable thickness (up to 7 m) are the most widespread components and distributary mouth bars make up a greater proportion of sandstone bodies. Coal seams are distinctly thinner than in the upper delta plain facies (e.g., section C, Fig. 2). Distally delta front sandstones can be replaced by orthoquartzite bodies which result from the reworking of delta front sands into beach barriers, washover, back-barrier, and tidal-delta deposits. Cyclic sequences including extensive orthoquartzites of beach barrier origin associated with coals have been described from the lower Pennsylvanian of Alabama by Hobday (1974). Their structure and origin seem to be similar to cyclothems known from the Cretaceous Western Interior Seaway.

Cyclothems described from the Carboniferous of Northern England, commonly referred to as Yoredale cycles, and from the Namurian to Westphalian of the Ruhr district in Germany are similar to those exemplified by the Illinois and Appalachian cyclothems and reflect strong deltaic influence. Closer sedimentological interpretation of the Ruhr cyclothems was previously impeded by the assumption of an ideal complete cyclothem, with all deviations from it being local irregularities (Jessen 1961). A trend from more distal deltaic to upper delta plain facies, however, can be recognized within the Westphalian.

3.2 Cyclothems of the Cretaceous Western Interior Seaway

A distinctly different type of coal cyclothem was first described in detail by Young (1957) from Late Cretaceous cyclic sequences exposed in the Book Cliffs, eastern Utah (USA). It consists of four lithologic units (compare section D in Fig. 2):

1. Marine shale unit, consisting of even-bedded gray mud- or claystone with calcareous ironstone concretions and rare macrofossils, but frequent foraminifera and ostracods, grading upward into
2. Littoral marine unit, consisting of medium- to fine-grained sandstone of wide lateral extent and showing bedding features typical of beach deposits.
3. Lagoonal unit, composed of a mixture of coals, silty/sandy carbonaceous shales and siltstones, poorly sorted sandstones with fresh- to brackish-water fossils and plant remains, occasionally also with mud- or sandstone-filled scours.
4. Coal unit with a seam thickness up to 7 m which locally rests directly on the littoral marine sandstone.

This cyclothem represents a mode of facies change occurring widely in the Upper Cretaceous of the Rocky Mountains area. Here, along the Western Interior Seaway, the clastic wedges prograding from the rising cordillera have been reworked into barrier beaches by longshore currents. The sequence represents a single progradational trend beginning with normal marine shales and grading into foreshore, shoreface, dune, lagoonal, and peat swamp deposits.

More recently, Ryer (1983) summarized results of several papers dealing with transgressive-regressive cyclic sequences in the Cretaceous of the Western Interior and, in expanding on Vail et al. (1977b), recognizes four orders of cycles in the Cretaceous succession of central and southwestern Utah. According to this, the Upper Cretaceous sequence shown in Fig. 3 represents one large transgressive-regressive cycle (second order cycle) which is part of the still larger Zuni Sequence (first order cycle). Three distinctive transgressive-regressive cycles of the third order can be recognized in the section shown in Fig. 3. They are marked by the transgression of the Tropic Shale, the Bluegate Shale, and the Masuk Shale. All three shale members are westward-protruding tongues of the continuous Mancos Shale sequence in the basin separated by sandstone units of delta front or barrier beach origin.

Superimposed upon the three third order cycles are many minor transgressive-regressive movements of the shoreline represented by the sawtooth edges of the delta-front and shoreline facies in Fig. 3. These fourth order cycles occur throughout the Cretaceous of the Western Interior (e.g., Peterson 1969; Ryer 1977) and result from the stacking of littoral units during both the transgressive and the regressive maximum of third order cycles.

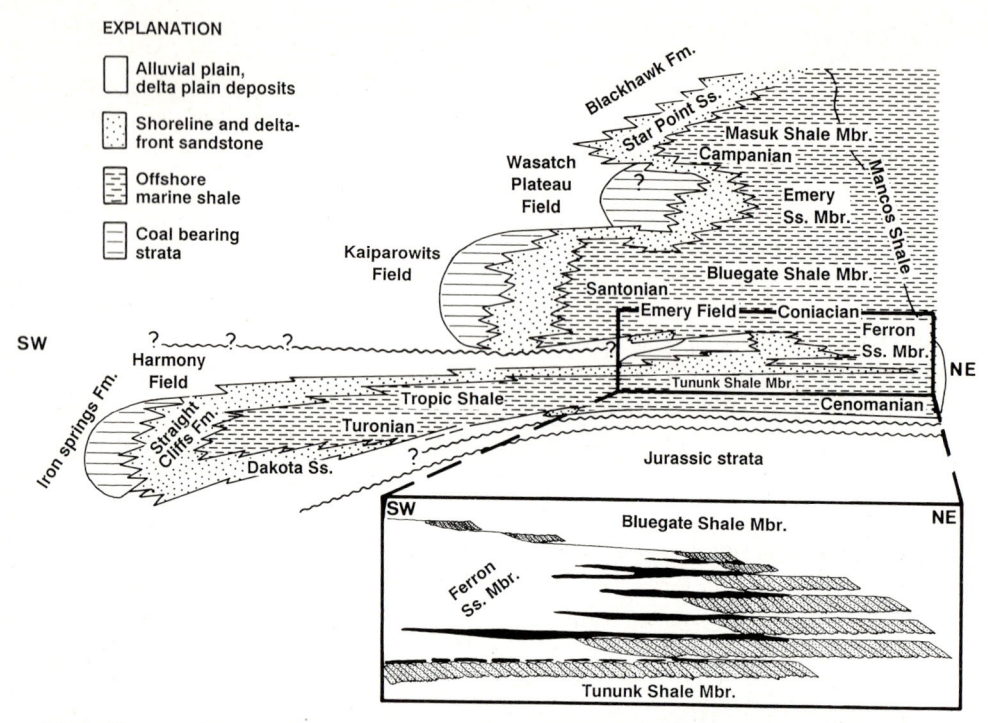

Fig. 3. Diagramatic cross-section of Cretaceous strata of central and southwestern Utah showing third and fourth order transgressive-regressive cycles of sedimentation (after Ryer 1983). *Insert* showing details of vertical stacking of fourth order cyclic shoreline deposits formed during the regressive maximum of the Ferron Sandstone. (Ryer 1981)

The Ferron Sandstone (Fig. 3), separating the lower and the middle of third order cycles, is a good example of vertical stacking of shoreline deposits at the regressive maximum. Here, Ryer (1981) has documented a one-to-one correspondence between fourth order cycles of the Ferron sandstone and thick coal seams of the Emery coal field in central Utah.

As Ryer (1983) points out, coal formation is conspicuously lacking during the major episodes of transgressions and regressions responsible for third order cycles in the Western Interior Seaway. Major transgression pulses apparently proceed at rates too fast to allow for peat buildups, while major regressions are accompanied by a general lowering of base level likewise detrimental to extensive peat formation. Since coal formation generally requires a gradual rise of water table and protection from erosion by channel cutting or wave action, Ryer's argument that major coal deposits are preferentially associated with prolonged periods of a slow relative rise of sea level and areas in which this is compensated by sedimentary progradation, is well grounded. Only in these situations can delta lobe switching or barrier beach progradation lead to the vertical stacking of sedimentary units and the development of coal cyclothems.

3.3 An Example from China

A very regular cyclic succession has recently been described from the Upper Carboniferous Taiyuan Formation at Lingchuan, southeastern Shanxi Province, China (Xu Hui-long et al. 1987). The cyclothems exposed here are noteworthy for the close coupling of fairly thick coals and open marine limestones and may serve to illustrate the problem of marine transgressions over peat-forming environments.

A typical example is represented in Fig. 4. Here, a low volatile bituminous coal of up to 2 m in thickness is immediately overlain by a thick limestone (K4 limestone, up to 5 m thick) which contains a diverse offshore marine fauna. The succession indicates a shallowing-upward trend continuing into the shale unit, where land plant remains appear together with silt and sand layers. In between the clastic layers, a thinner limestone without marine fossils occurs. Sandstones in the upper part of the cyclothem are commonly not massive but rather parallel-bedded, though sedimentary structures have not been described in detail. The root horizon of the succeeding coal is developed in a clay bed above the sandstone.

Several sedimentological aspects have to be considered when events and environmental conditions in these cycles are reconstructed. The compaction ratio of peat to low volatile bituminous coal may be as high as 1 to 10. Thus the 2 m of coal represent an original peat thickness of approximately 15 to 20 m. Since peat accumulation is dependent on a simultaneous rise of the water table, a nearly equal relative sea level rise seems to be a prerequisite for the formation of the seam beneath the K4 limestone. The beginning of relative sea level rise should therefore be placed near the base of the seam and distinguished from the actual transgression which is reflected by the contact between coal and limestone. At this contact no obvious signs of erosion can be detected. Our own observations in the Florida Everglades have shown, however, that even minute changes in hydrologic conditions are accompanied by extensive erosion. Nearer to the shoreline the peat is deeply dissected by fluvial and tidal channels and eroded along the surge zone (Spackman et al. 1969).

Since in the Lingchuan section the change from peat-forming environments to fully marine conditions is sudden and seems to take place without reworking or

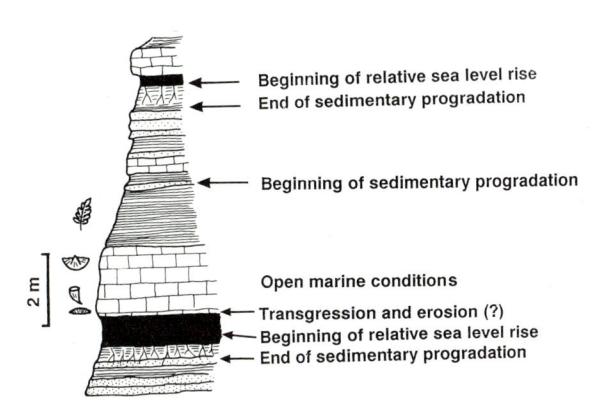

Fig. 4. Cyclothem from the Upper Carboniferous Taiyuan Formation at Lingchuan, southeastern Shanxi Province, China (from Xu Hui-long et al. 1987) showing open marine limestone (K4 limestone) immediately superseding thick coal seam. Position of events within the sequence as interpreted in this chapter

scouring, it may be suggested that the peat surface has grown in adjustment to a rising sea level in a protected embayment or lagoon. The peat surface may even have subsided below sea level prior to the infiltration of sea water. By such a mechanism the space required for the bathymetry of the overlying limestone could have been provided. The relative sea level rise observed in the transition from coal to overlying limestone may be calculated as the sum of original peat thickness and of the water depth required for the limestone. It may have reached tens of meters.

The assumption that major peat beds mark the beginning of relative sea level rise in marginal marine swamps has also been maintained for deltaic environments by several authors (e.g., Elliott 1974; Leeder and Strudwick 1987). In such situations the base of coal seams should be a plausible choice for the placement of cyclothem boundaries.

4 Some Lessons from Modern Peat-Forming Environments

Interpretation of Pennsylvanian cyclothems has been greatly influenced during the past three decades by studies on the modern Mississippi Delta (e.g., Fisk 1960; Ferm 1970; Coleman 1968). They clearly demonstrated that the switching of delta lobes is a simple and readily acceptable mode of cyclothem generation. The seven lobes that can be distinguished here were all formed within about the past 6000 years of retarded sea level rise (Fig. 5). Thus an average of 800 to 900 years may be calculated for the time between initiation and abandonment of a single delta lobe. Up to three large delta lobes, the Cocodrie, the St. Bernard, and the Plaquemines delta lobe were more or less superimposed upon one another within about 4000 years. On a smaller scale, the formation of individual subdeltas with diameters in excess of 10 km takes place in the order of 100 years. Consequently, in fluvial-dominated deltas the time intervals involved in the generation of deltaic cycles may be far below those associated with allocyclic forcing mechanisms, i.e., tectonic or Milankovitch cycles.

It should also be kept in mind that transgression events in a Mississippi-type delta situation are not due to events of sea level rise but rather a consequence of sediment starvation and internal compaction. It is typical of such fluvial-dominated deltas that abandoned delta plains gradually subside below sea level and conditions successively grade from fresh water to brackish and eventually normal marine, while barrier beaches or chenier islands are formed from the reworking of delta front sediments. Under this protection delta plains may be invaded by sea water through seepage or a "back-door" transgression that may be characterized by nearly continuous sedimentation from fluvial to marine conditions and a lack of erosional hiatuses.

This would present a contrast to transgressions which are generally associated with erosional surfaces and sharp facies breaks due to gaps in the sedimentary record of varying proportions. An example of this is provided by the mangrove swamp along the shore of the southwestern tip of the Florida Peninsula (Spackman et al. 1969). Here, despite the cohesiveness of mangrove peat, large portions of the 4.5 m thick peat blanket that was formed during the past 4500 years is being removed by wave action and dissected by deep tidal channels along unprotected shores. The minor peat occurrences detected beneath marine marls in the Gulf of Mexico and Florida Bay

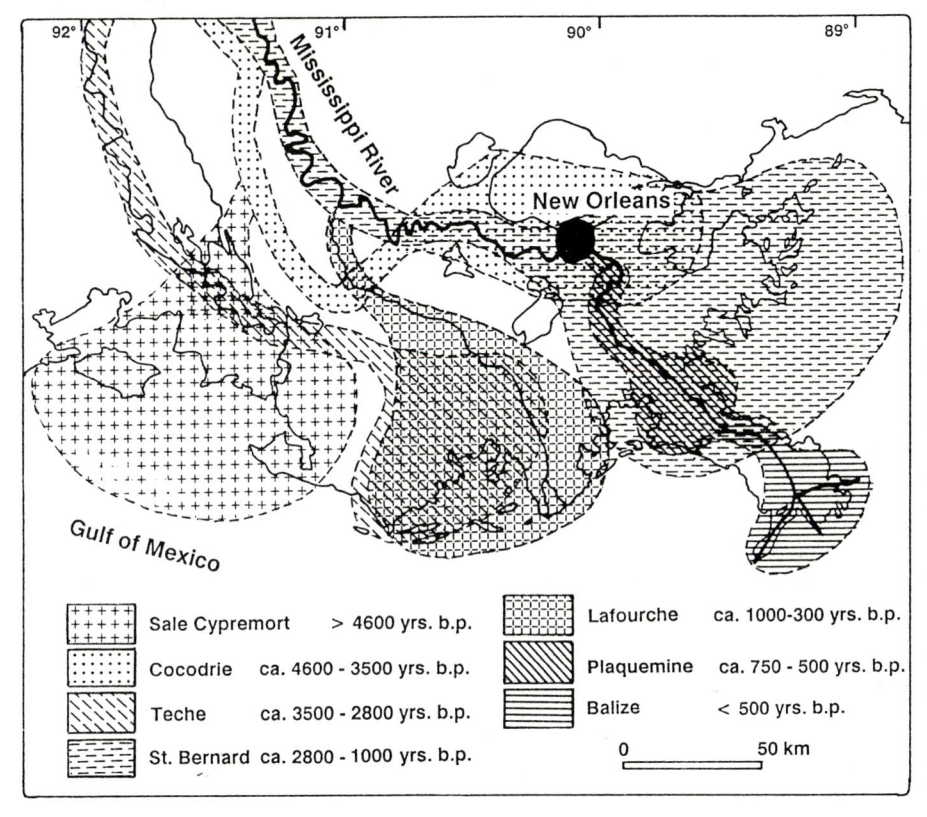

Fig. 5. Sequence of delta lobe development in Mississippi River delta complex during the last Holocene rise of sea level. (Coleman 1968)

represent merely erosional remnants of a previously more extensive and thicker peat blanket. Thus it remains doubtful whether coals of economic proportions and the orderly succession of coal cyclothems could ever have been formed under the mechanisms operating along an open shoreline receiving restricted sedimentary input. It also follows that the immediate superposition of major coal seams by marine units (limestones or black shales) as described in the previous chapter as well as by Heckel (1977) requires the backdoor marine invasion mechanism observed in fluvial dominated deltas (e.g., Mississippi delta) or lagoonal opening and indicated by arrows in Fig. 2.

5 Controls of Cyclothem Formation

The pervasive theme in the discussion of cyclothems is the interpretation of events reflected in their succession and the causes controlling their development. Comprehensive reviews mainly concerned with Upper Carboniferous cyclothems have been given by Duff et al. (1967), Westoll (1968), and Heckel (1984). Since it has been generally accepted that a variety of controls can produce coal cycles, the discussion has concentrated on the question of which control plays the dominant role in the various examples studied.

5.1 Sedimentary Control

The concept of sedimentary control of cyclothem formation implies that the vertical stacking of cyclic sequences can be accomplished under constant rates of basin subsidence or base level rise by periodic shifts in sediment transport. This applies in particular to fluvial-dominated deltas where relatively high rates of delta progradation and delta top accretion lead to instabilities between the topographically raised active delta plain and the adjacent sediment-starved bays. In this situation channel avulsion in the upper reaches of the delta plain eventually results in delta lobe switching. Providing continued gradual basin subsidence, this process can be repeated many times and may produce a more or less regular en echelon arrangement of sedimentary bodies which are wedge-shaped in longitudinal and lenticular in cross-section. It is obvious that the en echelon stacking of delta lobe units may produce more or less identical sequences in sections a few kilometers apart, though individual beds have no physical and chronological continuity since they have been formed during different sedimentary events.

The recognition of delta lobe switching and its effects in the Mississippi delta system by Fisk (1944, 1960) and others (e.g., Coleman 1968; Frazier and Osanik 1969) and its successful application to the coal-bearing Upper Carboniferous of the Appalachian region (e.g., Ferm 1974; Donaldson 1974) and the Lower Carboniferous of Britain (Moore 1959) soon focused the attention of most coal geologists on the great importance of sedimentary processes for the make-up of coal-bearing strata. As a consequence, the discontinuity of coal seams and associated sediments was stressed and the use of coal cyclothems as time-stratigraphic units abandoned.

Even under the assumption of dominant sedimentary control, it should be expected that the general geometry of delta lobe wedges and resulting cyclothems depends on the paleogeography and tectonic stability of the region. To account for this, Donaldson (1974) distinguished three types of deltas on the basis of the ratio between sediment supply and subsidence rate. The Mississippi delta serves as an example in which supply slightly exceeds subsidence and both are relatively great. As a result, fairly thick sequences are developed in a somewhat off-lapping succession. In the Rockdale-type delta (Lower Wilcox Group, Eocene, Texas), supply kept pace with subsidence, and facies units are stacked more or less vertically. Delta progradation and abandonment alternate most rapidly in the West Virginia-type delta (Conemaugh and Monongahela Groups, Pennsylvanian, West Virginia, Ohio, and Pennsylvania), which has rapidly and repeatedly prograded toward a stable platform. Resulting cyclothems are relatively thin.

Sedimentary mechanisms commonly exert the dominant control on the formation of cyclic sequences in strictly fluvial sediments. Here, lateral channel migration or switching and corresponding vertical accretion of overbank and flood basin deposits leads to the stacking of fining upward cycles. Such purely fluvial-controlled cycles can be highly differentiated into channel sandstones, overbank clay-siltstones, rooted flood plain clays, back river swamp deposits (coals and carbonaceous shales) and terminal dark lacustrine shales. In this respect they meet the requirement of a multi-elemental sequence inherent in the definition of cyclothems, but they often lack the thickness and lateral continuity to be of mappable proportions.

5.2 Climatic Control

Direct climatic control of cyclothem formation in the midcontinent Pennsylvanian has been proposed by Beerbower (1961) and Swann (1964). According to Swann, marine limestones represent periods of a warm humid climate. A change to drier climate caused deterioration of the vegetation cover and increased erosion and sediment transport, resulting in the deposition of sand in the coastal region. The return to more humid climates promoted plant growth in the depositional as well as in the source area, reducing clastic input and enhancing peat formation. Due to sediment starvation, the sea eventually transgressed and limestone was deposited again, starting the succeeding cycle. Similar ideas were previously expressed by Wanless and Shepard (1936). Opposite interpretations using coarser clastics as the result of increased runoff under humid climates and carbonates as deposits of more arid stages have likewise been proposed (Brough 1928). These models require a rather rapid alternation of humid and dry climates or extended periods of nondeposition. More important, there is no additional paleobotanical evidence in Pennsylvanian cyclic sequences supporting the periodic recurrence of dry climates.

5.3 Eustatic Control

Eustatic sea level fluctuations in the earth's history have essentially been referred to two different causes:
1. Fixing of sea water on the continents by glacial buildups or other climatic processes and later release of water by melting glaciers.
2. Varying volumes of ocean water displaced by rising mid-ocean ridges due to changes in spreading rates.

Since the existence of glacial buildups is well established for certain periods and can be related to the earth's orbital variations, glacio-eustatic effects on cyclic deposition are the most commonly discussed. While short-term glacio-eustatic fluctuations of sea level have periods approximating those of cyclothem formation, eustatic sea level changes induced by expanding ocean ridges appear to be too long-termed to be related to coal cyclothem formation. On the other hand, glacio-eustatic effects are not generally accepted to explain coal cyclothem formation in the Cretaceous of the Western Interior for climatic reasons. Among the many attempts to apply glacio-eustatic control to the causes for cyclothem formation in the Carboniferous, perhaps those of Heckel (1977, 1984, 1986) are the most advanced. During the Pennsylvanian, the cratonic margin of the North American Midcontinent area provides the necessary crustal stability, low topographic relief, and wide space for optimal sedimentary reflection of eustatic cycles.

It should be stressed that the assumption of crustal stability of the cratonic margin allowing for continuous gradual subsidence is a major prerequisite for identifying these cycles as the sedimentary response to glacio-eustatic sea level fluctuations and for constructing a sea level curve for the Pennsylvanian (Heckel 1986). The degree of marine inundation in any cycle may be determined by directly tracing the maximum landward extent of transgressions, or may be inferred from the degree to which the deepest facies is developed. The extent of regression, on the other hand, is determined by how far basinward a paleosol or terrestrial deposit capping the cycle can be traced. On the basis of extent and completeness, a sequence of major, intermediate, and minor cycles has been set up for the upper part of the Pennsylvanian (Heckel 1986), ranging from about the middle of the Desmoinesian through the Missourian and Virgilian. Attempts at sequential groupings of cycles in order to detect any hierarchy of cyclicity have not been conclusive.

With regard to relating sedimentary cycles to glacio-eustatic sea level fluctuations, all major and intermediate cycles are counted individually by Heckel. For the 10 Ma time interval calculated for the European Stephanian, closely corresponding to the North American Missourian and Virgilian, Heckel (1986) counted 25 cyclothems to obtain a 400 ka period for each. Using different time assumptions for the period length, cycle length, and different period ratios for minor and major cycles, Heckel shows that major Pennsylvanian cycles may have a range of periods from 235 to 393 ka, while minor cycles have periods ranging from 44 to 118 ka. These values correspond reasonably well with the dominant period of obliquity near 40 ka and the two dominant periods of eccentricity near 100 and 413 ka respectively (Imbrie and Imbrie 1980; Fischer Chap. 1.2, this Vol.). Similar periodicities of about 100 ka have been calculated on the basis of landward tracing of bentonite layers for the fourth order cycles of Ryer (1983) in the Cretaceous Western Interior Seaway (Kauffman pers. commun. 1989).

The fit of observed and theoretical values is quite remarkable, though it should be pointed out that no information is given concerning gaps in the record and rates of deposition which are highly variable for the different types of sediments involved.

5.4 Tectonic Control

The idea of diastrophic control has been advocated at the outset of the cyclothem concept by Weller (1931) and renewed by him later (Weller 1956). Diastrophism assumes periodic alternations of uplift and subsidence in the sedimentary basin as well as in the source area. Unconformities at the base of sandstones were taken as signals of widespread erosion during periods of uplift, while sedimentation during periods of subsidence leading from massive sandstone to shale, coal, and marine beds was considered to be the reflection of waning stream gradients during the completion of an erosional cycle.

Ideas concerning tectonic control may be classified into two categories:

1. those relating cyclothems to regionally limited periodic movements along fault scarps and
2. those advocating synchronous craton-wide or even global perturbations.

Modern basin analysis in the Carboniferous of the British Isles has revealed several examples in which thick cyclic sequences can be related to major faults bordering extensional basins (Bott and Johnson 1967; Belt 1984; Leeder and Strudwick 1987). In most of these cases, however, successive uplift episodes merely provide the space needed for the accommodation of additional sedimentary wedges while the actual formation of cyclothems is attributed to delta progradation. Leeder and Strudwick (1987) point out that purely tectonic subsidence along normal faults should cause rapid delta drowning. Resulting cyclothems should be characterized by the lack of sheet-like peat development, evidence of catastrophic delta front and delta plain drowning, and a long ensuing period of tectonic quiescence represented by deep to shallow carbonate facies. Increasing dominance of sedimentary over tectonic factors leads to delta lobe switching and the formation of sheet-like peat beds on abandoned delta lobes. With continuing compactional subsidence this in turn is followed by carbonate deposits whose facies may show a deepening-upward trend.

Purely tectonic control of cyclic successions, sometimes including coal cyclothems, is probably best displayed in intramontane basins. Here, as well as in foreland basins, deposition of coarse clastics is traditionally considered as the sedimentary response to rapid uplift along the boundary fault. However, Blair and Bilodeaux (1988) have suggested a reversal of arguments. They assume that fine-grained sediments represent the stage immediately following rapid basin subsidence prior to the development of a new erosional cycle and the progradation of a new generation of alluvial fans across the basin.

A new version of relating coal cyclothem formation to global tectonics has recently been proposed by de Klein and Willard (1989). They contrast the three different types of cyclothems represented in the North American Carboniferous:

1. The marine-dominated Kansas-type cyclothem.
2. The mixed marine-nonmarine Illinois-type cyclothem.
3. The nonmarine-dominated Appalachian-type cyclothem.

On the basis of regional tectonic history they recognize the Kansas-type cyclothem of the western Midcontinent as being exposed to minimal tectonic influence and being essentially controlled by glacio-eustatic sea level fluctuations. The other end member, the Appalachian-type cyclothem, is developed in a foreland basin influenced by episodic thrust loading.

In their explanation, de Klein and Willard (1989) refer to a model of periodic alternation of flexural deformation due to thrust loading and a viscoelastic response during a relaxation phase proposed by Tankard (1986). In this, overthrust loading in the fold belt is associated with shallowing in the foreland basin and thus sedimentary progradation and basin infill, while relaxation leads to downwarp and deepening of the basin. Each renewed overthrust loading can thus be linked with the onset of the progradational position of a cyclothem. The Tankard model can also be applied to cyclothem development in the Cretaceous Western Interior Seaway (Tankard 1986; Kauffman 1986).

The Illinois-type cyclothem, according to de Klein and Willard (1989), is typical for basins wedged between the foreland basins and the stable craton (Michigan, Illinois, and Forest City basins) undergoing moderate flexural foreland subsidence.

5.5 Distinction of Various Controls

The foregoing discussion shows that three major types of controls – sedimentary, climatic-eustatic, and tectonic – are capable of producing coal cyclothems. They most likely operate concurrently, with different intensities, to produce the great variety of cyclothems observed in different regions and at different times.

Few attempts have been made to look into the sedimentary record for specific signatures of these controls. Some recently published examples of sequence analysis have aimed specifically at the distinction between allocyclic cyclothems showing eustatic or tectonic control and autocyclic cyclothems caused by lobe switching within a prograding delta system (Belt 1984; Johnson 1984b; Leeder and Strudwick 1987).

A quantitative approach was chosen by Belt (1984), who studied a well-exposed Late Dinantian section along the Fife coast near the eastern end of the Midland Valley in Scotland. Here, as many as 76 complete cycles occur in 1180 m of section. Among other tests, Belt (1984) used thickness histograms of various units (e.g., frequencies of basin-phase or delta-top thickness) to demonstrate that there are two clearly separated and unequal modes of distribution. The great majority of cyclothems following a "normal" pattern and only a few being extraordinary. When separated in this way, cyclothems containing units with extraordinary thicknesses commonly reveal specific sedimentary structures or lateral facies gradations upon closer sedimentological reexamination.

Johnson (1984b) summarizes the distinction between cycles caused by tectonic subsidence and those caused by eustatic rise in sea level as developed by Bott and Johnson (1967): "In eustatically controlled cycles, deposition at the margin of the basin takes place during the early part of the cycle, after rise in sea level, and erosion should occur later as sea level returns to its initial value. At the basin margin, the early sediments of consecutive cycles should be present and there should be strong erosion between them. Tectonic control of cycles, on the other hand, requires that only the upper part of each cycle should be laid down at the basin margin and there should be no strong erosion between them."

From their study of the classical Yoredale cycles in northern England, Leeder and Strudwick (1987) proposed a number of criteria to be tested for their use in assessing the controls of cyclicity. According to them, the lack of shallowing-upward signals in the carbonate facies, the markedly diachronous onset of pro-delta mud deposition, the development of sheet-like peat beds on abandoned delta lobes and a transgressive coarsening-upward sand body above it are suggested as possible criteria for sedimentary control. In addition, limestone members should pass laterally into deposits of an active delta lobe on a regional scale.

Cycles controlled by eustatic mechanisms should show a shallowing-upward trend and eventually signs of emergence in their laterally persistent carbonate member. The clastic/carbonate boundary would be diachronous and accompanied by channel cutting. The lack of widespread peat formation is attributed to delta retreat dominating over lobe abandonment in response to relative sea level rise. Characteristics pointing to purely tectonic control of cycle formation are discussed in Section 5.4.

A combined tectonic-sedimentary mechanism is suggested by Johnson (1984b) as being most effective in active extensional basins such as the Northumberland Basin (Dinantian, northern England) in which the Yoredale-type cycles developed. Here, in contrast to purely sedimentary control, fault movements due to crustal extension cause upward-deepening trends at the tops of carbonate members while delta progradation proceeds from the margin to the basin interior. Continued fault movement produces a number of minor cycles due to local crevassing and distributary migration, and results in a close relationship between clastic member thickness and tectonic subsidence.

These attempts to distinguish the effects of various controls in coal cyclothems have been conducted in specific examples. Successions produced in different paleogeographic and tectonic settings may exhibit quite different effects. Thus, more studies regarding event and sequence analysis of coal-bearing cycles are needed before more consistent criteria for the distinction of controls can be recognized.

6 Concluding Remarks

A somewhat paradoxical situation has evolved at present regarding the geologists' concern with coal cyclothems. With the wave of recent developments in terrestrial sedimentology, most coal geologists have become enthusiastic sedimentologists, exploiting the newly acquired tools to reconstruct environments of deposition and to employ facies patterns for predictions in coal exploration. The concern with coal cyclothems, once a domain of coal geologists, was deliberately dismissed as misleading.

The renewed discussion of Milankovitch cycles and their effects on sedimentological and biological events has been stimulated by quite different fields, such as seismic stratigraphy and evolutionary paleontology, and has caught coal geologists rather by surprise. As a consequence, the current revival of coal cyclothems as an object for studying the effects of orbital, eustatic, or tectonic cycles is largely taking place without consultation of coal sedimentologists and, with notable exceptions, ignores their insights into sedimentary processes and events.

On the other hand, coal sedimentologists need to take a new look at coal-bearing sequences to identify specific signatures of potential allocyclic controls. Especially the various modes of transgression do not seem to have been analyzed in sufficient detail. In addition, some generally accepted assumptions, such as the notion that regressive phases are associated with widespread delta progradation, have to be critically revised in the light of the observed delta progradation during the postglacial sea level rise and the expected downcutting of channels due to lowered base levels at the peak of regression. Comparative sequence analyses and three-dimensional studies of lithobody geometries from coal cyclothems of selected paleotectonic and paleogeographic settings appear to be most useful.

The stable cratonic margin of the American midcontinent (Kansas) has already been recognized as an area well suited for the study of glacio-eustatic effects in coal cyclothems of the Upper Carboniferous, since clastic influx is low and a chronological link with the southern hemisphere glaciations exists. Rift and pull-apart basins should be considered as sites in which tectonic control of cyclothem formation is dominant, while mobile cratonic margins and foreland basins display a particularly strong sedimentary fingerprint. Eustatic and tectonic effects may also be significant in foreland basins characterized by strong longshore transport and barrier beach formation, since transgression of beach barriers requires that progradation rates are greatly outstripped by relative sea level rise.

6.8 Cycles, Rhythms, and Events on High Input and Low Input Glaciated Continental Margins

R. Henrich

1 Introduction

Today, oceans and shelves in the northern and southern hemispheres are strongly influenced by glacial processes. Together with mountain glaciations, about 10% of the Earth's surface is covered by ice and snow. During the Cenozoic, as well as in earlier glacial periods, variations in the magnitude and dimension of the Earth's ice and snow cover resulted in repetitive shifts of the global climate through changes in albedo and total ice volume. Development of the Antarctic ice cap in the early Cenozoic (Kennett 1977) and the onset of Northern Hemisphere glaciation in the late Cenozoic (Plafker and Addicot 1976; Armentrout 1983; Henrich et al. 1989 and references therein) mark a progressive global cooling that is well documented in the climatic records of the oceans (Miller et al. 1987). Superimposed on these long-term changes in Earth's climate – which correspond to specific plate tectonic and paleoceanographic settings – are short-term changes that are controlled primarily by cyclic variations in Earth's orbital parameters, e.g., Milankovitch cycles.

In addition to the Cenozoic, extensive glaciations are documented during other periods of Earth's history, such as the Proterozoic and the Permocarboniferous Gondwana glaciations (Frakes and Crowell 1975). During the Pleistocene, more than 40% of the Earth's continental shelves were glaciated (Climap 1976, 1981). Glaciations also influence continental margins through sea level fluctuations, with the consequent migration of facies belts, changes in width of the continental shelves (Beard et al. 1982), and variations in the quantity and type of continental erosional products and sediment delivery to slope and deep sea environments (Ruddiman 1977; Thiede et al. 1986; Henrich et al. 1989). Glaciations indirectly influence nonglaciated continental margins by perturbing high level atmospheric circulation, with its associated variations in aeolian dust supply and oceanic productivity (Sarnthein et al. 1982; Stein 1986a).

Ancient glaciomarine sections have been studied extensively by numerous workers (see bibliography in Eyles et al. 1985 and Edwards 1986). Studies of modern glaciomarine environments were comparatively few until about 20 years ago, but much progress has been achieved since then. A summary of our knowledge on glaciomarine environmental settings has been compiled in a special volume on *Glaciomarine Sedimentation*, edited by B. Molnia (1983), and in special issues of Marine Geology (Volume 57, 1984; Volume 85, 1989).

The objectives of this chapter are manifold:

1. To provide an overview of the principal controls on glaciomarine deposition in continental margin environments
2. To discuss depositional events, cycles and rhythms in Subarctic and Antarctic Pleistocene glaciomarine sequences, and their significance for paleoceanographic and paleoclimatic reconstructions.
3. To compare high input (Subarctic) and low input (Antarctic) glaciomarine records.

1.1 Terminology of Glaciomarine Deposits

The most commonly used definitions of the varieties and subtypes of till are not merely descriptive, but incorporate aspects derived from theoretical modeling. Eyles et al. (1983) propose that glacial facies analysis should apply a purely descriptive nomenclature, which can be easily applied during field work as well as during core logging. Use of the term till (and tillite) should be reserved for sediments directly deposited at the base of a glacier by sub- and supraglacial aggregation of englacial debris, without subsequent reworking (Eyles et al. 1983). The term glaciomarine diamicton (and diamictite) can be employed for any poorly sorted gravel-sand-mud mixture (Frakes 1978), deposited directly or indirectly from ice in glaciomarine environments.

1.2 The Basal Thermal Regime: a Major Control on Glacial Deposition

The volume of diamicton and lithofacies sequences that is deposited by a glacier largely depends on its basal thermal regime. Three thermal or glaciodynamic basal regimes are often recognized at modern ice margins (Fig. 1).

1.2.1 Temperate Glaciers

The glacier may be wet-based and slide over its bed (e.g., Iceland, Boulton 1972; Alaska, Powell 1983; Powell and Molnia 1989). A typical vertical profile (Fig. 1A) reveals a wide range of diamicton, including basal lodgement till, deposited during traction over the bed, and melt-out diamicton interbedded and channeled with glaciofluvial facies. Erosional contacts, multiple diamictons with undulating geometry, thin stratified units released by subglacial melt-out, as well as resedimented diamictons released from drumlin slumps, are common features in temperate glacier lithofacies profiles.

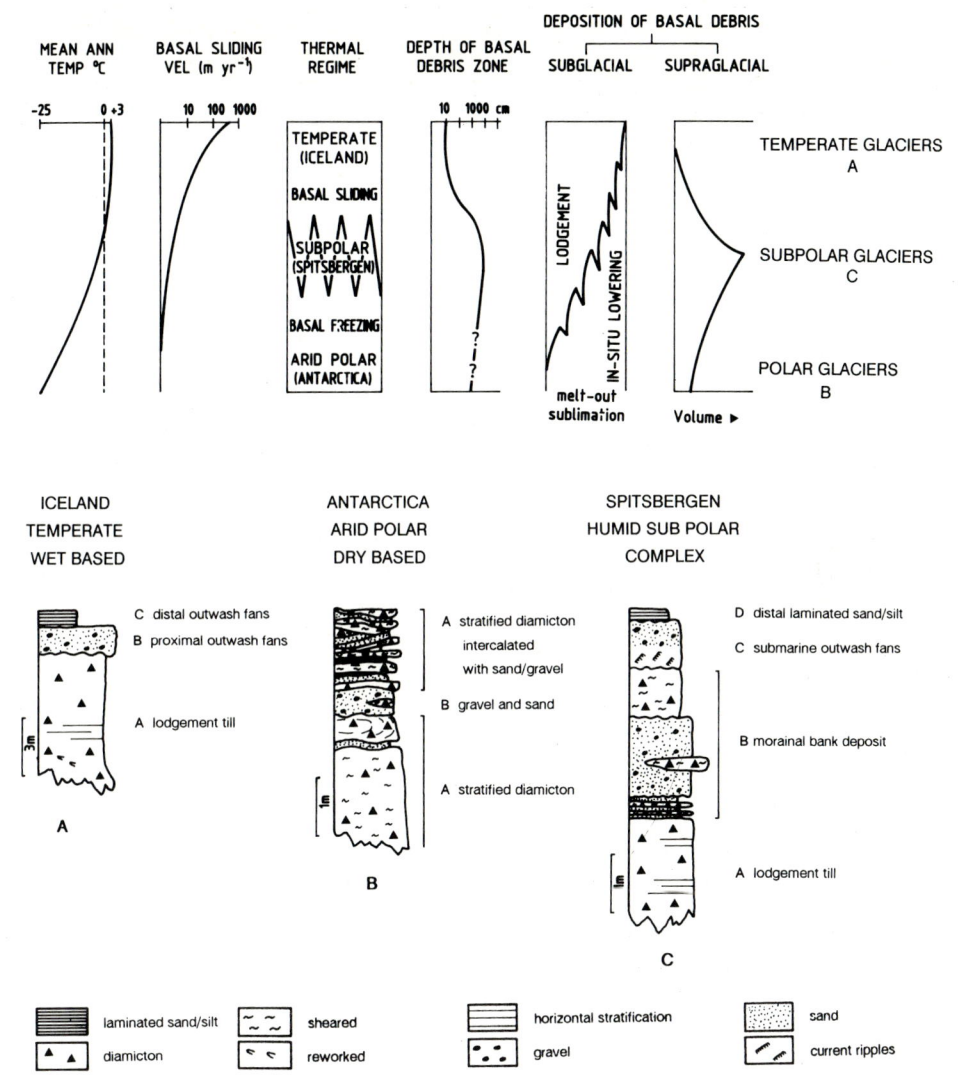

Fig. 1. Evolution of glacial diamicton sequences of differing basal thermal regimes. *A* temperate glaciers; *B* polar glaciers; *C* subpolar glaciers. (After Eyles et al. 1983)

1.2.2 Dry Polar Glaciers

Ice may also be dry and frozen to the substrate, such that the glacier margin moves by internal deformation at an order of magnitude slower than temperate glaciers (e.g., arid polar glaciers, Antarctic dry valleys; Anderson et al. 1980, 1983). During recession of such Antarctic glaciers, in situ stratified basal debris deposits and englacial debris are simply lowered on to the substrate. Interstitial ice is lost not by melting, but by sublimation, so that basal debris layers may be preserved intact. Vertical profiles of polar glaciers (Fig. 1B) reveal a thin accumulation of crudely stratified diamictons, most commonly with well-developed parallel clast orientation resulting from former foliation within the glacier.

1.2.3 Subpolar Glaciers

The third basal thermal regime reflects complex patterns of freezing and thawing zones. Commonly, these subpolar glaciers (Fig. 1C, Spitsbergen, Boulton 1972; Alaska, Lawson 1982; Powell 1983) are wet-based at their interior, sliding over their beds, but they also exhibit cold bases with freezing on of subglacial meltwaters.

Some of these subpolar glaciers tend to surge (Elverhøi 1984; Solheim and Pfirman 1985). Subpolar glaciers with complex freezing/thawing bases reveal thick basal debris accumulation at their margins (Fig. 1C, Lawson 1982), some of which are characterized by intense compressional folding. As they are affected by melt-out and sublimation, depending on the regional climate, no englacial structures are preserved, but generally massive, normally consolidated diamictons are deposited.

1.3 Modern, Glacially Influenced Continental Margin/Basin Configurations

Glaciomarine facies assemblages and bathymetric features of the modern, glacially influenced continental margins are a result of a variety of climatic, oceanographic, tectonic, and icemass-related constraints, and their close and complex interaction.

Climatic factors of importance are temperature, moisture supply, pathways of major storm tracks, and the direction and strength of katabatic winds. The vigor and circulation patterns of shelf surface and bottom currents determine meltwater dispersion, iceberg drift patterns, and sea floor sediment resuspension. Open-ocean surface current patterns and water mass parameters strongly influence iceberg melting rates, moisture supply to the continental ice, seasonal or perennial sea ice coverage, and oceanic productivity. Dense brines released during sea ice formation may erode shelf sediments (Elverhøi et al. 1989) and cascade into adjacent deep sea water masses (Quadfasel et al. 1988). Contour currents along the shelf break and slope may cause considerable erosion and winnowing of the sediments.

Ice mass related constraints are: the total volume of ice, ice sheet isostacy, ice sheet/glacier aerial extension and erosion, and advance/retreat cyclicity, which determine the dimension and morphology of the shelf/slope configuration. Other important ice mass related parameters are basal thermal regime (as discussed above), debris content in various parts of the ice, rate of meltwater supply, rate of iceberg calving, and surge activity.

The large number of possible combinations of limiting factors makes it obvious for a great variety of glacially influenced continental margin/basin configurations to be envisaged. The large variety of glacially influenced continental margin/basin configurations will be illustrated through discussions of modern extremes, e.g., the low input margin typical of Antarctica, and the high input margin typical of the Gulf of Alaska.

2 Subarctic and Antarctic Glaciomarine Sequences

2.1 Tidewater Ice Fronts (Alaska and Svalbard)

Today, temperate and subpolar tidewater glaciers are best exposed around the Gulf of Alaska and on Svalbard. Most of the glaciers in Alaska are valley-type glaciers that terminate either on land or at the head of the fjords, while only a few reach the open ocean (Powell 1983; Powell and Molnia 1989). On Svalbard also, many glaciers have fjord or land termini. On Nordaustlandet, the second largest island on Svalbard, the Austfonna ice cap has a 200-km-long, open marine tidewater ice front (Pfirman and Solheim 1989). Since recent models of the major Quaternary glaciations suggest frequent development of open marine tidewater ice fronts (Henrich 1989; Henrich et al. 1989), the Austfonna ice cap provides a unique opportunity for studying a modern analog. A second important feature of Svalbard's glaciers, their tendency to surge (Elverhøi 1984; Solheim and Pfirman 1985), is of importance in understanding Pleistocene glaciations.

The most important element of a wet-based tidewater ice front setting is the efflux of turbid meltwater suspensions either at the grounding line (Powell 1983; Powell and Molnia 1989; Solheim and Pfirman 1985; Pfirman and Solheim 1989) or at the sea surface (Elverhøi and Roaldset 1983). Dumping the bed load may deposit a mouth bar, or in combination with minor glacier oscillations, a submarine morainal bank (Powell 1983). Depending on the densities of turbid sediment plumes and sea water density stratification overflows, interflows, or in rare cases underflows, will be initiated (Fig. 2). Fine-grained sand and mud are deposited from low to medium density interflow and overflow sediment plumes (Powell 1983), while the formation of traction currents and underflows (= continuous turbidity currents) requires high sediment concentrations in subglacial streams and are most likely episodic events.

Proximal deposits from subglacial streams are likely to be coarse-grained, clast-supported gravels and sands. Sand and silt laminae with dropstone diamicton or dropstone mud form interlaminated mud, silt and sand (= laminites, Fig. 2), within about 0.5 to 1 km of a grounding line (Mackiewicz et al. 1984). Distal facies (Fig. 2) consist of dropstone diamictons and muds and, in the case of low iceberg drift, outwash muds (e.g., rock flour, Molnia 1983).

Depending on glacier movement, different thicknesses and successions of glaciomarine facies have been observed. Quasistable tidewater glaciers (Powell 1983) form thick submarine morainal banks and elongate ridges, or isolated mounds composed of grounding line melt-out till, dumped supraglacial debris, and gravity flow deposits down oversteepened slopes on morainal banks (Fig. 2). Gravity flows may also be triggered by ice push, iceberg calving and porewater pressure fluctuations by storm waves or tides. Morainal bank sediments affected by storm or tidal currents will have a higher ratio of sorted sediment to diamicton.

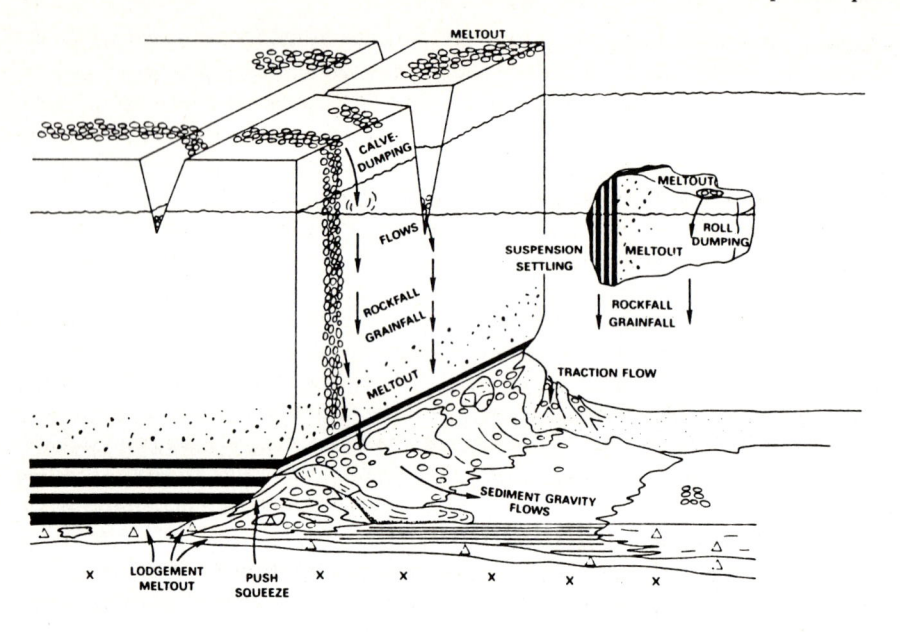

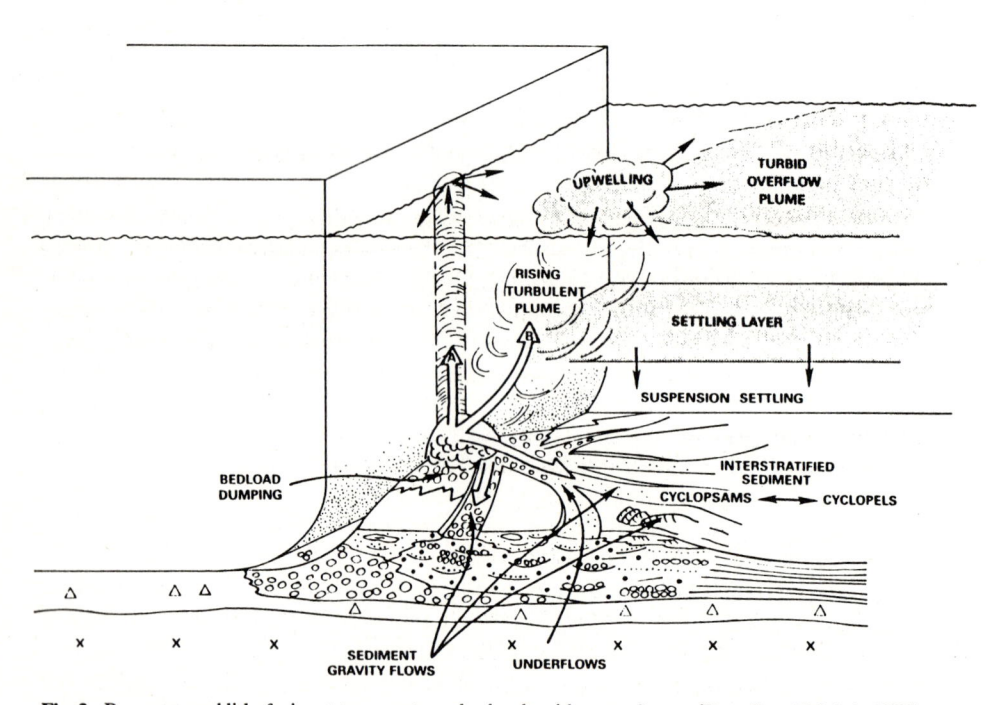

Fig. 2. Processes and lithofacies at temperate and subpolar tidewater fronts. (Powell and Molnia 1989)

During the gradual retreat of a tidewater front, subglacial streams may form eskers or erosional channels. Rapid retreat causes rapid iceberg calving, and thus rafts further offshore a higher proportion of englacial and basal debris. Glacier surges may transport large quantities of glacial erosional products to the ice front (Solheim and Pfirman 1985) and result in thick deposits deposited from meltwater, (e.g., in Kongsfjorden and Van Mjenfjorden on Svalbard, Elverhøi 1984).

Careful logging of advance and retreat successions of tidewater ice front associations is a powerful tool in glaciomarine basin analysis. Identification of rhythmic associations within a given stratigraphic framework might aid in deciphering the response of ice sheet growth and decay during orbitally induced climatic deteriorations. Modern tidewater glacier fronts on the inner and outer shelf of the Gulf of Alaska, where outwash mud dominates with only a minor contribution of ice rafted debris (IRD) (Molnia 1983; Powell and Molnia 1989), should not be considered as a major criterion for recognition of a tidewater ice front situation in ancient sequences. Spreading of turbid sediment plumes as well as transport and melting rates of icebergs are determined by water mass properties and current circulation. An advancing ice sheet, with a tidewater ice front on the outer shelf or at the shelf edge, might result in a completely different facies succession. Assuming weak coastal currents, sediment-laden meltwater plumes may spread further offshore. Deposition from suspension in addition to higher IRD rain out could result in dropstone diamicton deposition along an offshore open ocean depocenter (compare following sections, Henrich 1989; Henrich et al. 1989).

2.2 Depositional Rhythms in Subarctic Quaternary Shelf Deposits

Glaciomarine shelf records may have incomplete or disturbed stratigraphic sections, due to erosional events during repetitive glacial advances or disturbance by iceberg turbation (Elverhøi et al. 1989; Vorren et al. 1984). Particularly wide shallow shelves may be affected by glacial erosion in addition to oceanic current reworking during isostatic rebound. In contrast, thick, glaciomarine, prograding sediment wedges may form on continental slopes (Fig. 3; Vorren et al. 1989) or on deeper shelves during repetitive glacial advances terminating at or close to the shelf edge. Mass vasting, e.g., slumps, slides, and sediment gravity flows, frequently distort and remove sediment deposited on the slope (Jansen et al. 1987).

The following discussion will focus on:

1. Glacial/interglacial depositional rhythms on a typical high input passive margin, i.e., the Norwegian margin.
2. Regressive sedimentary rhythms deposited in response to isostatic rebound and current reworking on the northwestern and central Barents shelf.

The morphology of the Barents Sea, characterized by northeast-southwest-trending basins (300–500 m deep) and shallow banks (30–150 m deep), is most likely a response to repeated glaciations in the late Cenozoic (Elverhøi and Solheim 1987). The southwestern Barents shelf/slope (Fig. 3) reveals a variety of characteristic glaciogenic units, which are typical of Subarctic passive continental margins (Vorren

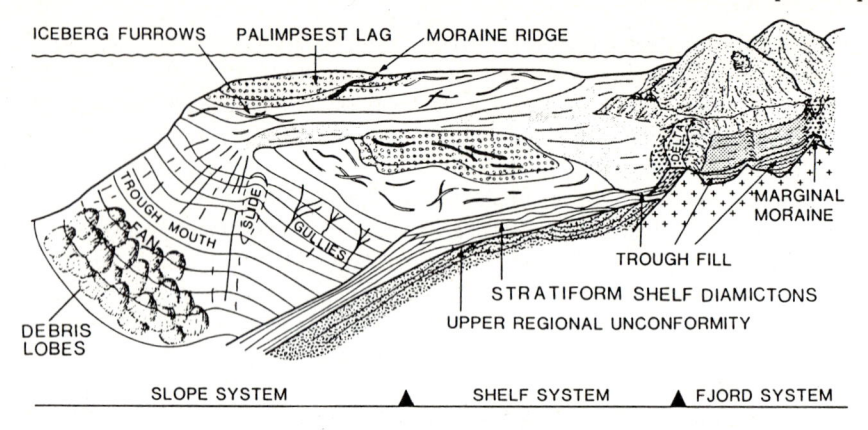

Fig. 3. Morphology, depocenters, and glaciogenic units of Subarctic passive margins. (Vorren et al. 1989)

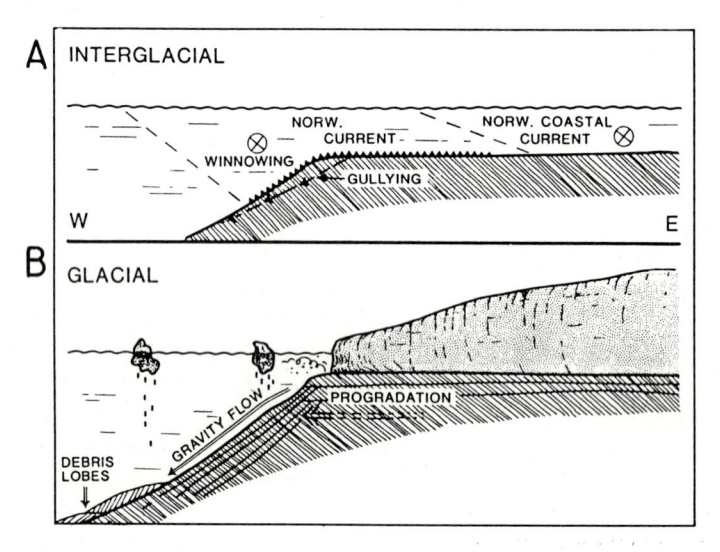

Fig. 4. Model of the formation of thick prograding sediment wedges on the shelf break and upper slope of Subarctic passive margins, **A** Wedge progradation during glacials. **B** Current-winnowing and gullying during interglacials. (Vorren et al. 1989)

et al. 1989). On the shelf, up to 300 m of stratiform glaciogenic sediments overlie an upper regional unconformity with a glacially eroded morphology. The shelf sediments comprise up to 150-m-thick seismic units, which are separated by glacial erosional surfaces. Various depositional environments have been inferred by different seismic signatures (Vorren et al. 1989), e.g., glaciomarine sediments (= semi-transparent signature), submarine outwash fans (= diverging stratified signature), glaciomarine/marine trough fills (= stratified onlap pattern), and till (= chaotic reflection). The greatest thicknesses of glaciogenic sediments (up to 1000 m) occur on the shelf break and upper slope. The depocenters are situated at the trough-mouth fans (Fig. 3). Slope sediments comprise prograding sequences with a complex, sigmoid-oblique character, indicating alternating upbuilding during glacial progradation (Fig. 4A) and depositional bypass/erosion in the topset during interglacial periods (Fig. 4B).

An alternative model, e.g., till-tongue stratigraphy (King and Fader 1986; King et al. 1987), has been considered for the up to 400-m-thick glaciomarine section on the Mid-Norwegian shelf. The till-tongue model comprises a modification of the floating ice shelf model by Carey and Ahmad 1961. The till tongues consist of wedge-shaped deposits of till (= seismically incoherent), interbedded with stratified glaciomarine sediments (= many coherent reflections). Within a till-tongue, accretion of a basal lodgement till is followed by deposition of till at the buoyancy line of a grounded ice sheet, and rain out of glaciomarine debris below a floating ice shelf in front of it (Fig. 5A). Seaward migration of the buoyancy line and subsequent retreat results in deposition of glaciomarine sediment over the till tongue (Fig. 5B), and finally glaciomarine deposition below a larger floating ice shelf (Fig. 5C). Frequently associated with the till tongue occurrences are transgressive and regressive till tongues, intraformational erosional surfaces, as well as a variety of buoyancy line moraines (Fig. 6). Recently, there has been much discussion on the applicability of the till tongue model on Quaternary Canadian and Norwegian shelf deposits. The main constraints are the existence of a floating ice shelf over wide open shelf areas (e.g., the problem of anchor points that would stabilize the floating ice shelf), and a lack of core data confirming the seismic interpretations of the till tongue model.

Due to glacio-isostatic rebound, a shallowing of 60 to 100 m has taken place in the central and northern Barents Sea during the Holocene (Salvigsen 1981). The Late Weichselian to Holocene section on Spitsbergenbanken reflects these conditions in a typical regressive facies succession (Fig. 7A, Bjørlykke et al. 1978; Elverhøi 1984). Deposition of Late Weichselian till and overconsolidated diamictons is succeeded by proximal glaciomarine sedimentation, due to gradual grounding line retreat of the Barents Sea ice sheet since about 10 ka. At about 8 to 9 ka, a sparse, marine, Arctic molluscan infauna is observed in the diamictons. Shallowing due to strong glacio-isostatic rebound after 8 ka and prior to 4 ka increased bottom current vigor. Excavation of molluscan shells and ice rafted pebbles formed a current-reworked, shell-rich, residual diamicton or a gravelly coquina lag deposit. Since about 3 to 2 ka a hard substrate epifauna, dominated by barnacles, encrusting bryozoans, and serpulides, used the gravelly shell-lag deposit as a secondary hardground. In addition, bottom currents reworked the rich benthic foraminiferal fauna, forming local sand lenses on top of the encrusted lag deposit.

Sampling of the trough and deeper bank sections (Fig. 7B,C) is incomplete. Here, environmental reconstructions are based mostly on specific features in seismic profiles (Elverhøi and Roaldset 1983; Elverhøi 1984; Elverhøi and Solheim 1987). The principal facies succession found on the shallow banks (Fig. 7A) occurs, but with increased thicknesses (Fig. 7B,C). The Holocene section comprises either local development of a thin, iron-encrusted lag deposit on deeper bank areas (Ingri 1985), or accumulation in the troughs of resuspended fine-grained sediment winnowed from diamictons on the shallow banks.

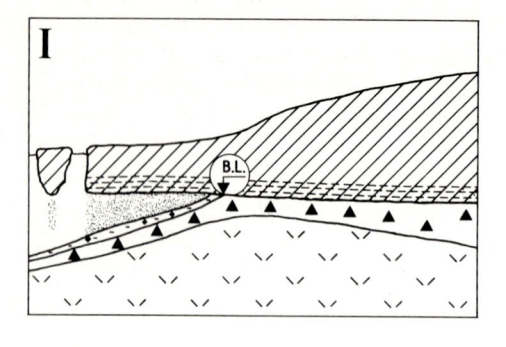

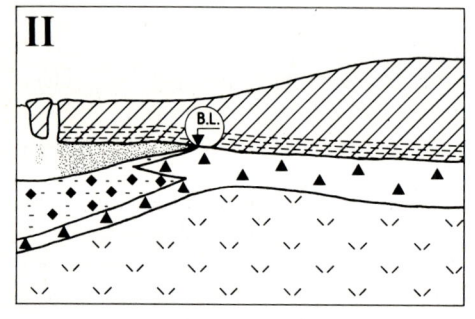

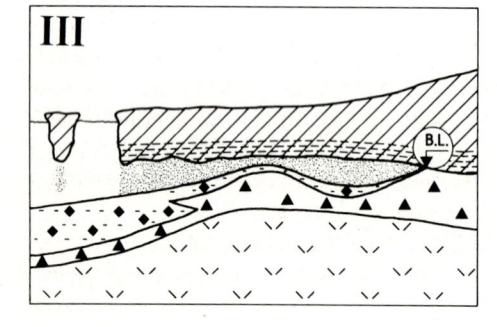

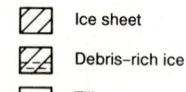

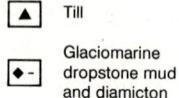

Fig. 5. The till tongue model. (After King and Fader 1986). **I** Deposition of lodgement till at the base of a grounded ice sheet, and glaciomarine sediment below a floating ice shelf. **II** Till tongue formation by seaward migration and subsequent retreat of the buoyancy line; till tongue is covered by melt-out debris from the floating ice shelf. **III** Lift off of ice from sea bed and formation of a second till tongue closer to the mainland

2.3 Cycles, Rhythms and Events in Norwegian Sea Deep Sea Cores

Offshore environments in the Norwegian-Greenland Sea have been strongly influenced by repetitive growth and decay of continental ice sheets over Greenland and Scandinavia, and by migration of the oceanic polar front. Sedimentary records from this area are excellent monitors of glacial/interglacial climatic shifts and changes in ocean circulation patterns. The two main surface current systems in the Norwegian-Greenland Sea are the inflow of warm saline North Atlantic surface water (NAW)

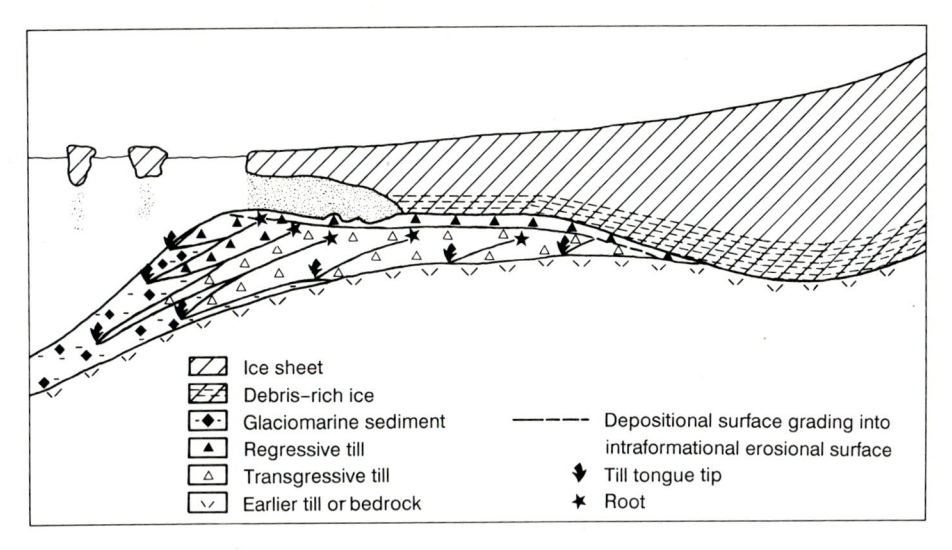

Fig. 6. Schematic representation of transgressive and regressive till tongue associations, intraformational surfaces, and buoyancy line moraine types. (After King et al. 1987)

between the Shetland and Faroe Islands (Norwegian Current), and the inflow of cold polar surface water and sea ice through the Fram Strait (East Greenland Current). Between these two major current systems, a wide area in the center of the Greenland and Iceland Seas is occupied by a mixed surface water layer called Arctic Surface Water (ASW).

Surface sediments on the floor of the Norwegian-Greenland Sea reflect the major surface water circulation pattern (Kellogg 1975). High carbonate shell productivity, e.g., high numbers of planktonic foraminifers with high percentages of subpolar species and coccolithophorids, is found under the Norwegian Current. Considerably lower carbonate fluxes dominated by the planktonic foraminifer *Neogloboquadrina pachyderma* sin., a cold-water species, indicate the presence of the Greenland Current or Arctic Surface Water in surface sediments. During glacial and deglacial periods, iceberg production and drift patterns, the density and extension of sea ice and its seasonal fluctuations, the discharge of meltwater, and the progressive intrusion or retreat of Atlantic waters in response to sea level fluctuations have enormously changed sediment characteristics in the Norwegian-Greenland Sea.

The most noticeable features of glacial/interglacial cycles in the Norwegian Sea sediments are pronounced color changes between light and dark layers. They appear both on a large scale (meters) and a small scale (centimeters to tens of centimeters). Light layers are usually thicker than dark layers and display three major lithofacies types (Henrich 1989; Henrich et al. 1989):

Facies A consists of typical interglacial deposits, consisting of brownish, strongly bioturbated foraminiferal muds and foram-nanno oozes with high carbonate, low organic carbon, and low terrigenous debris contents.

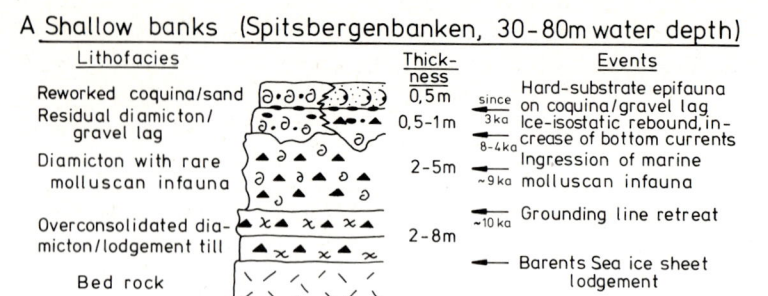

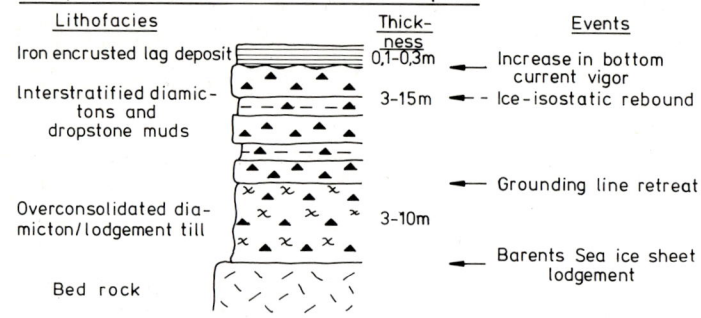

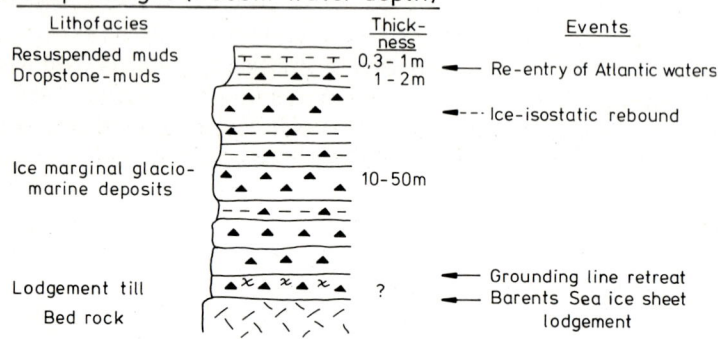

A Shallow banks (Spitsbergenbanken, 30–80 m water depth)

Lithofacies	Thickness	Events
Reworked coquina/sand	0,5 m	Hard-substrate epifauna on coquina/gravel lag (since 3 ka)
Residual diamicton/gravel lag	0,5–1 m	Ice-isostatic rebound, increase of bottom currents (8–4 ka)
Diamicton with rare molluscan infauna	2–5 m	Ingression of marine molluscan infauna (~9 ka)
Overconsolidated diamicton/lodgement till	2–8 m	Grounding line retreat (~10 ka)
Bed rock		Barents Sea ice sheet lodgement

B Deep banks (200–300 m water depth)

Lithofacies	Thickness	Events
Iron encrusted lag deposit	0,1–0,3 m	Increase in bottom current vigor
Interstratified diamictons and dropstone muds	3–15 m	Ice-isostatic rebound
		Grounding line retreat
Overconsolidated diamicton/lodgement till	3–10 m	Barents Sea ice sheet lodgement
Bed rock		

C Deep troughs (>300 m water depth)

Lithofacies	Thickness	Events
Resuspended muds	0,3–1 m	Re-entry of Atlantic waters
Dropstone-muds	1–2 m	
		Ice-isostatic rebound
Ice marginal glaciomarine deposits	10–50 m	
		Grounding line retreat
Lodgement till	?	Barents Sea ice sheet lodgement
Bed rock		

Fig. 7. Diagnostic regressive glaciomarine facies succession in response to glacio-isostatic rebound, central and northwestern Barents shelf. (Bjørlykke et al. 1978; Elverhøi and Solheim 1987)

Facies B consists of brownish, moderately bioturbated silty muds and sandy muds, with intermediate to low carbonate contents and low to moderate organic carbon values. Coarse terrigenous material is enriched, often within distinct layers. Type B sediments accumulated during early and late glacial periods, as well as during glacial periods.

Facies C, the typical glacial background sediment, consists of gray, moderately bioturbated silty muds with low carbonate, moderate organic carbon and moderate amounts of coarse, sand-sized terrigenous debris, scattered dropstones and few mud clasts.

Three major types of dark lithofacies (i.e., facies D, E, and F) are intercalated in these light sediment packages. These were deposited either during the transition from a glacial to an interglacial stage or within a glacial period with a very specific oceanographic configuration, as discussed below. All three dark diamictons consist of sandy to silty muds with very low carbonate contents (usually 0–0.3% $CaCO_3$). They show strong dissolution features on planktonic forams, high organic carbon values, high sand-sized terrigenous debris contents, as well as densely scattered dropstones and abundant mud clasts.

Facies D is characterized by its sharp top and base and its very dark gray color.

Facies E reveals a dark olive gray color and commonly has a sharp top and base, but sometimes contacts are blurred by bioturbation.

Facies F forms a complex layer consisting of a basal, very dark gray diamicton grading upward into dark olive gray diamicton. Higher parts Facies F commonly reveal brownish diagenetic laminations that cut sedimentary structures, but are themselves truncated by burrows filled with brownish sediment from above.

Mapping of facies distribution along an E-W transect across the Norwegian Sea (Vøring Plateau-Jan Mayen) reveals typical facies successions that correspond to glacial/interglacial climatic shifts. Dark diamictons occur in greatest thicknesses and highest frequency close to the continental margin (Fig. 8), decrease in number and thickness westward, and grade far offshore into oxic sediments of Facies B. This pattern indicates that the major source of the diamictons is the continental margin. Based on flux calculations, Henrich et al. 1989 assume a rapid deposition of diamictons.

Environmental parameters associated with rhythmic facies successions are displayed in Fig. 9. The most prominent rhythm is facies succession C-F-B-A (Fig. 9a), representing a complete shift from a full glacial to an interglacial climate. Other rhythmic facies successions, e.g., C-D-C, C-E-C (Fig. 9b), or C-F-B-C, commonly observed within certain glacial stages, reveal similar environmental changes but without return to or initiation of full interglacial conditions.

The environmental conditions during deposition of Facies C indicate a low carbonate shell production and some ice rafting in surface waters (Fig. 9a,b). Bottom waters are oxygen-depleted, but far from azoic, as evidenced by not very diverse benthic foraminiferal fauna and good carbonate preservation. Facies C (Fig. 10a) is deposited during progressive advance of the continental ice sheets onto the shelf. The offshore situation is characterized by a seasonally varying sea ice pack, interspersed with a few drifting icebergs.

Faciescode   Lithology

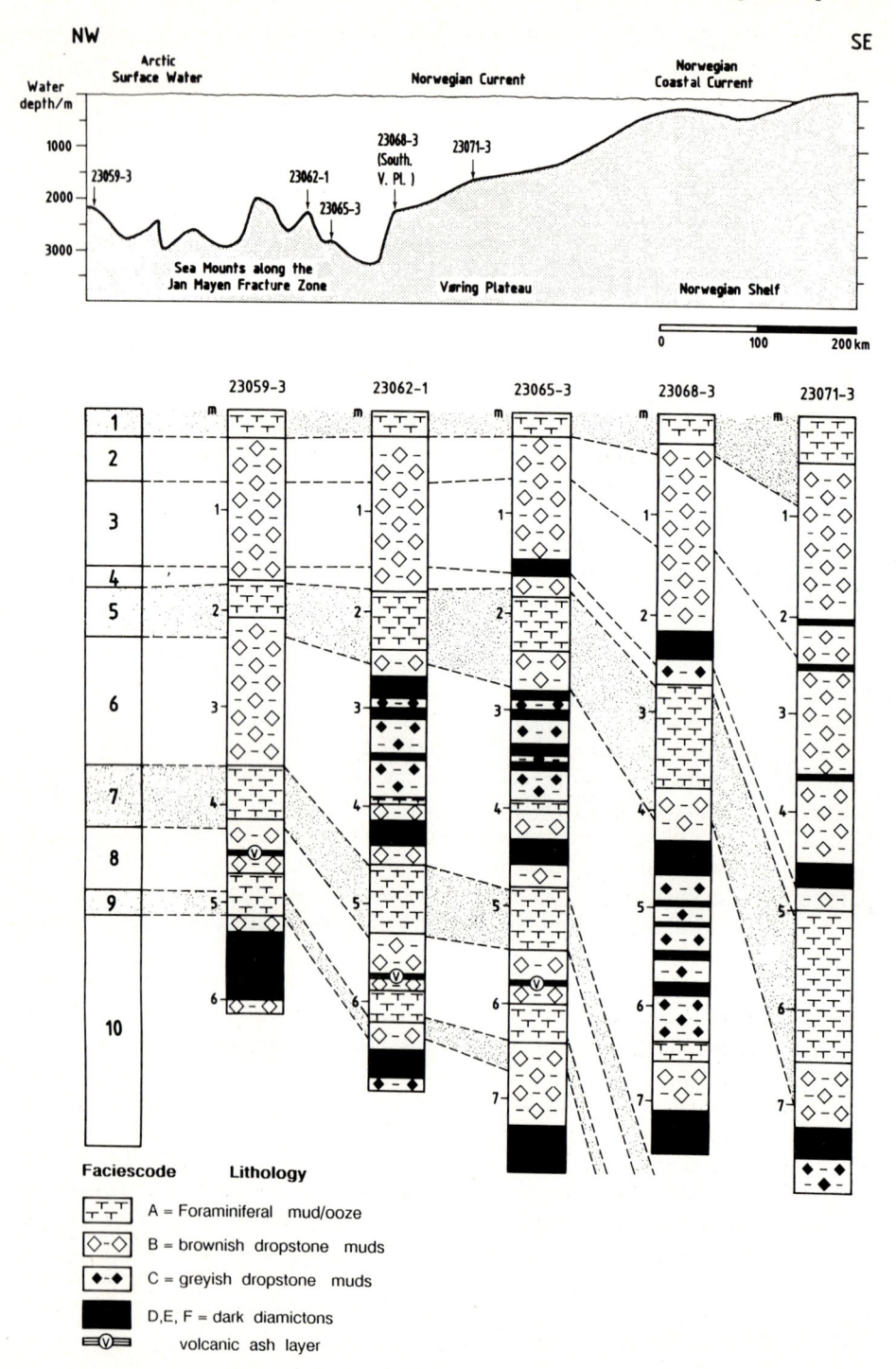

A = Foraminiferal mud/ooze

B = brownish dropstone muds

C = greyish dropstone muds

D, E, F = dark diamictons

volcanic ash layer

Facies succession C–F–B–A: Major deglaciation sequence

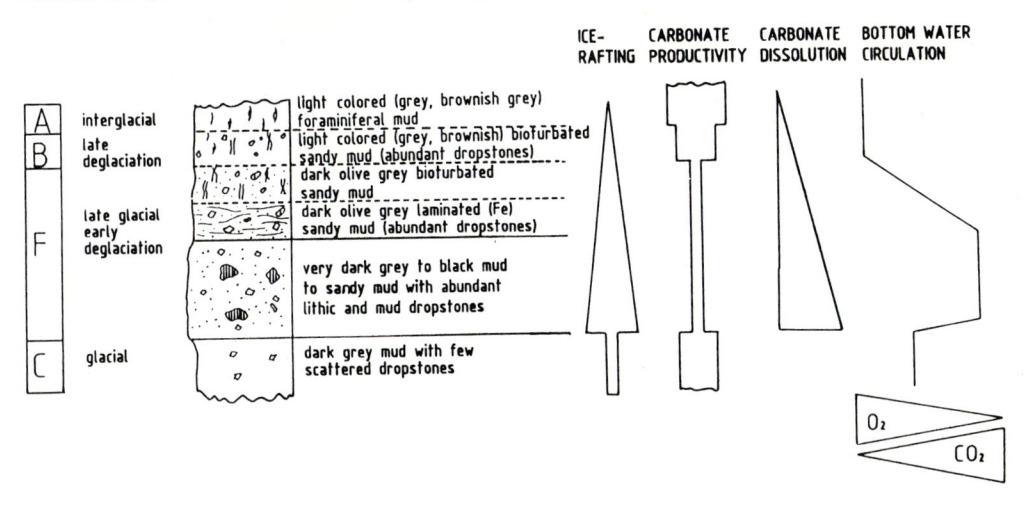

A	interglacial		light colored (grey, brownish grey) foraminiferal mud
B	late deglaciation		light colored (grey, brownish) bioturbated sandy mud (abundant dropstones)
			dark olive grey bioturbated sandy mud
F	late glacial early deglaciation		dark olive grey laminated (Fe) sandy mud (abundant dropstones)
			very dark grey to black mud to sandy mud with abundant lithic and mud dropstones
C	glacial		dark grey mud with few scattered dropstones

ICE-RAFTING   CARBONATE PRODUCTIVITY   CARBONATE DISSOLUTION   BOTTOM WATER CIRCULATION

O₂   CO₂

Facies succession C–E–C: Glacial period with continental ice sheet close to the shelf break

C	glacial		dark grey mud with few scattered dropstones
E	minor deglaciation		dark olive grey laminated (Fe) sandy mud with abundant dropstones
C	glacial		dark grey mud with few scattered dropstones

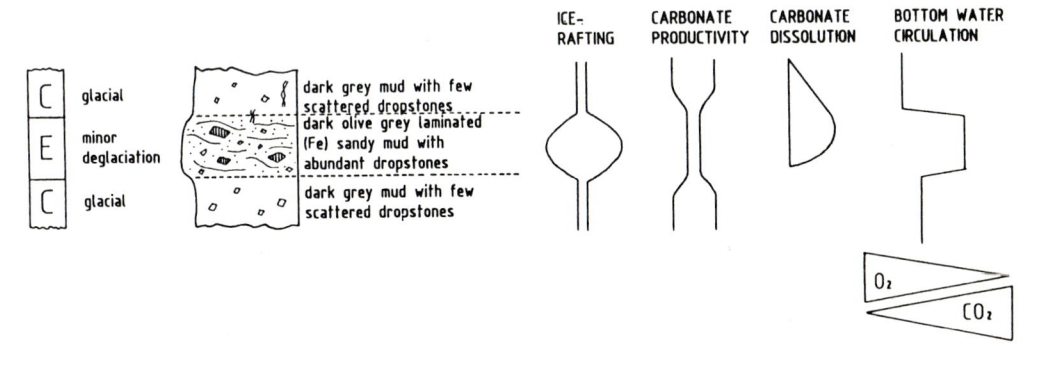

ICE-RAFTING   CARBONATE PRODUCTIVITY   CARBONATE DISSOLUTION   BOTTOM WATER CIRCULATION

O₂   CO₂

Fig. 9. Compilation of basic lithofacies successions in the Norwegian Sea deposited during the transition from glacial to interglacial periods, and during periods of maximum glacial advance and disintegration of the marine based parts of the continental ice sheets. Environmental changes in surface and bottom water properties are indicated on a semiquantitative scale. (After Henrich et al. 1989)

Fig. 8. Intercalated glaciogenic facies in Norwegian Sea pelagic environments. Temporal framework is based on oxygen isotope stratigraphy. *Dark lines* represent stratigraphic correlations along the transect. Note the E-W transitions of dark diamictons (Facies F, D and E) into dropstone muds of Facies B and C. (After Henrich et al. 1989)

Deposition of Facies F on top of Facies C reflects a drastic increase in ice rafting and a strong decrease in carbonate shell production in surface waters (Fig. 9a). Highly corrosive bottom waters, as seen in strong dissolution of carbonate tests and an interval barren of benthic organisms, suggest unfavorable bottom water conditions. Facies F (Fig. 10b) documents the maximum extension of the ice sheet close to, at, or even below the shelf edge, in addition to the early period of ice sheet retreat. During this time, large-scale calving along the tidewater front of the grounded ice sheet or frequent surges delivers huge amounts of debris-laden icebergs into the open sea. In addition, large meltwater discharges may have formed a low salinity surface lid and deposited fine-grained sediment from sediment-laden plumes. Decrease in bottom water renewal and oxidation of organic matter at this time would result in high corrosivity of bottom waters.

During deposition of Facies B on top of Facies F, a gradual decrease in ice rafting and an increase in carbonate shell production are observed (Fig. 9a). Bottom waters become progressively less corrosive, as dissolution decreases, and more oxygenated, seen in a change in redox potential of the sediment with the reappearance of benthic fauna. Facies B (Fig. 10c) represents ice sheet retreat on the shelf and the major deglaciation, characterized by rapid retreat of the low salinity surface water lid towards the coast. Rapid rise in sea level most probably caused sudden disintegration of the marine based parts of the continental ice sheets. Offshore, carbonate productivity increased, and bottom waters returned to oxygenated conditions.

The topmost facies, Facies A, records a high interglacial carbonate productivity and completely oxygenated bottom waters (rich benthic fauna).

The above outlined paleoceanographic model on the formation of dark diamictons might prove a useful tool for offshore paleoceanographic reconstructions. It recognizes glacial stages with strong advances of the continental ice sheets on the shelves and permits correlation of the open ocean records with the shelf records.

2.4 Sedimentary Cycles in Antarctic Continental Slope Deposits

The Antarctic continental margin is a high relief margin due to strong isostatic depression and glacial overdeepening by variations in the volume of the huge Antarctic ice cap. The Antarctic margin is characterized by its wide, extended ice shelf environments, developed under extreme polar aridity (Anderson et al. 1980). Under these conditions, very restricted meltwater production and supply of ice rafted debris results in generally low sedimentation rates on the shelf and slope (Anderson et al. 1980, 1983; Fütterer et al. 1988).

Calved icebergs drift predominantly within the Antarctic coastal current (Fütterer et al. 1988), a pattern which is reflected by an increase in IRD contents of surface samples. Furthermore, glacial debris is restricted almost entirely to the basal portions of shelf ice, with most of it deposited at the ice shelf front; rain out of the remainder on the shelf is restricted because of the very low temperature of the shelf waters. Most of it is released at the shelf-slope break, where upwelling of warmer water masses occurs. As a consequence, biogenic siliceous sedimentation, with only a minor contribution by IRD, dominates over wide areas.

Contour currents at the shelf break and upper slope as well as strong currents on the shelf cause intensive winnowing and form wide-spread residual lag diamictons. These sediments provide the substrate for a hardground fauna, with coldwater corals, bryozoans and barnacles contributing biogenic carbonate debris to the residual lag diamictons (Elverhøi and Roaldset 1983). The overdeepened slopes of the Antarctic continental margin are characterized by intensive downslope resedimentation processes, while the deep basins act as sediment traps.

Glacial/interglacial cyclic sedimentation patterns have been recognized on slope terraces of the continental margin off Cape Norvegia in the Weddell Sea (Grobe 1986, 1987; Fütterer et al. 1988). Shelf deposits in this region are characterized by a "paratill facies" (Grobe 1986), an intensively winnowed, gravelly diamicton, with an admixture of biogenic carbonate (mainly bryozoans) and opaline silica (sponge spicules). Sediments at the base of the upper continental slope consist of the so-called "morainic facies" (Grobe 1986), i.e., gravelly diamictons and sediment gravity flows deposited during the last glacial maximum when grounded ice extended to the shelf break and a floating ice shelf continued over the slope.

Cyclic sediments on the slope terrace off Cape Norvegia were deposited in response to advance and retreat of ice shelves and sea ice during glacial/interglacial climatic changes (Grobe 1986, 1987). Cyclicity is indicated by variations in grain size, IRD content, clay mineral assemblages, biogenic silica, pelagic carbonate content and preservation, as well as sedimentary structures and bioturbation features (cf. Fig. 11). Three principal facies, i.e., "cold", "transitional", and "warm" facies have been defined.

The "cold" facies consists of silty muds with only minor IRD, biogenic carbonate and silica admixtures, rare bioturbation, and a clay mineral assemblage dominated by smectite (Fig. 11). Reduced biogenic production and weak IRD contribution are thought to result from a permanent sea ice cover over the area during glacial periods (Fig. 12a). Carbonate preservation is rather poor, with the CCD located at a water depth of 2000 to 3000 m.

The "transitional" facies is comprised of silty muds with larger (but still low) IRD contents, slightly coarser grain sizes, a significant contribution of radiolarians, as well as a low abundance of planktonic foraminifers (Fig. 11). The increase in grain size is interpreted as reflecting an increased calving of icebergs from rapidly refloated ice shelves during late glacial time. High biogenic silica production may reflect enhanced upwelling conditions over the slope, in response to stronger katabatic winds on surface waters with less extensive sea ice cover (Fig. 12b).

The "warm" facies consists of strongly bioturbated foraminiferal muds dominated by the cold water planktonic foraminifer species *N. pachyderma* sin. and high IRD contents (Fig. 11). High carbonate fluxes are rather unusual in Antarctic surface waters. They are attributed to the presence of ice-free waters in the Weddell Sea Polynya (Fig. 12c). Carbonate preservation is generally good, indicated by a deep CCD at a water depth of 3500 to 4000 m. High IRD contribution results from rapidly calving glaciers and ice shelves during interglacial periods, enhanced in early stages by the global sea level rise (Fig. 12b/c). Icebergs drift mainly within the Antarctic Coastal Current, as evidenced by higher IRD contents in the surface sediments (Fütterer et a. 1988).

a) **Facies C:**

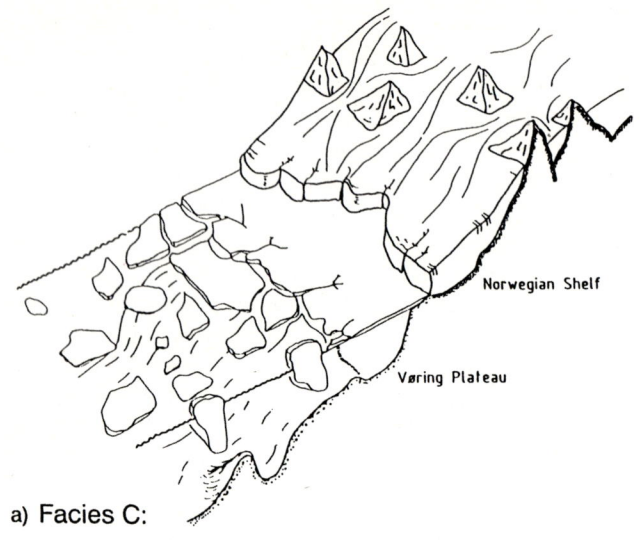

Advance of continental ice onto the shelf

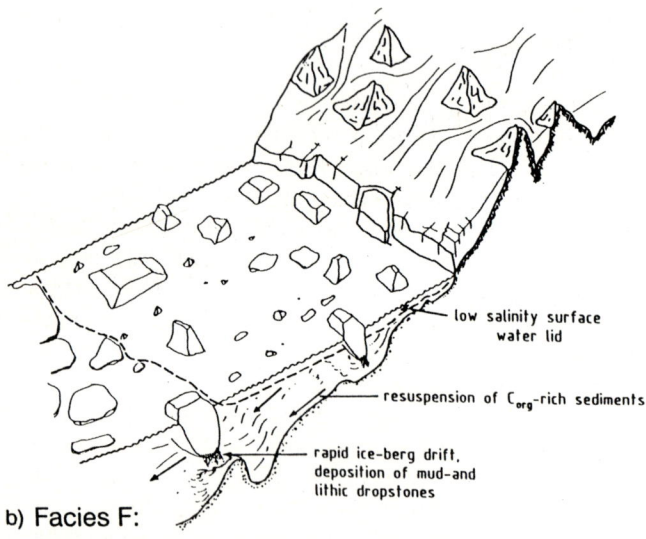

b) **Facies F:**

Ice sheet close to the shelf break during glacial maximum and its retreat in late glacial / early deglacial time

Fig. 10. Conceptual paleoenvironmental model on the formation of dark diamictons in the Norwegian Sea pelagic realm. **a** Strong glaciation, reflecting advance of continental ice on the Norwegian coastal shelf associated with offshore pack ice drift. **b** Late glacial/early deglacial strong iceberg drift, induced by surges of the marine-based parts of continental ice. **c** Intrusion of the Norwegian Current, with retreat of iceberg drift progressively closer to the coastal regions and offshore increase in carbonate productivity. (After Henrich et al. 1989)

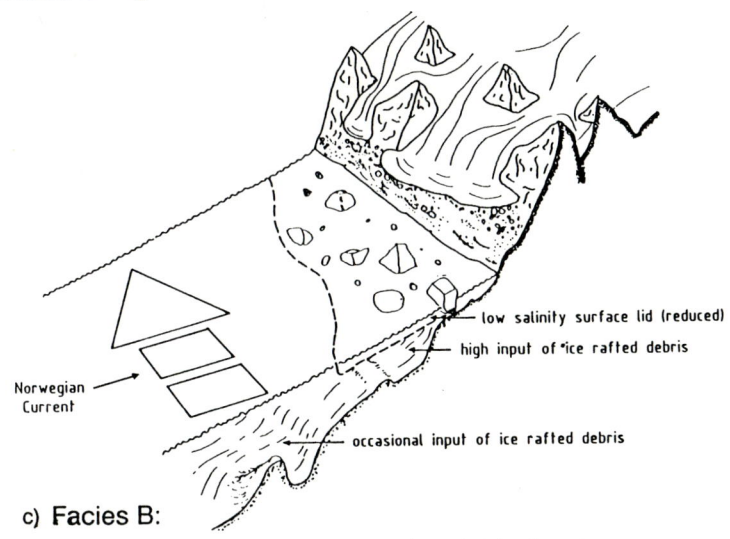

c) Facies B:
Rapid deglaciation on the shelf and intrusion of Atlantic water

3 Summary and Conclusions: a Comparison of High Input (Sub-Arctic) Versus Low Input (Antarctic) Glaciomarine Depositional Cycles, Rhythms, and Events

Subarctic glaciomarine deposits contain a large variety of facies, due to:

1. complex water mass configurations in the open ocean and on the shelves, e.g., Atlantic surface waters (Norwegian Current) versus cold polar waters (Greenland Current, Arctic Surface Water, Spitsbergen Current);
2. changes in sea ice cover, iceberg production, and ice rafting of debris; and
3. variable discharges of meltwater along ice fronts. Pronounced changes in water mass characteristics and configurations in response to glacial/interglacial climatic shifts result in deposition of cyclic and rhythmic facies successions, both in offshore and in shelf environments.

Antarctic glaciomarine environments are characterized by the prevalence of cold polar waters over the shelves, with seasonally variable sea ice coverage and negligible meltwater input. Glacial/interglacial shifts in water mass properties and configurations mainly influence the outer shelf and continental slope environments.

Advance/retreat cyclicity of the large glacial ice sheets reshaped the topography of the Antarctic and Arctic shelves. In response to strong isostatic depression, deep basins formed on the inner Antarctic shelves, while the Subarctic shelves are characterized by deep, glacially eroded troughs and intermittent shallow banks, arranged along the flow direction of the major ice streams.

Due to their rather shallow average depths and complex topographic structures, facies patterns on the Subarctic shelves have been strongly influenced by sea level fluctuations and isostatic rebound. As a result, regressive facies successions are observed which were deposited by grounded ice sheets and within proximal

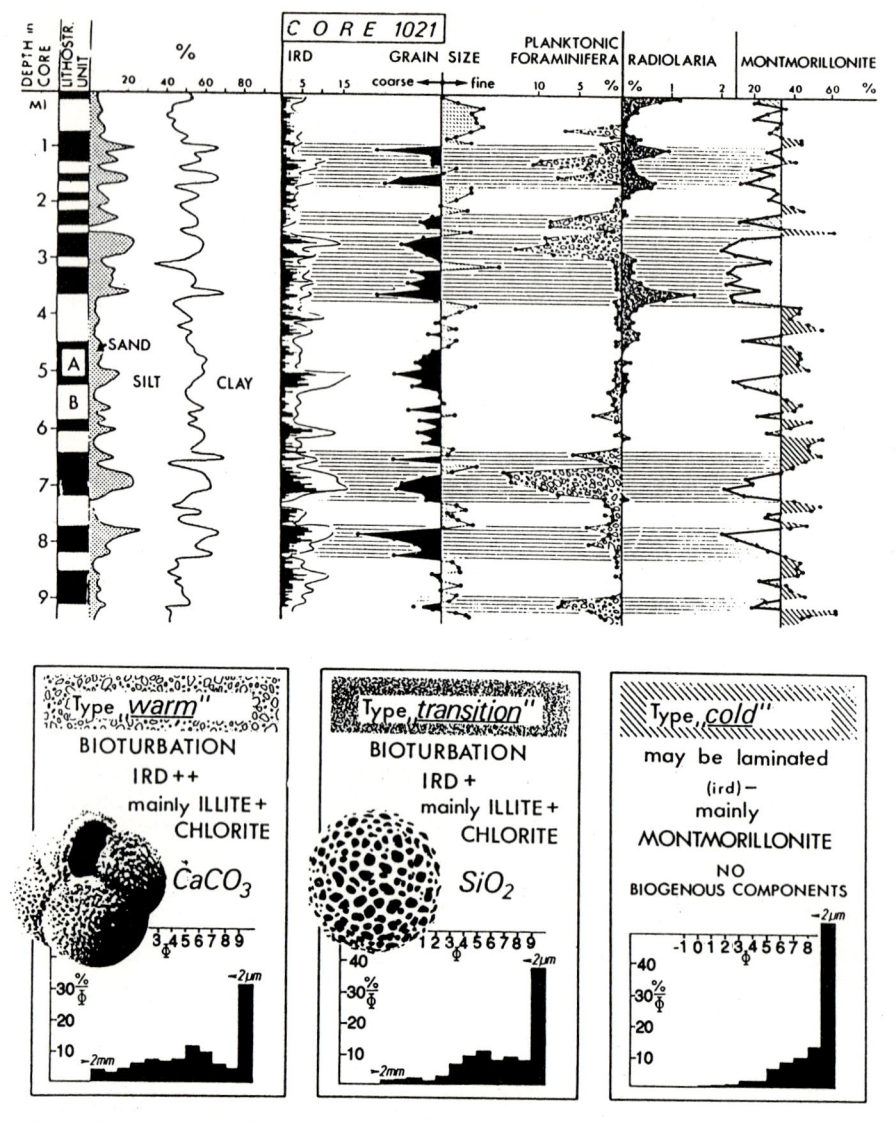

Fig. 11. Distribution of sedimentological parameters in Core 1021, Cape Norvegia, Weddell Sea, upper slope terrace. Cyclic pattern of "cold", "transitional", and "warm" facies indicated by cyclic shifts in IRD content, grain size, and biogenic components. (Grobe 1986)

glaciomarine sediments during glacial time, followed by a high energy reworked top sequence as eustatic sea level rose during late glacial isostatic uplift. The high energy deposits consist of lag deposits on shallow and deep banks and accumulations of fine-grained, winnowed sediment in the troughs. On shallow banks such as Spitsbergenbanken, a polygenetic biogenic carbonate lag deposit with a current-excaved, soft bottom molluscan infauna and a secondary, hard substrate, high energy epifauna

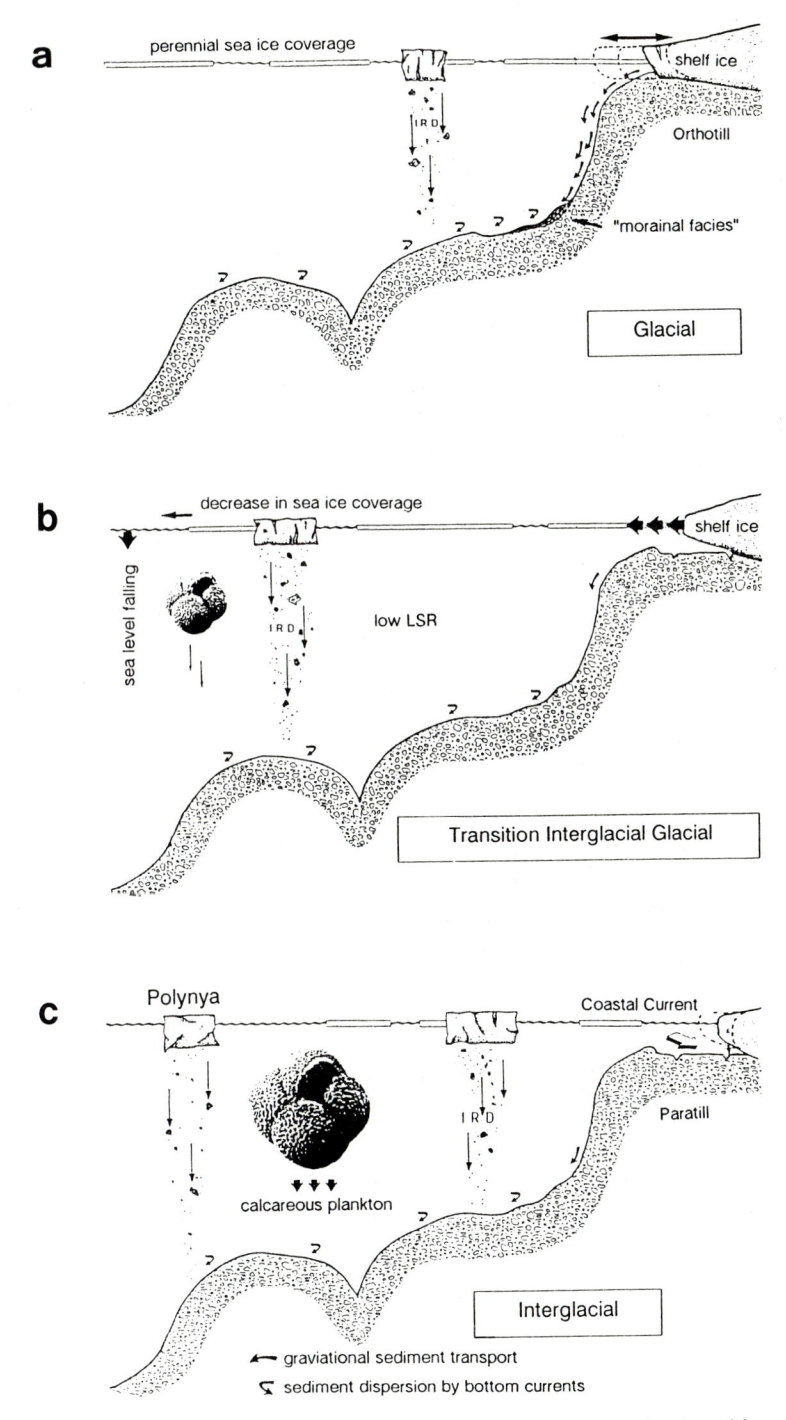

Fig. 12. a Environmental reconstruction of glacial facies with reduced biogenic productivity, deposition of fine-grained sediment by dowslope currents, and a shallow CCD at about 2000 m. **b** Environmental reconstruction of transitional glacial to interglacial facies, with increased calving of icebergs from rapidly refloated ice shelves during late glacial time and high biogenic silica production during early interglacial time. **c** Environmental reconstruction of interglacial facies with high biogenic calcareous productivity at the outer slope terrace and a deep CCD at about 3000 m. (After Grobe 1987)

(bryozoans, barnacles, and serpulids) forms the top unit. On deeper banks, lag deposits are frequently iron-encrusted.

So far, no similar rhythmic facies successions have been reported from the Antarctic shelves. Here, biogenic siliceous muds or gravelly diamictons with admixtures of a monogenetic, biogenic, hardground epifauna (e.g., bryozoans and siliceous sponges) have been deposited at low sedimentation rates. The low sedimentation rate and the overall small thickness of the glaciomarine unit is the result of a combination of various factors, e.g., intensive winnowing by the Antarctic Coastal Current, low melting rates of icebergs in cold shelf waters, low debris concentrations in the cold polar ice masses, and deposition of a thin blanket of lodgement till during glacial advance.

Subarctic shelf sections that were less affected by reworking during deglacial glacio-isostatic uplift record advance/retreat cycles of continental ice sheets grounded on the shelves; these are identified by thick, complex transgressive and regressive till and associated glaciomarine deposits. No such pattern has been recognized on the Antarctic shelves.

Glacial/interglacial cyclicity in open ocean environments of the Norwegian-Greenland Sea is best recorded by cyclic variations in carbonate productivity of surface waters (e.g., variations in the strength or absence of the Norwegian Current) and, to lesser degree by cyclicity in IRD fluxes. IRD fluxes, terrigenous, reworked organic carbon, and dissolution records reflect rhythmic patterns that correlate with major advances and retreats of the continental ice sheets close to the shelf break. Variations in meltwater discharge, surges, and melting rates of icebergs form oscillating low salinity surface water lids offshore, that may have decreased deep water renewal in the eastern Norwegian Sea and episodically resulted in corrosive bottom waters. A rather humid glacial climate would favor these configurations through a strong moisture supply and through alternating freeze/thawing conditions on the base of the grounded ice sheets.

Multiple glacial/interglacial cyclicity in the Antarctic has been reported from the upper continental slope off Cape Norvegia in the Weddell Sea. The cyclicity corresponds to advance/retreat cycles of the Antarctic ice sheet across the shelf, episodic development of floating ice shelves, variations in the extension of sea ice coverage, cyclic variations in surface water productivity induced by changes in upwelling conditions over the upper slope, as well as cyclic development of a polynya within the Weddell Sea sea ice cover.

Acknowledgments. I gratefully acknowledge comments on the manuscript by Dr. A. Elverhøi, Dr. D. Fütterer, Dr. S. Pfirman, Dr. J. Thiede and Dr. T. Vorren. I offer my special thanks to Mrs. S. Körsgen, Mrs. O. Runze, Mr. S. Rumohr and to my wife Claudia for assistance in preparing the final draft of the manuscript. This is publication No. 74 of the Special Research Project (SFB 313) at Kiel University.

Chapter 7 Timing and Correlation

7.1 Time Span Assessment – an Overview

W. Ricken

1 Introduction

The assessment of the amount of time inherent in sedimentary rocks (i.e., the "timing") provides the base for a detailed investigation of various depositional and biological processes addressed in this book. Without a precise time control, the depositional mechanisms forming beds and sequences cannot be sufficiently understood. Such limitation is obvious for rhythmic bedding and cyclic sequences on various scales, where the interpretation of the repetition (i.e., periodic, non-periodic, or stochastic) directly depends on the success of time span determinations. This is also true for detailed studies of event sequences and the assessment of erosional gaps. Although providing the base for a detailed investigation of many depositional and biological processes, timing has remained an elusive problem. Too many inaccuracies are involved in resolving stratigraphic durations, including a large range of error in radiometric age determination, poor biostratigraphic as well as magnetostratigraphic resolution, and an incompleteness of sedimentary sections. As a result, time estimates are commonly imprecise, and the range of error is often larger than the actual time span considered, especially when the time spans of small intervals, such as beds, have to be evaluated.

It seems therefore appropriate to find solutions to time span estimations other than those usually applied, using relative timing methods. The amount of time inherent in sedimentary rocks or stratigraphic gaps can be expressed to chemical changes by assuming particular patterns of depositional flux, or can be compared to thicknesses and compositions of laterally correlated sections. In addition, statistical methods for evaluation of bedding rhythms and time spans can be used. Unconventional, relative timing methods applied in this paper include

1. the determination of fractional sedimentation rates to derive relative depositional fluxes and time spans in vertical sections,
2. the lateral correlation of compositions and section thicknesses to calculate basin-wide relative input patterns as well as time spans,
3. the statistical evaluation of the time inherent in a particular bed using time series analyses as well as chemical parameters, and
4. the assessment of stratigraphic gaps on various scales, including gaps at the base of event beds. Such time estimation is performed using lag sediment concentrations, carbonate contents, lateral correlation of event sequences, as well as the time span dependency of sedimentation rates.

Einsele et al. (Eds.)
Cycles and Events in Stratigraphy
©Springer-Verlag Berlin Heidelberg 1991

All these timing methods are closely associated with the determination of the various depositional processes forming sequences and beds. Many of these relative methods can be applied in three dimensions with respect to basin fill or bed geometry (Fig. 1), thus providing a basic tool for analyzing many of the depositional processes, resulting bed types, and sequences addressed in this book.

2 Timing and Depositional Input Variation of Larger Stratigraphic Intervals

The determination of the time span inherent in any given stratigraphic interval is commonly performed by comparing two radiometric data at the bottom and top of a given interval, or by using biostratigraphic, magnetostratigraphic and isotopic data (e.g., $^{87}Sr/^{86}Sr$ ratio) that are tied to conventional radiometric time scales (e.g., Odin 1982; Harland et al. 1982; Palmer 1983; Salvador 1985; Kent and Gradstein 1985; Haq et al. 1987; Bayer 1987). There are two sources of error involved in such a conventional time span determination:

1. The radiometric ages and time scales have a considerable range of error themselves.
2. The stratigraphic distance between two radiometric dates or biostratigraphic zone boundaries is often much larger than the interval of the section which is to be dated. For the interpolation down to smaller intervals, it is usually assumed that time is linearly distributed (Fig. 2); in consequence, the precise timing of smaller stratigraphic intervals is a rather difficult procedure.

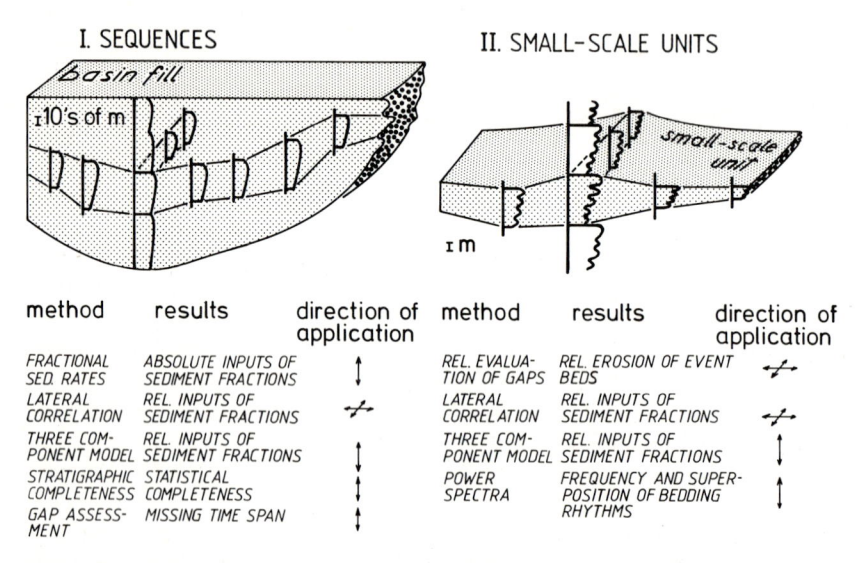

Fig. 1. Time span assessment and sediment input analysis for sequences and beds. Diagram presents basic methods and the direction of application with respect to sequence or bed geometry

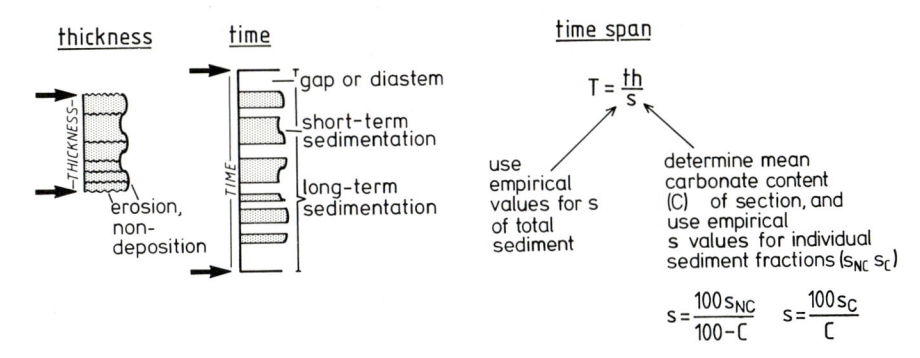

Fig. 2. Basic procedures for sedimentation rate and time span determinations. Relationships between stratigraphic thickness (*th*), time span (*T*), and sedimentation rate (*s*). Determination of sedimentation rates by using the individual sedimentation rates for the noncarbonate (*sNC*) and carbonate fraction (*sC*). Calculation of *sNC* and *sC* is based on the mean carbonate content (*C*)

2.1 Fractional Sedimentation Rates and the Duration of Stratigraphic Sections

Unlike the timing method explained above, the time span represented in stratigraphic intervals can be estimated through the employment of standard, long-term sedimentation rates. This is achieved by dividing the thickness of a given stratigraphic interval by the typical sedimentation rate for the sediment or rock type under consideration (Fig. 2). The time span (T) of a sedimentary interval equals

$$T = \frac{th}{s} \text{ with} \tag{1}$$

$$s = \frac{th}{T}, \tag{2}$$

where th is the rock or sediment thickness, and s the average long-term sedimentation rate for a given environment. In this simple approach to time span estimation, sedimentation rates are used that are derived from compacted rocks in various environments, averaged over large time intervals. The error from using average sedimentation rates for time estimations may be large, because long-term sedimentation rates are variable within most individual environments by 0.5 to 2 orders of magnitude (see Fig. 6a).

The assessment of time spans can be more accurately addressed by using fractional sedimentation rates, instead of the average, whole rock sedimentation rates as utilized above. Such fractional sedimentation rates are defined as representing the individual sedimentation rates of compositional sediment or rock fractions (Fig. 2). Fractional sedimentation rates are especially used for rocks composed of a carbonate and a noncarbonate fraction (e.g., in shales, marls, and limestones). Such fractional sedimentation rates can be transformed into conventional, whole rock sedimentation rates by employing the mean carbonate content of the considered stratigraphic interval. Thus, the whole rock sedimentation rate (s) is expressed as

$$s = \frac{100 \, s_{NC}}{100-C} \quad \text{and} \tag{3}$$

$$s = \frac{100 \, s_C}{C}, \tag{4}$$

where s_{NC} and s_C are the sedimentation rates of the noncarbonate and carbonate fractions, respectively, and C is the mean carbonate content. The time span inherent in a given stratigraphic interval can be derived when Eqs. (3) and (4) are substituted for the s value used in Eq. (1).

Under what conditions is it appropriate to apply fractional sedimentation rates? They can be used when the sedimentation rate of one fraction can be more precisely estimated than that of the other fraction; or when the sedimentation rate of one fraction is relatively constant, while that of the other fraction varies greatly. An example would be when the clay fraction in pelagic carbonates has a similar long-term sedimentation rate as that of deep sea clay; or, when the pelagic carbonate fraction of hemipelagic marls has a typical long-term sedimentation rate similar to that of pure chalks.

An instructive example (Fig. 3) is presented by modeling the Gubbio section in central Italy (see Fischer Chap. 1.2, Schwarzacher Chap. 7.5, both this Vol.). Mean carbonate contents in this entirely pelagic sequence range between 70 and 95% (Arthur 1979b); these values have been averaged for all stratigraphic stages, in order to smooth out diagenetically enhanced and lowered carbonate contents found in individual marl-limestone couplets. The whole rock sedimentation rates and the durations of the stratigraphic stages are calculated according to Eq. (3), by assuming that the noncarbonate content of the sequence is equivalent to typical deep sea clay with a long-term sedimentation rate of $s_{NC} = 1$ m/Ma (see Fig. 6a). These calculated sedimentation rates are then compared in Fig. 3 with actual sedimentation rates derived individually for the carbonate and noncarbonate fractions according to the time scale of Palmer (1983), and by solving Eqs. (3) and (4) for s_{NC} and s_C. The overall pattern of the two types of sedimentation rates (and calculated time spans) is for the most part similar, indicating that the carbonate input to the Gubbio sequence was rather variable, while the clay input remained relatively constant, especially for the lower and the upper part of the sequence.

The whole rock sedimentation rates and the associated time spans derived for equally thick stratigraphic intervals contained in the Gubbio sequence rely on the mean carbonate contents. The higher the carbonate contents averaged for such intervals, the larger were the whole rock sedimentation rates, and the shorter the related time spans. This is illustrated through a special presentation of the Albian in the Gubbio section (Fig. 3), depicting small-scale variations in carbonate contents, sedimentation rates, and time spans averaged for 10-m-thick intervals, by assuming a steady noncarbonate input. Essentially, this pattern also seems valid for smaller units down to the scale of bundles, when sediment inputs are derived by assuming that each bundle represents an equivalent time period (Herbert et al. 1986). Individual beds are commonly subjected to diagenetic overprints, and their input pattern is more difficult to evaluate.

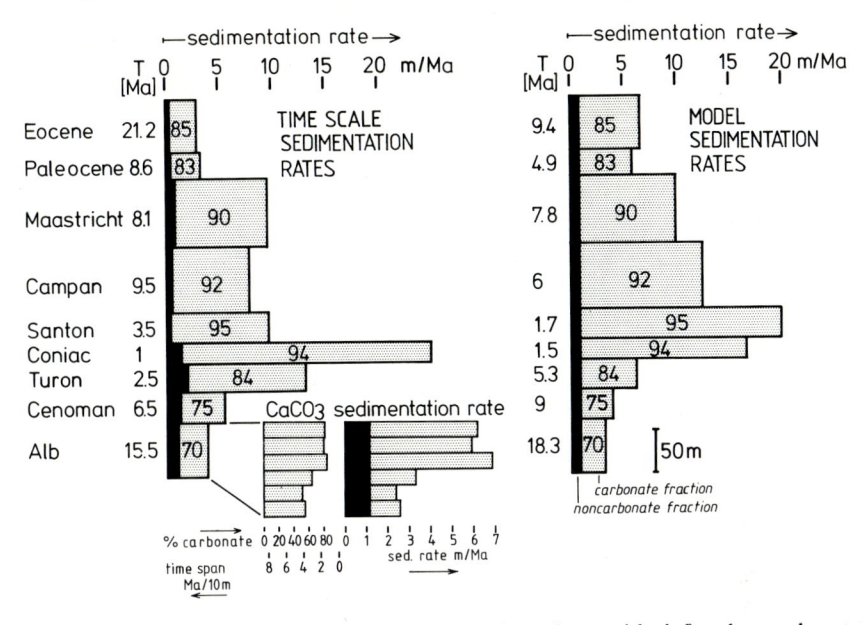

Fig. 3. Time scale-derived and model sedimentation rates for the carbonate (*shaded*) and noncarbonate fraction (*black*) of the Gubbio section, Italy, based on carbonate data from Arthur (1979b). *Numbers on the left side* of the two diagrams indicate time spans for stratigraphic stages (*T*, in Ma), while *numbers within shaded parts* of histogram indicate mean carbonate content (w%). Model sedimentation rates and time spans (*T*, in Ma) are determined by assuming a constant sedimentation rate for the noncarbonate fraction of 1 m/Ma. Inset presents the Albian with smaller-scale carbonate variations and supposed variations of sedimentation rates and time spans, based on an assumed constant noncarbonate input of 1.2 m/Ma

When carbonate input is largely constant, however, the above-mentioned relationships would be reversed. Theoretically, both types of sequence can develop, with deposition dominated by either carbonate or noncarbonate (i.e., terrigenous) variations, and with transitions between these end members. Large carbonate variations with only minor variations in clastic input are thought to occur in pelagic sequences far away from nearshore influences, or in areas of very high carbonate production (i.e., platforms and reefs), whereas dominant terrigenous variation is related to clastic shelf environments. Such end members form sequences where time spans and carbonate contents are easily determined. This aspect is further discussed in the following section, with respect to laterally correlated stratigraphic units.

2.2 Lateral Correlation: Relative Sedimentation Rates and Sediment Inputs

Lateral correlation procedures allow one to precisely determine relative values of stratigraphic time spans and sediment inputs. The methods are based on an accurate, lateral correlation of stratigraphic intervals, either by performing graphic correlations

(Shaw 1964) or common lithostratigraphic and biostratigraphic high-resolution correlations. Refined correlation methods presented by Kauffman et al. (Chap. 7.2, this Vol.) demonstrate that lateral correlations can be performed very accurately, spanning tens to thousands of kilometers. As follows from such precise lateral correlations, an equivalent amount of time must be represented by the various thicknesses of the correlated unit. Hence, when differences in porosity and compaction between sections are small enough to be negligible, variations in thicknesses are directly related to variations in sedimentation rates. Such variations can be expressed relative to a chosen time-equivalent reference section. As described in Fig. 4, the relative sedimentation rate of a given, laterally correlated section is defined by comparing its thickness (th_2) with that of such a reference section (th_1); the latter is assigned a relative sedimentation rate of $s_r = 1$. Hence, the relative sedimentation rate of a correlated section (s_r) is expressed by

$$s_r = \frac{th_2}{th_1}. \tag{5}$$

After such relative sedimentation rates are established for laterally correlated, isochronous intervals or subsections, individual sedimentation rates for the carbonate (s_C) and noncarbonate fraction (s_{NC}) can be calculated in order to document changes in depositional input. Such calculation is performed by substituting the relative sedimentation rate (s_r) obtained in Eq. (5) for the long-term sedimentation rate (s) expressed in Eqs. (3) and (4), and by solving these equations for s_{NC} and s_C (Fig. 4). Generally, for laterally correlated isochronous intervals, the carbonate or noncar-

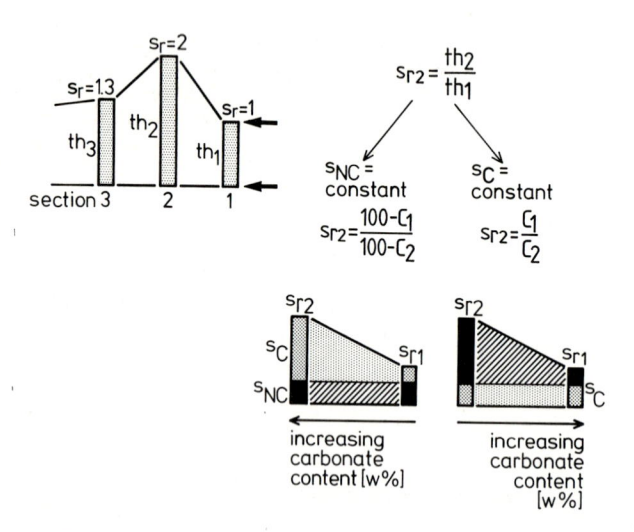

Fig. 4. Determination of sedimentation rates relative to a time-equivalent reference section. *Left side of diagram* shows lateral correlation of three sections with different thicknesses (th) and relative sedimentation rates (s_r; expressed relative to section 1). *Right side of diagram* indicates the relationship between relative sedimentation rate and carbonate content (C) for two idealized end members, with either laterally equal noncarbonate (s_{NC}, *black*) or carbonate sedimentation rate (s_C, *shaded*)

bonate fractions (or both) can change laterally, documenting various depositional regimes.

An excellent example for the promising tool of lateral correlation and related input determination is provided by an E-W transect through the Western Interior Basin, U.S.A. In Fig. 5, a 1500 km-long, time-equivalent cross section through the western part of this basin is presented, spanning the Cenomanian-Turonian boundary. Individual sections were precisely correlated, controlled by bentonite beds and biostratigraphic data, carried out by Elder (1987). In this transect, mid-basin pelagic chalks grade into marls and silty clays towards the western shore zone. In the chalk, correlated, time-equivalent sections are relatively thin, while sections become thicker towards the western shore zone. Such increasing thickness is reflected by raising relative sedimentation rates for the individual sections when a section in mid-basin position is chosen for the reference section (i.e., section Colorado 2). The depositional input pattern causing this nearshore related increase in sedimentation rate can be expressed by the determination of relative fractional sedimentation rates (s_C, s_N) using the mean carbonate content of the individual sections as explained above. The

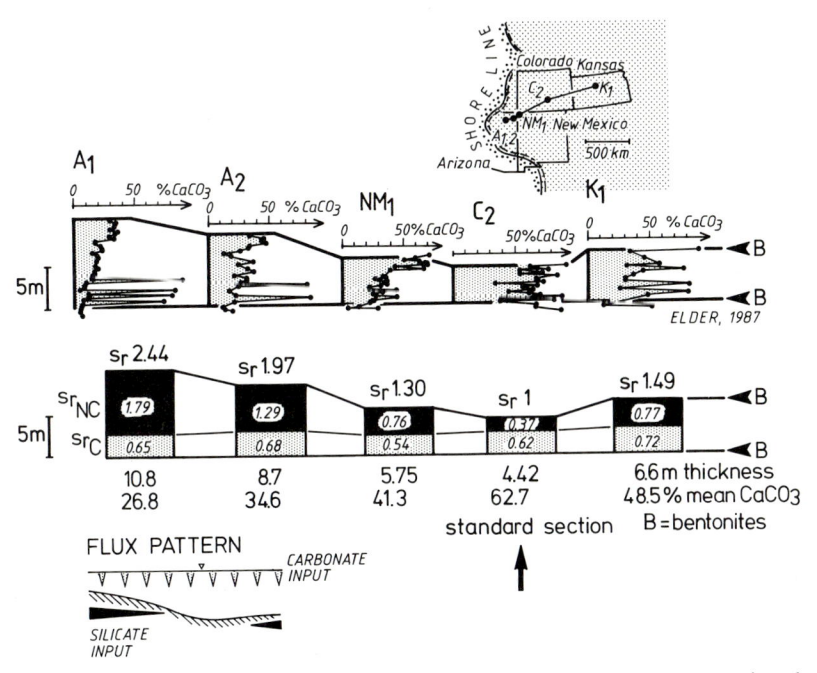

Fig. 5. Time-equivalent E-W transect through the Western Interior Basin, U.S.A., spanning the Cenomanian-Turonian Boundary. Sections (A_1 to K_1) show variations in carbonate content (w%). *Lower part of diagram* schematically depicts section thicknesses with relative sedimentation rates (s_r expressed relative to the standard section with $s_r = 1$), as well as relative sedimentation rates for the noncarbonate (s_{rNC}, *black*) and carbonate fractions (s_{rC}, *shaded*). B indicates bentonite marker bed. *Inset on the lower left side* exhibits schematic interpretation of sediment flux pattern. Carbonate data and section correlation according to Elder (1987)

noncarbonate input (i.e., terrigenous sediment) steadily increases as the western shore zone is approached, but the carbonate input (i.e., coccolithic calcareous ooze) is basically the same in all sections (Fig. 5). Hence, sections become thicker towards the west and more depleted in carbonate content solely due to increased terrigenous input; whereas equal carbonate input reflects laterally constant surface water productivity. Under such conditions, the entire 1500-km-long transect becomes an easily determinable system:

a) In isochronous, laterally correlated units, the various types of carbonate and clastic input may form two end members, with either laterally constant carbonate or terrigenous supply (see Fig. 4). The various depositional parameters of such end members can be easily determined, including section thicknesses, relative sedimentation rates and inputs of the carbonate and noncarbonate fluxes, carbonate contents, as well as time spans. In our example of a laterally constant carbonate input, the relative sedimentation rate of a given section is also expressed by dividing the carbonate content of the reference section by the carbonate content of that given section ($s_{r2} = C_1/C_2$), the carbonate content is $C_2 = C_1/s_{r2}$, the expected thickness of the given second section is $th_2 = C_1*th_1/C_2$, and the relative time span is $T_{r2} = 1/s_{r2}$ (for equally-thick units).

b) The question remains whether or not this epeiric, hemipelagic foreland basin (Fig. 5) represents unusual environmental conditions because the carbonate input is so uniform over a very large distance of 1500 km. It is thought that basins with laterally even carbonate input develop in temperate and boreal settings, because such basins commonly have low shallow water carbonate production and greater planktonic carbonate production; the latter condition is prevalent since the Upper Jurassic (see Fig. 11 in Ricken and Eder Chap. 3.1, this Vol.). On the other hand, tropical basins are dominated by greater shallow water carbonate than terrigenous deposition; the clastic input may even be laterally more constant than the carbonate supply. Additionally, there are also many basins with lateral simultaneous changes of the carbonate and clastic flux, such as mixed shelves and Paleozoic basins. The depicted transect therefore belongs to an end member of various basin types with more complex input configurations, where sedimentation rates and inputs are easily determined so that relative time assessment can be precisely performed.

3 Timing and Sediment Input of Individual Beds

In assessing the time span represented within individual beds, one is faced with serious difficulties, including the fact that beds are usually not resolvable for radiometric age determinations, and carbonate contents are often considerably altered by diagenetic cementation and dissolution processes that bias indirect input determinations (Bathurst Chap. 3.2; Ricken and Eder Chap. 3.1, both this Vol.). In principle, time span and input estimations of individual nonevent beds may be performed equivalent to that in larger sequences, when beds are not affected by secondary, differential compaction and carbonate redistribution. In addition, two methods are briefly mentioned; these include time series analyses, which quantify the frequency of bedding rhythms, and relative input determinations through the evaluation of carbonate-organic carbon contents.

3.1 Time Spans Inherent in Individual Beds in Various Environments

As for larger sequence, the timing of beds can be estimated using two procedures: (1) The time span of larger intervals between two radiometric dates or other datums is divided equally by the number of beds, or (2) the bed thickness is divided by typical values for long-term sedimentation rates. Both methods are restricted through the nonlinear distribution of sediment thickness and associated time spans, the occurrence of stratigraphic gaps, diagenetic overprints, and through scale effects in sedimentation rates; the latter are explained in Section 4.4.

In Fig. 6a, long-term sedimentation rates are compiled according to data presented by Schwab (1976), Sadler (1981), Seibold and Berger (1982), Scholle et al. (1983a), Stow et al. (1985), and Anders et al. (1987). Altogether, for various environments, the sedimentation rates span four orders of magnitude from 1 to 10 000 m/Ma. In contrast, most sedimentation rates for individual environments vary by only one to two orders of magnitude, except deep sea clay, siliceous ooze, and lacustrine sediments, which have smaller ranges spanning 0.3 to 0.8 orders of magnitude.

Taking a common bed thickness in most sedimentary rocks of 30 cm, the time span contained in such a bed varies substantially for the various environments, related to the various magnitudes of sedimentation rate. Such standard beds in deltaic environments represent, on the average, 0.02 to 10 ka of deposition, in turbidite environments 0.1 to 10 ka of deposition, and in continental slope, shelf, and fluvial environments anywhere from 1 to 40 ka (Fig. 6a). Only standard beds in deep sea clay, radiolarites, chalks, carbonate platforms, reefs, and cratonic basins are, on the average, within the Milankovitch frequency band with periodicities between 21 and 400 ka (see Fischer Chap. 1.2, this Vol.). Of these environments, deep sea clay, radiolarites, and chalks may reflect bedding rhythms related to orbital periodicities. Reefs, carbonate platforms, and cratonic basins, however, with long-term sedimentation rates similar to those of the deep sea environments, reflect sedimentation rates controlled by the low subsidence in these environments. In such shallow water environments, the instantaneous sedimentation rate is larger than the long-term subsidence rate, but the sedimentary record is discontinuous and frequently interrupted by stratigraphic gaps, restricting the depositional expression of Milankovitch rhythms to periods of rising and elevated sea level (see Einsele and Ricken Chap. 1.1, this Vol.).

Consequently, only in pelagic to hemipelagic environments may individual *beds* reflect orbital, Milankovitch-type variations. In most other environments with higher sedimentation rates, *sequences* which are a few to tens of meters thick (i.e., at the scale of bundles and parasequences) may be related to Milankovitch frequencies. However, such assumptions are not always valid, as such sequences can also be formed by various depositional processes not related to Milankovitch variations, but which are controlled through autocyclic, tectonic, and eustatic processes (see Einsele

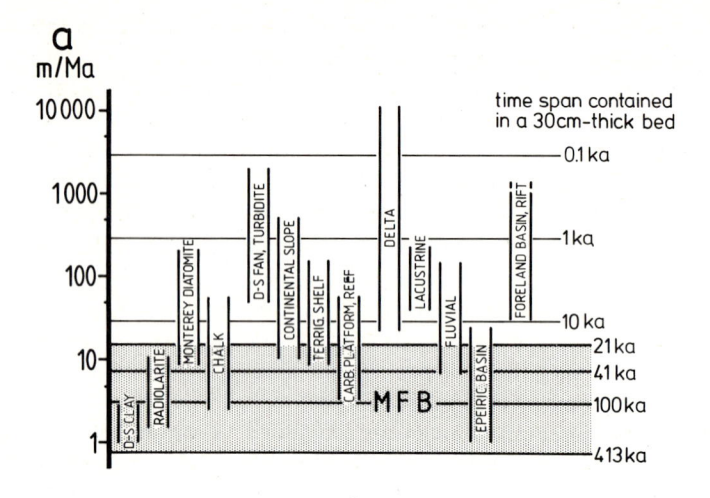

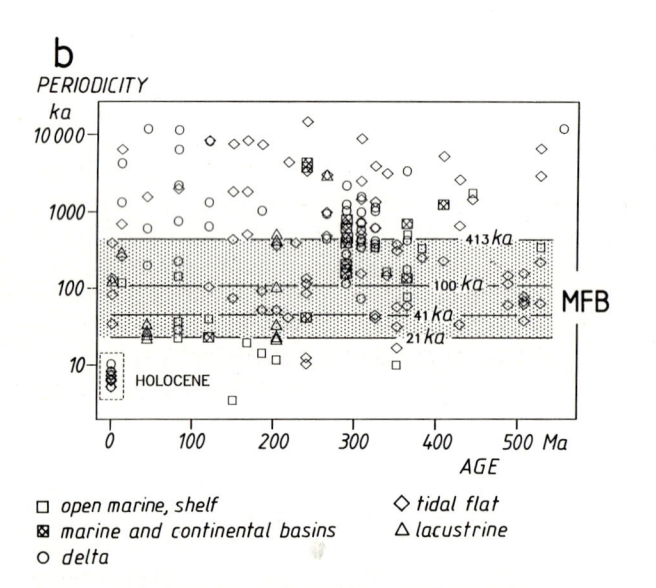

Fig. 6. a Major long-term sedimentation rates (in m/Ma) compiled after several authors as referred in the text. *Horizontal lines* indicate time spans (in ka) inherent in 30-cm-thick standard beds for environments with various sedimentation rates. Note that only a few environments have standard beds within the Milankovitch frequency band (*MFB, shaded*). **b** Periods of Phanerozoic smaller-scale cycles. Except for Paleozoic coal cycles, no significant clustering around the periods of the Milankovitch frequency band (*MFB*) is observed. (After Algeo and Wilkinson 1988)

et al., Introduction, this Vol.). Algeo and Wilkinson (1988) showed that the amount of time represented by depositional cycles found in various environments span three orders of magnitude with no significant clustering in the Milankovitch frequency band except Paleozoic coal cycles (Fig. 6b; see Riegel Chap. 6.7, this Vol.). The fact that the average duration of 30-cm-thick beds in pelagic environments is coincident with the broad frequency band of Milankovitch cycles, however, is no proof that their bedding is in fact due to orbital-climatic forcing. In the following, a reliable technique for testing for orbital frequencies, contained in the bedding rhythm, is briefly described.

3.2 Time Series Analysis

In order to test bedding rhythms for orbital forcing (i.e., Milankovitch cycles), one has to demonstrate that the distinctive superposition of orbital frequencies is contained in bedding rhythms above the level of depositional background noise (see Fig. 3 in Einsele and Ricken Chap. 1.1, this Vol.). Such testing tools are time series analyses, which are described in detail in articles by Weedon (Chap. 7.4) and Schwarzacher (Chap. 7.5, both this Vol.). Therefore, only a brief overview is presented herein. Time series analysis is based on treating lithologic variations as a superposition of different wave lengths; constant intervals of sediment or rock thickness are thereby assumed to represent equal amounts of time. As described by various authors, time series analysis is performed by digitizing lithology, and by numerically separating wave trains by using Fourier or Walsh transformations; the results are plotted as "power" spectra which show the significance of certain rock thicknesses for the whole spectrum (e.g., Schwarzacher 1964, 1975; Dunn 1974; Berger et al. 1984; Weedon 1986, 1989). Such thicknesses are then tied to time periods by essentially using the above described methods. A precise time span estimation may be performed when the power spectra indicate that the wave hierarchies reflect orbital ratios (such as 1:2:5:20) which represent the frequencies of precession, obliquity, and the two eccentricity cycles. However, there is evidence that the ratio of orbital frequencies has changed through Earth's history (Berger et al. 1987).

Because of gaps and varying sedimentation rates, time series analysis has to be performed individually for different but overlapping parts of a larger section (see Weedon Chap. 7.4, this Vol.). Power spectra are not free of producing artificial effects, depending on sampling methods, data processing, and filtering procedures. A further complication comes into play, because rhythmic bedding is typically related to sedimentation rate changes between succeeding carbonate-rich and carbonate-poor beds (see Arthur and Dean Chap. 1.7; Ricken Chap. 1.8, both this Vol.). As a consequence, equal intervals of sediment thickness in rhythmic sequences do not represent equal amounts of time, as assumed in time series analysis. Weedon (Chap. 7.4) and Schwarzacher (Chap. 7.5, both this Vol.) show that such sedimentation rate changes and subsequent processes of differential compaction influence the magnitude and the position of peaks in the frequency analysis (i.e., power spectra).

3.3 Time Spans of Beds Rich and Poor in Carbonate Content

An organic carbon-carbonate model for the assessment of depositional inputs, sedimentation rates, and time spans inherent in rhythmically bedded sediment is already presented in Part I (see Ricken Chap. 1.8, this Vol.). Therefore, only a few comments are made here. In this model, bedding rhythms are viewed as being generated by either depositional variations in individual sediment fractions (i.e., the carbonate, clastic, or organic fractions) or by combinations of these three types. Chalk-marl alternations formed by such depositional variations are related to distinctive patterns in carbonate and organic carbon contents; such patterns in turn can be utilized for the relative determination of depositional inputs and time spans.

Evaluation of the organic carbon and carbonate contents has shown that rhythmic bedding, related to either depositional variation of the carbonate or clastic fraction, is in fact closely associated with large variations in sedimentation rates. Such variations are expressed by rhythmically varying bed thicknesses, with thicker beds reflecting higher sedimentation rates. Productivity and carbonate dissolution rhythms have thicker limestone beds than carbonate-poor interbeds, whereas the opposite is observed for rhythms owing to terrigenous dilution (see Arthur and Dean Chap. 1.7; Decker Chap. 4.1; Einsele and Ricken Chap. 1.1 and Ricken Chap. 1.8, all this Vol.). Consequently, for the various types of rhythmic bedding, equally thick intervals in carbonate-rich and carbonate-poor interbeds represent different time spans.

4 How to Treat Stratigraphic Gaps

4.1 The Hierarchy Concept of Stratigraphic Gaps

It has long been known that many depositional environments show episodic sedimentation, and that most stratigraphic sequences contain some sort of temporal gaps (Barrell 1917; Reineck 1960; Ager 1973; van Andel 1981; Sadler 1981; Schindel 1982; Behrensmeyer 1983; Tipper 1983; Dott 1983, 1988; Mc Kinney 1985; Badgley et al. 1986; Wetzel and Aigner 1986; Algeo and Wilkinson 1988). In general, stratigraphic gaps are formed by various geological processes which represent various magnitudes of time span, including many types of sequence boundaries and smaller gaps associated with depositional processes. Stratigraphic gaps are schematically expressed through a hierarchy concept that classifies them according to their duration, genesis, and whether or not they are associated with event deposition (Fig. 7).

The hierarchy system of stratigraphic gaps must reflect the modern view on the organization of the sedimentary record, in which sediments are grouped into various orders of sedimentary sequences and subsequences (see Vail et al. Chap. 6.1, this Vol.). Commonly, such sequences contain considerable stratigraphic gaps at their bases and tops (e.g., Wheeler 1958; Sloss 1963, 1988; Vail et al. 1977b; Algeo and Wilkinson 1988). Major sequence-boundary gaps are related to larger unconformities of various origin (e.g., base-level variations, tectonic processes, etc.) and may comprise ca. 1 to 10 Ma (Salvador 1987; Sloss 1988). Within these sequences,

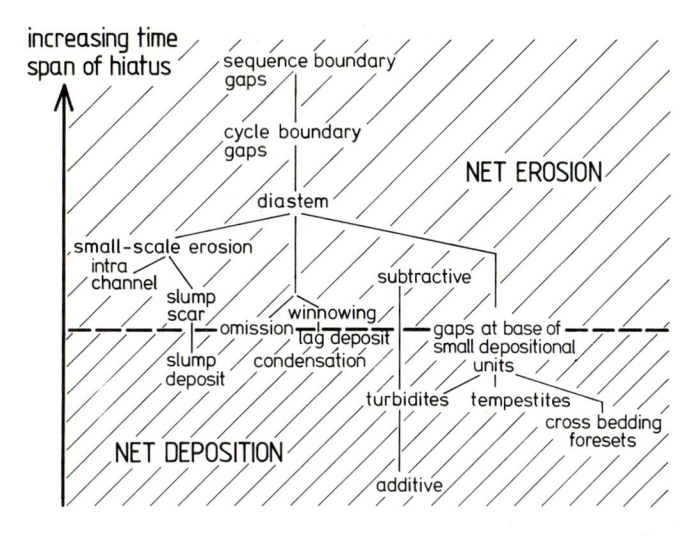

Fig. 7. Schematic representation of type and hierarchy of stratigraphic gaps. Gaps are grouped according to magnitude, genetic type, and net effect in terms of erosion and deposition. For detailed explanation see text

smaller, sedimentary units with boundary gaps of various magnitude are superimposed, as known from many offlap-onlap cycles representing supercycle sets, supercycles, and third order cycles according to the sea level curves presented by Haq et al. (1987) and Vail et al. (Chap. 6.1, this Vol.). In these units, third order offlap-onlap cycles are associated with cycle boundary gaps on the order of a 100 ka to a few Ma. The time span magnitude of such gaps encountered in a shelf-basin transect generally decreases from the margin towards the depocenter of a basin (e.g., Einsele and Bayer Chap. 6.2, this Vol.). Again, these offlap-onlap cycles may be grouped in smaller units or parasequences (Haq et al. 1987), including fining-upward and coarsening-upward cycles, and bundles (i.e., sequence-scale cycles, see Einsele et al. Introduction, this Vol.), which may also contain smaller cycle boundary gaps. An example would be the widely observed unconformities at the top of shallowing-upward carbonate cycles (e.g., Goodwin and Anderson 1985; Koerschner and Read 1989; see Strasser Chap. 6.5, and Haas Chap. 6.6, both this Vol.).

All these boundary gaps of sedimentary sequences and cycles are due to non-deposition or erosion, comprising relatively large durations. The various sequence boundary gaps related to the transgression-regression processes can be discriminated by various stratigraphic condensation patterns described by Kidwell (Chap. 6.3, this Vol.). On the scale of beds, however, erosion or small breaks in sedimentation, called diastems (Barrell 1917), are generated by many processes, such as erosion without subsequent deposition (e.g., in slumps or channels), or erosion which occurs at the base of small depositional units (e.g., ripple sets; Fig. 7). The latter includes typical event beds, which usually have, despite their preceding erosion, a net depositional effect.

4.2 Gaps Related to Nondeposition and Erosion

Gaps related to nondeposition and erosion are formed by various processes, which can be distinguished through some of the taphonomic processes and condensation types as shown by Kidwell (Chap. 6.3, this Vol.). Genetically, such hiatuses may be grouped into three categories: (1) Essentially nondeposition (i.e., starvation) over large time intervals, which occurs in sediment starved basins, and under backlap and downlap conditions (see Kidwell Chap. 6.3; Vail et al. Chap. 6.1, both this Vol.); (2) repeated deposition and erosion (i.e., winnowing and bypassing), found in areas of current activity, slopes, and onlap and toplap situations, causing multiple hardgrounds and lag deposits; and (3) net erosion and transportation of the fine-grained sediment away from the site of erosion. This third type may be found in discrete channels or in large areas of wave-induced erosion observed in transgressive lags. Also, slump scars may belong to this hiatus group. The missing sediment in these gaps and the represented time spans may be assessed through field observations, as demonstrated by the various methods listed below:

a) Time span estimation can be performed by the employment of detailed stratigraphic investigations, when the time gap is larger than at least one biostratigraphic or magnetostratigraphic zone (Gebhard 1982). A minimum value of erosion can be obtained through vertical determination of truncation magnitudes, for example, between beds contained in two succeeding and unconformity-bounded sequences, or between incised channel fills and the underlying strata (Fig. 8). The time span represented by the eroded sediment can be estimated according to eq. 1, by employing either average sedimentation rates for given environments or sedimentation rates derived from datable sections underlying the stratigraphic gap in question.

b) When hiatuses are represented by lag deposits, the amount of missing sediment can be assessed by utilizing the enrichment of relic constituents. Lag deposits are formed during intervals of essential nondeposition, repeated erosion and deposition, or nearly complete erosion. Solid sediment constituents (e.g., larger skeletal hard parts, clasts, and small-sized concretions), which are difficult to transport and break up, are incorporated in the forming lag deposit, while the fine-grained sediment matrix is removed; this process is called hiatal concentration (Kidwell Chap. 2.4, this Vol.). The degree of hiatal concentration can be used for assessing the amount of nondeposited or eroded sediment, under the assumption that the solid constituents contained in the eroded sediment previously had the equivalent concentration found in the preserved sediment (or sedimentary rock) below the lag deposit (Fig. 8). Thus, the thickness of the eroded or nondeposited sediment (e) can be estimated by comparing the number of concentrated constituents in the lag deposit with that contained in the underlying sediment or rock,

$$e = \frac{N_L}{N}, \tag{6}$$

where e is the thickness of eroded or nondeposited sediment in meters, N_L is the number of relic components that occur in the lag deposit for a chosen horizontal width, and N is the number of components contained in the reference rock below the lag deposit. This reference rock is assigned a thickness of one meter and the same width as that used for measuring the lag deposit (Fig. 8). The time span represented by this missing sediment can be assessed by employing essentially the same methods as listed above.

The assesment of erosion using hiatal concentration can be applied to stratigraphic gaps related to essential nondeposition and winnowing, repeated erosion and deposition, and to widespread, continuous erosion. Lag sediments forming channel fills should not be employed for this method, as the presence of channels indicates lateral transportation. The amount of missing sediment is calculated as being too high when relic constituents are transported from elsewhere into the lag, or when the breakage and additional production of skeletal parts by soft bottom and gravel dwellers remain undetected (i.e., taphonomic feedback; see Kidwell Chap. 2.4, this Vol.). All these restrictions augment the original number of constituents contained in the lag deposits and appear to enhance the required amount of erosion, while the transportation of shelly parts away from the lag will reduce the quantity of missing sediment.

4.3 Gaps Associated with Depositional Events

Erosion at the base of event beds (i.e., tempestites and turbidites) can be generally recognized by the presence of various sedimentary structures, such as sharp lithologic contacts, undulating lower surfaces, flute casts, burrow casts at the event bed base, and rip-up clasts (see Seilacher and Aigner Chap. 2.3; Eberli Chap. 2.8; and Einsele

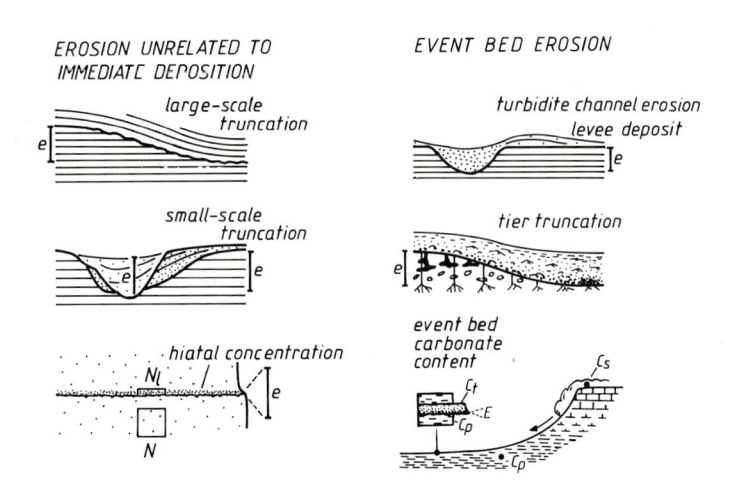

Fig. 8. Assessment of the missing sediment (i.e., erosion, e) for various types of stratigraphic gaps. *Left side of diagram*, erosion unrelated to subsequent deposition; *right side of diagram*, event bed erosion. N_l and N indicate number of constituents contained in the lag deposit (*stippled*) and in the underlying rock, respectively. C_t, C_s, and C_p symbolize carbonate contents in the turbidite, source area, and in the background sediment, respectively

Chap. 2.7, all this Vol.). Some basic methods may be employed to estimate the amount of event bed erosion, as listed below.

a) The amount of truncation of underlying beds, for example, below a turbiditic channel fill, provides a measurement of channel erosion relative to the interchannel areas (Fig. 8). In addition, the comparison of truncated trace fossil tiers with complete standard tiers (see Savrda et al. Chap. 5.2, this Vol.) allows an estimation of the minimum amount of event bed erosion (Wetzel and Aigner 1986). Furthermore, the magnitude of fauna mixing between source area and the area of deposition can be used as a measure of erosion. Valuable hints on source area composition and the degree of intra-basinal erosion may be obtained through exotic components and rip-up clasts (see Einsele Chap. 2.7, this Vol.).

b) The carbonate content of turbidite sediment may serve as an indicator of the amount of incorporated basinal sediment during turbidity current flow, when the source and deposition areas have different carbonate contents. This is true for many shelves to basin facies transitions, when reefs, carbonate platforms, and calcareous shelves interfinger with siliciclastic basinal sediments (see Eberli Chap. 2.8, this Vol.), or when siliciclastic shelves or shore lines pass into deeper water carbonate oozes. The amount of erosion and sediment incorporation during turbidity current flow, and the resulting carbonate changes in the subsequently deposited event bed, may be schematically expressed through various transitions between the following two end members:

1. The mean carbonate content of the event bed equals that of the source area, when no additional erosion occurs within the basinal area.
2. The carbonate content of the event bed may also be very similar to the basinal sediment when large quantities of it are incorporated. As a consequence, the amount of incorporated sediment can be estimated by comparing the mean carbonate content of the turbidite or tempestite sediment (C_t) with that of both the source area (C_s) and the nonevent background sediment (C_p, Fig. 8). The percentage of incorporated sediment (E) contained in a given event bed at a given site in the basin is then expressed as

$$E\,[\%] = \frac{100\,(C_s - C_t)}{C_s - C_p}. \tag{7}$$

The application of this carbonate method is limited by several geological constraints. Erosion and incorporation of background sediment can theoretically occur within the entire transport route between source area and the depositional site in question. In addition, shelf to basin carbonate changes may be gradual, and fractionation of the carbonate and noncarbonate portions can occur during turbidity current flow. Further limitations arise when the carbonate content of the source area (C_s) cannot be directly measured and must be indirectly derived. This may involve the study of components transported in the event beds, including clasts of source area sediments contained in tempestitic to turbiditic channel fills, and transported fossils which indicate the environment of the source area (e.g., reefal). For lithified sections, differential cementation between beds may cause additional complications. Methods to correct for such effects are discussed in Ricken (1986, 1987).

c) Under ideal conditions, the amount of event bed erosion may be ascertained by comparing the portion of background sediment of an event sequence with that of a time-equivalent, nearby reference section (Fig. 9a). This reference section, composed entirely of nonevent sediment, is more complete than the event section, where the background sediment is partly eroded. When event beds erode more sediment than they subsequently deposit, the event sequence may be termed "subtractive", because it becomes smaller in thickness than the reference section. On the other hand, a sequence (e.g., in a turbidite fan) may be termed "additive" when erosion is less than event bed deposition, making the event section thicker than the reference section. The transition between subtractive and additive sections of a sequence is characterized by "constant" relationships in terms of section thickness, found for a situation in which event erosion equals event bed deposition (Fig. 9a).

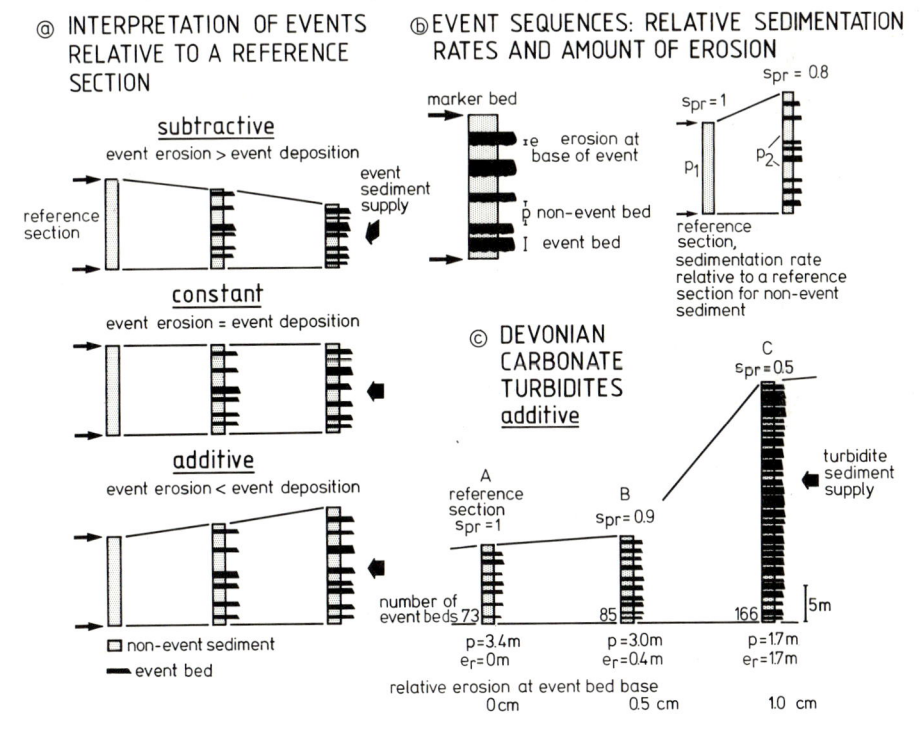

Fig. 9. Relative assessment of event bed erosion through lateral correlation. **a** Net effect of erosion and deposition in event sequences by comparison with isochronous, nonevent reference sections. **b** Turbidite sequences composed of event beds and pelagic background sediment (p); erosion (e) occurs at the event bed base. Relative sedimentation rates of the background sediment (s_{pr}) is determined through lateral correlation of time-equivalent sections; low values of background sedimentation indicate event bed erosion. **c** Example shows estimation of relative event bed erosion (e_r) associated with carbonate turbidites, Devonian of Rheno-Hercynian Sea, West Germany; based on sections described by Eder et al. (1983)

Thus, the three types of event sequences may hint at the amount of erosion associated with event deposition. Amalgamated event beds with only the lower members of the Bouma sequence deposited (see Einsele Chap. 2.7, this Vol.) point to subtractive sequences with a large amount of erosion, while thick intercalations of background sediment and the deposition of only the upper members of the Bouma sequence indicate little erosion and additive sections. A basic problem for the application of comparative methods is the difficulty of distinguishing precisely between the fine-grained upper part of the turbidite and the pelagic interbed (see Piper and Stow Chap. 2.9, this Vol.).

The amount of time represented in an event sequence is largely due to the presence of nonevent sediment (e.g., pelagic sediment) and the missing, eroded nonevent sediment at the event bed base. The time span inherent in the deposition of individual event beds can be neglected when evaluating the duration of the whole sequence (Fig. 9b). In other words, the sequence duration (T) is expressed as the thickness of the nonevent (e.g., pelagic) portion (p) plus the amount of erosion at the event bed base (e) divided by the nonevent sedimentation rate (s_p)

$$T = \frac{p+e}{s_p}, \tag{8}$$

where the amount of erosion at the event base (e) is

$$e = s_p * T - p. \tag{9}$$

For the determination of the nonevent sedimentation rate (s_p) one can use the previously-explained methods and data compilations (see Fig. 6a), while the sequence duration (T) has to be determined by independent means (e.g., biostratigraphy). This method only allows an unreliable estimate of the amount of event bed erosion, because of the errors involved in sedimentation rate and time span estimations. A more precise result is obtained when event bed erosion (e_r) is expressed relative to a time-equivalent, laterally correlated reference section, resulting in the equation

$$e_r = p_1 - (p_1 * s_{pr}) = p_1 - p_2; \text{ with } s_{pr} = \frac{p_2}{p_1}, \tag{10}$$

where p_1 and p_2 are the nonevent thicknesses of two time-equivalent sections, with p_1 as the reference section, and s_{pr} as the nonevent sedimentation rate (relative to that reference section). The application of Eq. (10) is essentially a comparison of the nonevent proportions of two or more laterally correlated sections. This is illustrated in the following example:

Figure 9c schematically shows an additive, time-equivalent sequence of carbonate turbidites which originated tens of kilometers from a Middle Devonian reef complex situated in the Rheno-Hercynian Sea (West Germany), after Eder et al. (1983). If the most distal turbidite section is defined as the reference section (section A) with a nonevent sedimentation rate of $s_{pr} = 1$, relative sedimentation rates for the nonevent portion of sections B and C are $s_{pr} = 0.9$ and 0.5, respectively. Such low sedimentation rates indicate, that a portion of background sediment was eroded and incorporated in the turbidite beds; this portion is equivalent to 0.4 and 1.7 m of

Fig. 10. Time span dependency of sedimentation rates in calcareous ooze. Sedimentation rates (in m/Ma, *vertical axis*) are plotted against the time span for which they were derived (*horizontal axis*). Data show median rates for 22 500 data points. *Triangles* and *circles* represent two ways of data processing. *Triangles* indicate decompacted sedimentation rate values for constant thickness intervals, while *circles* indicate decompacted values for constant time intervals, *Lower end* of "stalk" shows compacted sedimentation rates. Diagram also depicts lines of constant interval thickness in meters, and the periodicities of the Milankovitch frequency (*MFB*). (After Anders et al. 1987)

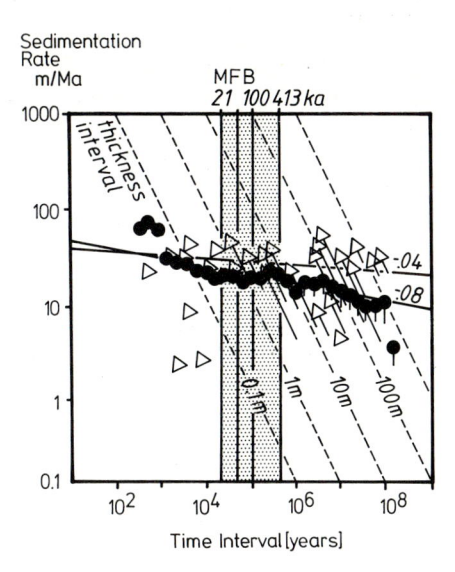

relative erosion for the two sequences, respectively. When dividing these figures by the number of event beds per section, the average, compacted amount of erosion at the event bed base relative to the reference section, equals 1 and 0.5 cm, respectively, which is approximately 5% of the turbidite thickness.

4.4 Testing Completeness of the Stratigraphic Record

Reineck (1960), Sadler (1981), and Schindel (1982) have introduced elegant stratigraphic completeness tests based on a statistical evaluation of large data sets of sedimentation rates. The authors made the observation that mean sedimentation rates averaged for sections comprising long time spans included many gaps of nondeposition and erosion. As a result, these long-term sedimentation rates were relatively low. However, sedimentation rates determined for small time spans (i.e., short-term sedimentation rates) were relatively large, because of a higher stratigraphic completeness. The comparison between such long-term and short-term sedimentation rates would then give a measurement of the stratigraphic completeness of the larger sections.

The success of this method depends highly on the quality of the utilized sedimentation rates and techniques used in data processing. Sadler (1981) plotted 25 000 sedimentation rates against the various time spans for which they were derived. For differing environments, the sedimentation rates drastically decreased with increasing time span over several orders of magnitude. A recent interpretation suggests that this extreme time span dependency was an artifact due to improper data processing (Anders et al. 1987). A new compilation of 22 500 sedinentation rates for calcareous ooze show that this environment is much more complete than previously thought, though not entirely free of gaps, as there is a slight decrease in the mean sedimentation rate with increasing time interval for which the sedimentation rates were derived (Anders et al. 1987; Fig. 10). For the timing questions addressed herein, these results would mean that carbonate ooze is incomplete by 10 to 40% for large stratigraphic intervals of roughly 10 Ma. Such values are obtained when short-term sedimentation rates on the time scale of beds (i.e., 20 ka) are used to determine the theoretical thickness of larger sections. An incompleteness on this order of magnitude was also estimated by Schwarzacher (1987a, and Chap. 7.5, this Vol.) for various pelagic sections in central Italy by employing a random walk simulation of the sedimentation process. These data, which are derived for sediments representing a relatively quiet, stable environment, may indicate the average magnitude of stratigraphic completeness contained in sedimentary sections.

5 Discussion and Conclusion

Time span estimations (i.e., timing) serve as basic instruments for describing and interpreting depositional processes forming sedimentary rhythms and events, as addressed by most of the articles in this book. Related to a better understanding of the sedimentary mechanisms involved, time span estimations have become more precise and methodically diversified, providing the scale on which various (i.e., cyclic and noncyclic) depositional, taphonomic, and evolutionary processes occur. As demonstrated herein, time span estimations are closely associated with the derivation of depositional inputs of sediment fractions, which provide the ultimate clues to depositional processes forming various sedimentary types and sequences. In this context, the amount of stratigraphic completeness of the rock record, as well as the origin and distribution of stratigraphic gaps, are major questions.

Sedimentary time span estimations are faced with two general problems. These are
1. the disagreement between sediment thicknesses and associated time spans, related to variations in sedimentation rates and the presence of stratigraphic gaps; and
2. the usually large errors involved in common stratigraphic age determinations, related to weak biostratigraphic and magnetostratigraphic resolution and substantial errors in radiometric age determination.

In order to avoid such constraints, time span determinations are addressed here by using promising, but indirect procedures. They include fractional sedimentation rates, relative depositional input determinations, numerical investigations of frequencies expressed in the bedding rhythms, and time span determinations for stratigraphic gaps. Some of these methods are semi-quantitative, and all the methods are limited by geological constraints. However, some of these indirect methods are more precise than conventional time span estimations. Such new tools in timing are based on the lateral correlation of sections and comparison of carbonate contents, the relative determination of sedimentation rates and depositional inputs, and the quantitative treatment of stratigraphic gaps based on lag sediment composition and event bed carbonate content. Most of these methods can be performed three-dimensionally with respect to sequence and basin geometry. In the following, the various time span estimation techniques are summarized for the two major stratigraphic units addressed in this book (i.e., smaller sequences and individual beds). In addition, the time spans represented by various types of stratigraphic gap can be assessed as they are related to sequence boundaries, hiatal concentration in lag deposits, and incorporation of basinal sediment prior to event bed deposition.

Time span estimations for smaller sedimentary sequences. Sedimentary sequences may comprise various time spans according to the various environments and sedimentation rates which they represent (Algeo and Wilkinson 1988). For sequences composed of a carbonate and noncarbonate fraction, a relatively precise time span estimation can be performed when the sedimentation rate of the carbonate or noncarbonate fraction can be derived individually (i.e., fractional sedimentation rates). Such fractional sedimentation rates are obtained for vertically succeeding stratigraphic intervals and laterally correlated isochronous sections; correlation techniques are summarized by Kauffman et al. (Chap. 7.2, this Vol.). As a consequence of lateral correlation, sedimentation rates can be accurately expressed relative to a reference section by comparing the thickness of time-equivalent stratigraphic units. When fractional sedimentation rates are combined with the mean carbonate contents, changes in depositional inputs (e.g., the carbonate and noncarbonate fractions) can be obtained in both vertical and lateral direction in respect to the basin fill. Fractional sedimentation rates therefore provide ultimate clues to the depositional input patterns and the involved styles of sedimentary basin fills.

Time span estimations for individual beds. Mean sedimentation rates and associated time spans for individual beds vary considerably from one environment to the other. For a standard, 30-cm-thick bed, only deep sea clay, siliceous sediment, and chalk seem to overlap with the Milankovitch frequency band, while the similarly low sedimentation rate of shallow water carbonates and continental basins reflects the low subsidence rate of these environments. The average time span inherent in 30-cm-thick beds of other environments (e.g., turbidite, slope, shelf, deltaic, fluvial, and lacustrine) is, for the most part, considerably shorter than orbital frequencies, ranging solely between 0.02 and 10 ka.

The significance of orbital frequencies above stochastic depositional background variation can be demonstrated through the application of time series analyses, demonstrating frequencies of bed thicknesses contained in the bedding rhythm (see Weedon Chap. 7.4, Schwarzacher Chap. 7.5, both this Vol.). Frequency ratios of 1:2:5:20 may indicate the Milankovitch hierarchy of orbital variation related to the periods in precession, obliquity and eccentricity. The use of time series analysis for the detection of orbital frequencies is limited, because of systematic as well as arbitrary variations in sedimentation rates and the existence of stratigraphic gaps. Diagenetic processes and differential compaction influence the frequency analysis of bedding rhythms (Weedon Chap. 7.4, Schwarzacher Chap. 7.5, both this Vol.). In addition, relative time spans and sediment inputs associated with individual beds contained in non-event sediment can be assessed through the evaluation of carbonate – organic carbon contents, as described by Ricken (Chap. 1.8, this Vol.).

Time span estimation for stratigraphic gaps. Statistical evaluation of many thousand sedimentation rates by Sadler (1981) and Anders et al. (1987) have provided new insight into the incompleteness of the stratigraphic record. According to these authors, sedimentation rates decrease systematically, when derived from sections comprising large time spans, because gaps with longer durations are included in such sections.

Such frequently occurring stratigraphic gaps may be grouped into two major types, including (1) gaps associated with sequence boundaries, which commonly represent larger time spans; and (2) shorter time gaps related to various depositional processes. The missing time spans represented by the various sequence boundary gaps may be ordered into a hierarchical pattern. Such a pattern is thought to reflect the hierarchical organization of various scales of offlap-onlap sequences documented in the depositional record of basin margins. The most common sequence boundaries observed in the field are related to the third order transgression-regression cycles according to Haq et al. (1987), and to smaller units representing parasequences or bundles of beds (see Vail et al., and Einsele et al. Introduction, both this Vol.). Such third order cycles are associated with maximum sequence boundary gaps of several Ma, while the smaller parasequences may represent maximum gaps on the order of several 100 ka, with the largest gaps occurring at the margins of the basin fills. The time span and the missing sediment contained in such gaps can be estimated through various field observations (e.g., truncations, rip-ups, etc.). When lag deposits are developed, the eroded or nondeposited sediment can be derived by comparing the number of hard parts concentrated in the lag deposits with that contained in the underlying sediment.

Smaller stratigraphic gaps are associated with various depositional processes; the most important types of this group are hiatuses occurring at the bases of event beds. The relative amount of event bed erosion can be evaluated based on precise lateral correlation and comparison with suitable reference sections. In this manner, event sequences are characterized as being subtractive, constant, or additive with respect to whether or not the amount of event bed erosion is greater, the same, or lesser than the accumulated event sediment. The comparison among the carbonate contents of event beds, source areas, and basin sediments may serve as an indicator for the amount of event bed erosion and incorporation of basinal sediments. This method can be controlled by employing the truncation of standard burrow tiers.

Acknowledgments. The author would like to acknowledge Gerhard Einsele, Linda Hobert, and Bradley Sageman, who all reviewed this manuscript and made many helpful comments and improvements. Also I would like to thank Graham Weedon and other colleagues for discussion on timing problems.

Chapter 7 Timing and Correlation

7.2 High-Resolution Correlation: a New Tool in Chronostratigraphy

E. G. Kauffman, W. P. Elder, and B. B. Sageman

1 Introduction

As traditional means of regional correlation reach their effective limits of resolution, we must seek new and more precise tools for dating and correlation in support of geological interpretation. Ideally, these tools would allow establishment of data-based regional to global time lines – chronostratigraphic surfaces or intervals – which could be calibrated to magnetostratigraphy and radiometric geochronology. Neither geochronology (because of analytical errors) nor magnetostratigraphy (because of widely spaced intervals between reversals) can, by themselves, provide the levels of stratigraphic resolution we seek. Biostratigraphic resolution can attain higher levels of refinement through more precise definition of taxa and temporal changes in populations or character states, and by further integration of multi-lineage data into composite assemblage biozones (Kauffman 1970); but even biostratigraphic systems are limited by ecological and preservational factors, and ultimately by the maximum evolutionary rates of component taxa. In addition, isochroneity of biozone boundaries cannot be demonstrated in most test cases (e.g., Kauffman 1988a).

A major focus on cycle and event stratigraphy during the last decade has led to the recognition of abundant, closely spaced, short-term (hours to 100 ka) cycles and episodic events in the rock record; these are preserved as isochronous to near-isochronous surfaces and thin stratal units, and are widely distributed in most modern and ancient sedimentary basins, especially in hemipelagic and pelagic fine-grained facies. Cyclic and event deposits include those representing: short-term physical events (e.g., meteorite impacts, volcanic ash falls, gravity flow deposits); chemical events (abrupt changes in atmospheric or aquatic chemistry as determined through elemental and isotopic geochemistry); biological events (punctuated evolutionary and extinction events, mass mortalities, abrupt ecological and paleobiogeographic changes); and composite events, e.g., interrelated chemical, physical, and biological changes. Initial chronostratigraphic testing also indicates that certain long-term (> 100 ka) physical, chemical, or biological phenomena (e.g., regional to global anoxic events, oceanic advection events, greenhousing intervals) may have very short-term boundary transitions with background environments. These boundaries, as well as magnetostratigraphic reversal boundaries, comprise valuable short-term event intervals for chronostratigraphic correlation.

Where marker beds are stratigraphically closely spaced, chronostratigraphic resolution for cycle and event-bounded units may reach 20–50 ka or less. This is

Einsele et al. (Eds.)
Cycles and Events in Stratigraphy
©Springer-Verlag Berlin Heidelberg 1991

10–50 times more refined than the average duration of most biostratigraphic zones. The potential is great, therefore, for a "revolution" in modern stratigraphy. This paper discusses concepts and methods of a high-resolution event stratigraphy (HIRES; Kauffman 1988a), with selected Phanerozoic examples and references.

1.1 Basic Concepts of High-Resolution Correlation

Some basic concepts behind high-resolution correlation are the following:

1. The Earth-atmosphere-life system is delicately perched on a series of environmental thresholds, and thus highly dynamic; this system responds to very short (daily, yearly) to long-period (ka-Ma) cyclic phenomena, and is punctuated by small to large-scale perturbational events in a manner that is commonly preserved in the rock record. Abrupt changes in this system, previously regarded as geologically rare events under Uniformitarian doctrines of past decades, are now seen as common elements of a dynamic Earth system.
2. The Earth-atmosphere-life system during the Quaternary, our Uniformitarian reference interval, is glacially influenced and relatively more stressed than it was during the nonglacial periods which have characterized most of geologic time. Much of the Phanerozoic was characterized by warmer climates, higher sea level, lower thermal gradients, and more sluggish ocean circulation than at present. Taxa adapted to more equable Phanerozoic systems would be expected to have had narrower environmental tolerances than modern analogs, and thus to have been more prone to rapid ecologic and evolutionary responses to perturbations and short-term cyclic changes than are modern biotas. This should have produced a greater number and diversity of biologic phenomena related to cyclic and event processes, over broader regions, than found in the Quaternary record.
3. Short-term cycles and events have been considered as independent phenomena, the first representing predictable patterns of change, the second, a series of random perturbational events. Yet, inasmuch as short-term cycles represent significant shifts in the rate and intensity of environmental change, the frequency and magnitude of event deposits should also increase in association with the development of short-term cyclic sedimentation.
4. Both local to regional and global levels of forcing of cyclic and event phenomena are evident in the rock record. These have been termed autocyclic and allocyclic by some authors (e.g., Kauffman 1988a; see Einsele et al. Introduction, this Vol.). Localized forcing mechanisms such as tectonic, volcanic, sedimentologic, oceanographic, and climatic phenomena produce intrabasinal responses. Regionally to globally expressed mechanisms include eustasy, plate tectonics, global oceanographic and climatic changes, large-scale volcanic explosions, and meteorite impacts, many of which are capable of producing very widespread near-isochronous stratigraphic signals. It is imperative that regional to global event and cycle deposits be isolated from local and background deposits by detailed stratigraphic testing. The methodology for these tests is subsequently discussed.

1.2 Methodology for Correlation

Methods of data collection and analysis in high-resolution event stratigraphy (HIRES) are described in detail by Kauffman (1986, 1988a) and only summarized here. Basinwide sections are selected to provide detailed tests for lateral continuity and synchroneity of short-term depositional cycles and events. Where possible these are aligned to transect facies belts or paleobiogeographic zones. Tens to hundreds of sections covering hundreds to thousands of square kilometers provide a desirable data base.

1.2.1 Establishing Integrated Event Units

Individual sections are cleaned to fresh bedrock and described at cm scale. Paleontologic and geochemical sampling is done at 10–50-cm intervals. Complete biostratigraphic, bioevent, and standard chemostratigraphic analyses (C_{org}, $\delta^{13}C$, $\delta^{18}O$, CaCO3, trace elements, etc.) are conducted on these samples in the laboratory. Physical, chemical, and biological data are then carefully plotted against the detailed stratigraphic section, and integrated into a local composite event stratigraphy (Fig. 1). The integrated event stratigraphy (IES; Fig. 1) for each section is a chronostratigraphic hypothesis; each proposed event surface or interval must subsequently be tested against numerous other sections to prove its regional extent and relative isochroneity. Only those that can be shown to occur at the same stratigraphic position in all or most tested sections are included into the basinal system of chronostratigraphy.

Some event and cycle intervals or surfaces are easily identified by unique physical and chemical characteristics, for example, the Cretaceous-Tertiary boundary bed with its characteristic shocked quartz and trace element signature (Bohor et al. 1987), or geochemically and geometrically unique volcanic ash beds of the Western Interior Cretaceous of North America (Kowallis et al. 1989). But the majority of the events and cycles are not as obvious in their correlation because of their similarity to other units of the same origin (e.g., multiple volcanic ash beds, storm beds, or turbidites from the same source areas, organic carbon-rich layers, mass mortality, or event community beds involving similar taxa, etc.). Not all such beds are isochronous and the differentiation of these units between sections constitutes the main challenge to event chronostratigraphy. Graphic correlation is one of the best tests of correlation and isochroneity available to meet this challenge (Miller 1977; Edwards 1984, 1989; Kauffman 1988a).

1.2.2 Graphic Correlation

Graphic correlation (Shaw 1964; Miller 1977; Edwards 1984, 1989; Kauffman 1988a) provides an analytical means for correlating all kinds of stratigraphic data within and between facies belts and basins. The graphic correlation method arose as a means of biostratigraphic correlation utilizing diverse taxa from various types of

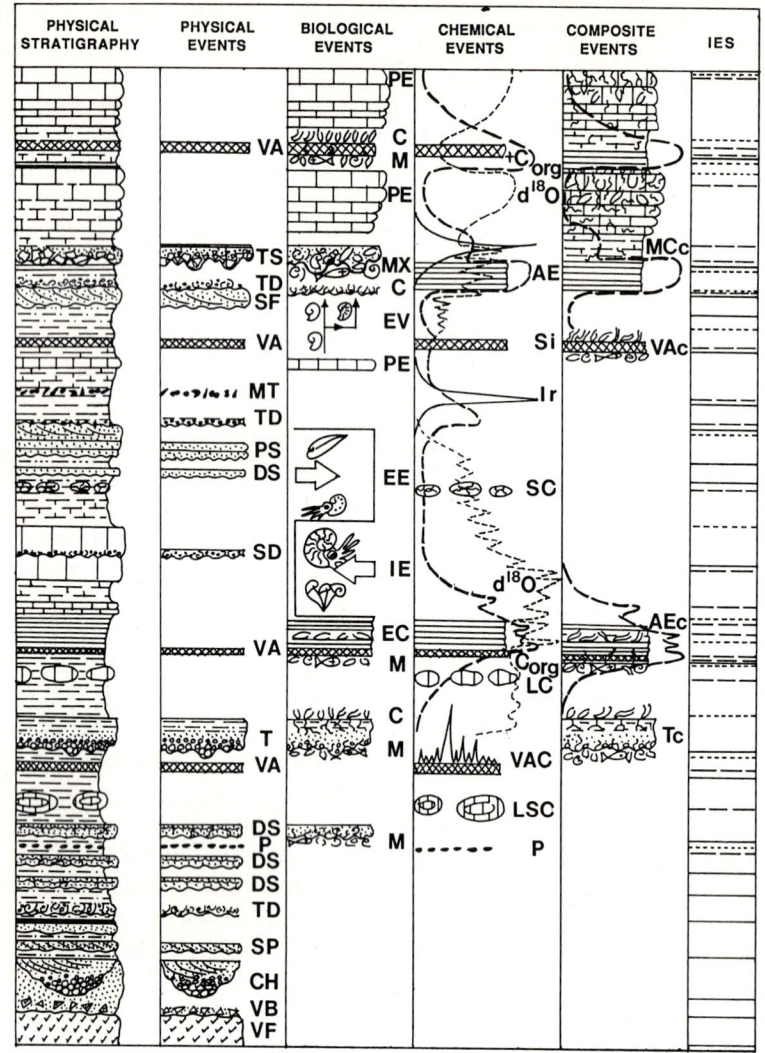

Fig. 1. a Hypothetical stratigraphic section of a complex marine transgressive hemicyclothem showing (*left to right columns*): physical stratigraphy; physical event units; biological event units; chemical event units; composite (related physical, chemical, and biological) event units; and integrated event chronostratigraphy (IES) collectively comprised of the physical (*horizontal lines*), biological (*rows of dots*), and chemical (*dashes*) events. Key to abbreviations (alphabetical): *AE* anoxic event; *AEc* composite physical, chemical, and biological events associated with AE; *C* colonization bioevent; *CH* channelization event; *C*$_{org}$ organic carbon enrichment interval or "spike"; $\delta^{18}O$ light stable isotope (^{18}O) spike or enrichment interval; *DS* distal storm bed; *EC* event community; *EE* emigration bioevent; *EV* punctuated evolutionary event; *IE* immigration event; *Ir* iridium or other trace metal spike; *LC* limestone concretion zone; *LSC* limestone-siderite concretion zone; *M* mass mortality bioevent; *MCc* composite physical, chemical, and biological events associated with Milankovitch climate cycle, *MT* microtektite layer; *MX* mass extinction bioevent; *P* phosphate pebble (condensed) zone; *PE* productivity bioevent; *PS* proximal storm bed; *SC* septarian limestone concretion bed; *SD* sediment starvation disconformity; *SF* submarine fan deposit; *Si* silica enrichment interval or "spike"; *SP* splay deposit; *T* turbidite deposit; *Tc* composite physical and biological events associated with turbidite deposit; *TD* transgressive disconformity with shell or pebble layers; *TS* tsunami deposit; *VA* volcanic ash-tuff-bentonite deposit; *VAC* chemical signature of individual volcanic ash; *VAc* composite physical, chemical, and biological events associated with a volcanic ash fall; *VB* volcanic breccia; *VF* volcanic flow

b   **KEY TO LITHOLOGIES**

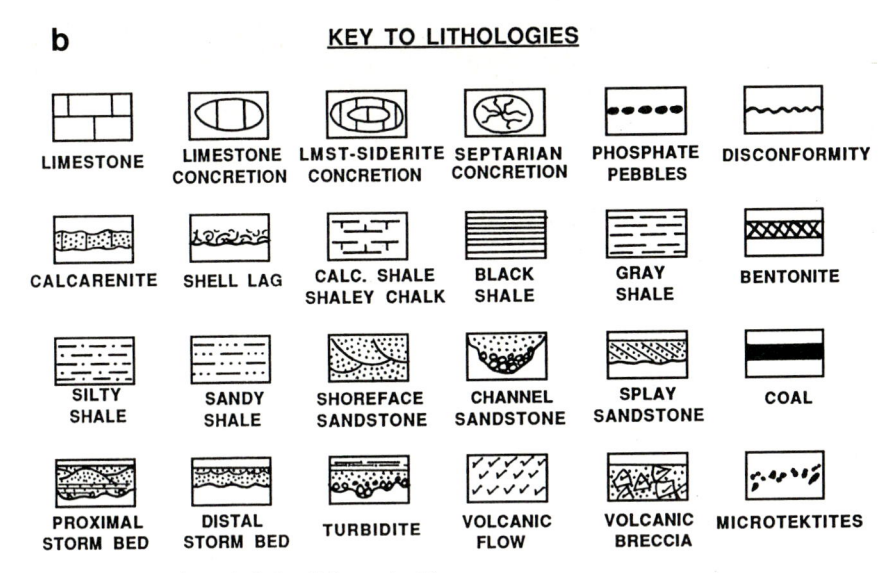

LIMESTONE	LIMESTONE CONCRETION	LMST-SIDERITE CONCRETION	SEPTARIAN CONCRETION	PHOSPHATE PEBBLES	DISCONFORMITY
CALCARENITE	SHELL LAG	CALC. SHALE SHALEY CHALK	BLACK SHALE	GRAY SHALE	BENTONITE
SILTY SHALE	SANDY SHALE	SHOREFACE SANDSTONE	CHANNEL SANDSTONE	SPLAY SANDSTONE	COAL
PROXIMAL STORM BED	DISTAL STORM BED	TURBIDITE	VOLCANIC FLOW	VOLCANIC BRECCIA	MICROTEKTITES

Fig. 1b. Key to lithologic symbols for all figures in this paper

facies (Shaw 1964; Miller 1977). Edwards (1984, 1989) provides the clearest modern review of the classic methodology; to summarize: first-and-last occurrences of taxa from the most and second-most complete and fossiliferous sections are plotted on the X and Y axes of a graphic plot, respectively. A line of correlation is then placed, either by eye or by regression analysis, between the intersects of these data points, dividing first- and last-occurrences to the maximum extent possible. Discrepant ranges for shared taxa between the sections are then adjusted to the line of correlation (see Edwards 1984, 1989), and resultant composite ranges projected onto the vertical graphic axis. This axis is rescaled to time units and becomes a composite reference section as additional sections are placed against the horizontal axis and their data similarly projected onto the vertical axis. A standard composite reference section incorporating all range data from within a basin is thus generated and can be used to correlate stratal sequences of the same age range into other basins.

Data based on isochronous or short-term cycle and event deposits also can be used to generate graphic correlation plots. The resultant line of correlation becomes a line of isochronous correlation, and any two points or intervals that intersect it are considered to have occurred at the same absolute point or interval of time. The line of isochronous correlation can further be calibrated to real time by tying it to high-temperature radiometric dating. The advantages of this system are:

1. It can be based on many independent data points other than biostratigraphy (i.e., event-stratigraphic units);
2. It does not have the same potential for error as biostratigraphic correlations in which ecological, evolutionary, and preservational controls can severely distort or modify the field data and correlation for any section;

3. It provides a real time matrix for basin analysis and for interpreting the short-term dynamics of any system; and
4. If the event units are carefully chosen (i.e., volcanic ash beds, Milankovitch cycles, geochemical spikes) this system has a more regional, facies-wide potential for correlation than biostratigraphy, the latter of which may be influenced by various ecological (habitat) controls.

2 Units of High-Resolution Event Stratigraphy

2.1 Units Representing Physical Events

Physical event units (Fig. 2) include deposits or surfaces formed by global to regional sedimentary processes within very short time intervals, e.g., magnetic polarity reversals, bolide impact ejecta layers, volcanic ash deposits, or giant storms, earthquakes, and gravity flow events. Cyclic short-term physical event deposits useful in high-resolution stratigraphy include thin sedimentary units reflecting different phases of climate cycles. Sediment bypass or starvation surfaces and disconformities may represent short-time durations that are useful in high-resolution correlation, but most are significantly diachronous and omit long intervals of sedimentary history.

2.1.1 Magnetostratigraphy

Abrupt changes in the polarity orientation of magnetic iron oxide minerals in sediments reflect fluctuations and reversals in the Earth's magnetic field (Fig. 2). Magnetic reversals are global events that can be integrated between igneous and sedimentary sequences, especially in Mesozoic and Tertiary rocks (Opdyke 1972; Cox 1973; review in Harland et al. 1982). The resolution of magnetostratigraphy is constrained by the duration of each polarity reversal, an average of 4–5 ka (Tarling 1983), and by blurring caused by bioturbation and other sedimentary mixing processes. Factors that limit the utilization of paleomagnetic reversals in HIRES are:

1. The wide spacing between reversal events, typically 10 ka to 2 Ma. Reversals spaced 50 to 100 ka apart, termed events or subchrons, are common only in certain Mesozoic and Tertiary intervals; many reversals are separated by long quiet zones that may last for over 30 Ma (e.g., mid-Cretaceous). Paleomagnetic "excursions" during which the Earth's polarity undergoes partial reversals lasting only ca. 1000 years have great potential in HIRES but are extremely difficult to correlate at present (Harland et al. 1982);
2. Reversals lack unique signals and are difficult to correlate unless they can be independently dated by geochronology or biostratigraphy.

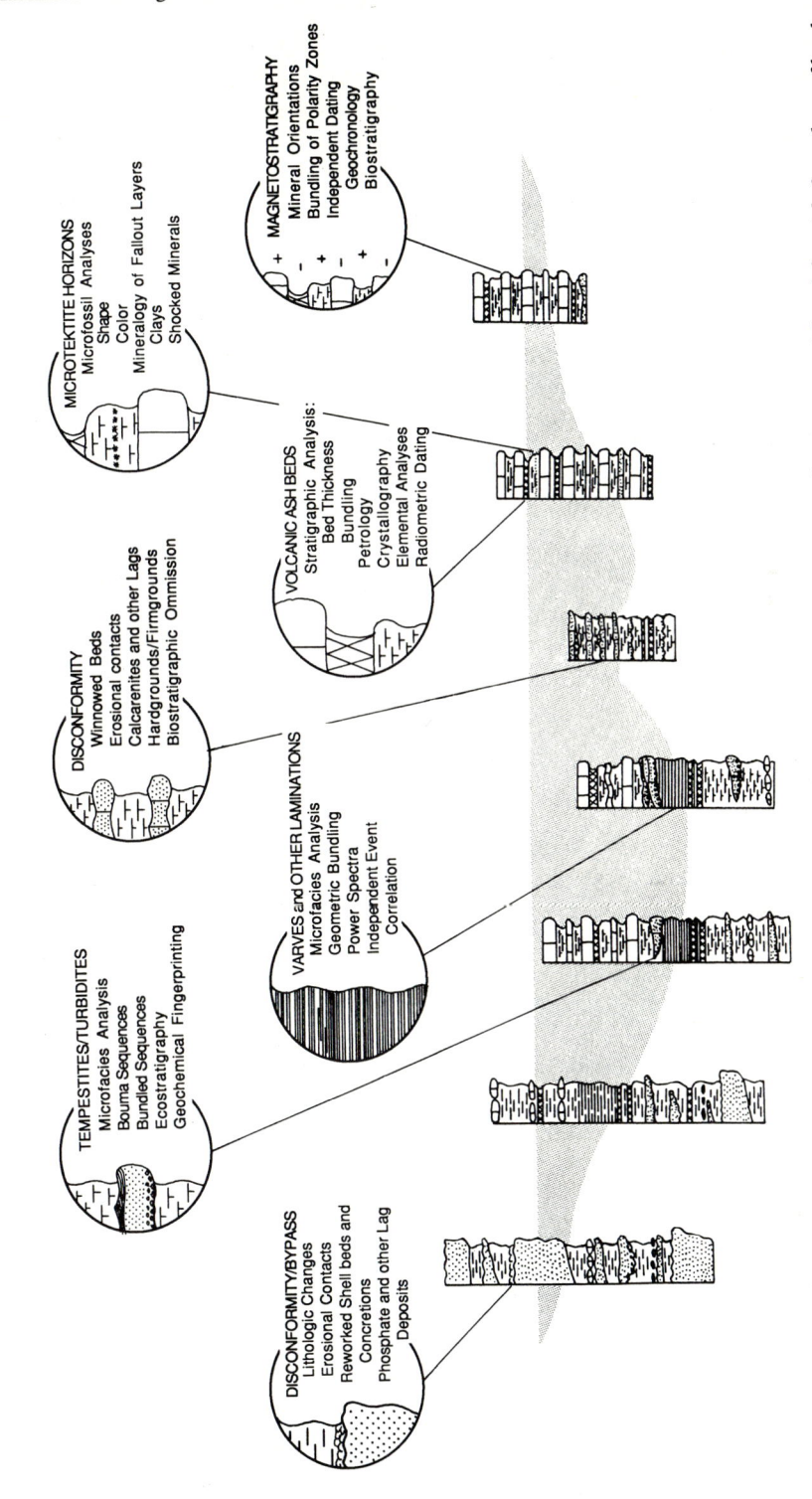

Fig. 2. Physical events are illustrated at sections on a transect across an idealized epicontinental basin. Event units are arranged *from left to right* from those of local correlation potential to those with global correlation potential. Many event units are regional in scale and can be traced throughout the basin. See Fig. 1 for explanation of symbols and event units

2.1.2 Impact-Related Beds

Second only to magnetostratigraphy in areal correlation potential, impact-related beds form rare, but extremely widespread stratigraphic markers. Impacts by large bolides on land or in shallow seas result in cratering and the ejection of great quantities of impact debris into the atmosphere. Ultimately this debris settles out over large areas, in some cases globally (Alvarez et al. 1980, 1984), as an isochronous, chemically and physically unique event deposit. The physical component of such deposits may contain shocked minerals, microtektites, and fine particulate fallout (see Sect. 2.2.1). The primary example of a global chronostratigraphic impact marker bed is the Cretaceous-Tertiary boundary clay, known from diverse marine and nonmarine facies (Alvarez et al. 1980, 1984; Izett 1987; Bohor et al. 1987). In addition, small impacts may have regional sedimentologic expression as laterally graded coarse to fine ejecta plumes (e.g., the Precambrian Acraman impact layer of South Australia; and the Cenozoic tektite strewn-fields; Glass 1982) and may also be detected in petrologic and microfossil analyses of pelagic marine strata as geographically widespread microtektite layers. These layers are known from several ages and areas (Glass 1969; Hut et al. 1987; Koeberl and Glass 1988). Finally, tsunamis associated with aquatic impact of large bolides, as proposed by Bourgeois et al. (1988) for the Cretaceous-Tertiary boundary in the Gulf and Caribbean region, are thought to have produced widespread and easily recognizable high-energy event beds in the ancient record.

2.1.3 Volcanic Ash Beds

Although typically less areally extensive than impact-related deposits, sedimentary ash, tuff, and bentonite beds are common to abundant in many sedimentary basins, particularly near sources of explosive, subduction-related volcanism (see Schmincke and van den Bogaard Chap. 2.12, this Vol.). These beds form ideal stratigraphic markers for high-resolution correlation due to their wide geographic distribution, ease of recognition on outcrop, and the isochronous nature of their bases (Fig. 2). Widely distributed bentonite and tuff beds form useful chronostratigraphic markers in Ordovician, Devonian, and Cretaceous marine strata of North America and in Devonian, Cretaceous, and Paleocene marine strata of western Europe. Some of these deposits exceed the thickness, number, and areal distribution of ash deposits from the largest historical eruptions (Krakatoa, Tambura, Vesuvius, etc.), which have laid down marine ash layers up to a few cm thick for distances of 100 to 3000 km within a few years. Terrestrially deposited ash beds or tonsteins are not as widely traceable as marine deposits, but in some cases may be correlated from brackish, estuarine, or coal swamp facies of ancient coastal plains, across the shoreface, and into more basinal marine facies.

Due to local omission, changes in thickness, mixing with surrounding sediments by bioturbation, mineralogical similarities, and diagenetic alteration, regional correlation of individual volcanic ash and bentonite beds between stratigraphic sections is commonly difficult, but can be substantiated by a number of independent physical and chemical methods (Fig. 2). These methods include: Analyses of the relative thickness, stratigraphic placement, and characteristic bundling of beds (e.g., doublets, triplets) between localities, and cross-referencing of the placement of bentonite beds relative to biostratigraphic and other event-stratigraphic markers. These methods allow bentonite correlations in outcrop and subsurface over 100 to 1000 of km^2 (Knechtel and Patterson 1962; Hattin 1971; Slaughter and Earley 1965; Amajor and Lerbekmo 1980; Wood et al. 1984; Dahmer and Ernst 1986; Elder 1988).

Compositional analyses ranging from bulk elemental and crystal chemistry to phenocryst morphology, as well as petrologic analysis of thin-sections (Droste and Vitaliano 1973), have all been successful in correlating individual bentonite and ash beds. Chemical fingerprinting using trace elements has proven to be most useful for correlation of the Ordovician K-bentonites of the upper Mississippi valley (Kolata et al. 1986). In other studies, geochemical analysis of elements such as Mn, Fe, K, Ca and Ti (Amajor and Lerbekmo 1980), and combined analyses of bulk geochemistry and phenocryst morphology and geochemistry (Winter 1977, 1982; Kowallis et al. 1989) have produced confident correlations among Ordovician, Devonian, and Cretaceous bentonites and tuffs.

2.1.4 Short-Term Depositional Events: Storm, Flood, and Gravity Flow Deposits

Major storms, floods, earthquakes, and slope failure events may produce unique short-term deposits that are ideally suited for local to regional HIRES correlation (see Seilacher and Aigner Chap. 2.3, this Vol.). Most such deposits are relatively limited in areal extent (<1000 km^2) and many are subsequently reworked. But confident local to regional correlation has been achieved using these deposits in some cases.

Storm deposits (tempestites; Ager 1974) represent day- to week-long storm events and may comprise isochronous markers beds (Fig. 2). Large modern tropical storm and hurricane deposits span 100 to 1000 km^2 in nearshore and shelf environments (Reineck and Singh 1980; Allen 1982; Gagen et al. 1988) and may be preserved intact when subsequent deposition is rapid enough to bury them below the zone of bioturbation (Gagen et al. 1988; see Sepkoski et al. Chap. 2.6, this Vol.). Many factors make the recognition and correlation of individual tempestites difficult, e.g., similar regional composition of beds, or removal of beds in some areas and amalgamation in others. The high frequency of major storm events (ranging from 20 to 3000 years; Reineck et al. 1968; Ager 1974) precludes the possibility of independent correlation using classical biostratigraphic methods. Various methods for their recognition and correlation are given in Seilacher (1982a,b) and Aigner (1982, 1985); these include the following: detailed microfacies analysis, correlation of unique individual and bundled coarsening-upward sequences, identification of unique ecostratigraphic signatures, and geochemical fingerprinting (Fig. 2). Although most tempestites are localized, spanning 10's of km^2, some can be correlated by normal or graphic means within certain basins for up to 200 km^2 (Kelling and Mullin 1975; Aigner 1982, 1985; Bloos 1982; Handford 1986; Glenister and Kauffman 1985; Kauffman 1988a).

Turbidite deposits hold great potential for high-resolution correlation on a local to regional scale, as turbidite fans often cover over 100 km^2. Correlation of single gravity flow beds or individual fining- or coarsening-upward sequences (several m in thickness) may be possible for areas of 10 to more than 100 km^2 (see Piper and Stow Chap. 2.9, and Einsele Chap. 2.7, both this Vol.). At present, many turbidite studies have achieved detailed vertical resolution, but lateral correlation is only local in scale (10–20 km) (Schlager and Schlager 1973; Crimes 1973; Ricken 1985); thus, the high-resolution correlation potential of turbidite deposits has not been fully exploited.

Other catastrophic event deposits with potential application to HIRES include terrestrial flood deposits (inundites), sedimentary features and mass flow deposits resulting from earthquakes (seismites), and beds or features resulting from rarely preserved phenomena such as tidal waves (tsunamites) (Seilacher 1969, 1982a, 1982b; Neff et al. 1985; Bourgeois et al. 1988). Correlation of any fluvial deposit is normally local at best and examples are scarce. Some authors have suggested that individual channeling events may be widely synchronous between drainages as a result of rapid regional shifts in base level. Paleosols and coals have been the most successfully employed short-term deposits for local to regional HIRES correlation in terrestrial facies (Retallack 1983; Flores 1979; Fastovsky 1987). Seismites provide important information for the reconstruction of basin history (Seilacher 1969), but HIRES correlations employing these features are lacking. Although probable tsunami orgin has been suggested for some high-energy event beds (e.g., Neff et al. 1985), extensive lateral correlation remains to be tested.

2.1.5 Varves and Other Fine Laminations

In stratified lakes and oxygen-deficient marine basins, fine mm-scale laminations of alternating grain-size or carbonate and organic content result from annual or longer period (10's of years) climate cycles. Varve-like features also may form from other small- to large-scale depositional events (e.g., productivity and mass mortality events among plankton), especially in association with organic-rich deposition. The correlation of individual laminae over distances as great as 1400 km has been achieved in modern Black Sea studies (Ross et al. 1970; Arthur et al. 1988) and has been suggested for some well-laminated black shale deposits, such as the Toarcian Posidonienschiefer (Seilacher pers. commun. 1987). Varve-type chronologies have been established in a number of lacustrine deposits, ranging from the Triassic to the Holocene (Olsen 1984; Halfman and Johnson 1988), and show resolution of climate cycles down to 10's of years. These types of physical event units may be geographically widespread (100–1000 km^2) and provide the most detailed levels of correlation known (Fig. 2). Individual varves are difficult to identify regionally without other event-stratigraphic markers, or the unique bundling of lamina that allow correlation by power spectrum statistical methods.

2.1.6 Disconformities and Bypass Surfaces

Disconformity, hardground, and sediment bypass surfaces are potential regional stratigraphic markers (Fig. 2). Most regional disconformities involve considerable loss of time and are significantly diachronous over large distances (see Kidwell Chap. 6.3, this Vol.). A combination of eustatic sea level fluctuations, basin uplift or subsidence, and variations in sediment input and transport rates control the development and characteristics of nondepositional and erosional events. The most conspicuous application of eustatically generated disconformities to stratigraphic correlation has been the development of sequence stratigraphy (Vail et al. 1977a, 1984; Haq et al. 1987; see Vail et al. Chap. 6.1, this Vol.). Integration of seismic sections with data from well logs and outcrops has allowed sequence stratigraphy to become an important tool for basin analysis and regional correlation (Haq et al. 1987). Sequence boundaries can set broad limits for more precise HIRES correlations, but claims that they represent true chronostratigraphic surfaces are unfounded.

Whereas regional regressive-transgressive disconformities are commonly diachronous on the scale of 100 ka to several Ma, condensed, sediment-starved basinal surfaces associated with times of maximum transgression are more nearly isochronous (Fig. 2). Studies have traced individual burrowed firmgrounds, hardgrounds, and omission or erosion surfaces over 100 to 10 000 km^2 within narrow time intervals (Jefferies 1963; Kennedy 1969; Kennedy and Garrison 1975b; Francis 1984; Kidwell 1984; Elder and Gustason 1984; Kauffman et al. 1987).

Small-scale, near-isochronous transgressive, nondepositional events corresponding to parasequence and punctuated aggradational cycle boundaries of Haq et al. (1987) and Goodwin and Anderson (1985), respectively, are of most interest to cycle and event stratigraphers. Such surfaces have been widely correlated in Devonian rocks of New York (Goodwin and Anderson 1985; Goodwin et al. 1986; Baird and Brett 1986), and in Cretaceous sequences of Europe and the Western Interior of North America (Kennedy and Garrison 1975b; Francis 1984; Kauffman et al. 1987); their apparent cyclic development has been attributed to orbitally forced climate cycles (Goodwin et al. 1986; Bottjer et al. 1986; Kauffman et al. 1987) (see also Sect. 2.4; Strasser Chap. 6.5, and Haas Chap. 6.6, both this Vol.).

2.2 Units Representing Chemical Events

Chemostratigraphic data valuable to HIRES correlation are obtained through closely spaced stratigraphic analyses of trace elements, organic and carbonate carbon, $\delta^{18}O$, $\delta^{13}C$, and other isotopes (Fig. 3). Major geochemical excursions ("spikes") of unusual magnitude and short duration (100 ka or less) have been successfully used in regional correlation (e.g., Scholle and Arthur 1980; Pratt and Threlkeld 1984; Pratt 1985; Kauffman 1988a). Longer-term, unusual chemostratigraphic intervals associated with protracted changes in oceanic or atmospheric processes can have widely correlative, short-term chemostratigraphic transition zones, reflecting rapid change from background to abnormal conditions and back again.

2.2.1 Elemental Chemostratigraphy

Analysis of bulk samples for trace elements (Fig. 3) has been applied in a number of detailed stratigraphic investigations (Alvarez et al. 1984; Walters et al. 1987; Orth et al. 1988). Iridium and a variety of other trace elements may be enriched in marine sediments by a number of means, including fallout from bolide impacts, giant deep-seated volcanic eruptions, deep mantle outgassing, and through condensation by sediment bypass or starvation, biological concentration, and chemical oceanographic or sediment recycling. Ratios of different rare elements within enrichment zones may reflect their collective origins and differentiate between deep mantle versus extraterrestrial or biological origins. The Ir-enrichment layer(s) associated with the Cretaceous-Tertiary boundary interval represent a unique chemical "spike" of extraterrestrial origin and comprise the best-known global chemostratigraphic event markers (Alvarez et al. 1980, 1984; Hut et al. 1987).

Phosphorus, manganese, uranium, and other trace elements may be enriched in marine strata by oxygen depletion in the water column (Fig. 3). Upwelling areas, typically on the western margins of continents, are characterized by nutrient excess, extremely high productivity, and depleted oxygen content, conditions resulting in sedimentary phosphorous enrichment (Manheim et al. 1975). Widespread thin phosphorite horizons may indicate short-term regional upwelling events in ancient black shales, and serve as important chronostratigraphic event markers (Heckel 1977). Under highly stratified conditions in relatively shallow marine basins, expansion of oxygen-minima zones into benthic sediments rich in trace elements commonly results in the remobilization of Mn and other elements into the water column and extensive benthic Mn reprecipitation at the oxic-anoxic interface during eustatic highstand or sea level fall (Frakes and Bolton 1984; Force and Cannon 1988). In basins where the bathymetric gradient is very low, short-term relative sea level fluctuations may produce rapid shifts in the zone of Mn precipitation, leading to deposition of thin, near-isochronous, Mn deposits over relatively large areas. Also, some black shales deposited under anoxic conditions are enriched in authigenic U, determined by measurement of the U:Th ratio (Myers and Wignall 1987). Correlation of U enrichment zones may provide a regional to global event-chemostratigraphic tool (Zelt 1985).

2.2.2 Light Stable Isotope Chemostratigraphy

Analyses of stable isotopes in carbonate and organic matter, especially $\delta^{18}O$ and $\delta^{13}C$ of foraminifer and mollusc shells, provide a powerful tool for sensing short-term paleoceanographic and paleoclimatic changes. Potential problems involving chemostratigraphic analyses and HIRES correlation are failure to recognize and sort out the effects of:

1. Diagenetic overprinting;
2. Local to regional changes in the expression of the isotopic signal and in sedimentation style; and
3. The varying effects of chemical, physical, and biological isotopic fractionation.

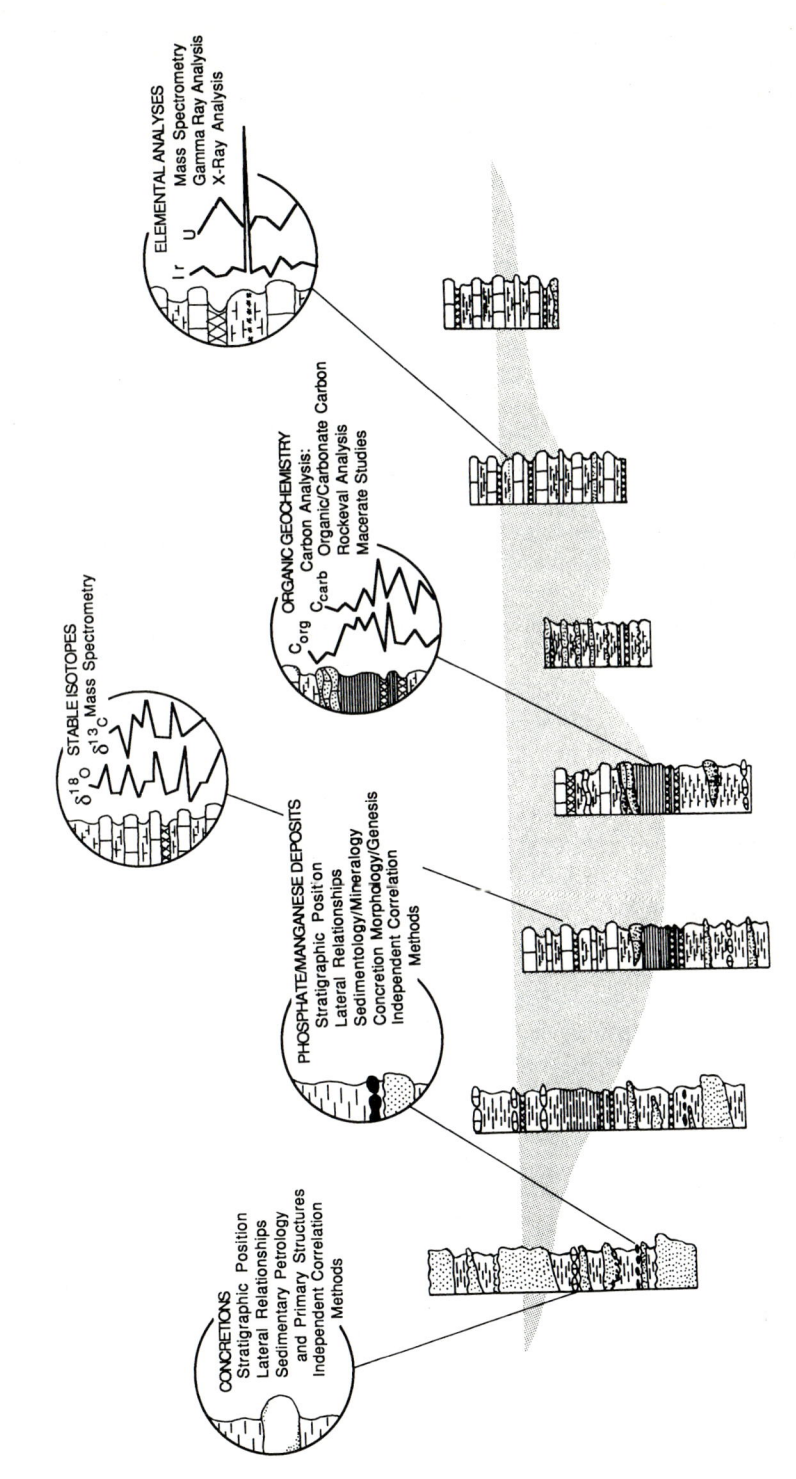

Fig. 3. Chemical events are illustrated as in Fig. 2. See Fig. 1 for explanation of symbols and event units

Abrupt shifts in $\delta^{18}O$ and $\delta^{13}C$ commonly result from rapid, short-term changes in water temperature, salinity, or the cycling and benthic storage of carbon. The Phanerozoic stratigraphic record contains numerous examples of abrupt positive and negative shifts in these stable isotopic values that comprise regionally, and possibly globally correlative chemoevent units (Kauffman 1988a; Berry et al. 1989). For example, correlation of short-term Pleistocene $\delta^{18}O$ isotopic shifts in cores from the Caribbean and equatorial Atlantic have established the global synchrony of 22 glacio-eustatically controlled isotope stages during the last 900 ka (Shackleton and Opdyke 1973, 1976); these stages reflect the 21–100 ka Milankovitch orbital cycles and yield resolutions of great value to HIRES chronostratigraphy. Similar $\delta^{18}O$ fluctuations reflecting Milankovitch climate cycles occur in Pliocene sequences (Shackleton 1982) and among $\delta^{18}O$ and $\delta^{13}C$ values in Miocene marine strata (Pisias et al. 1985). Regional chemostratigraphic correlations may therefore be possible in the middle Tertiary, but the values are altered by diagenetic overprints. Williams et al. (1988) presented quantitative methods for filtering, deconvoluting, and correlating isotopic and other time-series data.

High-resolution isotopic studies in Mesozoic and older rocks are not as common, and largely confined to specific intervals such as mass extinctions and oceanic anoxic events. Zachos and Arthur (1986) found an abrupt negative shift in $\delta^{13}C$ at the Cretaceous-Tertiary boundary in ocean cores that appears to form a global isochronous marker. Detailed $\delta^{13}C$ and $\delta^{18}O$ analyses of the anoxic event associated with the Cenomanian-Turonian boundary interval in the Western Interior Basin, U.S.A. (Pratt and Threlkeld 1984; Pratt 1985) indicate abrupt, large-scale near-isochronous fluctuations in both isotopic ratios around much of the basin (e.g., Kauffman 1988a) and in coeval European strata (Hilbrecht and Hoefs 1986; Jarvis et al. 1988). Detailed studies through carbonate-shale bedding rhythms, reflecting 21, 42, and 100 ka Milankovitch climate cycles (de Boer 1982a; Barron et al. 1985) indicate that $\delta^{18}O$ values track these cycles (Kauffman 1988a, Fig. 15, for model). If these fluctuations reflect primary temperature or salinity effects on isotopic ratios, rather than diagenetic overprint (e.g., Ricken 1986), then this isotopic signal may be useful for regional correlation from the center to the clastic margins of basins and between basins. Although the use of stable isotopes for regional to global high-resolution stratigraphic correlation of abrupt to short-term (<100 ka) oceanographic and climatic fluctuations is still in its formative stages, initial attempts at applying isotope chemostratigraphy to HIRES correlation show great promise.

2.2.3 Organic and Carbonate Carbon Chemostratigraphy

Variations in organic and carbonate carbon values may reflect primary short-term signals in upwelling, water stratification, carbon cycling, and productivity useful for local to regional HIRES correlation (Fig. 3). The use of these values for lateral correlation is limited by variability in productivity, sedimentation rate, terrigenous sediment supply, benthic recycling, and diagenetic remobilization of carbonate and organic carbon. In special circumstances (i.e., oceanic anoxic events), precise regional to global correlation of carbonate and organic carbon variations based on abrupt oceanographic changes appears possible. Large-scale, rapid fluctuations in organic and carbonate carbon are typically reflected by marked lithologic or color changes in mixed calcareous-siliciclastic facies, thus also enhancing physical correlation. Where such changes are lithologically subtle (e.g., in monotonous shale sequences), detailed geochemical analyses are necessary for the determination and correlation of carbon-based chemostratigraphic markers. Most detailed regional to global correlations of organic and carbonate carbon percentages are based on abrupt shifts resulting from a wide range of factors, including climatic, tectonic, and eustatic controls on sedimentation rates (Kauffman et al. 1987) and basin configurations, productivity fluctuations (Fischer et al. 1985; Eicher and Diner 1985; Zachos and Arthur 1986), water mass stagnation events (Cita and Grignani 1982), oceanic anoxic events (Schlanger and Jenkyns 1976), or abrupt shifts in the carbonate compensation depth (Berger and Mayer 1978). Many of these causal mechanisms may be interrelated.

More subtle but still abrupt changes in organic and carbonate carbon content may also be useful for regional correlation. For example, Prell (1978) correlated Quaternary strata in the Colombia Basin based on small-scale changes in carbonate content and Elder (1987) traced rapid shifts in carbonate and organic carbon content in Cenomanian-Turonian boundary shale sequences throughout the Colorado Plateau region. The resolution of these geochemical events can be very high, but they should be rigorously tested for chronostratigraphic relationships inasmuch as local depositional facies may shift in geographic distribution and through time.

2.2.4 Concretion and Nodule Zones

Persistent beds of concretions, lenses, and nodules of calcium carbonate and silica commonly parallel tested isochronous marker beds (e.g., volcanic ashes) and thus are apparently useful in HIRES. Concretions produced by early diagenetic precipitation of carbonate cements in thin stratigraphic intervals most commonly form regional stratigraphic marker beds (Fig. 3). Studies suggest several mechanisms for the production of persistent isochronous concretion zones of regional extent (see Ricken and Eder Chap. 3.1, this Vol.). Raiswell (1989) indicated that Jurassic carbonate concretion growth is the result of anaerobic methane oxidation promoted by breaks in sedimentation. Savrda and Bottjer (1988) suggested that early diagenetic concretion formation in the Jurassic Posidonienschiefer was produced by short-term oxygenation events. Kauffman (1965) suggested that carbonate concretions commonly form around organic nuclei and thus that concretion zones may reflect unusually high concentrations of shells and coarse organic detritus. Similarly, persistent chert layers are thought to reflect, in part, unusually high concentrations of biogenic silica (sponges, radiolarians, etc.).

Although the correlation of concretion beds over large geographic regions has rarely been attempted (see Hattin 1985; Kauffman et al. 1987), these beds are potential chronostratigraphic markers. Based on characteristic faunal constituents, Waage (1964) correlated concretion beds in the Fox Hills Formation of South Dakota for thousands of km^2 and many other beds on a local scale. Continuous limestone beds of the lower part of the Bridge Creek Limestone Member of the Greenhorn Formation in the central Western Interior Basin can be traced westward into concretion beds covering over 1000 km^2 (Hattin 1985; Elder 1985, 1988; Kauffman et al. 1987). Zones of septarian limestone and limestone-siderite concretions, which form later in early diagenesis (Kauffman 1965), also appear to be correlative throughout large regions of the central Western Interior Basin in the Blue Hill Shale Member of the Carlile Shale (Hattin 1962; Glenister and Kauffman 1985).

2.3 Units Representing Biological Events

Bioevents play an important role in both high-resolution event stratigraphy and biostratigraphy, although the two are not necessarily related. Bioevents include phenomena that may serve as biostratigraphic zone boundaries (Fig. 4), such as episodes of punctuated evolution, or steps and catastrophes within mass extinction intervals (Kauffman 1986, 1988b). Yet most bioevents do not mark species origination or extinction, but are ecological phenomena acting at the population and community level. These bioevents (Fig. 4) include short-term regional changes in community structure (ecostratigraphy), mass mortalities, rapid species level immigration and emigration events, productivity events, population bursts (acmezones), and rapid benthic colonization events (Kauffman 1986, 1988a). Whereas correlations based on biostratigraphic data are typically regional to global in scope, population and ecostratigraphic phenomena are normally local to regional in extent, except where associated with mass extinction or widespread climatic and oceanographic changes.

2.3.1 Evolution and Extinction Events

Evolutionary bioevents are based on punctuated and allopatric speciation, macromutation, and macroevolutionary phenomena (Gould and Eldridge 1977). These are common patterns for the rapid origination of new taxa. If followed by rapid regional to global dispersal (<100 ka) via planktonic or nektonic adult dispersal or planktotrophic larval drift (Kauffman 1975; Scheltema 1977), these origination events have important utility in both event chronostratigraphy and in biostratigraphy. Because different lineages within an ecosystem respond at different times to the same broad set of ecological stresses, successive punctuated evolutionary events may create stratigraphically closely spaced evolutionary bioevent boundaries broadly applicable to high-resolution event chronostratigraphy.

Extinction bioevents are those in which taxa or groups of taxa became extinct throughout their geographic range within a very short time interval (<100 ka). This differs from background extinction in which regional diachroneity of last occurrences typically results from survival of small local species populations within refugia or distal areas. Abrupt regional to global extinctions are known throughout the Phanerozoic, some of which are catastrophic (days to years; mm to cm of rock) and comprise recognized mass extinction intervals (Raup and Sepkoski 1986). Examples

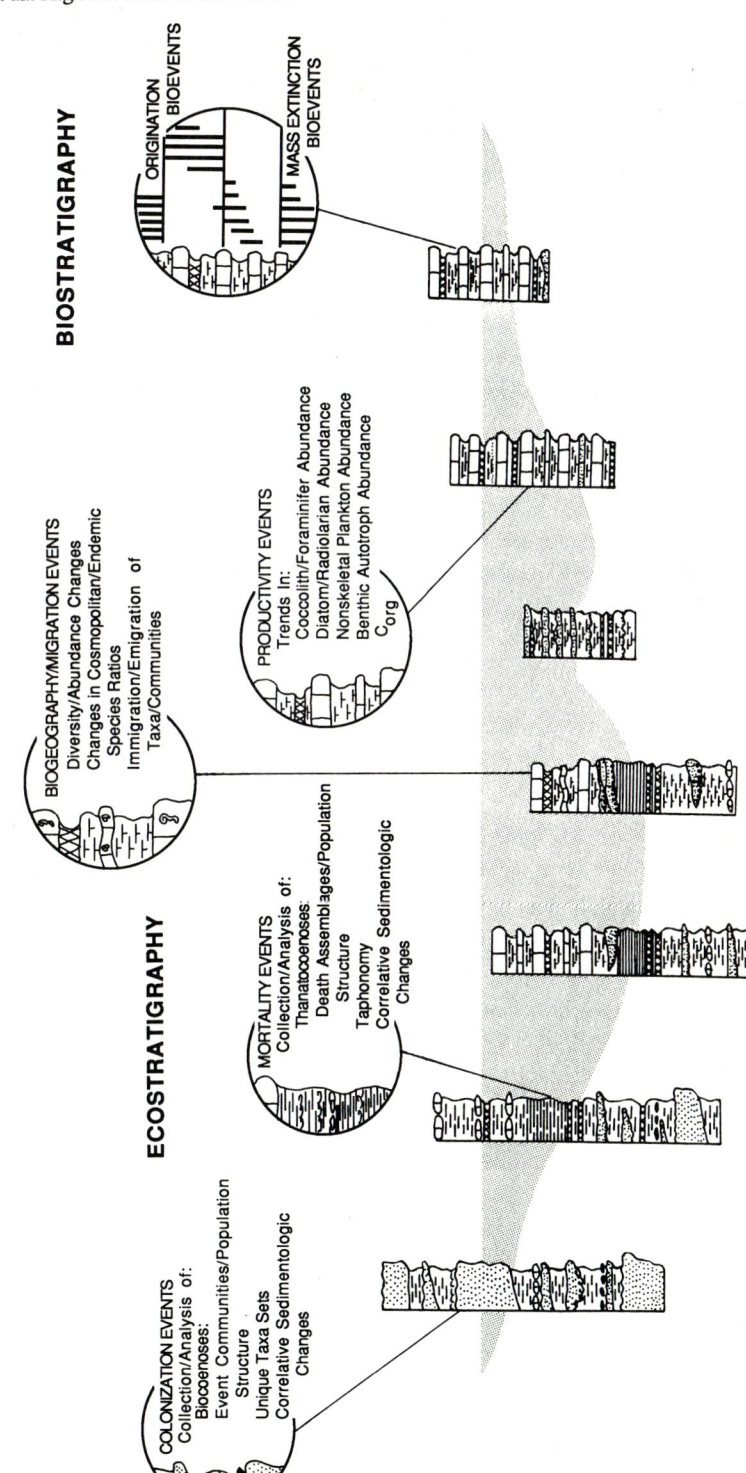

Fig. 4. Biological events are illustrated as in Figs. 2 and 3. See Fig. 1 for explanation of symbols and event units

are Cambrian biomere boundaries (Palmer 1982), Devonian carbonate platform taxa at the Frasnian-Famennian boundary (McLaren 1982) or diverse taxa at the Cretaceous-Tertiary boundary (Alvarez et al. 1984; Kauffman 1988b). Some mass extinctions appear to be characterized by discrete steps (Kauffman 1986, 1988b) each involving abrupt loss of ecologically or biogeographically related taxa, usually along an ecological gradient from more stenotopic to more eurytopic organisms. These steps (bioevents), each 100 ka or less in duration, have been demonstrated on a regional scale, e.g., across the Cenomanian-Turonian boundary in the Interior of North America (Elder 1987; Kauffman 1988b), across the Cretaceous-Tertiary boundary in Tethys and the North Atlantic regions (Alvarez et al. 1984; Kauffman 1988b), and at the Eocene-Oligocene boundary among oceanic foraminifers (Keller 1986) and Gulf Coast molluscs (Hut et al. 1987; Hanson 1988; Kauffman 1988b). Both catastrophic and stepwise extinction levels form valuable regional to global bioevent markers for HIRES correlations.

Biostratigraphy is commonly the primary method of establishing initial correlations between distant localities prior to other high-resolution stratigraphic studies. Abrupt origination and extinction bioevents comprise the major components of biostratigraphy (Fig. 4). Among these, events of origination or first occurrence are typically more reliable for biostratigraphic correlations than last occurrence levels, especially during times of low or background extinction rates (Birkelund et al. 1984; Kirkland 1987; Kauffman 1988a). Although biotic provincialism largely prohibits global correlations based on first occurrences, except in rare cases, precise regional correlation based on origination bioevents is possible, especially among groups like the ammonites (e.g., Kennedy and Cobban 1976). In contrast, the upper boundaries of biozones tend to be isochronous only in cases of widespread, abrupt habitat disruption associated with regional to global mass extinction.

2.3.2 Ecostratigraphic Events

Ecostratigraphic events represent abrupt changes in population or community structure resulting from changes in regional temperature, climate, oceanography, water chemistry, substrate character, etc. These events are recognizable through detailed bed-by-bed sampling, documentation of all fossil material, and reconstruction of paleocommunities. In some cases these events involve biostratigraphically important pelagic and benthic taxa and can be used to define local-regional acmezones (epiboles).

Most ecostratigraphic events involve rapid, short-term expansion of a species' population size beyond normal background levels as a result of environmental changes that favor proliferation and survival of the species (e.g., increases in trophic resources or ecospace; removal of competitors). Such population changes may occur rapidly over wide areas, and appear as thin intervals of exceptionally high population abundance compared to background levels. Many population bursts are among pelagic taxa, reflecting rapid nutrient or water mass changes, but benthic events are also well documented (e.g., *Ostrea beloiti* biostrome, Cenomanian, North America; Kauffman 1988a).

Productivity events (Fig. 4) refer to short-term proliferation of a few to diverse species within communities, usually among plankton, resulting from favorable changes in nutrient levels, temperature, salinity or reduced cropping levels. Pelagic productivity events are commonly short-lived, even seasonal, and may be recorded as abrupt changes in benthic sedimentation toward higher carbonate or silica content. Coccolith-rich laminae in black shales may represent an annual (or longer) cycle of this kind. Other examples include bedded shales and radiolarites (e.g., Monterey Shale, California), or the braarudosphere and thoracosphere peaks (disaster species) which immediately follow the Cretaceous-Tertiary mass extinction (Percival and Fischer 1977). Large-scale productivity events are associated with limestone-shale bedding couplets representing Milankovitch climate cycles (Barron et al. 1985; Kauffman 1988a). A portion of carbonate enrichment in these cycles is due to diagenetic enhancement (Ricken 1986 Chap. 7.1, this Vol.), but a primary carbonate signal is widely acknowledged and attributed in some way to the dynamics of plankton productivity and population size (Fischer et al. 1985; Eicher and Diner 1985).

Short-term benthic population bursts or colonization events (Fig. 4) involve abrupt introduction of new communities to a region in conjunction with short-term changes in benthic substrates, chemistry, or trophic resources. Rapid changes in substrate texture immediately following large volcanic ash falls, storm events, or other sedimentologic changes in basinal facies favor widespread population expansion of benthic species (Stel 1975; Aigner 1980; Elder 1985; Hattin 1986b; Kauffman 1988a; Sageman 1989). Short-term oxygenation events in normally dysaerobic to anaerobic benthic environments also can result in benthic bioevent horizons (Kauffman 1981; Hattin 1986; Sageman 1989) that are correlative over large distances (more than 1000 km^2).

Rapid paleobiogeographic immigration or emigration events (Fig. 4) are indicated by abrupt changes in total diversity and composition of regional biotas, and are confirmed by reconstruction of changing biogeographic patterns within narrow time intervals (see Kauffman 1984, 1988a). Normally these paleobiogeographic shifts reflect rapid, widespread environmental changes, such as the incursion or retreat of cool polar or warm tropical water masses into temperate marine basins (Kauffman 1984).

Mass mortalities may occur widely and very abruptly in response to rapid deterioration of marine life conditions (Fig. 4). Some mass mortality events are associated with short-term fluctuation of benthic oxygen levels or overturn (advection) of toxic bottom waters within a basin (Wilde et al. 1988; Berry et al. 1988). These events may result in widespread death among most or all benthic and pelagic organisms, or selective species death. Other mortality events are related to rapid burial by storm, flood, or turbidite deposits (Kauffman 1988a; Brett and Seilacher Chap. 2.5, this Vol.). Examples of mass mortality surfaces in middle Paleozoic, Jurassic, and Cretaceous epicontinental deposits are well known, and in some cases have been correlated over 1000 km^2 (e.g., Kauffman 1981, 1986, 1988a; Sageman 1985).

2.4 Composite Events

Many event-stratigraphic markers are closely related to abrupt changes in the magnitude of external factors that collectively regulate sedimentologic, oceanographic, climatic, and biotic processes. As physical and biological stress gradients increase, so does the probability of event deposition. Externally forced environmental stress, e.g., asteroid impacts, major glaciation, or orbitally driven climatic cycles, may result in increased complexity and frequency of stratigraphic event markers, producing closely linked or composite events.

Dramatic catastrophic events, such as the bolide impact at the Cretaceous-Tertiary boundary, result in widespread and diverse stratigraphic event units: physical event beds associated with impact fallout debris and giant tsunami waves; chemical events associated with stable isotope excursions and trace element enrichment; and biological events associated with dramatic marine extinctions and population bursts of disaster species (Alvarez et al. 1984; Hut et al. 1987; Kauffman 1988b). These types of composite events comprise a relatively small portion of the geologic record. Smaller-scale catastrophic perturbations (e.g., major volcanic ash falls, turbidites, storm beds) also result in composite event markers such as mass mortality beds overlain by colonization events, and abrupt chemical and sedimentological changes (Kauffman 1988a).

Of the mechanisms producing composite events, orbital forcing of Milankovitch cycles is probably the most ubiquitous and widely applicable to HIRES analyses (see Chap. 1, this Vol.). Short-term climatic cycles may regulate weather patterns, sea level (at least during glaciations), weathering and erosion, fluvial transport, as well as oceanic and lacustrine chemistry, circulation, nutrient input, productivity, and ecological structure. These parameters may change rapidly within individual climate cycles to form widespread, correlative lithologic, geochemical, and biotic event units, most of which are discussed in the previous sections of this paper.

Cyclic composite event markers apparently reflecting climatic cycles are most readily preserved in lacustrine and marine strata. These markers include varved sediments, cyclic evaporite deposits, and associated palynomorph fluctuations [e.g., Pennsylvanian Paradox basin deposits of Colorado (Anderson 1984)], cyclic isotopic curves (Imbrie et al. 1984), relatively large-scale cyclothems (ca. 400 ka) with associated chemical and biotic events [e.g., the North American mid-continent Pennsylvanian cyclothems (Heckel 1986)], small-scale cyclothems (ca. 21–100 ka) and associated faunal changes [e.g., the Devonian Hamilton and Helderberg Groups of New York (Brett 1986; Brett and Baird 1986a); the Cretaceous Greenhorn and Niobrara Formations of the U.S. Western Interior (Elder and Gustason 1984; Barron et al. 1985; Kauffman et al. 1987)), and finally, the small-scale bedding rhythms that are so widely developed in pelagic to hemipelagic carbonate facies (Fischer et al. 1985; Bottjer et al. 1986; see Chap. 1, this Vol.). The composite nature of these climate-regulated cycles potentially allows widespread, high-resolution correlation based not only on the cyclothem boundaries themselves, but on the numerous event-units developed within each cycle.

3 Integrated High-Resolution Correlation: Time Lines for Basin Analysis

The preceding sections provide examples of diverse cyclic and event data that can be collected and applied to the development of a high-resolution stratigraphic framework. Second only to the detailed documentation of these data is their integration; this provides the multiple cross-checks of individual correlations necessary for the development of a robust HIRES data base. The integrated data base represents the highest-known level of stratigraphic resolution and provides an extremely powerful tool for basin analysis.

Construction of the HIRES framework should start with highly detailed section description at localities spanning distinct facies tracts and biogeographic zones in order to document all potential event units (Kauffman 1988a). Initial broad correlation of sections within a basin usually involves partitioning them into relatively large-scale units such as biozones, magnetic reversal intervals, transgressive-regressive cycles, disconformity-bounded stratigraphic sequences, or a series of units displaying relatively stable, long-term geochemical signals separated by abrupt shifts. Many of these broad units are prone to some diachroneity, but their utilization is important in developing an initial stratigraphic framework for study and first-order regional correlations. Following this, correlation of cycle and event units can proceed within these large-scale intervals, utilizing at first those thought to be most easily identified, most widespread, and least diachronous.

Although somewhat diachronous, biostratigraphic units are usually superior to other methods of initial correlation because they are based on nonrepetitive evolutionary phenomena having broad geographic extent, yet relatively short duration. In addition, first occurrence data are reasonably time-correlative on a regional to global scale.

Where present, magnetostratigraphic reversals and volcanic ash beds comprise the second step in developing and refining correlations, especially when the latter are chemically or petrologically "fingerprinted" or either occur in definitive bundles. These units are the most trusted HIRES correlation markers and define a matrix of isochronous time lines against which all other data can be compared. The regional expression of Milankovitch cycles in both basinal and nearshore facies provides the next most pervasive means of regional event correlation. Once these basic chronostratigraphic correlations have been established, they can be used to develop a regional composite chronostratigraphic matrix against which geographically more restricted, less diagnostic event units (e.g., many bioevents, concretion zones, turbidite and storm beds, etc.) can be tested and integrated using graphic correlation methods (see Sect. 1.2.2). These more facies-restricted event units can play an important role in further refining regional correlations. Thus, it is important to consider all event data when developing a HIRES framework for each basin or region. Subsequently, more far-reaching interregional correlations can be made with a high degree of confidence by integration of diverse cyclic and event units with magnetostratigraphic, biostratigraphic, oceanic chemostratigraphic, and sequence stratigraphic data having global expression.

The establishment of a highly resolved (<100 ka) event and cycle chronostratigraphy allows detailed three-dimensional reconstructions of the environmental dynamics and stratigraphic relationships among facies containing short-term physical, chemical, biological, and composite event deposits. These strata and surfaces reflect dynamic changes in climate, oceanography, eustasy, sedimentation, tectonism, and ecosystems. Kauffman (1988a) provides many examples of how the HIRES framework established for Upper Cretaceous marine deposits in the Western Interior Basin of North America can be used in comprehensive basin analysis interrelating these short-term dynamic factors. Figure 5 schematically illustrates the integrated HIRES framework for a hypothetical basin and shows relationships established between principal events and cycles recorded in the basin. Changes in facies and sediment thicknesses between time lines can be used to analyze patterns of sedimentation through time and space; these may reflect eustatic changes, transgressive-regressive or progradational events, as well as subsidence and uplift histories for different parts of a basin. In addition, the distribution of lithofacies (particularly organic-rich facies) and biofacies in single basins can be reconstructed in space and time and used to interpret such basinal phenomena as productivity and circulation histories. All of the above are of great value in developing a basin history and in understanding the relationship between potential petroleum source rock and reservoir bodies, as well as in interpreting regional tectonic and oceanographic trends.

4 Conclusions

Detailed documentation of marine and lacustrine strata in diverse Phanerozoic basins indicates that cycle and event units of short duration (<100 ka) are abundant and closely spaced through most depositional sequences; these units are less common in terrestrial and marginal marine facies. These stratigraphic markers represent diverse physical, chemical, biological, and composite events and small-scale cycles that collectively form the basis for high-resolution event and cycle stratigraphy (HIRES) and its broad application to basin analysis.

Widespread application of the HIRES methodology described in this paper provides one means of further refining the resolution and accuracy of stratigraphic dating and correlation. With it we can answer challenging questions about the dynamic interrelationships between diverse geological, climatic, oceanographic, and biological phenomena within and between sedimentary basins, and apply these answers to new levels of geological interpretation.

High-resolution stratigraphy offers temporal resolutions 10's to 100's of times greater than magnetostratigraphy or biostratigraphy alone in tested examples. In basinal facies particularly conducive to HIRES analysis, essentially the entire section can be broken into small-scale cycle and event units, allowing the construction of a highly resolved, integrated chronostratigraphic correlation matrix (Fig. 1).

Some limitations to the HIRES methodology should be noted. These relate to the degree of development and preservation, and the ease of confident identification of widespread cyclic and event-stratigraphic markers in the sequences being studied. These limitations can be partially overcome by the collection of more detailed field data with potential for correlation, at more closely spaced sections, at high levels of resolution, and by the application of more sophisticated techniques to process and

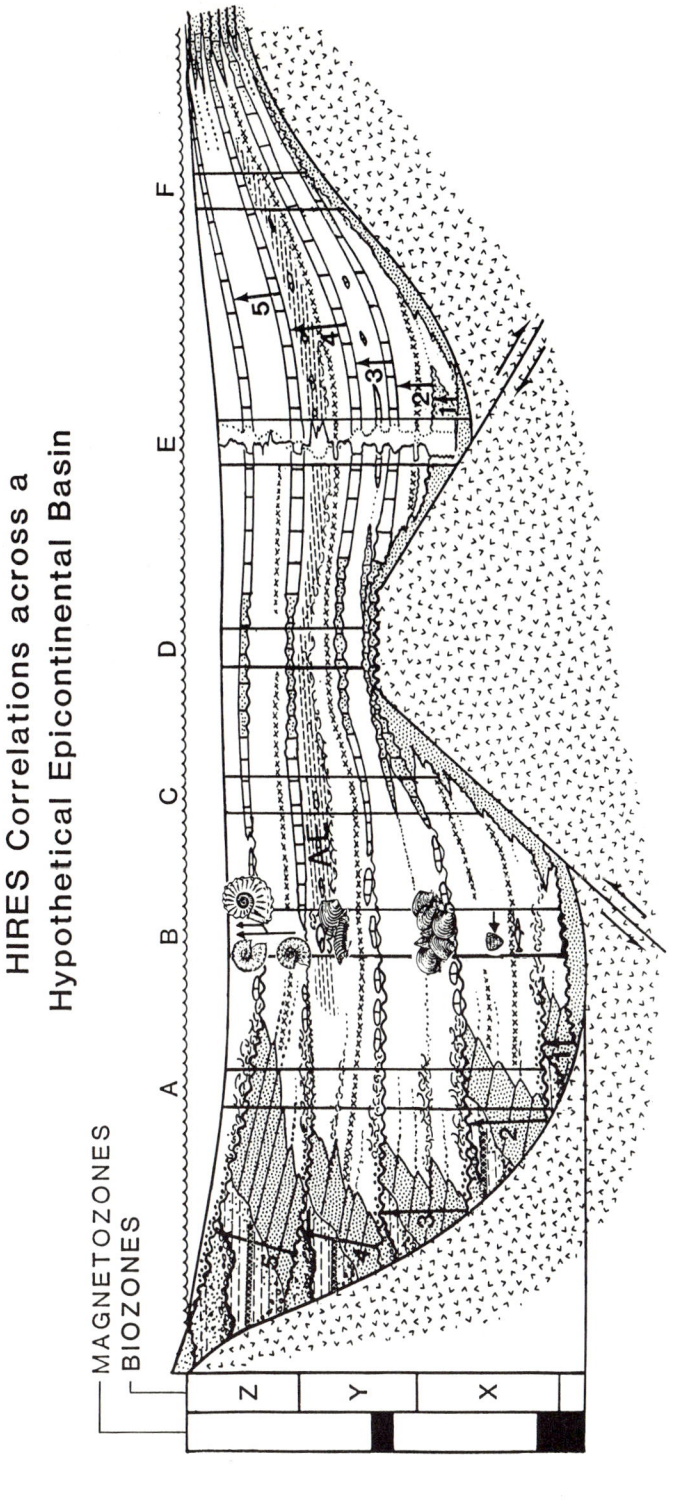

Fig. 5. Integration of physical, chemical, and biological event units in basin analysis is illustrated for an idealized epicontinental basin (not drawn to scale; scale changes within figure). The resolution of traditional correlation methods is represented at *left*. Predominate physical event units are shown at widely distributed localities (sections *A* through *F*) within a matrix of closely spaced sections. For simplicity, biological and chemical event data, though collected at every locality, are represented at sections *B* and *E*, respectively. When compared to traditional methods of correlation, the potential advantage of HIRES in recognition, reconstruction, and interpretation of cyclic and event stratigraphy is clear (e.g., note the relationship between the disconformity-progradational cycles on left hand margin of basin and shale-limestone cycles at the basin center). See Fig. 1 for explanation of lithologic symbols

correlate these data. Further, new geochemical and magnetostratigraphic methods of observation may provide more types of high-resolution stratigraphic data than are presently available. Even so, not all stratigraphic sections preserve cyclic and event deposits in recognizable form and some facies virtually lack them (e.g., highly bioturbated mudstones).

Thus, some geologic settings are predictably more conducive to the deposition and preservation of events and small-scale cycles than others, and should be selectively studied. For example, the high depositional rates present along active plate margins dilute most nonturbidite and tempestite marker units and produce sections of extreme stratigraphic thickness. Also, the predominance of turbidite deposition and contour current activity along continental slope and rise settings tends to both produce short-term mass flow events, and to erode or dilute most regional to global marker beds to the point where they are difficult or impossible to define. Likewise, multiple periods of erosion and redeposition in fluvial and strand plain settings generally inhibit preservation and recognition of regional HIRES correlation units. Paleosols, regional channelization events, coal-forming intervals, and coastal effects of small-scale, relative sea level changes comprise major HIRES units to be investigated in these facies.

Conversely, low-energy basinal environments having moderate to low sedimentation and erosional rates, such as those present in many lakes, starved marine basins, and mid- to outer-shelf (especially on passive margins) and deep-sea settings, are conducive to the deposition and preservation of abundant, diverse HIRES markers which may dominate these facies. Condensation and omission of sediments, however, present a problem in these environments.

Further, certain periods in the Earth's history appear to have been more conducive to deposition and preservation of short-term cycles and events than others, in particular those periods when the sedimentary reflection of orbitally driven climatic cyclicity (Milankovitch cycles) appears to have been enhanced. These time intervals include glacial periods when sea level fluctuated rapidly due to orbital forcing of ice formation, and periods of high global sea-level stands when sedimentation rates were low and the climatically forced salinity and density stratification, and planktonic productivity were apparently enhanced, such as during long intervals of the Mesozoic. The preservation of cyclic and event units through geological time is also an important component governing their presence in strata. Biogenic mixing and reworking of sediments is of primary importance in the destruction of event markers (see Sepkoski et al. Chap. 2.6, this Vol.). Therefore, marker beds deposited in strata of the earliest Paleozoic, prior to the extensive development of diverse, abundant, deep-burrowing organisms, should be more widely preserved than those deposited in younger strata. Also, sediments laid down in dysoxic to anoxic, hypersaline, or hyposaline environments where bioturbation is limited, typically have well-preserved marker beds. Preservation of these units is enhanced by their rapid burial, below bioturbating levels, and by low levels of subsequent diagenesis.

Thus, HIRES analysis must be somewhat tailored to the geologic setting under study. But the primary objectives and advantages of high-resolution event and cycle stratigraphy must not be overlooked; HIRES methods lead to development of an isochronous stratigraphic matrix that in modern stratigraphy is unsurpassed for dating and correlation. HIRES systems allow the maximum possible temporal resolution attainable in a stratigraphic sequence by the integration of diverse types of short-term processes, having local, regional, and even global expression, with existing systems of biostratigraphy, magnetostratigraphy, and geochronology. This, in turn, allows highly evolved interpretations of the relationships between physical, chemical, and biological data, and their causal factors, in the analysis of sedimentary basins.

Acknowledgments. The authors greatly appreciate the work of the editors and reviewers in Tubingen, especially Dr. Werner Ricken, whose comments significantly improved the manuscript. At the University of Colorado, the input of ideas and data from E. Gustason and J. Kirkland, and an early review of the manuscript by P. Harries aided in the development of the paper and are gratefully acknowledged.

7.3 Varves, Beds, and Bundles in Pelagic Sequences and Their Correlation (Mesozoic of SE France and Atlantic)

P. Cotillon

1 Introduction

Southeastern France corresponds partly to the subalpine ranges (Fig. 1) located on a last subsiding cratonic area close to the European margin represented by the Briançon and Piemont zones. Pelagic facies define a former basin referred to as Dauphinois throughout the Jurassic and Vocontian throughout the Cretaceous. As early as the upper Jurassic, this basin was surrounded by platforms marked by hemipelagic facies and temporarily by shallow carbonate banks. Until the Early Cretaceous, the tectonic regime was extensional, followed later by a compressive phase coinciding with the closure of the Tethyan ocean that led to the collision between the European and African plate margins.

Southeastern France is the most suitable area in the Tethyan domain for the study of Mosozoic cyclicity. Thick Jurassic and Cretaceous series are well exposed and little affected by subsequent tectonics and metamorphism, and they can be subdivided accurately on a biostratigraphic basis. The thickness from the Liassic to the Cretaceous sequence is up to 8 km in a few parts of the basin, showing almost continuous fine-grained calcareous deposits with a constant trend to form multiscaled limestone-marl alternations or cycles. Similarly, in the Atlantic ocean, DSDP cores from the Cretaceous and Upper Jurassic show a wide spectrum of high-frequency cycles, including bundles of beds and interbeds up to the finest submillimetric laminations.

As early as 1950, Gignoux interpreted bed-scale cycles as the result of a submarine climatic control. More recently, the same cyclicity has been studied through the Jurassic and Cretaceous sequences of the Vocontian basin and characterized by stratonomy, mineralogy, organic and inorganic geochemistry, and micropaleontology (Cotillon et al. 1980; Darmedru et al. 1982; Curial 1983; Tribovillard 1988; Dromart 1989). Comparisons between the Vocontian Cretaceous alternations and coeval marl-limestone couplets cored at DSDP sites in the Atlantic have led to the conclusion of a global, mainly climatic control of this pelagic deposition (Cotillon 1984; Cotillon and Rio 1984).

However, millimetric and submillimetric cycles were rather neglected in these studies. Indeed, these delicate structures are often disturbed or destroyed by bioturbation and/or compaction. In addition, the distinction between laminae originating from a vertical flux of materials and structures due to tractive currents is not very clear and often controversial. Finally, apart from Holocene series (e.g., Gulf of

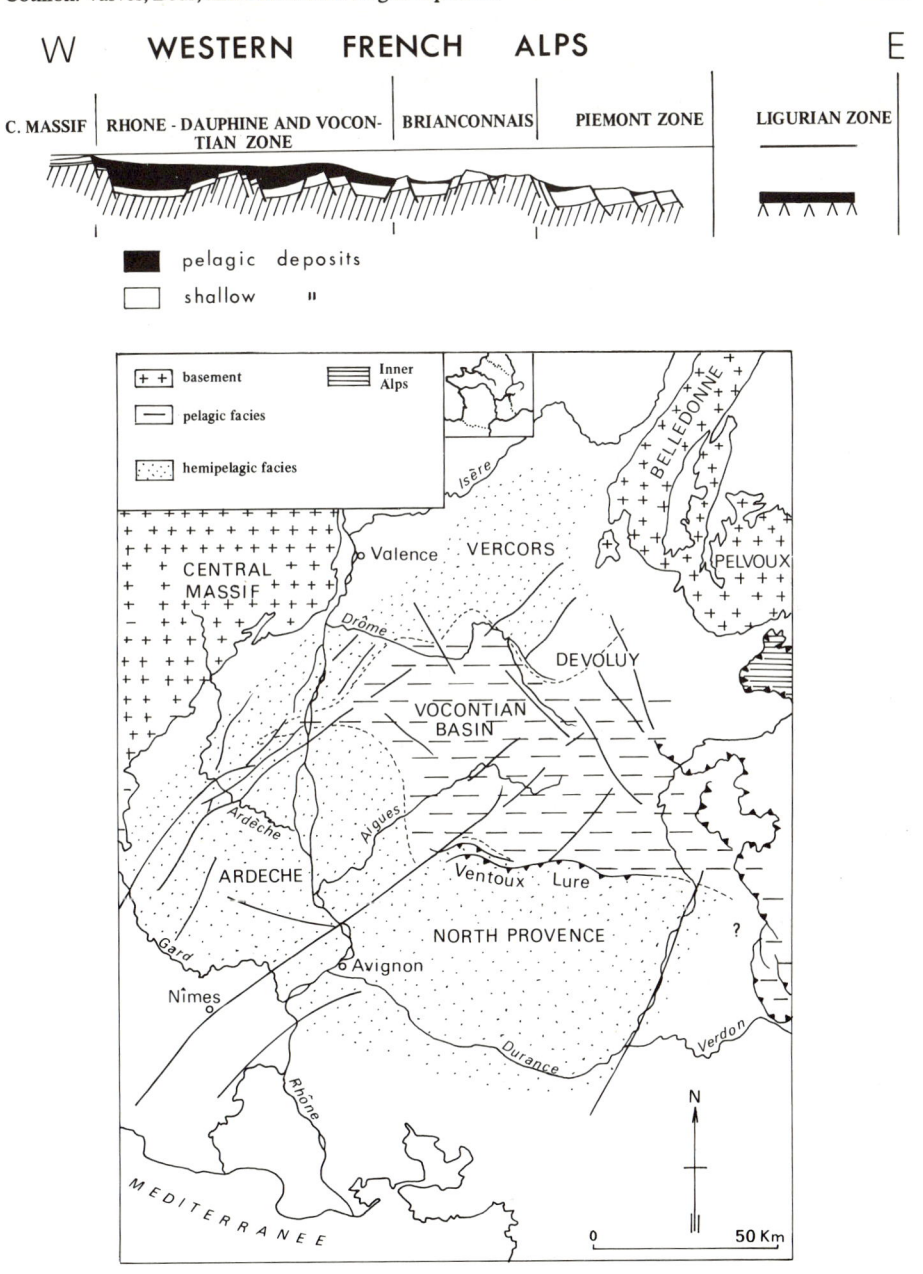

Fig. 1. Location of the Vocontian Basin and cross-section of Southeastern France during the Cretaceous showing special paleogeographic zones which later formed the Alpine and Subalpine chains. (After Argyriadis et al. 1980)

California and Black Sea (Donegan and Schrader 1982; Noel et al. 1987)), marine laminations, particularly within pre-Quaternary deposits, cannot be clearly linked to climatic causes which generally control seasonal planktonic production and/or terrestrial runoff. Here the attempt is made to identify and explain various scales of high-frequency cycles from a climatic point of view.

2 Bed-Scale Cycles

A number of formations within Mesozoic series are marked by thick marl-limestone alternation patterns, in which salient calcareous beds gradually pass into marly scoured interbeds and reversely, are illustrated by a sinusoidal fluctuation of the carbonate content (Fig. 2). Cycles may be binary (marl-limestone couplets) or ternary with an additional intermediate lithology. In this case, a more or less asymmetrical character of cycles may be represented and quantified. The Mesozoic series cored in the Atlantic and Gulf of Mexico display a similar pattern.

2.1 Vertical Successions of Cyclic Bedding

Typical alternations occur when marls and limestones are nearly equally represented, i.e., in sequences just beneath or above major carbonate formations. This pattern occurs in the Aalenian, Bathonian, Upper Oxfordian, Kimmeridgian, Berriasian, Lower Valanginian, Middle Hauterivian, Lower Barremian, Clansayesian, and Lower Cenomanian of the Vocontian basin (Fig. 2). The calcareous content of deposits becomes dominant or exclusive in some intervals, making important barriers in the Vocontian landscape: the Tithonian, Upper Hauterivian, Upper Barremian and Bedoulian, and Turonian to Santonian. Here, interbeds are often reduced to thin clayey seams. An opposite tendency leads to marl-dominated formations with scarce limestone beds or exfoliated argillaceous limestones. This is common in the Toarcian and Upper Valanginian, as well as in thinner intervals of the Upper Oxfordian, Middle Hauterivian, Lower Barremian, Albian, and Cenomanian. An extreme case of the same tendency is represented by some parts of the Oxfordian "Terres Noires" and the Apto-Albian blue marls. They appear to be homogeneous, but show in fact a weak cyclicity due to subtle changes in color. The darkest layers are more carbonate-rich (Tribovillard 1988). Elementary bed-interbed cycles are often grouped in bundles forming cycles of lower frequency (second order cycles). Bundles may be successions of cycles of still upper (first) order.

Fig. 2. Cyclic bedding in the Mesozoic of Southeastern France. *1* Total Jurassic-Cretaceous succession. *2* Cretaceous sequence in Vocontian basin. *3* Patterns of alternating layers (binary and ternary cycles)

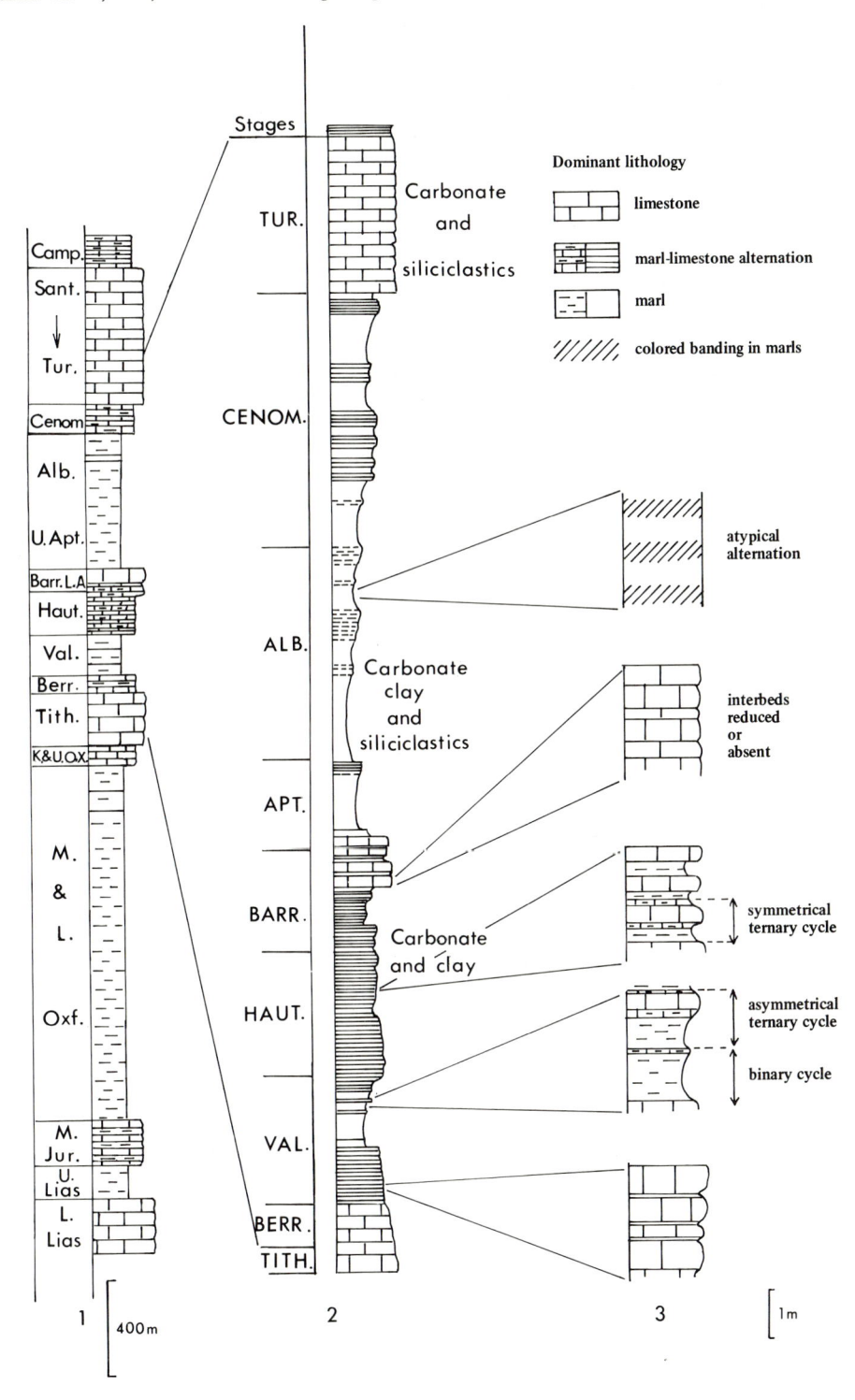

Cotillon: Varves, Beds, and Bundles in Pelagic Sequences   823

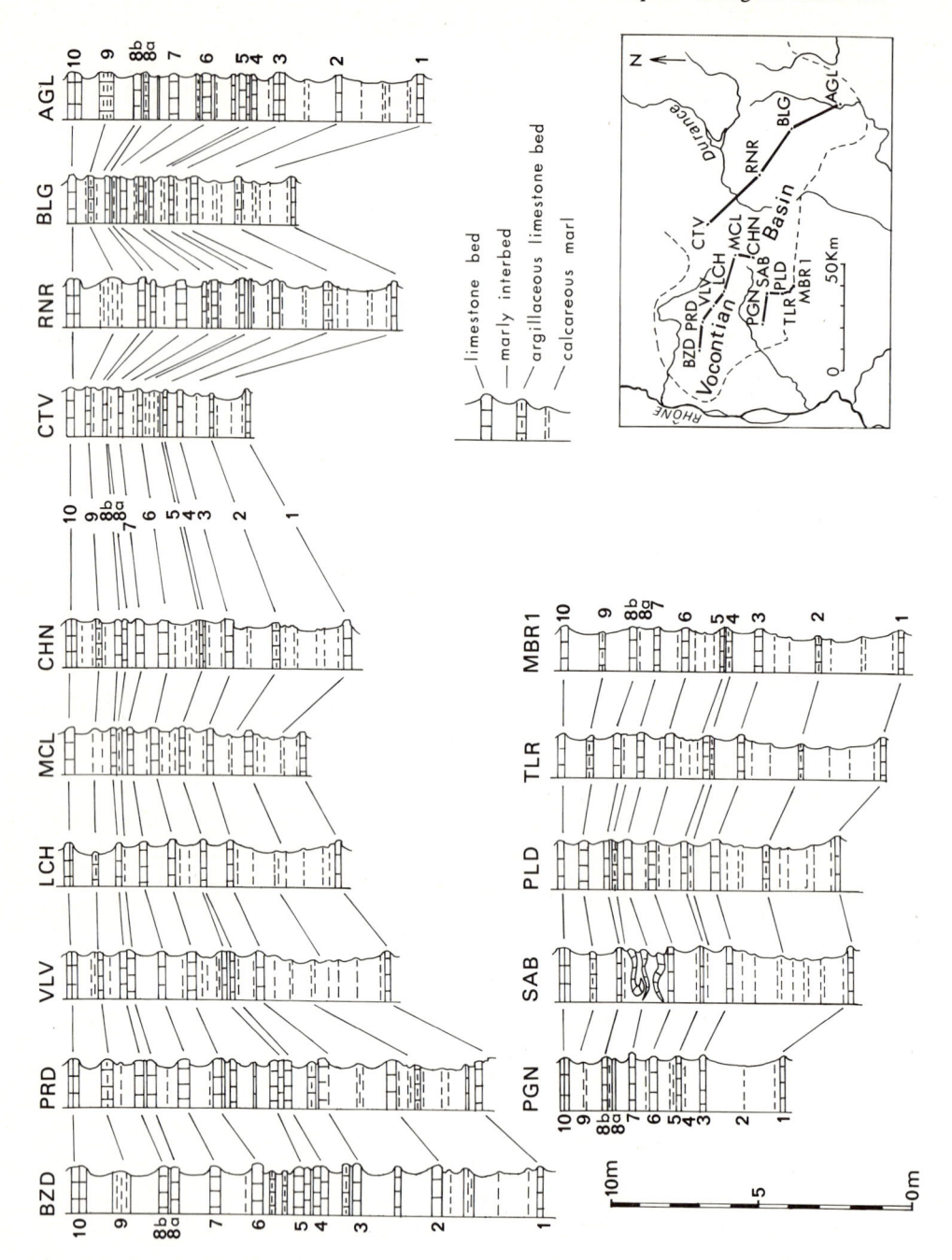

Fig. 3. Correlation of "The Toulourenc" marker sequence (Upper Valanginian) throughout the Vocontian basin. Some layers (2) may be affected by horizontal variations of CaCO3 content. Others (4–5) may subdivide when the global sequence thickens

2.2 Lateral Changes in Cyclic Bedding

Whatever their order, marl-limestone cycles are correlatable throughout the pelagic part of the Southeastern basin of France. Curial (1983), Tribovillard and Ducreux (1986), Tribovillard (1988), and Dromart (1989) demonstrated their constancy over large distances in the Upper and Middle Oxfordian respectively, when the pelagic domain dominated in Southeastern France. In the Upper Valanginian, a first order cycle (the marker sequence of "the Toulourenc", Cotillon et al. 1980), composed of ten beds and bundles, can be traced with only thickness variations over an area of 12 000 km^2 as long as the facies remain pelagic (Fig. 3). However, when passing into the hemipelagic zone surrounding the Vocontian basin (Vercors, Ardèche, Provence) the marker sequence is no longer recognizable. This disappearance may have two causes:

1. The alteration of a pelagic deposition, mainly resulting from vertical setting, to a composite deposition, the result of vertical and lateral fluxes. The latter is controlled by tractive currents evidenced by large channels clearly exposed in wide outcrops such as the high Hauterivian-Barremian cliffs of the Vercors and Ventoux-Lure chains, which correspond to the northern and southern boundaries of the Vocontian trough.
2. An increase in sedimentation rate from the basin to the marginal slopes, actively subsiding and fed directly by carbonate-producing platforms. Here, the sedimentation rate in the Lower Cretaceous is higher by a factor of 4 and in the Lower Barremian by a factor of up to 13. Therefore, the number of minor bed-interbed cycles increases (Cotillon 1985, 1987; Ferry and Monier 1987) (Fig. 4). Consequently, a pelagic bed may be represented by a bundle of beds in the hemipelagic zone.

2.3 Various Expressions of Cyclicity in Limestone-Marl Alternations

Lithological cyclicity, due to carbonate variations, is accompanied by sinusoidal fluctuations of biological, mineralogical, and geochemical characteristics which may change in phase or 180° out of phase.

2.3 Biological Characteristics

Darmedru (1982) demonstrated that the abundance and proportion of all microfaunal components (calcareous and arenaceous foraminifers, radiolarians) of the Toulourenc bundle-marker vary with the carbonate content according to equal or twofold and even threefold wave lengths. Limestones and marls contain the more specialized associations with principally benthic species for marls, planktonic species (and particularly radiolarias) for limestones. Argillaceous limestones exhibit more diversified populations and multiplication of planktonic foraminifers. On the other hand, coccolites are more abundant in marls and Nannoconus in limestones.

Briefly, the faunal distribution within marl-limestone cycles is as follows:

1. Limestones. spherical radiolarians, planktonic foraminifers – *Lenticulina* – *Nannoconus*.
2. Marly limestones. conical radiolarians – *Spirillina* – *Lenticulina* and *Dorothia* – Planktonic foraminifers.
3. Marls. calcareous and arenaceous benthonic foraminifers – *Gavelinella* – coccoliths.

These distributions are independent of differential preservation induced by dissolution or recrystallization. Consequently they are of paleoecological significance (Cotillon and Rio 1984).

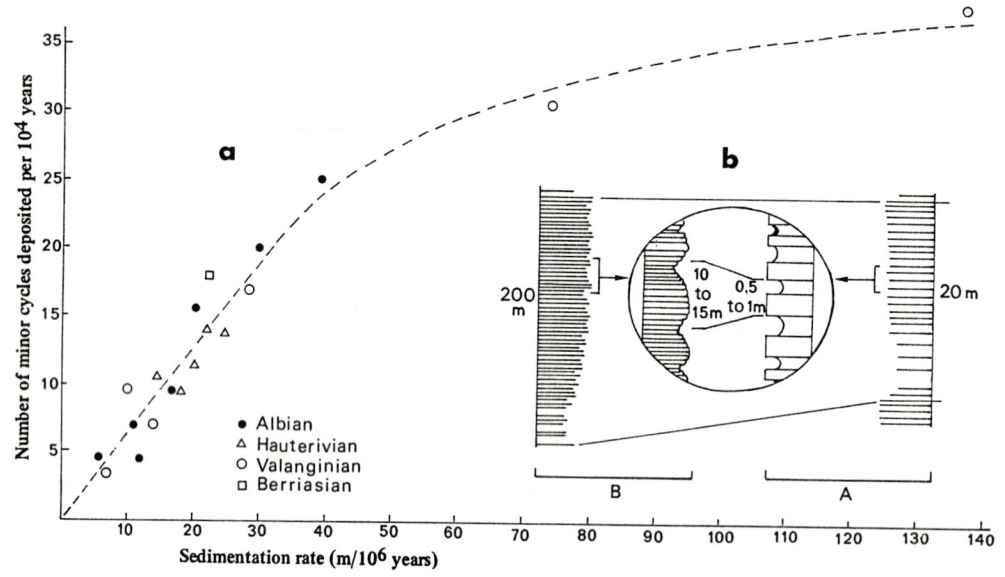

Fig. 4. a Sedimentation rate versus number of marl-limestone couplets deposited per 10^4 years. Lower Cretaceous sections drilled at DSDP sites and investigated in Southeastern France, Northern Italy, and Northern Tunisia. (After Cotillon 1987). **b** Correlation between *A* pelagic and *B* hemipelagic series: a bed-interbed cycle in the Vocontian basin (*A*) may correspond to a bundle of beds and interbeds in the hemipelagic zone (*B*). (After Ferry and Monier 1987, simplified)

2.3.2 Mineralogical Content

Illite, kaolinite, and silty quartz are more abundant in marls than in limestones and conversely for smectite. According to Ferry et al. (1983), Deconinck and Chamley (1983), and Levert (1989), this distribution of clay minerals in the marl-limestone couplets is original but may be strongly modified by diagenesis. Thus, in some bed-interbed cycles from Kimmeridgian, Upper Valanginian, and Clansavesian intervals, smectite is replaced by chlorite in an eastward direction toward alpine overthrusts (Levert 1989).

2.3.3 Geochemical Content

The Toulourenc marker sequence shows variations in many major or minor elements (e.g., Fe, Mg, Sr, K, Li, Zn, Nb, Rb, Ni) parallel or opposite to that of carbonate, of many elements which occur either in carbonate or in the insoluble residue (Jouchoux 1982). Organic matter is the higher concentrated in marls; limestones are richer in ^{18}O and ^{12}C than marls.

2.3.4 Bioturbation

Marl-limestone couplets are bioturbated throughout. Nevertheless, limestones are more intensely churned.

The cyclic distribution of various biological, mineralogical, and geochemical components in marl-limestone alternations on a decimetric scale is inconsistent with the interpretation of the limestone beds as distal turbidites (e.g., Beaudoin et al. 1974; Le Doeuff 1977) as well as of a diagenetic origin of limestones (Ricken 1986).

3 Lamination

3.1 Observed Features

Laminations were observed in Cretaceous successions of many offshore sites (Donnelly et al. 1980; Jansa et al. 1978, Thiede et al. 1981; Jenkyns 1976; Sheridan et al. 1983; Freeman and Enos 1978) and onshore sites (Kauffman 1982; Breheret 1985). Generally used for characterization of an oxygen-poor environment (e.g., Noel and Manivit 1978), such laminations were rarely used for a better understanding of a basic behavior of pelagic sedimentation. Their minimal qualitative and quantitative variations are particularly interesting.

In the Atlantic ocean, the Berriasian-Hauterivian interval exhibits a multi-scaled alternating pattern of marl-limestone rhythms with laminae, generally 1 mm or less in thickness. They occur in gray marls, black shales or argillaceous limestones. At DSDP sites 535 and 540, in the Straight of Florida, some laminae display riplets and result from current action. Others, showing a fine-grained matrix, are more regular. Some of them, 30 to 350 µm thick, are composed of an alternation of (1) light-colored micrite with abundant foraminifers, radiolarians, and pelagic pelecypods and (2) dark micrite with more clay, iron oxide, and organic matter. More detailed examination shows that these laminae represent at least three orders of cycles.

The third order (often destroyed and deformed by compaction) corresponds to alternations of 8 to 25 µm thickness with thin dark laminae (concentration of iron oxide and clay) and larger light-colored laminae consisting of a concentration of calcareous nannofossils (Fig. 5A-B). In some layers, the calcareous laminae grow into alignments of ovoid, well-outlined, white specks which are accumulations of well-preserved nannofossils (Fig. 5D-F). The pelletic nature of the specks has been assumed by several authors. They most likely correspond to the "stringers of white

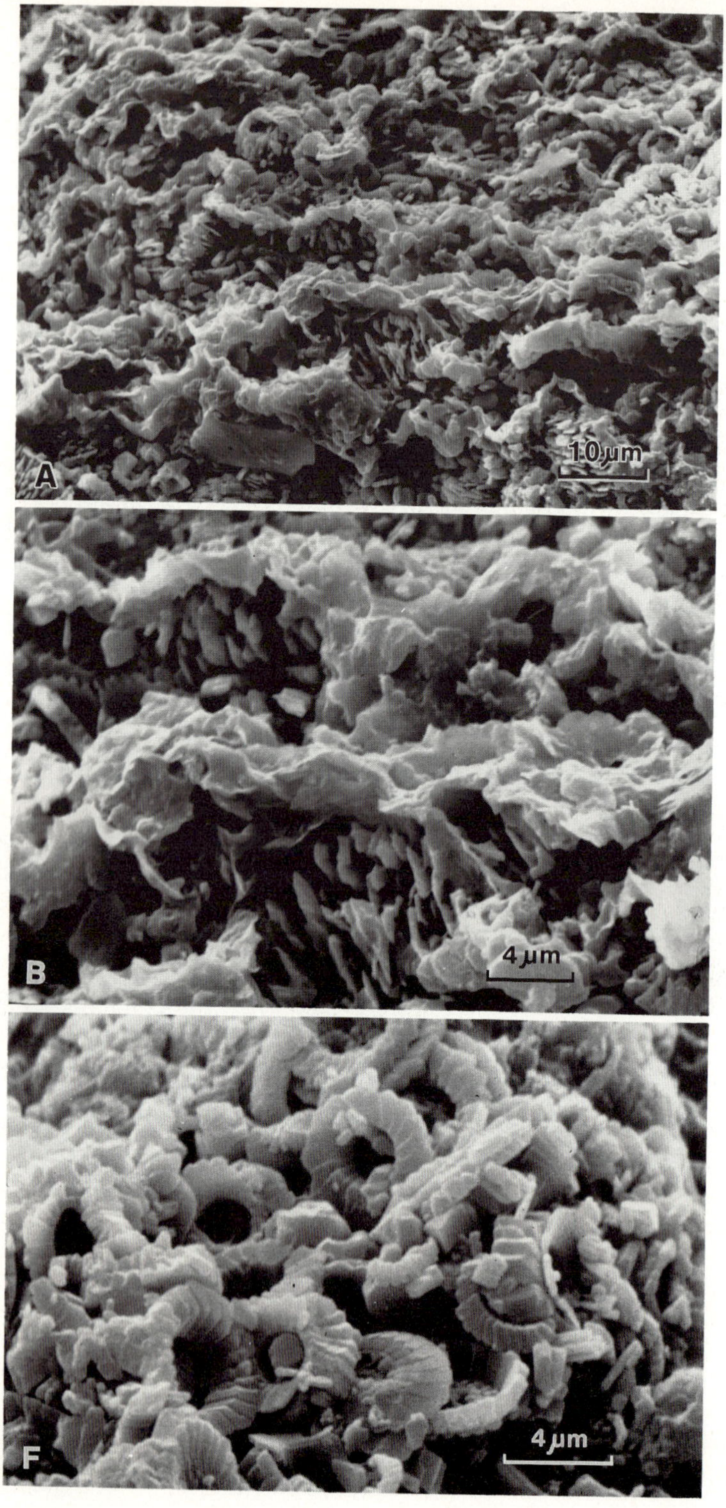
A
10 µm
B
4 µm
F
4 µm

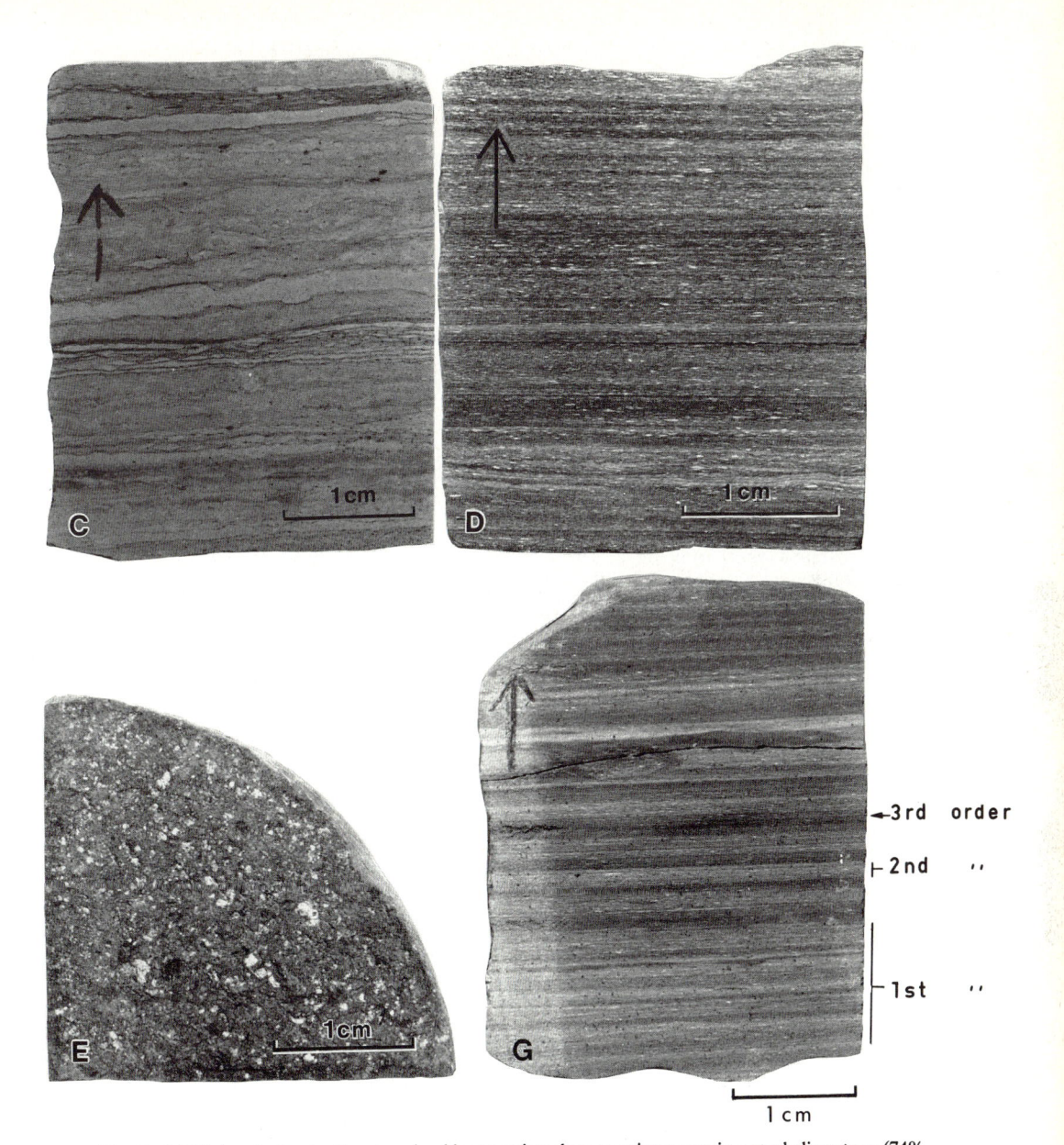

Fig. 5. A Third-order lamination examined by scanning electron microscopy in a marly limestone (74% CaCO$_3$). Nannoconid-bearing layers alternate with argillaceous ones (Upper Valanginian, Site 535 DSDP, sample 59–2, 74–79 cm). **B** Same sample, close-up view. **C** Effect of compaction and pressure-solution in a laminated limestone. Part of the laminae are deformed and constricted or disrupted into lens-shaped amygdaloid units (Albian, DSDP Site 535). **D-F** Alignments of calcareous white specks in a laminated marly limestone, corresponding to crushed fecal pellets bearing nannofossils (Lower Hauterivian, DSDP Site 535, sample 53–4, 41–44 cm). **D** Polished section. Specks are concentrated in second order and first order laminae. **E** same as **A** viewed on bedding plane. **F** White speck examined by scanning electron microscopy in a laminated marl: fecal pellet crowded with coccoliths (Upper Valanginian, DSDP Site 535, sample 57–4, 39–44 cm). **G** The three orders of laminae in a marly limestone. Each of them forms an alternation between light-colored units (rich in carbonate) and dark ones (richest in clay and organic matter). First order laminae form a banding (see a light band in the lower part of the section) (Lower Valanginian, DSDP Site 535, sample 67–4, 122–128 cm).

specks" observed in Aptian-Albian laminated marls of the Central Atlantic (Donnelly et al. 1980).

Light colored calcareous units 0.05–1 mm thick (Fig. 5G) representing a higher order of laminations, may be groups of flattened white pelletic specks alignments (in marls) (Fig. 5D-F) or groups of lens-shaped, amygdaloid, and unraveled micritic layers, separated by a thin, brown, argillaceous film (in marly and argillaceous limestones) (Fig. 6A-C).

Dark-clayey third order to first order units (0.010 to 1 mm thick or more) contain more organic matter, fish debris, iron oxide, detrital quartz, and mica. Coccolites are the most frequent among planktonic calcareous remains. In limestones, laminae may be represented by parallel alignments of alternatively large and small-sized radiolarians (Fig. 6D). Micritic specks and amygdaloid patches are either pellets, detrital elements, or the result of a disruption of continuous layers into microlenses by compaction. When specks are sparse and made of packed nannofossils, a pelletal origin is highly probable.

However, when these microlenses or specks are pressed on each other, deformed, and stretched, they may be affected by differential compaction and boudinage (Fig. 5C). Such structures, similar to a micro flaser bedding, were also noticed by Jansa et al. (1978) in the Cretaceous succession at site 367 (Central Atlantic) and by Thiede et al. (1981) in Aptian-Albian deposits of the Central Pacific. Compaction is also responsible for stylolitic seams. Sometimes occurring between the laminae in marly interbeds, these seams are outlined by black concentrations of organic matter.

3.2 Interpretation

Deep-sea laminated facies are often thought to be distal turbidite structures, particularly by Montadert et al. (1979), Thiede et al. (1981). However, the laminations described here were deposited in a way precluding an origin from current action. Instead, they have in millimetric or submillimetric scale the main characteristics of the bed-interbed alternations:

1. a grouping of elementary cycles into bundles (second and first order cycles);
2. a cyclic distribution of carbonate marked by the alternation, described above, of light calcareous and dark marly or clayey units;
3. a cyclic distribution of some elements of the biological content. Thus, coccolites and planktonic foraminifers are more abundant in marly laminae while *Nannoconus* are dominant in calcareous-rich ones.

Laminae containing abundant planktonics are mostly interpreted as being current-winnowed ooze (Ryan et al. 1973; Leclaire 1974; Cronan et al. 1974). However, in calcareous laminated beds cored in the Gulf of Mexico, laminae provided with abundant and mostly large-sized radiolarians alternate with laminae enclosing small and scattered ones. In each of these laminae, calcitized tests are unsorted and are matrix-supported. Occasionally their long and fine spines are still intact, which more or less excludes current transport. Fluctuations in components of the couplets of laminae (e.g., terrigenous products) may explain the changes in radiolarian size;

indeed, the size of spheric radiolarians has been correlated with $CaCO_3$ content of deposits in the Vocontian basin (Darmedru 1982).

In summary, the alternating laminae of the Lower Cretaceous sequences at sites 535 and 540 are interpreted as varve-like structures resulting mainly from a vertical supply of biocalcareous and terrigenous particles, although the distinction between deposits from pelagic settling and low-density distal turbidites is not easy (Stanley and Maldonado 1981). In fact, deep currents (although probably weak in the Tethyan ocean) could have contributed to the deposition of laminated sediments (Honjo et al. 1982) but their winnowing and reworking did not operate beyond the scale of a given lamina, thus preserving the original microcyclicity of deposition.

3.3 Lamination and Rate of Sedimentation

At sites 535 and 540, the average thickness of a given order of laminae in sections of different ages is correlated with the lithology and increases exponentially from marls to limestones up to be 18 to 50 times as thick as the thinnest unit (Fig. 7A). Two causes working in the same direction have effected this. The first is differential compaction between marls and limestones (Rieke and Chilingarian 1974) which normally leads to more flattened sedimentary structures (particularly the laminae) in marls than in limestones. The second cause is the higher rate of sedimentation for limestones than marls. After filtering the first cause with an independent calculation of the differential compaction rate, and assuming a constant time-span for the couplets of laminae, the compaction-corrected variation of laminae thickness should directly represent the relative sedimentation rates. In the Vocontian basin and at sites 535 and 540, the differential compaction rate is illustrated by the deformation of horizontal cylindrical structures of bioturbation (Fig. 6E,F). These observations may be used to calculate the compaction rate (C.R.) which is correlated with the carbonate content. A relation C.R. = f(% $CaCO_3$) set up for bioturbated Hauterivian Vocontian deposits (Cotillon 1985) has been used for the Neocomian and the Albian at sites 535 and 540 (Fig. 6E,F). This relation shows that, whatever the stratigraphic level, the variations in laminae thickness cannot be caused solely by compaction. Indeed, when its effect is ruled out, the amplitude of variation is divided by 1.5 to 6 only. Taking into account the absence of significant dissolution of carbonate in marls during and after its deposition (Cotillon and Rio 1984), it can be concluded that the rate of sedimentation increases from marl to limestone by a factor of from 3 to 20 in different stratigraphic levels.

Although still imprecise, this result permits calculation of the relative changes of the carbonate and clay fluxes (i.e., amounts of clay and carbonate deposited per unit area and time) through bed-interbed cycles. Figure 7C shows tentatively these variations for Valanginian, Hauterivian, and Albian deposits at sites 535 and 540. In these cases, the relation C.R. = f(% $CaCO_3$) is assumed to be the same for both the Vocontian Upper Hauterivian and the Cretaceous of the northeastern Gulf of Mexico, because stratonomy and facies are alike in the two areas. Flux variations are either out of phase (Hauterivian), or in phase (Valanginian, Albian). When peaks of terrigenous and biocalcareous inputs coincide, the latter show the largest variations.

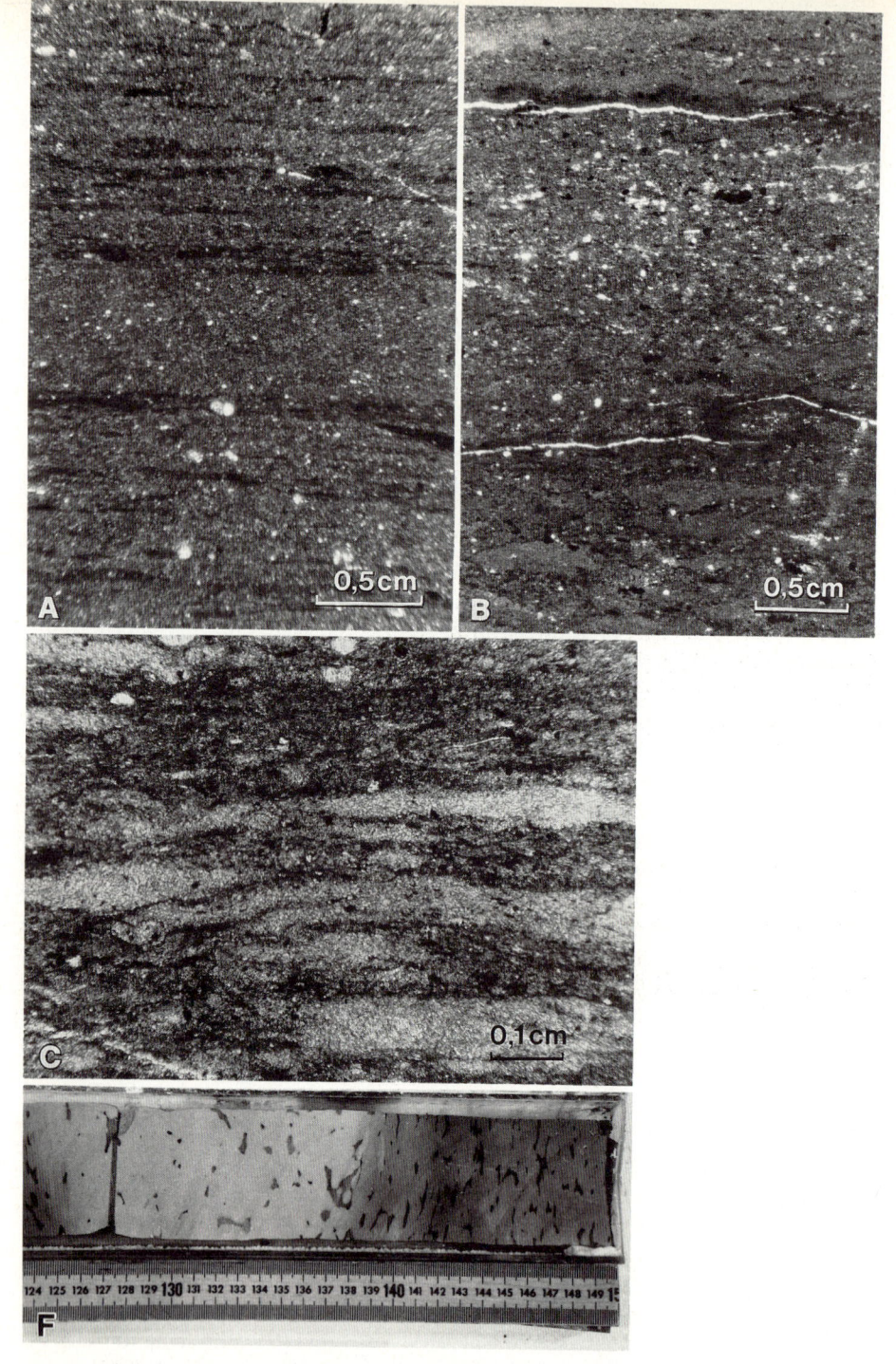

Fig. 6. A First order and second order laminae examined in thin sections. **A** Marly limestone (73% Ca CO$_3$): second-order alternation between light calcareous layers and dark ones, thinner, richer in clay and organic matter, deformed by compaction; first-order alternation between calcareous sets of laminae (in the center laminae are indistinct because joined) and clayey calcareous ones (Barremian at DSDP Site 535). **B** Marly limestone (67% CaCO$_3$): above and below large second-order calcareous laminae, deformed and

disrupted by differential compaction, form two calcareous first-order laminae. *Middle part of section* first order lamina, predominantly marly, and groups of thin second order laminae (Hauterivian, DSDP Site 535). **C** Argillaceous limestone (83% CaCO$_3$): portion of a calcareous first order lamination showing second order units separated by a thin dark clayey film, deformed and disrupted by compaction (Cenomanian, DSDP Site 535). **D** Second order lamination in a radiolarian-rich limestone (93% CaCO$_3$). Alternation between layers with numerous, large, spherical radiolarians and layers with small, scattered ones (Hauterivian, DSDP Site 535). **E, F** Differential compaction of limestone and clayey limestone evidenced by bioturbation showing various degrees of deformation. **E** Spotty burrows in Hauterivian beds at La Charce, Drôme, S-E France. Undeformed burrows in limestone (*top*), flattened ones in clayey limestones (*bottom*). *Bar* = 1 cm. **F** Spotty bioturbation in some Albian cores from DSDP 540 Hole. Closeness and flatness of bioturbations are function of clay content

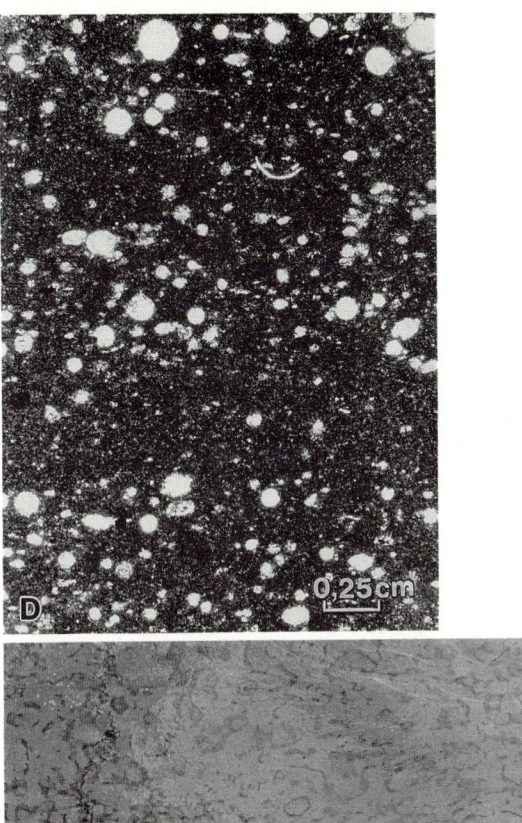

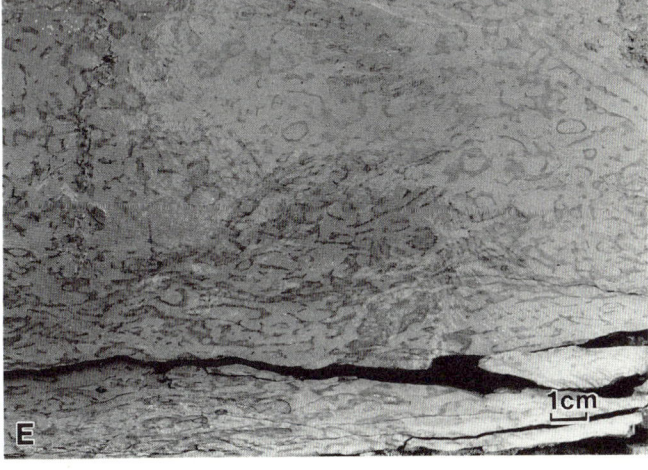

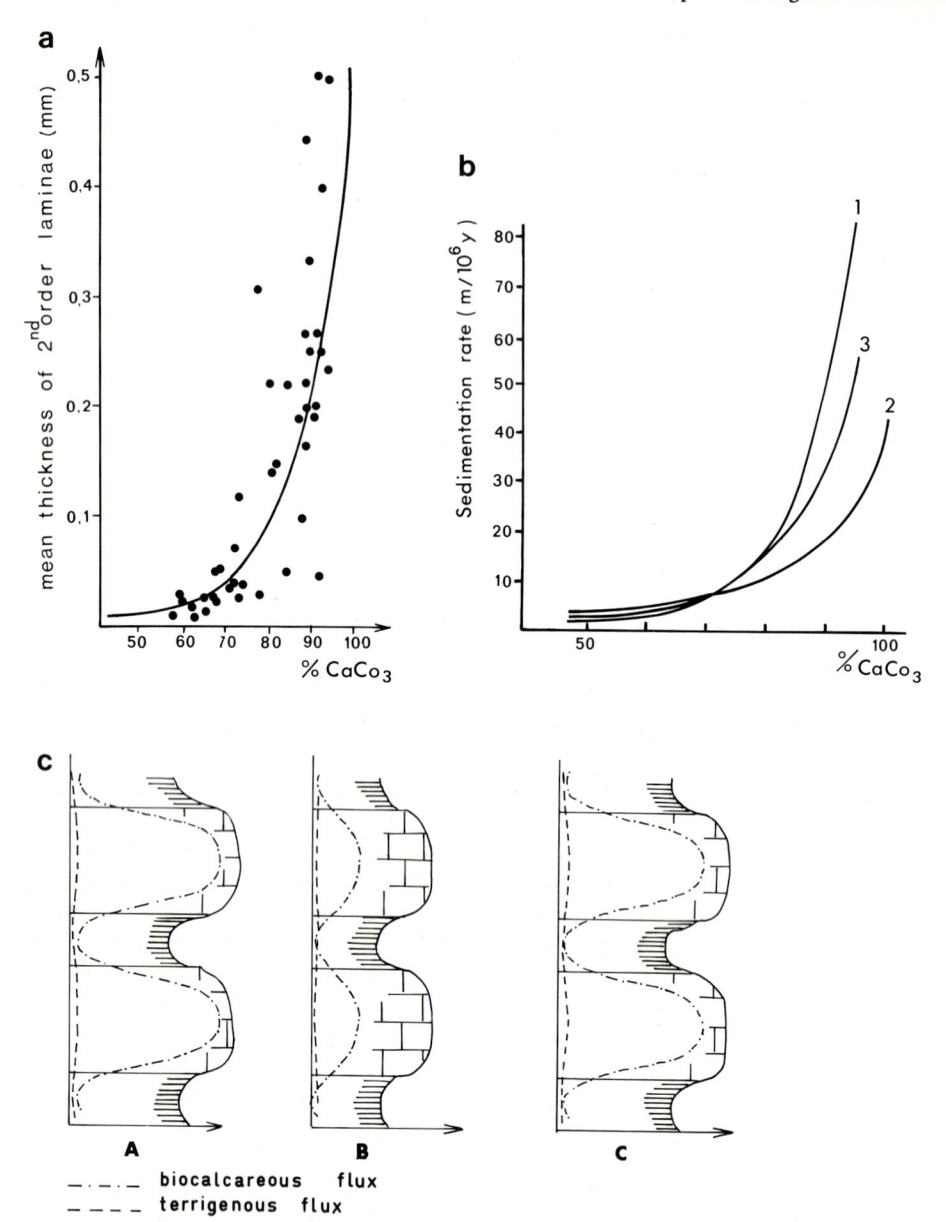

Fig. 7. a Mean thickness of laminae (mm) versus CaCO₃ content (Valanginian section, DSDP Site 535). Calculated exponential curve: Y = 0.147e 0.081x r = 0.87. **b** Mean sedimentation rate of laminae couplets versus CaCO₃ content of deposits for the Albian (*1*), the Hauterivian (*2*) and the Valanginian (*3*) at DSDP sites 535 and 540 (Gulf of Mexico). **c** Models of relative fluctuations of terrigenous and biocalcareous fluxes causing limestone-marl alternations. *A* Albian DSDP Site 540; *B* Hauterivian, DSDP Site 535; *C* Valanginian, DSDP Site 535. Relation between compaction rate and CaCO₃ content: CR = 4.4.10⁻²% CaCO₃ + 4.7 (based on Upper Valanginian, Vocontian Basin)

This example demonstrates that biocarbonate flux may be enhanced by influx of suspended and dissolved river load, if these also carry nutrients (Summerhayes 1981; Williams and Von Bodungen 1989). By other methods, Darmedru (1982) concluded that the deposition of thickest calcareous beds of the Vocontian Valanginian occurred simultaneously with the highest clay input.

4 Climatic Interpretation of Cycles

4.1 Bed-Scales Cycles

Some characters of the Vocontian alternations mentioned above, particularly the cyclic repartition of biological, mineralogical, and geochemical characteristics, were also found in similar Cretaceous on-shore and off-shore successions around the world. The variations in couplet thickness may be an additional element for comparison. One aspect of cyclostratigraphy consists in the tentative correlation between such series on the basis of cyclograms illustrating the cycle thickness variation (Cotillon and Rio 1984). Correlations between Neocomian Vocontian, Atlantic and Gulf of Mexico series are based upon similar kinds of thickness evolution in time and are controlled by the available biostratigraphic data. Obviously this method has allowed the introduction of stratigraphic subdivisions although index fossils were lacking. Gaps are evident, such as that of the Upper Hauterivian in the Straight of Florida. The correlations between oceanic series may also be based on the frequency variation of decimetric cycles per core, reflecting fluctuation of the sedimentation rate (Cotillon 1984). Figure 8 shows striking analogies for Albian and Hauterivian curves constructed for Atlantic and Central Pacific sites.

Similar characters shown by numerous alternating successions dispersed over large regions (Cotillon and Rio 1984) support the interpretation for a global control of cyclic sedimentation. It is generally assumed (see Einsele and Ricken Chap. 1.1, or Fischer Chap. 1.2, both this Vol.) that Milankovitch periodicities testify to a climatic control of marl-limestone successions. Harmonic analysis was applied to the Vocontian series of Angles from the Berriasian to the Bedoulian, after having calculated a mean cycle period of 21 000 years for elementary cycles (Rio et al. 1988). Some of the periods correspond to the 40 100 and 400 ka cycles, but the majority of them fall in the range between 40 and 100 ka and 100 to 400 ka (Fig. 9). The shortest periodicities are principally recorded in marly sections (Upper Valanginian) and the longest in calcareous formations (Berriasian, Lower Valanginian, Barremian). This difference may be explained by differences in sedimentation rate: slow and continuous for marls, rapid but discontinuous for limestones, commonly interrupted by diastems or clayey seams. Another possibility rests in two different responses of terrigenous and biocalcareous fluxes to orbital forcings. Variations of bioturbation and organic matter content through bed-interbed couplets suggest marine environmental changes affecting both surface and bottom waters. Lamination, present essentially in marls, and the absence of bioturbation traces indicate poorly oxygenated conditions at the sea floor. Water stratification due to poor oceanic circulation is mainly proposed (Bralower and Thierstein 1984; Noel et al. 1987). The marl-lime-

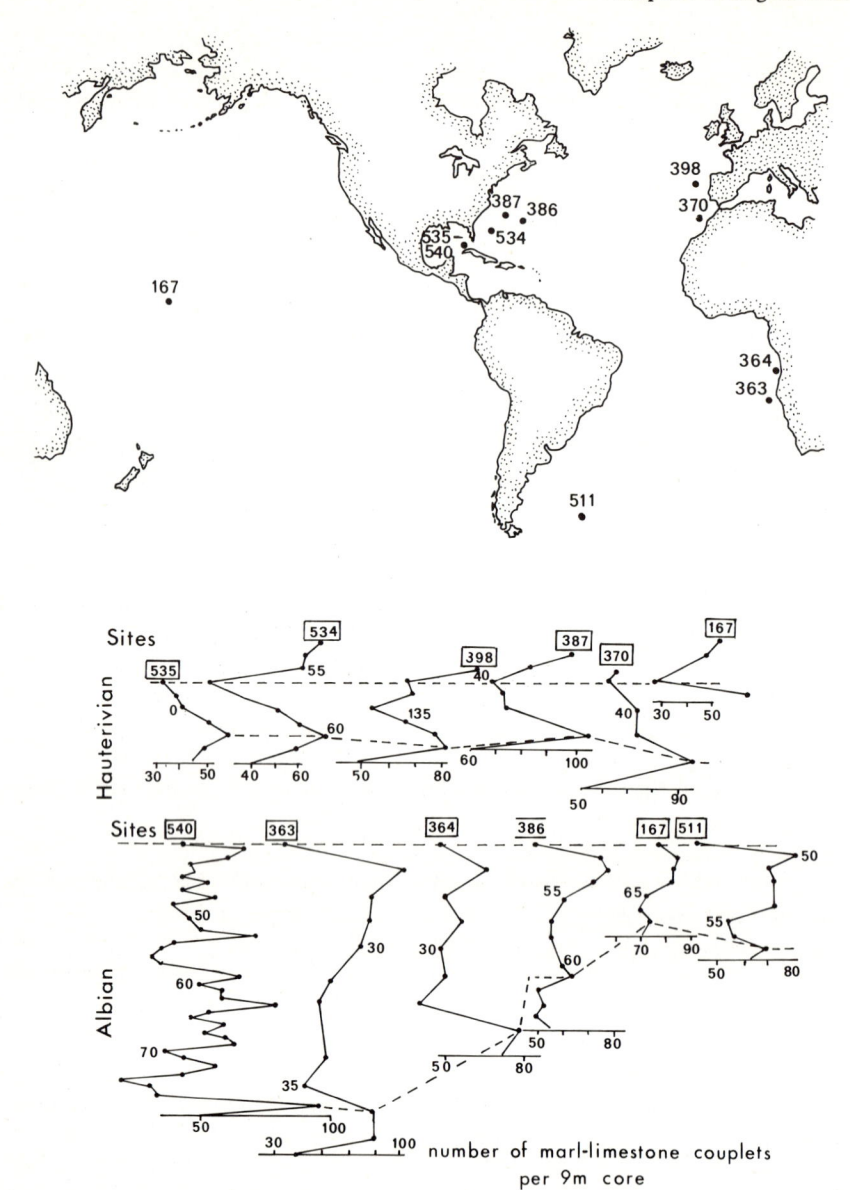

Fig. 8. Decimetric marl-limestone cycle frequency (number of cycles per 9 m core) in Hauterivian and Albian sections at various DSDP sites (see map.) *Dashed line* correlated levels. (After Cotillon 1987)

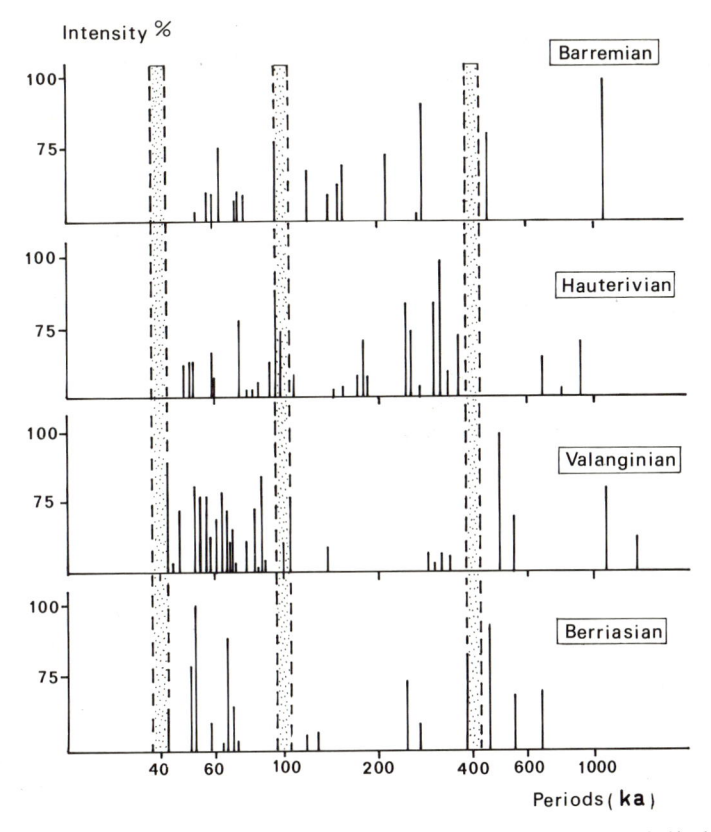

Fig. 9. Period spectra from Fourier analysis on bed-interbed cycles recorded in the Early Cretaceous stages of Angles (Vocontian basin). (After Rio et al. 1988)

stone passage is marked by a renewal of oxic conditions, causing a benthonic re-colonization and subsequent obliteration of lamination. From this point of view, the marly intervals may be regarded as a step towards black-shale facies, which is occasionally realized by centimetric layers (Buffler and Schlager 1984). If diagenetic effects are ruled out, variations in clay minerals as described above indicate fluctuating river transport and thus climatic changes on continents. Acting simultaneously on land and in the ocean, Milankovitch periodicities correspond to climatic fluctuations which may induce lithologic variations (Rossignol-Strick 1983) even during ice-free periods. Variations in planktonic productivity most likely are the most dominant response to these fluctuations.

4.2 Lamination

Only few attempts have been made to estimate the duration of laminae microcycles. In the Valanginian, at DSDP site 534, a duration of 10–50 a was proposed for laminae of unspecified thickness (Sheridan et al. 1983). At site 391, in the same basin and for coeval sediments, Freeman and Enos (1978) counted 67 laminae per cm and deduced a duration of 6 a for the deposition of each unit. At sites 535 and 540, because of the relatively poor influence of compaction upon the change of laminae thickness versus $CaCO_3$ content, it can be assumed that the sedimentation rate through bed-interbed couplets follows the same exponential law as described above. Curves were drawn for the three stages studied (Fig. 7B), taking into account their average sedimentation rate and the fact that the volume of limestone is nearly twice that of marls. From these curves, elementary rates of sedimentation are deduced for each laminae couplet with respect to its $CaCO_3$ content, and thus to its thickness. From these values, mean times for the deposition of couplets can be obtained: 1.2 a for the third, 4 a for the second, and 16.3 a for the first order couplets. Considering the approximation used for this calculation, I assume that the first value corresponds with an annual cycle. The others, corrected by the same factor, would be 3.3 and 13.6 a long, respectively. Although tentative, these periods may have some meaning. The shortest cycle reminds us of phytoplanktonic blooms which in the presentday oceans occur nearly every year when most favorable conditions are prevailing (e.g., Noel et al. 1987; Wefer 1989). The 13.6-a cycle is of the same magnitude as the cycles of solar activity but it is not known at present whether the solar cycles affect planktonic production. So far these is no explanation for the 3-a period.

5 Conclusions

Southeastern France and the Atlantic exemplify high-frequency cyclic sedimentation during the Mesozoic. Climatic control appears to be the most likely cause for reasons as follows. (1) The long-distance continuity of marl-limestone alternations of a 20-ka mean period, particularly in the Vocontian basin; (2) the detailed correlation, based on this cyclicity, between off-shore and on-shore sites; (3) environmental variation occurring simultaneously in oceans and on continents during a cycle, as is deduced from cyclic fluctuations of biological, mineralogical, and geochemical components of deposits.

Climatic forcing is supported by the recognition of Milankovitch periodicities in Vocontian alternating successions. Periodic variations in seasonality, coupled with other changes in climate and oceanic circulations, are required to explain pelagic bed-interbed cycles. The periodic trend towards anoxic conditions, indicated by an alternation of bioturbated limestone beds and laminated marly interbeds in the Atlantic series, was facilitated by a narrow Neocomian ocean, which was not affected by high latitudinal cool and oxygenated waters and was devoid of effective bottom currents.

Millimetric or submillimetric varve-like marl-limestone laminations occur in Atlantic marly sediments devoid of bioturbation. These laminae successions are also interpreted as bed-interbed rhythms but on a reduced scale. They show the main characteristics of larger cycles, particularly their grouping in bundles (second and first orders of laminae). The association of first order laminae may also constitute a banding. Thus it appears that a large and continuous spectrum of cycles exists in oceanic successions ranging from millimetric to metric scales and beyond. A tentative calculation of the time period for the three orders of laminae shows that they possibly reflect annual and multiannual climatic changes, causing variations in planktonic calcareous production.

High-frequency sedimentary cycles, preserved in pelagic environments far from massive detrital input and from flourishing reefal constructions, allow the reconstruction of orbital forcing, detection of small sedimentation rate variations, and accurate stratigraphic correlations.

Acknowledgments. I gratefully acknowledge the careful reviews of G. Einsele and G. Dromart. Since 1979, C. Darmedru, S. Ferry, C. Gaillard, E. Jautee, A. Jouchoux, † G. Latreille, and M. Rio were associated with the research work. Studies were supported by URA 11, CNRS (ATP IPOD, GGO) and INSU (DBT, message sédimentaire).

7.4 The Spectral Analysis of Stratigraphic Time Series
G. P. Weedon

1 Introduction

Spectral analysis is an objective, statistical method for detecting regular cyclicity in data called time series (Fig. 1). Many of the mathematical aspects of it have probably discouraged sedimentary geologists, who would otherwise have found a useful tool for their studies. This chapter is designed to explain why spectral analysis is useful for sedimentologists, it lists references to introductions and computer algorithms, tackles the geological limitations of the method, and describes some examples. As the principal application at present concerns sedimentary cyclicity related to orbital/climatic variations or Milankovitch cycles, the chapter concentrates on this field.

2 The Use and Limitations of Spectral Analysis

In many modern and ancient pelagic sequences it has been postulated that variations in sediment composition record orbital/climatic, or Milankovitch, cycles (Fischer 1986; see Part I, this Vol.). Spectral analysis cannot explain why there might have been a connection between climatic variations and sediment composition. However, it can provide supporting evidence that such a connection existed, as Pleistocene studies have shown that orbital and climatic variations on the 10 ka to 1 Ma time scale are dominated by a few regular components whose periods are known (Berger 1984; Imbrie et al. 1984). These components are related to variations in eccentricity (about 410 and 100 ka), tilt or obliquity (41 ka) and precession (about 21 ka) (Berger 1984). During the Phanerozoic, the components gradually changed their periods, but over the time intervals represented by most stratigraphic time series, their periods were essentially constant (Walker and Zahnle 1986).

Thus when sedimentologists speculate that the cyclicity they observe records orbital/climatic oscillations, they can test the idea by using spectral analysis to hunt for a few discrete regular cycles possessing the Milankovitch periods. Spectral analysis has also been used to look for regular sedimentary cyclicity related to sunspot cycles and tidal (i.e., lunar) cycles (e.g., Yang and Nio 1985; Sonnet and Williams 1985; Williams 1989).

Yet there are important limitations to the spectral analysis of Pre-Pliocene stratigraphic time series. Firstly it cannot be used to *disprove* a connection between climatic and sedimentary cycles. This is because spectral analysis of ancient sections

---

Einsele et al. (Eds.)
Cycles and Events in Stratigraphy
©Springer-Verlag Berlin Heidelberg 1991

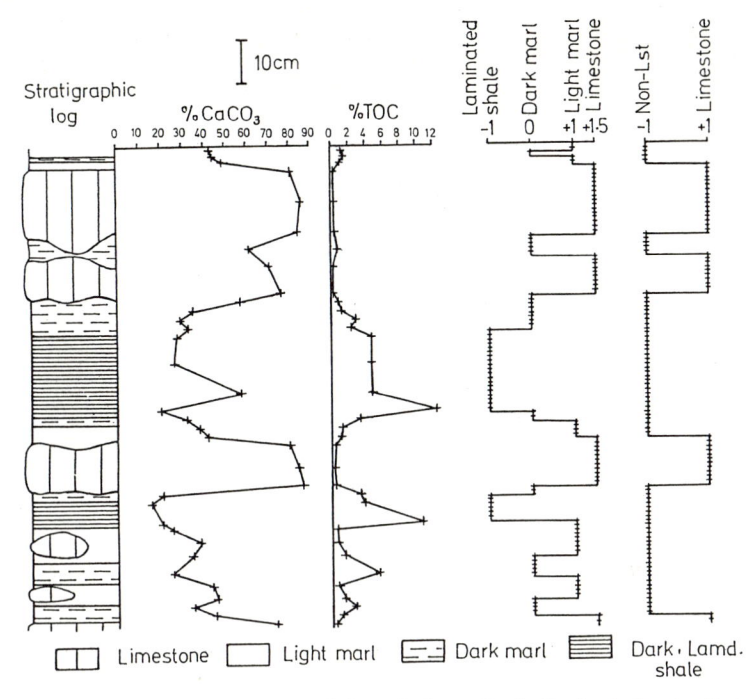

Fig. 1. A section through part of the *S. angulata* zone, *L* Jurassic, Lyme Regis, England, showing that many different types of time series can be generated from the same sequence. Sample interval for digitized logs = 1 cm, variable for geochemical data. *TOC* Total organic carbon. (Data from Weedon 1987 unpublished D. Phil. Thesis, University of Oxford)

uses a thickness scale instead of a time scale for the time series-yielding spectra with axes in "cycles per metre" rather than "cycles per thousand years". Many geological processes, such as variable sedimentation rates, can distort an originally linear relationship between time and thickness (Sect. 5.). Consequently, the lack of a regular cycle in thickness could result from these distorting factors acting during and after deposition or because there really was no connection between sediment composition and the orbital cycles in the original environment.

The other limitation concerns the assignment of periods in years to regular cycles in thickness. Aside from using rare varved sequences (Olsen 1986) the main solution has been to use the assumption that the orbital cycles are the only *regular* processes with periods between 10 ka and 1 Ma capable of affecting sedimentation (Schwarzacher 1975; Weedon 1986). Critical to this argument is the demonstration of regularity-requiring spectral analysis (cf. Algeo and Wilkinson 1988). The wavelength ratios of regular cycles can be used to assign particular sedimentary cycles to particular orbital cycles (Weedon 1986; Schwarzacher 1987a; but see Weedon 1989). Currently it is hoped that sedimentary cycles recording Milankovitch cyclicity could be used for refined dating in the Mesozoic although hiatuses will pose a major problem (House 1985; Weedon 1989; Fischer Chap. 1.2, and Schwarzacher Chap. 7.5, both this Vol.).

3 Time Series from Stratigraphic Sections

A time series, by definition, consists of observations of some parameter obtained at a constant interval of time or space (in this case thickness). Spectral analysis in this context is used to search for regular cycles in terms of thickness rather than in terms of time or in terms of order of rock types (Schwarzacher 1975). Here "regular cycle" is used to denote a sedimentary cycle detected using power spectra. A search for regular cyclicity in stratigraphic time series usually implies a test of the hypothesis that the time series results from an "oscillating random process" to use Schwarzacher's (1975) terminology. Such a process produces successive values that are determined by a combination of

1. one or several regular components,
2. an independent random component, and
3. a certain dependence on previous value or values (i.e., a Markov property).

A large variety of methods are available for producing lithostratigraphic time series. For instance on Fig. 1 the same rock sequence could yield at least four different types of time series. Two different ways of digitizing the rock types are illustrated. Both rely on rather sharp contacts between different lithologies (Weedon 1989 for further discussion of digitization). Other methods of time series generation could be used – the exact choice of method will depend on the nature of the rocks and the type of investigation involved.

It is stressed that before a time series is generated from a stratigraphic section, it should be clear what it signifies geologically. If regular cycles in time are sought, there must be some relationship, however complex, between the thickness of beds and the time involved for their deposition. Thus where a thickness scale is treated as a distorted time scale (the premise of this chapter), event beds must be absent from the section in question (Schwarzacher 1975). As well as investigations into the mode of deposition, diagenesis must also be considered. For instance, it has been suggested that some limestone/shale sequences were generated entirely diagenetically by migration of carbonate, after deposition, within sequences that were originally homogeneous (Eder 1982; Hallam 1986; Bathurst Chap. 3.2, this Vol.). Thus where limestone/shale sequences are to be investigated it is essential that a primary distinction of rock types is demonstrated using palaeontological, petrographic or geochemical evidence (Part I, this Vol.). Yet contrary to the pure diagenesis hypothesis, in many cases it has been argued that carbonate cementation during early diagenesis or stylolite formation was controlled by primary sediment composition (Weedon 1986; Ricken 1986).

An important aspect of time series generation involves the choice of sampling interval. When the sample spacing is too wide, high frequency variations of the parameter under investigation will not be resolved. The result is "aliasing", whereby the incorrect sampling produces a time series containing artificial or spurious low frequency variations (Pisias and Mix 1988). When digitizing a section, the sample interval should be equal to or less than the thickness of the thinnest bed present. If discrete samples are collected, perhaps because bed contacts are gradational, it is recommended that initially a variety of sample intervals are experimented with.

Digitization is obviously a fast method for time series generation by comparison with geochemical methods. The time series thus produced have a square-wave rather than a sine-wave shape (Fig. 1). Two main types of spectral analysis can be applied to any time series. As it is based on square waves, Walsh spectral analysis is more appropriate for analysing digitized section data (Weedon 1989). Fourier analysis is based on sine and cosine waves instead and is therefore more appropriate for geochemical time series. The shape of Walsh spectra is dependent on the position of "zero-crossings" in time series much more than amplitude values between zero-crossings (Weedon 1989). On the other hand, the shapes of Fourier spectra are determined by all the values in a time series, so the two spectra methods are not quite comparable. Given the choice, a continuous time series and Fourier analysis is perhaps preferable, but the effort involved in each method of time series generation needs consideration.

4 Spectral Analysis

4.1 Introduction

This section is merely designed to briefly introduce the most important concepts in spectral analysis. An exhaustive and clear account is provided by Priestley (1981) and his terminology has been applied here where synonyms are involved. The operations used and references for FORTRAN algorithms employed for the spectra in this paper are summarized in Table 1. As an alternative source of algorithms, Press et al. (1986) provide FORTRAN and PASCAL programs for Fourier spectral analysis, assuming prior trend removal, based on the Parzen spectral window. A disadvantage of this program is that the number of points in the time series must be an integer power of two (e.g., 1024, 2048, etc.).

Table 1. Operations and FORTRAN algorithms for spectral analysis.

Operation	Fourier	Walsh
Trend removal	Least-squares regression – Bloomfield 1976	–
Tapering	Split cosine bell applied to first and last 10% of time series – Bloomfield 1976	–
Transform	Fast Fourier Transform – Davis 1973	Fast Walsh Transform – Kanasewich 1981 (Limited to 2^k pts k = integer e.g., 1024, 2048, 4096 ... etc.)
Spectral estimates	Periodogram – Davis 1973	"Averaged Walsh power spectrum" – Beauchamp 1984
Normalization of estimates	By sum of squares of estimates – Nowroozi 1967	By sum of squares of estimates – Nowroozi 1967
Spectral window	3 pt Hanning, applied three times – Davis 1973, Bloomfield 1976	3 pt Hanning applied once – Davis 1973

This paragraph is devoted to a very basic explanation of spectral analysis. Any time series can be treated mathematically as the sum of many different frequency regular waves/components, each of which can have different amplitudes and phases (Fig. 2). The transform operation separates the different regular components of the time series. The components involved are sine and cosine waves in Fourier analysis and CAL and SAL functions in Walsh analysis. The "importance" of the different

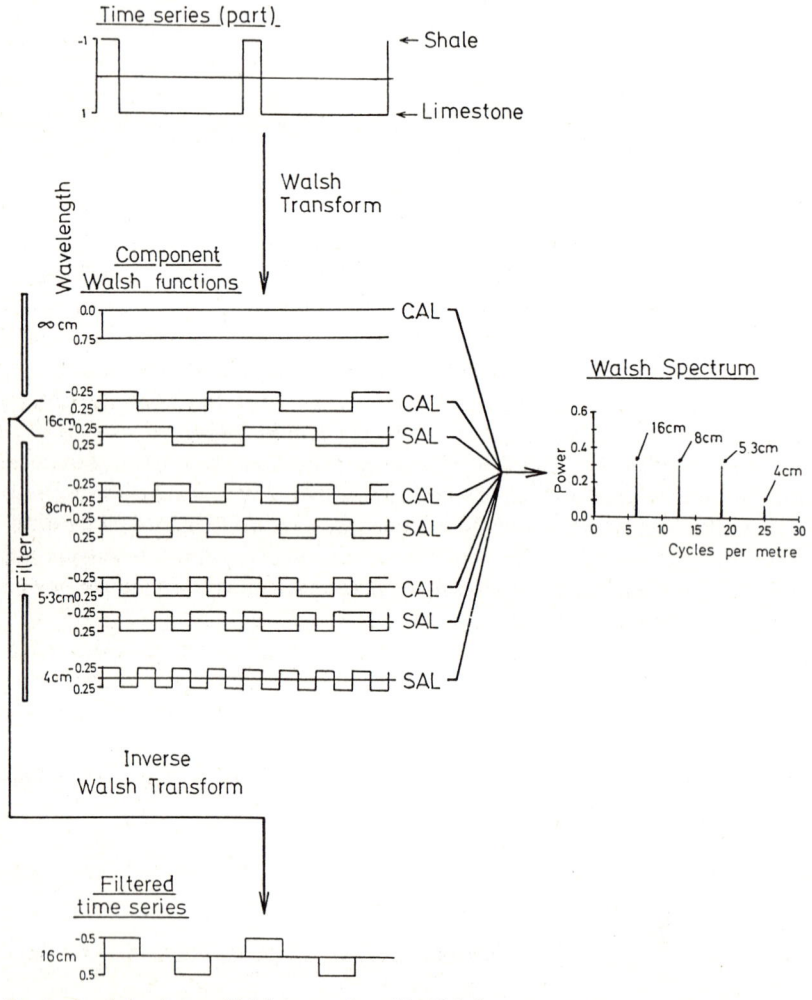

Fig. 2. Analysis of an artificial time series with thick limestone and thin shale beds forming repeating couplets 16 cm thick (only the first two couplets are shown). Only the component Walsh functions with nonzero amplitude are illustrated. In this time series all the components have constant amplitude along the length of the time series. The 8, 5.3 and 4 cm spectral peaks represent harmonics of the 16 cm cycle (i.e., 16/2, 16/3, 16/4 cm); they account for the asymmetry of the time series. Note: a point-by-point addition of the component waves recreates the original time series

components to the time series is determined by using the average amplitudes involved. In a simple form of "spectral estimation", the periodogram approach, squared amplitude (or power) is plotted against frequency – producing a power spectrum. For a particular frequency on a Fourier periodogram, the power is determined by the square of the average cosine amplitude plus the square of the average sine amplitude; an analogous procedure applied to CAL and SAL functions yields the Walsh periodogram (Fig. 2). All spectral values lie at or between 0 frequency (infinite wavelength) and the Nyquist frequency [= 1/(2 * sample interval)]. For a certain sample interval, the frequency resolution of a spectrum is proportional to the number of samples or points in the time series. Judging whether a peak is significant involves deciding whether it emerges from the general noise level; interpretation of a spectrum is similar to that of X-ray diffraction traces.

4.2 Trend Removal

Spectral analysis is designed for the analysis of time series that are "stationary", i.e., the mean and variance of the top halves are more or less the same as those of the bottom halves of the data. When this is not the case, the time series possesses a trend (e.g., progressively larger values on average up-section) which should be removed before analysis. One approach is to "detrend" the data by performing a linear regression and then subtracting the regression line point-by-point from the time series. However, often trends in average amplitude values are accompanied by changes in average wavelength (Schwarzacher 1975). For instance, a sequence of alternating sand and clay beds may not only become sandier on average up-section, but successive beds may become thicker on average as well (implying a trend in average sedimentation rate). A trend in average bed thickness is not removed by the linear regression method. Unfortunately, compensating for a trend in average bed thickness or average sedimentation rate requires a quantitative knowledge of the nature of the trend (Schwarzacher 1975).

In digitized sequence time series there may also be a progressive trend from more negative values at the bottom to more positive values at the top of the data (i.e., a greater proportion of one rock type at the base). Providing that overall sedimentation rates remain nearly constant, the shape of Walsh spectra are unaffected by detrending using linear regression as detrending square waves hardly affects the position of zero-crossings (Sect. 4.1).

4.3 Tapering

Smoothly varying time series, such as those composed of geochemical values, terminate abruptly at both ends. In Fourier analysis this leads to the Gibb's phenomenon where the step-like ends of the time series cannot be simply represented by the sum of different frequency sine and cosine waves. The consequence is that each genuine peak on the spectrum is associated with small side-lobes that interfer with side-lobes from other genuine peaks. Side-lobes can be suppressed by tapering the time series at each end so that the oscillations in the data are forced to gradually diminish to zero amplitude (Priestley 1981). A common taper, employed here, involves tapering the first and then the last ten percent of the time series using a cosine wave (Bloomfield 1976; Priestley 1981). Tapering is unnecessary in Walsh analysis, as the abrupt terminations to the time series are easily represented by the stepped Walsh functions (Beauchamp 1984).

4.4 Generating Smooth Spectra

There are many ways to generate power spectra; brief descriptions and comparisons of the main procedures are given by Pestiaux and Berger (1984a). For Fourier spectra there are two main methods based on

1. the Fast Fourier Transform and
2. the standard Blackman-Tukey method (or the classical truncated-lagged-autocovariance method) (Priestley 1981).

The Fast Fourier Transform is an efficient computational algorithn which transforms the data into a series of component regular waves. The periodogram is then calculated from the sum of the squares of the average amplitudes of paired sine and cosine waves. In the Blackman-Tukey method the autocorrelation function is calculated by finding the correlation coefficient between the time series and the time series offset by varying amounts or lags (the maximum number of lags equals the number of points). The periodogram is derived by calculating the autocovariance function from the autocorrelation function and transforming it using the Cosine Transform (Priestley 1981; Pestiaux and Berger 1984a).

However, the periodogram always varies erratically around the true spectral shape (Priestley 1981). Instead a smoothed version of the periodogram is needed. Adjacent periodogram power values can be smoothed by using some sort of weighted moving average known as a spectral window. One application of the three-point Hanning window, for instance, directly to the periodogram may not be enough to produce a sufficiently smooth spectrum, so the window can be reapplied several times or a series of different windows used (Bloomfield 1976).

Alternatively, in the Blackman-Tukey method spectral smoothing can be achieved by using the majority, but not all the available lags to produce the autocovariance function. A typical procedure is to use all the lags up to the "truncation point" but with diminishing weight attached to values from higher lags. The sequence of weights used is termed a lag window and each lag window has a corresponding spectral window or smoothing effect (Priestley 1981).

The smoother a spectrum, the more closely the calculated spectrum corresponds to the true spectrum. However, smoother spectra have a lower frequency resolution (viz. a larger "bandwidth"). Thus the investigator must choose a level of smoothness which eliminates a substantial fraction of the periodogram's erratic nature, but maintains an acceptable frequency resolution (Priestley 1981).

The Walsh spectrum can be derived using the Fast Walsh Transform and then periodogram values calculated as before. However, unlike the Fourier periodogram, the Walsh periodogram depends on the phase of the time series. This can be avoided by calculating the averaged Walsh power spectrum (Beauchamp 1984; Weedon 1989).

5 Geological Complications

Interpreting spectra based on stratigraphic time series using a thickness scale is complicated by geological processes. Schwarzacher (1975) discussed Sander's (1936) work which showed that, for time series recording varying environmental conditions but no event deposits, a regular cycle in stratal thickness cannot result from a random process in time. However, the absence of a regular cycle in thickness does not rule out the existence of a regular cycle in time.

The effects of these processes on spectra derived from time series with a thickness scale can be gauged in a general way from Fig. 3. This represents the results of a crude simulation of different geological processes acting on an initially perfect square wave 64 cm long repeated 32 times described elsewhere (Weedon 1989).

5.1 Processes Related to Rock Type

Sedimentation rates, the incidence of hiatuses and the degree of compaction may in some cases be systematically related to rock type, leading to consistently thicker beds of one rock type rather than another. When acting on an initially regular and symmetrical alternation of rock types these processes generate regular but asymmetric alternations (Fig. 2). The regularity, and thus the original spectral peaks involved, are still preserved, but the asymmetry leads to harmonic peaks with frequencies that are integer multiples of the original peak (Fig. 3). When the initial spectrum is complex, systematic processes such as these may generate harmonics which are indistinguishable from noise, as for instance in the case of systematic compaction used in Fig. 3. As well as affecting the asymmetry of the time series, compaction also leads to a reduction in the thickness of all beds and thus a shift of original spectral peaks to higher frequencies. Schiffelbein and Dorman (1986) describe a method for reducing time series asymmetry and thus eliminating harmonics from spectra.

5.2 Processes Unrelated to Rock Type

Much more complicated effects are encountered with processes that were uninfluenced by rock type or composition. Sedimentation rates, for example, can vary randomly about a mean value (Schwarzacher Chap. 7.5, this Vol.). In general, spectral peaks become diminished in amplitude and broader as a result (Pestiaux and Berger 1984b). Random variations in compaction caused by isolated stylolite formation for instance, and random measurement errors, have the same effect. In the case of Fig. 3 the relevant simulation of sedimentation rate variations also led to a nonlinear interaction of the 64-and 32-cm peaks – generating a combination tone (or beat).

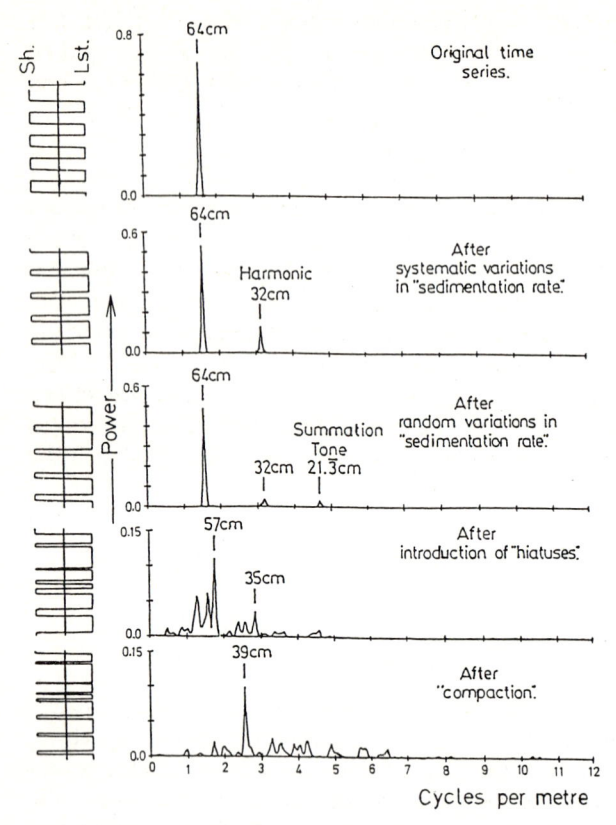

Fig. 3. Different geological processes cause different distortions of initial time-depth relationships in stratigraphic sections. Here their effects on Walsh spectra are illustrated based on time series generated by simulations described by Weedon (1989). Parts of the appropriate time series are illustrated next to each spectrum. Original time series – all beds 32 cm thick. After systematic sedimentation rate variations – all limestone beds 48 cm, all shale beds 16 cm thick. After random variations in sedimentation rate – couplets 52 to 76 cm thick. After introduction of hiatuses at random intervals with random amounts removed – section about 47% complete. After compaction – all limestone beds reduced to 90% and all shale beds reduced to 50% of former thicknesses

Combination tone peaks can be recognized on spectra as they have frequencies which are the sum or difference of the frequencies of the interacting peaks (e.g., on Fig. 3 1/21.3 cm = 1/64 cm + 1/32 cm). Park and Herbert (1987) have developed a way to remove some of the effects of random sedimentation rate variations.

When mean sedimentation rates gradually increase or decrease up-section, the resulting time series is not stationary (Sect. 4.2). If successive beds become, on average, thinner and thinner or thicker and thicker, then the spectrum for the top and bottom halves will be dissimilar. Thus splitting a time series into halves and comparing the resulting spectra acts as an important check for stationarity.

Bioturbation has the effect of smoothing-out time series oscillations. It also reduces the size of spectral peaks by an amount that increases towards high frequencies (Dalfes et al. 1984; Pestiaux and Berger 1984b).

Where the occurrence of hiatuses is not related to rock type or composition, they can seriously distort spectral shapes. The example of Fig. 3 demonstrates a substantial decrease in the signal-to-noise ratio, and additionally that wavelength ratios of major peaks can also be affected. This may partly explain why, quite commonly, wavelength ratios observed on spectra are not exactly those of the current orbital cycles even when orbital-forcing has been inferred (Schwarzacher 1987a; Weedon 1989). The alternative explanation is that the periods of the orbital cycles at the time studied differed from the periods observed today (Walker and Zahnle 1986). The size and distribution of hiatuses controls the effect they have on spectra (Weedon 1989).

Hiatuses in sections not only affect the structure of spectra, they also affect the approach taken to estimate the periods, in years, of regular cycles. Sadler (1981) has demonstrated that hiatuses, although usually undetected, can be expected at all scales in every stratigraphic section. Thus a section spanning one million years of deposition is statistically more likely to contain sediment from every 100-ka period than from every 21-ka period of deposition, therefore, mean sedimentation rates calculated from long periods of deposition (e.g., 1 Ma) should be regarded as underestimates of the rates applicable to much shorter periods of deposition (e.g., 21 ka) (Weedon 1986, 1989; Algeo and Wilkinson 1988). Unfortunately, several studies of cyclicity related to Mesozoic orbital cycles, have assumed 100% completeness at the 20-ka scale even when employing sedimentation rates calculated using radiometric dates that differ by millions of years.

6 Examining and Interpreting Power Spectra

6.1 Testing the Significance of Peaks

The time series produced by an independent random process (i.e., no dependence on previous values) produces a spectrum with a nearly constant noise level. This is termed white noise by analogy with optics: white light is made up of equal contributions of light of all frequencies. A real spectrum can be tested against a white noise model using Fisher's test (Nowroozi 1967; Priestley 1981). But eliminating a white noise model for a spectrum could mean that (1) the real spectrum contains a significant peak denoting regular cyclicity or (2) some other sort of noise model is applicable instead or (3) some sort of combination of (1) and (2) applies.

The majority of spectra for real stratigraphic time series have noise levels which gradually increase towards lower frequencies, that is they possess "red noise". Red noise spectra are produced by time series generated by processes with an independent random component plus some dependence of each value on previous value(s). This dependence (or Markov property) can arise because the environmental systems affecting sedimentation possess a certain degree of "inertia". This forces any rapid variation acting on the input to a system to affect the output over a significant period of time (e.g., major instantaneous temperature decreases cannot cause instantaneous growth of ice sheets). Additionally, bioturbation has the effect of smoothing-out time series oscillations and also leads to "reddened" spectra (Dalfes et al. 1984).

In Fig. 4, 256 values from a red noise process were used to create a time series. Both the Fourier and Walsh spectra produced (Table 1) have the same arbitrary frequency units. The LOG_{10} plots allow all the spectral values to be inspected simultaneously. In both cases noise levels increase gradually towards low frequencies, as expected. Note that the size of the peaks and troughs on the linear plots are roughly inversely proportional to the frequency.

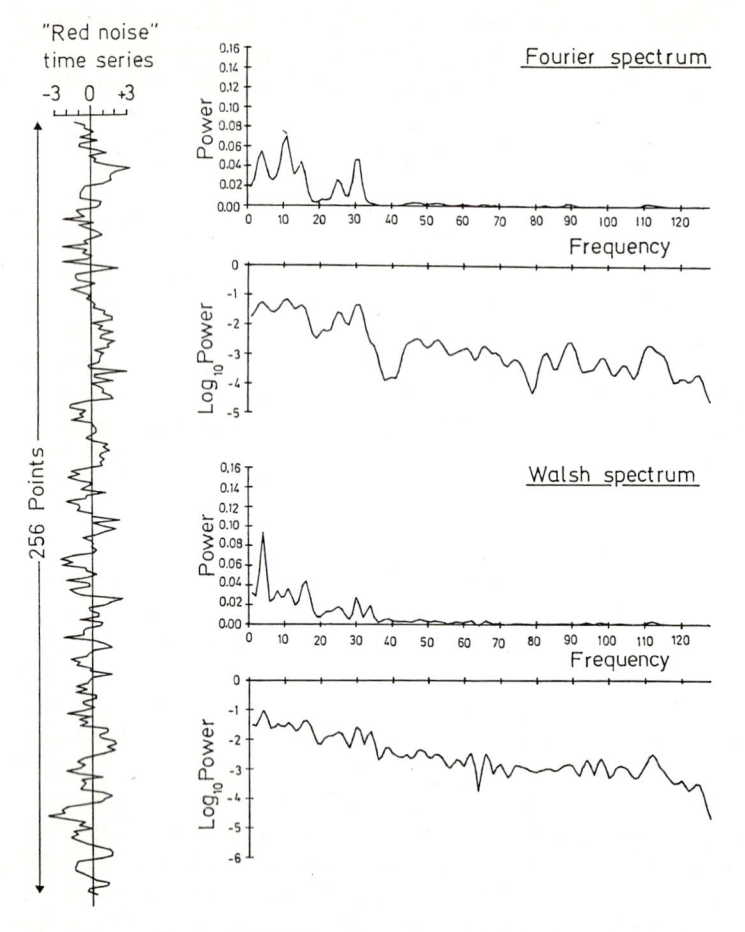

Fig. 4. Spectra for a red noise (first order autoregressive) time series. (Data from Priestley 1981)

The theoretical spectra for this time series rises smoothly from high to low frequencies without peaks and troughs. All the observed peaks and troughs could be eliminated by a greater degree of smoothing. However, if used for spectra from real data this would eliminate any peaks related to regular cyclicity that might be in the data, as all the narrow features would be smoothed out. Therefore, in order to have a reasonable chance of detecting regular cycles in the data, spectra must have a reasonable frequency resolution; the consequence is that spectra for time series *without* regular cyclicity still contain peaks (Priestley 1981). Thus a method is needed to distinguish between peaks related to variations of the spectral estimates about the theoretical noise level due to the erratic nature of the periodogram, and peaks related to regular cyclicity.

One method for this involves assuming that deviations from the theoretical red noise spectrum are distributed normally and can thus be approximated by a Chi-squared distribution (Schwarzacher 1975; Priestley 1981). The degrees of freedom used for the estimated distribution depend on the level of smoothing used (i.e., the exact type of spectral window) and the type of tapering applied to the data (Priestley 1981). This procedure is usually applied to the log plot of the spectrum so that the limits (e.g., 5 and 95% limits) form bands or confidence intervals of constant height above and below the theoretical noise level (Schwarzacher 1975). Peaks that emerge from this band are then distinguishable from the red noise model at a certain level of confidence and might then be attributed to a regular cyclicity in the time series. However, many different types of red noise models can be applied in each case; a particular peak may be distinguishable from one model with a certain level of confidence, but not from another (Pestiaux and Berger 1984a). Priestley (1981) discusses this problem in detail and reviews other tests of significance for spectral peaks.

6.2 Spectra for Artificial and Real Digitized Sections

In Fig. 5A an artificial digitized stratigraphic section has been generated using a random number table. One column was used to determine the rock type for each layer with limestone (code = +1) denoted by even numbers and shale (= –1) by odd numbers in the table. The other column (00–99) was used to define the thickness of each layer in centimetres, this value was divided by 4 and rounded to the nearest integer to produce a time series with an average bed thickness similar to that in Fig. 5B for real data. The number of limestone/shale couplets is 22 and 25.5 in Fig. 5A and B respectively. As random numbers have been used to generate the spectrum in Fig. 5A, one would not expect the spectra to contain significant spectral peaks. The linear plot of the Fourier spectrum suggests that peaks near two and three cycles per metre may be significant. On the log plot, these peaks emerge slightly from the red noise background. However, the sub-spectra from each half of the time series reveals that the major peaks cannot be detected throughout the time series and should therefore be attributed to "noise". This is supported by the Walsh spectra which yield no major peaks and again the sub-spectra reveal different frequency components in each half of the data.

In Fig. 5B the time series has the same number of points as in Fig. 5A, so the resulting spectra have an identical resolution and possess identical scaling. The data were obtained from a Middle Pliensbachian section in Breggia Gorge, southern Switzerland, and interpretation of their Walsh spectra was discussed elsewhere (Weedon 1989). The Fourier spectrum contains two peaks near two and four cycles per metre. On the log spectra these peaks appear to emerge from the red noise background. The size of these peaks is greater than those of Fig. 5A and it is clear from the sub-spectra that both halves of the data contain the same frequency components. The Walsh spectra also contain two major peaks with the same wavelengths as the peaks in the Fourier spectra (51 and about 26 cm); again the two sub-spectra have detected the same components. The spectra demonstrate that there are two

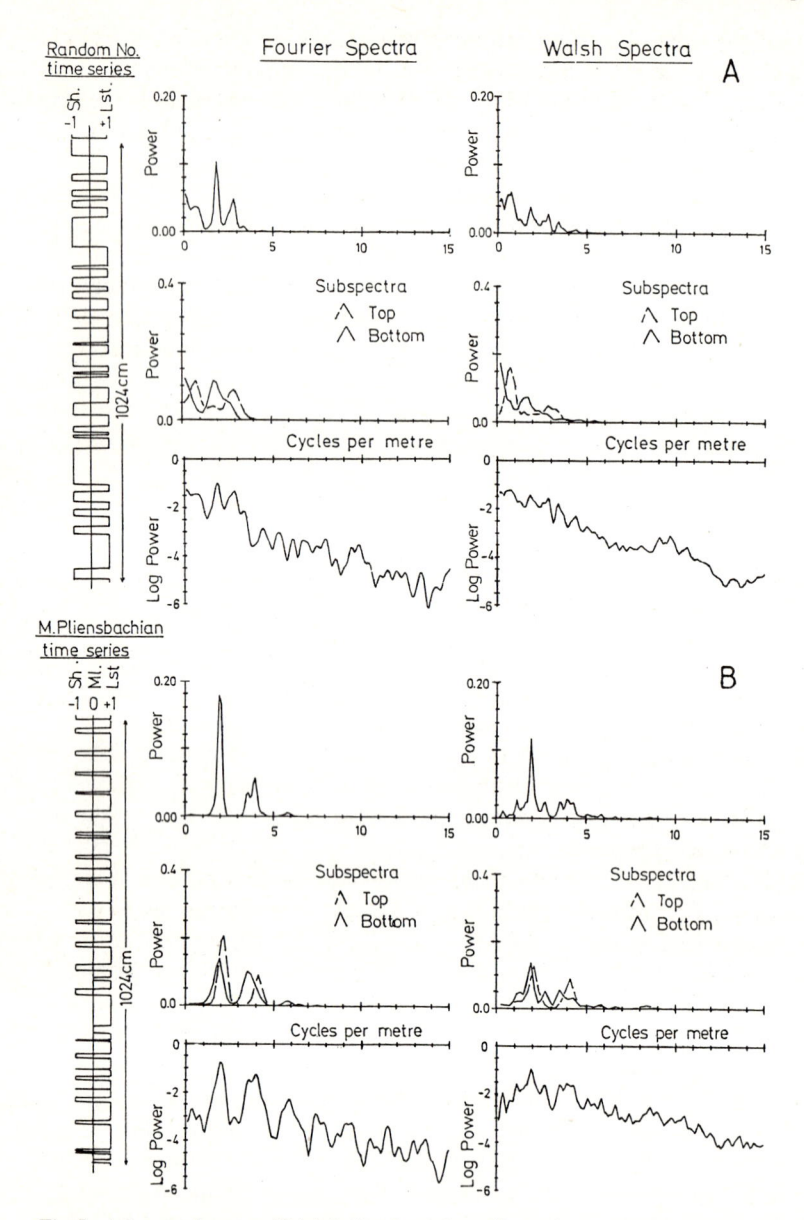

Fig. 5. A Spectra for an artificial digitized stratigraphic section produced using a random number table. **B** Spectra for a real digitized section from the M. Pliensbachian section in Breggia Gorge, Switzerland. (Data from Weedon 1989)

Fig. 6. Band-pass filtering based on the Walsh transform using the data from the M. Pliensbachian section of Breggia Gorge

regular cycles in the time series, but the cycle with a wavelength of about 26 cm appears to be a harmonic (51/2 cm) caused by the asymmetry of the time series (Weedon 1989). It is clear that although the time series in Fig. 5A and B appear similar visually, spectral analysis can be used to decide whether regular cyclicity is present.

7 Future Directions: Filtering and Cross-Spectra

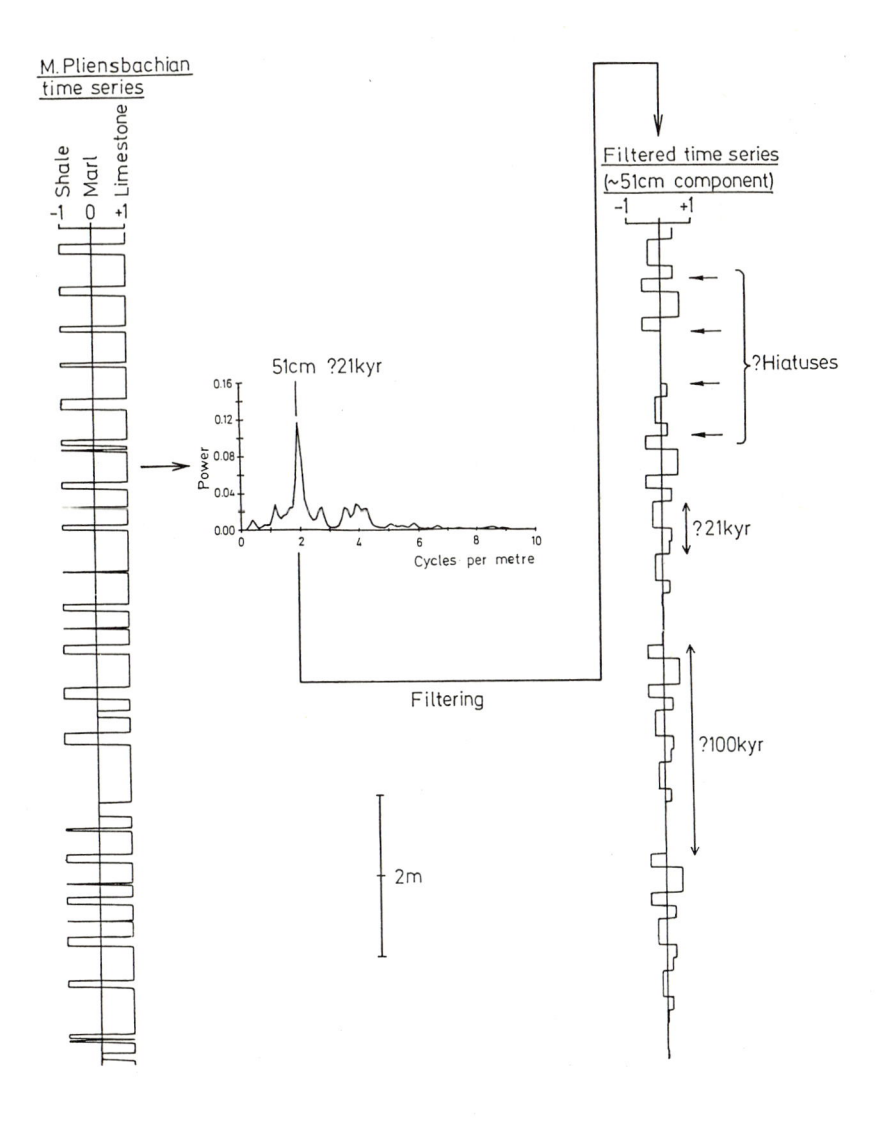

Once a regular cycle has been identified in a time series, it can be extracted and illustrated by band-pass filtering (Priestley 1981; Beauchamp 1984). An example of Walsh filtering is illustrated in Fig. 6 – the 51-cm cycle identified on the spectrum has been extracted from the data. Before filtering it was suspected that this cycle was related to the 21-ka orbital-precession cycle (Weedon 1989). The filtered time series shows amplitude variations or packaging with four or five 51-cm cycles per packet. Such amplitude variations are very similar to the modulations of the precession cycle by the 100-ka eccentricity cycle (Imbrie et al. 1984). Such variations in amplitude may be useful for refining lithostratigraphic correlation and could help in the detection of hiatuses (Fig. 6) (Weedon 1989).

Another way to extract information from stratigraphic data is to apply cross-spectra to parallel time series (e.g., %$CaCO_3$ and %TOC determined for the same samples) (Priestley 1981). Coherency spectra reveal the correlation of the amplitudes of two time series at particular frequencies (e.g., for 100-ka cycles). Significant coherence at the frequency of a regular cycle would imply that variations in the amplitude of this cycle in one time series are correlated with amplitude variations in the other and thus that the same environmental factors controlled both measured parameters. Phase spectra are used to measure the phase relationship between two time series for different frequencies. Currently no cross-spectral studies have been published for Pre-Cenozoic strata, but they would help build conceptual models of the links between different environmental processes and determine the relative order of environmental events.

8 Conclusions

Spectral analysis applied to stratigraphic sections provides an objective method for detecting regular cyclicity. Time series generation should start with a consideration of the nature of deposition and diagenesis. It is essential to choose an appropriate sample interval to avoid aliasing the signal under investigation. The choice of spectral methods employed depends on the nature of the time series, but where no particular procedure is required, a series of different spectral methods could be applied (Pestiaux and Berger 1984a). The stationarity of the time series should be checked by splitting the data into two and comparing the sub-spectra produced. The significance of a well developed peak can be checked by searching for peaks at the same frequency in the sub-spectra (Fig. 5A). If several significant peaks are present the possibility that harmonics and combination tones are present should be investigated.

Currently there are two major limitations to spectral analysis of Pre-Cenozoic time series: geological processes which affect the time-thickness relationship and direct dating of regular cycles in stratigraphic thickness. Recently, however, a variety of methods have been introduced to tackle time-thickness distortions caused by geological processes. It is clear that orbital-climatic cyclicity has been demonstrated objectively in many Pre-Cenozoic sections using spectral analysis. The possibility of refining the Mesozoic timescale based on information appears promising (Schwarzacher Chap. 7.5, this Vol.). Further time series analysis based on filtering and cross-spectral techniques is likely to provide a range of new lithostratigraphic and sedimentological insights.

7.5 Milankovitch Cycles and the Measurement of Time

W. Schwarzacher

1 Introduction

The first quantitative analysis of a relationship between climate and regular changes in the orbital elements of the earth was given by Milankovitch, who used the theory to explain the origin of ice ages.

The Milankovitch theory is well supported for the Pleistocene by evidence from deep sea sediments (See Grötsch et al. Chap. 1.6, this Vol.). The isotope signal which is preserved in the latter is directly related to the size of terrestrial ice caps, which provides one of the most powerful indicators of climate. Furthermore, Pleistocene sediments can be dated relatively accurately and the orbital elements are known fairly accurately.

This directly supportive evidence is missing for older sediments. In interpreting older stratification cycles in terms of the Milankovitch theory, one is left with three criteria: the persistency of the cycles, their spectral composition, and the climatic history which they are supposed to represent. The first two rely heavily on accurate timing, which is considerably less accurate in pre-Pleistocene times. The interpretation of the stratigraphic record in terms of climate is difficult and often ambiguous, particularly when the land-sea configurations of the planet are not known precisely. In addition, the calculation of orbital data becomes less accurate in the more remote past.

In this chapter, the effect of an irregular recording of cyclic signals is examined and an attempt is made to estimate errors in quantitative terms. It is hoped that a better knowledge of the inaccuracies of the stratigraphic record will lead to a more positive identification of Milankovitch cycles.

2 Models for Cyclic Sedimentation

The most elementary model of cyclic sedimentation is given by the periodic function:

$$y(t) = A \sin (\omega t). \tag{1}$$

In this equation, y is a function of time, t, which represents some environmental variable which ultimately will determine the nature of the sediment. The cycle has a fixed frequency, ω, and a constant amplitude, A.

Einsele et al. (Eds.)
Cycles and Events in Stratigraphy
©Springer-Verlag Berlin Heidelberg 1991

It is useful to think of this signal as being transmitted through a sequence of stages such as deposition, diagenesis, or tectonism. During this transmission, the signal will be modified and parts of it will be eliminated. The concept of transmission makes it possible to apply many results from the theory of communication to the problems of stratigraphic analysis.

In the language of radio communications, y(t) could represent the so-called carrier wave. Carrier waves can either be amplitude modulated (AM) or frequency modulated (FM). Modulation can be either by some signal or simply by noise. In the stratigraphic model, the carrier wave will be regarded as the signal and the modulation as interference. Thus for amplitude modulation to take place, the model becomes:

$$y(t) = A \sin(\omega t) + e(t), \tag{2}$$

in which e(t) is a time-dependent error term which causes the modulation. The error can be interpreted as an observational error, or it could be some bias introduced by the sedimentation process itself. Clearly, no environmental variable will be recorded with absolute accuracy in the stratigraphic section, and the error term will therefore always be present. The amplitude modulation does not change the frequency information of the record. Its effect on the power spectrum is only to add a constant background noise. The correlogram of this process is a cosine wave with a constant amplitude which is smaller than one. In contrast to amplitude modulation, frequency modulation changes the time scale which is used to measure the carrier wave. The resulting function can be written as:

$$y(t) = A \cos[\omega t - \phi(t)]. \tag{3}$$

In this equation, ω is the frequency of the carrier representing the stratigraphic signal, and $\phi(t)$ is an instantaneous phase shift which represents noise and which produces a distortion in the carrier. If $\phi(t)$ is directly proportional to the modulating signal, one talks about phase modulation. If $\phi(t)$ represents the integrated change of modulation over a short time period, it is referred to as frequency modulation.

To make the transition from the signal, which is a function of time, t, to the stratigraphic section as a function of stratigraphic position, s, the rules by which this transformation is achieved must be known. If the sedimentation rates are constant, no problem arises but normally the rates also undergo error-like fluctuations and this can be indicated by writing the stratigraphic position-time relationship as:

$$s(t) = c(t + e_t). \tag{4a}$$

In this equation, c is a constant and e_t an error affecting time. On a small scale in particular, sedimentation can be regarded as a sequence of deposition and nondeposition steps which is a process that can be modeled by a random walk. In its simplest form, the accumulating sediment may be represented by a sequence of jumps. The magnitude of each jump is governed by a probability distribution, and the jump may be either negative (erosion) or positive (deposition) or zero (nondeposition). The resulting picture is a step function which, providing that deposition is more likely than erosion, will have an upward trend.

The endpoint of this random walk gives the stratigraphic position, which is simply the sum of all individual jumps.

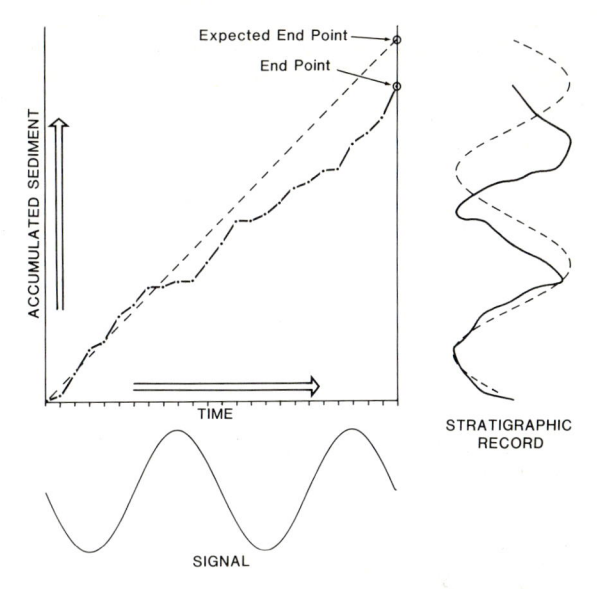

Fig. 1. Diagramatic representation of a sedimentation process. The straight line relationship between time and stratigraphic thickness is shown by *dashed lines*. If the sedimentation rate fluctuates, the shape of the signal will be distorted

$$s(n) = (c + e_1) + (c + e_2) + \ldots (c + e_n) = c\sum_{1}^{n} e_t. \tag{4b}$$

The position of the endpoint is predictable, and it is a normally distributed random variable. If the mean size of each jump is $E(p)$ and the variance is $var(p)$, then the endpoint position after n jumps will be given by:

$$E(endpoint) = nE(p) \text{ and } var(endpoint) = nvar(p). \tag{5}$$

Replacing $\phi(t)$ in Eq. (3) with either Eq. (4a) or (4b) results in phase and frequency modulation respectively. The effect of incorporating errors into the vertical scale is shown diagramatically in Fig. 1.

The system of Eq. (3) is nonlinear and if it is strongly nonlinear, the reconstruction of the time history becomes impossible. In weakly nonlinear systems, the fluctuations of the phase variable $\phi(t)$ are small compared with the cycle of the carrier wave, and such systems can be regarded as quasi-stationary.

If a sine wave is modulated by a normally distributed noise, its power spectrum can be calculated. Middleton's analysis (1960) has shown that the disturbed spectrum will be a normal distribution centred around the original frequency of the carrier wave. The power is reduced to:

$$W(\omega_0) = A^2{}_0/2 \exp(-\vartheta^2), \tag{6}$$

where ϑ^2 is the mean square phase deviation of the disturbance. The power decreases with increasing noise and the spectrum becomes flatter and widens out. The phase distortion process does not depend on the frequency of the signal and cycles of different lengths are equally affected. Furthermore, the time errors are not additive, and providing that the signal is not completely eliminated, it is possible to reconstruct the length of the cycle from the distorted record. In the more realistic case of frequency modulation, the time errors are additive and the reduction in power is strongly related to the frequency of the signal. It is found that:

$$W(\omega_0) = A^2_0/2 \exp(-2\vartheta^2\omega). \tag{7}$$

It follows that with increasing errors, the high frequencies are rapidly reduced and short cycles are the first to be eliminated. It is important for the purpose of stratigraphic analysis to note that random gaps or hiatuses in the sequence can reduce the power of higher frequencies, but they will not change the frequencies of the observed cycles.

As long as the hiatuses in the sequence happen statistically with equal density, the behavior of the spectrum is predictable. If the section contains possibly long unidentified gaps, then it is not suitable for spectral analysis.

Timing errors in the previous models were taken to be independent of the signal and are generally of a much higher frequency. In most situations, however, sedimentation rates will be related to the signal itself which leads to further nonlinearities and departures from a sinusoidal shape. The Fourier spectrum of non-sinusoidal functions invariably contains a large number of harmonics. This makes for complex spectra which are difficult to interpret. The complexity of the spectrum can be measured and a search can be made for the best correction which will produce a simple spectrum. The method was successfully used in estimating the sedimentation changes during glacial and interglacial periods of a Pleistocene deep sea core, for which it was assumed that the climatic changes can be represented by sine waves (Schiffelbein and Dorman 1986).

The method of simplyfing the power spectrum by introducing transformations of the measuring scale, can be applied to a variety of problems, and the complexity of both Fourier and Walsh spectra can be measured by the varimax method. However, it is important to have some reason for believing that one knows the original shape of the signal. In the case of the Pleistocene isotope data, the existence of regular climatic cycles is well established but such cyclicity may only be suspected in other sediments. Clearly, to make the record conform to such a preconceived model, proves very little. Problems of adjusting the time scale become particularly difficult when dealing not only with variable sedimentation rates, but also with often vastly varying degrees of compaction for different lithologies and possibly, with the diagenetic transfer of material.

3 The Evidence for Milankovitch Cyclicity

Milankovitch cycles, which are of particular interest in geology, are the 21 ka precession cycle, the 41 ka obliquity cycle, and the 100 ka eccentricity cycle. The latter has further important periodicities at 0.4, 1.23, 2.04, and 3.4 ma. All Milankovitch cycles are composed of a large number of harmonic components and their wavelength is not absolutely constant. Precession and obliquity, which are dependent on the earth's rotational speed and the earth-moon distance, are likely to have changed throughout geological time. The lengths of these cycles may have increased with time (Berger and Loutre 1989).

Most sedimentary cycles which are believed to be ultimately due to orbital variation, consist of single beds or small groups of beds. To prove that such cycles are, in fact, Milankovitch cycles is often impossible, but if one can demonstrate that the cyclicity is present and persistent and if the time represented by the cycles is approximately within the Milankovitch spectrum, the likelihood of astronomical control should be accepted. Groups of beds were originally called "bundles" but because they can vary laterally and in fact become a single bed, the term stratification cycle may be used as an alternative (Schwarzacher 1987b). The variability of such stratification cycles is shown in Fig. 2, displaying actually measured cycles from limestone-marl sequences (Fig. 2A,B) and from Lofer cycles of limestones separated by supratidal dolomites (Fig. 2C). Each row in this figure shows the development of well-differentiated into low-differentiated cycles. This change can occur within a very short stratigraphic interval, and in the examples represents genuine facies changes. The latter in example B can also be established according to a regional pattern (Schwarzacher 1982). Similar differentiation can, however, also be produced by different degrees of weathering. If measurements are taken at surface exposures, this must be taken into account.

If data on the lithological or faunal composition of such cycles are available, power-spectral analysis can be used to establish cyclicity (see Weedon Chap. 7.4, this

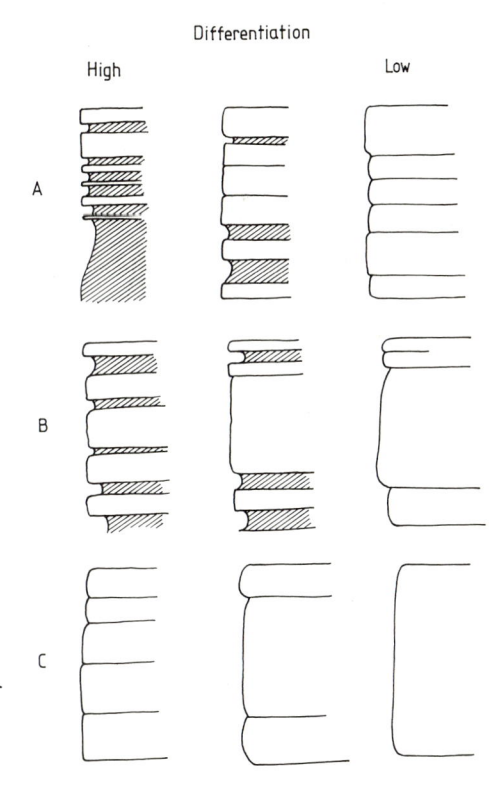

Fig. 2. Different facies developments of similar cycles. Marl is indicated by shading. **A** Cenomanian Scaglia from the Gubbio district. Well-differentiated cycles occur in the schisti fucuoidi and differentiation decreases in stratigraphically higher positions. **B** Carboniferous limestone from Co. Sligo (Ireland). Highly differentiated cycles are found on the margin of the basin and differentiation decreases towards the center. **C** Bedded Dachstein limestone from Lofer (Austria). Bedding planes are caused by supratidal dolomite horizons. Changes in development occur both laterally and vertically

Vol.). Because of the variability of cycles due to facies changes or weathering, the higher frequencies representing individual beds will often give unreliable results in spectral analysis.

Further complications can arise from the asymmetry of some cycles, for example the Cenomanian limestone marl cycle (Fig. 2A) is thickening upwards and the Triassic Lofer cycle is generally thinning upwards. However, exceptions occur in both cases, and these lead to phase shifts which will reduce the power of the basic cycle.

Spectral analysis is a statistical method which yields average frequencies for complete sections. A more detailed stratigraphic analysis can be obtained by mapping individual cycles and by locating the cycle boundaries precisely within the stratigraphic scale. This is possible when the cycles are well defined, but in many cases, locating individual cycles is very difficult, if not impossible. In such cases, spectral analysis can still give recognizable frequency peaks, and the persistency of cyclic behavior can be tested by calculating power spectra for shorter subsections.

Estimating the absolute time represented by cycles will always depend on the quality of the available time scale. In most situations it is possible only to demonstrate that a sedimentary cycle is within the Milankovitch spectrum. However, for the Pleistocene and parts of the Tertiary, such identifications are possible.

An alternative approach to direct timing is to deduce the Milankovitch nature of a group of cycles from the spectral composition of the sedimentary cycle. For example, one might suspect that a cycle represents eccentricity and precession. Since the frequency of the latter is about five times that of the eccentricity, one would expect a sedimentary cycle containing five subcycles. In using such arguments, one has to remember that the ratios between the Milankovitch frequencies may have changed in geological time.

4 The Stratigraphic Importance of Cycles

If it is possible to associate a sedimentary cycle with an identified Milankovitch cycle, absolute time measurements can be made. This has been successful for the Pleistocene and some parts of the Tertiary and possibly also the Cretaceous (see Fischer Chap. 1.2, this Vol.). Even if the definite identification of the time durations is impossible, cycles can be used for comparative studies of sedimentation rates, the completeness of sedimentation, and the variability of sedimentation rates. As an example, data from some very different carbonate environments are given in Table 1.

In each case, it is believed that the analysed cycle represents the 100 ka eccentricity cycle. The accumulation rates which can be deduced from this assumption fit the environments which the sediments represent. The Carboniferous limestone of Co.Sligo (Ireland) is a shallow water carbonate with relatively high sedimentation rates. The Triassic limestone from Lofer (Austria) is deposited on a rapidly sinking carbonate platform and has very high sedimentation rates. All Cretaceous and Tertiary examples are from the Gubbio district (Italy) and represent the low sedimentation of a pelagic environment.

Table 1. Data from different carbonate environments

	Mean thickness of cycles (cm)	Coefficient of variation	Number of cycles
Glencar limestone (Visean, Co.Sligo)	409.3	32.6	19
Dachstein limestone (Noric, Lofer)	1600.0	49.6	88
Maiolica (Barremian, Gubbio)	98.6	29.9	32
Schisti fucoidi (Aptian Albian, Gubbio)		44.1	21.9
Data from Herbert and Fischer (1986)			
Scaglia (Cenomanian, Gubbio)	72.9	36.0	108
Scaglia (Paleocene Eocene, Gubbio)	59.2	40.5	81

In order to be able to compare the variability of the cycles they are given as coefficients of variation (i.e., standard deviation expressed as percentage of the mean). The data show a quite clearly increasing variability with increasing length of cycles and increasing thickness of the measured sections. The Aptian-Albian data are measured on core material and are of a much better quality. This gives them the lowest variability. The general increase of variability with increasing thickness is to be expected from the random walk model and is expressed by Eq. (5). The applicability of this model can be tested by actually determining the variances of the thickness intervals provided by 1,2,3... cycles. A cycle in this analysis is accepted as a time unit which is assumed to be constant. Figure 3 gives the results of such analyses for the Glencar limestone of the Sligo area and for the Paleocene-Eocene Scaglia of the Gubbio district.

Both examples show a linear increase of variance with cycle thickness. Furthermore, the correlated points fall more or less unto the same regression line. This suggests that despite the different sedimentation rates, the random walk which generated the sequence was similar. An interpretation of this result in terms of sedimentation alone is, however, not indicated. This is because the random series incorporates not only all the errors of observation and misinterpretations but also all diagenetic and postdiagenetic processes which may very well have been similar in both cases.

The variance-time relationship can also be investigated by using sedimentary beds as "time units". This analysis shows a clear difference between the Carboniferous and Tertiary example. In the latter, beds behave in a similar way to the cycles, but in the Carboniferous example, beds no longer have a linear variance-thickness relationship. Indeed, their variance is much higher than one would expect from their thickness. This indicates that the Gubbio data are statistically homogeneous, but the Sligo data are not, and the Sligo beds contain units which are not time equivalent.

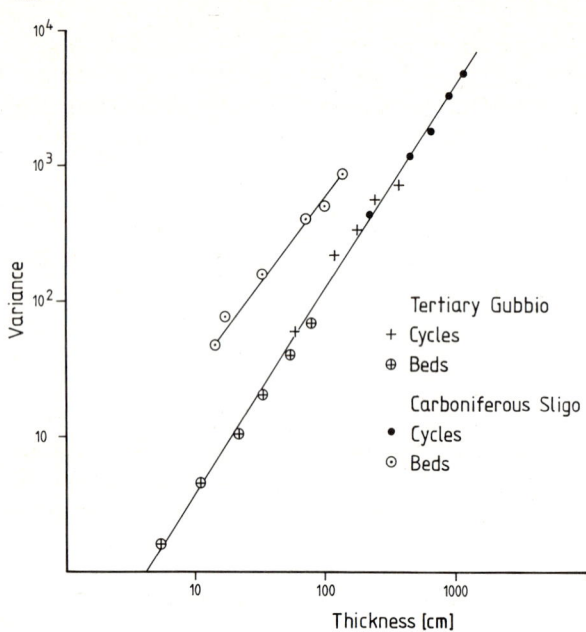

Fig. 3. Relationship between the mean thickness of stratigraphic intervals, provided by beds and groups of beds and their variance. Cycles and bedding are statistically homogeneous for the Gubbio data, but not for the Sligo data. The variance of beds is too high in both cases to measure time within the cycle

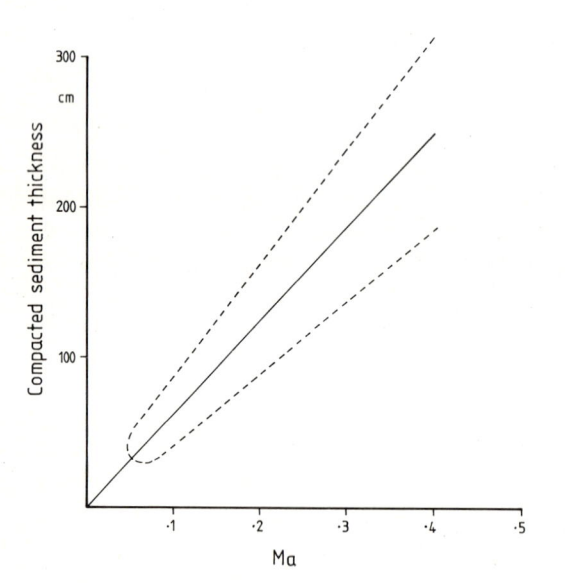

Fig. 4. Relationship between stratigraphic thickness and time for the Paleocene of the Gubbio district. The *dashed curve* gives the standard deviation to this regression. Note that thickness intervals below 30 cm cannot be dated by this regression

In the case of the Gubbio data, there is fairly good evidence that the measured cycle does in fact represent a time interval of 100 ka (Schwarzacher 1987a) and it is therefore possible to construct a time-thickness diagram (Fig. 4). The data from the variance analysis can be used to indicate the accuracy of this graph, and the dashed curve in Fig. 4 indicates the standard deviation of the time thickness regression. Note that the variance becomes zero at a thickness of 31 cm, corresponding to a time interval of 50 ka. This must be taken as the limit of stratigraphic resolution. Any analysis of shorter time intervals or cycles is meaningless in this particular material.

5 Conclusions

Milankovitch cycles promise to be a powerful tool in stratigraphy and sedimentology. Their full potential can only be realized, however, if the interpretation of cyclic sediments is done critically. The geologist only sees the result of a complicated recording mechanism which is full of errors. A study of such errors is necessary to establish the accuracy and limitations of stratigraphic records and thus to prevent the over-interpretation of cyclic processes. In the present situation, the collection of more data appears to be more valuable than speculating on processes about which no information is available as yet.

References

Acker KL, Risk MJ (1985) Substrate destruction and sediment production by the boring sponge Cliona caribbea on Grand Cayman Island. J Sediment Petrol 55:705-711

Adelseck CG, Anderson TF (1978) The Late Pleistocene record of productivity fluctuations in the eastern equatorial Pacific Ocean. Geology 6:388-391

Ager DV (1973) The nature of the stratigraphic record. Wiley, New York, pp 1-114

Ager DV (1974) Storm deposits in the Jurassic of the Morocco High Atlas. Palaeogeogr Palaeoclimatol Palaeoecol 15:83-93

Agterberg FP, Banerjee I (1969) Stochastic model for the deposition of varves in glacial Lake Barlow-Ojibway, Ontario, Canada. Can J Earth Sci 6:625-652

Aharon P (1984) Implication of the coral-reef record from New Guinea concerning the astronomical theory of ice ages. In: Berger A, Imbrie J, Hays J, Kukla G, Saltzman B (eds) Milankovitch and climate. Reidel, Dordrecht, pp 379-389

Aigner T (1980) Biofabrics and stratinomy of the Lower Kimmeridge Clay (U. Jurassic, Dorset, England). N Jb Palaeontol Abh 159:324-338

Aigner T (1982) Calcareous tempestites: storm-dominated stratification in Upper Muschelkalk limestones (Middle Trias, SW Germany). In: Einsele G, Seilacher A (eds) Cyclic and event stratification. Springer, Berlin Heidelberg New York, pp 180-198

Aigner T (1985) Storm depositional systems: dynamic stratigraphy in modern and ancient shallow-marine sequences. (Lect Notes Earth Sci 3), Springer, Berlin Heidelberg New York, 174 pp

Aigner T, Futterer (1978) Kalk-Töpfe und Rinnen (pot and gutter casts) im Muschelkalk. Anzeiger für Wattenmeer? N Jb Geol Palaeontol Abh 156:285-304

Aigner T, Reineck HE (1982) Proximality trends in modern storm sands from the Helgoland Bight (North Sea), and their implications for basin analysis. Senckenb Marit 14:183-215

Aigner T, Reineck HE (1983) Seasonal variation of wave-base in the shoreface of Barrier Island Norderney. Senckenb Marit 15 (1-3):87-92

Akou AE (1984) Subaqueous debris flow deposits in Baffin Bay. Geomar Lett 4/2:83-90

Akpan EB, Farro GE (1985) Shell bioerosion in high-latitude low-energy environments: firths of Clyde and Lorne, Scotland. Mar Geol 67:139-150

Aksu AE (1984) Subaqueous debris flows in Baffin Bay. Geomar Lett 4:83-90

Algeo TJ, Wilkinson BH (1988) Periodicity of mesoscale Phanerozoic sedimentary cycles and the role of Milankovitch orbital modulation. J Geol 96:313-322

Allen JRL (1972) A theoretical and experimental study of climbing-ripple cross-lamination, with a field application to the Uppsala esker. Geogr Ann A 53:157-187

Allen JRL (1982) Sedimentary structures, their character and physical basis, vol II. (Developments in Sedimentology 30 B), Elsevier, Amsterdam, 663 pp

Allen JRL (1985) Loose boundary hydraulics and mechanics: selected advances. Geol Soc Lond Spec Publ 18:7-30

Allen PA, Collinson JD (1986) Lakes. In: Reading HG (ed) Sedimentary environments and facies, 2nd edn. Blackwell, Oxford, pp 63-94

Aller RC (1982a) Carbonate dissolution in nearshore terrigenous muds: the role of physical and biological reworking. J Geol 90:79-95

Aller RC (1982b) The effects of macro-benthos on chemical properties of marine sediment and overlying water. In: McCall PL, Tevesz MJS (eds) Animal-sediment relations. Plenum, New York, pp 53-102

Allison PA (1988a) Konservat Lagerstaetten: cause and classification. Paleobiology 14:331-344

Allison PA (1988b) The decay and mineralization of proteinaceous macrofossils. Paleobiology 14:139-154

Alvarez LW (1986) Toward a theory of impact crises. EOS Trans Am Geophys Union 67:649-658

Alvarez LW (1987) Mass extinctions caused by large bolide impacts. Phys Today 40:24-33

Alvarez LW, Alvarez W, Asaro F, Michel HV (1980) Extraterrestrial cause for the Cretaceous-Tertiary extinction. Science 208:1095-1108

Alvarez W, Colacicchi R, Montanari A (1985) Synsedimentary slides and bedding formation in Apennine pelagic limestones. J Sediment Petrol 55:720-734

Alvarez W, Kauffman EG, Surlyk F, Alvarez LW, Asaro F, Michel HV (1984) Impact theory of mass extinctions and the invertebrate fossil record. Science 223:1135-1141

Amajor LC, Lerbekmo JF (1980) Subsurface correlation of bentonite beds in the Lower Cretaceous Viking Formation of south-central Alberta. Bull Can Petrol Geol 28:149-172

Anadón P, Cabrera L, Juliç R (1988) Anoxic-oxic cyclical lacustrine sedimentation in the Miocene Rubielos de Mora Basin, Spain. In: Fleet AJ, Kelts K, Talbot MR (eds) Lacustrine petroleum source rocks. Geol Soc Lond Spec Publ 40:353-367

Anders MH, Krueger SW, Sadler PM (1987) A new look at sedimentation rates and the completeness of the stratigraphic record. J Geol 95:1-14

Anderson JB, Brake C, Domack EW, Myers N, Wright R (1983) Development of a polar glacial-marine sedimentation model from Antarctic Quaternary deposits and glaciological information. In: Molnia BF (Glacial-marine sedimentation. Plenum, New York, pp 233-264

Anderson JB, Kurtz DD, Domack EW, Balshaw KM (1980) Glacial and glacial-marine sediments of the Antarctic continental shelf. J Geol 88:399-414

Anderson RO (1983) Radiolaria. Springer, Berlin Heidelberg New York

Anderson RY (1961) Solar-terrestrial climatic patterns in varved sediments. Annu NY Acad Sci 95:424-435

Anderson RY (1982) A long geoclimatic record from the Permian. J Geophys Res 87:7285-7294

Anderson RY (1984) Orbital forcing of evaporite sedimentation. In: Berger A, Imbrie J, Hays J, Kukla G, Saltzman B (eds) Milankovitch and climate. NATO ASI Ser C 126, pp 147-162

Anderson RY (1986) The varve microcosm: propagator of cyclic bedding. Paleoceanol 1:373-382

Anderson RY, Dean WE (1988) Lacustrine varve formation through time. Paleogeogr Palaeoclimatol Palaeoecol 62:215-235

Anderson RY, Kirkland DW (1960) Origin, varves, and cycles of Jurassic Todilto Formation, New Mexico. Am Assoc Petrol Geol Bull 44:37-52

Anderson RY, Kirkland DW (1966) Interbasin varve correlation. Geol Soc Am Bull 77:241-256

Anderson RY, Kirkland DW (1969a) Paleoecology of an early Pleistocene lake on the High Plains of Texas. Mem Geol Soc Am 113:pp 215

Anderson RY, Kirkland DW (1969b) Paleoecology of the Rita Blanca Lake area. Mem Geol Soc Am 113:141-157

Anderson RY, Koopmans LH (1969) Statistical analysis of the Rita Blanca varve time-series. Mem Geol Soc Am 113:59-75

Antevs E (1929) Cycles in variations of glaciers and ice sheets and in ice melting. Reports of the Conference on Cycles. Carnegie Institution Washington, 83 pp

Antevs E (1935) Telecorrelation of varve curves. Geol Foren Forh 57/1:47-58

Anthony RS (1977) Iron-rich rhythmically laminated sediments in Lake of the Clouds, northeastern Minnesota. Limnol Oceanogr 22:45-54

Arche A (1983) Coarse-grained meander lobe deposits in the Jarama River, Madrid, Spain. Spec Publ Int Assoc Sediment 6:313-321

Argyriadis I, de Graciansky PC, Marcoux J, Ricou LE (1980) The opening of the Mesozoic Tethys between Eurasia and Arabia-Africa. 26e Congr Géol Int Paris 1980, Coll C5, Géologie des chaines alpines issues de la Téthys, pp 199-214

Arkell WJ (1933) The Jurassic system in Great Britain. Clarendon, Oxford, 681 pp

Armentrout JT (1983) Glacial lithofacies of the Neogene Yakataga Formation, Robinson Mountains, southern Alaska Coast Range, Alaska. In: Molnia BF (ed) Glacial-marine sedimentation. Plenum, New York, pp 629-666

Armstrong RL (1968) Sevier orogenic belt in Nevada and Utah. Geol Soc Am Bull 79:429-458

Arrhenius G (1952) Sediment cores from the East Pacific. Rep Swedish Deep-Sea Expedition 1947-48, 5:1-228

Arrhenius G (1988) Rate of production, dissolution and accumulation of biogenic solids in the ocean. Palaeogeogr Palaeoclimatol Palaeoecol 67:119-146

Arthur MA (1979a) North Atlantic Cretaceous black shales: the record at Site 398 and a brief comparison with other occurrences. In: Sibuet JC, Ryan WB et al. Init Repts DSDP 47: Washington (US Govt Print Office), pp 719-751

Arthur MA (1979b) Sedimentologic and geochemical studies of Cretaceous and Paleocene pelagic sedimentary rocks: the Gubbio sequence. Diss Princeton University, Part 1, 174 pp

Arthur MA, Bottjer DJ, Dean WE, Fischer AG, Hattin DE, Kauffman EG, Pratt LM, Scholle PA (1986) Rhythmic bedding in Upper Cretaceous pelagic carbonate sequences: varying sedimentary response to climate forcing. Geology 14:153-156

Arthur MA, Dean WE (1986) Cretaceous paleoceanography of the western North Atlantic Ocean. In: Vogt PR, Tucholke BE (eds) The geology of North America, vol M. Geol Soc Am, pp 617-630

Arthur MA, Dean WE, Bottjer D, Scholle PA (1984b) Rhythmic bedding in Mesozoic-Cenozoic pelagic carbonate sequences: the primary and diagenetic origin of Milankovitch-like cycles. In: Berger A, Imbrie J, Hays J, Kukla G, Saltzman B (eds) Milankovitch and climate. Riedel, Hingham, pp 191-222

Arthur MA, Dean WE, Claypool GE (1985b) Anomalous ^{13}C enrichment in modern marine organic matter. Nature 315:216-218

Arthur MA, Dean WE, Hay BJ, Honjo S, Neff ED, Konuk TY (1988b) Late Holocene sedimentation and environmental history of the Black Sea: preliminary results from R/V Knorr 134-8 expedition. Geol Soc Am Cent Mtng, Denver CO, Abstr with Programs: A199

Arthur MA, Dean WE, Pollastro RM, Scholle PA, Claypool GE (1985a) A comparative geochemical and mineralogical study of two transgressive pelagic limestone units, Cretaceous Western Interior Basin U.S. In: Pratt P, Kauffman EG, Zelt F (eds) Fine-grained deposits and biofacies of the Cretaceous Western Interior Seaway: evidence of cyclic sedimentary processes. Soc Econ Paleontol Mineral, Tulsa, pp 16-27

Arthur MA, Dean WE, Pratt LM (1988a) Geochemical and climatic effects of increased marine organic carbon burial at the Cenomanian/Turonian boundary. Nature 335:714-717

Arthur MA, Dean WE, Schlanger SO (1985c) Variations in the global carbon cycle during the Cretaceous related to climate, volcanism, and the changes in atmospheric CO_2. In: Sundquist ET, Broecker WS (eds) The carbon cycle and atmospheric CO_2: natural variations Archean to Present. Am Geophys Union Geophys Monogr 32:504-529

Arthur MA, Dean WE, Stow DAV (1984a) Models for the deposition of Mesozoic-Cenozoic fine-grained organic-carbon-rich sediment in the deep sea. In: Stow DAV, Piper DJW (eds) Fine-grained sediments: deep-water processes and facies. Geol Soc Lond Spec Publ 15:527-559

Arthur MA, Fischer AG (1977) Upper Cretaceous-Paleocene magnetic stratigraphy at Gubbio, Italy. I. Lithostratigraphy and sedimentology. Geol Soc Am Bull 88:367-389

Arthur MA, Jenkyns HC (1981) Phosphorites and paleoceanography. 26th Int Geol Congr Paris 1980, Oceanol Acta Spec Publ, pp 83-96

Arthur MA, Natland JH (1979) Carbonaceous sediments in North and South Atlantic: the role of salinity in stable stratification of Early Cretaceous basins. In: Talwani M, Hay WW, Ryan WBF (eds) Deep drilling results in the Atlantic Ocean: continental margins and paleoenvironment. Washington, Am Geophys Union, Maurice Ewing Ser 3:297-344

Arthur MA, Premoli Silva I (1982) Development of widespread organic-carbon rich strata in the Mediterranean Tethys. In: Schlanger SO, Cita MB (eds) Nature and origin of Cretaceous carbon-rich facies. Academic Press, New York, pp 8-54

Arthur MA, Schlanger SO, Jenkyns HC (1987) The Cenomanian-Turonian oceanic anoxic event, II. Paleoceanographic controls on organic-matter production. In: Brooks J, Fleet AJ (eds) Marine petroleum source rocks. Geol Soc Lond Spec Publ 26:401-420

Ashley GM (1975) Rhythmic sedimentation in glacial Lake Hitchcock, Massachusetts-Connecticut. Spec Publ Soc Econ Paleontol Mineral 23:304-320

Astin TR, Scotchman IC (1988) The diagenetic history of some concretions from the Kimmeridge Clay, England. Sedimentology 35:349-368

Attolini MR, Galli M, Nanni T (1988) Long and short cycles in solar activity during the last millennia. In: Stephenson FR, Wolfendale AW (eds) Secular solar and geomagnetic variations in the last 10,000 years. Kluwer Academic, Dordrecht, pp 49-68

Austin JA, Schlager W et al. (1986) Proc ODP, Init Repts Leg 101. College Station, Texas (Ocean Drilling Program), 569 pp

Awramik SM (1981) The Pre-Phanerozoic biosphere - three billion years of crisis and opportunities. In: Nitecki MH (ed) Biotic crisis in ecological and evolutionary time. Academic Press, London, pp 83-102

Bacescu M (1963) Contribution a la biocoenologie de la Mer Noire. L'étage periozoique et le facies dreissenifère, leurs caractéristiques. Rapp Proc Verb Reun CIESM 17:107-122

Backman J, Duncan RA et al. (1988) Proc ODP, Init Repts Leg 115. College Station, TX (Ocean Drilling Program), 1085 pp

Bader RG (1954) The role of organic matter in determining the distribution of pelecypodes in marine sediments. J Mar Res 13:32-47

Badgley C, Tauxe L, Bookstein FL (1986) Estimating the error of age interpolation in sedimentary rocks. Nature 319:139-141

Bagnold RA (1962) Auto-suspension of transported sediment: turbidity currents. Proc R Soc (Lond), Ser A, 205:315-319

Bahrig B (1989) Stable isotope composition of siderite as an indicator of the paleoenvironmental history of oil shale lakes. Palaeogeogr Palaeoclimatol Palaeoecol 70:139-151

Bailard JA, Inman DL (1981) An energetics bedload model for a plane sloping beach: local transport. J Geophys Res 86:10938-10954

Bailey RA, Miller CD, Sieh K (1989) Long Valley caldera and Mono-Inyo Craters volcanic chain, eastern California. New Mexico Bur Mines Miner Res Mem 47:227-254

Baillie MGL, Munro MAR (1988) Irish tree rings, Santorini and volcanic dust veils. Nature 332:344-346

Baird GC, Brett CE (1986) Erosion on an anaerobic sea floor: significance of reworked pyrite deposits from the Devonian of New York State. Palaeogeogr Palaeoclimatol Palaeoecol 57:157-193

Baker PA, Kastner M, Beyerlee JD, Lockner DA (1980) Pressure solution and hydrothermal recrystallization of carbonate sediments - an experimental study. Mar Geol 38:185-203

Baker VR (1973) Paleohydrology and sedimentology of Lake Missoula flooding in eastern Washington. Spec Pap Geol Soc Am 144:1-79

Baldwin B, Butler CO (1985) Compaction curves. Am Assoc Paleontol Geol Bull 69:622-262

Bally AW (1987) Atlas of seismic stratigraphy. Am Assoc Paleontol Geol Stud Geol 27:125 pp

Bally AW (1989) Phanerozoic basins of North America. In: Bally AW, Palmer AR (eds) The geology of North America - an overview, vol A. Geol Soc Am, Boulder, CO, pp 397-446

Balsam WL (1981) Late Quaternary sedimentation in the western North Atlantic: stratigraphy and paleocanograpy. Palaeogeogr Palaeoclimatol Palaeoecol 35:215-240

Balsam WL (1983) Carbonate dissolution on the Muir Seamount (western North Atlantic): interglacial/glacial changes. J Sediment Petrol 53 (3):719-731

Balson PS, Taylor PD (1982) Palaeobiology and systematics of large cyclostome bryozoans from the Pliocene Coralline Crag of Suffolk. Palaeontology 25:529-554

Bambach RK (1969) Bivalvia of the Siluro-Devonian Arisaig Group, Nova Scotia. PhD Thesis, Yale University, Conn, 376 pp

Bambach RK (1983) Ecospace utilization and guilds in marine communities through the Phanerozoic. In: Tevesz MJS, McCall PL (eds) Biotic interactions in recent and fossil benthic communities. Plenum, New York, pp 719-746

Bambach RK (1985) Classes and adaptive variety: the ecology of diversification in marine faunas through the Phanerozoic. In: Valentine JW (ed) Phanerozoic diversity patterns: profiles in macroevolution. Princeton University Press and Pacific Div Am Assoc Adv Sci, Princeton, NJ, pp 191-253

Bambach RK, Sepkoski JJ Jr (1979) The increasing influence of biologic activity on sedimentary stratification through the Phanerozoic. Geol Soc Am Abstr with Program 11: 383

Banerjee I (1977) Experimental study of the effects of deceleration on the vertical sequence of sedimentary structures in silty sediments. J Sediment Petrol 47:771-783

Barale G, Blanc-Louvel C, Buffetaut E, Courtinat B, Peybernes B, Via Boada L, Wenz S (1984) Les gisements de calcaires lithographiques du Cretacé inferieur du Montsech (Province de Lerida, Espagne) considérations paléoécologiques. Geobios Mém Spéc 8:275-283

Barale G, Bourseau JP, Buffetaut E, Gaillard C, Gall C, Wenz S (1985) Cerin. Une lagune tropicale au temps des dinosaures. CNRS and Association des Amis du Muséum - Lyon, 136 pp, Lyon

Barber NF, Ursell F (1948) The generation and propagation of ocean waves and swell. I. Wave periods and velocity. Philos Trans R Soc (Lond) Ser A, 240:527-560

Barker PF, Carlson RL, Johnson DA et al. (1983) Init Repts DSDP 72: Washington (US Govt Print Office), 1O24 pp

Barlow LK, Kauffman EG (1985) Depositional cycles in the Niobrara Formation, Colorado Front Range. In: Pratt LM, Kauffman EG, Zelt FB (1985) Fine-grained deposits and biofacies of the Cretaceous Western Interior Seaway: evidence of cyclic sedimentary processes. Soc Econ Paleontol Mineral, Field Trip Guidebook 4:199-208

Barnola JM, Raynaud D, Korotkevich YS, Lorius C (1987) Vostok ice core provides 160,000-year record of atmospheric CO_2. Nature 329:408-414

Barrell J (1917) Rhythms and measurements of geologic time. Geol Soc Am Bull 28:745-904

Barrell J (1917) Rhythms and measurements of geologic time. Geol Soc Am Bull 28:745-904
Barrett PJ (1964) Residual seams and cementation in Oligocene shell calcarenites, Te Kuiti Group. J Sediment Petrol 34:524-531
Barrett TJ (1982) Stratigraphy and sedimentology of Jurassic bedded chert overlying ophiolites in the northern Apennines, Italy. Sedimentology 29:353-373
Barron EJ (1986) Physical paleoceanography: a status report. In: Hsü KJ (ed) Mesozoic and Cenozoic oceans. Am Geophys Union Geodyn Ser 15:1-9
Barron EJ, Arthur MA, Kauffman EG (1985) Cretaceous rhythmic bedding sequences: a plausible link between orbital variations and climate. Earth Planet Sci Lett 72:327-340
Barron EJ, Hay WW, Thompson S (1989) The hydrologic cycle: a major variable during earth history. Palaeogeogr Palaeoclimatol Palaeoecol 75:157-174
Barron EJ, Washington WM (1982) Cretaceous climate: a comparison of atmospheric simulations with the geologic record. Palaeogeogr Palaeoclimatol Palaeoecol 40:103-133
Barron EJ, Washington WM (1985) Warm Cretaceous climates: high atmospheric CO_2 as a plausible mechanism. In: Sundquist ET, Broecker WS (eds) The carbon cycle and atmospheric CO_2: natural variations Archean to Present. Am Geophys Union Mono 32:546-553
Barron EJ, Whitman JM (1981) Oceanic sediments in space and time. In: Emiliani C (ed) The sea4, vol 7. The oceanic lithosphere. Wiley-Interscience, New York, pp 689-731
Bartek L, Vail PR, Anderson JB, Emmet PA, Wu S (1990) The effect of Cenozoic ice sheet fluctuations in Antarctica on the stratigraphic signature of the Neogene. J Geophys Res (in press)
Barthel KW (1970) On the deposition of the Solnhofen lithographic limestone (Lower Tithonian, Bavaria, Germany). N Jb Geol Palaeontol Abh 135:1-18
Barthel KW (1972) The genesis of the Solnhofen Lithographic Limestone (Lower Tithonian): further data and comments. N Jb Geol Palaeontol Monatsh 3, Fossil-Lagerstätten 21:133-1452
Barthel KW (1978) Solnhofen. Ein Blick in die Erdgeschichte. Otto Verlag, Thun, 393 pp
Barthel KW, Swinburne NHM, Conway Morris S (1990) Solnhofen. A study in Mesozoic palaeontology. Cambridge University Press, Cambridge, 236 pp
Bathurst RGC (1975) Carbonate sediments and their diagenesis. Dev Sediment 12: 658
Bathurst RGC (1980a) Deep crustal diagenesis in limestones. Rev Inst Invest Geol 34:89-100
Bathurst RGC (1980b) Les rôles des compactions mécaniques et chimiques dans la déformation et la diagenèse des sédiments carbonatés. In: Humbert L (ed) Cristallisation, déformation, dissolution des carbonates. Inst Géodyn Bordeaux III, pp 25-32
Bathurst RGC (1980c) Lithification of carbonate sediments. Sci Prog Oxford 66:451-471
Bathurst RGC (1984) The integration of pressure solution with mechanical compaction and cementation. In: Yaha FA (ed) Stylolites and associated phenomena: relevance to hydrocarbon reservoirs. Abu Dhabi Nat Reserv Res Found, pp 41-55
Bathurst RGC (1986) Carbonate diagenesis and reservoir development: conservation, destruction and creation of pores. Q J Colo Sch Mines 81:1-25
Bathurst RGC (1987) Diagenetically enhanced bedding in argillaceous platform limestones: stratified cementation and selective compaction. Sedimentology 34:749-778
Bathurst RGC (1990) Thoughts on the growth of stratiform stylolites in buried limestone. In: Heling D, Rothe P, Förstner U, Stotters P (eds) Sediments and environmental geochemistry. Springer, Berlin Heidelberg New York, pp 3-15
Battarbee RW (1981) Diatom and Chrysophyceae microstratigraphy of the annually deposited sediments of a small meromictic lake. Striae 14:105-109
Baturin GN (1971) Stages of phosphorite formation on the ocean floor. Nat Phys Sci 232:61-62
Baturin GN (1982) Phosporites on the sea floor. Developments in sedimentology, vol 33. Elsevier, Amsterdam, 343 pp
Baum GR, Vail PR (1987) Sequence stratigraphy, allostratigraphy, isotope stratigraphy and biostratigraphy: putting it all together in the Atlantic and Gulf Paleogene. 8th Annu Res Conf, Gulf Coast Section, Soc Econ Pal Min Found, Earth Enterprises, Austin, pp 15-23
Baumgartner P (1987) Age and genesis of Tethyan Jurassic radiolarites. Eclogae Geol Helv 80:831-875
Bayer U (1987) Chronometric calibration of a comparative time scale for the Mesozoic and Paleozoic. Geol Rdsch 76:485-503
Bayer U (1989) Stratigraphic and environmental patterns of ironstone deposits. In: Young T, Taylor WEG (eds) Phanerozoic ironstones. Geol Soc Lond Spec Publ 46:105-117
Bayer U, Altheimer E, Deutschle W (1985) Environmental evolution in shallow epicontinental seas: sedimentary cycles and bed formation. In: Bayer U, Seilacher A (eds) Sedimentary and evolutionary cycles. (Lect Notes Earth Sci 1), Springer, Berlin, pp 347-381

Bayer U, McGhee GR (1984) Iterative evolution of Middle Jurassic ammonite faunas. Lethaia 17:1-16

Bayer U, McGhee GR (1985) Evolution in marginal epicontinental basins: the role of phylogenetic and ecological factors. In: Bayer U, Seilacher A (eds) Sedimentary and evolutionary cycles. (Lect Notes Earth Sci 1), Springer, Berlin Heidelberg New York, pp 164-220

Bayer U, McGhee GR (1986) Cyclic patterns in the Paleozoic and Mesozoic: implications for time scale calibrations. Paleoceanography 1:383-402

Bayer U, Seilacher A (eds) (1985) Sedimentary and evolutionary cycles. (Lect Notes Earth Sci 1), Springer, Berlin Heidelberg New York

Bé AWH (1977) An ecological, zoogeographic and taxonomic review of recent planktonic foraminifera. In: Ramsay ATS (ed) Oceanic micropaleontology, vol 1. Academic Press, New York, pp 1-100

Bé AWH, Damuth JE, Lott L, Free R (1976) Late Quaternary climatic record in western equatorial Atlantic sediment. In: Paleoceangraphy and paleoclimatology. Geol Soc Am Mem 145:165-200

Beard JH, Sangree JB, Smith LA (1982) Quaternary chronology. Paleoclimate depositional sequences and eustatic cycles. Am Assoc Paleontol Geol Bull 66:158-169

Beardsley RC, Butman B (1974) Circulation on the New England continental shelf: response to strong winter storms. Geophys Res Lett 1:181-184

Beauchamp KG (1984) Applications of walsh and related functions, with an introduction to sequence theory. Academic Press, New York

Beaudoin B, Bié J, Conard M, Guy B, Le Doeuff D (1974) Essai d'analyse des formations marno-calcaires alternantes. Bull Soc Géol Fr Paris 7 XVI:634-642

Bechstädt T (1979) The lead-zinc deposits of Bleiberg-Kreuth (Carinthia, Austria): palinspastic situation, paleogeography and ore mineralization. Verh Geol Bundesanst (3):221-235

Beckvar N, Kidwell SM (1988) Hiatal shell concentrations, sequence analysis and sea level history of a Pleistocene costal alluvial fan, Punta Chueca, Sonora. Lethaia 21:257-270

Beecher CE (1894) On the mode of occurrence, structure and development of the trilobite Triarthrus becki. Am Geol 13:38-43

Beer J, Siegenthaler U, Blinov A (1988) Temporal 10Be variations in ice: information on solar activity and geomagnetic field intensity. In: Stephenson FR, Wolfendale AW (eds) Secular solar and geomagnetic variations in the last 10,000 years. Kluwer Academic, Dordrecht, pp 297-313

Beer J, Siegenthaler U, Bonani G, Finkel RC, Oeschger H, Suter M, Wülfli W (1988) Information on past solar activity and geomagnetism from 10Be in the Camp Century ice core. Nature 331:675-679

Beerbower JR (1961) Origin of cyclothems of the Dunkard Group (Upper Pennsylvanian-Lower 2Permian) in Pennsylvania, West Virginia and Ohio. Bull Geol Soc Am 72:1029-1050

Beerbower JR (1964) Cyclothems and cyclic depositional mechanisms in alluvial plain sedimentation. In: Merriam DF (ed) Symposium on cyclic sedimentation. Kansas Geol Surv Bull 169 (1):31-42

Behrensmeyer AK (1983) Resolving time in paleobiology. Paleobiology 9:1-8

Beiersdorf H, Knitter H (1986) Diagenetic layering and lamination. Mitt Geol Paläontol Inst Hamburg 60:267-273

Belt ES (1984) Origin of Late Dinantian cyclothems, East Fife, Scotland. In: Neuvième Congrès Internat Stratigr Géologie Carbonifère, Washington and Campaign-Urbana, Compte Rendu 3:570-588

Bender F (1968) Geologie von Jordanien. Beiträge zur Regionalen Geologie der Erde 7, Borntraeger, Berlin

Bender ML (1984) On the relationship between ocean chemistry and atmospheric CO_2 during the Cenozoic. In: Hansen JE, Takahashi T (eds) Climate processes and climate sensitivity. Am Geophys Union, Geophys Monogr 29:352-359

Bentor YK (1980) Phosphorites - the unsolved problems. Soc Econ Paleontol Mineral Spec Publ 29:3-18

Berger A (1984) Accuracy and frequency stability of the Earth's orbital elements during the Quaternary. In: Berger A, Imbrie J, Hays J, Kukla G, Saltzman B (eds) Milankovitch and climate. NATO ASI Ser C 126:3-39

Berger A (1988) Milankovitch theory and climate. Rev Geophys 26:624-657

Berger A, Loutre MF (1989) Pre-Quaternary Milankovitch frequencies. Nature 342: 133

Berger A, Pestiaux P (1985) Modelling the astronomical theory of paleoclimates in the time and frequency domain. An example of the relationship between long term and short term climate changes. In: Fantechi R, Ghazi A (eds) Current issues in climate research. Reidel, Dordrecht, pp 77-95

Berger A, Imbrie J, Hays J, Kukla G, Saltzman B (eds) (1984) Milankovitch and climate. NATO ASI Ser C 126 (I,II): 895

Berger A, Loutre MF, Dehant V (1987) Influence of the variation of the lunar orbit on the astronomical frequencies of the pre-Quaternary paleo-insolation. Sci Rep 1987/15, Inst d'Astron Geophys, Louvain-la-Neuve

Berger A, Tricot C (1986) Global climatic changes and astronomical theory of paleoclimates. In: Cazenave A (ed) Earth rotation: solved and unsolved problems. Reidel, Dordrecht, pp 111-129

Berger AL (1978a) Long-term variations of caloric insolation resulting from the earth's orbital elements. Quat Res 9:139-167

Berger AL (1978b) Long-term variations of daily insolation and Quaternary climatic cycles. J Atmos Sci 35:2362-2367

Berger RA, Raiswell R (1983) Burial of organic carbon and pyrite sulfur in sediments over Phanerozoic time; a new theory. Geochim Cosmochim Acta 47:855-862

Berger WH (1968) Planktonic foraminifera: selective solution and paleoclimatic interpretation. Deep Sea Res 15:31-43

Berger WH (1970) Planktonic foraminifera: selective solution and the lysocline. Mar Geol 8:111-138

Berger WH (1971) Sedimentation of planktonic foraminifera. Mar Geol 11:325-358

Berger WH (1973) Deep sea carbonates: Pleistocene dissolution cycles. J Foraminiferal Res 3:187-195

Berger WH (1976) Biogenous deep-sea sediments: production, preservation and interpretation. In: Riley JP, Chester R (eds) Chemical oceanography. Academic Press, New York, pp 265-347

Berger WH (1977) Carbon dioxide excursions and the deep-sea record: aspects of the problem. In: Andersen NR, Malahoff A (eds) The fate of fossil fuel CO_2 in the oceans. Plenum, New York, pp 505-542

Berger WH (1978) Sedimentation of deep-sea carbonate: maps and models of variations and fluctuations. J Foraminiferal Res 8(4):286-302

Berger WH (1979a) Impact of deep sea drilling on paleoceanography. In: Talwani M, Hay W, Ryan WBF (eds) Deep drilling results in the Atlantic Ocean continental margins and paleoenvironment. Am Geophys Union, Maurice Ewing Ser 3:297-314

Berger WH (1979b) Preservation of foraminifera. In: Foraminiferal ecology and paleocology. Soc Econ Paleontol Mineral, Short Course 6:105-155

Berger WH (1981) Paleoceanography: the deep-sea record. In: Emiliani C (ed) The oceanic lithosphere. The Sea, vol 7. John Wiley, New York, pp 1437-1519

Berger WH (1982) Deep-sea stratigraphy: Cenozoic climate steps and the search for chemo-climatic feed-back. In: Einsele G, Seilacher A (eds) Cyclic and event stratification. Springer, Berlin Heidelberg New York, pp 121-157

Berger WH, Bonneau MC, Parker FL (1982) Foraminifera on the deep-sea floor: lysocline and dissolution rate. Oceanol Acta 5 (2):249-257

Berger WH, Diester-Haass L (1988) Paleoproductivity: the benthic/planktonic ratio in foraminifera as a productivity index. Mar Geol 79:15-25

Berger WH, Ekdale AA, Bryant PF (1979) Selective preservation of burrows in deep-sea carbonates. Mar Geol 32:205-230

Berger WH, Fischer K, Lai C, Wu G (1987) Ocean productivity and organic carbon flux. I. Overview and maps of primary production and export production. University of California, San Diego, SIO Ref 87-30, 67pp

Berger WH, Keir RS (1984) Glacial-Holocene changes in atmospheric CO_2 and the deep-sea record. In: Hansen JE, Takahashi T (eds) Climate processes and climate sensitivity. Am Geophys Union Monogr 29:337-351

Berger WH, Mayer CA (1978) Deep-sea carbonates: acustic reflectors and lysocline fluctuations. Geology 6:11-15

Berger WH, Piper DJW (1972) Planktonic foraminifera: differential settling, dissolution and redeposition. Limnol Oceanogr 17:275-287

Berger WH, Roth PH (1975) Oceanic micropaleontology: progress and prospect. Rev Geophys Space Phys 13:561-585

Berger WH, Smetacek VS, Wefer G (1989) Ocean productivity and paleoproductivity - an overview. In: Berger WH, Smetacek VS, Wefer G (eds) Productivity of the ocean: present and past. Dahlem Konferenzen. John Wiley, Chichester, pp 1-34

Berger WH, Soutar A (1970) Preservation of plankton shells in an anaerobic basin of California. Geol Soc Am Bull 81:275-282

Berger WH, Vincent E (1981) Chemostratigraphy and biostratigraphic correlation: exercises in systematic stratigraphy. Proc 26th Int Geol Congr, Paris. Geology of the oceans, Oceanol Acta Spec Publ 4:115-127

Berger WH, Vincent E (1986) Deep-sea carbonates: reading the carbon isotope signal. Geol Rdsch 75:249-269

Berger WH, von Rad U (1972) Cretaceous and Cenozoic sediments form the Atlantic ocean. In: Hayes DE, Pimm AC et al. Init Repts DSDP 14: Washington (US Govt Print Office), pp 787-954

Berger WH, Winterer EL (1974) Plate stratigraphy and fluctuating carbonate line. In: Hsü KJ, Jenkyns HC (eds) Pelagic sediments on land and under the sea. Int Assoc Sediment Spec Publ 1:11-48

Berggren WA, Kent DV, Flynn JJ, van Couvering JA (1985) Cenozoic geochronology. Geol Soc Am Bull 96:1407-1418

Berner RA (1980) Early diagenesis. A theoretical approach. Princeton Univ Press, Princeton, 237 pp

Berner RA, Raiswell R (1983) Burial of organic carbon and pyrite sulfur in sediments over Phanerozoic time; a new theory. Geochim Cosmochim Acta 47:855-862

Berniér P (1984) Les formations carbonatées du Kimmeridgien et du Portlandien dans le Jura méridional (stratigraphie, micropaléontologie, sédimentologie). Doc Lab Géol Lyon 92:802

Bernoulli D (1972) North Atlantic and Mediterranean Mesozoic facies: a comparison. In: Hollister CD, Ewing JI et al. Init Repts DSDP 11: Washington (US Govt Print Office), pp 801-871

Bernoulli D, Bichsel M, Bolli H, Häring M, Hochuli P, Kleboth P (1981) The Missaglia megabed, a catastrophic deposit in the Upper Cretaceous Bergamo flysch, northern Italy. Eclogae Geol Helv 74:421-442

Bernoulli D, Jenkyns HC (1974) Alpine, Mediterranean, and central Atlantic Mesozoic facies in relation to the early evolution of the Tethys. In: Dott RH Jr, Shaver RH (eds) Modern and ancient geosynclinal sedimentation. Soc Econ Paleontol Mineral Spec Publ 19:129-160

Bernoulli D, Lemoine M (1980) Birth and early evolution of the Tethys: the overall situation. In: Auboin J, Debelmas J, Latreille M (eds) Geology of the Alpine chains born of the Tethys. Mem BRGM 115:168-198

Berry WBN, Wilde P (1978) Progressive ventilation of the oceans - an explanation for the distribution of Lower Paleozoic black shales. Am J Sci 278:257-275

Berry WBN, Wilde P, Quinby-Hunt MS (1988) Ordovician-Silurian graptolite extinctions: rates, patterns and casual mechanisms. Abstr 3rd Int Conf Global Bioevents, Boulder, CO, University of Colorado, 10A

Berry WBN, Wilde P, Quinby-Hunt MS (1990) Late Ordovician graptolite mortality and subsequent Early Silurian radiation. In: Kauffman EG, Walliser OH (eds) Global bioevents: abrupt changes in the global biota. (Lect Notes Earth Sci 30), Springer, Berlin Heidelberg New York, pp 1115-123

Berthelin J (1983) Microbial weathering processes. In: Krumbein WE (ed) Microbial geochemistry. Blackwell Scientific, Oxford, pp 223-262

Betsey DD, Lyalikova NN, Murray D, Doyle M, Kolesov GM, Krumbein WE (1989) Role for microorganisms in the formation of iridium anomalies. Geology 17:1036-1039

Billi P, Magi M, Sagri M (1987) Coarse-grained low-sinuosity river deposits: example from the Plio-Pleistocene Valdarno Basin, Italy. In: Etheridge GF, Flores R, Harvey MD (eds) Recent developments in fluvial sedimentology. Soc Econ Paleontol Mineral Spec Publ 39:197-203

Birkelund T, Hancock JM, Hart MB, Rawson PF, Remane J, Robaszwnski F, Schmid F, Surlyk F (1984) Cretaceous stage boundaries - proposals. Geol Soc Denmark Bull 33:1-20

Bishop JDD (1988) Disarticulated bivalve shells as substrates for encrustation by the bryozoan Cribrilina puncturata in the Plio-Pleistocene Red Crag of eastern England. Palaeontology 31:237-253

Bishop JK (1987) The barite-opal-organic carbon association in oceanic particulate matter. Nature 332:341-343

Bitschene P, Schmincke HU (1989) Tephra layers: their properties and significance. In: Heling D (ed) Sediments and environmental geochemistry. Springer, Berlin Heidelberg New York, pp 48-82

Bjorlykke K (1979) Discussion - cementation of sandstones. J Sediment Petrol 49:1358-1359
Bjorlykke K, Bue B, Elverhoi A (1978) Quaternary sediments in the northwestern part of the Barents Sea and their relation to the underlying Mesozoic bedrock. Sedimentology 25:227-246
Blair TC (1987) Sedimentary processes, vertical stratification sequences, and geomorphology of the Roaring River alluvial fan, Rocky Mountain National Park, Colorado. J Sediment Petrol 57:1-18
Blair TC, Bilodeau WL (1988) Development of tectonic cyclothems in rift, pull-apart, and foreland basins: sedimentary response to episodic tectonism. Geology 16:517-520
Blanckenhorn M (1914) Syrien, Arabien, Mesopotamien. Handb Reg Geol Abt 5 (5):1-159
Blanpied C, Stanley DJ (1981) Uniform mud (unifite) deposition in the Hellenic Trench, eastern Mediterranean. Smithson Contrib Mar Sci 13, pp 40
Blatt H, Middleton G, Murray R (1980) Origin of sedimentary rocks, 2nd edn. Prentice Hall, Englewood Cliffs, 782 pp
Bloomfield P (1976) Fourier analysis of time series: an introduction. John Wiley, New York
Bloos G (1976) Untersuchungen über Bau und Entstehung der feinkörnigen Sandsteine des Schwarzen Jura alpha (Hettangium und tiefstes Sinemurium) im schwäbischen Sedimentationsbereich. Arb Inst Geol Pal Univ Stuttgart, NF 71:1-269
Bloos G (1982) Shell beds in the Lower Lias of South Germany - facies and origin. In: Einsele G, Seilacher A (eds) Cyclic and event stratification. Springer, Berlin Heidelberg New York, pp 223-239
Bluck BJ (1979) Structure of coarse grained braided stream alluvium. Trans R Soc Edinb 70:181-221
Boardman DR II, Mapes RH, Yancey TE, Malinky JM (1984) A new model for the depth-related allogenic community succession within North American Pennsylvanian cyclothems and implications on the black shale problem. In: Hyne NJ (ed) Limestones on the mid-continent. Tulsa Geol Soc Spec Publ 2:141-182
Boardman MR, Neumann CA (1984) Sources of periplatform carbonates: Northwest Providence Channel, Bahamas. J Sediment Petrol 54:1110-1123
Bohor BF, Triplehorn DM, Nichols DJ, Millard HT Jr (1987) Dinosaurs, spherules, and the "magic" layer: a new K-T boundary clay site in Wyoming. Geology 15:896-899
Boles JR, Franks SG (1979) Clay diagenesis in Wilcox sandstones of southwest Texas: implications of smectite diagenesis on sandstone cementation. J Sediment Petrol 49:55-70
Bolli HM, Loeblich AR Jr, Tappan H (1957) Planktonic foraminiferal families Hantkeninidae, Orbulinidae, Globorotaliidae and Globotruncanidae. US Natl Mus Bull 215:3-50
Bolli HM, Ryan WBF et al. (1978) Init Repts DSDP 40: Washington (US Govt Print Office), 1079 pp
Bolt BA (1978) Earthquakes. Freeman, San Francisco, 241 pp
Bond G (1976) Evidence for continental subsidence in North America during the Late Cretaceous global submergence. Geology 4:557-560
Bond GC, Kominz MA (1984) Construction of tectonic subsidence curves for the Early Paleozoic miogeocline, southern Canadian Rocky Mountains: implication for subsidence mechanisms, age of breakup, and crustal thinning. Geol Soc Am Bull 95:155-173
Bond GC, Kominz MA, Grotzinger JP (1988) Cambro-Ordovician eustasy: evidence from geophysical modeling of subsidence in Cordilleran and Appalachian passive margin. In: Klienspehn K, Paola C (eds) New perspective in basin analysis. Springer, Berlin Heidelberg New York, pp 129-161
Bond GC, Nickeson PA, Kominz MA (1984) Breakup of a supercontinent between 625 Ma and 555 Ma: new evidence and implications for continental histories. Earth Planet Sci Lett 70:325-345
Boon JJ, de Leeuw JW, Krumbein WE (1985) Biogeochemistry of Gavish Sabkha sediments. II. Pyrolysis mass spectrometry of the laminated microbial mat. In: Friedman GM, Krumbein WE (eds) Hypersaline ecosystems - the Gavish Sabkha. Ecol Stud 53:368-380
Bosellini A (1984) Progradation geometries of carbonate platforms: examples from the Triassic of the Dolomites, northern Italy. Sedimentology 31:1-24
Bosellini A, Hardie LA (1973) Depositional theme of a marginal marine evaporite. Sedimentology 20:5-27
Bosellini A, Masetti D, Sarti M (1981) A Jurassic "Tongue of the Ocean" infilled with oolitic sands: the Belluno trough, Venetian Alps, Italy. Mar Geol 44:59-95
Bosellini A, Rossi D (1974) Triassic buildups of the Dolomites, northern Italy. In: Laporte LF (ed) Reefs in time and space. Soc Econ Paleontol Mineral Spec Publ 18:209-233
Bosence DWJ, Rowlands RJ, Quine ML (1985) Sedimentology and budget of a recent carbonate mound, Florida Keys. Sedimentology 32:317-343

Boston WB, McComas GA, McGhee GR (1988) The fossil occurrence of epizoans on living coiled cephalopods in the Upper Paleozoic (Carboniferous). Geol Soc Am Abstr with Programs 20 (1):9

Bott MHP, Johnson GAL (1967) The controlling mechanism of Carboniferous cyclic sedimentation. Geol Soc Lond Q J 122:421-441

Bottjer DJ (1986) Campanian-Maastrichtian chalks of SW Arkansas: petrology, paleo-environments and comparison with other N American and European chalks. Cretaceous Res 7:161-196

Bottjer DJ, Arthur MA, Dean WE, Hattin DE, Sarda CE (1986) Rhythmic bedding produced in Cretaceous pelagic carbonate environments: sensitive recorder of climatic cycles. Paleoceanography 1:467-481

Bottjer DJ, Ausich WI (1986) Phanerozoic development of tiering soft substrata suspension - feeding communities. Paleobiology 12:400-420

Boudreau BP, Canfield DE (1988) A provisional diagenetic model for pH in anoxic porewaters. J Mar Res 46:429-455

Boulton GS (1972) The role of thermal regime in glacial sedimentation. Inst Brit Geogr Spec Publ 4:1-19

Bouma AH (1962) Sedimentology of some flysch deposits: a graphic approach to facies interpretation. Elsevier Scientific, Amsterdam, 167 pp

Bouma AH, Coleman JM, Meyer AW et al. (1986) Init Repts DSDP 96: Washington (US Govt Print Office), 824 pp

Bouma AH, Hollister CD (1973) Deep ocean basin sedimentation. In: Middleton GV, Bouma AH (eds) Turbidites and deep water sedimentation. Soc Econ Paleontol Mineral Pacific Sect Short Course, Anaheim 1973, pp 79-118

Bourgeois J, Hansen TA, Wiberg PL, Kauffman EG (1988) A tsunami deposit at the Cretaceous-Tertiary boundary in Texas. Science 241:567-569

Bourrouilh R (1987) Evolutionary mass flow-megaturbidites in an interplate basin: example of the North Pyrenean basin. Geomar Lett 7:69-81

Bova JA, Read JF (1987) Incipiently drowned facies within acyclic peritidal ramp sequence. Geol Soc Am Bull 98:214-227

Bowen AJ (1980) Simple models of nearshore sedimentation; beach profiles and longshore bars. In: McCann SB (ed) The coastline of Canada. Geol Surv Can 80-10:1-11

Bowen AJ, Inman DL (1969) Rip currents. 2. Laboratory and field observations. J Geophys Res 74:5479-5490

Bowen AJ, Inman DL (1971) Edge waves and crescentic bars. J Geophys Res 76:8662-8671

Bowen AJ, Normark WR, Piper DJW (1984) Modelling of turbidity currents on Navy deep sea fan, California Continental Borderland. Sedimentology 31:161-185

Bowles FA, Jack RN, Carmichael ISE (1973) Investigation of deep-sea volcanic ash layers from equatorial Pacific cores. Geol Soc Am Bull 84:2371-2388

Bowman SA, Glaser KS, Jordan JE, Jacquin T, Vail PR (1990) Balancing of stratigraphic cross-sections of the Delaware Basin and the New Jersey continental shelf to determine tectonic and eustatic controls. Am Assoc Petrol Geol, San Francisco Annu Meeting (in press)

Boyce RE (1981) Electrical resistivity, sound velosity, thermal conductivity, density-porosity, and temperature, obtained by laboratory techniques and well logs: site 462 in the Nauru Basin of the Pacific Ocean. In: Larson RL, Schlanger SO et al. Init Repts DSDP 61: Washington (US Govt Print Office), pp 743-761

Boyer BW (1982) Green River laminites: does the playa-lake model really invalidate the stratified-lake model? Geology 10:321-324

Boyle EA, Keigwin LD (1982) Deep circulation of the North Atlantic over the last 200,000 years: geochemical evidence. Science 218:784-787

Boyoko-Diakonow M (1979) The laminated sediments of Crawford Lake, southern Ontario, Canada. In: Schlüchter Ch (ed) Moraines and varves - origin/genesis/classification. Balkema, Rotterdam, pp 303-307

Bradley RS (1985) Quaternary paleoclimatology. Allen and Unwin, Boston, 472 pp

Bradley WH (1929) The varves and climate of the Green River Epoch. US Geol Surv Prof Pap 158-E:87-110

Bradley WH (1931) The origin of the oil shale and its microfossils of the Green River Formation in Colarado and Utah. US Geol Surv Prof Pap 168: 58 pp

Bralower TJ, Thierstein HR (1984) Low productivity and slow deep-water circulation in mid-Cretaceous oceans. Geology 12:614-618

Bralower TJ, Thierstein HR (1987) Organic carbon and metal accumulation rates in Holocene and mid-Cretaceous sediments: palaeoceanographic significance. In: Brooks J, Fleet AJ (eds) Marine petroleum source rocks. Geol Soc Spec Publ 26:345-369

Bramlette MN (1946) The Monterey Formation of California and the origin of its siliceous rocks. US Geol Surv Prof Pap 212:1-57

Brandt DS (1986) Preservation of event beds through time. Palaios 1:92-96

Brass GW, Southam JR, Peterson WH (1982) Warm saline bottom waters in the ancient ocean. Nature 296:620-623

Bréhéret JG (1985) Indice d'un événement anoxique étendu à la Téthys alpine à l'Albien inférieur (événement Paquier). CR Acad Sci Paris 300:355-358

Brenchley PJ (1985) Storm influenced sandstone beds. Mod Geol 9:369-396

Brenner K (1976) Ammoniten-Gehäuse als Anzeiger von Palaoströmungen. N Jb Geol Paläontol Abh 157:11-18

Brenner W (1988) Dinoflagellaten aus dem Unteren Malm (Oberer Jura) von Süddeutschland, Morphologie, Ökologie, Stratigraphie. Diss Inst Mus Geol Palaeontol Univ Tübingen, Tübingen, 115 pp

Bretsky PW (1969) Evolution of Paleozoic benthic marine invertebrate communities. Palaeogeogr Palaeoclimatol Palaeoecol 6:45-59

Brett CE (ed) (1986) Dynamic stratigraphy and depositional environments of the Hamilton Group (Middle Devonian) in New York State, Part I. NY State Mus Bull 457:156

Brett CE, Baird GC (1986a) Comparative taphonomy: a key to paleoenvironmental interpretation based on fossil preservation. Palaios 1:207-227

Brett CE, Baird GC (1986b) Symmetrical and upward shallowing cycles in the Middle Devonian of New York State and their implications for the punctuated aggradational cycle hypothesis. Paleoceanography 1:431-445

Brett CE, Eckert JD (1982) Palaeoecology of a well preserved crinoid colony from the Silurian Rochester Shale in Ontario. Life Sci Contrib R Ontario Mus 131: 20

Brett CE, Liddell WD (1978) Preservation and paleoecology of a Middle Ordovician hardground community. Paleobiology 4:329-348

Brett CE, Dick VB, Baird GC (1991) Comparative taphonomy and paleoecology of Middle Devonian dark gray and black shales facies from western New York. In: Landing E, Brett C (eds) Dynamic stratigraphy and depositional environments of the Hamilton Group (Middle Devonian) in New York State, Part 2. NY State Mus Bull, 469, pp 5-36

Brett CE, Speyer SE, Baird GC (1988) Storm-generated sedimentary units: tempestite proximality and event stratification in the Middle Devonian Hamilton Group of New York. In: Brett CE (ed) Dynamic stratigraphy and depositional environments of the Hamilton Group (Middle Devonian) in New York State, Part 1. NY State Mus Bull 457:129-155

Bretz JH, Smith HTU, Neff GE (1956) Channeled scablands of Washington: new data and interpretations. Geol Soc Am Bull 67:957-1049

Briggs DEG, Clarkson ENK, Aldridge RJ (1983) The conodont animal. Lethaia 16:1-14

Brink AH (1974) Petroleum geology of Gabon. Am Assoc Paleontol Geol Bull 58:216-235

Brock TD (1976) Environmental microbiology of living stromatolites. In: Walter MR (ed) Stromatolites. (Developments in sedimentology 20). Elsevier Scientific, Amsterdam, pp 141-148

Broecker WS (1971) A kinematic model for the chemical composition of sea water. Quat Res 1:188-207

Broecker WS (1982) Glacial to interglacial changes in ocean chemistry. Prog Oceanogr 11:151-197

Broecker WS, Peng TH (1982) Tracers in the sea. Palisades. Lamont-Doherty Geological Observatory. Eldigio, New York, 690 pp

Broecker WS, Peng TH (1986) Carbon cycle: 1985, glacial to interglacial changes in the operation of the global carbon cycle. Radiocarbon 28 (2A):309-327

Broecker WS, Peng TH (1987) The role of $CaCO_3$ compensation in the glacial to interglacial atmospheric CO_2 change. Global Biogeochem Cycles 1:15-29

Broecker WS, Takahashi T (1978) The relationship between lysocline depth and in situ carbonate ion concentration. Deep Sea Res 25:65-95

Broecker WS, van Donk J (1970) Insolation changes, ice volumes and the 18O record in deep-sea cores. Rev Geophys Space Phys 8:169-198

Bromley RG (1975) Trace fossils at omission surfaces. In: Frey RW (ed) The study of trace fossils. Springer, Berlin Heidelberg New York, pp 399-428

Bromley RG, Ekdale AA (1984a) Chondrites: a trace fossil indicator of anoxia in sediments. Science 224:872-874

Bromley RG, Ekdale AA (1984b) Trace fossil preservation in flint in the European chalk. J Paleontol 58:298-311

Bromley RG, Ekdale AA (1986) Composite ichnofabrics and tiering in burrows. Geol Mag 123:59-65

Bromley RG, Ekdale AA (1987) Mass transport in European chalk: fabric criteria for its recognition. Sedimentology 34:1079-1092

Brongersma-Sanders M (1957) Mass mortality in the sea. In: Hedgpeth JW (ed) Treatise on marine ecology and paleoecology. Geol Soc Am Mem 67 (1):941-1010

Brongersma-Sanders M (1971) Origin of major cyclicity of evaporites and bituminous rocks: an actualistic model. Mar Geol 11:123-144

Brooks J, Fleet AJ (eds) (1987) Marine petroleum source rocks. Geol Soc London Spec Publ 26: 444 pp

Brough J (1928) On rhythmic deposition in the Yoredale Series. Proc Univ Durham Philos Soc 8:116-126

Brown LF Jr (1969) Geometry and distribution of fluvial and deltaic sandstones (Pennsylvanian and Permian), north-central Texas. Gulf Coast Assoc Geol Soc Trans 19:23-47

Brumsack HJ (1986) The inorganic geochemistry of Cretaceous black shales (DSDP Leg 41) in comparison to modern sediments from the Gulf of California. In: Shackleton NJ, Summerhayes CP (eds) North Atlantic paleoceanography. Geol Soc Lond Spec Publ 21:447-462

Bryant ID, Kantorowicz JD, Love CF (1988) The origin and recognition of laterally continuous carbonate-cemented horizons in the Upper Lias Sands of southern England. Mar Petrol Geol 5:108-133

Bubenicek L (1971) Géologie du gisement de fer de Lorraine. Bull Cent Rech Pau SNPA 5:223-320

Buchheim HP, Surdam RC (1977) Fossil catfish band the depositional environment of the Green River Formation, Wyoming. Geology 5:196-198

Buchheim HP, Surdam RR (1981) Paleoenvironments and fossil fishes of the Lancy Member, Green River Formation, Wyoming. In Gray J, Boucot AJ, Berry WBN (eds) Communities of the past. Hutchinson Ross, Stroudsburg, PA, pp 415-452

Budd DA, Perkins RD (1980) Bathymetric zonation and paleoecological significance of microborings in Puerto Rican shelf and slope sediments. J Sediment Petrol 50:881-904

Budyko MI, Ronov AB (1979) Chemical evolution of the atmosphere in the Phanerozoic. Geochem Int 1979:1-9 (Geokhimiya 5:643-653)

Buffler RT, Schlager W et al. (1984) Sites 535, 539, and 540. Init Repts DSDP 77: Washington (US Govt Print Office), pp 25-217

Bullen SB, Sibley DF (1984) Dolomite selectivity and mimic replacement. Geology 12:655-658

Burbank DW, Grant MJ (1985) Plio-Pleistocene cyclic sedimentation in the Kashmir Basin, northwestern Himalaya: Milankovitch periodicities and grain-size data. Z Gletscherkd Glazialgeol 21:229-236

Burne RV, Ferguson J (1983) Contrasting marginal sediments of a seasonally flooded saline lake - Lake Eliza, S Australia: significance for oil shale genesis. BMR J Aust Geol Geophys 8:99-108

Burnett WC (1977) Geochemistry and origin of phosphorite deposits from off Peru and Chile. Geol Soc Am Bull 88:813-823

Burnett WC, Roe KK, Piper DZ (1983) Upwelling and phosphorite formation in the ocean. In: Suess E, Thiede J (eds) Coastal upwelling and its sediment record, Part A. Responses of the sedimentary regime to present coastal upwelling. Plenum, New York, pp 377-397

Burton R, Kendall CGStC, Lerche I (1987) Out of our depth: on the impossibility of fathoming eustacy from the stratigraphic record. Earth Sci Rev 24:237-277

Buxton TM, Sibley DF (1981) Pressure solution features in shallow buried limestone. J Sediment Petrol 51:19-26

Byers CW (1974) Shale fissility: relation to bioturbation. Sedimentology 21:479-484

Byers CW (1977) Biofacies patterns in euxinic basins: a general model. In: Cook HE, Enos P (eds) Deep water carbonate environments. Soc Econ Pal Min Spec Publ 25:5-17

Byers CW (1979) Biogenic structures of black shale paleoenvironments. Postilla 174:1-43

Byers CW, Stasko LE (1978) Trace fossils and sedimentologic interpretation - McGregor Member of Platterville Formation (Ordovician) of Wisconsin. J Sediment Petrol 48:1303-1309

Cacchione DA, Drake DE, Grant WD, Tate GB (1984) Rippled scour depressions on the inner continental shelf off central California. J Sediment Petrol 54:1280-1291

Calvert SE (1964) Factors affecting distribution of laminated diatomaceous sediments in Gulf of California. In: Van Andel TH, Schor JJ Jr (eds) Marine geology of the Gulf of California. Am Assoc Paleontol Geol Mem 3:311-330

Calvert SE (1974) Deposition and diagenesis of silica in marine sediments. In: Hsü KJ, Jenkyns HC (eds) Pelagic sediments on land and under the sea. Int Assoc Sedimentol Spec Publ 1:273-299

Calvert SE (1987) Oceanographic controls on the accumulation of organic matter in marine sediments. In: Brooks J, Fleets AJ (eds) Marine petroleum source rocks. Geol Soc Lond Spec Publ 26:137-151

Campbell CV (1967) Laminae, lamina set, bed and bedset. Sedimentology 8:7-26

Carey DL, Roy DC (1985) Deposition of laminated shale: a field and experimental study. Geomar Lett 5:3-9

Carey S, Sigurdsson H (1978) Deep-sea evidence for distribution of tephra from the mixed magma eruption of the Soufriere on St. Vincent 1902 ash turbidites and air fall. Geology 6:271-274

Carey S, Sigurdsson H (1980) The Roseau ash: deep-sea tephra deposits from a major eruption on Dominica, Lesser Antilles Arc. J Volcanol Geotherm Res 7:67-86

Carey S, Sigurdsson H (1986) The 1982 eruptions of El Chichon volcano, Mexico (2) Observations and numerical modelling of tephra-fall distribution. Bull Volcanol 48:127-141

Carey S, Sparks RSJ (1986) Quantitative models of the fallout and dispersal of tephra-fall from volcanic eruption columns. Bull Volcanol 48:109-126

Carey SW, Ahmad N (1961) Glacimarine sedimentation. In: Raasch GO (ed) Geology of the Arctic. Proc 1st Int Symp Arctic Geol, Univ Toronto Press, Toronto 2:865-894

Carozzi AV, Gerber MS (1978) Synsedimentary chert breccia: a Mississippian tempestite. J Sediment Petrol 48:705-708

Carranante G (1971) Ricerche sedimentologiche sulla successione ciclotemica dell'Infralias del Paso dell'Annunziata Lunga (Monti di Venafro). Boll Soc Nat Napoli 80:389-412

Cary SC, Fisher CH, Felbeck H (1988) Mussel growth supported by methan as sole carbon and energy source. Science 240:78-80

Cas RAF, Wright JV (1987) Volcanic rocks: modern and ancient. Allen and Unwin, Winchester, pp 1-528

Casanova J (1986) East African rift stromatolites. In: Frostick LE, Renaut RW, Reid I, Tiercelin JJ (eds) Sedimentation in the African rifts. Geol Soc Spec Publ 25:201-210

Casey RE, Spaw JM, Kunze FR (1982) Polycystine radiolarian distributions and enhancements related to oceanographic conditions in a hypothetical ocean. Trans Gulf Coast Assoc Geol Soc 32:319-322

Castenholz RW (1969) The thermophilic cyanophytes of Iceland and the upper temperature limit. J Phycol 5:360-368

Catalano R, d'Argenio B, Lo Cicero G (1974) I Ciclotemi Triassici di Capo Rama (Monti di Palermo). Geol Rom XIII:125-145

Cavanaugh CM (1985) Symbiosis of chemoautotrophic bacteria and marine invertebrates from hydrothermal vents and reducing sediments. Bull Biol Soc Wash 6:373-388

Chamley H (1979) North Atlantic clay sedimentation and paleoenvironments since the Late Jurassic. In: Talwani M, Hay WW, Ryan WBF (eds) Deep drilling results in the Atlantic Ocean: continental margins and paleoenvironment. Am Geophys Union, Maurice Ewing Ser 3:342-361

Chamley H, Debrabant P (1984) Paleoenvironmental history of the North Atlantic region from mineralogical and geochemical data. Sediment Geol 40:151

Chamley H, Debrabant P, Candillier AM, Foulou J (1983) Clay mineralogical and inorganic geochemical stratigraphy of Blake-Bahama Basin since the Callovian, Site 534, DSDP Leg 76. In: Sheridan RE, Gradstein FM et al. Init Repts DSDP 76: Washington (US Govt Print Office), pp 437-451

Chamley H, Debrabant P, Foulou J, Giroud d'Argoud G, Latouche C, Maillet N, Maillet H, Sommer F (1979) Mineralogy and quantitative carbonate analysis methods. J Sediment Petrol 50:631-6372

Chanda SK, Bhattacharyya A, Sarkar S (1976) Early diagenetic chert nodules in limestone, India. J Geol 84:213-224

Chanda SK, Bhattacharyya A, Sarkar S (1977) Deformation of ooids by compaction in the Precambrian Bhander limestone, India: implications for lithification. Geol Soc Am Bull 88:1577-1585

Chanda SK, Bhattacharyya A, Sarkar S (1983) Compaction of limestones: a reappraisal. J Geol Soc India 24:73-92

Choquette PW, James NP (1987) Diagenesis 12. Diagenesis in limestones 3. The deep burial environment. Geosci Can 14:3-35

Chough SK (1984) Fine-grained turbidites and associated mass flow deposits in th Ulleung (Tsushima) back-arc basin, East Sea (Sea of Japan). In: Stow DAV, Piper DJW (eds) Fine-grained sediments: deep-water processes and facies. Geol Soc Lond Spec Publ 15:185-196

Ciesielski PF, Kristoffersen Y et al. (1988) Preliminary results of subantarctic south Atlantic leg 114 of the Ocean Drilling Program. Proc Init Repts DSDP 114: Washington (US Govt Print Office), pp 797-803

Cisne JL (1973) Beecher's trilobite bed revisited: ecology of an Ordovician deep-water fauna. Postilla 160:9-34

Cisne JL (1986) Earthquakes recorded stratigraphically on carbonate platforms. Nature 323:320-322

Cita MB, Beghi C, Cammerleugni A, Kastens K, McCoy FW, Nosetto A, Parisi E, Scolari F, Tomadin L (1984) New findings of Bronze Age homogeneities in the Ionian Sea: geodynamic implications for the Mediterranean. Mar Geol 55:47-62

Cita MB, Grignani D (1982) Nature and origin of late Neogene Mediterranean sapropels. In: Schlanger SO, Cita MB (eds) Nature and origin of Cretaceous carbon-rich facies. Academic Press, London, pp 165-196

Cita MB, Ricci Lucchi F (eds) (1984) Seismicity and sedimentation. Mar Geol 55 1/2:161

Clark DL, Cheng-Yuan W, Orth CJ, Gilmore JS (1986) Conodont survival and low iridium abundances across the Permian-Triassic boundary in South China. Science 233:984-986

Claypool GE, Kaplan IR (1974) The origin and distribution of methan in marine sediments. In: Kaplan IR (ed) Natural gases in marine sediments. Plenum, New York, pp 99-139

Clifton HE (1981) Progradational sequences in Miocene shoreline deposits, southeastern Caliente Range, California. J Sediment Petrol 51:165-184

Clifton HE (ed) (1988) Sedimentologic consequences of convulsive geologic events. Geol Soc Am Spec Pap 229:157

CLIMAP project members (1976) The surface of the ice-age earth. Science 191:1131-1137

CLIMAP (1981) Maps of northern and southern hemisphere continental ice, sea ice, and sea surface temperatures in August for the modern and the last glacial maximum. Geol Soc Am Map and Chart Ser MC-36

Cloetingh S (1986) Intraplate stresses: a new tectonic mechanism for fluctuations of relative sea level. Geology 14:617-620

Cluff RM (1980) Paleoenvironment of the New Albany Shale Group (Devon-Mississippian) of Illinois. J Sediment Petrol 50:767-780

Coastal Engineering Research Center (CERC) (1984) Shore Protection Manual, 4th edn. Department of the Army, US Army Corps of Engineers (US Govt Print Office) 2 vols (not consecutively paginated)

Cobban WA, Scott GR (1972) Stratigraphy and ammonite fauna of the Graneros Shale and Greenhorn Limestone near Pueblo, Colorado. US Geol Surv Prof Pap 645:108

Cochrane JD, Kelley FJ (1986) Low-frequency circulation on the Texas-Louisiana continental shelf. J Geophys Res 91 (C9):10645-10659

Cohen AS (1989) Facies relationships and sedimentation in large rift lakes and implications for hydrocarbon exploration: examples from Lake Turkana and Tanganyika. Palaeog Palaeoclimatol Palaeoecol 70:65-80

Cohen Y, Castenholz RW, Halvorson HO (1984) Microbial mats: stromatolites. Alan Liss, New York, 498 pp

Cohen Y, Krumbein WE, Shilo M (1977a) Solar Lake (Sinai) 2 distribution of photosynthetic microorganisms and primary production. Limnol Oceanogr 22:609-620

Cohen Y, Krumbein WE, Goldberg M, Shilo M (1977b) Solar Lake (Sinai) 1 physical and chemical limnology. Limnol Oceanogr 22:597-608

Coleman JM (1968) Deltaic evolution. In: Fairbridge RW (ed) Encyclopedia of geomorphology. Reinhold, Washington, pp 255-260

Coleman JM, Gagliano SM (1964) Cyclic sedimentation in the Mississippi River deltaic plain. Gulf Coast Assoc Geol Soc Trans 14:67-80

Coleman JM, Roberts HH, Murray SP, Salama M (1981) Morphology and dynamic sedimentology of the eastern Nile Delta shelf. In: Nittrouer CA (ed) Sedimentary dynamics of continental shelves. (Developments in sedimentology 32) Elsevier, Amsterdam, pp 301-326

Coleman ML, Raiswell R (1981) Carbon, oxygen and sulphur isotope variations in concretions from the Upper Lias of NE England. Geochim Cosmochim Acta 45:329-340

Collins LB (1988) Sediments and history of the Rottnest Shelf, southwest Australia: a swell-dominated, non-tropical carbonate margin. Sediment Geol 60:15-49

Collinson JD, Thompson DB (1982) Sedimentary structures. Allen and Unwin, London, 194

Compston W, Williams ES, Meyer C (1984) U-Pb geochronology of zircons from lunar b4 breccia 73217 using a sensitive high mass-resolution ion microprobe. J Geophys Res Suppl B:525-534

Cook DO (1970) The occurrence and geologic work of rip currents off southern California. Mar Geol 9:173-186

Cook DO, Gorsline DS (1972) Field observations of sand transport by shoaling waves. Mar Geol 13:31-55

Cook HE (1972) Leg 9, Deep Sea Drilling Project: stratigraphy and sedimentation. In: Hays JD et al. Init Repts DSDP 9: Washington (US Govt Print Office), pp 933-943

Cook HE, Egbert RM (1981) Late Cambrian-Early Ordovician continental margin sedimentation, central Nevada. In: Taylor ME (ed) Procs 2nd Int Symp Cambrian System. US Geol Surv Open File Rept 81-743:50-56

Cook HE, Enos P (eds) (1977) Deep-water carbonate environments. Soc Econ Paleontol Mineral Spec Publ 25: 336

Cook HE, Jenkyns HC, Kelts K (1976) Redeposited sediments along the Line Islands, Equatorial Pacific. In: Schlanger SO, Jackson ED et al. Init Repts DSDP 33: Washington (US Govt Print Office), pp 837-846

Cook HE, McDaniel PN, Mountjoy EW, Pray LC (1972) Allochthonous carbonate debris flows at Devonian bank ("reef") margins Alberta, Canada. Bull Can Petrol Geol 20:439-497

Cook HE, Taylor ME (1977) Comparison of continental slope and shelf environments in the Upper Cambrian and lowest Ordovician of Nevada. In: Cook HE, Enos P (eds) Deep-water carbonate environments. Soc Econ Paleontol Mineral Spec Publ 25:51-81

Cornell W, Carey SN, Sigurdsson H (1983) Computer simulation of transport and deposition of the Campanian Y5 ash. J Volcanol Geotherm Res 17:89-109

Costa JE (1983) Paleohydraulic reconstruction of flash-flood peaks from boulder deposits in the Colorado Front Range. Geol Soc Am Bull 94:986-1004

Costa JE (1984) Physical geomorphology of debris flows. In: Costa JE, Fleisher PJ (eds) Developments and applications of geomorphology. Springer, Berlin Heidelberg New York, pp 268-317

Cotillon P (1974) Sédimentation rythmique et milieux de dépôt: données fournies par l'étude du Crétacé inférieur de l'arc subalpin de Castellane (France SE). Bull Soc Géol Fr (7) 16:583-592

Cotillon P (1984) Tentative world-wide correlation of Early Cretaceous strata by limestone-marl cyclicities in pelagic deposits. Bull Geol Soc Denmark 33:91-102

Cotillon P (1985) Les variations à différentes échelles du taux d'accumulation sédimentaire dans les séries pélagiques alternantes du Crétacé inférieur, conséquences de phénomènes globaux. Bull Soc Géol Fr 8:59-68

Cotillon P (1987) Bed-scale cyclicity of pelagic Cretaceous successions as a result of world-wide control. Mar Geol 78:109-123

Cotillon P, Ferry S, Gaillard C, Jautée E, Latreille G, Rio M (1980) Fluctuation des paramètres du milieu marin dans le domaine Vocontien (France Sud-Est) au Crétacé inférieur: mise en evidence par l'étude des formations marno-calcaires alternantes. Bull Soc Géol Fr (7) XXII 5:735-744

Cotillon P, Rio M (1984) Cyclic sedimentation in the Cretaceous of DSDP Sites 535 and 540 (Gulf of Mexico), 534 (central Atlantic), and in the Vocontian Basin (France). In: Buffler RT, Schlager W et al. Init Repts DSDP 77: Washington (US Govt Print Office), pp 339-376

Cottle R (1988) Orbitally induced climatic cycles in the chalk of southern England. Abstr CRER Workshop Perugia, Sept 1988

Courtillot V, Besse J, Vandamme D, Montigny R, Jaeger JJ, Cappetta H (1986) Deccan flood basalts at the Cretaceous/Tertiary boundary? Earth Planet Sci Lett 80:361-374

Cox A (ed) (1973) Plate tectonics and geomagnetic reversals. Freeman, San Francisco, 702 pp

Cox MA, Whitford-Stark JL (1987) Stylolites in the Caballos Novaculite, west Texas. Geology 15:439-442

Crevello PD, Patton JW, Oesleby TW, Schlager W, Droxler A (1984) Source rock potential of Bahamian trough carbonates. In: Stow DAV, Piper DJW (eds) Fine-grained sediments: deep-water processes and facies. Geol Soc Lond Spec Publ 15:469-480

Crevello PD, Schlager W (1980) Carbonate debris sheets and turbidites, Exuma Sound, Bahamas. J Sediment Petrol 50:1121-1148

Crimes TP (1973) From limestones to distal turbidites: a facies and trace fossil analysis in the Zumaya flysch (Paleocene-Eocene), North Spain. Sedimentology 20:105-131

Cronan DS, Damiani VV, Kinsman DJJ, Thiede J (1974) Sediments from the Gulf of Aden and Western Indian Ocean. In: Fisher L, Bunce ET et al. Init Repts DSDP 24: Washington (US Govt Print Office), pp 1047-1110

Cross TA (1988) Controls on coal distribution in transgressive-regressive cycles, Upper Cretaceous, Western Interior, USA. In: Sea-level changes - an integrated approach. Soc Econ Paleontol Mineral Spec Publ 42:371-380

Cross TA, Pilger RH Jr (1978) Tectonic controls of Late Cretaceous sedimentation, Western Interior, USA. Nature 274:653-675

Crossley R, Owen B (1988) Sand turbidites and organic-rich diatomaceous muds from Lake Malawi, central Africa. In: Fleet AJ, Kelts K, Talbot MR (eds) Lacustrine petroleum source rocks. Geol Soc Lond Spec Publ 40:369-374

Crout RL (1983) Wind-driven, near-bottom currents over the west Louisiana inner continental shelf. PhD Diss Louisiana State University, Baton Rouge, 117 pp

Crout RL, Murray SP (1978) Shelf and coastal boundary layer currents, Miskito Bank of Nicaragua. Proc 16th Coastal Engr Conf, Am Soc Civ Engr, Hamburg, Germany 3:2715-2729

Crowley TJ (1983) Calcium carbonate preservation patterns in the central North Atlantic during the last 150,000 years. Mar Geol 51:1-14

Crowley TJ (1985) Late Quaternary changes in the North Atlantic and Atlantic/Pacific - comparison. In: Sundquist ET, Broecker WS (eds) The carbon cycle and the atmospheric CO_2: natural variations Archean to Present. Am Geophys Unions Monogr 32:271-284

Csanady GT (1982) Circulation in the coastal ocean. Reidel, Dordrecht, 279 pp

Cuffey RJ (1985) Expanded reef-rock textural classification and the geologic history of bryozoan reefs. Geology 13:307-310

Curial A (1983) Etude sédimentologique des alternances marne-calcaire de l'Oxfordien supérieur de l'Ardèche. Rapp DEA Lyon, inédit, 29 pp

Curray JR, Moore DG (1974) Sedimentary and tectonic processes in the Bengal deep-sea fan and geosyncline. In: Burk CA, Drake CI (eds) The geology of continental margins. Springer, Berlin Heidelberg New York, pp 617-627

Curray JR, Moore DG et al. (1982) Guaymas Basin slope: Sites 479 and 480. In: Curray JR, Moore DG et al. Init Repts DSDP 64: Washington (US Govt Print Office), 417-504 pp

Currie RG (1983) Detection of 18.6 year nodal induced drought in the Patagonian Andes. Geophys Res Lett 10:1089-1092

Currie RG (1984) Periodic 18.6-year and cyclic 11-year induced drought and flood in western North America. J Geophys Res 89:7215-7230

Currie RG (1987) Examples and implications of 18.6- and 11-yr terms in world weather records. In: Rapino MR, Sanders JE, Newman WS, Königsson LK (eds) Climate-history, periodicity, and predictability. Van Nostrand Reinhold, New York, pp 378-403

Currie RG, Fairbridge RW (1985) Periodic 18.6-year and cyclic 11-year induced drought and flood in northeastern China and some global implications. Quat Sci Rev 4:109-134

Curtis CD (1980) Diagenetic alteration in black shales. J Geol Soc Lond 137:189-194

Dahanayake K, Gerdes G, Krumbein WE (1985): Stromatolites, oncolites and oolites biogenically formed in situ. Naturwissenschaften 72:513-518

Dahanayake K, Krumbein WE (1985) Ultrastructure of a microbial mat-generated phosphorite. Mineral Depos 20:260-265

Dahmer D, Ernst G (1986) Upper Cretaceous event-stratigraphy in Europe. In: Walliser OH (ed) Global bio-events. (Lect Notes Earth Sci 8), Springer, Berlin Heidelberg New York, pp 353-362

Dalfes HN, Schneider SH, Thompson SL (1984) Effects of bioturbation on climatic spectra inferred from deep sea cores. In: Berger A, Imbrie J, Hays J, Kukla G, Saltzman B (eds) Milankovitch and climate. NATO ASI Ser C126 (I):481-492

Dally WR (1987) Longshore bar formation - surf beat or undertow. Coast Seds '87, Am Soc Civ Engr, New York, pp 71-86

Dally WR, Dean RG (1984) Suspended sediment transport and beach profile evolution. J Waterway Port Coast Ocean Engr, Am Soc Civ Engr 110:15-33

Dalrymple GB, Duffield WA (1988) High precision 40Ar/39Ar dating of Oligocene rhyolites from the Mogollon-Datil volcanic field using a continuous laser system. Geophys Res Lett 15:463-466

Damuth JE (1980) Use of high-frequency (3.5-12 kHz) echograms in the study of near-bottom sedimentation processes in the deep sea: a review. Mar Geol 38:51-75

Damuth JE, Embley RW (1981) Mass-transport processes on the Amazon cone: western equatorial Alantic. Am Assoc Paleontol Geol Bull 65:629-643
Damuth JE, Kowsmann RO, Monteiro MC, Gorini MA, Palma JJC, Belderson RH (1983) Distributary channel meandering and bifurcation patterns on the Amazon deep sea fan as revealed by long-range side-scan (GLORIA). Geology 11:94-98
Dando PR, Southward AJ (1986) Chemoautotrophy in bivalve molluscs of the genus Thyasira. J Mar Biol Assoc UK 66:915-929
Dando PR, Southward AJ, Southward EC, Terwilliger NB, Terwilliger, RC (1985) Sulphur-oxidising bacteria and haemoglobin in gills of the bivalve mollusc Myrtea spinifera. Mar Ecol Prog Ser 23:85-98
Darmedru C (1982) La microfaune dans les alternances marne-calcaire pélagique du Crétacé inférieur vocontien (Sud-Est de la France). Mise en évidence d'oscillations climatiques. Thèse 3e cycle, Université de Lyon 1
Darmedru C, Cotillon P, Rio M (1982) Rhythmes climatiques et biologiques en milieu marin pélagique. Leurs relations dans les dépots Crétacés alternants du Basin Vocontien (SE France). Bull Soc Géol Fr 7:627-640
Davies AG, Soulsby RL, King H (1988) A numerical model of the combined wave and current boundary layer. J Geophys Res 93:491-508
Davies DJ, Powell EN, Stanton RJ Jr (1989) Relative rates of shell dissolution and net sediment accumulation - a commentary: can shell beds form by the gradual accumulation of biogenic debris on the seafloor? Lethaia 22:207-212
Davies GR (1977) Turbidites, debris sheets, and truncation structures in Upper Paleozoic deep-water carbonates of Sverdrup Basin, Arctic Archipelago. In: Cook HE, Enos P (eds) Deep-water carbonate environments. Soc Econ Paleontol Mineral Spec Publ 25:221-247
Davis JC (1973) Statistics and data analysis in geology, 1st edn. Wiley, New York
Dean RG (1977) Equilibrium beach profiles: US Atlantic and Gulf coasts. Dept Civ Engr, University of Delaware, Newark, Del, Tech Rept 12: 46
Dean RG, Perlin M (1986) Intercomparison of near-bottom kinematics by several wave theories and field and laboratory data. Coast Engr 9:399-437
Dean WE (1981) Carbonate minerals and organic matter in sediments of modern north temperate hardwater lakes. Soc Econ Paleontol Mineral Spec Publ 31:213-231
Dean WE, Arthur MA (1986) Origin and diagenesis of Cretaceous deep-sea, organic-carbon-rich lithofacies in the Atlantic Ocean. In: Mumpton FA (ed) Studies in diagenesis. US Geol Surv Bull 1578:97-128
Dean WE, Arthur MA (1987) Inorganic and organic geochemistry of Eocene to Cretaceous strata recovered from the lower continental rise, North American Basin, Site 603, DSDP Leg 93. In: Wise SW Jr, Van Hinte JE et al. Init Repts DSDP 93: Washington (US Govt Print Office), pp 1093-1197
Dean WE, Arthur MA (1989) Iron-sulfur-carbon relationships in organic-carbon-rich sequences. I: Cretaceous Western Interior Seaway. Am J Sci 289:708-743
Dean WE, Arthur MA, Claypool GE (1986) 13C/12C of organic carbon in Cretaceous black shales: influence of source, diagenesis, and environmental factors. Mar Geol 70:119-157
Dean WE, Arthur MA, Stow DAV (1984a) Origin and geochemistry of Cretaceous deep-sea black shales and multicolored claystones, with emphasis on DSDP Site 530, southern Angola Basin. In: Hay WW, Sibuet JC et al. Init Repts DSDP 75: Washington (US Govt Print Office), pp 819-844
Dean WE, Bradbury JP, Anderson RY, Barnowsky CW (1984b) The variability of Holocene climatic change: evidence from varved lake sediments. Science 116:1191-1194
Dean WE, Gardner JV (1982) Origin and geochemistry of redox cycles of Jurassic to Eocene age, Cape Verde Basin (DSDP Site 367), continental margin of North-west Africa. In: Schlanger SO, Cita MB (eds) Nature and origin of Cretaceous organic carbon-rich facies. Academic Press, London, pp 55-78
Dean WE, Gardner JV (1985) Cyclic variations in calcium carbonate and organic carbon in Miocene to Holocene sediments, Walvis Ridge, South Atlantic Ocean. In: Hsu KJ Weissert HJ (eds) South Atlantic paleoceanography. Cambridge University Press, Cambridge, pp 61-78
Dean WE, Gardner JV (1986) Milankovitch cycles in Neogene deep-sea sediments. Paleoceanography 1:539-553
Dean WE, Gardner JV, Cepek P (1981) Tertiary carbonate-dissolution cycles on the Sierra Leone Rise, eastern equatorial Atlantic Ocean. Mar Geol 39 (1/2):81-101
Dean WE, Gardner JV, Jansa LF, Cepek P, Seibold E (1977) Cyclic sedimentation along the continental margin of northwest Africa. In: Lancelot Y, Seibold E et al. Init Repts DSDP 41: Washington (US Govt Print Office), pp 965-986

Dean WE, Parduhn NL (1984) Inorganic geochemistry of sediments and rocks recovered from the southern Angola Basin and adjacent Walvis Ridge Site 530 and 532. In: Hay WW, Sibuet JC et al. Init Repts DSDP 75: Washington (US Govt Print Office), pp 923-958

de Boer PL (1980) The paleo-environment of mid-Cretaceous black shale deposition as deduced from stable carbon isotopes. Comparative Sedimentology Research Group, Institute of Earth Sciences, Utrecht, sep. 42, 19 pp

de Boer PL (1982a) Cyclicity and the storage of organic matter in Cretaceous pelagic sediments. In: Einsele G, Seilacher A (eds) Cyclic and event stratification. Springer, Berlin Heidelberg New York, pp 456-475

de Boer PL (1982b) Some remarks about stable isotope composition of cyclic pelagic sediments from the Cretaceous in the Apennines (Italy). In: Schlanger SO, Cita MB (eds) Nature and origin of Cretaceous carbon-rich facies. Academic Press, London, pp 129-143

de Boer PL (1983) Aspects of Middle Cretaceous pelagic sedimentation in S Europe. Geol Ultraiectina 31:112

de Boer PL (1986) Changes in the organic carbon burial during the Early Cretaceous. In: Summerhayes CP, Shackleton NJ (eds) North Atlantic paleoceanography. Geol Soc Lond Spec Publ 21:321-331

de Boer PL (1991) Astronomical cycles reflected in sediments. Zbl Geol Paläont, Teil I, 8:911-930, Stuttgart

de Boer PL, Wonders AAH (1984) Astronomically induced rhythmic bedding in Cretaceous pelagic sediments near Moria (Italy). In: Berger A, Imbrie J, Hays J, Kukla G, Saltzman B (eds) Milankovitch and climate. Reidel, Hingham, Mass, pp 177-190

de Boer RB (1977) On the thermodynamics of pressure solution - interaction between chemical and mechanical forces. Geochim Cosmochim Acta 41:249-256

Debrabant P, Chamley H (1982) Influences océaniques et continentales dans les premières dépots de l'Atlantique Nord. Bull Soc Géol Fr 7:473-486

Debrabant P, Chamley H, Foulou J (1984) Paleoenvironmental implications of mineralogic and geochemical data in the Western Florida Straits (DSDP Leg 77). In: Buffler RT, Schlager W et al. Init Repts DSDP 77: Washington (US Govt Print Office), pp 377-396

Decker K (1987) Faziesanalyse der Oberjura- und Neokomschichtfolge der Grestener- und Ybbsitzer Klippenzone im westlichen Niederösterreich. Thesis, University Vienna, Vienna

Decker K (1990) Plate tectonic control on facies development: Late Jurassic to Early Cretaceous deep-dea sediments in the Ybbsitz Klippen Belt, eastern Alps. Sediment Geol 67:85-99

Deconinck JF, Chamley H (1983) Héritage et diagenèse des minéraux argileux dans les alternances marno-calcaires du Crétacé inférieur du domaine subalpin. CR Acad Sci Paris 297:589-594

Deconinck JF, Strasser A (1987) Sedimentology, clay mineralogy and depositional environment of Purbeckian green marls (Swiss and French Jura). Eclogae Geol Helv 80:753-772

de Geer G (1912) A geochronology of the last 12,000 years. 11th Int Geol Congr Stockholm, 1910. Proc S Rep, vol 1, pp 241-253

Degens ET, Emeis KC, Mycke B, Wiesner MG (1986) Turbidites, the principle mechanism yielding black shales in the early deep Atlantic ocean. In: Summerhays CP, Shackleton NJ (eds) North Atlantic palaeoceanography. Geol Soc Spec Publ 21:361-376

Degens ET, Kurtman F (1978) The geology of Lake Van. Min Res Expl Inst Turkey Mem 169:158

Degens ET, Stoffers P (1980) Environmental events recorded in Quaternary sediments of the Black Sea. J Geol Soc Lond 137:131-138

Degens ET, Stoffers P, Golubic S, Dickmann MD (1978) Varve chronology: estimated rates of sedimentation in the Black Sea deep basin. In: Ross DA, Neprochnov YP et al. Init Repts DSDP 42: Washington (US Govt Print Office), pp 499-508

de Graciansky PC, Auffret GA, Dupeuble P, Montadert L, Müller C (1979) Interpretation of depositional environments of the Aptian/Albian black shales of the north margin of the Bay of Biscay (DSDP Sites 400 and 402). In: Montadert L, Roberts DG et al. Init Repts DSDP 48: Washington (US Govt Print Office), pp 877-907

de Graciansky PC, Brosse E, Deroo G, Herbin JP, Montadert L, Müller C, Sigal J, Schaaf A (1982) Les formations d'age Crétacé de l'Atlantique Nord et leur matière organique: paléogéographie et milieux de dépot. Rev Inst Fr Pet 37:275-337

de Graciansky PC, Brosse E, Deroo G, Herbin JP, Montadert L, Müller C, Sigal J, Schaaf A (1987) Organic-rich sediments and paleoenvironmental reconstructions of the Cretaceous North Atlantic. In: Brooks J, Fleet A (eds) Marine petroleum source rocks. Geol Soc Lond Spec Publ 26:317-344 de Graciansky PC, Chenet PY (1979) Sedimentological study of Cores 138 to 56 (Upper Hauterivian to Middle Cenomanian), an attempt at reconstruction of paleoenvironments. In: Sibuet JC, Ryan WBF et al. Init Repts DSDP 47: Washington (US Govt Print Office), pp 403-418

Dehaires F, Chesselet R, Jedwab J (1980) Discrete suspended particles of barite and the barium cycle in the open ocean. Earth Planet Sci Lett 49:528-550

Deines P (1980) The isotopic composition of reduced organic carbon. In: Fritz P, Fontes JC (eds) The isotopic composition of reduced organic carbon. Elsevier, Amsterdam, pp 329-406

de Klein G, V, Willard DA (1989) Origin of the Pennsylvanian coal-bearing cyclothems of North America. Geology 17:152-155

Delmas RJ, Ascesio JM, Legrand M (1980) Polar ice evidence that atmospheric CO_2 20,000 years B.P. was 50% of present. Nature 284:155-157

Demaison GJ, Moore GT (1980) Anoxic environments and oil source Bed Genesis. Am Assoc Petrol Geol Bull 64 (8):1179-1209

Dengler AT, Nota EK, Wilde P, Normark WR (1984) Slumping and related turbidity currents along proposed OTEC cold water pipe route resulting from hurricane Iwa. Proc 16th Offshore Tech Conf, Houston, Texas OTC 4702

Denis C (1986) On the change of kinetical parameters on the Earth during geological times. Geophys J R Astr Soc 87:559-568

Deroo G, Herbin JP, Huc AY (1984) Organic geochemistry of Cretaceous black shales from DSDP site 530, leg 75, eastern South Atlantic. In: Hay WW, Sibuet JC et al. Init Repts DSDP 75: Washington (US Govt Print Office), pp 933-999

Dethlefsen V, Westernhagen H (1983) Oxygen deficiency and effects on bottom fauna in the eastern German Bight 1982. Meeresforschung 30:42-53

Devine JD, Sigurdsson H, Davis AN (1984) Estimates of sulfur and chlorine yield to the atmosphere from volcanic eruptions and potential climatic effects. J Geophys Res 89:6309-6325

de Wever P (1987) Radiolarites rubanées et variations de l'orbite terrestre. Bull Soc Géól Fr 3 (4):957-960

Dewers T, Ortoleva PJ (1990) Interaction of reaction, mass transport, and rock deformation during diagenesis: Mathematical modeling of intergranular pressure solution, stylolites, and differential compaction/cementation. In: Meshri D, Ortoleva PJ (eds.) Prediction of reservoir quality through chemical modeling. Am Ass Petrol Geol Memoir 49: 147-160

Dia A, Manhes G, Dupre B, Allegre CJ (1989) The Cretaceous-Tertiary boundary problem: an assessment from lead isotope systematics. Chem Geol 75:291-304

Dickinson KA (1988) Paleolimnology of Lake Tubutulik, an iron-meromictic Eocene Lake, eastern Seward Peninsula, Alaska. Sediment Geol 54:303-320

Dickman M (1985) Seasonal succession and microlamina formation in a meromictic lake displaying varved sediments. Sedimentology 32:109-118

Diester L (1972) Zur spätpleistozänen und holozänen Sedimentation im zentralen und östlichen Persischen Golf. Meteor Forsch Ergeb C 8:37-83

Diester-Haass L (1975) Sedimentation and climate in the Late Quaternary between Senegal and the Cape Verde Islands. Meteor Forsch Ergeb C 20:1-32

Diester-Haass L (1976) Late Quaternary climatic variations in NW Africa deduced from East-Atlantic sediment cores. Quat Res 6:299-314

Diester-Haass L (1978) Sediments as indicators of upwelling. In: Boje R, Tomczak M (eds) Upwelling ecosystems. Springer, Berlin Heidelberg New York, pp 261-281

Diester-Haass L (1983) Late Quaternary sedimentation processes on the West-African continental margin and climatic history of West-Africa (12-18°N). Meteor Forsch Ergeb C 37:47-84

Diester-Haass L (1985) Late Quaternary sedimentation on the eastern Walvis Ridge, SE Atlantic (HPC 532 and four piston cores). Mar Geol 65:145-189

Diester-Haass L, Meyers PA, Rothe P (1986) Light-dark cycles in opal-rich sediments near the Plio-Pleistocene boundary, DSDP Site 532, Walvis Ridge continental terrace. Mar Geol 73:1-23

Diester-Haass L, Meyers PA, Rothe P (1990) The Benguela Current: glacial-interglacial fluctuations during its Miocene northward migration across the Walvis Ridge. Paleoceanography (in press)

Diester-Haass L, Rothe P (1988) Plio-Pleistocene sedimentation on the Walvis Ridge, southeast Atlantic (DSDP Leg 75, Site 532) - influence of surface currents, carbonate dissolution and climate. Mar Geol 77:53-85

Diester-Haass L, Schnitker D (1990) Plio-Pleistocene sedimentation regimes leading to chalk-marl cycles in the North Atlantic (DSDP Site 552-Hole 552 A). Geol Rdsch (in press)

Diester-Haass L, Schrader HJ, Thiede J (1973) Sedimentological and paleoclimatological investigations of two pelagic ooze cores of Cape Barbas, north-west Africa. Meteor Forsch Ergeb C 16:19-66

Dill RF, Shinn EA, Jones AT, Kelly K, Steiner RP (1986) Giant subtidal stromatolites forming in normal salinity waters. Nature 324:55-58

Ditchfield P, Marshall JD, (1989) Isotopic variation in rhythmically bedded chalks: paleotemperature variation in the Upper Cretaceous. Geology, 17:842-845

Dix GR, Mullins HT (1986) Shallow, subsurface growth and burial alteration of Middle Devonian calcite concretions. J Sediment Petrol 57:140-152

Dix GR, Mullins HT (1988) Rapid burial diagenesis of deep-water carbonates: Exuma Sound, Bahamas. Geology 16:673-768

Doeglas DJ (1962) The structure of sedimentary deposits of braided rivers. Sedimentology 1:167-190

Donaldson AC (1974) Pennsylvanian sedimentation of central Appalachians. In: Briggs G (ed) Carboniferous of the southeastern United States. Geol Soc Am Spec Pap 148:47-78

Donegan D, Schrader H (1982) Biogenic and abiogenic components of laminated hemipelagic sediments in the central Gulf of California. Mar Geol 48:215-237

Donn WL, Ninkovich D (1980) Rate of Cenozoic explosive volcanism in the North Atlantic Ocean inferred from deep sea cores. J Geophys Res 85:5455-5460

Donnelly TW, Francheteau J et al. (1980) Init Repts DSDP 51, 52, 53, Pts 1 and 2: Washington (US Govt Print Office), 1613 pp

Donovan DT, Jones EJW (1979) Causes of world-wide changes in sea level. J Geol Soc Lond 136:187-192

Dott RH (1983) Episodic sedimentation - how normal is average? How rare is rare? Does it matter? J Sediment Petrol 53:5-23

Dott RH (1988) An episodic view of shallow marine clastic sedimentation. In: de Boer PL, van Gelder A, Nio SD (eds) Tide-influenced sedimentary environments and facies. Reidel, Dordrecht, pp 3-12

Dott RH, Bourgeois J (1982) Hummocky stratification: significance of its variable bedding sequences. Geol Soc Am Bull 93:603-680

Douglas RG (1972) Paleozoogeography of Late Cretaceous planktonic foraminifera in North America. J Foraminiferal Res 2:14-34

Douglas RG, Savin SM (1975) Oxygen and carbon isotope analyses of Tertiary and Cretaceous microfossils from Shatsky Rise and other sites in the North Pacific Ocean. In: Larson RL, Moberly R et al. Init Repts DSDP 32: Washington (US Govt Print Office), pp 509-520

Dries RR, Theede H (1974) Sauerstoffmangelresistenz mariner Bodeninvertebraten aus der westlichen Ostsee. Mar Biol 25:327-333

Dromart G (1989) Deposition of Upper Jurassic fine-grained limestones in the Western Subalpine Basin, France. Palaeogeogr Palaeoclimatol Palaeoecol 69:23-43

Droser ML (1987) Trends in extent and depth of bioturbation in Great Basin Precambrian-Ordovician strata, California, Nevada and Utah. PhD Thesis, University of Southern California, Los Angeles, 365 pp

Droser ML, Bottjer DJ (1986) A semi quantitative field classification of ichnofabric. J Sediment Petrol 56:558-559

Droser ML, Bottjer DJ (1988) Trends in depth and extent of bioturbation in Cambrian carbonate marine environments, western United States. Geology 16:233-236

Droser ML, Bottjer DJ (1989) Ordovician increase in extent and depth of bioturbation: implications for understanding early Paleozoic ecospace utilization. Geology 17:850-852

Droste JB, Vitaliano CJ (1973) Tioga bentonite (Middle Devonian) of Indiana. Clays and Clay Minerals 21:9-13

Droxler AW, Schlager W (1985) Glacial versus interglacial sedimentation rates and turbidite frequency in the Bahamas. Geol Soc Am Geol 13:799-802

Droxler AW, Schlager W, Whallon CC (1983) Quaternary aragonite cycles and oxygen isotope record in Bahamian carbonate ooze. Geology 11:235-239

Duff KL (1975) Paleoecology of a bituminous shale - the lower Oxford Clay of central England. Palaeontology 18:443-482

Duff PM, Hallam A, Walton EK (eds) (1967) Cyclic sedimentation. (Developments in sedimentology 10), Elsevier, Amsterdam, 280 pp

Duke WL (1985) Hummocky cross-stratification, tropical hurricanes, and intense winter storms. Sedimentology 32:167-194

Duke WL (1990) Geostrophic circulation or shallow marine turbidity currents? The dilemma of paleoflow patterns in storm-influenced prograding shoreline systems. J Sediment Petrol 60:870-883

Duncan AD, Hamilton RFM (1988) Palaeolimnology and organic geochemistry of the Middle Devonian in the Orcadian Basin. In: Fleet AJ, Kelts K, Talbot MR (eds) Lacustrine petroleum source rocks. Geol Soc Spec Publ 40:173-201

Dunn CE (1974) Identification of sedimentary cycles through Fourier analysis of geochemical data. Chem Geol 13:217-232

Duplessy JC, Moyes J, Pujol C (1980) Deep water formation in the North Atlantic Ocean during the last ice age. Nature 286:479-482

Dzulynski S, Ksiazkiewicz M, Kuenen PH (1959) Turbidites in flysch of the Polish Carpathian Mountains. Geol Soc Am Bull 70:1089-1118

Eaton GP (1963) Volcanic ash deposits as a guide to atmospheric circulation in the geologic past. J Geophys Res 68:521-528

Eberli GP (1987) Carbonate turbidite sequences deposited in rift-basins of the Jurassic Tethys Ocean (eastern Alps, Switzerland). Sedimentology 34:363-388

Eberli GP (1988a) Physical properties of carbonate turbidite sequences surrounding the Bahamas - implications for slope stability and fluid movements. In: Austin JA, Schlager W et al. Proc Ocean Drilling Program, Scientific Results, 101:305-314

Eberli GP (1988b) The evolution of the southern continental margin of the Jurassic Tethys Ocean as recorded in the Allgäu Formation of the Austroalpine nappes of Graubünden (Switzerland). Eclogae Geol Helv 81:175-214

Eberli GP, Ginsburg RN (1989) Cenozoic progradation of northwestern Great Bahama Bank, a record of lateral platform growth and sea level fluctuations. In: Crevello PD, Sarg PD, Wilson JL, Read JF (eds) Controls on carbonate platform and basin evolution. Soc Econ Paleontol Mineral Spec Publ 44:339-355

Eddy JA (1988) Variability of the present and ancient Sun: a test of solar uniformitarianism. In: Stephenson FR, Wolfendale AW (eds) Secular solar geomagnetic variations in the last 10,000 years. Kluwer Academic, Dordrecht, pp 1-24

Eder FW (1971) Riff-nahe detritische Kalke bei Balve im Rheinischen Schiefergebirge. Gött Arb Geol-Palaeont 10:66

Eder FW (1982) Diagenetic redistribution of carbonate, a process in forming limestone-marl alternations (Devonian and Carboniferous, Rheinisches Schiefergebirge, W Germany). In: Einsele G, Seilacher A (eds) Cyclic and event stratification, Springer, Berlin Heidelberg New York, pp 98-112

Eder FW, Engel W, Franke W, Sadler PM (1983) Devonian and Carboniferous limestone-turbidites of the Rheinisches Schiefergebirge and their tectonic significance. In: Martin H, Eder FW (eds) Intracontinental fold belts. Springer, Berlin Heidelberg New York, pp 93-124

Edwards LE (1984) Insights on why graphic correlation (Shaw's method) works. J Geol 92:583-597

Edwards LE (1989) Supplemented graphic correlation: a powerful tool for paleontologists and nonpaleontologists. Palaios 4:127-143

Edwards M (1986) Glacial environments. In: Reading HG (ed) Sedimentary environments and facies. Blackwell Scientific, Oxford, pp 445-470

Eicher DL (1967) Foraminifera from Belle Fourche Shale and equivalents Wyoming and Montana. J Paleontol 41:167-188

Eicher DL (1969) Paleobathymetry of Cretaceous Greenhorn Sea in eastern Colorado. Am Assoc Petrol Geol Bull 53:1075-1090

Eicher DL, Diner R (1985) Foraminifera as indicators of water mass in the Cretaceous Greenhorn Sea, Western Interior. In: Pratt LM, Kauffman EG, Zelt FB (eds) Fine-grained deposits and biofacies of the Cretaceous Western Interior Seaway: evidence of cyclic sedimentary processes. Soc Econ Paleontol Mineral Fieldtrip Guidebook 4:60-71

Eicher DL, Diner R (1989) Origin of the Cretaceous Bridge Creek cycles in the Western Interior, United States. Paleogeogr Paleoclimatol Paleoecol 74:127-146

Eicher DL, Worstell P (1970) Cenomanian and Turonian Foraminifera from the Great Plains, United States. Micropaleontology 16:269-324

Einsele G (1982) Limestone-marl cycles (periodites): diagnosis, significance, causes - a review. In: Einsele G, Seilacher A (eds) Cyclic and event stratification. Springer, Berlin Heidelberg New York, pp 8-53

Einsele G (1985) Response of sediment to sea level changes in differing subsiding storm-dominated marginal and epeiric basins. In: Bayer U, Seilacher A (eds) Sedimentary and evolutionary cycles. (Lect Notes Earth Sci 1), Springer, Berlin, pp 68-97

Einsele G (1989) In situ water contents, liquid limits, and submarine mass flows due to a high liquifaction potential of slope sediments (results from DSDP and subaerial counterparts). Geol Rdsch 78:821-840

Einsele G, Elouard P, Herm D, Kögler FC, Schwarz HU (1977) Source and biofacies of Late Quaternary sediments in relation to sea level on the shelf off Mauritania, West Africa. Meteor Forsch Ergeb C 26:1-43

Einsele G, Kelts K (1982) Pliocene and Quaternary mud turbidites in the Gulf of California: sedimentology, mass physical properties and significance. In: Curray JR, Moore DG et al. Init Repts DSDP 64: Washington (US Govt Print Office), pp 511-528

Einsele G, Seilacher A (eds) (1982) Cyclic and event stratification. Springer, Berlin Heidelberg New York, 536 pp

Einsele G, Wiedmann J (1982) Turonian black shales in the Moroccan Coastal Basins: first upwelling in the Atlantic Ocean? In: von Rad U, Hinz K, Sarnthein M, Seibold E (eds) Geology of the northwest African continental margin. Springer, Berlin Heidelberg New York, pp 396-414

Ekdale AA, Bromley RG, Pemberton SG (1984a) Ichnology (Short Course Notes 15). Soc Econ Paleontol Mineral, 317 pp

Ekdale AA, Muller LN, Novack MT (1984b) Quantitative ichnology of modern pelagic deposits in the abyssal Atlantic. Paleogeogr Palaeoclimatol Palaeoecol 45:189-223

Elam JC, Chuber S (eds) (1972) Cyclic sedimentation in the Permian basin, 2nd edn. West Texas Geol Soc, Midland Texas

Elder WP (1985) Biotic patterns across the Cenomanian-Turonian extinction boundary near Pueblo, Colorado. In: Pratt LM, Kauffman EG, Zelt FB (eds) Fine-grained deposits and biofacies of the Cretaceous Western Interior Seaway: evidence of cyclic sedimentary processes. Soc Econ Paleontol Mineral Fieldtrip Guidebook 4:157-169

Elder WP (1987) The Cenomanian-Turonian (Cretaceous) stage boundary extinctions in the Western Interior of the United States. PhD Thesis, University of Colorado, 621 pp

Elder WP (1988) Geometry of Upper Cretaceous bentonite beds: Implications about volcanic source areas and paleowind patterns, Western Interior, United States. Geology 16:835-838

Elder WP, Gustason ER (1984) Regionally correlated fourth-order cyclothems of Upper Cretaceous strata, Colorado Plateau. Abstr with Programs, Soc Econ Paleontol Mineral Annu Midyr Mtng, 31 pp

Elder RL, Smith GR (1988) Fish taphonomy and environmental inference in paleolimnology. In: Gray J (eds) Aspects of freshwater paleoecology and biogeography. Palaeogeogr Palaeoclimatol Palaeoecol 62:577-592

Eldholm O, Thiede J, Taylor E et al. (1987) Proc Init Repts ODP 104. (PT A):1-783 College Station, Texas (Ocean Drilling Program)

Eller MG (1981) The Red Chalk of eastern England: a Cretaceous analogue of Rosso Ammonitco. In: Farinacci A, Elmi S (eds) Rosso Ammonitico Symp Proc. Edizioni Tecnoscienza, Roma, pp 207-231

Elliott T (1974) Abandonment facies of high-constructive lobate deltas, with an example from the Yoredale Series. Proc Geol Assoc 85:359-365

Elmi S (1981) Classification typologique et genétique des Ammonitico-Rosso et des facies noduleux ou grumeleux: essai de synthèse. In: Farinacci A and Elmi S (eds) Rosso Ammonitico Symp Proc. Edizioni Tecnoscienza, Roma, pp 233-249

Elmore RD, Pilkey OH, Cleary WJ, Curran HA (1979) Black shell turbidite, Hatteras abyssal plain, western Atlantic Ocean. Geol Soc Am Bull 90:1165-1176

Elverhoi A (1984) Glacigenic and associated marine sediments in the Weddell Sea, fjords of Spitsbergen and the Barents Sea: a review. Mar Geol 57:53-88

Elverhoi A, Liestol O, Nagy J (1980) Glacial erosion, sedimentation and microfauna in the inner part of Kongsfjorden, Spitsbergen. Norsk Polarinst Skrift 172:33-61

Elverhoi A, Pfirman S, Solheim A, Larsen BB (1989) Glaciomarine sedimentation on epicontinental seas - exemplified by the northern Barents Sea. Mar Geol 85: 225-250

Elverhoi A, Roaldset E (1983) Glaciomarine sediments and suspended particulate matter, Weddell Sea shelf, Antarctica. Polar Res 1:1-21

Elverhoi A, Solheim A (1987) Shallow geology and geophysics of the Barents Sea. Norsk Polarinst Rapp 37:71

Emerson S (1985) Organic carbon preservation in marine sediments. In: The carbon cycle and atmospheric CO_2. Am Geophys Union, Geophys Monogr 32:78-87

Emerson S, Bender M (1981) Carbon fluxes at the sediment-water interface of the deep sea: calcium carbonate preservation. J Mar Res 39:139-162

Emiliani C (1955) Pleistocene temperatures. J Geol 63:538-578

Enos P (1977) Tamabra limestone of the Pozo Rico trend, Retaceous, Mexico. In: Cook HE, Enos P (eds) Deep-water carbonate environments. Soc Econ Paleontol Mineral Spec Publ 25:273-314

Enos P (1985) Carbonate debris reservoirs, Poza Rica field, Veracruz, Mexico. In: Roehl PO, Choquette PW (eds) Carbonate petroleum reservoirs. Springer, Berlin Heidelberg New York, pp 455-489

Erba E (1988) Aptian-Albian calcareous nannofossil biostratigraphy of the Scisti a Fucoidi cored at Piobbico (central Italy). Riv Ital Paleont Stratigr 94:36 pp

Ericson DB, Ewing M, Heezen BC (1952) Turbidity currents and sediments in the North Atlantic. Am Assoc Paleontol Geol Bull 36:489-511

Ernst G, Schmidt F, Klisches G (1979) Multistratigraphische Untersuchungen in der Oberkreide des Raumes Braunschweig-Hannover. In: Wiedmann J (ed) Aspekte der Kreide Europas. Int Union Geol Sci Ser A 6:11-46

Ernst WG, Calvert SE (1969) An experimental study of the recrystallization of porcellanite and its bearing on the origin of some bedded cherts. Am J Sci 267A:114-133

Espitalié J, La Porte JL, Madex M, Marquis F, Leplat P, Paulet J, Boutefeu A (1977) Méthode rapide de charactérisation des roches mères de leu potentiel pétrolier et de leur degré d'évolution. Rev Inst Fr Pet 32:23-42

Esteban M, Klappa CF (1983) Subaerial exposure environments. In: Scholle PA, Bebout DG, Moore CH (eds) Carbonate depositional environments. Am Assoc Paleontol Geol Mem 33:1-54

Eugster HP, Hardie LA (1978) Sedimentation in an ancient playa-lake complex: the Wilkins Peak Member of the Green River Formation of Wyoming. Geol Soc Am Bull 86:319-334

Eugster HP, Kelts K (1983) Lacustrine chemical sediments. In: Goudie A, Pye K (eds) Chemical geomorphology. Academic Press, New York, pp 321-368

Evans CC, Ginsburg RN (1987) Fabric-selective diagenesis in the late Pleistocene Miami Limestone. J Sediment Petrol 57:311-318

Evans J, Kendall CGH (1977) An interpretation of the depositional setting of some deep-water Jurassic carbonates of the central High Atlas mountains, Morrocco. In: Cook HE, Enos P (eds) Deep-water carbonate environments. Soc Econ Palaeontol Mineral Spec Publ 25:249-261

Eyles CH, Eyles N, Miall AD (1985) Models of glaciomarine sedimentation and their application to the interpretation of ancient glacial sequences. Palaegeogr Palaeoclimatol Palaeoecol 51:15-84

Eyles N, Eyles CH, Miall AD (1983) Lithofacies types and vertical profile models; an alternative approach to the description and environmental interpretation of glacial diamict and diamictite sequences. Sedimentology 30:393-410

Eyles N, Miall AD (1984) Glacial facies. In: Walker RG (ed) Facies models. Geosci Can Reprint Ser 1, Toronto, pp 15-38

Fabre J (1981) Les Rhynchocphales et les Pterosauriens crètes paritales du Kimridgien superieur-Barriasien d'Europe occidentale. Les gisements de Canjuers (Var-France) et ses abords. Edit Fond Singer-Polignac, Paris, pp 188

Fairbridge RW (1976) Convergence of evidence on climatic change and ice ages. Ann NY Acad Sci 91:542-579

Fairbridge RW (1984) Planetary periodicities and terrestrial climate stress. In: Mürner NA, Karlen W (eds) Climate changes on a yearly to millenial basis. Reidel, Boston, pp 509-520

Fairbridge RW, Sanders JE (1987) The Sun's orbit, A.D. 750-2050: basis for new perspectives on planetary dynamics and the Earth-Moon linkage. In: Rampino MS, Sanders JE, Newman WS, Künigsson LK (eds) Climate - history, periodicity, predictability. Van Nostrand Reinhold, New York, pp 446-471

Falvey DA, Middletow MF (1981) Passive continental margins: evidence for a pre-breakup deep crustal metamorphism. In: Colloquium on geology of continental margins, C3, Paris 7-17 July 1980, Oceanol Acta Suppl 4:103-114

Farell, Prell WL (1987) Climate forcing of calcium carbonate sedimentation: a 4.0 my record from the central equatorial Pacific Ocean. EOS Trans, Am Geophys Union 68:333

Farinacci A, Elmi S (eds) (1981) Rosso Ammonitico Symp Proc. Edizioni Tecnoscienza, Roma, pp 1-602

Fastovsky DE (1987) Paleoenvironments of vertebrate-bearing strata during the Cretaceous-Paleogene transition, eastern Montana and western North Dakota. Palaios 2:282-295

Fedonkin MA (1977) Precambrian-Cambrian ichnocoenoses of the East European platform. In: Crimes TP, Harper JC (eds) Trace fossils 2. Seel House Press, Liverpool, pp 183-194

Felbeck H (1983) Sulfide oxidation and carbon fixation by the gutless clam Solemya reidi: an animal-bacteria symbiosis. J Comp Physiol B 152:3-11

Felder WM (1974) Lithostratigraphische Gliederung der oberen Kreide in Süd-Limburg (Niederlande) und den Nachbargebieten. Publicaties Natuurhistorisch Genootschap Limburg, Reeks 24, 43 pp

Feldhausen PH, Stanley DJ, Knight RJ, Maldonado A (1981) Homogenization of gravity-emplaced muds and unifites: models from the Hellenic Trench. In: Wezel FC (ed) Sedimentary basins of Mediterranean margins. CNR, Italian Project of Oceanography, Tecnoprint, Bologna, pp 203-226

Fenchel TM, Riedl RJ (1970) The sulfide system: a new biotic community underneath the oxidized layer of marine sand bottoms. Mar Biol 7:255-268

Ferm JC (1970) Allegheny deltaic deposits. In: Morgan JP (ed) Deltaic sedimentation, modern and ancient. Soc Econ Palaeontol Mineral Spec Publ 15:246-255

Ferm JC (1974) Carboniferous environmental models in eastern United States and their significance. In: Briggs G (ed) Carboniferous of the southeastern United States. Geol Soc Am Spec Pap 148:79-95

Ferm JC, Cavaroc VV Jr (1968) A nonmarine sedimentary model for the Allegheny rocks of West Virginia. Geol Soc Am Spec Pap 106:1-20

Ferry S, Cotillon P, Rio M (1983) Digenèse croissante des argiles dans les niveaux isochrones de l'alternance calcaire-marne Valanginienne du bassin vocontien. Zonation géographique. CR Acad Sci Paris 297:51-56

Ferry S, Monier PH (1987) Correspondance entre alternances marno-calcaires de bassin et de plateforme (Crétacé du SE de la France). Bull Soc Géol Fr 8 III:961-964

Ferry S, Schaaf A (1981) The early Cretaceous environment at DSDP Site 463 (Mid Pacific Mountains), with reference to the Vocontian Trough (French Supalpine Ranges) In: Thiede J, Vallier TL et al. Init Repts DSDP 62: Washington (US Govt Print Office), pp 669-682

Field ME, Gardner JV, Jennings AE, Edwards BD (1982) Earthquake-induced sediment failures on a 0.25 slope, Klamath River delta, California. Geology 10:542-546

Fischer AG (1964) The Lofer cyclothems of the Alpine Triassic. Kansas Geol Surv Bull 169:107-149

Fischer AG (1965) Eine Lateralverschiebung in den Salzburger Kalkalpen. Verh Geol Bundesanstalt Wien Heft 1/2:20-33

Fischer AG (1980) Gilbert-bedding rhythms and geochronology. In: Yochelson EI (ed) The scientific ideas of G.K. Gilbert. Geol Soc Am Spec Pap 183:93-104

Fischer AG (1981) Climatic oscillations in the biosphere. In: Nitecki MH (ed) Biotic crises in ecological and evolutionary time. Elsevier, New York, pp 103-131

Fischer AG (1982) Long-term climatic oscillations recorded in stratigraphy. In: Berger WH, Crowell JC (eds) Climate in earth history. Studies in geophysics. National Academy Press, Washington, pp 25-57

Fischer AG (1984) Biological innovation and the sedimentary record. In: Holland HD, Trendall AF (eds) Patterns and change in Earth evolution. Springer, Berlin Heidelberg New York Tokyo, pp 145-157

Fischer AG (1986) Climatic rhythms recorded in strata. Annu Rev Earth Planet Sci 14:351-376

Fischer AG, de Boer PL, Premoli-Silva I (1989) Cyclostratigraphy. In: Ginsburg RN, Beaudoin B (eds) Cretaceous resources, rhythms, and events. NATO ASI Ser C, Kluwer, Dordrecht, 304:139-172

Fischer AG, Herbert TD, Premoli-Silva I (1985) Carbonate bedding cycles in Cretaceous pelagic and hemipelagic sediments. In: Pratt LM, Kauffman EG, Zelt FB (eds) Fine-grained deposits and biofacies of the Cretaceous Western Interior Seaway: evidence of cyclic sedimentary processes. Soc Econ Paleontol Mineral, pp 1-10

Fischer AG, Schwarzacher W (1984) Cretaceous bedding rhythms under orbital control? In: Berger A, Imbrie J, Hays J, Kukla G, Saltzman B (eds) Milankovitch and climate. NATO ASI Series C 126, Reidel, Dordrecht, pp 163-176

Fisher MR, Hand SC (1984) Chemoautotrophic symbionts in the bivalve Lucina floridana from eelgrass beds. Biol Bull 167:445-459

Fisher RV, Glicken HX, Hoblitt RP (1987) May 18th 1980, Mount St. Helens deposits in South Coldwater Creek, Washington. J Geophys Res 92:10267-10283

Fisher RV, Schmincke HU (1984, 1990) Pyroclastic rocks, 1st and 2nd edn Springer, Berlin Heidelberg New York, pp 1-472

Fisher RV, Schmincke HU, Bogaard P (1983) Origin and emplacement of a pyroclastic flow and surge unit at Laacher See, Germany. J Volcanol Geotherm Res 17:375-392

Fisk HN (1944) Geological investigation of the alluvial valley of the lower Mississippi River. US Army Corps Eng, Mississippi Riv Comm, Vicksburg, Miss, 78 pp

Fisk HN (1960) Recent Mississippi River sedimentation and peat accumulation. Congr Int Stratigr Géol Carbonif, Heerlen 1958, CR 4, 1:187-199

Fiske RS (1963) Subaqueous pyroclastic flows in the Ohanapecosh Formation Washington. Geol Soc Am Bull 74:391-406

Fiske RS, Matsuda T (1964) Submarine equivalents of the ash flows in the Tokiwa Formation, Japan. Am J Sci 262:76-106

Fleet AJ, Brooks J (1987) Introduction. In: Brooks J, Fleet AJ (eds) Marine petroleum source rocks. Geol Soc Spec Publ 26:1-14

Flessa KW, Erben HK, Hallam A, Hsu KJ, Hussner HM, Jablonski D, Raup DM, Sepkoski JJ, Soule ME, Sousa W, Stinnesbeck W, Vermeij GJ (1986) Causes and consequences of extinction: group report. In: Raup DM, Jablonski D (eds) Pattern and process in the history of life. Springer, Berlin Heidelberg New York, pp 235-257

Flood RD (1978) X-ray mineralogy of DSDP Legs 44 and 44A, western North Atlantic: lower continental rise hills, Blake Nose, and Blake-Bahama Basin. In: Benson WE, Sheridan RE et al. Init Repts DSDP 44: Washington (US Govt Print Office), pp 463-476

Flores RM (1979) Restored stratigraphic cross-sections and coal correlations in the Tongue River Member of the Fort Union Formation, Powder River area, Montana. US Geol Surv Map MF-1127

Flügel E (1982) Microfacies analysis of limestones. Springer, Berlin Heidelberg New York, 633 pp

Föllmi KB (1989) Evolution of the mid-Cretaceous platform triad: carbonates - phosphatic sediments and pelagic carbonates along the northern Tethys margin. Lecture Notes Earth Sci 23, 153 pp, Springer-Verlag, Berlin Heidelberg New York

Foland KA, Faul H (1977) Ages of the White Mountain intrusives - New Hampshire, Vermont, and Maine, USA. Am J Sci 277:888-904

Folk RL, Siedlecka A (1974) The 'schizohaline' environment: its sedimentary and diagenetic fabrics as exemplified by Late Paleozoic rocks of Bear Island, Svalbard. Sediment Geol 11:1-15

Force ER, Cannon WF (1988) Depositional model for shallow-marine manganese deposits around black shale basins. Econ Geol 83:93-117

Forster W (1809) A treatise on a section of strata from Newcastle-upon-Tyne, to the mountain of Cross Fell, in Cumberland; with remarks on mineral veins in general, 1st edn. Preston & Heaton, Newcastle

Foucault A, Fang N (1987) Contr$_o$le climatique de la sedimentation quaternaire dans le Golfe du Bengale. CR Acad Sci Paris 305 II:1383-1388

Foucault A, Renard M (1987) Controle climatique de la sedimentation marno-calcaire dans le Mésozoique d'Espagne (Sierra de Fontcalent, Prov Alicante): arguments isotopiques. CR Acad Sci Paris 305 II:517-521

Foucault A, Powichrowski L, Prud'homme A (1987) Le controle astronomique de la sedimentation turbiditique: exemple du flysch à Helminthoides des Alpes Ligures (Italie). CR Acad Sci Paris 305 II:1007-1011

Frakes LA (1978) Diamictite. In: Fairbridge RW, Bourgeois (eds) The encyclopedia of sedimentology. Dowden, Hutchinson and Ross, Stroudsburg, pp 262-263

Frakes LA (1979) Climates throughout geologic time. Elsevier, Amsterdam, 310 pp

Frakes LA, Bolton BR (1984) Origin of manganese giants: sea-level change and anoxic-oxic history. Geology 12:83-86

Frakes LA, Crowell JC (1975) Characteristics of modern glacial marine sediments: application to the Gondwana Glacials. In: Campell KSW (ed) Gondwana geology. Aust Natl Univ Press, Canberra, pp 373-380

Frakes LA, Francis JE (1988) A guide to Phanerozoic cold polar climates from high-latitude ice-rafting in the Cretaceous. Nature 333:547-549

Francis EH (1984) Correlation between the north temperate and Tethyan realms in the Cenomanian of western France and the significance of hardground horizons. Cretaceous Res 5:259-269

Francis EH (1988) Mid-Devonian to early Permian volcanism: old world. In: Harris AL, Fettes DJ (eds) The Caledonian-Appalachian orogen. Geol Soc Spec Publ 38:573-584

Franke W, Walliser OH (1983) "Pelagic" carbonates in the Variscian Belt -their sedimentary and tectonic environments. In: Martin H, Eder FW (eds) Intracontinental fold belts. Springer, Berlin Heidelberg New York, pp 77-92

Frazier DE, Osanik A (1969) Recent peat deposits - Louisiana coastal plain. In: Dapples EC, Hopkins ME (eds) Environments of coal deposition. Geol Soc Am Spec Pap 114:63-85

Freeman T, Enos P (1978) Petrology of Upper Jurassic-Lower Cretaceous limestones, DSDP Site 391. In: Benson WE, Sheridan RE et al. Init Repts DSDP 44: Washington (US Govt Print Office), pp 463-475

Freundt A, Schmincke HU (1985a) Hierarchy of facies of pyroclastic flow deposits generated by Laacher See-type eruptions. Geology 13:278-281

Freundt A, Schmincke HU (1985b) Lithic-enriched segregation bodies in pyroclastic flow deposits of Laacher See Volcano (East Eifel, Germany). J Volcanol Geotherm Res 25:193-224

Freundt A, Schmincke HU (1986) Emplacement of small-volume pyroclastic flows at Laacher See Volcano (East Eifel, Germany). Bull Volcanol 48:39-60

Frey RW, Seilacher A (1980) Uniformity in marine invertebrate ichnology. Lethaia 13:183-207

Friedman GM, Amiel AJ, Braun M, Miller DS (1973) Generation of carbonate particles and laminites in algal mats - example from sea-marginal hypersaline pool, Gulf of Aqaba, Red Sea. Am Assoc Paleontol Geol Bull 57:541-557

Froelich PN, Arthur MA, Burnett WC, Deakin M, Hensley V, Jahnke R, Kaul L, Kim KH, Roe K, Soutar A, Vathakanon C (1988) Early diagenesis of organic matter in Peru continental margin sediments: phosphorite precipitation. Mar Geol 80:309-343

Froelich PN, Bender ML, Luedtke NA, Heath GR, DeVries J (1982) The marine phosphorus cycle. Am J Sci 282:474-511

Froelich PN, Kim KH, Jahnke R, Burnett WC, Soutar A, Deakin M (1983) Pore water fluoride in Peru continental margin sediments. Geochim Cosmochim Acta 47:1605-1612

Froitzheim N (1988) Synsedimentary and synorogenic faults within a thrust sheet of the eastern Alps (Ortler Zone, Graubünden, Switzerland). Eclogae Geol Helv 81:593-610

Frush MP, Eicher DL (1975) Cenomanian and Turonian foraminifera and paleoenvironments in the Big Bend region of Texas and Mexico. In: Caldwell WGE (ed) The Cretaceous system in North America. Geol Assoc Canada Spec Pap 13:277-301

Frydl P, Stearn CW (1978) Rate of bioerosion by parrotfish in Barbados reef environments. J Sediment Petrol 48:1149-1158

Füchtbauer H (ed) (1988) Sedimente und Sedimentgesteine. Schweizerbart, Stuttgart, 1141 pp

Füchtbauer H, Richter DK (1988) Karbonatgesteine. In: Füchtbauer H (ed) Sedimente und Sedimentgesteine. Schweizerbart, Stuttgart, pp 233-434

Führböter A, Manzenrieder H (1987) Biostabilisierung von Sandwatten durch Mikroorganismen. In: Gerdes G, Krumbein WE, Reineck HE (eds) Mellum - Portrait einer Insel. Kramer, Frankfurt, pp 123-138

Fulthorpe CS, Melillo AJ (1988) Middle Miocene carbonate gravity flows in the Straits of Florida at Site 626. In: Austin JA Jr, Schlager W et al. Proc ODP, Sci Results 101: (Ocean Drilling Program) College Station, Texas, pp 179-191

Fürsich H (1971) Hartgründe und Kondensation im Dogger von Calvados. N Jb Geol Palaeontol Abh 138:313-342

Fürsich FT (1973) Thalassinoides and the origin of nodular limestone in the Corallian Beds (Upper Jurassic) of southern England. N Jb Geol Palaeontol Mh 140:136-156

Fürsich FT (1978) The influence of faunal condensation and mixing on the preservation of fossil benthic communities. Lethaia 11:243-250

Fütterer DK, Grobe H, Grüning S (1988) Quaternary sediment patterns in the Weddell Sea: relations and environmental conditions. Paleoceanography 3 (5):551-561

Futterer E (1982) Experiments on the distinction of wave and current influenced shell accumulations. In: Einsele G, Seilacher A (eds) Cyclic and event stratification. Springer, Berlin Heidelberg New York, pp 175-179

Gagen MK, Johnson DP, Carter RM (1988) The cyclone Winifred stormbed, central Great Barrier Reef, Australia. J Sediment Petrol 58:845-856

Gall J, Bernier P, Gaillard C, Barale G, Bourseau J, Buffetaut E, Wenz S (1985) Influence du développement d'un voile algaire sur la sédimentation et la taphonomie des calcaires des lithographiques. Example du gisement de Cerin (Kimmridgien Supérieur, Jura Méridional francais). CR Acad Sci Paris 301:547-552

Gall J, Larsonneur L (1972) Séquences et environements sedimentaires dans la Baie des Veys (Manche). Rev Geogr Geol Dyn 14:189-204

Gallardo VA (1977) Large benthic microbial communities in sulphide biota under the Peru-Chile subsurface countercurrent. Nature 268:331-332

Garber RA, Levy Y, Friedman GM (1987) The sedimentology of the Dead Sea. Carbonates and Evaporites 2:43-57

Gardner JV (1975) Late Pleistocene dissolution cycles in the eastern equatorial Atlantic. In: Silter WV, Bé AWH, Berger WH (eds) Dissolution of deep-sea carbonates. Cushman Found Foraminifer Res Spec Publ 13:129-141

Gardner JV (1982) High-resolution carbonate and organic-carbon stratigraphies for the late Neogene and Quaternary from the western Carribean and eastern equatorial Pacific. In: Prell WL, Gardner JV et al. Init Repts DSDP 68: Washington (US Govt Print Office), pp 347-364

Gardner JV, Dean WE, Wilson CR (1984) Carbonate and organic carbon cycles and the history of upwelling at DSDP Site 532, Walvis Ridge, South Atlantic Ocean. In: Hay WW, Sibuet JC et al. Init Repts DSDP 75: Washington (US Govt Print Office), pp 905-921

Garrett P (1970) Phanerozoic stromatolites: noncompetitive ecologic restriction by grazing and burrowing animals. Science 169:171-173

Garrison RE (1967) Pelagic limestones of the Oberalm Beds (Upper Jurassic-Lower Cretaceous), Austrian Alps. Bull Can Petrol Geol 15:21-49

Garrison RE (1981) Diagenesis of oceanic carbonate sediments: a review of the DSDP perspective. Soc Econ Paleontol Mineral Spec Publ 32:181-207

Garrison RE, Fischer AG (1969) Deep-water limestone and radiolarites of the alpine Jurassic. Soc Econ Paleontol Mineral Spec Publ 14:20-56

Garrison RE, Kastner M, Zenger DH (eds) (1984) Dolomites of the Monterey Formation and other organic-rich units. Pacific Section SEMP, Los Angeles, 215 pp

Garrison RE, Kastner M, Koldny Y (1987) Phosphorites and phosphatic rocks in the Monterey Formation and related Miocene units, coastal California. In: Ingersoll RV, Ernst WG (eds) Cenozoic basin development in coastal California, vol 6. Prentice-Hall, New Jersey, pp 348-381

Garrison RE, Kastner M, Reimers CE (1990) Miocene phosphogenesis in California. In: Burnett WC, Riggs S (eds) Genesis of Neogene to modern phosphorites. Cambridge University Press, Cambridge (in press)

Garrison RE, Kennedy WJ (1977) Origin of solution seams and flaser structures in the Upper Cretaceous chalks of southern England. Sediment Geol 19:107-137

Gautier DL (1982) Siderite concretions: indicators of early diagenesis in the Gammon Shale (Cretaceous). J Sediment Petrol 52:859-871

Gautier DL, Claypool GE (1984) Interpretation of methanic diagenesis in ancient sediments by analogy with processes in modern diagenetic environments. In: McDonald DA, Surdam RC (eds) Clastic diagenesis. Am Assoc Petrol Geol Mem 37:111-123

Gebhard G (1982) Glauconitic condensation through high energy events in the Albian near Clars (SE France). In: Einsele G, Seilacher A (eds) Cyclic and event stratification. Springer, Berlin Heidelberg New York, pp 286-298

Geeslin JH, Chafetz HS (1982) Ordovician Aleman ribbon cherts: an example of silification prior to carbonate lithification. J Sediment Petrol 52:1283-1293

Genthon C, Barnola JM, Raynaud D, Lorius C, Jouzel J, Barkov NI, Korotkevich YS, Kotlyakov VM (1987) Vostok ice core: climatic response to CO_2 and orbital forcing changes over the last climatic cycle. Nature 329:414-418

Gerdes G, Krumbein WE (1987) Biolaminated deposits. (Lect Notes Earth Sci 9), Springer, Berlin Heidelberg New York, 183 pp

Gerdes G, Krumbein WE, Reineck HE (1985a) The depositional record of sandy, versicolored tidal flats (Mellum Island, southern North Sea). J Sediment Petrol 55:265-278

Gerdes G, Krumbein WE, Reineck HE (1985b) Verbreitung und aktuogeologische Bedeutung mariner mikrobieller Matten im Gezeitenbereich der Nordsee. Facies 12:75-96

Gerrard AJ (1988) Rocks and landforms. Unwin Hyman, London, 319 pp

Geyh M, Merkt J, Müller H (1971) Sediment, Pollen und Isotopenanalysen an jahreszeitlich geschichteten Ablagerungen im zentralen Teil des Schleinsees. Arch Hydrobiol 69 (3):366-399

Gieskes JM (1981) Deep sea drilling intersitial-water studies: implications for chemical alteration of the ocean crust, layers I and II. In: Warme JE, Douglas RG, Winterer EL (eds) The Deep Sea Drilling Project: a decade of progress. Soc Econ Paleontol Mineral Spec Publ 32:149-167

Gignoux M (1950) Sédimentation rythmique dans les plaines maritimes et au fond des mers. CR Acad Sci D Paris 230 (8):693-698

Gilbert GK (1895) Sedimentary measurement of geologic time. J Geol 3:121-127

Ginsburg RN (1971) Landward movement of carbonate mud: new model for regressive cycles in carbonates. Bull Am Assoc Petrol Geol 55:340-340

Giraudi C (1989) Lake levels and climate for the last 30,000 years in the Fucino area (Abruzzo-Central Italy) - a review. Palaeogeogr Palaeoclimatol Palaeoecol 70:249-260

Glaser KS, Jordan JE, Vail PR (1990) Tectonic and eustatic controls on the carbonate stratigraphy of the Leonardian-Guadalupian (Permian) section, NW Delaware Basin, New Mexico and Texas. Am Assoc Petrol Geol Annu Meet Abstr, San Francisco (in press)

Glass BP (1969) Reworking of deep-sea sediments as indicated by the vertical dispersion of the Australasian and Ivory Coast microtektite horizons. Earth Planet Lett 6:409-415

Glass BP (1982) Possible correlations between tektite events and climatic changes? In: Silver LT, Schultz PH (eds) Geological implications of impacts of large asteroids and comets on the earth. Geol Soc Am Spec Pap 190:251-256

Gleick J (1987) Chaos, making a new science. Viking-Penguin, New York, 352 pp

Glenister LM, Kauffman EG (1985) High resolution stratigraphy and depositional history of the Greenhorn hemicyclothem, Rock Canyon anticline, Pueblo, Colorado. In: Fine-grained deposits and biofacies of the Cretaceous Western Interior Seaway. Soc Econ Paleontol and Mineral, Tulsa, pp 170-183

Glicken H (1986) Rockslide-debris avalanches of May 18, 1980 Mt. St. Helens Volcano. Thesis, University of California, Santa Barbara

Glikson H, Taylor GH (1985) Cyanobacterial mats: major contributors to the organic matter in Toolebuc Formation oil shales. Geol Soc Aust Spec Publ 12:273-286

Gluyas JG (1984) Early carbonate diagenesis within Phanerozoic shales and sandstones of the NW European shelf. Clay Min 19:309-321

Gocht H (1973) Einbettungslage und Erhaltung von Ostracoden-Gehäusen im Solnhofener Plattenkalk (Unter-Tithon, SW-Deutschland). N Jb Geol Paläontol Monatsh 1973:189-206

Goldhaber MB, Kaplan IR (1974) The sulfur cycle in the sea. In: Goldberg (ed) The sea, marine chemistry, vol 5, John Wiley, New York, pp 569-655

Goldhammer RK, Dunn DA, Hardie LA (1987) High-frequency glacio-eustatic sea-level oscillations with Milankovitch characteristics recorded in Middle Triassic platform carbonates in northern Italy. Am J Sci 287:853-892

Goldsmith IR, King P (1987) Hydrodynamic modelling of cementation patterns in modern reefs. In: Marshall JD (ed) Diagenesis of sedimentary sequences. Blackwell Scientific, London, pp 1-13

Goll RM, Bjorklund KR (1971) Radiolaria in surface sediments of the North Atlantic Ocean. Micropaleontology 17:434-454

Golubic S, Krumbein W, Schneider J (1979) The carbon cycle. In: Trudinger PA, Swaine, DJ (eds) Biogeochemical cycling of mineral-forming elements (Studies in environmental science 3) Elsevier, Amsterdam, pp 29-45

Golubic S, Schneider J (1979) Carbonate dissolution. In: Trudinger PA, Swaine DJ (eds) Biogeochemical cycling of mineral-forming elements. (Studies in environmental science 3), Elsevier, Amsterdam, pp 107-129

Goodwin PW, Anderson EJ (1985) Punctuated aggradational cycles: a general hypothesis of episodic stratigraphic accumulation. J Geol 93:515-533

Goodwin PW, Anderson EJ, Goodman WM, Saraka LJ (1986) Punctuated aggradational cycles: implications for stratigraphic analysis. Paleoceanography 1:417-429

Gore PJW (1988) Lacustrine sequences in an Early Mesozoic rift basin: Cilpeper Basin, Virginia. In: Fleet AJ, Kelts K, Talbot MR (eds) Lacustrine petroleum source rocks. Geol Soc Spec Publ 40:247-278

Gore PJW (1989) Toward a model for open- and closed-basin deposition in ancient lacustrine sequences: the Newark supergroup (Triassic-Jurassic), eastern North America. Palaeogeogr Palaeoclimatol Palaeoecol 70:29-51

Görler K, Reutter KJ (1968) Entstehung und Merkmale der Olisthostrome. Geol Rdsch 57:484-514

Gornitz V, Lebedeff S, Hansen J (1982) Global sea level trend in the past century. Science 215:1611-1614

Gould SJ, Eldridge N (1977) Punctuated equilibria: the tempo and mode of evolution reconsidered. Paleobiology 3 (2):115-151

Govean FM, Garrison RE (1981) Significance of laminated and massive diatomites in the upper part of the Monterey Formation, California. In: Garrison RE, Pisciotto KA (eds) The Monterey Formation and related siliceous rocks in California. Soc Econ Paleontol Mineral Pacific Sect Spec Publ 15:181-198

Graham NE, White WB (1988) The El Niño cycle: a natural oscillator of the Pacific Ocean-atmosphere system. Science 240:1293-1302

Grande L (1980) Paleontology of the Green River Formation, with a review of the fish fauna. Geol Surv Wyoming Bull 63:1-325

Grant WD (1986) The continental shelf bottom boundary layer. Ann Rev Fluid Mech 18:265-305

Grant WD, Madsen OS (1979) Combined wave and current interaction with a rough bottom. J Geophys Res 84 (C4):1797-1808

Gray DI, Benton MJ (1982) Multidirectional paleocurrents as indicators of shelf storm beds. In: Einsele G, Seilacher A (eds) Cyclic and event stratification. Springer, Berlin Heidelberg New York, pp 350-353

Greenlee SM, Schroeder FM, Vail PR (1988) Seismic stratigraphy and geohistory analysis of Tertiary strata from the continental shelf off New Jersey: calculation of eustatic fluctuations from stratigraphic data. In: Sheridan RE, Grow JA (eds) The Atlantic continental margin. US Geol Soc Am Bull I-2:437-444

Grobe H (1986) Sedimentation processes on the Antarctic continental margin at Cape Norvegia during the late Pleistocene. Geol Rdsch 75:97-104

Grobe H (1987) Facies classification of glacio-marine sediments in the Antarctic. Facies 17:99-109

Groiß JTh (1967) Mikropaläontologische Untersuchungen der Solnhofener Schichten im Gebiet um Eichstätt (südliche Frankenalb). Erlanger Geol Abh 66:75-93

Grotzinger JP (1986a) Cyclicity and paleoenvironmental dynamics, Rocknest Platform, northwest Canada. Bull Geol Soc Am 97:1208-1231

Grotzinger JP (1986b) Upward shallowing platform cycles: a response to 2.2 billion years of low-amplitude, high-frequency (Milankovitch band) sea level oscillations. Paleoceanography 1:403-416

Group ROCC (Arthur MA, Bottjer DJ, Dean WE, Fischer AG, Hattin DE, Kauffman EG, Pratt LM, Scholle PA) (1986) Rhythmic bedding in Upper Cretaceous pelagic carbonate sequences: varying sedimentary response to climatic forcing. Geology 14:153-156

Gründel J, Rösler HJ (1963) Zur Entstehung der oberdevonischen Kalkknollengesteine Thüringens. Geologie 12:1009-1037

Guidish TM, Lerche I, Kendall CGSTC, O'Brien JJ (1984) Relationship between eustatic sea level changes and basement subsidence. Bull Am Assoc Petrol Geol 68:164-177

Gustavson TC (1975) Sedimentation and physical limnology in proglacial Malaspina Lake, southeast Alaska. In: Jopling AV, McDonald BC (eds) Glaciofluvial and glaciolacustrine sedimentation. Soc Econ Paleontol Mineral Spec Publ 23:249-263

Gustavson TC (1978) Bed forms and stratification types of modern gravel meander lobes, Nueces River, Texas. Sedimentology 25:401-426

Guzzetta G (1984) Kinematics of stylolite formation and physics of the pressure solution process. Tectonophysics 101:383-394

Gygi RA (1981) Oolitic iron formations: marine or not marine? Eclogae Geol Helv 74:233-254

Haak AB, Schlager W (1989) Compositional variations in calciturbidites due to sea-level fluctuations, Late Quaternary, Bahamas. Geol Rdsch 78:477-486

Haas J (1982) Facies analysis of the cyclic Dachstein Limestone Formation (Upper Triassic) in the Bakony Mountains, Hungary. Facies 6:75-84

Haas J, Dobosi K (1982) Investigation of Upper Triassic cyclic carbonate rocks in key sections in the Bakony. Annu Rep Hung Geol Inst 1980:135-168

Habib D (1979) Sedimentary origin of North Atlantic Cretaceous palynofacies. In: Talwani M, Hay WW, Ryan WBF (eds) Deep drilling results in the Atlantic Ocean: continental margins and paleoenvironments. Am Geophys Union, Maurice Ewing Ser 3:420-437

Habib D (1982) Sedimentary supply origin of Cretaceous black shales. In: Schlanger SO, Cita MB (eds) Nature and origin of Cretaceous carbon-rich facies. Academic Press, London, pp 113-127

Habib D (1983) Sedimentation-rate-dependent distribution of organic matter in the North Atlantic Jurassic-Cretaceous. In: Sheridan RE, Gradstein FM et al. Init Repts DSDP 76: Washington (US Govt Printing Office), pp 781-794

Hacket WR, Houghton BF (1989) A facies model for a Quaternary andesitic composite volcano: Ruapehu, New Zealand. Bull Volcanol 51:51-68

Haddon M, Wear RG, Packer HA (1987) Depth and density of burial by the bivalve *Paphies ventricosa* as refuges from predation by the crab *Ovalipes catharus*. Mar Biol 94:25-30

Haff E ten (1956) Significance of convolute lamination. Geol Mignbouw 18:188-194

Hagdorn H (1982) The "Bank der kleinen Terebrateln" (Upper Muschelkalk, Triassic) near Schwäbisch Hall (SW-Germany) - a tempestite condensation horizon. In: Einsele G, Seilacher A (eds) Cyclic and event stratification. Springer, Berlin Heidelberg New York

Hagdorn H, Simon T (1988) Geologie und Landschaft des Hohenloher Landes. Jan Thorbecke, Sigmaringen, 192 pp

Halfman JD (1987) High-resolution sedimentology and paleoclimatology of Lake Turkana, Kenya. PhD Diss, Duke University, Durham, North Carolina, 188 pp

Halfman JD, Johnson TC (1988) High-resolution record of cyclic climatic change during the past 4 ka from Lake Turkana, Kenya. Geology 16:496-500

Hallam A (1964) Origin of the limestone-shale rhythms in the Blue Lias of England: a composite theory. J Geol 72:157-168

Hallam A (1977) Secular changes in marine inundation of USSR and North America through the Phanerozoic. Nature 269:769-772

Hallam A (1981a) A revised sea-level curve for the early Jurassic. J Geol Soc Lond 138:735-743

Hallam A (1981b) Facies interpretation and the stratigraphic record. Freeman, Oxford, 291 pp

Hallam A (1984) Pre-Quaternary sea-level changes. Annu Rev Earth Planet Sci 12:205-244

Hallam A (1985) A review of Mesozoic climates. J Geol Soc Lond 142:433-445

Hallam A (1986) Origin of minor limestone-shale cycles: climatically induced or diagenetic? Geology 14:609-612

Hallam A (1987) Mesozoic marine organic-rich shales. In: Brooks J, Fleet AJ (eds) Marine petroleum source rocks. Geol Soc Am Spec Publ 26, pp 251-261

Hallermeister RJ (1981) A profile zonation for seasonal sand beaches from wave climate. Coast Engr 4:253-277

Halley RB (1987) Burial diagenesis of carbonate rocks. Colo Sch Mines 82:1-15

Halley RB, Harris PM (1979) Fresh-water cementation of a 1,000-year-old oolite. J Sediment Petrol 49:969-988

Hallock P (1988) The role of nutrient availability in bioerosion: consequences to carbonate buildups. Palaeogeogr Palaeoclimatol Palaeoecol 63:275-291

Hallock P, Schlager W (1986) Nutrient excess and the demise of coral reefs and carbonate platforms. Palaios 1:389-398

Hamilton EL (1976) Variations of density and porosity with depth in deep sea sediments. J Sediment Petrol 46:280-300

Hammer CU, Clausen HB, Dansgaard W (1980) Greenland ice sheet evidence of post-glacial volcanism and its climatic impact. Nature 288:230-235

Hammer CU, Clausen HB, Friedrich WL, Tauber H (1987) The Minoan eruption of Santorini in Greece dated to 1645 BC? Nature 328:517-519

Hampton MA (1972) The role of subaqueous debris flow in generating turbidity currents. J Sediment Petrol 42:775-793

Hampton MA (1979) Buoyancy in debris flows. J Sediment Petrol 49:753-758

Hancock JM, Scholle PA (1975) Chalk of the North Sea. In: Woodland AW (ed) Petroleum and continental shelf of north-west Europe, I. Applied Science Publishers, Barking, Essex, England, pp 413-425

Hand SC, Somero GN (1983) Energy metabolism pathways of hydrothermal vent animals: adaptions to a food-rich and sulfide-rich deep-sea environment. Biol Bull 165:167-181

Handford CR (1986) Facies and bedding sequences in shelfstorm-deposited carbonates - Fayetteville Shale and Pitkin Limestone (Mississippian), Arkansas. J Sediment Petrol 56:123-137

Hanson TA (1988) Early Tertiary radiation of marine molluscs and the long-term effects of the Cretaceous-Tertiary extinction. Paleobiology 14:37-51

Haq BU (1990) Sequence stratigraphy and sea level change. Int Sediment Assoc Spec Publ (in press)

Haq BU, Hardenbohl J, Vail P (1987a) Chronology of fluctuating sea levels since the Triassic. Science 235:1156-1167

Haq BU, Hardenbol J, Vail P (1987b) The chronology of fluctuating sea level since the Triassic. Science 269:483-489

Haq BU, Hardenbol J, Vail PR (1988) Mesozoic and Cenozoic chronostratigraphy and eustatic cycles. In: Wilgus C, Hastings B, Ross C, Posamentier H, Van Wagoner J, Kendall CGStC (eds) Sea-level changes: an integrated approach. Soc Econ Paleontol Mineral Spec Publ 42:71-108

Hardenbol J, Vail PR, Ferrer J (1981) Interpreting paleoenvironments, subsidence history and sea-level changes on passive margins from seismic biostratigraphy. 26th Int. Geol Congr, Geology of Continental Margins. Oceanol Acta (Suppl 4):33-44

Hardie LA, Bosellini A, Goldhammer RK (1986) Repeated subaerial exposure of subtidal carbonate platforms, Triassic, northern Italy: evidence for high frequency sea level oscillations on a 104 year scale. Paleoceanography 1:447-457

Hardie LA, Ginsburg RN (1977) Layering: the origin and environmental significance of lamination and thin bedding. In: Hardie LA (ed) Sedimentation on the modern carbonate tidal flats of northwest Andros Island, Bahamas. Johns Hopkins University Press, Baltimore, pp 50-123

Harland WB, Cox AV, Llewellyn PG, Pickton CAG, Smith AG, Walters R (1982) A geologic time scale. Cambridge University Press, Cambridge, 131 pp

Harmon RS, Schwarcz HP, Ford DC (1978) Late Pleistocene sea-level history of Bermuda. Quat Res 9:205-218

Harms JC, Southard JB, Walker RG (1982) Structures and sequences in clastic rocks. Soc Econ Paleontol Mineral Short Course Notes 9

Harris MT (1988) Margin and foreslope deposits of the Latemar carbonate buildup (Middle Triassic), the Dolomites, northern Italy. PhD Thesis, Johns Hopkins University, Baltimore, 473 pp

Harris PM (1979) Facies anatomy and diagenesis of a Bahamian ooid shoal. Sedimenta 7:163 pp

Hart MB (1987) Orbitally induced cycles in the chalk facies of the United Kingdom. Cretac Res 8:335-348

Hart MB, Bigg PJ (1981) Anoxic events in the Late Cretaceous chalk seas of northwest Europe. In: Neale JW, Brasier MD (eds) Microfossils from recent and fossil shelf seas. Ellis Horwood, London, pp 177-185

Hartman O, Barnard JL (1958) The benthic fauna of the deep basins off southern California, Parts I and II. Allen Hancock Pacific Exp 22:297 pp

Hattin DE (1962) Stratigraphy of the Carlile Shale (Upper Cretaceous) in Kansas. Kansas State Geol Surv Bull 156:155

Hattin DE (1971) Widespread, synchronously deposited, burrow-mottled limestone beds in Greenhorn Limestone (Upper Cretaceous) of Kansas and central Colorado. Am Assoc Paleontol Geol Bull 55:412-431

Hattin DE (1975) Stratigraphy and depositional environment of Greenhorn Limestone (Upper Cretaceous) Kansas. Kansas Geol Surv Bull 209:128

Hattin DE (1982) Statigraphy and depositional environments of the Smoky Hill Chalk Member of the Niobara Formation (Upper Cretaceous) of the type area, Western Kansas. Kansas Geol Surv Bull 225:108

Hattin DE (1985) Distribution and significance of widespread, time-parallel pelagic limestone beds in the Greenhorn Limestone (Upper Cretaceous) of the central Great Plains and southern Rocky Mountains. In: Pratt L, Kauffman E, Zelt F (eds) Fine-grained deposits and biofacies of the Cretaceous Western Interior Seaway-evidence of cyclic sedimentary processes. Soc Econ Paleontol Mineral Fieldtrip Guidebook 4:28-37

Hattin DE (1986a) Interregional model for deposition of Upper Cretaceous pelagic rhythmites, U.S. Western Interior. Paleoceanography 1:483-494

Hattin DE (1986b) Carbonate substrates of the Late Cretaceous sea, central Great Plains and southern Rocky Mountains. Palaios 1:347-367

Hay R (1959) Formation of the crystal-rich glowing avalanche deposits of St Vincent. J Geol 67:540-562

Hay WW (1983) The global significance of regional Mediterranean Neogene paleoenvironmental studies. In: Mulenkamp JE (ed) Reconstruction of marine paleoenvironments. Utrecht Micropa Bull 30:9-23

Hay WW (1988) Paleoceanography: a review fo the GSA centennial. Geol Soc Am Bull 100:1934-1956

Hay WW, Sibuet JC et al. (1984) Init Repts DSDP 75: Washington (US Govt Print Office), 1303 pp

Hayes MO (1967) Hurricanes as geological agents; case studies of hurricane Carla, 1961, and Cindy, 1963. Bur Econ Geol, Univ Texas Rept Invest 61: 56

Hayes MO (1979) Barrier island morphology as a function of tidal and wave regime. In: Leatherman S (ed) Barrier islands from the Gulf of St. Lawrence to the Gulf of Mexico, Academic Press, San Diego, pp 1-27

Hays JD, Imbrie J, Shackelton NJ (1976) Variations in the Earth's orbit - pacemaker of the ice ages. Science 194:1121-1132

Hays JD, Saito T, Opdyke ND, Burckle LH (1969) Pliocene-Pleistocene sediments of the equatorial Pacific: their paleomagnetic, biostratigraphic, and climatic record. Geol Soc Am Bull 80:1481-1514

Heald MT (1955) Stylolites in sandstones. J Geol 63:101-114

Heath GR, Moberly R (1971) Cherts from the western Pacific, Leg 7, Deep Sea Drilling Project. In: Winterer EL, Ewing JI et al. Init Repts DSDP 7: Washington (US Govt Printing Office), pp 991-1007

Heath GR, Moore TC, Dauphin JP (1977) Organic carbon in deep-sea sediments. In: Anderson NR, Malahoff A (eds) The fate of fossil fuel CO_2 in the oceans. Plenum, New York, 605-625

Heath KC, Mullins HT (1984) Open-ocean off-bank transport of fine-grained carbonate sediment in the northern Bahamas. In: Stow DAV, Piper DJW (eds) Fine-grained sediments: deep-water processes and facies. Geol Soc Lond Spec Publ 14:199-208

Heckel PH (1974) Carbonate buildups in the geological record: a review. In: Laporte LF (ed) Reefs in time and space. Soc Econ Paleontol Mineral Spec Publ 18:90-154

Heckel PH (1977) Origin of black phosphatic shale facies in Pennsylvanian cyclothems of the midcontinent, North America. Am Assoc Paleontol Geol Bull 61:1045-1068

Heckel PH (1983) Diagenetic model for carbonate rocks in Midcontinent Pennsylvanian eustatic cyclothems. J Sediment Petrol 53:733-759

Heckel PH (1984) Changing concepts of Midcontinent Pennsylvanian cyclothems, North America. Neuvième Congrès International de Stratigraphie et de Géologie Carbonifère, Washington and Champaign-Urbana, Compte Rendu 3:535-553

Heckel PH (1986) Sea level curve for Pennsylvanian eustatic marine transgressive-regressive depostional cycles along midcontinent outcrop belt, North America. Geology 14:330-334

Hecker R (1948) Der jurassische Faunen- und Florenfundpunkt im Karatau. In: Hecker R, Rjabinin, Rammelmayer, Filipova (eds) Ein fossiler jurassischer See im Karatau-Gebirge. Trudy Pal Inst, Moskau Leningrad 15:1-115

Hedberg HD (ed) (1976) International stratigraphic guide. Wiley, New York, 200 pp

Hein JR, Scholl DW, Barron JA, Jones MG, Miller J (1978) Diagenesis of Late Cenozoic diatomaceous deposits and formation of the bottom simulating reflector in the southern Bering Sea. Sedimentology 25:155-181

Hein JR, Vallier TL, Allan MA (1981) Chert petrology and geochemistry, Mid-Pacific Mountains and Hess Rise, DSDP Leg 62. In: Thiede J, Vallier TL et al. Init Repts DSDP 62: Washington (US Govt Print Office), pp 711-748

Hemleben C (1977a) Autochthone und allochthone Sedimentanteile in den Solnhofener Plattenkalken. N Jb Geol Palaeontol Monatsh 1977:257-271

Hemleben C (1977b) Rote Tiden und die oberkretazischen Plattenkalke im Libanon. N Jb Geol Palaeontol Monatsh 1977:239-255

Hemleben C, Freels D (1977a) Algen-laminierte und gradierte Plattenkalke in der Oberkreide Dalmatiens (Jugoslawien). N Jb Geol Palaeontol Abh 154:61-93

Hemleben C, Freels D (1977b) Fossilführende dolomitisierte Plattenkalke aus dem "Muschelkalk superior" bei Montral (Provinz Tarragona, Spanien). N Jb Geol Palaeontol Abh 154:186-212

Hennessy J, Knauth LP (1985) Isotopic variations in dolomite concretions from the Monterey Formation, California. J Sediment Petrol 55:120-130

Henningsmoen G (1974) A comment. Origin of limestone nodules in the Lower Paleozoic of the Oslo region. Nor Geol Tidsskr 54:401-412

Henrich R (1989): Glacial/interglacial cycles in the Norwegian Sea: sedimentology, paleoceanography and evolution of Late Pliocene to Quaternary Northern Hemisphere climate. In: Eldholm O, Thiede J, Taylor E (eds) Proc. ODP Sci Results 104, College Station, Texas, pp 189-217

Henrich R, Kassens H, Vogelsang E, Thiede J (1989) Sedimentary facies of glacial-interglacial cycles in the Norwegian Sea during the last 350 ka. Mar Geol 86:283-319

Henterich K, Sarnthein M (1984) Brunhes time scale: tuning by rates of calcium-carbonate dissolution and cross spectral analysis with solar insolation. In: Berger A, Imbrie J, Hays J, Kukla G, Saltzman B (eds) Milankovitch and climate. NATO Series C, Reidel, Dordrecht, 126(I): pp 447-466

Herbert TD, Fischer AG (1986) Milankovitch climatic origin of mid-Cretaceous black shale rhythms in central Italy. Nature 321:739-743

Herbert TD, Stallard RF, Fischer AG (1986) Anoxic events, productivity rhythms and the orbital signature in a mid-Cretaceous deep-sea sequence from central Italy. Paleoceanography 1:495-506

Herbin JP, Deroo G (1982) Sédimentologie de la matière organique dans les formations du Mésozoic de l'Atlantique Nord. Bull Soc Géol Fr 7:497-510

Herbin JP, Montadert L, Muller C, Gomez R, Thurow J, Wiedmann J (1986) Organic-rich sedimentation at the Cenomanian-Turonian boundary in oceanic and coastal basins in the North Atlantic and Tethys. In: Summerhayes CP, Shackleton NJ (eds) North Atlantic paleoceanography. Geol Soc Lond Spec Publ 22:389-422

Herterich K, Sarnthein M (1984) Brunhes time scale: tuning by rates of calcium carbonate dissolution and cross spectral analyses with solar insolation. In: Berger A, Imbrie J, Hays J, Kukla G, Saltzman B (eds) Milankovitch and climate. Reidel, Dordrecht, pp 446-466

Hesse R (1974) Long-distance continuity of turbidites: possible evidence for an Early-Cretaceous trench-abyssal plain in the East Alps. Bull Geol Soc Am 85:859-870

Hesse R (1975) Turbiditic and non-turbiditic mudstones of Cretaceous flysch sections of the East Alps and other basins. Sedimentology 22:387-416

Hesse R (1986) Early diagenetic porewater-sediment interaction. Geosci Can 13:165-196
Hesse R (1989) Diagenetic origin of chert; diagenesis of biogenic siliceous sediments. Geosci Can 15(3):171-192
Hesse R, Chough SK (1980) The Northwest Atlantic mid-ocean channel of the Labrador Sea II. Deposition of parallel laminated levee-muds from the viscous sublayer of low density turbidity currents. Sedimentology 27:697-711
Hickman CS (1983) Radular patterns, systematics, diversity, and ecology of deep-sea limpets. Veliger 26:73-92
Hieke W (1984) A thick Holocene homogenite from the Ionian Abyssal Plain (eastern Mediterranean). Mar Geol 55:63-78
Highsmith RC (1980) Geographic patterns of coral bioerosion: a productivity hypothesis. J Exp Mar Biol Ecol 46:177-196
Highsmith RC (1981) Coral bioerosion at Enewetak: agents and dynamics. Int Rev Ges Hydrobiol 66:335-375
Hilbrecht H, Hoefs J (1986) Geochemical and palaeontological studies of the δ 13 C anomaly in boreal and north Tethyan Cenomanian-Turonian sediments in Germany and adjacent areas. Palaeogeogr Palaeoclimatol Palaeoecol 53:169-189
Hildreth W (1983) The compositionally zoned eruption of 1912 in the Valley of Ten Thousand Smokes, Katmai National Park, Alaska. J Volcanol Geotherm Res 18:1-56
Hill PR (1984) Facies and sequence analysis of Nova Scotian slope muds: turbidite vs. "hemipelagic" deposition. In: Stow DAV, Piper DJW (eds) Fine-grained sediments: deep-water processes and facies. Geol Soc, Spec Publ 15:311-318
Hine AC, Wilber RJ, Bane JM, Neuman AC, Lorenson KR (1981b) Offbank transport of carbonate sands along open, leeward bank margins: northern Bahamas. Mar Geol 42:327-348
Hine AC, Wilber RJ, Neumann AC (1981a) Carbonate sand bodies along contrasting shallow bank margins facing open seaways in Northern Bahamas. Am Assoc Petrol Geol Bull 65:261-290 Part I
Hird K, Tucker ME (1988) Contrasting diagenesis of two Carboniferous oolites from South Wales: a tale of climatic influence. Sedimentology 35:587-602
Hiscott RN (1981) Deep-sea fan deposits in the Macigno Formation (middle-upper Oligocene) of the Gordana Valley, northern Appalachians, Italy Discussion. J Sediment Petrol 51:1015-1021
Hiscott RN, Pickering KT (1984) Reflected turbidity currents on an Ordovician basin floor, Canadian Appalachians. Nature 411:143-145
Hobday DK (1974) Beach- and barrier-island facies in the Upper Carboniferous of Northern Alabama. In: Briggs G (ed) Carboniferous of the Southeastern United States. Geol Soc Am Spec Pap 148:209-223
Hoefs J (1970) Kohlenstoff- und Sauerstoff-Isotopenuntersuchungen an Karbonatkonkretionen und umgebendem Gestein. Contrib Mineral Petrol 27:66-79
Hoffman PF (1989) Precambrian geology and tectonic history of North America. In: Bally AW, Palmer AR (eds) The geology of North America - an overview, vol A. Geol Soc Am, Boulder, Co, pp 447-512
Hogg SE (1982) Sheetfloods, sheetwash, sheetflow, or...? Earth Sci Rev 18:59-76
Hollander DJ (1989) Carbon and nitrogen isotopic cycling and organic geochemistry of eutrophic Lake Greifen: implications for preservation and accumulation of ancient organic carbon-rich sediments. PhD Thesis, ETH-Zürich, 315 pp
Holman RA (1981) Interfragravity energy in the surf zone. J Geophys Res 86 (C7):6442-6450
Holman RA, Bowen AJ (1982) Bars, bumps, and holes: models for the generation of complex beach topography. J Geophys Res 87 (C1):457-468
Holman RA, Sallenger AH Jr (1985) Setup and swash on a natural beach. J Geophys Res 90 (C1):945-953
Holser WT (1984) Gradual and abrupt shifts in ocean chemistry during Phanerozoic time. In: Holland HD, Trendall AF (eds) Patterns of change in Earth evolution. Springer, Berlin Heidelberg New York, pp 123-143
Holtkamp E (1985) The microbial mats of the Gavish Sabkha (Sinai). Diss Univ Oldenburg, 151 pp
Honjo S, Maganini SJ, Poppe LJ (1982) Sedimentation of lithogenic particles in the deep oceans. Mar Geol 50:199-220
Hooke RL (1967) Processes on arid region fluvial fans. J Geol 75:438-460
Horodyski RJ (1977) Lyngbya mats at Laguna Mormona, Baja California, Mexico: comparison with Proterozoic stromatolites. J Sediment Petrol 47:1305-1320

House MR (1985a) A new approach to an absolute time scale from measurements of orbital cycles and sedimentary microrhythms. Nature 316:721-725

House MR (1985b) Correlation of mid-Palaeozoic ammonoid evolutionary events with global sedimentary perturbations. Nature 313:17-22

Hsü KJ (1977) Studies of Ventura field, California. I Facies geometry and genesis of Lower Pliocene turbidites. Am Assoc Petrol Geol Bull 61:137-168

Hsü KJ, Jenkyns HC (eds) (1974) Pelagic sediments: on land and under the sea. Int Assoc Sediment Spec Publ 1, Blackwell, Oxford, 447 pp

Hsü KJ, Kelts KR (eds) (1984) Quaternary geology of Lake Zürich: an interdisciplinary investigation by deep-lake drilling. Contrib Sediment 13:210

Huang TC, Watkins ND, Shaw DM (1975) Atmospherically transported volcanic glass in deep-sea sediments: volcanism in sub-antarctic latitudes of the south Pacific during Late Pleistocene time. Geol Soc Am Bull 86:1305-1315

Hubbard RJ (1988) Age and significance of sequence boundaries on Jurassic and Early Cretaceous rifted continental margins. Am Assoc Petrol Geol Bull 72:49-72

Hudson JD (1977) Stable isotopes and limestone lithification. Geol Soc Lond Q J 133:637-660

Hudson JD (1978) Concretions, isotopes, and the diagenetic history of the Oxford Clay (Jurassic) of central England. Sedimentology 25:339-370

Hückel U (1970) Die Fischschiefer von Haqel und Hjoula in der Oberkreide des Libanon. N Jb Geol Paläontol Abh 135:113-149

Hückel U (1974) Vergleich des Mineralbestandes der Plattenkalke Solnhofens und des Libanon mit anderen Kalken. N Jb Geol Paläontol Abh 145:153-182

Hulsemann J, Emery KO (1961) Stratification in recent sediments of Santa Barbara Basin as controlled by organisms and water character. J Geol 69:279-290

Hurd DC (1972) Factors affecting solution rate of biogenous opal in seawater. Earth Planet Sci Lett 15:411-417

Hurd DC (1973) Interactions of biogenous opal sediment and seawater in the Central Equatorial Pacific. Geochim Cosmochim Acta 37:2257-2282

Hurd DC, Wernkam C, Pankratz HS, Fugate J (1979) Variable porosity in siliceous skeletons: determinations and importance. Science 203:1340-1343

Hut P, Alvarez W, Elder WP, Hansen T, Kauffman EG, Keller G, Shoemaker EM, Weissman PR (1987) Comet showers as a cause of mass extinctions. Nature 329:118-126

Hutschinson GE (1957) A treatise on limnology. Wiley, New York, 1015 pp

Iijima A, Kakuwa Y, Matsuda H (1988) Silicified wood from the Adoyama Chert, Kuzuh, central Honshu, and its bearing on compaction and depositional environment of radiolarian bedded chert. In: Hein JR, Obradovic J (eds) Siliceous deposits of the Tethys and Pacific regions. Springer, Berlin Heidelberg New York, pp 151-168

Iijima A, Kakuwa Y, Yanagimoto Y (1978) Shallow-sea organic origin of the Triassic bedded chert in central Japan. J Fac Sci, Univ Tokyo Sec II, 19:369-400

Iijima A, Matsumoto R, Tada R (1985) Mechanism of sedimentation of rhythmically bedded chert. Sediment Geol 41:221-233

Iijima A, Matsumoto R, Watanabe Y (1981) Geology and siliceous deposits in the Tertiary Setogawa Terrain of Shizuoka, central Honshu. J Fac Sci, Univ Tokyo Sec II, 20:241-276

Iijima A, Tada R (1981) Silica diagenesis of Neogene diatomaceous and volcanoclastic sediments in northern Japan. Sedimentology 28:185-200

Imbrie J (1985) A theoretical framework for the Pleistocene ice ages. J Geol Soc 142; 3:417-432

Imbrie J, Hays JD, Martinson DG, McIntyre A, Mix AC, Morley JJ, Pisias NG, Prell WL, Shackleton NJ (1984) The orbital theory of Pleistocene climate: support from a revised chronology of the marine $\delta^{18}O$ record. In: Berger A, Imbrie J, Hays J, Kukla G, Saltzman B (eds) Milankovitch and climate. NATO Ser C126 (I):269-305

Imbrie J, Imbrie JZ (1980) Modeling the climatic response to orbital variations. Science 207:943-953

Imbrie J, Imbrie KP (1979) Ice ages: solving the mystery, 2nd edn. Harvard Univ Press, Cambridge

Imoto N (1983) Sedimentary structures of Permian-Triassic cherts in the Tamba District, southwest Japan. In: Iijima A, Hein JR, Siever R (eds) Siliceous deposits in the Pacific region. (Development in sedimentology 36), Elsevier, Amsterdam, pp 377-393

Ingri J (1985) Geochemistry of ferromanganese concretions in the Barents Sea. Mar Geol 67:101-119

Inman DL, Nordstrom CE, Flick RE (1976) Currents in submarine canyons: an air-sea-land interaction. Annu Rev Fluid Mech 1976:275-310

Irwin H, Curtis C, Coleman M (1977) Isotopic evidence for the source of diagenetic carbonates formed during burial of organic-rich sediments. Nature 269:209-213

Isaacs CM (1981) Porosity reduction during diagenesis of the Monterey Formation, Santa Barbara Coastal Area, California. In: Garrison RE, Douglas RG, et al. (eds) The Monterey Formation and related siliceous rocks of California. Soc Econ Paleontol Mineral Pacific Section Spec Publ, pp 257-271

Izett GA (1987) The Cretaceous-Tertiary (K-T) boundary interval, Raton Basin, Colorado and New Mexico, and its content of shock-metamorphosed minerals: implications concerning the K-T boundary impact theory. US Geol Surv, Fed Center Denver, Open-file Rept 87-606:125 pp

Jablonski D (1986) Evolutionary consequences of mass extinction. In: Raup DM, Jablonski D (eds) Pattern and process in the history of life. Springer, Berlin Heidelberg New York, pp 313-329

Jackson MJ (1985) Mid-Proterozoic dolomitic varves and microcycles from the McArthur Basin, northern Australia. Sediment Geol 44:301-326

Jacobs JA (1984) Reversals of the Earth's magnetic field. Adam Hilger, Bristol, 230 pp

Jacquin T, de Gracianski PC (1988) Cyclic fluctuations of anoxia during Cretaceous time in the South Atlantic Ocean. Mar Petrol Geol 5:359-369

Jacquin T, Ravenne C, Vail PR (1990) Systems tracts and depositional sequences in a carbonate setting: study of continuous outcrops from platform to basin at the scale of seismic sections (in press)

Jahnke RA, Emerson SR, Roe KK, Burnett WC (1983) The present day formation of apatite in Mexican continental margin sediments. Geochim Cosmochim Acta 47:259-266

James NP (1983) Reef environment. In: Scholle PA, Bebout DG, Moore CH (eds) Carbonate depositional environments. Am Assoc Petrol Geol Mem 33:346-440

James NP (1984) Shallowing-upward sequences in carbonates. In: Walker RG (ed) Facies models, 2nd edn. Geosci Can Reprint Ser 1, Geol Assoc Can, Toronto, pp 213-228

James NP, Bone Y (1989) Petrogenesis of Cenozoic, temperate water calcarenites, South Australia: a model for meteoric/shallow burial diagenesis of shallow water calcite cements. J Sediment Petrol 59:191-203

Janecek TR, Rea DK (1984) Pleistocene fluctuations in northern hemisphere tradewinds and westerlies. In: Berger A, Imbrie J, Hays J, Kukla G, Saltzman B (eds) Milankovitsch and climate. NATO ASI Ser C, Reidel, Dordrecht, 126:331-348

Jannasch HW (1984a) Chemosymbiosis: the nutritional basis for life at deep-sea vents. Oceanus 27(3):73-78

Jannasch HW (1984b) Chemosynthetic microbial mats of deep-sea hydrothermal vents. In: Cohen Y, Castenholz RW, Halvorson HO (eds) Microbial mats: stromatolites. Alan Liss, New York, 498 pp

Jansa LF, Gardner JV, Dean WE (1978) Mesozoic sequences in the central North Atlantic. Init Repts DSDP 141: Washington (US Govt Print Office), pp 991-1031

Jansa LF, Enos P, Tucholke B, Gradstein FM, Sheridan RE (1979) Mesozoic Cenozoic sedimtary formations of the North Atlantic basin, Western North Atlantic. In: Talwani M, Hay WW, Ryan WBF (eds) Deep drilling results in the Atlantic Ocean: continental margin and paleoenvironment. Am Geophys Union, (Maurice Ewing Ser 3) pp 1-57

Jansen E, Befring S, Bugge T, Holtedahl H, Sejrup HP (1987) Large submarine slide on the Norwegian continental margin: sediments, transport and timing. Mar Geol 78(1/2):77-108

Jarrett RD, Costa JE (1985) Hydrology, geomorphology, and dam-break modeling of the July 15, 1982 Lawn Lake Dam and Cascade Lake Dam failures, Larimer County, Colorado. US Geol Surv Open-File Rep 84-612, 109 pp

Jarrett RD, Malde HE (1987) Paleodischarge of the Late Pleistocene Bonneville Flood, Snake River, Idaho, computed from new evidence. Geol Soc Am Bull 99:127-134

Jarvis I, Carson GA, Cooper MKE, Hart MB, Leary PN, Tocher BA, Horne D, Rosenfeld A (1988) Microfossil assemblages and the Cenomanian-Turonian (Late Cretaceous) oceanic anoxic event. Cretac Res 9:3-103

Javoy M, Courtillot V (1989) Intense acidic volcanism at the Cretaceous-Tertiary boundary. Earth Planet Sci Lett 94:409-416

Jeans CV (1980) Early submarine lithification in the Red Chalk and Lower Chalk of eastern England: a bacterial control model and its implications. Proc Yorkshire Geol Soc 43:81-157

Jefferies RPS (1963) The stratigraphy of the Actinocamax plenus subzone (Turonian) in the Anglo-Paris Basin. Proc Geol Assoc 74:1-33

Jefferies RPS, Minton P (1965) The mode of life of two Jurassic species of "Posidonia" (Bivalvia). Palaeontology 8:156-185

Jenkyns HC (1971) The genesis of condensed sequences in the Tethyan Jurassic. Lethaia 4:327-352

Jenkyns HC (1974) Origin of red nodular limestones (Ammonitico Rosso, Knollenkalke) in the Mediterranean Jurassic: a diagenetic model. Spec Publ Int Assoc Sediment 1:249-271

Jenkyns HC (1976) Sediments and sedimentary history of the Manihiki Plateau, South Pacific Ocean. In: Schlanger SO, Jackson ED et al. Init Repts DSDP 33: Washington (US Govt Print Office), pp 873-890

Jenkyns HC (1980) Cretaceous anoxic events: from continents to oceans. J Geol Soc Lond 137:171-188

Jenkyns HC, Clayton CJ (1986) Black shales and carbon isotopes in pelagic sediments from the Tethyan Lower Jurassic. Sedimentology 33:87-106

Jenkyns HC, Winterer EL (1982) Paleooceanography of Mesozoic ribbon radiolarites. Earth Planet Sci Lett 60:351-375

Jervey MT (1988) Quantitative geological modeling of siliciclastic rock sequences and their seismic expression. In: Wilgus C, Hastings S, Ross C, Posamentier H, Van Wagoner J, Kendall CGStC (eds) Sea-level changes: an integrated approach. Soc Econ Paleontol Mineral Spec Publ 42:47-69

Jessen W (1956a) Allgemeine Erkenntnisse aus feinstratigraphisch erarbeiteten Faunen- und Sediment-Zyklen des Ruhrkarbons. Geol Rdsch 45:119-128

Jessen W (1956b) Die marinen Sonderhorizonte unter Flöz Mausegatt (Unteres Westfal A) im Ruhrgebiet. Z Dtsch Geol Ges 107:73-82

Jessen W (1961) Zur Sedimentologie des Karbon mit Ausnahme seiner festländischen Gebiete. In: 4th Congr Int Stratigr Géol Carbonif, Heerlen 1958, CR 2:307-322

Johansson CE (1976) Structural studies of frictional sediments. Geogr Ann 58:201-300

Johnson AM (1984) Debris flows. In: Brunsden D, Prior DB (eds) Slope instability. Wiley, Chichester, pp 257-361

Johnson GAL (1984) Carboniferous sedimentary cycles in Britain controlled by plate movements. Neuvième Congrès Internat Stratigr Géologie Carbonifère, Washington and Champaign-Urbana, Compte Rendu 3:367-371

Johnson HD, Baldwin CT (1986) Shallow siliciclastic seas. In: Reading HG (ed) Sedimentary environments and facies, 2nd edn, Blackwell, Oxford, pp 229-282

Johnson JG, Klapper G, Sandberg CA (1985) Devonian eustatic fluctuations in Euramerica. Geol Soc Am Bull 96:567-587

Johnson ME (1989) Tempestites recorded as variable Pentamerus layers in the Lower Silurian of southern Norway. J Paleontol 63:195-205

Johnson TC, Halfman JD, Showers WJ (in press) A recent, anomalous departure in climate recorded in sediments of Lake Turkana, Kenya

Johnson TC, Hamilton EL, Berger WH (1977) Physical properties of calcareous ooze: control by dissolution at depth. Mar Geol 24:259-277

Jorgensen BB, Revsbech NP (1985) Diffusive boundary layers and the oxygen uptake of sediments and detritus. Limnol Oceanogr 30(1):111-122

Jouchoux A (1982) Les alternances pélagiques marne-calcaire du bassin vocontien (SE France): essai de caractérisation géochimique des bancs et interbancs dans le Valanginien supérieur. 9e Réunion Ann Sci Terre, Soc Géol Fr, Paris, 330 pp (Abstr)

Jouzel J, Lorius C, Petit JR, Genthon C, Barkov NJ, Kotlyakov VM, Petrov VM (1987) Vostok ice core: a continuous isotope temperature record over the last climatic cycle (160.000 years). Nature 329 (1):403-414

Jumars PA (1976) Deep sea diversity. J Mar Res 34:217-246

Kaiser NFJ (1979) Ein späteiszeitlicher Wald im Dättnau bei Winterthur, Schweiz. Ph D Thesis Univ Zürich, Ziegler Druck- und Verlags-AG, Winterthur, 90 pp

Kakuwa Y (1988) Geochemical study of Triassic to Jurassic bedded cherts in the Ashio, Mino and Tamba Terranes in Japan. Sci Pap Coll Arts Sci Univ Tokyo 38:19-41

Kalkowsky E (1908) Oolith und Stromatolith im norddeutschen Buntsandstein. Z Dtsch Geol Ges 60:68-125

Kamatani A, Ejiri N, Treguer P (1988) The dissolution kinetics of diatom ooze from the Antarctic area. Deep-Sea Res 35:1195-1203

Kammer TW, Brett CE, Boardman DR, Mapes RH (1986) Ecological stability of the dysaerobic biofacies during the Late Paleozoic. Lethaia 19:109-121

Kanari S, Fuji N, Horie S (1984) The paleoclimatological constituents of paleotemperature in Lake Biwa. In: Berger AL, Imbrie J, Hays J, Kukla G, Saltzman B (eds) Milankovitch and climate, Part 1. NATO ASI Ser C, Reidel, Dordrecht, 126:405-414

Kanasewich ER (1981) Time sequence analysis in geophysics. Univ Alberta Press, Alberta

Kaneps A (1973) Carbonate chronology for Pliocene deep-sea sediments. In: van Andel TH, Heath GR et al. Init Repts DSDP 16: Washington (US Govt Print Office), pp 873-881

Kantorowicz JD, Bryant ID, Dawans JM (1987) Controls on the geometry and distribution of carbonate cements in Jurassic sandstones: Bridgeport Sands, southern England and Viking Group, Troll Field, Norway. In: Marshall JD (ed) Diagenesis of sedimentary sequences. Blackwell Scientific, Oxford, pp 103-122

Karcz I (1972) Sedimentary structures formed by flash floods in southern Israel. Sediment Geol 7:161-182

Kashiwaya K, Yamoto A, Fukuyama K (1987) Time variations of erosional force and grain size in Pleistocene lake sediments. Quat Res 28:61-68

Kasper DC, Larue DK, Meeks YJ (1987) Fine-grained terrigenous turbidites in Barbados. J Sediment Petrol 57:440-448

Kastens KA, Cita MB (1981) Tsunami-induced sediment transport in the abyssal Mediterranean Sea. Geol Soc Am Bull 92:845-857

Kastner M (1981) Authigenic silicates in deep-sea sediments: formation and diagenesis. In: Emiliani C (ed) The sea, vol 7. Wiley Interscience, pp 915-980

Kastner M, Gieskes JM (1983) Opal-A to opal-CT transformation: a kinetic study. In: Iljima A, Hein JR, Siever R (eds) Siliceous deposits in the Pacific region. (Developments in sedimentology 36), Elsevier, Amsterdam, pp 211-227

Kastner M, Keene JB, Gieskes JM (1977) Diagenesis of siliceous oozes - I. Chemical controls on the rate of opal-A to opal-CT transformation - an experimental study. Geochim Cosmochim Acta 41:1041-1059

Katz A, Kolodny Y (1977) The geochemical evolution of the Pleistocene Lake Lisan-Dead Sea system. Geochim Cosmochim Acta 41:1609-1626

Katz BJ (1983) Limitations of Rock-Eval pyrolysis for typing organic matter. Org Geochem 4:195-199

Kauffman EG (1965) Collecting in concretions, nodules, and septaria. In: Raup DM, Kummel B (eds) Handbook of paleontologic techniques. Freeman, San Francisco, pp 175-194

Kauffman EG (1967) Coloradan macroinvertebrate assemblages, central Western Interior, United States. In: Kauffman EG, Kent HC (eds) Paleoenvironments of the Cretaceous Seaway in the Western Interior, a symposium. Golden, Colorado School of Mines, Spec Publ, pp 67-143

Kauffman EG (1970) Population systematics, radiometrics, and zonation: a new biostratigraphy. Proc N Am Paleontol Conv 1 F:612-666

Kauffman EG (1975) Dispersal and biostratigraphic potential of Cretaceous benthonic bivalvia in the Western Interior. In: Caldwell WGE (ed) The Cretaceous system in the Western Interior of North America. Geol Assoc Can Spec Pap 13:163-194

Kauffman EG (1976) British Middle Cretaceous inoceramid biostratigraphy. Ann Hist Nat Nice 4:1-11

Kauffman EG (1977a) Geological and biological overview: Western Interior Cretaceous basin. Mount Geol 14:75-100

Kauffman EG (1977b) Upper Cretaceous cyclothems, biotas, and environments, Rock Canyon anticline, Pueblo, Colorado. Mount Geol 14:129-152

Kauffman EG (1978) Benthic environments and paleoecology of the Posidonienschiefer (Toarcien). N Jb Geol Paleontol Abh 157:18-36

Kauffman EG (1981) Ecological reappraisal of the German Posidonienschiefer (Toarcian) and the stagnant basin model. In: Gray J, Boucot AJ, Berry WBN (eds) Communities of the past. Hutchinson Ross, Stroudsburg, pp 311-381

Kauffman EG (1982) Ecology and depositional environments of chalk-marl and limestone-shale rhythms in the Cretaceous of North America. In: Einsele G, Seilacher A (eds) Cyclic and event stratification. Springer, Berlin Heidelberg New York, p 97 (Abstr)

Kauffman EG (1983) A geological and paleoceanographic overview of Cretaceous history in the Western Interior Seaway of North America. In: Kauffman EG (ed) Depositional environments and paleoclimates of the Greenhorn tectonoeustatic cycle, Rock Canyon anticline, Pueblo, Colorado. Geol Soc Am, Penrose Conf Guidebook, pp 3-22

Kauffman EG (1984) Paleobiogeography and evolutionary response dynamic in the Cretaceous Western Interior Seaway of North America. In: Westermann GEG (ed) Jurassic-Cretaceous biochronology and paleogeography of North America. Geol Assoc Can Spec Pap 27:273-306

Kauffman EG (1986) High-resolution event stratigraphy: regional and global Cretaceous bioevents. In: Walliser OH (ed) Global bioevents. (Lect Notes Earth Sci 8), Springer, Berlin Heidelberg New York, pp 279-335

Kauffman EG (1988a) Concepts and methods of high-resolution event stratigraphy. Annu Rev Earth Planet Sci 16:605-654

Kauffman EG (1988b) The case of the missing community: low-oxygen adapted Paleozoic and Mesozoic bivalves ("flat clams") and bacterial symbioses in typical Phanerozoic seas. Geol Soc Am Centen Mtng, Denver, CO, Abstr with Programs:A48

Kauffman EG (1988c) The dynamics of marine stepwise mass extinction. In: Lamolda MA, Kauffman EG, Walliser OH (eds) Paleontology and evolution: extinction events. Rev Espanola Paleontol, No. Extraord, pp 57-71

Kauffman EG, Sageman BB, Gustason ER, Elder WP (1987) A field trip guidebook - high-resolution event stratigraphy Greenhorn cyclothem (Cretaceous: Cenomanian-Turonian), Western Interior basin of Colorado and Utah. Geol Soc Am, Rocky Mtn Sect Mtng, Boulder, CO, 198 pp

Kazmierczak J, Goldring R (1978) Subtidal flat-pebble conglomerates from the Upper Devonian of Poland: a multiprovenant high-energy product. Geol Mag 115:359-366

Keene JB (1975) Cherts and prolecanites from the North Pacific, Leg 32. In: Larson RL, Moberly R et al. Init Reps DSDP 32: Washington (US Govt Print Office), pp 553-570

Keigwin L (1986) Pliocene stable-isotope record of Site 606: sequential events of ^{18}O enrichment at 3.1 Ma. In: Ruddiman WF, Kidd RB, Thomas E et al. Init Repts DSDP 94: Washington (US Govt Print Office), pp 911-920

Keir RS (1983) Reduction of thermohaline circulation during deglaciation: the effect on atmospheric radiocarbon and CO_2. Earth Planet Sci Lett 64:445-456

Keir RS (1988) On the Late Pleistocene ocean geochemistry and circulation. Paleoceanography 3:413-445

Keir RS, Berger WH (1983) Atmospheric CO_2 content in the last 120,000 years: the phosphate-extraction model. J Geophys Res 88 (C10):6027-6038

Keir RS, Berger WH (1985) Late Holocene carbonate dissolution in the equatorial Pacific: reef growth or neoglaciation? In: Sundquist E, Broecker WS (eds) The carbon cycle and atmospheric CO_2: natural variations Archean to Present. Am Geophys Union Geophys Monogr 32:208-219

Keller G (1980) Middle to Late Miocene planktonic foraminiferal datum levels and paleoceanography of the north and southeastern Pacific Ocean. Mar Mikropaleontol 5:249-281

Keller G (1986) Stepwise mass extinction and impact events: Late Eocene to Early Oligocene. Mar Micropaleontol 10:267-293

Keller G (1989) Extended period of extinctions across the Cretaceous/Tertiary boundary in planktonic foraminifera of continental-shelf sections: implications for impact and volcanism theories. Geol Soc Am Bull 101:1408-1419

Keller J, Ryan WBF, Ninkovich D, Altherr R (1978) Explosive volcanic activity in the Mediterranean over the past 200.000 years as recorded in deep-sea sediments. Geol Soc Am Bull 89:591-604

Kelling G, Mullin PR (1975) Graded limestones and limestone-quartzite couplets: possible storm deposits from the Moroccan Carboniferous. Sediment Geol 13:161-190

Kellogg TB (1975) Late Quaternary climatic changes in the Norwegian-Greenland sea. In: Bowling SA, Weller G (eds) Climate of the Arctic. Univ Alaska, Fairbanks, pp 3-36

Kelts K (1988) Environments of deposition of lacustrine source rocks. In: Fleet AJ, Kelts K, Talbot MR (eds) Lacustrine petroleum source rocks. Geol Soc Lond Spec Publ 40:3-26

Kelts K, Arthur MA (1981) Turbidites after ten years of deep-sea drilling - wringing out the mop? In: Warme JE, Douglas RG, Winterer EL (eds) The Deep Sea Drilling Project: a decade of progress. Soc Econ Paleontol Mineral Spec Publ 32:91-127

Kelts K, McKenzie JA (1982) Diagenetic dolomite formation in Quaternary anoxic diatomaceous muds of Leg 64, Gulf of California. In: Curry JR, Moore DG et al. Init Repts DSDP 64: Washington (US Govt Print Office), pp 553-569

Kelts K, McKenzie J (1984) A comparison of anoxic dolomite from deep-sea sediments: Quaternary Gulf of California and Messinian Tripoli Formation of Silicy. In: Garrison RE, Kastner M, Zenger DH (eds) Dolomites of the Monterey Formation and other organic-rich units. Soc Econ Paleontol Mineral Pacific Section, pp 19-28

Kelts K, Talbot M (1990) Lacustrine carbonates as geochemical archives of environmental change and biotic-abiotic interactions. In: Tilzer MM, Serruya C (eds) Ecological structure and function in large lakes. Sci Tech, Madison (in press)

Kelts KR, Hsü KJ (1978) Freshwater carbonate sedimentation. In: Lerman A (ed) Lakes: chemistry geology physics. Springer, Berlin Heidelberg New York, pp 295-323

Kelts KR, Shahrabi M (1986) Holocene sedimentology of Hypersaline Lake Urmia, northwestern Iran. Palaeogeogr Palaeoclimatol Palaeoecol 54:105-130

Kempe S, Degens ET (1979) Varves in the Black Sea and in Lake Van (Turkey). In: Schlüchter CH (ed) Moraines and varves - origin/genesis/classification. A.A. Balkema, Rotterdam, pp 309-318

Kempe S, Liebezeit G, Dethlefsen V, Harms U (eds) (1988) Biochemistry and distribution of suspended matter in the North Sea and implications to fisheries biology. Mitt Geol Paläontol Inst Univ Hamburg 65:1-547

Kemper E (1987) Das Klima der Kreide-Zeit. Geol Jb Reihe A Heft 96, Geologische Bundesanstalt, Hannover, 402 pp

Kendall CGStC, Schlager W (1981) Carbonates and relative changes in sea level. Mar Geol 44:181-212

Kendall CGStC, Warren J (1987) A review of the origin and setting of tepees and their associated fabrics. Sedimentology 34:1007-1027

Kennedy WJ (1969) The correlation of the Lower Chalk of southeast England. Proc Geol Assoc 80:459-560

Kennedy WJ, Cobban WA (1976) Aspects of ammonite biology, biogeography, and biostratigraphy. Spec Pap Palaeontol 17, 94 pp

Kennedy WJ, Garrison RE (1975a) Morphology and genesis of nodular phosphates in the Cenomanian glauconite marl of south east England. Lethaia 8:339-360

Kennedy WJ, Garrison RE (1975b) Morphology and genesis of nodular chalks and hardgrounds in the Upper Cretaceous of southern England. Sedimentology 22:311-386

Kennedy WJ, Juignet P (1974) Carbonate banks and slump beds in the Upper Cretaceous (Upper Turonian-Santonian) of Haute Normandie, France. Sedimentology 21:1-42

Kennett JP (1977) Cenozoic evolution of Antarctic glaciation, the circum Antarctic Ocean, and their impact on global paleoceanography. J Geophys Res 82:3843-3859

Kennett JP (1982) Marine geology. Prentice-Hall, London, 813 pp

Kennett JP, Thunell RC (1975) Global increase in Quaternary volcanism. Science 187:497-503

Kennett JP, Thunell RC (1977) On explosive Cenozoic volcanism and climatic implications. Science 196:1231-1234

Kennett JP, von der Borch CC et al. (1986) Init Rept DSDP 90: Washington (US Govt Print Office), 1517 pp

Kent DV, Gradstein FM (1985) A Cretaceous and Jurassic geochronology. Bull Geol Soc Am 96:1419-1427

Keränen R (1984) Certain relationships between lake level variations and some climatic factors in Finland. In: Mürner NA, Karlen W (eds) Climatic changes on a yearly to millennial basis. Reidel, Amsterdam, pp 381-389

Kern JP (1980) Origin of trace fossils in Polish Carpathian flysch. Lethaia 13:347-362

Kerr RA (1981) Milankovitch climate cycles: old and unsteady. Science 213:1095-1096

Kerr RA (1988) Sunspot-weather link holding up. Science 242:1124-1125

Keupp H (1977) Ultrafazies und Genese der Solnhofener Plattenkalke (Oberer Malm, Südliche Frankenalb). Abh Nat Ges Nürnberg 37:1-128

Kidwell SM (1983) Stratigraphic taphonomy: predicting occurrence, time scales and modes of formation of fossil deposits. Geol Soc Am Abstr Prog 14:529

Kidwell SM (1984) Outcrop features and origin of basin margin unconformities in the lower Chesapeake Group (Miocene), Atlantic Coastal Plain. In: Schlee JS (ed) Interregional unconformities and hydrocarbon accumulation. Am Assoc Paleontol Geol Mem 36:37-58

Kidwell SM (1986a) Models for fossil concentrations: paleobiological implications. Paleobiology 12:6-24

Kidwell SM (1986b) Taphonomic feedback in the fossil record: testing the role of dead hardparts in benthic communities. Palaios 1:239-255

Kidwell SM (1988a) Reciprocal sedimentation and non-correlative hiatuses in marine-paralic siliciclastics: Miocene outcrop evidence. Geology 16:609-612

Kidwell SM (1988b) Taphonomic comparison of passive and active continental margins: Neogene shell beds of the Atlantic coastal plain and northern Gulf of California. Palaeogeogr Palaeoclimatol Palaeoecol 63:201-224

Kidwell SM (1989) Stratigraphic condensation of marine transgressive records: origin of major shell deposits in the Miocene of Maryland. J Geol 97:1-24

Kidwell SM (1990) Phanerozoic evolution of macroinvertebrate shell accumulations: preliminary data from the Jurassic of Britain. In: Miller WE III (ed) Paleocommunity temporal dynamics. Paleontol Soc Spec Publ (in press)

Kidwell SM, Aigner T (1985) Sedimentary dynamics of complex shell beds: implications for ecologic and evolutionary patterns. In: Bayer U, Seilacher A (eds) Sedimentary and evolutionary cycles. (Lect Notes Earth Sci 1), Springer, Berlin Heidelberg New York, pp 382-395

Kidwell SM, Jablonski D (1983) Taphonomic feedback: ecological consequences of shell accumulation. In: Tevesz MJS, McCall PL (eds) Biotic interactions in recent and fossil benthic communites. (Topic Geobiol 3), Plenum, New York, pp 195-248

Kier JS, Pilkey OH (1971) The influence of sea-level changes on sediment carbonate mineralogy, Tongue of the Ocean, Bahamas. Mar Geol 11:189-200

King LH, Fader G (1986) Wisconsinan glaciation of the continental shelf - southeast Atlantic Canada. Geol Surv Can Bull 363: 72

King LH, Rokoengen K, Gunsleiksrud T (1987) Quaternary seismostratigraphy of the Mid Norwegian Shelf, 650-670 30'N. - A till tongue stratigraphy. Inst Kontinent Sokkel Unders, Publ 114: 58

Kinsey DW (1985) Metabolism, calcification and carbon production. I. Systems level studies. Proc 5th Int Coral Reef Congr Tahiti 4, pp 505-526

Kirkland JI (1987) Integrated high-resolution event and biotic data using graphic correlation: examples from the mid-Cretaceous of Arizona and Colorado. Geol Soc Am Abstr with Program 19: 287

Klein GD deV (1985) The frequency and periodicity of preserved turbidites in submarine fans as a quantitative record of tectonic uplift in collision zones. Tectonophysics 119:181-193

Klöden KF (1828) Beiträge zur Mineralogischen und Geognostischen Kenntnis der Mark Brandenburg, I. Dieterici, Berlin

Knechtel MM, Patterson SH (1962) Bentonite deposits of the Northern Black Hills District Wyoming, Montana, and South Dakota. US Geol Surv Bull 1082-M:893-1030

Knight RJ, McLean JR (eds) (1986) Shelf sands and sandstones. Can Soc Petrol Geol Mem II: 347

Knoll AH (1985) A paleobiological perspective on sabkhas. In: Friedman GM, Krumbein WE (eds) Hypersaline ecosystems: the Gavish Sabkha (Ecological Studies 53). Springer, Berlin Heidelberg New York, pp 407-425

Knoll AH, Awramik SM (1983) Ancient microbial ecosystems. In: Krumbein WE (ed) Microbial geochemistry. Blackwell Scientific, Oxford, pp 287-315

Koch DL, Strimple HL (1968) A new Upper Devonian cystoid attached to a discontinuity surface. Iowa Geol Surv Rept Invest 5:1-49

Koeberl C, Glass BP (1988) Chemical composition of North American microtektites and tektite fragments from Barbados and DSDP site 612 on the continental slope off New Jersey. Earth Planet Sci Lett 87:286-292

Koepnick RB (1984) Distribution and vertical permeability of stylolites within a Lower Cretaceous carbonate reservoir, Abu Dhabi, United Arab Emirates. In: Yaha FA (ed) Stylolites and associated phenomena: relevance to hydrocarbon reservoirs. Abu Dhabi Nat Reser Res Found, Abu Dhabi, pp 261-277

Koerschner WF, Read JF (1989) Field and modelling studies of Cambrian carbonate cycles, Virginia Appalachians. J Sediment Petrol 59:654-687

Kolata DR, Frost JK, Huff WD (1986) K-bentonites of the Ordovician Decorah Subgroup, upper Mississippi Valley: correlation by chemical fingerprinting. Illinois Geol Surv Circ 537:30

Kolla V, Coumes F (1987) Morphology, internal structure, seismic stratigraphy, and sedimentation of the Indus fan. Am Assoc Petrol Geol Bull 71:650-677

Komar PD (1969) The channelized flow of turbidity currents with application to Monterey deep sea fan channel. J Geophys Res 74:4544-4558

Komar PD (1970) The competence of turbidity current flow. Geol Soc Amer Bull 81:1555-1562

Komar PD (1975) Nearshore currents: generation by obliquely incident waves and longshore variations in breaker heights. In: Hails J, Carr A (eds) Nearshore sediment dynamics and sedimentation. Wiley, London, pp 17-45

Komar PD (1976) Beach processes and sedimentation. Prentice-Hall, Englewood Cliffs, NJ, 429 pp

Komar PD (1983) Nearshore currents and sand transport on baches. In: Johns B (ed) Physical oceanography of coastal and shelf seas. Oceanography Series 35, Elsevier, Amsterdam, pp 67-109

Komar PD (1985) The hydraulic interpretation of turbidites from their grain sizes and sedimentary structures. Sedimentology 32:395-408

Kominz MA (1984) Oceanic ridge volumes and sea-level change - an error analysis. Interregional unconformities and hydrocarbon accumulation. Am Assoc Petrol Geol Mem 36:109-127

Kondo Y (1989) Bivalve orientation analysis as a method of biostratinomic interpretation of shallow sea fossiliferous facies. Abstr 28th Int Geol Congr 2:210

Kowallis BJ, Christiansen EH, Daino A (1989) Multi-characteristic correlation of Upper Cretaceous volcanic ash beds from southwestern Utah and central Colorado. Utah Geol Mineral Surv Misc Publ 89-5, 22 pp

Kranz PM (1974) The anastrophic burial of bivalves and its paleoecological significance. J Geol 82:238-265

Kraus NC, Dean JL (1987) Longshore sediment transport rate distribution measured by trap. Coast Seds '87, Am Soc Civ Engrs 1:881-896

Kreisa RD (1981) Storm-generated sedimentary structures in subtidal marine facies with examples from the Middle and Upper Ordovician of southwestern Virginia. J Sediment Petrol 51:823-848

Krone RV (1978) Aggregation of suspended particles in estuaries. In: Kjerfve B (ed) Estuarine transport processes. Univ South Carolina Press, Columbia, SC, pp 177-190

Krumbein WE (1983) Stromatolites - the challenge of a term in space and time. Precambrian Res 20:493-531

Krumbein WE (1987) Das Farbstreifensandwatt: Bau, Struktur und Erdgeschichte von Mikrobenmatten. In: Gerdes G, Krumbein WE, Reineck HE (eds) Mellum - Portrait einer Insel. Kramer, Frankfurt/Main, pp 170-187

Krumbein WE, Buchholz H, Franke P, Giani D, Giele C, Wonneberger K (1979) O_2 and H_2S coexistence in stromatolites. A model for the origin of mineralogical lamination in stromatolites and banded iron formations. Naturwissenschaften 66:381-389

Krumbein WE, Cohen Y, Shilo M (1977) Solar Lake (Sinai) 4 stromatolitic cyanobacterial mats. Limnol Oceanogr 22:635-656

Krumbein WE, Swart PK (1983) The microbial carbon cycle. In: Krumbein WE (ed) Microbial geochemistry. Blackwell Scientific Publ, Oxford, pp 5-62

Krusat G (1966) Beitrag zur Geologie und Paläontologie der Sierra del Monsech (Provinz L Lerida, Spanien). Thesis, Dept Geosciences, Berlin, 118 pp

Kuenen PH, Migliorini CI(1950) Turbidity currents as a cause of graded bedding. J Geol 58:91-127

Kuhn G, Meischner D (1988) Quaternary and Pliocene turbidites in the Bahamas, Leg 101. Sites 628, 632, and 635. In: Austin JA Jr, Schlager W et al. Proc ODP, Sci Results, 101, (Ocean Drilling Program) College Station, Texas, pp 203-212

Kuhnt W, Thurow J, Wiedmann J, Herbin JP (1986) Oceanic anoxic conditions around the Cenomanian/Turonian boundary and the response of the biota. Mitt Geol Pal Inst Univ Hamburg 60:205-246

Kutzbach JE, Otto-Bliesner BL (1982) The sensitivity of the African-Asian monsoonal climate to orbital parameter changes for 9000 years B.P. in a low-resolution general circulation model. J Atmos Sci 39:1177-1188

Kutzbach JE, Street-Perrott FA (1985) Milankovitch forcing of fluctuations in the level of tropical lakes from 18 to 0 ka BP. Nature 317:130-134

Kyle PR, Seward D (1984) Dispersed rhyolitic tephra from New Zealand in deep-sea sediments of the southern ocean. Geology 12:487-490

Labaume P, Mutti E, Seguret M, Rosell J (1983) Megaturbidites carbonatees du bassin turbiditique de l'Eocene inférieur et moyen sud-pyrénéen. Bull Soc Geol Fr 25:927-941

Lacroix A (1904) La Montagne Pelée et ses éruptions. Masson et Cie, Paris, 662 pp

Laferriere AP, Hattin DE, Archer AW (1987) Effects of climate, tectonics and sea-level changes on rhythmic bedding patterns in the Niobrara Formation (Upper Cretaceous), U.S. Western Interior. Geology 15:233-236

Lahann RW (1980) Smectite diagenesis and sandstone cement: the effect of reaction temperature. J Sediment Petrol 50:755-760

LaMarche VC, Hirschboeck KK (1984) Frost rings in trees as records of major volcanic eruptions. Nature 307:121-126

Lamb HH (1970) Volcanic dust in the atmosphere; with a chronology and assessment of its meteorological significance. Philos Trans R Soc Lond Ser A 266:425-533

Lamb HH (1984) Some studies of the Little Ice Age of recent centuries and its great storms. In: Mürner NA, Karlän W (eds) Climate changes on a yearly to millennial basis. Reidel, Dordrecht, pp 309-329

Lambert A, Hsü KJ (1979) Non-annual cycles of varve-like sedimentation in Walensee, Switzerland. Sedimentology 26:453-461

Lambert AM, Kelts KR, Marshall NF (1976) Measurements of density underflows from Walensee, Switzerland. Sedimentology 23:87-105

Lancelot Y (1973) Chert silica diagenesis in sediments from the Central Pacific. In: Winterer EL, Ewing JI et al. Init Repts DSDP 17: Washington (US Govt Print Office), pp 377-405

Lancelot Y, Hathaway JC, Hollister CD (1972) Lithology of sediments from the western North Atlantic. In: Hollister CD, Ewing JI et al. Init Repts DSDP 11: Washington (US Govt Print Office), pp 901-950

Land LS, Dutton SP (1978) Cementation of a Pennsylvanian deltaic sandstone: isotopic data. J Sediment Petrol 48:1167-1176

Land LS, Milliken KL, McBride EF (1987) Diagenetic evolution of Cenozoic sandstones, Gulf of Mexico sedimentary basin. Sediment Geol 50:195-225

Lane NG (1973) Paleontology and paleoecology of the Crawfordsville fossil site (Upper Osagian, Indiana). University Calif Publ Geol Sci 99:141

Larson DW, Rhoads DC (1983) The evolution of infaunal communities and sedimentary fabrics. In: Tevesz MJS, McCall PL (eds) Biotic interactions in Recent and fossil benthic communities. Plenum, New York, pp 627-648

Larson RL, Schlanger SO et al. (1981) Init Repts DSDP 61: Washington (US Govt Print Office), 885 pp

Lash GG (1988) Sedimentology and evolution of the Martinsburg Formation (Upper Ordovician) fine-grained turbidite depositional system, central Appalachians. Sedimentology 35:429-447

Lawrence DT, Doyle M, Aigner T (1990) Stratigraphic simulation of sedimentary basins: concepts and calibration. Am Assoc Petrol Geol Bull 74:273-295

Lawrence JR, Dreyer JI, Anderson TF, Brueckner HK (1979) Importance of alteration of volcanic material in the sediments of Deep Sea Drilling Site 323: chemistry, 18O/16O and 87Sr/86Sr. Geochim Cosmochim Acta 47:573-588

Lawson DE (1982) Mobilisation, movement and deposition of active subaerial sediment flow, Matanuska Glacier, Alaska. J Geol 90:279-300

Layer PW, Hall CM, York D (1987) The derivation of 40Ar/39Ar age spectra of single grains of hornblende and biotite by laser step-heating. Geophys Res Lett 14:757-760

Leary PN, Cottle RA, Ditchfield P (1989) Milankovitch control of foraminiferal assemblages from the Cenomanian of southern England. Terra Nova, 1: 416-419, Oxford

Leckie DA, Krystinik LF (1989) Is there evidence for geostrophic currents preserved in the sedimentary record of inner to middle shelf deposits? J Sediment Petrol 59:862-870

Leckie M (1985) Foraminifera of the Cenomanian-Turonian boundary interval, Greenhorn Formation, Rock Canyon Anticline, Pueblo, Colorado. In: Pratt LM, Kauffman EG, Zelt FB (eds) Fine-grained deposits and biofacies of the Cretaceous Western Interior Seaway: evidence of cyclic sedimentary processes. Soc Econ Pal Min Fieldtrip Guidebook 4:139-150

Leclaire L (1974) Late Cretaceous and Cenozoic pelagic deposits - paleoenvironment and paleoceanography of the Central Western, Indian Ocean. In: Simpson ESW, Schlich R et al. Init Repts DSDP 25: Washington (US Govt Print Office), pp 481-505

Ledbetter MT (1985) Tephrochronology of marine tephra adjacent to Central America. Geol Soc Am Bull 96:77-82

Ledbetter MT, Sparks RSJ (1979) Duration of large-magnitude explosive eruptions deduced from graded bedding in deep-sea ash layers. Geology 7:240-244

Le Doeuff D (1977) Rythmes et contournements synsédimentaires en série carbonatée alternante. Reconstitution paléomorphologique au Crétacé inférieur dans les chaines subalpine méridionales. Thèse 3e cycle, Univ Paris

Leeder MR (1982) Sedimentology: process and product. Allen and Unwin, London, 344 pp

Leeder MR, Strudwick AE (1987) Delta-marine interactions: a discussion of sedimentary models for Yoredale-type cyclicity in the Dinantian of northern England. In: Miller J, Adams AE, Wright VP (eds) European Dinantian environments. Wiley, New York, pp 115-129

Leinen M, Cwienk D, Heath GR, Biscaye PE, Kolla V, Thiede J, Dauphin JP (1986) Distribution of biogenic silica and quartz in recent deep-sea sediments. Geology 14:199-203

Leonard EM (1985) Glaciological and climatic controls on lakes sedimentation, Canadian Rocky Mountains. Ztsch Gletscherkd Glazialgeol 21:35-42

Leppakoski E (1969) Transitory return of the benthic fauna of Bornholm Basin after extermination by oxygen insufficiency. Cah Biol Mar 10:163-172

Lerman A (ed) (1978) Lakes: chemistry geology physics. Springer, Berlin Heidelberg New York, 356 pp

Levert J (1989) Répartition géographique des mineraux argileux dans les sédiments mésozoiques du bassin subalpin: mise en évidence d'une diagenèse complexe. Thèse Univ Lyon, 143 pp

Levinton JS (1970) The paleoecological significance of opportunistic species. Lethaia 3:69-78

Linck O (1965) Stratigraphische, stratinomische und ökologische Betrachtung zu *Encrinus lilliformis* Lamarck. Jb Geol Landesamt, Bad-Württ 7:123-148

Linsley RM, Yochelson EL (1973) Devonian carrier shells (Euomphalidae) from North America and Germany. US Geol Surv Prof Pap 824:26

Lipman PW (1984) The roots of ash flow calderas in western North America: windows into the tops of granitic batholiths. J Geophys Res 89:8801-8841

Lipman PW, Mullineaux DR (eds) (1981) The 1980 eruptions of Mount St. Helens Washington. US Geol Survey Prof Pap 1250:844

Lipman PW, Normark WR, Moore JG, Wilson JB, Gutmacher C (1988) The giant submarine Alika debris slide, Mauna Loa, Hawaii. J Geophys Res 93:4279-4299

Lippmann F (1955) Ton, Geoden und Minerale des Barrème von Hoheneggelsen. Geol Rdsch 43:475-502

Lippmann F (1973) Sedimentary carbonate minerals. Springer, Berlin Heidelberg New York, 228 pp

Lippman TC, Holman RA (1989) Wave dissipation on a barred beach: a method for determining sand bar morphology. Contract Report CERC-89-1 Department of the Army, US Army Corps of Engineers 54

Lippolt HJ, Fuhrmann U, Hradetzky H (1986) 40Ar/39Ar age determinations on sanidines of the Eifel volcanic field (FRG): constraints on age and duration of a Middle Pleistocene cold period. Chem Geol 59:187-204

Lisitzin AP (1972) Sedimentation in the World Ocean. Soc Econ Paleontol Mineral Spec Publ 17:218

Lisitzin AP (1985) The silica cycle during the last ice age. Palaeogeogr Palaeoclimatol Palaeoecol 50:241-270

Lister GS (1984) Lithostratigraphy of Zübo sediments. In: Hsü KJ, Kelts KR (eds) Quaternary geology of Lake Zürich: an interdisciplinary investigation by deep-lake drilling. Contrib Sediment 13:31-58

Lloyd RM (1977) Porosity reduction by chemical compaction - stable isotope model. Abstr Am Assoc Petrol Geol Bull 61:809

Lo Bello P, Feraud G, Hall CM, York D, Lavina P, Bernat M (1987) 40Ar/39Ar step-heating and laser fusion dating of a Quaternary pumice from Neschers, Massif Central, France: the defeat of xenocrystic contamination. Chem Geol 66:61-71

Logan BW, Rezak R, Ginsburg RN (1964) Classification and environmental significance of algal stromatolites. J Geol 72:68-83

Logan BW, Semeniuk V (1976) Dynamic metamorphism: processes and products in Devonian carbonate rocks, Canning Basin, Western Australia. Geol Soc Aust Spec Publ 6:138

Loh H, Maul B, Prauss M, Riegel W (1986) Primary production, maceral formation and carbonate species in the Posidonia Shale of NW Germany. In: Degens ET, Meyers PA, Brassel SC (eds) Biogeochemistry of black shales. Mitt Geol Palaeontol Inst Univ Hamburg 60:307-421

Lohse J, Kempe S, Liebezeit G (1989) Weiträumige Belastung der Nordsee. Geowissenschaften 7:151-156

Lorenz V (1974) On the formation of maars. Bull Volcanol 37:183-204

Lotter AF (1989) Evidence of annual layering in Holocene sediments of Soppensee, Switzerland. Aquat Sci 51:19-30

Loutit TS, Hardenbol J, Vail PR, Baum P (1988) Condensed sections: the key to age dating and correlation of continental margin sequences. In: Wilgus C, Hastings B, Ross C, Posamentier H, Van Wagoner J, Kendall CGStC (eds) Sea-level changes: an integrated approach. Soc Econ Paleontol Mineral Spec Publ 42:183-213

Lowe DR (1975) Water escape structures in coarse-grained sediments. Sedimentology 22:157-204

Lowe DR (1979) Sediment gravity flows: their classification and some problems of application to natural flows and deposits. Soc Econ Paleontol Mineral Spec Publ 27:75-82

Lowe DR (1982) Sediment gravity flows. II. Depositional models with special reference to the deposits of high-density turbidity currents. J Sediment Petrol 52:279-297

Lucas J, Prévôt L (1985) The synthesis of apatite by bacterial activity: mechanism. Sci Géol Mém 77:83-92

Ludlam SD (1979) Rhythmite deposition in lakes of the northeastern United States. In: Schlüchter C (ed) Moraines and varves - origin/genesis/classification. Balkema, Rotterdam, pp 295-302

Lutit TS, Hardenbol J, Vail PR, Baum GR (1988) Condensed sections: the key to age determination and correlation of continental margin sequences. In: Sea-level changes - an integrated approach. Soc Econ Paleontol Mineral Spec Publ 42:183-213

Lutz TM (1987) Limitations to the stastical analysis of episodic and periodic models of geologic time series. Geology 15:1115-1117

Luz B, Shackleton NJ (1975) CaCO$_3$ solution in the tropical east Pacific during the past 130,000 years. In: Sliter WV, Bé AWH, Berger WH (eds) Dissolution in deep-sea carbonates. Cushman Found Foraminiferal Res Spec Publ, US Natl Mus, Washington DC, 13:142-150

Lynts GW, Judd JB, Stehman CF (1973) Late Pleistocene history of Tongue of the Ocean, Bahamas. Geol Soc Am Bull 84:2665-2684

MacDonald DIM (1986) Proximal to distal sedimentological variation in a linear turbidite trough: implications for the fan model. Sedimentology 33:243-259

Machida H (1981) Tephrochronology and Quaternary studies in Japan. In: Self S, Sparks RJS (eds) Tephra studies. Reidel, Dordrecht, pp 161-192

Mack GH, James WC (1986) Cyclic sedimentation in the mixed siliciclastic-carbonate Abo-Hueco transitional zone (Lower Permian), southwestern New Mexico. J Sediment Petrol 56:635-647

Mackiewicz NE, Powell RD, Carlson PR, Molnia BF (1984) Interlaminated ice-proximal glacimarine sediments in Muir Inlet, Alaska. Mar Geol 57: 113-147

Maiklem WR (1968) Some hydraulic properties of bioclastic carbonate grains. Sedimentology 10:101-109

Maldonado A (1979) Upper Cretaceous and Cenozoic depositional processes and facies in the distal North Atlantic continental margin off Portugal, Site 398. In: Sibuet JC, Ryan WB et al. Init Repts DSDP 47(II): Washington (US Govt Print Office), pp 373-392

Malinverno A, Ryan WBF, Auffret G, Pautot G (1988) Sonar images of the path of recent failure events on the continental margin off Nice, France. In: Clifton HE (ed) Sedimentologic consequences of convulsive geologic events. Geol Soc Am Spec Pap 229:59-75

Maliva RG, Siever R (1989) Nodular chert formation in carbonate rocks. J Geol 97:421-433

Malouta DN, Gorsline DS, Thornten SE (1981) Processes and rates of recent (Holocene) basin filling in an active transform margin: Santa Monica Basin, California continental borderland. J Sediment Petrol 51:1077-1096

Manheim FT, Rowe GT, Jipa D (1975) Marine phosphorite formation off Peru. J Sediment Petrol 45:243-251

Manley PL, Flood RD (1988) Cyclic deposition within Amazon deep-sea fan. Am Assoc Paleontol Geol Bull 72:912-925

Mantz PA (1978) Bedforms produced by fine cohesionless, granular and flakey sediments under subcritical water flows. Sedimentology 25:83-103

Margaleff R (1973) Fitoplancton marino de la region de affloramiento del NW de Africa. Res Exp Cient B/O Cornide 2:65-94

Marsaglia KM, Klein G (1983) The paleogeography of Mesozoic storm depositional systems. J Geol 91:117-142

Marsh B (1981) On the crystallinity, probability of occurrence, and rheology of lava and magma. Contrib Mineral Petrol 78:85-98

Marsh OC (1867) On the origin of the so-called lignilitites or epsomites. Proc Am Assoc Adv Sci 16:135-143

Marshak S, Engelder T (1985) Development of cleavage in limestones of a fold-thrust belt in eastern New York. J Struct Geol 7:3-14

Marshall BA (1987) Osteopeltidae (Mollusca: Gastropoda): a new family of limpets associated with whale bone in the deep sea. J Molluscan Stud 53:121-127

Marshall JD (ed) (1987) Diagenesis in sedimentary sequences. Geol Soc Lond, Blackwell, Oxford, pp 41-54

Marshall JF, Davies PJ (1988) Halimeda bioherms of the northern Great Barrier Reef. Coral Reefs 6:139-148

Martini IP, Sagri M, Doveton JH (1978) Lithologic transition and bed thickness periodicities in turbidite successions of the Antola Formation, northern Apennines, Italy. Sedimentology 25:605-624

Martinson DG, Nicklas G, Pisias JD, Hays J, Imbrie TC, Moore JR, Nicholas J, Shackleton NJ (1987) Age dating and the orbital theory of the Ice Ages: development of a high resolution 0 to 300,000 year chronostratigraphy. Quat Res 27:1-29

Massari F (1983) Tabular cross-bedding in Messinian fluvial channel conglomerates, southern Alps, Italy. Spec Publ Int Assoc Sediment 6:287-300

Massari F, Sorbini L (1975) Aspects sédimentologiques des couches à poissons de l'Eocène de Bolca (Vérone - Nord Italie). IXe Congrès Internat Sédimentol, Nice 1975, Internat Assoc Sedimentologists, pp 55-61

Mather JA (1982a) Choice and competition: their effects on occupancy of shell homes by Octopus joubini. Mar Behav Physiol 8:285-293

Mather JA (1982b) Factors affecting the spatial distribution of natural populations of Octopus joubini Robson. Anim Behav 30:1166-1170

Matsumoto R, Matsuda H (1987) Occurrence, chemistry and isotopic composition of carbonate concretions in the Miocene to Pliocene siliceous sediments of Aomori, northeast Japan. J Fac Sci, Univ Tokyo Sec II, 21:351-377

Matter A (1974) Burial diagenesis of pelitic and carbonate deep-sea sediments from the Arabian Sea. In: Whitmarsh RB, Weser OE, Ross DA et al. Init Repts DSDP 23: Washington (US Govt Print Office), pp 421-470

Matter A, Douglas RG, Perch-Nielsen K (1975) Fossil preservation, geochemistry and diagenesis of pelagic carbonates from Shatsky Rise, northwest Pacific. In: Larson RL, Moberly R et al. Init Repts DSDP 32: Washington (US Govt Print Office), pp 891-922

Matti JC, McKee EH (1976) Stable eustacy, regional subsidence, and a carbonate factory: a self-generating model for onlap-offlap cycles in shallow-water carbonate sequences. Geol Soc Am Abstr 8:1000-1001

May RM (1989) Detecting density dependence in imaginary worlds. Nature 338:16-17

McCall PL, Tevesz MJS (eds) (1982) Animal-sediment relations. The biogenic alteration of sediment. Plenum, New York, 336 pp

McCave IN (1979) Depositional features of organic-carbon-rich black and green mudstones at Sites 386 and 387, Western North Atlantic. In: Tucholke B, Vogt P et al. Init Repts DSDP 43: Washington (US Govt Print Office), pp 411-416

McCave IN (1984) Erosion, transport and deposition of fine-grained marine sediments. Geol Soc Lond Spec Publ 16:35-69

McCave IN (1988) Biological pumping upwards of the coarse fraction of deep-sea sediments. J Sediment Petrol 58:148-158

McCave IN, Jones KPN (1988) Deposition of ungraded muds from high-density non-turbulent turbidity currents. Nature 333:250-252

McCave IN, Swift SA (1976) A physical model for the rate of deposition of fine-grained sediments in the deep sea. Bull Geol Soc Am 87:541-546

McClintock TS (1985) Effects of shell condition and size upon the shell choice behavior of a hermit crab. J Exp Mar Biol Ecol 88:271-285

McDougall I, Harris TM (1988) Geochronology and thermochronology by the method. Oxford Univ Press, New York, 212 pp

McDougall I, Schmincke HU (1976) Geochronology of Gran Canaria Canary Islands: age of shield-building volcanism and other magmatic phases. Bull Volcanol 40:1-21

McGhee GR (1981) Evolutionary replacement of ecological equivalents in Late Devonian benthic marine communities. Palaeogeogr Palaeoclimatol Palaeoecol 34:267-283

McGhee GR (1989) Catastrophes in the history of life. In: Allen KC, Briggs DEG (eds) Evolution and the fossil record. Belhaven, London, pp 26-50

McGhee GR, Bayer U (1985) The local signature of sea-level changes. In: Bayer U, Seilacher A (eds) Sedimentary and evolutionary cycles. (Lect Notes Earth Sci 1), Springer, Berlin Heidelberg New York, pp 98-112

McIlreath JA, James NP (1984) Carbonate slopes. In: Walker RG (ed) Facies models, 2nd edn. Geol Assoc Can, Geosci Reprint Ser 1, pp 245-257

McKee ED (1945) Stratigraphy and ecology of the Grand Canyon Cambrian, Part I. In: McKee ED, Resser CE (eds) Cambrian history of the Grand Canyon region. Carn Inst Wash Publ 563:5-168

McKee ED, Weir GW (1953) Terminology for stratification and cross stratification in sedimentary rocks. Am Assoc Geol Bull 64:381-390

McKelvey VE, Williams JS, Sheldon RP, Cressman ER, Cheney TM, Swanson RW (1959) The Phosphoria and Shedhorn formations in the western phosphate field. US Geol Surv Prof Pap 313-A:47

McKenzie JA, Bernouilli D, Garrison RE (1978) Lithification of pelagic-hemipelagic sediments at site 372: oxygen isotope alteration with diagenesis. In: Ross DA, Neprochnov YP et al. Init Repts DSDP 42 A: Washington (US Govt Print Office), pp 473-478

McKenzie JA, Hsü KJ, Schneider JF (1980) Movement of subsurface waters under the sabkha, Abu Dhabi, UAE, and its relation to evaporative dolomite genesis. In: Zenger DH, Dunham JB, Ethington RL (eds) Concepts and models of dolomitization. Soc Econ Paleontol Mineral Spec Publ 28:11-30

Mc Kinney ML (1985) Distinguish patterns of evolution from patterns of deposition. J Paleobiol 59:561-567

McLaren DJ (1982) Frasnian-Famennian extinctions. In: Silver LT, Schultz PH (eds) Geological implications of impacts of large asteroids and comets on the earth. Geol Soc Am Spec Pap 190:477-484

McLean R (1983) Gastropod shells: a dynamic resource that helps shape benthic community structure. J Exp Mar Biol Ecol 69:151-174
Meischner KD (1964) Allodapische Kalke, Turbidite in Riff-nahen Sedimentationsbecken. In: Bouma AH, Brouwer A (eds) Developments in sedimentology 3, Turbidites. Elsevier Scientific, Amsterdam, pp 156-191
Meischner KD (1967) Palökologische Untersuchungen an gebankten Kalken - ein Diskussionsbeitrag. Geol Fören Stockholm 89:465-469
Meldahl KH (1987) Sedimentologic and taphonomic implications of biogenic stratification. Palaios 2:350-358
Melvin J (1986) Upper Carboniferous fine-grained turbidite sandstones from SW England. J Sediment Petrol 56:19-34
Menard HW (1956) Archipelagic aprons. Am Assoc Petrol Geol Bull 40:2195-2210
Merino E, Ortoleva P, Strickholm P (1983) Generation of evenly-spaced pressure solution seams during (late) diagenesis. Contrib Mineral Petrol 82:360-370
Merriam DF (ed) (1964) Symposium on cyclic sedimentation. Kansas Geol Surv Bull 169(1/2), 636 pp
Meyers PA (1987) Synthesis of organic geochemical studies, Leg 93, North American continental margin. In: Wise SW Jr, Van Hinte JE et al. Init Repts DSDP 93: Washington (US Govt Print Office), pp 1333-1342
Middleton D (1960) Introduction to statistical communication theory. McGraw-Hill, New York, pp 604-610
Middleton GV, Hampton MA (1976) Subaqueous sediment transport and deposition by sediment gravity flows. In: Stanley DJ, Swift DJP (eds) Marine sediment transport and environmental management. Wiley, New York, pp 197-218
Milankovitch M (1941) Kanon der Erdbestahlung und seine Anwendung auf das Eiszeitproblem. Akad R Serbe 133, 633 pp
Miller FM, Byers CW (1984) Abundant and diverse early Paleozoicinfauna indicated by the stratigraphic record. Geology 12:40-43
Miller FX (1977) The graphic correlation method in biostratigraphy. In: Kauffman E, Hazel J (eds) Concepts and methods of biostratigraphy. Dowden, Hutchinson & Ross, Stroudsberg, pp 165-186
Miller KG, Fairbanks RG, Mountain GS (1987) Tertiary oxygen isotope synthesis, sea level history, and continental margin erosion. Paleoceanography 2(1):1-19
Miller MC, Komar PD (1977) The development of sediment threshold curves for unusual environments (Mars) and for inadequately studied materials (foram sands). Sedimentology 24:709-721
Mitchell JM Jr (1976) An overview of climatic variability and its causal mechanism. Quat Res 6:481-493
Mitchell JM Jr, Stockton CW, Meko DM (1979) Evidence of a 22-year rhythm of drought in the western United States related to the Hale solar cycle since the 17th century. In: McCormac BM, Seliga TA (eds) Solar-terrestrial influences on weather and climate. Reidel, Dordrecht, pp 125-144
Mitchum JM Jr, Vail PR, Thompson S III (1977) Part II. The depositional sequence as a basic unit for stratigraphic analysis. In: Payton CE (ed) Seismic stratigraphy - application to hydrocarbon exploration. Am Assoc Petrol Geol Mem 26:53-62
Mitsui K, Taguchi K (1977) Silica mineral diagenesis in Neogene Tertiary shales in the Tempoku district Hokkaido, Japan. J Sediment Petrol 47:158-167
Mizutani S (1970) Silica minerals in the early stage of diagenesis. Sedimentology 15:419-436
Möller NK, Kvingan K (1988) The genesis of nodular limestones in the Ordovician and Silurian of the Oslo Region (Norway). Sedimentology 35:405-420
Mörner NA (1984) Eustacy, geoid changes, and multiple geophysical interaction. In: Berggren WA, Van Couvering JA (eds) Catastrophes and Earth history. Princeton Univ Press, Princeton, N5, pp 395-415
Molina-Cruz A, Price P (1977) Distribution of opal and quartz on the ocean floor of the subtropical southeastern Pacific. Geology 5:81-84
Molnia BF (1983) Subarctic glacial-marine sedimentation: a model. In: Molnia BF (ed) Glacial-marine sedimentation. Plenum, New York, pp 95-144
Montadert L, Roberts DG et al. (1979) Init Repts DSDP 48: Washington (US Govt Print Office) 1183 pp
Monty C (1970) An autoecological approach of intertidal and deep water stromatolites. Ann Soc Geol Belg 94:265-276
Monty C (1981) Phanerozoic stromatolites. Springer, Berlin Heidelberg New York, 249 pp

Mooers CNK (1976) Introduction to the physical oceanography and fluid dynamics of continental margins. In: Stanley DJ, Swift DJP (eds) Marine sediment transport and environmental management. Wiley, New York, pp 7-22

Moon CF, Hurst CW (1984) Fabrics of muds and shales: an overview. In: Stow DAV, Piper DJW (eds) Fine-grained sediments: deep water processes and facies. Geol Soc Lond Spec Publ 15:579-593

Moore CH (1989) Carbonate diagenesis and porosity. (Developments in sedimentology 46), Elsevier, Amsterdam, 338 pp

Moore D (1959) Role of deltas in the formation of some British Lower Carboniferous cyclothems. J Geol 67:522-539

Moore DG (1969) Reflection profiling studies of the California continental borderland: structure and Quaternary turbidite basins. Geol Soc Am Spec Pap 107:142

Moore DG, Curray JR, Einsele G (1982) Salado-Vinorama submarine slide and turbidity current off southeast tip of Baja California. In: Curray JR, Moore DG et al. Init Repts DSDP 64: Washington (US Govt Print Office), pp 1071-1082

Moore DG, Scrutton PC (1957) Minor internal structures in some Recent unconsolidated sediments. Am Assoc Paleontol Geol Bull 91:2723-2751

Moore JG, Moore GW (1984) Deposits from a giant wave on the island of Lanai, Hawaii. Science 226:1312-1315

Moore JG, Nakamura K, Alcaraz A (1966) The 1965 eruption of Taal Volcano. Science 151:955-960

Moore RC (1936) Stratigraphic classification of the Pennsylvanian rocks of Kansas. Kansas Geol Surv Bull 22:1-256

Moore TC Jr (1969) Radiolaria: Change in skeletal weight and resistance to solution. Geol Soc Am Bull 80:2103-2108

Moore TC Jr, Pisias NG, Heath GR (1977) Climate changes and lags in Pacific carbonate preservation, sea surface temperature and global ice volume. In: Andersen NR, Malahoff A (eds) The fate of fossil fuel CO_2 in the oceans. Plenum, New York, pp 145-165

Moore TC Jr, Rabinowitz PD et al. (1984) Init Rept DSDP 74: Washington (US Govt Print Office), 894 pp

Morgenstern NR (1967) Submarine slumping and the initiation of turbidity currents. In: Richards AF (ed) Marine geotechnique. Univ Illinois Press, Urbana, pp 189-220

Morley JJ, Hays JD (1979) Comparision of glacial and interglacial oceanographic conditions in the South Atlantic from variations in calcium carbonate and radiolarian distribution. Quat Res 12:396-408

Morris KA (1979) A classification of Jurassic marine sequences: an example from the Toarcian (Lower Jurassic) of Great Britain. Palaeogeogr Palaeoclimatol Palaeoecol 26:117-126

Morris KA (1980) Comparison of major sequences of organic-rich mud deposition in the British Jurassic. J Geol Soc Lond 137:157-170

Morton RA (1981) Formation of storm deposits by wind-forced currents in the Gulf of Mexico and the North Sea. In: Nio S-D, Shuttenhelm RTE, van Weering Tj CE (eds) Holocene marine sedimentation in the North Sea Basin. Int Assoc Sediment Spec Publ 5:385-396

Morton RA, Price WA (1987) Late Quaternary sea-level fluctuations and sedimentary phases of the Texas coastal plain and shelf. In: Nummedal D, Pilkey OH (eds) Sea-level fluctuations and coastal evolution. Soc Econ Paleontol Mineral Spec Publ 41:181-198

Mossop GD (1972) Origin of the peripheral rim, Redwater Reef, Alberta. Bull Can Petrol Geol 20:238-280

Mossop GD (1979) The evaporites of the Ordovician Baumann Fiord Formation, Ellesmere Island, arctic Canada. Bull Geol Surv Can 298:52

Müller G, Irion G, Förstner U (1972) Formation and diagenesis of inorganic Ca-Mg carbonates in the lacustrine environment. Naturwissenschaften 59:158-164

Müller J, Fabricius F (1974) Magnesian-calcite nodules in the Ionian deep sea: an actualistic model for the formation of some nodular limestones. In: Hsü KJ, Jenkyns C (eds) Pelagic sediments. Int Assoc Sediment Spec Publ 1:235-247

Müller PJ, Erlenkeuser H, von Grafenstein R (1983) Glacial-interglacial cycles in oceanic productivity inferred from organic carbon centents in eastern North Atlantic sediment cores. In: Thiede J, Suess E (eds) Coastal upwelling, Part B. Plenum, New York, pp 365-398

Müller PJ, Suess E (1979) Productivity, sedimentation rate, and sedimentary organic matter in the oceans. I. Organic carbon preservation. Deep-Sea Res 26A:1347-1362

Mullins HT (1983) Comment to: "Eustatic control of turbidites and winnowed turbidites". Geology 11:57-58

Mullins HT, Cook HE (1986) Carbonate apron models: alternatives to submarine fan model for paleoenvironmental analysis and hydrocarbon exploration. J Sediment Geol 48:37-79

Mullins HT, Gardulski AF, Hine AC (1986) Catastrophic collapse of the west Florida carbonate platform margin. Geology 14:167-170

Mullins HT, Neumann AC, Wilber RJ, Boardman MR (1980) Nodular carbonate sediment on Bahamian slopes: possible precursors to nodular limestones. J Sediment Petrol 50:117-131

Mullins HT, Rasch RF (1985) Sea-floor phosphorites along the central California continental margin. Econ Geol 80:696-715

Murata KJ, Friedman I, Gleason JD (1977) Oxygen isotope relations between diagenetic silica minerals in Monterey Shale, Temblor Range, California. Am J Sci 277:259-272

Murata KJ, Larson RR (1975) Diagenesis of Miocene siliceous shale, Temblor Range, California. J Res US Geol Surv 3:553-566

Murata KJ, Nakata JK (1974) Cristobalitic stage in the diagenesis of diatomaceous shale. Science 184:567-568

Murata KJ, Norman MB (1976) An index of crystallinity for quartz. Am J Sci 276:1120-1130

Murray SP (1975) Trajectories and speeds of wind-driven currents near the coast. J Phys Ocean 5:347-360

Murray SP, Coleman JM, Roberts HH, Salama M (1980) Eddy currents and sediment transport off the Damietta Nile. 17 Coast Engr Conf Proc Am Soc Civil Engrs, Sydney, Australia II:1680-1699

Murray SP, Young M (1985) The nearshore current along a high-rainfall, trade-wind coast - Nicaragua. Estuarine Coastal Shelf Sci 21:687-699

Mutti E (1977) Distinctive thin-bedded turbidite faces and related depositional environments in the Eocene Hecho Group (south-central Pyrenees, Spain). Sedimentology 24:107-131

Mutti E (1985) Turbidite systems and their relation to depositional sequences. In: Zuffa GG (ed) Provenance in arenites. Reidel, Dordrecht, pp 65-93

Mutti E (1989) Submarine sand mounds and their relations to turbidite systems reworked by bottom currents. Giornale di Geologia, Bologna, 51, Suppl 4

Mutti E, Normark WR (1987) Comparing examples of modern and ancient turbidite systems: problems and concepts. In: Legett JK, Zuffa GG (eds) Marine clastic sedimentology. Graham and Trotman, London, pp 1-38

Mutti E, Ricci Lucchi F (1975) Turbidite facies and facies associations. IX Int Congr Sed Nice Fieldtrip Guidebook A-11:21-36

Mutti E, Ricci Lucchi F (1978) Turbidites of the northern Apennines: introduction to facies analysis. In: Nilsen T (ed) Am Geol Inst Reprint Ser 3:127-166 (translation of 1972 article in Italian)

Mutti E, Ricci Lucchi F, Seguret M, Zanzucchi G (1984) Seismoturbidites: a new group of resedimented deposits. Mar Geol 55:103-116

Mutti E, Sonnino M (1981) Compensation cycles: a diagnostic feature of sandstone lobes. Int Assoc Sed 2 Europ Mtg Bologna Abstr, pp 120-123

Myers KJ, Wignall PB (1987) Understanding Jurassic organic-rich mudrocks - new concepts using gamma-ray spectrometry and paleoecology: examples from the Kimmeridge Clay of Dorset and the Jet Rock of Yorkshire. In: Legget JK, Zuffa GG (eds) Marine clastic sedimentology. Graham and Trotman, London, pp 172-189

Naeser CW, Naeser ND (1988) Fission-track dating of Quaternary events. In: Easterbrook DJ (ed) Dating Quaternary sediments. Geol Soc Am Spec Pap 227:1-12

Nathan Y (1984) The mineralogy and geochemistry of phosphorites. In: Niagru JO, Moore PB (eds) Phosphate minerals. Springer, Berlin Heidelberg New York, pp 275-291

Neev D, Emery KO (1967) The Dead Sea, depositional processes and environments of evaporites. Geol Surv Israel Bull 41:147

Neff G, Pilmer IR, Botrill RS (1985) Submarine-fan deposited sandstone and rudite in a mid-Cenozoic interarc basin in Maewo, Vanuatu (New Hebrides). Sedimentology 32(4):519-542

Neftel A, Oeschger H, Schwander J, Stauffer B, Zumbrunn R (1982) Ice core sample measurements give atmospheric CO_2 content during the past 40,000 years. Nature 295:220-223

Negendank JFW (1989) Pleistozäne und holozäne Maarsedimente der Eifel. Z Dtsch Geol Gesell 140:13-24

Nelson CH (1982) Modern shallow-water graded sand layers from storm surges, Bering Shelf: a mimic of Bouma Sequences and turbidite systems. J Sediment Petrol 52 (2):537-545

Nelson CH, Maldonado A (1988) Factors controlling depositional patterns of Ebro turbidite system, Mediterranean Sea. Am Assoc Paleontol Geol Bull 72:698-716

Nelson CH, Normark WR, Bouma AH, Carlson PR (1978) Thin-bedded turbidites in modern submarine canyons and fans. In: Stanley DJ, Kelling G (eds) Sedimentation in submarine canyons, fans and trenches. Dowden, Hutchinson & Ross, Stroudsburg, pp 177-189

Nelson CS, Keane SL, Head PS (1988) Non-tropical carbonate deposits on the modern New Zealand shelf. Sediment Geol 60:71-94

Nelson DM, Gordon LI (1982) Production and pelagic dissolution of biogenic silica in the Southern Ocean. Geochim Cosmochim Acta 46:491-501

Neumann AC, Land LS (1975) Lime mud deposition and calcareous algae in the Bight of Abaco, Bahamas: a budget. J Sediment Petrol 45:763-786

Newell CR, Hidu H (1982) The effects of sediment type on growth rate and shell allometry in the soft shelled clam Mya arenaria L. J Exp Mar Biol Ecol 65:285-295

Newell RE (1985) Volcanism and climate. In: 1985 yearbook of science and the future. The new Encyclopaedia Britannica, Encycl Brit Inc, Chicago, pp 206-225

Nichols JA (1976) The effect of stable dissolved-oxygen stress on marine benthic invertebrate community diversity. Int Rev Gesamten Hydrobiol 61:747-760

Niedoroda AW, Swift DJP, Hopkins TS (1985) The shoreface. In: Davis RA Jr (ed) Coastal sedimentary environments, 2nd edn. Springer, Berlin Heidelberg New York, pp 533-624

Nielsen P (1979) Some basic concepts of wave sediment transport. Technical Univ Denmark, Inst Hydro Hydraul Engr, Ser Pap 20: 160

Niessen F, Kelts K (1989) The deglaciation and Holocene sedimentary evolution of southern perialpine Lake Lugano - implications for Alpine paleoclimate. Eclogae Geol Helv 82:235-263

Nilsen TH (1980) Modern and ancient submarine fans: discussion of papers by R.G. Walker and W.R. Normark. Am Assoc Paleontol Geol Bull 64:1094-1101

Ninkovich D, Sparks RSJ, Ledbetter MT (1978) The exceptional magnitude and intensity of the Toba eruption, Sumatra: an example of the use of deep-sea tephra layers as a geological tool. Bull Volcanol 41:286-298

Nipkow F (1927) Über das Verhalten der Skelette planktischer Kieselalgen im geschichteten Tiefenschlamm des Zürich- und Baldeggersee. PhD Thesis, ETH-Zürich 455, 49 pp

Nisbet E, Price I (1974) Siliceous turbidites: bedded cherts as redeposited ocean ridge derived sediments. In: Hsü KJ, Jenkyns HC (eds) Pelagic sediments on land and under the sea. Spec Publ Int Assoc Sediment 1:351-366

Noel D, Bréheret JG, Lambert B (1987) Enregistrement sédimentaire de floraisons phytoplantoniques calcaires en milieu confiné. Synthèse de données sur l'Actuel et observation géologiques. Bull Soc Géol Fr 8 III:1097-1106

Noel D, Manivit H (1978) Nannofaciès de "black shales" Aptiennes et Albiennes d'Atlantique sud (Legs 36 et 40). Intéret sédimentologique. Bull Soc Géol Fr 7; 20, 4:491-502

Non-metallic Mineral Recources Research Committee (1986) Basic research on the utility of cristobalitic rocks of Higashidouri Village, Aomori Prefecture. Sendai, 51 pp (in Japanese)

Norris RD (1986) Taphonomic gradients in shelf fossil assemblages: Pliocene Purisima Formation, California. Palaios 1:256-270

Notholt AJG (1980) Economic phosphatic sediments: mode of occurrence and stratigraphical distribution. J Geol Soc Lond 137:793-805

Nottrecht A, Kreisa RD (1987) Model for the combined flow origin of hummocky cross section. Geology 15:357-361

Nowroozi AA (1967) Table for Fisher's test of significance in harmonic analysis. Geophys J R Astro Soc 12:517-520

Nummedal D, Sonnenfeld DL, Taylor K (1984) Sediment transport and morphology at the surf zone of Presque Isle, Lake Erie, PA. Mar Geol 60:99-122

Nummedal D, Swift DJP (1987) Transgressive stratigraphy at sequence-bounding unconformities: some principles derived from Holocene and Cretaceous examples. In: Nummedal D, Pilkey OH (eds) Sea-level fluctuations and coastal evolution. Soc Econ Pal Min Spec Publ 41:241-260

O'Brien GW, Heggie D (1988) East Australian continental margin phosphorites. EOS Am Geophys Union Trans 69:2

O'Brien NR, Nakazawa K, Tokuhashi S (1980) Use of clay fabric to distinguish turbiditic and hemipelagic siltstones and silts. Sedimentology 27:47-61

O'Sullivan PE (1983) Annually-laminated lake sediments and the study of Quaternary environmental changes-a review. Quat Sci Rev 1:245-313

Oba T (1969) Biostratigraphy and isotopic paleotemperatures of some deep-sea cores from the Indian Ocean. Sci Rep Tohoku University Ser 2, 41:129-195

Oberhänsli H, Heinze P, Diester-Haass L, Wefer G (1990) Up-welling off Peru during the last 430.000 years and its relationship to the bottom water environment as deduced from coarse grain size distributions and analyses of benthic foraminifers at Sites 679 D, 680 B, and 681 B (ODP Leg 112)

Oberli F, Fischer H, Meier M (1989) High-resolution zircon dating of Tertiary pyroclastic sediments by low-level U-Pb techniques. Int Geol Congr Wash 9 Int Union Geol Sciences, Abstr 2: 536

Odin GS (ed) (1982) Numerical dating in stratigraphy. Wiley, Chichester

Oechsle E (1958) Stratigraphie und Ammoniten-Fauna der Sonninien-Schichten des Filsgebietes unter besonderer Berücksichtigung der sowerbyi-Zone. Paleontographica 111 A:47-129

Oertsen JA, Schlungbaum G (1972) Experimentell-ökologische Untersuchungen über O2-Mangel und H2S-Resistenz an marinen Evertebraten der westlichen Ostsee. Beitr Meereskd 29:79-91

Oeschger H, Beer J, Siegenthaler W, Stauffer W, Dansgaard W, Langway CC (1984) Late glacial climate history from ice cores. In: Hansen J, Takahashi T (eds) Climate processes and climate sensitivity. Geophys Monogr 29:299-306

Ogg JG, Haggerty J, Sarti M, von Rad U (1987) Lower Cretaceous pelagic sediments of Deep Sea Drilling Project Site 603, western North Atlantic: a synthesis. In: van Hinte JE, Wise SW et al. Init Repts DSDP 93: Washington (US Govt Print Office), pp 1305-1331

Ogura K (1987) Organic compounds in the 1400 m core sample of Lake Biwa. In: Horie S (ed) History of Lake Biwa. Inst Paleolimnol Paleoenviron Lake Biwa, Kyoto Univ, Contrib 553:157-169

Olausson E (1960) Sediment cores from the Indian Ocean. Swedish Deep-Sea Exped (1947-1948) Rept 9:53-88

Oliver JE, Fairbridge RW (eds) (1987) The encyclopedia of climatology. Van Nostrand Reinhold, New York

Olsen PE (1984) Periodicity of lake-level cycles in the Late Triassic Lockatong Formation of the Newark Basin, Newark Supergroup, New Jersey and Pennsylvania. In: Berger AL, Imbrie J, Hays J, Kukla G, Salzman B (eds) Milankovitch and climate, Part 1. Reidel, Dordrecht, pp 1129-1146

Olsen PE (1986) A 40-million-year lake record of Early Mesozoic orbital climatic forcing. Science 234:842-848

Olsson RK (1988) Foraminiferal modeling of sea-level change in the Late Cretaceous of New Jersey. In: Wilgus CK, Hastings BS, Posamentier H, Van Wagoner J, Ross CA, Kendall CGSC (eds) Sea level changes: an integrated approach. Soc Econ Paleontol Mineral Spec Publ 42:289-287

Opdyke NB (1972) Paleomagnetism of deep-sea cores. Rev Geophys Space Phys 10:213-249

Orth CJ, Attrep M Jr, Mao XY, Kauffman EG, Diner R, Elder WP (1988) Iridium abundance maxima in the Upper Cenomanian extinction interval. Geophys Res Lett 15:346-349

Oschmann W (1985) Faziesentwicklung und Provinzialismus in Nordfrankreich und Südengland zur Zeit des obersten Jura (Oberkimmeridge und Portland). Münch Geowiss Abh A 2:1-119

Oschmann W (1988a) Kimmeridge Clay sedimentation-a new cyclic model. Palaeogeogr Palaeoclimatol Palaeoecol 65:217-251

Oschmann W (1988b) Upper Kimmeridgian and Portlandian marine macrobenthic associations from southern England and northern France. Facies 18:49-82

Osgood RG (1970) Trace fossils of the Cincinnati area. Palaeontogr Am 6:281-444

Palmer AR (1971) The Cambrian of the Great Basin and adjacent areas, western United States. In: Holland CH (ed) Cambrian of the New World. Wiley, London, pp 1-78

Palmer AR (1982) Biomere boundaries: a possible test for extraterrestrial perturbation of the biosphere. In: Silver LT, Schultz PH (eds) Geological implications of impacts of large asteroids and comets on the earth. Geol Soc Am Spec Pap 190:469-476

Palmer AR (1983) Geologic time scale. The decade of North American geology. Geol Soc Am, Boulder, CO, 2 pp

Park J, Herbert TD (1987) Hunting for palaeoclimatic periodicities in a geologic time series with an uncertain time scale. J Geophys Res 92:14027-14040

Park R (1976) A note on the significance of lamination in stromatolites. Sedimentology 23:379-393

Parker FL (1954) Distribution of the foraminifera in the northeastern Gulf of Mexico. Bull Mus Comp Zool 111(10):453-588

Parker FL, Berger WH (1971) Faunal and solution patterns of planktonic foraminifera in surface sediments of the South Pacific. Deep-Sea Res 18:73-107

Parker G, Fukushima Y, Pantin HM (1986) Self accelerating turbidity currents. J Fluid Mech 171:145-181

Parker WE, Arthur MA, Wise SW, Wenkam CR (1983) Carbonate and organic carbon cycles in Aptian-Albian black shales at Deep Sea Drilling Project Site 511, Falkland Plateau. In: Ludwig WJ, Krasheninikov VA et al. Init Repts DSDP 71: Washington (US Govt Print Office), pp 1051-1070

Parkin DW (1974) Trade-winds during the glacial cycles. Proc R Soc Lond 337:73-100

Parnell P (1988) Significance of lacustrine cherts for the environment of source-rock deposition in the Orcadian Basin, Scotland. In: Fleet AJ, Kelts K, Talbot MR (eds) Lacustrine petroleum source rocks. Geol Soc Spec Publ 40:205-217

Parrish JT, Curtis RL (1982) Atmospheric circulation, upwelling, and organic-rich rocks in the Mesozoic and Cenozoic eras. Palaeogeogr Palaeoclimatol Palaeoecol 40:31-66

Parrish JT, Ziegler AM, Scotese CR (1982) Rainfall patterns and the distribution of coals and evaporites in the Mesozoic and Cenozoic. Palaeogeogr Palaeoclimatol Palaeoecol 40:67-101

Parrish JT, Ziegler AM, Humphreville RG (1983) Upwelling in the Paleozoic era. In: Thiede J, Suess E (eds) Coastal upwelling and its sediment record, Part B. Sedimentary records of ancient coastal upwelling. Plenum, New York, pp 553-578

Partheniades E (1972) Results of recent investigations on erosion and deposition of cohesive sediment. In: Shen HW (ed) Sedimentation, a symposium to honor Professor H.A. Einstein. Fort Collins, Colorado, pp 1-39

Paull KC, Hecker B, Commeau R, Freeman-Lynde RF, Neumann C, Corso WP, Golubic S, Hook JE, Sikes E, Curray J (1984) Biological communities at the Florida escarpment resemble hydrothermal vent taxa. Science 226:965-967

Paull KC, Hills SJ, Thierstein HR (1988) Progressive dissolution of fine carbonate particles in pelagic sediments. Mar Geol 81:27-40

Pearce JB (1965) On the distribution of *Tresus nuttalli* and *Tresus capax* (Pelecypoda: Mactridae) in the waters of Puget Sound and the San Juan Archipelago. Veliger 7:166-170

Pedersen TF (1983) Increased productivity in the eastern equatorial Pacific during the last glacial maximum (19.000 to 14.000 B.P.). Geology 11:16-19

Perch-Nielsen K, Supko PR et al. (1977) Init Repts DSDP 39: Washington (US Govt Print Office), 1139 pp

Percival SF, Fischer AG (1977) Changes in calcareous nannoplankton in the Cretaceous - Tertiary biotic crisis at Zumaya, Spain. Evol Theory 2:1-35

Perkins JA, Sims JD (1983) Correlation of Alaskan varve thickness with climatic parameters and use in paleoclimatic reconstruction. Quat Res 20:308-321

Perlmutter MA, Matthews MD (1990) Global cyclostratigraphy: a model. In: Cross TA (ed) Quantitative dynamic stratigraphy. Prentice Hall, Englewood Cliffs, New Jersey, pp 233-260

Pestiaux P, Berger A (1984a) An optimal approach to the spectral characterization of deep-sea climatic records. In: Berger A, Imbrie J, Hays J, Kukla G, Saltzman B (eds) Milankovitch and climate. Reidel, Dordrecht, NATO ASI Ser C 126:417-445

Pestiaux P, Berger A (1984b) Impacts of deep-sea processes on palaeoclimatic spectra. In: Berger A, Imbrie J, Hays J, Kukla G, Saltzman B (eds) Milankovitch and climate. Reidel, Dordrecht, NATO ASI Ser C 126: 493-510

Pestiaux P, Van der Mersch I, Berger A (1988) Paleoclimatic variability at frequencies ranging from 1 cycle per 10 000 years to 1 cycle per 1000 years: evidence for nonlinear behaviour of the climate system. Clim Change 12:9-37

Peterson F (1969) Four new members of the Upper Cretaceous Straight Cliffs Formation in the southeastern Kaiparowits region, Kane County, Utah. US Geol Surv Bull 1274-J:1-28

Peterson LC (1984) Late Quaternary deep-water paleoceanography of the eastern equatorial Indian Ocean: evidence from benthic Foraminifera, carbonate dissolution and stable isotopes. PhD Thesis, Brown University, Providence, RI, 429 pp

Peterson LC, Prell WL (1985a) Carbonate dissolution in recent sediments of the eastern equatorial Indian Ocean: preservation patterns and carbonate loss above the lysocline. Mar Geol 64:259-290

Peterson LC, Prell WL (1985b) Carbonate preservation and rates of climate change: an 800 kyr record from the Indian Ocean. In: Sundquist ET, Broecker WS (eds) The carbon cycle and atmospheric CO_2: natural variations Archean to Present. Am Geophys Union Geophys Monogr 32:251-284

Peybernes B, Oertli H (1972) La série de passage du Jurassique au Crétacé dans le Basin sud-pyrénéen (Espagne). CR Acad Sci Paris 274:33-48

Pfirman S, Solheim A (1989) Subglacial meltwater discharge in the open marine tidewater glacier environment: observations from Nordaustlandet, Svalbard Archipelago. Mar Geol 86:265-281

Picard MD, High LR Jr (1972) Criteria for recognizing lacustrine rocks. In: Rigby JK, Hamblin WK (eds) Recognition of ancient sedimentary environments, Soc Econ Paleontol Mineral Spec Publ 16:108-145

Picard MD, High LR Jr (1981) Physical stratigraphy of ancient lacustrine deposits. Soc Econ Paleontol Mineral Spec Publ 31:233-259

Pickering KT (1981) Two types of outer fan lobe sequence, from the Late Precambrian Kongsfjord Formation Submarine Fan, Finnmark, northern Norway. J Sediment Petrol 51:1277-1286

Pickering KT, Coleman J, Cremer M, Droz L, Kohl B, Normark WR, O'Connell S, Stow DAV, Meyer-Wright A (1986) A high-sinuosity, laterally migrating submarine fan channel-levee-overbank: results from DSDP Leg 96 on the Mississippi Fan, Gulf of Mexico. Mar Petrol Geol 3:3-18

Pierson TC (1981) Dominant particle support mechanisms in debris flows at Mt Thomas, New Zealand, and implications for flow mobility. Sedimentology 28:49-60

Pierson TC, Scott KM (1985) Downstream dilution of a Lahar: transition from debris flow to hyperconcentrated streamflow. Water Resour Res 21:1511-1524

Pilkey OH, Locker SD, Cleary WJ (1980) Comparison of sand layer geometry on flat floors of 10 modern depositional basins. Am Assoc Petrol Geol Bull 64:841-856

Piper DJW (1972) Turbidite origin of some laminated mudstones. Geol Mag 109:115-126

Piper DJW (1978) Turbidite muds and silts on deep sea fans and abyssal plains. In: Stanley DJ, Kelling G (eds) Sedimentation in submarine canyons, fans and trenches. Dowden, Hutchinson & Ross, Stroudsburg, PA, pp 163-176

Piper DJW, Brisco CD (1975) Deep water continental margin sedimentation, DSDP Leg 28, Antarctica. In: Hayes DE, Frakes LA etal. Init Repts DSDP 28: Washington (US Govt Print Office), pp 727-755

Piper DJW, Normark WR (1982) Effects of the 1929 Grand Banks earthquake on the continental slope off eastern Canada. Geol Surv Can Curr Res B Pap 82-1B

Piper DJW, Normark WR (1983) Turbidite depositional patterns and flow characteristics, Navy submarine fan, California borderland. Sedimentology 30:681-694

Piper DJW, Shor AN (1988) The 1929 "Grand Banks" earthquake, slump, and turbidity current. In: Clifton HE (ed) Sedimentologic consequences of convulsive geologic events. Geol Soc Am Spec Pap 229:77-92

Pisciotto KA (1981a) Diagenetic trends in the siliceous facies of the Monterey Shale in the Santa Maria region, California. Sedimentology 28:547-571

Pisciotto KA (1981b) Distribution, thermal histories, isotopic compositions, and reflection characteristics of siliceous rocks recovered by the Deep Sea Drilling Project. In: Warme JE, Douglas RG, Winterer EL (eds) The Deep Sea Drilling Project: a decade of progress. Soc Econ Paleontol Mineral Spec Publ 32:129-147

Pisciotto KA, Garrison RE (1981) Lithofacies and depositional environments of the Monterey Formation, California. In: Garrison RG, Pisciotto KA (eds) The Monterey Formation and related siliceous rocks in California Los Angeles. Soc Econ Paleontol Mineral Pacific Sect Spec Publ 15:97-122

Pisciotto KA, Mahoney JJ (1981) Isotopic survey of diagenetic carbonates, DSDP Leg 63. In: Yeats B, Haq B et al. Init Repts DSDP 63: Washington (US Govt Print Office), pp 595-609

Pisias NG (1976) Late Quaternary sediment of the Panama Basin: sedimentation rates, periodicities, and controls of carbonate and opal accumulation. Geol Soc Am Mem 145:375-391

Pisias NG, Imbrie J (1986/1987) Orbital geometry, CO_2, and Pleistocene climate. Oceanus 29:43-49

Pisias NG, Mix AC (1988) Aliasing of the geologic record and the search for long-period Milankovitch cycles. Paleoceanography 3:613-619

Pisias NG, Shackleton NJ, Hall MA (1985) Stable isotope and calcium carbonate records from hydraulic piston cored hole 574A: high-resolution record from the Middle Miocene. In: Moberly R, Schlanger SO et al. Init Repts DSDP 85: Washington (US Govt Print Office), pp 735-748

Pitman WC (1978) Relationship between eustacy and stratigraphic sequences of passive margins. Geol Soc Am Bull 89:1389-1403

Pitman WC (1989) The interpretation of sedimentary sequences. Symp Controvers in Modern Geol, ETH Zürich (Abstr)

Pitman WC, Golovchenko X (1983) The effect of sea level change on the shelf edge and slope of passive margins. In: Stanley DJ, Moore GT (eds) The shelf break: critical interface on continental margins. Soc Econ Paleontol Mineral Spec Publ 33:41-58

Plafker G, Addicott WO (1976) Glaciomarine deposits of Miocene through Holocene age in the Yakataga Formation along the Gulf of Alaska margin, Alaska. In: Miller TP (ed) Recent and Ancient sedimentary environments in Alaska. Alaska Geol Soc, Anchorage, pp 1-12

Poag CW, Valentine PC (1988) Mesozoic and Cenozoic stratigraphy of the United States, Atlantic continental shelf slope. In: Sheridan RE, Grow JA (eds) The Atlantic continental margin. US Geol Soc Am Bull I-2:67-85

Pokras EM, Winter A (1987) Variability of holocene diatom assemblages in laminated sediments near Walvis Bay, southwest Africa. Mar Geol 76:185-194

Pollastro RM, Martinez CJ (1985) Whole-rock, insoluble residue, and clay mineralogies of marl, chalk, and bentonite, Smoky Hill Shale Member, Niobrara Formation near Pueblo, Colorado - depositional and diagenetic implications. In: Pratt LM, Kauffman EG, Zelt FB (eds) Fine-grained deposits and biofacies of the Cretaceous Western Interior Seaway: evidence of cyclic sedimentary processes. Soc Econ Paleontol Mineral Fieldtrip Guidebook 4:215-222

Popenoe P (1985) Cenozoic depositional and structural history of North Carolina margin from seismic-stratigraphic analysis. In: Poag CW (ed) Geological evolution of the U.S. Atlantic Margin. Van Nostrand Reinhold, New York, pp 125-187

Posamentier HW, Jervey MT, Vail PR (1988a) Eustatic controls on clastic deposition II - conceptual framework. In: Wilgus C, Hastings B, Ross C, Posamentier H, Van Wagoner J, Kendall CGStC (eds) Sea-level changes: an integrated approach. Soc Econ Paleontol Mineral Spec Publ 42:109-124

Posamentier HW, Vail PR (1988b) Eustatic controls on clastic deposition II - sequence and systems tract models. In: Wilgus C, Hastings B, Ross C, Posamentier H, Van Wagoner J, Kendall CGStC (eds) Sea-level changes: an integrated approach. Soc Econ Paleontol Mineral Spec Publ 42:125-154

Postma G (1986) Classification of sediment gravity-flow deposits based on flow conditions during sedimentation. Geology 14:291-294

Postma G, Wojciehn N, Kleinspehn K (1988) Large floating clasts in the turbidites - a mechanism for their emplacement. Sediment Geol 55:47-61

Potter P, Maynard JB, Pryor W (1980) Sedimentology of shale, study guide and reference source. Springer, Berlin Heidelberg New York, 306 pp

Powell RD (1983) Glacial-marine sedimentation processes and lithofacies of temperate tidewater glaciers, Glacier Bay, Alaska. In: Molnia BF (ed) Glacial-marine sedimentation. Plenum, New York, pp 185-231

Powell RD (1984) Glaciomarine processes and inductive lithofacies modelling of ice shelf and tidewater glacier sediments based on Quaternary examples. Mar Geol 57:1-52

Powell RD, Molnia BF (1989) Glaciomarine sedimentary processes, facies and morphology of the south-southeast Alaska shelf and fjords. Mar Geol 85:359-390

Pratt BR (1979) Early cementation and lithification in intertidal cryptalgal structures, Boca Jewfish, Bonaire, Netherlands Antilles. J Sediment Petrol 49:379-386

Pratt BR (1982) Stromatolite decline - a reconsideration. Geology 10:512-515

Pratt BR, James NP (1986) The St George Group (Lower Ordovician) of western Newfoundland: tidal flat island model for carbonate sedimentation in shallow epeiric seas. Sedimentology 33:313-343

Pratt LM (1981) A paleo-oceanographic interpretation of the sedimentary structures, clay minerals, and organic matter in a core of the Middle Cretaceous Greenhorn Formation near Pueblo, Colorado. PhD Thesis, Princeton Univ, Princeton, 176 pp

Pratt LM (1984) Influence of paleoenvironmental factors on preservation of organic matter in Middle Cretaceous Greenhorn Formation, Pueblo, Colorado. Am Assoc Petrol Geol Bull 68:1146-1159

Pratt LM (1985) Isotopic studies of organic matter and carbonate in rocks of the Greenhorn Marine Cycle. In: Pratt LM, Kauffman EG, Zelt FB (eds) Fine-grained deposits and biofacies of the Cretaceous Western Interior Seaway: evidence for cyclic sedimentary processes. Soc Econ Paleontol Mineral Fieldtrip Guidebook 4:38-48

Pratt LM, Arthur MA, Dean WE, Scholle PA (1991) Paleoceanographic cycles and events during the Late Cretaceous in the Western Interior Seaway of North America. In: Caldwell WGE, Kauffman EG (eds) Evolution of Western Interior Basin. Geol Assoc Can (in press)

Pratt LM, Kauffman EG, Zelt F (eds) (1985) Fine-grained deposits and biofacies of the Cretaceous Western Interior Seaway: evidence for cyclic sedimentary processes. Soc Econ Paleontol Mineral Fieldtrip Guidebook 4:249 pp

Pratt LM, King JD (1986) Variable marine productivity and high eolian input recorded by rhythmic black shales in mid-Cretaceous pelagic deposits from central Italy. Paleoceanography 1:507-522

Pratt LM, Threlkeld CN (1984) Stratigraphic significance of C-13/C-12 ratios in mid-Cretaceous strata of the Western Interior Basin. In: Stott DF, Glass DJ (eds) Mesozoic of middle North America. Can Soc Petrol Geol Mem 9:305-312

Prell WL (1978) Upper Quaternary sediments of the Colombia basin: spatial and stratigraphic variation. Geol Soc Am Bull 89:1241-1255

Prell WL (1982) Oxygen and carbonate isotope stratigraphy for the Quaternary of hole 502B: evidence for two modes of isotopic variability. In: Prell WL, Gardner JV et al. Init Repts DSDP 68: Washington (US Govt Print Office), pp 455-466

Prell WL, Hays JD (1976) Late Pleistocene faunal and temperature patterns of the Colombia Basin, Caribbean Sea. In: Cline RM, Hays JD (eds) Investigation of Late Quaternary paleoceanography and paleoclimatology. Geol Soc Am Mem 145:201-220

Prell WL, Kutzbach JE (1987) Monsoon variability over the past 150 000 years. J Geophys Res 92:8411-8425

Prell WL, Niitsuma N et al. (1989) Proc ODP, Init Repts 117: College Station, Texas (Ocean Drilling Program), 1236 pp

Premoli-Silva J (1986) A new biostratigraphic interpretation of the sedimentary record recovered at Site 462, Leg 61, Nauru Basin, Western Equatorial Pacific. In: Moberly R, Schlanger SO et al. Init Repts DSDP 89: Washington (US Govt Print Office), pp 311-320

Premoli-Silva J, Brusa C (1981) Shallow-water skeletal debris and larger foraminifers from Deep Sea Drilling Project Site 462, Nauru Basin, western equatorial Pacific. In: Larson RL, Schlanger SO et al. Init Repts DSDP 61: Washington (US Govt Print Office), pp 439-453

Press WH, Flannery BP, Teukolsky SA, Vetterling WT (1986) Numerical recipes, the art of scientific computing. Cambridge Univ Press, Cambridge

Price J (1977) Deposition and derivation of clastic carbonates on a Mesozoic continental margin, Othris, Greece. Sedimentology 24:529-546

Priestley MB (1981) Spectral analysis and time series. vols 1 and 2. Academic Press, New York

Prior DB, Bornhold BD, Wisemand WJ, Lowe DR (1987) Turbidity current activity in a British Columbia fjord. Science 237:1330-1333

Prior DB, Coleman JM (1984) Submarine slope instability. In: Brunsden D, Prior DB (eds) Slope instability. Wiley, New York, 455 pp

Purser BH (1972) Subdivision et interprétation des séquences carbonatées. Mém Bur Rech Geol Min 77:679-698

Purser BH (ed) (1973) The Persian Gulf. Springer, Berlin Heidelberg New York, 471 pp

Purser BH (1978) Early diagenesis and the preservation of porosity in Jurassic limestones. J Petrol Geol 1:83-94

Purser BH (1984) Stratiform stylolites and the distribution of porosity: examples from the Middle Jurassic limestones of the Paris Basin. In: Yaha FA (ed) Stylolites and associated phenomena: relevance to hydrocarbon reservoirs. Abu Dhabi Nat Reserv Res Found, Abu Dhabi, pp 203-216

Putnam JA, Munk WH, Traylor MA (1949) The prediction of longshore currents. Trans Am Geophys Union 30:337-345

Pyle DM (1988) The thickness, volume and grainsize of tephra fall deposits. Bull Volcanol 51:1-15

Quadfasel D, Rudels B, Kurz K (1988) Outflow of dense water from a Svalbard fjord into the Fram Strait. Deep-Sea Res 35:1143-1150

Quinn WH, Neal VT, Antunez de Mayolo SE (1987) El Nino occurrences over the past four and half centuries. J Geophys Res 92:14449-14461

Quinn WH, Zopf DO, Short KS, Yang RTW (1978) Historical trends and statistics of the southern oscillation, El Nino, and Indonesian droughts. Fish Bull 76:663-678

Rachor E (1980) The inner German Bight - an ecologically sensitive area as indicated by the bottom fauna. Helgoländer Wissensch Meeresuntersuch 33:522-530

Rachor E (1982) Indikatorkarten für die Umweltbelastung im Meer. Decheniana Beih 26:128-137

Rachor E, Albrecht H (1983) Sauerstoffmangel im Bodenwasser der Deutschen Bucht. Veröffent Inst Meeresforsch Bremerhaven 19:209-227

Rahmani RA, Flores RM (1984) Sedimentology of coal and coal-bearing sequences of North America: a historical review. In: Rahmani RA, Flores RM (eds) Sedimentology of coal and coal-bearing sequences. Int Assoc Sediment Spec Publ 7:3-10

Raiswell R (1971) The growth of Cambrian and Liassic concretions. Sedimentology 17:147-171

Raiswell R (1987) Non-steady state microbiological diagenesis and the origin of concretions and nodular limestones. In: Marshall JD (ed) Diagenesis in sedimentary sequences. Geol Soc Lond, Blackwell, Oxford, pp 41-54

Raiswell R (1988a) Chemical model for the origin of minor limestone-shale cycles by anaerobic methane oxidation. Geology 16:641-644

Raiswell R (1988b) Evidence for surface reaction-controlled growth of carbonate concretions in shales. Sedimentology 35:571-575

Rampino MR, Self S (1982) Historic eruptions of Tambora (1815), Krakatau (1883), and Agung (1963), their stratospheric aerosols, and climatic impact. Quat Res 18:127-143

Rampino MR, Self S, Stothers RB (1988) Volcanic winters. Annu Rev Earth Planet Sci 16:73-99

Rasmusson EM, Carpenter TH (1982) Variations in tropical sea surface temperatures and wind fields associated with the Southern Oscillation/El Nino. Mon Weather Rev 111:517-528

Rau GH, Arthur MA, Dean WE (1987) 15N/14N variations in Cretaceous Atlantic sedimentary sequences: implication for the past changes in marine nitrogen biogeochemistry. Earth Planet Sci Lett 82:269-279

Raup DM, Sepkoski JJ (1986) Periodic extinction of families and genera. Science 231:833-836

Rea DK, Leinen M, Janecek TR (1985) Geologic approach to long-term history of atmospheric circulation. Science 227:721-725

Read JF, Goldhammer RK (1988) Use of Fischer plots to define third-order sea-level curves in Ordovician peritidal cyclic carbonates, Appalachians. Geology 16:895-899

Read JF, Grotzinger JP, Bova JA, Koerschner WF (1986) Models for generation of carbonate cycles. Geology 14:107-110

Reading HD (ed) (1982) Sedimentary environments and facies. Elsevier, New York, 569 pp

Reeder RJ (ed) (1983) Carbonates: mineralogy and chemistry. Mineral Soc Am, Rev Mineral 11: 394 pp

Reeside JB Jr (1944) Map showing thickness and general character of the Cretaceous deposits in the Western Interior of the United States, US Geol Surv Oil Gas Invest Map OM-10, Denver

Reid RGB, Brand DG (1986) Sulfide-oxidizing symbiosis in Lucinaceans: implications for bivalve evolution. Veliger 29:3-24

Reimers CE, Kastner M, Garrison RE (1990) The role of bacterial mats in phosphatic mineralisation with particular reference to the Monterey Formation. In: Burnett WC, Riggs S (eds) Genesis of Neogene to modern phosphorites. Cambridge Univ Press, Cambridge

Reimnitz E, Toimil LJ, Shepard FP, Gutierrez-Estrada M (1976) Possible rip current origin for bottom ripple zones to 30-m depth. Geology 4:395-400

Reineck HE (1960) Über Zeitlücken in rezenten Flachsee-Sedimenten. Geol Rdsch 49:149-161

Reineck HE (1978) Das Watt. Waldemar Kramer, Frankfurt, 185 pp

Reineck HE (1979) Rezente und fossile Algenmatten und Wurzelhorizonte. Nat Mus 109:290-296

Reineck HE, Dorjes J, Gadow S, Hertweck G (1968) Sedimentologie, Faunenzonierung und Faziesabfolge vor der Ostküste der inneren Deutschen Bucht. Senckenb Lethaea 49:261-309

Reineck HE, Singh IB (1980) Depositional sedimentary environments. Springer, Berlin Heidelberg New York, 549 pp

Remane J (1960) Les formations bréchiques dans le Tithonique du sud-est de la France. Trav Lab Géol, Fac Sci Grenoble 36:6-114

Renberg I, Segerstroem U, Wallin JE (1984) Climatic reflection in varved lake sediments. In: Mürner NA, Karlen W (eds) Climatic changes on a yearly to millennial basis. Reidel, Amsterdam, pp 249-256

Research on Cretaceous Cycles (ROCC) Group (1986) Rhythmic bedding in Upper Cretaceous pelagic carbonate sequences: varying sedimentary response to climatic forcing. Geology 14:153-156

Retallack GJ (1983) Late Eocene and Oligocene paleosols from Badlands National Park, South Dakota. Geol Soc Am Bull 94:823-840

Reymer JJG, Schlager W, Droxler AW (1988) Site 632: Pliocene-Pleistocene sedimentation cycles in a Bahamian basin. In: Austin JA Jr, Schlager W et al. (eds) Proc ODP, Sci Results 101, (Ocean Drilling Program) College Station, Texas, pp 213-220

Rhoads DC, Boyer LF (1982) The effects of marine benthos on physical properties of sediments: a successional perspective. In: McCall PL, Tevesz MJS (eds) Animal-sediment relations. Plenum, New York, pp 3-52

Rhoads DC, Morse JM (1971) Evolutionary and ecologic significance of oxygen-deficient marine basins. Lethaia 4:413-428

Rhoads DC, Speden IG, Waage KM (1972) Trophic group analysis of Upper Cretaceous (Maastrichtian) bivalve assemblages from South Dakota. Am Assoc Petrol Geol Bull 56(6):1100-1113

Rhoads DC, Young DK (1970) The influence of deposit-feeding organisms on sediment stability and community trophic structure. J Mar Res 28:150-178

Ricci Lucchi F (1977) Depositional cycles in two turbidite formations of northern Apennines. J Sediment Petrol 45:1-43

Ricci Lucchi F, Valmori E (1980) Basin-wide turbidites in a Miocene "over-supplied" deep-sea plain: a geometrical analysis. Sedimentology 27:241-270

Richmond BM, Sallenger AH Jr (1984) Cross-shore transport of bimodal sands. 19th Coast Engr Conf Proc, Am Soc Civil Engrs, Houston, Texas, pp 1997-2008

Richter (1985) Mikrodolomite in Crinoiden des Trochitenkalks (mo1) und die Wärmeanomalie von Vlotho. N Jb Geol Palaeontol Monatsh 11:681-690

Richter-Bernburg G (1955) Über salinäre Sedimentation. Z Dtsch Geol Ges 105:593-645

Ricken W (1985) Epicontinental marl-limestone alternations: Event deposition and diagenetic bedding (Upper Jurassic, southwest Germany). In: Bayer U, Seilacher A (eds) Sedimentary and evolutionary cycles. (Lect Notes Earth Sci 1), Springer, Berlin Heidelberg New York, pp 127-162

Ricken W (1986) Diagenetic bedding: a model for marl-limestone alternations. (Lect Notes Earth Sci 6), Springer, Berlin Heidelberg New York, 210 pp

Ricken W (1987) The carbonate compaction law: a new tool. Sedimentology 34:571-584

Ricken W (1991) A volume and mass approach to carbonate diagenesis: the role of compaction and cementation. In: Wolf KH, Chilingar GV (eds) Diagenesis, vol III (in press)

Ricken W, Hemleben C (1982) Origin of marl-limestone alternations in southwest Germany. In: Einsele G, Seilacher A (eds) Cyclic and event stratification. Springer, Berlin Heidelberg New York, pp 63-71

Riech V, von Rad U (1979) Silica diagenesis in the Atlantic Ocean: diagenetic potential and transformations. In: Talwani M, Harrison CG, Hayes DE (eds) Deep drilling results in the Atlantic Ocean: continental margins and paleoenvironment. Am Geophys Union, Washington, pp 315-340

Riegel W, Loh H, Maul B, Prauss M (1986) Effects and causes in a black shale event - the Toarcian Posidonia Shale of NW Germany. In: Walliser OH (ed) Global bioevents. (Lect Notes Earth Sci 8), Springer, Berlin Heidelberg New York, pp 267-276

Riegraf W (1977) Goniomya rhombifera (Goldfuss) in the Posidonia Shales (Lias epsilon). N Jb Geol Palaeontol Monatsh 1977 (7):446-448

Riegraf W (1985) Mikrofauna, Biostratigraphie und Fazies im unteren Toarcium Südwestdeutschlands und Vergleiche mit benachbarten Gebieten. Tübinger Mikropal Mitt 3:233

Rieke HH, Chilingarian GV (1974) Compaction of arillaceous sediments. (Developments in sedimentology 16), Elsevier Scientific, Amsterdam, 424 pp

Riggs SR (1984) Paleoceanographic model of Neogene phosphorite deposition, U.S. Atlantic Continental Margin. Science 223:123-131

Rio M, Ferry S, Cotillon P (1988) La périodicité dans les séries alternantes. Exemple du Crétacé inférieur de la région d'Angles-Saint-André-les-Alpes (Sud-Est de la France). Géotrope 1:14-21

Ripepe M (1988) Strata base: a stratigraphical and processing program for microcomputers. Comput Geosci 14:369-375

Ripepe M, Fischer AG (1989) Stratigraphic rhythms synthesized from orbital variations. Terra Abstr 1:241

Roberts DG, Schnitker D et al. (1984) Init Repts DSDP 81: Washington (US Govt Print Office), 923 pp

Robertson AHF (1984) Origin of varve-type lamination, graded claystones and limestone-shale couplets in the Lower Cretaceous of the western North Atlantic. In: Stow DAV, Piper DJW (eds) Fine-grained sediments: deep water processes and facies. Blackwell, Oxford, pp 437-452

Robertson AHF, Bliefnick DM (1983) Sedimentology and origin of Lower Cretaceous pelagic carbonates and redeposited clastics, Blake-Bahama Formation, DSDP Site 534, western equatorial Atlantic. In: Sheridan RE, Gradstein FM et al. Init Repts DSDP 74: Washington (US Govt Print Office), pp 795-828

Robin PYF (1978) Pressure solution at grain-to-grain contacts. Geochim Cosmochim Acta 42:1383-1389

Rodine JD, Johnson AM (1976) The ability of debris, heavily freighted with coarse clastic materials, to flow on gentle slopes. Sedimentology 23:213-234

Roehl PO (1967) Stony Mountain (Ordovician) and Interlake (Silurian) facies analogs of recent low-energy marine and subaerial carbonates, Bahamas. Am Assoc Petrol Geol Bull 51:1979-2032

Röhl U (1990) Parallelisierung des norddeutschen oberen Muschelkalks mit dem süddeutschen Hauptmuschelkalk anhand von Sedimentationszyklen. Geol Rdsch 79:13-26

Rona A (1973) Relations between rates of sediment accumulation on continental shelves, sea-floor spreading, and eustasy inferred from the central North Atlantic. Geol Soc Am Bull 84:2851-2872

Rose WI, Bonis SB, Stoiber RE, Keller M, Bickford T (1973) Studies of volcanic ash from two recent Central American eruptions. Bull Volcanol 37:338-364

Rose WI, Chesner CA (1987) Dispersal of ash in the great Toba eruption, 75 ka. Geology 15:913-917

Rosenberg R (1977) Benthic macrofaunal dynamics, production, and dispersion in an oxygen-deficient estuary of west Sweden. J Exp Mar Biol Ecol 26:107-133

Rosenkranz D (1971) Zur Sedimentologie und Ökologie von Echinodermen-Lagerstätten. N Jb Geol Paleontol Abh 138:221-258

Ross DA, Degens ET (1974) Recent sediments of the Black Sea. Am Assoc Petrol Geol Mem 20:183-199

Ross DA, Degens ET, MacIlvaine J (1970) Black Sea: recent sedimentary history. Science 170:163-165

Rossignol-Strick M (1983) African monsoons, an immediate climate response to orbital insolation. Nature 303:46-49

Roth PH (1986) Mesozoic paleoceanography of the North Atlantic and Tethys Oceans. In: Shackleton NJ, Summerhayes CP (eds) North Atlantic paleoceanography. Geol Soc Lond Spec Publ 21:299-320

Roth PH (1987) Mesozoic calcareous nannofossil evolution: relation to paleoceanographic events. Paleoceanography 2:601-611

Roth PH, Bowdler JL (1981) Middle Cretaceous calcareous nannoplankton biogeography and oceanography of the Atlantic Ocean. Soc Econ Paleontol Mineral Spec Publ 32:517-546

Ruddiman WF (1971) Pleistocene sedimentation in the equatorial Atlantic: stratigraphy and faunal paleoclimatology. Geol Soc Am Bull 82:283-302

Ruddiman WF (1977) Late Quaternary deposition of ice-rafted sand in the subpolar North Atlantic (lat 40° to 65° N). Geol Soc Am Bull 88:1813-1827

Ruddiman WF, Glover LK (1972) Vertical mixing of ice-rafted volcanic ash in North Atlantic sediments. Geol Soc Am Bull 83:2817-2836

Ruddiman WF, Kidd RB, Thomas E et al. (1987) Init Repts DSDP 94: Washington (US Govt Print Office), 1261 pp

Ruddiman WF, McIntyre A (1976) Northeast Atlantic paleoclimate changes over the past 600.000 years. In: Cline RM, Hays JD (eds) Investigation of Late Quaternary paleoceanography and paleoclimatology. Geol Soc Am Mem 145:111-146

Ruddiman WF, McIntyre A (1984) Ice-age thermal response and climatic role of the surface Atlantic Ocean, 40°N to 63°N. Geol Soc Am Bull 95:381-396

Ruddiman WF, Raymo M, Backman J, Clement BM (1990) The Mid-Pleistocene change in northern hemisphere climate (in prep)

Ruhrmann G (1971) Riff-ferne Sedimentation unterdevonischer Krinoidenkalke im Kantabrischen Gebirge (Spanien). N Jb Geol Palaeontol Monatsh 1971:231-248

Rupke NA (1976) Sedimentology of very thick calcarenite-marlstone beds in a flysch succession, southwestern Pyrenees. Sedimentology 23:43-6

Rupke NA, Stanley DJ (1974) Distinctive properties of turbiditic and hemipelagic mud layers in the Algero-Balearic Basin, western Mediterranean Sea. Smiths Contrib Earth Sci 13:40

Rusnak GA, Nesteroff WD (1964) Modern turbidites: terrigenous abyssal plain versus bioclastic basin. In: Miller RL (ed) Papers in marine geology. MacMillan, New York, pp 488-507

Russo AR (1980) Bioerosion by two rock-boring echinoids (Echinometra mathaei and Echinostrephus aciculatus) on Enewetak Atoll, Marshall Islands. J Mar Res 38:99-110

Rust BR (1972) Pebble orientation in fluviatile sediments. J Sediment Petrol 42:384-388

Rust BR (1984) Proximal braidplain deposits in the Middle Devonian Malbaie Formation of eastern Gaspé, Quebec, Canada. Sedimentology 31:675-695

Ryan WBF, Cita MB (1977) Ignorance concerning episodes of ocean wide stagnation. Mar Geol 23:197-215

Ryan WBF, Hsü, KJ et al. (1973) Init Repts DSDP 13 (Pt 1): Washington (US Govt Print Office), 949 pp

Ryer TA (1977) Patterns of Cretaceous shallow-marine sedimentation, Coalville and Rockport areas, Utah. Geol Soc Am Bull 88:177-188

Ryer TA (1981) Deltaic coals of Ferron Sandstone Member of Mancos Shale - predictive model for Cretaceous coal-bearing strata of Western Interior. Am Assoc Petrol Geol Bull 65:2323-2340

Ryer TA (1983) Transgressive-regressive cycles and the occurrence of coal in some Upper Cretaceous strata of Utah. Geology 11:207-210

Sadler PM (1981) Sediment accumulation rates and the completeness of stratigraphic sections. J Geol 89:569-584

Sageman BB (1985) High-resolution stratigraphy and paleobiology of the Hartland Shale Member: analysis of an oxygen-deficient epicontinental sea. In: Pratt LM, Kauffman EG, Zelt FB (eds) Fine-grained deposits and biofacies of the Cretaceous Western Interior Seaway: evidence of cyclic sedimentary processes. Soc Econ Paleontol Mineral Fieldtrip Guidebook 4:110-121

Sageman BB (1989) The benthic boundary biofacies model: Hartland Shale Member, Greenhorn Formation (Cenomanian), Western Interior, North America. Palaeogeogr Palaeoclimatol Palaeoecol 74:87-110

Sahagian DL (1987) Epeirogeny and eustatic sea level changes as inferred from Cretaceous shoreline deposits: applications to the central and western United States. J Geophys Res 92:4895-4904

Salinas I (1984) Thesis, Univ Copenhagen, Geol Inst

Sallenger AH Jr, Holman RA, Birkemeier WA (1985) Storm-induced response of a nearshore-bar system. Mar Geol 64:237-257

Sallenger AH Jr, Howd PA (1989) Nearshore bars and the break-point hypothesis. Coast Engr 12:301-313

Salvador A (1985) Chronostratigraphic and geochronometric scales in COSUNA stratigraphic correlation charts of the US. Am Assoc Petrol Geol Bull 69:181-189

Salvador A (1987) Unconformity-bounded stratigraphic units. Bull Geol Soc Am 98:232-237

Salvigsen O (1981) Radiocarbon dated raised beaches in Kongs Karls Land, Svalbard, and their consequences for the glacial history of the Barents Sea area. Geogr Ann 63A:283-291

Salzman ES, Barron EJ (1982) Deep circulation in the Late Cretaceous: oxygen isotope paleotemperatures from Inoceramus remains in DSDP cores. Palaeogeogr Palaeoclimatol Palaeoecol 40:167-181

Sandberg PA (1984) Recognition criteria for calcitized skeletal and non-skeletal aragonites. Palaeontologr Am 54:272-281

Sandberg PA (1985) Nonskeletal aragonite and pCO_2 in the Phanerozoic and Proterozoic. In: Sunquist ET, Broecker WS (eds) The carbon cycle and atmospheric CO_2. Geophys Monogr 32:585-594

Sander B (1936) Beiträge zur Kenntnis der Anlagerungsgefüge rhythmischer Kalke und Dolomite aus der Trias. Min Petrogr Mitt 48:27-209

Sangree JB, Wiedmier JM (1977) Seismic interpretation of clastic depositional facies. In: Paytow CE (ed) Seismic stratigraphy - applications to hydrocarbon exploration. Am Assoc Petrol Geol Mem 26:165-184

Sarg JF (1988) Carbonate sequence stratigraphy. In: Wilgus C, Hastings B, Ross C, Posamentier H, Van Wagoner J, Kendall CGStC (eds) Sea-level changes: an integrated approach. Soc Econ Paleontol Mineral Spec Publ 42:155-181

Sarna-Wojcicki AM (1985) Ages of tuff beds at East African early hominid sites and sediments in the Gulf of Aden. Nature 313:306-308

Sarna-Wojcicki AM, Morrison SD, Meyer CE, Hillhouse JW (1987) Correlation of Upper Cenozoic tephra layers between sediments of the western United States and eastern Pacific Ocean and comparison with biostratigraphic and magnetostratigraphic age data. Geol Soc Am Bull 98:207-203

Sarna-Wojcicki AM, Shiply S, Waitt R, Dzurisin D, Wood S (1981) Areal distribution, thickness, mass, volume and grain size of airfall ash from the six major eruptions of 1980. In: Lipman PW, Mullineaux DR (eds) The 1980 eruptions of Mount St. Helens, Washington. US Geol Surv Prof Pap 1250:577-600

Sarnthein M (1978) Sand deserts during glacial maximum and climatic optimum. Nature 271:43-46

Sarnthein M, Bartolini C (1973) Grain size studies on turbidite components from Tyrrhenian deep sea cores. Sedimentology 20:425-436

Sarnthein M, Diester-Haass L (1977) Eolian sand turbidites. J Sediment Petrol 47:868-890

Sarnthein M, Erlenkeuser H, von Grafenstein R, Schröder C (1984) Stable - isotope stratigraphy for the last 750.000 years: "Meteor" core 13519 from the eastern equatorial Atlantic. Meteor Forsch Ergeb C 38:9-24

Sarnthein M, Fenner J (1988) Global wind-induced change of deep-sea sediment budgets, new ocean production and CO_2 reservoirs ca. 3.3-3.25 Ma BP. Philos Trans R Soc Lond B 318:487-504

Sarnthein M, Thiede J, Erlenkeuser H, Fütterer D, Koopmann B, Lange H, Seibold E (1982) Atmospheric and oceanic circulation patterns off northwest Africa during the past 25 million years. In: von Rad U, Hinz K, Sarnthein M, Seibold E (eds) Geology of the northern African continental margin. Springer, Berlin Heidelberg New York, pp 545-604

Sarnthein M, Winn K, Duplessy JC, Fontugne MR (1988) Global variations of surface ocean productivity in low and mid latitudes: influences on CO_2 reservoirs of the deep ocean and atmosphere during the last 21.000 years. Paleoceanography 3:361-399

Sarnthein M, Winn K, Zahn R (1987) Paleoproductivity of oceanic upwelling and the effect on atmospheric CO_2 and climatic change during deglaciation times. In: Berger HW, Labeyrie LD (eds) Abrupt climatic change. Reidel, Dordrecht, pp 311-337

Sass E, Kolodny Y (1972) Stable isotopes, chemistry and petrology of carbonate concretions (Mishash Formation, Israel). Chem Geol 10:261-286

Sastry AN (1979) Pelecypoda (excluding Ostreidae). In: Giese AC, Pearse JS (eds) Reproduction of marine invertebrates 5. Academic Press, New York, pp 113-292

Savrda CE, Bottjer DJ (1986) Trace fossil model for reconstruction of paleo-oxygenation in bottom waters. Geology 14:3-6

Savrda CE, Bottjer DJ (1987a) The exaerobic zone, a new oxygen-deficient marine biofacies. Nature 327:54-56

Savrda CE, Bottjer DJ (1987b) Trace fossils as indicators of bottom-water redox conditions in ancient marine environments. In: Bottjer DJ (ed) New concepts in the use of biogenic sedimentary structures for paleoenvironmental interpretation. Soc Econ Paleontol Mineral Pacific Sect 52:3-26

Savrda CE, Bottjer DJ (1988) Limestone concretion growth documented by trace-fossil relations. Geology 16:908-911

Savrda CE, Bottjer DJ (1989) Development of a trace fossil model for the reconstruction of paleo-bottom-water redox conditions: evaluation and application to Upper Cretaceous Niobrara Formation, Colorado. Palaeogeogr Palaeoclimatol Palaeoecol 74:49-74

Savrda CE, Bottjer DJ (1989a) Anatomy and implications of bioturbated beds in "black shale" sequences: examples from the Jurassic Posidonienschiefer (southern Germany). Palaios, 4:330-342

Savrda CE, Bottjer DJ (1989b) Trace fossil model for reconstructing oxygenation histories of ancient marine bottom waters: application to Upper Cretaceous Niobrara Formation, Colorado. Palaeogeography, Palaeoclimatology, Palaeoecology, 74:49-74

Savrda CE, Bottjer DJ (in press) Trace fossil assemblages in fine-grained strata of the Cretaceous western interior. In: Caldwell WGE, Kauffman EG (eds) Evolution of the western interior foreland basin. Geol Assoc Canada Spec Paper

Savrda CE, Bottjer DJ, Gorsline DS (1984) Development of a comprehensive oxygen-deficient marine biofacies model: evidence from Santa Monica, San Pedro and Santa Barbara Basins, California continental borderland. Am Assoc Petrol Geol Bull 68:1179-1192

Sawyer DS (1986) Effects of basement topography on subsidence history analysis. Earth Planet Sci Lett 78:427-434

Saxov S, Nieuwenhuis JK (eds) (1982) Marine slides and other mass movements. NATO Conf Ser IV, Marine Sciences. Plenum, New York, 353 pp

Schäfer W (1972) Ecology and palaeoecology of marine environments. Univ Chicago Press, Chicago, Ill, 568 pp

Schairer G, Janicke V (1970) Sedimentologisch-paläontologische Untersuchungen an den Plattenkalken der Sierra de Montsech (Provinz Lérida, NE-Spanien). N Jb Geol Palaeontol Abh 135:171-189

Scheltema RS (1977) Dispersal of marine invertebrate organisms: paleobiogeographic and biostratigraphic implications. In: Kauffman E, Hazel J (eds) Concepts and methods of biostratigraphy. Dowden, Hutchinson & Ross, Stroudsberg, pp 73-108

Schieber J (1989) Facies and origin of shales from the mid-Proterozoic Newland Formation, Belt Basin, Montana, USA. Sedimentology 36:203-219

Schiffelbein P, Dorman L (1986) Spectral effects of time-depth non-linearities in deep-sea sediment records: a demodulation technique for reanalysing time and depth scales. J Geophys Res 91:3821-3835

Schindel DE (1982) Resolution analysis: a new approach to the gaps in the fossil record. Paleobiology 8:340-353

Schlager W (1981) The paradox of drowned reefs and carbonate platforms. Bull Geol Soc Am 92:197-211

Schlager W, Camber O (1986) Submarine slope angles, drowning unconformities, and shelf-erosion on limestone escarpments. Geology 14:762-765

Schlager W, Chermak A (1979) Sediment facies of platform-basin transition, Tongue of the Ocean, Bahamas. In: Doyle LJ, Pilkey OH (eds) Geology of continental slopes. Soc Econ Paleontol Mineral Spec Publ 27:193-208

Schlager W, Ginsburg RN (1981) Bahamas carbonate platforms - the deep and the past. Mar Geol 44:1-24

Schlager W, James NP (1978) Low-magnesian calcite limestones forming at the deep-sea floor, Tongue of the Ocean, Bahamas. Sedimentology 25:675-702

Schlager W, Schlager M (1973) Clastic sediments associated with radiolarites (Tauglboden-Schichten), Upper Jurassic, eastern Alps. Sedimentology 20:65-89

Schlanger SO, Arthur MA, Jenkyns HC, Scholle PA (1987) The Cenomanian-Turonian ocaeanic anoxic event. I. Stratigraphy and distribution of organic carbon-rich beds and the marine $\delta^{13}C$ excursion. In: Brooks J, Fleet AJ (eds) Marine petroleum source rocks. Geol Soc Lond Spec Publ 26:371-399

Schlanger SO, Douglas RG (1974) The pelagic ooze-chalk-limestone transition and its implications for marine stratigraphy. In: Hsü KJ, Jenkyns C (eds) Pelagic sediments. Int Assoc Sediment Spec Publ 1:117-148

Schlanger SO, Jenkyns HC (1976) Cretaceous oceanic anoxic events - causes and consequences. Geol Mijnbouw 55:179-184

Schlanger SO, Premoli-Silva I (1986) Oligocene sea-level falls recorded in mid-Pacific atoll and archipelagic apron settings. Geology 14:392-395

Schlee JS (ed) (1984) Interregional unconformities and hydrocarbon accumulation. Am Assoc Petrol Geol Mem 36:184 pp

Schmincke HU (1967) Graded lahars in the type section of the Ellensburg Formation, south-central Washington. J Sediment Petrol 37:438-448

Schmincke HU (1970) "Base surge"-Ablagerungen des Laacher-See-Vulkans. Aufschluß 21:350-364

Schmincke HU (1976) Geology of the Canary Islands. In: Kunkel G (ed) Biogeography and ecology in the Canary Islands. Junk, The Hague, pp 67-184

Schmincke HU (1977) Phreatomagmatische Phasen in quartären Vulkanen der Osteifel. Geol Jb A 39:3-45

Schmincke HU (1982) Volcanic and chemical evolution of the Canary Islands. In: von Rad U, Hinz K, Sarnthein M, Seibold E (eds) Geology of the northwest African Continental Margin. Springer, Berlin Heidelberg New York, pp 273-306

Schmincke HU (1987) Geological field guide of Gran Canaria, 2nd edn. Pluto, Witten, pp 1-179

Schmincke HU (1988) Pyroklastische Gesteine. In: Füchtbauer H (ed) Sedimentgesteine. Schweizerbart, Stuttgart, pp 720-785

Schmincke HU (1990) Die quartären Vulkanfelder der Eifel. Schweizerbart, Stuttgart, pp 1-220

Schmincke HU, Bednarz U (1990) Pillow-, sheet flow- and breccia-volcanoes and volcano-tectonic-hydrothermal cycles in the extrusive series of the northwestern Troodos ophiolite (Cyprus). In: Malpas J, Moores F, Panayiolou A, Xenophantos C (eds) Ophiolites - oceanic crustal analogues. Proc Symp "Troodos 1987", Geol Survey Deptm, Nicosia, pp 185-206

Schmincke HU, Fisher RV, Waters AC (1973) Antidune and chute and pool structures in the base surge deposits of the Laacher See area, Germany. Sedimentology 20:553-574

Schmincke HU, von Rad U (1979) Neogene evolution of Canary Island volcanism inferred from ash layers and volcanoclastic sandstones of DSDP site 397 (Leg 47 A). In: von Rad U, Ryan WBF et al. Init Repts DSDP 47 A: Washington (US Govt Print Office), pp 703-725

Schmitz B (1987) Barium equatorial high productivity and the northward wandering of the Indian continent. Paleoceanography 2:63-78

Schneider FK (1964) Erscheinungsbild und Entstehung der rhythmischen Bankung der altkretazischen Tongesteine Nordwestfalens und der Braunschweiger Bucht. Ein Beitrag zur Absolutchronologie von Hauterive bis Alb. Fortschr Geol Rheinl Westf 7:353-382

Schneiderman N, Harris PM (eds) (1985) Carbonate cements. Soc Econ Paleontol Mineral Spec Publ 36:1-379

Schofield K (1984) Are pressure solution, neomorphism and dolomitization generally related? In: Yaha FA (ed) Stylolites and associated phenomena: relevance to hydrocarbon reservoirs. Abu Dhabi Nat Reserv Res Found, Abu Dhabi, pp 183-201

Scholle PA (1971) Sedimentology of fine-grained deep-water carbonate turbidites, Monte Antola Flysch (Upper Cretaceous), northern Apennines, Italy. Bull Geol Soc Am 82:629-658

Scholle PA (1977) Chalk diagenesis and its relation to petroleum exploration - oil from chalks, a modern miracle? Am Assoc Petrol Geol Bull 61:982-1009

Scholle PA (1978) A color illustrated guide to carbonate rock constituents, textures, cements, and porosities. Mem Assoc Petrol Geol Mem 27:pp 241

Scholle PA, Arthur MA (1980) Carbon-isotopic fluctuations in Cretaceous pelagic limestones: potential stratigraphic and petroleum exploration tool. Am Assoc Petrol Geol Bull 64:67-87

Scholle PA, Arthur MA, Ekdale AA (1983b) Pelagic environment. In: Scholle PA, Bebout CH, Moore CH (eds) Carbonate depositional environments. Am Assoc Petrol Geol Mem 33:620-691

Scholle PA, Bebout DG, Moore CH (eds) (1983a) Carbonate depositional environments. Am Assoc Petrol Geol Mem 33:345-440

Scholle PA, Halley RB (1985) Burial diagenesis: out of sight, out of mind. In: Schneidermann N, Harris PM (eds) Carbonate cements. Soc Econ Paleontol Mineral Spec Publ 36:309-334

Schove DJ (1983) (ed) Sunspot cycles. Benchmark Pap Geol 68, Hutchinson Ross, Stroudsburg, 397 pp

Schrader HJ (1972) Kieselsäure-Skelette in Sedimenten des ibero-marokkanischen Kontinentalrandes und angrenzender Tiefsee-Ebenen. Meteor Forsch Ergeb 8:10-36

Schroeder J (1988) Spatial variations in the porosity development of carbonate sediments and rocks. Facies 18:181-204

Schroeder JH, Purser BH (1986) Reef diagenesis. Springer, Berlin Heidelberg New York, 455

Schultz LG, Tourtelot HA, Gill JR, Boerngen JG (1980) Comparison of the Pierre Shale and equivalent rocks, Northern Great Plains region. US Geol Surv Prof Pap 1064-B, 114 pp

Schumacher R, Schmincke HU (1990) The lateral facies of ignimbrites at Laacher See volcano. Bull Volcanol 52:271-285

Schumacher R, Schmincke HU (1991) Structure of accretionary lapilli and cluster sedimentation of fine-grained volcanic ashes. Bull Volcanol 54 (submitted)

Schwab FL (1976) Modern and ancient sedimentary basins: comparative accumulation rates. Geology 4:723-727

Schwarz HU (1982) Subaqueous slope failures - experiments and modern occurrences. (Contributions to sedimentology 11), Schweizerbart, Stuttgart, 116 pp

Schwarzacher W (1947) Über die sedimentäre Rhythmik des Dachsteinkalkes bei Lofer. Verh Geol Bundesanst Wien (10-12):176-188

Schwarzacher W (1954) Die Großrhythmik des Dachsteinkalkes von Lofer. Tschermaks Mineral Petrogr Mitt 4:44-54

Schwarzacher W (1958) The stratification of the Great Scar Limestone in the settle district of Yorkshire. Liverpool Manchester Geol J 2:124-142

Schwarzacher W (1964) An application of statistical time-series analysis of a limestone-shale sequence. J Geol 72:195-213

Schwarzacher W (1975) Sedimentation models and quantitative stratigraphy. (Developments in sedimentology 19), Elsevier, Amsterdam, 382 pp

Schwarzacher W (1982) Quantitative correlation of a cyclic limestone-shale formation. In: Cubitt JM, Reyment RA (eds) Quantitative stratigraphic correlation. Wiley, Chichester, pp 275-286

Schwarzacher W (1987a) Astronomically controlled cycles in the Lower Tertiary of Gubbio. Earth Planet Sci Lett 84:22-26

Schwarzacher W (1987b) The analysis and interpretation of stratification cycles. Paleoceanography 2:79-95

Schwarzacher W, Fischer AG (1982) Limestone-shale bedding and perturbations of the Earth's orbit. In: Einsele G, Seilacher A (eds) Cyclic and event stratification. Springer, Berlin Heidelberg New York, pp 72-95

Schwarzacher W, Haas J (1986) Comparative statistical analysis of some Hungarian and Austrian Upper Triassic peritidal carbonate sequences. Acta Geol Hung 29:175-196

Sclater JG, Hellinger S, Tapscott C (1977) The paleobathymetry of the Atlantic Ocean from the Jurassic to the Present. J Geol 85:509-552

Scoffin TP (1987) An introduction to carbonate sediments and rocks. Blackie, Glasgow, 274

Scoffin TP, Henry MD (1984) Shallow-water sclerosponges on Jamaican reefs and a criterion for recognition of hurricane deposits. Nature 307:728-729

Scotese CR, Denham CR (1987) User's guide to Terra Mobilis: a plate tectonics program for the Macintosh. Earth in Motion Technologies, Houston, Texas, 55 pp

Scott GR (1969) General and engineering geology of the northern part of Pueblo, Colorado. US Geol Surv Bull 1262:131 pp

Scott GR, Cobban WA (1964) Stratigraphy of the Niobrara Formation at Pueblo, Colorado. US Geol Surv Prof Pap 454-L:30 pp

Scott JT, Csanady GT (1976) Nearshore currents off Long Island. J Geophys Res 81 (30):5401-5409

Seibertz E (1979) Stratigraphisch-fazielle Entwicklung des Turon im südöstlichen Münsterland (Oberkreide, NW-Deutschland) Newsl Stratigr 8:3-60

Seibold E (1952) Chemische Untersuchungen zur Bankung im unteren Malm Schwabens. N Jb Geol Palaeontol Abh 95:337-370

Seibold E (1962) Kalk-Konkretionen und karbonatisch gebundenes Magnesium. Geochim Cosmochim Acta 26:899-909

Seibold E, Berger WH (1982) The sea floor. An introduction to marine geology. Springer, Berlin Heidelberg New York, 288 pp

Seibold E, Seibold I (1953) Foraminiferenfauna und Kalkgehalt eines Profils im gebankten unteren Malm Schwabens. N Jb Geol Palaeontol Abh 98:28-86

Seilacher A (1953) Studien zur Palichnologie. I. Über die Methoden der Palichnologie. N Jb Geol Palaeontol Abh 96:421-452

Seilacher A (1962) Paleontological studies on turbidite sedimentation and erosion. J Geol 70:227-234

Seilacher A (1968) Origin and diagenesis of the Oriskany Sandstone (Lower Devonian, Appalachians) as reflected in its fossil shells. Recent developments in sedimentology in central Europe. Springer, Berlin Heidelberg New York, pp 175-185

Seilacher A (1969) Fault graded beds interpreted as seismites. Sedimentology 13:155-159

Seilacher A (1971) Preservational history of ceratite shells. Palaeontology 14:16-21

Seilacher A (1978) Evolution of trace fossil communities in the deep sea. In: Seilacher A, Westphal F (eds) Paleoecology. N Jb Geol Palaeontol Abh 157:251-255

Seilacher A (1982a) Ammonite shells as habitats in the Posidonia Shales of Holzmaden: floats or benthic islands?. N Jb Geol Palaeontol Monatsh 1982:98-114

Seilacher A (1982b) Distinctive features of sandy tempestites. In: Einsele G, Seilacher A (eds) Cyclic and event stratification. Springer, Berlin Heidelberg New York, pp 333-349

Seilacher A (1982c) Posidonia Shales (Toarcian, S Germany) - stagnant basin model revalidated. In: Gallitelli EM (ed) Proc 1st Int Meet Paleontology, essential of historical geology, Venezia 1981, Modena, pp 25-55

Seilacher A (1982d) General remarks about event deposits. In: Einsele G, Seilacher A (eds) Cyclic and event stratification. Springer, Berlin Heidelberg New York, pp 161-174

Seilacher A (1983) Palökologie - Wechselwirkung zwischen geologischen und biologischen Prozessen. In: Seibold (ed) Forschung in der Bundesrepublik Deutschland, DFG. Verlag Chemie, Weinheim, pp 689-696

Seilacher A (1984a) Constructional morphology of bivalves: evolutionary pathways in primary versus secondary soft-bottom dwellers. Palaeontology 27:207-237

Seilacher A (1984b) Sedimentary structures tentatively attributed to seismic events. Mar Geol 55:1-12

Seilacher A (1985) The Jeram model: event condensation in a modern intertidal environment. In: Bayer U, Seilacher A (eds) Sedimentary and evolutionary cycles. (Lect Notes Earth Sci 1) Springer, Berlin Heidelberg New York, pp 336-342

Seilacher A (1988a) Schlangensterne (Aspidura) als Schlüssel zur Entstehungsgeschichte des Muschelkalks. In: Hagdorn H (ed) Neue Forschungen zur Erdgeschichte von Crailsheim. Goldschneck, Stuttgart

Seilacher A (1988b) Why are nautiloid and ammonite sutures so different? N Jb Geol Palaeontol Abh 177:41-69

Seilacher A (1989) Photo- and chemosymbiotic bivalves: the constructional morphology approach to evolution. In: Ross R, Allmon W (eds) Biotic and abiotic factors in evolution. Univ Chicago Press, 25 pp

Seilacher A (1990a) Aberrations in bivalve evolution related to photo- and chemosymbiosis. Historical Biology, 3:289-311

Seilacher A (1990b) Paleozoic trace fossils in Egypt. In: Said R (ed) Geology of Egypt, new edition. Balkema, Rotterdam

Seilacher A, Andalib F, Dietl G, Gocht H (1976) Preservational history of compressed Jurassic ammonites from southern Germany. N Jb Geol Palaeontol Abh 152:307-336

Seilacher A, Hemleben C (1966) Spurenfauna und Bildungstiefe der Hunsrückschiefer (Unterdevon). Notizbl Landesamt Bodenforsch 94:40-53

Seilacher A, Matyja BA, Wierzbowski A (1985a) Oyster beds: morphologic response to changing substrate conditions. In: Bayer U, Seilacher A (eds) Sedimentary and evolutionary cycles. Springer, Berlin Heidelberg New York pp 421-435

Seilacher A, Reif WE, Westphal F (1985b) Sedimentological, ecological and temporal patterns of fossil Lagerstaetten. In: Whittington HB, Conway Morris S (eds) Extraordinary fossil biotas: their ecological and evolutionary significance. Philos Trans R Soc Lond B 311:5-23

Seilacher A, Seilacher-Drexler E (1986) Sekundäre Weichbodenbewohner unter den Cirripediern. Palaeontol Z 60:75-92

Self S, Rampino MR (1981) The 1883 eruption of Krakatau. Nature 294:699-704

Self S, Sparks RJS (eds) (1981) Tephra studies. Reidel, Dordrecht, pp 1-481

Selg M (1988) Origin of peritidal carbonate cycles: Early Cambrian, Sardinia. Sediment Geol 59:115-124

Selley RC (1972) Diagnosis of marine and non-marine environments from the Cambro-Ordovician sandstones of Jordan. Q J Geol Soc Lond 128:135-150

Sellwood BW, Jenkyns HC (1975) Basins and swells and the evolution of an eperic sea (Pliensbachian-Bajocian of Great Britain). J Geol Soc Lond 131:373-388

Sepkoski JJ Jr (1977) Dresbachian (Upper Cambrian) stratigraphy in Montana, Wyoming, and South Dakota. PhD Thesis, Harvard Univ, Cambridge, Mass, 563 pp

Sepkoski JJ Jr (1978) Taphonomic factors influencing the lithologic occurrence of fossils in Dresbachian (Upper Cambrian) shaley facies. Geol Soc Am Abstr with Program 10:490

Sepkoski JJ Jr (1981) A factor analytic description of the marine fossil record. Paleobiology 5:222-251

Sepkoski JJ Jr (1982) Flat-pebble conglomerates, storm deposits, and the Cambrian bottom fauna. In: Einsele G, Seilacher A (eds) Cyclic and event stratification. Springer, Berlin Heidelberg New York, pp 371-385

Sepkoski JJ Jr, Miller AI (1985) Evolutionary faunas and the distribution of Paleozoic benthic communities in space andtime. In: Valentine JW (ed) Phanerozoic diversity patterns: profiles in macroevolution. Princeton Univ Press and Pacific Div Am Assoc Adv Sci, Princeton, NJ, pp 153-190

Sepkoski JJ Jr, Sheehan PM (1983) Diversification, faunal change, and community replacement during the Ordovician radiations. In: Tevesz MJS, McCall PL (eds) Biotic interactions in recent and fossil benthic communities. Plenum, New Yrok, pp 673-717

Seymour RJ (1980) Longshore sediment transport by tidal currents. J Geophys Res 85:1899-1904

Seymour RJ (1986) Nearshore auto-suspending turbidity flows. Ocean Engr 13:435-447

Seymour RJ (1990) Autosuspending turbidity flows. In: LeMehaute B, Hanes M (eds) The Sea 9. Ocean Engr Sci, Wiley, New York, pp 919-940

Shackleton NJ (1977) Carbon-13 in Uvigerina: tropical rainforest history and the equatorial Pacific carbonate dissolution cycles. In: Andersen NR, Malakoff A (eds) The fate of fossil fuel CO_2 in the oceans. Plenum, New York, pp 401-428

Shackleton NJ (1982) The deep-sea record of climate variability. Prog Oceanogr 11:199-218

Shackleton NJ, Hall MA (1984) Oxygen and carbon isotope stratigraphyof Deep-Sea Drilling Project Hole 552 A: Plio-Pleistocene glacial history. In: Roberts DG, Schnitker D et al. Init Repts DSDP 81: Washington (US Govt Print Office), pp 599-609

Shackleton NJ, Hall MA, Line J, Cang Shuxi (1983) Carbon isotope data in core V19-30 confirm reduced carbon dioxide concentration in the ice age atmosphere. Nature 306:319-322

Shackleton NJ, Opdyke ND (1973) Oxygen isotope and paleomagnetic stratigraphy of equatorial Pacific Core V28-238: oxygen isotope temperatures and ice volume on a 105 year and 106 year scale. Quat Res 3:39-55

Shackleton NJ, Opdyke ND (1976) Oxygen-isotope and paleomagnetic stratigraphy of Pacific core V 28-239: Late Pliocene to latest Pleistocene. In: Cline RM, Hays JD (eds) Investigation of Late Quaternary paleoceanography and paleoclimatology. Geol Soc Am Mem 145:449-464

Shackleton NJ, Pisias NG (1985) Atmospheric CO_2, orbital forcing and climate. In: Sundquist ET, Broecker WS (eds) The carbon cycle and atmospheric CO_2. Geophys Monogr 32:303-318

Shaffer G (1986) Phosphate pumps and shuttles in the Black Sea. Nature 321:515-517

Shanmugam G (1980) Rhythms in deep sea, fine-grained turbidite and debris flow sequences, Middle Ordovician, eastern Tennessee. Sedimentology 27:419-432

Shanmugam G, Moiola RJ (1982) Eustatic control of turbidites and winnowed turbidites. Geology 10:231-235

Shanmugam G, Moiola RJ (1984) Eustatic control of calciclastic turbidites. Mar Geol 56:273-278

Shanmugam G, Moiola RJ (1985) Submarine fan models: problems and solutions. In: Bouma AH, Normark WR, Barnes NE (eds) Submarine fans and related turbidite systems. Springer, Berlin Heidelberg New York, pp 29-34

Shaw AB (1964) Time in stratigraphy. McGraw-Hill, New York, 365 pp

Shaw DM, Watkins ND, Huang TC (1974) Atmospherically transported volcanic glass in deep-sea sediments: theoretical considerations. J Geophys Res 79:3087-3094

Sheehan PM, Schiefelbein DRJ (1984) The trace fossil Thalassinoides from the Upper Ordovician of the eastern Great Basin: deep burrowing in the Early Paleozoic. J Paleontol 58:440-448

Sheldon RP (1980) Episodicity of phosphate deposition and deep ocean circulation - a hypothesis. In: Bentor YK (ed) Marine phosphorites - geochemistry, occurrence, genesis. Soc Econ Paleontol Mineral Spec Publ 29:239-247

Sheridan MF, Wohletz KH (1983) Hydrovolcanism: basic considerations and review. J Volcanol Geotherm Res 17:1-29

Sheridan RE (1987) Pulsation tectonics as the control of long-term stratigraphic cycles. Paleoceanography 2:97-118

Sheridan RE, Gradstein FM et al. (1983) Init Repts DSDP 76: Washington (US Govt Print Office), 947 pp

Sherwood BA, Sager SL, Holland HD (1987) Phosphorus in foraminiferal sediments from North Atlantic Ridge cores and in pure limestones. Geochim Cosmochim Acta 51:1861-1866

Shepard FP, Inman DL (1950) Nearshore water circulation related to bottom topography and wave refraction. Trans Am Geophys Union 31:555-565

Shi NC, Larsen LH (1984) Reverse sediment transport induced by amplitude modulated waves. Mar Geol 54:181-200

Shimkus KM, Trimonis ES (1974) Modern sedimentation in Black Sea. In: Degens ET, Ross DA (eds) The Black Sea - geology, chemistry and biology. Am Assoc Petrol Geol Mem 20:249-278

Shinn EA (1969) Submarine lithification of Holocene carbonate sediments in the Persian Gulf. Sedimentology 12:109-144

Shinn EA (1983) Tidal flat environment. In: Scholle PA, Bebout DG, Moore CH (eds) Carbonate depositional environments. Am Assoc Petrol Geol Mem 33:171-210

Shinn EA, Lloyd RM, Ginsburg RN (1969) Anatomy of a modern carbonate tidal-flat, Andros Island, Bahamas. J Sediment Petrol 39:1202-1228

Shinn EA, Robbin DM (1983) Mechanical and chemical compaction in fine-grained shallow-water limestones. J Sediment Petrol 53:595-618

Shoemaker EM (1984) Large body impacts through geologic time. In: Holland HD, Trendall AF (eds) Patterns of change in Earth evolution. Springer, Berlin Heidelberg New York, pp 15-40

Short AD (1984) Beach and nearshore facies: southeast Australia. Mar Geol 60:261-282

Short AD (1985) Rip-current type, spacing and persistence, Narrabeen Beach, Australia. Mar Geol 65:47-71

Shrock RR (1948) Sequence in layered rocks. McGraw-Hill, New York, 507 pp

Shukla V, Baker PA (1988) Sedimentology and geochemistry of dolostones. Soc Econ Paleontol Mineral Spec Publ 43:266 pp

Siebert L (1984) Large volcanic debris avalanches: characteristics of source areas, deposits, and associated eruptions. J Volcanol Geotherm Res 22:163-197

Siegenthaler U, Wenk T (1984) Rapid atmospheric CO_2 variations and oceanic circulation. Nature 308:624-626

Siegfried P (1954) Die Fischfauna des westfälischen Oberbenons. Palaeontographica A 106:1-36

Sigurdsson H, Carey S (1989) Plinian and co-ignimbrite tephra fall from the 1815 eruption of Tambora volcano. Bull Volcanol 51:243-270

Sigurdsson H, Loebner B (1981) Deep sea record of Cenozoic explosive volcanism in the North Atlantic. In: Self S, Sparks RSJ (eds) Tephra studies, NATO Adv Stud Inst. Reidel, Dordrecht, pp 289-316

Sigurdsson H, Sparks RSJ, Carey SN, Huang TC (1980) Volcanogenic sedimentation in the Lesser Antilles Arc. J Geol 88:523-540

Simm RW, Kidd RB (1984) Submarine debris flow deposits detected by large-range side-scan sonar 1000 km from source. Geo-Mar Lett 3/1:13-16

Simon M, Schmincke HU (1983) Late Cretaceous volcaniclastic rocks from the Walvis Ridge, southeast Atlantic, Leg 74. In: Sheridan RE, Gradstein FM et al. Init Repts DSDP 74: Washington (US Govt Print Office), pp 765-791

Simpson J (1985) Stylolite-controlled layering in an homogeneous limestone: pseudo-bedding produced by burial diagenesis. Sedimentology 32:495-505

Sirocko F (1989) Accumulation of eolian sediments in the northern Indian ocean; record of the climatic history of Arabia and India. Rep Geol Palaeontol Inst Univ Kiel 27:185

Siscoe GL (1978) Solar-terrestrial influences on weather and climate. Nature 276:348-351

Slaughter M, Earley JW (1965) Mineralogy and geological significance of the Mowry bentonites, Wyoming. Geol Soc Am Spec Pap 83:116

Sloss LL (1963) Sequences in the cratonic interior of North America. Geol Soc Am Bull 74:93-114

Sloss LL (1988) Forty years of sequence stratigraphy. Geol Soc Am Bull 100:1661-1665

Sloss LL, Krumbein WC, Dapples EC (1949) Integrated facies analysis. In: Longwell CR (chairman) Sedimentary facies in geologic history. Geol Soc Am Mem 39:91-124

Sloss LL, Speed RC (1974) Relationship of cratonic and continental margin tectonic episodes. In: Dickinson WR (ed) Tectonics and sedimentation. Soc Econ Paleontol Mineral Spec Publ 22:38-55

Smith CR, Hamilton SC (1983) Epibenthic megafauna of a bathyal basin off southern California: patterns of abundance, biomass, and dispersion. Deep-Sea Res 30:907-928

Smith GA (1988) Sedimentology of proximal to distal volcaniclastics dispersed across an active foldbelt: Ellensburg Formation (Late Miocene), central Washington. Sedimentology 35:953-977

Smith GI (1984) Paleohydrological regimes in the southwestern Great Basin, 0-3.2 my ago, compared with other long records of "global" climate. Quat Res 22:1-17

Smith ND (1978) Sedimentation processes and patterns in a glacier-fed lake with low sediment input. Can J Earth Sci 15:741-756

Smith RA (1984) The lithostratigraphy of the Karoo Supergroup in Botswana. Bull Geol Surv Dep 26:239

Smith RL (1960a) Ash flows. Geol Soc Am Bull 71:795-842

Smith RL (1960b) Zones and zonal variations in welded ash flows. US Geol Surv Prof Pap 354-F:149-159

Smith RL (1979) Ash-flow magmatism. Geol Soc Am Spec Pap 180:5-27

Smith RL, Bailey RA (1968) Resurgent cauldrons. Geol Soc Am Mem 116:613-662

Snavely PD (1981) Early diagenetic controls on allochthonous carbonate debris flows - examples from Egyptian lower Eocene platform-slope. Am Assoc Petrol Geol Bull 65:995 (Abstr)

Snedden JW, Nummedal D (1990) Coherence of surf zone and shelf current flow along the Texas coast and its implications for interpretation of paleocurrent measurements in ancient shelf to shoreline sequences. Sediment Geol (in press)

Snedden JW, Nummedal D, Amos AF (1988) Storm and fair-weather combined flow on the central Texas continental shelf. J Sediment Petrol 58:580 595

Snodgrass D, Groves DW, Hasselman KF, Miller GR, Munk WH, Powers WH (1966) Propagation of ocean swell across the Pacific. Philos Trans R Soc Lond Ser A 259:431-497

Snyder SW, Hine AC, Riggs SR (1990) Seismic stratigraphic record of shifting Gulf Stream flow paths in response to Miocene glacio-eustasy. In: Burnett WC, Riggs SR (eds) Phosphate deposits of the world, vol 3. Cambridge Univ Press, Cambridge (in press)

Solheim A, Pfirman S (1985) Sea-floor morphology outside a grounded surging glacier-Brasvellbreen, Svalbard. Mar Geol 65: 127-143

Somero GN (1984) Physiology and biochemistry of the hydrothermal vent animals. Oceanus 27:67-72

Sonett CP (1984) Very long solar periods and the radiocarbon record. Rev Geophys Space Phys 22:239-254

Sonett CP, Finney SA, Williams CR (1988) The lunar orbit in the Late Precambrian and the Elatina sandstone laminae. Nature 335:806-808

Sonett CP, Williams GE (1985) Solar periodicities expressed in varves from glacial Skilak Lake, southern Alaska. J Geophys Res 90A:12019-12026

Sonnenfeld DL, Nummedal D (1987) Morphodynamics and sediment dispersal in a tideless surf zone. Coast Seds '87, Am Soc Civ Engr II:1938-1949

Sorby HC (1908) On the application of quantitative methods to the study of the structure and history of rocks. Q J Geol Soc Lond 64:171-233

Soudry D (1987) Ultra-fine structures and genesis of the Campanian Negev high-grade phosphorites (southern Israel). Sedimentology 34:641-660

Soudry D, Champetier Y (1983) Microbial processes in the Negev phosphorites (southern Israel). Sedimentology 30:411-423

Soudry D, Lewy Z (1988) Microbially influenced formation of phosphate nodules and megafossil moulds (Negev, southern Israel). Palaeogeogr Palaeoclimatol Palaeoecol 64:15-34

Soudry D, Southgate PN (1989) Ultrastructure of a Middle Cambrian primary nonpelletal phosphorite and its early transformation into phosphate vadoids: Georgina Basin, Australia. J Sediment Petrol 59:53-64

Souquet P, Eschard R, Lods H (1987) Facies sequences in large-volume debris- and turbidity-flow deposits from the pyrenees (Cretaceous; France, Spain). Geo-Mar Lett 7:83-90

Southam JR, Peterson WH, Brass GW (1982) Dynamics of anoxia. Palaeogeogr Palaeoclimatol Palaeoecol 40:183-198

Southard JB, Young RA, Hollister CD (1971) Experimental erosion of calcareous ooze. J Geophys Res 76 (24):5903-5909

Southgate PN (1986) Cambrian phoscrete profiles, coated grains, and microbial processes in phosphogenesis: Georgina Basin, Australia. J Sediment Petrol 56:429-441

Southgate PN (1989) Relationships between cyclicity and stromatolite form in the Late Proterozoic Bitter Springs Formation, Australia. Sedimentology 36:323-339

Spackman W, Riegel WL, Dolsen CP (1969) Geological and biological interactions in the swamp-marsh complex of southern Florida. In: Dapples EC, Hopkins ME (eds) Environments of coal depostion. Geol Soc Am Spec Pap 114:1-35

Sparks RJS (1976) Grain-size variations in ignimbrites and implications for the transport of pyroclastic flows. Sedimentology 23:147-188

Sparks RJS (1986) The dimensions and dynamics of eruption columns. Bull Volcanol 48:3-15

Sparks RJS, Huang TC (1980) The volcanological significance of deep-sea ash layers associated with ignimbrites. Geol Mag 117:425-436

Sparks RJS, Self S, Walker GPL (1973) Products of ignimbrite eruption. Geology 1:115-118

Sparks RJS, Walker GPL (1977) The significance of vitric-enriched air-fall ashes associated with crystal-enriched ignimbrites. J Volcanol Geotherm Res 2:239-341

Sparks RJS, Wilson L (1976) A model for the formation of ignimbrite by gravitational column collapse. J Geol Soc Lond 132:441-451

Spencer RJ, Baedecker MJ, Eugster HP, Forester RM, Goldhaber MB, Jones BF, Kelts K, Mckenzie J, Madsen DB, Rettig SL, Rubin M, Bowser CJ (1984) Great Salt Lake, and precursors, Utah: the last 30,000 years. Contrib Min Petrol 86:321-334

Speyer SE (1985) Moulting in phacopid trilobites: Trans R Ser Edinb 76:239-254

Speyer SE (1987) Comparative taphonomy and paleoecology of trilobite Lagerstätten. Alcheringa, Sydney, Australia 11 (3-4):205-232

Speyer SE, Brett CE (1986) Trilobite taphonomy and Middle Devonian Hamilton Group. Lethaia 18:85-103

Speyer SE, Brett CE (1987) Trilobite taphonomy and Middle Devonian taphofacies. Palaios 1:312-327

Speyer SE, Brett CE (1988) Taphofacies models for epeiric sea environments: Middle Paleozoic examples. Palaeogeogr Palaeoclimatol Palaeoecol 63:225-262

Spies RB, Davis PH (1979) The infaunal benthos of a natural oil seep in the Santa Barbara Channel. Mar Biol 50:227-237

Stabell B (1986a) A diatom maximum horizon in Upper Quaternary deposits. Geol Rdsch 75:175-184

Stabell B (1986b) Variations of diatom flux in the eastern equatorial Atlantic during the last 400,000 years ("Meteor" cores 13519 and 13521). Mar Geol 72:305-329

Stal LJ, Gemerden H, Krumbein WE (1985) Structure and development of a benthic marine microbial mat. FEMS Microbiol Ecol 31:111-125

Stanley DJ (1982) Welded slump-graded sand couplets: evidence for slide generated turbidity currents. Geo-Mar Lett 2:149-155

Stanley DJ (1985) Mud depositional processes as a major influence on Mediterranean margin-basin sedimentation. In: Stanley DJ, Wezel FC (eds) Geological evolution of the Mediterranean Basin. Springer, Berlin Heidelberg New York, pp 377-410

Stanley DJ, Kelling G (eds) (1978) Sedimentation in submarine canyons, fans, and trenches. Dowden, Hutchinson & Ross, Stroudsburg, 382 pp

Stanley DJ, Maldonado A (1981) Depositional models for fine-grained sediments in the Western Hellenic Trench, eastern Mediterranean. Sedimentology 28(2):273-290

Stanley SM (1970) Relations of shell form to life habits in the bivalvia. Geol Soc Am Mem 125:296

Stanley SM (1987) Extinction. Scientific American Books, New York, 242 pp

Start GG, Prell WL (1984) Evidence for two Pleistocene climatic modes: data from DSDP site 502. In: Berger A, Nicolis C (eds) New perspectives in climate modeling. (Develop Atmos Sci 16), Elsevier, Amsterdam, pp 3-22

Staudigel H, Schmincke HU (1984) The Pliocene seamount series of La Palma (Canary Islands). J Geophys Res 89:11195-11215

Steckler MS, Watts AB (1978) Subsidence of the Atlantic-type continental margins. Earth Planet Sci Lett 41:1-13

Steckler MS, Watts AB, Thorne JA (1988) Subsidence and basin modelling at the U.S. Atlantic passive margin. In: Sheridan RE, Grow JA (eds) The Atlantic continental margin. US Geol Soc Am, The Geology of North America I-2, pp 399-416

Steel RJ, Thompson DB (1983) Structures and textures in Triassic braided stream conglomerates ('Bunter' Pebble Beds) in the Sherwood Sandstone Group, North Staffordshire, England. Sedimentology 30:341-367

Steimle FW, Sindermann CJ (1978) Review of oxygen depletion and associated mass mortalities of shellfish in the Middle Atlantic Bight in 1976. Mar Fish Rev 40:17-26

Stein CL, Kirkpatrick RJ (1976) Experimental porcelanite recrystallization kinetics: a nucleation and growth model. J Sediment Petrol 46:430-435

Stein R (1986a) Late Neogene evolution of paleoclimate and paleoceanic circulation in the northern and southern hemispheres - a comparison. Geol Rdsch 75(1):125-138

Stein R (1986b) Organic carbon and sedimentation rate - further evidence for anoxic deep-water conditions in the Cenomanian/Turonian Atlantic Ocean. Mar Geol 72:199-209

Stel JH (1975) The influence of hurricanes upon the quiet depositional conditions in the Lower Emsian La Vid Shales of Colle (N.W. Spain). Leidse Geol Meded 49:475-486

Stewart RW (1967) Mechanics of the air-sea interface. Phys Fluids Suppl 10:547-555

Stoffers P, Hecky RE (1978) Late Pleistocene-Holocene evolution of the Kivu-Tanganyika Basin. In: Matter A, Tucker ME (eds) Modern and ancient lake sediments. Spec Publ Int Assoc Sediment Spec Publ 2:43-55

Stolz JF (1984) Fine structure of the stratified microbial community at Laguna Figueroa, Baja California, Mexico: II. Transmission electron microscopy as a diagnostic tool in studying microbial communities in situ. In: Cohen Y, Castenholz RW, Halvorson HO (eds) Microbial mats: stromatolites. Liss, New York, pp 23-38

Stothers RB (1984) The great Tambora eruption and its aftermath. Science 224:1191-1198

Stothers RB, Rampino MR, Self S, Wolff JA (1989) Volcanic winter? Climatic effects of the largest volcanic eruptions. Latter JH (ed) Volcanic hazards. IAVCEI Proc Volcanol 1:3-9

Stow DAV (1979) Distinguishing between fine-grained turbidites and contourites on the Nova Scotian deep water margin. Sedimentology 26:371-387

Stow DAV (1980) A physical model for the transport and sorting of fine-grained sediment by turbidity currents. Sedimentology 27:31-46

Stow DAV (1981) Laurentian fan: morphology, sediments, processes, and growth pattern. Am Assoc Petrol Geol Bull 65 I:375-393

Stow DAV (1984a) Anatomy of debris-flow deposits. In: Hay WW, Sibuet JC et al. (eds) Init Repts DSDP 75: Washington (US Govt Print Office), pp 801-807

Stow DAV (1984b) Turbidite facies, associations and sequences in the southeastern Angola Basin. In: Hay WW, Sibuet JC et al. (eds) Init Repts DSDP 75: Washington (US Govt Print Office), pp 785-799

Stow DAV (1986) Deep clastic seas. In: Reading HD (ed) Sedimentary environments and facies, 2nd edn. Blackwell, Oxford, pp 399-444

Stow DAV (1987) South Atlantic organic-rich sediments: facies processes and environments of deposition. In: Brooks J, Fleet AJ (eds) Marine petroleum source rocks. Geol Soc Spec Publ 26:287-299

Stow DAV, Alam M, Piper DJW (1984a) Sedimentology of the Halifax Formation, Nova Scotia: Lower Paleozoic fine-grained turbidites. Geol Soc Lond Spec Publ 16:127-144

Stow DAV, Bishop CD, Mills SJ (1982) Sedimentology of the Brae Oil Field, North Sea: fan models and controls. J Petrol Geol 5:129-148

Stow DAV, Bowen AJ (1980) A physical model for the transport and sorting of fine-grained sediments by turbidity currents. Sedimentology 27:31-46

Stow DAV, Cochran JR (1989) The Bengal Fan: some preliminary results from ODP drilling. Geo-Marine Lett 9:19-29

Stow DAV, Cremer M, et al. (1986) Facies, composition and texture of Mississippi Fan sediments, DSDP Leg 96, Gulf of Mexico. In: Bouma AH, Coleman JM, Meyer AW et al. Init Repts DSDP 96: Washington (US Govt Print Office), pp 475-487

Stow DAV, Dean WE (1984) Middle Cretaceous black shale at Sites 530 in the southeastern Angola Basin. In: Hay WW, Sibuet JC et al. Init Repts DSDP 75 II: Washington (US Govt Print Office), pp 809-817

Stow DAV, Howell DG, Nelson CH (1985) Sedimentary, tectonic, and sea-level controls. In: Bouma AH, Normark WR, Barnes NE (eds) Submarine fans and related turbidite systems. Springer, Berlin Heidelberg New York pp 15-22

Stow DAV, Piper DJW (1984a) Deep-water fine-grained sediments: facies models. In: Stow DAV, Piper DJW (eds) Fine-grained sediments: deep-water processes and facies. Blackwell, Oxford, Geol Soc Lond Spec Publ 15: 611-646

Stow DAV, Piper DJW (1984b) Fine-grained sediments: deep-water processes and facies. Blackwell, Oxford, Geol Soc Spec Publ 15:659 pp 659

Stow DAV, Shanmugam G (1980) Sequence of structures in fine-grained turbidites: comparison of recent deep-sea and ancient flysch sediments. Sediment Geol 25:23-42

Stow DAV, Wezel FC, Savelli D, Rainey SCR, Angell G (1984b) Depositional model for calcilutites: Scaglia Rossa limestones, Umbro Marchean, Apennines. In: Stow DAV, Piper DJW (eds) Fine-grained sediments: deep-water processes and facies. Geol Soc Lond Spec Publ 14:223-240

Strasser A (1988) Shallowing-upward sequences in Purbeckian peritidal carbonates (lowermost Cretaceous, Swiss and French Jura Mountains). Sedimentology 35:369-383

Strasser A, Davaud E (1986) Formation of Holocene limestone sequences by progradation, cementation, and erosion: two examples from the Bahamas. J Sediment Petrol 56:422-428

Street-Perrott A, Harrison SP (1984) Temporal variations in lake levels since 30,000 yr BP - an index of the global hydrological cycle. In: Hansen JE, Takahashi T (eds) Climate processes and climate sensitivity. (Maurice Ewing vol 5) Am Geophys Union Monogr 29:118-129

Strobel J, Soewito F, Kendall CGStC, Biswas G, Bezdek J, Cannon R (1990) Interactive simulation (SED-pak) of clastic and carbonate sedimentation in shelf to basin settings. In: Cross TA (ed) Quantitative dynamic stratigraphy. Prentice Hall, Englewood Cliffs, New Jersey, pp 433-455

Stuiver M, Braziunas TF (1988) The solar component of the atmospheric 14 C record. In: Stephenson FR, Wolfendale AW (eds) Secular solar and geomagnetic variations in the last 10,000 years. Kluwer Academic, Dordrecht, pp 245-266

Stuiver M, Braziunas TF (1989) Atmospheric 14C and century-scale solar oscillations. Nature 338:405-408

Stuiver M, Grootes PM (1980) Trees and the ancient record of heliomagnetic cosmic ray flux modulation. In: Pepin RO, Eddy JA, Merrill RB (eds) The ancient sun. Pergamon, New York, pp 165-173

Stuiver M, Quay PD (1980) Changes in atmospheric carbon-14 attributed to a variable sun. Science 207:11-19

Sturm M, Matter A (1978) Turbidites and varves in Lake Brienz (Switzerland): deposition of clastic detritus by density currents. Modern and ancient lake sediments. Spec Publ Int Assoc Sediment 2:147-168

Suess E (1906) The face of the earth, vol 2. Clarendon Press, Oxford, 556 pp

Suess E (1981) Phosphate regeneration from sediments of the Peru continental margin by dissolution of fish debris. Geochim Cosmochim Acta 45:577-588

Suess E, Kulm LD, Killingley JS (1987) Coastal upwelling and a history of organic-rich mudstone deposition of Peru. In: Brooks J, Fleet AJ (eds) Marine petroleum source rocks. Geol Soc Spec Publ 26:181-197

Suess E, von Huene R et al. (1988) Proc ODP Init Repts 112: College Station, Texas (Ocean Drilling Program) 1015 pp

Sujkowski ZL (1958) Diagenesis. Am Assoc Petrol Geol Bull 42:2692-2717

Summerhayes CP (1981) Organic facies of Middle Cretaceous black shales in deep North Atlantic. Am Assoc Petrol Geol Bull 65:2364-2380

Summerhayes CP (1987) Organic-rich Cretaceous sediments from the North Atlantic. In: Brooks J, Fleet AJ (eds) Marine petroleum source rocks. Geol Soc Lond Spec Publ 26:301-316

Summerhayes CP, Masran TC (1983) Organic facies of Cretaceous and Jurassic sediments from DSDP Site 534 in the Blake-Bahama Basin, Western North Atlantic. In: Sheridan RE, Gradstein FM et al. Init Repts DSDP 76: Washington (US Govt Print Office), pp 469-480

Sunamura T, Maruyama K (1987) Wave-induced geomorphic response of eroding beaches - with special reference to seaward migrating bars. Coast Seds '87, Am Soc Civ Engr 1:788-801

Sundquist ET, Broecker WS (eds) (1985) The carbon cycle and atmospheric CO_2. Geophys Monogr 32:627 pp

Supko PR (1963) A quantitative X-ray diffractometer-method for the mineralogical analysis of carbonate sediment from Tongue of the Ocean. MS Thesis, University of Miami, 158 pp

Surdam RC, Stanley KO (1979) Lacustrine sedimentation during the culminating phase of Eocene Lake Gosiute, Wyoming (Green River Formation). Geol Soc Am Bull 90:93-110

Surdam RC, Wolfbauer CA (1975) Green River Formation, Wyoming: a playa-lake complex. Geol Soc Am Bull 86:335-345

Surlyk F (1987) Slope and deep shelf gully sandstones, Upper Jurassic, East Greenland. Am Assoc Petrol Geol Bull 71:464-475

Sutton RG, McGhee GR (1985) The evolution of Frasnian marine "community-types" of south-central New York. Geol Soc Am Spec Pap 201:211-224

Swann DH (1964) Late Mississippian rhythmic sediments of Mississippi valley. Bull Am Assoc Petrol Geol 48:637-658

Swanson RL, Sindermann CJ, Han G (1979) Oxygen depletion and the future: an evaluation. In: Swanson RL, Sindermann CJ (eds) Oxygen depletion and associated benthic mortalities in New York Bight, 1976. Nat Ocean Atmos Admin Prof Pap 11:1-345

Swift DJP (1978) Continental shelf sedimentation. In: Fairbridge RW, Bourgeois J (eds) The encyclopedia of sedimentology. Dowden, Hutchinson & Ross, Stroudsburg, pp 190-196

Swift DJP, Figueiredo AG Jr, Freeland GL, Oertel GF (1983) Hummocky cross-stratification and megaripples: a geological double standard? J Sediment Petrol 53:1295-1317

Swift DJP, Han G, Vincent CE (1986) Fluid processes and sea-floor response on a modern storm-dominated shelf: middle Atlantic shelf of North America. Part I The storm-current regime. In: Knight RJ, McLean JR (eds) Shelf sands and sandstones. Can Soc Petrol Geol Mem II:99-119

Swift DJP, Hudelson PM, Brenner RL, Thomson P (1987) Shelf construction in a foreland basin: storm beds, shelf sandbodies and shelf-slope depositional sequences in the Upper Cretaceous Mesaverde Group, Book Cliffs, Utah. Sedimentology 34:423-457

Swift DJP, Niedoroda AW, Vincent CE, Hopkins TS (1985) Barrier island evolution, middle Atlantic shelf, U.S.A. Part I Shoreface dynamics. Mar Geol 63:331-361

Swift DJP, Nummedal D (1987) Hummocky cross-stratification, tropical hurricanes and intense winter storms: a discussion. Sedimentology 34:338-344

Swift SA, Wenkam C (1978) Holocene accumulation rates of calcitein the Panama Basin: lateral and vertical variations in calcite dissolution. Mar Geol 27:67-77

Swirydczuk K (1988) Mineralogical control on porosity type in Upper Jurassic Smackover ooid grainstones, southern Arkansas and northern Louisiana. J Sediment Petrol 58:339-347

Tada R, Iijyima A (1983) Petrology and diagenetic changes of Neogene siliceous rocks in northern Japan. J Sediment Petrol 53:911-930

Tada R, Maliva R, Siever R (1987) A new mechanism for pressure solution in porous quartzose sandstone. Geochim Cosmochim Acta 51:2295-2301

Tada R, Siever R (1986) Experimental knife-edge pressure solution of halite. Geochim Cosmochim Acta 50:29-36

Tada R, Siever R (1989) Pressure solution during diagenesis: a review. Annu Rev Earth Planet Sci 17:89-118

Talbot MR (1988) The origin of lacustrine oil source rocks: evidence from the lakes of tropical Africa. In: Fleet AJ, Kelts K, Talbot MR (eds) Lacustrine petroleum source rocks. Geol Soc Spec Publ 40:29-43

Talbot MR, Kelts K (1986) Primary and diagenetic carbonates in the anoxic sediments of Lake Bosumtwi, Ghana. Geology 14:912-916

Talwani H, Udintsev G et al. (eds) (1976) Init Repts DSDP 38, 39, 40, 41: Washington (US Govt Print Office), 1256 pp

Tankard AJ (1986) On the depositional response to thrusting and lithospheric flexure: examples from the Appalachian and Rocky Mountain basins. In: Allen PA, Homewood P (eds) Foreland basins. Int Assoc Sediment Spec Publ 8:369-392

Tarling DH (1983) Paleomagnetism. Chapman & Hall, London, 379 pp

Taub FB (1984) Ecosystem processes. In: Taub FB (ed) Ecosystems of the world, 23. Lakes and reservoirs. Elsevier, Amsterdam, pp 9-42

Taylor FJ, Taylor NJ, Walsby JR (1985) A bloom of planktonic diatom, Cerataulina pelagica, off the coast of northeastern New Zealand in 1983, and its contribution to an associated mortality of fish and benthic fauna. Int Rev Ges Hydrobiol 70:773-795

Taylor ME (1976) Indigenous and redeposited trilobites from Late Cambrian basinal environments of central Nevada. J Paleontol 50:668-700

Temmler H (1966) Über die Nusplinger Fazies des Weißen Jura der Schwäbischen Alb. Z Dtsch Geol Ges 116: 891-907

Ten Brink US, Brocher TM (1987) Multichannel seismic evidence for a subcrustal intrusive complex under Oahu and a model for Hawaiian volcanism. J Geophys Res 92:13687-13707

Ten Haaf E (1956) Significance of convolute lamination. Geol Mijnbouw 18:188-194

Ten Kate WG, Sprenger A (1989) On the periodicity in a calcilutite-marl succession (SE Spain). Cret Res 10:1-31

Thayer CW (1983) Sediment-mediated biological distrubance and the evolution of marine benthos. In: Tevesz MJS, McCall PL (eds) Biotic interactions in Recent and fossil benthic communities. Plenum, New York, pp 480-626

Theede H (1973) Comparative studies on the influence of oxygen deficiency and hydrogen sulphide on marine invertebrates. Neth J Sea Res 7:244-252

Theede H, Ponat A, Hiroki K, Schlieper C (1969) Studies on the resistance of marine bottom invertebrates to oxygen deficiency and hydrogen sulphide. Mar Biol 2:325-337

Thein J, von Rad U (1987) Silica diagenesis in continental rise and slope sediments off eastern North America (Sites 603 and 605, Leg 93; Sites 612 and 613, Leg 95). In: Poag CW, Watts AB, et al. Init Repts DSDP 95: Washington (US Govt Print Office), pp 501-525

Theyer F, Mayer LA, Barron JA, Thmas E (1985) The equatorial Pacific high-productivity belt: elements for a synthesis of Deep Sea Drilling Project Leg 85 results. In: Moberly R, Schlanger SO et al. Init Repts DSDP 85: Washington (US Govt Print Office), pp 971-985

Thickpenny A, Leggett JK (1987) Stratigraphic distribution and paleo-oceanographic significance of European organic-rich sediments. In: Brooks J, Fleet AJ (eds) Marine petroleum source rocks. Geol Soc Lond Spec Publ 26:231-247

Thiede J (1981) Reworked neritic fossils in Upper Mesozoic and Cenozoic central Pacific deep-sea sediments monitor sea-level changes. Science 211:1422-1424

Thiede J, Dean WE, Claypool GE (1982a) Oxygen-deficient depositional paleo-environments in the mid-Cretaceous tropical and subtropical Pacific Ocean. In: Schlanger SO, Cita MB (eds) Nature and origin of Cretaceous organic-rich facies. Academic Press, New York, pp 79-100

Thiede J, Diesen GW, Knudsen BE, Snare T (1986) Patterns of Cenozoic sedimentation in the Norwegian-Greenland Sea. Mar Geol 69:323-352

Thiede J, Ehrmann WU (1986) Late Mesozoic and Cenozoic sediment flux to the central North Atlantic Ocean. In: Summerhayes CP, Shackleton NJ (eds) North Atlantic paleoceanography. Geol Soc Lond Spec Publ 21:3-15

Thiede J, Suess E (eds) (1983) Coastal upwelling and its sediment record, part B, Sedimentary records of ancient coastal upwelling. Plenum, New York, 610 pp

Thiede J, Suess E, Müller PJ (1982b) Late Quaternary fluxes of major sediment components to the sea-floor at the northwest African continental slope. In: von Rad U, Hinz K, Sarnthein M, Seibold E (eds) Geology of the northwest African continental margin. Springer, Berlin Heidelberg New York, pp 605-631

Thiede J, Vallier TL et al. (1981) Init Repts DSDP 62: Washington (US Govt Print Office), 1120 pp

Thiede J, Van Andel TH (1977) The paleoenvironment of anaerobic sediments in the Late Mesozoic South Atlantic Ocean. Earth Planet Sci Lett 33:301-309

Thierstein HR (1989) Inventory of paleoproductivity records: the mid-Cretaceous enigma. In: Berger WH, Smetacek VS, Wefer G (eds) Productivity of the oceans: present and past. Wiley, New York, pp 355-375

Thierstein HR, Berger WH (1978) Injection events in Earth history. Nature 276:461-466

Thompson I, Johnes DS, Ropes JW (1980) Advanced age for sexual maturity in the ocean, quahog *Arctica islandica* (Mollusca: Bivalvia). Mar Biol 57:35-39

Thompson JB, Ferris FG (1988) Microbial ultrastructure of a thrombolitic microbialite and bacterial mat from the chemocline in Fayetteville Green Lake. Geol Soc Am Centen Mtng, Denver, CO, 1988, Geol Soc Am, Abstr with Programs:A226

Thompson JB, Mullins HT, Newton CR, Vercoutere TL (1985) Alternative biofacies model for dysaerobic communities. Lethaia 18:167-179

Thompson JB, Newton CR (1987) Ecological reinterpretation of the dysaerobic Leiorhyncus fauna: Upper Devonian Geneseo black shale, central New York. Palaios 2(3):274-281

Thompson PR (1976) Planktonic foraminifera dissolution and the progress towards a Pleistocene equatorial Pacific transfer function. J Foraminiferal Res 6(3):208-22

Thompson PR, Saito T (1974) Pacific Pleistocene sediments: planktonic foraminifera dissolution cycles and geochronology. Geology 2:333-335

Thornburg TM, Kulm LD (1987) Sedimentation in the Chile Trench: depositional morphologies, lithofacies, and stratigraphy. Geol Soc Am Bull 98:33-52

Thorson G (1957) Bottom communities (sublittoral or shallow shelf). In: Hedgpeth J (ed) Treatise on marine ecology and paleoecology 1. Mem Geol Soc Am 67:461-534

Thunell RC (1976) Optimum indices of calcium carbonate dissolution in deep-sea sediments. Geology 4:525-528

Thunell RC, Williams DF (1989) Glacial-Holocene salinity changes in the Mediterranean Sea: hydrographic and depositional effects. Nature 338:493-496

Thurow J, Kuhnt W (1986) Mid-Cretaceous of the Gibraltar Arch Area. In: Summerhayes CP, Shackleton NJ (eds) North Atlantic. Paleoceanography, Geol Soc Lond Spec Publ 22:423-445

Tillmann RW, Swift DJP, Walker RG (eds) (1985) Shelf sands and sandstone reservoirs. Soc Econ Paleontol Mineral Short Course Notes 13, Tulsa

Tipper JC (1983) Rates of sedimentation, and stratigraphical completeness. Nature 302:696-698

Tissot BP (1979) Effects of prolific petroleum source rocks and major coals deposits caused by sea level changes. Nature 277:463-465

Tissot BP, Demaison G, Masson P, Delteil JR, Combaz A (1980) Paleoenvironment and petroleum potential of Middle Cretaceous black shales in Atlantic basins. Am Assoc Petrol Geol Bull 64:2051-2063

Tissot BP, Deroo G, Herbin JP (1979) Organic matter in Cretaceous sediments of the North Atlantic: contribution to sedimentology and paleogeography. In: Talwani M, Hay WW, Ryan WBF (eds) Deep drilling results in the Atlantic Ocean: continental margins and paleoenvironment. Am Geophys Union, Washington, Maurice Ewing Ser 3:362-374

Tornaghi ME, Premoli Silva I, Ripepe M (1989) Lithostratigraphy and planktonic foraminiferal biostratigraphy of the Aptian-Albian "Scisti a Fucoidi" in the Piobbico area: background for cyclostratigraphy. Riv Ital Paleont Stratigr, 95,3:223-264

Tornaghi ME, Ripepe M, Premoli Silva I (1989) Planktonic foraminiferal distribution records productivity cycles: evidence from the Aptian-Albian of central Italy. Terra Nova I:443-448

Tribovillard NP (1988) Controles de la sédimentation marneuse en milieu pélagique semi-anoxique. Exemples dans le Mésozoïque du Sud-Est de la France et de l'Atlantique. Thèse Univ Lyon, 116 pp

Tribovillard NP, Ducreux JL (1986) Mise en évidence de cycles de 100,000 et 400,000 ans dans les Terres Noires du Callovien et de l'Oxfordien de la région de Buis-les-Baronnies (SE France). CR Acad Sci Paris 303:1508-1512

Trurnit P (1968) Analysis of pressure solution contacts and classification of pressure solution phenomena. In: Müller G, Friedman GM (eds) Recent developments in carbonate sedimentology in Central Europe. Springer, Berlin Heidelberg New York, pp 75-84

Tucholke BE, Vogt PR (1979) Western North Atlantic: sedimentary evolution and aspects of tectonic history. In: Tucholke BE, Vogt PR et al. Init Repts DSDP 43: Washington (US Govt Print Office), pp 791-825

Tucholke BE, Hollister CD, Weaver FM, Vennum WR (1976) Continental rise and abyssal plain sedimentation in the Southeast Pacific basin - Leg 35 Deep Sea Drilling Project. In: Hollister CD, Craddoc C et al. Init Repts DSDP 35: Washington (US Govt Print Office), pp 359-400

Tucker ME (1969) Crinoidal turbidites from the Devonian of Cornwall and their paleogeographic significance. Sedimentology 13:281-290

Tucker ME (1974) Sedimentology of Palaeozoic pelagic limestones: the Devonian Griotte (southern France) and Cephalopodenkalk (Germany). In: Hsü KJ, Jenkyns HC (eds) Pelagic sediments. Int Assoc Sediment Spec Publ 1:71-92

Tucker ME (1982) Storm-surge sandstones and the deposition of interbedded limestones: Late Precambrian, southern Norway. In: Einsele G, Seilacher A (eds) Cyclic and event stratification. Springer, Berlin Heidelberg New York, pp 354-362

Tucker ME, Wright VP (1990) Carbonate sedimentology. Blackwell Scientific Publications, Oxford, London, 482 pp

Tuttle JH (1985) The role of sulfur-oxidizing bacteria at deep-sea hydrothermal vents. Bull Biol Soc Wash 6:335-343

Tyson RV (1987) The genesis and palynofacies characteristics of marine petroleum source rocks. In: Brooks J, Fleet AJ (eds) Marine petroleum source rocks. Geol Soc Spec Publ 26:47-67

Tyson RV, Wilson RC, Downie C (1979) A stratified water column environmental model for the type Kimmeridge Clay. Nature 277:377-380

Udden JA (1912) Geology and mineral resources of the Peorisa quadrangle, Illinois. US Geol Surv Bull 506:1-103

Ui T (1983) Volcanic dry avalanche deposits - identification and comparison with nonvolcanic debris stream deposits. J Volcanol Geotherm Res 18:135-150

Vail PR (1987a) Sequence stratigraphy workbook. Fundamentals of sequence stratigraphy. 1988 Am Assoc Petrol Geol Annu Conv Short Course Notes, Tulsa

Vail PR (1987b) Seismic stratigraphy interpretation procedure. In: Bally AW (ed) Atlas of seismic stratigraphy. Am Assoc Petrol Geol. Stud Geol 27:1-10

Vail PR, Audemard F, Bartek LR, Bowman SA, Coterill K, Emmet PA, Liu C, Perez-Cruz G, Ross MI, Wu S (1991) The global stratigraphic signature of the Neogene. In: Macdonald DJM (ed) Sedimentation, tectonics and enstacy. Spec Pub Int Ass Sedimentol 12 (Blackwell, Oxford), in press

Vail R, Hardenbol J (1979) Sea-level changes during the Tertiary. Oceanus 22:71-79

Vail PR, Hardenbol J, Todd RG (1984) Jurassic unconformities, chronostratigraphy, and sea level changes from seismic stratigraphy and biostratigraphy. In: Schlee JS (ed) Interregional unconformities and hydrocarbon accumulation. Am Assoc Petrol Geol Mem 36:129-144

Vail PR, Mitchum RM Jr, Thompson S (1977a) Seismic stratigraphy and global changes of sea level. Part 3 Relative changes of sea level from costal onlap. In: Payton CE (ed) Seismic stratigraphy. Am Assoc Petrol Geol Mem 26:63-81

Vail PR, Mitchum RM Jr, Thompson S (1977b) Seismic stratigraphy and global changes of sea level. Part 4 Global cycles of relative changes of sea levels. Am Assoc Petrol Geol Mem 26:83-97

Vail PR, Mitchum RM Jr, Todd RG, Widmier JW, Thompson S, Sangree JB, Bubb JN, Hatlelid WG (1977c) Seismic stratigraphy and global changes of sea level. In: Payton CE (ed) Seismic stratigraphy - application to hydrocarbon exploration. Am Assoc Petrol Geol Mem 26:49-212

Van Andel TH (1981) Consider the incompleteness of the geological record. Nature 294:397-398

Van Andel TH (1985) New views on an old planet. Cambridge Univ Press, Cambridge, 324 pp

Van Bennekom AJ, Jansen JH, Van der Gaast SJ, Van Iperen JM, Pieters J (1989) Aluminium-rich opal: an intermediate in the preservation of biogenetic silica in the Zaire (Congo) deep-sea fan. Deep-Sea Res 36(2):173-190

Van den Bogaard P, Hall CM, Schmincke HU, York D (1987) $^{40}Ar/^{39}Ar$ Laser dating of single grains. Ages of Quaternary tephra from the East Eifel Volcanic Field FRG. Geophys Res Lett 14:1211-1214

Van den Bogaard P, Hall CM, Schmincke HU, York D (1989b) Precise single-grain $^{40}Ar/^{39}Ar$ dating of a cold to warm climate transition in Central Europe. Nature 342:523-525

Van den Bogaard P, Schmincke HU (1984) The eruptive center of the Late Quaternary Laacher See Tephra. Geol Rdsch 73:935-982

Van den Bogaard P, Schmincke HU (1985) Laacher See tephra: a widespread isochronous Late Quaternary tephra layer in central and northern Europe. Geol Soc Am Bull 96:1554-1571

Van den Bogaard P, Schmincke HU (1988) Aschenlagen als quartäre Zeitmarken in Mitteleuropa. Geowissenschaften 6:75-84

Van den Bogaard P, Schmincke HU, Freundt A, Hall CM, York D (1988) Eruption ages and magma supply rates during the Miocene evolution of Gran Canaria. Naturwissenschaften 75:616-617

Van den Bogaard P, Schmincke HU, Freundt A, Park C (1990) Evolution of complex Plinian eruptions: the Late Quaternary Laacher See case history. In. Harvey DA (ed) Thera and the Aegean world III/2, The Thera Foundation, London, pp 463-485

Van den Bogaard C, Van den Bogaard P, Schmincke H-U (1989a) Quartärgeologisch - tephrostratigraphische Neuaufnahme und Interpretation des Pleistozänprofils Kärlich. Eiszeitalter Gegenwart 39:62-86

Van Hinte JE (1978) Geohistory analysis - application of micro-paleontology in exploration geology. Am Assoc Petrol Geol Bull 62:201-222

Van Hinte JE, Wiese SW et al (1987) Init Repts DSDP 93: Washington (US Govt Print Office), 469 pp

Van Houten FB (1962) Cyclic sedimentation and the origin of analcime-rich Upper Triassic Lockatong Formation, west-central New Jersey and adjacent Pennsylvania. Am J Sci 260:561-576

Van Houten FB (1964) Cyclic lacustrine sedimentation, Upper Triassic Lockatong Formation, central New Jersey and adjacent Pennsylvania. Kansas Geol Surv Bull 169:497-531

Van Leckwijck W (1948) Quelques observations sur les variations verticales des caractères lithologiques et fauniques de divers horizons marins du terrain houillier de Belgique. Ann Soc Géol Belg 71:377-406

Van Loon H, Labitzke K (1988) Association between the 11-year solar cycle, the QBO, and the atmosphere. Part II Surface and 700 mb on the northern hemisphere in winter. J Clim 1:905

Van Wagoner JC, Mitchum RM, Campion KM, Rahmanian VD (1990) Siliciclastic sequence stratigraphy in well logs, core and outcrop: concepts for high-resolution correlation of time and facies. Am Assoc Petrol Geol, Methods Explor 7:55 pp

Van Wagoner JC, Mitchum RM Jr, Posamentier HW, Vail PR (1987) Seismic stratigraphy interpretation using sequence stratigraphy. Part II Key definitions of sequence stratigraphy. In: Bally AW (ed) Atlas of seismic stratigraphy 1. Am Assoc Petrol Geol Stud Geol 27:11-14

Van Wagoner JC, Posamentier HW, Mitchum RM, Vail PR, Sarg JF, Loutit TS, Hardenbol J (1988) An overview of the fundamentals of sequence stratigraphy and key definitions. In: Wilgus C, Hastings B, Ross C, Posamentier H, Van Wagoner J, Kendall LGStC (eds) Sea-level changes - an integrated approach. Soc Econ Paleontol Mineral Spec Publ 42:39-45

Van Weering TCE, van Iperen J (1984) Fine-grained sediments of the Zaire deep-sea fan, southern Atlantic Ocean. Geol Soc Lond Spec Publ 15:95-114

Van Woerkom AJJ (1953) The astronomical theory of climatic changes. In: Shapley H (ed) Climatic change: evidence, causes, and effects. Harvard Univ Press, Cambridge, pp 147-157

Vecsei A, Frisch W, Pirzer M, Wetzel A (1988) Origin and tectonic significance of radiolarian chert in the Austroalpine rifted continental margin In: Hein JR, Obradovic H (eds) Siliceous deposits of the Tethys and Pacific regions. Springer, Berlin Heidelberg New York, pp 65-80

Veizer J (1983a) Chemical diagenesis of carbonates: theory and application of trace element technique. In: Stable isotopes in sedimentary geology. Soc Econ Paleontol Mineral Short Course Notes 10:3-1-3-100

Veizer J (1983b) Trace elements and isotopes in sedimentary carbonates. In: Reeder RJ (ed) Carbonates: mineralogy and chemistry. Min Soc Am Rev Min 11:265-299

Vermeij GJ (1977) The Mesozoic marine revolution: evidence from snails, predators, and grazers. Paleobiology 3:245-258

Via Boada L, Villalta JF, Esteban Cerda M (1977) Paleontologia y Paleoecologia de los Yacimientos Fosiliferos del Muschelkalk Superior entre Alcover y Mont-Ral (Montanas de Prades, Provincia de Tarragona). Cuad Geol Iberica 4:247-256

Viereck LG, Simon M, Schmincke HU (1985) Primary composition, alteration, and origin of Cretaceous volcaniclastic rocks, East Mariana Basin, site 585, Leg 89. In: Moberly R, Schlanger SO et al. Init Repts DSDP 89: Washington (US Govt Print Office), pp 529-553

Vincent E, Berger WH (1985) Carbon dioxide and polar cooling in the Miocene: the Monterey hypothesis. In: The carbon cycle and atmospheric CO_2. Geophys Monogr 32:455-468

Vincent E, Killingley JS, Berger WH (1981) Miocene stable isotope composition of benthic foraminifera from the equatorial Pacific. Nature 289:639-643

Viohl G (1983) Eichstätts Patenort Bolca als Fossillagerstätte. Archaeopteryx, Eichstätt (Jura-Museum), pp 42-69

Viohl G (1985) Geology of the Solnhofen lithographic limestone and the habitat of Archaeopteryx. In: Hecht MK, Ostrom JH, Viohl G, Wellnhofer P (eds) The beginnings of the birds. Proc Int Archaeopteryx Conf 1984, Eichstätt, pp 31-44

Viohl G (1987) Raubfische der Solnhofener Plattenkalke mit erhaltenen Beutefischen. Archaeopteryx 5:33-64

Viohl G (1989) Die Plattenkalke der Sierra de Monsetch (Katalonien) - eine bedeutsame Fossillagerstätte. Archaeopteryx 7:13-29

Vischer H (1985) Report of subcommission on Triassic stratigraphy. Albertiana 3:1-2

Vistelius AB (1980) Osznovü matematicseszkoj geologii. Nauka, Leningrad, pp 1-389

Vogt PR (1979) Global magmatic episodes: new evidence and implications for the steady-state mid ocean ridge. Geology 7:93-98

Voight B, Glicken H, Janda RJ, Douglass PM (1981) Catastrophic rockslide avalanche of May 18. In: Lipman PW, Mullineaux DR (eds) The 1980 eruptions of Mount St. Helens, Washington. US Geol Surv Prof Pap 1250:347-377

Voigt E (1968) Über Hiatus-Konkretionen, dargestellt an Beispielen aus dem Lias. Geol Rdsch 58:281-296

Voigt E (1979) Wann haben sich die Feuersteine gebildet? Nachrichten Akad Wissensch, Math-Phys Kl 6:75-127

Volat JL, Pastouret L, Vergnaud-Grazzini C (1980) Dissolution and carbonate fluctuations in Pleistocene deep-sea cores: a review. Mar Geol 34:1-28

Von der Borch CC, Galehouse J, Nesteroff WD (1971) Silicified limestone-chert sequences cored during Leg 8 of the Deep Sea Drilling Project: a petrologic study. In: Tracey JI Jr et al. Init Repts DSDP 8: Washington (US Govt Print Office), pp 819-827

von Freyberg B (1966) Der Faziesverband im Unteren Malm Frankens. Ergebnisse der Stromatometrie. Erlanger Geol Abh II 62:1-92

von Stackelberg U (1972) Faziesverteilung in Sedimenten des indisch-pakistanischen Kontinentalrandes (Arabisches Meer). Meteor Forsch Ergebn C 9:1-73

Vorren TO, Hald M, Thomsen E (1984) Quaternary sediments and environments on the continental shelf off northern Norway. Mar Geol 57:229-257

Vorren TO, Lebesbye E, Andreassen K, Larsen KB (1989) Glacigenic sediments on a passive continental margin as exemplified by the Barents Sea. Mar Geol 85:251-272

Waage KM (1964) Origin of repeated fossiliferous concretion layers in the Fox Hills Formation. Bull State Geol Surv Kansas 169(2):541-563

Wächter J (1987) Jurassische Massflow- und Internbreccien und ihr sediment-tektonisches Umfeld im mittleren Abschnitt der nördlichen Kalkalpen. Bochumer Geol Geotech Arb 27:239

Wagner G (1913) Stylolithen und Drucksuturen. Geol Palaeontol Abh (NF) 11:101-128

Waitt RB Jr (1984) Periodic jükulhlaups from Pleistocene glacial Lake Missoula - new evidence from varved sediment in northern Idaho and Washington. Quat Res 22:46-58

Walanus A (1989) Periodicities in sequence of laminae thicknesses in laminated sediments from the Gosciaz Lake. Zeszyty Naukowe Politechniki Slaskiej, Ser Mat-Fizy, Z 57, Geochronomet 5(989):53-62

Walker EH (1967) Varved lake beds in northern Idaho and northeastern Washington. US Geol Surv Prof Pap 575-B:83-87

Walker GPL (1973) Explosive volcanic eruptions - a new classification scheme. Geol Rdsch 62:431-466

Walker GPL (1979) A volcanic ash generated by explosions where ignimbrite entered the sea. Nature 281:642-646

Walker GPL (1981) Plinian eruptions and their products. Bull Volcanol 444:223-240

Walker GPL, Croasdale R (1971) Two Plinian-type eruptions in the Azores. J Geol Soc Lond 127:17-55

Walker GPL, Wilson CJN, Froggatt PC (1981) An ignimbrite veneer deposit: the trail marker of a pyroclastic flow. J Volcanol Geotherm Res 9:409-421

Walker JCG, Zahnle KJ (1986) Lunar nodal tide and the distance to the Moon during the Precambrian. Nature 330:600-602

Walker RG (1973) Mopping up the turbidite mess. In: Ginsburg RN (ed) Evolving concepts in sedimentology. Johns Hopkins Univ Press, Baltimore, pp 1-37

Walker RG (1978) Deep-water sandstone facies and ancient submarine fans: models for exploration for stratigrphic traps. Am Assoc Petrol Geol Bull 62:932-966

Walker RG (1984a) Shelf and shallow-marine sands. Geosci Can Rep Ser 1:141-170

Walker RG (1984b) Turbidites and associated coarse-grained clastic deposits. In: Walker R (ed) Facies models, 2nd edn. Geol Assoc Can, Geosci Reprint Ser 1:171-188

Walker RG (1985) Mudstones and thin-bedded turbidites associated with the Upper Cretaceous Wheeler Gorge conglomerate, California: a possible channel-levee complex. J Sediment Petrol 55:279-290

Walker RG, Mutti E (1973) Turbidite facies and facies associations. In: Middleton GV, Bouma AH, (eds) Turbidites and deep water sedimentation. Soc Econ Paleontol Mineral Pacific Section Short Course, Anaheim 1973, pp 119-158

Walker SE (1988) Taphonomic significance of hermit crabs (Anomura: Paguridea): epifaunal hermit crab - infaunal gastropod example. Palaeogeogr Palaeoclimatol Palaeoecol 63:45-71

Walling DE, Webb BW (1986) Solutes in river systems. In: Trudgill ST (ed) Solute processes. Wiley, Chichester, pp 251-327

Walliser OH (1986) Global bio-events: a critical approach. Springer, Berlin Heidelberg New York, 442 pp

Walsh I, Dymond J, Collier R (1988) Rates of recycling of biogenic components of settling particles in the ocean derived from sediment trap experiments. Deep-Sea Res 35:43-58

Walters LJ Jr, Owen DE, Henley AL, Winsten MS, Valek KW (1987) Depositional environments of the Dakota Sandstone and adjacent units in the San Juan Basin utilizing discriminant analysis of trace elements in shales. J Sediment Petrol 57(2):265-277

Walther JV, Helgeson HC (1977) Calculation of the thermodynamic properties of aqueous silica and the solubility of quartz and its polymorphs at high pressures and temperatures. Am J Sci 277:1315-1351

Walther M (1982) A contribution to the origin of limestone-shale sequences. In: Einsele G, Seilacher A (eds) Cyclic and event stratification. Springer, Berlin Heidelberg New York, pp 113-120

Wanless HR (1979) Limestone response to stress: pressure solution and dolomitization. J Sediment Petrol 49:437-462

Wanless HR, Baroffio JR, Trescott PC (1969) Conditions of deposition of Pennsylvanian coal beds. In: Dapples EC, Hopkins ME (eds) Environments of coal deposition. Geol Soc Am Spec Pap 114:105-142

Wanless HR, Patterson J (1952) Cyclic sedimentation in the marine Pennsylvanian of the southwestern United States. Congr Int Strat Géol Carbonif, Heerlen 1951, CR 3(2):655-664

Wanless HR, Shepard FB (1936) Sea level and climatic changes related to Late Paleozoic cycles. Bull Geol Soc Am 47:1177-1206

Wanless HR, Tedesco LP, Tyrrell KM (1988) Production of subtidal tubular and surficial tempestites by Hurricane Kate, Caicos Platform, British West Indies. J Sediment Petrol 58:739-750

Wanless HR, Weller JM (1932) Correlation and extent of Pennsylvanian cyclothems. Bull Geol Soc Am 43:1003-1016

Watkins ND, Sparks RSJ, Sigurdsson H, Huang TC, Federman A, Carey S, Ninkovich D (1978) Volume and extent of the Minoan tephra from Santorini: new evidence from deep-sea sediment cores. Nature 271:122-126

Watkins R (1979) Benthic community organization in the Ludlow Series of the Welsh borderland. Br Mus Nat Hist Bull Geol 31:175-280

Watts AB (1982) Tectonic subsidence, flexure and global changes of sea level. Nature 297:469-475

Watts AB, Karner GD, Steckler MS (1982) Lithospheric flexure and the evolution of sedimentary basins. Philos Trans R Soc A 305:249-281

Watts AB, Ten Brink US, Buhl P, Brocher TM (1985) A multichannel seismic study of lithospheric flexure across the Hawaiian-Emperor seamount chain. Nature 315:105-111

Watts AB, Thorne J (1984) Tectonics, global changes in sea level and their relationship to stratigraphical sequences at the US Atlantic continental margin. Mar Petrol Geol 1:319-339

Watts KF (1988) Triassic carbonate submarine fans along the Arabian platform margin, Sumeini group, Oman. Sedimentology 35:43-72

Webb JA (1986) Ordovician bedded cherts of eastern Victoria, Australia. 12th Int Sediment Congr, Int Assoc Sediment, 24th-30th August 1986, Canberra, Australia, Abstracts, Theme A 8

Weber HP (1981) Sedimentologische und geochemische Untersuchungen im Greifensee (Kanton ZH/Schweiz). PhD Thesis, ETH-Zürich 6811

Weber HS (1967) Zur Westgrenze der ostschwäbisch-fränkischen Fazies des Braunjura (Dogger) beta in der Schwäbischen Alb (Württemberg). Jahresber Mitt Oberrhein Geol Ver NF 49:47-54

Weber P (1969) Bildung und Regelung von Knollengefügen im Oberdevon des Rheinischen Schiefergebirges. Fortsch Geol Rheinl Westf 17:81-94

Weedon GP (1986) Hemipelagic shelf sedimentation and climatic cycles: the basal Jurassic (Blue Lias) of South Britain. Earth Planet Sci Lett 76:321-335

Weedon GP (1989) The detection and illustration of regular sedimentary cycles using Walsh power spectra and filtering, with examples from the lias of Switzerland. J Geol Soc Lond 146:133-144

Wefer G (1989) Particle flux in the ocean: effects of episodic production. In: Berger WH, Smetacek VS, Wefer G (eds) Productivity of the ocean: present and past. Wiley, New York, pp 139-153

Weimer RJ (1960) Upper Cretaceous stratigraphy, Rocky Mountain area. Am Assoc Petrol Geol Bull 44:1-20

Weimer RJ (1970) Rates of deltaic sedimentation and intrabasin deformation, Upper Cretaceous of Rocky Mountain region. In: Morgan JP (ed) Deltaic sedimentation, modern and ancient. Soc Econ Paleontol Mineral Spec Publ 15:211-222

Weimer RJ (1984) Relation of unconformities, tectonics, and sea-level changes in the Cretaceous of Western Interior, USA. In: Schlee S (ed) Interregional unconformities and hydrocarbon accumulation. Am Assoc Petrol Geol Mem 36:7-35

Weissert H (1979) Die Paläoozeanographie der südwestlichen Tethys in der Unterkreide. Mitt Geol Inst ETH Zürich 226

Weissert H (1981a) Depositional processes in ancient pelagic environment: the Lower Cretaceous Maiolica of the southern Alps. Eclogae Geol Helv 74:339-352

Weissert H (1981b) The environment of deposition of black shales in the early Cretaceous: an ongoing controversy. Soc Econ Paleontol Mineral Spec Publ 32:547-560

Weissert H, McKenzie J, Hochuli P (1979) Cyclic anoxic events in the early Cretaceous Tethys Ocean. Geology 7:147-151

Weller JM (1930) Cyclic sedimentation of the Pennsylvanian period and its significance. J Geol 38:97-135

Weller JM (1931) The conception of cyclical sedimentation during the Pennsylvanian period. Illinois State Geol Surv Bull 60:163-177

Weller JM (1956) Arguments for diastrophic control of Late Paleozoic cyclothems. Bull Am Assoc Petrol Geol 40:17-50

Weller JM (1964) Development of the concept and interpretation of cyclic sedimentation. In: Merriam DF (ed) Symposium on cyclic sedimentation. Kansas Geol Surv Bull 169:607-621

Wendt J, Aigner T (1985) Facies patterns and depositional environments of Paleozoic cephalopod limestones. Sediment Geol 44:263-300

Wendt J, Aigner T, Neugebauer J (1984) Cephalopod limestone deposition on a shallow ridge: the Tafilalt Platform (Upper Devonian, eastern Anti-Atlas, Morocco). Sedimentology 31:601-625

Wepfer E (1926) Die Auslaugungsdiagenese, ihre Wirkung auf Gestein und Fossilinhalt. N Jb Mineral Beil 54:17-94

Westernhagen H, Dethlefsen V (1983) North Sea oxygen deficiency 1982 and its effects on bottom fauna. Ambio 12:264-266

Westgate JA, Gorton MP (1981) Correlaton techniques in tephra studies. In: Self S, Sparks RJS (eds) Tephra studies. Reidel, Dordrecht, pp 73-94

Westgate JA, Walter RC, Pearce GW, Gorton MP (1985) Distribution, stratigraphy, petrochemistry, and paleomagnetism of the Late Pleistocene Old Crow tephra in Alaska and the Yukon. Can J Earth Sci 22:893-906

Westoll TS (1968) Sedimentary rhythms in coal-bearing strat. In: Murchison D, Westoll TS (eds) Coal and coal-bearing strata. Oliver & Boyd, Edinburgh, pp 71-103

Wetzel A (1983) Biogenic sedimentary structures in a modern upwelling regime: northwest Africa. In: Thiede J, Suess E (eds) Coastal upwelling and its sediment record. Part B Sedimentary records of ancient coastal upwelling. Plenum, New York, pp 123-144

Wetzel A, Aigner T (1986) Stratigraphic completeness: tiered trace fossils provide a measuring stick. Geology 14:234-237

Weyl PK (1959) Pressure solution and the force of crystallization - a phenomenological theory. J Geophys Res 64:2001-2025

Wheeler HE (1958) Time-stratigraphy. Am Assoc Petrol Geol Bull 42:1047-1063

Whitham AG (1989) The behaviour of subaerially produced pyroclastic flows in a subaqueous environment: evidence from the Roseau eruption, Dominica, West Indies. Mar Geol 86:27-40

Whittaker A (ed) (1985) Atlas of onshore sedimentary basins in England and Wales: post-Carboniferous tectonics and stratigraphy. Blackie, Glasgow

Wignall PB (1989) Sedimentary dynamics of the Kimmeridge Clay: tempestites and earthquakes. J Geol Soc Lond 145:273-284

Wignall PB (in press) Observations on the evolution and classification of dysaerobic communities. In: Miller W (ed) Paleocommunity temporal dynamics: the long-term development of multispecies assemblies. Spec Publ Palaeontology 10

Wignall PB, Myers KJ (1988) Interpreting benthic oxygen levels in mudrocks: a new approach. Geology 16:452-455

Wignall PB, Simms MJ (1990) Pseudoplankton. Palaeontology (in press)

Wilde P, Berry WBN (1982) Progressive ventilation of the oceans - potential for return to anoxic conditions in the post-Paleozoic. In: Schlanger SO, Cita MB (eds) Nature and origin of Cretaceous carbon-rich facies. Academic Press, London, pp 209-224

Wilde P, Quinby-Hunt MS, Berry WBN (1988) Vertical advection from oxic or anoxic waters from the main pycnocline as a cause of rapid extinction or rapid radiations. Abstr 3rd Int Conf Global Bioevents, Boulder, CO, Univ Colorado, 36 pp

Wilkinson BH (1979) Biomineralization, paleoceanography, and the evolution of calcareous marine organisms. Geology 7:524-527

Wilkinson BH, Owen RM, Carroll AR (1985) Submarine hydrothermal weathering, global eustacy, and carbonate polymorphism in Phanerozoic marine oolites. J Sediment Petrol 55:171-183

Willett HC (1987) Climate responses to variable solar activity - past, present and predicted. In: Rampino MS, Sanders JE, Newman WS, Künigsson LK (eds) Climate-history, periodicity, and predictability. Van Nostrand Reinhold, New York, pp 404-414

Williams DF, Lerche I, Full WE (1988) Isotope chronostratigraphy: theory and methods. Academic Press, San Diego, 345 pp

Williams EG, Bragonier WA (1974) Controls of Early Pennsylvanian sedimentation in western Pennsylvania. In: Briggs G (ed) Carboniferous of the southeastern United States. Geol Soc Am Spec Pap 148:135-152

Williams GE (1988) Late Precambrian tidal rhythmites in South Australia and the history of the Earth's rotation. J Geol Soc Lond 146:97-111

Williams GE (1989) Precambrian tidal sedimentary cycles and Earth's paleorotation EOS 70:33-41

Williams GE, Sonett CP (1985) Solar signature in sedimentary cycles from the Late Precambrian Elatina Formation, Australia. Nature 318:523-527

Williams LA (1983) Deposition of the Bear Gulch limestone: a Carboniferous Plattenkalk from central Montana. Sedimentology 30:843-860

Williams LA, Crerar DA (1984) Silica diagenesis, II. General mechanisms. J Sediment Petrol 55:312-321

Williams LA, Parks GA, Crerar DA (1985) Silica diagenesis: solubility controls. J Sediment Petrol 55:301-311

Williams LA, Reimers C (1983) Role of bacterial mats in oxygen-deficient marine basins and coastal upwelling regimes: preliminary report. Geology 11:267-269

Williams PJ, Von Bodungen B (1989) Group report "export productivity from the photic zone". In: Berger WH, Smetacek VS, Wefer G (eds) Productivity of the ocean: present and past. Wiley, New York, pp 99-115

Wilson AO (1985) Depositional and diagenetic facies in the Jurassic Arab-C and -D reservoirs, Qatif Field, Saudi Arabia. In: Roehl PO, Choquette PW (eds) Carbonate petroleum reservoirs. Springer, Berlin Heidelberg New York, pp 321-340

Wilson CJN (1985) The Taupo eruption, New Zealand II. The Taupo ignimbrite. Philos Trans R Soc Lond A 314:229-310

Wilson CJN, Walker GPL (1985) The Taupo eruption, New Zealand I. General aspects. Philos Trans R Soc Lond A 314:199-228

Wilson JL (1975) Carbonate facies in geologic history. Springer, Berlin Heidelberg New York, 471 pp

Wilson L, Sparks RJS, Huang TC, Watkins ND (1978) The control of volcanic column eruption heights by eruption energetics and dynamics. J Geophys Res 83:1829-1836

Wilson RC (1980) Changing sea-levels: a Jurassic case study. Open Univ Press, Milton Keynes, pp 49-65

Wilson V, Welch FBA, Robbie JA, Green GW (1958) Geology of the country around Bridgeport and Yeovil. Geol Surv Great Brit Mem, 232 pp

Winter J (1977) "Stabile" Spurenelemente als Leit-Indikatoren einer tephrostratigraphischen Korrelation (Grenzbereich Unter-/Mitteldevon, Eifel-Belgien). Newsl Stratigr 6:152-170

Winter J (1982) Habits of zircon as a tool for precise tephrostratigraphic correlation. In: Einsele G, Seilacher A (eds) Cyclic and event stratification. Springer, Berlin Heidelberg New York, pp 423-428

Winterer EL, Bosellini A (1981) Subsidence and sedimentation on Jurassic passive continental margin, southern Alps, Italy. Am Assoc Petrol Geol Bull 65:394-421

Wise SW, Kelts KR (1972) Inferred diagenetic history of a weakly silicified deep sea chalk. Trans Gulf-Coast Assoc Geol Soc 22:177-203

Wissmann G (1979) Cape Bojador slope, an example for potential pitfalls in seismic interpretation without the information of outer margin drilling. In: von Rad U, Ryan WBF et al. Init Repts DSDP 47-1: Washington (US Govt Print Office), pp 491-500

Wohletz KH (1983) Mechanism of hydrovolcanic pyroclast formation: grain size, scanning microscopy, and experimental studies. J Volcanol Geotherm Res 17:31-63

Wollast R, Garrels RM (1971) Diffusion coefficient of silica in seawater. Nature 229:94

Wonders AAH (1980) Middle and Late Cretaceous planktonic foraminifera of the western Mediterranean area. Utrecht Micropaleontol Bull 24:1-157

Wong PK, Oldershaw AE (1980) Causes of cyclicity in reef interior sediments, Kaybob Reef, Alberta. Bull Can Petrol Geol 28:411-424

Wong PK, Oldershaw AE (1981) Burial cementation in the Devonian, Kaybob Reef Complex, Alberta, Canada. J Sediment Petrol 51:507-520

Wood CJ, Ernst G, Rasemann G (1984) The Turonian-Coniacian stage boundary in Lower Saxony (Germany) and adjacent areas: the Salzgitter-Salder quarry as a proposed international standard section. Bull Geol Soc Denmark 33:225-238

Wright LD (1981) Nearshore tidal currents and sand transport in a macrotidal environment. Geo-Mar Lett 1:173-179

Wright LD (1987) Shelf-surf zone coupling: diabatic shoreface transport. Coast Seds '87, Am Soc Civ Engr 1:25-40

Wright LD, Boon JD III, Green MO, List JH (1986a) Response of the mid shoreface of the southern mid-Atlantic bight to a "northeaster". Geo-Mar Lett 6:153-160

Wright LD, Guza RT, Short AD (1982) Dynamics of a high-energy dissipative surf zone. Mar Geol 45:41-62

Wright LD, Short AD (1983) Morphodynamics of beaches and surf zones in Australia. In: Komar PD (ed) CRC handbook of coastal processes and erosion. CRC Press, Boca Raton, pp 35-64

Wright LD, Yang Z-S, Bornhold BD, Keller GH, Prior DB, Wiseman WJ Jr (1986b) Hyperpycnal plumes and plume fronts over the Huanghe (Yellow River) delta front. Geo-Mar Lett 6:97-105

Wright VP (1984) Peritidal carbonate facies models: a review. Geol J 19:309-325

Wright VP (1986) Facies sequences on a carbonate ramp: the Carboniferous limestone of South Wales. Sedimentology 33:221-241

Wu G, Berger WH (1989) Planktonic foraminifera: differential dissolution and the Quaternary stable isotope record in the west-equatorial Pacific. Paleoceanography 4(2):181-198

Wyrtki K (1962) The oxygen minima in relation to ocean circulation. Deep-Sea Res 9:11-23

Xu Hui-long, Sun Wei-han, Hou Ji-hui, Pan Sui-xian, Liu Yu, Zhao Xiu-hu, Liu Lu-jan, Rui Lin (1987) Carboniferous and Permian stratigraphy in Shanxi. 11th Int Congr Carbonif Stratigr Geol Beijing, Guidebook 1:1-66

Yamamoto S, Tokuyama H, Fujioka K, Taakeuchi A, Ujiié H (1988) Carbonate turbidites deposited on the floor of the Palau Trench. Mar Geol 82:217-233

Yang CS, Nio SD (1985) The estimation of palaeohydrodynamic processes from subtidal deposits using time series analysis methods. Sedimentology 32:41-58

Yemane K, Siegenthaler C, Kelts K (1989) Lacustrine environment during Lower Beaufort (Upper Permian) Karoo deposition in northern Malawi. Palaeogeogr Palaeoclimatol Palaeoecol 70:165-178

Yokoyama T, Horie S (1974) Lithofacies of a 200 m core sample from Lake Biwa. Paleolimnol Lake Biwa Jpn Pleistocene 2:31-37

York D, Hall CM, Yanase Y, Hanes JA, Kenyon WJ (1981) 40 Ar/39 Ar dating of terrestrial minerals with a continuous laser. Geophys Res Lett 8:1136-1138

Yose LA, Heller PL (1989) Sea level control of mixed-carbonate-siliciclastic gravity flow deposition: lower part of the Keeler Canyon Formation (Pennsylvanian), southeastern California. Geol Soc Am Bull 101:427-439

Young HR, Nelson CS (1988) Endolithic biodegradation of cool-water skeletal carbonates on Scott Shelf, northwestern Vancouver Island, Canada. Sediment Geol 60:251-267

Young RG (1957) Late Cretaceous cyclic deposits, Book Cliffs, eastern Utah. Am Assoc Petrol Geol Bull 41:1760-1774

Yuretich RF (1979) Modern sediments and sedimentary processes in Lake Rudolf (Lake Turkana), East Rift Valley, Kenya. Sedimentology 26:313-331

Zachos JC, Arthur MA (1986) Paleoceanography of the Cretaceous, Tertiary boundary event: inferences from stable isotopic and other data. Paleoceanography 1:5-26

Zangerl R, Richardson ES (1963) The paleoecological history of two Pennsylvanian black shales. Fieldiana Geol Mem 4:352

Zankl H (1967) Die Karbonatsedimente der Obertrias in den nördlichen Kalkalpen. Geol Rdsch 56:128-139

Zankl H (1971) Upper Triassic carbonate facies in the Northern Limestone Alps. In: Müller G (ed) Sedimentology of parts of Central Europe. 8th Int Sediment Congr, Int Assoc Sediment, Heidelberg 1971, Guidebook, pp 147-185

Zelt FB (1985) Natural gamma-ray spectrometry, lithofacies, and depositional environments of selected Upper Cretaceous marine mudrocks, western United States, including Tropic Shale and Tununk Member of Mancos Shale. PhD Thesis, Princeton Univ, Princeton, NJ, 340 pp

Zenger DH, Dunham JB (ed) (1980) Concepts and models of dolomitization. Soc Econ Paleontol Mineral Spec Publ 28:320

Zhao XF, Hsü KJ, Kelts K (1984) Varves and other laminated sediments of Zübo. In: Hsü KJ, Kelts KR (eds) Quaternary geology of Lake Zürich: an interdisciplinary investigation by deep-lake drilling. Contrib Sediment 13:2161-2177

Zhou L, Kyte FT (1988) The Permian-Triassic boundary event: a geochemical study of three Chinese sections. Earth Planet Sci Lett 90:411-421

Ziegler B (1958) Feinstratigraphische Untersuchungen im Oberjura Südwestdeutschlands - ihre Bedeutung für Paläontologie und Paläogeographie. Eclogae Geol Helv 58:265-278

Ziegler PA (1982) Geologic atlas of Western and Central Europe. Shell Int Petrol Maatschappij, The Hagne, 130 pp

Zolitschka B (1990) Spätquartäre Sedimentationsgeschichte des Meerfelder Maares (Westeifel). Mikrostratigraphie jahreszeitlich geschichteter Seesedimente. Eiszeitalter Gegenwart 38 (in press)

Zwarts L, Wanink J (1989) Siphon size and burying depth in deposit- and suspension-feeding benthic bivalves. Mar Biol 100:227-240

Subject Index

accomodation space (see sediment accomodation)
anachronistic facies 310
Antarctic continental slope deposits 766 767
allocyclic, allogenetic processes 7
allodapic limestones 328-330, 340
Alpine Triassic 57, 58, 722-732
apparent saturation level (ASL) 113, 124
ash layers (see tephra layers)
atmospheric CO_2 (see carbon cycle)
attached carbonate shelves 260
autocyclic, autogenetic processes 7, 712, 713, 742-745

backlap 678
backset beds 387
Bahama 262-63
Baltimore Canyon Trough 626, 645
Barents shelf 762
basin evolution 352-356
bedded cherts
    diagenesis (see diagenetic overprinting)
    events of redeposition 474-479
    global opal cycle 465-467, 471-473
    opal - carbonate rhythms 470, 471
    opal dilution rhythms 467-469, 473-474
    opal productivity rhythms 467-469, 471-473
    radiolarians, hydrodynamic properties 475, 476
    radiolarian turbidites 476, 477
    siliceous varves 473, 474
bentonite layers (see tephra layers)

bioclasts 13, 14
bioerosion 272
biogenic varvites 598, 599
biolamination 592
biological response to sedimentological events 13
    benthic events 14
    pre-event community 13
    post-event community 13
bioturbation 237, 543, 557
    diagenetic effects 300
    horizons 525
    secular changes 298
    tiering 301
biozonal boundaries 14
bivalve stratinomy 569
    chemosymbiosis 535
black shales 296, 500-523, 524, 527,
    adaptive strategies of fauna 554-556
    bed-scale variations 512, 513
    benthic faunas 546, 551
    bioturbation 543, 557
    chemosymbiosis 555
    depositional environments 508, 509, 711-717
    dysaerobic biofacies 556-561
        in regions of upwelling 511
    lithologic characteristics 508
    oxygenation sediment-water interface 552
    redox cycles (see limestone-marl rhythms)
    related to transgressions 511, 521
    sequence-scale phenomena 510, 501, 523

shell islands 556, 557
time periods of black shales 509-511
varve-scale laminations 34, 521-522
bone beds 690, 592
Bridge Creek Limestone 84-89, 130-134, 144-45, 559-563
Bubnoff units 50
bundles (see nomenclature)

calcite compensation depth, CCD 74-77, 110-115, 328-331, 344
California, Neogene 692
Cambrian, USA 304
carbonate content (see limestone-marl rhythms)
   bedding rhythms, prediagenetic 180
   in sediments of glacial periods 33, 765, 766
carbonate dissolution (see dissolution cycles) 33, 344, 765, 766
carbonate platforms 57-61, 260, 655-657
   lagoonal-peritidal cycles 709-721, 722-732
carbonate preservation 113-116
carbonate production rates 711
carbonate ramps 260
carbonate reservoirs 118
carbonate saturation 113-116
carbon cycle, global 27, 45
   atmospheric CO_2, last glacial 29
Catskill delta 701
Cenonanian-Turonian Event 89-93
cementation (see diagenetic overprinting)
chalk, chalk rhythms 56, 57, 70, 130-134, 446
channel incision 664, 665
chemical compaction 36
chemical cycles 52
chemosymbiosis 535, 555
chert stratification (see bedded chert)
Chondrites 534
chronostratigraphic sequences, diagrams 620-622, 624-626, 630, 664-671, 815-817
classification of cyclic sequences (see nomenclature) 4-7, 618

according to time periods 5, 619
descriptive 4
eustatic cycles, order 643-644
genetic, to depositional types 131, 168, 171-173
glacio-eustatic cycles 620
quasi-periodic processes 11, 13
Milankovitch frequency band (see orbital paramters) 5
tephra layers 392-429
third-order sequence cycles 644-645
Vail-Haq nomenclature 5, 61, 618
climatic change
   cause of limestone-marl rhythms 46, 835-837
   effect on high-frequency sea-level changes 710
coal cycles
   climatic control 745
   cyclothem concept, evolution 733-738
   delta lobe switching 742-743
   depositional control, autocyclicity 744-745
   eustatic control 745, 746
   Pennsylvanian, N.America 735, 738-739
   Shanxi Province, China, Carboniferous 741-742
   tectonic controls 746-748
   types, environments 735, 737, 738 742
   Western Interior Seaway, Cretaceous 739, 740
coal deposits 652
coastal jet 237
combined flow 242, 378
compensation currents 250
composite dissolution index (CDI) 113
concretions 437, 441
   correlation 809, 810
condensation 652
condensed sections 652-653, 668-671 (see sequence stratigraphy)
continental flooding cycles 618-640
coquinas 689

correlation, sideward 18, 795-819
  bypass surfaces 804, 805
  carbonate, organic carbon contents 808, 809
  chemostratigraphy, elements 806
  chemostratigraphy, isotopes 806-808
  composite events 814
  concretions 809
  ecological events 812, 813
  extinctions 810-812
  high resolution event stratigraphy (HIRES) 797-800, 815-819
  impacts 802
  lateral, flux determination 777-789
  magnetostratigraphy 800,
  tempestites 803, 804
  varves 804
  volcanic ash beds 802, 803
Cretaceous seas 34, 46, 65-67, 88, 518, 519
  epicontinental seas 91-92
  France, southeastern 820-830
  Italy, central 83-84
  North Atlantic Ocean 128
  Western Interior Seaway, 739, 740, 779, 780
cross-spectral analysis (see time series analysis)
cycle asymmetry, Jurassic Germany 671-680
cycle chart 628
cycle hierarchy 48-50, 352, 353
cycle modelling 713-715
cycles versus events 222
cyclograms 835-837
cyclothem (see coal cycles)

Dachstein Limestone, Hauptdolomit 58, 727-732
debris flows and mud flows (see also sequence stratigraphy, lowstand)
  amalgamation 321
  carbonate debris flows 656
  cohesive and frictional freezing 321
  matrix 319
  megaturbidites 321
  olistostromes 648

deep-sea carbonate cycles 110-25
  Atlantic 121-123, 128
  carbonate contents 180-184
  Indian Ocean, Pacific 121-123
deep-sea fans 317, 332-336, 374
  (see also sequence stratigraphy)
  channel fills 332-3
  channel-levee complexes 332
  channel, lobe switching 334, 335
  channel-lobe transition 334
  overbank deposits 333-334
  sedimentation rates 336
delta switching, Mississippi 743
depositional sequence, systems (see nomenclature)
Devonian faunas 701
diaclasts 14
diagenetic overprinting 15-7, 71, 344-5, 431-463, 480-491
  burial diagenesis 138-143, 446, 489, 490
  carbonate cementation 431, 454-457
  carbonate concretions 16, 17, 437-441
  carbonate compaction law 433
  carbonate dissolution 431
  chalk, succession of overprints 446
  changes through Earth history 446-449
  compaction, porosity reduction siliceous sediment 483-485
  diagenetic bedding, differential cementation and dissolution 36, 432-437, 454-463
  differential compaction, 16, 432, 433-435
  differential compaction, siliceous 485-486
  dissolution seams 451-454
  dolomitization 442, 443, 725-727
  early diagenesis 134-138
  event bed cementation 17, 444, 445
  fitted fabric 451-453
  flint nodule bedding 441, 442
  hardground cementation 16, 443, 444
  influence on bedding rhythm 434-437, 457-463
  isotope shifts 139-143, 439, 440
  nodular limestones 445, 446

pore water, diffusion and advection 456, 457, 487, 488
pressure dissolution, carbonate 15, 344, 450-457, 457-463
pressure dissolution, siliceous 489, 490
selective cementation 15, 17
silica phase transformations 480-483
silica redistribution between beds 469, 470, 487-490
silification 344
stylolitic structures 344, 452-454
underbeds 17, 444, 445
diastems (see stratigraphic gaps)
dilution cycles (see limestone-marl rhythms)
discontinuity surfaces (see erosional unconformities)
dissolution cycles (see limestone marl-rhythms)
dolomitization (see diagenetic overprinting)
downlap 678
downwelling 246
dropstone muds 764, 765

echinoderm graveyards 289, 293
encrustation 271
entrophization 566
epibenthos, epifauna 13
erosional unconformites 622-623, 665-681
    dating 672
    hiatus timing 784-792
eustatic sea level changes (see sea level changes)
events versus cycles 222
exaerobic facies 538

facies models
    carbonate platforms 716, 725, 729
    deltaic, coastal plain 737
    glacial-interglacial shelf-slope setting 768-769, 771
    turbidites 374, 354
fallout deposits (see tephra layers)
false bedding 386
feedback systems climatic-oceanic 24-27, 45

fish graveyards 290
flash flood deposits 383
flat-pebble conglomerates 305, 306
floating ice shelf 760
fossil lagerstaetten 257, 283
    Solnhofen, Germany 581-584
Francfort Shales 293

Gavish Sabkha 605
geotrophic flow 235
glacial-interglacial cycles 760-772
    glaciomarine cycles 769-772
    isotpe curve, last glacial 28
    lysocline variation 110-125
glacial rhythms, varves 198
glacier types 752-754
glacio-isostatic rebound 762
glaciomarine deposits 752-755
glaciomarine facies assemblages 754, 755
global silica cycle 465-467, 471-473
gravity mass movements 315-318, 346, 347
    collapse events 319
    compound mass movements 315
    frequency 336-338, 351
    grain flows 316
    liquefaction, liquefied mass 316
    slide-debris flows 316
    slides 319
    slumping 314
    transport distances 317
    viscoplastic flows 316
    volcaniclastic (see tephra layers)
    volume of mass movements 317
greenhouse and icehouse effects 45
Green River Shales 29
Gubbio sequence 777

Hamilton Shales 293-94
Hauptrogenstein 291
hiatal accumulations 279
hierarchical sequences (see cycle hierarchy)
high-resolution event stratigraphy (see correlation)
Hunsrück Shales 293

ichnofabric 307, 525
ignimbrites (see tephra layers)
impacts, cosmic 802
infragravity waves 230-231
integrated event stratigraphy (see correlation, high resolution)
interaction subsidence, eustacy, sedimentation 617-20, 713, 715
inundites 224
isochrones 372-372, 798, 815, 816
    landward merging 372
isotope signals
    concretions 440
    diagenesis of beds 138-143
    rhythmic sedimentation 25-29, 85-88

Jordan 384
Jurassic, southern Germany 671-80

Kimmeridge Clay 568, 569
Klüpfel cycles 696

lacustrine life cycle 217
lacustrine-playa facies 50, 51
lacustrine sedimentation (see lake sediment) 189-91
lag deposits 665, 770
    timing 786, 787
    transgressive lags 626, 671
lagoonal-peritidal carbonate cycles 709-732
    allocyclic processes 714-720
    autocyclic processes 712, 713
    climatic-eustatic influence 714
    controlling factors 710, 771
    Dachstein Limestone, Hauptdolomit 727-729
    dolomitisation 725-727
    facies successions 712, 716, 724, 725, 729
    Lofer cycles 722-732
    models 715-717
    preservation potential 720
    Prubeckian, French Jura 718-720
    stacking pattern 731
    tectonic influence 720

Lake Biwa 213
    Bonneville 218-219
    Gosiute, Green River Form. 209-211, 574-576
    Greifen 201
    Rita Blanca 202, 209
    Searles Lake 212-213
    Silak 207
    Turkana 209-210
    Urmia 197, 201
    Van 197, 201, 208, 576-578
    Zürich 197-199, 217
lake sediments
    algal, microbial mats 199, 219
    allochthonous events 198
    biological productivity 199
    black shales, sapropels 195, 202, 203, 218, 219, 512
    chemical processes 199
    closed and open systems 217-219
    El Nino events 210
    extraterrestrial forcing 203
    glacial rhythms 192-198, 220, 221
    high-latitude lakes 207
    low-latitude lakes 208
    lunar-solar tidal periods 203
    Milankovitch rhythms 211-217
    solar rhythms, sun spot control 203-205
    transgressive-regressive cycles 217
larval development 570
limestone-marl rythms 30, 68, 69, 822-825
    bedding rythhms, distinction 170-173
    bed thickness ratio 31, 32, 34, 131, 163, 171
    bioturbation 37, 155
    calcareous redox cycles 34, 46, 68, 69
    carbonate difference, prediagenetic 180-184
    carbonate-organic carbon ratio 169, 171, 177, 179
    Cenomanian-Turonian Boundary Event 89-93
    composition, biota 825, 88-9
    diagenesis (see diagenetic overprinting)

differentiation 859
dilution cycles 31-33, 81, 131, 144-149
dissolution cycles 33, 34, 110-112, 131
depositional environment 42
depositional models, distinction 158-162, 168-170
fauna and preservation 37
geochemical element ratios 134-138, 144-54
input variation, single and composite 173-178
isotopes 37, 69-71, 85-88
lateral correlation 18, 38, 824, 825, 636
light-dark layers 761-763
organic carbon - carbonate relationship 137, 159, 169
organic matter type 134-135, 154-158
productivity cycles 30-31, 74-74, 81-83, 110-112, 131, 149-59, 519
pyrite content 135
redox cycles 34, 134-138
relation to sea level changes 44, 615
sedimentation rate variation 39-40, 89, 167-187
siliceous and calcareous flux 83, 84
thickness limestone-marl successions 43
time series analysis 840-854
timing of bedding couplets 38-40, 167-187
trace elements 36, 827
weathering limits, field recognition 27-29
lithographic limestones (see plattenkalk facies)
lithostratigraphy, lith. units 621, 622
live/dead interactions 270
Lofer cycles (see lagoonal-peritidal cycles)
Long Island 239
longshore currents 232
lysocline (see calcite comp. depth) 113-116

marker beds 14, 319, 621-622, 668, 688, 779, 795-819
diachronous, reworked horizons 671
Maryland, Miocene 688
mass extinctions 706
mass flows (see gravity mass movements)
mass mortality 286
maturation 381
compositional 381
textural 381
Mesozoic cyclicity 821
microbial mats 532
Milankovitch oscillations 23, 48, 644 (see orbital parameters)
Miocene, Maryland 688
Monterey Formation 497, 505
mud blanketing 288
mud flows (see debris flows)
mud turbidites 330-332, 361-366, 381
aggregates of mud particles 372
criteria for recognition 366-368
cyclicity in turbidite successions 368-371
facies distribution 374
flow processes 375
internal structures 361-367
lamina deposition 373
organic-rich 34
transport and deposition 371-373
Muschelkalk 292

nearshore bars 231, 233
Neogene
Baltimore Canyon 643-644
California 493, 692
Neogene sequence cycles 628, 643-645
Newark supergroup 50-53, 214-217
Niobrara Formation 56-7, 134, 528
nodular limestones 17, 445, 446 (see diagenetic overprinting)
nomenclature, cycles, rhythms (see classification, sequence stratigraphy)
bundles 4, 7, 50, 822, 859
cyclic sequences 4
cyclothem 5, 733-738

depositional sequences, systems 5, 620-621
discyclic, nonperiodic sequences 7
episodic phenomena 10, 618-620
event beds, event deposits 10, 377, 492-415
high-energy episodes 377
parasequence 5, 620
periodic sequences 7
rhythmic bedding, sequences 4
sedimentary cycle 5
stacking of cycles 630
systems tracts 623
non-actualistic effects 222
North Sea 684
Norwegian Sea 760-765, 768, 769
numerical model of cyclic sedimentation 855-863
random walk 856, 857
stratification cycle, differentiation 859
time-thickness relationship 860-863

obrution deposits 275, 283
oceanic circulation 45, 75, 89, 90
sensitivity 46, 75
Oman 261
omission (see stratigraphic gaps) 279-281
onlap 686
opal production and dissolution (see bedded cherts)
orbital cycles, parameters 23-25, 48-49, 63, 72-73, 714, 835
atmospheric-climatic feedback 24-27, 710
Earth's axial obliquity 24, 48, 49,
Earth's axial precession 23, 48, 49, 72-76
orbital excentricity 24, 48, 49, 72
translation into sediments 25-27, 82, 170-173
Ordovician, USA 304
organic carbon oxidation (Emerson-Bender effect) 116
organic matter, organic carbon 74-75
compactional enrichment 36
relation to carbonate 169, 129, 136, 137, 159, 171, 174, 177, 179
metabolism of organic matter 36
terrestrial, in black shales 34
in mud turbidites 34
Oriskany Sandstone 29
overbeds 256
oxygen deficiency (see black shales)
Oxygen isotope record 69-70, 119-21
oxygen minimum zone 10
redox cycles 46, 134-138, 514, 517

paleocurrent indicators 242
Paris Basin 267
Pennsylvanian, N.America 735, 738-739
peritidal carbonates (see lagoonal-peritidal cycles)
Permian, Texas 532
Petra 383
phosphatic sediments, phosphorites 492-507, 513-516
allochthonous 503
apatite, carbonate fluor apatite 492
classification 506-507
condensed layers 498-505
depositional environments 492, 505-506
laminae, layers 494-499
phosphatic coatings 495, 498, 506
phophogenesis 492-493
pristine phosphatic sediments 495-498
rhythmic stratification 496-497
winnowing events 501, 505
phytoplankton blooms 566, 836
plattenkalk facies, lithographic limestones
Bear Gulch, USA 584-586
diversification 591
Green River formation, USA 574-576
Gürün formation, Turkey 576-578
Haqel and Hjoula, Lebanon 587-589
Hvar, Yugoslavia 580-581
lagoonal environments 578-584
lake environments 573-578
Montsech 578-580
shelf environment 584-589
Solnhofen, Germany 581-584
poikiloaerobic facies 565
Posidonia Shales 296, 531, 532
power spectra (see time series analysis)

preservation spikes 117-118, 810-813
pressure dissolution (see diagenetic overprinting)
productivity cycles (see limestone-marl rhythms, bedded cherts)
provenance of calcareous turbidites 347-348 (see turbidites)
Purbeckian, French Jura 718, 719
pyroclastic flow (see tephra)

ravinement surfaces 627, 630, 652 (see sequence stratigraphy)
recurrenc time interval 8, 10-12
   black shales 509-511
   turbidites 336-338, 348-351
   storm, flooding events 8
   volcanic eruptions 415-420
redox cycles (see oxygen minimum, limestone marl cycle) 46, 514, 517, 524, 540, 565
regression (see sequence stratigraphy, lowstand progradation) 660
   prograding peritidal carbonates 709-721
   prograding sands, ooids 665, 668
rhythmic bedding, rythmic sequences (see limestone-marl rhythms) 4
   random walk model 855-863
   time series analysis 840-854
   types, reflecting deposition 30-33, 131, 170-173
rip currents 231, 245
Rochester Shales 294

Saramouj Conglomerate 384
Saudi Arabia 267
Scisti a fucoidi 52-55, 83-84
sea level changes, sea level curves 660-681
   amplitude 661
   control on coal cycles 745, 746
   control on lagoonal-peritidal cycles 715
   diversity fluctuations 703-705
   ecological response 700
   eustatic 620
   evolutionary response 698
   high frequency, factors 710, 771
   inflection point 666
   period 669
   rates of sea level change 620, 629, 711
   relative sea level changes 617, 660
   shape of sea level curve 662
   turbidites related to s.l.ch. 348-354
sea level highstand (see sequence stratigraphy)
   highstand deposits 66
sea level lowstand (see sequence stratigraphy)
   lowstand deposits 666
   lowstand fan 664
seasonal rhythms 191-197
sediment accomodation 617, 662, 713, 715
sedimentary cycles (see limestone-marl rhythms)
   amalgamated, stacked 668, 672, 674-78, 731
   asymmetric 663-81, 715, 717
   coal cycles 737
   glaciomarine cycles 768-769
   symmetric, Lofer cycles 724
sedimentary structures 253-255
sedimentation rate 40, 678-79, 782, 825-826, 834
   alternating beds 40, 167-187
   carbonate-organic carbon interrelation 169, 175-179
   fractional sedimentation rates 775-777
   relative sedimentation rates 167-187, 777-780
   time span dependency 791, 792
sedimentation thresholds 200, 202
sediment bypassing 662, 665
   timing of stratigraphic condensation 786, 787
sediment supply (see sedimentation rate) 661, 665
seismic episodes 355
seismic stratigraphy 620-622
seismites 224-226
sensitivity of depositional systems 73
sequence cycle chart (Cenozoic) 628

sequence boundaries (see sequence
    stratigraphy)
sequences (see sedimentary cycles)
    coarsening, shallowing-upward 630,
        665, 678, 717
    fining, deepening-upward 665, 678
    symmetric 724
sequence stratigraphy 262, 620-658, 645,
    682
    backstepping 625
    condensed sections 624, 653, 655
    downlap, downlap surface 625
    first flooding surface 625
    highstand deposits, highstand systems
        624-627, 630, 653-654, 666
    highstand progradation 629
    incised valleys 624, 629, 647-649
    lowstand deposits, lowstand systems
        623-627, 630, 646-651, 666
    lowstand fan, basin floor fan 624,
        628, 648-649, 656, 664
    lowstand progradation 625-629, 650,
        657
    maximum flooding surface 625, 653
    offlap 625
    onlap 623-625
    parasequences 623, 630, 644, 717
    retrogradation 629
    sequence boundaries 623-627, 668
    shelf margin surface 627
    shelf margin systems tract 623-624,
        629, 646-648
    systems tracts 623-30
    top lowstand surface 625
    toplap 623
sequence stratigraphy in various settings
    646-658
    carbonate environments 655-657
    ramp settings 647, 651
    siliciclastic environments 646-655
    transgressive systems tract 624-627,
        651-65
shallow sea, epicontinental sea (see
    Cretaceous seas)
    models 662-72
    time lines 817

shallow water carbonate cycles (see
    lagoonal-peritidal carbonate cycles)
Shanxi Province, China, Carboniferous
    741-742
shelf anoxia 566
shelf diamicton 758
shelf prograding (see sequence
    stratigraphy) 758
shelf-slope-deep sea setting 672-673
shell beds 14, 271, 689
shelly tempestites 257
shoreface, equilibrium profile 669
shoreface morphodynamics 235
silica diagenesis (see diagenetic
    overprinting)
siliceous sediments (see bedded chert)
siliciclastics 652
Silurian, USA 308
Solnhofen 290, 295-96
spectral analysis (see time series analysis)
starvation 652, 678
storm deposition 249
storm-generated currents 229
storm-generated suspension currents 246
storms, energy transfer 227
storm waves 220
storm wave base 662, 669
    exhumed concretions 671
    storm wave erosion 671
stratified cementation (see diagenetic
    overprinting)
stratigraphic correlation (see
    correlation)
stratigraphic gaps (hiatuses) 665-681, 775
    base of depositional events 787-791
    hierarchy 784, 785
    omission 279-281
    stratigraphic completeness 791, 792
    time span estimation 784-787
submarine erosion (see erosional
    unconformities) 66
subsidence
    analysis 631, 713, 715
    curves 637-638
    effective subsidence 661
    differential subsidence 669

953

interaction subsidence-sea level (see interaction of ...)
rates 637, 662-664711
subsidence curves, history 558, 637-639
superposition of cyclic phenomena 8, 678
surface water salinity 70
suspension sedimentation 327
systems tracts (see sequence stratigraphy)

taphofacies 284
taphonomic feedback 13, 268, 686
taphonomy 283
tectonic events 636-637
tectono-stratigraphic analysis 631-634
tempestites, storm bed 249, 304
   basin scale 261
   biological feedback 250
   bipolar sole marks 377
   body fossils 379
   calcareous tempestites 381
   correlation 803, 804
   diagenetic effects 256
   distinction from turbidites 377-382
   general characteristics 253
   gutter casts 378
   hummocky cross stratification 377-378
   ichnofauna 379
   modern examples 227
   mud tempestites 381
   proximality 258
   sedimentary structures 377-378
   sequences 259
temporary burial 275
tephra layers and events 392-492
   bentonite beds 399
   convective plume 393
   correlation 802, 803
   debris avalanche deposits 406-407
   duration and periodicity of eruptions 415-420
   elutriation cloud, co-ignimbrite fallout 400
   impacts on climate and biota 425-428
   fallout ash 393-396
   Laacher See ash layer 398
   marine ash layers 396-399
   Mount St. Helens eruption cloud 396
   overbank ash 400, 401
   pyroclastic flows, ignimbrites 402-405
   subaereal volcaniclastic mass flows, lahars 394, 405, 406
   submarine eruption deposits 409, 410
   submarine redeposition 412-413
   submarine volcaniclastic mass flows 407-415
   tephrochonology 421-424
   tephrofacies, tephrostratigraphy 420, 421
   volcaniclastic aprons 413, 414
Tethyan ocean 820-30
thickness of glaciomarine sediments 755-759
tidewater ice fronts 755-760
tier migration 531
tiering of trace fossils 34, 527
till tongue association 760-761
time averaging 690
time series analysis 59-60, 783, 840-854
   cross spectra analysis 119, 854
   digitization of lithology 841, 843
   filtering 853, 854
   Fourier spectra 843, 850, 852
   implications by sedimentation, compaction 847, 849
   red noise 849-851
   smoothing sectra 846
   tapering 845, 846
   trend removal 845
   vertical overlapping 852
   Walsh spectra 59, 843, 844, 850, 852, 853
time span asessment, timing
   beds 40, 780-784, 167-187
   Gubbio sequence, Italy 777, 862, 863
   lateral correlation 777-780, 789, 798, 815, 816
   methods, overview 774
   power spectra (see time series analysis)
   sequences 774-780
   stratification cycles 860-863
   stratigraphic completeness 791, 792

stratigraphic gaps, base of event beds 787-790
stratigraphic gaps, hierarchy 784, 785
stratigraphic gaps, nondeposition 786, 787
using sedimentation rates 178, 775-782, 791
Western Interior Basin transsect 779, 780
Todilto Formation (Jurassic) 209
toplap 688, 691
transgression (definition) 660
transgression-regression cycles 621, 622, 658, 659, 662-681, 717, 724, 737
transgressive lag (see also stratigraphic gaps) 686
trilobite graveyards 288
turbidites 322-332
   body fossils 379
   calcareous turbidites 328-330, 381, 340-359
   composition of calc. turb. 341
   current directions, paleocurrents 326
   distinction from tempestites 377-382
   frequency of turbidite events 336-368, 348-351
   graded cross-bedding and lamination 328
   megaturbidites 321
   nomenclature (Bouma divisions) 323, 326
   organic carbon rich 518-520
   post-turbidite, pre-turbidite infauna 379
   proximal facies association 379

radiolarian turbidites 476, 477
sedimentary structures 341-346, 377-378
silt turbidites 361
turbidite sequences 351-354
turbidites in basin evolution 354-359
turbidity currents 322-332
   autosuspension 326
   density 324
   hydraulic behavior of grains 328
   hydraulic jump 333
   low-velocity, high-velocity 324-327, 344

unconformities (see erosional unconformities)
underbeds 256

Van Houten cycles 52, 214-217
varve-scale laminations 10-12, 521, 827-834
   correlation 804
   bundled varves 830
   siliceous varves 473, 474
   varve counts 49
   varves in lake sediments 191-197
varvites, biogenic 598
Vocontian basin 821-822
volcanic events (see tephra layers)

wadi sedimentation 383
water depth curves 666-677

Zechstein 210

Printing: Druckerei Zechner, Speyer
Binding: Buchbinderei Schäffer, Grünstadt